Earthquake Prediction
An International Review

Edited by

David W. Simpson
Paul G. Richards

American Geophysical Union
Washington, D.C.

Earthquake Prediction
An International Review

Library of Congress Cataloging in Publication Data
Main entry under title:

Earthquake predictions.

   (Maurice Ewing series; 4)
    1. Earthquake prediction—Addresses, essays, lectures.
I. Simpson, David W. II. Richards, Paul G.,
III. American Geophysical Union. IV. Series.
QE538.8.E38   551.2'2   81-7941
ISBN 0-87590-403-3   AACR2

Maurice Ewing Series

CONTENTS

Short-Term and Immediate Precursors to Earthquakes

Fundamental Studies, Laboratory Investigations and Models

Reviews of National Programs

From May 12 to May 16, 1980, eighty-eight scientists from eleven countries attended a Symposium on Earthquake Prediction at Mohonk Mountain House, Mohonk, New York. This was the third in a biennial series honoring Maurice Ewing, first director of Lamont-Doherty Geological Observatory. The Symposium was one of several events that were held in 1980 to celebrate the 100th anniversary of the Graduate School of Arts and Sciences at Columbia University. The two earlier Ewing Symposia, on island arcs and deep sea drilling, reflected Ewing's lifelong interest in the structure and evolution of the ocean floor. In the Third Ewing Symposium we touch another area - earthquake seismology - that played an important part in Ewing's career. Work on surface waves and long-period seismology under Ewing's direction during the 1950's and 1960's, along with his exploration of the earth beneath the oceans, provided much of the framework on which current ideas on earthquake generation and plate tectonics are based.

To see how far we have come in understanding the earthquake process and the predictability of earthquakes, it is helpful to look back over the past two decades. In 1960 there occurred in Chile one of the greatest earthquakes of modern times. Free oscillations of the earth were first detected unambiguously from the records of that event. One aspect of the Chilean shock was particularly impressive, the great length of the rupture zone, nearly 1000 km. Nevertheless, our knowledge at that time of tectonics and earthquake mechanisms was so poor that there was considerable debate whether the mechanism was one of strike-slip or thrust faulting. In fact, there was still controversy as to whether movements along faults were the direct cause of earthquakes or if they were merely a secondary manifestation of some other process. Several years were to pass before the large displacements in that shock were appreciated in the context of the plate tectonic model or before the observational data were described in terms of the seismic moment, stress drop, corner frequency and moment tensor. Those concepts, which are so widely used today, were not developed until well after 1960.

In 1964, the first United States-Japanese conference on earthquake prediction was held in Tokyo. Prior to that conference the subject of earthquake prediction was not taken very seriously in the United States. The great Alaskan earthquake of 1964 occurred about a week after that conference. Again, the event involved a huge rupture zone and large vertical motions. Nevertheless, whether the fault plane of that shock was of a shallow or a steep dip was debated for more than a year. Geodetic measurements were to indicate horizontal movements in excess of 10 m. By 1966 the distribution of aftershocks, the focal mechanism, the observed free oscillations and the static displacements all seemed to be compatible with an average displacement of about 10 to 20 m on a thrust fault that dipped about 15° to the northwest.

The discovery of plate tectonics in the late 1960's revolutionized many of our concepts about earthquakes, volcanoes, earth deformation and the tectonic setting of major earthquakes like those in Chile and Alaska. In just the last 13 years the plate tectonic hypothesis has become a central part of many earthquake studies. That hypothesis provides a rationale for the seismic gap concept, i.e., the notion that segments of major plate boundaries that have not been the sites of large earthquakes for many decades should be considered the likely sites of future great shocks. The gap concept has been remarkably successful in forecasting the locations and magnitudes of many of the large shocks of the past ten years. It does not, however, provide more than a very rough estimate of the time of occurrence of future earthquakes.

Plate tectonics also underlies the strain-buildup and release model that H. F. Reid hypothesized about 70 years ago. Reid and others proposed that buildup of strain occurred over a period of many years before large shocks and that successful prediction would likely involve the careful monitoring of the strain accumulation process. Space technology may eventually provide rapid and frequent monitoring of the strain field in a large region.

It was only 13 years ago that definitive evidence for subduction became available. Since then we have been gradually refining our knowledge of the upper 100 km of plate boundaries of the convergent type. One may anticipate that detailed studies during the next decade, combining techniques from seismology, geology and marine geophysics, will provide a much greater understanding of the subduction process and of the occurrence of large earthquakes and seismic sea waves that are associated with subduction.

During the past five years exciting new evidence on the occurrence of prehistoric earthquakes has come from geologic studies of fault zones, particularly trenching and the dating of offset geologic units. These studies are beginning to provide more extensive information about the repeat times of large earthquakes at a given point along major fault zones. Only a few areas of the world, such as China and parts of Japan, have an historic record of earthquakes that is longer than 1000 years. Hence, for most areas, geologic studies are needed to cover more than one cycle of strain build-up to a great shock and then its sudden release.

Japan began an official program of earthquake prediction in 1965. Studies of seismicity in Central Asia in the 1950's evolved into the program of earthquake prediction in the U.S.S.R. China also mounted a large effort in earthquake prediction following a damaging shock in 1966. Work on earthquake prediction intensified in the United States during the 1960's and 1970's. Following the Alaskan earthquake of 1964, a committee chaired by Frank Press recommended that the U.S. start a major program of prediction. An official government program in that area was not authorized, however, until 1977.

Fortunately, seismology has had a long history of international cooperation and exchange of data. In organizing the symposium we purposely sought participants who were working in a variety of tectonic settings and who were employing many different techniques and hypotheses for prediction.

It is evident that there is a wide range of opinions about the prospects for earthquake prediction. In the United States there was, from about 1971 to 1975, a surge of interest in forerunning effects, and several physical mechanisms were put forward by American, Soviet and Japanese scientists to account for these observed effects, and to explain why they were precursory to earthquakes. Several U.S. scientists expected in the early 1970's, probably with too much optimism, that routine prediction of

earthquakes was just around the corner. But most experience in the last few years has been more equivocal, leading some seismologists to react and become pessimistic about the prediction of earthquakes.

However, an overview of large earthquakes in several countries during the past five years does show that a number of forerunning effects have been clearly and quite consistently observed. One of the goals of the symposium reported in this volume was to obtain such an overview, and included here are a number of case histories of recent large events in China, Japan, Mexico, the U.S.S.R., and the U.S.A. More than half a million people have been killed in earthquakes during the last twenty years. It is our belief that this social problem can be mitigated in the future by the progress that may be expected in earthquake prediction. From what was presented at the symposium, and from what is now contained in this volume, we sense a renewed optimism about these prospects over the next decade.

Acknowledgments. In addition to ourselves, other members of the Organizing Committee were Keiiti Aki, Massachusetts Institute of Technology; Carl Kisslinger, University of Colorado; Christopher Scholz, Columbia University; Manik Talwani, Columbia University; and Robert Wesson, U.S. Geological Survey. Many people from Lamont-Doherty helped to make the Third Ewing Symposium a success. Ellie Wellmon and Margaret Swan helped with much of the initial correspondence and with other arrangements for the meeting. Linda Murphy organized all of the correspondence and did much of the editing related to the preparation of this volume of papers. We are grateful to the G. Unger Vetlesen Foundation, the National Science Foundation (EAR 79-26376), and the U.S. Geological Survey (14-08-0001-G-686) for their financial support of the symposium.

Lynn R. Sykes, Chairman, Organizing Committee

David W. Simpson and Paul G. Richards, Editors

THE NATURE OF SEISMICITY PATTERNS BEFORE LARGE EARTHQUAKES

Hiroo Kanamori

Seismological Laboratory

California Institute of Technology, Pasadena, California 91125

Abstract. Various seismicity patterns before major earthquakes have been reported in the literature. They include foreshocks (broad sense), preseismic quiescence, precursory swarms, and doughnut patterns. Although many earthquakes are preceded by all, or some, of these patterns, their detail differ significantly from event to event. In order to examine the details of seismicity patterns on as uniform a basis as possible, we made space-time plots of seismicity for many large earthquakes by using the NOAA and JMA catalogs. Among various seismicity patterns, preseismic quiescence appears most common, the case for the 1978 Oaxaca earthquake being the most prominent.

Although the nature of other patterns varies from event to event, a common physical mechanism may be responsible for these patterns; details of the pattern are probably controlled by the tectonic environment (fault geometry, strain rate) and the heterogeneity of the fault plane. Here a simple asperity model is introduced to explain these seismicity patterns. In this model, a fault plane with an asperity is divided into a number of subfaults. The subfaults within the asperity are, on the average, stronger than those in the surrounding weak zone. As the tectonic stress increases, the subfaults in the weak zone break in the form of background small earthquakes. If the frequency distribution of the strength of the subfaults has a sharp peak, a precursory swarm occurs. By this time, most of the subfaults in the weak zone are broken and the fault plane becomes seismically quiet. As the tectonic stress increases further, eventually the asperity breaks and sympathetic displacement occurs on the entire fault zone in the form of the main shock. Foreshocks do or do not occur depending upon the distribution of the strength of the subfaults within the asperity. Since the spatio-temporal change in the stress on the fault plane is most likely to dictate the change in seismicity patterns, detailed analysis of seismicity patterns would provide a most direct clue to the state of stress in the fault zone. However, because of the large variation from event to event, seismicity pattern alone is not a definitive tool for earthquake prediction; measurements of other physical parameters such as the spectra, the mechanism and the wave forms of the background events should be made concurrently.

Introduction

Spatio-temporal variations of seismicity before major earthquakes have been studied by many investigators in an attempt to understand the physical mechanism of earthquakes and to use them as a tool for earthquake prediction. In this paper, we review the recent progress in this field, add some new data, and propose a simple model which facilitates the understanding of the nature of these seismicity patterns.

Since these patterns have not been defined unambiguously, we first discuss some representative patterns by using a schematic diagram shown by Figure 1. This figure includes, following Mogi (1976), the pattern of foreshocks, precursory swarms, precursory quiescence and doughnut patterns.

Foreshocks

Although there is no widely accepted definition of foreshocks, some earthquakes (e.g., 1974 Haicheng, China earthquake; 1963 Kurile Islands earthquake) were preceded by a very remarkable short-term increase in the number of small events in the epicentral area so that little ambiguity exists in calling them the foreshocks. In other cases, however, ambiguity arises because of either too small number of events, too spreadout time interval, or both. Yet these events may be causally and/or physically related to the mainshock, and may be called the foreshocks.

Sometimes small events which preceded a mainshock and occurred in, or in the neighborhood of, the mainshock rupture zone are called preshocks. In this paper, we will use the term foreshocks in a rather loose sense of the word to include both "obvious" foreshocks and preshocks.

According to Jones and Molnar (1976), about 44% of large shallow earthquakes in the world were

Seismicity Pattern

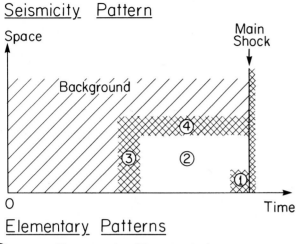

Fig. 1. Schematic space-time diagram showing various seismicity patterns. (Modified from Mogi, 1977).

preceded by foreshocks of their definition. A very useful summary of foreshock activity in Japan can be found in Mogi (1963). Among the best documented foreshock sequences are those of the 1974 Haicheng, China earthquake (Wu et al., 1978), the 1978 Oaxaca, Mexico earthquake (Ponce, et al., 1977-1978) and the 1963 Kurile Islands eqrthquake (Santô, 1964).

Occasionally, a very tight clustering of activity occurs before the mainshock. Mogi (1968b) and Kelleher and Savino (1975) demonstrated that seismic activity prior to a great earthquake tends to cluster around the epicenter of the eventual mainshock. More recently, Ishida and Kanamori (1978) and Fuis and Lindh (1979) found a very tight clustering of activity before the 1971 San Fernando, California and the 1975 Galway Lake, California earthquakes, respectively. Engdahl and Kisslinger (1977) found a clustering of small events before a magnitude 5 earthquake in the Central Aleutians. Although these events are not usually called typical foreshocks, they can be considered to be foreshocks in a broader sense, or preshocks.

Gap

Imamura (1928) investigated historical data on large earthquakes in southwest Japan (Tokaido-Nankaido region), and found that large earthquakes in this region had occurred repeatedly at approximately the same location with a repeat time of about 100 to 150 years. He pointed out that the area southeast of the Kii peninsula and Shikoku

island had not experienced a large earthquake for 70 years, and suggested that a large earthquake was imminent there. The Tonankai ($M_S = 8.0$) and the Nankaido ($M_S = 8.2$) earthquakes indeed occurred there in 1944 and 1946 respectively. Fedotov (1965) and Mogi (1968a) studied seismicity in the Kamchatka, Kurile and Japan regions and identified several zones which had not experienced a large earthquake for a long time. These zones were considered to be candidate sites of major earthquakes in the future. In fact, several major earthquakes including the 1968 Tokachi-Oki, Japan earthquake ($M_W = 8.2$) occurred in these zones subsequently. These results were furhter extended to the concept of seismic gaps, and have been used more globally by many investigators (Kelleher, 1970; Utsu, 1970; Kelleher et al., 1973; Sykes, 1971; Ohtake et al., 1977). Kelleher and Savino (1975) demonstrated that gaps in seismicity for great earthquakes are also gaps for smaller magnitude activity and such gaps commonly persist until the time of the mainshock. A very comprehensive review can be found in McCann et al. (1980). Usually these seismic gaps refer to a spatial gap of seismic activity, particularly of large earthquakes.

Quiescence

Inouye (1965) found that seismicity in the epicentral area of several large earthquakes in Japan (e.g., 1952 Tokachi-Oki and 1964 Niigata earthquakes) became very low before the mainshock. This quiescence was followed by increased activity for several years before the mainshock. Mogi (1968a) showed that before several large earthquakes (e.g., 1944 Tonankai and the 1946 Nankaido earthquakes), the focal region became very calm. These studies suggested that seismic activity in the eventual rupture zone of a large earthquake decreases more or less abruptly sometime before the mainshock. In this regard this pattern may be called a temporal gap. Perhaps the most pronounced of the temporal gaps is the one before the 1978 Oaxaca, Mexico earthquake ($M_W = 7.6$) reported by Ohtake et al. (1977). Mogi (1979) called the spatial gap and the temporal gap, the gap of the 1st and 2nd kind respectively. In any case, a preseismic quiescence of seismic activity in the epicentral area of a large earthquake appears very common to many large earthquakes.

Precursory Swarm

McNally (1977) found that distinct clusters of small earthquakes occurred in the near-source region of several moderate size earthquakes in Central California 2 to 10 years before the mainshock. Sekiya (1977) and Ohtake (1976) reported that anomalous seismic activity occurred about 10 years before the 1974 Izu-Hanto-Oki earthquake in the epicentral area which had generally been quiet before the earthquake. Sekiya (1977) reported further examples for about ten other Japanese earthquakes. Evison (1977a,b) found such precursory activities before the 1968 Borrego Mountain, California earthquake and several earthquakes in New Zealand. Evison considered that a burst of seismic activity marks the start of a precursory sequence, and called it the precursory earthquake swarm. Brady (1976) found a clustering of seismic activity before the 1971 San Fernando, California earthquake and interpreted it as a "primary inclusion zone" of the impending failure.

Doughnut Pattern

Mogi (1969) found that before several large earthquakes in Japan, the region surrounding the focal region became very active, while the focal region was quiet at the same time. This pattern is often called a doughnut pattern. A similar doughnut pattern has been reported for a magnitude 6 earthquake in Kyushu, Japan by Mitsunami and Kubotera (1977) and for a magnitude 6.1 earthquake in the Shimane prefecture, Japan by Yamashina and Inoue (1979).

For many earthquakes, these elementary seismicity patterns described above appear either by itself or as combinations so that the actual pattern often becomes very complex. Furthermore, identification and classification of the patterns depend upon the catalog used for the study, the magnitude threshold, the time period, the depth range, and the judgement of the investigators, so that entirely different patterns have often been identified for the same event. Because of this ambiguity, the reported seismicity pattern should not be regarded as a unique feature of the earthquake, but should rather be regarded as a manifestation of the physical process leading to an earthquake.

In general, two approaches are possible in using seismicity patterns for earthquake prediction. The first is represented by the works of Keilis Borok et al. (e.g., 1980), Wyss et al. (1978) and Habermann (1980). In this method, various seismicity patterns are treated as rigorously as possible in a statistical framework to establish an empirical algorithm for earthquake prediction. Essential to this approach are the uniformity of earthquake catalogs and rigorous definition of the seismicity patterns.

In the second approach, various seismicity patterns are used as a clue to the physical mechanism of earthquake failure process. Although the observed seismicity patterns are very complex, the fundamental physical mechanism may be simple. The complex structures of the fault zone may be primarily responsible for the variation of the observed seismicity patterns. In this approach, rigorous definition of seismicity patterns and uniformity of the catalogs are less important than in the first approach. Although seismicity patterns thus somewhat loosely defined may not be used directly for prediction purposes, they can be

used to identify a possible physical mechanism. Once a physical mechanism is identified, other means such as monitoring temporal variations of source mechanism, spectra and wave forms may be used for prediction purposes. In view of the large degree of non-uniformity of the presently available seismicity catalogs and of the methods used by various investigators, we will be primarily concerned with the second approach in this paper.

Examples of Seismicity Patterns

Various seismicity patterns which have been reported so far are summarized in Table 1 in terms of the elementary patterns described in the Introduction. Although these results provide a fundamental data base for the present study, we made a global survey of seismicity patterns associated with large earthquakes by using space-time plots of seismicity in order to clarify the nature and regional variation of seismicity patterns.

We used the NOAA catalog for all of the regions except for Japan. Since the uncertainty in the location is probably about 30 to 50 km, we will be primarily concerned with the patterns for earthquakes larger than magnitude 7.5 (The long-period magnitude, M_w, is preferred whenever available. If it is not available, the surface-wave magnitude, M_s, is used.) whose rupture length is 70 km or larger.

The non-uniformity of the catalogues results from combinations of many factors which include: 1) temporal variations in the number and distribution of stations, 2) changes in the practice of the magnitude determination both at the individual stations and the central agency, 3) changes in the instruments, 4) changes in the personnel and operation practice at the individual stations, and 5) changes in the location procedure. It is not clear how we can remove the effect of all of these factors to extract the real spatio-temporal variations of seismicity. Hence, we will plot the raw data from routinely available catalogues without heavy processing. Symbols with different sizes are used for different magnitude ranges to facilitate visual inspection of the patterns. As will be shown later, it is encouraging that some patterns are discernible from the space-time plots produced in this manner from routinely available catalogues. We first illustrate the method for Mexico.

Mexico

We extracted all the events shallower than 60 km which occurred within the box shown in the index map (Figure 2). Then the distance to the individual epicenter from the pole shown in Figure 2 was measured and the events were plotted as a function of time in the form of a space-time plot as shown in Figure 3. The pole is placed on the approximate extention of the strike of the region considered, and at a distance comparable to the total length

TABLE 1.   OBSERVED SEISMICITY PATTERNS*

EVENT	PRECURSORY SWARM	QUIESCENCE	FORESHOCK (PRESHOCK)	DOUGHNUT PATTERN	REFERENCE	NOTE
Cape Kamchatka,1971 ($M_s = 7.8$)	1969 during the quiet period	1962-1971	No	No	Wyss et al. (1978)	See alao Fedotov et. al. (1977)
Friuli, 1976 ($M_s = 6.4$)	See the column under Foreshock	No activity prior to 1960	Clustered activity in March 1975	No	Wyss et al. (1978)	
Assam, 1897 ($M_s = 8.2$)		During 28 years before the main shock			Khattri & Wyss (1978)	
Kalapana,Hawaii 1975 ($M_s = 7.2$)	No	About 2 years from May,1972			Wyss et al. (1978)	
San Fernando, California,1971 ($M_s = 6.5$)	1961 ∿ 1964	1965 to 1969	From Sept.,1970	No	Ishida & Kanamori(1978)	See also Brady (1976,1977)
		Quiet period began 597 days before the main shock.	211 days before the main shock	No	Ohtake et al. (1978	
Kamchatka,1952 ($M_w = 9.0$)		From 1920	During 3 years just before the main shock	Not obvious	Kelleher & Savino (1975)	
Alaska,1964 ($M_w = 9.2$)		From 1944	1954 to 1964 increased activity near the edge of the rupture zone	Not obvious	Kelleher & Savino (1975)	
Sitka,Alaska 1972 ($M_s = 7.2$)	No	Quiet everywhere near the rupture zone	No	No	Kelleher & Savino (1975)	
Alaska,1958 ($M_s = 7.9$)	No		Increased activity during about 10 years before the main shock	No	Kelleher & Savino (1975)	
Kern County, California,1952 ($M_s = 7.7$)	1939∿1941	Much of the rupture zone quiet for at leat 20 years	Clustered activity near the epicenter	No	Wesson & Ellsworth(1973), Kelleher & Savino(1975), Ishida & Kanamori(1980)	
Parkfield, California,1966 ($M_L = 5.5$)	No	Quiet in the rupture zone about 9 months	Small foreshocks 8 days before the main shock	No	McEvilly et al. (1967), Kelleher & Savino(1975) Ohtake et al. (1978)	

Table 1. cont'd.

EVENT	PRECURSORY SWARM	QUIESCENCE	FORESHOCK (PRESHOCK)	DOUGHNUT PATTERN	REFERENCE	NOTE
Chile, 1960 ($M_w = 9.5$)		Quiet for at least about 5 to 8 years	Immediate foreshocks (33 hours before the main shock)	No	Kelleher & Savino(1975)	
Tokachi-Oki,1952 ($M_w = 8.1$)	No	1943 to 1951 1926 to 1951 1934 to 1947	Increased activity during 1948 to 1952	Yes	Mogi (1969) Utsu (1968) Inouye (1965)	
		1937 to 1949	2 years before the main shock		Katsumata (1973), Katsumata & Yoshida (1980)	
Tonankai,1944 ($M_w = 8.1$) and Nankaido,1946 ($M_w = 8.1$)	No	20 years	No	Yes	Mogi (1969)	
Sanriku,1933 ($M_w = 8.4$)		12 years	No	Yes	Mogi (1969)	
Tokachi-Oki,1968 ($M_w = 8.2$)	No	1961 to main shock except 1965		Yes	Mogi (1969)	
		1948 to 1963	Increased activity near the epicenter 1964 to 1968		Katsumata & Yoshida (1980)	
		1962 to May 1968			Habermann(1980)	
Shimane, Japan 1978($M_{JMA} = 6.1$)	No	5 months before the main shock	No	Yes	Yamashina & Inoue(1979)	
Oaxaca, Mexico 1978($M_w = 7.6$)	No	About 5 years before the main shock	Yes	No	Ohtake et al. (1977), Ponce et al.(1977-78)	
Oaxaca, Mexico 1968($M_s = 7.1$)	No	1966 to 1968	Increased activity just before main shock	No	Ohtake et al. (1977)	
Oaxaca, Mexico 1965($M_s = 7.6$)	No	1964	Preshocks 1 year before main shock	No	Ohtake et al. (1977)	
Milford Sound, New Zealand, 1976($M_L = 7.0$)	1968	1969 to 1975			Evison (1977a)	
Borrego Mountain, California,1968 ($M_L = 6.4$)	1965	1965 to 1967 About 1 year	Yes Increased activity for 395 days	No	Evison (1977a) Ohtake et al. (1978)	
Mendocino Ridge, California,1960 ($M_L = 6.2$)	1959	1959 to 1960	No	No	Evison (1977b)	

Table 1. cont'd.

EVENT	PRECURSORY SWARM	QUIESCENCE	FORESHOCK (PRESHOCK)	DOUGHNUT PATTERN	REFERENCE	NOTE
Gulf of California 1966 (M_L = 6.3)	1963	1963 to 1966	No	No	Evison (1977b)	
Gisborne, New Zealand,1966 (M_L = 6.2)	August 1964	Sept 1964 to Jan 1966	No	No	Evison (1977c)	
Seddon, New Zealand,1966 (M_L = 6.0)	October 1964	Nov 1964 to March 1966	No	No	Evison (1977c)	
Inangahua, New Zealand,1968 (M_L = 7.1)	1962	1963 to 1967	No	No	Evison (1977c)	
Hastings, New Zealand,1973 (M_L = 5.7)	March 1972	April 1972 to Jan 1973	No	No	Evison (1977c)	
Off Izu Peninsula Japan,1974 (M_{JMA} = 6.9)	1963 to 1965 / 1961 to 1966	1965 to 1974 / 1967 to 1974		No / No	Sekiya (1977) / Ohtake (1976)	
Central Gifu, Japan,1969 (M_{JMA} = 6.6)		About 5 years before the main shock			Sekiya (1977)	
Choshi, Japan 1974 (M_{JMA} = 6.1)	3 years and 5 months before main shock		Some foreshocks	No	Sekiya (1977)	
Fukui, Japan 1948 (M_{JMA} = 7.3)	19 years and 3 months before the main shock				Sekiya (1977)	
N. Miyagi, Japan 1962(M_{JMA} = 6.5)	1956 to 1958				Sekiya (1977)	
Shizuoka, Japan 1965(M_{JMA} = 6.1)	4 years before the main shock				Sekiya (1977)	
Ebino, Japan 1975(M_{JMA} = 4.1)	About 15 days before the main shock				Sekiya (1977)	
Kanto, Japan 1923(M_W = 7.9)			About 82 years before the main shock		Sekiya (1977)	

Table 1. cont'd.

EVENT	PRECURSORY SWARM	QUIESCENCE	FORESHOCK (PRESHOCK)	DOUGHNUT PATTERN	REFERENCE	NOTE
Central Aleutian, 1976 ($m_b = 5$)		4-1/2 months prior to the main shock	6 foreshocks during 5 week period		Engdahl & Kisslinger (1977)	
Markansu, Central Asia 1974 ($M_s = 7.4$)	About 7 years before main shock	1968 to 1974			Kristy & Simpson (1980)	
Zaalai, Central Asia 1978 ($M_s = 6.7$)	About 2 years after main shock				Kristy & Simpson (1980)	
Imperial Valley, California 1979 ($M_s = 6.9$)	About 4 months before the main shock	About 3-1/2 months before the main shock	Yes		Johnson & Hutton (1980)	
Off Fukushima, Japan, 1938 ($M_{JMA} = 7.7$)		Quiet at least from 1926 to 1933	Increased activity from 1934 to 1938		Inouye (1965)	
Niigata, Japan 1964 ($M_{JMA} = 7.6$)		1946 to 1961	Slight increase in activity from 1962		Inouye (1965)	
Aso, Japan 1975 ($M_{JMA} = 5.9$)	3 days before the main shock	1-1/2 days before the main shock	During 30 hours before the main shock		Mitsunami & Kubotera (1977)	3-dimensional feature
Assam, 1950 ($M_W = 8.6$)		> 30 years			Khattri & Wyss (1978)	Similar results for 4 other events in same area
Kurile Is., 1963 ($M_W = 8.5$)	1961				Katsumata* & Yoshida	
		2.2 years before the main shock			Tanaka et al.	
Kurile Is., 1969 ($M_W = 8.2$)	1967 to 1969 Precursory activity	1961 to 1971			Katsumata and Yoshida (1980)	
Nemuro-Oki			1972 to main shock		Katsumata Yoshida (1980)	
Wakayama, 1968 & 1977 (M = 4.8, 4.7)		2 to 3 years	A precursory event		Mizoue et al. (1978)	

Table 1. cont'd.

EVENT	PRECURSORY SWARM	QUIESCENCE	FORESHOCK (PRESHOCK)	DOUGHNUT PATTERN	REFERENCE	NOTE
Garm, 1969 (M = 5.7)		About 1.5 years			Nersesov et al. (1973)	
Central California (M_L = 4.0 to 5.1)	2 to 10 years before the main shock				McNally (1977)	
Kamchatka, 1973 (M_s = 7.2)		5 1/2 years			Wyss and Habermann (1979)	

* Details of the definition of the patterns differ from event to event. Refer to the original references for details.

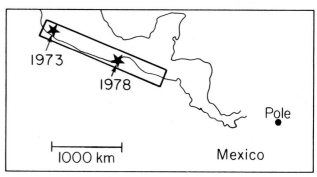

Fig. 2. Index map for Mexico. All the events shallower than 60 km which occurred in the box are shown in Fig. 3. The asterisks show the location of the 1973 Colima earthquake and the 1978 Oaxaca earthquake. The location of the pole is arbitrary, and the scale refers to the middle of the figure.

of the region. The pole position would not have a drastic effect on the overall pattern. The magnitude ranges for the larger symbols are indicated in the figure; the magnitude ranges for the events smaller than 6 are not indicated but the size of the symbols is approximately proportional to the magnitude. The dashed curves indicate the period of quiescence before the 1973 Colima and 1978 Oaxaca earthquakes. No rigid criterion is used for drawing these curves; they are drawn mainly to indicate the region being discussed rather than to define it. Therefore the estimate of the length of the quiet period and the size

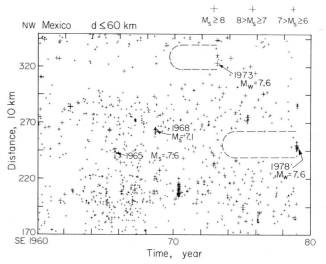

Fig. 3. Space-time plot of seismicity for Mexico obtained from the NOAA catalog (See Fig. 2 for the location). Dotted curves encircle the period of quiescence. The magnitude ranges for the larger symbols are indicated in the figures. The magnitude ranges for the events smaller than 6 are not indicated but the size of the symbols is approximately proportional to the magnitude.

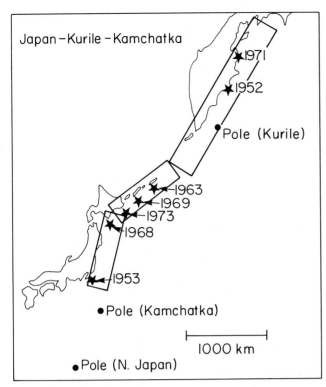

Fig. 4. Index map for the Kurile Is., Kamchatka and Northern Japan. For details, see caption for Fig. 2.

of the zone indicated by the dashed curves should not be given too much significance; no attempt is made here to use it for statistical arguments.

Since the detection capability, the location accuracy, the reporting procedure for the individual station, and the data reduction procedure are not uniform during this time period, the result shown here is inevitably nonuniform. For example, the sudden increase in the number of small events in 1963 is probably due to the establishment of the World Wide Seismographic Station Network (WWSSN). Also, the number of small events seems to have decreased abruptly in 1969. This sudden change could be an artifact of the reporting and the cataloguing procedures. Despite this nonuniformity, the quiescence before these two earthquakes, particularly the 1978 Oaxaca earthquake, appears very obvious. The quiescence before the Oaxaca earthquake was first noted by Ohtake et al. (1977) and was one of the basis of their forecast of this earthquake. The quiescence before the Colima earthquake is less pronounced, but the activity in the encircled area seems to be lower than that during the preceding time period. As seen in Figure 3, the Oaxaca earthquake was preceded by several foreshocks which were located by the world-wide network. At a smaller magnitude level, more than 10 foreshocks were located by a local network (Ponce et al., 1977-1978). For both the Oaxaca and

Colima earthquakes, no obvious precursory swarms or doughnut patterns are seen.

Ohtake et al. (1977) reported a pattern of quiescence before the 1965 (M_S = 7.6) and 1968 (M_S = 7.1) Oaxaca earthquakes. Although there is some indication of reduced seismicity prior to these earthquakes, the pattern is not obvious on the scale of this plot (see Figure 3).

As seen in Figure 3, there are many quiet zones which are not followed by a large earthquake. It is important to note that while large earthquakes appear to be preceded by a period of quiescence, the mere existence of a quiet period does not necessarily point to an impending large earthquake.

We will proceed with a similar analysis for other seismic zones. The analysis method and the basic philosophy of interpretation will be the same unless noted otherwise.

Kurile

Figure 4 shows the index map for the Kurile, Kamchatka and Northern Japan regions. The result for the Kurile Islands is shown in Figure 5. During the period from 1960 to 1978, there were three events larger than 7.5, the 1963 event being the largest.

The 1963 event (M_W = 8.5) was preceded by a distinct period of quiescence. During 1961, the seismic activity became very high along a substantial length of the arc. This increase may be an artifact of increased number of reports from a regional network. However, it consists of many events with a magnitude larger than 6, and is unlikely to be entirely due to nonuniformity of the catalog. For example, the numbers of events with $m_b \geq 6.0$ which occurred within this box are 2, 3, 15 and 2 for 1959, 1960, 1961 and 1962 respectively. For events with $m_b \geq 6.3$, the corresponding numbers are 1, 2, 8 and 2. The 1963 earthquake was preceded by remarkable foreshock activity (e.g., Santô, 1964). Also, an increased activity during about a one year period before the mainshock is seen in Figure 5.

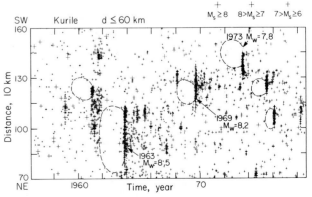

Fig. 5. Space-time plot of seismicity for the Kurile Is. region. See the caption for Fig. 3 for details.

A similar pattern is seen for the 1969 Kurile Islands earthquake ($M_w = 8.2$). This earthquake was preceded by several foreshocks during the 30 minute period before the mainshock. The pattern for the 1973 Nemuro-Oki earthquake ($M_w = 7.8$) is more or less similar to those for the 1963 and 1969 events, though the precursory swarm is not very distinct. The 1973 event was preceded by several small foreshocks.

Thus, for all three major earthquakes which occurred in the Kurile Islands since 1963, the swarm-quiescence-foreshocks pattern can be seen. Although the location accuracy is not good enough to investigate such patterns for smaller events, there is an indication that such patterns preceded smaller events in 1961, 1975 and 1976, as shown by dashed curves in Figure 5.

South America

Figure 6 shows the index map for South America. Figure 7 shows the space-time plot for Chile. Although the data prior to 1960 are, in general, rather poor, it is very clear that seismic activity in the rupture zone of the 1960 Chilean earthquake ($M_w = 9.5$), the largest event in this century, had been lower than in the adjacent segment to the north during the preseismic period. This low activity has already been pointed out by Kelleher and Savino (1975). Because of the poor quality of the data, it is unclear when the quiet

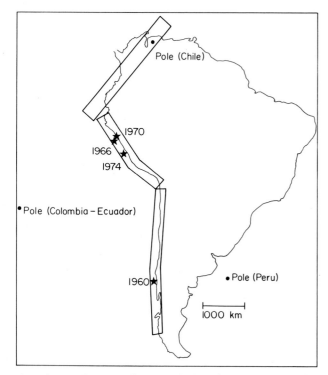

Fig. 6. Index map for South America. See caption for Fig. 2 for details.

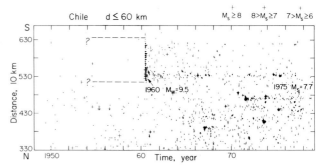

Fig. 7. Space-time plot of seismicity for Chile. See caption for Fig. 3 for details.

period started. The 1960 Chilean earthquake was preceded by remarkable foreshock activity during the 33 hour period just before the mainshock.

For the Peruvian subduction zone (see Figure 8), three events larger than 7.7 occurred during the period since 1960. The 1970 event ($M_w = 7.9$) occurred at a depth of about 70 km and probably represents failure within the down-going slab (Abe, 1972; Isacks and Barazangi, 1977). Since the location of this event is very close to that of the 1966 event, the pattern of seismicity for this event cannot be studied very well with the space-time plot used here. For both the 1966 ($M_w = 8.1$) and 1974 ($M_w = 8.1$) events, a period of seismic quiescence seems to have preceded the mainshock, as shown in Figure 8. However, since the total number of events is relatively small, the statistical significance of these patterns is considered marginal. The pattern of precursory swarms and doughnut patterns are not evident. Neither of these events had foreshock activity detectable by the world-wide network.

For the Colombia-Ecuador subduction zone, the data are too sparse to study seismicity patterns.

Alaska-Aleutians

Figure 9 shows the locations of the seismic zones in the Alaska-Aleutians region studied here. Figure 10 shows the space-time plot for the 1964 Alaska earthquake ($M_w = 9.2$) and the 1957 Fox Island earthquake ($M_w = 9.1$).

In contrast to the examples shown above, both of these earthquakes were preceded by a distinct increase in seismic activity which may be called a preshock activity during about 10 years before the mainshock. Although this increase may be due to increased detection capability, the fact that the commencement of the increased activity for the 1964 event differs from that of the 1957 event suggests that it represents a real seismicity change. This kind of increase in activity before large earthquakes concentrated near either end of the rupture zone, has already been pointed out by Kelleher and Savino (1975). This preshock activity was preceded by a relatively quiet period, although it is not very distinct compared with the

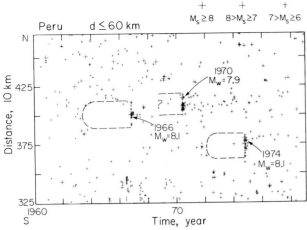

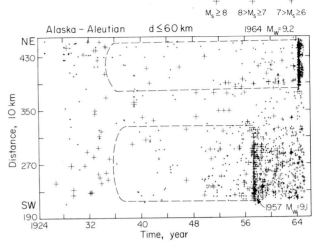

Fig. 8. Space-time plot of seismicity for Peru. See caption for Fig. 3 for details.

Fig. 10. Space-time plot of seismicity for the rupture zones of the 1964 Alaskan and the 1957 Fox Is. earthquake. For details, see caption for Fig. 3.

adjacent regions. During the period from 1928 to 1936, an increased level of seismic activity is seen in the rupture zone of both the 1964 and 1957 events. This activity may be considered to be a precursory swarm.

Figure 11 shows the pattern for the 1965 Rat Island earthquake ($M_W = 8.7$). The main feature of the seismicity pattern is similar to that of the 1964 and 1957 earthquakes. A relatively quiet period from 1957 to 1961 was followed by an increased level of activity for about 4 years, and there is some indication of an increased activity which may be considered to be a precursory swarm around 1956.

Thus, all three major earthquakes in the Alaska-Aleutian region have a common feature which is not observed in the other regions discussed above.

Kamchatka

The area to be considered is shown in Figure 4, and the results are shown in Figures 12 and 13. Figure 12 shows the result for the 1952 Kamchatka earthquake ($M_W = 9.0$). Unfortunately, the data are too incomplete to investigate the pattern. As Kelleher and Savino (1975) pointed out, a higher seismic activity than during the previous period is seen near the epicenter for about 3 years. Although the quiescence is not very clear, the seismic activity during the 15 year period from 1935 to 1950 appears lower than the preceding period.

Figure 13 shows the space-time plot for the period 1960 to 1978. During this period only one event larger than 7.6 occurred ($M_S = 7.8$, 1971). However, the rupture zone for this event appears very small and no obvious pattern is seen on the scale shown here. Wyss et al. (1978) made a detailed analysis of this event and found a quiet period from 1962 to 1971 with a short period of increased activity in 1969. For a smaller event ($M_S = 7.2$) which occured in 1973, a period of

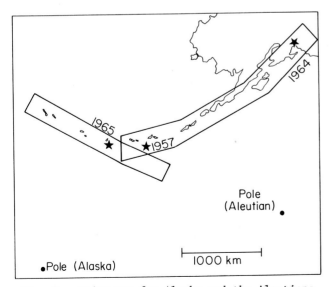

Fig. 9. Index map for Alaska and the Aleutians. See caption for Fig. 2 for details.

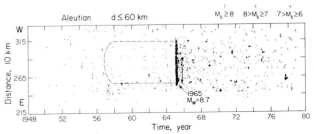

Fig. 11. Space-time plot of seismicity for the rupture zone of the 1965 Rat Is. earthquake. For details, see caption for Fig. 3.

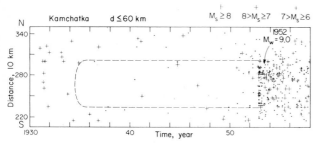

Fig. 12. Space-time plot of seismicity for the rupture zone of the 1952 Kamchatka earthquake. For details, see caption for Fig. 3.

quiescence may be identified (Figure 13). Wyss and Habermann (1979) examined seismicity within 100 km radius from the epicenter of this event, and concluded that a 50% decrease in seismicity rate began in mid 1967. This period of quiescence can be identified also in the space-time plot shown in Figure 13.

Northern Japan

Figure 14 shows the seismicity pattern for Northern Japan (see Figure 4 for the location). For this plot, the earthquake catalog compiled by the Japan Meteorological Agency (JMA) is used, and the earthquakes with $M_{JMA} > 5$ are shown. The largest earthquake during this time period is the 1968 Tokachi-Oki earthquake ($M_w = 8.2$). The activity during about 3 years just before the mainshock is considerably lower than during the preceding period. No obvious foreshocks were reported for this earthquake, although Nagumo et al. (1968) recorded a number of very small events during several days before the mainshock by a ocean-bottom seismograph which had been deployed in the epicentral area.

Mogi (1969), Katsumata and Yoshida (1980) and Habermann (1980) have made detailed analyses of seismicity patterns associated with this earthquake.

Summary

Both the results summarized in Table 1 and those described above show that many large earthquakes were preceded by a period of quiescence. Some events have a pronounced foreshock activity during a period of hours to weeks before the mainshock. Examples include earthquakes in the Kurile Islands (1963 and 1969), Nemuro-Oki (1973), and Chile (1960). Many large events in the Alaska-Aleutians and the Kamchatka regions tend to have an increased seismic activity during several years before the mainshock. However, some events do not have obvious forshocks or preshocks; examples are the large earthquakes in Peru and Northern Japan. Large earthquakes along the Mid-America Trench were often preceded by moderate foreshocks (e.g., 1978 Oaxaca earthquake, 1979

Petatlan earthquake, Meyer et al., 1980). Thus there appears to be a fairly systematic regional variation in the pattern for large earthquakes along various subduction zones. Precursory swarms and doughnut patterns are not always obvious in the space-time plot on regional scales, but detailed studies by other investigators have identified such patterns for some of the events.

Asperity Model

The coseismic motion on earthquake faults is often irregular as evidenced by complex wave forms of seismic waves generated by large earthquakes. This observation suggests that the fault plane is irregular either geometrically or in its physical or mechanical properties. The strength of the contact zone between the two sides of the fault is larger at some places than elsewhere. Such places of increased strength, either of geometrical origin or of some other causes, are generally called asperities. The importance of asperities in various failure processes was recognized in laboratory studies (Byerlee, 1970; Scholz and Engelder, 1976) and the concept of asperity has been frequently used in seismology, either explicitly or implicitly, to explain non-uniform seismicity along fault zones (Wesson and Ellsworth, 1973; Bakun et al., 1980) complex events (Wyss and Brune, 1967; Nagamune, 1971, 1978; Kanamori and Stewart, 1978; Lay and Kanamori, 1980a; Das and Aki, 1977; Aki, 1979), seismic clustering (Ishida and Kanamori, 1978, 1980), and certain aspects of seismicity patterns (Mogi, 1977, Tsumura, 1979; Katsumata and Yoshida, 1980; Lay and Kanamori, 1980b).

Kanamori (1978) interpreted preseismic clustering of events near the main shock epicenter in terms of stress concentration around a strong asperity due to failure of weaker asperities surrounding it. Jones and Molnar (1979) explained observed time dependence of foreshocks by using a fault model with inhomogeneous contact planes on which asperities fail by static fatigue. Ebel

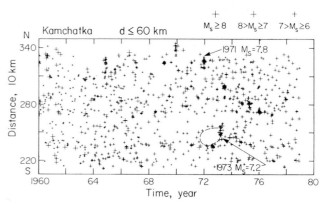

Fig. 13. Space-time plot of seismicity for the Kamchatka region for the period 1960 to 1978. For details, see caption for Fig. 3.

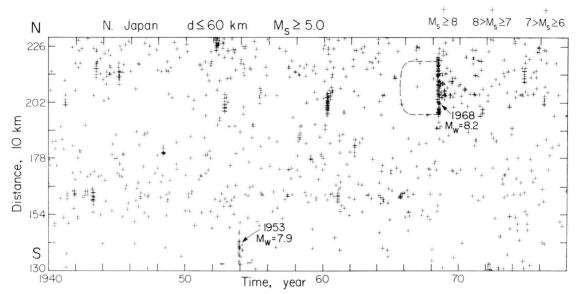

Fig. 14. Space-time plot of seismicity for Northern Japan obtained from the JMA catalog. Earthquakes smaller than 5 are not plotted.

(1980) interpreted a foreshock-main shock-aftershock sequence in the New Hebrides Islands in terms of loading and subsequent failure of asperities on the fault plane.

Mogi (1977) explained the pattern of temporal seismic quiescence and doughnut patterns in terms of a heterogeneous stress distribution on the focal zone. As the tectonic stress increases, small earthquakes occur at high-stress spots in the focal region of an impending large earthquake. When all the high stress spots are broken, the focal region becomes seismically quiet, but the activity in the surrounding region increases.

Tsumura (1979) argued that various seismicity patterns including foreshock patterns can be explained by introducing fault surfaces with variable strength. Katsumata and Yoshida (1980) proposed a model in which the coupling conditions between the lithospheric plates control the temporal variation of seismicity patterns.

Brune (1979) discussed the importance of heterogeneous stress distribution (asperity) on the fault plane as a controlling factor of various premonitory phenomena.

In order to explain the nature of various seismicity patterns presented in the previous section, we propose a very simple asperity model. As mentioned above, this type of asperity model has been used by various investigators in the past; the main emphasis here is to parameterize the model and relate it to the variation of observed seismicity patterns.

Figure 15 illustrates the model. The rectangular box represents all or part of the rupture surface of a large earthquake, and hereafter is called a unit fault. The unit fault is divided into smaller subfaults. Let δ be the strength of the subfaults. In general, the strength of the fault surface is not uniform. Here we use, for simplic-

ity, a Gaussian distribution with the average $\bar{\delta}$ and the standard deviation Σ to represent the variation of the strength (see Figure 15b). The details of the form of this distribution are unimportant for the present purpose.

An asperity is introduced as a region within this unit fault where the strength is higher than in the surrounding region (Figure 15a). Let δ_a be the strength of the subfaults located in the asperity. We assume that δ_a follows another Gaussian distribution with the average $\bar{\delta}_a$ ($\bar{\delta}_a > \bar{\delta}$) and the standard deviation Σa. Thus the overall distribution of the strength of the subfaults on the fault surface is given by a bimodal distribution as shown by Figure 15b. Although the actual fault is more complex and may be more adequately represented by a multi-modal distribution, the bimodal distribution shown by Figure 15b is introduced to isolate the effect of an asperity, and represents the simplest case.

We consider a loading stress σ_o, which varies linearly in time:

$$\sigma_o = \sigma_{oo} + \alpha t \qquad (1)$$

where t is the time and σ_{oo} and α are constants. When the stress at a grid point (i,j) exceeds the strength of the subfault there, the subfault fails and the stress there drops to 0. For simplicity, we assume that once a subfault fails, the fault surface there is decoupled (i.e., no healing takes place), and the loading stress σ_o is held uniformly by the remaining subfaults. Thus, under this assumption, the stress at the subfault at (i,j) is given by:

$$\sigma(i,j) = \sigma_o/[1-(\ell/N)] \qquad (2)$$

where N is the total number of subfaults in the

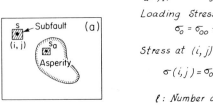

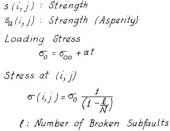

$s(i,j)$: Strength

$s_a(i,j)$: Strength (Asperity)

Loading Stress

$$\sigma_0 = \sigma_{00} + \alpha t$$

Stress at (i,j)

$$\sigma(i,j) = \sigma_0 \, \frac{1}{\left(1 - \frac{\ell}{N}\right)}$$

ℓ : Number of Broken Subfaults

N : Total Number of Subfaults

Modified:

$$\tilde{\sigma}(i,j) = \sigma(i,j) \, \frac{1}{1 + C\left(\frac{\ell/N}{1 - \ell/N}\right)}$$

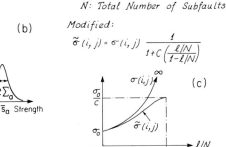

Fig. 15. Schematic figure showing a unit fault containing an asperity (a), the distribution of strength of subfaults within the asperity and the surrounding area (b), and the stress $\tilde{\sigma}$ on subfaults (c).

unit fault and ℓ is the number of broken subfaults. Thus, although the loading stress is linear in time, the rate of stress increase on unbroken faults is accelerated as the number of broken subfaults increases. This kind of accelerated instability due to failure of asperities has been used by Jones and Molnar (1979) to explain temporal variations of foreshocks. As ℓ approaches N, σ tends to infinity. However, this is physically inplausible. Since the unit fault considered here (the rectangular box in Figure 15a) is not an isolated system, the loading stress will be partially held by the fault zones adjacent to the unit fault, as a large part of the unit fault is broken. Therefore the stress on the subfault will approach a finite value instead of becoming infinity. In order to accommodate this situation we introduce a constant C and represent the stress on the subfault at (i,j) by:

$$\tilde{\sigma}(i,j) = \sigma(i,j)/[1 + C \left(\frac{\ell/N}{1 - \ell/N}\right)] \qquad (3)$$

which is shown in Figure 15c together with $\sigma(i,j)$. This modification is made in order to accommodate a physically more reasonable situation, but it does not affect our conclusion qualitatively.

According to this asperity model, the failure sequence on the unit fault would be schematically shown by Figure 16. When the tectonic loading stress is relatively low $\tilde{\sigma}(i,j)$ is substantially lower than $\bar{s}$ so that a relatively small number of subfaults break as a scattered background activity, as shown by stage 1 in Figure 16. As $\tilde{\sigma}$ approaches $\bar{s}$, a large number of subfaults fail and the process is accelerated resulting in a

swarm-like activity, and $\tilde{\sigma}$ exceeds $\bar{s}$ in a relatively short time. This stage (stage 2, Figure 16) corresponds to a precursory swarm. When this stage is passed, most subfaults outside the asperity are broken so that few subfaults fail as the tectonic stress increases, resulting in seismic quiescence (stage 3, Figure 16). At this stage, stress is concentrated on the asperity and the area surrounding the asperity is essentially decoupled. This situation may result in loading of the fault zones adjacent to the unit fault, and a doughnut pattern may develop. Finally, when $\tilde{\sigma}$ approaches $\bar{s}_a$, the asperity begins to break in an accelerated sequence in the form of a foreshocks-main shock sequence (stage 4, Figure 16). When the entire asperity breaks, sympathetic slip occurs in the area surrounding it, causing coseismic overall fault movement. If the effect of the sudden stress drop is large enough to trigger the adjacent fault zones, the event will develop into a more complex multiple event. After this entire sequence is completed, a new episode of stress loading begins, producing another earthquake cycle.

In order to see the above sequence in more detail, we made a simple numerical simulation. As shown by Figure 17b, a square unit fault is divided into 100 subfaults and nine of them near the center are designated as an asperity. A Gaussian random number table is used to assign variable strengths to the subfaults. By choosing the parameters as shown in Figure 17b and 17c, an earthquake sequence shown by Figure 17c can be generated. This sequence may be compared to one of the representative seismicity patterns, such as the one for the 1963 Kurile Islands earthquake shown in Figure 17a. The pattern shown in Figure 17c has a precursory swarm, seismic quiescence and foreshocks. The difference in time scale between Figure 17a and Figure 17c could be adjusted very easily by changing the model parameters and is not important. Among the most important parameters in this model are Σ_a, $\bar{s}_a/\bar{s}$ and Σ which control the foreshock activity, quiescence and precursory swarm respectively. When Σ_a is reduced, the asperity tends to fail in a single event without foreshocks as shown by Figure 17d. When the difference in the strength of the

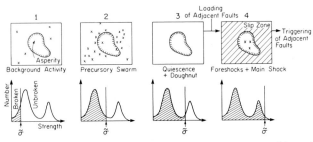

Fig. 16. Sequence of seismicity pattern predicted by the asperity model shown in Fig. 15. $\tilde{\sigma}$ is the stress on subfaults.

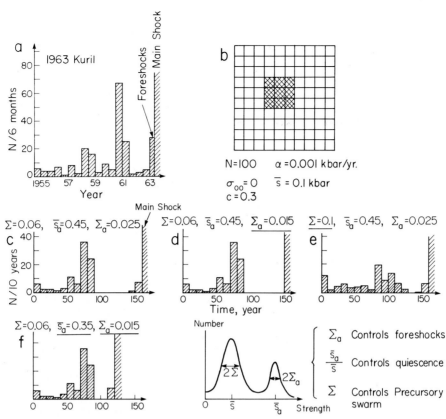

Fig. 17. Comparison of temporal variation of number of events between the 1963 Kurile Is. sequence and the asperity models.

asperity and the surrounding area is small (i.e., small $\bar{s}_a/\bar{s}$), the duration of the quiet period decreases (Figure 17f). This situation is similar to that discussed by Tsumura (1979). A large Σ results in a spread-out precursory swarm activity as shown in Figure 17e.

In the model presented here, we did not consider any physical failure criterion, healing mechanism, anelastic time dependent mechanism or dynamic response of the medium. The primary purpose of this model is to provide the simplest possible model with which the variation and complexity of observed seismicity patterns can be reproduced. At present, our knowledge of the nature, distribution and regional variation of asperities on the fault is too limited to fully test this model. However, if this model proves useful for interpreting seismicity patterns, more physical and dynamic models such as the one developed by Mikumo and Miyatake (1979) need to be introduced to study further details of seismicity patterns.

Discussion

Although it is not possible at present to test this model directly, it is desirable to investigate whether this model is reasonable in the light of available seismological data other than seismicity patterns.

Among the important consequences of this model in terms of observable seismological parameters are the clustering of events and increasing (in time) stress drops as schematically shown in Figure 18. The overall loading stress increases linearly with time. When a weaker subfault breaks, a stepwise increase in $\check{\sigma}$ occurs on other subfaults. Thus the stress drop of small earthquakes would increase as a function of time. Since the in situ condition is far more complex than is modeled here, one would expect a considerable variation in stress drops at a given time. Nevertheless, if the asperity model is correct, the stress drop should increase, on the average, as the final failure of the asperity approaches.

Since foreshocks and preshocks occur as a result of local stress concentration within the asperity, they would be tightly clustered in space and probably have the same mechanism. As a result, they would have approximately the same wave form, if the magnitude of the events is about the same. The mechanism of the preshocks may be different from the background mechanism.

Hamaguchi and Hasegawa (1975) studied a large number of aftershocks of the 1968 Tokachi-Oki earthquake having similar wave forms, and concluded that these similar events occurred at approximately the same location under the same mechanical condition. Geller and Mueller (1980)

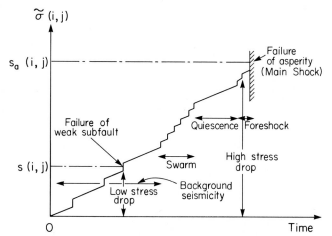

Fig. 18. Temporal variation of stress $\tilde{\sigma}$ at a subfault. A step-wise increase is caused by failure of another weaker subfault.

found four small earthquakes with similar wave forms on the San Andreas Fault in Central California, and suggested that they may represent breakings of an asperity where stress is repeatedly concentrated and released.

Waveform

There are not many high-quality data sets of wave forms for foreshocks or preshocks. Ishida and Kanamori (1978) found that the wave forms of all of the five events which occurred in the proximity of the epicenter of the 1971 San Fernando earthquake during about 2 years before the main shock were nearly identical to each other. Ishida and Kanamori interpreted this clustering and similarity of the wave forms in terms of stress concentration around an asperity whose failure led to the San Fernando earthquake.

For the 1974 Haicheng earthquake, Jones et al. (1980) found that most foreshocks can be classified into two groups, each having approximately the same wave form. For the 1979 Imperial Valley, California earthquake, the main shock was preceded by three foreshocks whose wave forms are very similar to each other, as shown in Figure 19 (James Pechmann, Personal communication, 1980).

These examples strongly indicate that a very tight clustering of activity which occurred more or less under the same stress preceded the main shock in the close proximity of the main shock rupture zone.

However, Tsujiura (1979) found that, while the wave forms of small events which preceded several earthquake swarms were very similar, those which occurred before several distinct main shock aftershock sequences varied considerably. Although the distinction between swarms and main shock aftershock sequences is not very obvious and whether the wave form is similar or not depends on the frequency band used, Tsujiura's observations

suggest that the stress distribution near the epicenter of an impending earthquake can be very heterogeneous.

Spectrum and Stress Drops

Whether foreshocks and preshocks are higher stress drop events than the earlier events is as yet unresolved. Although several examples indicate that foreshocks had higher frequency content than other events, the quality of the data was somewhat limited.

Ishida and Kanamori (1980) analyzed the wave forms of small events which occurred near the epicentral area of the 1971 San Fernando and the 1952 Kern County earthquakes and found that foreshocks had, on the average, more high-frequency energy than the earlier events. Although these results are obtained with one-station data, they are based on very uniform broad-band data (Wood-Anderson seismograms) collected over a very long period of time, 10 and 18 years for the San Fernando and the Kern County earthquakes respectively. Since the characteristic time scale of earthquake loading cycles is at least 10 to 100 years, it is extremely important to have a long-term data base for this kind of study.

Some studies indicated, however, that the identification of foreshocks is not very straight-forward. Tsujiura (1977) observed that some foreshocks had higher stress drops, but it was not always the case. Bakun and McEvilly (1979) examined several foreshocks and aftershocks of the 1966 Parkfield, California earthquake and concluded that the difference between the foreshocks and the aftershocks in terms of their frequency content is extremely subtle.

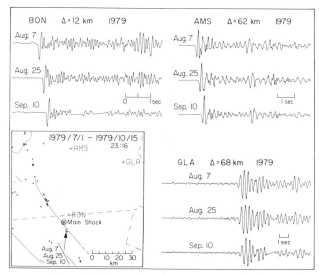

Fig. 19. Wave forms of three foreshocks of the 1979 Imperial Valley earthquake recorded at three stations.

According to the numerical experiment described above, a ratio of 4 of the stress drop of foreshocks to other events would be enough to yield the observed seismicity patterns. In view of the large error in the measurements of stress drops, particularly for small events, detection of possible temporal variation of stress drops of this magnitude would be very difficult. Nevertheless, with a better (wider dynamic range, broader frequency band with digital recording) instrumentation, it will eventually be possible to make more accurate stress drop measurements for monitoring stress variation on a fault plane. In this regard, Archambeau's (1978) approach (M_S/m_b ratio), House and Boatwright's (1978) analysis (use of local strong-motion record), and Mori's (1980) study (use of short-period WWSSN data) would have a very important potential for monitoring seismic gaps.

Conclusion

Although various seismicity patterns have been reported for many earthquakes, the nature of the patterns varies substantially from event to event. A global survey of seismicity patterns before major subduction-zone events indicates significant regional variations of the nature of seismicity patterns. It appears that the heterogeneity and the complexity of the individual fault zones are responsible for the observed variations. The fundamental physical process leading to an earthquake may be common to most events, but its manifestation as seismicity patterns may vary significantly depending upon the regional and local variations of the fault-zone structure. A very simple asperity model is presented in this paper to reproduce this situation. The basic physical process in this model is gradual stress concentration at an asperity on the fault zone. This stress concentration followed by failure of the asperity manifests itself as a variety of seismicity patterns depending upon the strength and heterogeneity of both the asperity and the surrounding area. Although it is not presently possible to test this model directly, it serves as a useful working model for a better understanding of earthquake precursors.

Although seismicity patterns provide important information on the earthquake preparatory process, its usefulness for prediction purposes is somewhat limited because of the substantial variations from event to event. However, the asperity model would suggest use of other seismological data such as wave forms, spectra and mechanism of preshocks for monitoring the state of stress on the fault plane, the key information for earthquake prediction.

Acknowledgments. I thank Carl Johnson and Bernard Minster for teaching me how to plot earthquakes by computer. I also thank Larry Ruff, Thorne Lay, Don Anderson and James Pechmann for many helpful comments on the manuscript. I benefited from comments by anonymous reviewers concerning several problems arising from non-uniformity of the catalog.

This work was partially supported by U.S. Geological Survey Contract No. 14-08-0001-18371. Contribution No. 3508, Seismological Laboratory, California Institute of Technology, Pasadena, California 91125.

References

Abe, K., Mechanisms and tectonic implications of the 1966 and 1970 Peru earthquakes, Phys. Earth Planet. Interiors, 5, 367-379, 1972.

Aki, K., Characterization of barriers on an earthquake fault, J. Geophys. Res., 84, 6140-6148, 1979.

Archambeau, C., Estimation of non-hydrostatic stress in the earth, by seismic methods: Lithospheric stress levels along Pacific and Nazca plate subduction zones, Proc. of Conference VI; Methodology for identifying seismic gap and soon-to-break gaps, U.S. Geological Survey Open-File Report 78-943, p. 47-138, 1978.

Bakun, W. H., and T. V. McEvilly, Are foreshocks distinctive? Evidence from the 1966 Parkfield and the 1975 Oroville, California sequence, Bull. Seismol. Soc. Am., 69, 1027-1038, 1979.

Bakun, W. H., R. M. Stewart, C. G. Bufe, and S. M. Marks, Implication of seismicity for failure of a section of the San Andreas fault, Bull. Seismol. Soc. Am., 70, 185-201, 1980.

Brady, B. T., Theory of earthquakes, Pageoph., 114, 1031-1082, 1976.

Brady, B. T., Anomalous seismicity prior to rock bursts: Implications for earthquake prediction, Pageoph., 115, 357-374, 1977.

Brune, J. N., Implications of earthquake triggering and rupture propagation for earthquake prediction based on premonitory phenomena, J. Geophys. Res., 84, 2195-2198, 1979.

Byerlee, J. D., Static and kinetic friction of granite under high stress, Int. J. Rock Mech. Min. Sci., 7, 577-582, 1970.

Das, S., and K. Aki, Fault planes with barriers: A versatile earthquake model, J. Geophys. Res., 82, 5658-5670, 1977.

Ebel, J. E., Source processes of the 1965 New Hebrides Islands earthquakes inferred from teleseismic waveforms, Geophys. J., in press, 1980.

Engdahl, E. R., and C. Kisslinger, Seismological precursors to a magnitude 5 earthquake in the central Aleutian Islands, J. Phys. Earth, 25, 5243-5250, 1977.

Evison, F. F., Fluctuations of seismicity before major earthquakes, Nature, 266, 710-712, 1977a.

Evison, F. F., The precursory earthquake swarm, Phys. Earth Planet. Int., 15, 19-23, 1977b.

Evison, F. F., Precursory seismic sequences in New Zealand, N.Z. Journal of Geology and Geophysics, 20, 129-141, 1977c.

Fedotov, S. A., Regularities of the distribution

of strong earthquakes in Kamchatka, the Kurile islands and northeastern Japan, Trans. Acad. Sci. USSR, Inst. Phys. Earth, 36, (203), 66-93, 1965 (in Russian).

Fuis, G. S. and A. G. Lindh, A change in fault-plane orientation between foreshocks and aftershocks of the Galway Lake earthquake, M_L = 5.2, 1975, Mojave Desert, California (abstract), Tectonophysics, 52, 601-602, 1979.

Geller, R. J., and C. S. Mueller, Four similar earthquakes in central California, Geophys. Res. Lett., 7, 821-824, 1980.

Habermann, R. E., Precursory seismicity patterns: Stalking the mature seismic gap, This volume, 1980.

Hamaguchi, H., and A. Hasegawa, Recurrent occurrence of the earthquakes with similar wave forms and its related problems, J. Seismol. Soc. Japan, 28, 153-169, 1975 (in Japanese).

House, L., and J. Boatwright, Investigation of two high stress-drop earthquakes in the Shumagin seismic gap, Alaska (abstract), EOS, 59, 1124, 1978.

Imamura, A., On the seismic activity of central Japan, Jap. J. Astron. Geophys., 6, 119-137, 1928.

Inouye, W., On the seismicity in the epicentral region and its neighborhood before the Niigata earthquake, Kenshin Jiho, 29, 31-36, 1965 (in Japanese).

Isacks, B. L., and M. Barazangi, Geometry of Benioff zones: Lateral segmentation and downward bending of the subducted lithosphere, in Talwani, M. and W. C. Pittman (eds.), Island Arcs, Deep Sea Trenches and Back-Arc Basins, Maurice Ewing Series I, 99-114, 1977.

Ishida, M., and H. Kanamori, The foreshock activity of the 1971 San Fernando earthquake, California, Bull. Seismol. Soc. Am., 68, 1265-1279, 1978.

Ishida, M., and H. Kanamori, Temporal variation of seismicity and spectrum of small earthquakes preceding the 1952 Kern County, California earthquake, Bull. Seismol. Soc. Am., 70, 509-527, 1980.

Johnson, C. and L. K. Hutton, The 15 October, 1979 Imperial Valley earthquake: A study of aftershocks and prior seismicity, to appear in the U.S. Geological Survey Professional Paper on the 1979 Imperial Valley Earthquake, 1980.

Jones, L., and P. Molnar, Frequency of foreshocks, Nature, 262, 677-679, 1976.

Jones, L. M., and P. Molnar, Some characteristics of foreshocks and their possible relationship to earthquake prediction and premonitory slip on faults, J. Geophys. Res., 84, 3596-3608, 1979.

Jones, L., Wang Biquan, and Xu Shaoxie, The Haicheng foreshock sequence, This volume, 1980.

Kanamori, H., Use of seismic radiation to infer source parameters, Proc. of Conference III; Fault mechanics and its relation to earthquake prediction, U.S. Geological Survey Open-File Report 78-380, p. 283-317, 1978.

Kanamori, H. and G. S. Stewart, Seismological

aspects of the Guatemala earthquake of February 21, 1976, J. Geophys. Res., 83, 3427-3434, 1978.

Katsumata, M., and A. Yoshida, Change in seismicity and development of the focal region, Papers in Meteorology and Geophysics, 31, 15-32, 1980.

Keilis-Borok, V. I., L. Knopoff and I. M. Rotvain, Burst of aftershocks, long-term precursors of strong earthquakes, Nature, 283, 259-263, 1980.

Kelleher, J. A., Space-time seismicity of the Alaska-Aleutian seismic zone, J. Geophys. Res., 75, 5745-5756, 1970.

Kelleher, J., and J. Savino, Distribution of seismicity before large strike slip and thrust-type earthquakes, J. Geophys. Res., 80, 260-271, 1975.

Kelleher, J., L. Sykes, and J. Oliver, Possible criteria for predicting earthquake locations and their application to major plate boundaries of the Pacific and the Caribbean, J. Geophys. Res., 78, 2547-2585, 1973.

Khattri, I., and M. Wyss, Precursory variation of seismicity rate in the Assam area, India, Geology, 6, 685-688, 1978.

Kristy, M. J., and D. W. Simpson, Seismicity changes preceding two recent central Asian earthquakes, Submitted to J. Geophys. Res., 1980.

Lay, T., and H. Kanamori, Earthquake doublets in the Solomon Islands, Phys. Earth Planet. Int., 21, 283-304, 1980a.

Lay, T., and H. Kanamori, An asperity model of great earthquake sequences, in this volume, 1980b.

McCann, W., R. S. P. Nishenko, L. R. Sykes, and J. Kraus, Seismic gaps and plate tectonics: Seismic potential for major boundaries, Pageoph., 117, 1087-1147, 1980.

McEvilly, T., W. Bakun, and K. Casady, The Parkfield, California earthquake of 1966, Bull. Seismol. Soc. Amer., 57, 1221, 1967.

McNally, K., Patterns of earthquake clustering preceding moderate earthquakes, central and southern California, abstract, EOS, 58, 1195, 1977 (full text in preprint form).

Meyer, R. P., W. D. Pennington, L. A. Powell, W. L. Unger, M. Guzman, J. Havskov, S. K. Singh, C. Valdes, and J. Yamamoto, A first report on the Petatlán Guerrero, Mexico earthquake of 14 March 1979, Geophys. Res. Lett., 7, 97-100, 1980.

Mikumo, T., and T. Miyatake, Earthquake sequences on a frictional fault model with non-uniform strengths and relaxation times, Geophys. J. R. Astron. Soc., 59, 497-522, 1979.

Mitsunami, T., and K. Kubotera, On the activity of the earthquake swarm in the northern part of Aso caldera, 1975 - Interpretation of hypocentral migration, J. Seismol. Soc. Japan, 30, 73-90, 1977 (in Japanese).

Mizoue, M., M. Nakamura, Y. Ishiketa, and N. Seto, Earthquake prediction from micro-earthquake observation in the vicinity of Wakayama city,

Northwestern part of the Kii peninsula, central Japan, J. Phys. Earth, 26, 397-416, 1978.

Mogi, K., Some discussions on aftershocks, foreshocks and earthquake swarms - The fracture of a semi-infinite body caused by an inner stress origin and its relation to the earthquake phenomena, Bull. Earthquake Res. Inst. Tokyo Univ., 41, 615-658, 1963.

Mogi, K., Some features of recent seismic activity in and near Japan, (1), Bull. Earthquake Res. Inst. Tokyo Univ., 46, 1225-1236, 1968a.

Mogi, K., Source locations of elastic shocks in the fracturing process in rocks (1), Bull. Earthquake Res. Inst. Tokyo Univ., 46, 1103-1125, 1968b.

Mogi, K., Some features of recent seismic activity in and near Japan, (2) Activity before and after great earthquakes, Bull. Earthquake Res. Inst. Tokyo Univ., 47, 395-417, 1969.

Mogi, K., Seismic activity and earthquake predictions, in Proceedings of the Symposium on Earthquake Prediction Research, 203-214, 1977 (in Japanese).

Mogi, K., Two kinds of seismic gaps, Pageoph, 117, 1-15, 1979.

Mori, J., Effective stress drops of moderate earthquakes in the eastern Aleutians (abstract), EOS, 61, 294, 1980.

Nagamune, T., Source regions of great earthquakes, Geophys. Mag., 35, 333-399, 1971.

Nagamune, T., Tectonic structures and multiple earthquakes, Zisin: J. Seismol. Soc. Japan, 31, 457-468, 1978 (in Japanese).

Nagumo, S., H. Kobayashi and S. Koresawa, Foreshock phenomena of the 1968 Tokachi-Oki earthquake observed by ocean-bottom seismographs off Sanriku, Bull. Earthquake Res. Inst. Tokyo Univ., 46, 1355-1368, 1968.

Nersesov, I. L., A. A. Lukk, V. S. Ponomarev, T. G. Rautian, B. G. Rulev, A. N. Semonov, and I. G. Simbireva, Possibilities of earthquake prediction, exemplified by the Garm area of the Tadzlink SSR, in Earthquake Precursors, Acad. Sciences, USSG, eds. M. A. Sadovsky, I. L. Nersesov, and L. A. Latynina, 72-99, 1973.

Ohtake, M., Search for precursors of the 1974 Izu-Hanto-Oki earthquake, Japan, Pageoph, 114, 1083-1093, 1976.

Ohtake, M., T. Matumoto, and G. V. Latham, Seismicity gap near Oaxaca, Southern Mexico as a probable precursor to a large earthquake, Pageoph, 115, 375-385, 1977.

Ohtake, M., T. Matumoto, and G. V. Latham, Patterns of seismicity preceding earthquakes in Central America, Mexico, and California, Proc. Conference Methodology for identifying seismic gaps and soon-to-break gaps, U. S. Geological Survey Open-File Report 78-943, 585-610, 1978.

Ponce, L., K. C. McNally, V. Sumin de Portilla, J. Gonzalez, A. Del Castillo, L. Gonzalez, E. Chael, and M. French, Oaxaca Mexico earthquake of 29 November 1979: A preliminary report

on spatio-temporal pattern of preceding seismic activity and main shock relocation, Geofísica Internacional, 17, 109-126, 1977-1978.

Santô, T., Shock sequences of the southern Kurile Islands from October 9 to December 31, 1963, Bull. Int. Inst. Seismol. Earthquake Eng., 1, 33, 1964.

Scholz, C. H. and J. T. Engelder, The role of asperity indention and ploughing in rock friction, I. Asperity creep and stick slip, Int. J. Rock Mech. Min. Sci., 13, 149-154, 1976.

Sekiya, H., Anomalous seismic activity and earthquake prediction, J. Phys.Earth, 25, 585-593, 1977.

Sykes, L. R., Aftershock zones of great earthquakes, seismicity gaps, and earthquake prediction for Alaska and the Aleutians, J. Geophys. Res., 76, 8021-8041, 1971.

Tanaka, K., T. Ohuchi and T. Santô, Formation patterns of seismic gaps before and after great shallow earthquakes, Program and abstracts, Seismol. Soc. Jap. Meeting, No. 1, 24, 1979 (in Japanese).

Tsujiura, M., Spectral features of foreshocks, Bull. Earthquake Res. Inst. Tokyo Univ., 52 357-371, 1977.

Tsujiura, M., The difference between foreshocks and earthquake swarms, as inferred from the similarity of seismic waveform, Bull. Earthquake Res. Inst. Tokyo Univ., 54, 309-315, 1979 (in Japanese).

Tsumura, K., A model for temporal variation of seismicity and precursors, abstract of the paper presented at the Spring meeting of the Seismological Society of Japan, Tokyo, p. 159, 1979.

Utsu, T., Seismic activity and seismic observation in Hokkaido in recent years, Rep. Coord. Committee Earthquake Pred., 2, 1-2, 1970 (in Japanese).

Wesson, R. L. and W. L. Ellsworth, Seismicity preceding moderate earthquakes in California, J. Geophys. Res., 78, 8527-8546, 1973.

Wu, Kai-tong, Ming-sheng Yue, Huan-ying Wu, Shin-ling Chao, Hai-tong Chen, Wei-Qong Huang, Kang-yuan Tien and Shou-de Lu, Certain characteristics of the Haicheng earthquake (M = 7.3) sequence, Chinese Geophysics, 1, 289-308, 1978.

Wyss, M. and J. Brune, The Alaska earthquake of 28 March 1964: A complex multiple rupture, Bull. Seismol. Soc. Am., 57, 1017-1023, 1967.

Wyss, M. and R. E. Habermann, Seismic quiescence precursory to a past and a future Kurile Island earthquakes, Pageoph, 117, 1195-1211, 1979.

Wyss, M., R. E. Habermann, and A. C. Johnston, Long term precursory seismicity fluctuations, in Methodology for Identifying Seismic Gaps and Soon-to-Break Gaps, Conference VI Nat. Eq. Haz. Red. Prog., 869-894, 1978.

Yamashina, K., and Y. Inoue, A doughnut-shaped pattern of seismic activity preceding the Shimane earthquake of 1978, Nature, 278, 48-50, 1979.

SEISMIC POTENTIAL FOR THE WORLD'S MAJOR PLATE BOUNDARIES: 1981

S.P. Nishenko[1,2] and W.R. McCann[1]

1. Lamont-Doherty Geological Observatory of Columbia University
Palisades, New York 10964
2. Also, Department of Geological Sciences of Columbia University

Abstract. Specific temporal and tectonic criteria are useful for estimating the potential for large interplate earthquakes to occur along segments of the world's major plate boundaries. Thirteen shocks of $M_s \geq 7.5$ have occurred in previously identified seismic gaps (i.e. regions of relatively high seismic potential) since 1968. This report updates the map of seismic potential for the Pacific, Caribbean, Indonesian and South Sandwich regions for January 1, 1981. It also introduces our estimates of the seismic potential for the New Zealand region.

Introduction

The theory of plate tectonics provides a basic framework for evaluating the potential for future great earthquakes to occur along major plate boundaries. Along most of the world's major convergent and transform plate boundaries, the majority of seismic slip occurs during large earthquakes, i.e. those of magnitude 7 or greater. This paper summarizes the potential for certain segments of major plate boundaries to rupture in large earthquakes using the technique of seismic gaps. The term seismic gap is taken to refer to any region along an active plate boundary that has not experienced a large thrust or strike slip earthquake for more than 30 years. Spatially, seismic gaps are identified using the concepts that rupture zones, as delineated by aftershocks or the region experiencing intensities of IX or more on the Modified Mercalli (M.M.) scale, tend to abut rather than overlap and that large events tend to occur in regions with histories of long and short-term seismic quiescence. Segments of plate boundaries that have not been the site of large earthquakes for tens to hundreds of years (i.e. have been seismic gaps for large shocks) are more likely to be the sites of future large shocks than segments that have experienced rupture during, say, the last 30 years.

In detail, however, the distribution of large shallow earthquakes along convergent plate margins is not always consistent with a simple model derived from plate tectonics. Certain plate boun-daries, for example, appear in the long-term to be nearly aseismic with respect to large earthquakes. The identification of specific tectonic regimes, as defined by the dip of the inclined seismic zone [Kelleher et al., 1974], the presence or absence of aseismic ridges and seamounts on the downgoing lithospheric plate [Kelleher and McCann, 1976], the age contrast between the overthrust and underthrust plates [McCann et al., 1978] and the presence or absence of backarc spreading [Uyeda and Kanamori, 1979] have led to a refinement in the application of plate tectonic theory to the evaluation of seismic potential.

Using these observations and earlier criteria as a guide, McCann et al. [1978, 1979] developed a set of six categories to reflect their assessment of 1) the relative seismic potential of the region, 2) the completeness of the seismic history, and 3) the understanding of the tectonic regime. The term seismic potential is qualitatively defined as the likelihood of a region to have a large earthquake within a specified period of time. A region of high seismic potential is a seismic gap, that for historic or tectonic reasons is considered likely to produce a large shock during the next few decades. The areas thought to be of the highest potential are assigned to category 1, those of lower or less certain potential are assigned higher numbers up to 6. Each segment of the plate boundaries studied is assigned to one of the six following categories:

1. The region (portion of a plate boundary) has experienced at least one large shock in the historic past with the most recent event occurring more than 100 years ago, i.e., prior to 1881. This category represents the highest seismic potential.
2. The region has experienced at least one large shock in the past with the most recent event occurring more than 30 years ago, but less than 100 years ago, i.e., between 1881 and 1951.
3. The region has an incomplete history of large earthquakes. No historic event is clearly documented as having ruptured the plate boundary. There is no evidence, how-

ever, that would indicate that the region may not be the site of a future large earthquake. A comparison of the tectonic framework with that of other areas known to be sites of historic large shocks may also suggest that the region is capable of being the site of a future large shock.

4. Motion between the plates is parallel or nearly parallel to the local strike of the subduction zone (trench). This category applies to the Puerto Rico-Virgin Island region, the Commander Islands in the westernmost Aleutians, and the Andaman-Nicobar region in the Indian Ocean. All appear to have a similar tectonic setting. A resolution of the question of seismic potential for one area may be useful in assessing the potential for the other two.

5. The region does not have a history of great earthquakes. Several tectonic hypotheses, which are proposed by various investigators, suggest that these regions will not be the sites of great shocks in the future.

6. The region has been ruptured by a large earthquake during the last 30 years (since 1 January 1951). This category is considered to represent the lowest seismic potential for the next few decades.

The identification of an area as a seismic gap provides estimates or forecasts for 1) the location and magnitude of future events, and 2) the origin time to a few tens of years. The term prediction is reserved for estimates that involve a more precise calculation of the time and probability of occurrence as well as the magnitude and location.

Seismic Potential, 1981

Figure 1 displays our assessment of the relative seismic potential for the major plate boundaries of the Pacific, Caribbean, Indonesian and South Sandwich areas. This map updates those of McCann et al. [1978, 1979] to reflect the occurrence of more recent gap-filling earthquakes and category changes associated with the temporal criteria. Recent studies of the seismic potential of various regions, including the New Zealand (this study) and the Alaskan (this volume) seismic zones have also been incorporated. It is important to remember that the forecasts are made subject to several assumptions and limitations which are as follows:

1. only shallow earthquakes of magnitude 7 or greater are considered or forecast;

2. some future events of magnitude near 7 may fail to be forecast because the error in mapping aftershock zones and rupture zones is comparable to the size of the rupture zone of an event near magnitude 7;

3. only simple plate boundaries of the thrust (convergent) or transform type are considered;

4. events near a major plate boundary as in central Alaska or along the grabens and volcanic zone of Central America are excluded;

5. shocks involving normal faulting of the down-bent lithosphere along the outer walls of trenches or in the bottoms of deep-sea trenches are excluded;

6. zones of multibranched deformation (i.e. 2 or more major subparallel faults) are not considered;

7. large thrust earthquakes are assumed to rupture the plate boundary from about 0 to 40 km in depth, large strike-slip earthquakes are assumed to rupture from 0 to 15 km in depth;

8. a second large event cannot occur along the same plate interface for many decades after a large earthquake until the stress is slowly built up again by plate movements;

9. the rupture zone is accurately reflected by the extent of the aftershock zone, by the extent of the area of intense shaking or damage, or by the area of tsunami generation. One case is known (Nankaido, Japan, 1944) in which the rupture zone is larger than the area inferred from aftershocks;

10. the forecasts are valid only for the next few tens of years;

11. the designation of high seismic potential is not a prediction in the sense that a precise origin time of a future shock is estimated.

Here it is appropriate for a review of how the seismic potential has changed for certain regions since the map was first published in 1978. Table 1 lists all earthquakes of $M_s \geq 7.5$ that occurred along segments of the world's major plate boundaries after the seismic potential for the region was estimated. Those shocks from 1968 to 1974 (seven) occurred in "seismic gaps" as identified by various investigators. Those events since 1978 (six) have occurred in regions assigned to various levels of seismic potential by McCann et al. [1978]. Shocks in Mexico (two), the Solomon Islands and Colombia occurred in regions placed in category 2 (last shock between 30 and 100 years ago). The event in the Santa Cruz islands ruptured a portion of a gap in category 3. The shock in Southern Alaska ruptured a small portion of a gap placed in category 3. More recent studies of this region, however, indicate that the ruptured portion and the remaining gap were in category 2 [McCann et al., 1980; Perez and Jacob, 1980].

No shocks have occurred in those segments designated category 1. This is not too surprising as these regions produce very great ($M_w \sim 9$) shocks with long repeat times and are few in number when compared with regions in category 2. Similarly, no large earthquakes ($M \geq 7.5$) have occurred in those segments designated categories 5 or 6. The great Sumba earthquake of 1977 is not included in Table 1 as it was a normal faulting event seaward of the Sunda trench [Newcomb and McCann, in preparation].

Studies of large shocks in the Mexican seismic zone have vastly improved our knowledge of the

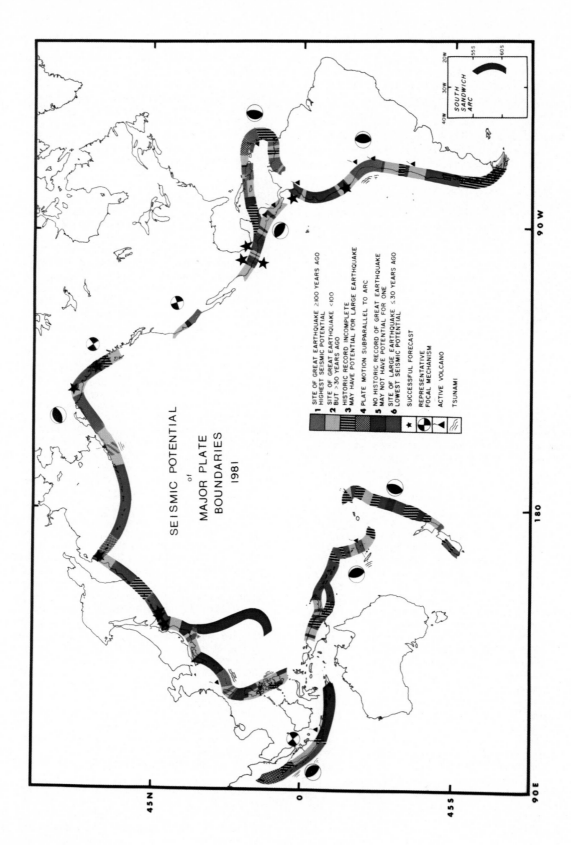

SEISMIC POTENTIAL
of
MAJOR PLATE
BOUNDARIES
1981

1 SITE OF GREAT EARTHQUAKE ≥100 YEARS AGO
HIGHEST SEISMIC POTENTIAL
2 SITE OF GREAT EARTHQUAKE <100
BUT >30 YEARS AGO
3 HISTORIC RECORD INCOMPLETE
MAY HAVE POTENTIAL FOR LARGE EARTHQUAKE
4 PLATE MOTION SUBPARALLEL TO ARC
5 NO HISTORIC RECORD OF GREAT EARTHQUAKE
MAY NOT HAVE POTENTIAL FOR ONE
6 SITE OF LARGE EARTHQUAKE ≤30 YEARS AGO
LOWEST SEISMIC POTENTIAL

SUCCESSFUL FORECAST

REPRESENTATIVE
FOCAL MECHANISM

ACTIVE VOLCANO

TSUNAMI

SOUTH
SANDWICH
ARC

40W   30W   20W
55 S
60 S

90 W

180

45 N

0

45 S

90 S

Fig. 1. Seismic potential for events M > 7 for the next few decades along certain major plate boundaries. The six categories presented are based on historic and tectonic criteria. Colored areas are those portions of plate boundaries about which we have the most data, historic and/or tectonic, and hence the most confidence in our evaluation. Red areas (category 1) have not ruptured in a great earthquake in over 100 years, and are considered likely candidates for major or great shocks within the next few decades. Similarly, green areas (category 6) have ruptured in a large earthquake within the last 30 years and are considered to have the lowest seismic potential for the present time. Other regions (categories 3 and 5, black and yellow and blue respectively) are areas where more information is needed to more accurately assess their seismic potential. Areas that are crossed hatched (category 4) are characterized by plate motion subparallel to the arc, and appear to be tectonically similar. The seismic potentials presented on this map are meant as general forecasts, not specific predictions of the time of occurrence. Stars are placed on those segments of plate boundaries that experienced large shocks after the region was cited in the literature as a major seismic gap. Typical focal mechanisms (lower hemisphere projection, compressional quadrant shaded) indicate the probable sense of motion during the next large earthquake on a particular plate boundary. Triangles are shown in gaps where active volcanoes occur in conjunction with the subduction process. Symbols indicating a tsunami (wavy lines) are those regions in category 1 that have experienced an historic, destructive tsunami.

seismic history of that region [e.g., Singh et al., 1981]. These results and the two recent shocks have left five seismic gaps in the Mexican region. The Michoacan and Tehuantepec gaps may not have the potential for producing large shocks as none are known in the history for those regions and tectonic factors (i.e. subduction of an aseismic ridge) also imply relative quiescence for large shocks [Kelleher and McCann 1976; Singh et al., 1980]. McNally et al. [1980] show evidence of uplift (i.e., coastal terraces) along the Michoacan gap that is in accordance with the model of modified subduction proposed by Kelleher and McCann [1976]. The uplift of these terraces could indicate the occurrence of large earthquakes with repeat times of hundreds of years.

New Zealand

The following discussion attempts to identify prominent seismic gaps in the New Zealand seismic zone, using the temporal and tectonic criteria outlined in the introduction. As indicated in the previous sections and Figure 1, we are examining the occurrence of large and great earthquakes on a global scale, and are primarily cataloging the time elapsed since the last major earthquake at any plate boundary. Hence, many of the fine details of seismicity and tectonic structure have been purposely excluded from our discussion. More thorough discussions of New Zealand tectonics and seismicity can be found in the following studies: Clark et al. [1965], Eiby [1968a,b, 1971, 1975], Hamilton and Gale [1968], Hatherton [1970], Evison [1971], Scholz et al. [1973], Caldwell and Frohlich [1975], Robinson and Arabasz [1975], Arabasz and Robinson [1976], Rynn and Scholz [1978], Walcott [1978a, b] and Reyners [1980].

The New Zealand seismic zone, extending through both North and South Island, marks a broad zone of contact between the Indian and Pacific plates. The relative motion between the two plates in this region ranges from 4 to 5 cm/yr [Minster and Jordan, 1978]. The Pacific plate subducts beneath the North Island at the Hikurangi Trench, which marks the southern continuation of the Tonga-Kermedec trench system before it shoals and abuts against the Chatham Rise. On land, the Taupo rift zone, an area of extensive andesitic volcanism, appears to be a zone of back arc rifting which continues north into the Havre Trough [Karig, 1970]. The Indian plate subducts beneath the Fiordland region of South Island. This zone of subduction continues south to the Puysegur Trench and north along Fiordland until its termination at the Lord Howe Rise (see Figure 2).

The tectonic evolution of New Zealand is complicated by its close proximity to the Indian-Pacific pole of rotation. Prior to the Miocene, the necessary transform motion between the Hikurangi and Fiordland Trenches is thought to have been taken up by the Alpine fault. In post-Miocene times, motion on the Alpine fault has changed from transform to oblique convergence as a consequence

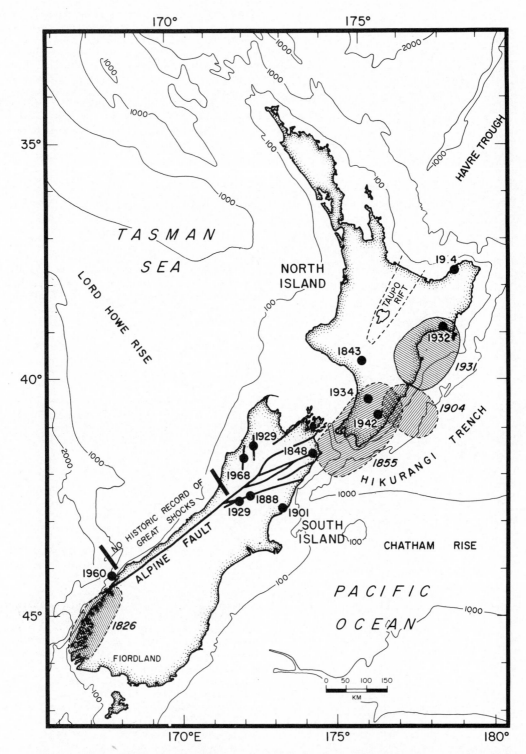

Fig. 2. Large earthquakes and rupture zones along the New Zealand seismic zone.
Epicenters are from Eiby [1968a, b]; rupture zones (dashed where inferred) are
based on aftershocks and intensity reports [Richter, 1958; Hogben, 1904; Adams et
al., 1933; Henderson, 1933; Ongley, 1943; Eiby, 1968a, b].   Submarine topography
is from Mammerickx et al. [1974].   Contours are in fathoms (1 fathom =
1.8 meters).

of a shift in the location of the Indian-Pacific pole of rotation [Scholz et al., 1973]. The initiation of convergence along the Alpine fault resulted in the rapid uplift of the Southern Alps since the Pliocene [Suggate, 1963; Adams, 1980a]. Transform motion between the two trenches is presently being accommodated by a wide zone of deformation that includes the Marlborough-East Nelson fault system in northern South Island as well as the Alpine fault [Scholz et al., 1973; Arabasz and Robinson, 1976; Walcott, 1978b].

The zone of highest earthquake risk for New Zealand [Clark et al., 1965] encompasses the entire broad zone of deformation between the Indian and Pacific plates, extending from the Hikurangi margin and Taupo region in North Island, through the Marlborough-West Nelson area to Fiordland, South Island. Within this broad zone, however, there are a number of separate tectonic elements (e.g., Hikurangi and Fiordland subduction zones) that have been the sites of large historic earthquakes. We are applying the temporal and tectonic criteria of seismic gaps only to the basic or fundamental tectonic structures within the New Zealand seismic zone.

A complete seismic history for large shocks in New Zealand extends back to the mid 1800's for North Island, only until the early 1900's for South Island [Eiby, 1968a]. Prior to these times the low population density of New Zealand results in an incomplete catalog of earthquake activity. Adequate seismograph coverage of New Zealand was not realized until the 1940's, after most of the great earthquakes discussed in this paper had occurred. As a result, the rupture zones shown by dashed lines in Figure 2 are based on published isoseismals, felt reports and descriptions of coseismic changes of topography [Eiby, 1968a,b; Hogben, 1904; Richter, 1958; Henderson, 1933; Adams et al., 1933]. Kelleher [1972] found that the area enclosed by the Modified Mercalli (MM) VIII-IX intensity contour is a good estimate of the rupture zone for many shallow, interplate events. We follow this convention for the rupture zones shown in Figure 2, i.e. the dashed lines represent the MM IX contour, when known. The repeat times for great shocks along the New Zealand seismic zone appears to be greater than 100-150 years, as no great shocks are known to have occurred in the same area more than once. The rupture zones for events shown in Figure 2 are typically 100-200 km long. Only the 1931 rupture zone is known with any certainty. The rupture zones of the 1826, 1855 and 1904 events are poorly known. Nevertheless, these rupture zones appear to be similar in size to the ruptures along other southwest Pacific margins [McCann, 1981].

A major seismic gap is situated near Wellington and the northern end of South Island, near where the Hikurangi Trench shoals and eventually terminates. The last great earthquake to occur in this region was in 1855 [M ∿8; Eiby, 1968a], and was accompanied by vertical offsets of 1 to 3 meters for ∿150 km along strike of the Wairarapa fault

[Ongley, 1943]. Extensive horizontal displacements also occur along strike of the Wairarapa fault. Lensen and Vella [1971] calculate a Holocene slip-rate of 3.4 to 6.0 mm/yr, by correlating river terraces that have been dextrally offset by the Wairarapa fault, and estimate the recurrence interval for a shock the size of the 1855 event to be 500 to 900 years. Eiby [1968a] mentions legend of a large event occurring near Wellington area about 1460, which was also accompanied by vertical offset. This ∿400 year repeat time is in reasonable agreement with the reoccurrence times estimated by Lensen and Vella [1971].

Large events in 1934 and 1942 may have reruptured the northern portion of the 1855 gap (see Figure 2). The 1904 event is located seaward of the 1855 and 1931 events, in the vicinity of another large shock in 1863 [Hogben, 1904]. None of these earthquakes can be classified as great, hence a significant portion of the 1855 gap still remains unbroken. Walcott [1978a] cites geodetic evidence that directions of principal strain along the 1855 gap changed orientation during the 1920's, which he interprets as signifying the onset of strain accumulation prior to the next great shock. Focal mechanisms in this region also indicate compressive axes normal to the arc [Reyners, 1980]. Geodetic observations suggest repeat times of 100-200 years for earthquakes like the 1931 Hawkes Bay and 1855 events [Walcott, 1978a]. Both estimates of recurrence times (100-200 years and 500-900 years) vary considerably and reflect the choice of location for the 1855 event (i.e., as a reverse or strike-slip fault, respectively). Low population densities in this region between the 1400 and 1800's suggest that other large events in the area have been missed, i.e., the repeat times for great shocks may be shorter than ∿400 years. The historic record alone cannot resolve this discrepancy in repeat times for the Wellington region. In terms of the criteria adopted for recognizing seismic gaps, however, it is clear that the Wellington region does have the potential for a future great earthquake. We assign the southern segment of the 1855 gap to category 1 and the northern segment to category 2 (see Figure 3).

The 1931 Hawkes Bay earthquake (M ∿8; Richter, 1958] broke the segment of plate boundary north of the 1855 event, and was well documented [Adams et al., 1933]. Major zones of uplift, subsidence and secondary faulting [Henderson, 1933; Walcott, 1978a] are consistent with an underthrusting model and are contained within the instrumentally located aftershock zone [Adams et al., 1933] (see Figure 2). The directions of principal strain following the 1931 event have changed orientation, from compressional to extensional strain normal to the plate boundary, implying the underthrust zone in this region is presently unlocked [Walcott, 1978a]. This unlocked state is also manifest by low levels of microseismicity in the overlying plate [Reyners, 1980]. The southern boundary of the 1931 rupture zone abuts against part of the 1904 zone, and appears to be coincident with a major discontinuity in the underthrust plate [Reyners, 1978]. This discontinuity is also within the transition region for the directions of principal strain associated with the 1855 and 1931 zones.

Extensional strain, normal to the plate boundary, is also observed north of the 1931 zone [Walcott, 1978a], however, there is no historic record of great shocks in the region. The 1914 East Cape event (M = 6.5 [Richter, 1958]) may have ruptured a portion of the plate boundary, however, data are sparse (see Figure 2). This segment of the Hikurangi margin lies adjacent to that part of the Kermedec trench which has been aseismic for large and moderate size shocks during the last several decades [McCann et al., 1979; McCann, 1981]. Back arc spreading in the Havre trough appears to have modified the subduction process for great events along the southern Kermedec margin, similarly, incipient rifting in the Taupo Basin may be modifying strain release associated with subduction in North Island. The transition zone between areas capable of great shocks and those areas that move aseismically is poorly defined, and we assign this segment to category 3.

The northern end of South Island is a region of complex tectonics. At shallow crustal levels, the Marlborough-East Nelson faults (Hope, Clarence, Awatere and Wairau faults) appear to be accommodating some of the necessary transform motion between the Hikurangi and Fiordland subduction zones. Focal mechanisms of microearthquake activity in the area exhibit right-lateral, strike-slip motion [Scholz et al., 1973; Rynn and Scholz, 1978; Arabasz and Robinson, 1976] as are ground offsets associated with great shocks (e.g., 1888 Amuri event). At deeper levels, the lower limit of the Benioff zone associated with the Hikurangi trench shoals and terminates beneath the Marlborough region. The shoaling of the leading edge of the subducted Pacific plate in this region is thought to represent the southern migration of the trench to meet the newly formed splay faults [Scholz et al., 1973].

The Marlborough-East Nelson area has been compared to the Tranverse Ranges of southern California [Richter, 1958, Scholz, 1977], where a master fault breaks into a series of splay faults with the frequent occurrence of large shocks. The Marlborough region does have the potential for large shocks (see Figure 2), however, because of the complex nature of deformation in the area (e.g., motion taken up by a series of splay faults rather than one master fault) we refrain from discussing the area further.

The central segment of the Alpine fault is notable for a lack of great shocks in the historic record [Evison, 1971]. Microseismicity in the area, while present, is of a diminished level compared to other portions of the plate boundary [Scholz et al., 1973]. Present day motions in the vicinity of the Alpine fault are of thrust type and define a zone of oblique continental convergence which has resulted in the rapid post-Miocene uplift of the southern Alps. The high strain rates inferred by Walcott [1978b] east of the Alpine fault suggest that the convergence is being accommodated by a sizeable component of aseismic slip. Nevertheless, while this area is historically aseismic with respect to great shocks, there is geologic evidence of pre-historic great earthquakes [Adams, 1980b]. Recurrence intervals for great shocks along this segment of the Alpine fault are about 500 years, based on ^{14}C dating of offset river terraces. The last pre-historic event is at 550 ybp, with older events at 1000 and 1500 ybp. Hence, while the area has no historic record, the geologic record provides compelling evidence to place this gap in category 1. Segments of the central Alpine fault where offsets of river terraces have not been dated are placed in category 3 (see Figure 3).

Along the Fiordland region of South Island, subduction of the Indian plate is manifested by seismic activity to a depth of 150 km [Hamilton and Evison, 1967; Smith, 1971; Scholz et al.,

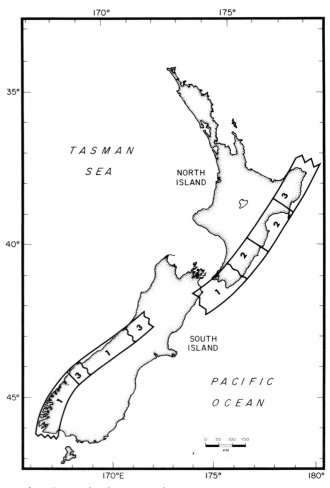

Fig. 3. Seismic potential for segments of the New Zealand seismic zone. Numbered regions refer to categories listed in Figure 1 and text.

1973]. This region was apparently the site of a great shallow earthquake in 1826 [Eiby, 1968a]. While poorly documented, the area from Milford Sound to Dusky Sound (Figure 2) was marked by regional uplift and subsidence of up to 3-4 meters. Thus, the area has the potential for great shocks and is placed in category 1 (Figure 3).

Summary

Plate tectonics has given us a fuller understanding of the mechanism by which strain energy is stored along plate boundaries and then suddenly released in large earthquakes. In this paper, major plate boundaries of the Pacific, Caribbean, Sunda and South Sandwich regions are classified into six categories of seismic potential for large earthquakes for the next few decades. These categories are meant to reflect our knowledge and assessment of the historic record, and our understanding and speculation about tectonic regimes and the length of time since the last large earthquake. These estimates of seismic potential for various segments of the world's major plate boundaries have successfully forecast the location of 13 shocks of $M_s \geq 7.5$; six of these occurring in the last few years.

We have applied these six categories of potential to portions of the New Zealand seismic zone. Three regions are delineated in category 1 (i.e., the highest seismic potential): Wellington-northern South Island, the Central Alpine fault and the Fiordland region of South Island.

Acknowledgments. We wish to thank John Adams and Dick Walcott for prodding us to write the New Zealand section. Conversations with, and advice from Martin Reyners are gratefully appreciated. The paper was critically reviewed by Tracy Johnson and Lynn Sykes. L. Murphy typed the manuscript and K. Nagao drafted the figures. This study was supported by the National Science Foundation under grants EAR 79-26350 and EAR 78-22770. Lamont-Doherty Geological Observatory Contribution Number 3117.

References

Adams, C. E., M. A. F. Barnett, and R. C. Hayes, Seismological report of the Hawke's Bay earthquake of 3rd February 1931, New Zealand J. Sci. Technol., 15, 93-107, 1933.

Adams, J., Contemporary uplift and erosion of the southern Alps, New Zealand, Geol. Soc. Amer. Bull., 91, Part I pp. 2-4, Part II pp. 1-114, 1980a.

Adams, J., Paleoseismicity of the Alpine fault seismic gap, New Zealand, Geology, 8, 72-76, 1980b.

Arabasz, W. J., and R. Robinson, Microseismicity and geologic structure in the northern South Island, New Zealand, New Zealand J. Geol. Geophys., 19, 569-601, 1976.

Caldwell, J. G., and C. Frohlich, Microearthquake study of the Alpine fault zone near Haast, South Island, New Zealand, Bull. Seismol. Soc. Amer., 65, 1097-1104, 1975.

Clark, R. H., R. R. Dibble, H.E. Fyfe, G. J. Lensen, and R. P. Suggate, Tectonic and earthquake zoning, Trans. Roy. Soc. New Zealand, General 1, 10, 113-126, 1965.

Eiby, G. A., A descriptive catalogue of New Zealand earthquakes, Part I - Shocks felt before the end of 1845, New Zealand J. Geol. Geophys., 11, 16-40, 1968a.

Eiby, G. A., An annotated list of New Zealand earthquakes, 1460-1905, New Zealand J. Geol. Geophys., 11, 630-647, 1968b.

Eiby, G. A., Seismic regions of New Zealand, in Recent Crustal Movements, Collins, B. W., and R. Frasier, eds., Roy. Soc. New Zealand Bull., 9, 161-165, 1971.

Eiby, G. A., Seismology in New Zealand, Geophys. Surv., 2, 55-72, 1975.

Evison, F. F., Seismicity of the Alpine fault, New Zealand, in Recent Crustal Movements, Collins B. W., and R. Fraisier, eds., Roy. Soc. New Zealand Bull., 9, 161-165, 1971.

Hamilton, R. M., and F. F. Evison, Earthquakes of intermediate depth in southwest New Zealand, N. Z. J. Geol. Geophys., 10, 1319-1329, 1967.

Hamilton, R. M., and A. W. Gale, Seismicity and structure of North Island, New Zealand, J. Geophys. Res., 73, 3859-3876, 1968.

Hatherton, T., Gravity, seismicity and tectonics of North Island, New Zealand, New Zealand J. Geol. Geophys., 13, 126-144, 1970.

Henderson, J., The geological aspects of the Hawke's Bay earthquakes, New Zealand J. Techn., 15, 38-75, 1933.

Hogben, G., Notes on the East Coast earthquake of 9th August 1904, Trans. New Zealand Inst., 37, 421-424, 1904.

Karig, D. E., Kermadec arc - New Zealand tectonic confluence, New Zealand J. Geol. Geophys., 13, 21-29, 1970.

Kelleher, J. A., Rupture zones of large South American earthquakes and some predictions, J. Geophys. Res., 77, 2087-2103, 1972.

Kelleher, J. A., and W. R. McCann, Buoyant zones, great earthquakes, and unstable boundaries of subduction, J. Geophys. Res., 81, 4885-4908, 1976.

Kelleher, J. A., J. Savino, H. Rowlett, and W. McCann, Why and where great thrust earthquakes occur along island arcs, J. Geophys. Res., 79, 4889-4899, 1974.

Lensen, G., and P. Vella, The Waiohine River faulted terrace sequence, in Recent Crustal Movements, Collins, B. W., and R. Fraiser, eds., Roy. Soc. New Zealand Bull., 9, 117-119, 1971.

Mammerickx, J., T. E. Chase, S. M. Smith, and J. L. Taylor, Bathymetry of the South Pacific charts 11-21, Scripps Inst. of Oceanogr., La Jolla, California, 1974.

McCann, W. R., Seismic potential and seismic

regimes of the Southwest Pacific, *J. Geophys. Res.*, 1981, in press.

McCann, W. R., S. P. Nishenko, L. R. Sykes, and J. Krause, Seismic gaps and plate tectonics: seismic potential for major plate boundaries, *Proc. Conf. VI: Methodology for Identifying Seismic Gaps and Soon-to-Break Gaps*, U.S. Geological Survey Open File Report 78-943, 441-584, 1978.

McCann, W. R., S. P. Nishenko, L. R. Sykes, and J. Krause, Seismic gaps and plate tectonics: seismic potential for major boundaries, *Pure Appl. Geophys.*, 117, 1082-1147, 1979.

McCann, W. R., O. J. Perez, and L. R. Sykes, Yakataga gap, Alaska: seismic history and earthquake potential, *Science*, 207, 1309-1314, 1980.

McNally, K., J. Alt, D. Helmberger, C. Lomnitz, and K. Sieh, New observations of coastal uplift patterns near the Middle America Trench, Mexico, preprint, 1980.

Minster, J. B., and T. H. Jordan, Present day plate motions, *J. Geophys. Res.*, 83, 5331-5354, 1978.

Ongley, M., The surface trace of the 1855 earthquake, *Trans. Roy. Soc. New Zealand*, 73, 84-89, 1943.

Perez, O. J., and K. H. Jacob, St. Elias, Alaska, earthquake of February 28, 1979: tectonic setting and precursory seismic pattern, *Bull. Seismol. Soc. Amer.*, 70, 1595-1606, 1980.

Reyners, M. E., A microearthquake study of the plate boundary, North Island, New Zealand, Ph.D. dissertation: Institute of Geophysics, Victoria University of Wellington, New Zealand, 1978.

Reyners, M. E., A microearthquake study of the plate boundary, North Island, New Zealand,

Geophys. J. Roy. astr. Soc., 63, 1-22, 1980.

Richter, C. F., *Elementary Seismology*, W. H. Freeman and Company, San Francisco, California, 768 pp., 1958.

Robinson, R., and W. J. Arabasz, Microearthquakes in the north-west Nelson region, New Zealand, *New Zealand J. Geol. Geophys.*, 18, 83-91, 1975.

Rynn, S. M. W., and C. H. Scholz, Seismotectonics of the Arthur's Pass region, South Island, New Zealand, *Bull. Seismol. Soc. Amer.*, 89, 1373-1388, 1978.

Scholz, C. H., Transform fault systems of California and New Zealand: similarities in their tectonic and seismic styles, *J. Geol. Soc. Lond.*, 133, 215-229, 1977.

Scholz, C. H., J. M. W. Rynn, R. W. Weed, and C. Frohlich, Detailed seismicity of the Alpine fault zone and Fiordland region, New Zealand, *Geol. Soc. Amer. Bull.*, 84, 3297-3316, 1973.

Singh, S. K., L. Astiz, and J. Havskov, Seismic gaps and recurrence periods of large earthquakes along the Mexican subduction zone: a reexamination, *Bull. Seismol. Soc. Amer.*, in press, 1981.

Smith, W. D., Earthquakes at shallow and intermediate depths in Fiordland, New Zealand, *J. Geophys. Res.*, 76, 4901-4907, 1971.

Suggate, R. P., The Alpine Fault, *Trans. Roy. Soc. New Zealand Geol.*, 2, 105-129, 1963.

Uyeda, S., and H. Kanamori, Back arc opening and mode of subduction, *J. Geophys. Res.*, 84, 1049-1062, 1979.

Walcott, R. I., Geodetic strains and large earthquakes in the axial tectonic belt, New Zealand, *J. Geophys. Res.*, 83, 4419-4429, 1978.

Walcott, R. I., Present day tectonics and Late Cenozoic evolution of New Zealand, *Geophys. J. Roy. astro. Soc.*, 52, 137-164, 1978b.

PRECURSORY SEISMICITY PATTERNS:
STALKING THE MATURE SEISMIC GAP

R.E. Habermann

Cooperative Institute for Research In Environmental Sciences
University of Colorado / NOAA Boulder, Colorado 80309

Abstract

Mature seismic gaps are defined here as gaps in which a possible precursor has been identified. The seismic gap hypothesis by itself provides an excellent means of narrowing spatial limits along plate boundaries within which the great earthquakes of the next several decades are expected. To narrow the temporal limits on these forecast events, identified gaps must be monitored for signs of maturity. Decreased seismicity rates (number of events reported/unit time) have been reported before over fifty large earthquakes. Seismicity rates measured teleseismically are tested as a possible indicator of maturity by examining data from the NOAA HDF in the regions of eleven recent large events. These events were chosen as test cases because they occurred in previously recognized seismic gaps.

Many previous seismicity studies lack quantitative evaluation of patterns or changes of patterns which are proposed as precursors. This makes many of the results unconvincing. In this work seismicity data is quantified by studying seismicity rates, the number of events per month. Changes in rates are evaluated using the normal deviate (z) test for a difference between two means. Several techniques for appling this test to seismicity data are proposed. These techniques offer various approaches to quantitatively defining seismicity rate changes and evaluating their temporal and spatial uniqueness.

Two of the eleven events examined (Sitka, 1972 and St. Elias,1978) occurred in regions for which the teleseismic seismicity was too sparse to allow study of seismicity rates. Rates preceding two events in the Hokkaido corner, South Kuriles, 1969 and Nemuro-oki 1973, could not be evaluated because of the complex seismicity of the region. Of the remaining seven events, the Tokachi-oki 1968, Colima 1973, and the Oaxaca 1978 earthquakes were preceded by temporally and spatially unique seismic quiescence. The Kamchatka 1971 event was preceded by a precursory cluster of events. The Lima 1974 and Solomon Islands 1978 events were preceded by clusters which may reflect similar phenomena but neither case was as clear as the Kamchatka 1971 case. Only the Guerrero 1979 event was not preceded by a recognizable seismicity anomaly.

The multiple asperity model for earthquake precursors predicts that two types of precursors will be observed in upcoming rupture zones. High-stress precursors (dilatancy related velocity anomalies and increased seismicity) occur in locked portions of the upcoming rupture (asperities). Seismic quiescence occurs in regions of the rupture zone which experience precursory displacement and decreased stresses. The ratio between locked and unlocked areas of the future rupture (Rl) may be important in determining what type of seismicity precursors will occur. Possible differences in Rl for different ruptures offers a speculative explanation for the occurrence of different types of precursory anomalies.

Introduction

Predicting earthquakes successfully depends on narrowing temporal and spatial limits within which we expect an event to occur. One of the most effective tools proposed to date for narrowing these limits is provided by the seismic gap hypothesis. According to this hypothesis, major events are expected along sections of plate boundaries which have a history of great earthquakes and have not ruptured during the last thirty years. The application of this concept provides a first order estimate of where the long-term (decades) seismic-potential is greatest. This is an excellent first step towards narrowing the spatial limits mentioned above, however, only a very rough estimate of the time of occurrence of the anticipated earthquake is possible. Narrowing the temporal limits for the expected event depends on monitoring recognized gaps for changes in some crustal parameter possibly related to the earthquake preparation process. In most seismic gaps, teleseismically located earthquake hypocenters are the only continuous data available which may provide a means to detect changes in processes occurring in the crust. In this paper the study of seismicity rates measured teleseismically is tested as a possible technique for evaluating how close a seismic gap may be to rupture. The first stage of this test, reported on here, involves studying seismicity patterns preceding past events which have occurred in seismic gaps. Eleven such events have occurred recently.

Decreased seismicity in the rupture zone of an upcoming earthquake has been discussed by many authors (Mogi [1968], [1969], Kelleher et al. [1973], Kelleher and Savino [1975], Engdahl and Kisslinger [1977], Ohtake et al. [1977a,b], Khattri and Wyss [1978], Wyss et al. [1978], Wyss and Habermann [1979], Wyss et al. [1981]). Most of these reports have been qualitative in nature which makes many of the results difficult to evaluate. Confidently recognizing precursory seismicity rate anomalies depends on *measuring* characteristics of seismicity relevant to these anomalies. Techniques for applying standard statistical tools to seismicity data which make quantitative evaluation of rate changes possible are proposed in this paper.

The seismic gap concept, as developed by Fedotov [1965], Mogi [1968] and Sykes [1971], has its origins in the theory of plate tectonics. When a region is recognized as a seismic gap, no statement is made about preparation for an earthquake. The concept of "maturity" is linked to the physical preparation for an earthquake. A mature seismic gap is one in which a change in some physical parameter which may signal the onset of the preparation for rupture (a possible precursor) has been recognized. At the time of such an occurrence, the gap becomes mature, and the seismic–potential of the gap becomes increased relative to immature seismic gaps.

Data

Detecting changes in seismicity rates which reflect real changes of the processes occurring in the crust and not changes in detection of events or reporting depends on a homogeneous data set. Homogeneous, in this sense, means that a constant portion of all events in any magnitude band are recorded through time. The homogeneity of the data is more crucial than the completeness because changes in rates, not absolute rates, are the object of study. The NOAA Hypocenter Data File (HDF) is the data source for this work. All events reported by NOAA since 1963 with depths less than 100 km are included. Habermann (in preparation) has shown that detection capability decreases since 1963 have severely affected the homogeneity of the NOAA data set. The detection capability decreases are apparent as substantial (>50%) decreases in the rate of reporting of smaller events ($mb \leq 4.5$) which occur at times of constant rates of reporting of larger events. In each area studied the homogeneity of the data is assessed by examining the seismicity rates for the entire area in several magnitude bands. Central

America is the only region in which evidence for inhomogeneity which affects the results presented here is found. In this case magnitude cut-offs are applied as larger events are not affected by changes in detection or reporting.

Techniques

Several tools are used for evaluating seismicity rate changes in some region. First, the cumulative number of events as a function of time is plotted for visual evaluation of seismicity rates. When an apparent quiet period is noted, it is tested for significance. Presumably a change in seismicity rate reflects a change in processes which are occurring in the crust. This hypothesized change in process makes the earthquake time series non–stationary on the time scale of interest, five to twenty years. The appropriate statistical test in this case is not a test for the probability of such a quiet period occurring in some stationary process, but a test for a change in process. To make this test we rely on the standard normal deviate (z) test for a difference between two means c.f. [Meyer, 1975, p.235]. In this test the hypothesis that two samples (seismicity rates during two periods) come from the same population (or represent the same process) is tested. Each sample is characterized by a mean, a standard deviation and a number of cases. The z–test is based on the fact that means of large samples drawn from the same population are normally distributed, a feature generally independent of the actual distribution of the parent population c.f. (Hogg and Tanis, [1977], p.155). Because of this, values of z have the same statistical interpretation as the number of standard deviations from the mean of a normal distribution, ie. 90, 95 and 99% significance levels correspond to z values of 1.64, 1.96 and 2.57.

In order to apply this test to the data, a period of background activity must be chosen for comparison with the hypothesized quiet period. The time between the beginning of the data set and the onset of the anomaly is used as the background whenever possible. In some regions, aftershock sequences or swarms make the determination of a meaningful rate difficult. If this is the case two data sets are studied. First, the entire data set and second, a set of independent events extracted from all events. The independent events are recognized using the algorithm of McNally [1977], which assumes that the earthquake time series is made up of a random (poissonian) and a dependent component. Events which occur within a given time period of one another are assumed to be dependent. A series of dependence times are tested by removing dependent events and comparing the resulting series with a stationary poissonian series. The dependence time which results in the closest fit to a poisson process is termed the cluster length for that data set. Sequential events separated by less than the cluster length are represented in the independent time series by the first event of the cluster. The cluster lengths determined for the regions studied here are generally less than five days, sometimes as long as ten days.

When a background period has been chosen the anomaly is tested by forming a function termed $AS(t)$. $AS(t)$ gives the z value resulting from comparisons of the rate between the beginning of the background period and t to the rate between t and the mainshock. A change in rate with a sharp onset results in a sharp peak in $AS(t)$ at the onset. The sign convention is positive AS =rate decrease, only positive values of $AS(t)$ are shown in this paper. The beginning of the anomaly (and therefore the anomaly duration) can be picked from this series of comparisons as the time at which the significance of the difference between the two means surpasses some level.

Several steps are involved in testing the temporal uniqueness of an anomaly of duration A. First, seismicity rates are calculated for time periods of length A before and after each month of the data. The rates for the periods before and after each month, t, are then compared using a z-test. This yields a function, $z(t)$, which gives the significance of rate changes for each month of the data (except for months within A of either end of the data set). As with $AS(t)$, positive values of $z(t)$ indicate rate decreases. If the anomaly is unique in time, the highest peak in $z(t)$ will occur at its beginning. The function $z(t)$ differs from $AS(t)$ in the period used to define the background rate. $z(t)$ carries its constant duration background estimate along with it while $AS(t)$ expands the background period as time increases. The advantage of $z(t)$ is that aftershock sequences or swarms cease to affect the background estimate after they leave the following window. This makes the application of $z(t)$ to a general data set easier as the beginning of the background period need not be adjusted as it must be with $AS(t)$.

Once a temporally unique rate decrease is established it must be evaluated for spatial extent. This is done using a z-map. Seismicity rates during the background and anomalous periods are calculated in a series of adjacent spatial slices (usually $50-150$ km wide), and compared, again using a z-test. The resulting z values are examined as a function of space, $z(x)$. The slices experiencing the most significant differences between the background and anomalous periods appear as peaks in $z(x)$. This technique assumes that the portion of the plate boundary surrounding an anomalous zone remains at the background rate throughout the anomaly.

Earthquakes in the west and north Pacific

Solomon Islands; November 5, 1978; Ms=7.25

This event ruptured 200 km of the Solomon Arc between $10-11°$S and $161-162.5°$E. The region occurred in, but did not fill, a region recognized as a seismic gap by McCann et al. [1979]. The cumu-

lative seismicity for this gap and $z(t)$ for a thirteen month window are shown in Figure 1. The thirteen month window was used to test the uniqueness of a short quiet period preceding a group of events in 1972 discussed below. The t−test is used in this case for the calculation of $z(t)$ because of the short window length. Note the noisy character of the curve, also an artifact of the short window length. This demonstrates a problem with recognizing short anomalies, the noise and the false alarm rates are likely to be high. Several peaks are apparent in this Figure. The first, at July, 1967, occurs one window length after a mainshock (Ms=7.5) and associated aft-

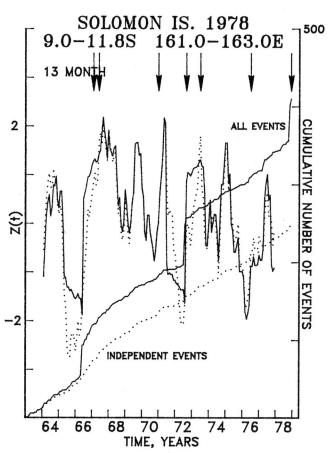

Fig. 1. Cumulative seismicity and $z(t)$ for all (solid) and independent (dotted) events in the region of the Solomon Islands gap. Note the noisy character of the $z(t)$ functions resulting from the short (13 month) window length. Two peaks in $z(t)$ are apparent, the first following an event with Ms=7.5 in June 1966 which occurred adjacent to this gap. The second peak is associated with a 13 month quiet period preceding a group of events during 1972. No rate decrease is detected in this gap before the large (Ms=7.25) event of November 5, 1978. Arrows show the occurrence times of events with Ms≥6.5.

TOKACHI-OKI 1968

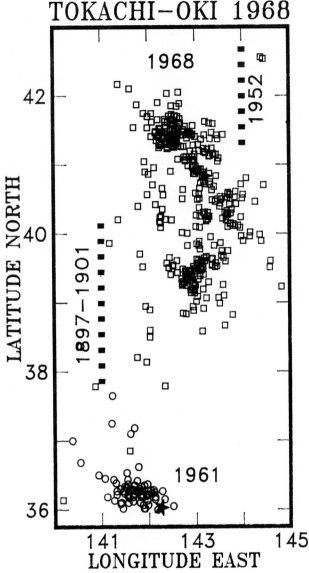

Fig. 2. Seismicity map showing the locations of the 1968 Tokachi-oki aftershocks (squares) in relation to the 1952 and 1897–1901 events. One month aftershocks of a Ms=6.9 event during 1961 are also shown (circles). The mainshocks of both sequences are shown by stars. The anomaly preceding the Tokachi-oki 1968 event was confined primarily to the region between 39–41°N, it was present to a lesser extent between 41–43°N.

group of events in August, 1972 and reflects a thirteen month quiet period which preceded these events. This series of events exhibits several unusual characteristics. It consisted of thirty-three events which occurred between August 2–6, 1972. The series included one 6.0, three 6.1's and one 6.5 (Ms). The largest of the shocks occurred at the end of the sequence and was followed by only seven aftershocks. The combined energy release of these events is equivalent to that of an earthquake with Ms=6.7, yet they covered an area of 5000 square kilometers, a factor of 14 larger than the area expected for a rupture of Ms=6.7 according to the formula of Wyss [1979].

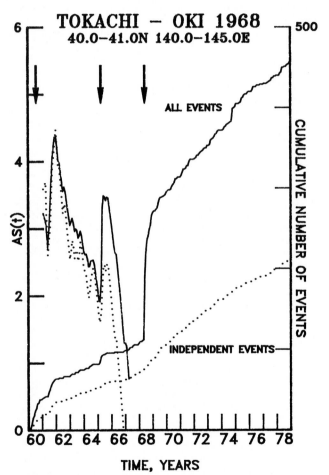

Fig. 3. Superimposed on the cumulative number of events (solid) and independent (dotted) events for the region of the gap in which the Tokachi-oki event occurred are the $AS(t)$ functions with January 1960 as the first month of the background and April 1968 as the final month. Note the strong peak in $AS(t)$ during 1961 for both all data (solid) and independent (dotted) data. The later peak, during 1965, reflects the change seen as most significant in the JMA data, Figure 4.

ershocks in June, 1966. The high rate during and shortly following this aftershock sequence is responsible for this first peak. Note that the peak remains in the independent case, this indicates that the high rate following this event was not entirely due to short term aftershocks.

The only other peak in $z(t)$ occurs before a

The occurrence of a quiescent period before this "event" suggests that the process reflected by seismic quiescence may be terminated in more than one way. At present the quality of the data for this series of events is too low to permit conclusive results. The series may be an analog to multiple ruptures observed in many great earthquakes. The time scale of this series is, of course, much longer than that of conventional multiple ruptures.

No peak in $z(t)$ exists in this gap before the GFEQ of November, 1978, which occurred in the same location as the events of 1972. Smaller regions of the gap were also studied to see if any part of the gap was quiet before this event. No precursory quiescence was recognized in these smaller regions. Nothing in our interpretation of the seismicity data suggests a connection between the 1972 and the 1978 events. The unusual characteristics of the 1972 events, their tight spatial and temporal clustering, and their proximity to the 1978 shock are, however, strikingly similar to the cluster type anomaly before the Kamchatka event discussed below. Possibly they represent similar phenomena.

Tokachi-oki, Japan; May 16, 1968; Ms=7.9, Mw=8.2

This event occurred at 40.8°N–143.2°E in the Japanese trench. The aftershock zone (May 16–July 16, 1968) extended from 39 to 42°N, a distance

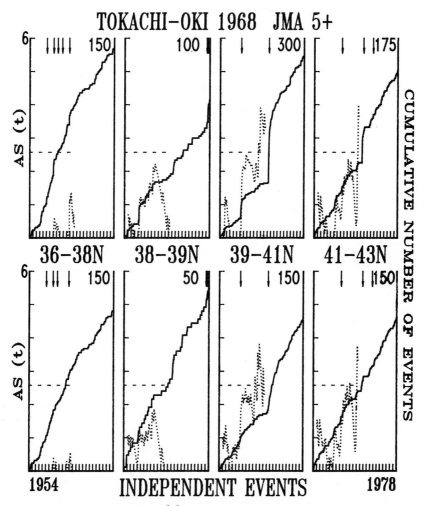

Fig. 4. Cumulative seismicity and $AS(t)$ for four east–west slices of the Japanese trench. All data is shown above and independent events below. First note the lack of high peaks in the $AS(t)$ functions in the set of all events. This is due to the increase in the variance of the rates calculated for periods including large aftershock sequences. Second, note the high peak in $AS(t)$ in the region of the mainshock (39–41°N) during 1961. This peak reflects a decrease in seismicity rate which is interpreted as a precursor to the Tokachi-oki event. The similar peak in the southernmost region is due to the aftershocks of the 2961 event shown in Figure 1.

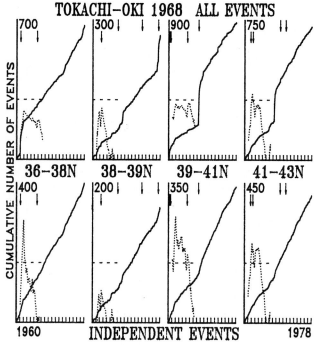

TOKACHI-OKI 1968 ALL EVENTS

700 | 300 | | | 900 | | 750 | |

36-38N   **38-39N**   **39-41N**   **41-43N**

400 | 200 | | | 350 | 450 | |

CUMULATIVE NUMBER OF EVENTS

1960   **INDEPENDENT EVENTS**   1978

Fig. 5. Cumulative seismicity for the same regions as in Figure 4 from the JMA catalog. All data is shown above and independent events below. Only events with Magnitudes above 5.0 are included as this was the level of completeness since 1954. Again peaks in the $AS(t)$ are apparent in the region of the Tokachi-oki event of 1968 and not in other regions. Both the decrease during 1961 and during 1965 show up as peaks in these functions.

of 330 km, see Figure 2. The event filled a seismic gap between the 1952 Tokachi-oki event and events of 1897-1901. This area was recognized as a gap by Fedotov [1965]. Mogi [1968, 1969] showed that the aftershock zone of this event was quiet at the $Ms \geq 7.0$ level between 1961 and the time of the mainshock and presented evidence that the focal region experienced low seismic energy release between 1961 and 1967.

The cumulative seismicity for the gap (40-41°N, 140-145°W) is shown in Figure 3 along with $AS(t)$ for all (solid) and independent events (dotted). Because this event occurred rather early, data from 1960 instead of 1963 was considered. Peaks in December 1961 in both AS functions reflect rate decreases from 2.67±2.20 to .64±1.04 earthquakes/month (eq/m) and 1.46±.93 to .53±.72 independent earthquakes/month (ieq/m).

In order to determine the spatial extent of this decrease, rates for the background (Jan., 1960-Dec., 1961) and quiet (Jan. 1962-Apr. 1968) periods were calculated and compared using a z-test in one degree latitude slices of the trench between 36-43°N. On the basis of these comparis-

ons the one degree slices were grouped into four regions with seismicity histories distinct from one another. The cumulative seismicity and $AS(t)$ are shown for these regions in Figure 4. Data for all events is shown above, data for independent events below. The removal of dependent events results in a much clearer picture of seismicity

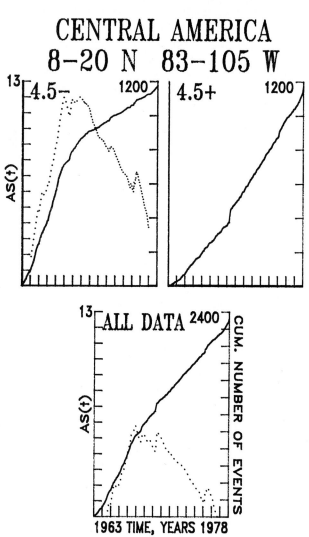

CENTRAL AMERICA
8-20 N   83-105 W

13 | 4.5-   1200   4.5+   1200

AS(t)

13 | **ALL DATA** 2400

AS(t)   **CUM. NUMBER OF EVENTS**

1963 TIME, YEARS 1978

Fig. 6. Cumulative seismicity and $AS(t)$ for the Central American trench in two magnitude bands, and for all events. $AS(t)$ is calculated for the entire data set. Note the large rate decreases apparent in the lower magnitude band during 1967, 1969 and 1970. This decrease is also evident in the set of all events for this region. Magnitude bands above 4.5, in contrast to those below, show constant or slightly increasing rates. This pattern of rate decreases in the smaller events and constant or increasing rates in the larger events is interpreted as an indication of a detection capability decrease in this region.

rate changes. This is mostly because of the resulting decrease in the variance of the rates determined for periods which include aftershock sequences. Peaks in $AS(t)$ in the first and third regions reflect decreases in seismicity rates which are apparent in the cumulative number curves. Several observations point to a difference between the phenomena which caused these two peaks.

First, the high rate in the southern region during 1961 reflects a post-seismic rate following an event with Ms=6.9 which occurred at 36°N on January 16, 1961. The one month aftershocks of this event are shown in Figure 2. Aftershocks, apparent in Figure 4, continued until the end of 1961. They are responsible for the peak in AS in

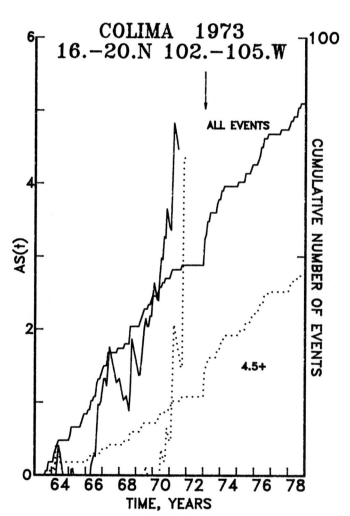

Fig. 7. Cumulative seismicity and $AS(t)$ for all (solid) and Independent (dotted) events in the region of the Colima gap. January 1963 is the beginning of the background and December 1972 is the last month for the calculation of $AS(t)$. Peaks in $AS(t)$ in both data sets reflect a quiescence which preceded the Colima event.

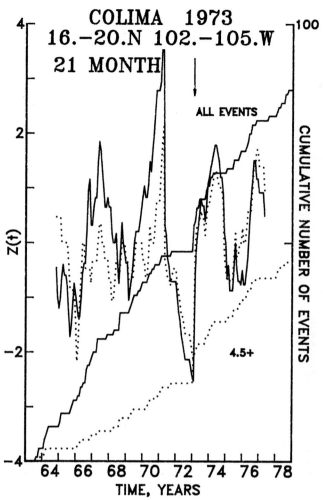

Fig. 8. $z(t)$ for a 21 month window superimposed on the cumulative seismicity for the region of the Colima gap. Note that the highest peaks in $z(t)$ are associated with a quiescent period preceding the Colima event of 1973. This quiet period is the most significant of duration 21 months or longer in both data sets. Arrows show the occurrence times of events with Ms≥6.5.

the southern region. A mainshock-aftershock sequence also occurred early in the data in the region of the Tokachi-oki event (39-41°N). These aftershocks were all dependent events and play no role in the AS peak in this region (see Figure 4). In the slice between 39-41°N the peak in $AS(t)$ is due to a rate decrease which may be a precursor to the Tokachi-oki event of 1968. Second, in the southern region (36-38°N) the rate remains constant between 1962 and 1978. This contrasts sharply with the mainshock region (39-41°N) in which the rate between 1962 and April 1968 is clearly lower than the rate following the Tokachi-oki aftershocks (1969-1978). The rate following the

Tokachi–oki aftershocks is identical to the rate between 1962–1978 in the southern region, 1.66±1.03 vs. 1.63±1.04 ieq/m. This rate may represent a regional background rate. Finally, the region between these two experienced no significant rate change near the end of 1961.

In order to increase the time period available for a background rate estimate data from the Japan Meteorological Agency (JMA) for this region was studied. In this case a complete data set was used to avoid potential problems with increasing resolution in the lower magnitude bands during the period studied (1954-1978). The data were found to be complete above 5.0 for the entire period. These data are shown in Figure 5. A significant rate decrease is again apparent in the region between 39–41°N. In these data, however, the most significant change occurs during October, 1965, several years after the rate decrease in the NOAA data. The rate decrease near 1962 found in the NOAA data is visible in the AS functions for the region between 39–41°N as a broad peak reaching 2.47 and 2.51 in the complete and independent data sets respectively. The absence of peaks in the other regions confirms the spatial limitation of this decrease to the region of the Tokachi–oki event of 1968.

These data demonstrate that a seismicity rate decrease existed in the rupture zone of the Tokachi–oki earthquake between 1962 and the time of the event, a period of 76 months. The anomaly became significant at the 99% level (peak) in the NOAA data during January, 1962, and during October, 1965 in the JMA data. The joint anomaly duration is 50±20 months. This anomalous rate is defined relative to a background rate beginning in January,1960 for the NOAA data and January, 1954 for the JMA data. The spatial coincidence of the anomaly and the rupture zone of the 1968 event strongly suggest a link between the quiescence and the physical preparation for this event. This study extends the work of Mogi [1969] (Figure 12) who showed that between 1961 and 1967 the upcoming rupture zone experienced few events with magnitudes above 6.0.

South Kuriles; Aug. 11, 1969; Ms=7.8, Mw=8.2
Nemuro–oki; June 17, 1973; Ms=7.7

These events occurred in a region recognized by Fedotov [1965] as a seismic gap. The seismicity of the region, known as the Hokkaido corner, is among the most complex in the world. Since 1960, 14 events with Ms≥7.0 have occurred there. The spatial relationships of the aftershock zones of these events are very complicated. Some events appear to have overlapping aftershock zones, making the application of the seismic gap hypothesis difficult in this region. For example, the western half of the 1969 South Kuriles rupture broke in January, 1968 during the Shikotan–oki earthquake (Ms=7.1). The epicenter of the 1969 quake was in the aftershock zone of the 1968 event. During 1973 and 1975, regions of the 1969 aftershock zone ruptured again in the Nemuro-oki

earthquakes of June 17, 1973 (Ms=7.7) and January 10, 1975 (Ms=7.0). While some of these events are not large enough to be considered great earthquakes, they certainly play a role in the tectonics of this region. The overlapping of these aftershock zones makes the study of seismicity rates in this region extremely difficult. Possibly detailed study of the three–dimensional seismicity of this region would make the recognition of seismicity precursors to these events possible, but no clear seismicity anomalies were found before either the 1969 or the 1973 events using techniques discussed above. This is not surprising considering the simplicity of the techniques, the quality of the data, and the complexity of the region. Recently Katsumata and Yoshida [1980] argued that small regions within the aftershock zones of these events were quiet before and after the events. They did not demonstrate that any change occurred in these regions before the mainshocks.

Kamchatka; Dec. 15, 1971; Ms=7.8

This event ruptured about 100 km of the Kamchatka trench near its intersection with the Aleutian trench. It occurred in a seismic gap recognized by Fedotov [1965]. Seismicity before this event was discussed by Wyss et al. [1978]. They proposed that a seismicity precursor consisting of three stages preceded this event. The first and third stages of the precursor consisted of quiescence in the entire rupture zone. The second stage consisted of quiescence in most of the rupture zone and a cluster of events which occurred between January 21 and 30, 1969 in the epicentral region. This earthquake occurred in a region of low teleseismically reported activity so the quiescent features of the precursor could not be reasonably tested using the techniques proposed here. A computer algorithm which analyses spatial and temporal clustering as well as the number of events in a sliding time window (Wyss et al. [1978]) picked the events of January 21–30, 1969 as the most clustered group of events which occurred in the region of the mainshock between 1960 and 1977. Fedotov et al. [1977] showed that the events of this cluster had S to P–wave energy ratios ($\log(E_s/E_p) = K_s - K_p$ where K_s and K_p are energy classes) more than two standard deviations below the mean for the Kamchatka region between 1968 and 1970. The combination of anomalous clustering and anomalous radiation characteristics for these events, and their proximity to the 1971 mainshock (40 km) strongly suggest that these events were related to the preparation for the mainshock.

Cluster type anomalies similar to this one have been reported before other large events (Habermann and Wyss [1977]), moderate events in California [McNally 1977], and small events in South Carolina (Talwani [1979]). This type of anomaly seems to be relatively rare compared to the pure quiescence type anomalies demonstrated before other events in this paper.

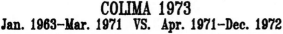

COLIMA 1973
Jan. 1963–Mar. 1971 VS. Apr. 1971–Dec. 1972

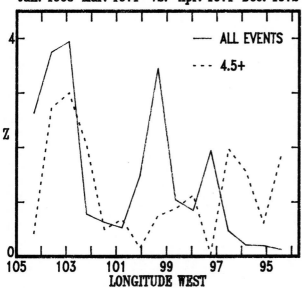

Fig. 9. z–map comparing the rates between January 1963–March 1971 to rates between April 1971–December 1972 for 150 km north–south slices of the Central American trench. Note that the greatest difference between these two rates occurs in the region of the Colima 1973 rupture. This is true both for all events (solid) and for events with magnitudes above 4.5 (dotted) which are a homogeneous data set.

St. Elias, Alaska; February 28, 1979; Ms=7.25

This event occurred near the junction of the Fairweather and southeast Alaska seismic zones. It had a rupture length of about 120 km, partially filling a gap pointed out by Sykes [1971]. The earthquake occurred in a region of very low teleseismically recorded activity (Tobin and Sykes [1968]). The activity is too low to allow a meaningful study of the seismicity rates.

Sitka, Alaska; July 30, 1972; Ms=7.6

Seismic preconditions for this event, which ruptured 100–150 km of the Fairweather fault system between 56–57°N, have been discussed by Kelleher and Savino [1975]. The event occurred in a seismic gap described by Sykes [1971]). The small number of events reported in this gap between 1960 and the time of the mainshock (six) renders a study of seismicity rates impossible. Four of these six events occurred between March, 1966 and April,1967, forming an unusual burst of seismic activity. Kelleher and Savino [1975] relocated these events and showed that they occurred near the boundaries of the future rupture zone. This activity may have been related to the 1972 mainshock but the low amount of teleseismic activity makes any case unconvincing.

Earthquakes in Central and South America

Central American Seismicity

The seismicity of the Central American trench between 8–20°N and 83–105°W was examined for homogeneity of the data set. The cumulative seismicity for two magnitude bands studied is shown in Figure 6. The most obvious feature of these data is a large rate decrease in the magnitude band below 4.5 between 1967 and 1970. This magnitude band contains 1175 earthquakes (50% of the total for this region), 45 of which have no mb given. The rate of reporting of events with $mb \leq 4.5$ decreased from 12.02 ± 5.05 eq/mo between January 1963 and November 1967 to 2.74 ± 2.02 eq/mo between October 1970 and December 1979. The effect of these decreases is

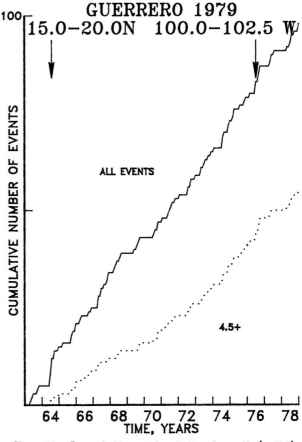

Fig. 10. Cumulative seismicity for all (solid) and 4.5+ (dotted) events for the region of the Guerrero gap. Note the lack of any seismicity rate decrease preceding this event which occurred three months off of the end of the data. The long term rate decrease in the set of all events during the late 1060's reflects the decrease in the rate of reporting of small events in this region. Arrows show the occurrence times of events with $Ms \geq 6.5$.

clearly felt in the set of all events (see Figure 6). The total rate of reported earthquakes in Central America decreases from 16.12±6.44 to 10.33±6.37 in October 1967, a decrease of 36%. The comparison of these two rates yields a z value of 5.74 (99+%) significance. The magnitude bands above 4.5, in contrast to those below, show constant or slightly increasing rates throughout the data period (see Figure 6).

This pattern, a decrease in the rate of reporting of smaller events in a large region during a period of constant or increasing rates of reporting of larger events, has been interpreted by Habermann (in preparation) as an indication of decreasing detection capability. The fact that detection capability in Central America decreases during the late 1960's is probably due to the closure of the Vela arrays in the Western United States (North [1977], Habermann, [in preparation]). As discussed

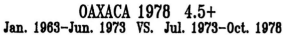

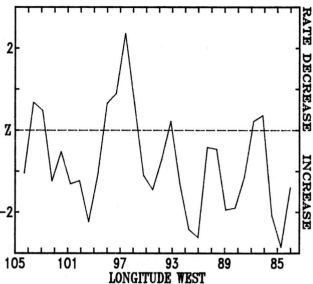

Fig. 12. z–map comparing the rates between January 1963–June 1973 to rates between July 1973–October 1978 for 150 km north–south slices of the Central American trench. Only events with mb≥4.5 are included. Points below the dashed line indicate that the second rate was higher than the first. Note that the region of the Oaxaca rupture (96–97°W) is the only region in which the second rate is significantly lower than the first.

above, a homogeneous data set, or at least one in which detection is constant or increasing, is crucial for this study. These data show that the seismicity data for Central America provided by the NOAA HDF is non-homogeneous. The results of North [1977] indicate that this is also true for the data provided by the International Seismological Center (ISC). All data, as well as events with $mb \geq 4.5$ will be examined in each case.

Colima, Mexico; January 30, 1973; Ms=7.4

The northwestern end of the Central American trench was described as a seismic gap by Kelleher et al (1973). A portion of this gap ruptured in the Colima event of January 30, 1973. The exact extent of the rupture is difficult to determine from teleseismic data as only two aftershocks are reported by NOAA. Reyes et al [1979] concluded from a local study that the Colima event re-ruptured the region of the 1941 earthquake rather than the gap to the southeast of it. This interpretation is also given by McCann et al [1979]. Ohtake et al [1977b] argued that the region surrounding this event was quiet between October 1971 and December 1972.

The cumulative seismicity and $AS(t)$ for the region of the gap (16–20°N, 102–105°W) are shown

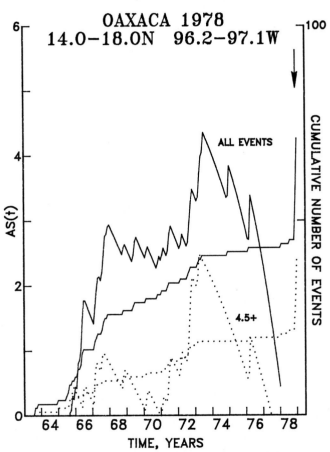

Fig. 11. Cumulative seismicity and $AS(t)$ for the region of the Oaxaca gap. $AS(t)$ is calculated for the period between January 1963 and October 1978. Note outstanding peaks in the $AS(t)$ functions for both all events (solid) and events with magnitudes above 4.5 (dotted). Arrows show the occurrence times of events with Ms≥6.5.

in Figure 7. Peaks in $AS(t)$ are clear in both data sets before this event, at March 1971 for all events and October 1971 for the larger events. The period between November 1971 and December 1972 is the α phase recognized by Ohtake et al [1977a,b]. The decrease in reporting of small events near 1967 has no observable effect in this region.

Several other periods of relatively low activity are apparent in the region. The relative significance of these periods was examined using $z(t)$, a function described above. The cumulative seismicity along with $z(t)$ for a 21 month window (the duration of the anomaly determined by $AS(t)$) are shown in Figure 8. In both data sets the decrease preceding the Colima event forms the highest peak in $z(t)$. This means that this anomaly is the most significant rate decrease with a period of 21 months or longer in this 192 month sample. This is true in both data sets.

In order to evaluate the spatial extent of this decrease, rates during the background and quiet periods were compared in 150 km wide north−south slices of the trench between 94−105°W. The resulting values of z are plotted as a function of longitude in Figure 9. It is clear that the most significant difference between rates in these two time periods is confined to the region of the Colima rupture, 102.9−103.4°W. It is the only region with localized peaks in both data sets.

This data indicates that a decrease in seismicity rate preceded the Colima event of 1973. The decrease was the most significant of 21 month or longer duration the region where it occurred. The decrease was tightly confined in space to the region of the rupture even though the rupture did not fill the entire gap in which it occurred. In this case a recognized gap could be divided into a mature and an immature segment before the time of the mainshock which filled the mature segment.

Guerrero, Mexico; March 14, 1979; Ms=7.6

This event ruptured the Middle America trench near its intersection with the Orozco fracture zone. This area was recognized as a seismic gap by Kelleher et al. [1973]. McCann et al. [1979], also discussed the zone. The Guerrero event ruptured the region between 100.6 and 101.6°W, it did not fill the entire gap (100−102.5°W).

The cumulative seismicity for the gap is shown in Figure 10. The NOAA data is compiled only until December, 1978. The seismicity rate in the gap shows no variation which is recognizable as a possible precursor. The seismicity rate in the rupture zone was also constant throughout the data period. Any seismicity rate variation which occurred before this event lasted a maximum of three months. Even if we had the data, the recognition of a change over such a short time period would be difficult if not impossible. This event then could not have been foreseen on the basis of our analysis of the seismicity data.

Oaxaca, Mexico; November 29, 1978. Ms=7.8

This event ruptured the Central American trench between 96.5− 97.0°W, filling the gap between the 1968 and 1965 Oaxaca earthquakes. Ohtake et. al [1977a,b] recognized a quiescent region roughly three times the size of this gap which they proposed as a precursor to a future event. The event occurred during November 1978 in the gap, filling only a third of the quiescent zone. This is the only case, at present, in which a sign of maturity in a seismic gap has been recognized before the gap is filled.

The cumulative seismicity and $AS(t)$ for the gap are shown in Figure 11. Peaks in $AS(t)$ are apparent in both data sets, during March 1973 for the larger events and June 1973 for all events. This quiet period, between June, 1973 and the mainshock is the α phase of Ohtake et al's [1977a,b] model for precursory seismicity.

Because this event occurred in a region strongly affected by the decrease in rate of small events discussed above, (see Figure 6), only events with magnitudes above 4.5 were considered when determining the extent of this anomaly. Rates during the background and quiet periods were compared in thirty 150 km north−south slices of the trench, each overlapping the next by 75 km. The z−values resulting from these comparisons are shown in Figure 12. The rate during the second period was higher than during the first in twenty-two of the thirty regions. Against this background of increasing activity the Oaxaca region stands out as the only region in which the second rate is significantly lower than the first. On the basis of these data it is clear that a temporally and spatially unique seismicity rate decrease preceded the Oaxaca event. This result is in agreement with the results of Ohtake et al [1977a,b].

Lima, Peru; October 10, 1974; Ms=7.5, Mw=8.1

This event ruptured the interplate−thrust zone off of the coast of Peru between 12−14°S, partially filling a gap recognized by Kelleher [1972], see Figure 13 inset. The seismicity in the region of this event has been discussed by Brady [1976] and Dewey and Spence [1979]. Brady pointed out the peak in the number of events per year which occurred in 1971 in the immediate epicentral zone of the 1974 event. He postulated that the events making up this peak represented the formation of a primary inclusion zone (PIZ) related to the preparation for the 1974 event. Dewey and Spence [1979] relocated about 350 events in this region and showed that the events of Brady's PIZ were actually 60 km or so from the mainshock and on the landward edge of the rupture zone. They point out that these events could have been related to the future mainshock according to several precursor hypotheses. They do not discuss how these events would be recognized as precursory at the time of their occurrence.

The cumulative seismicity for the gap (12−14°S, 75−79°W), is shown in Figure 13. A quiet period is evident between September, 1969 and July, 1971.

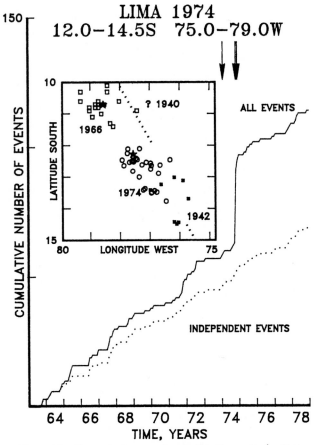

Fig. 13. Cumulative seismicity for all (solid) and independent (dotted) events in the region of the Lima gap. Note the lack of any apparent quiescence preceding this event. Arrows show the occurrence times of events with Ms≧6.5. Inset is seismicity map showing aftershocks of the 1966 (open boxes) and 1974 (open circles) events. Approximate rupture lengths are shown for the 1940 and 1942 events in this region (From Kelleher [1972]). Solid boxes indicate events which occurred in the southern portion of the gap between September 1971 and July 1972. These events form the slow swarm apparent in the cumulative seismicity curve in Figure 14.

This quiescence is separated from the mainshock by a period of high activity, (August, 1971–July, 1972) and a period of background rate, (August, 1972– September, 1974). This suggests that the quiet period was not related to the mainshock. No clearly anomalous quiescence is seen in this gap before the 1974 mainshock.

The algorithm of Wyss et al [1978] was applied to these data to determine if anomalous clustering occurred in this region before the 1974 mainshock. Two groups of clustered events were identified. The first was the same group as Brady

[1976] pointed out, this group appears to be a foreshock–mainshock sequence. The second was a group of events which occurred on July 17–18, 1972 near the southern end of the gap. The area of these events and the southern half of the rupture zone (13–14.5°S) were quiet between early 1969 and the time of the mainshock, except for events which occurred between September 20, 1971 and the time of the cluster, see Figure 14, which shows the cumulative seismicity for this region. These events are shown as solid boxes in Figure 13 (insert). This pattern of quiescence–swarm–quiescence has been proposed

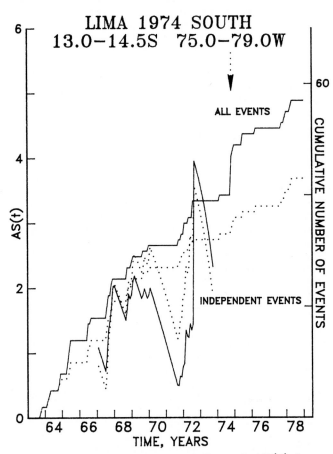

Fig. 14. Cumulative seismicity and $AS(t)$ for the southern portion of the Lima gap. January 1963 is the first month of the background and September 1974 is the last month for the $AS(t)$ calculation. Note quiet period with a slow swarm in the center apparent in both data sets between 1969 and the time of the mainshock. Events in the slow swarm are shown as solid boxes in Figure 10 inset. This pattern is very similar to the precursor to the 1971 Chili event discussed by Habermann [1979]. Dotted arrow indicates time of occurrence of Lima 1974 event slightly to the north.

as a type of seismicity precursor by Wyss et al [1978]. The pattern in the region between 13–14.5°S between 1969 and October 1974 is very similar to the pattern observed before the Kamchatka event discussed above and before an event with Ms=7.8 (Chili, July 9, 1971) discussed by Habermann [1979]. On the basis of the pattern in this southern region an event would be expected at 13.7°S–76.4°W (the average position of the swarm events) between January 1973 and October 1974 (time window from empirical formula of Wyss et al [1978]). The event in 1974 occurred in this time window but the location was 225 km from the expected location. This distance is small (<2 source dimensions) compared to distances between mainshocks and clusters proposed as precursors by Keilis-Borok et al [1980], but still to large to be considered as a clear precursor.

Discussion and Conclusions

Seismicity rates preceding eleven recent seismic gap filling earthquakes has been examined using new quantitative techniques. Data included in the NOAA HDF is too sparse for determining if precursory changes in seismicity rates occurred in two of the regions studied (Sitka 1972, and St. Elias 1979). In the Hokkaido corner, site of the South Kuriles 1969 and the Nemuro-oki 1973 events, neither the seismic gap hypothesis or the techniques proposed here work well. The region is just too complex. Of the remaining seven events, three were preceded by temporally and spatially unique seismic quiescence, (Tokachi-oki 1968, Colima 1973, Oaxaca 1978) and one was preceded by a clear cluster type anomaly (Kamchatka 1971). Possible cluster type anomalies preceded the Lima 1974 and Solomon Islands 1978 events. Only the Guerrero 1979 event was not preceded by some type of anomaly.

The minimum magnitude of completeness of all data sets is roughly the same (5.0±.25), therefore there is no reason to believe that differences in observed anomalies are due to differences in completeness of the data. The statistical significance and the tight spatial coincidence of the observed anomalies and the rupture zones strongly suggests that the anomalies are real and are related to the preparation for the mainshocks. Whether or not an earthquake will be preceded by an observable seismicity anomaly appears to be impossible to assess *a priori* on the basis of simple parameters, ie., region, tectonic setting, data completeness, background seismicity rate.

Data presented by Wyss et al. [1981] demonstrates that the rupture zone of the Kalapana, Hawaii earthquake (Nov. 29, 1975) can be divided into two parts on the basis of precursory phenomena. One part, termed low-stress, experienced significantly decreased seismicity. The other part, somewhat smaller, experienced constant or slightly increased seismicity. This region is termed the high-stress region. Wyss et al. [1981] and Wyss and Habermann [1979] have inter-

preted seismic quiescence in terms of precursory displacement and aseismic stress release on some portion of the future rupture surface(s). Constant or increased seismicity reflect locked portions of the rupture surface(s), asperities. The ratio of locked to unlocked portions of the rupture surfaces (Rl) may be an important factor in determining the type of precursory seismicity which may be observed. If Rl is large, no precursory quiescence would be expected, on the contrary, one might expect slightly increased seismicity in the upcoming rupture zone. If Rl is small, a large portion of the future rupture zone would experience precursory stress drop and decreased seismicity.

The application of this concept to the events discussed above is premature at this time. It does offer a speculative explanation for why some earthquakes are preceded by quiescence, some by clustering, and some by no observable quiescence or clustering. Of course the occurrence of seismicity precursors may be controlled by any number of other parameters, ie., fault zone composition, pore pressure, ambient stress level, temperature, homogeneity of the fault surface, time elapsed since the last major event etc. Rl is also dependent on these factors and may, in the long run, represent some measure of the combined effect of all of them.

The study of seismicity rates falls into the same category as the study of most possible precursors, it works well in some cases. Seismicity rates are not the "master key" which will open the door to predicting all major earthquakes. Probably no single such key exists. Seismicity rates will, however, open some doors, as demonstrated by Ohtake et al.'s specific forecast (based on the observation of a possible precursor) of the Oaxaca, 1978 event. Data presented here shows that this forecast was not an isolated success. Four of seven GFEQ's which occurred in regions with reasonable data could have been specifically forecast on the basis of teleseismic seismicity patterns. We have presented techniques which make the evaluation of the uniqueness of these patterns in space and time possible. The use of some such techniques is essential in the study of possible precursory seismicity changes.

Acknowledgements

I would like to thank Max Wyss, Steve Ihnen, Selena Billington, Bob Engdahl, and Duncan Agnew for their hours of helpful discussion. Thanks also to Bill Rinehart without whom I could not have had such easy access to the data. Danny Harvey, Rich Gross, Ernie Harkins, and Martin Smith answered many of my innocent questions about the computer. This work was funded by U. S. Geological Survey contract number 14–08–0001–18386 .

References

Brady, B.T., Theory of earthquakes IV; General

implications for earthquake prediction, *PAGEOPH, 115*, 357, 1976.

Dewey, J.W. and W. Spence, Seismic gaps and source zones of recent large earthquakes in coastal Peru, *PAGEOPH, 117*, 1148, 1979.

Engdahl, E.R. and C. Kisslinger, Seismological precursors to a magnitude 5 earthquake in the Central Aleutian Islands, *Jour. Phys. Earth, 25, Suppl.*, 243, 1977.

Fedotov, S.A., Regularities of the distribution of strong earthquakes in Kamchatka, the Kurile islands, and northeast Japan, *Trudy Inst. Fiz. Zemli., Acad. Nauk. SSSR, 36*, 66, 1965.

Fedotov, S.A., G.A. Sobolev, A.A. Boldyrev, A.M. Gusev, A.M. Kondratenko, O.V. Potapova, L.B. Slavina, V.D. Theophylaktov, A.A. Khramov, and V.A. Shirokov, Long and short term earthquake prediction in Kamchatka, *Tectonophysics, 37*, 305, 1977.

Habermann, R.E. and M. Wyss, Seismicity patterns before five major earthquakes (abstr.), *EOS, 58*, 884, 1977.

Habermann, R.E., Precursory seismicity: quiescence and clusters (abstr.), *EOS, 60*, 884, 1979.

Habermann, R.E. and R.E. Habermann, Teleseismic detection capability decreases since 1963, *to be submitted to B.S.S.A.*, 1981a.

Hogg, R.V. and E.A. Tanis, *Probability and statistical inference*, 450pp, Macmillan Publishing Co. Inc., New York, 1977.

Katsumata, M. and A. Yoshida, Change in seismicity and development of the focal region, *Meteorlogical Research Institute, Japan, Papers in Meteorology and Geophysics, 31*, 15, 1980.

Keilis-Borok, V.I., L. Knopoff, I.M. Rotvain, and T.M. Sidorenko, Bursts of seismicity as long-term precursors of strong earthquakes, *J.G.R., 85*, 803, 1980.

Kelleher, J.A., Rupture zones of large South American earthquakes and some predictions, *J. Geophys. R., 77*, 2087, 1972.

Kelleher, J.A., L.R. Sykes, and J. Oliver, Possible criteria for predicting earthquake locations and their applications to major plate boundaries of the Pacific and Caribbean, *J. Geophys. Res, 78*, 2547, 1973.

Kelleher, J.A. and J. Savino, Distribution of seismicity before large strike-slip and thrust type earthqaukes, *J. Geophys. Res, 80*, 260, 1975.

Khattri, K. and M. Wyss, Precursory variation of seismicity in the Assam area, India, *Geology, 6*, 685, 1978.

McCann, W.R., S.P. Nishenko, L.R. Sykes, and J. Krause, Seismic gaps and plate tectonics: Seismic potential for major boundaries, *PAGEOPH, 117*, 1082, 1979.

McNally, K., Patterns of earthquake clustering preceding moderate earthquakes, central and southern California (abstr.), *EOS, Trans. AGU, 58*, 1195, 1977.

Meyer, S.L., *Data Analysis for scientists and engineers*, 513pp, John Wiley & Sons, New York, 1975.

Mogi, K., Some features of recent seismic activity in and near Japan (1), *Bull. EQ. Res. Inst., Tokyo Univ., 46*, 1225, 1968.

Mogi, K., Some features of recent seismic activity in and near Japan (2) Activity before and after great earthquakes, *Bull. EQ. Res. Inst., Tokyo Univ., 47*, 395, 1969.

North, R.G., Station magnitude bias - Its determination, causes, and effects, in *Technical Note # 1977-24, Lincoln Lab., MIT*, 1977.

Ohtake, M., T. Matumoto, and G. Latham, Seismicity gap near Oaxaca, southern Mexico as a probable precursor to a large earthquake, *PAGEOPH, 115*, 375, 1977a.

Ohtake, M., T. Matumoto, and G. Latham, Temporal changes in seismicity preceding some shallow earthquakes in Mexico and Central America, *Bull. Inter. Inst. of Seismology and EQ. Engineering, 15*, 105, 1977b.

Reyes, A., J.N. Brune, and C. Lomnitz, Source mechanism and aftershock study of the Colima, Mexico earthquake of January 30, 1973, *B.S.S.A, 69*, 1819, 1979.

Scholz, C., The frequency-magnitude relation of microfracturing in rocks and its relation to earthquakes, *B.S.S.A., 58*, 399, 1968.

Sykes, L.R., Aftershock zones of great earthquakes, seismicity gaps, and earthquake prediction for Alaska and the Aleutians, *J. Geophys. Res, 76*, 8021, 1971.

Talwani, P., An emperical model for earthquake prediction, *Phys. Earth Planet. Int., 18*, 288, 1979.

Tobin, D.G. and L.R. Sykes, Seismicity and tectonics of the northeast Pacific Ocean, *J. Geophys. Res, 73*, 3821, 1968.

Wier, S., A change in seismicity near the Mid-America trench following the Oaxaca earthquake, *in press Phys. of Earth and Planet. Int.*, 1980.

Wyss, M., R.E. Habermann, and A.C. Johnston, Long term precursory seismicity fluctuations, p.869 in *U.S.G.S. open file report no. 78-943, Methodology for identifying seismic gaps and soon-to-break gaps*, 1978.

Wyss, M., Estimating maximum expectable magnitude of earthquakes from fault dimensions, *Geology, 7*, 336, 1979.

Wyss, M. and R.E. Habermann, Seismic quiescence precursory to a past and a future Kuriles islands earthquake, *PAGEOPH, 117*, 1195, 1979.

Wyss, M., F.W. Klein, and A.C. Johnston, Precursors to the Kalapana M=7.2 earthquake, *submitted to J. Geophys Res.*, 1981.

SEISMICITY IN WESTERN JAPAN AND LOMG-TERM EARTHQUAKE FORECASTING

Kiyoo Mogi

Earthquake Research Institute, Tokyo University, Tokyo 113

Abstract. Western Japan, along the Nankai trough is a typical subduction zone where great shallow earthquakes have frequently occurred. In this region, records of seismic activity and ground movement have been accumulated for a long period of time. Because of the excellent quality of the data, this region is one of the most suitable sites for studying the long-term spatio-temporal variation of seismicity in the crust. In this paper, the time of occurrence of great shallow earthquakes and the space-time distribution of seismicity between successive great earthquakes are discussed. On the basis of the seismicity and the crustal movement data, it is concluded that a great earthquake occurs when the stress in this region reaches to the ultimate stress level. These results are useful for long-term forecasting of great shallow earthquakes along subduction zones.

Introduction

Western Japan, along the Nankai trough, is a typical subduction zone where great shallow earthquakes of thrust type have occurred with the recurrence time of 100~200 years (Fig. 1). In this region, both recent instrumental data and historical records of seismicity and ground movements have been accumulated over a long period. With this excellent data, crustal activity in this region has previously been discussed by a number of investigators (Mogi, 1969, 1970; Fitch and Scholz, 1971; Ando, 1975; Shimazaki, 1976; Seno, 1979). Recently, reliable earthquake data for the period (1885-1925) has been reported by Utsu (1979). Because of the large amount of data available, this region is one of the most suitable sites for studying the space-time variation of seismicity between successive great shallow earthquakes along subduction zones.

In this paper, two main subjects are considered. First, the time of occurrence of great shallow earthquakes in this region is discussed. On the basis of the seismic data, it is concluded that a great earthquake occurs when the stress in this region reaches to the ultimate stress level. Secondly, the space-time variation of seismic activity between successive great shallow earthquakes is discussed. These results are useful for long-term forecasting of great shallow earthquakes along subduction zones.

Time of Occurrence of Great Shallow Earthquakes Along The Nankai Trough

Previous Works

Although the occurrence of great shallow earthquakes in this region is rather regular, the time interval between successive great shallow earth-

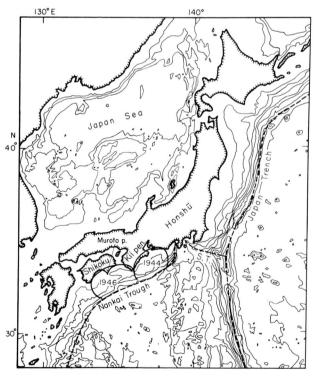

Fig. 1. Submarine topography around the Japanese islands (Hydrographic Department, Marine Safety Agency, Japan). The Japan trench, the Sagami trough (S.T.) and the Nankai trough are shown by thick broken curves. The source regions of the 1944 Tonankai and the 1946 Nankaido earthquakes are shown by thick curves.

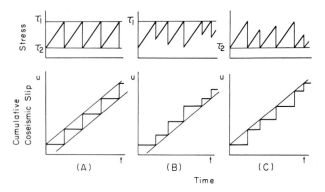

Fig. 2. Three typical models of the temporal variation of the cumulative coseismic fault slip in a fault (bottom). The top figure shows the temporal variation of stress in the source region (after Shimazaki and Nakata, 1980).

quakes varies in the range from one hundred to two hundreds years (e.g. Usami, 1975). It is important to address the problem of how and why the recurrence time interval changes under a constant rate of plate motion. Figure 2 shows the three typical models presented by Shimazaki and Nakata (1980). The bottom figures in Fig. 2 show the cumulative coseismic fault slip and the top figures show the deduced stress variation as functions of time.

The model A is the case that the amount of fault slip and the time interval are all equal. Evidently, this is not the actual case. The model B is the case that a time interval is proportional to the amount of fault slip in the preceding earthquake. The model C is the case that a time interval is proportional to the amount of the fault slip in the following earthquake.

Several investigators have tried to apply these latter two different models for prediction of the time of occurrence of earthquakes in different cases (e.g. Bufe et al., 1977; Shimazaki and Nakata, 1980; Acharya, 1979). In the case of the Nankaido earthquakes, Shimazaki and Nakata (1980) showed that the model B is consistent with historical data of coseismic uplift at Murotsu near Muroto point, Shikoku. However, their evidence is limited only to the three uplift data at a station and they could not explain why the model B is applicable. In this section, it is shown that the temporal variation of seismicity in western Japan also supports model B. It is also shown that the model B can be explained reasonably from the standpoint of fault mechanics.

Time of Occurrence of Great Shallow Earthquakes

In the model B, great earthquakes occur at a constant ultimate stress and the stress drop is not constant. In the model C, rupture strength markedly fluctuates and the stress is nearly completely released by the occurrence of each great earthquake. Observation of the spatial and temporal variation of seismicity can be used to examine

which model is correct. Previous studies show that the seismicity of the land area of western Japan increases appreciably before great shallow earthquakes along the Nankai trough (Mogi, 1969; Ozawa, 1973; Utsu, 1974; Shimazaki, 1976; Seno, 1979). This seismicity may be caused by the increase of stress, similar to stress dependent acoustic emission activity (c.f. Mogi, 1962). The above-mentioned land area is not the source region of the great shallow earthquakes, but rather the surrounding wide area of the source region. Therefore, this activation of seismicity in the land area indicates a stress increase encompassing the wide area of western Japan along the Nankai trough.

According to this view, it is expected that in model B, the period of increased seismicity preceding the great shallow earthquakes is constant for different recurrence time intervals between great earthquakes, as shown in Fig. 3. In this case, the length of the calm period (following the great earthquakes) is different for recurrence time interval of different length. In this figure, the active and the calm periods are shown by bars at top and bottom, respectively. On the other hand, in the model C, the length of the ac-

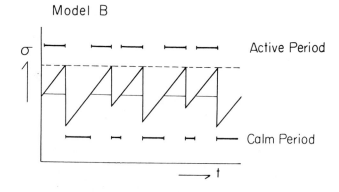

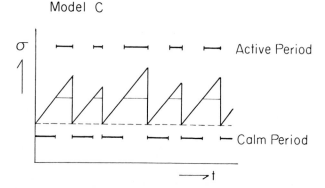

Fig. 3. Seismic active period and temporal variation of stress in the models B and C. Top bar: seismic active period; bottom bar: quiet period. The length of seismic active periods is constant in the model B and that of the quiet periods is constant in the model C.

tive period preceding great earthquakes varies according to the recurrence time intervals and the calm period is constant.

Therefore, we can examine which model is consistent with actual earthquake data. In the following analysis, earthquake data are taken from the catalogue of historical destructive earthquakes by Usami (1975). However, for recent years, only earthquakes of magnitude 7 and over are taken from the catalogues of Japan Meteorological Agency (or JMA) and Utsu (1979).

Figure 4 shows the temporal variation of seismicity preceding great shallow earthquakes along the Nankai trough. In this figure, great earthquakes are divided into two groups by the time intervals preceding each great earthquake. The upper figure is the case of relatively short time intervals, such as about one hundred years. The number of earthquakes (N) before great earthquakes are the sum of earthquakes in the case of the 1605, the 1707 and the 1944 great earthquakes, and the preceding small arrows show the preceding each great earthquakes. The lower figure is the case of long time intervals, such as one hundred fifty years and over. The number of earthquakes (N) in this case is the sum of earthquakes before the 1096, the 1361, the 1498 and the 1854 great earthquakes.

This results show that seismicity is significantly high before great shallow earthquakes and the length of the active period is nearly similar in the cases of different time intervals. The active

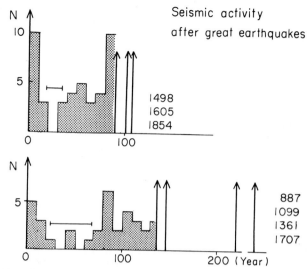

Fig. 5. Seismic activity after great earthquakes. Upper and lower figures show the case of the relatively short and the long time intervals respectively. In each figure, the number of earthquakes (N) is the sum of the earthquakes before three or four great earthquakes. Numerals in the right side show the time of these great earthquakes. The small arrows show the following great earthquakes. The length of the calm period shown by a horizontal bar in the upper figure is shorter than that in the lower figure.

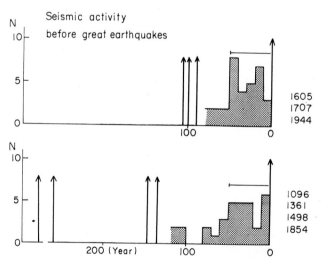

Fig. 4. Seismic activity before great earthquakes. Upper and lower figures show the cases of the relatively short and the long-time intervals respectively. In each figure, the number of earthquakes (N) is the sum of the earthquakes before three or four great earthquakes. Numerals in the right side show the time of these great earthquakes. The small arrows show the preceding great earthquakes. The lengths of the active period in both cases shown by horizontal bars are nearly similar (about 50 years).

period is about 50 years. This result supports the model B.

Figure 5 shows the temporal variation of the seismic activity after great shallow earthquakes. In this figure, great earthquakes are divided into two groups by the time intervals after each great earthquake. The upper and the lower figures show the cases of relatively short and long time intervals. The number of earthquakes (N) after great earthquakes in the case of the short time interval is the sum of earthquakes after the 1498, the 1605 and the 1854 great earthquakes. The number of earthquakes (N) in the case of long time interval is the sum of earthquakes in the case of the 887, the 1099, the 1361 and the 1707. According to the result, the region becomes quiet after the great earthquakes except for aftershocks in a strict sense. The length of the calm period shown in Fig. 5 is not constant and is nearly proportional to the time interval. This result also supports the model B.

The above-mentioned results suggest that, when the stress level in this region reaches a certain value, a great shallow earthquake occurs. If it is assumed that the stress increases linearly by the plate motion of a constant rate, the time of occurrence of a great earthquake can be predicted approximately from the amount of the fault slip of the preceding great earthquake, as pointed out by Shimazaki and Nakata (1980).

Relationship between The Recurrence Time and The Fault Length

As mentioned above, the time interval between successive great shallow earthquakes is not constant, but varies significantly. If the constant rate of plate motion is assumed, this means that the amount of fault slip varies significantly. This is consistent with the historical data of the coseismic uplift at Muroto point, Shikoku. In this section, it is discussed why the amount of fault slip was not constant in this region.

Figure 6 shows the past successive three great rupture zone along the Nankai trough. The top figure shows the temporal variation of stress in the land area deduced from the seismic activity in the western Japan and the vertical movement at Muroto point.

In 1707, the whole region ruptured simultaneously and the coseismic uplift at Muroto point was about 2m. On the other hand, in 1854 and 1944, first the eastern part ruptured, and then the western part ruptured in 1854 and 1946. In these latter cases, the ruptured regions of a great earthquakes are only a half of the 1707 earthquake and the coseismic uplift at Muroto point was about 1.2m. Therefore, the difference in the coseismic uplift or the amount of fault slip between the 1707 and the latter two earthquakes correspond to the difference in rupture lengths. It is reasonable that the amount of fault slip is proportional to the length of rupture zone. Thus, the physical basis of the above-mentioned model B, that the ultimate strength is constant and the amount of fault slip fluctuate significantly, can be reasonably understood by consideration of the difference in rupture lengths.

It is also an interesting problem why the whole region ruptured in 1707 only. In 1703, the great Genroku earthquake occurred along the Sagami trough, as shown in Fig. 6. This earthquake is the greatest rupture along the Sagami trough during the last thousand years. Therefore, it is very probable that the 1707 great rupture in a part of the northern boundary of the Philippine sea plate (the Sagami trough) strongly affected the great rupture of the adjacent part of the plate boundary (the Nankai trough) in 1707.

Relationship between "The Tokai Earthquake" and The 1891 Nobi Earthquake

The Tokai region, an area in central Japan between Tokyo and Nagoya, has been pointed out as a potential region for a future great earthquake (Mogi, 1970; Ishibashi, 1977; Utsu, 1977). This region is indicated in the bottom figure in Fig. 6. According to historical data, the rupture of the Tokai region occurred together with the rupture of the Tonankai-Nankaido region, except for the case of the 1944 Tonankai earthquake. Various data shows that the main rupture zone of the 1944 Tonankai earthquake did not extend eastward to the Suruga trough, and so the Tokai region remains a

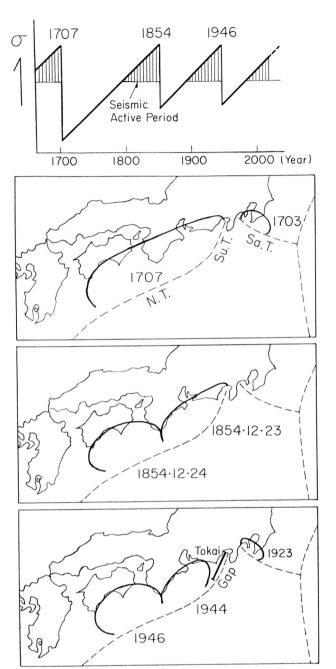

Fig. 6. Temporal variation of the crustal stress in western Japan (top) and the rupture zones along the Nankai trough and the Sagami trough (the northern boundary of the Philippine sea plate) during the last three hundreds years. Great earthquakes occurred at a constant ultimate stress, but the amount of slip differed between the 1707 and the other earthquakes. It is noted that the 1707 greatest earthquake along the Nankai trough occurred four years after the 1703 greatest earthquakes along the Sagami trough. The Tokai region along the Suruga trough is a seismic gap of the first kind (bottom figure). N.T.: Nankai trough; Su.T.: Suruga trough; Sa.T.: Sagami trough.

seismic gap. Now, it is important questions why the Tokai region remains as a gap only in the case of the 1944, rupture, and whether or not "the Tokai earthquake" can occur independently from the next Tonankai-Nankaido earthquake which are not expected in near future.

In relation to this problem, it is suggested that the 1891 Nobi earthquake (M=8.4) might have played an important role. The Nobi earthquake is a great shallow earthquake of left lateral strike-slip fault with NW-SE direction. Such a great intra-plate earthquake with NW-SE direction occurred once several hundred years in the central part of Honshu because of low strain accumulation in this region (Figure. 7).

As shown in Fig. 7, the strain accumulation in the Tokai region may be released slightly by the left lateral movement in the 1891 Nobi earthquake, and so the rupture zone of the 1944 Tonankai earthquake did not extend to the Tokai region. However, recent geodetic and seismic data suggest the appreciable strain accumulation in this region. Therefore, "the Tokai earthquake" may be expected as a delayed rupture independent from the next Tonankai earthquake.

Space-Time Pattern of Seismic Activity between
Two Successive Great Earthquakes

Previous Works and Data Used

In 1969, the author discussed the space-time distribution of earthquakes of magnitude 6 and over before and after the 1944 Tonankai - the 1946

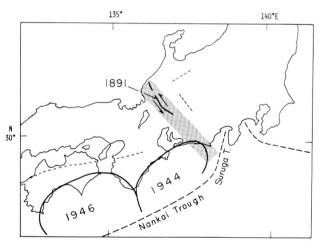

Fig. 7. Possible effect of the 1891 Nobi earthquake (M=8.4) to the 1944 Tonankai earthquake. The arrows show the left lateral displacement by the 1891 Nobi earthquake. This figure suggests that the rupture zone of the 1944 Tonankai earthquake did not extend to the Suruga trough by the interaction between the great intraplate earthquake and the great interplate earthquake, and so the Tokai region along the Suruga trough remained as a seismic gap.

Nankaido earthquakes, together with other great shallow earthquakes along the Japan - Kurile trench (Mogi, 1969). In this previous paper, the spatial and temporal variation in seismic activity before great shallow earthquakes has been found not only in the source region, but also in the wide surrounding region. In other words, the doughnut-shaped pattern has been found. Since the accuracy of earthquake data in the earlier period was low, the reliable discussion was limited to the period from 1926. Recently, Utsu (1979) published a reliable list of major earthquakes in and near Japan for the period from 1885 to 1925. Abe (1979) also reported the instrumental magnitudes of major earthquakes from 1901 to 1925. Now, we can discuss the space-time distribution of major earthquakes for about one hundred years which is nearly a recurrence time interval of the great shallow earthquakes in this region. In this paper, earthquake data were taken from Utsu (1979) for the period (1885-1925) and from JMA catalogue for the period (1926-1979).

Space-Time Pattern

Figure 8 shows the spatial distribution of shallow earthquakes of magnitude 6 and over for the successive six periods before and after the 1944 and the 1946 great shallow earthquakes.
(1) 1885-1898 Figure 8(1) shows the calm stage after the preceding great earthquake. The source regions of the future great earthquakes are shown by broken curves.
(2) 1899-1923 In this period, the seismic belt along the Nankai trough became markedly active (Fig. 8(2)). This activity began about 50 years before the following 1944 Tonankai - 1946 Nankaido earthquakes.
(3) 1924-1943 In this stage, the source regions of future great shallow earthquakes were quiet except for the Kii peninsula, and the surrounding region became active (Fig. 8(3)). That is, a seismic gap of the second kind (Mogi, 1979b) and the doughnut pattern appeared. This stage began about 20 years before the great earthquakes. Three major earthquakes occurred near the boundary between the 1944 and the 1946 rupture zones. Figure 9 shows the temporal variation in seismic activity within the source region. The above-mentioned three earthquakes which are indicated by closed circles seem to be the long-term foreshock activity. It should be noted that such foreshocks occurred near the boundary of the source regions of the 1944 and the 1946 earthquakes.
(4) 1944-1947 In 1944 and 1946, the Tonankai and the Nankaido earthquakes occurred. In this period, a number of aftershocks occurred in the source regions of these great earthquakes (Fig. 8(4)).
(5) 1948-1956 The seismic activity became quiet in the wide surrounding region, but a number of earthquakes including large ones of magnitude 7 and over continued to occur along the boundary between the source regions of the 1944 Tonankai and the 1946 Nankaido earthquakes (Fig. 8(5)). By this

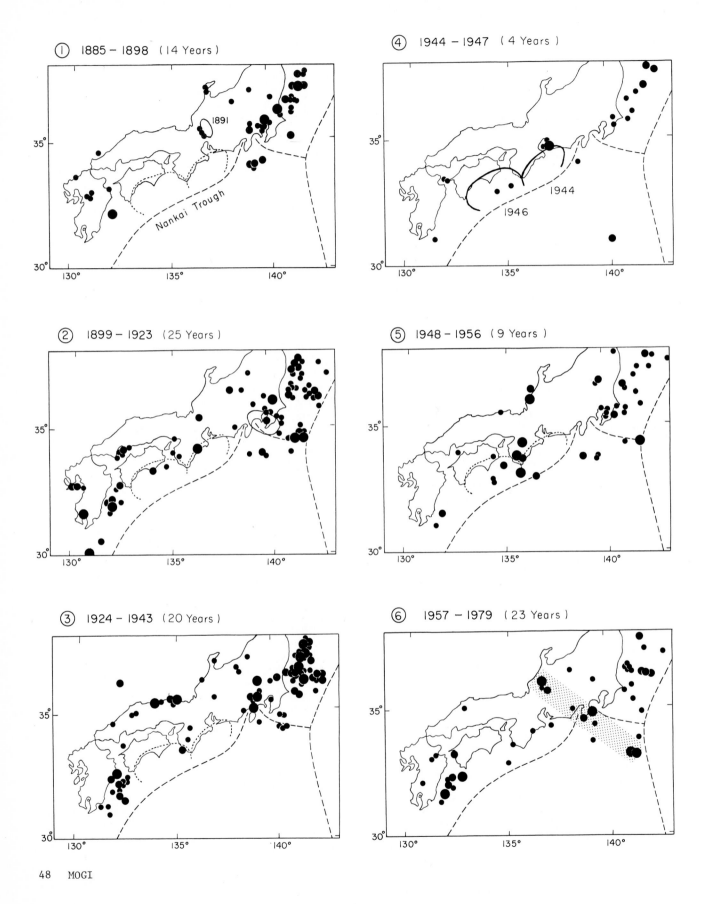

① 1885 – 1898 (14 Years)

1891

Nankai Trough

② 1899 – 1923 (25 Years)

③ 1924 – 1943 (20 Years)

④ 1944 – 1947 (4 Years)

1944

1946

⑤ 1948 – 1956 (9 Years)

⑥ 1957 – 1979 (23 Years)

Fig. 8. Spatial distribution of shallow earthquakes of magnitude 6 and over for the successive periods before and after the 1944 and the 1946 great shallow earthquakes. Large closed circle: M ≥ 7.0; small closed circle: 7.0 > M ≥ 6.0. The source regions of the 1944 Tonankai and the 1946 Nankaido earthquakes are shown by broken curves or solid curves. (1) 1885-1898 Western Japan is calm, except for the occurrence of the 1891 Nobi earthquake. (2) 1899-1923 The seismic belt along the Nankai trough became active. (3) 1924-1943 The source region of future great shallow earthquakes were quiet and the surrounding region became active. Three foreshocks occurred near the boundary between the two future great earthquakes. (4) 1944-1947 The 1944 Tonankai and the 1946 Nankaido earthquakes occurred. Their aftershocks occurred within the source regions. (5) 1948-1956 The wide surrounding region is quiet, but the activity is abnormally high along the boundary between the 1944 and the 1946 great earthquakes. By this seismic activity, the amount of slip of the subduction zone along the Nankai trough might be unified. (6) 1957-1979 Western Japan is calm, but the zone of NW-SE direction which passes through the Tokai region is active in the recent years.

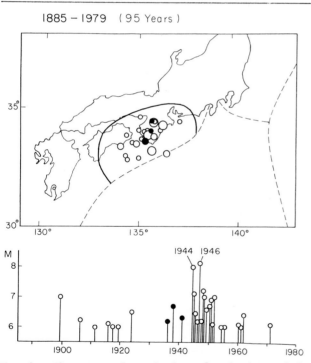

Fig. 9. The temporal variation of seismic activity within the source regions of the 1944 Tonankai and the 1946 Nankaido earthquakes during the period from 1885 to 1979. The top figure shows the location of major earthquakes of magnitude 6 and over and bottom figure shows the magnitude of these earthquakes as a function of time. Closed circles are foreshocks in a broad sense. The quiescence before the great earthquakes, except for the foreshocks, is noted.

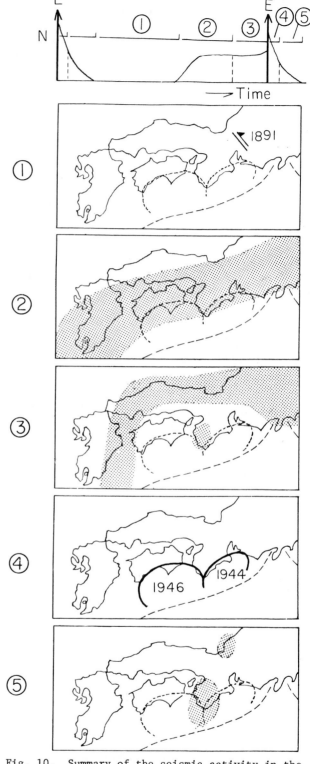

Fig. 10. Summary of the seismic activity in the successive five periods before and after the 1944 and the 1946 great earthquakes. Top figure shows the temporal variation of seismic activity in western Japan between two successive great earthquakes.

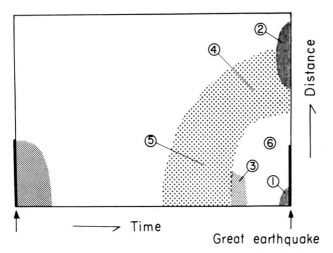

Fig. 11. The space-time distribution of seismic
activity between two successive great earthquakes
(revised Mogi (1977)). Various kinds of foreshock
activities, the first kind of seismic gap and the
aftershock activity are shown in the space-time
graph. Distance is from the epicenter of the great
earthquake. (1) is foreshocks in a strict sense
and (2) (3) (4) and (5) are different kinds of
foreshocks in a broad sense. (6) is the seismic
gap of the first kind.

remarkable aftershock occurrence along the boundary, the amount of slip of the subduction zone of
the Philippine sea plate might be unified along the
Nankai trough. The occurrence of the above-
mentioned foreshocks and large aftershocks along
the boundary between the two great rupture zones
suggests that this boundary zone is the important
mechanical singularity along the Nankai trough.
(6) 1957-1979 Figure 8(6) shows the following calm
stage. However, the zone of NW-SE direction which
passes through the Tokai region is markedly active
in the recent years (Utsu, 1977; Seno, 1979; Mogi,
1979a). This direction is that of the plate motion. This recent high activity including Izu peninsula suggests the high stress state in the Tokai
region.
    The main pattern of seismic activity in the
above-mentioned successive stages are summarized in
Fig. 10. The space-time pattern between two successive great earthquakes is schematically shown in
the space-time graph in Fig. 11.

Conclusion

    Western Japan along the Nankai trough is one of
the most suitable sites for studying the long-term
variation of the crustal activity in subduction
zones, because both recent and historical records
of seismic activity and ground movements have been
accumulated for a long period.
    First, the time of occurrence of great shallow
earthquakes along the Nankai trough is discussed
on the basis of the seismic data. The seismic

activity in western Japan increases appreciably
before the great earthquakes and decreases after
those events. The length of the preceding seismically active period (about 50 year) is independent of the time intervals between the successive
great shallow earthquakes and that of the quiet
period is nearly proportional to the time interval.
Since the seismic activity in the wide area is an
indication of stress level, this result shows that,
when the stress level reaches to a certain constant
value, a great shallow earthquake occurs. If it is
assumed that the stress increases linearly by the
plate motion of a constant rate, the time of occurrence of a great earthquake can be predicted approximately from the amount of the fault slip of
the preceding great earthquake, as pointed out by
Shimazaki and Nakata (1980). The difference in the
amount of the fault-slip appears to be related to
the rupture length.
    Secondly, the space-time distributions of major
shallow earthquakes between two successive great
earthquakes are discussed on the basis of seismic
data of about one hundred years. The result suggests the following stages before and after a great
shallow earthquake:

stage 1: quiet in the wide region.
stage 2: active in the belt along the trough.
stage 3: quiet in a source region of a future
         great shallow earthquake and active in
         the surrounding region (doughnut
         pattern).
stage 4: a great shallow earthquake with fore-
         shocks and aftershocks in a strict sense.
stage 5: quiet in a wide area, but active in a
         limited region (aftershocks in a broad
         sense).
stage 6: stage 1 in the next round.
This temporal variation obtained in the seismic
belt along the Nankai trough may be useful for
long-term earthquake forecasting in subduction
zones.

Acknowledgement

    I thank Norio Yamakawa for critical reading the
manuscript and for his helpful suggestions.

References

Abe, K., Instrumental magnitudes of Japanese
    earthquake, 1901-1925, J. Seismol. Soc. Japan,
    II, 32, 341-353, 1979.
Acharya, H. K., A method to determine the duration
    of quiescence in a seismic gap, Geophys. Res.
    Lett., 6, 681-684, 1979.
Ando, M., Source mechanisms and tectonic signifi-
    cance of historical earthquakes along the
    Nankai trough, Japan, Tectonophysics, 27, 119-
    140, 1975.
Bufe, C. G., P. W. Harsh and R. O. Burford,
    Steady-state seismic slip - a precise recurrence
    model, Geophys. Res. Lett., 4, 91-94, 1977.
Fitch, T. J. and C. H. Scholz, Mechanism of

underthrusting in southwest Japan: a model of convergent plate interactions, J. Geophys. Res., 76, 7260-7292, 1971.

Ishibashi, K., Re-examination of a great earthquake expected in the Tokai district, central Japan - Possibility of the "Suruga Bay Earthquake", Rep. Coord. Comm. Earthquake Prediction, 17, 126-132, 1977 (in Japanese).

Mogi, K., Study of elastic shocks caused by the fracture of heterogeneous materials and its relation to earthquake phenomena, Bull. Earthq. Res. Inst., 40, 125-173, 1962.

Mogi, K., Some features of recent seismic activity in and near Japan, 2. Activity before and after great earthquakes, Bull. Earthq. Res. Inst., 47, 395-417, 1969.

Mogi, K., Recent horizontal deformation of the earth's crust and tectonic activity in Japan, 1. Bull. Earthq. Res. Inst., 48, 413-430, 1970.

Mogi, K., Dilatancy of rocks under general triaxial stress states with special references to earthquake precursors, J. Phys. Earth, 25, Suppl., S203-S217, 1977.

Mogi, K., A feature of recent seismic activity in the Kanto-Tokai district, Rep. Coord. Comm. Earthquake Prediction, 21, 91-92, 1979a (in Japanese).

Mogi, K., Two kinds of seismic gaps, Pageoph, 117, 1172-1186, 1979b.

Ozawa, I., Forecast of occurrence of earthquakes in the northwestern part of Kinki district, Contrib. Geophys. Inst., Kyoto Univ., 13, 147-161, 1973.

Seno, T., Pattern of intraplate seismicity in southwest Japan before and after great interplate earthquakes, Tectonophysics, 57, 267-283, 1979.

Shimazaki, K., Intraplate seismicity and interplate earthquakes - historical activity in southwest Japan, Tectonophysics, 33, 33-42, 1976.

Shimazaki, K. and T. Nakata, Time-predictable recurrence model for large earthquakes, Geophys. Res. Lett., 7, 279-282, 1980.

Usami, T. Descriptive Catalogue of Disaster Earthquakes in Japan, Tokyo, 327pp, 1975 (in Japanese).

Utsu, T., Correlation between great earthquakes along the Nankai trough and destructive earthquakes in western Japan, Rep. Coord. Comm. Earthquake Prediction, 12, 120-122, 1974 (in Japanese).

Utsu, T., Possibility of a great earthquake in the Tokai district, central Japan, J. Phys. Earth, 25, Suppl., S219-S230, 1977.

Utsu, T., Seismicity of Japan from 1885 through 1925 - A new catalogue of earthquakes of $M \geq 6$ felt in Japan and smaller earthquakes which caused damage in Japan -, Bull. Earthq. Res. Inst., 54, 253-308, 1979 (in Japanese).

EVALUATION OF THE FORECAST OF THE 1978 OAXACA, SOUTHERN MEXICO EARTHQUAKE BASED ON A PRECURSORY SEISMIC QUIESCENCE

Masakazu Ohtake

National Research Center for Disaster Prevention
Tennodai, Sakura-mura, Niihari-gun, Ibaraki-ken, Japan

Tosimatu Matumoto and Gary V. Latham

Marine Science Institute of the University of Texas
700 The Strand, Galveston, Texas 77550

Abstract. The rupture zone and rupture type of the 1978 Oaxaca, southern Mexico earthquake (M_s = 7.7) were successfully predicted based on the premonitory quiescence of seismic activity and the spatial and temporal relationships of recent large earthquakes. The magnitude and epicenter location of the Oaxaca earthquake were well within the range of the most probable predicted values. Although the forecast was published based upon the observational data available as of the end of 1975, any modification of the initial conclusions was not brought about by adding new data down to the occurrence of the main shock. The additional data revealed that a renewed seismic activity, signaling the imminent stage of the earthquake, occurred in the quiescent area as expected.

Introduction

An abnormal quiescence of seismic activity preceding a large earthquake has been reported for a number of major earthquakes since the phenomenon was first reported by Inouye [1965] and was intensively investigated by Mogi [1968, 1969] and others. Based upon the observed characteristics of the seismic quiescence, Ohtake et al. [1977a] forecasted occurrence of a large earthquake near Oaxaca, southern Mexico in the near future. The anticipated earthquake took place on November 29 of 1978 (M_s=7.7).

The forecast, which was based upon the observational data available as of the end of 1975, included predictions of the following items; (1) rupture zone and rupture type, (2) magnitude and epicenter location, and (3) the pattern of seismicity change expected to precede the main shock. The present paper attempts to evaluate the results of the predictions by examining seismic data from the World-Wide Standardized Seismograph Stations. The source of data is the Monthly Listing of the Preliminary Determination of Epicenters (PDE) published by the National Earthquake Information Service of the United States Geological Survey.

The Oaxaca Quiescent Zone

Seismicity in the Mexico-Central America region is characterized by the active seismic belt along the Middle America Trench. Notable gaps of epicentral distribution are not found in the seismic belt so far as a seismicity map covering a long period is referred to (see Fig. 1). Systematically investigating the space-time pattern of shallow seismicity in the region, Ohtake et al. [1977a, b] found a remarkable seismic quiescence near Oaxaca, southern Mexico, beginning in mid-1973. Figure 2 compares the epicentral distribution of shallow earthquakes (H<60 km) for the initial two-year period of seismic quiescence and the preceding two years of normal seismicity. As delineated by a dashed box in the figure, a 270 km segment of the seismic belt between 95.5°W and 98.0°W, became aseismic in June 1973. The last earthquake took place on June 8 (m_b=4.5). All the earthquakes reported in PDE are plotted in Fig. 2. However, a frequency versus magnitude plot of earthquakes (Fig. 3) suggests that the PDE file is not complete for shallow earthquakes with body wave magnitude less than 4.5 in the Mexico-Guatemala region.

The seismic quiescence, which covered the entire state of Oaxaca, has cusomarily been called the Oaxaca gap [e.g., Garza and Lomnitz, 1978]. However, the terminology of the seismic gap is usually taken to refer to a portion of a seismic belt which does not experience a large rupture during the recent seismic sequence [e.g., Sykes, 1971]. In order to avoid confusion, we take the term of the "quiescent zone" for the area of precursory low seismicity.

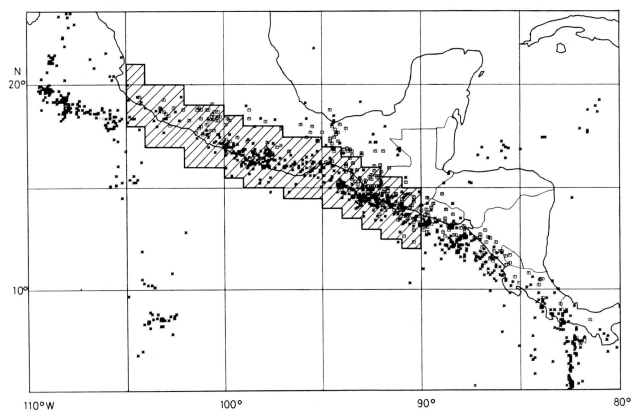

Fig. 1. Seismicity of the Mexico-Central America region for 1962-1969 (taken from the Seismicity of Middle America published by the National Earthquake Information Center, 1970). See the legend of Fig. 11 for the shaded zone.

The Oaxaca quiescent zone had lasted 2.6 years as of the end of 1975. The absence of earthquakes for such a prolonged period was concluded to be non-accidental by assuming a stationary random occurrence of earthquakes within the region. Under this assumption, the probability that x shocks take place during a time interval of t is expressed by a Poisson's distribution,

$$p(x) = \exp(-kt) \cdot (kt)^x / x! \qquad (1)$$

where k is the average frequency of shocks per unit time. Therefore, the probability that no shock occurs in a period equal to or longer than t is

$$P(0) = \int_t^\infty \exp(-kt) \, dt$$
$$= \exp(-kt)/k \qquad (2)$$

For shallow earthquakes (H<60 km) of $m_b \geq 4.5$, the most probable estimate of k for the Oaxaca quiescent zone is 3.09 per year because 34 such earthquakes, excluding five aftershocks, took place during the eleven years from 1963 to 1973. Substituting t=2.6 and k=3.09 into (2), P(0)=1.0 $\times 10^{-4}$ is obtained. The seismic quiescence could

scarcely have occurred by chance if the random model is an adequate description of the process.

Figure 4 illustrates frequency distribution of the time interval between the successive earthquakes. The distribution fairly well fits the exponential curve which corresponds to the stationary Poisson process so that the random assumption represented by the formula (1) will provide a good expression for the first approximation.

Forecast of the Pending Earthquake

In the Mexico-Guatemala region, five large shallow earthquakes (M_s>7.0, H<60 km) took place during the eleven years from 1964 to 1974. Investigating spatio-temporal changes in seismic activity, Ohtake et al. [1977b] found that without exception all of the five earthquakes were preceded by a period of seismic quiescence. Those cases provided empirical data by which to judge the seismic quiescence of Oaxaca as a possible precursor to a large earthquake.
Ohtake et al. [1977a] published a forecast of the possible Oaxaca large earthquake by analyzing the past seismicity of the region. Parameters of the pending earthquake were predicted to be φ=

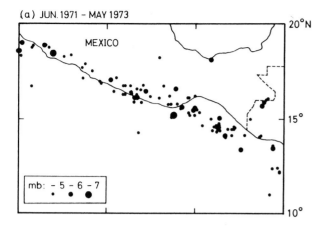

(a) JUN. 1971 – MAY 1973

MEXICO

mb: – 5 – 6 – 7

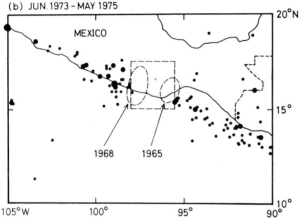

(b) JUN. 1973 – MAY 1975

MEXICO

1968   1965

105°W   100°   95°   90°

Fig. 2. Epicentral distribution of shallow
earthquakes (H<60 km) which took place in
the southern Mexico-Guatemala region during
the period of (a) June 1971–May 1973, and
(b) June 1973–May 1975. The quiescent zone
is outlined by a box. Two ellipses show
aftershock areas of the Oaxaca earthquakes
of 1965 and 1968, respectively [after Ohtake
et al., 1977a].

$16.5°\pm0.5°N$, $\lambda=96.5°\pm0.5°W$ for the epicenter
location, and $M_s=7\ 1/2\pm1/4$ for the magnitude,
respectively.

If a quiescent zone delineates the aftershock
area of the eventual main shock, as was the case
of the 1973 Nemuro-Hanto-oki, Japan earthquake
(M=7.4) for example, an estimate of the magnitude
can be obtained from the empirical relationships
of Utsu and Seki [1955], and Utsu [1961];

$$\log A = 1.02\ M - 4.0 \qquad (3)$$

$$\log L = 0.5\ M - 1.8 \qquad (4)$$

where M is the magnitude of the main shock, and
A and L are the area (km^2) and linear dimension
(km) of the aftershock zone. The Oaxaca quiescent zone was 270 km in linear dimension, and

roughly $7\times10^4\ km^2$ in area so that direct application of the formulae results in M=8.5–8.7.
However, the eastern and western parts of the
quiescent zone had been ruptured by recent
earthquakes in 1965 ($M_s=7\ 1/2$-7 3/4) and 1968 (M_s
=7.5). The central part of the quiescent zone
remained as a gap in the recent sequence of large
earthquakes.

For that reason, it was concluded that only the
central part of the Oaxaca quiescent zone (80–90
km in linear dimension) would be ruptured by the
next earthquake. The formula (4) predicts M=7.4
-7.5 for L=80–90 km. On the other hand, the
recent Oaxaca earthquakes of 1965 and 1968 which
ruptured the eastern third and western third of
the present quiescent zone were 7 1/2-7 3/4 in
the surface wave magnitude. Judging from the
size of the potential rupture zone, the similar
magnitude was expected for the pending earthquake. The consistency of the magnitude estimates from different sources added confidence to
the reliability of the magnitude prediction, $M_s=$
7 1/2±1/4.

Rupture of both the 1965 and 1968 earthquakes
atarted near the landward end of the final rupture zone, and propagated seaward along the plate
interface. The prediction of the epicenter was
made by assuming that the pending earthquake

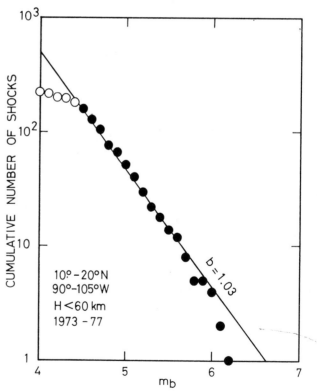

Fig. 3. Frequency distribution of body wave
magnitudes for shallow earthquakes in the
southern Mexico-Guatemala region. The open
circles are not used for computing the b
value. Data source is PDE for 1973-1977.

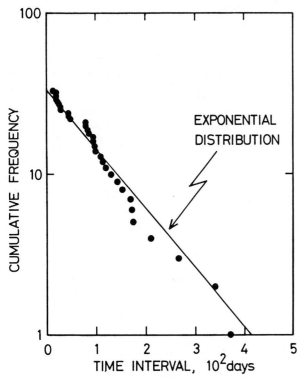

Fig. 4. Frequency distribution of the time interval between the successive earthquakes, and the corresponding exponential curve which is expected for the stationary Poisson process.

would also follow the similar rupture pattern.

A firm prediction of the occurrence time was not made because available relations between magnitude of the main shock and precursor time [Whitcomb et al., 1973; Scholz et al., 1973; Rikitake, 1976, 1978] give a large variation in the precursor time; 6-52 years for the magnitude 7 1/2±1/4 earthquake. However, it was remarked that a renewal of seismic activity in the quiescent zone would signal the imminence of the large earthquake. Ohtake et al. [1977a] proposed the following: "Systematic monitoring of microearthquakes may be the most effective and simplest method of obtaining a short-term warning in the Oaxaca case, since a renewal of activity preceding the main event is expected."

The 1978 Oaxaca Earthquake

On November 29, 1978, the forecasted earthquake took place near the Pacific coast of the State of Oaxaca, southern Mexico. It was 5.5 years since the seismic quiescence began. Parameters of the earthquake are listed in TABLE 1 in comparison with those of the prediction. As illustrated in Fig. 5, the focal mechanism of the earthquake was of the typical thrust type expected.

Figure 6 shows the time sequence of the shallow earthquakes which were located inside the Oaxaca

quiescent zone (15.0°-17.5°N, 95.5°-98.0°W) for the period of 1963-1978. The figure demonstrates the almost complete absence of earthquakes during the 4.6 years from June 1973 to December 1977. The event which occurred on July 23 of 1976 (m_b = 4.5) was located near the western margin of the quiescent zone. It would be eliminated from the figure if a slightly smaller area was defined as the Oaxaca quiescent zone (see Fig. 7). Another notable feature of the time sequence is the renewal of seismicity beginning on January 11, 1978. The period of quiescence and renewal of seismic activity are referred to as the α-stage and the β-stage of premonitory seismicity hereafter. Duration time of the α- and β-stages was 4.6 years and 0.9 year, respectively. Figures 7 and 8 are epicenter plots for the periods of the α-stage and the β-stage, respectively, for the Oaxaca case.

The probability that the α-stage, with a total duration of 4.6 years, appears by chance can be obtained following the same argument as given in the second section. The probability that only one shock occurs in a period equal to or longer than t is

$$P(1) = (t + 1/k) \cdot \exp(-kt) \qquad (5)$$

Therfore, the probability that one earthquake at most takes place in the duration time of 4.6 years or longer is computed to be P(0)+P(1)=3.5× 10^{-6} by substituting t=4.6 and k=3.09 into the formulae (2) and (5).

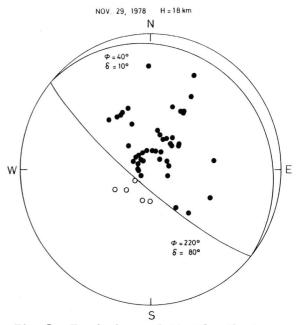

Fig. 5. Focal plane solution for the Oaxaca earthquake of November 29, 1978 (equal area projection on the lower hemisphere). Solid and open circles indicate compressional and dilatational initial motions of the P wave, respectively.

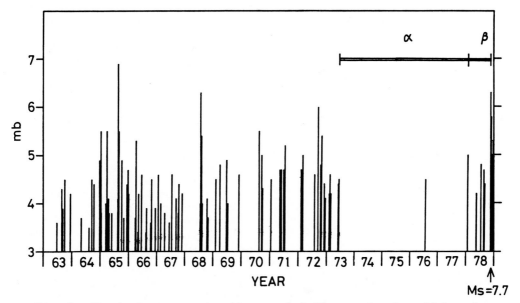

Fig. 6. Magnitude versus time diagram of shallow earthquakes which took place within the Oaxaca quiescent zone since 1963. Each event reported in PDE is plotted by a vertical segment with length proportional to the body wave magnitude. See the text for the periods designated as α and β.

Aftershocks that occurred during about one month following the main shock are plotted in Fig. 9. The aftershock area was roughly 90 km × 60 km. It is clearly seen that the aftershock area abuts those of the 1965 and 1968 events.

Discussions

The epicenter location and magnitude of the 1978 Oaxaca earthquake were well within the range of uncertainty in the forecast as shown in TABLE 1. The success of the forecast was a consequence of the validity of the basic concept that the

central part of the Oaxaca quiescent zone should be the source region of the coming earthquake. The spatial distribution of the aftershocks (Fig. 9) confirms that the main shock ruptured only the central portion of the zone.

Based upon the aftershock distribution taken from the quick report of PDE, Singh et al. [1980] suggested a possibility that the gap between the rupture zones of the 1965 and 1968 events might not be entirely broken at the time of the 1978 Oaxaca earthquake. However, most of the epicenters reported in the quick report are considerably, more than 30 km in some cases, deviated

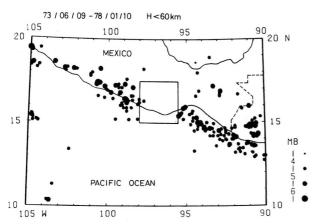

Fig. 7. Epicentral distribution of shallow earthquakes for the period of the α-stage. The box shows the Oaxaca quiescent zone that was defined in Fig. 2.

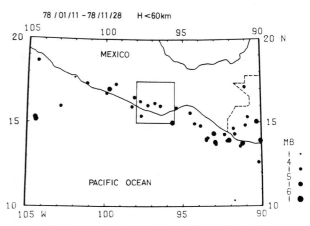

Fig. 8. Epicentral distribution of shallow earthquakes for the period of the β-stage. The box shows the Oaxaca quiescent zone that was defined in Fig. 2.

OHTAKE ET AL.   57

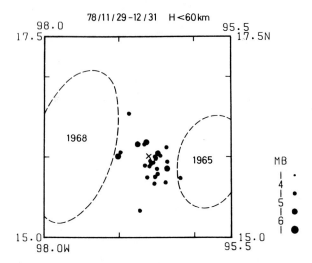

78/11/29-12/31 H<60km

MB
4
5
6

Fig. 9. Epicentral distribution of the aftershocks of the 1978 Oaxaca earthquake for the period of about one month following the main shock (cross mark). The ellipses are the aftershock areas of the large earthquakes of 1965 and 1968.

TABLE 1. Parameters of the 1978 Oaxaca Earthquake in Comparison with the Prediction

Parameter		Prediction	Observed*
Origin Time (UT)			Nov. 29, 1978 19h 52m 47.6s
Epicenter	Lat.	16.5°±0.5°N	16.010°N
	Lon.	96.5°±0.5°W	96.591°W
Depth		shallow	18 km
M_s		7 1/2±1/4	7.7

* after PDE

from those of the final version of PDE (Monthly Listing) because of limited number of stations. Figure 9, of which data were taken from the Monthly Listing of PDE, seems to be consistent with the idea that the gap between the past large earthquakes was entirely ruptured by the 1978 Oaxaca earthquake.

Figure 9 does not completely match the assumption that the rupture will start near the landward end of the rupture zone. The epicenter of the main shock was located about 20 km to the south of the northern boundary of the aftershock

1973 / 06 / 09 — 1978 / 01 /10

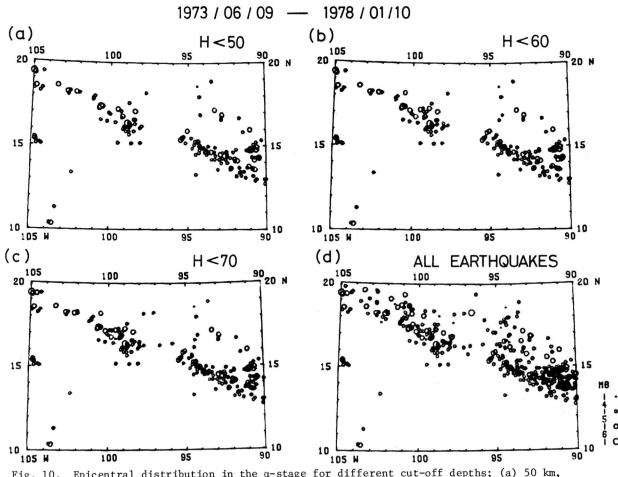

(a) H<50 (b) H<60 (c) H<70 (d) ALL EARTHQUAKES

MB
4
5
6

Fig. 10. Epicentral distribution in the α-stage for different cut-off depths; (a) 50 km, (b) 60 km, (c) 70 km, and (d) infinity.

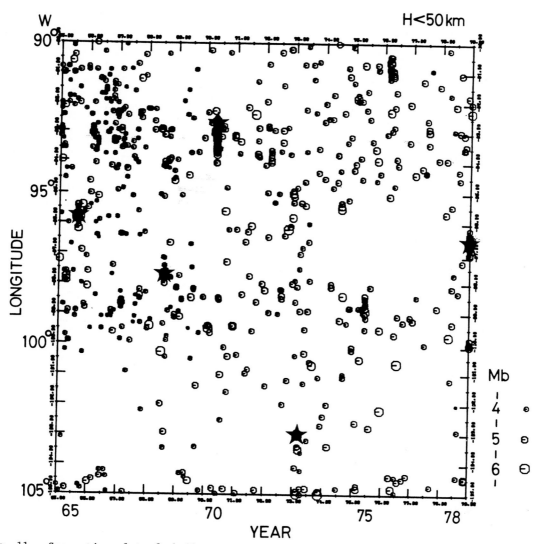

Fig. 11. Space-time plot of shallow earthquakes which occurred in the shaded zone of Fig. 1. Earthquakes are plotted on the plane where ordinate and abscissa are epicenter longitude and occurrence time, respectively. A star mark indicates a large earthquake with surface wave magnitude equal to or larger than 7.0.

area. The unsatisfactory assumption resulted in the southward deviation of the observed epicenter by about 50 km compared with the predicted epicenter. The deviation, however, was within the range of the uncertainty that was specified in the forecast.

It has been reported that seismically quiescent areas were completely filled by aftershocks for some of the typical cases in and near Japan [e.g., Mogi, 1968; Sekiya et al., 1974]. But, it is not always the case with the quiescence phenomenon. Compiling past research works on the precursory seismic quiescence, Ohtake [1980] pointed out that a quiescent zone is not always comparable with the aftershock area but scatters in a wide range of from one to ten times the aftershock area in linear dimension. The Oaxaca earthquake provides a typical example of a case

in which the quiescent zone is remarkably larger than the aftershock area. In such a case, a simple assumption that the area of seismic quiescence is equal to the rupture zone of the main shock would result in overestimation of the size of the pending earthquake. The present study demonstrates that the combined investigation of seismic quiescence and the history of recent large earthquakes in the quiescent zone, can be effective for successful forecast of a large earthquake.

A renewal of seismic activity in the quiescent zone (β-stage) did occur as expected (see Figs. 6 and 8). The β-stage started 0.9 year prior to the main shock. The renewal of seismic activity is not a rare phenomenon. According to Ohtake [1980], such a renewed seismicity was found at least for 28 cases (35 %) among the 81 seismic quiescences so far reported. For the Mexico-Central America region particularly, all of the large shallow earthquakes ($M_s>7.0$, H<60 km) which occurred during 1964-1974 are reported to have been preceded by the β-stage of premonitory seismicity change [Ohtake et al., 1977b].

We delineated the Oaxaca quiescent zone based on the seismicity map for shallow earthquakes with the focal depth less than 60 km (Fig. 7). The cut-off depth was selected following the previous studies for the Mexico-Central America region [Ohtake et al., 1977a, b]. Figure 10, comparing the epicentral distribution for different cut-off depths, demonstrates that selection of the cut-off depth is not crucial for the present case. As shown in Fig. 10 (d), the Oaxaca quiescent zone can still be recognized even if the focal depth is not limited. It is interesting that the large intermediate-depth earthquake (m_b=6.8) plotted at 18.3°N, 96.6°W in Fig. 10 (d) took place on August 28 of 1973, coinciding with the start of the seismic quiescence by a margin of three months.

Figure 11 is a space-time plot of seismicity for shallow earthquakes in the Mexico-Guatemala region. The aseismic character of the Oaxaca region is clearly shown. Temporal low seismicity of comparable size is also seen in the western region (100°-105°W) for the period of 1965-1968, 1970-1972, and 1976-1978. A large earthquake corresponding to the first gap in epicentral distribution was not found while the second one was followed by the Michoacan earthquake (M_s=7.5) of January 30, 1973. The third seismic quiescence may be related to the Guerrero earthquake (M_s=7.6) of March 14, 1979. The epicenter was 17.813°N, 101.276°W according to PDE. Thus, three of the four cases of the most remarkable seismic quiescences in the region are associated with large earthquakes of magnitude larger than 7.0.

Conclusions

The epicenter location and magnitude of the 1978 Oaxaca earthquake were successfully predicted based on the premonitory seismic quiescence and the seismic gap of the past large earthquakes. From a scientific point of view, it is important that the rupture zone of the main shock was accurately predicted.

Seismicity in the Oaxaca area followed the predicted pattern; (1) seismic quiescence for a prolonged period (α-stage), (2) renewal of seismic activity (β-stage), and (3) occurrence of the main shock. If the β-stage is common for all large earthquakes, highly sensitive observation of earthquakes will be quite effective for improving the time accuracy of a long-term prediction. Detection of a significant increase in seismic activity will not pinpoint the time-of-occurrence, but it will serve as a warning that the time remaining before the earthquake is not long. Further investigations including precise

location of hypocenters and focal mechanism studies are encouraged for revealing the physical mechanism of the β-stage.

The present study demonstrates that seismic quiescence is one of the most promising precursory phenomena for long-term prediction of a large earthquke in an active seismic belt. The phenomenon is quite useful to delineate the area where intensified observations for the short-term prediction should be concentrated.

Acknowledgements. We wish to thank Dr. Ray E. Habermann for helpful comments on the manuscript, which resulted in substantial improvements of the paper.

References

Garza, T., and C. Lomnitz, The Oaxaca gap: A case history, U. S. Geological Survey Open-File Report, 78-943, 173-188, 1978.

Inouye, W., On the seismicity in the epicentral region and its neighbourhood before the Niigata earthquake, Kenshin-Jiho, 29, 139-144, 1965 (in Japanese).

Mogi, K., Some features of recent seismic activity in and near Japan (1), Bull. Earthq. Res. Inst., 46, 1225-1236, 1968.

Mogi, K., Some features of recent seismic activity in and near Japan (2). Activity before and after great earthquakes, Bull. Earthq. Res. Inst., 47, 395-417, 1969.

Ohtake, M., Earthquake prediction based on the seismic gap with special reference to the 1978 Oaxaca, Mexico earthquake, Rep. National Res. Center for Disaster Prevention, 23, 65-110, 1980 (in Japanese).

Ohtake, M., T. Matumoto, and G. V. Latham, Seismicity gap near Oaxaca, southern Mexico as a probable precursor to a large earthquake, Pure Appl. Geophys., 115, 375-385, 1977a.

Ohtake, M., T. Matumoto, and G. V. Latham, Temporal changes in seismicity preceding some shallow earthquakes in Mexico and Central America, Bull. Internat. Inst. Seism. Earthq. Eng., 15, 105-123, 1977b.

Rikitake, T., Earthquake Prediction, Elsevier, Amsterdam, 357 pp, 1976.

Rikitake, T., Earthquake prediction, Physics of the Earth (ed. H. Kanamori), Iwanami-Shoten, Tokyo, 169-210, 1978 (in Japanese).

Scholz, C. H., L. R. Sykes, and Y. P. Aggarwal, Earthquake prediction: A physical basis, Science, 181, 803-810, 1973.

Sekiya, H., S. Hisamoto, E. Mochizuki, E. Kobayashi, T. Kurihara, K. Tokunaga, and M. Kishio, The off-Nemuro Peninsula earthquake of 1973 and the large earthquakes off southern part of Hokkaido, Kenshin-Jiho, 39, 33-39, 1974 (in Japanese).

Singh, S. K., J. Havskov, K. McNally, L. Ponce, T. Hearn, and M. Vassiliou, The Oaxaca, Mexico, earthquake of 29 November 1978: A preliminary report on aftershocks, Science, 207, 1211-1213, 1980.

Sykes, L. R., Aftershock zones of great earthquakes, seismicity gap, and earthquake prediction for Alaska and the Aleutians, J. Geophys. Res., 76, 8021-8041, 1971.

Utsu, T., A statistical study on the occurrence of aftershocks, Geophys. Mag., 30, 521-605, 1961.

Utsu, T., and A. Seki, A relation between the area of after-shock region and the energy of main-shock, Zisin, Ser. 2, 7, 233-240, 1955 (in Japanese).

Whitcomb, J. H., J. H. Garmany, and D. L. Anderson, Earthquake prediction: Variation of seismic velocity before the San Fernando earthquake, Science, 180, 632-635, 1973.

PLATE SUBDUCTION AND PREDICTION OF EARTHQUAKES ALONG THE MIDDLE AMERICA TRENCH

Karen C. McNally

Seismological Laboratory, California Institute of Technology, Pasadena, California 91125

Abstract. Results of detailed studies of seismic
slip, tectonic structures, fault mechanisms, and
rupture patterns along various segments of the
Middle America Trench provide a tectonic
framework for analyzing long, intermediate, and
short term patterns of seismicity preceding large
($M_s \geqslant 7$), shallow earthquakes.
Long term. Average repeat times of 33 ± 8 yrs for
earthquakes $M_s \geqslant 7.5$ since 1898 and 35 ± 24 yrs
since 1540 indicate that the 81 yr history (1898
to 1979) may be adequate for estimating temporal
deficiencies in seismic slip. Seismic slip rates
are spatially nonuniform on a scale of 100 km and
forecasts are only valid when local conditions
(eg. subduction of seafloor topography) are
considered. The current temporal deficiency in
seismic slip offshore Mexico and Guatemala is
equivalent to 5 earthquakes of $M_s = 8$ in the next
5 years, or 6 in the next 10 years. The most
likely areas for the next large shallow
earthquakes ($M_s \geqslant 7.5$) are between longitudes
$98.5°$–$99.7°$W, $97.2°$–$97.7°$W, and $90°$ – $92°$ W. A
major episode of 6 earthquakes with $7.9 \leqslant M_s \leqslant
8.2$ occurred between 1899 and 1907. At least 3
of these events were in the regions listed above,
where average repeat times have been 31.7 ± 9.2
yrs ($M_s \geqslant 7.5$).
Intermediate term. Five of the six largest
earthquakes ($M_s \geqslant 7$) since 1965 were preceded by
between 19 to 46.5 months of relative seismic
quiescence ($m_b \geqslant 4$), with probabilities of 0.8%
to 2.6% of this quiescence having occurred due to
chance. Between $98.5°$ to $99.7°$W, a conspicuous
seismic quiescence has persisted for 48 months;
this quiescence has a 0.05% probability of
occurring due to chance.
Short term. A local field array of seismographs
documented the following patterns during the 3
weeks prior to the 29 November 1978 earthquake
($M_s = 7.8$) in Oaxaca, Mexico:
(1) The mainshock location was surrounded by a
seismically quiet area of 2800 km², which was
broken by only a single cluster of activity 2
weeks before the final foreshock sequence.
(2) At 1.8 days before the mainshock, the
foreshock sequence began migrating NE to SW
through the quiescent zone toward the eventual

rupture point; activity then subsided for 12
hours until failure.
(3) The foreshock sequence was identifiable, on
a statistical basis, relative to previous
activity during the 3-week period. A significant
result of this study was the documentation of a
foreshock sequence at energy levels below
standard WWSSN detection thresholds, but
otherwise similar to other sequences observed
worldwide. This observation suggests the
possibility that many large earthquakes may have
foreshock activity, which has not been detected
at teleseismic distances, and that small
earthquakes are at least as sensitive as
indicators of the level of stress as are larger
($m_b \geqslant 4$) earthquakes.

Introduction

A comprehensive study of the Middle America
Trench has been undertaken for the purpose of
identifying, and determining the physical basis
for, seismicity patterns associated with the
occurrence of large earthquakes. This region was
selected for its relatively uniform faulting
mechanism (underthrusting of the Cocos plate),
simple tectonic history, and frequent recurrence
of large earthquakes compared with the available
seismic record. This paper summarizes the
conclusions of these studies to date and is
organized in accordance with the concept of long,
intermediate and short term behavior prior to a
major throughgoing rupture.

Long-Term Behavior

Kelleher et al. [1973] and McCann et al.
[1980] have contributed substantial information
regarding seismic gaps on a worldwide basis.
These studies provide a foundation for further
research regarding long term earthquake
forecasts. McNally and Minster, who compiled a
revised catalog of earthquakes along the Middle
America Trench [1981], found that seismic slip
rates are lower than plate convergence rates
(Minster and Jordan [1978]) on the average but
match these rates locally. Along the Cocos-North

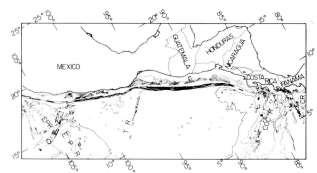

Fig. 1. Shallow earthquakes (h ≤ 60 km) since 1898; "x": M_s ≥ 7, "*": M_s ≥ 7.5, "star": M_s ≥ 8.0 (McNally and Minster, [1981]). Bathymetry of the Cocos plate, Chase et al [1970]. Solid lines offshore indicate depths of the Middle America Trench ≥ 2400 fathoms (≃ 4.3 km); seafloor topographic highs are indicated by dotted lines (depths ≤ 1800 fathoms [≃ 3.3 km]). Features of the seafloor shown are (left to right): East Pacific Rise (EPR), Orozco Fracture Zone (OFZ), Tehuantepec Ridge (TR), Cocos Ridge (COC R), Coiba Ridge (CR).

American plate boundary, this discrepancy can be explained by nonuniformities in slip at points of aseismic ridge or fracture zone subduction (Figure 1) (McNally and Minster [1981]; McNally et al. [1981a]). In this region recurrence periods for large (M_s ≥ 7.5) earthquakes are 33 to 35 yrs . However, a larger discrepancy between plate convergence and seismic slip rates exists along the Cocos-Caribbean plate boundary; this is more likely caused by decoupling and downbending of the subducted plate. Cifuentes [1981] has found that repeat times for large earthquakes may reach 50-100 yrs in this region. Using these observations, the release of seismic moment during the last 18 years is compared with that for the last 50 years along the trench in order to infer sites of present day deficiencies in seismic slip (Figure 2). Areas with relatively high moment release during the 50-year period are not shaded gray in Figure 2 and are being investigated for their present seismic potential. Several conspicuous seismic gaps left during the 18-year period (1960-1978) had been the sites of large events between 1928 and 1960. Among the most obvious is one near Jalisco (distance [d] ≃ 650 km in Figure 2) with a last significant sequence in 1932-1934. The gap near Petatlan (d ≃ 950 km) was the site of a M_s = 7.6 event in 1979; this event is not included in this graph. Another obvious gap, having no significant activity since 1950-1953, lies between 90° and 92°W along the Guatemala coastal region (d ≃ 2050 to 2250 km). In addition, a broad seismic gap between 98.5° and 99.7°W appears in the Acapulco region (d ≃ 1180 to 1300 km) and a narrow gap from 97.2° to 97.7°W appears on the NW side of the 1978 Oaxaca event (d ≃ 1450

km), with the last significant ruptures in 1957 and 1928, respectively. A special study of the Jalisco region (Eissler and McNally [1981]) indicates that the 1932 earthquake occurred north of the Rivera plate/Cocos plate boundary; relative plate convergence rates predict that the repeat time should double across this boundary. We therefore conclude that the regions of Acapulco and Oaxaca, Mexico, and Guatemala are sufficiently anomalous to deserve special attention.

The seismic moments for the 6 largest events in this region from 1965 through 1979 (Stewart et al. [1981]; Chael et al. [1981]) and the five largest events 1928-1964 (Wang and McNally [1981]) have been determined using surface- and body-wave data. The resulting cumulative seismic slip for the offshore Mexico region (90°W to 105°W) is shown in Figure 3, where slip is assumed to be uniform along the 1700-km-long trench, dipping at 45° with depth ≤ 60 km (Davies and Brune [1971]; McNally and Minster [1981]; Wang and McNally [1981]). Based on the 35-year period 1907 to 1942, the average slip rate is only 3.4 cm/yr, which is approximately equal to one recurrence period in this region. The average slip rate is 3.7 cm/yr from 1898 to 1979, where the relation Log M_O = 1.5 M_s + 16.0 and the magnitudes of all instrumentally recorded earthquakes are used. However the seismograms for earthquakes at the turn of the century have

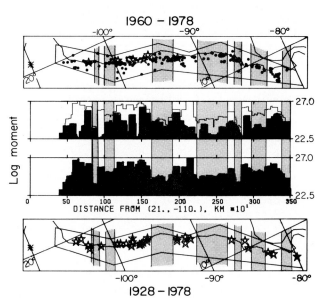

Fig. 2. Comparison of seismicity maps and cumulative moment release for 1960-78 (2 top frames) and 1928-78 (2 bottom frames). Only events with M ≥ 7 are shown on bottom frame. Outline of moment release for 1928-78 is repeated on second frame for comparison. Shaded bands outline areas without moment release in earthquakes of M_s ≥ 7.0 during the 50 yr period (after McNally and Minster, [1981]).

not been reanalyzed to confirm the seismic moment. A temporal deficiency in seismic slip is currently indicated (arrows, Figure 3). This deficiency is equivalent in moment release to five earthquakes of $M_s = 8.0$ in the next 5 years or six in the next 10 years. Historically, a major episode of six earthquakes with $7.9 \leqslant M_s \leqslant 8.2$ did occur during a 9-year period at the turn of the century (1899-1907)(McNally and Minster [1981]). At least three of these events occurred in the regions of Guatemala, Acapulco, and Oaxaca (discussed above), where the average repeat times are 31.7 ± 9.2 yrs. A fourth occurred in Jalisco.

Intermediate-Term Behavior

Fault mechanisms and rupture patterns for the six largest earthquakes since 1965 have been studied using surface- and body-wave synthetic seismograms: 1973 ($M_s = 7.5$, Colima), 1978 ($M_s = 7.8$, Oaxaca), 1979 ($M_s = 7.6$, Petatlan), 1970 ($M_s = 7.3$, Chiapas), 1965 ($M_s = 7.6$, Oaxaca), and 1968 ($M_s = 7.1$, Oaxaca)(Stewart et al. [1981]; Chael et al. [1981]). These studies determined that all events are (1) predominantly thrust type with dip angle of $12°$ to $20°$, consistent with subduction to the NE of the Cocos plate; (2) shallow, with focal depths of 15 to 20 km; and (3) characterized by low stress drops on the order of 10 bars. The events have seismic moments ranging from 1.0×10^{27} to 3.2×10^{27} dyne-cm. The long-period body waves all indicate a simple faulting process except for the 1970 and 1973 events, which lie in close proximity to plate triple-junction points. Presumably these triple-junctions result in fault zone complexities.

Having ascertained the basic similarity of faulting in these events, the patterns of moderate seismicity ($m_b \geqslant 4$) in each rupture zone prior to the mainshock and the seismicity in the remaining areas along the Cocos-North America plate boundary that have had earthquakes with $M_s \geqslant 7.5$ during historic time are investigated. The well-determined aftershock zones from special studies of each event since 1965 are shown in Figure 4. The areas selected for seismicity analyses, however, are controlled by the locations of aftershocks obtained from standard catalogs, except in the case of the 3 Oaxaca events. In this case all seismicity between 1964-1978 has been relocated (Tajima and McNally [1980]). (The width of the seismically active zone increases significantly from west to east in Oaxaca and may result from bending of the trench; the boundaries for the seismicity study are taken accordingly [Figure 4]). The boundaries for the Acapulco area are based on the aftershock zone of the 1957 ($M_s = 7.5$) earthquake (Kelleher et al. [1973]). NOAA and ISC catalogs (as available) have been combined for completeness ($m_b \geqslant 4$); no events are excluded by depth because it is not reliably determined (Tajima and McNally [1980]).

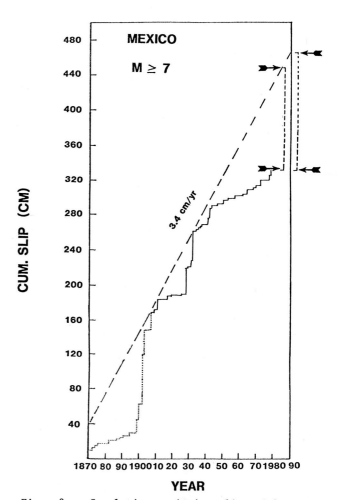

Fig. 3. Cumulative seismic slip with time, assuming slip is uniform along the trench offshore Mexico ($90°W$ - $105°W$)(after Wang and McNally, [1981]). Seismic moments for all earthquakes with $M_s \geqslant 7.8$ since 1928 and other earthquakes in 1965, 1968, 1970, 1973 and 1979 are from seismogram analyses (Wang and McNally, [1981]; Stewart et al. [1981]; Chael et al, [1981]); all others are estimated from $\log M_o = 1.5 M_s + 16.0$ (McNally and Minster [1981]). Data prior to 1898 are from Singh et al, [1981]. The arrows show the amount of slip needed to "catch up" in the next 5 or 10 years.

Prior to 1964, the catalogs are much less reliable for a threshold of $m_b \geqslant 4$. Other problems of catalog homogeneity are discussed by Tajima and McNally [1981c] and Habermann [1981].

The cumulative number of events associated with each mainshock-aftershock region is shown as a function of time in Figure 5. Since most earthquakes- excepting the mainshocks in 1965 and 1968- are about m_b 4-5, these graphs can also be taken to represent moment release or seismic strain-energy release locally with time. A drop in seismic activity, relative to the rates during

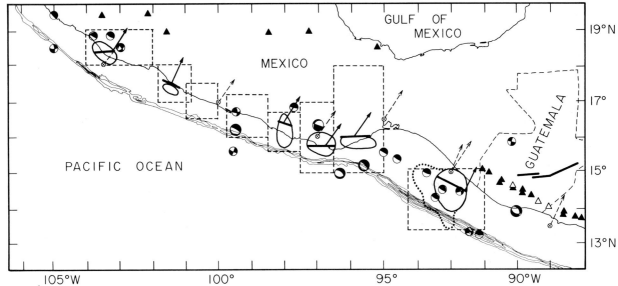

Fig. 4. Aftershock zones from special studies and areas delimited for seismicity analyses in Figures 5, 6 and 7 (dashed outlines). From left to right: Colima 1973 (Ms = 7.5)(Reyes et al, [1979]); Petatlan 1979 (Ms = 7.6)(Meyer et al, [1980]; Gettrust et al, [1981]); Tecpan; Acapulco; Oaxaca 1968 (Ms = 7.1)(Tajima and McNally, [1981]); Oaxaca 1978 (Ms = 7.8) (Singh et al, [1980]); Oaxaca 1965 (Ms = 7.6)(Tajima and McNally, [1981]); Chiapas 1970 (Ms = 7.3)(Yamamoto, [1978]). The fault strike and slip direction on the fault plane for each mainshock are shown as bars with arrows (Stewart et al, [1981]; Chael et al, [1981]). Dashed arrows indicate relative plate motions (Minster and Jordan, [1979]; McNally and Minster, [1981]); 2 arrows are shown at 93°W due to uncertainty in the North America- Caribbean plate boundary position. Small fault mechanisms are from Dean and Drake (1978): in some areas fault complexity with strike-slip and normal faulting are suggested. Volcanoes are shown as triangles.

the preceding 70 to 92 months, is apparent beginning 24, 43, and 30-46.5 months before the mainshocks in, respectively, 1973 (Colima), 1978 (Oaxaca) and 1979 (Petatlan). For a Poisson process, the probability of observing x events in a time interval t, where (λ) is the mean rate, is

$$\text{prob } (N_t = x) = [(\lambda t)^x e^{-\lambda t}]/x! \quad (x = 0,1...).$$

Based on the preceding background rates, the probabilities of observing one event in 24 months, five events in 43 months, and five events in 44 months due to chance are 0.9%, 2.6% and 1.2%, respectively. Forty-seven days of significant foreshock activity followed the 44 months of relative quiescence before the 1979 Petatlan event and are not included in the probability calculation. A drop in activity during the 15 months before the 1970 Chiapas earthquake does not appear significant on a visual basis, but the probability of observing 18 events in 15 months due to chance is only 0.8%, relative to the previous rate. A significant foreshock sequence occurred within 18 days preceding the mainshock and is not included in the probability calculation. Due to large mainshock-aftershock sequences in 1966, the previous rate is based on only 25 months of observation, rather than the preferred longer

time average (≥ 70 months) available for the events discussed previously. Activity was low (three events) for at least 19 months before the 1965 Oaxaca earthquake. (Prior to 1964 the catalog is clearly incomplete, and the onset of seismic quiescence cannot be determined with reliability.) If the activity from 1966-1975 is taken as a "background" rate, the probability of observing only three events in 19 months due to chance is 2.2%. (Note that the seismic quiescence beginning in 1975 which was associated with the 1978 Oaxaca earthquake also affected the adjoining areas of the 1965 and 1968 earthquakes. This was also observed by Ohtake et al. [1977]; Kanamori [1981]; and Habermann [1981]). In summary, the five earthquakes --in 1965 (Oaxaca), 1970 (Chiapas), 1973 (Colima), 1978 (Oaxaca), and 1979 (Petatlan)-- were all preceded by > 19 months of relative seismic quiescence locally, with probabilities of ≤ 2.6% of occurring due to chance. In contrast, the 22 months prior to the 1968 (Oaxaca) earthquake were marked by an increase in local activity (Figure 5). This pattern is different from that reported by Ohtake et al. [1977] as a result of additional events listed by ISC (Tajima and McNally [1981]). Fourteen events concentrated in the rupture zone during this time period, with a 4.8% chance probability relative to the "background" rate

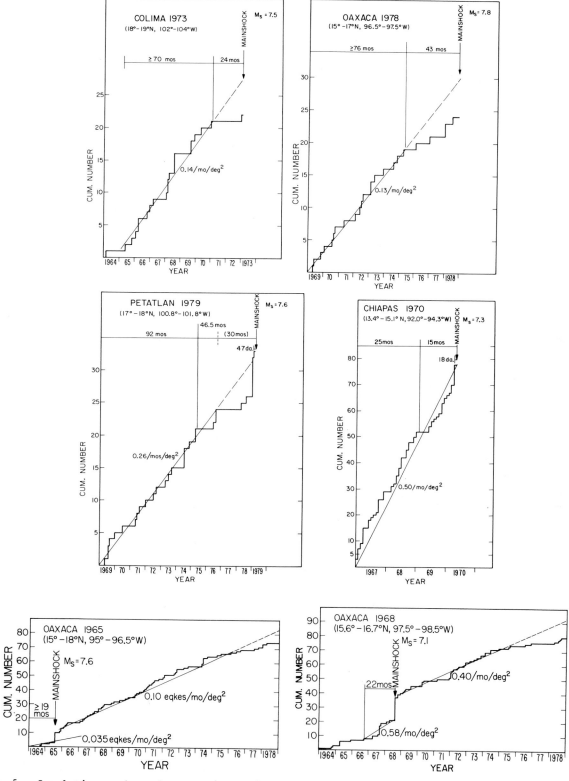

Fig. 5. Cumulative number of events ($M_b > 4$) with time for regions in Figure 4. Arrows indicate times of mainshocks. Rates normalized by area are also shown and are based on the time periods for which the rates are shown as solid lines.

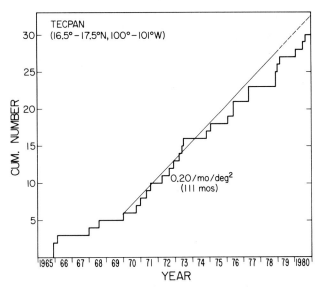

Fig. 6. As in Figure 5.

large ($M_s \geq 7.5$) earthquakes since 1911 (?) (uncertain location) and 1957, respectively. A possible reduction in activity began near Tecpan in June 1977 (Figure 6). The chance probability of seven earthquakes in 44 months relative to the preceding rate from 1970- May 1977 (where we require ≥ 70 months to establish the prior "background" rate) is 12.0%. This probability indicates that no significant quiescence exists in this region. A conspicuous quiescence began near Acapulco in January 1977 and has persisted for 48 months (through December 1980). Relative to the rate observed for 72 months from 1971-1976, the probability of observing only seven events in 48 months due to chance is only 0.05%, a value which is comparable with that for the patterns observed preceding the large earthquakes described above. Maps of seismicity in the Acapulco region for equal time periods are shown in Figure 8. Only three events, and one cluster of four events in a single day, are observed during the last 4 years; the last similar clustering occurred in 1967. The recent change in activity from the previous period, 1964-1976, is clear. The last major earthquakes ($M_s \geq 7.5$) in this general region (± 100 km) occurred in 1957 ($M_s = 7.5$), 1937 ($M_s = 7.5$), and 1907 ($M_s = 8.0$). Although the locations of the 1907 and 1937 events are not very well determined, it does appear that this region is characterized by frequent large events. A

between 1969 and 1974. The physical basis for the deviation of this pattern from those preceding all other mainshocks is the focus of current studies. The 1968 earthquake is not distinguished by fault dip or strike, stress drop, hypocentral depth, or long period body waveform complexity. However, it is the smallest of the events studied and the background seismicity rate is high relative to that in all other areas except the 1970 event. The latter observations are consistent with the model of Kanamori [1981] and Lay and Kanamori [1981], who propose that differences in premonitory seismicity patterns result from relative differences in the distributions of asperity sizes and strengths producing the "background" seismicity and the mainshock rupture. We suggest the possibility that background seismicity rates might be used to predict the type of precursory pattern expected for each local region. Current studies suggest that the relative complexity of rupture as indicated by the mainshock short-period body waveforms at teleseismic distances can be correlated with the pattern of seismic quiescence or foreshock type activity (Tajima et al.[1980]). The characteristic wavelengths measured by short-period ($T_o = 1$ sec) seismometers are probably more appropriate for studying local asperities or heterogeneities on the length scale of the m_b 4 - 6 events (1-15 km) than are the wavelengths of long-period ($T_o = 15$ sec) seismometers used in the studies by Stewart et al. [1981] and Chael et al. [1981]. Long-period seismometers are more likely to reflect larger asperities or heterogeneities on a scale of 50-100 km.

The seismicity patterns are investigated for the remaining coastal regions, near Tecpan and Acapulco (Figures 4, 6 and 7), which have had no

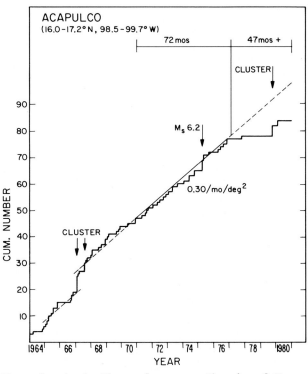

Fig. 7. As in Figure 5; an earthquake of $M_s = 6.2$ and earthquake clusters are also indicated.

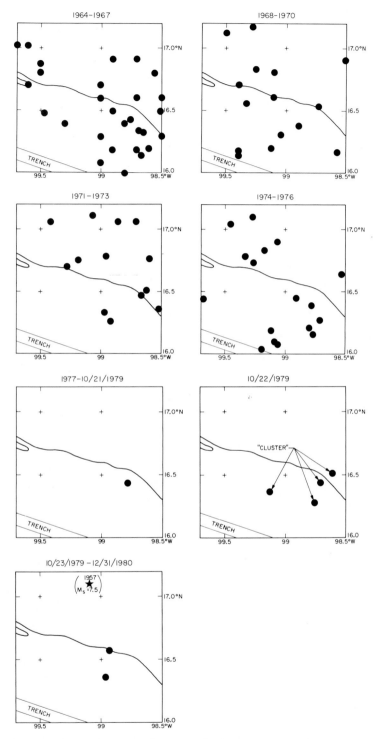

Fig. 8. Maps of seismicity near Acapulco (Figure 7) during equal time periods. The locations used are primarily those given by ISC (for greater reliability with more reporting stations). Otherwise, NOAA/PDE locations are used. The location of the 1957 earthquake (M_s = 7.5) is shown in the last frame.

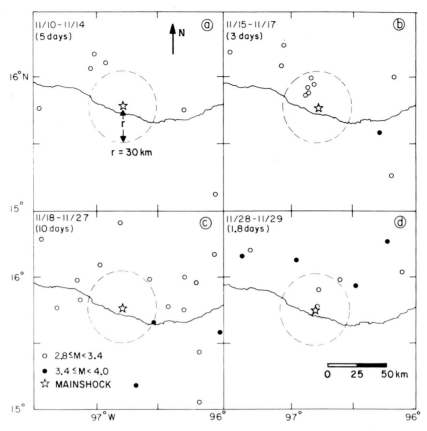

Fig. 9. Foreshocks (M $\geq$ 2.8) to the Oaxaca (M$_s$ = 7.8) earthquake for the time periods indicated, up to the mainshock on 29 November 1978, 19h. 52 m. GMT.

cluster of earthquakes preceded the 1957 earthquake by 18 months, but no short-term foreshock sequence was observed. Due to large uncertainties in earthquake locations and reporting, special studies are required to establish the detailed seismicity patterns during this earlier period.

It is concluded that while the seismicity patterns near Acapulco are similar to others that are observed prior to large earthquakes in this region, our current knowledge is not sufficient to predict the exact time, magnitude, and location of future earthquakes; a special study of this area is warranted, however. Out of the six case histories discussed above, all four of the mainshocks with M$_s$ $\geq$ 7.5 were preceded by a relative seismic quiescence (m$_b$ $\geq$ 4) in the rupture area, which began between 19 and 46.5 months before the mainshock. A short-term (47 days) foreshock sequence was also observed for one of the four events at the same magnitude threshold. At least 70 months of local activity should be studied to establish the background rate for determining the statistical significance of a change in activity; a probability of $\leq$ 3-5% for chance occurrence may indicate significant quiescence. All seismicity (m$_b$ $\geq$ 4) listed by

NOAA/PDE and ISC that coincides with local rupture areas should be considered because depth is not well determined.

Short-Term Behavior

As previously discussed, only two of the six largest earthquakes since 1965 were preceded by short-term (18 and 47 days) foreshock activity (m$_b$ $\geq$ 4) that could be detected at teleseismic distances. Data regarding foreshock activity at smaller magnitude thresholds are extremely rare. In a special field seismograph study by Caltech and the Universidad Nacional Autonoma de Mexico (UNAM) (McNally et al. [1979]; Ponce et al. [1977-78]; McNally et al. [1981b]), we were able to document the detailed seismicity patterns during the 3 weeks prior to the 29 November 1978 (M$_s$ = 7.8) earthquake in Oaxaca, Mexico. All seismicity (M $\geq$ 2.8) is shown for time periods from 10 November 1978 up to the mainshock (large star) in Figure 9. The mainshock location was surrounded by a seismically quiet area of 2800 km^2, which was broken by only a single cluster of activity 2 weeks before the final foreshock sequence. Most activity surrounding the quiet area was landward (away from the trench) from the mainshock epicenter. The foreshocks of all magnitudes occurring in the 1.8-day period before the mainshock are shown in Figure 10. The sequence began with an event in the central area (marked "1") and subsequently migrated from NE to SW toward the eventual rupture point and then outwards following the dashed line; activity then subsided for 12 hours until failure. (The aftershock sequence of this event has been described by Singh et al. [1980]).

The number of earthquakes of $M \geq 2.8$ observed was 43, which is equivalent to a rate of 47 earthquakes/3 weeks. However, a "background" rate of 3 to 20 earthquakes/3 weeks ($M \geq 2.8$) is estimated from the rate of earthquakes $m_b \geq 4$ (Figure 10) locally (or the rate of earthquakes $M_s \geq 7$ along the trench) from Log $N = a - bM$ with $b = 1$ (McNally and Minster [1981]; McNally and Hearn [1981]). This suggests that the activity during the 3 weeks before the mainshock was at least twice as high as the background rate.

The inverse of the interoccurrence times (1/T) between consecutive events of $M \geq 2.8$ is shown as a function of time in Figure 11. The values of 1/T can be thought of as the "instantaneous frequencies" of earthquakes. Dashed lines indicate times by which consecutive events are separated with Poisson probability of 0.1 due to chance: line "A" represents $1/T = 0.4$ hrs^{-1} for a rate of 20 earthquakes/3 weeks (or similarly, $1/T = 0.6$ hr^{-1} for a rate of 3 earthquakes/3

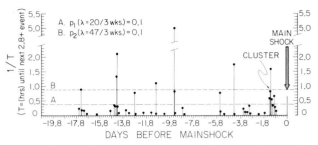

Fig. 11. Interoccurrence times (inverse) between consecutive events ($M \geq 2.8$) as a function of time until the mainshock (see text for further explanation).

weeks, with a probability of 0.01 due to chance). Line "B" represents $1/T = 0.9$ for the rate of 47 earthquakes/ 3 weeks actually observed. In this type of graph it is the groups of relatively large values of 1/T, rather than the single very large values, that may indicate earthquake clustering, such as a foreshock sequence. The temporal clustering of earthquakes on November 28 is seen in Figure 11 and is likely to indicate a true clustering , rather than an apparent clustering due to chance, based on the Poisson probability of .01-.1 for the background rate (line "A"). From this graph the final stage in activity before the mainshock can be distinguished from the previous activity during the 3 week period. The subsidence in activity until the mainshock (29 November, 19 h. 52 m. GMT) is also apparent.

None of the activity described above was sufficiently large for detection by WWSSN worldwide. However, the pattern of temporal clustering of foreshock activity followed by a subsidence in activity until the mainshock is typical of foreshock patterns observed worldwide. This observation suggests the possibility that many large earthquakes may have foreshock activity that cannot be detected at teleseismic distances.

Acknowledgments. The author thanks H. Kanamori, H. Eissler, and W. Savage for helpful review and suggestions; Bette Sheppard and Sue Depta for assistance in manuscript preparation.

This work was partially supported by U.S. Geological Survey Contract No. 14-08-0001-18371 and 14-08-0001-19265. Contribution No. 3571, Division of Geological and Planetary Sciences, California Institute of Technology, Pasadena, California 91125.

References

Chael, E., G. Stewart, and K.C. McNally, Recent large earthquakes along the Middle America Trench, J. Geophys. Res., 1981, to be submitted.

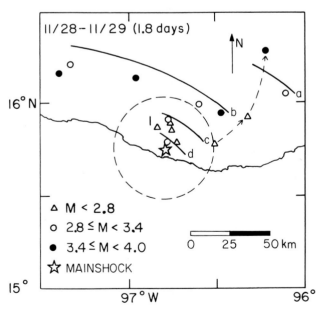

Fig. 10. Foreshocks of all magnitudes occurring in the 1.8 day period prior to the Oaxaca mainshock. "1" represents the initial foreshock in this time period with the sequence subsequently migrating from NE to SW, from a to b to c to d, toward the mainshock epicenter and then outwards following the dashed line.

Chase, T. F., H. W. Menard and J. Mammerick, Scripps Inst. of Oceanography and Inst. of Marine Resources, Univ. of California, 1970.

Cifuentes, I., Historica Sismica de Centroamerica, Proceedings of Simposium Los Riesgos Sismicos y los Asentamientos Humanos (Mexico, Centroamerica y el Caribe), SAHOP, 1981, in press.

Davies, G. and J. Brune, Regional and global fault slip rate from seismicity, Nature Phys. Sci., 229, 101-107, 1971.

Dean, B. W. and C. L. Drake, Focal mechanism solutions and tectonics of the middle America arc, J. Geology, 86, 111-128, 1978.

Eissler, K., and K. C. McNally, Seismicity of the Rivera Fracture Zone and the great Jalisco earthquake of 1932, Proceedings of Simposium los Riesgos Sismicos y los Asentamientos Humanos (Mexico, Centroamerica y el Caribe), SAHOP, 1981, in press.

Gettrust, J., V. Hsu, C. Helsley, E. Herrero and T. Jordan, Patterns of local seismicity preceding the Petatlan earthquake of March 14, 1979, 1980, in preparation.

Habermann, R. E., Precursory seismicity patterns: Stalking the mature seismic gap, This volume, 1981.

Kanamori, H., The nature of seismicity patterns before large earthquakes, This volume, 1981.

Kelleher, J., L. Sykes and J. Oliver, Possible criteria for predicting earthquake locations and their application to major plate boundaries of the Pacific and Caribbean, J. Geophys. Res., 78, 2547-2585, 1973.

Lay, T. and H. Kanamori, An asperity model of great earthquake sequences, This volume, 1981.

McCann, W., R. S. P. Nishenko, L. R. Sykes, and J. Kraus, Seismic gaps and plate tectonics: Seismic potential for major boundaries, Pageoph., 117, 1087-1147, 1980.

McNally, K. C., E. Chael, and L. Ponce, The Oaxaca Mexico earthquake (M_s = 7.8) of 29 November 1978: New "pre-failure" observations, EOS Trans. AGU, 60 332, 1979.

McNally, K. C. and T. Hearn, A "master catalog" of large shallow earthquakes along the Middle America Trench (1898-1979), Bull. Seism. Soc. Am., 1981, in preparation.

McNally, K. C. and J. B. Minster, Non-uniform seismic slip rates along the Middle America Trench, J. Geophys. Res., 1981, in press.

McNally, K., J. Alt, D. V. Helmberger, C. Lomnitz, A. Nava, C. Sanders and K. Sieh, New observations of coastal uplift patterns near the Middle America trench, Mexico, Science, 1981a, submitted.

McNally, K., L. Ponce, T. Hearn and M. Vassilious, The 1978 November 29, Oaxaca, Mexico earthquake-foreshock, mainshock aftershock sequence, J. Geophys. Res., 1981b, in preparation.

Meyer, R. P., W. D. Pennington, L. A. Powell, W. L. Unger, M. Guzman, J. Havskov, S. K. Singh, C. Valdes, and J. Yamamoto, A first report on the Petatlan, Guerrero, Mexico earthquake of 14 March 1979, Geophys. Res. Lett., 7, 97-100, 1980.

Minster, J. B. and T. H. Jordan, Present-day plate motions, J. Geophys. Res., 83, 5331-5334, 1978.

Ohtake, J., T. Matumoto, and G. V. Lathanm, Seismicity gap near Oaxaca, southern Mexico as a probable precursor to a large earthquake, Pageoph., 115, 375-385, 1977.

Ponce, L., K. McNally, V. Sumin de Portilla, J. Gonzalez, A. del Castillo, L. Gonzales, E. Chael, and M. French, Oaxaca, Mexico, earthquake of 29 November 1978: a preliminary report on spatio-temporal pattern of preceding seismic activity and mainshock relocation, Geofis. Intern. 17, 109-126, 1977-78.

Reyes, A., J. N. Brune, and C. Lomnitz, Source mechanism and aftershock study of the Colima, Mexico earthquake of January 30, 1973, Bull. Seism. Soc. Am., 69, 1819-1840, 1979.

Singh, S. K., J. Havskov, K. McNally, L. Ponce, T. Hearn, and M. Vassilious, The Oaxaca, Mexico earthquake of 29 November 1978: a preliminary report on aftershocks, Science, 207, 1211-1213, 1980.

Singh, S. K., M. Guzman, R. Castro, and D. Novelo, A catalog of major nineteenth century earthquakes of Mexico, Bull. Seism. Soc. Am., 1981, submitted.

Stewart, G. S., E. C. Chael and K. C. McNally, The 1978 November 28 Oaxaca Mexico earthquake--a large simple event, J. Geophys. Res., 1981, in press.

Tajima, F. and K. C. McNally, Seismic rupture patterns in Oaxaca, Mexico, EOS Trans., AGU, 61, 288, 1980.

Tajima, F., and K. C. McNally, Seismic rupture patterns in Oaxaca, Mexico, J. Geophys. Res., 1981, submitted.

Tajima, F. M. Kikuchi and K. McNally, Seismic rupture patterns in Oaxaca, Mexico (Part II), EOS Trans., AGU, 61, 1044, 1980.

Wang, S., and K. McNally, Seismic moment release in large earthquakes since 1928 offshore Mexico, Earthquke Notes, 1981, in press.

Yamamoto, J., Rupture processes of some complex earthquakes in southern Mexico, Ph.D. Thesis, Saint Louis University, 1978.

RUPTURE ZONES AND REPEAT TIMES OF GREAT EARTHQUAKES
ALONG THE ALASKA-ALEUTIAN ARC, 1784-1980

Lynn R. Sykes[1], Jerome B. Kisslinger[2], Leigh House[1],
John N. Davies, and Klaus H. Jacob

Lamont-Doherty Geological Observatory of Columbia University
Palisades, New York 10964

Abstract. The dimensions of the rupture zones
of known great earthquakes along the plate
boundary in Alaska and the Aleutians are inferred
from the distribution of aftershocks, the intensity and duration of strong shaking, seismic sea
waves and ground deformation. The historic record
of great shocks for a segment of the arc is
extended back nearly 200 years using Russian
documents prior to 1867 as well as more recent
descriptions. A great earthquake off the Alaska
Peninsula in 1847 re-ruptured most of a 600-km
long zone that broke 59 years earlier in 1788. At
least half of a major seismic gap in the Shumagin
Islands ruptured during those shocks but has not
been the site of a great earthquake for at least
77 years. The rupture zone of an earthquake in
1938 also broke in 1788 and 1847 and may have
broken between 1899 and 1903. Average repeat
times for that zone are 50 to 75 years. The
longest repeat time, 91 years, is obtained
assuming the zone did not rupture between 1899 and
1903. It is likely that the Shumagin gap will be
the site of a great earthquake within the next 10
to 20 years. Large parts of the plate boundary
along southern Alaska and the Aleutians ruptured
in sequences of major and great shocks from 1898
to 1907 and from 1938 to 1965. This and some other
plate boundaries appear to be nearly quiescent for
large earthquakes for long periods of time and
then to rupture in a series of major shocks.
Almost all of the plate boundary along the Alaska-
Aleutian arc is now known to have ruptured
previously in large or great earthquakes. Hence,
presently-existing seismic gaps should be
considered to be probable sites of future large
shocks and not regions where the plate motion is
relieved mainly by aseismic slip.

[1]Also with the Department of Geological Sciences,
Columbia University.

[2]Also with the Department of Slavic Languages,
Columbia University.

Introduction

The Alaska-Aleutian island arc is one of the
world's most active zones of earthquake activity,
volcanism and subduction. From 1938 to 1979 nine
earthquakes (Figure 1) of magnitude (M) 7.4 or
larger have ruptured much of the zone of contact
between the North American and Pacific plates from
offshore British Columbia to southern Alaska and
thence along the Aleutian arc. Plate motion is
largely strike slip off British Columbia and
southeast Alaska and occurs mainly by thrust
faulting along the Aleutians. The rupture zones,
magnitudes and seismic moments of several of these
shocks are among the largest known anywhere in the
world. Even for great earthquakes (M > 7.7) the
historic record for the Alaska-Aleutian zone was
thought to be very short and incomplete. Hence,
previous authors have debated whether the repeat
time of large earthquakes at a given place on the
plate boundary is about 30 to 60 years [Sykes,
1971; Kanamori, 1977a] or as much as 500 to 1350
years [Plafker and Rubin, 1978]. A better knowledge of repeat times of large earthquakes is
obviously needed to estimate seismic hazards more
accurately, to understand the subduction process
in more detail and to make realistic progress
toward the prediction of earthquakes.

We have translated a number of documents from
Russian that contain material on earthquakes and
volcanism in the Aleutians and in coastal Alaska
prior to the advent of a global instrumental
record in 1897. These descriptions along with
more recent instrumental data now allow us to map
(Figure 1) the rupture zones of great earthquakes
for nearly the past 200 years in a 600 to 1000-km
long segment of the arc adjacent to the Alaska
Peninsula and to Kodiak Island. We show that two
great earthquakes ruptured and re-ruptured at
least a 500 km segment of this plate boundary in
1788 and 1847. These and more recent great earthquakes in that area indicate an average repeat
time of about 50 to 75 years between great shocks
that involve thrust faulting at shallow depths
along the same segment of the plate boundary. A

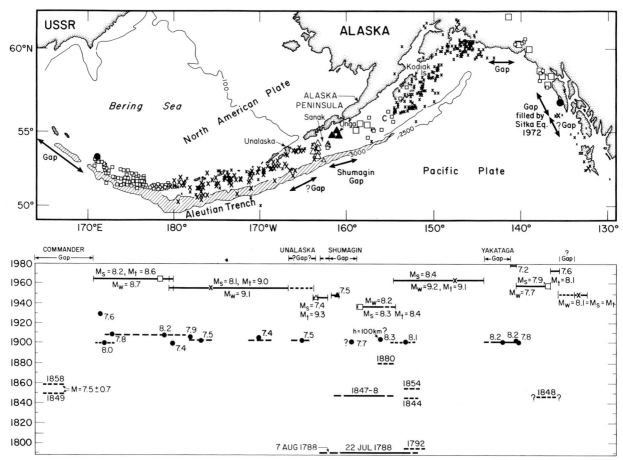

Fig. 1. <u>Above</u>: Rupture zones of earthquakes of magnitude M ≥ 7.4 from 1925-1971 as delineated by their aftershocks along plate boundary in Aleutians, southern Alaska and offshore British Columbia [after Sykes, 1971]. Contours in fathoms. Various symbols denote individual aftershock sequences as follows: crosses, 1949, 1957 and 1964; squares, 1938, 1958 and 1965; open triangles, 1946; solid triangles, 1948; solid circles, 1929, 1972. Larger symbols denote more precise locations. C = Chirikof Island. <u>Below</u>: Space-time diagram showing lengths of rupture zones, magnitudes [Richter, 1958; Kanamori, 1977<u>b</u>; Kondorskaya and Shebalin, 1977; Kanamori and Abe, 1979; Perez and Jacob, 1980] and locations of mainshocks for known events of M ≥ 7.4 from 1784 to 1980. Dashes denote uncertainties in size of rupture zones. Magnitudes pertain to surface wave scale, M_s unless otherwise indicated. M_w is ultra-long period magnitude of Kanamori [1977b]; M_t is tsunami magnitude of Abe [1979]. Large shocks in 1929 and 1965 that involve normal faulting in trench and were not located along plate interface are omitted. Absence of shocks before 1898 along several portions of plate boundary reflects lack of an historic record of earthquakes for those areas.

250-km long segment of the arc in this area, the Shumagin seismic gap, has not been the site of a great earthquake for at least 77 years. Davies et al. [1980] discuss earthquake prediction and earthquake hazards for the Shumagin gap in more detail. An abbreviated version of this paper is published by Sykes et al. [1980].

Rupture Zones of Large Earthquakes

<u>1929-1980</u>. Sykes [1971] mapped the rupture zones of large (M ≥ 7.0) earthquakes of the past 50 years along the plate boundary in southern Alaska and the Aleutians by relocating aftershocks

of those events and by assuming that the latter are a good measure of the area of the rupture surface. Aftershocks of events of M ≥ 7.4 are shown in the upper half of Figure 1 from 1925 to 1971. The inferred rupture zones of all known shocks of M ≥ 7.4 from 1784 to 1980 are indicated in the lower half of Figure 1. Rupture zones of events of M ≤ 7.4 do not exceed about 70 km [Sykes, 1971; Kelleher et al., 1973], a dimension small compared with the rupture dimensions of great events, which for this arc extend 200 to 1000 km along strike. Shocks of M ≥ 7.4 account for most of the slip that occurs seismically between large lithospheric plates. Small shocks

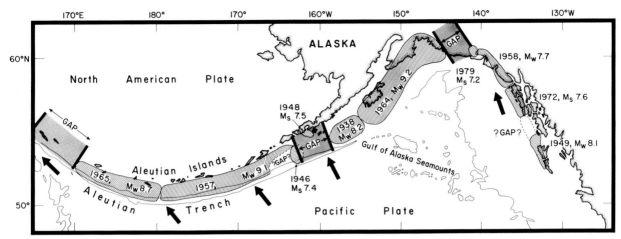

Fig. 2. Aftershock areas of earthquakes of magnitude ≥ 7.4 in the Aleutians, southern Alaska and offshore British Columbia from 1938 to 1979, after Sykes [1971] and McCann et al. [1979]. Heavy arrows denote motion of Pacific plate with respect to North American plate as calculated by Chase [1978]. Two thousand fathom contour is shown for Aleutian trench. M_s and M_w denote magnitude scales described by Kanamori [1977b].

generally account for only a few percent of either the cumulative seismic moment, seismic energy release or total plate movement. From 1938 to 1971, nine large earthquakes ruptured much of the length of the plate boundary of Figures 1 and 2. Aftershock zones of several of these earthquakes abut without significant overlap, which indicates that the rupture pattern of sequences of large shocks is rather simple.

Most of the large earthquakes of the past 15 years that have occurred along simple plate boundaries of the world ruptured areas that had not been the sites of large shocks for many decades. Sykes [1971] pointed out that three segments of the plate boundary in Figure 1 had not broken in large earthquakes for several decades and called them seismic gaps. One of these gaps was ruptured by the Sitka earthquake of 1972. Only a small portion of a major seismic gap near 60°N, 142°W was ruptured by the St. Elias earthquake (M = 7.2) of February 28, 1979 [Lahr et al., 1980; McCann et al., 1980; Perez and Jacob, 1980]. The remaining unruptured area, the Yakataga gap, extends for about 175 km along the coast of southern Alaska between the rupture zones of the 1964 and 1979 shocks. Another gap in the Commander Islands of the westernmost Aleutians has not experienced a great earthquake in the 20th century and may not have been the site of such a shock since 1858 (Figures 1 and 2).

Sykes [1971] was uncertain about the dimensions of the rupture zone of a great earthquake that occurred near the Alaska Peninsula in 1938. Recent work [Davies et al., 1980] on the aftershocks of that event, the source region of the seismic sea wave (tsunami) generated by the mainshock, and the rupture area inferred from the seismic moment all indicate that the 1938 event did not rupture west of about 158.5°W into what is

called the Shumagin gap in Figure 1. Although the 1946 earthquake generated a great tsunami that affected many areas in the Pacific, the generating area of the sea wave appears to have been largely confined to the rather small aftershock area. Hence, that event neither ruptured into the Shumagin gap to the east nor into the possible Unalaska gap to the west. The possible Unalaska gap is described in the accompanying paper by House et al. [1981].

The rupture zones and seismic gaps in Figure 1 only pertain to large earthquakes of the thrust type that occur along the plate interface at shallow depth. Large shocks do occur less frequently beneath the deeper part of the trench, at intermediate depths and to the north of the plate boundary in south-central Alaska. No attempt is made here to apply gap theory to events of those types. Two large earthquakes involving normal faulting [Kanamori, 1972; Abe, 1972] within the deeper part of the Aleutian trench in March 1929 and March 1965 are omitted from Figure 1 since they occurred within the Pacific plate and were not situated along the plate interface.

The rupture zone of the 1948 shock of magnitude 7.5 near 161°W is small and is located farther from the trench than most shallow events of the thrust type. Its main shock is located in the midst of a band of high activity along the northwest side of the gap. Data from a recently-installed seismic network in the area indicate that the region of high activity is centered at a depth of about 40 to 50 km near the downdip end of the zone of shallow thrusting [Davies et al., 1980]. These observations and the greater depths (44 and 48 km) computed from the times of pP-P for the two largest aftershocks of the 1948 event suggest that it may not have ruptured the plate boundary at shallow depths. The magnitude and size of the aftershock zone of the 1948 event indicate it could not have ruptured more than about 15% of the Shumagin gap. It is not clear if the great Aleutian earthquake of 1957 ruptured the shallow, thrust portion of the plate boundary in the area near Unalaska [Davies et al., 1980; House et al., 1981]. Hence, that area may also be a seismic gap.

1907-1929. Since aftershocks of large events in the Alaska-Aleutian zone before about 1929 generally cannot be located very precisely, the dimensions of rupture zones for older shocks can only be estimated qualitatively. Nevertheless, the record of great (M > 7.7) shocks appears to be complete or nearly complete since 1898. Hence, it is clear that no great shock ruptured any of the plate boundary in Figure 1 from 1907 to 1938. This quiescence for great shocks is remarkable since numerous events of that class occurred there from 1938 to 1965 and from 1898 to 1907. Only one shock of M $\geq$ 7.4, an event of M = 7.6 in the westernmost Aleutians, occurred along the entire plate boundary from 1907 to 1938.

1898-1907. A remarkable sequence of 14 shocks of M $\geq$ 7.4 ruptured large parts of the plate boundary in the decade starting soon after the beginning of a world-wide instrumental record in 1897. The distribution of felt reports of aftershocks, the occurrence of landslides and reports of strong shaking [Tarr and Martin, 1912] led McCann et al. [1980] to conclude that two great earthquakes in 1899 ruptured all or most of the Yakataga gap. Similar information appears to be very scarce, however, for the other large events in Figure 1 that occurred near the turn of the century. An event of M = 8.1 on October 9, 1900 was felt with decreasing intensities proceeding in a northeasterly direction away from Kodiak Island [Tarr and Martin, 1912; McCann et al., 1980]. Hence, its rupture zone appears to have been centered either to the south or southwest of Kodiak Island as indicated by dashes in Figure 1. Richter [1958] lists a shock of M = 8.3 on June 2, 1903 near 57°N, 156°W with a questionable depth of 100 km. That event may have ruptured either the Shumagin gap or the zone that subsequently broke in 1938 or it may, in fact, have been of deeper focus. A shock on July 14, 1899 of M = 7.7, for which the epicentral location [Kanamori and Abe, 1979] is very poor, was felt on Unalaska in the eastern Aleutians and on Unga in the Shumagin Islands [Tarr and Martin, 1912]. It also may have occurred in the Shumagin gap.

The occurrence of five earthquakes of M $\geq$ 7.4 between about 177°W and 170°E from 1898 to 1907 strongly suggests that at least that portion of the plate boundary that ruptured in the great shock of February 4, 1965 previously broke about 60 years earlier. Two shocks of magnitude 7.5 occurred in the central Aleutians near 177°W and 165°W on December 31, 1901 and January 1, 1902, only 20 hours apart. While the two events may be related in some way, their relatively modest magnitudes suggest that they ruptured only two

small portions of the zone that broke in 1957. It is difficult to estimate the amount of displacement or the extent of rupture in the series of events near the turn of the century in the western and central Aleutians since their seismic moments have not been measured and felt reports are almost lacking. The apparent absence of large tsunamis accompanying these events suggests, however, that individually they were not as large as the great earthquakes of 1957, 1964 and 1965 [Abe, 1979; H. Kanamori, written communication, 1979]. Nevertheless, that absence may be attributed to rupture of a considerable portion of the arc being distributed among several large shocks from 1898 to 1907 rather than in a few very great events. Hence, large portions of the plate boundary from about 153°W to 170°E appear to have broken in a series of at least 11 large shocks from 1898 to 1907. Whether the Shumagin gap, the questionable Unalaska gap or the rupture zones of the 1938 and 1957 earthquakes definitely broke between 1899 and 1903, however, cannot be resolved at present.

1784-1867. When we examined several catalogues in the English language of earthquakes in Alaska and the Aleutians for the period of permanent Russian settlement from 1784 to 1867, we soon realized that they contain mainly third and fourth-hand accounts. We translated a number of descriptions from Russian in an attempt to obtain as original documentation as possible for earthquakes during that period. The material that follows is a summary of more extensive descriptions published elsewhere [Davies et al., 1980; Kisslinger et al., 1980]. To our knowledge, the important catalogue by Doroshin [1870] for the period 1840 to 1866 was not consulted in the preparation of catalogues of Alaskan earthquakes that were published subsequently. That paper and other early sources [Merkul'ev, 1789 in Solov'iev, 1968; Davydov, 1812; Veniaminov, 1840, 1888; Grewingk, 1850; Mushketov and Orlov, 1893] contain invaluable qualitative discriptions of strong shaking, seismic sea waves, landslides, changes in sea level and aftershock sequences from which we infer the approximate rupture zones of previous major and great earthquakes. In fact, several of these reports of earlier earthquakes are more complete and more useful in this regard than information available to us for many of the large shocks that occurred from 1898 to 1907. Many of the seismic events listed in early reports appear to be clearly local in their effects and to have been associated with volcanic eruptions. Hence, they are not included in Figure 1 as large events.

The first permanent Russian settlement was established on Unalaska Island near 167°W in 1784. By the time of the great earthquake of 1788 reports of the shock and its tsunami were available (Figure 1) from Unga in the Shumagin Islands, Sanak Island, Kodiak Island, the Alaska Peninsula and Unimak Island (55°N, 164°W). Most of the descriptions of earthquakes during the period of Russian control come from either that 1000-km long segment of the arc, the Commander

Islands in the westernmost Aleutians or the region near Sitka in southeast Alaska. The absence of shocks in other areas in Figure 1 is attributed to the lack of an historic record prior to 1897.

We take two or more of the following as indicative of an earthquake that ruptured a considerable (> 100 km) portion of the plate boundary: extremely strong shaking or extensive damage of intensity IX or greater on the Modified Mercalli scale at two or more separated localities [Kelleher, 1972], shaking lasting a minute or more, permanent changes in sea level, a large tsunami associated with the shock, ground breakage, landslides, or aftershocks lasting for weeks to months. Places experiencing such effects are taken to have been located near the rupture zone.

Descriptions of these types for an earthquake on July 22, 1788 lead us to interpret it as a great shock that ruptured at least a 600 km segment of the plate boundary from Kodiak Island to Unga in the Shumagin Islands (Figure 1). [Note that considerable confusion arises in the literature about date of events since some are reported on the old (Julian) calendar and some on the modern (Gregorian). For the first event of 1788, July 11 (old) = July 22 (modern); for second, July 27 (old) = August 7 (modern)]. In a letter of 1789 to G. I. Shelikov, Merkul'ev [in Solov'iev, 1968, p. 235] describes very intense shaking on Kodiak Island, an intense flood (tsunami) consisting of two large waves and other smaller waves, aftershocks every day for a month or longer and a permanent change in sea level. Davydov [1812, p. 154] also mentions landslides on Kodiak Island and that the sea first withdrew from shore, surged onshore and carried a vessel onto the top of a cabin. Veniaminov [1840, 1888] describes strong shaking, landslides and a "horrible flood" on Unga on the same date. He refers to another flood on Unga 16 days later in which the water rose to 50 sazhens, about 91 m. It is not clear, however, if 91 m is the vertical height of water or the distance the waves ran up the beach. Veniaminov [1888] states, "The deluge or flood which took place on Unga and on the southern side of Alaska (Peninsula) in 1788 did not have any effect on the northern side of Unimak." "The tradition of Aleuts...reports that during the flood which took place on Sanak around the year 1790 the water preceded as strong and infrequent large waves."

Solov'iev [1968] interprets the second flood on Unga 16 days later (i.e. on or about August 7) as a great tsunami generated by a second earthquake that ruptured the area from Sanak Island near the western end of the Shumagin gap in Figure 1 to Unga. He takes the vertical height of that tsunami to be greater than 30 m at those two islands. A 30-m sea wave generated by an earthquake on April 1, 1946 to the west of the Shumagin gap destroyed the nearby Scotch Cap lighthouse on Unimak Island. If the second flood was, in fact, generated by a second earthquake, its source area would seem to have been close to Unga. Neverthe-

less, neither the date of the flood on Sanak nor the occurrence of an earthquake on August 7 is mentioned in any of the older Russian documents we examined. Dall [1870, pp. 310, 467], a secondary source, reports tsunami damage to Sanak Island on the date of the second flood on Unga. Solov'iev [1968] proposes that Dall used some other primary source in addition to Veniaminov [1840] for the events of 1788. Hence, while the evidence that an earthquke ruptured the zone from Kodiak Island to Unga is quite strong, we are forced to rely on Dall's account to infer that a second large event appears to have ruptured the western half of the Shumagin gap 16 days later.

Doroshin [1870] describes ground cracking, landslides, shaking continuing for about 4 hours, and aftershocks lasting about 5 weeks on Ukamok (Chirikof Island, C in Figure 1) in association with a large earthquake in 1847. He states that it was impossible to remain standing on Unga during the earthquake and that shocks were felt several times on the Alaska Peninsula. He also describes another large shock accompanied by its own aftershocks on Chirikof Island in 1848. From these reports we infer that at least a 500 km segment of the plate boundary between Chirikof Island and Unga ruptured in 1847 and 1848. The sequence of 1847-1848 is particularly significant since that portion of the plate boundary appears to have also broken 60 years earlier in 1788. This is probably the best documented repeat time obtained thus far for the plate boundary in Alaska and the Aleutians. Although the effects described by Doroshin [1870] are similar to those reported for the 1788 event he does not mention a tsunami accompanying the shock of 1847.

From similar reports [Davydov, 1812; Doroshin, 1870] of strong shaking and aftershocks we infer that large earthquakes also ruptured the plate boundary somewhere near Kodiak Island in 1792, 1844 and 1854. Local tsunamis are also described in association with the events of 1792 and 1854. Since reports of these three shocks are only available from Kodiak Island, the dimensions of their rupture zones cannot be ascertained. While their rupture zones in Figure 1 are indicated as being off Kodiak Island, the events may well not have ruptured the same segment of the plate boundary. The portion of the 1964 rupture zone opposite Kodiak Island does appear to have moved several times in the last 200 years.

An earthquake on March 18, 1848 [Doroshin, 1870; Veniaminov, 1897, p. 210] may have ruptured either that portion of the Fairweather fault that broke in 1958 or the part that ruptured near Sitka in 1972 (Figure 1). That shock and one listed for an unknown date in 1847 by Dall [1870] may be identical. The evidence for inferring either the size, dimension of the rupture zone or the causative fault zone, however, is poor and fragmentary. Aftershocks were felt at Sitka for nearly a month after the event of 1848 [Doroshin, 1870].

Two large earthquakes that occurred in the

Commander Islands in the westernmost Aleutians on September 28, 1849 and January 22, 1858 are assigned magnitudes of 7.5 + 0.7 by Kondorskaya and Shebalin [1977]. They incorrectly list the date of the 1849 event as October 28. The shock of 1849 generated a local sea wave that affected the Commander Islands [Doroshin, 1870; Kondorskaya and Shebalin, 1977] and islands of the South Pacific [Ella, 1890]. Since the configuration of its rupture zone or that of the poorly documented shock of 1858 cannot be ascertained, the respective rupture zones are merely indicated in Figure 1 to be somewhere off the southern coast of the Commander Islands. The generation of a tsunami along the southern coast is in accord with the observation that thrust faulting and vertical motion movements along that coast while strike-slip motion occurs along the north coast [Cormier, 1975].

1867-1897. We infer that a large earthquake occurred off the Alaska Peninsula near Chirikof Island on September 28, 1880 based on the following effects [Secretary of War, 1883, p. 120] aftershocks continuing for 19 days, numerous deep fissures, strong shaking lasting about 20 minutes, extensive damage to a log house, several sea waves that travelled about 55 m onshore, and permanent changes in sea level. Moore [1962] concludes that a vertical displacement of 2 m, which is still preserved in dammed streams and uplifted wave-cut terraces, occurred along a northeast-striking fault on Chirikof Island during the earthquake of 1880. Hence, the earthquake at least ruptured that fault, a secondary imbricate fault within the upper plate. While the full extent of rupture is not clear, the earthquake of 1880 also may have broken a segment of the main plate boundary to the southwest of the 1964 rupture zone. The evidence is unclear whether that segment also ruptured in 1938. One or more imbricate faults also appear to have ruptured in conjunction with movement on shallow dipping thrust faults during the great earthquakes of 1899 and 1964 [Tarr and Martin, 1912; Plafker and Rubin, 1978; McCann et al., 1980].

Rupture Sequences Along Plate Boundaries

As noted earlier, large portions of the plate boundary in Figures 1 and 2 ruptured in a series of major and great earthquakes from 1898 to 1907 and then again from 1938 to 1965. Each sequence encompassed a time interval that was short compared to the repeat time of great shocks which appears to be at least 60 years for individual segments of this arc. Similarly, much of the plate boundary along the North Anatolian fault in Turkey ruptured in a series of large earthquakes from 1939 to 1943 [Richter, 1958, p. 612]. Also, the entire plate boundary off northern Japan and the southern Kuril Islands ruptured in a series of large thrust earthquakes from 1952 to 1973, a time interval that is short compared to repeat times of about 50 to 100 years [Abe, 1977]. Thus, a strong

temporal clustering of large events appears to be a common feature of several simple plate boundaries. In some instances, rupture along one segment (i.e. 1957 in Figure 1) is followed (with a time delay) by rupture along an adjacent part of the plate boundary (i.e. 1965 zone). Other events in a sequence, however, do not appear to be triggered in such a simple manner.

Almost the entire plate boundary in Figure 1 is known to have ruptured in a previous large earthquake. If any of the boundary moves dominantly in a non-seismic manner, i.e. by slow slip or creep, the total lengths of such segments cannot exceed more than a few percent of the length of the plate boundary. A conservative approach to seismic zonation seems to demand that all of the remaining gaps in Figure 1 be considered probable sites of future large earthquakes.

Repeat Times of Great Earthquakes

A repeat time of 59 years is indicated by the occurrences of shocks in 1788 and 1847 for a segment of the plate boundary near the Alaska Peninsula (Figure 1). Both of those shocks appear to have ruptured the entire portion of the arc that broke in 1938. Since it is not clear if that segment also ruptured between 1899 and 1903, an average repeat time of 50 to 75 years is obtained by dividing 150 years (1938-1788) by either two or three earthquake cycles. The longest possible repeat time, 91 years (1938-1847), is obtained if that zone did not rupture in 1880 or between 1897 and 1903.

At least the eastern half of the Shumagin gap also broke in 1788 and 1847. The entire Shumagin gap has not been the site of a great shock since at least 1903 and possibly even since 1847. Thus, the interval that has elapsed since 1903 is somewhat more than the average repeat time of 50 to 75 years estimated for the 1938 rupture zone, but it is still somewhat less than the upper limit of 91 years. Thus, if the observations from the 1938 zone can be applied to the Shumagin gap, it seems likely that one or more large earthquakes will rupture the Shumagin gap sometime in the next 15 years. The history of tsunamis from that region, including waves locally reaching heights of more than 30 m, indicates that a large future earthquake in the Shumagin gap could be expected to generate a sizable seismic sea wave.

Repeat times of historical earthquakes at a given place along the Nankai trough of southwest Japan vary by a factor of about two [Ando, 1975]. Shimazaki and Nakata [1980] find that the longer repeat times in that area are associated with larger seismic displacements and larger rupture zones. The experience from Japan suggests that the interval between the last major shock and a future large event will be in the shorter end of this spectrum of repeat times for both the Shumagin gap and the 1938 rupture zone since the 1938 shock and events between 1899 and 1903 appear to have had shorter rupture dimensions that the shocks of 1788 and 1847.

Sykes and Quittmeyer [1981] find that the average repeat times of great earthquakes along simple plate boundaries of the world are governed by three factors: the relative velocity of the interacting plates, the ratio of seismic to aseismic motion and the geometry of the zone of plate contact, particularly the down-dip width. Using the observed repeat times of about 60 years for the 1938 and 1965 rupture zones (Figure 1), they conclude that most of the slip in those areas occurs seismically in great earthquakes and that the amount of aseismic slip is small. They calculate a longer repeat time, about 180 years, for the 1964 rupture zone, where the zone of plate contact has a greater downdip width, and a shorter repeat time for the central and western Aleutians, where the width is less. The actual repeat time of great shocks in the 1964 zone, however, is not known. The calculated repeat time of 180 years, however, is much smaller than that estimated by Plafker and Rubin [1978] from uplifted marine terraces, 500 to 1350 years. Those very long repeat times, however, appear to us to reflect the movement of a secondary imbricate fault within the upper portion of the North American plate rather than slip along the main plate boundary. This is understandable since such splay faults may not move during every major thrust event along the main plate boundary. Hence, we conclude that the actual repeat times of great earthquakes at a given place along the Alaska-Aleutian zone are likely to be much shorter than those estimated from those marine terraces.

The Commander gap in the westernmost Aleutians is not known to have been the site of a large earthquake since 1858, i.e. 122 years ago. Since plate movement along the southern side of the Commander Islands is nearly parallel to the plate boundary and occurs along shallow-dipping thrust faults [Cormier, 1975], the repeat time of large earthquakes in that area may well differ from that farther east where a large component of plate convergence is present. Also, it is not clear that the historic record of large shocks is complete for the last 122 years.

A repeat time of 110 to 125 years is obtained for the rupture zones of either the 1958 or the 1972 earthquakes in Southeast Alaska if the shock of 1848 (Figure 1) did, in fact, rupture one of those zones and the zone did not rupture during that interval. The 1972 zone may have broken in the magnitude 7.1 shock of 1927. From the measured seismic moments [Thatcher and Plafker, 1977] of two great earthquakes in southern Alaska during September 1899 and the areas of the inferred rupture zones, McCann et al. [1980] estimate an average displacement of about 5 m. Approximately 5 m of potential slip could have been built up as strain by plate movements in the 81 years since 1899 if aseismic slip is negligible. Hence, since there is no historical record of large earthquakes in that region prior

to 1899, estimates of repeat time are uncertain to the extent that the seismic moment is uncertain for the 1899 sequence (by at least 50%) and to the extent that the amount of aseismic slip is unknown. A small portion of the zone that ruptured in 1899 appears to have broken again in 1979.

Hence, repeat times appear to be about 50 to 100 years for several portions of the Alaska-Aleutian arc and to exceed 100 years for much of the rupture zone of the 1964 earthquake, the Commander gap and possibly for the strike-slip portion of the plate boundary in southeast Alaska. Nevertheless, knowledge of repeat times for this plate boundary must still be regarded as incomplete.

Acknowledgments. We thank R. Bilham and C. Scholz for critically reading the manuscript. G. W. Moore kindly provided translations of several documents from Russian. This work was supported by contracts with the U.S. Departments of Energy (DE-AS02-76 ERO-3134) and Commerce (NOAA 03-05-022-70) and by grants EAR 78-22770 and EAR 79-26350 from the National Science Foundation. Lamont-Doherty Geological Observatory Contribution Number 3112.

References

Abe, K., Lithospheric normal faulting beneath the Aleutian trench, Phys. Earth Planet. Int., 6, 346-359, 1972.

Abe, K., Some problems in the prediction of the Nemuro-Oki earthquake, J. Phys. Earth, 25, (supplement), S261-S271, 1977.

Abe, K., Size of great earthquakes of 1837-1974 inferred from tsunami data, J. Geophys. Res., 84, 1561-1568, 1979.

Ando, M., Source mechanisms and tectonic significance of historical earthquakes along the Nankai trough, Japan, Tectonophysics, 27, 119-140, 1975.

Chase, C. G., Plate kinematics: The Americas, East Africa and the rest of the world, Earth Planet. Sci. Lett., 37, 355-368, 1978.

Cormier, V., Tectonics near the junction of the Aleutian and Kuril-Kamchatka arcs, and a mechanism for middle Tertiary magmatism in the Kamchatka Basin, Geol. Soc. Amer. Bull., 86, 443-453, 1975.

Dall, W. H., Alaska and Its Resources, Boston, Lee and Shepard, 627 p., (reprinted by Arno Press and New York Times), 1870.

Davies, J., L. Sykes, L. House, and K. Jacob, Shumagin seismic gap, Alaska Peninsula: History of great earthquakes, tectonic setting, and evidence for high seismic potential, J. Geophys. Res., in press, 1980.

Davydov, G. I., Dvukratnoe puteshestvie v Ameriku [Double Voyage to America] translated by W. L. Klawe, provided by G. Moore, October 1979, St. Petersburg, Marskaia Tip., v. 2, 224 p., 1812.

Doroshin, P., Onekotorykh vulkanakh, ikh izverzheniyakhi zemletryaseniyakh v byvshikh Amerikanskikh vladeniyakh Rossii, "Some volcanoes, their eruptions, and earthquakes in the former Russian holdings in America", Verh. Russ. Kais. Min. Ges., ser. 2, part 5, 25-44, St. Petersburg, 1870.

Ella, S., Some physical phenomena of the South Pacific Islancs, Rpt. 2nd Meeting Australian Assoc. Advancem. Sci., Melbourne, Australia, 559-572, 1890.

Grewingk, C., Beitrag zur kentniss der orographischen und geognostischen Beschaffenheit der Nordwest Kuste Amerikas, mitden anliegenden Inselin, Russ. K. min. Gesell Verh., J. 1848-49, 50-53, 204-273, St. Petersburg, 1850.

House, L., L. R. Sykes, J. N. Davies, and K. H. Jacob, Identification of a possible seismic gap near Unalaska Island, eastern Aleutians, Alaska, this volume, 1981.

Kanamori, H., Mechanism of tsunami earthquakes, Phys. Earth Planet. Int., 6, 346-359, 1972.

Kanamori, H., Seismicity and aseismic slip along subduction zones and their tectonic implications, in Island Arcs, Deep Sea Trenches and Back-Arc Basins, M. Talwani and W. C. Pitman III (eds.), Amer. Geophys. Union, Washington, D. C., Maurice Ewing Series 1, 163-174, 1977a.

Kanamori, H., The energy release in great earthquakes, J. Geophys. Res., 82, 2981-2987, 1977b.

Kanamori, H., and K. Abe, Reevaluation of the turn-of-the-century seismicity peak, J. Geophys. Res., 84, 6131-6139, 1979.

Kelleher, J., Rupture zones of large south American earthquakes and some predictions, J. Geophys. Res., 77, 2087-2103, 1972.

Kelleher, J., L. R. Sykes, and J. Oliver, Possible criteria for predicting earthquake locations and their applications to major plate boundaries of the Pacific and Caribbean, J. Geophys. Res., 78, 2547-2585, 1973.

Kisslinger, J., J. Davies, L. Sykes, L. House, and K. Jacob, Historical earthquakes of Alaska and the Aleutian Islands: A compilation of original references, some newly translated, in preparation, 1981.

Kondorskaya, N. V., and N. V. Shebalin (editors), New Catalog of Strong Earthquakes in the Territory of the U.S.S.R., Nauka, Moscow, U.S.S.R., 347-433, 1977.

Lahr, J. C., C. D. Stephens, H. S. Hasagawa, and J. Boatwright, Alaskan seismic gap only partially filled by 28 February 1979 earthquake, Science, 207, 1351-1353, 1980.

McCann, W. R., S. P. Nishenko, L. R. Sykes, and J. Krause, Seismic gaps and plate tectonics: Seismic potential for major plate boundaries, Pure Appl. Geophys., 117, 1082-1147, 1979.

McCann, W. R., O. J. Perez, and L. R. Sykes, Yakataga Gap, Alaska: Seismic history and earthquake potential, Science, 207, 1309-1314, 1980.

Moore, G. W., Geology of Chirikof Island, Alaska, U.S. Geol. Surv. Techn. Lett., Aleut 1, 10 p., 1962.

Mushketov, I. V., and A. P. Orlov, Catalogue of earthquakes in the Russian Empire, Proc. Imperial Russian Geographical Soc., 26, 1893.

Perez, O., and K. H. Jacob, Tectonic model and seismic potential of the eastern Gulf of Alaska and Yakataga seismic gap, J. Geophys. Res., in press, 1980.

Plafker, G., and M. Rubin, Uplift history and earthquake recurrence as deduced from marine terraces on Middleton Island, Alaska, U.S. Geol. Surv. Open-File Report 78-943, 687-721, 1978.

Richter, C. F., Elementary Seismology, W. H. Freeman and Company, San Francisco, California, 768 p., 1958.

Secretary of War, U.S. War Dept., Miscellaneous phenomena: U.S. Signal Corp. Ann. Rept. for 1882, 4, part 2, p. 120, Washington, D. C., 1883.

Shimazaki, K., and T. Nakata, Time-predictable recurrence model for large earthquakes, Geophys. Res. Lett., 7, 279-282, 1980.

Solov'iev, S. L., Sanak-Kodiak Tsunami 1788, Problema Tsunami, Moscow, 232-237, 1968.

Sykes, L. R., Aftershock zones of great earthquakes, seismicity gaps, earthquake prediction for Alaska and the Aleutians, J. Geophys. Res., 76, 8021-8041, 1971.

Sykes, L. R., J. B. Kisslinger, L. House, J. Davies, and K. H. Jacob, Rupture zones of great earthquakes, Alaska-Aleutian arc, 1784-1980, in Science, 210, 1343-1345, 1980.

Sykes, L. R., and R. C. Quittmeyer, Repeat times of great earthquakes along simple plate boundaries, this volume, 1981.

Tarr, R. S., and L. Martin, The earthquake at Yakutat, Alaska, in September 1899, U.S. Geol. Surv. Prof. Paper 69, 1912.

Thatcher, W., and G. Plafker, The 1899 Yakutat Bay, Alaska earthquakes, Intern. Union Geod. Geophys., IASPEI/IAVCEI Assembly, (abstracts), p. 54, Durham, England, 1977.

Veniaminov, I., (Zapiski ob Ostrovakh Unalashkago Otdiela): Parts 1-3, Published at the Expense of the Russian American Company, St. Petersburg, 1840.

Veniaminov, I., Zapiski ob Ostrovakh Unalashkago Otdeliela - vol. 2 of Tvoreniya Innokentiya, Ivan Barsukov (ed.), Sinodal'naya Tipografiya, Moscow, 1888.

Veniaminov, I., Letter to N. and A. Lozhechnikov of May 8, 1848, in Pis'ma Innokentiya, Mitropolita Moskovskago i Kolomenskago 1828-1855, Ivan Barsukov (ed.), book 1, Sinodal'naya Tipografiya, St. Petersburg (in Russian), 1897.

IDENTIFICATION OF A POSSIBLE SEISMIC GAP NEAR
UNALASKA ISLAND, EASTERN ALEUTIANS, ALASKA

L. S. House[1], L. R. Sykes[1], J. N. Davies, and K. H. Jacob

Lamont-Doherty Geological Observatory of Columbia University
Palisades, New York 10964

Abstract. A portion of the eastern Aleutians, between about 168°W and 160°W, appears to be a major tectonic transition between the oceanic arc structure to the west and the continental arc structure to the east. The 1200 km long aftershock zone of the 1957 Andreanof-Fox Islands earthquake, Mw = 9.1, extends 200 km into the western portion of this zone. This 200 km long segment, which is located near Unalaska Island, underwent deformation quite different from that of the remaining 1000 km long zone. It experienced aftershocks along only a narrow (20 km width) zone at its arcward (northern) edge, in remarkable contrast to the 80 km width of the aftershock zone to the west. Since 1957, the interior of this segment has produced only two earthquakes of magnitude 5 or larger. Two clusters of events occurred prior to the 1957 main shock, one near the western end of the entire aftershock zone (180°W), the other, a more dense cluster, occurred near the western edge of the Unalaska segment (168°W). Travel times of the 1957 tsunami to tide gauges in western North and South America indicate that the eastern extent of the tsunami generating area was situated at or near the western boundary of the Unalaska segment. If the Unalaska segment slipped in 1957 it must either have undergone delayed rupture, so that seismic and tsunami energy were concealed in the coda of the main shock, or have slipped in a rupture process so slow as to be an inefficient tsunami source. Another possibility clearly exists, however, namely that the Unalaska segment did not rupture in 1957. If it did not, the Unalaska segment is a seismic gap, and could also have a high potential for producing an earthquake as large as magnitude 8. This possibility has serious implications for the evaluation of seismic and tsunami hazards in the eastern Aleutians.

---

[1]Also with the Department of Geological Sciences, Columbia University.

Introduction

The concept of seismic gaps has been highly successful in identifying those portions of simple plate boundaries that are likely sites of future large (M $\geq$ 7) earthquakes [Fedotov, 1965; Mogi, 1968a; Sykes, 1971; Kelleher et al., 1973]. Evaluating the seismic potential of segments of a plate boundary represents a refinement of the seismic gap concept and provides the impetus to focus earthquake prediction studies on areas that seem most likely to produce large earthquakes in the near future (one to several decades) [McCann et al., 1979]. McCann et al. [1979] consider an area to be a seismic gap if it is part of an active plate boundary and has not experienced a large earthquake in the past 30 years. In order to assign a high seismic potential to a gap, they add the additional requirement that the area must have broken in a large earthquake at least once in the historic past.

To identify a portion of a plate margin as a seismic gap requires delineating the rupture areas of previous large earthquakes along it. This is usually done by assuming that the aftershock area coincides with the rupture area of an earthquake. This assumption is supported by studies demonstrating that aftershock areas of great (M $\geq$ 7.8) earthquakes tend to abut without significant overlap [e.g. Fedotov, 1965; Mogi, 1968a; Kelleher, 1970; Sykes, 1971]. Felt areas of intensity VIII-IX are also used to define rupture areas [Kelleher, 1972] if aftershock data are inadequate.

Aftershock data from an earthquake in 1957 that broke a large portion of the Aleutian plate margin show an unusual distribution. This earthquake is one of the largest events ever recorded, with a magnitude, Mw, of 9.1. Its aftershock zone stretches 1200 km along the central and eastern Aleutians, from 180°W to 163°W. In the western 1000 km of this zone, the main segment, aftershocks distribute over a zone about 80 km wide, measured normal to the arc. In contrast, aftershocks in the eastern 200 km segment, the Unalaska segment, define a narrow zone less than 20 km wide. This narrow zone of activity extends the northern edge of the aftershock zone defined by the main aftershock segment. To better understand the behavior of the Unalaska segment of the 1957 aftershock zone, we studied earthquake data in the area of the 1957 earthquake from the past 33 years and tide gauge recordings of the tsunami excited by the 1957 earthquake.

We find that the Unalaska segment, if it underwent deformation in 1957, did so in a manner quite different from the manner of the main segment of the 1957 aftershock zone. Either it ruptured in a delayed event, whose seismic and tsunami signatures were masked in the coda of the main rupture or, alternatively, it ruptured with a very slow source process. The seismic energy of a very slow (longer than several tens of minutes) rupture process in the Unalaska segment might not be distinguishable from the coda of the main rupture and would be inefficient in generating a tsunami.

A third possibility also exists, however. The Unalaska segment may not have ruptured at all in 1957. If it did not, it cannot have broken since about 1902 [Sykes et al., 1981] and would therefore be a seismic gap. It could also have a high seismic potential [McCann et al., 1979] and would produce an earthquake as large as magnitude 8 if it were to rupture completely in a single earthquake. Thus, the evaluation of seismic and tsunami hazards in the eastern Aleutians could be substantially affected if the Unalaska segment did not rupture in 1957.

Figure 1 (top) shows the location of the 1957 earthquake rupture zone and its relation to the rest of the Alaska-Aleutian plate margin. It also shows the locations of three seismic gaps identified by previous studies [Kelleher, 1970; Sykes, 1971; Davies et al., 1981]. From the west, these are: the western Aleutians, an area where relative motion is highly oblique [Sykes, 1971]; the Shumagin Gap, discussed by Davies et al. [1981]; the Yakataga Gap, described by McCann et al. [1980] and Lahr et al. [1980]. The possible Unalaska Gap is the area at the eastern end of the 1957 aftershock zone, between about 164°W and 167°W, labelled as a queried gap.

Tectonic Setting

Geologic evidence indicates that the region of the Alaska-Aleutian arc between 168°W and 160°W is a transitional region between an oceanic type arc structure to the west and a thicker, more continental type arc structure to the east. The Bering Sea Shelf, indicated by the 100 fathom bathymetric contour in Figure 1 (bottom) is believed to be an extinct plate margin that was active during the Mesozoic [Marlow et al., 1977]. At that time, the active plate margin trended subparallel to the present margin westward to about 162°W then began to curve northwestward to follow the trend of the Bering Sea Shelf [Moore, 1972; Scholl et al., 1975; Moore and Connelly, 1977]. The active plate margin jumped to its

present position along the Aleutian arc prior to the early Tertiary [Scholl et al., 1975]. The trend of the Bering Sea shelf edge intersects the Aleutian arc at about 165°W, near the eastern edge of the Unalaska segment.

Geophysical data also indicate that the Unalaska region is a transitional region along the Alaska-Aleutian arc. The volcano-trench separation is nearly constant at about 160 km along the entire portion of the Aleutians from Kiska Island (about 178°E) to Akutan Island, at about 166°W [Davies, 1975]. To the east of Akutan Island, it systematically and smoothly widens, until at Augustine Volcano, in Cook Inlet (about 154°W) it measures nearly 400 km. The Aleutian Terrace, a flat, well defined forearc basin about 70 km wide, extends along the entire central and eastern Aleutians east of about 180°W, but begins to narrow near Akutan Island and has disappeared entirely by 160°W [Nishenko and McCann, 1979]. Similarly, the trend of a prominent 100-150 mgal gravity high, which follows the volcanic line in the central and eastern Aleutians, diverges from this trend at Akutan Island where it turns eastward and follows the trend of the shelf edge break [Watts et al., 1976].

The transitional nature of the Alaska-Aleutian arc between about Unalaska and Unimak Islands (for locations, see Figure 4) may influence the rupture process of earthquakes whose ruptures approach or extend into it. The unusual behavior of the Unalaska segment during and since the 1957 earthquake may be an example that merits special attention.

Seismicity Data

Mogi [1968b], Kelleher [1970] and Sykes [1971] delineated the rupture zone of the 1957 earthquake from its aftershocks, which extend 1200 km eastward along the central and eastern Aleutians from about 180°W. This is the longest aftershock zone ever identified. The earthquake involved bilateral rupture, since the epicenter of the main shock at 51.6°N, 175.4°W, is located well within the aftershock zone (Figure 2). Good quality long period seismograms of this event are sparse [H. Kanamori, written communication, 1979] so its moment, 6×10^{29} dyne cm, and magnitude, $M_w = 9.1$, are estimated from the aftershock area [Kanamori, 1977].

The bottom half of Figure 1 [after Sykes, 1971] shows, as large X's, well-located aftershocks of the 1957 earthquake that occurred during March 1957 (about 20 days) and includes shocks of all magnitudes. Several features of the aftershock distribution are notable. First, along most of the length of the aftershock zone, aftershocks are scattered between the island arc and the Aleutian Trench. This is the main segment of the aftershock zone. Aftershocks in the eastern 200 km, the Unalaska segment, occur only along the northern (arcward) edge of the aftershock zone. Second, a large number of epicenters plot within

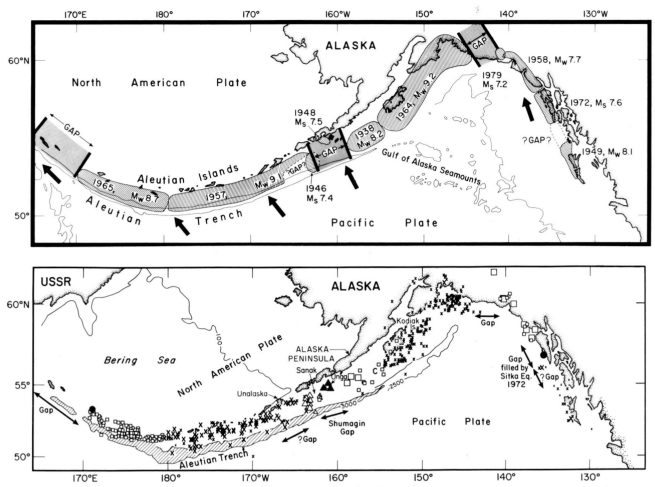

Fig. 1. Top: Location map and identification of aftershock zones of major earthquakes and previously identified seismic gaps in Alaska and the Aleutians. The possible Unalaska Gap is the area labelled as a queried gap near 165°W. Note the proximity of the 1946 tsunamigenic earthquake to the possible Unalaska Gap, and also that the Shumagin Gap nearly abuts the eastern edge of the possible Unalaska Gap. Arrows indicate direction of relative convergence. [After Davies et al., 1981]. Bottom: Map of relocated aftershocks of recent major earthquakes in Alaska and the Aleutians. Only those aftershocks of the 1957 earthquake that occurred in March of 1957 are included; these are plotted as large X's. Note how far trenchward the aftershocks extend between 180°W and 166°W as compared to the very narrow band between 166°W and 163°W. Also note that 3 aftershocks plot within the aftershock zone of the 1946 earthquake, which is identified by triangles (see text for discussion of these events). Bathymetry is in fathoms (1 fathom = 1.83 m). [After Sykes, 1971].

the Aleutian Trench. We shall demonstrate that these events resulted from normal faulting triggered or stimulated by the large underthrust event, and are not aftershocks in the strict sense of defining the rupture zone. Third, several aftershocks of the 1957 earthquake plot within the aftershock zone of the 1946 tsunamigenic earthquake (identified by triangles in Figure 1). No magnitudes are assigned to these events in the International Seismological Summary for 1957, but from the small number of stations reporting, they are likely to be magnitude 6 or smaller.

A more detailed plot of aftershocks during the first year after the 1957 earthquake is shown in Figure 2 and contains only shallow (depth $\leq$ 60 km) earthquakes with assigned magnitude of 5.0 and larger. Epicenters of these events are from the relocations of Sykes [1971] and from the locations of the International Seismological Summary and the U.S. Coast and Geodetic Survey. From a frequency-magnitude plot, we estimate that the data set is complete only for earthquakes larger than about magnitude 6 1/4. The distribution of seismicity in this figure is somewhat more heterogeneous than that plotted in Figure 1, with much clustering evident. Within

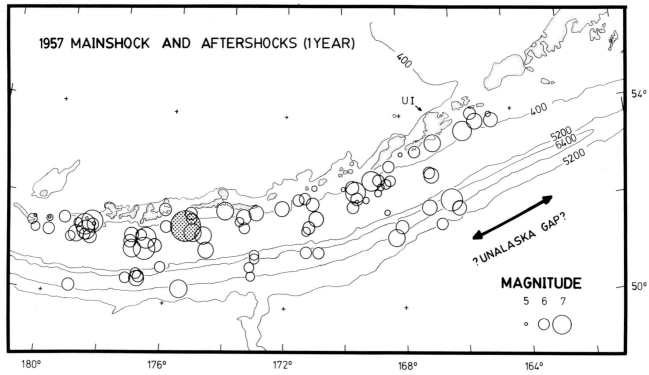

Fig. 2. Detailed plot of the 1957 main shock and aftershocks over a period of 1 year. Aftershocks of March 1957 relocated by Sykes [1971] are supplemented by aftershocks located by the International Seismological Summary and the U.S. Coast and Geodetic Survey. Size of symbols is scaled to magnitude; only events assigned a magnitude of 5 or larger are plotted. The 1957 main shock is shaded and is plotted with a symbol size appropriate for its surface wave magnitude (M_s) of 8.2, rather than its M_w of 9.1. We infer that earthquakes located beneath the Aleutian Trench between the 5200 m contours are of normal faulting type although such mechanisms have actually been obtained for only two of the events [Stauder and Udias, 1963]. Earthquakes at depths shallower than 80 km are included although none occurred that were deeper than 60 km. UI indicates the location of Unalaska Island. Bathymetry is in meters.

the main portion of the aftershock zone, events scatter over a width normal to the arc of about 80 km. Note the large number of aftershocks with magnitudes about 7. For comparison, no aftershocks larger than magnitude 6.5 occurred in the aftershock zone of another great earthquake nearby, the 1964 Alaska earthquake [Davies et al., 1981] (Figure 1).

Activity within the Unalaska segment of the aftershock zone defines an eastward extension of the northern margin of activity to the west of about 167°W longitude. Depths of these 5 events, based on pP-P times reported in the International Seismological Summary, range from 45 to 60 km. This depth range is appropriate for the deeper, or down dip, edge of the main thrust zone or seismic portion of the plate interface in the Aleutians [Davies and House, 1979]. Stauder and Udias [1963] obtained an underthrust-type focal mechanism for the westernmost event of the cluster of five (53.6°N, 165.8°W, M = 7). Although they are not well constrained, both nodal planes dip steeply and strike subparallel to the trend of the

arc. It is plausible, therefore, that the aftershock activity located at the northern edge of the Unalaska segment represents underthrust-type motion along the deeper edge of the plate interface. The remaining shallower portion of the interface was aseismic, at least for earthquakes larger than about magnitude 6 1/4.

Another characteristic of the aftershock activity is the large number of events whose epicenters plot within the Aleutian Trench (Figure 2). Focal mechanisms of two of these events [Stauder and Udias, 1963] show normal faulting, which is a typical focal mechanism for earthquakes beneath trenches. During the 15 years from 1959 to 1974 (see Table 1), earthquakes of magnitude 6 and larger occurred beneath the trench in areas offshore of the 1957 earthquake on an average of less than once every 2 years. None were located within this area during the 12 years prior to the main shock. Fully 13 events occurred in this area during the first year after the 1957 main shock and a total of 19 occurred there in the 2 years following the main shock. These numbers

Table 1. Normal Faulting Following the
1957 Earthquake

Numbers of events of magnitude ≥ 6.0 with
epicenters within the Aleutian Trench between
longitude 160°W and 180°W

Time Interval	Number of Events
Before the main shock (12 years)	0
The first year after the main shock	13
The second year after the main shock	6
Average number of events per year from 3/9/59 to 3/9/74 (15 years)	0.40
Total Number of Events Considered from 3/9/57 to 3/9/74	25

contrast sharply with the quiescence before and near quiescence after. It seems clear that the 1957 mainshock stimulated these large numbers of events. Similar phenomena have been noted by Sykes [1971], Spence [1977], and Hanks [1979] for several other major underthrust-type events.

Interestingly, none of the many normal fault-type events occurred seaward of the possible Unalaska Gap in the first year after the main shock. Only one did in the second year, and none have since then (1959-1979). If the Unalaska segment did not rupture in 1957 we would not expect induced activity to occur beneath the adjacent portion of the trench. The distribution of such activity within the rest of the 1957 rupture zone is quite heterogeneous, however, so the lack of it seaward of the Unalaska segment may be only fortuitous.

Figure 3 illustrates earthquake activity during 11 years before the 1957 main shock. This figure clearly shows two distinct clusters. One is at the western end of the aftershock zone, and developed over about 3 years before the main shock. The other plots at the eastern end of the main portion of the aftershock zone (compare Figures 2 and 3). The latter cluster primarily developed as a swarm over a period of about 1 week in early January 1957, although it also includes several events from as early as 1950. Mogi [1968b] noted the occurrence and location of the swarm of January 1957 and suggested that it represented an area weakened by partial failure prior to the main rupture. If the area of the 1957 aftershock zone had been identified as a seismic gap prior to 1957, the occurrence of these two very clear earthquake clusters might have prompted the issuance of an intermediate-term earthquake prediction for the area.

The two clusters seem highly suggestive of

pre main shock failure processes at the margins of the eventual rupture zone. Kelleher and Savino [1975] noted the occurrence of such clusters of activity prior to several great (magnitude ≥ 7.8) underthrust type earthquakes. By this interpretation, the eastern cluster suggests that rupture during the main shock may not have extended east of about 168°W longitude.

An alternative interpretation of the eastern cluster of activity is that it represents failure of a distinct stress concentration (or asperity) prior to the main shock. Thus, the events in 1950, 1952 and 1955 may have resulted from a build up of stress in the area, while the swarm of January 1957 could have resulted from the failure of that highly stressed region. Since the area had already failed prior to the main rupture of March 1957, it probably would not produce aftershocks associated with the March main shock. Note that in fact no aftershocks occurred in the region of this cluster in Figure 2. In this interpretation, the location of the January 1957 swarm would not have particular significance for the dimensions of the main 1957 rupture, which could have continued eastward beyond it.

Seismicity data since 1957 also show remarkable quiescence within the Unalaska segment compared to seismicity within the main portion of the 1957 aftershock zone. During the entire 21 year period from 1957 to 1979, the Unalaska segment experienced only two events (both m_b = 5.0) with assigned magnitude of 5 or greater [Davies et al., 1981]. Numerous events occurred along the northern margin of this segment during this time. The main segment of the 1957 aftershock zone experienced a large number of earthquakes larger than magnitude 5 during this time. These events could represent aftershock activity in a general sense. Thus, the Unalaska segment does not seem to have experienced any significant activity that could be interpreted as aftershocks in any sense.

Study of the Source Area of the 1957 Tsunami

Method. Large underthrust-type earthquakes commonly generate sizable tsunamis [Cox, 1963; Hatori, 1970; Ando, 1975; Abe, 1979; Nishenko and McCann, 1979], and the 1957 earthquake was no exception [Salsman, 1959; Abe, 1979; Nishenko and McCann, 1979]. The size of the tsunami source area generally corresponds well to the aftershock areas of large earthquakes [Hatori, 1970; Nishenko and McCann, 1979].

As noted by Salsman [1959] the tsunami from the 1957 earthquake was generated over a region of sizeable dimensions, rather than at a point. Tsunami energy excited by the eastern portion of the 1957 rupture zone would be recorded by tide gauges along the western coasts of North and South America. Therefore, the travel times to gauges in these locations would constrain the eastern extent of the tsunami generating area.

The speed of tsunami waves can be approximated quite closely by the simple formula: $S = (gd)^{1/2}$,

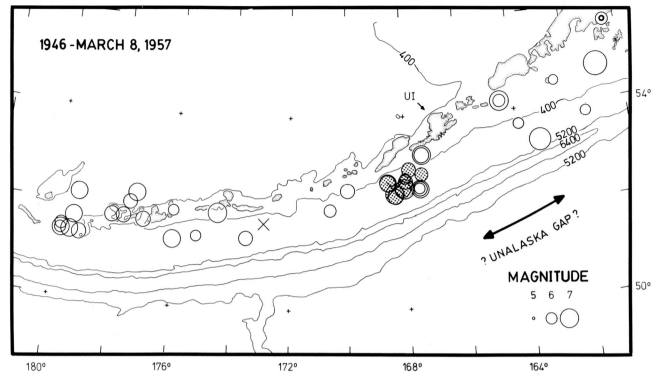

Fig. 3. Plot similar to Figure 2, but of earthquakes from 1946 to just before the 1957 main shock. Earthquakes at depths shallower than 60 km are plotted as circles; those at depths of 60 to 80 km as X's. Note the clustering of epicenters near 180°W and near 168°W. Earthquakes that are part of a swarm that occurred in early January 1957 are shaded. Note also the location of the 1946 tsunamigenic earthquake, immediately to the left of the 5200 meter label.

where g is the acceleration due to gravity ($\sim$ 9.8 m/s^2) and d is the water depth [Cox, 1963; Li and Lam, 1964]. From this formula, travel-time diagrams can be used to locate the extent of the source. This approach, however, yields estimates of the eastern extent of the source of the 1957 tsunami that are so scattered it is not possible to distinguish whether the source area extends eastward to 164°W or only extends to 168°W [L. R. Sykes, unpublished data, 1979]. One possible source of relatively large errors in this approach is the calculation of tsunami speed in the very shallow water portions of the path, where small depth errors can produce large travel time errors.

We use the relative location technique to locate the eastern extent of the source area of the 1957 tsunami more precisely. The nearby 1946 Aleutian earthquake provides a well-located tsunami source for comparison.

As noted by Green [1946] and Sheperd et al. [1950] the 1946 Aleutian earthquake generated an extremely large tsunami that was destructive over much of the Pacific Coast. This large tsunami is especially notable because of the moderate surface-wave magnitude of the earthquake (Ms = 7.4, Sykes [1971]; Kanamori [1972a]; Fukao [1979]). The aftershock zone of this event covers

an area of only about 100-150 km in diameter (Figures 1 and 4) that extends to the NE of the epicenter of the main shock at 53.3°N, 163.2°W [Sykes, 1971]. Green [1946] and Bodle [1946] studied the tsunami of this event and concluded that its source area was confined very near the epicenter. Davies et al. [1981] find that the aftershock and tsunami generating areas are nearly identical for the 1946 shock. Since many tide gauges recorded the tsunami of both this event and the 1957 earthquake, it is a good reference event.

Results. The results of the relative relocation are compiled in Table 2. Travel times of the 1957 tsunami to eastern Pacific stations are 18 to 27 minutes longer than those of the 1946 event to the same tide gauges. If the tsunami source of the 1957 event had extended over the entire after-shock zone indicated in Figure 1, the travel time differential for the two events would have been small, about 6 minutes or less.

Several assumptions were made in our analysis. These were: 1) the tsunamis of 1946 and 1957 followed nearly identical paths, except in the immediate area of the 1957 source; 2) the arrival times of the 1957 tsunami correspond to the first arriving tsunami energy; 3) rupture during the 1957 earthquake propagated smoothly between the hypocenter and the easternmost extent of the

Table 2. Comparison of Travel Times from the 1957 and 1946 Tsunamis

Tide Gauge	1957 Tsunami Arrival time[1]	1957 Tsunami Travel time[1]	1946 Tsunami Arrival time[1]	1946 Tsunami Travel time[1]	Δ (1957-1946) (min)	From Tsunami Source to Tide Gauge Azimuth	From Tsunami Source to Tide Gauge Distance (km)	Tsunami Speed Along Differential Path Between 1946 and 1957 Source Areas (km/min)	Distance West from 1946 Tsunami to 1957 Tsunami (km)
Neah Bay, Washington	19:20[2]	4:57	17:00[3]	4:31	+26	82°	3,000	14.2	327
Crescent City, California	19:30[4]	5:07	17:09[4]	4:40	+27	93°	3,400	13.6	327
San Francisco	20:18[2]	5:55	18:00[3]	5:31	+24	98°	3,789	13.6	283
La Jolla	20:57[4]	6:34	18:43[4]	6:14	+20	99°	4,500	13.6	226
Antofagasta	32:04[2]	17:41	29:50[3]	17:21	+20	100°	12,477	13.6	226
Valparaiso	32:48[2]	18:25	30:36[3]	18:07	+18	108°	13,165	13.5	210

[1]Units are hour:minute.
[2]Published arrival time, Salsman [1959].
[3]Published arrival time, Green [1946].
[4]Reread from records published in Green [1946] or Salsman [1957]. Reread arrivals agreed with published arrivals within 3 minutes.

tsunami source area. Assumption 1 probably is quite good, since the distance between source and receiver is never less than 10 times the distance between the two sources. Assumptions 2 and 3 may not be valid; the discussion section considers them in more detail.

For the individual differential travel paths between the 1946 and 1957 tsunami source areas we calculate depths that are distance weighted averages of the depths along each travel path. We obtain average tsunami speeds (listed in Table 2) from the average depths by using the formula above, and reduce differential travel times of the two tsunamis to relative locations of the eastern extent of the 1957 tsunami source area compared to that of the 1946 tsunami. Our calculations corrected the differential travel times for the finite rupture velocity of the 1957 earthquake, which we chose as 3.5 km/s. Our results do not depend strongly on the actual velocity chosen since the rupture velocity of earthquakes generally is more than 10 times the velocity of tsunamis.

The last column of Table 2 lists the relative locations obtained, which are plotted in Figure 4. The solid lines in the figure are determinations from travel times reread from published records, dashed lines are from published travel times.

Uncertainties in the determination of the eastern extent of the 1957 tsunami arise

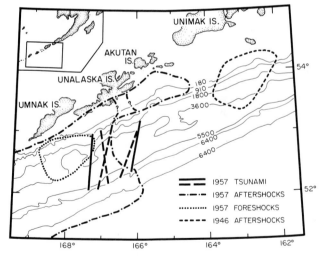

Fig. 4. Detailed map and summary of information about the eastern portion of the aftershock zone of the 1957 earthquake. The eastern extent of the 1957 tsunami is indicated by a solid line for stations that the authors could reread arrival times and by a dashed line when the determination is based on published arrival times (see text for sources). Other lines enclose the aftershock zones of the 1957 and 1946 earthquakes, and a swarm that occurred 2 months before the 1957 main shock. Bathymetry is in meters.

principally from two sources - error in reading the arrival times, which we estimate to be about ±2 minutes, and error in the tsunami speeds along the differential travel path, which we estimate to be at most ±20%. Error limits of 20% in the tsunami speed were chosen arbitrarily; actual errors seem unlikely to exceed this, since depth errors greater than about 25% would be required. The uncertainty resulting from errors in tsunami speed are ±65 km or less, and dominate the uncertainty from travel time errors (about 30 km), so we take ±65 km as a reasonable uncertainty in our estimates of the source location. This corresponds to about 1° of longitude in Figure 4. Individual estimates fall within this uncertainty and determine the eastern extent of the 1957 tsunami source at about 166.5°W to 167°W. This location is nearly identical to that of the eastern end of the main portion of the aftershock zone (see Figure 4).

We note also that the tsunami data excludes the possibility of significant tsunami generation within the aftershock zone at the northern edge of the Unalaska segment. Consider, for example, if seismic rupture continued eastward into the Unalaska segment along a steeply dipping imbricate fault. Walcott [1978] inferred that the 1931 Hawke's Bay earthquake in New Zealand may have broken such a fault. This type of fault would outcrop substantially arcward of the outcrop of the more shallowly dipping fault zone that we presume ruptured within the main segment of the 1957 aftershock zone. A steeply dipping rupture geometry might produce the pattern of aftershocks observed within the Unalaska segment, as well as the focal mechanism of one of these [Stauder and Udias, 1963] which was described earlier. The travel times of a tsunami produced by this process, however, would be substantially shorter, by as much as 10 minutes or more, than the observed travel times.

Reports of near source tsunami effects, while not definitive, seem consistent with our identification of the source area. A published report that the 1957 tsunami reached 40 feet high (12 m) at Scotch Cap [Brazee and Cloud, 1959] at

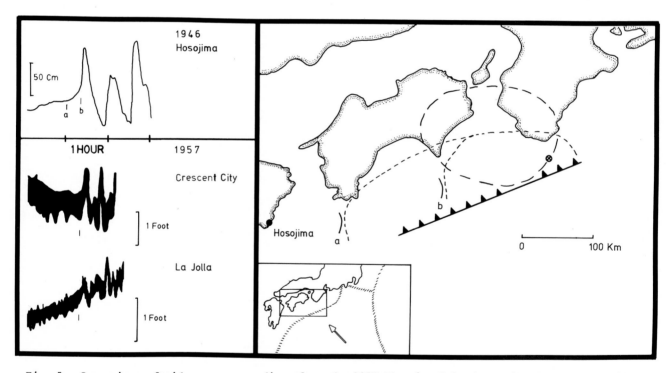

Fig. 5. Comparison of tide gauge recordings from the 1957 Aleutian Islands earthquake and the well-studied 1946 Nankaido earthquake, southwest Japan. Right: map of southwest Japan, showing the area of the 1946 Nankaido earthquake. The location of the main shock is indicated by a star and the after-shock zone after 1 month by a coarse dashed line. Hypothesized location of tsunami source area is indicated by the fine dashed line. Two arcs, labelled 'a' and 'b' are the locations of two phases identified on tide gauge record from Hosojima, to the southwest [after Ando, 1975]. Toothed line indicates surface trace of rupture zone obtained from geodetic data [after Fitch and Scholz, 1971]. Left: tracings of tsunami from 1946 Nankaido earthquake recorded at Hosojima (top) and of a tsunami from 1957 Aleutian earthquake recorded at Crescent City and La Jolla, California (below). Tracings are at the same time scale; the time between tick marks is 1 hour. Arrival times are indicated by tick marks under records; 'a' and 'b' are identifications of Ando [1975]. Tracings are from Ando [1975] and Salsman [1959].

the southwest end of Unimak Island (see Figure 4) might suggest that the source of such a large wave must be closer to Scotch Cap than our determination of the source is. For comparison, however, the tsunami ran up to at least 30 m above sea level at the southwest end of Umnak Island [R. Black, written communication, 1980]. This island is directly adjacent to the main portion of the rupture. The observation from Umnak Island clearly establishes a tsunami amplitude at least 3 times larger near the main source than the tsunami amplitude about 200 km away.

A decrease of amplitude of a factor of about 2 over a distance of 200 km seems reasonable, since tsunami amplitude decreases as:

$$h \approx h_o/(d)^{1/2}$$

where h_o is the amplitude at the source and d is distance from the source [Solov'ev, 1965]. Tsunami run-up associated with the 1968 Tokachi-oki earthquake decayed by a factor of about 2 over a distance of about 200 km [Kajiura et al., 1968].

We note, also, that local bathymetry can strongly influence tsunami amplitudes [e.g. Hatori, et al., 1973]. Therefore, observations of near source tsunami heights alone do not allow us to distinguish whether tsunami excitation occurred in the Unalaska segment. Our only conclusion in this regard is that near source observations do not contradict the previous result that the Unalaska segment did not excite significant tsunami energy.

Comparison with Nankaido Earthquake of 1946

The coincidence of the easternmost extent of the 1957 tsunami source area and the main portion of the aftershock zone is remarkable and could lead to the premature conclusion that the 1957 earthquake did not rupture the Unalaska segment of the aftershock zone. It is instructive, therefore, to compare the observations of the 1957 earthquake with similar observations from an unusual and well-studied earthquake in Japan, the 1946 Nankaido earthquake. One month after the Nankaido main shock, aftershocks covered an area with a length of 210 km along strike [Kanamori, 1972b; Ando, 1975], whereas geodetic data suggest a rupture length of about 320 km [Fitch and Scholz, 1971]. Of this extended zone, the western 90 km lacked aftershocks, but probably slipped more (by about a factor of 2) than did the rest of the rupture zone [Fitch and Scholz, 1971]. Figure 5 illustrates the region near the Nankaido event and the main features of the earthquake.

Ando [1975] studied the records from two tide gauges located near the aftershock zone of the Nankaido earthquake. One, located near the northern edge of the aftershock zone, shows a simple motion at the start of the tsunami. The other, located at Hosojima, about 250 km southwest of the aftershock zone, shows a more complex onset. Ando identified two distinct arrivals within the initial portion of the tsunami (see Figure 5). The main energy of the tsunami, arriving at 'b' in the record, was generated at the location labelled 'b', which coincides with the outline of the aftershock area. An earlier, smaller and less distinct arrival that Ando identifies at 'a' seems to originate from a source that extends somewhat to the west of the geodetically determined rupture zone. In addition to lacking aftershocks, the westernmost portion (90 km in length) of the rupture zone was inefficient in generating a tsunami. The absence of a strong tsunami source in the western portion strongly suggests this area ruptured with a very long time constant of at least several minutes in duration [Kanamori, 1972a; 1972b]. A rupture with such a long time constant could involve either a very slow rupture propagation combined with rapid displacement [Fukao, 1979; Das and Scholz, 1981], or a fast rupture propagation with very slow displacements on the fault, possibly as in an episode of creep [Kanamori, 1972b]. The identification of a small tsunami phase excited by the westernmost portion of the rupture zone suggests, however, that the rupture propagated rapidly (compared to tsunami speeds) across the western portion of the zone but because of slow strain release was inefficient at generating a tsunami. Otherwise, the small tsunami excited by the western area would have been concealed by the later and much larger main tsunami generated by the eastern region.

In the lower corner of Figure 5 we compare two California tide gauge records from the 1957 tsunami with that of Hosojima from the 1946 Nankaido tsunami. A precursor comparable to 'a' produced by the 200 km long Unalaska segment would arrive approximately 20 minutes before the main tsunami phase. There is no obvious arrival prior to the main tsunami in the two records of 1957. We conclude that any such phase, if present, is very small and that if any tsunami was generated at all by the Unalaska segment, generation was inefficient.

Discussion

The behavior of the Unalaska segment at the time of the 1957 earthquake seems to have been distinctly different from that of the rest of the aftershock zone. An unusual type of rupture in the Unalaska segment, either as a delayed event or as a very slow source process, might explain the observations. Alternatively, the lack of rupture within the Unalaska segment could, also.

Consider first that the Unalaska segment ruptured in an unusual manner in 1957. The possibility that it broke in a delayed seismic event with a normal-type rupture seems unlikely since it requires a fortuitous complexity of rupture that is not supported by the lack of aftershock activity within the Unalaska segment. Timing of the Unalaska rupture would be critical. Rupture would have to be delayed by at least 25 minutes in order not to interfere with the initial tsunami energy. Rupture would also have to occur quickly enough after the main shock that the seismic and tsunami energy that resulted would be lost in the coda of the main shock. In addition, the smaller tsunami observed at Scotch Cap compared to that at Umnak Island is inconsistent with this delayed normal rupture interpretation.

If the Unalaska segment did rupture in 1957 it seems more likely that it did so in a very slow and smooth rupture process that would leave stress heterogeneities too small to produce aftershocks. The result of such a rupture would be a slow earthquake [Kanamori and Stewart, 1979] whose seismological effects might not be distinguishable from the coda of the main shock. In this interpretation, however, the occurence of aftershocks at the northern and deeper edge of the plate interface is puzzling. If material at shallower depths undergoes non brittle deformation, why would the deeper material, which is subjected to higher temperature and pressure, undergo brittle deformation? Since the Unalaska area seems to be a tectonic transition zone, perhaps coupling of the plates is poor at shallow depths, but is strong enough along the deeper edge of the interface to rupture seismically.

The alternative possibility, that no rupture occurred within the Unalaska segment, could explain the aftershocks at 40-60 km depths as activity along the deeper edge of a strongly coupled segment of the plate interface. This activity may have been triggered by rupture of the main segment. The smaller magnitude events that occurred to the east of the Unalaska segment (and within the aftershock zone of the 1946 Aleutian earthquake) could be explained in a similar manner. Since most of the plate interface did not rupture, there would be no aftershocks within the segment. Also, activity would not be stimulated in the adjacent trench region. Finally, the precursory swarm near the eastern edge of the main portion of the aftershock zone could have resulted from preparatory failure at this end of the eventual rupture zone. Such failure may also have occurred near the western end of the aftershock zone.

Few data are available to quantify the behavior of this Unalaska segment during the 1957 event. Long and ultra-long period seismic or strain records might help to resolve if it ruptured in a slow event. No suitable geodetic data are known to the authors from this area. A tide gauge operated at the town of Unalaska, near the eastern side of Unalaska Island, showed somewhat less coseismic subsidence than did a tide gauge operated on Adak Island, near the epicenter of the main shock (12 cm compared to 18) [Wahr and Wyss, 1980]. It is not clear, however, that the amount of subsidence at Unalaska could distinguish whether rupture continued into the Unalaska segment, since Unalaska is so near the western boundary of the segment. Data from several temporary tide gauges that were operated for short

periods of time both before and after the 1957 earthquake may help to resolve whether the Unalaska segment ruptured. At least one gauge was located near the northeastern edge of the segment and presumably would not be influenced by slip on the main portion of the 1957 rupture zone [R. Stein, personal communication, 1980].

It is difficult to identify historic earthquakes that definitely ruptured into the Unalaska segment as opposed to earthquakes that may have ruptured only the main segment of the 1957 aftershock zone or the Shumagin Gap. Nevertheless, it is clear that great earthquakes have broken adjacent portions of the Alaska-Aleutian plate margin on both sides of the Unalaska segment. Therefore it would seem unwise to wait until finding definitive evidence that the Unalaska segment has ruptured in a great earthquake before considering that this area is capable of producing one.

The question of whether the Unalaska segment ruptured in 1957 has serious implications for the evaluation of seismic and tsunami hazards in the eastern Aleutians. If it did not break then, it could not have ruptured more recently than 1902, 78 years ago [Sykes et al., 1981]. In this case, not only would the Unalaska segment be a seismic gap, but it could also fall into a category of high potential for producing a large or great earthquake, as described by McCann et al. [1979]. Davies et al. [1981] found a range of repeat times of about 50 to 90 years for great earthquakes within 400 km to the east of the possible Unalaska Gap. So, from a recurrence time consideration it would seem that if it is a seismic gap it may also be a mature seismic gap.

We conclude that at the time of the earthquake of 1957, the 200 km long Unalaska segment behaved in a manner quite different from the rest of the 1200 km long aftershock zone. Although the behavior of this segment may not be resolvable with available data, one of two scenarios seems likely. The Unalaska segment may have ruptured in a slow event in 1957, and as a result would now have a low seismic potential [McCann et al., 1979]. On the other hand, if the segment did not rupture in 1957, it would be a seismic gap with a high seismic potential and hence, could pose significant seismic and tsunami hazards for the eastern Aleutians in the near future.

The dimensions of the unbroken arc segment would be about 210 km in length along the plate margin and 90 km in down dip width. If a single earthquake broke the entire area of the Unalaska segment and had an average stress drop ($\Delta\sigma$) of 20 bars, the displacement across the fault surface would be about 4 m, the moment, M_o, about 3×10^{28} dyne cm, and the magnitude, M_w, about 8.3 [Kanamori and Anderson, 1975; Kanamori, 1977].

Acknowledgments. Discussions with W. R. McCann provided the motivation for this study and were helpful to its development. L. Zappa typed the manuscript and K. Nagao assisted with drafting. We thank R. Bilham and W. R. McCann for critically reviewing this manuscript. We also thank M. Ohtake for helpful comments. This study was supported by the U.S. Department of Energy (Contract EY-76-02-3134C) and by the U.S. Bureau of Land Management through interagency agreement with the National Oceanic and Atmospheric Administration (Contract NOAA 03-5-022-70), under which a multi-year program responding to the needs of petroleum development of the Alaska Continental Shelf is managed by the Outer Continental Shelf Environmental Assessment Program (OCSEAP) office. The Department of Energy supports the broad seismotectonics aspects of the reported studies; OCSEAP supports the hazards evaluation aspects. Lamont-Doherty Geological Observatory Contribution Number 3111.

References

Abe, K., Size of great earthquakes of 1837-1974 inferred from tsunami data, J. Geophys. Res., 84, 1561-1568, 1979.

Ando, M., Source mechanisms and tectonic significance of historical earthquakes along the Nankai Trough, Japan, Tectonophysics, 27, 119-140, 1975.

Bodle, R. R., Note on the earthquake and seismic sea wave of April 1, 1946, Trans. AGU, 27, 464-465, 1946.

Brazee, R. J., and W. K. Cloud, United States earthquakes, 1957, U.S. Dept. of Commerce, Washington, D. C., 1959.

Cox, D. C., Status of tsunami knowledge, in Cox, D. C. (ed.), Proceedings of Tsunami Meetings Associated with 10th Pacific Science Congress, IUGG Monograph 24, 1963.

Das, S., and C. H. Scholz, Theory of time-dependent rupture in the earth, submitted to J. Geophys. Res., 1981.

Davies, J. N., Seismological investigations of plate tectonics in southcentral Alaska, Ph.D. thesis, University of Alaska, Fairbanks, 1975.

Davies, J. N., and L. House, Aleutian subduction zone seismicity, volcano-trench separation, and their relation to great thrust-type earthquakes, J. Geophys. Res., 84, 4583-4591, 1979.

Davies, J., L. Sykes, L. House, and K. Jacob, Shumagin seismic gap, Alaska Peninsula: History of great earthquakes, tectonic setting, and evidence for high seismic potential, J. Geophys. Res., in press, 1981.

Fedotov, S. A., Regularities of the distribution of strong earthquakes of Kamchatka, the Kuril Islands, and northeastern Japan (in Russian), Tr. Inst. Fiz. Zemli Akad. Nauk SSSR, 36, 66-93, 1965.

Fitch, T. J., and C. H. Scholz, Mechanism of underthrusting in southwest Japan: a model of convergent plate interactions, J. Geophys. Res., 76, 7260-7292, 1971.

Fukao, Y., Tsunami earthquakes and subduction processes near deep-sea trenches, J. Geophys. Res., 84, 2303-2314, 1979.

Green, C. K., Seismic sea wave of April 1, 1946, as recorded on tide gauges, Trans. Am. Geophys. Un., 27, 490-500, 1946.

Hanks, T. C., Deviatoric stresses and earthquake occurrence at the outer rise, J. Geophys. Res., 84, 2343-2347, 1979.

Hatori, T., Dimensions and geographic distribution of tsunami sources near Japan, in: Adams, W. M. (ed.), Tsunamis in the Pacific Ocean, East-West Center Press, Honolulu, Hawaii, 69-83, 1970.

Kajiura, K., T. Hatori, I. Aida, and M. Koyama, A survey of a tsunami accompanying the Tokachi-oki earthquake of May 1968, Bull. Earthq. Res. Inst., 46, 1369-1396, 1968.

Kanamori, H., Mechanism of tsunami earthquakes, Phys. Earth Planet. Int., 6, 346-359, 1972a.

Kanamori, H., Tectonic implications of the 1944 Tonankai and the 1946 Nankaido earthquakes, Phys. Earth Planet. Int., 5, 129-139, 1972b.

Kanamori, H., and D. L. Anderson, Theoretical basis of some empirical relations in seismology, Bull. Seismol. Soc. Am., 65, 1073-1095, 1975.

Kanamori, H., The energy release in great earthquakes, J. Geophys. Res., 82, 2981-2987, 1977.

Kanamori, H., and G. S. Stewart, A slow earthquake, Phys. Earth Planet. Int., 18, 167-175, 1979.

Kelleher, J. A., Space-time seismicity of the Alaska-Aleutian seismic zone, J. Geophys. Res., 75, 5745-5756, 1970.

Kelleher, J. A., Rupture zones of large South American earthquakes and some predictions, J. Geophys. Res., 77, 2087-2103, 1972.

Kelleher, J., L. Sykes, and J. Oliver, Possible criteria for predicting earthquake locations and their applications to major plate boundaries of the Pacific and the Caribbean, J. Geophys. Res., 78, 2547-2585, 1973.

Kelleher, J., and J. Savino, Distribution of seismicity before large strike slip and thrust-type earthquakes, J. Geophys. Res., 80, 260-271, 1975.

Lahr, J. C., C. D. Stephens, H. S. Hasegawa, and J. Boatwright, Alaskan seismic gap only partially filled by 28 February 1979 earthquake, Science, 207, 1351-1353, 1980.

Li, W-H, and S-H Lam, Principles of Fluid Mechanics, Addison-Wesley, p. 175-180, Reading, Mass., 1964.

Marlow, M. S., D. W. Scholl, and A. K. Cooper, St. George basin, Bering Sea shelf: a collapsed Mesozoic margin, in Talwani, M., and W. C. Pitman III (eds.), Island Arcs, Deep Sea Trenches and Back Arc Basins, Maurice Ewing Ser. vol. 1, 211-220, AGU, Washington, D. C., 1977.

McCann, W. R., S. P. Nishenko, L. R. Sykes, and J. Krause, Seismic gaps and plate boundaries: seismic potential for major boundaries, Pure Appl. Geophys., 117, 1082-1147, 1979.

HOUSE ET AL.   91

McCann, W. R., O. J. Perez, and L. R. Sykes, Yakataga gap, Alaska: seismic history and earthquake potential, Science, 207, 1309-1314, 1980.

Mogi, K., Some features of recent seismic activity in and near Japan (1), Bull. Earthq. Res. Inst., Tokyo University, 46, 1225-1235, 1968a.

Mogi, K., Development of aftershock areas of great earthquakes, Bull. Earthq. Res. Inst., Tokyo Univ., 46, 175-203, 1968b.

Moore, J. C., Uplifted trench sediments: southwestern Alaska-Bering Shelf edge, Science, 175, 1103-1105, 1972.

Moore, J. C., and W. Connelly, Mesozoic tectonics of the southern Alaska margin, in Talwani, M., and W. C. Pitman III (eds.), Island Arcs, Deep Sea Trenches and Back-Arc Basins, Maurice Ewing Series, vol. 1, 71-82, AGU, Washington, D.C., 1977.

Nishenko, S., and W. McCann, Large thrust earthquakes and tsunamis: implications for the development of fore arc basins, J. Geophys. Res., 84, 573-584, 1979.

Salsman, G. G., The tsunami of March 9, 1957, as recorded at tide stations, Technical Bulletin No. 6, U.S. Dept. of Commerce, Washington, D. C., 1959.

Scholl, D. W., E. C. Buffington, and M. S. Marlow, Plate tectonics and the structural evolution of the Aleutian-Bering Sea region, Geol. Soc. Amer. Spec. Paper 151, 1-31, 1975.

Shepard, F. P., G. A. Macdonald, and D. C. Cox, The tsunami of April 1, 1946, Bull. Scripps Inst. Ocean., 5, 391-528, 1950.

Solov'ev, S. L., The Urup earthquake and associated tsunami of 1963, Bull. Earthq. Res. Inst., 43, 103-109, 1965.

Spence, W., The Aleutian arc: tectonic blocks, episodic subduction, strain diffusion, and magma generation, J. Geophys. Res., 82, 213-230, 1977.

Stauder, W. J., and A. Udias, S-wave studies of earthquakes of the north Pacific, Part II: Aleutian Islands, Bull. Seismol. Soc. Am., 53, 59-77, 1963.

Sykes, L. R., Aftershock zones of great earthquakes, seismicity gaps, and earthquake prediction for Alaska and the Aleutians, J. Geophys. Res., 76, 8021-8041, 1971.

Sykes, L. R., J. B. Kisslinger, L. House, J. N. Davies, and K. H. Jacob, Rupture zones and repeat times of great earthquakes along the Alaska-Aleutian arc, 1784-1980, this volume, 1981.

Wahr, J., and M. Wyss, Interpretation of post seismic deformation with a viscoelastic relaxation model, J. Geophys. Res., 85, 6471-6477, 1980.

Walcott, R. I., Geodetic strains and large earthquakes in the axial tectonic belt of North Island, New Zealand, J. Geophys. Res., 83, 4419-4429, 1978.

Watts, A. B., M. Talwani, and J. R. Cochran, Gravity field of the northwest Pacific Ocean basin and its margin, in The Geophysics of the Pacific Ocean Basin and its Margin, Geophysical Monograph 19, 17-34, AGU, Washington, D. C., 1976.

SEISMICITY AND TECTONICS OF THE CENTRAL NEW HEBRIDES ISLAND ARC

Bryan L. Isacks(1), Richard K. Cardwell(1), Jean-Luc Chatelain(2), Muawia Barazangi(1), Jean-Michel Marthelot(1), Douglas Chinn(1), and Remy Louat(2)

(1)Department of Geological Sciences, Cornell University, Ithaca, New York 14853.
(2) Office de la Recherche Scientifique et Technique Outre-Mer (ORSTOM), Noumea Centre, B.P.A5, Noumea, New Caledonia.

Abstract. The seismicity of the central New Hebrides convergent plate boundary is investigated with three sets of data: (1) large earthquakes (Ms > 6.9) for the past 75 years, (2) moderate-sized earthquakes (mb > 4.5 and Ms < 7.0) during the past 20 years, and (3) small earthquakes (mb = 2.5 to 4.5) located by local networks for several intervals of 1-2 months each since 1975 and continuously since mid-1978. The second set includes a nearly complete collection of focal mechanism solutions for events with Ms > 5 3/4 and new determinations of accurate focal depths based on analyses of P waveforms recorded by WWSSN long-period seismographs. On a regional scale the geometry of the Benioff zone is relatively uniform. Within the resolution of the data there are no major disruptions of the descending plate nor changes in direction of plate convergence along the length of the arc. However, the three data sets all indicate marked variations in the seismicity patterns along the strike of the arc. These variations together with major bathymetric and structural complexities of the interacting plates divide the interplate boundary of the central New Hebrides into segments of about 100 km in length. The segments delimit episodes of seismic rupture but may also differ significantly in the long term balance of seismic versus aseismic slippage. In the segments near Santo and northern Malekula islands seismic rupture of the interplate boundary occurred in complex sequences of large earthquakes during the period 1965-1974, whereas in the segments near southern Malekula and Efate islands seismic rupture of the boundary may not have occurred during the past 75 years. The interaction of a subducted ridge with a pre-existing seaward protrusion of the upper plate may result in an increased coupling of the converging plates in the Santo-Malekula segments. The orientation of horizontal compressive stress within the upper plate inferred from focal mechanism solutions and geological data is perpendicular to the arc in the recently ruptured segments but changes in the region between southern Malekula and Efate islands to a more complex and variable pattern found in the southern New Hebrides arc. In addition, the diffuse spatial distribution of small earthquakes in the southern Malekula and Efate segments contrasts with the concentration of small events along the presumed interplate boundary of the recently ruptured Santo segment. The unusual concentration of events at shallow depths within the upper plate in the southern Malekula segment may be evidence for loading of a locked segment of the plate boundary. Unusual features of seismicity suggest that in the Efate segment a significant component of creep may accommodate interplate slippage. A persistently high rate of occurrence of small and moderate-size events (Ms < 6.5) in the Efate segment contrasts with the large fluctuations in activity associated with the major events in the recently ruptured segments. The persistent nest of activity in the Efate segment also contrasts with the relative quiescence in the adjacent segments. The most recent (1978-1979) sequence of three moderate-sized shocks (Ms=6) located in the Efate segment was caught by the local networks of seismographs and tilt measurements. Well-documented features of the temporal and spatial development of foreshocks and aftershocks include a clear migration of foreshock activity towards the epicenter of one of the mainshocks. In two cases the area of the aftershock zone expands during a period of several days to a size significantly larger than that expected for the surface wave magnitudes and body wave seismic moments of the mainshocks. For these cases tiltmeter stations on Efate Island, located at distances of 40 to 60 km from the mainshock hypocenters, did not record tilt indicative of post-seismic creep. Nevertheless, periodic releveling of a 1 km benchmark array on Efate reveals that a significant tilt of 3-4 microradians has accumulated during the four year interval between 1976 and 1980. This tilt signal could be

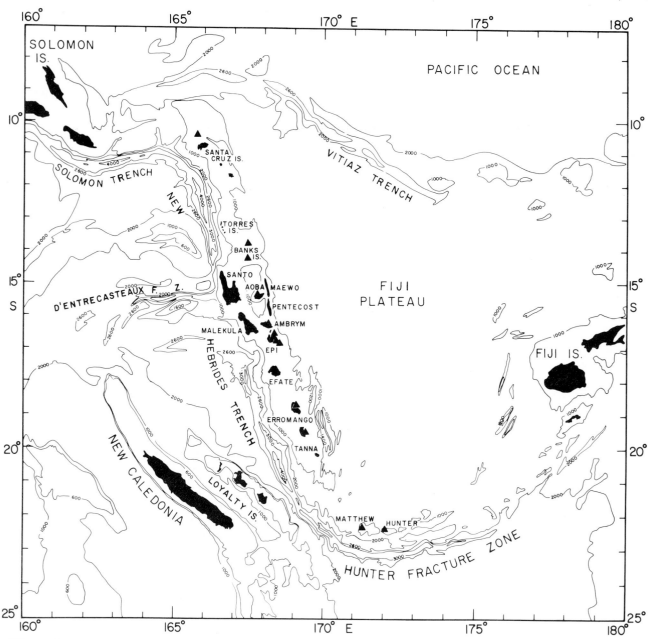

Figure 1. Bathymetric map of the New Hebrides Island arc and surrounding region, taken from Mammerickx et al., 1971. The filled triangles denote Quaternary volcanoes. Contours are in fathoms.

indicative of deformation near a transition between a creeping Efate segment and an adjacent locked segment.

Introduction

The New Hebrides island arc has a very distinctive style of seismicity. Although no earthquake has had a magnitude greater than about 8, numerous large events have occurred in remarkably clustered sequences. This style of seismicity contrasts with that of the Chilean or the Aleutian-Alaskan convergent margins where great earthquakes more or less regularly rupture well-defined segments of the plate boundary. Utsu (1974) considered similar contrasts in the seismicity in the region of Japan, and pointed out the difficulty of resolving seismic "gaps" in a region characterized by a style of seismicity like that in the New Hebrides. Major unresolved questions concern the role of aseismic creep in accommodating some fraction of interplate slippage; the nature of the loading process and its behavior over time scales less than the million year averages provided by analysis of sea-floor spreading data; and the role of structural and geometrical complexities, acting as "barriers" or "asperities" along the plate boundary, in governing the location, size and other characteristics of episodes of seismic slippage.

The inference that variations in the characteristics of seismicity among different convergent plate boundaries represent long-term properties rather than transient features depends largely on the association of seismicity characteristics with structural features of the plate boundary (e.g., Kelleher et al., 1974; Kelleher and McCann, 1976; Uyeda and Kanamori, 1979). The New Hebrides island arc offers a good opportunity to study this problem. In the class of the earth's convergent zones the New Hebrides arc is an extreme member in several respects and may thus exaggerate effects that are more difficult to see in other areas. While having an overall uniformity of configuration on a regional scale, the arc displays remarkable variations in detailed structure along strike, variations that seem to correlate with different characteristics of the seismicity.

The upper plate has a very complex structure that has resulted from the peculiar tectonic history of the region. After a Late Miocene disruption of an ancestral Solomons-New Hebrides-Fiji-Tonga subduction zone, the New Hebrides arc reversed subduction polarity, moved to its present position and left the newly created North Fiji Basin (Fiji Plateau) in its wake (see Figure 1 and Chase, 1971; Karig and Mammerickx, 1972; Gill and Gorton, 1973; Falvey, 1978; and Carney and Macfarlane, 1978). Thus, what is now the leading edge of the plate has played this role for a short time, probably less than 6-8 MY. The most outstanding feature of the New Hebrides arc is the anomalous morphology of the central region. There, the normal elements of physiography including the trench, island arc, and back arc rifts found in the northern and southern parts of the New Hebrides arc are replaced, respectively, by the island blocks of Santo and Malekula, the Aoba Basin, and the uplifted horst-like ridge upon which the islands of Maewo and Pentecost emerge (see Figures 1 and 2). The D'Entrecasteaux "Fracture Zone" (Mammerickx et al., 1971 and Daniel et al., 1977; referred to hereafter as the DFZ) intersects the arc in this region but appears to be subducted along with the rest of the oceanic plate (Isacks and Barazangi, 1977; Pascal et al., 1978; Chung and Kanamori, 1978a).

In our view the striking morphological anomalies of the central New Hebrides are best explained by (1) a Late Miocene episode of intra-arc rifting which produced a major seaward protrusion of the upper plate as the Santo and Malekula blocks moved outboard of the main arc and in so doing created the Aoba Basin (Karig and Mammerickx, 1972), and (2) a Quaternary interaction with the DFZ which affected the uplift and tilting of Santo and Malekula as well as the seismicity (Taylor et al., 1978) but which does not alone account for the main morphological anomalies as proposed by Ravenne et al. (1977) and Chung and Kanamori (1978b). Although unusual, intra-arc rifting of a type analogous to the central New Hebrides may also account for the anomalous morphology of the Bonin arc (Karig and Moore, 1975). The Quaternary subduction of a ridge beneath a pre-existing protrusion of the the upper plate is supported by the fit of the Santo and Malekula blocks into the Aoba Basin; the mismatch between the northern limits of the DFZ and the Santo block; the thickness of sediment in the basin (Ravenne et al., 1977; Luyendyk et al., 1974); and the geological history of the central New Hebrides (Mitchell and Warden, 1971; Mallick, 1973; Mallick and Greenbaum, 1977; and Carney and Macfarlane, 1978).

The complex structure of the upper plate does not appear to be associated with major disruptions or contortions of the descending plate. Pascal et al. (1978) show that on a regional scale the overall configuration of the subduction zone is relatively uniform as found for most other areas (Isacks and Barazangi, 1977). New results of this paper further support this result. The cross-sectional shape of the Benioff zone is clearly among the most sharply downbent and steep ones on earth, and the width of the arc-trench gap is among the smallest. These regional scale geometrical features imply that the width of the interplate boundary -- i.e., the width of the gently dipping thrust fault along which the major interplate earthquakes are generated -- is small compared to most other arcs, and again represents an extreme case in the spectrum of convergent zone variations.

Since 1975 Cornell University, the Office de la Recherche Scientifique et Technique Outre-Mer (ORSTOM), and the New Hebrides Mines Service have collaborated in studies of seismicity, tilting and uplift of the central New Hebrides. The program has included the operation of temporary seismograph networks, two of which included ocean bottom instruments deployed in cooperation with the University of Texas, Marine Sciences Institute (York, 1977; Stephens, 1978; Louat et al., 1979; Ibrahim et al., 1980; Coudert et al., 1981); a permanent network of 19 telemetered stations installed in 1978; measurements of tilt using bubble-level tiltmeters and relevelings of benchmark arrays (Isacks et al., 1978; Marthelot et al., 1980; Bevis and Isacks, 1981); and studies of the pattern of Late Quaternary uplift and tilting of coral terraces (Taylor et al., 1980; Jouannic et al., 1980). This study

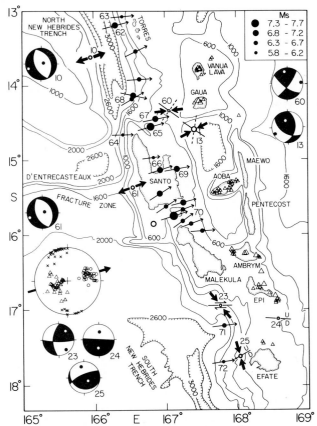

Figure 2. Focal mechanism solutions in the central New Hebrides for earthquakes during 1963 through 1976 with magnitudes (Ms>5 3/4) large enough for reliable focal mechanism determination. Intra-plate epicenters are shown as open circles and the mechanisms are shown in lower hemisphere equal area projections with quadrants of compressional first motions blackened, compressional axes as filled circles, and tensional axes as open circles. Large arrows show horizontal projections of stress axes and lines through epicenters show strike of selected nodal planes, with sense of motion indicated by strike-slip couple or by up (U) and down (D) for dip-slip motion. The arrows through the events interpreted as having interplate thrust-type mechanisms (epicenters shown by filled circles) give the horizontal projection of the slip vector. All interplate mechanisms for the arc are summarized on the larger lower hemisphere equal area plot in the lower left hand side of the figure. Open circles are poles interpreted as the slip vectors, triangles are the fault plane poles, and the X's the "B" or null axes. Numbers are from Pascal et al. (1978) or Table 1; interplate events from Pascal et al. (1978) are not numbered here. Bathymetric contours in fathoms are from Mammerickx et al., 1971. Open triangles on

map show volcanoes: large triangles show recently active centers, small triangles show Quaternary centers (Geological Map of the New Hebrides Condominium, 1975).

synthesizes results from these and other published studies plus the following : (1) new focal mechanism solutions for the central New Hebrides updating the collection of Pascal et al. (1978) through 1980 for much of the central region and including new data for events that occurred between 1960 and 1962; (2) focal depths and seismic moments determined by the method of matching observed and synthetic long-period $\underline{P}$ waveforms (Chinn and Isacks, 1979); (3) preliminary results of a study of seismicity based on data in the Preliminary Determination of Epicenters (PDE) for the past 20 years (Marthelot and Isacks, 1980); and (4) results from the new seismograph network relevant to the seismicity of the region of Malekula and Efate islands. In this study, we focus on the area of Santo through Efate islands where most of the observations have been made.

The northern part of the study area experienced two major sequences of earthquakes in 1965 and 1973-1974 which accommodated seismic slippage along a segment of the convergent plate boundary about 250 km in length. In contrast, much of the plate boundary in the southern part of the study area may not have experienced major seismic slippage during the past 75 years. Several interesting features of the southern area stand out. These include (1) persistent nests and quiescent zones of seismic activity; (2) an unusually diffuse distribution of small shallow earthquakes near the inferred location of the inclined plate boundary; and (3) high rates of tilting measured by releveling of benchmark arrays during a 5 year period since 1975. One interpretation is that these features are related to a cycle of strain accumulation leading to a major episode of seismic slippage of the plate boundary similar to that which occurred recently in the northern part of the study area. Alternatively, the mode of slippage of the plate boundary may vary significantly along the arc and may be associated with specific structural features of the complex plate boundary.

Without further information on the past seismicity the choice between these alternatives remains open, but an opportunity to examine in detail an important episode of activity in the area of anomalous seismicity near Efate Island was provided by the capture of a sequence of three moderate-sized (Ms = 6) earthquakes in 1978 and 1979. In this paper we present preliminary determinations of the source parameters of the mainshocks and the space-time pattern of occurrence of the associated small shocks located by the seismograph network for a 1.5 year period. In addition, we report tilt measurements made before and after the sequence.

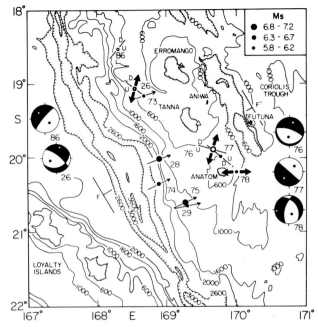

Figure 3. Same as Figure 2 but for the area south of the figure (after Coudert et al., 1981). Numbers are from Coudert et al. (1981), Pascal et al. (1978), and Table 1. Bathymetry interpolated from new results synthesized by the French Office de la Recherche Scientifique et Technique Outre-Mer (ORSTOM; Daniel, 1978) for 18°S to 20°S and Mammerickx et al. (1971) from 20°S to 22°S. X's show locations of seismograph stations on land and on the ocean bottom operated during August and September, 1977. Section F-F' is shown in Figure 5.

Tectonic Features of the Central New Hebrides

Complex Upper Plate

The upper plate has a major westward protrusion formed by the block-like morphological features upon which are located the islands of Santo and Malekula. The northern limit of the Santo block is clearly defined by the bathymetry. Focal mechanisms of two events (events 13 and 60, Figure 2) possibly located on the boundary between the Santo and Banks Islands (Vanua Lava and Gaua, Figure 2) blocks have large strike-slip components, but the sense of displacement is opposite that expected for the westward translation of the Santo block. The simplest interpretation is that these events represent a reactivation of the block boundary and reflect compressive stress now probably resulting from the interaction with the DFZ.

To the south the boundary of the upper plate protrusion is more complicated. The bathymetry suggests a distribution between a Santo-northern Malekula block and a southern Malekula block.

The region between southern Malekula and Efate Island has a complex bathymetry and widespread distribution of volcanic centers, both features indicative of major transverse structures (Figures 2 and 4). The northern termination of the South New Hebrides Trench is adjacent to a pronounced embayment in the shape of the leading edge of the upper plate and is also located just west of the northern termination of the system of back-arc rifts (the northern continuation of the Coriolis Trough). To the south, Efate Island is part of a westward salient of the upper plate located between the major embayment to the north and the less pronounced one which separates the Efate salient from the Tanna-Erromango salient.

Focal mechanism solutions for events in the upper plate for earthquakes located south of the Santo-Malekula area are interpretable as block faulting on nearly vertical nodal planes (events 24,86,26,76,77, Figures 2 and 3). Only one mechanism indicates a significant component of horizontal extensional stress for an event (event 78) located near the western side of the Coriolis Trough. The strong transverse trends between Malekula and Efate are reflected by the east-west striking nodal planes of the solutions for events 23 and 24 (Figure 2).

Based on the pattern of horizontal compressive stress trajectories inferred from the distribution of volcanic centers (following Nakamura, 1977) and other volcanic and structural features on the islands, Roca (1978) proposes

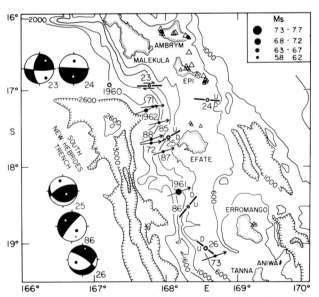

Figure 4. Earthquakes with focal mechanism solutions for the period 1960 through 1979 with symbols the same as in Figure 2 and bathymetry references of Figure 3, with the area of 16°S to 18°S taken from Mammerickx et al. (1971). Only the largest events for the period 1960-1962 are included.

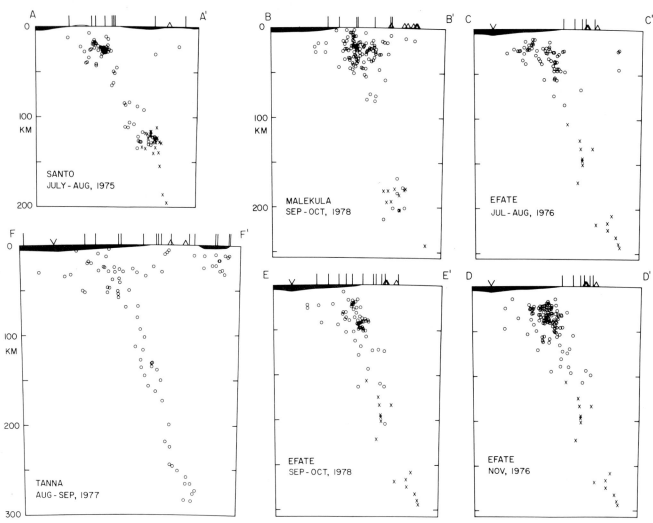

Figure 5. Cross sections of locations by temporary local networks operated for intervals of 1-2 months each. The horizontal and vertical scales are equal. Section locations are shown in Figure 3,6 and 7. The locations of sections C-C' and D-D' are the same as E-E', but only epicenters for E-E' are shown in Figure 6. The stations are shown by vertical lines on the surface, active or Quaternary volcanoes are shown by triangles, and the axis of the trench by a V. Results (open circles) are relocations based on data from Stephens (1978), York (1977) and Coudert et al. (1981) using the same flat layered model and the Hypo71 program of Lee and Lahr (1975). The model parameters include layer thicknesses (from the surface) of 15,10,175 and 100 km with $\underline{P}$ wave velocities, respectively, of 5.55, 6.5, 8.1 and 8.2 km/sec, and a Vp/Vs ratio of 1.75. The X's are reliable ISC locations for the Efate and Malekula sections and are relocations from Pascal et al. (1978) for the Santo section (all for depths greater than 70 km).

that subduction of the DFZ acts as a concentrated source of stress within the upper plate. The main evidence is a tendency for the inferred directions of maximum compressive stress to radiate out away (in a map view) from the DFZ. This effect is most strikingly illustrated by the variations in the alignments of volcanic centers from Aoba to Ambrym, as can be seen in Figure 2. Note also that the direction of the axis of compression in the focal mechanism solution of event 23 located south of Malekula agrees with the pattern. The focal mechanism solutions for

upper plate events located farther south along the arc show a rather complex and variable pattern without a clear trend in the orientation or type of the horizontal stress component.

A further argument supporting this change in upper plate stress along the strike of the arc can be made by considering the narrow ridge along which the islands of Maewo and Pentecost emerge (see Figure 2). These narrow, linear islands show clear evidence of rapid Quaternary uplift and appear to be located on horst blocks (Mallick and Neef, 1974; Luyendyck et al., 1974; Carney and Macfarlane, 1978). They occupy a location in the morphology of the arc normally occupied by the back-arc rifts found along both the northern and southern parts of the New Hebrides arc (Dubois et al., 1978). It is conceivable that a previously existing rift structure similar to that found in the northern and southern New Hebrides has been reactivated, but is now under compression rather than extension, and what were originally grabens have now become uplifted horst blocks. In a similar line of interpretation, the reactivation of faulting in a sense opposite to that involved in the original structure was suggested above as the interpretation of focal mechanisms of events in the northern part of the Aoba Basin. Mallick and Greenbaum (1977) also report reactivation of a Miocene normal fault (located in western Santo) as a reverse fault.

It is thus possible that the central New Hebrides, with its protruding upper plate interacting with major bathymetric features of the subducted plate, is the most tightly coupled part of the plate boundary and the region of major transverse features located near the northern end of the South New Hebrides Trench may be a transition zone between two different stress regimes along the arc.

Overall Geometry of the Subducted Plate and the Interplate Boundary

The relatively uniform configuration of the inclined zone of intermediate depth earthquakes demonstrated by Dubois (1971), Isacks and Molnar (1971) and Pascal et al. (1978) is simply related to the regional scale uniformity of the arc in map view as shown by the alignments of the North and South New Hebrides trenches and by the overall alignments of active and late Quaternary volcanic centers.

New data from local seismograph networks operated in the region support this result (Figure 5). In all sections the distribution of events with depths greater than about 50 km suggests a downbent form similar to that so well exhibited in the Tanna section. Although the activity is not uniformly distributed, comparisons of the four sections taking into account the 20-35 km thickness of the intermediate depth zone do not indicate any significant change in overall dip among the four areas. This result and the locations of events at depths between 50 and 125 km in the Santo, Malekula, and Efate sections provide fairly strong evidence against the "flap and gap" structure discussed by Choudhury et al. (1975) and Pascal et al. (1978).

It is remarkable that for three of the four areas, Efate, Malekula and Tanna, the shallow hypocenters located by the temporary networks fail to define a single thin zone that can be identified as the interplate boundary. We infer the existence of such a boundary from the abundant thrust-type focal mechanism solutions of

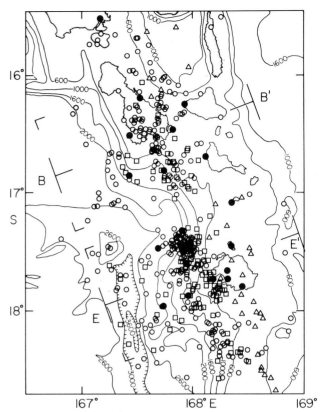

Figure 6. Epicenters of events located during September and October, 1978 by the network shown by filled circles. The ocean bottom stations operated during two weeks in September. Sections B-B′ and E-E′ are shown in Figure 5. Open circles are for depths less than 35 km, squares for depths between 36 and 69 km, and triangles for depths greater than 69 km. Bathymetry as in Figure 4.

earthquakes with epicenters in the arc-trench gap. Published and new solutions updating the results of Pascal et al. (1978) are summarized in Figures 2-4 and Table 1.

In the collections of data presented by Johnson and Molnar (1972) and Pascal et al. (1978) very few thrust-type solutions are found for events located in the southern half of the New Hebrides arc. A major result of the new data is to add thrust-type solutions for the southern half of the arc and thus largely remove the discrepancy between the two halves of the arc. As shown in Figure 8 the directions of slip plotted as a function of latitude show no significant change across the anomalous central region. The average direction of slip determined from these data, N76° ± 11°E, is essentially the same direction as that determined by Pascal et al. (1978).

The exact position and shape of the interplate boundary is difficult to determine. Relevant data include the dips of the slip vectors for the thrust-type focal mechanism solutions, the depths

TABLE 1 Focal Mechanism Solutions (see Fig. 20)

No.[1]	Date Mo/Dy/Yr	Origin Time Hr:Min	Location[2] Lat. S	Long. E	Depth, km	Pole[3][1] Tr.	Pl.	Pole 2 Tr.	Pl.	P Axis Tr.	Pl.
60	10/09/73	07:58	14.34	167.06	27	038	20	305	10	083	07
61	12/30/73	16:39	15.37	166.54	10	070	34	250	56	070	79
62	01/23/72	21:18	13.18	166.32	33	(4)					
63	01/24/72	03:56	13.07	166.40	38	(4)					
64	01/06/73	15:53	14.66	166.41	24	090	10	280	80	271	35
65	12/28/73	13:42	14.56	166.80	13	075	46	255	44	075	01
66	12/29/73	00:19	15.13	166.92	43	073	40	253	50	253	05
67	01/10/74	08:51	14.45	166.87	36	079	45	258	45	258	00
68	01/11/74	05:37	14.19	166.54	37	060	30	240	60	240	15
69	11/20/74	04:14	15.10	167.16	62	080	60	244	30	077	15
70	04/08/73	12:41	15.81	167.24	38	055	33	218	57	227	12
71	06/05/73	03:12	17.22	167.81	5	084	26	264	64	264	19
72	05/26/74	01:32	17.69	167.80	13	080	16	260	74	260	29
85[6]	08/01/78	04:16	17.38	167.88	20	077	18	257	72	257	27
86	01/27/79	18:15	18.54	168.21	25	311	84	131	06	141	51
87	08/17/79	12:59	17.73	167.87	25	090	15	270	75	270	30
88	08/26/79	11:47	17.63	167.71	22	090	15	270	75	270	30
	03/29/60	06:30	16.93	167.22	0	(5)					
	07/23/61	21:51	18.33	168.18	0	070	20	250	70	250	25
	10/06/62	04:23	17.26	167.72	0	070	15	250	75	250	30

1 Numbers continue numbering system of Pascal et al. (1978).

2 Locations are from Bulletins of the International Seismological Centre (ISC) or from the International Seismological Summary (ISS) except for 86, which is taken from the Monthly Bulletins of the PDE, and 85, 87 and 88, which are relocated with all available local and teleseismic data at the depth fixed by matching observed and synthetic long period P waveforms.

3 Trends measured clockwise from north; plunges measured from horizontal.

4 Solutions published by Kim and Nuttli (1975).

5 Solution not well determined but appears inconsistent with typical interplate thrust-type solutions.

6 Focal mechanism solutions for events 85-88, and the 1960-1962 events are preliminary and are not shown in Figure 20. The solutions are based on our readings of first motions recorded by the long-period seismographs of the Lamont International Geophysical Year (IGY) network plus first motions reported in the International Seismological Summary.

of some events, and, in the case of data from the operation of a temporary array beneath Santo Island, the location of a cluster of hypocenters whose common focal mechanism solution is of the thrust-type similar to that for larger events (see Figure 7). Better control on depth of moderate-sized events is provided by Chinn and Isacks' (1979) determination of accurate focal depths by matching observed and synthetic long period P, pP, and sP waveforms. The results for three areas beneath Santo, Efate, and Tanna islands are shown in Figure 9 together with data from local networks. A curve with the same shape was fitted to each of the sections as a crude approximation to the location of the plate boundary and as a reference for inter-comparison of the sections. At shallow depths the curve is an estimate of the interplate boundary, but at intermediate depths it approximates the upper envelope of the inclined seismic zone and thus

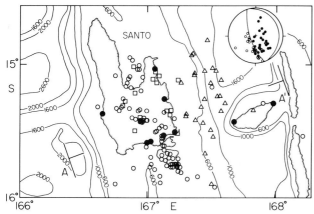

Figure 7. Epicenters of events located beneath Santo Island during July and part of August, 1975 (symbols are as in Figure 6). Not all of the stations operated at the same time. The lower hemisphere equal area plot in the upper right corner shows first motions (solid circles-compressions, open circles-dilatations) for events located in the concentrated cluster of events at depths near 20-25 km shown in section A-A' of Figure 5. Bathymetry as in Figure 2.

may pass slightly beneath the actual upper surface of the subducting plate.

Figure 9 shows that in comparison to most other subduction zones the sharply downbent configuration of the New Hebrides subducted plate results in a very narrow interface zone with relatively steeply dipping slip vectors. The maximum depth of interplate seismic slippage obtained from the accurate focal depths is 36 km. Unfortunately, the depths of the events located farthest west and with the steepest slip vectors were not determined due to complexity of the source process, and the depth of seismic slippage along the plate boundary may in one case reach 60 km (e.g., event 69, Table 1 and Figure 2). Nevertheless, if we take a depth of 40 km as that of the down-dip edge of the most seismically active part of the plate boundary, then the Efate and Tanna sections have widths of about 90-95 km as measured to the trench axis. The seaward protrusion of the Santo-northern Malekula block in the upper plate appears to add 10-20 km to this width, as would be expected if the intermediate depth seismic zone has a uniform configuration through the central New Hebrides.

The thickness of the seismically active parts of the subducted plate beneath the shallow plate boundary as determined by well-located events is about 25-35 km, in agreement with the thickness of the intermediate depth seismic zone found by Pascal et al. (1978). No clear evidence for distinct double Benioff zones similar to those described by Hasegawa et al. (1978) have been observed.

Recent History of Seismic Plate Boundary Slippage

The 1965 and 1973-1974 Sequences of Large Earthquakes: Plate Boundary Slippage Beneath the Santo-N. Malekula Segments

Two major sequences of large earthquakes occurred in the region of Santo Island during August, 1965 and December, 1973-January, 1974 (see Figure 10). The 1965 sequence has been studied in some detail by Taylor et al. (1980) and Ebel (1980). The boundaries of the aftershock zones shown in Figure 10 are based on locations of events occurring within two days of the mainshocks. Well-located shocks are selected according to data reported in the Bulletins of the International Seismological Centre.

Segment Boundaries. The estimate of the southern limit of the aftershock zone of the 1965 sequence is somewhat ambiguous because possible aftershocks occurred in the region of the southern Malekula and Epi islands during the month following the main sequence. However, the immediate aftershock activity is concentrated in the region shown. Further evidence of the southern extent of rupture are provided by the determinations by Taylor et al. (1978) of the uplift associated with the 1965 sequence. The decrease in uplift southwards along the west coast of Malekula indicates that the rupture of the 1965 sequence probably did not extend beneath southern Malekula. Furthermore, the rupture boundary may coincide with the remarkable discontinuity in the tilting of the late Quaternary coral terraces described by Taylor et al. (1978). This feature is located along the

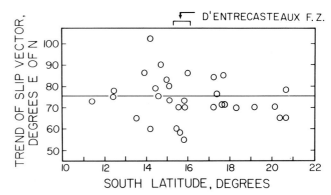

Figure 8. Trends of the nodal plane poles taken as the direction of the slip vector of all interplate thrust-type focal mechanism solutions for the New Hebrides, including data from Pascal et al. (1978), Coudert et al. (1981), and Table 1. The line gives the average azimuth of N76°E. (The trends are plotted against latitudes of the epicenters of the event.)

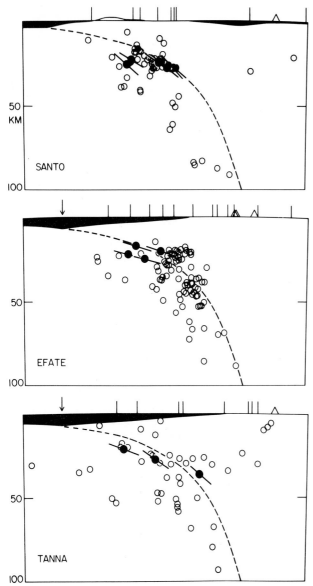

Figure 9. Detail of three of the sections, A-A′, E-E′, and F-F′ shown in Figures 5 through 7 plus earthquakes (large filled circles) with focal mechanisms and depths determined by analysis of long-period P records. The lines through the large filled circles are projections of those nodal plane poles chosen as the directions of the slip vectors. The continuous line has the same shape for the three sections and is adjusted in position to give the best fit to the locations of the interplate events and the dips of the slip vectors; at intermediate depths the line is an approximation to the upper envelope of the inclined seismic zone. The dashed part of this line is of variable length and is drawn to the assumed surface trace of the plate boundary at the trench (marked by an arrow).

eastward projection of the southern scarp and ridge of the DFZ (see Figure 2). We hypothesize that the main plate boundary rupture associated with the 1965 sequence terminates near this boundary.

The northern limits of the rupture are not very well defined. The very young uplift along the south coast of Santo (Taylor et al., 1980) could be associated with the 1965 earthquake, although the association is not clearly established. As pointed out by Pascal et al. (1978) and Chung and Kanamori (1978b) the boundary between the aftershock zones of the 1965 and 1973-1974 sequences closely corresponds to the eastward projection of the northern ridge and scarp of the DFZ. Careful examination of the well located aftershocks suggests a gap between the 1965 and 1973-1974 aftershock sequences as shown in Figure 10. It is possible that the earthquakes of October, 1971 and November, 1974 (see Figure 10) contributed to filling this gap, but the areas of rupture for these events are not known. Well-located aftershocks of the 1971 events are too few to define a clear aftershock zone and no aftershocks are reported by the ISC for the two days following the 1974 shock.

The boundary between the December 1973 and the January 1974 sequences is well defined by the aftershock zones and corresponds to the northern end of the main Santo block. The January 1974 sequence occurs beneath a triangular area of the upper plate which forms the southern part of the embayment between the Santo block and the Torres Island salient. The January sequence is further delineated by a zone of strike-slip faulting inferred from the locations and focal mechanisms of the earthquakes of October, 1973 (event 60) and May, 1965 (event 13).

In summary, the boundaries of the 1965 and 1973-1974 sequences appear to be strongly controlled by clear features in both the subducted and upper plates, including the northern boundary of the Santo block, the northern and southern scarps of the DFZ, and a boundary between the Santo-N. Malekula block and the S. Malekula block.

Development in time and space. The two-day long sequence in 1965 shows a clear southwards migration of the successive major events as shown in Figure 10. Significant slip apparently continued in the 8 years following this sequence as indicated by the events in 1966, 1971, 1972 and 1973. Similarly, the November, 1974 earthquake (no. 69) occurred nearly a year after the December, 1973-January, 1974 sequence. These "late aftershocks" occur on the eastern edges of the inferred rupture zones and could thus represent a down-dip expansion of the ruptured area. These observations could be accommodated by the model of Thatcher and Rundle (1979) in which post-seismic slippage of the interplate boundary occurs near a transition between a shallow region of stick-slip behavior and a

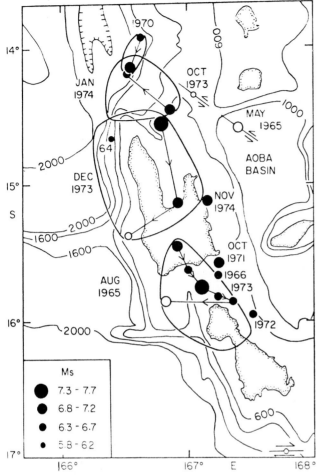

Figure 10. The temporal development of the three large sequences in 1965, 1973 and 1974 and the smaller sequence in 1970. The remaining activity is seen to be closely associated with these sequences. The irregular outlines are estimates of the boundaries of the aftershock zones associated with the four sequences, and the arrows show the sequence of events in time. The X shows the location of the leveling array on Santo Island. Earthquakes with thrusting focal mechanism solutions are shown as filled circles and other focal mechanism solutions are shown as open circles.

deeper region of creep. Further evidence for a transient effect lasting several years after the 1965 sequence is the exponential like decay in the rate of occurrence of earthquakes during the 3 year period following the sequence (see Figure 14). The 1973-1974 sequence occurs at the end of this transient, and Marthelot and Isacks (1980) describe a clear increase in the rate of occurrence of intermediate depth events during the transient.

The second sequence, a more complex one, occurs

in two periods of about two days each separated by a hiatus of 10 days. In the first period, December 28-30, 1973, the rupture appears to propagate southwards as shown in Figure 10. One of the larger shocks (event 61, Figure 2) of the sequence had a normal faulting mechanism and is located within the sub-oceanic plate. A significant number of the aftershocks of the first, largest event are also located beneath and seaward of the western edge of the Santo block and probably represent deformations within the sub-oceanic plate. The second part of the sequence was initiated on January 10, 1974 by an earthquake located near the epicenter of the first mainshock in December, but the rupture then appeared to proceed northwards.

The 1965 and the 1973-1974 sequences were in each case preceded by one of the upper plate shocks along the northern boundary of the Santo-northern Malekula block. Event 13 occurred nearly 83 days prior to the 1965 sequence, while event 60 occurred 70 days prior to the beginning of the 1973-1974 sequence.

Thus, nearly all the events located near Santo in Figure 2 can be associated with the 1965 and 1973-1974 sequences. The close coordination in time and space among the inter- and intraplate events can be taken as further evidence of the degree of coupling between the convergent plates in the Santo-northern Malekula segments.

Seismic Moments. Ebel's (1980) moments for the 1965 sequence determined from amplitudes of long-period surface waves total 4.3×10^{27} dyne-cm for the four interplate events (excluding the probable intraplate event of August 13). A crude estimate of the fault width and slip can be obtained from the moment if the fault length is known and the stress or strain drop assumed. We take a fault length of 100 km. If the stress drop is 30 bars (Kanamori and Anderson, 1975; Kanamori, 1977; Ebel, 1980), then, with respect to simple rectangular dislocation models, the fault width would be 30-40 km and the fault slip 3-4 meters. In support of these crude estimates Taylor et al. (1980) models the co-seismic uplift with a width of 40 km and a fault slip of 5.3 meters. This is less than half the width of the boundary as measured to the estimated surface outcrop (see Figure 9). The smaller width is further supported by the concentration of interplate seismicity (Figures 2-7) along a band located at depths greater than about 10-15 km and distances of about 30-40 km down-dip from the surface trace of the plate boundary. With an estimated slip of about 3-5 meters and a rate of convergence of about 10 cm/yr (Dubois et al., 1977) the "average" repeat time for the 1965 sequence would be 30-50 years.

Moments have not been determined for the 1973-1974 sequences. Estimates based on the surface wave magnitudes ($\log M_O = 1.5$ Ms + 16.1, taken from Kanamori and Anderson, 1975; Pucaru and Berckhemer, 1978; and Hanks and Kanamori,

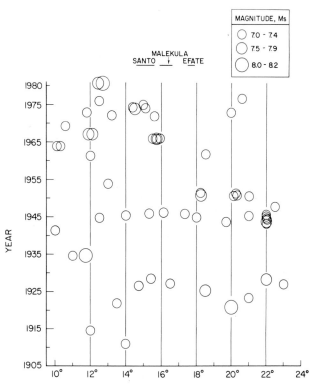

Figure 11. Large earthquakes (Ms > 6.9) plotted as a function of time and of the latitude of the epicenter. Data from Gutenberg and Richter (1954, with relocations of McCann, 1980), Rothe (1965), the "Seismological Notes" of the Bulletin of the Seismological Society of America, and the monthly listings of the Preliminary Determination of Epicenters (PDE).

1979) yield 3.7 x 10^{27} dyne-cm for the December, 1973 sequence and 1 x 10^{27} dyne-cm for the January, 1974 sequence, or a total of nearly 5 x 10^{27} dyne-cm. These data in combination with the aftershock zones indicate rupture of the plate boundary of approximately similar amount and extent to that in the 1965 sequence.

A Gap in the South Malekula-Efate Region?

In a time-space plot (Figure 11) of large earthquakes (Ms ≥7) occurring during the past 75 years, two nearly arc-wide periods of activation occur in the 1920's and again in the 1940's. Since then, seismicity has been most notable in the northern half of the arc. The southern Malekula-Efate region (16°-18°S) has not had any events with magnitudes as large as 7 since the 1940's. With respect to the estimated 30-50 year repeat time for the 1965 sequence, the Southern Malekula-Efate region may be considered a seismic gap. An important question is to what extent the pre-1960 earthquakes involved seismic rupture of

the plate boundary comparable to that accomplished in the 1965-1974 episodes.

In Figure 12 we have used McCann's (1980) relocations of some of the events listed by Gutenberg and Richter (1954). The differences between these locations are in certain cases quite substantial. For example, the largest discrepancy is nearly 200 km for the January 20, 1946 earthquake located by McCann beneath northern Malekula but located by Gutenberg and Richter near the northern end of the South New Hebrides trench (northwest of Efate Island). However, the locations of the 1944 event agree closely. Serious uncertainties in the magnitudes of the older events are also a problem. For

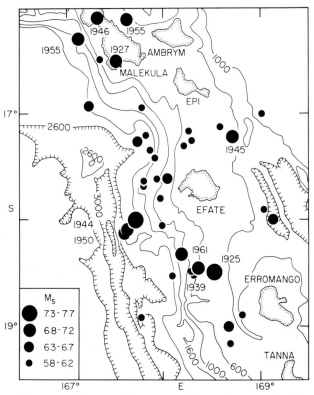

Figure 12. Large and moderate-sized earthquakes in the Malekula-Erromango region. The data for large earthquakes (Ms > 6.9) is the same as that for Figure 11 except that the December 2, 1950 earthquake is plotted with the magnitude revised by Geller and Kanamori (1977). McCann's relocation of the January 20, 1946 earthquake is beneath Santo, and located out of the figure. The data for moderate-sized earthquakes (Ms = 5.8 to 6.7) are shown for the period 1955 through 1979, with locations taken from Rothe (1965), the International Seismological Summary (ISS), the Bulletins of the International Seismological Centre (ISC), and the Preliminary Determination of Epicenters (PDE).

example, one of the largest events in the Efate region reported by Gutenberg and Richter is the December 2, 1950 shock with Ms = 7 3/4 (Richter, 1958 reports a magnitude, M, of 8.1). However, in the revisions of Geller and Kanamori (1977) the magnitude of the 1950 event is decreased to Ms = 7.2 (as plotted in Figure 12). Thus, any conclusions must be highly tentative until further analysis is made of the locations, focal mechanisms and moments of the older events.

The area from Santo through Erromango, a segment of the plate boundary of about 600 km in length, includes two groups of large, probably interplate earthquakes. Five events are located in the Santo-Malekula area and six events are located between Efate and Erromango. The 1945 earthquake is probably an intraplate event. The northernmost earthquake of the Santo-Malekula group, the January 20, 1946 earthquake, is located beneath Santo and out of the map of Figure 12. For comparative purposes we use the same moment-magnitude relationship used above in the analysis of the 1973-1974 episode. The four events beneath Santo and Malekula together contribute a net moment of about 2.6×10^{27}, while the six events located between Efate and Erromango contribute 7.1×10^{27} dyne-cm. The sum of the two groups is 9.8×10^{27} dyne-cm, a total which is close to net moment of 8.9×10^{27} dyne-cm estimated for the 1965-1974 sequence described in the last section. The pre-1965 events occur over a length of the arc nearly three times that ruptured during the 1965-1974 episode. Thus only about a third of the plate boundary from Santo through Erromango may have ruptured seismically during the 60 years before the 1965-1974 episode.

The locations in Figure 12 suggest that perhaps much of the area between Efate and Erromango may have slipped via the 1925, 1939, 1944, 1950 and 1961 events, in addition to southern Malekula via the 1927, 1946, and 1955 events. This would leave a prominent gap in the area of Efate and Epi islands and would imply that the 1965-1974 episode was the only rupture of the Santo-northern Malekula area during the past 75 years. Alternatively, the pre-1960 events beneath Malekula and Santo may have ruptured a part of the Santo-northern Malekula segment (with the 1965-1974 sequence including a repeat of such rupture) and a prominent gap remains beneath southern Malekula. In either case, the area of Efate and Epi islands appears deficient in large earthquakes during the past 75 years.

Unusual Seismicity Near Efate and Southern Malekula

Seismicity During the Past 20 Years

The largest interplate events in the southern Malekula-Erromango region during the past 20 years include the magnitude (Ms) 7.2 earthquake in 1961 and a striking spatial concentration of moderate-sized events in the Efate salient. The

1961 event has a well-determined thrust-type focal mechanism solution (Table 1 and Figure 4). The spatially concentrated activity beneath the Efate salient is developed mainly in three clusters of events which occurred in 1960-1962, 1973-1974, and 1978-1979 (Table 1 lists these events). These clusters are characterized by events with magnitudes (Ms) near 6. The largest one in 1962 had a magnitude of 6.6. In addition, an intraplate event occurred in 1966 (event 25, Figures 2 and 4) with a magnitude of 6.5.

The events reported by the PDE for the period 1961 through 1979 provide a sample of seismicity that covers magnitudes (mb) above about 4.5. The data for the New Hebrides arc are summarized in Figures 13 and 14 (Marthelot and Isacks, 1980). Although seismic quiescence preceding large earthquakes can sometimes be seen in the data of Figure 14 it is not very obvious in other cases. In particular, the activity in segment 23 in relation to the large earthquake (Ms = 8.0) of August, 1980 reveals no obvious precursory activity. The detection of such phenomena is difficult with the limited time sample and poorly defined background levels. What seems more striking in the data is the evidence for substantial variations from segment to segment in the character of the curves, variations which appear persistent at least over the time sampled.

The segments which experienced major earthquakes (e.g., segments 16-20 for the 1965-1974 episode) show very strong fluctuations in the cumulative number of events as a function of time. These fluctuations are dominated by the aftershock sequences and in one case includes a longer, exponential-like decay in the rate of occurrence during a period of about 3 years after the mainshocks.

In striking contrast to this strongly fluctuating activity, and also to all other segments, the Efate salient (mainly segment 13 but including part of 12) exhibits a very high rate of occurrence. This nest of seismic activity, a persistent feature throughout the 20 year period, coincides approximately with the concentration of magnitude 6 events described above. North of the nest the southern Malekula-Epi region (segments 14 and 15) exhibits a slight perturbation associated with the 1965 sequence, but is otherwise characterized by a moderate level of activity. South of the Efate nest the area between Efate and Erromango (including segments 10 and 11) has had a relatively low level of activity after the magnitude 7.2 event in 1961.

Small Earthquakes Located by Temporary Networks: Anomalous Cross Sections Beneath South Malekula and Efate

The data of Figure 5 (see also Figure 9) define the plate boundary only beneath Santo where the most important recent episode of interplate boundary slippage occurred 10 years prior to the

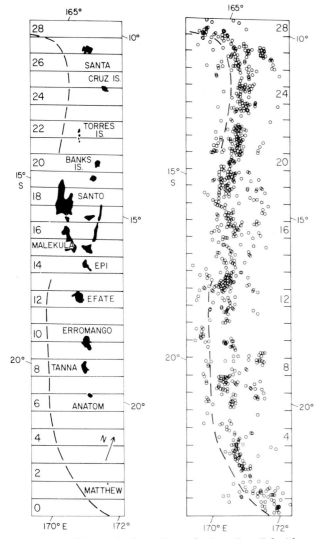

Figure 13.   Left: Map of the New Hebrides
island arc, divided in 29 contiguous strips
each a half degree wide and perpendicular to
the arc. The dashed line represents the axis
of the trench.   The numbers are used to
identify each strip in Figure 14.
Right: Spatial distribution of   shallow
earthquakes (depth < 70 km) from 1961 through
1979,   selected   from the   PDE bulletins
according to the reported magnitudes and the
number of stations used in the locations and
includes events with magnitudes (mb)
estimated to be greater than or equal to 5.

operation of the network. In contrast the Efate
sections (for three different periods of sampling
in Figure 5) show a very active shallow section
distributed in a rather broad zone near the
inferred plate boundary. The locations indicate
abundant lower plate and some probable upper
plate activity. The characteristically high rate

of occurrence of the Efate nest is manifested in
these sections.
   The thrust-type events in the Efate section
compared to those in the Santo and Tanna sections
(see Figure 9) are located at shallower depths
and closer to the surface trace of the plate
boundary and have more gently dipping slip
vectors.   The events appear to be located at
shallower depths along the interplate boundary
than those in the other two sections.   If this
were a general feature it would indicate that the
seismically tectonic part of the interplate
boundary is shallower and perhaps smaller in
width than that beneath Santo.   Further,   the
central plate activity located farther east might
thus be associated with a down-dip zone of creep
along the interplate boundary.
   The Malekula section (Figure 5) exhibits an
especially diffuse distribution of shallow events
near the inferred position of the interplate
boundary.   Many of the events are shallower than
those in the Efate section and are likely to
represent a concentration of activity in the
upper plate.   Also in contrast to the Efate
segment,   no   thrust-type   earthquakes   with
magnitudes greater than about 6 have occurred
beneath southern Malekula during the past 20
years.   Thus,   the high level of intra-plate
activity in the southern Malekula-Efate area
might be ascribed to loading of a locked section
of the plate boundary.   Alternatively,   the
features may be associated with the complex upper
plate structures which characterize the area.

Magnitude 6 Earthquakes in the Efate Nest:
1978-1979 Sequence

Time-Space Development of the Sequence

   The largest interplate events (Figure 4)
located near Efate since the early 1960's
occurred in 1973 (event 71), 1974 (event 72), and
a sequence of three events (events 85, 87, and
88) which occurred within a year during 1978 and
1979.   Two weeks after the first earthquake
(September 1, 1978) of the 1978-1979 sequence the
new local network commenced operation and for the
first two weeks of operation (the last half of
September, 1978) was augmented by a network of
ocean bottom seismographs as shown in Figure 6.
Since then, the network on land has continued to
monitor the activity at a threshold level of
about magnitude (mb) 3.0.   The locations were
done with a flat layered model and the Hypo 71
computer program (Lee and Lahr, 1975)   and by
methods similar to those described by Coudert et
al. (1981).   The hypocenter locations with OBS
readings are accurate for much of the shallow
zone of activity beneath the network,   but the
land network alone has good accuracy only for
hypocenters relatively near the islands.   In the
New Hebrides arc there does not appear to be a
large difference in locations determined with
teleseismic from those determined with local data

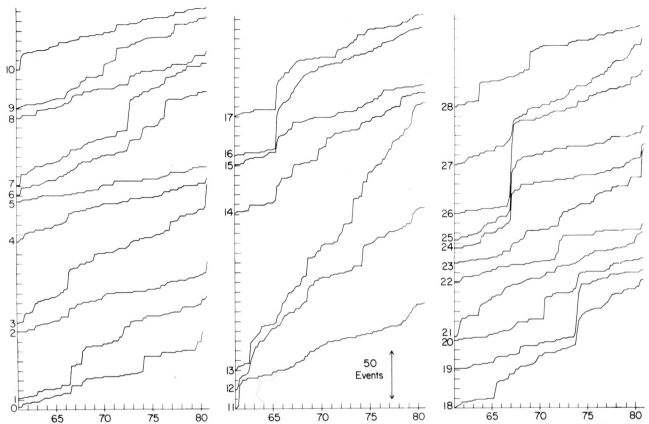

Figure 14. Curves of cumulative number versus time for all shallow earthquakes (depths < 70 km) reported in the PDE listings from 1961 through 1979 in each of the contiguous strips defined in Figure 13. The number at the origin of each curve identifies the strip along the arc from south (0) to north (28). The origin of each curve is arbitrary. The increment of time is 2 months.

(see also Coudert et al., 1981). The results are illustrated in Figure 6 and Figures 15-18. Although preliminary, these first results show several striking features outlined below which are not likely to be substantially modified by further analysis.

Foreshock and Aftershock Sequences. Although no foreshock sequence was detected by the first few stations of the network which began to operate several weeks before the September 1, 1978 event, an interesting foreshock sequence preceded the August 17, 1979 mainshock as illustrated in Figures 16 and 18. Several periods of increased activity occurred during the nine months preceding the August sequence. These include a cluster located beneath the trench (the December, 1978 cluster shown in Figure 15 but not included in the area covered by Figure 18), the striking cluster of March, 1979 (Figures 15 and 18), and increased activity during April and June, 1979. The March, 1979 cluster occurred near the boundary between the aftershock zones of the August 17 and August 26 events. During the nine days preceding the August 17 event seismic activity again increased in a series of clusters

(Figure 18). The locations show a clear time-space migration in a northeast direction from the trench to the epicenter of the mainshock. The final clusters (Figures 15,16,18) occurred close to the epicenter of the mainshock. During the 3.5 hours preceding the mainshock no events were detected. The cascading or accelerating activity found by Jones and Molnar (1979) in their "stacking" of many foreshock sequences is not apparent in this particular foreshock sequence.

The zone of aftershocks of each of the August 17 and August 26 events expands in area during a period of several days as shown in Figure 17. The expanded areas overlap to some extent, but are mainly separate and contiguous. The areas of the expanded aftershock zones are, however, quite large relative to the magnitude and seismic moment of the events, as will be discussed in a following section.

Spatial Trends in Earthquake Locations and Structure of the Upper Plate. The spatial distribution of the aftershock zones and other events in the Efate nest seem closely related to predominant morphological features of the Efate salient, notably the east-west embayment of the

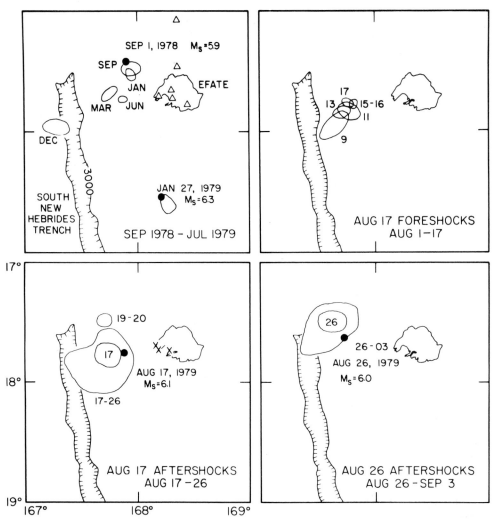

Figure 15.   Maps of the Efate area summarizing the regions of concentrated earthquake activity during the period from September 1978 to September 1979.   The event which initiated each cluster of earthquakes is shown by a solid circle whose diameter increases with the magnitude of the event. The extent of each cluster of earthquakes is shown by an irregular outline, and the dates are shown when the cluster was active.   The X's show the location of the tiltmeter stations on Efate (lower left hand frame).   The leveling array is near the westernmost two stations.

upper plate located north of Efate and the marked southwest-northeast trends defined by the bathymetry of the sea floor and the physiography and shape of northwest Efate Island.   The aftershock zones of the 1978-1979 sequence are bounded to the north by the embayment feature, as shown in Figures 4,6 and 15.   In Figure 6 the embayment is seen as a remarkable gap in activity.   The southern limit of the aftershock zone of the August 17 event is approximately limited by the southwest-northeast trend in the bathymetry.   The Pleistocene volcanoes of northern Efate (see Figure 2) are also located along that trend, as are the epicenters of the large 1944 and 1950 earthquakes (see Figure 12). Two other sets of features have the same

southwest-northeast trend but are offset to the north of the morphological feature.   One set (Figure 15) includes the epicenter of the September 1, 1978 event,   the clusters of December, 1978 and March and June, 1979, and the boundary between the aftershock zones of the two August events.   The second set includes the foreshocks of the August 17 event, the epicenters of the August 17 mainshock, the epicenters of the May 26, 1974 and February 16, 1966 (events 72 and 25, Figures 3 and 4), and the strike of the nearly vertical nodal plane of the focal mechanism of the February 16, 1966 event.   The southwest-northeast trends and the east-west trends in the morphology thus seem to reflect important structures in the interacting plates,

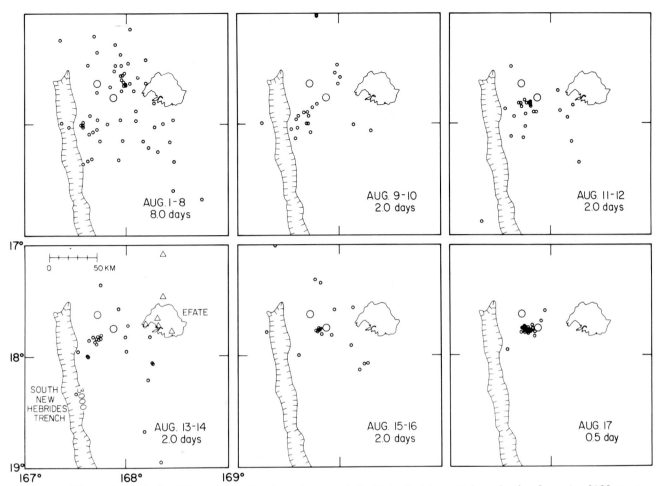

Figure 16. Maps of the Efate area showing the spatial distribution of foreshocks for six different intervals of time preceding the earthquake of August 17, 1979. Earthquake epicenters are representd by circles. The two large circles represent the locations for the mainshocks (Ms = 6) which occurred on August 17, 1979 and August 26, 1979. Triangles represent the locations of the six closest seismograph stations.

especially in the upper plate, which affect the seismicity.

The nest of seismicity near Efate is thus localized within the roughly triangular area of the Efate salient. It is likely that the salient constitutes a seismotectonic unit distinct from the southern Malekula block located to the north and possibly from the embayment of the upper plate between the Efate and Erromango salients (see Figure 3) located to the south.

Unusually Large Aftershock Areas for the August 1979 Earthquakes

Preliminary source parameters for five interplate events in the Efate nest (events 71,72,85,87 and 88) are listed in Table 2 together with those for two of the intraplate earthquakes in the area. The interplate earthquakes have body wave magnitudes (mb)

between 5.5 and 5.9, surface wave magnitudes (Ms) between 5.9 and 6.1, and seismic moments (Mo) between 1×10^{27} and 3×10^{27} dyne-cm. The moments are determined by matching observed and synthetic long period $\underline{P}$ waveforms (Chinn and Isacks, 1979). The relationships between the USGS mb and Ms values, and between the Ms and Mo values are close to those found for many other interplate earthquakes (Nagamune, 1972; Geller, 1976; Kanamori and Anderson, 1975; Pucaru and Berckhemer, 1978) and indicate no obvious anomaly.

What does appear anomalous is the large size of the aftershock areas of the two events in August. Even the areas of the zones that developed only several hours after the mainshocks appear rather large. For example, if Abe's (1975) results are used to estimate an area from the $\underline{P}$ wave moment, the initial aftershock zones are too large by about a factor of 2 to 3 in area or about 10 km

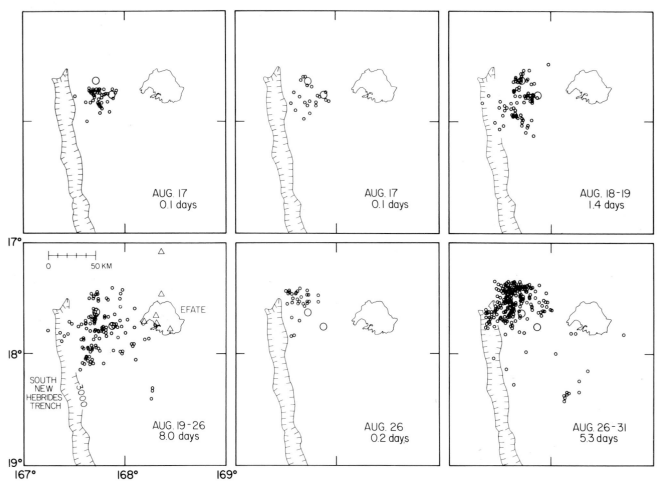

Figure 17. Maps of the Efate area showing the spatial distribution of aftershocks for six different time intervals following the earthquake of August 17, 1979. All symbols are the same as in Figure 16.

in linear dimensions. The expanded area developed in the days following the mainshock in each of the two cases is clearly anomalous. Some of this activity occurs within the sub-oceanic plate, as shown by the reliable location of hypocenters beneath and west of the axis of the trench (Figure 17), but many of the aftershocks are located near the interplate boundary.

Several limiting hypotheses can be considered to explain expanding aftershock zones. If the large area ruptured during the mainshock then, given the seismic moment, an anomalously low stress drop and small fault displacement are required. However, the 4 to 6 second durations of the source functions used in the syntheses of the long-period $\underline{P}$ waves are in better agreement with the dimensions of the small initial aftershock areas.

Alternatively, the small initial aftershock areas may coincide with the seismically ruptured areas in agreement with area-moment-magnitude relationships discussed above. If the larger

aftershock area represents an area of creep between the smaller patches of stick-slip deformation responsible for the earthquakes, two possibilities can be considered: (1) creep is an ongoing phenomena that loads the isolated patches or "asperities" to seismic failure or, (2) the rupture of the patches is closely associated with major episodes of creep in time in the form of "slow" earthquakes. In particular, a post-seismic episode of creep might be reflected by the temporal expansion of the aftershock zone. The persistently high rate of occurrence of small earthquakes in the Efate nest (Figure 14) for the past 20 years could be taken to support a more continuous loading process wherein various sized isolated patches loaded by the surrounding creep fail seismically to account for the continuously high level of seismicity. The moderate-sized events of the 1978-1979 and earlier sequences may represent repetitive failure of the largest patches in the area. In this model, the expansion of the aftershock sequence would

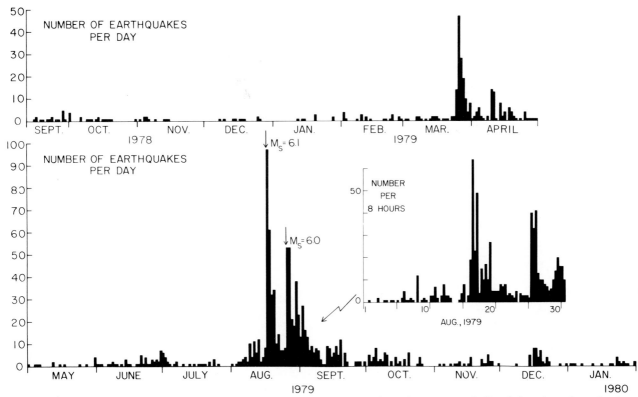

Figure 18. Histogram of earthquake activity versus time for the region defined by the aftershocks of the two August 1979 earthquakes. The insert shows clearly the small foreshock sequences which occurred on August 9, August 11, August 13, August 15, and August 17.

represent post-seismic diffusion of stress away from the patches of stick-slip behavior into the adjacent areas of creep surrounding the seismic rupture. The intervening areas of creep behavior also gives a mechanism to account for the time delays between the multiple events so characteristic of the seismicity.

Observations of surface deformations near the earthquakes might provide further evidence bearing on the question of whether a substantial component of creep occurs in episodes closely associated in time with the larger earthquakes or occurs as a longer term loading process. In the next section pertinent measurements of tilt obtained of Efate and Santo islands are described.

Tilt Measurements on Efate and Santo Islands

Tilt measurements in the New Hebrides have included the operation of a network of Kinemetrics bubble level sensors and releveling of two arrays of benchmarks located in southern Santo Island and western Efate Island (Isacks et al., 1978; Marthelot et al., 1980; Bevis and Isacks, 1981). The benchmark arrays were releveled from 2 to 4 times per year during the period since 1975 and provide measurements of tilt over baselines of nearly 1 km. The bubble

level tiltmeters have proven rather noisy, with noise increasing with increasing period, but are effective as "strong-motion" instruments for signals with time scales of several hours to weeks. In the short period end of the spectrum, at periods of minutes to hours, the resolution of the recordings approaches about 0.2 microradians. The sensitivity of the detection is estimated to be about a microradian at the longer periods.

The largest events recorded near the tiltmeter network are the three events of the 1978–1979 sequence. The August 17, 1979 earthquake was the largest and closest to three tiltmeter stations on Efate Island. The straight-line distance from that hypocenter to the nearest station is about 40 km. The tiltmeter records of the three Efate stations were searched carefully for tilt changes before and after the earthquake. Except for offsets during the arrival of the seismic waves no pre- or post-seismic signals with time scales in the range of about ten minutes to a week were detected by visual inspection. Comparisons of the records at the three stations which are located at distances of 3 and 11 km from one another, provides a good means for identifying spurious drifts and other local noise sources. The same results were found for the other two more distant events of the 1978–1979 sequence. The coseismic offsets (none greater than about 1

Table 2: Source Parameters of Earthquakes Near
Efate Island, 1973-1979

EQ No.	Mo/Dy/Yr	Depth[1] km	Magnitude mb	Magnitude Ms	Seismic Moment[1], X 10^{25} dyne-cm	Source Duration[1], sec
25	02/16/66	28	6.1	6.5	8.6	2
71	06/05/73	14	5.6	6.1	2.4	5
72	05/26/74	17	5.8	6.0	2.8	5
85	09/01/78	20	5.6	5.9	1.1	4
86	01/27/79	23	5.8	6.3	2.1	4
87	08/17/79	25	5.7	6.1	3.1	6
88	08/26/79	22	5.5	6.0	2.5	5

1 Determined by comparison of observed and synthetic long period P waveforms.

microradian) are not consistent in amplitude and polarity among the three stations and appear to be strongly influenced by the effects of seismic shaking of the ground at the tiltmeter site. This characteristic, also observed for smaller nearby and larger more distant events, discourages attempts to analyze co-seismic deformations.

Reliable long-term tilt measurements obtained by releveling the benchmark arrays are summarized in Figure 19. These measurements are shown by Isacks et al. (1978) and Bevis and Isacks (1981) to have a resolution of about 1 microradian. The Ratard array on southern Santo has accumulated little or no net tilt during the 5 years of measurements, but has exhibited two marginally significant excursions in 1976-1977 and 1979-1980. These excursions are possibly related to large earthquakes occurring several hundreds of kilometers north of Santo, as indicated in Figure 19.

The Efate array shows an accumulation of tilt at an overall average rate of about 1.5 microradians/year during the three year period from about 1977 to 1980. The tilt is downward toward the south, or in a direction more parallel than perpendicular to the strike of the arc. This progressive tilting is very well established by the consistency and redundancy of the array measurements as shown by Bevis and Isacks (1981). Thus, the accumulated tilt near the anomalous Efate segment is large whereas the tilt measured above the recently ruptured Santo' segment shows little accumulation of tilt.

The detailed temporal development of tilt at Efate, although close to or within the noise level of the measurements, exhibits some noteworthy features that may be significant. These include a ramp-like signal in 1977 (Figure 19) which occurs during the year preceding the 1978-1979 sequence of moderately large earthquakes, a possible coseismic tilting of

about 1 microradian upwards towards the west during a two month period spanning the August, 1979 earthquake sequence, and a recent reversal in the overall down-to-the-south trend of the tilt field. This last reversal is coincidental with the second large excursion seen at Ratard.

The data suggest that at a time scale of weeks surrounding the August, 1979 events (and the September, 1978 event) the tilt changes measured on Efate are not as important as the long term accumulation of tilt over the 4 years of measurements. If creep plays an important role in the interplate slippage of the Efate segment, and if the tilt signal is of tectonic origin, then the tilt data favor a long term development of creep rather than a type of slow earthquake where a large component of creep accompanies an episode of seismic slippage.

Conclusions

This study highlights spatial variations in the seismicity of the central New Hebrides that seem to be closely associated with the morphology and structure of each of the converging plates. Bathymetric irregularities of the sub-oceanic plate, notably the northern and southern bounding ridges of the DFZ, may act as asperities in the development of fault zone slippage and may also be responsible for uplift and block-like deformations of the upper plate. Irregularities in the shape and structure of the upper plate are probably indicative of variations in the width, physical state, and/or lithology of the interplate boundary. A notable example is the seaward protrusion of the upper plate in the area of Santo and Malekula islands, a protrusion that appears to increase the width of the interplate boundary. The interaction of this protrusion with the subducted topography of the DFZ may produce an area of increased coupling along the interplate boundary and increased compressional deformation in the upper plate.

In the central New Hebrides four main seismotectonic segments of the interplate boundary can be tentatively distinguished on the basis of the morphological features and characteristics of seismicity. The Santo-northern Malekula segment, including a major part of the protruding upper plate and its interaction with the DFZ, is the only segment for which seismic rupture of the interplate boundary by large earthquakes is well-established. The southern Malekula and Erromango segments have been very quiet during the past 20 years except for a magnitude 7 interplate event occurring in the Erromango segment in 1961. In contrast the segment comprising the Efate salient, located between the two quiet segments, has exhibited a persistently high rate of occurrence of small and moderate-sized earthquakes. The extent to which seismic rupture of these segments has taken place by large earthquakes occurring before 1960 is still highly uncertain, but the record suggests

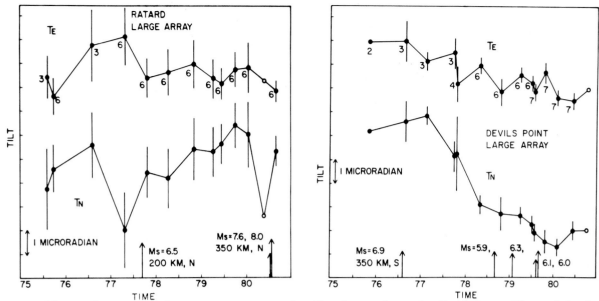

Figure 19. The north and east components of tilt observed at Devils Point, Efate Island and Ratard, Santo Island as determined by first-order relevelings of the arrays of benchmarks. Each array is clearly 1 km in dimension. The Devils Point array is located near the westernmost edge of Efate Island and the Ratard array is located near the central part of the southern coast of Santo Island (see Bevis and Isacks, 1981). The tilt components were calculated using an unweighted least squares analysis, with the September 1979 levelings as references. Also shown is the seismicity within 100 km of each array (open circles) and major events within 400 km of either array (arrows).

significant deficiencies in the number, magnitudes, and moments of events during the past 75 years in comparison with those which ruptured the Santo-northern Malekula segments in 1965-1974. The basic question is whether the variations in seismicity during the past 20 years represent different parts of a common earthquake cycle or different means by which slippage between the converging plates is accommodated.

The persistence of the anomalous seismicity of the Efate salient over 20 years suggests a long term anomaly related to the mode of plate boundary slippage. Approximate calculations show that the seismic activity, though high in respect to rate of occurrence, does not accumulate enough total moment to accommodate plate boundary slippage of 10 cm/yr in the entire Efate segment. Thus either the activity will substantially increase, or larger earthquakes will eventually occur, or a substantial fraction of the slippage occurs by creep. The last possibility is supported by the unusually large aftershock areas found for the recent sequence of moderately large earthquakes in the Efate salient in 1979. The data for this sequence suggest that relatively small, separate areas or patches of the plate boundary fail seismically and activate aftershocks in larger intervening areas. The time delays between mainshocks in the sequence and the time-space development of the aftershock sequence presumably reflect the non-elastic redistribution of stress. However, tilt

measurements do not reveal evidence for large pre- or post-seismic creep movements at time scales of days to months around the times of the moderately large mainshocks of the sequence. The tilt measurements do, however, indicate significant accumulation of tilt at an average rate of 1.5 microradians per year during a 3-4 year period.

These observations can be qualitatively accounted for (somewhat speculatively) by a model in which the interplate boundary in the Efate salient includes a significant area along which slippage occurs largely by creep but within which seismic slippage occurs on relatively small separated patches. The faulting of these patches, after being loaded to failure by the surrounding creep, accounts for the high level of moderately large and moderately small earthquakes. The area of seismic slippage may thus be smaller than in other segments. There is also a suggestion that the seismically active portion of the interplate boundary may be shallower and smaller in width than in other segments. The along-strike direction of tilting observed on Efate could be accounted for in terms of deformation near a transition from the creeping Efate segment to a locked Efate-Erromango segment and/or to a locked southern Malekula segment. Thus the Efate segment may be analogous to the central California part of the San Andreas fault system (Allen, 1968) in respect to being a zone of predominantly creep behavior

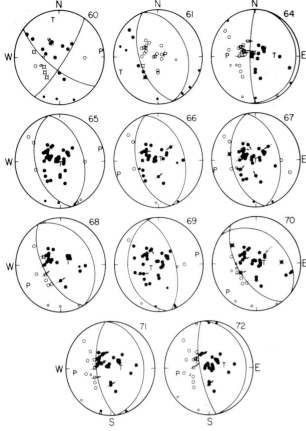

Figure 20. Focal mechanism solutions for earthquakes in the central New Hebrides, listed in Table 1. Compressions are filled circles, dilatations open circles, small symbols indicate less reliable readings, X's are readings judged to be near a nodal plane, and the arrows show first motions for S waves.

located between zones of predominantly stick-slip behavior. In particular, the moderate-sized earthquakes in the Efate nest may be comparable to the Parkfield earthquakes as recently described by Bakun and McEvilly (1979) and Lindh and Boore, 1981.

Acknowledgements. This paper would not have been possible without the support of the Office de la Recherche Scientifique et Technique Outre-Mer (ORSTOM), the New Hebrides Mines Service, the New Hebrides Geological Survey and the New Hebrides Topographic Survey. J. Dubois, J. Recy, and J.-L. Saos played particularly crucial roles in implementing the research. G. Hade, B. Campillo, R. Decourt, M. Bevis, B. Pontoise, C. Stephens, and J. York made especially valuable contributions to the field work. We also are grateful to numerous individuals and villages of New Hebrides (now the Republic of Vanuatu) for their friendly help during the extensive field operations. We thank G. Latham and A. Chen of the Marine Sciences Institute (University of Texas) for making the OBS data available to us. E. Coudert and M. Bevis offered valuable advice and assistance. We thank P. Bulack, D. Citron, and E. Farkas for typing and drafting. This research was supported by grants from the U.S. Geological Survey and the National Science Foundation.

References

Abe, K., 1975, Reliable estimation of the seismic moment of large earthquakes, J. Phys. Earth, 23, 381-390.

Allen, C.R., 1968, The tectonic environments of seismically active and inactive areas along the San Andreas fault system, in Proceedings of Conference on Geologic Profiles of San Andreas Fault System, Stanford University Publ., Geol. Sci., v. 11, 70-80.

Bakun, W.H., and T.V. McEvilly, 1979, Earthquakes near Parkfield, California: Comparing the 1934 and 1966 sequences, Science, 205, 1375-1377.

Bloom, A.L., and F.W. Taylor, 1977, Emerging Pleistocene islands in the New Hebrides island arc, Abstracts with Programs, 1977 Annual Meeting, 9, #7, Geol. Soc. Amer., p. 902.

Carney, J.N., and A. Macfarlane, 1978, Lower to Middle Miocene sediments on Maewo, New Hebrides, and their relevance to the development of the outer Melanesian arc system, Bull. Aust. Explor. Geophys., 9, 123-130.

Chase, C.G., 1971, Tectonic history of the Fiji Plateau, Geol. Soc. Am. Bull., 82, 3087-3110.

Chinn, D.S., and B.L. Isacks, 1979, Accurate depths of shallow earthquakes in the New Hebrides island arc, (Abstr.), Earthquake Notes, 49, 84.

Choudhury, M.A., G. Poupinet and G. Perrier, 1975, Shear velocity from differential travel times of short-period ScS-P in New Hebrides, Fiji-Tonga, and Banda Sea regions, Bull. Seism. Soc. Am., 65, 1787-1796.

Chung, W.-Y., and H. Kanamori, 1978a, A mechanical model for plate deformation associated with aseismic ridge subduction in the New Hebrides arc, Tectonophysics, 50, 29-40.

Chung, W.-Y., and H. Kanamori, 1978b, Subduction process of a fracture zone and aseismic ridges -- The focal mechanism and source characteristics of the New Hebrides earthquake of January 19, 1969 and some related events, Geophys. J.R. Astr. Soc., 54, 221-240.

Coudert, E., B.L. Isacks, M. Barazangi, R. Louat, R. Cardwell, A. Chen, J. Dubois, G. Latham, and B. Pontoise, 1981, Spatial distribution and mechanisms of earthquakes in the southern New Hebrides arc from a temporary land-OBS seismic network and world-wide observations, J. Geophys. Res., (in press).

Daniel, J., 1978, Morphology and structure of the southern part of the New Hebrides island arc system, J. Phys. Earth, 26, suppl., S181–S190.

Daniel, J., C. Jouannic, B. Larue, and J. Recy, 1977, Interpretation of D'entrecasteaux zone (north of New Caledonia), in International Symposium on Geodynamics in South-West Pacific, Noumea, New Caledonia, Editions Technip, Paris, France, pp. 117–124.

Dubois, J., 1971, Propagation of P waves and Rayleigh waves in Melanesia: Structural implications, J. Geophys. Res., 76, 7217–7240.

Dubois, J., J. Launay, J. Recy, and J. Marshall, 1977, New Hebrides trench: Subduction rate from associated lithospheric bulge, Canadian J. of Earth Sci., 14, 250–255.

Dubois, J., F. Dugas, A. Lapouille, and R. Louat, 1978, The troughs at the rear of the New Hebrides island arc: Possible mechanisms of formation, Canad. Journ. of Earth Sciences, 15, 351–360.

Ebel, J.E., 1980, Source processes of the 1965 New Hebrides Islands earthquakes inferred from teleseismic wave forms, Geophys. J.R. Astr. Soc., 63, 381–403.

Falvey, D.A., 1978, Analysis of paleomagnetic data from New Hebrides, Bull. Aust. Explor. Geophys., 9, 117–123.

Frankel, A., and W. McCann, 1979, Moderate and large earthquakes in the South Sandwich arc: Indicators of tectonic variation along a subduction zone, J. Geophys. Res., 84, 5571–5577.

Geller, R.J., 1976, Scaling relations for earthquake source parameters and magnitudes, Bull. Seism. Soc. Am., 66, 1501–1523.

Geller, R.J., and H. Kanamori, 1977, Magnitudes of great shallow earthquakes from 1904 to 1952, Bull. Seism. Soc. Am., 67, 587–598.

Geological Map of the New Hebrides Condominium, scale 1:1,000,000, Ministry of Overseas Development (Directorate of Overseas Surveys), D.O.S. 1196, British Residency, Vila, New Hebrides.

Gill, J., and M. Gorton, 1973, A proposed geological and geochemical history of eastern Melanesia, in The Western Pacific, Island Arcs, Marginal Seas, Geochemistry, P.J. Coleman, ed., Crane, Russals and Co., New York, p. 543–566.

Gutenberg, B., and C.F. Richter, 1954, Seismicity of the Earth and Associated Phenomena, Princeton Univ. Press, 310 pp.

Hanks, T.C., and H. Kanamori, 1979, A moment magnitude scale, J. Geophys. Res., 84, 2348–2350.

Hasegawa, A., N. Umino, and A. Takagi, 1978, Double-planed deep seismic zone and upper mantle structure in the northeastern Japan arc, Geophys. J.R. Astr. Soc., 54, 281–296.

Ibrahim, A., B. Pontoise, G. Latham, B. Larue, T. Chen, B. Isacks, J. Recy, and R. Louat, 1980, Structure of the New Hebrides Arc - Trench system, J. Geophys. Res., 85, 253–266.

Isacks, B.L., and M. Barazangi, 1977, Geometry of Benioff zones: Lateral segmentation and downwards bending of the subducted lithosphere

in M. Talwani and W.C. Pitman, III (editors), Island Arcs, Deep Sea Trenches, and Back-Arc Basins, Maurice Ewing Ser., 1, AGU, Washington, D.C., 99–114.

Isacks, B.L., G. Hade, R. Campillo, M. Bevis, D. Chinn, J. Dubois, J. Recy, and J.L. Saos, 1978, Measurements of tilt in the New Hebrides Island Arc, in Proceedings of Conference VII, Stress and Strain Measurements Related to Earthquake Prediction, U.S. Geological Survey Open-File Report 79-370, 176–221.

Isacks, B.L., and P. Molnar, 1971, Distribution of stresses in the descending lithosphere from a global survey of focal mechanism solutions of mantle earthquakes, Rev. Geophys. Space Phys., 9, 103–174.

Johnson, T., and P. Molnar, 1972, Focal mechanisms and plate tectonics of southwest Pacific, J. Geophys. Res., 77, 5000–5032.

Jones, L.M., and P. Molnar, 1979, Some characteristics of foreshocks and their possible relationship to earthquake prediction and premonitory slip on faults, J. Geophys. Res., 84, 3596–3608.

Jouannic, C., F.W. Taylor, A.L. Bloom, and M. Bernat, 1980, Late Quaternary uplift history from emerged reef terraces on Santo and Malekula islands, central New Hebrides island arc, United Nations E.S.C.A.P., CCOP/SOPAC, Tech. Bull. no. 3, 91–108.

Kanamori, H., and D.L. Anderson, 1975, Theoretical basis of some empirical relations in seismology, Bull. Seism. Soc. Am., 65, 1073–1095.

Kanamori, H., 1977, The energy release in great earthquakes, J. Geophys. Res., 82, 2981–2987.

Karig, D.E., and J. Mammerickx, 1972, Tectonic framework of the New Hebrides island arc, Marine Geology, 12, 187–205.

Karig, D.E., and G.F. Moore, 1975, Tectonic complexities in the Bonin Arc system, Tectonophysics, 27, 97–118.

Kelleher, J., J. Savino, H. Rowlett, and W. McCann, 1974, Why and where great thrust earthquakes occur along island arcs, J. Geophys. Res., 79, 4889–4898.

Kelleher, J., and W. McCann, 1976, Buoyant zones, great earthquakes and unstable boundaries of subduction, J. Geophys. Res., 81, 4885–4896.

Kim, S.G., and O.W. Nuttli, 1975, Surface-wave magnitudes of Eurasian earthquakes and explosions, Bull. Seism. Soc. Am., 65, 693–709.

Lee, W.H.K., and J.C. Lahr, 1975, HYPO-71 (Revised): A computer program for determining hypocenters, magnitudes, and first motion patterns of local earthquakes, U.S. Geol. Survey, Open File Report 75-311, Menlo Park, California, 113 p.

Lindh, A.G., and D.M. Boore, 1981, Control of rupture by fault geometry during the 1966 Parkfield Earthquakes, Bull. Seism. Soc. Am., 71, 75–116.

Louat, R., J. Dubois, and B. Isacks, 1979, Evidence for anomalous propagation of seismic waves within the shallow zone of shearing between the converging plates of the New Hebrides subduction zone, Nature, 281, 293-295.

Luyendyck, B., W.B. Bryan, and P.A. Jezek, 1974, Shallow structure of the New Hebrides island arc, Geol. Soc. Am. Bull., 85, 1287-1300.

Mallick, D.I.J., 1973, Some petrological and structural variations in the New Hebrides, in P.J. Coleman (ed.), The Western Pacific: Island Arcs, Marginal Seas, Geochemistry, Crane, Russals, and Co., New York, p. 193-211.

Mallick, D.I.J., and G. Neef, 1974, Geology of Pentecost, New Hebrides Geological Survey, Regional Report, British Residency, Vila, New Hebrides, 103 pp.

Mallick, D.I.J., and D. Greenbaum, 1977, Geology of Southern Santo, New Hebrides Geological Survey, Regional Report, 84 p.

Mammerickx, J., T.E. Chase, S.M. Smith, and I.L. Taylor, 1971, Bathymetry of the South Pacific, Chart 12, Scripps Inst. of Oceanography, La Jolla, California.

Marthelot, J.-M., E. Coudert, and B.L. Isacks, 1980, Tidal tilting from localized ocean loading in the New Hebrides island arc, Bull. Seism. Soc. Am., 70, 283-292.

Marthelot, J.-M., and B.L. Isacks, 1980, Space-time distribution of shallow and intermediate earthquakes in the New Hebrides island arc, (Abstr.), EOS, Trans. Am. Geophys. Union, 61, 288.

McCann, W.R., 1980, Seismic potential and seismic regimes of the Southwest Pacific, preprint.

Mitchell, A.H.G., and A.J. Warden, 1971, Geological evolution of the New Hebrides island arc, J. Geol. Soc. London, 127, 501-529.

Mogi, K., 1969, Relationship between the occurrence of great earthquakes and tectonic structure, Bull. Earthquake Res. Inst., Tokyo Univ., 47, 429-451.

Nagamune, T., 1972, Magnitudes estimated from body waves of great earthquakes (in Japanese), Quarterly J. of Seismol., 47, 1-8.

Nakamura, K., 1977, Volcanoes as possible indicators of tectonic stress orientation: Principle and proposal, J. Volc. Geoth. Res., 2, 1-16.

Pascal, G., B.L. Isacks, M. Barazangi, and J. Dubois, 1978, Precise relocations of earthquakes and seismotectonics of the New Hebrides island arc, J. Geophys. Res., 83, 4957-4973.

Pucaru, G., and H. Berckhemer, 1978, A magnitude scale for very large earthquakes, Tectonophysics, 49, 189-198.

Ravenne, C., G. Pascal, J. Dubois, F. Dugas, and L. Montadert, 1977, Model of a young inter-oceanic arc: The New Hebrides island arc, in International Symposium on Geodynamics in South-West Pacific, Noumea, New Caledonia, Editions Technips, Paris, France, 63-79.

Roca, J.-L., 1978, Essai de determination du champ de contraintes dans l'archipel des Nouvelles Hebrides, Bull. Soc. Geol. France, 20, 511-519.

Rothe, J.P., 1969, The Seismicity of the Earth, UNESCO, Paris, 336 p.

Santo, T., 1970, Regional study on the characteristic seismicity of the world, Part III, New Hebrides island region, Bull. Earthq. Res. Inst., 48, 1-18.

Stephens, C.D., 1978, A short-term microearthquake survey in the central New Hebrides island arc: Closing the gap, M.S. Thesis, Cornell University, 46 pp.

Taylor, F.W., B.L. Isacks, C. Jouannic, A.L. Bloom, and J. Dubois, 1980, Coseismic and Quaternary vertical tectonic movements, Santo and Malekula Islands, New Hebrides Island Arc, J. Geophys. Res., 85, 5367-5381.

Thatcher, W., and J.B. Rundle, 1979, A model for the earthquake cycle in underthrust zones, J. Geophys. Res., 10, 5540-5556.

Utsu, T., 1974, Space-time pattern of large earthquakes occurring off the Pacific coast of the Japanese Islands, J. Phys. Earth, 22, 325-342.

Uyeda, S., and H. Kanamori, 1979, Back-arc opening and the mode of subduction, J. Geophys. Res., 84, 1049-1061.

York, J.E., 1977, Seismotectonics in intraplate and interplate regions: Eastern North America, eastern Taiwan, China, and the New Hebrides, Ph.D. Thesis, Cornell University, 152 pp.

SEISMICITY PATTERNS IN CHINA

Shaoxie Xu and Peiwin Shen

Institute of Geophysics, State Seismological Bureau
Beijing (Peking), The People's Republic of China

Abstract. The pattern of seismicity in China, especially that seen in northern China is described in this paper. Strong earthquakes with magnitudes of $M \geq 7.5$ are found to occur at certain definite intervals of distance. In northern China, recent earthquakes with magnitudes M 7.0 occurred within a belt in the form of a stripe (along the two boundary lines of a stripe). The seismicity immediately before the strong earthquakes shows a similar distribution pattern. The strong earthquakes of given separate regions are obviously thus correlatable. Such a correlation can neither be shown to be due to a propagation process nor be explained by the assumption that correlatable earthquakes are linked by faults. It is thought to possibly be the result of the action of an ambient stress field over a broad region. Medium and small shocks, whose numbers maintain certain constant ratios, are usually observed as preludes to strong earthquakes.

All of these facts may be considered to be the results of buckling of crustal layers under an ambient stress field. According to this viewpoint, we are able to calculate the values of b: equalling 0.75 for average earthquakes; 1.5 for volcanic earthquakes, reservoir earthquakes and some earthquake swarms; and 0.5 for earthquakes on the same fault system. These figures agree rather well with actual observations, so the discussion presented here might have some general significance.

Introduction

This paper describes some seismicity patterns occurring before and after strong earthquakes which were observed in China, especially in northern China. The repeatability of these patterns is reasonably good. As described in the last part of this paper, the patterns of seismicity may have some general significance.

Pattern in Spatial Distribution

We found that before or after each strong earthquake ($M \geq 8$), another strong earthquake ($M \geq 7.5$) generally occurred at a definite distance away. Figure 1 shows the locations of many pairs of strong earthquakes ($M \geq 7.5$) that occurred at different time periods in China. Before 1900, only historical data existed, and records for earthquakes in western China were rather scarce. Therefore, Figures 1a-1f only show earthquakes occurring in the eastern part of China. From these figures it can be seen that separations between earthquakes fall into two categories: small separations and large ones. Instrumental data have been available since 1900, and strong earthquakes recorded from western China are shown in Figure 1h. Seismic activity is high in western China and the distribution of epicenters is more complicated, but separations between the epicenters are not irregular. If the time intervals between the earthquakes are plotted versus the distances between epicenters (Figure 2) it can be seen that the earthquakes fall into two groups; one group of earthquakes being separated by 480 to 550 km; the other group by about 1100 km. The cause may be due to stratum buckling as discussed in detail by Xu and Shen [1979].

The distribution of five strong earthquakes ($M \geq 7$) which have occurred since 1900 in northern China forms a stripe of blocks (along the two boundary lines of a stripe) as shown in Figure 3. Other characteristics of this stripe of blocks were discussed in Xu and Shen [1979].

Four of the five strong earthquakes mentioned above occurred after the establishment of the Beijing network of seismic stations. The seismic activity around the epicentral region prior to each of these strong earthquakes was rather similar to the activity preceding each of the others. In each case, approximately one year (the exact length of time period is shown in the lower left corner of Figure 4) before the occurrence of the strong earthquake, a $M \geq 4$ event occurred about 360 km away from it. This is shown in Figure 4 where all of the main shocks are superimposed onto the same location and plotted together with their foreshocks of $M \geq 4$ (and any $M \geq 5$ aftershocks of these foreshocks). The foreshocks occurred in different directions from the main shocks, but to emphasize the similarity in separation from the main shocks, the north-south orientation of each

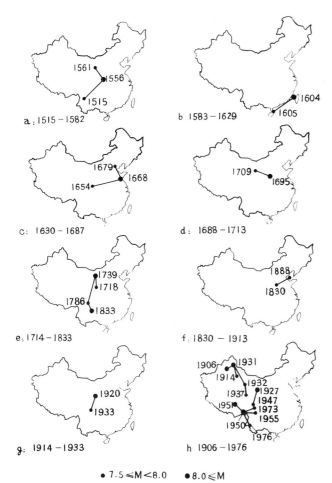

a: 1515-1582   1561 1556 1515

b 1583-1629   1604 1605

c: 1630-1687   1679 1668 1654

d: 1688-1713   1709 1695

e: 1714-1833   1739 1718 1786 1833

f: 1830-1913   1888 1830

g: 1914-1933   1920 1933

h 1906-1976   1906 1931 1914 1932 1937 1927 1951 1947 1973 1955 1950 1976

● 7.5 ≤ M < 8.0   ● 8.0 ≤ M

Fig. 1. Map of strong earthquakes (M ≥ 7.5) in China at different periods.

above phenomena is enhanced if we consider the fact that the epicentral distribution of the different regions are quite different.

Correlation of Seismic Activities of Given Separate Regions

It is often seen that the seismic activities of several regions increase and decrease almost simultaneously [Xu and Shen, 1979]. In Figure 5, the most important earthquakes of this area which occurred in the time period denoted in Figure 6 are shown. They seem to concentrate in the three regions shown by dashed lines: the central region A (Beijing - Tangshan region), the western region B (Liangcheng - Hunyuan region) and the eastern region C (Chaoyang - Haicheng region). Taking all of the earthquakes above certain magnitudes (magnitude 3-4 for the western region; 4 for Beijing, 5 for Tangshan, 4.5 for Chaoyan, 4-5 for Haicheng), the M(t) diagram for the three regions are plotted (Figure 6). It can be seen that, in many cases, when an earthquake occurred in region A, an earthquake above the magnitudes mentioned previously occurred very close in time (before or after) in the adjacent regions B or C. Some typical examples are: the M = 6.9 event of Ninghe (21^h, 15 November 1976) which was followed, within three days, by the M = 4.9 earthquake of Haicheng (10^h, 18 November) in the eastern region; and the M = 5.7 earthquake at Tangshan (08^h, 2 December 1976) which was followed, within 2 1/2 days, by the M = 5.2 earthquake of Haicheng (19^h, 4 December); the M = 5.0 event of Liangchen in the western region (13^h, 2 February 1977) which was followed, within 1 1/2 days, by the M = 4.5 earthquake of Majuqiao near Beijing (01^h, 4 February); the M = 5.0 event of Tangshan (07^h, 1 June 1978)

foreshock-main shock group plot has been rotated (through an angle listed in Figure 4) to place the foreshocks close together. This 360 km separation is seen as well for two M = 6 events (the Hejian earthquake and the Helin earthquake) which are also shown in Figure 4. The signficance of the

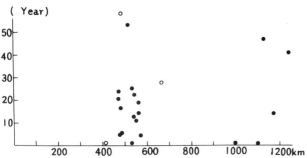

(Year)

Fig. 2. Distance span and time interval between strong earthquakes (M ≥ 7.5) occurring in China.

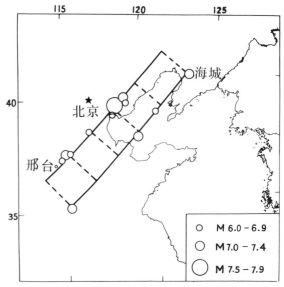

海城 北京 邢台

M 6.0 - 6.9
M 7.0 - 7.4
M 7.5 - 7.9

Fig. 3. The stripe block of recent strong earthquakes in northern China.

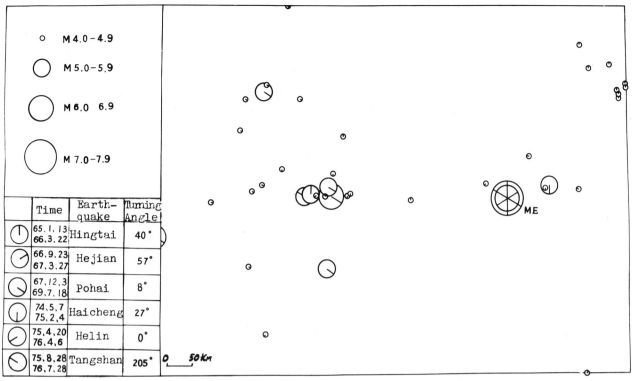

Fig. 4. Comparative map of seismicity distribution before strong earthquakes.

which was followed, within only 6 hours, by the M = 4.8 event of Hunyuan (13^h, 1 June) in the west; and the M = 4.8 earthquake at Huairou near Beijing (18^h, 3 October 1978) which was followed, only 16 hours later, by the M = 4.9 earthquake of Hunyuan (4 October). Since earthquakes above these magnitudes are rather rare in these regions, the occurrence of an earthquake above that magnitude in neighboring regions so soon after the

first may not be due to chance alone. Table 1 lists the time intervals between these sequential earthquakes occurring in neighboring regions and the durations of the quiescent periods between the pairs as well as their ratios (quiescent period/time interval). This ratio is always much larger than unity except in one case (ratio = 0.6). About 80% of the ratios are greater than 2, 50% of the ratios are greater than

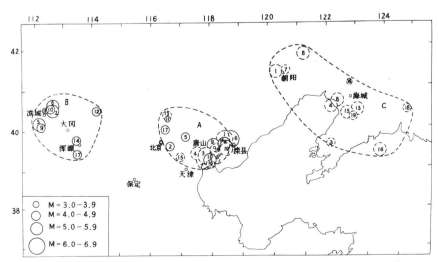

Fig. 5. Correlated regions of Beijing and areas to the east and west.

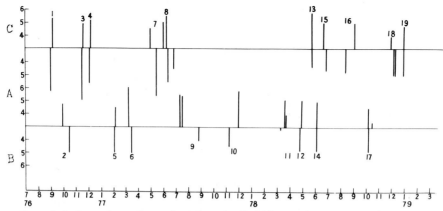

Fig. 6. Correlated series of earthquakes in Beijing and areas to the east and west.

6 and 25% of the ratios are greater than 20. This shows that the small temporal separation of earthquakes in neighboring regions is most probably not just a coincidence.

Correlation of earthquakes in different regions has also been seen in Huairou, Baijiatuan and in the northeastern and northwestern suburban areas of Beijing where earthquakes are relatively rare. Since the establishment of the Beijing network of seismic stations in 1966, only 5 earthquakes of about magnitude 4 in Huairou, and also only 5 earthquakes of about magnitude 3.5 in Baijiatuan have been recorded. It is interesting to note that each of the 5 events in Huairou was quickly followed by one in Baijiatuan. The most remarkable is the pair which took place on March 25-27, 1972, in which the M = 4.5 earthquake of Huairou ($22^h 22^m$, March 25) was followed 32 hours later by the M = 3.5 earthquake in Baijiatuan. Before this pair, no earthquakes of magnitude 3-4 had occurred for over four years in these two areas. The ratios of the durations of the quiescent periods to the lengths of the intervals between the members of each pair is much larger than those in Table 1, amounting to even 1000 in one case (see Table 2). Thus the correlation is quite remarkable.

Such a correlation exists not only between adjacent regions, but also for regions much further apart: for instance between northern China and southern China; between Sichuan and Yunnan; between northern China and the lower reaches of the Yangtze River; and between the Dengkou region and the Helin region. Closer study of such a correlatiion reveals that:

1. This correlation phenomena is not the result of a propagation process. The correlated earthquakes are divided into 3 groups according to their approximate separation, namely: a) short range (SD) of about 60 km for earthquakes occurring in Huairou and Baijiatuan near Beijing; b) medium range (MD) of about 350 km for earthquakes occurring in the Beijing area and its western and eastern neighboring areas; and c) long range (LD) of

about 900 km for earthquakes occurring in the northern and southern parts of northern China. In the lower part of Figure 7 these three groups of earthquakes are plotted according to the time interval between the sequential earthquake pairs. No dependence of the time intervals on the average distance between the earthquakes can be seen, and therefore this correlation cannot be explained by a propagation process.

2. This correlation may be the result of a force applied simultaneously to the different regions. By plotting the frequency of the earthquakes versus the time interval between the sequential earthquake pairs as shown in the upper part of Figure 7, it can be seen that earthquake pairs separated by short time intervals occurred much more frequently, and that the frequency decreases gradually with increase of time separation. Perhaps the two regions are subjected to a force simultaneously, and the temporal separation is only a manifestation of differences in rheologic responses.

3. This correlation cannot be explained by the effects of faults. Often it is assumed that earthquakes occur at the two extremities of a fault. Figure 8 shows straight lines drawn to connect sequential pairs of earthquake: one occurring in the Beijing region and the other in the eastern or western neighboring regions. The directions of the straight lines vary a great deal and are scattered over a considerable area, but the lengths of the lines do not change, and especially do not increase with time. Therefore, the earthquake pairs cannot be taken as the extremities of faults.

Furthermore, earthquake activity is usually distributed in the form of blocks [Xu and Shen, 1979]. The correlated earthquakes shown in Figure 5 are located at the corners of such blocks. An earthquake occurring in one corner of a block is followed shortly afterwards by an earthquake in the opposite corner of the same

TABLE 1

	B		A		C		b	a	a/b
	Liangcheng	Hunyuan	Beijing	Tangshan	Chaoyang	Haicheng	days		
1				76.08.31 11:25 6.3 11:27 6.3	76.09.05 08:14 5.3		4.9		4.8
								23.4	1.4
2	76.10.14 22:35 5.0		76.09.28 18:43 4.7				16.2		2.0
								32.0	11.4
3				76.11.15 21:53 6.9		76.11.18 18:10 4.9	2.8		4.9
								13.6	5.6
4			76.12.02 08:42 5.7			76.12.04 19:19 5.2	2.4		24.5
								59.8	40.7
5	77.02.02 13:44 5.0		77.02.04 01:02 4.5				1.5		21.3
								31.3	4.5
6	77.03.14 06:16 5.1			77.03.07 08:28 6.1			6.9		6.5
								45.0	3.1
7				77.05.12 19:17 6.7	77.04.28 06:46 4.6		14.5		1.6
								23.7	4.9
8			77.06.24* 00:53 4.6	77.06.10 08:40 5.6	77.06.05 12:37 5.5	77.05.29* 23:24 5.0	4.8		6.2
								29.8	0.6
9	77.08.25 13:29 4.1			77.07.10 04:27 5.5 77.07.15* 04:36 5.4			46.4		1.6
								72.6	3.4
10	77.11.06 03:33 4.5			77.11.27 06:46 5.8			21.1		4.7
								99.4	9.8
11	78.03.06 16:00 3.2		78.03.18* 18:33 3.9	78.03.16 20:27 5.1			10.2		3.5
								35.2	7.4
12		78.04.21 02:08 4.8		78.04.25 20:07 5.1			4.8		4.9
								23.0	32.4
13			78.05.19 13:33 4.4			78.05.18 20:33 5.8	0.7		17.9
								12.7	49.0
14		78.06.01 13:26 4.8		78.06.01 07:08 5.0			0.3		54.2
								14.1	1.9
15	78.06.21* 04:43 3.4		78.06.22 21:38 4.7			78.06.15 15:19 5.0	7.3		6.3
								45.9	2.1
16				78.08.07 20:03 4.8		78.08.30 01:28 5.0	22.2		1.6
								34.7	51.9
17	78.10.12* 00:12 3.4	78.10.04 10:57 4.9	78.10.03 18:50 4.5				0.7		77.2
								51.7	8.6
18	see Table 2 for explanation * – indicates earthquakes not used in calculation			78.12.01 02:39 5.0 78.12.06* 15:30 5.0		78.11.25 02:27 4.0	6.0		4.0
								23.9	79.8
19				78.12.25 08:00 5.0		78.12.25 00:44 4.8	0.3		

TABLE 2

	Huairou	Baijiatuan	b days	a days	a/b
1	67.11.18 18:01 4.2	68.01.01 16:07 3.0	43.92		35.2
				1545.3	1145.0
2	72.03.25 22:22 4.5	72.03.27 06:49 3.5	1.35		422.0
				570.3	15.1
3	73.10.18 14:39 3.4	73.11.25 11.09 3.4	37.85		41.6
				1574.3	151.0
4	78.03.18 18:33 3.9	78.03.29 05:36 3.5	10.46		4.9
				51.3	5.2
5	78.05.19 13:33 4.4	78.05.29 11:00 3.1	9.9		

Notes:
b=time interval between events in one sequence
a=quiescent time interval between sequences
 earthquake parameters given as:
  yr.mo.da
  hr:min mag

block. Perhaps this is due to the fact that when a block is under compression, similar stress patterns develop at the opposite corners of this block.

Pattern of Magnitude Distribution

We have noticed that for earthquake swarms and man-induced earthquakes at dam sites, large earthquakes are usually preceded by a series of small earthquakes with increasing frequency and magnitude. But what is the case in general for strong earthquakes? Are they also preceded by a series of smaller earthquakes?

Figure 9 shows a b-value graph for earthquakes before the M = 8.5 earthquake of Haiyuan (December 1920). This includes earthquakes back to January 1739 when another M = 8 event took place in the Yinchuan - Pingluo region. Figure 10 shows the b-value graph for earthquakes preceding the M = 7.9 earthquake of Luhuo (Feburary 1973). Figure 11 shows the b-value graph for earthquakes before the M = Songming - Yanglin earthquake of 6 September 1833 beginning with 1500.

In the above three cases, for each group of earthquakes there is an earthquake of approximately magnitude 8 at the end. We may venture to say that by projecting the b-value graph, strong earthquake (M = 8) could be predicted. For northern China, earthquakes since 1400 can be divided into two groups: one group from 1400 to 1699 and the second group from 1700 to the 28 July 1976 Tangshan earthquake. The "b-lines" of these two groups are shown in Figure 12. However, if a straight line joining the small circles of Figure 12 (those of the second group) is extended, it would cut the M-axis at about M = 8.5. Does this mean that there should be another strong earthquake of this family after the Tangshan earthquake (M = 8) of July 1976?

Considering the foregoing good relation between medium and small shocks before a large earthquake, perhaps the model which was suggested by Xu et al. [1980] as shown in Figure 13 may be reasonable. This model postulates that those medium and small shocks not only release energy themselves, but also help to concentrate stress for a future larger earthquake. Therefore, the largest earthquakes must have a series of medium and small shocks preceding it.

Discussion

As mentioned previously, we have observed that seismic activity in northern China is distributed

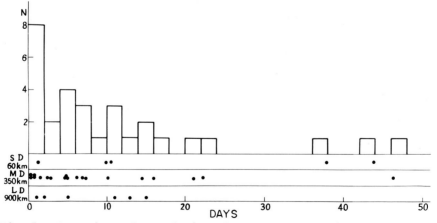

Fig. 7. Comparison of correlation of earthquakes at different scales.

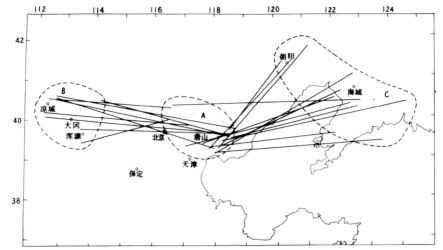

Fig. 8. Differences in the lines linking correlated pairs of earthquakes.

along the edges of blocks and that earthquakes occur at certain distances from one another. Strong earthquakes are always located at one corner of the block.   The correlation between earthquakes suggests that the stratum is simultaneously subjected to a large scale force and in our opinion, these phenomena are due to buckling of the stratum [Xu and Shen, 1979]. In addition, stratrum buckling may explain why before a large earthquake there must be several medium and small shocks in a certain ratio.

The distance between the earthquakes is equivalent to a half wavelength of buckling as shown by Xu and Shen [1979].   When the stratum is compressed and buckled, the half wavelength is determined by only the thickness of the stratum;

$$\lambda = \pi\,4\sqrt{\frac{R^2 h^2}{12(1-\mu^2)}} = C\sqrt{h} \qquad (1)$$

where λ is the half wavelength, R is the radius of the earth, μ is the Poisson coefficient, h is the thickness of the stratum, and C is a constant corresponding to a given R and μ.

The number, N, of earthquakes in an area, A, can be calculated by;

$$N = \frac{A}{\lambda^2}$$
$$= \frac{A}{C^2 h} \qquad (2)$$

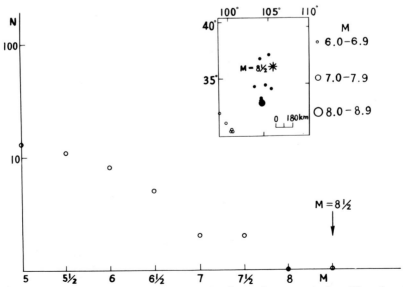

Fig. 9.   Relation between magnitude and cumulative frequency in Ningxia Haiyuan region (1765.5-1920.12.16).

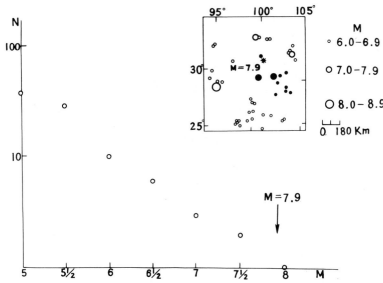

Fig. 10. Relation between magnitude and cumulative frequency in Xianshuihe region (1926.8-1973.2.6).

$$\therefore \log N = (\log A - \log C^2) - \log h. \qquad (3)$$

On the other hand, the energy, E, of the earthquake is proportional to the volume, v, which is involved in the earthquake activity. That is;

$$E = \varepsilon v \qquad (4)$$

where ε is a constant. According to (2), the number of earthquakes being N in the area A, the volume of every earthquake must be

$$v = \lambda^2 h = C^2 h^2 \qquad (5)$$

$$\therefore = E = \varepsilon C^2 h^2. \qquad (6)$$

Since the relation between energy E and magnitude M is;

$$\log E = \alpha + \beta M \qquad (7)$$
$$= 11.8 + 1.5M,$$

the magnitude M of those earthquakes are related by

$$\log \varepsilon C^2 h^2 = \alpha + \beta M \qquad (8)$$

Cancelling h in (3) and (8), we obtain

$$\log N = \left(\log \frac{A\sqrt{\varepsilon}}{C} - \frac{\alpha}{2}\right) - \frac{\beta}{2} M. \qquad (9)$$

Usually, the relation between large and small shocks is;

$$\log N = a - bM \qquad (10)$$

$$\therefore b = \frac{\beta}{2} \qquad (11)$$
$$= \frac{1.5}{2} = 0.75.$$

This is the b-value usually used for general earthquakes.

According to this analysis, we find small earthquakes corresponding to the buckling of a thin stratum and large earthquakes corresponding to the buckling of a thick stratum. The ratio between the numbers of large and small events depends only on this particular geometrical relation. There-

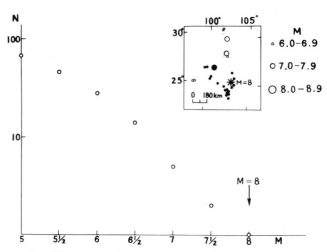

Fig. 11. Relation between magnitude and cumulative frequency in Songming - Yanglin region (1500-1833.9.6).

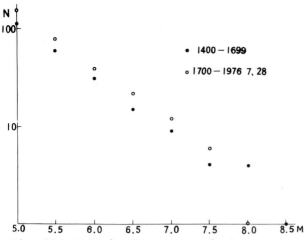

Fig. 12. Relation between magnitude an cumulative frequency in northern China.

fore, the b-value is very nearly the same for large and small earthquakes over the entire world [Utsu, 1967].

If the occurrence of such earthquakes is not controlled by stratum buckling over a large area, but only concentrated in a certain small volume, V, such as in the cases of volcanic earthquakes [Suzuki, 1967] or reservoir earthquakes and certain earthquake swarms, etc., then the number of earthquakes in equation (2) must be changed to N';

$$N' = \frac{V}{\frac{4\pi}{3} r^3} \quad (12)$$

where r is the equivalent radius of an earthquake. Similarly (5) must be changed to;

$$v' = \frac{4\pi}{3} r^3 \quad (13)$$

$$\therefore E' = \varepsilon \cdot \frac{4\pi}{3} r^3 \quad (14)$$
$$= \varepsilon \frac{v}{N'} \cdot$$

From (7), we get

$$\log N' = (\log \varepsilon v - \alpha) - \beta M \quad (15)$$

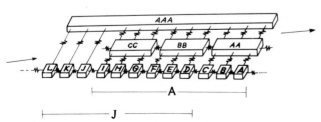

Fig. 13. Model of relation between large and small earthquakes.

$$\therefore b' = \beta \quad (16)$$
$$= 1.5 .$$

This explains the rather higher b-value found with volcanic earthquakes, reservoir earthquakes and certain earthquake swarms.

If the earthquakes only occur along a single fault, it is very simple to calculate the b-value from the above analysis;

$$b'' = \frac{\beta}{3} \quad (17)$$
$$= 0.5 .$$

There has been much discussion about b-values in papers on seismomolgy. Some scientists connect the b-value with the state of stress; the higher the stress, the lower the b-value; and vice versa. Other scientists connect the b-value with the homogeneity of the medium: the more heterogeneous the medium, the higher the b-value; the more uniform the medium, the lower the b-value. Especially in earthquake prediction, some say that the b-value decreases before large earthquakes while others say that the b-value increases before a large earthquake. It seems rather difficult to explain this discrepancy. In this paper, another point of view is suggested, that is, that the b-value is related to the seismicity pattern. It is proposed that at least part of the seismicity may be controlled by the buckling of a stratum - and this explains the b-value. We hope that this seismicity pattern which is observed in China, especially northern China, may have general meaning and can also be verified in other regions.

References

Gutenberg, B., and C. Richter, Magnitude and energy of earthquakes, Ann. Geofis., 9, 1956.

Ku, K., Scientific research and administration of earthquake prediction in China, Lecture in IUGG Meeting, 1979.

Suzuki, Z., Statistics of earthquakes, J. Seismol. Soc. Japan, 20, 138, 1967.

Utsu, T., Some problems of the frequency distribution of earthquakes in respect to magnitude (1), Geophys. Bull. Hokkaido Univ., 17, 106, 1967.

Xu, S., and P. Shen, Some features of earthquake distribution around Peking and crustal buckling, in Terrestrial and Space Techniques in Earthquake Prediction Research, Vogel, A., ed., Vieweg, pp. 585-601, 1979.

Xu, S., B. Wang, L. Jones, X. Ma, and P. Shen, The Haicheng foreshock sequence and earthquake swarm - a function of foreshock sequences in earthquake prediction, Acta Seismologica Sinica, 1980 (in press) (in Chinese).

THE 1906 SAN FRANCISCO EARTHQUAKE AND THE SEISMIC CYCLE

W. L. Ellsworth, A. G. Lindh, W. H. Prescott, and D. G. Herd

U.S. Geological Survey Menlo Park, California 94025

Abstract. The cyclic nature of strain accumulation and release in great earthquakes on the San Andreas fault appears to have modulated the historic pattern of seismicity in northern coastal California. The region experienced many more strong earthquakes in the half century preceding the 1906 earthquake than in the half century following it. Fedotov and Mogi recognized a similar seismic cycle accompanying great earthquakes in the subduction zones of the Western Pacific. The cycle consists of an extended period of quiescence after a major earthquake, followed by a period of increased activity that leads to the cycle-controlling earthquake and its foreshocks and aftershocks. The timing of the next great earthquake cannot be forecast with precision at present, although it appears to be decades away. The reemergence of $M \geq 5$ earthquakes since 1955 within the latitudes of the 1906 surface rupture to the north of San Jose, after an extended period of quiescence following 1906, lead us to conclude that the region is entering the active stage of the cycle in which events as large as M 6, to perhaps M 7 can be expected.

Introduction

On April 18, 1906 a M_s $8\frac{1}{4}$ earthquake occurred along a 430 km-long segment of the San Andreas fault in central and northern California. The ground breakage extended from San Juan Bautista to at least as far north as Point Arena (Figure 1). Right-lateral surface offsets averaged 4 m north of San Francisco and less then 2 m south of San Francisco to San Juan Bautista (Figure 1). Changes in the angles between geodetic monuments, measured after the earthquake, indicated that the slip at the time of the earthquake occurred to a depth of about 10 km and averaged more than $4\frac{1}{2}$ m north of San Francisco and about 3 m south of San Francisco (Thatcher, 1975a).

This earthquake represents the most recent event in a long sequence of very large earthquakes that have ruptured the San Andreas fault in northern coastal California. That another such earthquake will again rupture the fault at some future time is a foregone conclusion (Allen, 1981). However, the timing of that earthquake and the specification of the sequence of events (if any) that will precede it are virtually unknown to us at the present time. It is the purpose of this paper to review the observational evidence relating to the long-term behavior of the San Andreas fault system along the 1906 rupture and to relate these data to simple models of the cyclic recurrence of great earthquakes.

Framework for Prediction of Great Plate Boundary Earthquakes

Work on the problem of the prediction of the next 1906 earthquake is conducted in the face of a null hypothesis that earthquake occurrence is random in space and time. And clearly at this time most seismic activity can only be described stochastically. Randomness, however, is not a physical property of a process, but rather reflects only our level of information about that process. Our task then is to find patterns, regularities and correlations that refine our physical model of the earthquake process.

The most significant progress to date in localizing the occurrence of a future earthquake in space and time stems from the recognition by Imamura that in a region where large earthquakes are known to have occurred in the past, the longer the time since the last earthquake, the greater the risk of another (Imamura, 1924; Aki, 1980). This concept has been formalized into what has come to be known as seismic gap theory by Fedotov (1965) and Mogi (1968). They showed that along the Benioff zone beneath Japan, the Kurile Islands, and the Kamchatka Peninsula, over a long period of time (several hundred years), the aftershock zones of large earthquakes entirely fill in the arc with very little overlap. Repetition of large earthquakes at any position along the arc is now understood to be the direct result of the continuing motion between the Eurasian and Pacific plates. Thus, the average recurrence interval at a single site is, to first order, the ratio of the average

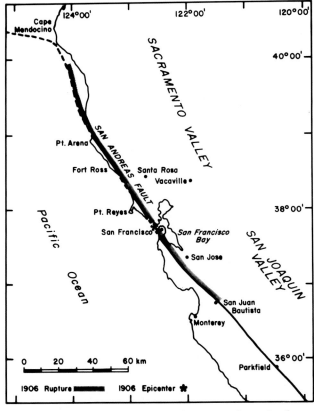

FIG. 1. Geographic location map of central and northern California. The average surface displacement in the 1906 earthquake was greater north of the epicenter than to its south.

displacement in a single earthquake to the average plate velocity at the site.

While the seismic gap theory has proven of great value in predicting where large earthquakes will occur (McCann et al., 1978), irregularity of the time intervals between successive earthquakes at a given site is such that the time of occurence cannot be confidently predicted to within a decade. Another limitation on its use in many areas, including California, is a short recorded history of earthquake occurrence that often leaves the average interevent time and/or the date of the last earthquake unknown.

The Seismic Cycle

An attempt to circumvent these difficulties and better define the time of occurence of a large earthquake on a known seismic gap was the concept of a seismic cycle introduced by Fedotov (1968). His analysis of the seismicity between the recurrence of a large gap filling earthquake suggested that once the aftershock activity had subsided, a relative lull in seismic activity

was observed, followed by an increase in activity to the time of the next event. Mogi (this volume) has formalized this model and characterized the activity during each period.

The cycle (Figure 2) begins with the seismically quiet period (I) which occupies approximately the first half of the interevent period. Activity is low in the immediate source region as well as in the surrounding area. The second period (II) is characterized by increased activity throughout the region. Activity in the eventual source region may subside during the later stages of this period (the "Mogi doughnut" pattern). The third period (III) includes the main event as well as its foreshocks and aftershocks. As the aftershocks in the rupture zone decay in frequency, regional activity may increase through diffusion of the aftershock zone. This leads into the quiescent period of the next cycle.

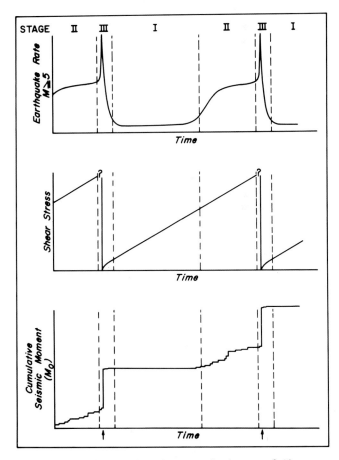

FIG. 2. Schematic diagram of stages of the seismic cycle. Hypothesized behavior of earthquake rate (upper), regional stress level (middle) and cumulative seismic moment (lower) are illustrated.

Time-Predictable Model

A further refinement of the quasi-periodic model of great earthquake occurrence has recently been proposed by Shimazaki and Nakata (1980). They found at three locations in Japan that the time between successive great gap-filling earthquakes was proportional to the displacement in the first event. This "time-predictable" model suggests that great earthquakes recur when the strain released in the last event recovers. In principle, this allows the estimation of the time of the future occurrence of a given great earthquake.

Such a calcuation could be made in several different ways. If one could measure the slip in the last event, then the local rate of plate motion could be used to estimate the time of occurrence of the next event (Sykes and Quittmeyer, this volume). Similarly, if one knew the strain drop in the last event, knowledge of the post-earthquake strain recovery and the current strain rate could also be used to give such an estimate.

Recurrence Time of the Next Great San Francisco Earthquake

Our current understanding of regional tectonics suggests that the segment that broke in the 1906 earthquake is presently locked and accumulating elastic strain but that it has not yet fully recovered the strain released in 1906 (Thatcher, 1975b). It is reasonable, then, to estimate the recurrence time of a M 8 earthquake for this section of the fault from the plate motion rate and the slip in the 1906 earthquake, using the time-predictable hypothesis of Shimazaki and Nakata (1980).

Geologic Slip Rate

Plate tectonic calcuations indicate that the long term slip rate between the North American and Pacific plates averages 5½ cm/yr (Minster and Jordan, 1978). However, geologic reconstructions of San Andreas fault offsets indicate a long-term displacement rate of only about 2 cm/yr along the San Francisco peninsula and 3 cm/yr to the north (Herd, 1979). Most of the remainder of the plate motion that is not transmitted by the San Andreas fault appears to be distributed along other strike slip faults in the coast ranges (Figure 3). The long-term slip rates for the San Andreas imply an average recurrence time of roughly 150 years for a 3 m event on the San Francisco peninsula segment of the fault and 150 years for a 4½ m event on the segment of the fault north of San Francisco. While these estimates of recurrence time coincidentally agree, it should be noted that we do not know that the next earthquake will necessarily rupture both segments of the fault.

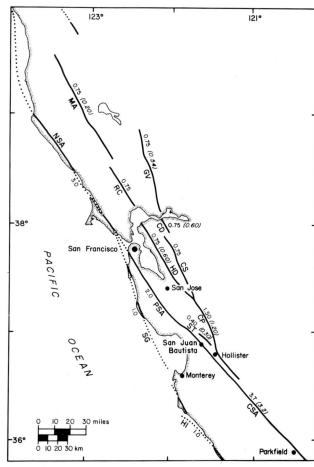

FIG. 3. Principal active faults in the greater San Francisco Bay region. Long-term slip rate (cm/yr) and creep rate (parens) appear next to fault name code. NSA-northern San Andreas, PSA-peninsular San Andreas, CSA-central San Andreas, SG-San Gregorio, HI-Hosgri, CP-Calaveras-Paicines, ST-Sargent, CS-Calaveras-Sunor, HD-Hayward, CD-Concord, GV-Green Valley, RC-Rogers Creek, MA-Maacama. Faults with slip rates lower than 0.1 cm/yr omitted.

Geodetic Strain Rate

Contemporary strain data are also consistent with a roughly 150 year recurrence interval. Strain rates across the San Andreas fault on the southern San Francisco peninsula between 1970 and 1980 were about 0.60 microstrain/yr (Prescott et al., 1980; Prescott, 1980). The near-fault coseismic strain drop determined by Thatcher (1975a) for this segment of the fault was -115 microstrain, which gives a recurrence time of about 190 years.

These calculations of recurrence time may be significantly in error if the strain rate is not constant in time. Thatcher (1975a, and

Table 2) suggested that the near fault strain rate was accelerated for roughly 30 years following the 1906 earthquake. If true, this could significantly reduce the 190 year value calculated above. However, the evidence for the high post-earthquake strain rate comes from the segment of the fault to the north of San Francisco, where the coseismic strain drop was substantially greater. A recurrence calculation for the Point Reyes region that includes the effect of the rapid post-seismic strain recovery gives a recurrence time of 220 ± 40 years.

An important point that must be emphasized is the uncertainty that must be attached to all these estimates. The best case is for the San Francisco peninsula where the two independent estimates agree reasonably well. But, the assumption that the next earthquake will occur when the strain recovers or reaches a predetermined threshold has not been demonstrated conclusively. The probability of a large earthquake in the San Francisco Bay area would be much more clearly defined if it could be established that the seismic cycle described above applied to large strike-slip earthquakes, and if it could be established where in that cycle we are at the present time.

Seismicity and Deformation Rates

In a discussion of the 1957 San Francisco earthquake (M 5.3), Tocher (1959) anticipated some of the key concepts in what was later formalized as the seismic cycle hypothesis. In that paper, he noted, as others had before (Willis, 1924; Gutenberg and Richter, 1954), that the seismicity rate in the San Francisco Bay area was higher during the decades preceeding the 1906 earthquake than in the 50 years after. He further suggested that this quiescence ended with a sequence of moderate earthquakes (M 5½) that occurred in the mid-1950's.

Recently published research into the pre-instrumental earthquake history of California (Toppozada et al., 1979, 1980), taken together with an additional quarter-century of instrumental data, permit us to re-examine Tocher's suggestion, and compare these data with the seismic cycle hypothesis. To that end, we have assembled a catalog of M ≥ 5 earthquakes for the region, covering the period from 1855 through 1980 (Appendix).

Regional Distribution of Earthquakes

Identification of meaningful variations in seismicity data presumes the existence of stable background patterns in those data. A comparison of the historic pattern of moderate-to-large earthquakes in the Coast Range with microearthquakes from 1969-1980, shows that spatial distribution has remained relatively unchanged for at least the past 125 years

(Figure 4). The fault systems that produced the strong earthquakes during the entire historic period are the same ones that account for contemporary background pattern of micro-earthquakes, with the single exception of that portion of the 1906 break north of San Francisco. In that region very few events can be associated with the San Andreas fault. In fact, most of the events locate well to the east of the fault with very few epicenters in the 30-km-wide coastal zone between the San Andreas fault and the Rodgers Creek - Maacama fault zone or seaward of the San Andreas.

Within the San Francisco Bay region, the overall distribution of events is similar to that observed to the north, with most events associated with strike-slip faults to the east of San Francisco Bay. A few isolated zones of epicenters are associated with the San Andreas, including activity in the epicentral region of the 1906 earthquake, to the west of San Francisco. South of the junction of the San Andreas and Calaveras-Paicines faults near San Juan Bautista, microearthquake activity is concentrated along the San Andreas fault, and is distributed in a broad zone across the entire breadth of the Coast Ranges.

Focal mechanism studies (Bolt et al., 1968; Ellsworth and Marks, 1980; Olson et al., 1980) show that throughout the entire region most events have strike-slip focal mechanisms. Nodal planes for right-lateral slip have strike directions that range from northwest to north, in good agreement with the orientation and sense of movement on geologically active faults. Some thrust and normal faulting focal mechanisms are observed, as are rare events that correspond to left-lateral (reversed!) movement on north to northwest striking faults.

Variations in Seismicity Rates

The historic record of moderate and large earthquakes (Figure 5) suggests that there have been significant temporal variations in the earthquake occurrence rate over the past 125 years. The region within the latitudes of the 1906 surface rupture experienced many more strong earthquakes in the half century preceding the 1906 earthquake than in the half century following it (Tocher, 1959; Kelleher and Savino, 1975; Turcotte, 1977). The fragmentary catalog for the period from 1808 to 1854 (Table 1) further suggests that the period of high activity extends at least to the beginning of the historic record. But the absence of any earthquake reports in the first 32 years of the records of missions in the San Francisco Bay area (1776-1807) might be taken as weak evidence for an earlier period of quiesence.

Since about 1955, the region has been more active at the M 5 level than it was in the preceding 50 years. Epicentral maps covering the last five quarter-centuries clearly illus-

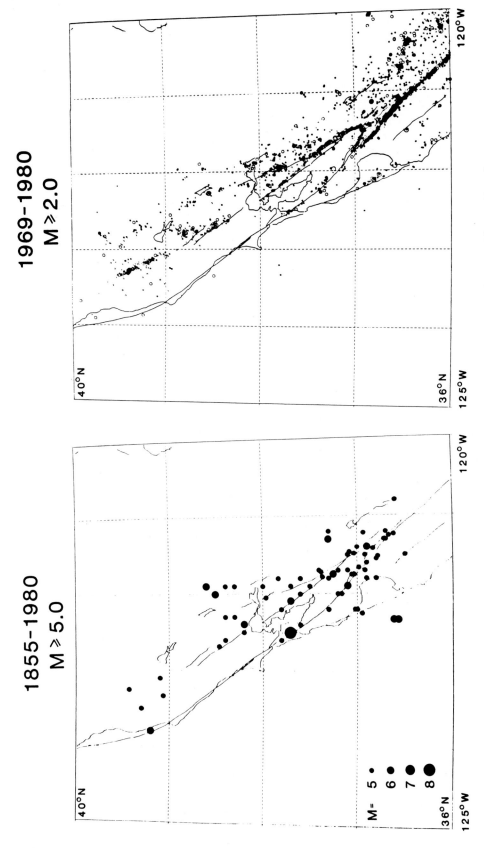

1969-1980
M ≥ 2.0

1855-1980
M ≥ 5.0

M= 5 •
6 ●
7 ●
8 ●

FIG. 4. Seismicity of central and northern California. Epicenters in the Sierra Nevada are not shown. Left: historic M≥5 earthquakes, 1855-1980. Right: microearthquake locations, M≥2, 1969-1980.

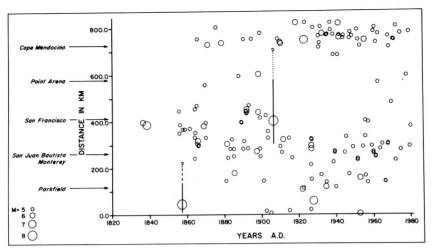

FIG. 5. Space-time plot of M>5 seismicity along the San Andreas fault system. Epicenters are from Real et al. (1978) and Toppozada et al. (1979, 1980). Surface faulting in great earthquakes of 1857 and 1906 shown by solid vertical lines. Surface faulting dotted offshore and dashed where uncertain.

trate the change in the seismicity rate that followed the 1906 earthquake and suggest the return to a higher level of activity during the past 25 years (Figure 6). Given the short historic record, the apparent variations in the earthquake rate must be subjected to some test of their significance.

We have chosen to test the historic catalog against the Poisson hypothesis of a constant average event occurrence rate, λ, for a population of unrelated events using the Kolmogorov test on the interevent times. According to the Poisson model, the cumulative fraction of the population with interevent times less than t is

$$F(t) = 1 - e^{-\lambda t}$$

The maximum deviation between the observed distribution and F(t) is the calculated Kolmogorov test statistic. When this test is applied to the catalog covering the entire region, with aftershocks removed, we cannot reject the hypothesis that the distribution is Poissonian at the 95% confidence level. But when we restrict our attention to the portion of the plate boundary that ruptured in the 1906 earthquake we reject the null hypothesis at the 95% confidence level (Figure 7).

The non-Poissonian behavior of the catalog for the region along the 1906 rupture is largely the result of one very long interevent time. The 473-month interval from 1914 to 1955 greatly exceeds the average interval of 44±79 months. If we hypothesize that this long interval reflects a change in the underlying event rate λ, possibly as a result of the strain drop of the 1906 earthquake, and recompute the Kolmogorov test with it excluded, we find that the null hypothesis cannot be rejected. Thus, we con-

clude that a first order feature of the data is that event rate, λ, precipitously declined following the 1906 earthquake.

Deformation Data

Variations in the rate of earthquake production and their possible association with the 1906 earthquake raise questions about the existence of concurrent variations in the rate or distribution of strain accumulation. Relative motion across the San Francisco Bay area is distributed on at least four major faults spanning a zone more than 80 km wide. Geologic determinations of slip rate indicate that all four transmit a part of the relative motion between the Pacific and North American plates (Figure 3). Offset geologic features indicate a long term (millions of years) average slip rate of 10 mm/yr on the San Gregorio fault (Webber and Lajoie, 1977; Webber and Cotton, 1980) and 20 mm/yr on the San Andreas fault (Cummings, 1968; Addicot, 1969). Uncertainties in age and displacement are such that both of these rates could vary by as much as 50%. Along the Hayward and Calaveras-Sunol faults the geologic rate is unknown, but south of San Jose the geologic rate on the Calaveras-Paicines fault is about 15 mm/yr and presumably this is distributed between the Hayward and Calaveras-Sunol faults (Herd, 1979).

Geodetic determination of the slip rate during the period 1970 to 1980 generally agrees with the geologic rate. Prescott et al. (1980; Prescott, 1980) found rates of 10±3 mm/yr slip on the San Andreas fault, 6±1 mm/yr on the Hayward fault, and 6±1 mm/yr on the Calaveras-Sunol fault. The slip associated with the San Andreas fault does not appear at the surface,

TABLE 1. Strong Earthquakes of the Greater San Francisco Bay Region, 1836-1980.

Date	Location[1]		Magnitude[1]	Comments	Reference[1]
1808/6/21	San Francisco?		?	Destroyed all of the buildings, in particular barracks and other houses of the Presidio of San Francisco	2
1836/6/10	37.8 N	122.2 °W	6.7	Ground breakage on the Hayward fault	3
1838/6	37.6	122.4°	7.0	Probable ground breakage on Peninsula San Andreas	3
1868/10/21	37.7	122.1°	6.7	Ground breakage on the Hayward fault	4
1892/4/21	38.5	122.0°	6.6	Largest (?) in series of earthquakes near the town of Vacaville that included a M 6.4 shock two days earlier	
1906/4/18	37.8	122.6°[5]	$8\frac{1}{4}$ [6]	Felt foreshock assigned epicenter off Golden Gate by Reid (1910)	4
1911/7/1	37.25	121.75°[7]	6.6[7]	Intensity data indicate an earthquake of comparable size to M 5.9 earthquake of 1979/8/6	8

1. All earthquakes are described in Townley and Allen, (1938). Locations and magnitudes from 1836-1900 are from Toppozada et al., (1979, 1980).
2. Quoted from Gaceta de Mexico (892, 1808) by Orozco y Berra (1887)
3. Louderback, (1947).
4. Lawson et al. (1908).
5. Boore (1977).
6. Bolt (1968) demonstrates that an M value of $8\frac{1}{4}$ is consistent with the teleseismic records. A moment-magnitude (Hanks and Kanamori, 1978) of 7.7 is obtained from the seismic and geodetic moment determined by Thatcher (1975a).
7. This magnitude is listed in Gutenberg and Richter (1954). We calcuate an intensity magnitude of 5.6 using relations of Toppozada (1975) and intensity data from Templeton (1911).
8. Templeton (1911) and Wood (1912).

but rather is distributed across some 40 km normal to the fault and presumably reflects (elastic) strain accumulation. North of San Francisco both geologic and geodetic determinations of slip rate are less certain.

Table 2 summarizes the strain rate since 1906 for the geodetic nets shown in Figure 8. Because of the spatial inhomogeneity of deformation in the San Francisco Bay area, care is required to avoid having spatial variations in strain rate masquerade as temporal variations. For this reason in Table 2 we have compared only the strain rates obtained over different time periods from a constant set of angles or distances.

The first three cases, Primary Arc, Hayward Net and CDWR Net (Table 2), compare geodetically determined strain rates in the San Francisco Bay area. Cases 1 and 2 were obtained from angle observations. In neither case is the change in strain rate (lines 1c and 2c, Table 2) significant at the two standard deviation level. Case 3 was derived from repeated measurements of the length of 11 lines from the California Department of Water Resources (CDWR). The difference between the 1960-1967 and the 1970-1980 strain rate (line 3c, Table 2) is apparently significant. Unfortunately a change of procedures and instrumentation occurred in 1968-1969 and Savage (1975) found that there

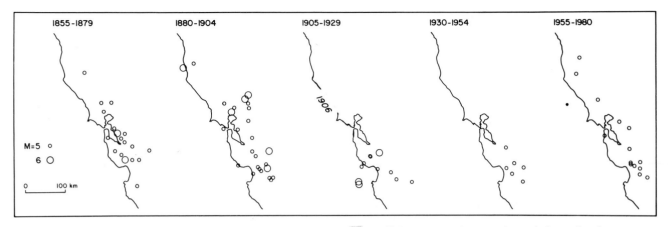

FIG. 6. Seismicity of the San Francisco Bay region, M≥5, by quarter-century intervals from 1855-1980.

are systematic differences between the two techniques. Thus the observed change is probably not reliable. The fact that all three cases indicate a decrease in strain rate may be significant.

North of San Francisco Bay the data of both Fort Ross and Point Reyes indicate a significantly higher rate prior to about 1940 (lines 5c and 6c, Table 2; Thatcher, 1975a). The data at Point Arena (line 4a, Table 2) are consistent with the higher early rate, but the second time period is too short to yield meaningful results.

South of the San Francisco Bay region, along the central segment of the San Andreas fault, the data indicate that plate motion is accommodated by rigid block motion across the San Andreas fault (Savage and Burford, 1971; Thatcher, 1979). Here, the geodetic data

clearly demonstrate that there have been no gross changes in the slip rate since 1885 (Thatcher, 1979). Seismicity along this reach of the fault similarly shows no significant temporal variation either before or after 1906 (Figure 5). Apparently, this segment of the fault is effectively buffered from the end-effects of great earthquakes to the north.

To summarize the deformation data: north of San Francisco the near-fault strain appears to have been elevated for a period of 25 to 30 years following 1906; in the San Francisco Bay area there is no hard evidence of any changes in the rate of strain accumulation or slip, but there is marginal evidence of a general decline in strain rate with time. No data exist to indicate whether the strain rates near the San Andreas fault along the San Francisco peninsula were high after 1906 as were the rates to the north. South of the 1906 break, the strain rate appears to have been constant since the mid-1880's. Finally it is unlikely that any significant increase in the rate of deformation accompanied the increase in M 5 and greater earthquakes that occurred about 1955.

Discussion

The hypothesis of a seismic cycle (Figure 2) as advanced by Fedotov (1968) and Mogi (this volume) is based principally upon observations of the subduction zones of the Western Pacific. While we must guard against premature acceptance of its universality, the principal features of the cycle do appear to be represented in the historic record of seismicity discussed above. Long-term variations in the production rate of M 5 earthquakes in the San Francisco Bay region indicate an active period (stage II) lasting for at least 50 years before the 1906 earthquake, followed by a marked quieting of the region for nearly 50 years after the earthquake (stage I). If, as Tocher (1959) suggested, the

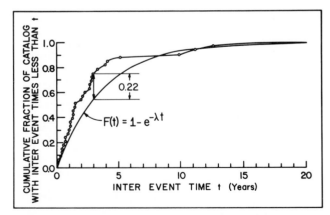

FIG. 7. Kolmogorov test of the interval time distribution of earthquakes in the San Francisco Bay region with a Poisson model (smooth curve). Maximum deviation of observed distribution from Poisson model, at vertical arrow, exceeds the expected range for a Poisson process at the 95% confidence level.

Table 2. Ψ is the direction across which right lateral shear is a maximum and γ is the rate of engineering shear strain across this direction

	Period	γ (ppm/yr)	Ψ	Source
Primary Arc				
1a	1907-1922	0.79 + 0.23	N 41 + 10	Thatcher (1975b)
1b	1922-1948	0.46 + 0.17	N 36 + 13	Thatcher (1975b)
1c	Difference	-0.33 + 0.29	-5 + 16	
Hayward Net				
2a	1951-1963	0.60 + 0.12	N 27 W + 6	Lisowski (unpublished)
2b	1963-1978	0.40 + 0.09	N 28 W + 7	Lisowski (unpublished)
2c	Difference	-0.20 + 0.15	1 + 9	
CDWR Net				
3a	1960-1967	0.87 + 0.10	N 24 W + 4	This paper
3b	1970-1980	0.42 + 0.02	N 26 W + 2	This paper
3c	Difference	-0.45 + 0.10	2 + 4	
Pt Arena				
4a	1906-1925	2.28 + 0.48	N 72 W + 6	Thatcher (1975a)
4b	1925-1930	16.75 + 7.96	N 59 E + 14	Thatcher (1975a)
4c	Difference	Not meaningful		
Fort Ross				
5a	1906-1930	2.47 + 0.80	N 57 W + 8	Thatcher (1975a)
5b	1930-1969	0.33 + 1.07	N 49 W + 91	Thatcher (1975a)
5c	Difference	-2.14 + 1.34	-8 + 91	
Pt Reyes				
6a	1930-1938	2.27 + 0.74	N 57 W + 10	Thatcher (1975a)
6b	1938-1961	0.75 + 0.16	N 52 W + 9	Thatcher (1975a)
6c	Difference	-1.52 + 0.75	-5 + 13	

quiet period following the earthquake ended in the mid 1950's, then the region can expect the continued occurrence of moderate earthquakes in the coming decades, culminating in another great earthquake. This may also imply that M 6-7 earthquakes (such as the 1836 and 1868 earthquakes on the Hayward Fault), capable of inflicting serious damage and significant loss of life, should be expected in the San Francisco Bay area at the rate experienced in the 19th century, or at an average rate of about one each decade.

Our analysis of the seismicity of the San Francisco Bay area leads us to the same general conclusions as reached by Tocher. Seismicity at the M 5 level since 1955 generally resembles that from the 19th century in both their spatial distribution and frequency, although the number of events in the sample (nine) is too small to conclusively establish the correspondence. The available evidence, taken literally, suggests that while the seismicty has increased, it has not fully returned to the level seen before 1906 (Figure 9).

At present, the very general descriptive model of a seismic cycle presented here is of little predictive value as to the time of either large earthquakes on any one of a number of faults, or of the next great gap filling earthquake along the 1906 break. The long duration of stage II before the 1906 earthquake would, however, suggest that the repeat of that earthquake is not imminent. The potential for these ideas in improving our estimate of the timing of the next great earthquake lies in the development of a quantitative physical model that predicts the increase in seismic activity, while satisfying the constraints provided by the geodetic measurements of the strain field.

Certainly the simplest model to account for

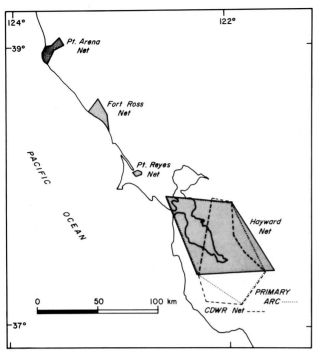

FIG. 8. Areal coverage of Geodetic Figures listed in Table 3.

the principle features of the seismic cycle is that the stress drop that accompanies the main event recovers in a linear fashion until the time of the next event (Figure 2). The regional seismicity is modulated by the stress relaxation of the main event and its gradual recovery. Work on possible laboratory analogs of these conditions have shown that increases in acoustic emissions prior to failure (Mogi, 1962; Scholz, 1968), or stick-slip (Weeks et al., 1978) do occur, even while the sample is in the linear stress-strain regime. Similar results have also been obtained however, under constant stress conditions, where the increase in acoustic emissions prior to failure accompanies accelerating strain (Mogi, 1962).

In detail, the regional extent of the M 5 quieting following 1906 is difficult to understand simply in terms of the coseismic displacement of the earthquake. The affected region is too broad, about 100 km, to be the result of the shallow, 10-km-deep, coseismic displacement alone, if the entire lithosphere responds elastically. The possibility that the earthquake triggered after-slip at depth, as the lower crust "caught up" with the surface displacement (Thatcher, 1975b) offers a resolution to this paradox. Recently presented models of earthquake cycles on transform faults (Lehner and Li, 1980; Mavko, 1980) display this behavior, which results in the gradual reduction of stress at distances from the fault that are

comparable to the lithospheric thickness. An alternative explanation for the quieting of the regional system of faults could be that the brittle crust decouples horizontally from the lower crust and mantle at a depth comparable to that of the coseismic slip (Lachenbruch and Sass, 1980).

The increase in seismicity that occurs during stage II of the cycle could result simply from the gradual recovery of stress, or it might be related to other factors, such as an accelerated strain rate. Thatcher (1975a) has argued that the strain rate was abnormally high in the late 19th century, during the period that we associate with stage II of the cycle. While his evidence for very large strains in the late 1800's has been challenged (Savage, 1978; Thatcher, 1978), the fact that, if anything, strain rates have decreased in the Bay area since 1950 essentially rules out the possibility that the apparent increase in magnitude 5 activity documented in this paper is the result of such an increase in strain rates in the region. As the strain, and by inference the stress, in the region is almost certainly increasing, the return of M 5 activity would appear to be caused by a stress threshold having been reached (Turcotte, 1977).

Paradoxically, the production rate of smaller earthquakes has apparently remained virtually constant since at least the 1930's. Except for short-term fluctuations related to aftershock sequences, the mean frequency of occurrence of M 3-4 earthquakes has not changed significantly since instrumental magnitudes have been routinely calculated for local earthquakes (Figure 9). Frequency-magnitude statistics for microearthquakes (M 1.5-3.0) that occurred during the 1970's give the same Gutenberg-Richter a and b-values as are obtained from the M 3-4 earthquakes from 1942-1969. Thus, by inference, the microearthquake production rate has also remained constant for 50 years.

The spatial distribution of earthquakes during stage II of the cycle may provide a clue as to the mechanism that connects the return of M 5 earthquakes with stress recovery. Activity during the second half of the 19th century, which we have associated with stage II, located preferentially to the east of the San Andreas fault, apparently along the faults that are, at the present time, creeping aseismically (Figure 3), have microearthquake activity associated with them (Figure 4), and produce significant local deflections in the surface strain field (Prescott, 1980).

The striking similarity between the geodetically determined strain pattern and the geologically determined pattern of slip on faults, taken together with the continuity of the spatial distribution of earthquakes through the time of the 1906 earthquake, strongly suggests that the entire coastal fault system maintains itself in a delicate equilibrium that is con-

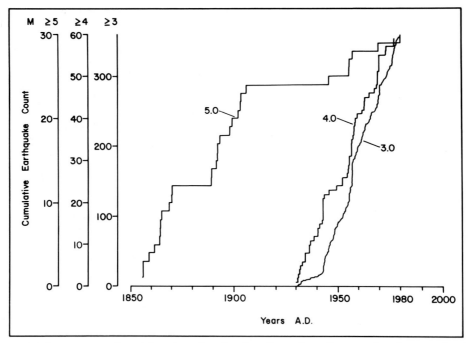

FIG. 9. Cumulative count of earthquakes in the San Francisco Bay region by magnitude. Counts of
M 4 and M 3 earthquakes are incomplete before 1930 and 1942, respectively.

trolled by long-wavelength forces. The occurrence of great earthquakes on the San Andreas
fault, and steady, deep-seated movement between
the plates in the lower lithosphere modulates
the stress level acting on the system without
disturbing the details of the balance between
its parts.

Concerning what directions future work on
this problem might take, we feel compelled to
point out that the most significant conclusion
of this paper, that the Bay area may now be in
stage II of the seismic cycle, is also the
least certain. The seismic cycle concept itself
is yet but a working hypothesis, and the
significance of the increase in M 5 activity
has not been established beyond any doubt. Yet
for earthquake prediction and hazard reduction,
the potential implications of the cycle are
great. Previous estimates of the annual probability of M 6-7 earthquakes on individual
faults in the Bay area have been of the order
of 10-2. A cumulative estimate of 10-1 for
the region involves a probability gain of two
to five (Aki, this volume), if we are indeed in
stage II of the cycle.

The task now is to reduce the uncertainties
in the factors that enter into this calculation.
Detailed geologic work on the major faults in
the area might improve our knowledge of average
recurrence intervals and the time of the most
recent events. Geodetic experiments now in
progress are likely to provide a better model
of the strain recovery process, further refining
the geodetic estimates of recurrence time. Care-

ful monitoring of microearthquakes may help
refine our knowledge of the stresses acting at
depth on the fault, and might lead to identification of long-term and/or short-term precursors to the next major earthquake.

Summary

1. Seismicity at the M 5 level was low in the
San Francisco Bay for the 50 years following
the 1906 earthquake, when compared to activity
during the previous 50 years.
2. Since about 1955, activity appears to have
increased, although it has not yet returned to
pre-1906 levels.
3. Offsets of geologic markers place
constraints on the long-term average slip rates
on a number of major Bay area faults. Faults
which accommodate a significant portion of the
North American-Pacific plate boundary motion
(5½ cm/yr) include the San Gregorio (1 cm/yr),
San Andreas (2 cm/yr), Calaveras-Paicines (1.5
cm/yr), Hayward (3/4 cm/yr), Calaveras-Sunol
(3/4 cm/yr), Rodgers Creek (3/4 cm/yr), Maacama
(3/4 cm/yr) and Green Valley (3/4 cm/yr) faults.
4. 3 m of offset was observed on the San
Francisco peninsula section of the San Andreas
during the 1906 earthquake, and no significant
aseismic slip has been observed since then.
Combined with the long-term geologic rate on
this fault, this implies a long term average
recurrence time of 150 years for this section
of the San Andreas fault.
5. Geodetic data give an strain drop of 115

microstrain at the San Andreas fault on the San Francisco peninsula during the 1906 earthquake. Current measurements of the strain recovery in the same region give 0.6 microstrain/yr, implying an _average_ recurrence time of 190 years.

6. Studies of the seismicity accompanying the recurrence of great trench earthquakes in the western Pacific have led to the concept of a seismic cycle, the principal features of which is quiesence following a great earthquake, with increased activity during the second half of the interevent period. Sesimicity data for the San Francisco Bay area summarized above fit this pattern, although they are not adequate to confirm its applicability to strike slip earthquakes on this transform boundary.

7. _If_ the seismic cycle is applicable and _if_ the estimates of 150-190 years for the average recurrence interval are correct, then it is likely that large (M 6-7) earthquakes are more probable in the next 70 years than their absence in the last 70 years might suggest.

8. However, even if all of these arguments are correct, nothing is implied as to where or when these earthquakes might occur, except that we believe from the historic record and geologic evidence that the faults named above are the most likely candidates.

9. One can arrive at the same conclusion more directly by simply observing that large earthquakes were far more common in the Bay area in the 19th century than they have been in the 20th. They occurred at a rate of about one event per decade in the 70 years prior to 1906.

10. Thus, we believe that a repeat of the 1906 earthquake on the San Andreas fault does not seem likely in the next few decades, but that one or more large (M 6-7) earthquakes on any number of faults (including the segment of the San Andreas fault on the San Francisco peninsula) are possible, or even likely, in the coming decades.

Appendix. Historic Record of Damaging Earthquakes

The brief pre-instrumental record of damaging earthquakes in the coastal ranges of central and northern California covers less than two centuries and is strongly influenced by the density and distribution of settlement in the region. Even in the immediate San Francisco Bay region the record does not approach an acceptable level of completeness until after the 1849 gold rush. Despite these difficulties, several strong earthquakes from before 1849 are known, and reports suggest others (Table 1). It should be stressed that the list of events in Table 1 is fragmentary. No earthquakes were reported by the inhabitants of either San Francisco and Mission Santa Clara (near San Jose) during the first 32 years of their establishment (1776-1807). No earth-

quakes are mentioned at all in the available records from the Russian settlement at Fort Ross (1812-1841). The absence of earthquake reports in the records of these settlements does not necessarily imply that there were no severe earthquakes during this period, as the records are principally concerned with economic matters.

The quantity and quality of written records of earthquakes dramatically improved with the influx of settlers in the 1850's. Toppozada et al. (1979, 1980) have systematically examined 19th-century newspapers and other sources in compiling their catalog of pre-1900 seismicity. We have adopted their catalog in our analysis of this period. We estimate that the catalog it is essentially complete down to M 5½ after 1855 in the San Francisco Bay region. However, the catalog contains only about half of the events to be expected from the Gutenberg-Richter frequency-magnitude law in the interval from M 5 to M 5½. Data on earthquakes from 1900 onward are taken from several sources, including Real et al. (1978); Bulletins of the Seismographic Stations, University of California, Berkeley; Savage and McNally (1974); and U.S.G.S. Catalogs of Earthquakes along the San Andreas Fault System. The compiled catalog appears in Table A1.

No aftershocks of the 1906 San Francisco earthquake are included in the catalog that we assembled, unless the 1911 event on the Calaveras-Paicines fault is considered as one. This is because no M 5 aftershocks of the earthquake have yet been documented, although, Lawson et al. (1908) tabluated felt reports for two years following the earthquake. Among the reports that they list, those from Berkeley, California were considered most reliable, due to the efforts there of trained observers who attempted to compile complete lists of felt events. We have appended to their list the listings of felt earthquakes at Berkeley for the period 1910-1949 from Bolt and Miller (1975), and have plotted the event rates versus time to assess the impact on the catalog of 1906 aftershocks not being included (Figure A1). The event rate decays in accordance with Omori's law (t-1) until about 1910, and appears to reach a constant value by about 1915. This suggests that the omission of 1906 aftershocks does not influence the conclusions reached in this paper.

Earthquake magnitudes and epicentral coordinates that we have used generally correspond to those listed in the cited reference. The only exceptions are for those events occurring between 1900 and 1927 for which we have computed local magnitudes from published felt area data using relations given by Toppozada (1975). When multiple sources listed different magnitude values, we have averaged the available data. Instrumental magnitudes determined from multiple records (Savage and McNally, 1974) have been

TABLE A1. Magnitude 5 and Larger Earthquakes
in Coastal California from 36½°N to 39½°N,
1855-1980

Year	Date		Time	Latitude		Longitude		Magnitude
1855	8	27	23: 0	38N	12	122W	30	5.1
1856	1	2	0: 0	37N	18	122W	12	5.3
1856	2	15	13:25	37N	36	122W	24	5.9
1858	11	26	8:35	37N	36	122W	0	5.9
1861	7	4	0:11	37N	42	121W	54	5.3
1864	2	26	8:40	36N	30	121W	30	5.8
1864	3	5	16:49	37N	24	121W	42	5.3
1864	5	21	2: 1	38N	24	122W	18	5.2
1864	7	22	6:41	37N	30	121W	54	5.4
1865	3	8	14:30	38N	24	122W	36	5.3
1865	5	24	11:21	37N	6	121W	42	5.0
1865	10	8	20:46	37N	6	121W	54	6.2
1866	3	26	20:12	37N	6	121W	30	5.8
1866	7	15	6:30	37N	18	121W	12	5.7
1868	10	21	15:53	37N	42	122W	6	6.7
1869	10	8	9:30	39N	6	123W	6	5.2
1870	2	17	20:12	37N	12	122W	0	5.5
1870	4	2	19:48	37N	48	122W	12	5.1
1881	4	10	10: 0	37N	18	121W	18	6.0
1882	3	6	21:45	36N	42	121W	12	5.5
1883	3	30	15:45	36N	54	121W	36	5.7
1884	3	26	0:40	37N	0	121W	12	5.3
1885	3	31	7:56	36N	42	121W	18	5.6
1885	4	2	15:25	37N	0	121W	24	5.5
1887	12	3	18:55	39N	18	123W	30	5.2
1889	5	19	11:10	38N	0	121W	54	5.6
1889	7	31	12:47	37N	48	122W	12	5.1
1890	4	24	11:36	36N	54	121W	36	5.9
1891	1	2	20: 0	37N	12	121W	42	5.1
1891	10	12	6:28	38N	18	122W	18	5.6
1892	4	19	10:50	38N	30	122W	0	6.8
1892	4	21	17:43	38N	36	121W	54	6.3
1892	4	30	0: 9	38N	24	121W	54	5.0
1892	11	13	12:45	36N	48	121W	48	5.2
1893	8	9	9:15	38N	24	122W	36	5.0
1897	6	20	20:14	36N	54	121W	24	6.0
1898	3	31	7:43	38N	12	122W	24	6.3
1898	4	15	7: 7	39N	12	123W	48	6.7
1899	4	30	22:41	36N	54	121W	36	5.7
1899	6	2	7:19	37N	48	122W	36	5.5
1899	7	6	20:10	36N	42	121W	18	5.5
1902	5	19	18:31	38N	18	121W	54	5.5
1903	6	11	13:12	37N	36	121W	48	5.6
1903	8	3	6:49	37N	18	121W	48	5.6
1906	4	18	13:12	37N	42	122W	30	8.3
1910	3	11	6:52	36N	54	121W	48	5.7
1910	12	31	12:11	36N	50	121W	25	5.0
1911	7	1	22: 0	37N	15	121W	45	6.0
1914	11	9	2:31	37N	10	122W	0	5.5
1916	8	6	19:38	36N	40	121W	15	5.4
1926	7	25	17:57	36N	36	120W	48	5.2
1926	10	22	12:35	36N	37	122W	20	6.2
1926	10	22	13:35	36N	34	122W	20	6.2
1926	10	24	22:51	37N	1	122W	12	5.6
1927	2	15	23:54	36N	57	122W	15	5.6
1939	6	24	13: 2	36N	48	121W	33	5.2
1945	1	7	22:25	36N	46	121W	13	5.0
1945	8	27	9:13	37N	23	121W	43	5.0
1949	3	9	12:28	37N	2	121W	29	5.2
1951	7	29	10:53	36N	37	121W	14	5.0
1954	4	25	20:33	36N	55	121W	42	5.3
1955	9	5	2: 1	37N	22	121W	47	5.6
1955	10	24	4:10	37N	58	122W	3	5.4
1957	3	22	19:44	37N	40	122W	29	5.3
1959	3	2	23:27	36N	59	121W	40	5.2
1960	1	20	3:25	36N	47	121W	31	5.1
1961	4	9	7:23	36N	43	121W	18	5.3
1961	4	9	7:25	36N	45	121W	22	5.2
1962	6	6	17:50	39N	4	123W	20	5.2
1964	11	16	2:46	37N	2	121W	45	5.1
1967	12	18	17:24	37N	3	121W	45	5.2
1969	10	2	4:56	38N	28	122W	41	5.5
1969	10	2	6:19	38N	28	122W	41	5.4
1974	11	28	23: 1	36N	55	121W	30	5.1
1977	11	22	21:15	39N	26	123W	15	5.0
1979	8	6	17: 5	37N	5	121W	28	5.9
1980	1	24	19: 0	37N	50	121W	48	5.8
1980	1	27	2:33	37N	51	121W	47	5.5

preferentially favored over single-station
magnitudes, especially during the period from
1936-1973. This has occasionally resulted in a
change in magnitude value of ½ unit relative to
those values listed in Bolt and Miller (1975)
and Real et al. (1978).

As the uniformity of the listed magnitudes
with time is essential for some aspects of the
analysis that follows, it is critical that the

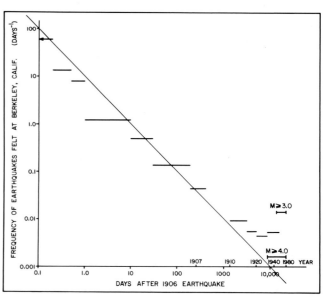

FIG. A1. Frequency of earthquakes felt in
Berkeley, California. Data from 1906-1907 from
Lawson et al. (1908). Data from 1910-1949 from
Bolt and Miller (1975). Reference line is proportional to t^{-1}. Earliest time interval (arrow)
extends to time of 1906 mainshock. Mean production rate of M$\geq$3 and M$\geq$4 earthquakes occurring
within 50 km of Berkeley area also shown.

intensity magnitudes (M_I) are not greatly biased relative to the instrumental magnitudes (M_L). A direct comparison of M_I with M_L for 19 events that occurred between 1945 and 1969 shows that $M_I - M_L = -0.2 \pm 0.2$. Thus, there is a suggestion that magnitude values based exclusively on intensities are marginally smaller than M_L. There is also some evidence that M_I values determined from the total felt area are systematically larger than those determined from the M. M. intensity V area (0.25 ± 0.25). As all computed intensity magnitudes were averaged by Toppozada et al. (1979) in preparing their catalog, we suspect that magnitudes from the pre-instrumental period may be underestimated by about 0.3. However, these magnitudes would have to be overestimated by 1.0 magnitude units to affect our conclusions. This appears to be highly unlikely.

The spatial uniformity of the earthquake catalog is of similar concern to us. We believe that it is reasonable to assume that the catalog attains its threshold reporting level about the time when local newspapers were established. This would correspond to about 1855 for the region from San Jose to Santa Rosa, 1860 for the region from Monterey to Point Arena, and about 1900 for the region north to Cape Mendocino. Magnitude 5 earthquakes are present in the catalog by 1865 from south of Monterey to Santa Rosa, and as far north as the latitude of Point Arena by 1870. Earthquakes located at the western edges of the San Joaquin Valley are also present in the catalog by the early 1860's.

Acknowledgements. We thank Mike Lisowski for the use of his unpublished analysis of the Hayward net. The critical comments of Tom Hanks, Wayne Thatcher and Dave Boore were of great value to us in preparing this paper. Graphics by C.R. McMasters.

References

Addicot, W.A., Late Pliocene mollusks from San Francisco Peninsula, California, and their paleographic significance, Proc. California Acad. Sci, Fourth Ser, 37, 57-93, 1969.

Aki, K., Possibilities of seismology in the 1980's, Bull. Seismol. Soc. Amer., 70, 1969-1976, 1980.

Allen. C.R., The modern San Andreas fault, in The Geotectonic Development of California, Ernst, W.G., ed., Prentice-Hall Inc., New Jersey, 511-534, 1981.

Bolt, B.A., The focus of the 1906 California Earthquake, Bull Seismol. Soc. Amer., 58, 457-471, 1968.

Bolt, B.A., and Miller, R.D., Catalogue of earthquakes in northern California and adjoining areas 1 January 1910 - 31 December 1972, Seismographic Stations Univ. Calif., Berkeley, California, 567 pp, 1975.

Bolt, B.A., Lomnitz, C., and McEvilly, T.V., Seismological evidence on the tectonics of central and northern California and the Mendocino escarpment, Bull Seismol. Soc. Amer., 58, 1725-1767, 1968.

Boore, D.M., Strong-motion recording of the California earthquake of April 18, 1906, Bull Seismol. Soc. Amer., 67, 561-579, 1977.

Cummings, J.C., The Santa Clara formation and possible post-Pliocene slip on the San Andreas fault in central California, in Dickinson, W. R., and Grantz, A., eds., Proc. Conf. Geol. Prob. San Andreas Fault System, Stanford Univ. Publ. Geol. Sci., 11, 191-207, 1968.

Ellsworth, W.L., and Marks, S.M., Seismicity of the Livermore Valley, California region 1969-1979, U.S. Geol. Surv. Open-File Report 80-515, 41 pp, 1980.

Fedotov, S.A., Regularites in the distribution of strong earthquakes in Kamchatka, the Kuril Islands and northeastern Japan, Akad. Nauk. SSSR Inst. Fiziki Zeml:Trudy, 36, 66-93, 1965.

Fedotov, S.A., The seismic cycle, quantiative seismic zoning, and long-term seismic forecasting, in Seismic Zoning of the USSR, S. Medvedev, ed., Izdatel'stvo "Nauka", Moscow, 1968.

Gutenberg, B., and Richter, C.F., Seismicity of the Earth and Associated Phenomena, Hafner Publ. Co., New York, 310 pp, 1954.

Hanks, T.C., and Kanamori, H., A moment magnitude scale. J. Geophys. Res, 2348-2350, 1978.

Herd, D.G., Neotectonic framework of central coastal California and its implications to microzonation of the San Francisco Bay Region, in Brabb, E. E., ed., Progress on Seismic Zonation in the San Francisco Bay Region, U. S. Geol. Surv. Circ. 807, 3-12, 1979.

Imamura, A., Preliminary note on the great earthquake of southeastern Japan on September 1, 1923, Bull. Seismol. Soc. Amer., 14, 136-149, 1924.

Kelleher, J., and Savino J., Distribution of seismicity before large strike slip and thrust-type earthquakes, J. Geophys. Res., 80, 260-271, 1975.

Lachenbruch, A.H., and Sass, J.H., Heat flow and energetics of the San Andreas fault zone, J. Geophys. Res., 85, 6185-6222, 1980.

Lawson, A.C., Chairman, California Earthquake of April 18, 1906, Report of the State Earthquake Investigation Commission, Vol. I., Carnegie Institution of Washington, 451 pp., 1908.

Lehner, F.K., and Li, V.C.F., On the stressing of the lithosphere in an earthquake cycle (abs), EOS, Trans. Amer. Geophys. Union, 61, 1051, 1980.

Louderback, G.D., Central California earthquakes of the 1830's, Bull. Seisim. Soc. Amer, 37, 33-74, 1947.

Mavko, G.M., Simulation of creep events and

earthquakes on a spatially variable model (abs), EOS, Trans. Amer. Geophys. Union, 61, 1120-1121, 1980.

McCann, W.R., Nishenko, S.P., Sykes, L.R., and Krause, J., Seismic gaps and plate tectonics: seismic potential for major plate boundaries, U.S. Geol. Surv. Open-File Report 78-943, 441-584, 1978.

Minster, J.B., and Jordan, T.H., Present day plate motions, J. Geophys. Res., 83, 5331-5354, 1978.

Mogi, K., Study of elastic shocks caused by the fracture of heterogenous materials and its relations to earthquake phenomena, Bull. Earthq. Res. Inst., 40, 125-173, 1962.

Mogi, K., Some features of recent seismic activity in and near Japan (1), Bull. Earthq. Res. Inst., 46, 1225-1236, 1968.

Olson, J.A., Lindh, A.G., and Ellsworth W.L., Seismicity and crustal structure of the Santa Cruz Mountains, California (abs), EOS, Trans. Amer. Geophys. Union, 61, 1042, 1980

Orozco y Berra, D.J., Seismologia: efemerides seismicas Mexicanas, Memorias de la Sociedad Cientifica, 1, 258, 1887.

Prescott, W.H., The accommodation of relative motion along the San Andreas fault system in California, Ph. D. Thesis, Stanford Univ., 195pp, 1980.

Prescott, W.H., Lisowski, M., and Savage, J.C., Geodetic measurement of crustal deformation on the San Andreas, Hayward and Calaveras faults near San Francisco, California (abs), EOS, Trans. Amer. Geophys. Union, 61, 1042, 1980.

Real, C.R., Toppozada, T.R., and Parke, D.L., Earthquake Catalog of California January 1, 1900 - December 31, 1974, Calif. Div. Mines and Geol. Special Pub 52, 15 pp, 1978.

Reid, H.F., The Mechanics of the Earthquake, Vol. II of The California Earthquake of April 18, 1906, Carnegie Institution of Washington, 192 pp, 1910.

Savage, J.C., A possible bias in the California state geodimeter data, J. Geophys. Res., 80, 4078-4088, 1975.

Savage, J.C., Comments on "Strain accummulation and release mechanism of the 1906 San Francisco earthquake by Wayne Thatcher, J. Geophys. Res., 83, 5487-5489, 1978.

Savage, J.C., and Burford, R.O., Discussion of paper by C. H. Scholz and T. J. Fitch, 'Strain accumulation along the San Andreas fault', Jour. Geophys. Res., 78, 832-845,1971.

Savage, W.U., and McNally, K.C., Moderate earthquake seismicity in central California 1936-1973 (abs), Earthquake Notes, 45, 33, 1974.

Scholz, C.H., The frequency-magnitude relation of microfracturing in rocks and its relation to earthquakes, Bull. Seism. Soc. Amer., 58, 399-415, 1968.

Shimazaki, K., and Nakata, T., Time predictabel recurrence model for large earthquakes,

Geophys. Res. Lett., 7, 279-282, 1980.

Templeton, E.C., The central California earthquakes of July 1, 1911., Bull. Seismol. Soc. Amer., 1, 167-169, 1911.

Thatcher, W., Strain accumulation and release mechanism of the 1906 San Francisco earthquake, J. Geophys. Res., 80, 4862-4872, 1975a.

Thatcher, W., Strain accumulation on the northern San Andreas fault zone since 1906, J. Geophys. Res., 80, 4873-4880, 1975b.

Thatcher, W., Reply to comments by J.C. Savage, J. Geophys. Res., 83, 5490-5492, 1978.

Thatcher, W., Systematic inversion of geodetic data in central California, J. Geophys. Res., 84, 2283-2295, 1979.

Tocher, D., Seismic history of the San Francisco region, Calif. Div. Mines Special Rep 57, 39-48, 1959.

Toppozada, T.R., Earthquake magnitudes as a function of intensity data in California and western Nevada, Bull. Seism. Soc. Amer., 65, 1223-1238, 1975.

Toppozada, T.R., Real, C.R., Bezore, S.P., and Parke, D.L., Compilation of pre-1900 California earthquake history, Calif. Div. Mines Geol., Open File Release 79-6, 271 pp, 1979.

Toppozada, T.R., Real, C.R., Bezore, S.P., and Parke, D.L., Preparation of isoseismal maps and summaries of reported effects for pre-1900 California earthquakes, Calif. Div. Mines Geol. Open File Release 80-15, 78pp, 1980.

Townley, S.D., and Allen, M.W., Descriptive Catalog of earthquakes of the United States 1769 to 1928, I. Earthquakes in California 1769-1928. Bull. Seism. Soc. Amer., 29, 21-252, 1938.

Turcotte, D.L., Stress accumulation and release on The San Andreas fault, Pageoph, 115, 413-427, 1977.

Weeks, J., Lockner, D., and Byerlyee, J., Change in b-value during movement on cut surfaces in granite, Bull. Seism. Soc. Amer., 68, 333-341, 1978.

Webber, G.E., and Cotton, W.R., Geologic investigation of recurrence intervals and recency of faulting along the San Gregorio fault zone, Final Tech. Rep., U.S.G.S. Grant 14-08-0001-16822, 135 pp, 1980.

Webber, G.E., and Lajoie, K.R., Late Pleistocene and Holocene tectonics of the San Gregorio fault zone between Moss Beach and Point Ano Nuevo, San Mateo County, California (abs), Geol. Soc. Amer. Abst. with Prog., 9, 524, 1977.

Willis, B., Earthquake risk in California 8. earthquake districts, Bull. Seismol. Soc. Amer., 14, 9-25, 1924.

Wood, H.O., On the region of origin of the central California earthquakes of July, August and September, 1911, Bull. Seism. Soc. Amer., 2, 31-39, 1912.

SEISMICITY PATTERN IN THE SOUTH ICELAND SEISMIC ZONE

Páll Einarsson, Sveinbjörn Björnsson, and Gillian Foulger.

Science Institute, University of Iceland, Reykjavík.

Ragnar Stefánsson and Thórunn Skaftadóttir.

Icelandic Meteorological Office, Reykjavík.

Abstract. The South Iceland Seismic Zone is an E-W trending zone of destructive, historic earthquakes, that takes up transform motion between the submarine Reykjanes Ridge and the Eastern Volcanic Zone in Southern Iceland. Major earthquake sequences affecting most of the 70 km long seismic zone recur at intervals ranging between 45 and 112 years. The sequences often begin with a large event $(M \geq 7)$ in the eastern part of the zone followed by similar or slightly smaller events farther west. Single, damaging shocks occur more frequently, and are usually located at the ends of the zone. In spite of the clear E-W alignment of the epicenters, no surface evidence can be found for a major E-W striking fault. En echelon arrays of faults and fissures are found, however, indicating right-lateral strike-slip along northerly striking faults. Destruction zones of individual earthquakes also tend to be elongated in the N-S direction. The distribution of recent microearthquakes and the surface faulting during past large earthquakes seem to indicate that the seismicity is associated with brittle deformation of a 10-20 km wide zone located above an E-W trending zone of aseismic deformation in the lower crust or upper mantle.

No major earthquake sequence has occurred since 1896, and the whole zone has been very quiet for the last 50 years of instrumental observation. Judging from past history there is a high probability of a large earthquake occurring in the South Iceland Seismic Zone within the next few decades. A modest effort of earthquake prediction has been initiated, including radon monitoring of eight geothermal wells and the operation of volumetric strainmeters at seven sites.

Introduction

The majority of destructive earthquakes in Iceland since its settlement in the ninth century A.D. occurred within the South Iceland Seismic Zone. In this paper we review the historical record of seismic activity in this zone, and describe its present state and type of faulting. We then give a short account of current efforts of earthquake prediction. From the historical record and the length of the present quiet period we conclude that a major earthquake sequence is likely to occur in this zone within the next few decades. The sequence is expected to begin with an event of magnitude about 7 in the eastern part of the zone with subsequent events of somewhat smaller magnitude farther west.

Tectonic setting

The South Iceland Seismic Zone is a part of the mid-Atlantic plate boundary that crosses Iceland. The boundary follows the crest of the Reykjanes Ridge southwest of Iceland and the crest of the Kolbeinsey Ridge to the north. Within Iceland the boundary is displaced to the east by two features that resemble transform faults in many respects, one in South Iceland, the other near the north coast. The transform fault characteristics are the high seismicity, occurrence of large earthquakes, the geometric relationship to the spreading axes, and the type and sense of faulting (Fig. 1). Both zones, on the other hand, lack the clear topographic expression characteristic of fracture zones on the ocean floor, and in neither zone is the transform motion taken up by a single, major fault. The diverging plate boundaries in Iceland are similarly complex. In South Iceland, for example, spreading appears to occur in two parallel rift zones, the Western and the Eastern Volcanic Zones (Fig. 1 and 2). On the Reykjanes Peninsula the plate boundary is oblique, and the vector of relative plate motion probably has a component of separation as well as a strike-slip component (Klein et al., 1973, 1977). The peninsula is an area of high seismic as well as volcanic activity, which is not usual for other

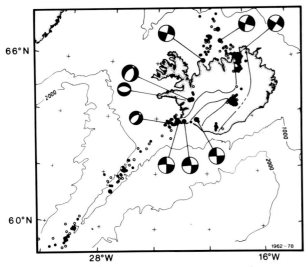

Fig. 1. Epicenters and focal mechanisms of
earthquakes in the Iceland region. Epicenters
are from the PDE lists of USCGS, later NOAA and
USGS, for the time period 1962-1978. Open circles denote epicenters determined with fewer
than 10 P-wave readings or epicenters of earthquakes smaller than m_b=4.5 Dots are events of
m_b=4.5 and larger, located with 10 or more readings. Large dots are events of m_b=5.0 and
larger. The focal mechanisms are shown schematically as lower hemisphere equal-area projections, compressional quadrants black. The volcanic rift zones of Iceland are shown.

active portions of the plate boundary in Iceland.
Near 21°W the boundary divides into the Western
Volcanic Zone, where the seismicity is only moderate, and the South Iceland Seismic Zone where
volcanic activity is nearly absent. The South
Iceland Seismic Zone bridges the gap between the
two volcanic rift zones in South Iceland, and
takes up the transform motion between the Reykjanes Ridge and the Eastern Volcanic Zone.
Stefánsson (1967) suggested that the large earthquakes in South Iceland were associated with
shear movements, which was later supported by
the distribution of microearthquakes (Ward et
al., 1969) and two focal mechanism solutions
(Ward, 1971). Ward et al. (1969) and Ward (1971)
used the term "Reykjanes Fracture Zone" for the
whole zone from the western tip of the Reykjanes
Peninsula to the Eastern Volcanic Zone. It has
since been customary to divide this zone according to tectonic characteristics into the Reykjanes Peninsula and the South Iceland Seismic Zone
(Tryggvason, 1973, Einarsson, 1979), which extends eastwards from the junction with the Western Volcanic Zone near Hengill (Fig. 2).
    Volcanism in the Eastern Volcanic Zone contin-
ues south of its junction with the South Iceland
Seismic Zone, but it does not seem to be associated with substantial rifting. Features normal-

ly associated with the axial rift zones, such
as fissure swarms, normal faults and long erup-
tive fissures, are lacking, and the volcanic
products are chemically different from the
tholeiites of the axial zones (Jakobsson 1972,
1979, Saemundsson, 1978).
    North of the transform zone recent rifting
structures occur in both the Eastern and the
Western Volcanic Zones indicating that crustal
spreading takes place in both zones. Heat-flow
and structural data suggest that most of the
crustal material in this region is produced by
the Western Zone and that the Eastern Zone may
be a relatively recent feature (Pálmason and
Saemundsson, 1974). In the postglacial time both
zones have produced equal amounts of lava
(Jakobsson, 1972), and in historical time only
the Eastern Zone has been volcanically active.
This may be taken to indicate that a major change
in the configuration of the plate boundaries in
Southern Iceland is occurring. The locus of
plate divergence seems to be shifting from the
Western to the Eastern Volcanic Zone. This interpretation implies that the South Iceland Seismic Zone is a young feature, even on the time
scale of tens of thousands of years.

The historical record

    Written accounts of earthquakes exist for most
of the time since Iceland was settled in the
ninth century A.D., although not contemporary
before the late twelfth century. The quality
and reliability of these accounts is quite variable. In some cases the writer himself witnessed the earthquake, in other cases he lived in a
different part of the country. Some of the reports were written a few hundred years after the
event. Most of the reports are very short and
unspecific, especially for the first centuries.
Thorvaldur Thoroddsen collected all reports
available to him in his books on Icelandic earthquakes and volcanoes (Thoroddsen 1899, 1905,
1925). Sigurdur Thorarinsson made a further
study of the written documents and assembled a
table of earthquakes strong enough to cause the
collapse of houses (in Tryggvason et al., 1958),
considered by him to indicate shaking of intensity VIII or more. Further studies by Sveinbjörn
Björnsson (1975, 1978) have revealed more information on the destruction zones of individual
earthquakes (see Fig. 2), and reports of an
earthquake in 1828 were found to be incorrect.
    Table 1 lists destructive earthquakes in the
South Iceland Seismic Zone and shows where damage and faulting occurred. It is largely based
on Thorarinsson's table, but only the districts
along the main epicentral belt are shown. The
event of 1828 is omitted, an event in 1389 is
added. The latter event is mentioned in one
account of questionable reliability.
    Annals and reports before 1700 are considered
to be incomplete with respect to destructive
earthquakes. It is noteworthy, in particular,

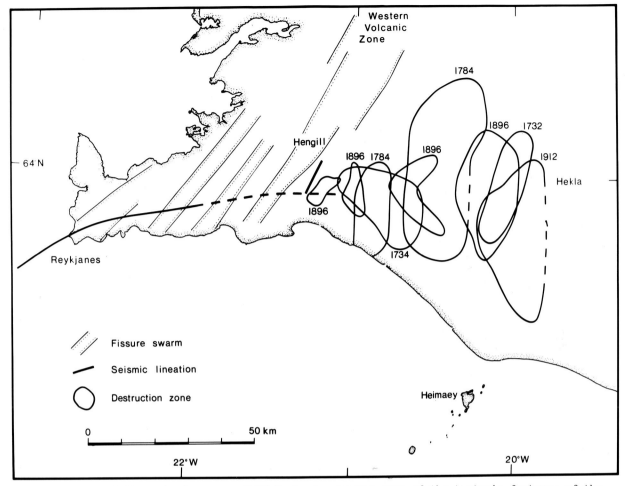

Fig. 2. Map of the seismic zones of SW-Iceland, showing some of the tectonic features of the Reykjanes Peninsula and the South Iceland Zone. The destruction zones of the earthquakes of 1732-34, 1784, 1896 and 1912 are shown. Within these zones more than 50% of houses at each farm were totally ruined. Corresponding intensity is MM VIII - IX.

that no earthquakes are mentioned in the fifteenth century, for which reports of other kinds of natural disasters are also conspicuously lacking (Tryggvason et al., 1958). There is no reason to believe that this was an unusually quiet period.

The events of 1784 and 1896 stand out in Table 1 because of the widespread destruction they caused. In both cases the damage was caused by a series of earthquakes and not by a single shock. The first and largest shock of 1784 occurred on August 14, and caused damage in an extensive zone centered on the districts Skeid and Holt. Two days later another shock, somewhat smaller than the first one, caused damage mostly in Flói. In 1896 the first earthquake struck in the evening of August 26 and caused destruction in a zone centered on the Land district. Another shock, probably about the same size, occurred the following morning. The highest intensities appear to have been in the same areas as during

the first shock and the damage zone was extended slightly to the north. On September 5 damage was caused farther west by what appears to have been a double earthquake. The destruction zones are clearly divided; one is centered on Skeid and Holt, the other on Flói and Ölfus. Felt reports (Thoroddsen, 1899) and descriptions of the seismograms from Strassburg (Rudolph, 1903) support the same. Further damage was caused by shocks on September 6 in Ölfus and on September 10 in Flói.

The magnitude of the South Iceland earthquakes has been a matter of some dispute. The only instrumentally determined magnitude is that of the 1912 event. It is assigned a surface wave magnitude of 7 by Kárník (1969). Tryggvason (1973) used the radius of destruction to estimate the magnitude of other earthquakes, and concluded that the largest event of 1784 probably exceeded magnitude 7.5. Similarly he found a magnitude of 7-7 1/2 for the largest event of 1896. Other

Table 1. Destruction and faulting during large earthquakes in Southern Iceland.

Year	:Olfus	Flói	Grimsnes	Skeid	Holt	Hreppar	Land	Rangárvellir
1164	X	X	+	..	..	..	..	o o
1182	?	?	?	..	..	..	?	?
1211	?	?	?	..	..	..	?	?
1294	..	..	..	0	..	..	X	⊕
1300	..	..	..	..	..	..	..	+
1308	?	?	?	..	..	..	?	?
1339	..	+	..	+	⊕	..	..	+
1370	+	..	..	..	..	..	..	..
1389	..	..	..	..	..	..	X	X
1391	+	⊕	+	..	..	..	..	..
1510	..	..	..	..	..	..	X	X
1546	+	..	..	..	..	..	X	X
1581	..	..	..	..	..	..	..	+
1597	+	..	..	..	..	..	..	..
1614	X	X	X	+	..	..	..	..
1624	..	+	..	..	..	..	..	..
1630	..	..	..	..	X	+	⊕	..
1633	+	..	..	..	..	..	..	..
1657	..	+	..	..	..	..	..	..
1671	+	..	+	..	..	..	..	..
1706	+	+	..	..	..	..	..	..
1725	..	..	..	..	..	..	..	+
1726	..	..	..	..	..	..	..	+
1732	..	..	..	..	..	+	+	+
1734	+	+	+	..	..	..	..	..
1749	+	..	..	..	..	..	..	..
1752	+	..	..	..	..	..	..	..
1766	+	..	..	..	+	..	+	..
1784	+	+	+	+	⊕	+	+	+
1789	+	..	..	..	..	..	..	..
1829	..	..	..	..	..	..	+	+
1896	+	+	+	⊕	+	+	⊕	+
1912	..	..	..	..	..	..	⊕	⊕

+ Annals state that houses fell in this district.
X It is inferred from annals that houses fell in the district.
? Casualties indicate that houses collapsed, most likely in these districts.
O Surface faulting mentioned or known.

The districts are shown on a map in Fig. 6.

events were assigned smaller magnitudes. Comparing the destruction zones in Fig. 2 one may find Tryggvason's estimates a little high. Stefánsson (1979) estimated a magnitude of 7.1 for the largest 1784 earthquake.

After 1700 the quality of the earthquake accounts improves, and it becomes possible to map the extent of the destruction zones. The zones in Fig. 2 are areas of more than 50% destruction, i.e. within their boundaries more than 50% of all buildings at each farm were ruined. There are several points to be emphasized about Fig. 2:
1. The destruction zones are arranged in an E-W trending zone.
2. Each destruction zone is elongated, with the long axis usually oriented transversely to the main seismic zone.
3. If the area of the destruction zone is a measure of the magnitude of the earthquake, the largest event is the one of 1784. Most other events are equal to or smaller than the 1912 earthquake.

4. The largest earthquakes tend to occur in the eastern half of the seismic zone.

The large, historic earthquakes can be tentatively grouped into three categories according to location and time sequence.

Category I: Earthquakes at the eastern end of the zone (districts Rangárvellir and Land), near the junction with the Eastern Volcanic Zone. The earthquakes of 1912, 1829, 1726, 1581 belong to this category. One may also want to count here earthquakes associated with the eruptions in the vicinity of Hekla volcano in 1913, 1878, 1725 and 1554. Earthquakes directly associated with eruptions of Hekla itself generally tend to be smaller (Thórarinsson, 1967). The largest earthquake associated with the beginning of the Hekla eruption in 1947 was of magnitude 5 1/4 and did not cause any damage (Tryggvason, 1978b). Earthquakes during the eruptions of 1970 and 1980 were much smaller.

Category II: Earthquakes at the western end of the zone (districts Ölfus and Flói), near the junction with the Western Volcanic Zone, such as the events of 1789, 1766, 1752, 1749, 1706, 1671, 1597, 1546, 1370. One may also want to count here the earthquakes of Oct. 9, 1935 (M=6) and April 1, 1955 (M=5.5), which caused only minor damage in Ölfus (Tryggvason, 1978a, 1979). It is noteworthy that the earthquakes of 1766 and 1597 occurred during eruptions of Hekla at the other end of the seismic zone. Similarly during the 1947 Hekla eruption an earthquake swarm (maximum magnitude 4 1/4) occurred in Ölfus (Tryggvason, 1978b). In spite of the small magnitude of the events, they caused some damage to houses and were associated with surface faulting and changes in hot springs. The earthquakes of 1789 were accompanied by appreciable rifting of the Hengill fault swarm in the Western Volcanic Zone.

Category III: Sequences of earthquakes that seem to affect most of the zone. The sequence starts with a large earthquake in the eastern part of the zone followed by large, but slightly smaller events farther west. These later events are not aftershocks in a strict sense, since they occur outside of the area affected by the first large event. The sequences of 1896, 1784 and 1732-34 belong to this category, possibly also the earthquakes in 1630-33 and 1389-1391. The duration of the major sequence may be quite variable, 2 days in 1784, 2 weeks in 1896 and 2 years in 1732-34. The earthquakes of 1294 were accompanied by faulting and damage in an area extending from the district Skeid to Rangárvellir. Similarly, the earthquake in 1339 appears to have affected the large area between Flói and Rangárvellir. These large macroseismic areas may indicate that more than one earthquake occurred in each case, and that these events belong to category III.

This categorization, which includes most earthquakes for which enough information is available, has some important implications in terms of earthquake forecasting.

1. Major earthquake sequences, affecting most of the seismic zone (category III), occur at the average rate of once per century. The intervals between sequences are 112, 50, 99, 239, 50 and 45 years. One should note, however, that the 239 years interval includes the fifteenth century, for which the historical record is known to be incomplete with respect to other kinds of events. The average rate found is therefore more likely to be too low than too high. Excluding the 239 years interval, the longest one is 112 years. With all the statistical limitations of 7 samples, one may conclude that there is more than 80% probability of a major earthquake sequence occurring within 112 years of the last one, i.e. before the year 2008.

2. A large earthquake in the eastern half of the seismic zone is likely to be followed by large earthquakes in the western part a few days to a few years later.

3. A few (2-4) earthquakes may occur at either end of the seismic zone (categories I and II) between two successive major sequences.

4. Prior to a major sequence premonitory effects of long and intermediate duration are only expected to occur in the eastern part of the zone. In the western part of the zone short-term effects are only to be expected after a large event has already occurred in the eastern part.

Recent activity

Continuous seismic recording began in Iceland in 1925 (Tryggvason, 1979). The first station was in Reykjavík and had two horizontal instruments with a magnification of about 100. The instrumentation was improved in 1951, when a vertical short-period Sprengnether pendulum with a magnification of up to 4000 was added. The old horizontal instruments were installed in Akureyri in North Iceland in 1954 and in Vík in South Iceland in 1955. A short-period station was added in Sída (SID) in SE-Iceland in 1958 and in Egilsstadir (EGI) in 1967. A WWSSN station has been in operation in Akureyri (AKU) since 1964. None of these stations was located within 40 km of the South Iceland Seismic Zone. Since 1974 there has been a great increase in the number of short-period seismograph stations distributed throughout the active zones of Iceland. The stations shown in Fig. 3 were established in 1972 (LV), 1974 (AR), 1976 (SL), 1977 (IR) and 1978 (HL). The station ST has been in intermittent operation since 1976.

Epicentral locations for the last 5 years are shown in Fig. 3. The largest earthquake during this period reached magnitude 4, but earthquakes as small as magnitude 1.5 are included in the map. The most prominent concentration of seismicity occurs in the Hengill area near the western edge of the map. This activity, described in some detail by Foulger and Einarsson (1980), is related to the high temperature geothermal

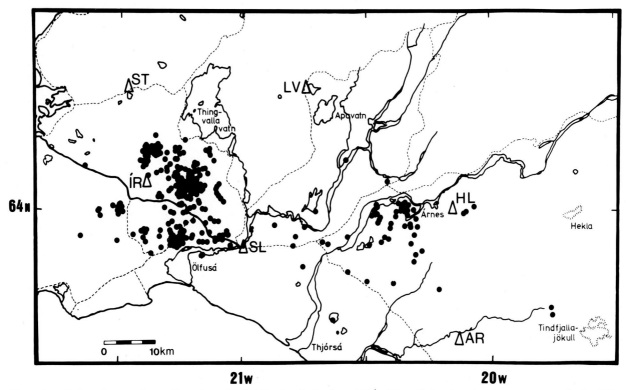

Fig. 3. Epicenters of earthquakes in SW-Iceland during 1974-79. Seismic stations are marked with triangles. Roads (dashed lines), lakes (-vatn) and rivers (-á) are shown for orientation.

area and the tectonic complexities near the Hengill central volcano. Its connection with the South Iceland Seismic Zone is unclear. Farther east the seismicity is very diffuse and only broadly delineates the seismic zone. The epicenters are clearly not associated with a single fault. It is noteworthy that large parts of the seismic zone are very quiet, even at the microearthquake level. A cluster of epicenters is found near Árnes in the central part of the zone. Most of these events, including the largest event in this map, occurred in several small sequences in 1978. The distribution of epicenters within the cluster vaguely suggests a N-S structure.

The seismicity of the South Iceland Seismic Zone since 1930 is summarized in a time-space plot in Fig. 4. The lowering of the detection threshold with time is quite apparent in this figure. It is close to magnitude 3 1/2 for the first decades but approaches magnitude 2 for the last years. In spite of this, the main features of the seismicity are demonstrated. In general the seismicity is very low. Only 5 events barely reach magnitude 5. Earthquakes that occurred in July 1967 are the only ones in the zone for which a focal mechanism solution is available. The solution is poorly constrained (Ward, 1971) but shows strike-slip faulting (see Fig. 1). The relatively high background activity of the area around Hengill stands out in this figure as in

Fig. 3. A remarkable feature of the activity in the central part of the zone is its tendency to come in bursts. Such bursts occurred in 1970 and 1978, for example. The activity is not limited to one spot. Earthquakes occur over a wide area within each burst of activity. The source volumes of the earthquakes are too small for them to directly influence the occurrence of each other. It seems more likely that the events are triggered by a regional strain pulse that affects a large part of the seismic zone. Some of the events of 1978 were preceded by a detected radon anomaly (Hauksson and Goddard, 1981).

Faulting

Evidence of recent faulting is found in numerous localities in the seismic zone. Historical accounts mention surface breakage associated with a few earthquakes, but the descriptions are generally short and in only very few cases are they specific enough to allow identification of the faults in the field. Most accounts prior to 1900 are collected in Thoroddsen's (1899) book, where faulting is mentioned in association with the earthquakes of 1294, 1308, 1339, 1391, 1630, 1784 and 1896. In addition, fairly detailed descriptions of faulting are available in newspaper reports of the earthquake of 1912.

Bedrock outcrops are relatively scarce in the

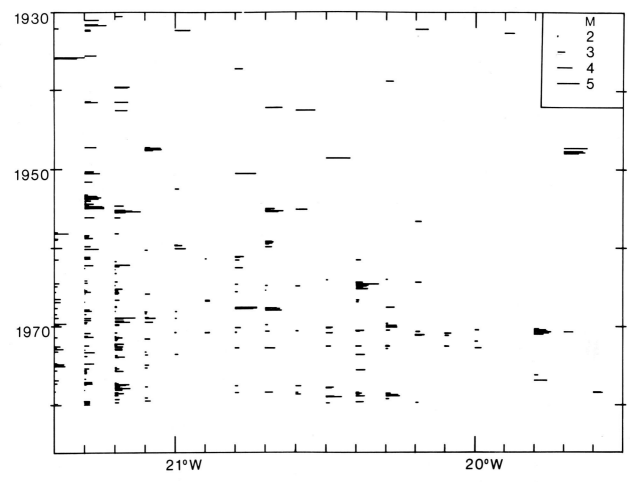

Fig. 4. Time-space diagram of seismic activity in the South Iceland Seismic Zone in the time interval 1930-80. The time of occurrence of the earthquake is given as a function of longitude of the epicenter. The length of the bar gives the magnitude of the event.

South Iceland Seismic Zone. Most of the area is covered by thick soil, glacial moraines, aa lavas and glaciofluviatile material. The appearance of the faults depends strongly on the nature of the surface material where they are exposed. They are most easily identified where 1-3 m of soil lies on top of lava, which is a common condition in the western part of the zone, in the districts Flói and Skeid. The faults become obscured where the loose material is very thick as in the moraines in the district Holt, in the central part of the zone. Farther east, in the district Land, most of the faults are exposed in soil-covered lava. In one place a fault cuts through the smooth surface of a pahoehoe lava.

In spite of their variable appearance, there are a few features characteristic of most of the faults, that make them identifiable in the field. Cases of doubt admittedly arise, especially where the fault traces merge with brooks, cattle tracks and old bridle paths. Flow structures in the lavas can also be deceptive. The most prominent

features of the faults are tension gashes, that are arranged in en echelon arrays (Fig. 5). The gashes often appear as fissures at the surface. In many instances the turf layer has not ruptured but has sagged down to form an elongated depression. Individual gashes may be tens of centimeters to tens of meters long. The en echelon pattern can be seen on many scales, ranging from meters to a kilometer, even along the same fault and superimposed on each other. The 1912 earthquake, for example was accompanied by faulting along several fault segments of a few hundred meters to over a kilometer in length. All the segments except one are arranged en echelon in a north trending zone (Fig. 6). Each segment is composed of shorter en echelon segments, some of which also consist of en echelon gashes down to a scale of meters. In other cases the en echelon pattern is relatively uniform as is shown near the center of Fig. 5, probably reflecting uniform conditions near the surface, in this case vegetated lava. These faults may be the faults

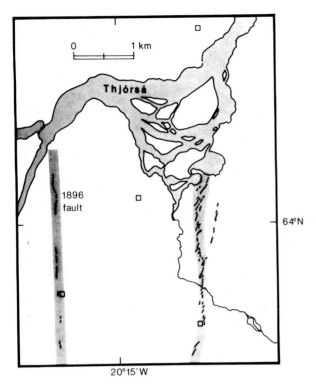

Fig. 5. Surface faulting during two earthquakes. One occurred on August 26 or 27, 1896, the other fault was probably active in the earthquake of 1630.

formed in the earthquake of 1630, cited by Thor-oddsen (1899).

The fault traces are in many cases marked by small mounds of soil and broken rock. The mounds are usually located between the tips of two en echelon tension gashes, and have apperently been pushed up during the fault movement. The top of some of the mounds has been cut into slices sub-parallel to the direction of the neighboring gashes. In a few areas, where the tension gash-es are obscured due to local conditions, the row of mounds is the primary indicator of the fault trace in the field.

The age of the faults is uncertain, with a few exceptions. The 1912 faults can be identified, as well as the faults in the Land district asso-ciated with the earthquakes of the 1896 sequence. Faults also moved in the Skeid district, probably during the event on Sept. 5th, 1896, but these are now hard to distinguish from the numerous older faults in the same area. The faults pre-sumed to be from the 1630 earthquake were men-tioned earlier. Interestingly, these seem to be as well preserved as the ones formed more than 250 years later. It is of great importance to date the faults, but this may prove to be diffi-cult. Tephrochronology is probably the most promising tool for this purpose. One may hope to obtain a seismic history for a good part of

the postglacial time, and thus be able to deter-mine more reliably the recurrence times of earth-quakes. An important question is also, whether each fault moves more than once. So far no evi-dence has been found to indicate repeated move-ment on the same fault.

All faults that have been mapped so far are shown in Fig. 6, as far as the scale of the map allows. More faults will undoubtedly be added to this map later since only parts of the zone have been systematically searched.

All the en echelon arrays identified so far are oriented in a northerly direction, a few slightly towards the NNE. Individual tension gashes have a NE trend, sometimes quite variable between N and N60°E. The en echelon arrange-ment is nearly always in the same sense, and indicates right-lateral displacement along an underlying fault. Vertical displacements are very rarely seen, except locally within an array of tension gashes where blocks have been tilted, downthrown or pushed up in the fault mounds described above. In one case the sense of motion along the fault can be determined directly from a manmade struc-ture. A wall built of turf and stones crosses the trace of the fault active in the earthquakes of August 26-27, 1896 near the farm Lunansholt in the district Land. The wall is displaced 75-80 cm right-laterally where it crosses the fault.

The structure and the appearance of the frac-tures described here is different from that of the fissures and faults of the volcanic rift zones of Iceland. There the fractures are grouped into swarms that have trends similar to that of the fractures themselves. En echelon arrangements can be found within the swarms, but the sense is not always the same and often depends on the strike of the fault with respect to the trend of the swarm. Each swarm has the structure of a graben, and vertical displacements across the fissures are common. The fractures are active during rifting episodes similar to the current events in the Krafla fault swarm in NE-Iceland, and the displacements have been shown to be in the form of widening of the fissures and subsi-dence of the graben floor (Gerke et al., 1978, Björnsson et al., 1979, Möller and Ritter 1980, Sigurdsson, 1980).

Faults of the South Iceland Seismic Zone were previously described by Tr. Einarsson (1967, 1968) and Tryggvason (1973) who both interpret them as right-lateral strike-slip faults. Tr. Einarsson, however, only looked at individual segments, which have a more north-easterly trend than the overall fracture pattern, and generalized his interpretation to the whole of the Icelandic rift zones, where individual fissures have similar trends.

The faults of the South Iceland Seismic Zone shown in Fig. 6 reveal a pattern which resembles that of the earthquake destruction zones in Fig. 2: The faults delineate an E-W trending zone, whereas each individual fault array has a N-S trend. The earthquakes seem to be associated

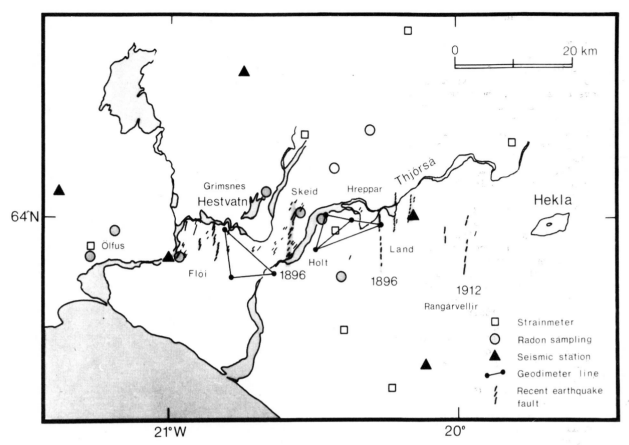

Fig. 6. A summary map of the South Iceland Seismic Zone, showing the surface fault breaks,
seismic stations, strainmeter locations, radon sampling sites and the names of the major
earthquake districts. Rivers and lakes are shown for orientation. Faults in the western
part of the zone are from an unpublished map by Helgi Torfason.

with brittle deformation of a 10-20 km wide zone
overlying an E-W trending zone of aseismic deformation in the lower crust or upper mantle. The
displacement along the zone is left-lateral
transform motion to accommodate the spreading in
the Eastern Volcanic Zone, but the brittle crust
responds by right-lateral faulting along the conjugate planes. The reason for this behaviour is
not clear, but it might be related to the apparent youth of this zone. It is possible that the
finite displacement along the zone is not yet
sufficient to break the whole crust and produce
a major throughgoing fault.

Current prediction research

Recent advances in earthquake prediction have
motivated effort to collect data that might be
useful in future prediction work in South Iceland.
Earthquake prediction has high significance in
this main farming area of Iceland. Although the
loss of life has not been great in past earthquakes, they have caused extensive damage and
economic loss. A large earthquake would doubt-

less have a serious impact on the modern community now in this area.

The possibility mentioned above, that the seismicity might be related to strain pulses of regional extent, gives hope that the earthquakes are
preceded by precursory effects detectable over
considerable areas. Effects such as foreshocks
and abnormal animal behaviour, given much attention in China (Raleigh et. al., 1977, Wallace and
Teng, 1980), are probably not of much value here.
No abnormal foreshock activity is reported prior
to any of the large historic earthquakes, and
stories of abnormal animal behaviour prior to
earthquakes are rare.

Conditions for geophysical monitoring work are
fairly favourable in South Iceland. The area is
flat, and all parts of it are easily accessible
by roads. Bedrock outcrops can be found in most
parts, and a large number of abandoned drill
holes is available.

The current prediction program is centered
around the monitoring of seismicity, radon and
strain. The instruments and sampling are mostly
taken care of by local people, thus forming a

link between the scientists and the local population.

Seismicity. In addition to the 6 seismic stations shown in Fig. 3 there are 9 stations immediately outside of the map area. With the exception of the station in Reykjavik, all the stations have a vertical, short-period seismometer. The signals are recorded with pen and ink on paper. The visible recording is an important feature of a monitoring system in a volcanic area. Radio time is used on all the stations, thus eliminating clock corrections. The set of locatable earthquakes is complete above magnitude 2, but events down to at least magnitude 1 are detected.

Radon. Discrete sampling of geothermal fluid for radon determination was initiated at 8 sites in South Iceland in 1978. A description of this program is given by Hauksson and Goddard (1981). Samples are taken every 1-2 weeks and sent to Reykjavik for analysis. The present sampling sites are shown in Fig. 6. Several anomalies have been observed prior to earthquakes, and the northernmost station (Flúdir) appears to be most sensitive to radon changes. A continuous radon meter has now been installed at that station. This project is run in cooperation between the Science Institute and Lamont-Doherty Geological Observatory. The very promising results of this work have provided stimulus for further prediction research.

Strain. Sacks-Evertson volumetric strainmeters were installed in 1979 at 7 sites (Fig. 6). After a settling period of about a month all instruments showed strain build-up at a more or less constant rate. This project is run in cooperation between the Icelandic Meteorological Office and the Carnegie Institution in Washington.

Other projects. The lake Hestvatn (Fig. 6) is used as a reference level for the monitoring of tilt changes in the central part of the seismic zone. Several benchmarks have been installed around the lake in cliffs directly above the water. Changes in relative lake level can be measured directly with a ruler. The measurements have to be made on calm days. A tilt resolution of 10^{-6} can be obtained in this way. Water level gauges are also operated around the lake by Egill Hauksson and Roger Bilham of Lamont-Doherty Geological Observatory for the same purpose.

Several geodimeter lines have been measured. The intent is to remeasure these lines and extend the geodimeter network over the whole zone in coming years.

Acknowledgements. Numerous individuals have contributed to the research described in this paper, designed instruments, operated monitoring stations, mapped faults, analyzed seismic records etc. Future work in South Iceland depends critically on continuing cooperation with these people. Dr. Helgi Torfason allowed the use of his unpublished map of faults in the western part of the seismic zone.

The construction and operation of some of the seismic stations was financed by the National Energy Authority of Iceland and by NATO Research Grant No. 715.

Sigurdur Thorarinsson, Eysteinn Tryggvason, and two anonymous reviewers made helpful suggestions for the improvement of the manuscript.

References

Björnsson, A., G. Johnsen, S. Sigurdsson, G. Thorbergsson and E. Tryggvason, Rifting of the plate boundary in north Iceland 1975-1978, J. Geophys. Res., 84, 3029-3038, 1979.

Björnsson, S., Earthquakes in Iceland (in Icelandic with English abstract), Náttúrufraedingurinn, 45, 110-133, 1975.

Björnsson, S., In: Large Earthquake in South Iceland, Report of a working group to the Civil Defense of Iceland (in Icelandic), 54pp., Reykjavik, 1978.

Einarsson, Páll, Seismicity and earthquake focal mechanisms along the mid-Atlantic plate boundary between Iceland and the Azores, Tectonophysics, 55, 127-153, 1979.

Einarsson, Tr., The Icelandic fracture system and the inferred causal stress field, In: Iceland and Mid-Ocean Ridges (ed. S. Björnsson), Soc. Sci. Islandica, Publ. 38, 128-141, 1967.

Einarsson, Tr., Submarine ridges as an effect of stress fields, J. Geophys. Res., 73, 7561-7575, 1968.

Foulger, G., and P. Einarsson, Recent earthquakes in the Hengill-Hellisheidi area in SW-Iceland, J. Geophys., 47, 171-175, 1980.

Gerke, K., D. Möller, B. Ritter, Geodätische Lagemessungen zur Bestimmung horizontaler Krustenbewegungen in Nordost-Island, Wissenschaftliche Arbeiten der Lehrstühle für Geodäsie, Photogrammetrie und Kartographie an der Technischen Universität Hannover, Nr. 83, pp. 23-33, 1978.

Hauksson, E., and J.G. Goddard, Radon earthquake precursor studies in Iceland, submitted to J. Geophys. Res., 1981.

Jakobsson, S., Chemistry and distribution pattern of Recent basaltic rocks in Iceland, Lithos, 5, 365-386, 1972.

Jakobsson, S., Petrology of Recent basalts of the eastern volcanic zone of Iceland, Acta Nat. Islandica 26, 103pp., 1979.

Kárník, V., Seismicity of the European Area, Part 1, Reidel, Dordrecht, Holland, 1969.

Klein, F.W., Páll Einarsson and M. Wyss, Microearthquakes on the mid-Atlantic plate boundary on the Reykjanes Peninsula in Iceland, J. Geophys. Res., 78, p. 5084-5099, 1973.

Klein, F.W., P. Einarsson and M. Wyss, The Reykjanes Peninsula, Iceland, earthquake swarm of September 1972 and its tectonic significance, J. Geophys. Res., 82, 865-888, 1977.

Möller, D., B. Ritter, Geodetic measurements in the rift zone of NE-Iceland, J. Geophys., 47, 110-119, 1980.

Pálmason, G., and K. Saemundsson, Iceland in relation to the Mid-Atlantic Ridge, Annual Rev. Earth Planet. Sci., 2, 25-50, 1974.

Raleigh, B., G. Bennett, H. Craig, T. Hanks, P. Molnar, A. Nur, J. Savage, C. Scholz, R. Turner, F. Wu, Prediction of the Haicheng earthquake, Eos, Trans. Am. Geophys. Union, 58, 236-272, 1977.

Rudolph, E., Seismometrische Beobachtungen (1889-1896), Beiträge zur Geophysik, 5, 94-170, 1903.

Saemundsson, K., Fissure swarms and central volcanoes of the neovolcanic zones of Iceland, Geol. J. Special Issue No. 10, 415-432, 1978.

Sigurdsson, O., Surface deformation of the Krafla fissure swarm in two rifting events, J. Geophys., 47, 154-159, 1980.

Stefánsson, R., Some problems of seismological studies on the Mid-Atlantic Ridge, In: Iceland and Mid-Ocean Ridges (ed. S. Björnsson), Soc. Sci. Islandica, Publ. 38, 80-89, 1967.

Stefánsson, R., Catastrophic earthquakes in Iceland, Tectonophysics, 53, 273-278, 1979.

Thorarinsson, Sigurdur, The eruptions of Hekla in historical times. The eruption of Hekla 1947-1948, I, p. 1-170, Soc. Sci. Islandica, 1967.

Thoroddsen, Th., Large earthquakes in Iceland (in Icelandic), The Icelandic Lit. Soc., Copenhagen 269pp., 1899 and 1905.

Thoroddsen, Th., Die Geschichte der isländischen Vulkane, Mem. Royal Acad. Sciences, Denmark, Natural Sci. Div., Series 8, IX. 458pp., Copenhagen, 1925.

Tryggvason, Eysteinn, Seismicity, earthquake swarms and plate boundaries in the Iceland region, Bull. Seismol. Soc. Am., 63, 1327-1348, 1973.

Tryggvason, E., Earthquakes in Iceland 1930-39 (in Icelandic), Science Institute, University of Iceland, Report RH-78-21, 92pp., 1978a.

Tryggvason, E., Earthquakes in Iceland 1940-49 (in Icelandic), Science Institute, University of Iceland, Report RH-78-22, 51pp., 1978b.

Tryggvason, E., Earthquakes in Iceland 1950-59 (in Icelandic), Science Institute, University of Iceland, Report RH-79-06, 90pp., 1979.

Tryggvason, E., S. Thoroddsen, S. Thorarinsson, Report on earthquake risk in Iceland (in Icelandic), J. Engineer's Soc. Iceland, 43, 1-9, 1958.

Wallace, R.E., T.L. Teng, Prediction of the Sungpan-Pingwu earthquakes, August 1976, Bull. Seismol. Soc. Amer., 70, 1199-1223, 1980.

Ward, P.L., New interpretation of the geology of Iceland. Geol. Soc. Am. Bull., 82, 2991-3012, 1971.

Ward, P.L., G. Pálmason, and C.L. Drake, Micro-earthquakes and the mid-Atlantic ridge in Iceland, J. Geophys. Res., 74, 665-684, 1969.

Earthquake Hazard in the Hellenic Arc[1]

M. WYSS

CIRES and Department of Geological Sciences,
University of Colorado, Boulder, Colorado 80309

M. BAER

Institute for Geophysics, ETH-Hoenggerberg
CH-8093 Zuerich, Switzerland

Abstract. The seismicity of the Hellenic arc was studied as a function of space and time by two methods. First data on large and great historic earthquakes were reinterpreted with the purpose of identifying seismic gaps along this plate boundary. We found that the 8 great and 3 large earthquakes which occurred along the Hellenic arc between 1805 and 1926 can be interpreted as shallow sources along the plate boundary. Based on macroseismic data we propose approximate locations and size of these earthquakes. Individual rupture lengths of the largest events were approximately 100 to 150 km. It appears that the entire Hellenic plate boundary can be considered to be a seismic gap of class 1,2, and 3 depending on the segment. Second a systematic analysis of the shallow seismicity (< 100 km) showed that the rate decreased by about 80% in the western third and portions of the eastern third of the arc. These anomalies started in 1962 and 1966 respectively, they lasted to the present, and they can be interpreted as earthquake precursors. Combining the seismic gap and the seismicity rate data we suggest that ruptures of about 120 ± 40 km lengths (M = 7 3/4 ± 1/2) may be expected to occur along the Hellenic plate boundary near 22.5 to 23.5° E and 26.5 to 27.5° E between now (1980) and 1990. Along the other segments of the Hellenic arc we were not able to define seismicity anomalies. Nevertheless, based on estimates of the stored strain derived from the approximate slip-rate one may conclude that large earthquakes may be expected there also. At the present stage of rudimentary understanding of earthquake preparatory processes, predictions like the above may be changed or rendered obsolete by new discoveries which may show that assumptions or hypotheses used may not be valid.

[1] Contribution No. 304, Institut fuer Geophysik, ETH-Zuerich

INTRODUCTION

For earthquake prediction and hazard research in Europe the area surrounding the Aegean Sea should be a prime target, because this area has the highest seismicity rate in Europe. Therefore the chances of gathering data relevant to the prediction of earthquakes are highest there, and the need for earthquake hazard estimates is perhaps greatest. The tectonic framework in Greece and western Turkey is complex. This makes earthquake prediction research in the area difficult, because the starting point for such research should be an understanding of the tectonic processes causing the earthquakes there. We thought it might be best to start with studying the most clearly defined tectonic feature in the area.

The Hellenic arc-trench system is a subduction zone of about 1000 km length [e.g. Papazachos, 1973]. At its northwestern end this plate boundary is formed by a trench which runs 50 to 100 km off the Messinian coast of the Peloponnesus, south of which it continues to strike SE until south of Crete where it takes a sharp bend to the northeast (Figure 1). Off of eastern Crete and Carpathos two parallel narrow trenches, Pliny and Stabro, exist [e.g. LePichon et al., 1979] and a deep ocean bottom depression terminates the trenches off of Rodhos and the Turkish coast. The slip vector along this plate boundary as determined by McKenzie [1978] is shown by an arrow in Figure 1. The present pole of rotation between the Southern Aegean and the African plates is located near the heel of the Italian boot formed by the Apennines [LePichon and Angelier, 1979].

The purpose of this paper is to extract from historic seismicity data as much information as possible to allow inferences about the location, size and occurrence time of future large earthquakes along the Hellenic trench. Two types of seismicity studies are particularly useful for our

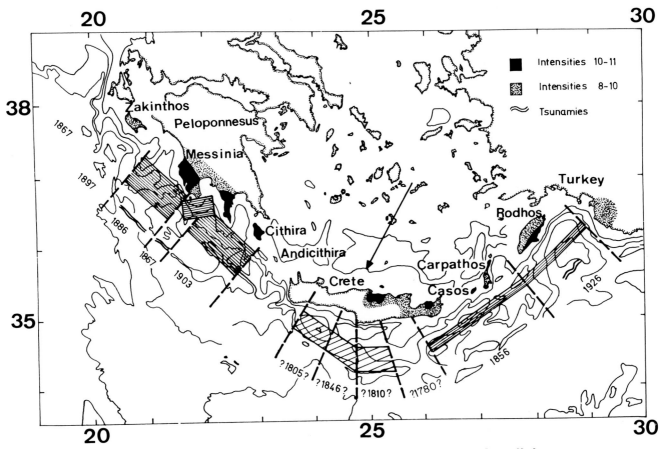

Figure 1. Map of the Hellenic arc showing the extent of fairly well documented ruptures by dark shading, and poorly documented ruptures by light shading. Question marks next to the dates indicate uncertainty of location. The slip direction (arrow) of the Aegean plate w.r. to Africa and the bathymetric contours are from McKenzie [1978]. The modified Mercalli intensities are from Sieberg [1932].

purpose: identification of seismic gaps and of periods of seismic quiescence. These two types of studies can jointly lead to an understanding of the seismicity regime, but they do not directly depend on each other. This paper can therefore be considered as consisting of two independent parts.

A seismic gap can be defined as a portion of a plate boundary which has not ruptured for a relatively long time in a large earthquake. We will follow the nomenclature of McCann et al [1979] and define a class 1 gap as a segment which has not broken for more than 100 years, and a class 3 gap as a segment for which evidence for historic rupture is not available. Class 2 gaps will represent segments which have not ruptured for more than 50 years (30 in McCann et al.). The expectation that seismic gaps should be filled in by great ruptures in the relatively near future [Fedotov, 1965] has been fulfilled in several cases during the last 15 years [McCann et al., 1979].

Periods of seismic quiescence can be defined by the numerous small earthquakes which occur con-

tinuously along most plate boundaries, within and outside of gaps. Mogi [1969] noticed this pattern for shocks of m ≥ 6, and recently seismic quiescence (m ≥ 4.5) has been demonstrated to last for several years within the source volumes before many large earthquakes [Ohtake et al., 1977; Wyss and Habermann, 1979; Habermann, 1981a]. When the background seismicity rate is high enough for statistical studies one can show that the rate drops by about 50% during the precursor time [Wyss and Habermann, 1979], and in one case it was shown that this anomaly occurred synchronously with two other independent precursory anomalies [Wyss et al., 1980]. Below we will attempt to define seismic gaps and seismicity quiescence if they exist in the Hellenic arc.

THE DEPTHS OF GREAT EARTHQUAKES IN THE HELLENIC ARC

Eight great earthquakes (M ≥ 7.8), and three large ones, of intermediate depths are commonly

Table 1. Macroseismic Data Summary for Great (M $\geq$7.8) and Large Earthquakes of the Hellenic Trench

Date	Location (severe damage)	I_o	Evidence for large size	Length[+]	Evidence for shallow depth	M*	Ref.
1805 July 3	Crete, Chania, Rethymnon	VIII	Felt in Sicily		Sequence of 4 strong shocks	7.6	S
1810 Feb 16	Crete, Heraklion, probably convent Asomatos	IX				8.2	S
1846 Mar 28	Crete, Heraklion, Chania	VII	Felt in Sicily, Italy, Syria, Egypt		Aftershocks until July, seaquake	8.1	S
1856 Oct 12	Crete, Casos, Carpathos, Heraklion, Sitia, Rodhos	XI	Felt in Italy, Syria, Palestine, Egypt	150	Steep isoseismal gradient, seaquake	8.6	S
1863 Apr 22	Rodhos, Massari	XI	Felt in Egypt			8.5	S
1867 Feb 4	Off Kefallinia					7.9	G67
1867 Sep 20	Messinia, Maina	IX		50	Large tsunami	7.6	G60,S
1886 Aug 27	Messinia Kyparissa Gargaliani	XI	Felt in Sicily Italy, Libya Egypt	100	Daily many aftershocks until Sep. 6, steep isoseismal gradient, large tsunami	8.4	G60,S
1897 May 28	Off Zakintos			100		7.6	G67
1903 Aug 11	Cythera, Peloponnesus, Mitata, Viaradica, Platsa	X	Felt in Sicily, Italy, Egypt Istanbul	100	Steep isoseismal gradient, 200m long cracks in ground striking NW-SE	8.3	S
1926 Jun 26	Rodhos, SW-Turkey, Archangelo, Fetich, Carpathos, Heraklion	XI	Felt in Sicily, Italy, Switz., Syria, Palestine, Egypt	100	Foreshock swarms since Jan. 13, inhabitants warned by foreshocks Jun. 26, many aftershocks, ISS location is shallow, multiple event rupture, steep isoseismal gradient, tsunami	8.3	G53,S

*Galanopoulos 1967
[+]Estimated from Figure 1

Reference code: S - Sieberg, 1932
                  G60 - Galanopoulos, 1960
                  G67 - Galanopoulos, 1967
                  G53 - Galanopoulos, 1953

quoted in the literature for the area of southern Greece [e.g. Galanopoulos, 1967]. These events occurred between 1805 and 1926 (Table 1). No great shallow earthquakes are reported for any time for which the information is detailed enough to allow a depth estimate. During the 25 years of relatively good depth determination capability from 1953 to 1977 [Rothe, 1969; ISC Bulletins] only two earthquakes with M ≥ 7.8 and no events with M > 8.0 were found to have occurred at intermediate depths in the entire world. Both of these large intermediate depth earthquakes were located in South America. It would be remarkable if Greece had such a concentration of intermediate depth great earthquakes, 8 known events compared to 2 known events in the rest of the world during 25 years. And further it is remarkable that in the Hellenic subduction zone no great shallow earthquakes are known. Because this seems unlikely we will examine the evidence for hypocentral depth of the eight great Hellenic earthquakes.

The large and great shallow thrusts which occur in subduction zones have rupture width dimensions of 50 to 200 km perpendicular to the trenches because they are rupturing along the megathrust between two colliding plates. Intermediate earthquakes occur within the cool part of the descending lithosphere, with either the greatest or least principal stress direction aligned with the plunge of the slab [Isacks and Molnar, 1971]. Therefore these rupture dimensions are usually limited to about 20 km, the thickness of the brittle part of the descending slab divided by cos(45°) [Wyss, 1973]. Based on the above facts one would expect to find similar conditions in the Hellenic consumption zone: the largest earthquakes should be shallow.

The degree of similarity between the Hellenic trench and areas of ocean floor consumption has been debated. Most authors believe that there are more similarities than differences [e.g. McKenzie, 1970; Karig, 1971; Comninakis and Papazachos, 1972; Berckhemer, 1977; McKenzie, 1978] some emphasize the differences [e.g. LePichon and Angelier, 1979]. The most significant question seems to be that of the composition of the subducted lithosphere. All authors agree that consumption of the Mediterranean crust occurs. The underthrusting in the western Hellenic trench is established by focal mechanisms [Papazachos and Delibasis, 1969; McKenzie, 1970; 1978] and deep sea drilling [Hsu and Ryan, 1973].

Seismicity cross sections show the typical pattern of subduction zones [Papazachos and Comninakis, 1971; Papazachos, 1973]. Where temporary stations were added for better hypocentral depth control [Leydecker, 1975] the Benioff zone becomes more clearly defined. In the western Hellenic arc the slip vector is perpendicular to the arc, in the eastern part it is oriented mostly in a strike-slip direction (parallel to the trench topography). This is the case in many island arcs (e.g. West Indies, South Sandwich, Aleutians), and the great 1965 Rat Island earthquake was an example

of a rupture with a slip vector at about 45° with respect to the dip and the strike (i.e. intermediate between down-dip and strike-slip) [Wu and Kanamori, 1973].

We conclude that from the seismotectonic point of view there is no significant difference between the Hellenic arc and other island arcs. In particular one should expect the largest earthquakes in the Hellenic trench to be shallow, because this is observed to be the case in areas of oceanic as well as continental (India - Asia) lithosphere consumption.

The chief argument for intermediate depths of the great earthquakes of Table 1 appears to be the large dimension of the macroseismically determined felt-areas. Because the macroseismic data are important for depth and source length determination we reproduce in Figures 2-5 the four detailed examples presented by Sieberg [1932]. These figures show several remarkable features: (1) These ruptures are clearly associated with the Hellenic arc. (2) The felt-areas are large. (3) The isoseismal contours are asymmetric. Fairly high intensities are reported at large distances in directions outside of the Hellenic arc, whereas intensities on the back-arc side decrease rapidly. (4) The sources appear to be located offshore south of the island arc. Nevertheless, the reported intensities reach XI in all cases.

The size of the felt-areas (Figures 2-5) were taken to imply intermediate depth sources. We therefore compare the macroseismic data of the best documented case (1926) with felt-areas of other great earthquakes. Table 2 shows the areas estimated within the intensity IV curve for the New Madrid [Nuttli, 1973] and for the San Francisco [Richter, 1958] earthquakes. For the Charleston earthquake the entire felt-area is given by Gordon et al. [1970]. We see that the Hellenic felt-area is about average. It is approximately the same as that of the Charleston quake (assuming that the Charleston intensity IV area was half of its entire felt-area), it is about three times larger than in the San Francisco, but three times smaller than in the New Madrid case. The San Francisco and New Madrid felt-areas can be taken as minimum and maximum estimates for expected felt-areas because of the known inefficient and efficient wave propagation characteristics in the respective crust and upper mantle. We conclude that the size of the felt-areas of the Hellenic great earthquakes is not excessive in comparison to other great shallow earthquakes. The focal depth of felt earthquakes can be determined from the isoseismal radii following the method derived by Sponheuer [1960]. A maximum depth of 38 km is obtained by this method for the 1926 earthquake when measuring the radii in a western direction where the macroseismic information is the best documented. This is also the direction of the largest isoseismal radii wherefrom the larger depth is inferred. Measuring the least extension of the isoseismal radii, i.e. perpendicular to the trench axis, a depth of 16 km is obtained. Since the

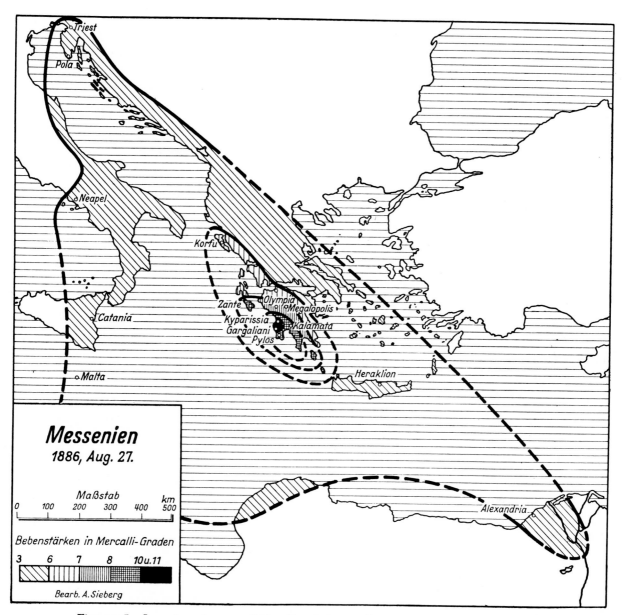

The map legend reads:

Messenien
1886, Aug. 27.

Maßstab
0 100 200 300 400 500 km

Bebenstärken in Mercalli-Graden
3   6   7   8   10u.11

Figure 2. Isoseismal map of the Messinia earthquake of 27 August 1886 [from Sieberg, 1932].

isoseismal maps of other large earthquakes are similar to those of the earthquake discussed above (Figures 2-5), we conclude that the isoseismal data can be interpreted to indicate that these large events were of shallow origin.

The pattern of asymmetrical distribution of intensities (Figures 2 through 5) is shown even more strongly by intermediate depth earthquakes. The macroseismic data of three earthquakes with epicenters near Corinth, Salamis and Delphi which occurred in 1962, 1964 and 1965 at respective depths of 120, 160 and 100 km are shown by Delibasis [1968] and Papazachos and Comninakis [1971]. The magnitudes of these events were given

as 6.75, 5.8 and 6.7. For these shocks the area of intensity >III to IV shaking reached half way up the Italian boot (2 larger events only) and northwest along the Yugoslavian coast up to between latitude 43° and 44°N (corresponding to the locations labelled Orebus and Sebenico in Figure 3). This means that intermediate depth earthquakes with $5.8 \leq M \leq 6.8$ and the $M \approx 8$ earthquakes of 1886 and 1903 caused intensity $\geq$ III at comparable distances in northwestern directions. This comparison suggests to us that the great earthquakes discussed here (Figures 2 to 5) must have been shallow ruptures. If they had been deep earthquakes, large enough to cause maximum

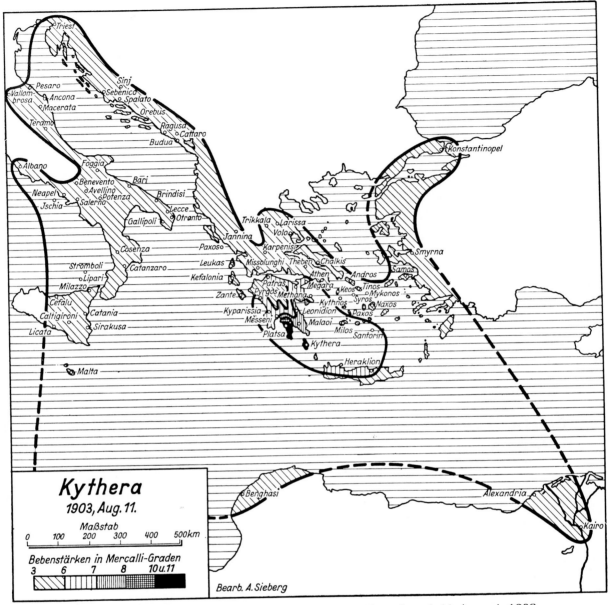

Figure 3. Isoseismal map of the Cithira earthquake of 11 August 1903 [from Sieberg, 1932].

intensities of XI, their intensity III radius should have been much larger than those of the smaller shocks reported by Delibasis [1968].

A further dissimilarity between the intensity patterns of the intermediate depth earthquakes and the examples of Figures 2 and 3 is the following: For the intermediate depth sources the highest intensities lie in the western to central Peloponnesus with values clearly decreasing towards the Messinian coast. In our cases the intensities along the coast are highest decreasing inland (Figures 2 and 3). From these differences we conclude that the 1886 and 1903 earthquake

damage could not have been due to intermediate depth sources located somewhere between Corinth and Milos.

A similarity in the intensity patterns of all events discussed is the strong attenuation in the back arc basin. Papazachos and Comninakis [1971] have further demonstrated the strong contrast between good transmission properties in the up-slab compared to other back arc paths, by showing strongly contrasting seismograms of stations equidistant to a deep earthquake. We agree with Papazachos and Comninakis [1971] that the strong attenuation in the Aegean area must be due

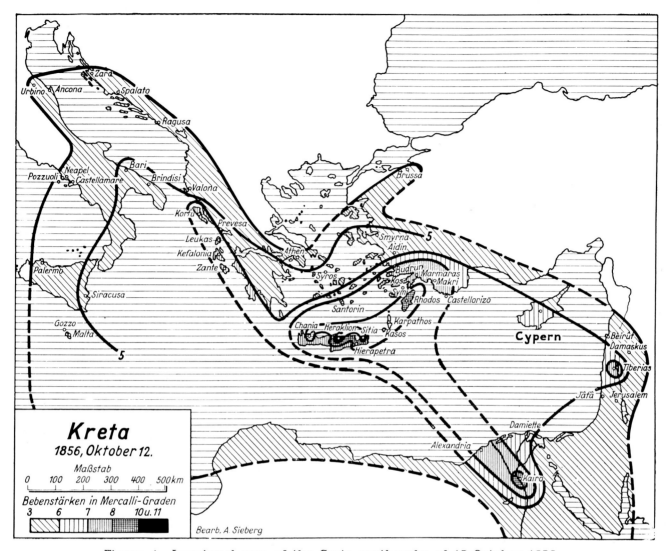

Figure 4. Isoseismal map of the Crete earthquake of 12 October 1956
[from Sieberg, 1932].

to low Q-values as they are usually observed for the upper mantle in back-arc basins [Molnar and Oliver, 1969; Barazangi et al., 1975].

Fore- and aftershocks have been reported for some of the mainshocks of Table 1. Generally, intermediate depth earthquakes have no aftershocks or very few. In a world-wide search for non-shallow aftershock sequences in island arcs we could find only a few cases with two to four aftershocks. There are some notable instances of intermediate depth swarm and aftershock activity in the Hindu-Kush, Colombia and Rumania. Since the intermediate seismicity in these areas is recognized as anomalous high activity clusters of small dimensions, and since we have no indications that the Hellenic arc contains such anomalies, we will accept fore- and aftershock activity as one of the criteria suggesting shallow

depth of the main shock. For three of the earthquakes listed in Table 1 many aftershocks were reportedly felt [Sieberg, 1932], in one case four strong shocks (including mainshock) were reported. It seems unlikely to us that the three events with "many aftershocks" could have been intermediate depth ruptures.

Tsunamis, some of which may have been small, were reported [Galanopoulos, 1960; Sieberg, 1932] for three of the events in Table 1. In addition, "seaquakes" were reported as felt for two shocks. It could have been that these latter observations also indicate tsunamis. The origin of tsunamis is now generally recognized as due to several meters of permanent coseismic displacement of the sea floor [e.g. Hatori, 1966; Nishenko and McCann, 1979]. These observations therefore suggest that three, perhaps five, of the events listed in Table 1

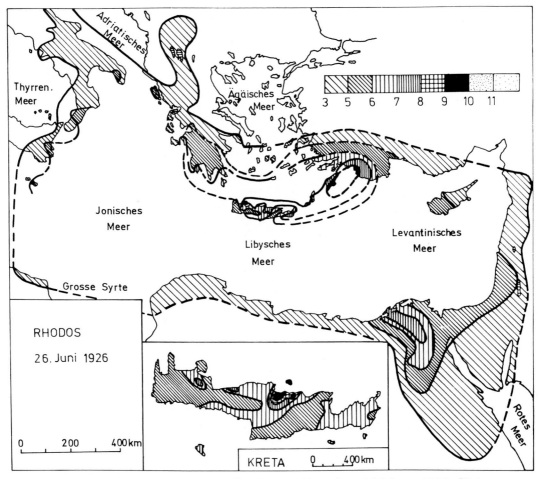

Figure 5. Isoseismal map of the Rodhos earthquake of 26 June 1926. Note that the epicenter is located east of Rodhos and near the Turkish coast. The high intensities at Heraklion (Crete) are clearly anomalous (see inset). [After Sieberg, 1932.]

were shallow earthquakes associated with surface deformations of permanent nature on the sea floor.

After the 1903 earthquake "200 m long cracks in the ground striking NW-SE" were reported [Sieberg, 1932]. While we do not propose that these represented the surface rupture of the earthquake, we do propose that this evidence also suggests a shallow source of the event.

An instrumental location exists only for the most recent event, that of 1926. In the ISS bulletin this earthquake is listed as shallow, and located at 36°N 28°E, just off the SE coast of Rodhos where the macroseismic evidence places the source (Figure 5). The ISS solution was largely ignored in the literature in favor of the 36.5°N 27.5°E location proposed by Critikos [1926] and used by Gutenberg and Richter [1948]. Sieberg [1932] pointed out that the latter solution could not be trusted because it was based mainly on an S-P time of 31 sec observed at Athens. Further he

remarked that the P-arrival time at Athens fits the ISS solution, however the S-arrival time did not. We found that at teleseismic distances the 1926 great earthquake is recorded as a complex multiple event (Figure 6 and Table 3) with records

Table 2. Dimensions of Felt Areas of Some Great Earthquakes

Year	Location	Area of Intensity IV 10^6 km^2
1926	Rodhos	1.8
1811	New Madrid	6.0
1895	Charleston	(1.3)*
1906	San Francisco	0.54

*half of entire felt area

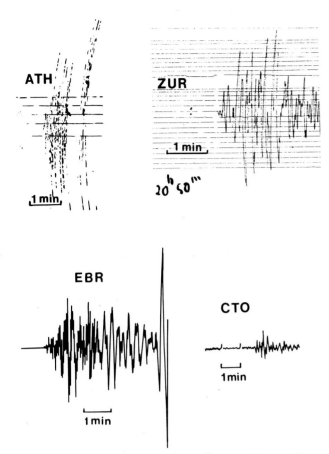

ATH   ZUR

1 min

20ʰ 50ᵐ

1 min

EBR

CTO

1 min

1min

1 min

Figure 6. Records of the 26 June 1926 earthquake. At least three clear arrivals can be
identified: The small amplitude first arrival,
then a five times larger arrival about 7 sec
later, and a ten times larger arrival 30 sec
later. At the recording distances in question
these phases could not be reflected or
refracted phases, they are interpreted as
multiple events. ATH (Athens), ZUR (Zuerich)
and EBR (Tarragona) are P-waves, CTO (Cape
Town) is an S-wave.

similar to other great multiple ruptures, like for
instance the 1964 Alaska earthquake [Wyss and
Brune, 1967]. The P- and S-waves show several
packages of energy with the largest pulse arriving
about 30 seconds after the initial arrival. This
delay time is independent of distance and
therefore this phase cannot be a reflected or
refracted phase. While the 30 second delayed
pulse could not be mistaken for the S-wave at
teleseismic distances it was apparently assumed
to be S in Athens. Also Galanopoulos [1953] examined the seismograms at Athens for Greek earthquakes which were located at intermediate depths
by Gutenberg and Richter [1948]. He compared
these signals with seismograms from sources
which were known to have been located at shallow
depths. He concluded that his observations "leave

little doubt that the 26 June 1926 earthquake was
a normal one".

The evidence which shows that the 1926 earthquake was a shallow event is summarized as follows: (1) The instrumental location places it at
shallow depth below the inner wall of the Hellenic
trench. (2) The detailed macroseismic isoseismal
map places it in the same location (Figure 5). (3)
The steepness of the isoseismal gradient near the
source suggests shallow origin depth, while the
dimensions of the felt-area do not suggest a deep
focus. (4) Foreshocks were reported for six
months. Foreshocks on the same day sent the
inhabitants of Archangelo in flight out of town,
with the result that in this totally destroyed town
no one was injured. Only two aftershocks were
large enough (M = 4.8 and 5.0) to have been
located. The ISS solutions indicate normal depth,
Papazachos and Comninakis [1971] obtained 60
km depth for one of them. Numerous aftershocks
were reported to have been felt [Sieberg, 1932].
We conclude that many fore- and aftershocks
occurred, but that none of them exceeded magnitude 5. (5) In the harbor of Heraklion boats
were reported to have collided. This suggests that a
tsunami was caused by this event.

We conclude that the evidence strongly suggests that the great 1926 earthquake was located
SE of Rodhos and that the source was shallow. For
most of the other large and great earthquakes in
Table 1 there is some evidence which also points
to shallowness of these sources. Where evidence is
lacking we will assume that the ruptures were
shallow, because their positions at the plate boundary suggest that they are of a tectonic type similar to the great Hellenic sources where evidence
for shallowness exists. We propose therefore that
the Hellenic arc experienced a sequence of great
ruptures between 1805 and 1926. Below, an
attempt is made to map the extent of breakage
along the arc as far as the data allow it.

THE SOURCE DIMENSIONS AND LOCATIONS
OF GREAT EARTHQUAKES
IN THE HELLENIC ARC

The locations and extent of the ruptures were
estimated from the macroseismic data, assuming
that the ends of the ruptures along the arc were
approximately at the isoseismal boundary between
intensity X-XI and VIII-X. A further constraint
used is that ruptures may not overlap, because
abutting aftershock-areas are generally observed
in other island arcs [e.g. Sykes, 1971]. Also the
relative size of the ruptures has to agree with the
relative size of the maximum intensity, and of the
felt area dimensions. Finally, as a weak constraint, subordinate to the others, we let ruptures
terminate at locations where the trench topography suggests a change of strike or a topographical
transverse feature that might indicate the presence of a tectonic element which could have
stopped an earthquake rupture.

Table 3. Reported Arrival Times of Seismic Energy Pulses.
Second Line of Each Station Indicates the Time Difference Between
that Phase and the First Arrival Time.

Station	Distance [deg]	Azimuth [deg]	t_1	t_2	t_3	t_4
ATH	3.9	302	19:47:24			19:47:55
						+31
ZUR*	18.3	314	19:50:40	19:50:42	19:50:50	19:51:07
				+2	+10	+27
EBR*	22.0	291	19:51:16		19:51:25	19:51:40
					+9	+25
TOL	25.4	289	19:51:50	19:51:52	19:51:57	19:52:22
				+2	+7	+32
CTO*[1]	70.5	188	20:06:53			20:07:13
						+20
HKC	74.1	75	19:58:11			19:58:40
						+29

*Readings by the authors
[1] S-phase data

Major shortcomings of intensity data are their dependence on the quality of local construction and on the soil conditions. Since the type of buildings is the same along the entire arc, building strength will not be an error source for the estimate of location and extent of ruptures. However, the strong shaking on unconsolidated sediments will cause high intensities and one will be tempted to place the source close to such locations. Examples of two such places are the Nile delta (Figures 2-5) and Heraklion (Figures 4 and 5) where intensities always far exceeded those observed at other locations at similar distances. The high intensities in the Nile delta are no problem for us since it is located too far (700 km) from the sources to contribute in the earthquake location procedure. The high intensities at Heraklion, however, are a problem. Sieberg [1932] noticed that relatively high intensities resulted at Heraklion from "shocks outside of the country". This town is built in a graben filled with alluvium [Sieberg, 1932]. We believe that many epicenters of large 19th century earthquakes were placed too close to Heraklion because of the excessive intensities reported from there. Accounting for Heraklions sensitivity we propose that the earthquakes of 1805, 1810, 1846 and 1856 took place south and east of Crete, strung out along the plate boundary, and not in one tight cluster close to and north of Heraklion.

For the 1886 rupture the SE end is fairly well defined by the intensities (Figure 2) as located between Gargaliani and Pylos. Along the Northwestern part of the rupture the coast line recedes and is located about 100 km NE of the trench. For this reason intensity X was not recorded there. Because of the intensities VIII-X on Zakintos (Zante), and because the felt area requires that the source had a length roughly equal to the 1903 and 1926 events, we chose the NW termination as shown in Figure 1. The approximate source length was 100 km.

The September 1867 event was clearly a much smaller earthquake than the one discussed above, but it caused a large tsunami from which a location in the gulf of Messinia was derived. This event just fits in between its larger neighbors (Figure 1).

The 1903 rupture has a fairly well defined northwestern end (Figure 3). One might argue that the rupture continued farther SE of Cithira than shown by us in Figure 1. However, there are two reasons for which we believe that the rupture probably did not extend much farther SE. First, the length of this earthquake was about the same as that of the 1886 and 1926 events according to the felt-areas (Figures 2, 3 and 5), and smaller than that of the 1856 event (Figure 4). In addition, the comparatively low intensities in Crete, especially in the highly sensitive area of Heraklion, which responded strongly to the 1926 event, suggest that the rupture did not come close to Crete.

The 1856 earthquake appears to have caused the strongest shaking at large distances, and the area of intensity VIII shaking is fairly extensive (Figure 4). It may have been the strongest of the great earthquakes in the Hellenic trench. The extensive destruction at Heraklion in central Crete (Figure 4) has suggested to some authors that the source of this earthquake was located north of Crete. We do not agree with this interpretation, because the area around Heraklion experiences high intensities even for relatively distant events [Sieberg, 1932]. For example, intensities up to X were reported at Heraklion due to the 1926 rupture (Figure 5). Accounting for this anomaly we placed the SW end of the 1856 rupture off southeastern Crete. The northeastern end is also difficult to determine. The rupture probably reached close to Rodhos, because intensities from VIII to X were observed there.

The length of the 1926 earthquake has to be about equal to the other great events, and its position cannot be shifted very much from that chosen in Figure 1 because of the detailed macroseismic data given by Sieberg [1932] and because of the ISS epicentre location. Judging from the relatively high intensities at the Turkish coast the rupture must have extended close to it.

The earthquakes of 1805, 1810 and 1846 are commonly plotted with locations close to Heraklion. However, because of the anomalously high sensitivity of this locale (e.g. Figure 5) we believe that these events could not have been located at Heraklion and yet have been large enough to be felt at the considerable distances indicated in Table 1. The macroseismic data shows that these events were fairly large but clearly smaller than the great earthquakes discussed above. According to the little information available (mostly from Sieberg, 1932) it is reasonable to assume that these events occurred along the subduction zone south of Crete. The positions in Figure 1 are chosen approximately according to the locations of the largest damage. The rupture lengths are assumed to be roughly half of the great earthquakes. If they were smaller the effects would probably not have been felt at the distant locations given in Table 1, and if they were larger the macroseismic data would be similar to those for the great earthquakes. The lightly shaded rupture areas and the question marks on Figure 1 express the uncertainty of our conclusions regarding these sources.

Two large earthquakes are reported to have occurred at the northwestern end of the Hellenic arc in 1867 and 1897 (Figure 1). Since this is a tectonically complicated area, and since the effects of these ruptures are poorly documented we will not attempt to study this end of the Hellenic plate boundary.

SEISMIC GAPS OF THE HELLENIC ARC

Based on the above discussion we propose that great and large shallow earthquakes have broken most of the Hellenic arc plate boundary in a sequence between 1805 and 1926. All of these earthquakes have occurred more than 50 years ago, many of them more than 100 years ago. Two portions of the arc seem not to have been broken since about 1800. According to the definitions of McCann et al. [1979] the entire Hellenic arc has to be considered in a stage of seismic gap of one category or another. Considering the uncertainties in determining location and size of large earthquakes during the 19th century we believe that 50 km under- or overestimations of the dimensions of seismic gaps are possible, while 100 km errors are unlikely.

The Andicithira gap (Figure 13) is of category 3 because we have not found a recent historic report of high intensities in western Crete and Andicithira. There is no reason to believe that creep displacement is occurring along this segment while it is not occurring along the rest of the arc. Therefore, we suggest that the seismic potential is highest here at the present. The gap is estimated to be about 100 km long, and it is located off western Crete and Andicithira. While the exact location and extent of this gap is open to question, we believe that the existence of it is fairly well documented. The greatest uncertainties are associated with the southeastern end of the gap. The gap could have been longer, if the 1805, 1846, 1810 events were located elsewhere, or if their source dimensions were smaller. On the other hand, the earthquake of 1805 could have been located farther west, in which case the Andicithira gap would be smaller, but an additional gap along southwestern Crete would exist. However, in whichever way we place the known past ruptures, we believe that there exists a seismic gap of category 3 in the area of Andicithira and western Crete. The dimensions of the gap indicate that the potential for another great earthquake with $M_s \leq 8$ and with effects similar to those during the great ruptures of 1856, 1886, 1903 and 1926 may exist there.

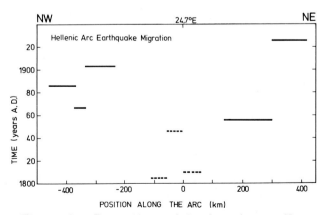

Figure 7. Space-time plot of major earthquakes along the Hellenic arc. Horizontal bars indicate the approximate extent of major ruptures along the arc, earthquakes with uncertain locations are shown as dashed lines. A possible migration pattern is not very clearly defined.

The Crete gap is a category 1 gap since most or all of it had ruptured more than 125 years ago. It extends from western Crete towards the east past Carpathos (Figure 13). As we have pointed out above, the evidence is weak for placing the earthquakes south of Crete as we did in Figure 1. It could be argued that the Hellenic arc segment between 23.5 and about 26 ° E should be considered a gap of class 3, like the Andicithira gap. We believe that the entire Crete gap has a strong potential for large earthquakes.

The Coast of Messinia gap is a category 2 gap (Figure 13). A small central part ruptured 114 years ago and the western part 95 years ago (Figure 1). Nevertheless, class 2 may be appropriate for this arc segment because the consumption rate is lowest here according to LePichon and Angelier [1979].

The Rodhos category 2 gap may have the lowest seismic potential of this island arc, because it ruptured most recently (Figures 1,13).

MIGRATION OF LARGE EARTHQUAKES ALONG THE HELLENIC TRENCH

A migration of ruptures from western Crete to Rodhos is suggested by the distribution of dates on Figure 1. To evaluate this possibility we constructed the time-distance plot shown in Figure 7. Here the suggestion of a systematic migration is less strong. Given the existing scatter and the uncertainties in the data, one might conclude that a possible northeastward migration with an average velocity of 4 to 6 km/year is suggested. These velocities are of the order of those inferred for the North Anatolian rupture migration velocity which was estimated at less or equal to 10 km/yr [Toksoz et al., 1979].

CHANGES OF SEISMICITY RATE AS A FUNCTION OF SPACE AND TIME

Seismic quiescence precursors have been defined at statistically highly significant levels for several large main shocks. Decrease by 50% of the background seismicity rate was measured with respect to the "normal" rate within the main shock volume, as well as with respect to neighboring segments of the same plate boundary [Wyss and Habermann, 1979, Habermann, 1981b]. The occurrence of seismicity quiescence along with strain softening before the M = 7.2 Hawaii earthquake suggests that a seismicity decrease indicates initiation of the precursory period by stress release in most of the future main-shock volume, with high-stress anomalies occurring near a few master asperities [Wyss et al., 1980]. We studied the seismicity rate as a function of space and time in the Hellenic arc in order to determine whether rate changes have occurred. This second part of our study is not dependent on the seismotectonic model derived in the first part. The overall assess-

ment of the seismic hazard is derived jointly from parts one and two. However, the implications of a seismicity quiescence anomaly does not depend critically on the exact depth, location and size of particular earthquakes in Table 1.

The data used consisted of all earthquakes with depth ≤ 100 km contained in the NOAA catalogue from 1900-1977, supplemented by the hypocenters listed by Galanopoulos [1963] for the period 1710-1959, by the data of Comninakis and Papazachos [1978] for the period 1901-1975, and by those in the bulletins of the National Observatory of Athens [e.g. Drakopoulos et al., 1978] for July 1977 to February 1979.

The completeness of the data set as a function of time was our first concern. As time and seismology progressed, more small earthquakes were located, therefore the rate appears to increase with time. The upper curve in Figure 8 shows the cumulative number of earthquakes as a function of time in the Hellenic arc based on the

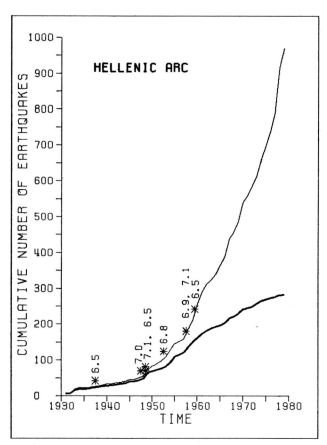

Figure 8. Cumulative number of earthquakes as a function of time from 1930 to 1979 for the Hellenic arc. The upper curve shows all events reported, the lower curve only those with M≥ 5.0. Hypocenters in the top 100 km of the earth are included. Stars mark shocks with M ≥ 6.5.

entire data set described above. This curve steepens gradually as more smaller earthquakes are located with increasing time. Before 1930, the origin of Figure 8, only few earthquakes are listed. In 1950 a fairly strong increase in reports occurs. We therefore will use in the further analysis the data from 1950 onward. The smallest magnitude for which all occurring events are located from 1950 on lies between 4 3/4 [Galanopoulos, 1967] and 5 [Comninakis and Papazachos, 1978]. We verified this by fitting a straight line to the frequency magnitude curve. Our conclusions are the same whether we take $M_{min} = 4.75$ or $M_{min} = 5.0$. We chose $M_{min} = 5$ in order to be certain to study a complete and homogeneous data set.

The data set with $M \geq 5.0$ is also plotted in Figure 8. This curve suggests that the seismicity in the Hellenic arc as a whole decreases after 1960. Below we will determine which part of the arc is mainly responsible for this decrease, and we will examine the statistical significance of the change.

For the systematic seismicity study we divided the arc into 12 arbitrary, overlapping segments (Figure 9) of a size roughly equal to the source length of the largest historic earthquakes in the arc (Figure 1). The width across the arc for the study volume was fixed at 150 km. For graphical representation of the seismicity rate we chose cumulative number versus time curves (Figures 10a,b,c). These have the advantage that they act as a low pass filter and allow visual identification of long term rate changes. Figure 10 shows that the total number of events ($M \geq 5.0$) per unit arc length (100 km) ranges between 20 and 45 over 29 years, corresponding to overall rates of approximately 0.7 to 1.6 events/year. At these low rates we need to combine volumes of similar rates and tectonics in order to arrive at statistically meaningful conclusions. Volumes 1, 2, 3 and 12 form clearly a group with the lowest total number of events (20 ± 3 during 29 years) and with their rates decreasing as a function of time. Volumes 5 through 10 show approximately constant seismicity rates and their total numbers are larger (33 ± 4 events during 29 years). The seismicity in volume 4 shows a behavior intermediate between the two groups, and volume 11 contains more earthquakes than all others because of swarm activity in it.

The numbers of events within individual volumes are too small to allow firm conclusions about rate changes. We therefore will merge neighboring volumes with similar behavior into super-volumes. Clearly, volumes 1 through 3 (and partially volume 4) define the super-volume I with a seismicity pattern distinctly different from the adjacent segments south of Crete. Volumes 5 through 11 seem to have similar seismicity characteristics and might therefore be studied jointly. However, for comparison of seismicity rate along the plate boundary we will need to consider volumes of equal dimensions. Therefore we have chosen volume II and III (defined in Figure 9) for comparison with volume I.

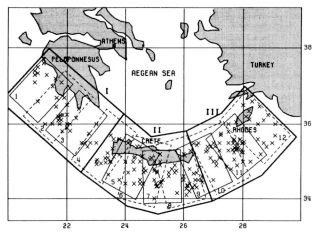

Figure 9. Epicenter map for all earthquakes used in this study. The seismicity rates of volumes 7 through 12 are shown in Figure 10. The boundaries of the even numbered volumes are shown wider and dashed for identification purposes only. The seismicity rate of the super-volumes I, II and III are shown in Figure 11.

The cumulative seismicity as a function of time in volume I, II and III is shown for $M \geq 5.0$ in Figure 11. We note the following main facts: (1) The three volumes had the same seismicity rate during the first 13 years. During the second 16 years their rates differ by more than a factor of 4. (2) Volume II shows a constant rate of about 3.3 events/year during the 29 year sample period. (3) Volume I shows a dramatic change of rate from 3.2 events/year to 0.5 events/year in 1962. This corresponds to a 80% drop in seismicity rate. According to the t-test the pre-1962 sample is different from the post-1962 sample at the 99.99% probability level. (4) The seismicity rate in volume III seems also lower during recent years. We will eliminate swarms in two ways and study this volume separately below.

The conclusions we draw from Figure 11 are the following: Volume I has entered a stage of seismic quiescence in 1962. This volume is now seismically far less active than the same volume was before 1962, and it is far less active than other segments of the Hellenic arc. The latter fact was noted by McKenzie [1978] based on the hypocenters with depth less than 50 km for the period of 1961-1975. He further suggested that a series of large earthquakes might occur here. We interpret the period of seismic quiescence which began in 1962 to represent the preparatory process to large Hellenic earthquakes in analogy to the precursors observed for the Oaxaca, the Hawaii and the Kurile island and other earthquakes [Ohtake et al., 1975; Wyss et al., 1980; Wyss and Habermann, 1979; Habermann, 1981a, 1981b]. This is a reasonable hypothesis even though no suitably large recent earthquake has occurred in the Hellenic arc,

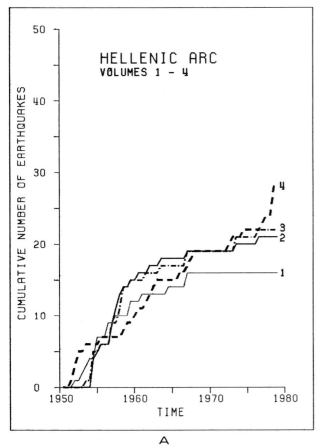

Figure 10. Cumulative number of earthquakes as a function of time for the 12 small
volumes defined in Figure 9. The magnitudes
used are M≥ 5.0. To avoid confusion of the
lines the data are presented in three figures
with the same scale. For rate comparison the
figures can be superposed.

events in the following way: If two or more events
occurred within 48 hours and at distances less
than 60 km from each other the sequence was
considered a cluster and only one event was
retained. (A time limit of 14 days gave essentially
the same results). The cumulative seismicity
without clusters (dashed lines in Figure 12) shows
that the rate change persists if clusters are
removed.

In the asperity model for precursors we expect
that the asperity portions of the source volume
produce earthquakes at a constant rate while the
rate decreases around the asperities in the rest of
the source volume [Wyss et al., 1980]. Volume III
contains a seismically especially active spot
(center of volume III, Figure 9) which contains
most of the clusters discussed above. If the
seismicity within a volume of 40 km radius around
35.42° N/27.74° E is removed the rest of the
volume III seismicity shows a clear seismicity
quiescence (fat solid curve, in Figure 12) starting
in 1966. The seismicity within the cluster is shown
as the lowest curve in Figure 12. We conclude that
the seismicity in volume III appears to show a
quiescence anomaly when all recorded earthquakes are considered (Figure 11). However, this
anomaly becomes fully convincing only after the
small volume with the highest seismicity is

where we could see whether quiescence precedes
large main shocks in the Hellenic island arc as it
seems to do in other island arcs.

A conclusion that may not be drawn from Figure 11 is that volume II is in a normal state.
Before some earthquakes in other island arcs constant seismicity was observed [Habermann,
1981b]. Since volume II is a seismic gap of class 1
it must be considered to have the potential for
great earthquakes at the present.

The analysis of the seismicity in volume III is
unsatisfactory, because strike-slip and thrust tectonism are lumped together, and because earthquake swarms obscure the rate estimates. In Figure 11 the rate in volume III appears to be lower at
least from 1970 onward. The pre-1965 rate in this
volume may appear high because the steep portions of the curve (1957, 1961) may be due to
swarms like the 1969 step. In order to study the
more randomly occurring background seismicity
in volume III we excluded obviously dependent

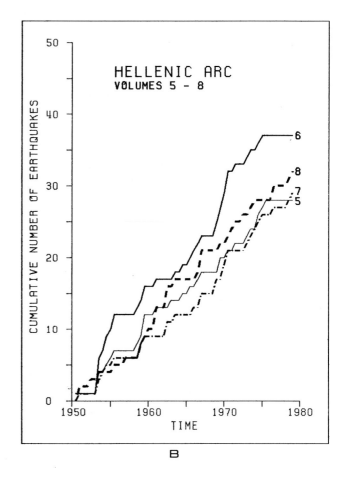

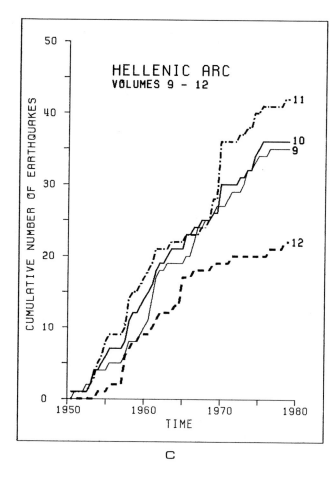

C

the slip rate should increase by a factor of about 2 from west to east along the Hellenic arc, because the rotation pole is located relatively close to the plate boundary. Arguing that subduction began 13.5 m.y. ago they obtained slip rates of 2 cm/yr (western part) to 4.5 cm/year (eastern part) by dividing the length of the subducted slab by the subduction time.

For our purpose, the rough estimation of the strain stored at the plate boundary since the last large earthquake, we will take a slip rate of 5 cm/year for the eastern arc and 2.5 cm/year for the western arc, allowing for an error of approximately 40%. The time elapsed since the last great earthquakes in the Hellenic arc is summarized in Table 4. Multiplying the time elapsed to 1980 since the last large stress-drop by 5 cm/year and 2.5 cm/year, as appropriate, we obtain the poten-

excluded from consideration (Figure 12). We feel that our procedure to exclude a high seismicity volume from the analysis is valid because it is based on a model explicitly stated before the analysis was carried out. Nevertheless, we recognize that the case for an anomaly in the eastern Hellenic arc is less convincing than that for the western arc anomaly, because manipulation of the data was necessary to clearly reveal the anomaly in volume III.

EARTHQUAKE POTENTIAL ESTIMATED
FROM THE SLIP-RATE

The rate of plate convergence across the Hellenic trench has been estimated by different methods and under different assumptions. McKenzie [1978] obtained a rate of 7 cm/year knowing the directions of the slip vectors Turkey-Eurasia, Turkey-Aegean and Aegean-Africa, and assuming 4 cm/year slip rate at the North Anatolian plate boundary. From the creep rate on this fault at Izmet Pasa, which is between 1 and 1.5 cm/year [Aytun, 1980], one may argue that McKenzie's estimate should be reduced by approximately a factor of 2, i.e. to 3.5 cm/year.

LePichon and Angelier [1979] estimated that

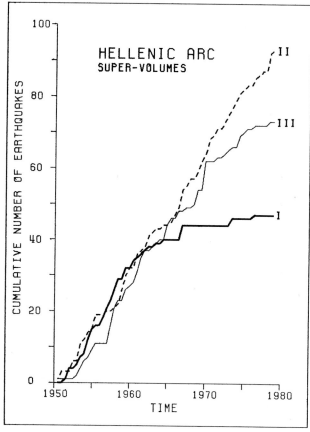

Figure 11. Cumulative number of earthquakes as a function of time for the super-volumes I, II and III defined in Figure 9. From 1950 to 1962 all volumes show the same seismicity rate. From 1963 to the present volume I exhibits remarkable quiescence, while earthquakes in volume II occur undiminished at a constant rate. In volume III the seismicity rate seems to be decreased from 1970 to the present.

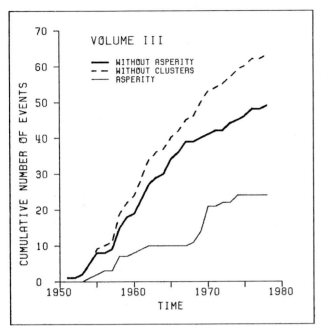

Figure 12. Cumulative number of earthquakes as a function of time for subsets of the volume III data. The heavy solid curve shows the seismicity without the contribution from the active central part of volume 11, which is shown by itself as the light solid curve (bottom). The dashed curve shows the seismicity of volume III with clusters removed by the method described in the text. The heavy solid curve shows a strong decrease of seismicity rate starting in 1966. This quiescence is very similar to the Hawaii (1975) observation where a rupture volume minus central asperity volumes showed such an anomaly.

tial slip stored along the the Hellenic arc (Table 4). In this calculation we assumed that underthrusting is not accommodated by aseismic creep. The approximate potential stress-drop available for the next sequence of large ruptures (Table 4) was calculated by assuming rupture widths of about 50 km for thrust and 20 km for strike-slip events, and a shear modulus of $3 \cdot 10^{11}$ dyne cm^2.

The worldwide average stress-drop for large interplate earthquakes is 30 bars [Kanamori and Anderson, 1975]. Given the large variations around this average, the possible errors in the slip rate estimate, plus the uncertainties in the assumption of fault width, one can only say that the approximate potential stress-drop estimates of Table 4 are roughly equal to the world average stress-drop.

SEISMIC HAZARD ALONG THE HELLENIC ARC

Identifying the locations of seismic gaps is a first step in a search for plate boundary segments of increased seismic potential. By categorizing the gaps one can make a further step towards pinpointing the segments most likely to rupture next. According to the definition of McCann et al. [1979] all of the Hellenic arc is to be considered a seismic gap. This assessment is supported by the estimates of stress stored along the arc due to plate motions. By shading dark the Andicithira gap (category 3) in the summary Figure 13, we express our opinion that this is the segment most likely to rupture next, based on the seismic gap analysis alone. The category 1 gap extending from south of Crete to the East past Carpathos (Figure 13) is a close second, unless slip rates are given more weight (Table 4), in which case the Crete-Carpathos gap may be expected to break first.

By our parallel investigation of the seismicity rates as a function of time and space we found segments of the Hellenic arc which turned quiet

Table 4. Approximate Potential Slip Accumulated in the Hellenic Arc

Location	Year (last rupture)	Time to date (years)	Slip accumulated (m)	Stress-drop possible (bars)
Western Hellenic Arc Quiet Volume	1886 to 1903	94 to 77	2.4 to 1.9	20 to 15
Andicithira Gap	<1800	>180	>4.5	>35
South of Crete	1805, 1810, 1846	175 to 134	4.4 to 3.4	30
Carpathos	1856	124	6.2	60
Rodhos	1926	54	2.7	30

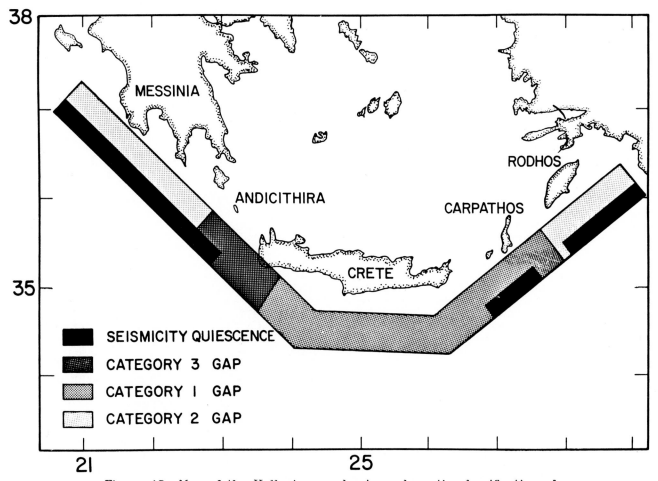

Figure 13. Map of the Hellenic arc showing schematic classification of seismic gaps along this plate boundary. The categories used are those defined by McCann et al. [1979] except that for category 2 gaps the time elapsed since the last large rupture is larger than 50 years. The segments where seismicity quiescence exists at present are delineated by black bars.

recently (black bars in Figure 13). According to our hypothesis a seismicity rate drop of more than 40% signals the beginning of the preparatory process to one or several earthquakes with combined dimensions roughly comparable to the quiescent volume.

The seismic hazard assessment should be based on the joint gap and quiescence data of Figure 13. If the evidence of the two independent methods is weighed equally, the darkest segments in Figure 1 are presently the most dangerous ones. In the segments southwest of Andicithira, and south of Carpathos both methods point to a very high seismic potential. Along the southern coast of Crete only the gap analysis points to a high seismic hazard, but it must be remembered that constant seismicity rate does not mean that the area is safe.

The fact that most parts of the western arc appears to be in a state of quiescence suggests that these segments are mature gaps [Habermann, 1981a] while the gap and strain analysis suggest that these segments are gaps of category 2 only. The cause of this difference in degree could be shortcomings of either method.

Two weaknesses of the quiescence method are: (1) The seismicity rate in the Hellenic arc is relatively low, so that the extent of quiescent segments is uncertain. (2) The hypothesis is still poorly tested and the phenomenon of quiescence is not fully understood yet. The categorizing of seismic gaps, on the other hand, is based on some assumptions which may not be correct. (1) If the slip rate does not decrease towards the west, then the northern half of the Messinia category 2 gap would be a category 1 gap. (2) Since we don't know the amount of slip during past and future large ruptures, we don't know the recurrence time for any of the segments of the arc.

Since both methods have shortcomings, we will

give about equal weight to their contributions of estimating the seismic hazard. From Figure 13 we conclude that the entire Hellenic arc has high seismic potential. The two segments where quiescence coincides with a seismic gap of category 3 and 1 are in our opinion the most likely places for large earthquakes in the near future.

Based on the source dimension estimates of past ruptures (Figure 1) we expect the future large earthquakes to also have lengths of about 120 ± 40 km. This length corresponds to a magnitude of approximately $M = 7\ 3/4 \pm 1/2$. The two most likely points of rupture initiation are along the plate boundary near 22.5 to 23.5° E, and near 26.5 to 27.5° E.

The time of the occurrence of these two earthquakes is difficult to estimate because the extent of the quiescence anomalies is poorly known, and because their relation to the seismic gaps is not clear. However, we believe that these ruptures may occur soon. Quiescence precursor times in other cases [Ohtake et al., 1975; Wyss and Habermann, 1979] have ranged from a few years to a couple of decades. Since the anomalies in the western and eastern segments started in 1962 and 1966 respectively, we would therefore expect that they will be terminated by main shocks no later than 10 years from now.

CONCLUSIONS AND RECOMMENDATIONS

We conclude that the Hellenic plate boundary ruptured in a series of large to great earthquakes, mostly during the 19th century with the last two earthquakes occurring in 1903 and 1926. Based on the approximate slip rate of the Aegean plate with respect to Africa we propose that enough strain has been accumulated at this plate boundary for the sequence of ruptures to reoccur. We showed that a clear period of seismicity quiescence (80% drop of seismicity rate) started in 1962 in the western Hellenic arc. In the eastern Hellenic trench seismic quiescence existed certainly post 1970, and probably post 1966. With the hypothesis that these anomalies represent precursors, and with the assumption that the precursor-time scales with anomaly dimension we suggest that large earthquakes should be expected to occur between 1980 and 1990 in the Hellenic trench segment near Andicithira and near Carpathos. These predictions should not be quoted without emphasizing the rudimentary state of the understanding of the processes which lead to earthquakes. Even though the number of well documented seismicity quiescence precursors is rapidly increasing, we still know only a couple of dozen such occurrences in the past. For several locations around the world, seismicity quiescence at highly significant confidence levels has been documented to exist at the present. However, we do not know whether it is possible that the seismicity rate is decreased in a crustal volume and no main-shock follows the anomaly. In the Hawaii case the concurrence of quiescence with two other independent precursor anomalies lends strong support to the hypothesis that quiescence is a precursory anomaly [Wyss et al., 1981]. Further the geodetically observed strain relaxation supports the interpretation that quiescence indicates a drop of stress in the crustal volume by strain softening, a model which is attractive from the rock mechanics point of view [e.g. Stuart, 1974]. Nevertheless, it may be possible that other explanations for seismicity quiescence can be found, and that the data shown in Figure 12 do not mean that large earthquakes are imminent in the Hellenic arc. But even if the above prediction is not accepted as significant, we would still argue that the seismic hazard is large in the Hellenic arc, given its seismic history and the estimated strain rate across it.

We believe that precautionary measures are warranted, even though we are aware of the negative effects the broadcasting of a so-called earthquake prediction can have [e.g. Garza and Lomnitz, 1979]. About two years before the $M = 7.8$ Oaxaca 1979 earthquake, Ohtake et al. [1977] showed that quiescence had started in the Oaxaca gap in mid-1973. Their publication in a scientific journal did not produce any good or bad effects: They were at first not able to mobilize funds to study the area further, nor did panic develop. More than a year later a panic was caused by the interference of non-professionals and by the broadcasting of non-facts which led to a financial loss for the region [Garza and Lomnitz, 1979]. We hope that these mistakes will not be repeated in the case of the Hellenic trench, namely non-facts will not be spread and a scientific surveillance of the area will be launched.

An earthquake prediction research program along the Hellenic arc should have two goals. Obviously, it should be aimed at measuring further precursors which would allow for a sharper prediction in space and time of the first earthquake in the sequence to come. But also it should be designed to define the normal values of crustal parameters in volumes where precursory anomalies and their main shocks may occur decades from now. In particular, we recommend the establishment of dense seismograph, tide gauge and gravity observation networks. With the tide gauge and gravity observation networks vertical crustal deformations of long term as well as short term precursory character may be observed. We believe that this area should be number one on the list of joint earthquake prediction efforts because it will take eight to ten large earthquakes to break the entire plate boundary again.

Acknowledgments. This research was supported by funds of the Institute of Geophysics, ETH-Zurich, by the U.S. Geological Survey Contract no. 14-08-001-1836, and by NASA order S-65000-B. We thank C. Kisslinger for criticism of the manuscript.

References

Aytun, A., *The creep measurements of the North Anatolian fault zone in the Ismet Pasa region*, presented at Interdisciplinary Conference on Earthquake Prediction Research in the North Anatolian Fault Zone, 1980.

Barazangi, M., W. Pennington, and B. Isacks, Global study of seismic wave attenuation in the upper mantle behind island arcs using pP waves, *J. Geophys. Res., 80*, 1079, 1975.

Berckhemer, H., Some aspects of the evolution of marginal seas deduced from observations in the Aegean region, *International Symposium on the Structural History of the Mediterranean Basins, 303*, 1977.

Comninakis, P. E. and B. C. Papazachos, Seismicity of the eastern Mediterranean and some tectonic features of the Mediterranean ridge, *Geol. Soc. Amer. Bull., 83*, 1093, 1972.

Comninakis, P. E. and B. C. Papazachos, *A catalogue of earthquakes in the Mediterranean and the surrounding area for the period 1901-1975*, University of Thessaloniki Geophysical Laboratory, 1978.

Critikos, N., Sur la sismicite des Cyclades et de la Crete, *Ann. de l'Observatoire National d'Athenes, 9*, 1926.

Delibasis, N., Focal mechanisms of the intermediate earthquakes of the area of Greece and their intensity distribution (in Greek), Sci. D. thesis, University of Athens, Athens, Greece, 1968.

Drakopoulos, J., P. Comninakis, A. Economides, S. Tassos, J. Latousakis, and J. Papis, *Bull. Nat. Obs. Athens, Seism. Inst.*, 1978.

Fedotov, S. A., Regularities of the distribution of strong earthquakes of Kamchatka, the Kurile Islands, and northeastern Japan, *Tr. Inst. Fiz. Zemli Akad. Nauk SSR, 66*(36), 1965.

Galanopoulos, A., On the intermediate earthquakes in Greece, *Bull. Seism. Soc. Amer., 43*, 159, 1953.

Galanopoulos, A., Tsunamis observed on the coasts of Greece from antiquity to the present time, *Ann. di Geofis., 8*, 369, 1960.

Galanopoulos, A., On mapping of seismic activity in Greece, *Ann. di Geofis., l6*, 37, 1963.

Galanopoulos, A., The seismotectonic regime in Greece, *Ann. di Geofis., 20*, 109, 1967.

Garza, T. and C. Lomnitz, The Oaxaca gap: A case history, *Pure and Appl. Geophys., 117*, 1187, 1979.

Gordon, D. W., T. Y. Bennett, and R. B. Herrman, The south-central Illinois earthquake of November 9, 1968: Macroseismic studies, *Bull. Seism. Soc. Amer., 60*, 953, 1970.

Gutenberg, B. and C. F. Richter, Deep-focus earthquakes in the Mediterranean region, *Geofisica pura e applicata, XII*(Fasc. 3-4), 1948.

Habermann, R. E., *Precursory seismicity patterns: Stalking the mature seismic gap*, this volume, 1981a.

Habermann, R. E., Seismicity rates in the central Aleutians: A quantitative study of precursors to moderate earthquakes, *J. Geophys. Res.*, 1981b.

Hatori, T., Vertical displacement in a tsunami source area and the topography of the sea bottom, *Bull. Earthq. Res. Inst., 44*, 1449, 1966.

Hsu, K. J. and W.B.F. Ryan, Deep-sea drilling in the Hellenic trench, *Bull. Geol. Soc. Greece, 81*, 1973.

Isacks, B. and P. Molnar, Distribution of stresses in the descending lithosphere from a global survey of focal-mechanism solutions of mantle earthquakes, *Rev. Geophys., 9*, 103, 1971.

Kanamori, H. and D. L. Anderson, Theoretical basis of some empirical relations in seismology, *Bull. Seism. Soc. Amer., 65*, 1073, 1975.

Karig, D. E., Origin and development of marginal basins in the Western Pacific, *J. Geophys. Res., 76*, 2542, 1971.

LePichon, X. and J. Angelier, The Hellenic arc and trench system: A key to the neotectonic evolution of the eastern Mediterranean area, *Tectonophysics, 60*, 1, 1979.

LePichon, X., J. Angelier, J. Aubouin, N. Lyberis, S. Monti, V. Renard, H. Got, K. Hsu, Y. Mart, J. Mascle, D. Matthews, D. Mitropoulos, P. Tsoflias, and G. Chronis, From subduction to transform motion; A seabeam survey of the Hellenic trench system, *Earth Planet. Sci. Letters, 44*, 441, 1979.

Lydecker, G., Seismizitaetsstudien im Bereich des Peloponnes auf Grund von Praezisionsherdbestimmungen, Dissertation, Geowissenschaften, Universitaet Frankfurt, Deutschland, 1975.

McCann, W., S. Nishenko, L. R. Sykes, and J. Krause, Seismic gaps and plate tectonics: Seismic potential for major boundaries, *Pure and Appl. Geophys., 117*, 1082, 1979.

McKenzie, D. P., Plate tectonics of the Mediterranean region, *Nature, 226*, 239, 1970.

McKenzie, D. P., Active tectonics of the Alpine-Himalayan belt: the Aegean Sea and surrounding regions, *Geophys. J. R. Astron. Soc., 55*, 217, 1978.

Mogi, K., Some features of recent seismic activity in and near Japan (2), Activity before and after great earthquakes, *Bull. Earthq. Res. Inst., 47*, 1172, Univ. of Tokyo, 1969.

Molnar, P. and J. Oliver, Lateral variations of attenuation in the upper mantle and discontinuities in the lithosphere, *J. Geophys. Res., 74*, 2648, 1969.

Nishenko, S. and W. McCann, Large thrust earthquakes and tsunamis: Implications for the development of fore arc basins, *J. Geophys. Res., 84*, 573, 1979.

Nuttli, O. W., The Mississippi valley earthquakes of 1811 and 1812: intensities, ground motion and magnitudes, *Bull. Seism. Soc. Amer., 63*, 227, 1973.

Ohtake, M., T. Matumoto, and G. V. Latham,

Seismicity gap near Oaxaca, Southern Mexico as a probable precursor to a large earthquake, *Pure and Appl. Geophys.*, *115*, 375, 1977.

Papazachos, B. C., Distribution of seismic foci in the Mediterranean and surrounding area and its tectonic implications, *Geophys. J. R. Astron. Soc.*, *33*, 421, 1973.

Papazachos, B. C. and P. E. Comninakis, Geophysical and tectonic features of the Aegean Arc, *J. Geophys. Res.*, *76*, 8517, 1971.

Papazachos, B. C. and N. D. Delibasis, Tectonic stress field and seismic faulting in the area of Greece, *Tectonophysics*, *7*, 321, 1969.

Richter, C. F., in *Elementary Seismology*, edited by J. Grithely and A. O. Woodford, W. H. Freeman and Company, San Francisco, 1958.

Rothe, J. P., The seismicity of the earth 1953-1965, *UNESCO*, 336 , 1969.

Sieberg, A., Untersuchungen ueber Erdbeben und Bruchschollenbau im oestlichen Mittelmeergebiet, *Denkschriften der med.-naturw. Ges. zu Jena*, *18. Band*, 2. Lief, 1932.

Sponheuer, W., Methoden zur Herdtiefenbestimmung in der Makroseismik, *Freiburger Forschungshefte*, *C88*, Akademic Verlag, 1960.

Stuart, W. D., Diffusionless dilatancy model for earthquake precursors, *Geophys. Res. Lett.*, *1*, 261, 1974.

Sykes, L. R., Aftershock zones of great earthquakes, seismicity gaps, and earthquake prediction for Alaska and the Aleutians, *J. Geophys. Res.*, *76*, 8021, 1971.

Toksoz, M. N., A. F. Shakal, and A. J. Michael, Space time migration of earthquakes along the North Anatolian fault zone and seismic gaps, *Pure and Appl. Geophys.*, *117*, 1258, 1979.

Wu, F. T. and H. Kanamori, Souce mechanism of February 4, 1965, Rat Island earthquake, *J. Geophys. Res.*, *78*, 6082, 1973.

Wyss, M., The thickness of deep seismic zones, *Nature*, *242*, 255, 1973.

Wyss, M. and J. N. Brune, The Alaska earthquake of 28 March 1964: A complex multiple rupture, *Bull. Seism. Soc. Amer.*, *57*, 1017, 1967.

Wyss, M. and R. E. Habermann, Seismic quiescence precusory to a past and a future Kurile Island earthquake, *Pure and Appl. Geophys.*, *117*, 1195, 1979.

Wyss, M., F. W. Klein, and A. C. Johnston, Precursors to the Kalapana M=7.2 earthquake, *Nature*, *289*, 231, 1981.

SHORT- AND LONG-TERM PATTERNS OF SEISMICITY IN KHORĀSĀN, IRAN

Mansour Niazi

TERA Corporation, 2150 Shattuck Avenue,
Berkeley, California 94704

Abstract. The Khorāsān province of north-eastern Iran has suffered scores of destructive earthquakes since 1968, including three major ones in 1968, 1978 and 1979. Three sets of data regarding the seismic activity in this province, at three time scales and magnitude levels, are available. They include the list of major historical earthquakes covering at least eight centuries, the strong regional earthquakes of the present century located by means of teleseismic information, and small regional and local earthquakes of the last few years located by the readings of the local seismographic network currently in operation. The distribution pattern of the instrumental epicenters suggests a tendency for clustering and migration across the tectonically controlled features of the region. A comparison of the three sets of data is made in an attempt to find answers to the following questions: (1) How representative the short-term (3+ years) seismic activity is of a) the 70-year seismicity picture constructed from the teleseismic coverage of larger events, and b) the 800-year activity as represented by the known major historical earth-quakes? (2) How characteristic are the cluster-ing and migration patterns of the regional events prior to major earthquakes? (3) How reasonably the regression parameters of the magnitude-fre-quency curve of the short-term activity can pre-dict or justify the recurrence of the major intermediate range and historical earthquakes? It is suggested that the short-term activity of the few years duration in active intraplate regions may be extrapolated to project long-term seismicity of several decades reasonably well.

INTRODUCTION

Because of the slow nature of tectonic pro-cesses, short-term observations are often pro-jected to make long-term conclusions. Such an extrapolation involves assumptions of an homo-geneous Poisson model for the earthquake occur-rence process (Cornell, 1968). It is, however, important to be able to confirm the validity of this projection whenever possible, questioning particularly the legitimacy of applying obser-vations in one magnitude range to assess expec-tations in another, often expanded, magnitude range. Here, a comparison is made between the regional seismicity of the Khorāsān province in northeastern Iran as viewed at three different time scales. The data base consists of the locally monitored events of recent years, world-wide observations of several decades and histori-cal records of several centuries.

REGIONAL TECTONICS

Historically, the name Khorāsān has been used for a vast (geographical) region covering north-eastern Iran, western Afghanistan, extensive parts of the Soviet Central Asian Republics of Turkmenia, Uzbekistan and Tadjikestan. In this study, however, the discussions are confined to a nearly 10^6 km^2 region bound from 30 to 40 de-grees north latitude and 52 to 62 degrees east longitude. The major tectonic provinces of the region are the Kopeh Dagh folded belt (Tchalenko, 1975), the eastern Alborz and the central and eastern Iran province, an area of complex block movement. Near its southwest corner, the region approaches the Zagros main thrust line. Along its eastern edge, the region is bound by the north-south trending Harirud fault which, de-spite its current aseismic character and pre-Jurassic age (Stöcklin, 1974), serves as a bound-ary between the aseismic zone of western Afghanistan and the highly seismic region of northeastern Iran (Taheri and Niazi, 1980). The prevailing tectonic environment in east-central Iran (the southern half of the region covered by Figures 1 and 3), as part of the so-called East-Central Iran Microcontinent (Takin, 1972), is that of complex horst and graben structure.

This truly intraplate environment extends over dimensions of several hundred kilometers in each direction and is bounded by an ophiolite ring, the southernmost extent of which coincides with the inner ranges of the Makran coast, bound-ing the Jaz-Murian depression from the south.

There is sufficient field evidence to suggest that the whole region was reconstituted from smaller continental pieces compressed together since the Cretaceous period by the compressive forces which have persisted in the region (Stöcklin, 1974). The present compressive axis is directed NE-SW.

SHORT-TERM SEISMICITY

The first elements of a regional network of single component vertical seismographs were installed in the epicentral area of the 1968 Dasht-e Bayāz earthquake in 1970 on a temporary basis for monitoring the aftershock activity (McEvilly and Niazi, 1975). By the mid-70s, the network was expanded to include half a dozen semi-permanent stations (Figure 1) and began providing sufficient coverage for regional monitoring of seismicity at lower magnitude threshold than provided by teleseismic observations. The establishment of the regional network, furthermore, improved teleseismic coverage through incorporation of arrival time data of selected stations of the network. Starting from April, 1975, the hypocentral parameters of the regional earthquakes, determined locally, have been published semi-annually in the Bulletin of Seismographic Network of Ferdowsi University (BSNFU). Although the information of other regional stations including Shiraz, Tehran, Quetta and Kermanshah is selectively incorporated for hypocentral determination, the coverage is by no means expected to be uniform below magnitude 3.

The available list of locally monitored earthquakes from April, 1975 through July, 1978, include 378 events with magnitudes ranging between 2 and 5.5. The data displayed in Figure 1 represent over three years of observations prior to the Tabas and Qainat earthquakes of August, 1978 and November, 1979, both having magnitudes above 7, and may serve to illustrate specific features of regional seismicity preceding major earthquakes. Also in Figure 1, the regional pattern of geologic faults and distribution of ophiolitic rocks are shown. The triangles represent the location of seismographic stations of the regional network (SNFU). Besides, the conspicuous concentration of activity near the Dasht-e Bayāz fault around 34 degrees north and 59 degrees east and those occurring along the nearly east-west trends of Kopeh-Dagh and Alborz, the seismicity pattern is diffuse as is the case with most of the intraplate seismic zones. As will be noted in the following sections, the Dasht-e Bayāz structure has remained the most active feature of the region since August, 1968.

The histogram of the locally monitored events for which a local magnitude is assigned is presented in Figure 2 at three magnitude levels. Considering that the data for 1978 only cover the first six months, the yearly variations appear significant only at magnitude levels 5

and above. The magnitude frequency relationship of the data is shown in Figure 5 and will be discussed in the following sections.

SEISMICITY SINCE 1910

The hypocentral coordinates of this century's earthquakes (since 1910), which were located by teleseismic information, are taken from the USGS Hypocenter Data File and the epicenters are plotted in Figure 3. Although some of the foci are reported as deep as 60 km, more recent observations suggest that all the earthquakes in the region are shallow, crustal events with focal depths seldom exceeding 30 km. The magnitude information for teleseismic data is mixed. Prior to 1963, nearly a third of the listed hypocenters have surface wave magnitudes, often assigned by Pasadena. Nearly all these earthquakes have a listed magnitude larger than 5, as would be expected from the sparse distribution of regional stations. The minimum reported magnitude for the first quarter of this century is 6.5, and decreases to 5.6 in the second quarter. Accordingly, a magnitude threshold of 6.0 and 5.5 is assigned to the events of these two periods for which no magnitude is reported, respectively. By a similar consideration, the threshold magnitude is lowered to 5.0 and 4.5 for 1951-1960, and 1961-1970 decades, respectively. There are relatively few events for which magnitude is not reported after 1970. Since during this period a number of stations of the regional network were reporting their observations of arrival times, the threshold magnitude is expected to have been further lowered by 0.5 magnitude unit to ∿ 4. The epicentral map of all the events which were located on the basis of teleseismic observations during the seventy-year period 1911-1980 is displayed in Figure 3. When two or more earthquakes within 0.5 magnitude units of each other have identical locations, only one box is plotted, while a count is maintained by repeated vertical bars within the box. As can be readily observed, except for the concentration of the epicenters near 33.5 degrees north, 57.0 degrees east, which belongs to the Tabas sequence of 1978 (Berberian, et al, 1979), Figures 1 and 3 representing the short- and intermediate-term seismicity of the region, bear a striking resemblance. The Tabas earthquake of September 16, 1978, Ms 7.8 (BRK), is probably the strongest regional earthquake of this century which interrupted several centuries of quiescence (Berberian, 1979), and as seen in the short-term seismicity map (Figure 1) covering over three years of data up to two months prior to its occurrence, with little or no apparent seismological warning. There is, however, a slight suggestion of Mogi's (1969) "donut" pattern of moderate to large earthquake epicenters to have developed in the preceding 25 years around the Tabas area.

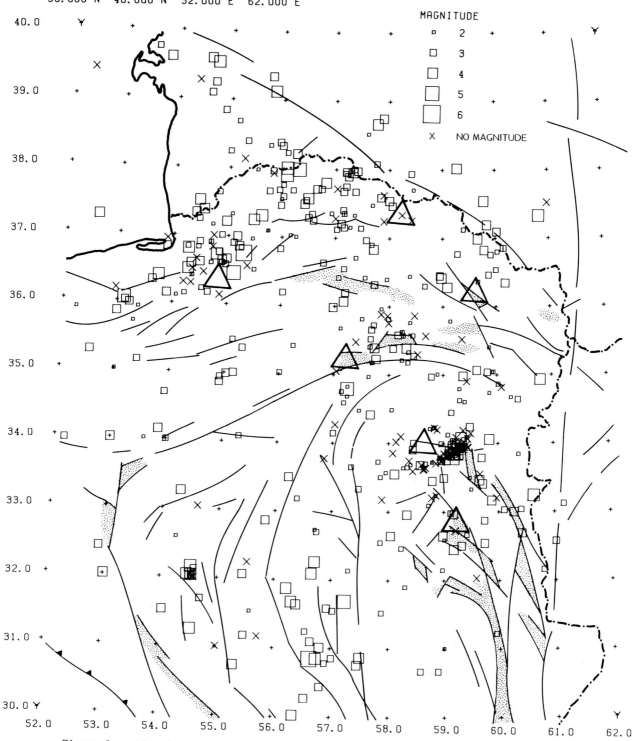

Figure 1. The short-term seismicity of the region as monitored by the regional network. The triangles show the location of recording stations. The shaded areas show the distribution of ophiolitic rocks. The background tectonic features are from Stöcklin (1974).

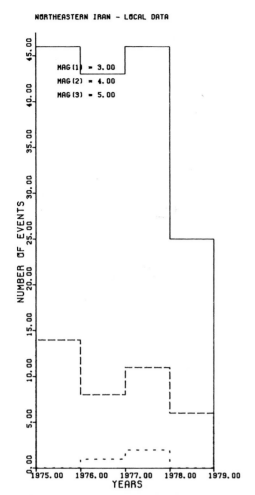

NORTHEASTERN IRAN - LOCAL DATA

MAG (1) = 3.00
MAG (2) = 4.00
MAG (3) = 5.00

Figure 2. The histogram of short-term seismicity at local magnitude levels 3, 4 and 5.

The histogram of earthquakes constituting the intermediate range instrumental data in successive decades since 1911 at three magnitude levels is given in Figure 4. All the reported pre-1960 data have been assigned a magnitude 5 and above. As seen in Figure 4, at this magnitude level, except for 1910s and 1940s, during which either due to insufficient coverage or because of the lack of significant activity, the reported seismicity is relatively low, overall regional activity does not vary substantially. An important observation concerning the intermediate-term instrumental seismicity is a tendency of large earthquakes to cluster in space and time. Most of the pre-1950 major earthquakes of the region including the destructive earthquakes of 1871, 1893, 1895, 1925, 1929, 1946 and 1948, occurred within the structural boundaries of Kopeh Dagh. However, with the exception of the July 30, 1970 earthquake (Ms 6.6), this zone has produced no destructive earthquakes since 1950. Instead, all

the strong earthquakes of the region since then have their epicenters in the Alborz and east-central Iran provinces. They include February 12, 1953; July 2, 1957; September 1, 1962; August 31 and September 1, 1968; November 7, 1976; September 16, 1978; January 16, 1979; and finally November 14 and 27, 1979. The 1968-1979 Dasht-e Bayāz - Tabas cluster will be discussed separately below.

DASHT-E BAYĀZ - TABAS CLUSTER

In a time window of just over ten years, within a radius of 150 km in eastern Iran, an unprecedented intensive seismic activity has produced 8 earthquakes of magnitude 6 and above including 3 major earthquakes of August 31, 1968, September 16, 1978, and November 27, 1979. The 1968 and 1979 events have been associated with similar movements on different segments of the same fault (Niazi, 1969; Haghipour and Amidi, 1980). Curiously, the causative Dasht-e Bayāz fault was identified among 28 potential structures as likely sites for future strong earthquakes in May, 1978 (Iwan, 1978, p. 13) with a 20% probability of the occurrence for an earthquake of M>6.5 in the following ten years. Within less than two years, two earthquakes M> 6.5 occurred near the eastern end of the east-west structure, one of which activating a north-south trending conjugate structure which bounds the Dasht-e Bayāz fault to the east (Haghipour and Amidi, 1980). The recent clustering of activity in the Dasht-e Bayāz - Tabas region delineates an active zone along the northern boundary of otherwise relatively rigid Lut Block, which despite their diverse fault geometry, suggest source mechanisms in general agreement with a NE compression.

HISTORICAL SEISMICITY

A list of historical earthquakes in the region is given in Table 1. The period spans nearly eleven centuries, from 810 through 1910, A.D. The July 2, 1895 earthquake with a maximum intensity IX has been assigned $M_{LH}=8.2$ by Shebalin (1974). Five events have been classified as major earthquakes with M > 7 by Ambraseys and Melville (1977). There are two notable gaps between 850 - 1050 A.D. and 1687 - 1871 A.D., each lasting nearly two centuries, for which no earthquake is listed. The lack of reported seismicity is likely to be due to the incomplete record. Despite the questions regarding its completeness, the data suggest an apparent clustering in time and space. The Neyshabur area, for instance, appears to be most active between 1208 and 1405 A.D., resulting in repeated destruction of the towns and villages around that cultural center. The approximate macroseismic centers of historical activity are also shown in Figure 3 with simple and multi-

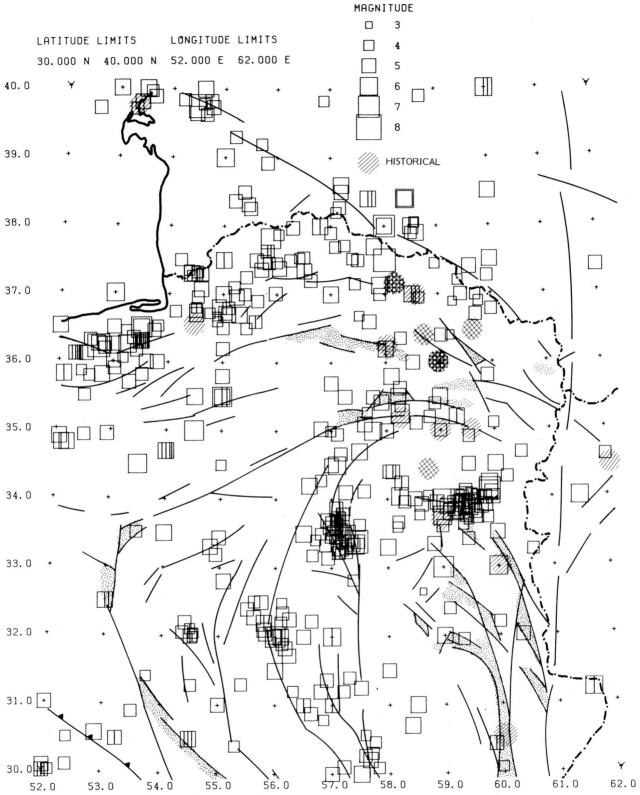

Figure 3.   The distribution of teleseismically located events from 1911 through 1979.   The reported pre-1910 activity as far back as the 9th century A.D. is shown by the hatched circles.   Multiple events are identified by vertical bars or cross hatching.

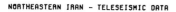

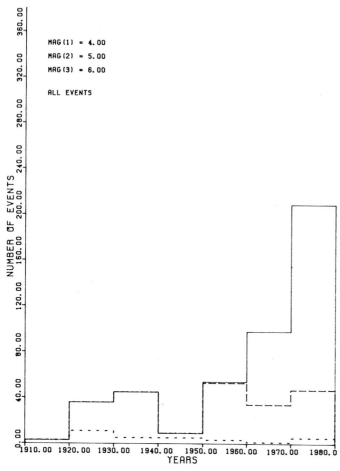

Figure 4. The histogram showing intermediate-range teleseismically determined regional activity in the last seventy years at three magnitude levels. The data also include the events with no assigned magnitude to which a threshold magnitude based on the smallest reported magnitude in each period is assigned.

hatched circular patches, designating centers of single or multiple occurrence, respectively. By and large, the reported historical seismicity appears to be sketchy, coinciding with the contemporaneous centers of population and would require further search and analysis for meaningful correlation with geological structures of the region. Further evidence for their incomplete coverage is provided in their recurrence analysis discussed below.

MAGNITUDE-FREQUENCY RELATIONSHIP

The magnitude frequency plots of both short-term local data and intermediate-term teleseismic data are presented in Figure 5. The teleseismic data below magnitude 6 are shown in two forms. The squares show only the events for which a magnitude had already been assigned in the original USGS list. The asterisks (*'s) represent all the events after a threshold magnitude is assigned to various periods based on the minimum reported magnitude in each period. Also, because of structural complexities in east-central Iran, major earthquakes of this region display a complicated multiple source character (Niazi, 1969) resulting in a pronounced saturation of body wave magnitude scale with a possible ceiling at $m_b = 6.5$. The Tabas earthquake of September 16, 1978, possibly the strongest regional earthquake of this century (Berberian, 1980), for instance, has an estimated body wave magnitude of 6.5. For the early earthquakes on USGS file which occurred prior to 1960, however, often the 20-second surface wave magnitude estimated at Pasadena (Gutenberg and Richter, 1954), is listed. Because of the growing difference between M_s and m_b beyond $M_s = 6.5$, and for the sake of uniformity with earlier magnitudes, the reported M_s was used for recent earthquakes whenever $M_s \gtrsim 6.5$.

The regression lines thus fitted to the linear portion of frequency magnitude plots of Figure 5 to the two sets of data have the following forms

$$\log N = 5.011 \pm 0.104 - (0.879 \pm 0.025)M_L; \text{for} M_L \gtrsim 3.0 \quad (1)$$

$$\log N = 6.330 \pm 0.253 - (0.802 \pm 0.041)M_s; \text{for} M_s \gtrsim 5.0 \quad (2)$$

Extension of formula (1) to a nearly 22 fold observation time, corresponding to the duration of teleseismic observation, will increase the coefficient a from 5.011 to 6.361 nearly identical to the corresponding coefficient in (2). Considering also the closeness of the b coefficients in the two formulae, it may be safe to say that the short time observations can be extrapolated to predict recurrence of strong regional earthquakes at an expanded time and magnitude scales of several decades and few magnitude units. This argument, however, cannot be extended to historical record, perhaps due to its nonuniformity and incompleteness. In an analysis of magnitude/frequency relationship of recent strong earthquakes in Kuhistan, corresponding to nearly 1/3 of the areal extent of Figures 1 and 3 and to the SE corner of the map, during this century, Ambraseys and Melville (1977) give a 39-year return period for magnitude 7 and 6.6 years for magnitude 6. On this basis alone, the occurrence of three major earthquakes within an 11-year period, 1968-1979, in this subregion indicated that the current seismicity is well above normal. The recent localization of activity in east-central Iran, however, should not be considered a true representative of long-term seismicity, partially due to clustering and internal migration of activity

Table 1. List of Historical Earthquakes and Their Approximate Location

Date (A.D.)	N. Lat. (°)	L. Long. (°)	Intensity/Magnitude		Location	Reference *
810	30 1/2	60		> 7		AM
848-49	34.4	62.2	VII-IX		Herat	QJ
1052, Jun 2	36 1/4	58			Sabzevar	B
1208	36 1/2	58 3/4		> 7	Neyshabur	W/AM
1238	34 1/2	58 3/4			Gonabad	B
1251	36 1/2	58 3/4			Neyshabur	B
1270	36	59			Neyshabur	B
1366, Oct 21	35	59 1/2		> 7	Khwaf	AM
1389	36	59			Neyshabur	B
1405	36	59			Neyshabur	W
1493, Jan 10	33	60		> 7	Nozad	AM/B
1549	33 3/4	59			Qaen	W/AM
1619, May	35	59			{ Torbat-e Heydarieh	W
1673, May 14	36	59			Neyshabur	W
1673, July 30	36 1/2	59 3/4		> 7	Mashad	AM/B
1678	34 1/2	58 3/4			Gonabad	B
1687, April	36 1/2	59 3/4			Mashad	W
1852, Feb 22 (?)	37 1/4	58 1/4			Quchan	W
1871, Dec 23	37 1/4	58 1/4			Quchan	W
1872, Jan 6	37 1/4	58 1/4			Quchan	W
1890, July 11	36.5	54.6			Tash	A
1893, Nov 17	37.1	58.4			Quchan	S
1895, Jan 17	37.1	58.5			Quchan	S
1895, July 9	39.8	53.3	IX	(8.2)	Krosnovodsk	SH
1903, Sept 25	35.2	58.2		6.4	Kashmar	AM

```
*    A = Ambraseys (1976)
    AM = Ambraseys and Melville (1977)
     B = Berberian (1979)
    QJ = Quittmeyer and Jacob (1979)
     S = Savarensky et al (1962)
    SH = Shebalin (1974)
     W = Wilson (1930)
```

within the region. We feel, however, in an active area such as northeast Iran periods of about a century should provide a reasonably stable seismicity pattern. Referring to the available historical records of Table 1, however, the inevitable conclusion is that they underestimate the regional seismic activity. Even if all the listed earthquakes of this table are considered equivalent to magnitude 7 and above, which is a clear exaggeration, allowing for the nearly 400-year gap in the data, it would lead to a 30-year recurrence period for major earthquakes. This predicted time interval still appears too long in veiw of this century's observations and thus, reconfirms the expressed belief that many of the historical earthquakes of the region are not yet accounted for.

CONCLUSIONS

Comparison of the seismicity pattern in the Khorāsān province, NE Iran, viewed from three different time scales of a few years, several decades and several centuries suggests: (1) the short-term local observation of small earthquakes may provide a reasonable representation of the long-term seismicity of the region, both for recurrence estimate and for delineation of active features. As for other intraplate regions, the projection may fail to foresee unexpected activity along seemingly dormant tectonic features with unusually long return periods. For the Tabas earthquake of September 16, 1978, for instance, no direct indication of the pending catastrophy could have been inferred, even by considering the available historical record; (2) in retrospect, however, it appears that the frequency of strong earthquakes around Tabas showed a marked increase, perhaps as early as two decades prior to its occurrence, in a so-called "donut" pattern as suggested by Mogi (1969); (3) there is strong evidence of clustering and migration of strong earthquakes across tectonic boundaries with a time scale of a few decades.

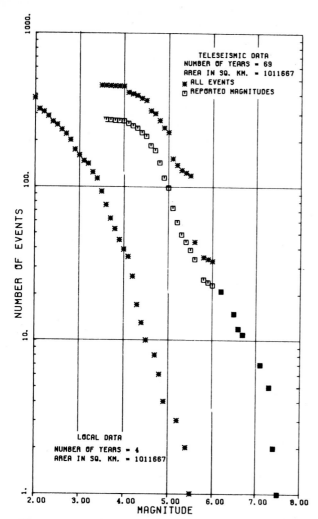

Figure 5. The magnitude-frequency plot for short-and intermediate-range seismicity. The double-valued part of the teleseismic data below magnitude 6 is caused by assigning a threshold magnitude to the events for which no magnitude was reported.

Acknowledgements. The teleseismic data was supplied by the National Geophysical and Solar Terrestrial Data Center of NOAA. The acquisition of the first three portable seismographs and the initial field investigations which led to the establishment of the regional seismographic network in Khorāsān was supported by the National Academy of Science through a grant from the Arthur L. Day Bequest and the Carnegie Institution of Washington, through a Harry Oscar Wood award. The efforts of the staff of the Geophysics Department of the University of Mashad, in maintaining the network, and in collecting, analyzing and publishing the local data, are acknowledged.

REFERENCES

Ambraseys, N.N., Studies in historical seismicity and tectonics, Geodynamics, Roy. Soc. London, 7-16, 1976.

Ambraseys, N.N. and C.P. Melville, The seismicity of Kuhistan, Iran, Geog. J., part II, 179-199, 1977.

Berberian, M., Earthquake faulting and bedding thrust associated with the Tabas-e Golshan (Iran) earthquake of September 16, 1978, Bull. Seism. Soc. Am., 69, 1861-1887, 1979.

Berberian, M., I. Asudeh, R. G. Bilham, C.H. Scholz, and C. Soufleris, Mechanism of the main shock and the aftershock study of the Tabas-e Golshan (Iran) earthquake of Sept. 16, 1978, Bull. Seism. Soc. Am., 69, 1851-1859, 1979.

Cornell, C.A., Engineering seismic risk analysis, Bull. Seism. Soc. Am., 58, 1583-1606, 1968.

Gutenberg, B. and C. F. Richter, Seismicity of the Earth, Princeton University Press, 2nd ed. 1954.

Haghipour, A. and M. Amidi, Geotectonics of Ghaenat earthquakes of NE Iran (November 14 to December 9, 1979) submitted for publication, Bull. Seism. Soc. Am. 1980.

Iwan, W.D., Strong-motion earthquake instrument arrays, Proceedings of the International Workshop, Honolulu, May 2-5, 1978.

McEvilly, T.V. and M. Niazi, Post earthquake observations at Dasht-e Bayāz, Iran, Tectonophysics, 26, 267-279, 1975.

Mogi, K., Relation between the occurrence of great earthquakes and tectonic structures, Bull. Earthq. Res. Inst., 47, 429-451, 1969.

Niazi, M., Source dynamics of the Dasht-e Bayāz earthquake of August 31, 1968, Bull. Seism. Soc. Am., 59, 1843-1861, 1969.

Quittmayer, R.C. and K. H. Jacob, Historical and modern seismicity of Pakistan, Afghanistan, northwestern India and southeastern Iran, Bull. Seism. Soc. Am., 69, 773-823, 1979.

Savarensky, S.P., S. L. Soloviev and D. A. Kharin (eds.), Seismologic Atlas of the USSR, 1962.

Shebalin, N.V., Strong earthquakes in the USSR Territories, Izdatel'stvo "Nauk", M., 1-54, 1974.

Stöcklin, J., Possible ancient continental margins in Iran, Geology of Continental Margins, C.A. Burk and C.L. Drake, eds., Springer-Verlag, New York, 1974.

Taheri, S.J. and M. Niazi, Seismicity of the Iranian Plateau and the bordering regions, in print, Bull. Seism. Soc. Am., 1980.

Takin, M., Iranian geology and continental drift in the Middle East, Nature, 235, 147-150, 1972.

Tchalenko, J.S., Seismicity and structure of the Kopeh Dagh (Iran, U.S.S.R.), Phil. Trans. Roy. Soc. London, 278 (1275), 1-25, 1975.

Wilson, A.T., Earthquakes in Persia, Bull. School Orient. Studies, London Univ., 6, 103-131, 1930.

A REVIEW OF GEOLOGICAL EVIDENCE
FOR RECURRENCE TIMES OF LARGE EARTHQUAKES

Kerry E. Sieh

Division of Geological and Planetary Sciences
California Institute of Technology
Pasadena, CA 91125

Abstract. The geological record of the past several thousand years contains valuable information for evaluating the earthquake potential of the earth's major fault systems. Geologists have begun to characterize past and, presumably, future behavior of active faults and recurrence intervals for large earthquakes by studying 1) uplifted marine terraces, 2) fault-scarp morphology, 3) physiographic features offset along faults, and 4) faulted or otherwise deformed young sediments.

Along the convergent plate margins of Alaska and Japan, for example, studies of uplifted marine terraces have aided in evaluating the likelihood of imminent rupture of faults in two seismic gaps. In Nevada, Utah, and eastern California, detailed studies of scarp morphology along normal faults of the Basin and Range Province are beginning to reveal the recurrence intervals, sizes, and patterns of prehistoric earthquakes. Studies of offset stream channels along the San Andreas fault have shown that right-lateral events of as much as 10 m have occurred repeatedly in the past with an average frequency of about two hundred years. Elsewhere along the San Andreas and on other faults in California and Japan, studies of faulted and deformed young sediments have enabled dating of specific prehistoric earthquakes or, at least, a determination of the minimum number of events that occurred during the deposition of the strata.

In the western U.S. and Japan and perhaps in other seismically active regions as well, there is good reason to believe that within the decade we will know the average recurrence intervals, regularity, and sizes of past large seismic events at several localities. Hopefully, this will enable forecasts of some future large earthquakes with uncertainties measured in decades rather than centuries and will provide a sound basis for hazard mitigation and for directing short-term predictive efforts to those fault segments in imminent danger of rupture.

Introduction

Purpose and Organization

An important element in the development of earthquake prediction and mitigation capabilities is knowledge of the long-term behavior of seismogenic faults. In a given earthquake-prone region the duration of the dormant period between large earthquakes, the regularity of that period, and the date of the latest event all aid in assessing whether or not an earthquake will occur in the near future.

Unfortunately, in most seismic regions, the period of historical and instrumental record is far too short to allow determination of the long-term seismic behavior for that region. However, the geologic record of the Holocene and late Pleistocene epochs is proving to be quite valuable in understanding long-term patterns of earthquake recurrence.

This paper calls attention to numerous studies of the geologic record that have aided in understanding the relationships of slip events on major faults in space and time. Although work in several tectonic settings around the globe is discussed, this is not an exhaustive review; rather, my intention has been to illustrate realized contributions and to suggest future contributions of geologic studies in evaluating long-term seismic behavior.

The first three sections of the main text describe and discuss studies of oceanic megathrusts, continental dip-slip faults, and strike-slip faults, in that order. Each of these three types of fault is associated with a distinctive geological environment, and each environment has fostered distinctly different approaches in studying long-term seismic behavior.

The fourth section of the main text outlines studies of the San Andreas fault. I have segregated these from discussion of other strike-slip faults because I wish to summarize and evaluate current understanding of the geologically recent behavior of that great fault.

The fifth major section discusses my perceptions of current problems and limitations and my impressions of new directions in which this field may be headed.

Importance of Past Seismic Behavior

Samplings from a broad spectrum of recent papers dealing with the evaluation of seismic potential [e.g. Allen, 1975; Kelleher et al., 1973; and Shimazaki and Nakata, 1980] reveal the general belief in a reconstituted uniformitarian principle. Simply stated, that guiding principle is that the historical and more ancient seismic past is a key to the seismic future. The degree to which this is true determines the predictive value of studying historical and more ancient seismicity.

Concept of average recurrence interval. The historical record does indicate that in several areas large seismic events are, to a remarkable degree, repeat performances. Large subduction events off the southwestern coast of Japan and southern coast of Chile are celebrated historic examples.

History records large Chilean earthquakes in 1575, 1737, and 1837 that were very similar to the great earthquake of 1960 (Fig. 1A) [Lomnitz, 1970; Kelleher, 1972]. If one assumes that the historical record for this region is complete, three periods of 162, 100, and 123 years separate these four great Chilean events. History also records a remarkable series of great Japanese earthquakes between 684 and 1946 A.D. (Fig. 1B) [Ando, 1975; Ishibashi, 1980]. Failure of a 200-km-long segment of the Nankai Trough megathrust occurred in 684, 887, 1099, 1361, 1605, 1707, 1854, and 1946. A similar set of dates applies to the adjoining 200-km-long segment, although there are significant ambiguities in interpretating the historical records for that segment. The range in intervals between events along both segments may be exaggerated by the exclusion of undiscovered events between 1707 and 684 A.D. Nevertheless, the historically recorded events affecting the western 200-km-long segment have occurred 92 to 262 years apart.

These two examples from Chile and Japan and others lend credence and usefulness to the concept of an average recurrence interval -- a number indicating the average period between large events for a given region or source. The average recurrence interval is 180 yrs in the Japanese example and 128 yrs in the Chilean case. In these two historical cases actual recurrence intervals stray from the average value by as much as 45%, if one assumes both records are complete. Using only the records of the latest 2-1/2 centuries, which are almost certainly complete, no actual interval varies from the average by more than 23%.

The historical record strongly suggests that neither Chile nor Japan need fear a repetition of the events discussed above for at least half a century and perhaps longer. The variation in actual recurrence intervals, however, precludes a more precise forecast of the date of the next great earthquake at either locality. Nevertheless, knowing these historical recurrence intervals may prove to be critical in motivating efforts at hazard mitigation and in predicting future great events by geodetic or geophysical methods. Likewise, geological studies that augment the historical earthquake record by providing average recurrence intervals increase the possibility of forecasting or predicting future seismic activity.

Concept of repeated earthquakes. The degree to which large fault ruptures physically mimic or duplicate their "predecessors" is poorly known, since documentation of the displacements, rupture lengths, and other physical characteristics of large events is sparse. The few examples that do seem to be adequately documented suggest that the physical characteristics of sequential events can be remarkably similar or appreciably different. In the Japanese example discussed above, for example, segments C and D tend to lead A and B by several hours or years, but in 1707 and possibly in 887 A.D. they broke at once. Also, the rupture of segments C and D in 1854 extended to the east, into Suruga Bay. This extension did not occur when C and D last broke, in 1944 [Ishibashi, 1980] -- thus, the current concern that the heavily industrialized Suruga Bay area (Tokai District) may soon experience a large earthquake.

The concept that earthquake and fault ruptures repeat and have repeat times that cluster around an average recurrence interval may not be valid in some regions. The time of occurrence and characteristics of a fault rupture in a region containing many faults of similar size and degree of activity may be influenced by far more complex cycles of stress application, unloading, and redistribution than are envisioned for "simple" plate boundaries characterized by one predominant, master fault. The deviations from the average recurrence interval for a fault in a region of multiple large seismic sources may be so great that knowing the average recurrence interval has little predictive value.

The above cautions notwithstanding, the availability and completeness of the known record of past earthquakes governs to a large degree our ability to understand, forecast, and thus, to predict future large seismic events. As the historical period lengthens, well-documented spatial and temporal relationships of large earthquakes will emerge in much greater detail than is presently available. Until the historical/instrumental period is several average recurrence intervals long, however, the need for augmentation of the historical record and discovery of the more ancient record will remain acute.

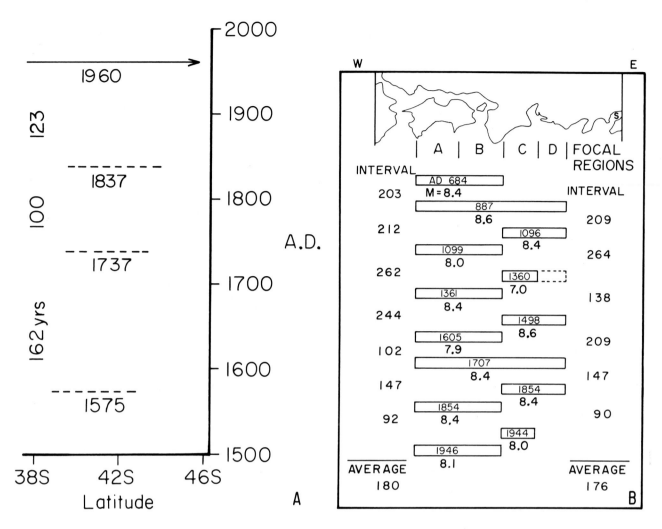

Fig. 1. A: Three large Chilean earthquakes in 1837, 1737, and 1575 preceded the great 1960 earthquake (from Kelleher, 1972). B: A remarkable series of large earthquakes has occurred between
684 and 1946 A.D. in coastal southwestern Japan [from Yonekura, 1975, based on Ando, 1975 and
other sources]. Both the Chilean and Japanese localities demonstrate a degree of regularity in
style and timing of large earthquakes. For a slightly different interpretation of the historical
documents, see Ishibashi [1980, Fig. 1]. S = Suruga Bay.

Convention Regarding the Reporting of ^{14}C Dates

In the many studies discussed below, radiocarbon
dating of materials is the principal means by
which actual ages can be assigned to ancient
earthquakes. In this paper I have adopted the
following convention for reporting these radiocarbon dates: Dates based on ^{14}C analyses
which have not been corrected for isotopic
fractionation or for known temporal variations in
atmospheric abundance of ^{14}C are given as "^{14}C
yrs B.P." ("B.P." means "before 1950 A.D.").
Corrected or calendar ages are given as "yrs
B.P." or "A.D."

Emergent Shorelines and Great
Subduction-Zone Earthquakes

Boso/Oiso Area, Japan

Studies of the history of coastal uplift as
recorded in emergent shorelines and related marine
terraces are currently the major avenue for
determining the long-term spatial and temporal
characteristics of large subduction zone earthquakes. Recent studies of the coastal terraces
south of Tokyo, Japan [Matsuda et al., 1978 and
Nakata et al., 1979] are exemplary. Not only
are marine terraces providing information on the

repeat time of great earthquakes there, but they are also enabling a relatively detailed under-standing of mechanisms and spatial patterns. Geologically, the region is characterized by complex (and still somewhat enigmatic) active shallow-dipping structures related to subduction of the Philippine Sea plate (Figs. 2 and 3). Forecasting future events is critically important in this heavily populated, urbanized region which suffered devastation in the great earth-quakes of 1923 and 1703.

The 1703 earthquake was accompanied by sudden emergence from the sea of a wave-cut platform and shoreline. Additional uplift of the shoreline and nearby region was measured geodetically following the 1923 event. The present elevation of the pre-1703 (Genroku) shoreline is the sum of the 1703 and 1923 uplifts minus subsidence that occurred between the events and after 1923 (Fig. 3). From these uplifts and other information, Ando [1974] deduced geometries for the main faults on which displacement occurred at the time of the earthquakes (Fig. 3). The 1703 earthquake may have been produced principally by dextral strike slip on segments B and C, and the southeastern part of segment A. The 1923 earthquake may have resulted from slip on segment A alone.

Three progressively older and higher shorelines and platforms exist progressively landward of the Genroku shoreline and platform [Matsuda et al., 1978, Fig. 6]. Each shoreline displays a pattern of deformation very similar to, but of progres-

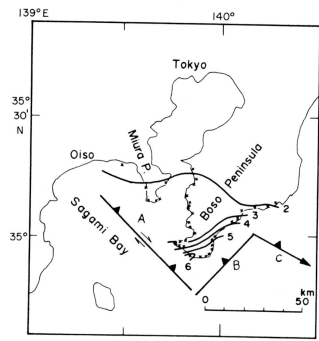

Fig. 3. Boso/Oiso area south of Tokyo. The contours (in meters) represent uplift asso-ciated with the great earthquakes of 1707 and 1923. They are based on the measured height above present sea level of the Genroku (pre-1703) shoreline. Crosses indicate localities of measurement [from Matsuda et al., 1978]. Uplift patterns for each earthquake were important in determining the rupture planes (A, B, and C) associated with the earthquakes [Ando, 1974].

sively greater magnitude than, the Genroku shore-line. Figure 4 illustrates the deformation of the oldest shoreline (Numa I). The similarity between this and the deformation of the youngest shoreline on the Boso Peninsula (Fig. 3) is remarkable. The dates of emergence from the sea of these older shorelines has been determined by Nakata et al. [1979]. Samples collected from the marine (pre-emergent) and non-marine (post-emergent) portions of the terraces yielded the radiocarbon ages plotted in Fig. 5A. These sug-gest emergence of Numa I ~5,500-6,150 ^{14}C yrs B.P., Numa II ~4,300 ^{14}C yrs B.P., and Numa III ~2,900 ^{14}C yrs B.P. Thus, three prehistoric great uplift events, or uplift sequences, very similar to the 1703/1923 sequence are documented in the marine terrace record on the Boso Penin-sula. As the periods between events, or event sequences, are ~1,000 to ~2,000 ^{14}C yrs long, the occurrence another event like the 1703 or 1923 earthquake is considered remote [Matsuda et al., 1978, p. 1616].

However, one significant difference between the uplift pattern of the Genroku shoreline and the older shorelines has caused great concern [Matsuda

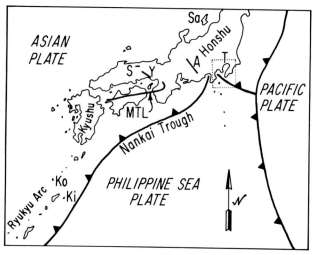

Fig. 2. Many geological studies pertinent to the assessment of seismic potential are being conducted in Japan and in the Ryuku Islands (adapted from Shimazaki and Nakata, 1980). A = Atera fault; Ki = Kikaijima; Ko = Kodakarajima; MTL = Median Tectonic Line; Sa = Sado Island; S = Shikano fault; T = Tokyo; Y = Yamasaki fault. Dot pattern indicates Boso/Oiso area south of Tokyo (Fig. 3); sawteeth are on upper plates of subduction megathrusts.

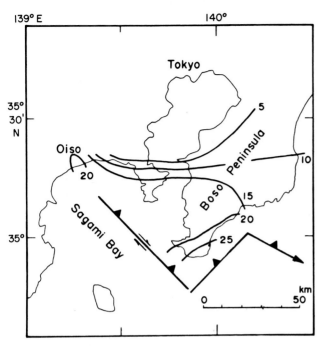

Fig. 4. The elevation of the Numa I shoreline above sea level ranges from a few meters
to >25 m. The similarity of this pattern of
deformation on the Boso Peninsula to that
of the Genroku shoreline indicates that the
1707 and 1923 uplifts mimicked earlier uplifts [from Matsuda et al., 1978]. The
dissimilarity of patterns in the Oiso region
lead Matsuda et al. [1978] to forecast a
great earthquake for that region.

et al., 1978, p. 1616-1617]. The 1703 and 1923
uplift in the Oiso region is much less than the
uplift on the Boso Peninsula (Fig. 3), whereas
the older terraces are at comparable elevations
in both regions (Fig. 4). If the past is a key
to the future, the Oiso area "is a candidate for
a future large earthquake", involving several
meters of uplift and rupture of nearby submarine
and subaerial faults [Matsuda et al., 1978].

Middleton Island, Alaska

The coastal terraces of Middleton Island, in
the northern Gulf of Alaska (M on Fig. 6), have
also provided valuable long-term information
concerning the recurrence of great earthquakes
[Plafker and Rubin, 1978]. Here also is a flight
of young terraces, the youngest of which was
lifted 3-1/2 m above the present shoreline
elevation during the great Alaskan earthquake of
1964. Mapping of the terraces and stratigraphic
studies of the terrace deposits are yielding a
detailed picture of several sudden uplifts of
the island during the past ~5,000 yrs. Current
estimates are that one or more closelytimed events
are represented by each terrace and that a new

terrace at least six to nine meters high has
emerged about every 850 ^{14}C yrs, on the average
(Fig. 7). Individual intervals shown in Fig. 7
deviate from this average by about 50%. The
lesser intervals are similar to estimates of ~480-
640 years made by dividing the "instantaneous"
(i.e. several-million-year average) plate rate of
62 mm/yr [Minster and Jordan, 1978] into the
maximum slip of ~30 to 40 m [Miyashita and
Matsu'ura, 1978, Fig. 6] on the Alaskan-Aleutian
megathrust associated with the 1964 earthquake.
Plafker and Rubin note that the 1964 uplift is
only one-third to one-half of the uplift values
estimated for previous events. Because of this,
they suggest, with certain reservations, that
another large event may produce additional uplift
within a period much shorter than the average
recurrence interval. They suggest that this
event may be the same event expected to rupture
the ~200-km-long Yakataga seismic gap between
the 1964 rupture and the Fairweather fault
[Sykes, 1971]. The occurrence within a few

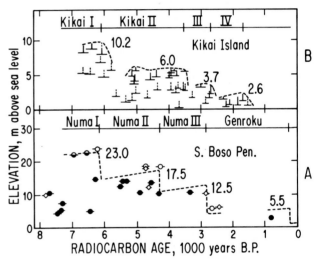

Fig. 5. A: Graph of ^{14}C ages of materials in
the Numa I, Genroku, and two intermediate
terraces indicates four sudden changes in
sea levels, which are inferred to represent
a great earthquake or great earthquake sequence. The open, half-closed and closed
circles indicate the samples above, near and
below the former sea level, respectively. No
information is available for the sample
shown by the diamond. B: Graph of coral
terrace elevations and ages indicate large
earthquakes on Kikaijima in the Ryuku
Islands. Horizontal bars indicate the sampling elevation and the uncertainty in age.
Top of the solid bar shows the estimated
height of the former sea level. The broken
curves show estimated uplift history assuming
some interseismic subsidence. The numeral
shows the present elevation of the former
shoreline mainly determined from beach angle
elevations [from Shimazaki and Nakata, 1980].

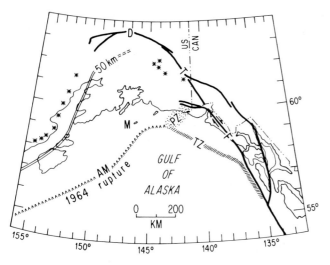

Fig. 6. In Alaska, geological studies have provided data pertinent to assessment of earthquake potential for the labelled structures: AM = Alaska-Aleutian megathrust; D = Denali fault; M = Middleton Island; PZ = Pamplona zone; T = Totschunda fault; TZ = Transition Zone; * = volcanos; double line is 50-km isobath of Benioff Zone [from Lahr and Plafker, 1980]

decades of such an earthquake similar in size to the 1964 event would even make 1000-yr-long recurrence intervals between terrace-producing earthquake sequences consistent with intervals calculated from plate rates.

Other Studies of Marine Terraces

On Kodakarajima, off southern Kyushu, Japan (Ko on Fig. 2), recent work by Nakata et al. [1979, p. 196-197] has revealed that a sudden uplift of ≳8 m occurred about 2,450 ^{14}C yrs B.P. A prior interval of gradual interseismic submergence seems to have lasted a few thousand years without interruption, and post-seismic uplift over a period of ~1000 yrs may have occurred also. They estimate an average recurrence interval of about 10,000 yrs for such seismic uplifts.

Nakata et al. [1979, p. 197-199] have also documented several periods of sudden uplift farther south, on Kikaijima, one of the northernmost of the Ryukyu Islands (Ki on Fig. 2, and Fig. 5B). These uplifts are inferred to have been seismic and occurred ~5,200-6,065; ~3,520; ~2,700; and ~1,700 ^{14}C yrs B.P. which yields recurrence intervals of ~1700-2500, ~800, and ~1000 ^{14}C yrs. (Average recurrence interval is between ~1200 and ~1400 ^{14}C yrs.) Inasmuch as no uplift has occurred in the past ~1700 ^{14}C yrs, this region should be considered well advanced in its current interseismic cycle. McCann et al. [1979, Fig. 1, p. 1125-1126] have stated that the megathrust in this region may not have the

potential for a great earthquake, judging from the lack of great earthquakes during the short historical record. Here, then, is a situation in which geological extension of the seismic record may play a critical role in reassessing earthquake potential.

Ota et al. [1976] estimated that the average recurrence interval for major earthquakes like that which struck Sado Island in the Japan Sea off central Honshu, Japan (Fig. 1) in 1802 to be 5000 to 9000 years during at least the past 100,000 years and suggest that the interval before the 1802 event was ≳6,000 ^{14}C yrs. The estimate of average interval is based on the ratio of tilting observed in 1802 to tilt of a Pleistocene shoreline. Tilt of the older surface is similar in style to that of the younger, but much greater in magnitude. This type of analysis provides no information on deviations of actual intervals from the average.

Studies of elevated shorelines in other parts of the world are as yet few in number, but include a study by Taylor et al. [1980] of the Santo and Malekula Islands segment of the New Hebrides island arc. These authors related recent seismic patterns and bathymetry of the subducting oceanic

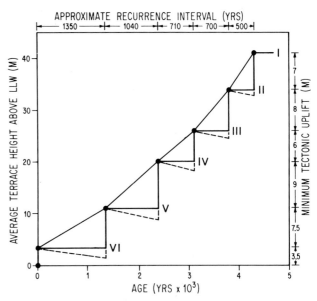

Fig. 7. Generalized diagram showing average terrace height (minimum tectonic uplift per event(s)) versus terrace age and approximate recurrence interval for Middleton Island. The solid curve shows an inferred uplift sequence assuming no interseismic vertical movement; the dashed line assumes some interseismic subsidence. The light solid line indicates the average uplift rate between terraces [from Plafker and Rubin, 1978]. The relatively small uplift of the youngest shoreline (3.4 m) in the great 1964 earthquake may indicate that a second large event will occur soon.

plate to patterns of terrace deformation on these islands which sit on the apex of the hanging wall of the New Hebrides megathrust.

Discussion

Megathrusts pose particular difficulties in the study of earthquake recurrence. One of the most severe problems has been an almost complete lack of exposure of megathrust traces on land. This has effectively prevented direct stratigraphic, geomorphic, or structural studies of the surface trace of the fault at appropriate scales and in appropriate detail.

Geologic determinations of the recurrence characteristics of great megathrust earthquakes necessarily are based upon study of accessible secondary features, principally elevated and tilted former shorelines and wave-cut surfaces. The raised shorelines and associated marine terraces of the Nankai Trough, Japan and Middleton Island, Alaska are situated just inboard of major imbricate thrust faults that branch upward from the megathrust. Yonekura and Shimazaki [1980] suggest that these terraces owe their existence not to slip on the megathrust but to slip on these lesser imbricate thrusts. Thus the dates of emergence of the terraces may represent the dates of seismic slip along the imbricate fault, not necessarily the megathrust.

Terrace uplifts associated with the 1944 and 1946 Nankai Trough earthquakes and the 1964 Alaskan earthquake indicate that slip occurred along the appropriate imbricate thrusts in conjuction with slip along the megathrust during those events. The uplift history of the marine terraces of the Nankai Trough suggest that during some great megathrust earthquakes slip along the imbricate fault has not occurred [Yonekura and Shimazaki, 1980]. The degree to which this is also true of other megathrusts will influence the extent to which a complete record can be obtained from emergent shorelines. Perhaps in some regions only a maximum average recurrence interval will be recoverable from the terrace record.

Complete elimination of a seismically produced marine terrace by wave action is a possibility that may be difficult to recognize. In such a situation, one emergent wave-cut surface, shoreline angle, and terrace would represent more than one seismic upheaval and the average recurrence interval would be overestimated. Similarly difficult to assess may be the possibility that some recurrence intervals between large earthquakes are shorter than the time required to cut a new wave-cut platform and shoreline angle. Such a situation would lead to underestimation of the average recurrence interval. In both cases above, the upheaval produced by more than one earthquake would be attributed to a solitary event.

Another factor that complicates the use of marine terraces to characterize and date large earthquakes is that major changes in sea level have occurred during the past several thousand years. Thus, the elevation of uplifted wavecut platforms and shorelines relative to modern sea level may not be equivalent to the amount of tectonic uplift. This is especially true for those features more than a couple thousand years old. In fact, emergent shorelines only exist where the rate of tectonic uplft is greater than the average rate of sea-level rise.

Although reliable curves of sea level vs. time have been established on some coasts [e.g. in southern Florida, Scholl et al., 1969] recent work has demonstrated that melting of the late Pleistocene continental glaciers produced Holocene coastal responses that varied between points on the globe [Clark et al., 1978]. Thus, the sea-level curve from a tectonically stable coast cannot necessarily be applied to correct for sea level changes affecting marine terraces in an area of tectonic interest. Modeling of the effects of the melting icecaps by Clark et al. [1978] does suggest some general characteristics of the Holocene sea-level curve for many areas of interest, however. Corrections involving a monotonic Holocene rise in sea level may pertain to coastal Japan and southeastern Alaska, for example. Thus, the elevation of older terraces may progressively underestimate true tectonic uplift. Further refinement of models of sea-level change will enable greater confidence in detailed comparisons of earthquake size based on amount of uplift.

Some emergent shorelines are probably not tectonic. Along some coasts widespread mid-Holocene wave-cut surfaces and shorelines are slightly elevated above sea level. These may be the result solely of sea-level fluctuations unrelated to tectonic activity [Clark et al., 1978]. To the best of my knowledge, the possibility that continuous or episodic aseismic subduction could produce emergent shorelines has not been investigated.

One promising new area of research involves the application of regional deformational patterns deduced from marine terraces [e.g. Ota and Hori, 1980, and Ota and Yoshikawa, 1978] to forecasting the nature of future earthquakes. The wave-length of deformations in Japan, for example (Fig. 8), might be roughly equal to the rupture length of related active faults. Many of the studies mentioned above support this suggestion.

Dip-Slip Faults On Land

Geomorphic Studies in the Great Basin by Wallace

Wallace [1977] pioneered recent attempts to use fault-scarp morphology to assess the repeat times of large earthquakes in the Great Basin of the western United States. This pilot study in north-central Nevada [see Fig. 1 of Wallace, this volume] demonstrated that fault scarps several meters in height in alluvial materials degraded systematically over hundreds, thousands, and

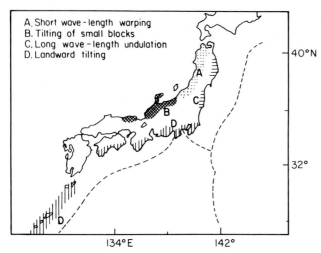

Fig. 8. Several styles and wavelengths of deformation are apparent from studies of the heights of former shorelines [from Ota and Yoshikawa, 1978]. Large historical earthquakes have mimicked these patterns, indicating that the deformational record preserved in marine terraces is an important key to the seismic future.

tens of thousands of years. With a few historical fault scarps and lakecliff scarps eroded by ancient Lake Lahonton about 12,000 yrs B.P. as calibrations, he was able to crudely date several pre-historic fault scarps in this arid region. Fig. 9 illustrates the distributions of principal (i.e. dominant) slope angles measured on several sets of fault scarps and the inferred approximate ages of the four scarp sets. He estimates average recurrence intervals of several thousand years for large events [Wallace, 1977, and this volume].

Along the fault scarp of the 1915 Pleasant Valley earthquake (M = 7.6, Richter, 1958) Wallace found geomorphic evidence for earlier events. Abrupt breaks in scarp profiles indicate at least two and possibly three older events. The principal slope of the youngest prehistoric event implies an age of about 10,000 yrs. Wallace [1978b] also estimated the magnitudes of the individual earthquakes represented by some of his scarps.

In addition, he observed that events cluster both in time and space [Wallace, 1977, 1978a]. "Most ranges shown [in Fig. 1 of Wallace, this volume] have been progressively tilted dominantly eastward for approximately the last 10-14 million years... Individual clusters of scarps, such as on the west flank of the Humboldt Range or the west flank of the Tobin Range, display repeated movement during periods of time when other clusters of scarps nearby were inactive. For example, the faults on the west flank of the East Range have remained inactive for possibly several tens of thousands of years, while the faults along the Tobin and Humboldt Ranges,

which bracket the East Range, were repeatedly active." The temporal and spatial distribution of large earthquakes as manifested by fault scarps in northern Nevada can be used for forecasting the sites of the next large earthquakes in this region. The observations also indicate that the number of large earthquakes accompanied by surface faulting during the 20th century is remarkably higher than usual. A careful study of the scarps of the penultimate event along each historic scarp might enable determination of whether this burst of recent activity has a prehistoric precedent in the late Pleistocene or earliest Holocene. Perhaps some faults ruptured during this postulated earlier burst of activity but have not ruptured recently. These might be candidates for rupture in the near future.

Owens Valley, Eastern California

On the western edge of the Great Basin, in eastern California, is the Owens Valley. Lubetkin [1980] has applied Wallace's method of geomorphic analysis to the Lone Pine fault, a small normal oblique-slip fault at the southern end of the valley, and argued that three events are probably represented by the 6-m-high scarp. The most recent event, associated with the great earthquake of 1872, is evidenced by the steepest 1-2 m of the scarp. The previous events occurred after ~10,000 to 21,000 ^{14}C yrs B.P., based upon the

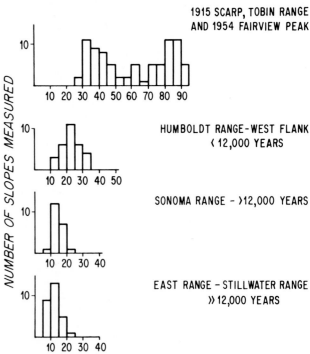

Fig. 9. Histograms of principal (i.e. dominant) slope angles on fault scarps, north-central Nevada. The slope angles indicate the age of the fault scarp and its associated earthquake [from Wallace, 1977].

inferred age of the faulted alluvial surface. This suggests an average recurrence of great earthquakes between ~3,300 and ~10,000 yrs. Lubetkin and Clark [1980] maintain that this interval is greater than that of the entire ~60 km length of the nearby mid-Owen's Valley fault zone, which last ruptured in 1872. Carver [1970, p. 74-88] studied a set of progressively deformed ancient shorelines farther south along the mid-Owen's Valley fault zone, and argued that the recurrence of events along the zone is between 700 and 1000 years.

The Wasatch Fault Zone, Utah

The ~370-km-long Wasatch fault zone delimits the eastern boundary of the Great Basin (Fig. 10) and poses a great threat to the major population centers of Utah, although it has not produced a large earthquake during the 133 years of historical record. Swan et al. [1980] have made an attempt to evaluate the recurrence intervals and sizes of earthquakes that occurred along this discontinuous zone of normal faulting during the Holocene. Their elegant study combined detailed geologic mapping, geomorphic analysis, and exploratory trenching at two localities -- Kaysville and Hobble Creek. Unfortunately, absolute chronologic control at both sites was very limited, but ~12,000 to ~15,000 yr ages for the youngest sediments of ancient Lake Bonneville and one ^{14}C date on a late Holocene unit at the Kaysville site enabled meaningful evaluations of earthquake recurrence.

At the Kaysville site, stratigraphic units and relationships in excavations clearly indicate two, and probably only two, large earthquakes since about 1,580 ± 150 ^{14}C yrs B.P. The authors suspect that the latest event occurred several hundred years ago and propose a 500- to 1000-yr average recurrence interval. Fault scarps more than 3 meters in height accompanied each event, although net tectonic movement between hanging- and foot-wall blocks was probably just under 2 m [Swan et al., 1980, p. 1441]. They suggest that scarps of this height along a normal fault are compatible with a M ≈ 7+ earthquake.

At the Hobble Creek site, the authors present evidence for six or seven faulting events since Lake Bonneville retreated from its Provo level

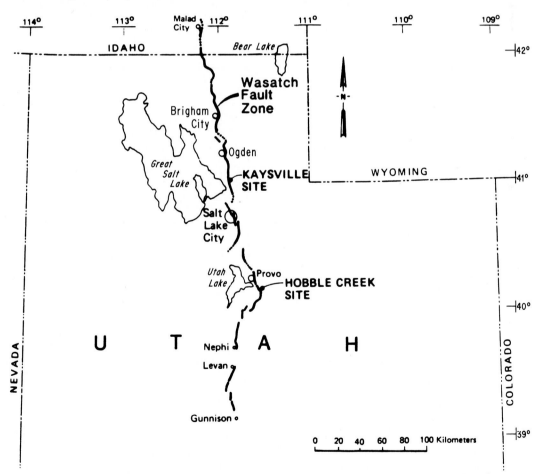

Fig. 10. The Kaysville and Hobble Creek sites have provided information on earthquake recurrence along the Wasatch fault zone [from Swan et al., 1980].

12,000 to 13,000 yrs B.P. Various considerations bracket the average recurrence interval between 1,500 and 2,600 years. The calculated average scarp height suggests that each offset may correspond with earthquakes of $M \approx 7$.

The Wasatch fault zone (Fig. 10) consists of several distinct segments. Because recurrence intervals at the two study sites are different, Swan et al. [1980] suggest that some or all of the fault segments produce events at different times. Estimating that there might be 6 to 10 individual seismogenic segments, they suggest that an event somewhere along the zone ought to occur every 50 to 430 years, on the average. Since the current seismic hiatus of >133 years is well within this range, they state that "a moderate-to-large magnitude earthquake (magnitude 6-1/2 to 7-1/2) may be due or past due somewhere along the Wasatch fault zone."

Cordillera Blanca Fault, Peru

The Cordillera Blanca fault is a 200-km-long normal fault whose trace lies parallel to and ~100 km inland from the Peruvian coast between 8-1/2° and 10°S latitude (Fig. 11A). The structure spectacularly cuts ancient alpine glacial moraines that protrude from the western front of the high

Cordillera Blanca of the Peruvian Andes. At two sites studied by Yonekura et al. [1979], rates of slip are estimated to be ~2 and ~3 mm/yr. Younger moraines display lesser offsets than older moraines and indicate to the investigators that 2 and 3 m offsets may occur with each event, implying an average recurrence interval of about 1000 years. Whether the fault breaks in unison or in segments is unknown. If it does break in unison, a moment, M_O, of about 6×10^{27} dyne-cm is likely. Yonekura et al. [1979] associate no large historical earthquake with slip along the fault, although four large earthquakes have occurred recently along or near the subduction zone to the west (1940, M = 8; 1966, M = 7.5; 1967, M = 7.0; 1970, M = 7.6; Fig. 11B).

The recognition and initial characterization of the Cordillera Blanca fault illustrates the need for careful geologic studies in refining synoptic maps of earthquake potential. Here is one place where the likelihood of a large earthquake may be considered low on the basis of a seismological analysis of recent large events along the subduction zone [McCann, 1979]. Nevertheless, the village of Yungay (Fig. 11A), devastated by a landslide loosened from the steep western fault escarpment of the Cordillera Blanca by the large 1970 event may have yet another

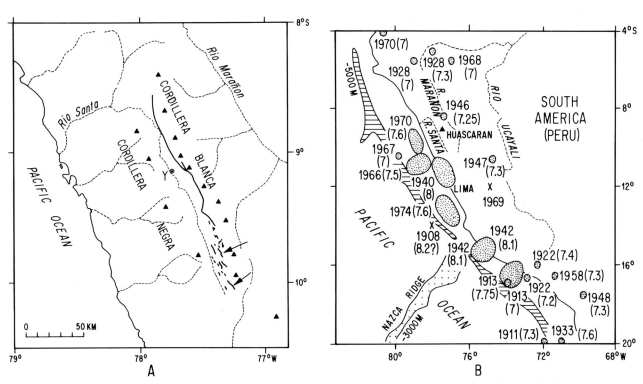

Fig. 11. The Cordillera Blanca fault is an active ~200-km-long normal fault about 100 km east of the coast of Peru. A: Geologic studies at two sites (arrows) indicate that the trace cuts young glacial moraines and may generate great earthquakes about every thousand years. Y = Yungay. B: Focal regions for earthquakes ($M_S \geq 7.0$ in Peru, 1904-1975 [from Yonekura et al., 1979]. The existence of the Cordillera Blanca fault indicates that the subduction megathrust is not the only source region for large earthquakes in this region.

source for large events only a few kilometers to the east.

Reverse Faults in California

Several reverse faults in California have been studied to determine earthquake recurrence intervals. Three studies illustrate the use of small fault-related stratigraphic units in identifying prehistoric earthquakes. These are discussed below. An excavation placed across the surface trace of one of the faults on which slip occurred during the San Fernando earthquake of 1971 (M_L = 6.6) (S on Fig. 14) revealed a wedge-shaped rubbly stratigraphic unit that may have been derived, at least in part, from a fault scarp produced before the 1971 event [Bonilla, 1973, p. 179-181]. Wood near the base of the wedge yielded a ^{14}C age of ~200 ± 100 ^{14}C yrs B.P. The age of the sample provides a maximal age of about three centuries for the scarp and the earthquake that accompanied its formation.

Near Ventura, California (V on Fig. 14), late Quaternary rates of uplift, folding, and faulting are quite rapid [Lajoie et al., 1979, p. 9-10]. Analysis of the Javon Canyon fault, a local high-angle reverse fault, where it is exposed in a natural outcrop, reveals 3.3 m of slip since ~2500 ^{14}C yrs B.P. [Sarna-Wojcicki et al., 1979]. Three superposed colluvial wedges in the sediment below the fault plane indicate that three slip events account for this movement. Similar relationships along a small reverse fault zone near Santa Cruz, California (SC on Fig. 14), have been interpreted by Weber et al., [1979, p. 117] as indicating 4 to 5 faulting events within the past 10^5 years.

Strike-Slip Faults

Investigations of the recent behavior of active strike-slip faults are numerous, especially those dealing with the determination of geologically recent slip-rates using offset geomorphic features. Also, the number of workers using trenching to expose subsurface evidence is rapidly increasing. Both approaches are discussed below.

Determination of the geologically recent rate of slip along a strike-slip fault can be useful in evaluating earthquake recurrence intervals. This is especially true if the amount of offset along the fault during one or more large earthquakes is known. In the simplest case, where slip associated with each earthquake is uniform along the fault and the slip value associated with each earthquake is identical to that of previous and later events, the average long-term slip-rate divided into the slip per event yields the average recurrence interval. Since no one has yet identified such a simple case, the interpretation of geologically determined rates of slip never yields such an unambiguous value for the average recurrence interval. Nevertheless,

estimates of the average interval can be meaningful in assessing earthquake potential.

Denali Fault, Alaska

Studies of the Denali fault [Sieh and Cluff, 1975] illustrate how the determination of slip rates and slip per event can be utilized. During the latest part of the Pleistocene Epoch glacial ice covered most of the 320-km long segment of the Denali fault between 144°W and 150°W. Subsequent to final deglaciation between ~10,000 and ~14,000 yrs B.P., various features of glacial origin (e.g. Fig. 12A) have been offset in a right-lateral sense along the fault. Most of these and other offsets, which were measured from vertical aerial photographs, are between 117 m and 143 m. If all of the features were produced during the latest deglaciation sometime between 10,000 and 14,000 yrs B.P., the fault-slip rate must be between 8 and 14 mm/yr. The oldest portion of a neoglacial moraine at 149°27'W is offset 35 ± 2 m (Fig. 12B). At 8-14 mm/yr, such an offset would accumulate in 2400 to 4600 yrs. This is compatible with Denton and Karlén's [1973] observations that a period of glacial expansion in Alaska occurred between 2400 and 3200 yrs B.P.

Coupled with estimates of the average displacement per earthquake, the slip rate along the Denali fault has been used to derive an average recurrence interval [Sieh and Cluff, 1975]. The most reasonable assessments of slip per event are those derived from measurements of small offsets along the fault. At 144°08'W, aerial photographic and field studies indicate a 15-m offset of two channels (Fig. 13A). The westernmost channel also has an older channel segment offset about 30 m. The lack at this locality of channels with offsets of intermediate value strongly suggests that the latest two events resulted in 15 m offsets. Walls of small channels draining northward into Augustana Creek (~146°50'W) display offsets as small as 11 m at one site and 7-1/2 to 9 m, 15 to 20 m, and 32 m, at another (Figs. 13 B and C). (These values were measured from very-low-altitude vertical aerial photographs, and are approximate). In summary, the range of offset values for the latest event at the few measured sites is 7-1/2 to 15 m. Division of this range of values by the 8 - 14 mm/yr slip rate yields an average recurrence interval somewhere between 540 and 1880 yrs. Trees that are about 150 yrs old are growing on an unfaulted moraine constructed across the fault at about ~145°45'. Thus, no appreciable slip has occurred there in at least the past century and a half [Stout et al., 1973]. Clearly, more precise and more numerous offset values determined by field measurements and direct radiocarbon dating of offset features would enable more precise and useful evaluations of the likelihood of a great earthquake along the Denali fault. Precise dating of individual events, a

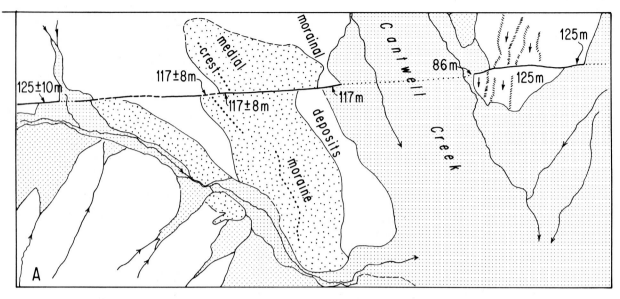

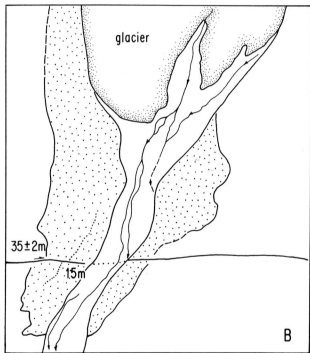

Fig. 12. A: Along the Denali fault, many glacial features created in latest Pleistocene time
have been offset between 117 and 143 m [Sieh and Cluff, 1975]. At this site at 149°20'W, offsets
of moraines and bedrock surfaces are between 117 and 125 m. A long-term slip rate of 8 to 14
mm/yr is derived from these types of data. Irregular dot pattern indicates well-defined glacial
deposits; regular pattern indicates well-defined Holocene alluvial and colluvial deposits; hachures
define walls of grooves eroded into bedrock. B: The oldest part of a neoglacial morainal complex
is offset ~35 m at 149°27'W and thus may be ~2400 to ~4600 yrs old.

topic discussed below, would enable testing
whether great events on the Denali fault might
be closely related in time to great events on
the Alaskan-Aleutian megathrust.

Fairweather Fault, Alaska

Plafker et al. [1978] estimated the average
recurrence interval for the Fairweather fault in

southeastern Alaska (Fig. 6) using a stream channel that bends right-laterally ~55 m along the fault and a nearby offset of 3-1/2 m that accompanied the 1958 (M_S = 7.9) Lituya Bay earthquake. Faulting at this locality has produced an uphill-facing scarp and diversion of the stream around the scarp may be responsible in part for the 55-m value. Nevertheless, the 55 m may be due entirely to displacement along the fault. A maximum age for the stream is based on a radiocarbon date of 940 ± 200 ^{14}C yrs B.P. on wood in the most recent neoglacial moraine in the area. The wood is in a soil buried by the moraine and thus gives a maximal age for the moraine and associated glacier. The glacier that formed the moraine was 300 m thick at the site of the offset stream and had to melt before the stream could form. Thus, an even longer period occurred between wood formation and creation of the the channel. The minimum slip rate for the fault would thus be 55 m/(940 + 200) yrs = 48 mm/yr, and the actual rate can be expected to be significantly higher.

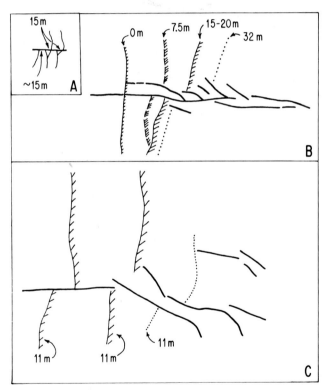

Fig. 13. Youngest offsets along the Denali fault range from 7-1/2 m to 15 m [Sieh and Cluff, 1975]. At a rate of 8 - 14 mm/yr, the range of average recurrence intervals for events of this size is 540 to 1880 years. A: Small tributaries to Jack Creek, 144°08'W. B and C: Small tributaries to Augustana Creek, ~145°52'W. Thick lines are faults; hachures are on channel banks whose bases are indicated by lines; dotted lines indicate offset linear ridge.

Plafker et al., using their best guess of about 58 mm/yr, derive an average recurrence interval of about 60 yrs for events producing 3-1/2 m offsets such as occurred nearby in 1958. Suggestions by Sykes [this volume] that slip along the Fairweather fault occurred in 1899 and 1847, as well as in 1958, are consistent with this estimate of average recurrence interval.

Median Tectonic Line, Japan

Extensive analyses of offset channels and fluvial terraces have yielded a complex but consistent record of recent right-lateral slip along the several active faults of the Median Tectonic Line fault system in southwestern Japan (Fig. 2) [Okada, 1968, 1970, 1973, 1980; Okada and Sangawa, 1978; Sangawa, 1978a, 1978b; Sangawa and Okada, 1977]. Rates of slip are highest (5 to 10 mm/yr) along the central segment of the fault system and decrease toward the ends of the ~300-km-long system. The most recent great earthquake probably occurred between 0 A.D. and the beginning of the historical period (~700 A.D.), as judged by a lack of great earthquakes during the historical period and a faulted alluvial fan containing logs with an age of 1860 ± 90 ^{14}C yrs B.P. in the central region of the fault [Okada, 1980, p. 99, 101]. Elastic strain accumulation at 5 to 10 mm/yr would produce a potential offset of 10 to 13 m during this most recent 1300- to 2000-year period of dormancy. Okada [1980, p. 101] concludes by forecasting a "severe" earthquake produced by slip along the faults of the Median Tectonic Line within a few hundred years. Therefore, a current priority is a more complete and detailed determination of past recurrence intervals [Okada, pers. comm., 1980 and 1981].

Atera Fault, Japan

The current status of knowledge concerning the active 80-km-long Atera fault in central Japan (Fig. 2) is similar to that for the Median Tectonic Line. Sugimura and Matsuda [1965] estimated the ages of terraces progressively offset ~70 to ~140 m left-laterally along the fault and published an estimated slip rate of 2-4 mm/yr. Okada [1975] confirmed this estimate by determining a 2-5 mm/yr late Pleistocene/Holocene slip rate using radiocarbon ages of offset river terraces at a different locality. The judgement of Okada [1975] is that an historical earthquake (M = 7.4) in 762 A.D. was the latest event associated with rupture of the Atera fault. At the determined rates, 2-1/2 to 6 m of potential fault slip may have accumulated between then and the present.

Recent studies of the Atera fault have also focussed on dating of individual seismic events. One fresh artificial exposure of the fault revealed a downwardpointing wedge of irregular, faulted, peat and rubble layers that probably

accumulated as colluvial fill and marsh deposits
in a swale along the fault [Hirano and Nakata,
1980]. The investigators interpreted contacts
with gravel rubble resting upon peat to indicate
prehistoric seismic events. The age of these
contacts are ~9300, ~5500, and ~3100 ^{14}C yrs B.P.
Perhaps the gravel represents caving of scarps
that were refreshened during earthquakes at
these times. Non-seismic causes for the formation of the rubble at these times has not been
ruled out, however. At least one of the earlier
events can be defended on the basis of more
severe disturbance of units below a layer that
formed about 4800 ^{14}C yrs B.P. Substantial disruption of a 3000-yr-old peat in this section
also indicates an event more recent than the three
proposed above. Even more events than those four
discussed above could be represented in the exposure, because the complexity of disturbances and
relationships in the exposure might very well
mask other events.

Okada and Matsuda [1976] studied a road cut
along the Atera fault which exposed sand-gravel
beds interbedded with black organic beds. They
argued that the sand and gravel beds represented
caving of the scarp formed after earthquakes
about 13,000, 11,400, 8,300 and <4,300 ^{14}C yrs
B.P. Okada [written comm., 1981] explains that
each of the coarse beds on the downthrown block
is derived from the upthrown block across the
fault, which is as one would expect for a scarp-
derived deposit. Additional support for the
four-earthquake interpretation is that each
organic layer is more severely deformed than
superjacent layers [Okada, 1981, written comm.],
although this is readily apparent only for the
layer ~8,300 ^{14}C yrs old.

Synthesis of the above mentioned studies and
further refinement of the dates and characteristics of Holocene seismic events along the Atera
fault may very well demonstrate that large earthquakes occur along the Atera fault every one
thousand years or so. Depending upon how greatly
actual recurrence intervals deviate from the
average value, a forecast of the behavior of
this fault during the next one hundred years or
so might be attainable.

Yamasaki and Shikano Faults, Japan

In excavations at a site along the Yamasaki
fault in western Honshu, Japan (Fig. 2), Okada
et al. [1980] have found evidence of an earthquake
that occurred between the 8th and 12th centuries
A.D. A stratum assigned an 8th-century age on
the basis of pottery fragments and a date of
1170 ^{14}C yrs B.P. is broken by faults, whereas
a 12th-century bed, dated using radiocarbon and
pottery, is not. The source of a M = 7.1 earthquake in 868 A.D. has been placed nearby on the
basis of historical records, and Okada et al.
[1980] consider this to be a likely candidate for
association with the faulting event recognized in
the excavation.

Other excavations, along the Shikano fault, in
western Honshu, Japan (Fig. 2), indicate two
events during the Holocene [Ando et al., 1980].
The latest event slightly disrupts strata deposited ~1,500 ^{14}C yrs B.P. and is believed to be
faulting related to the 1943 Tottori event. An
older event more severely disturbs strata with ^{14}C
ages of 8,000 and 9,000 yrs B.P. These results
on the Yamasaki and Shikano faults are encouraging and suggest that further work could lead to
knowledge of the average recurrence interval.

Totschunda, Boconó, Clarence, Wairau, and Wairarapa Faults

Other faults for which geologically recent
slip-rates have been estimated include the Totschunda in southeastern Alaska (Fig.6) [Richter
and Matson, 1971, ˜9-33 mm/yr], the Boconó
in Venezuela [Schubert and Sifontes, 1970, ~7
mm/yr], and the Clarence in New Zealand [Kieck-
hefer, 1979, ˜4 mm/yr]. All of these values
are based on offsets of latest Pleistocene or
Holocene features that have not been directly
dated and are roughly correlated with features
dated by radiocarbon elsewhere in these regions.
None of these authors venture an estimate of
average recurrence interval, though hazardous
guesses of slip per event, based on fault length,
could be combined with the slip rates to derive
intervals.

A flight of eight river terraces on the Waichine
River in New Zealand is progressively offset 12
to 99 m right-laterally along the Wairarapa fault
[Lensen and Vella, 1971]. Indirect age assignments
for the offset features imply latest Pleistocene/
Holocene slip-rates of 3.4 to 6.0 mm/yr. Assuming
the historical 1855 earthquake was associated
with about 3 m of slip, Lensen and Vella [1971,
p. 119] propose that the average recurrence
interval is between 500 and 900 years.

A very similar set of circumstances to that
outlined above occurs along the Wairau fault
where it crosses the Branch River in New Zealand.
Lensen [1968] used the same approach to derive a
fault-slip rate at this locality. He then interpreted several small offsets as indicating individual slip events of 5 to 7 m. The average
recurrence interval thus derived is between 500
and 900 years.

Raymond Fault, California

The 15-km-long Raymond fault, in the heavily
populated Los Angeles metropolitan area (R on
Fig. 14), has been the subject of recent studies
by Crook et al. [1978]. In addition to geologic
mapping and geomorphic and pedologic approaches
in assessing the fault's longer-term geologic
history, they utilized trench excavations across
the fault to assess its recent history of movement. They identified in late Pleistocene/Holo-
cene deposits evidence of several prehistoric
earthquakes -- at least 5, and perhaps 8, between

~1,630 and ~36,000 ^{14}C yrs B.P. [Crook et al., 1978, p. 78-82]. If these represent all the events that occurred in this span of time, the average recurrence interval is between 4500 and 9200 ^{14}C yrs. If, as the authors suspect, not all events are represented, these would be maximum values for the average recurrence interval. They suggest that 3,000 years might be a more accurate value. Although the events already identified by Crook et al. [1978] may represent the majority of the late Pleistocene and Holocene earthquakes along the Raymond fault, a stratigraphic record with greater sensitivity to closely-timed earthquakes needs to be found. The fault's location within a populous urban area mandates acquisition of more precise knowledge concerning its most recent behavior.

Garlock Fault, California

Two independent studies of Pleistocene features offset along the Garlock fault (G on Fig. 14) indicate an average slip rate of about 10 mm/yr [Carter, 1980; Clark and LaJoie, 1974]. Clark [1973] estimated offsets of about 3 to 5 m for several small channels that cross the fault. If these were offset during the latest large event and are representative of slip values for earlier events as well, the average recurrence interval

would be between 300 and 500 years. Burke [1979] has recognized 9 to 17 prehistoric earthquakes in an excavation that exposed the Garlock fault breaking sediments younger than ~15,000 ^{14}C yrs. He feels that this may not represent all the earthquakes that have occurred along the Garlock fault during this period, so the maximum average recurrence interval is about 1-1/2 millenia.

Studies of the Geologically Recent Behavior of the San Andreas Fault

Geomorphic and stratigraphic studies at several localities along the San Andreas fault system have provided a partial understanding of the long-term (i.e. millenial) behavior of that major plate boundary, including recurrence intervals of large earthquakes and fault slip-rates. This section consists of an outline and evaluation of these studies.

Based on its historical behavior, the San Andreas fault can be divided into four distinct segments (Fig. 14). The northern segment produced a great earthquake in 1906 [Lawson et al., 1908]. The south-central segment produced a great earthquake in 1857 [Agnew and Sieh, 1978; Sieh, 1978b]. Both of these segments have been characterized by extraordinarily low levels of microearthquake activity since their respective great earthquakes

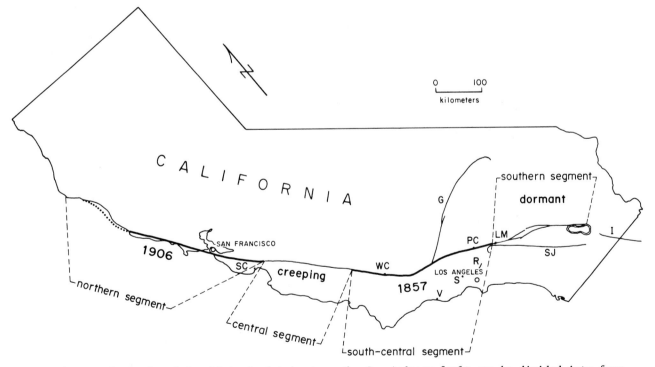

Fig. 14. On the basis of its historical behavior, the San Andreas fault can be divided into four segments. Forecasting the location and size of future great earthquakes generated along the fault requires understanding the degree to which this historical behavior characterizes the long-term behavior of the fault. WC, PC, and LM indicate Wallace Creek, Pallett Creek, and Lost Marsh. G = Garlock fault, I = Imperial fault, R = Raymond fault, SC = Santa Cruz, SF = San Fernando fault, SJ = San Jacinto fault zone, V = Ventura.

[see, for example, Carlson et al., 1979]. The central segment, in contrast, has been undergoing creep continuously throughout the twentieth century [Burford and Harsh, 1980] and is characterized by a high level of activity below $M_L = 6$ [Brown and Wallace, 1968]. The southern segment of the fault has not produced a great earthquake during the ~210 years of historical record, and creep, though suspected or recognized locally [Keller et al., 1978; Allen et al., 1972] is very minor.

To what degree does this historical pattern represent the behavior of the geologically recent past and the future? Allen [1968] has proposed, on the basis of long-standing geological and geometric characteristics of and contrasts between the creeping and dormant segments, that the historical behavior is representative of the long-term behavior. Irwin and Barnes [1975] have observed that the fault suffers creep only where a stratigraphic section of Franciscan Formation capped by relatively impermeable Great Valley Formation abuts it. They suggest that capping of the Franciscan may cause buildup of high pore pressures in the fault zone which greatly reduces the shear strength of the fault and enables creep. Exactly to what extent the historical behavior of the fault represents its long-term behavior is unknown. Nevertheless, the studies outlined below indicate that, at least in part, historic and ancient behavior are similar.

Geomorphic Evidence of Individual Large Events

Geomorphic features indicate that during at least the past thousand years or so great earthquakes have been the principal mode of strain relief along the south-central segment of the San Andreas fault. In 1857 right-lateral slip along this stretch varied along strike from ~3 to ~10 m (Fig. 15). These values were determined principally by measuring the offset of small gullies cut during periods of heavy rainfall in the decades before the earthquake [Sieh, 1978b]. The smallest offsets at Van Matre Ranch are ~8 m (Fig. 15 and Table 1). These are attributed to the 1857 event. Younger channels display no offsets, implying that there has been no appreciable slippage since the 1857 earthquake and associated aftercreep occurred. Three older channels are separated by about 16 m along the fault (Table 1). These probably record both the 1857 and a previous slip event of ~8 m. Offsets of ~26 m (Table 1) probably accumulated in three increments: ~8 m in 1857, ~8 m during a previous large earthquake, and ~10 m during the next previous earthquake. The geomorphic data along this portion of the fault seem to indicate that sudden, large, 8- to 10-m slip events characterize this segment of the fault. Similarly, though less convincingly, where slip in 1857 was ~3-4 m, previous events seem to have been of similar size [Sieh, 1977, Fig. 23].

These observations could fit at least two hypothetical models of large earthquake behavior (Fig. 16). In a uniform-slip model, large earthquakes must be twice as frequent where the slip is half as great as at Van Matre Ranch. In the uniform-earthquake model the nonuniform slip function of the latest earthquake repeats. Such a model requires substantial deformation in one or both of the blocks separated by the fault. Determination of the actual long-term slip rate at several carefully selected localities along the south-central segment enables a test of which, if either, model is appropriate.

Long-Term Slip Rates

Investigations at two localities along the south-central segment are providing information about the long-term rate of fault slip. These localities are Wallace Creek and Lost Marsh (WC and LM in Fig. 14).

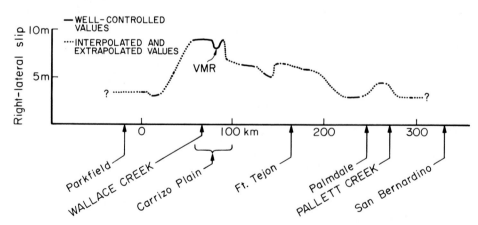

Fig. 15. Small channels and other features indicate that the great 1857 earthquake was associated with the values of fault slip shown here [from Sieh, 1978b]. Is this pattern characteristic of earlier earthquakes along the south-central segment as well? VMR indicates the location of Van Matre Ranch.

TABLE 1. Offset gullies at the Van Matre Ranch indicate that the magnitude of fault slip accompanying the last two prehistoric events there was very similar to the ~8 m of slip in 1857 [from Sieh, 1977].

Site #	Offsets (meters)		
50	7.6 ± 1.5		27.9 ± 1.8
49	8.5 ± 1.5		
48	7.8 ± 0.6 (48-F)	15.8 ± 1.2 (48-E)	
47	8.2 $^{+1.5}_{-0.3}$ (47-C)		24.4 ± 1.0 (47-A)
46	8.2 $^{+0.6}_{-1.5}$	16.3 ± 0.9	25.1 ± 0.9
45		15.2 ± 0.9	
44			27.7 ± 2.7

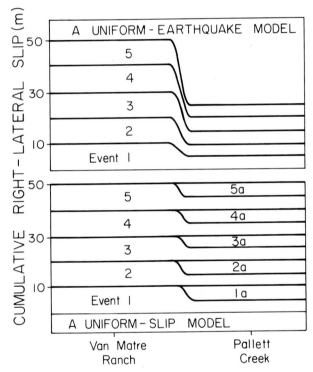

South-central segment, San Andreas fault

Fig. 16. The suggestion that slip at any one locality does not vary greatly from one earthquake to the next leads to two plausible models relating great earthquakes. In the uniform-earthquake model, all great earthquakes (1 through 5) have non-uniform slip functions like that of the 1857 earthquake. In the uniform-slip model, alternation of 1857-like events (1 through 5) with other large events (1a through 5a) accomplishes uniform slip along this 400-km-long stretch of the fault.

Wallace Creek is an ephemeral stream which flows perpendicular to the fault in a channel incised into an old alluvial fan (Fig. 17). Where it first reaches the fault, it turns northwestward and flows 130 m along the fault. It then leaves the fault and flows out into a valley. An older channel, created long ago by Wallace Creek, leaves the fault 250 m farther northwest. Detailed surficial and subsurficial studies of the geology associated with creek development [Sieh and Jahns, 1980] suggest the history illustrated in Fig. 18. Trenches at several locations revealed stratigraphic relationships critical to understanding the rate of slip of the fault. Trench #5 (Fig. 19) exposed late Pleistocene-early Holocene(?) alluvium (solid black). This alluvium was incised sometime after 19,000 ^{14}C yrs B.P. and before 5900 B.P. The channel then accumulated slope debris and fluvial gravels (unstippled and stippled, respectively) until sometime after about 3900 B.P. Some time after 3900 B.P., incision again occurred and the modern channel began to fill with slope debris and fluvial gravel (unstippled and hachured, respectively). The 130-m offset has occurred since deposition of the youngest of the stippled units and overlying slope debris, that is, since incision of the modern channel. The age of the youngest stippled deposits (<3900 yrs B.P.) is a maximum time for accumulation of the 130 m offset of the modern channel. The 380 m offset began to accumulate much earlier, after incision of the alluvium (solid black). Therefore, the 380 m offset is older than or approximately equal to the age of the oldest stippled deposits and slope debris (i.e. ≥5900 yrs old). Because 130 m has accumulated in no more than 3900 yrs, the minimum slip rate is 33 mm/yr; and because 380 m has accumulated in no less than 5900 yrs, the maximum slip rate is 64 mm/yr.

The average long-term slip rate at Wallace Creek is appreciably greater than it is about 250 km to the southeast, at Lost Marsh (LM in

Fig. 17. Models relating great earthquakes along the San Andreas fault can be tested by comparing the long-term slip rates of the fault at localities such as Wallace Creek, which is shown here. The modern channel is offset ~130 m and an older one is offset ~380 m. Photo courtesy of R. E. Wallace, U. S. Geol. Survey.

Fig. 14). There a study of offset fluvial terrace risers has yielded a tentative average slip rate of 25 mm/yr for the Holocene epoch [Weldon and Sieh, 1980]. These values favor the uniform earthquake model, in which the 1857 earthquake is the characteristic event along the south-central segment. The non-uniform slip associated with the 1857-earthquake is consistent with the non-uniform long-term slip along that stretch of the fault. If the uniform-earthquake model is valid, the next event to occur along the south-

central segment will be similar in static charac-teristics to the 1857 earthquake.

The slip-rate determination at Wallace Creek enables calculation of an average recurrence interval. Repeated nine- to ten-meter offsets seem to characterize the segment of the fault near Wallace Creek, so the 33–64 mm/yr limits on slip rate imply an average recurrence interval between 140 and 300 years. These data lead to the conclusion that a great event involving the Wallace Creek site within the next few decades is

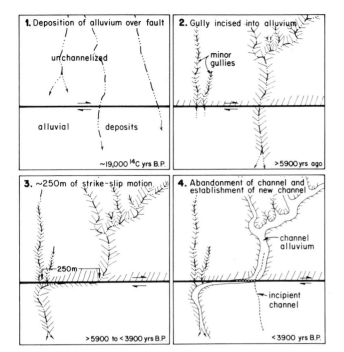

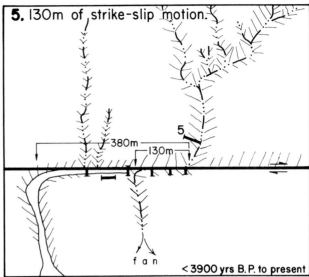

Fig. 18. Extensive surficial and subsurficial studies of Wallace Creek suggest the history indicated above. Thick bars indicate locations of trenches; trench #5 is numbered.

certainly well within the realm of possibilities.

We might also suspect from the Wallace Creek data that the central (creeping) reach of the fault is capable of being involved in great earthquakes. This suspicion stems from the probability that the long-term slip rate at Wallace Creek (33-64 mm/yr) is in excess of the historical creep rate along the central segment [$\gtrsim$32 mm/yr; Burford and Harsh, 1980]. A 1 to 32 mm/yr deficit in creep would result in a 100 to 3200 mm slip deficit per 100 years, which might well be relieved during great earthquakes on either the northern or south-central segments. This also raises the possibility of occasional future "superquakes" involving rupture of the northern, central, and south-central segments, since major seismic rupture through the central segment would provide physical linkage of its neighboring great-earthquake-generating segments.

Dating of Individual Prehistoric Events

At Pallett Creek (PC in Fig. 14) individual prehistoric earthquakes are recorded in the datable deposits of a late Holocene marsh [Sieh, 1978a]. The fault has broken these deposits repeatedly during the past 2000 yrs and rapid depositional rates have ensured that all or most earthquake ruptures and related phenomena are buried and preserved before the next large event.

Four types of evidence for prehistoric earthquakes exist here: 1) sandblows and other evidence of seismically induced liquefaction (Figs.

20 and 21), 2) fissures (Figs. 22 and 23), 3) fault scarps along the main fault (Figs. 24 and 25), and 4) lateral offsets (Fig. 26). Together, there is evidence for nine large events since about 500 A.D.

Sieh [1978a] did not convincingly demonstrate that displacements during each of the recognized events were as large as in 1857, but suggested that patterns and magnitudes of vertical deformation for at least six of the prehistoric events were similar to those of 1857. Whether all are as large as the 1857 event or not, all of the nine youngest events do have direct and indirect evidence that movement along the San Andreas was involved. The direct evidence is in the form of buried scarps at the proper horizons (e.g. Fig. 25). The indirect evidence is in the form of anticlines and other coseismic deformation. Davis [1981] reports evidence from excavations 125 km northwest of Pallett Creek for three episodes of slip along the San Andreas since the 16th century. These could well correlate with the latest three events at Pallett Creek. If so, the latest three events at Pallett Creek should be regarded as great earthquakes.

The data from Pallett Creek indicate an average recurrence of large earthquakes on the San Andreas fault of about 160 yrs [Sieh, 1978a]. Individual recurrence intervals vary by up to several decades, although this may be more a function of the uncertainty in ^{14}C age determinations than an indication of actual irregularities. Variation of several tens of percent in recurrence inter-

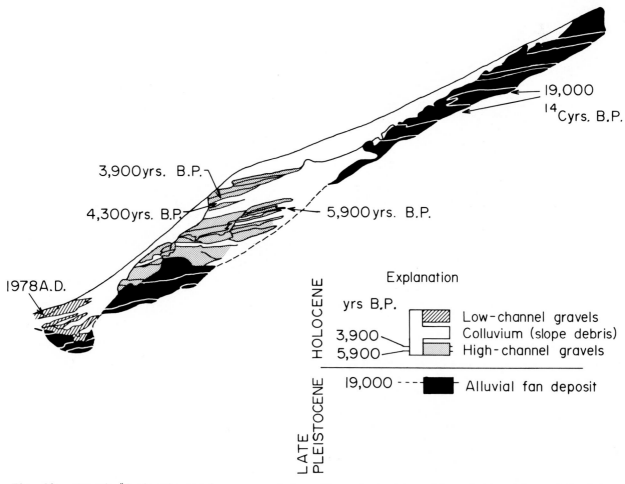

Fig. 19. Trench #5 clearly displays several deposits and stratigraphic relationships critical to determining a fault slip rate at Wallace Creek. Solid black = late Pleistocene alluvial fan deposit; stipple = old channel deposit; hachures = modern channel deposit; no pattern = slope debris.

vals, however, is certainly a possibility, especially in view of the known historical variability in recurrence of large earthquakes in Japan and Chile discussed above. I conclude that the probability of a large event involving all or part of the south-central reach of the San Andreas fault within the next several decades is high.

Problems and Directions

Absolute Dating

The power of the geological approaches illustrated above is limited greatly by the resolution of the dating techniques currently employed. At the present time, conventional ^{14}C analyses produce ages with uncertainties ($\pm 2\sigma$) of nearly a century, at best. Such large uncertainties preclude resolution of several types of problems, especially in regions characterized by average recurrence intervals of only a few centuries or less.

(1) Ancient seismic disturbances recognized in excavations at widely separated localities along a major fault might have identical radiocarbon ages and yet be separated in real age by a century. The converse situation is also possible -- seismic disturbances at different localities, with radiocarbon ages differing by a century, could represent the same earthquake.

(2) Recognition of the clustering of large seismic events within a decade or a century is prohibited by present levels of analytical uncertainty.

(3) Meaningful assessments of the variation of actual recurrence intervals about an average value is impossible at the present time for faults with average recurrence intervals of a few centuries or less.

Application of dendrochronologic and paleomagnetic methods and new radiocarbon techniques may eventually help overcome these problems to some extent. Along the San Andreas fault, for example, the youngest prehistoric events might someday

Fig. 20. Sandblow cone in plowed field after recent earthquake illustrates the type of feature preserved among the strata of Pallett Creek. The sand was deposited as a fountain of water and sand exited the ground from a small central crater. Feature is about 1 meter across.

Fig. 22. Secondary fissure related to ground failure resulting from a recent earthquake in Mexico illustrates another type of feature preserved in the strata of Pallett Creek.

Fig. 21. Cross-sectional view of a sandblow (between arrows) in the stratigraphic record at Pallett Creek. This feature was created during an earthquake in about 1740 A.D. The sand was ejected from a fissure along the San Andreas fault.

be dated to within half a year by analysis of the annual growth rings of trees affected by heavy shaking during these events. Meisling and Sieh [1980] have shown that several conifers near the fault display effects of the 1857 earthquake; older trees might record similar effects from large events in the 17th or 18th centuries. La Marche and Wallace [1972] may, in fact, have identified a predecessor of the 1906 earthquake in the ring record of an ancient redwood north of San Francisco.

New methods of radiocarbon analysis may indirectly enable reduction of uncertainties to about a decade. Short-term irregularities in the relationship of analytical ^{14}C age to real, or calendar, age have been found by determining the ^{14}C age of individual tree rings whose ages span the past 500 years [Stuiver, 1978]. The record of such fluctuations could be determined for the past several thousand years by analyzing the rings of bristlecone pines. This will be facilitated by the advent of new direct-counting

Fig. 23. This fissure (arrow) preserved in the geologic record at Pallett Creek is one of many features that indicate an earthquake occurred in about 1500 A.D. The string outlines a one-meter square.

production in the atmosphere, the concentration of ^{14}C in the atmosphere and in organic matter has not been constant. Fractionation of ^{14}C between organisms and the atmosphere also occurs. Thus ^{14}C ages must usually be corrected to obtain calendar ages. An extreme example would be a radiocarbon age of 6000 yrs. B.P. This would correspond to a calendar age of about 6700 yrs. Conventions in reporting dates determined by radiocarbon analysis vary among users. Some report ^{14}C ages corrected for these factors, others do not. Routine specification of official laboratory sample number, ^{14}C age, $\delta^{13}C$ value, real age, and analytical uncertainty (1σ or 2σ) would greatly aid in eliminating ambiguities that exist currently in the literature. Equations, tables or graphs for these corrections are given by Broeker and Olson [1961], Damon et al. [1972], and Yang and Fairhall [1972].

Seismic Areas Without Surface Faulting

In many regions large earthquakes have been accompanied by little or no surface faulting. Notable examples in the United States are the subduction events of the Aleutians and southern coastal Alaska, the great 1811-12 earthquakes of the upper Mississippi embayment, and the Charleston, S.C., and Boston, Mass., earthquakes of 1886 and 1755. Many great subduction earthquakes (and the great decollement earthquakes of the Ganges plain on the southern flank of the Himalayas) have also not been associated with surface fault rupture. Direct stratigraphic, geomorphic, and structural studies of the faults that caused these earthquakes are precluded by their lack of surface exposure. Only in the unusual coastal areas discussed above, where emergent beaches

Fig. 24. Slip on this scarp along the Imperial fault was about 20 cm of dip-slip and 20 cm of dextral strike-slip when this photo was taken about 3 days after the Imperial Valley earthquake of 1979. Scarps of similar height, but with appreciably greater components of horizontal slip, are buried in the strata at Pallett Creek.

methods that do not require large samples [Maugh, 1978]. In critically important studies where appropriate carbonaceous material spans several decades or centuries, the ^{14}C age vs. real age curve of the sample or samples could be matched against the dendrochronologically established curve and a more refined age determined.

Enormous benefits may soon be realized by the development of direct radiocarbon dating [Maugh, 1978]. This method will require only milligrams of carbon, whereas present conventional methods require grams. Innumerable attempts to date strata have failed for lack of enough carbonaceous material. This new method promises to enable determination of average recurrence intervals and earthquake dates where it is now impossible.

Another dating problem is more readily corrected than the problem of precision discussed above. Because of variations in the rate of ^{14}C

have been utilized in determining recent seismic behavior, have indirect methods of investigating the causative faults been applied extensively with great success.

Sims' [1975] study of seismically deformed soft-sediments has raised the possibility that liquefaction and other non-tectonic surficial seismic phenomena might also be recoverable indicators of ancient earthquakes whose causative faults are inaccessible. Sandblows and liquefaction-related ground failures were certainly widespread during the 1811-12 earthquakes in the Upper Mississippi Valley [Penick, 1976], the 1886 Charleston earthquake [Dutton, 1889], and the great 1897 and 1934 Indian earthquakes [Oldham, 1899 and Geol. Survey India, 1939], and similar features from older events are probably preserved in Holocene strata in these areas. Buried sandblows in the regions of the 1964 Alaskan earthquake or 1703 and 1923 Japanese earthquakes might provide evidence for or against the hypothesis that not all megathrust events there are represented by emergent shorelines.

Recognition of More Complex Patterns

Gaining even a simple knowledge of the average recurrence intervals of the earth's major faults will require enormous efforts. As new and more precise data become available, however, more complex patterns could emerge. Perhaps in some regions systematic relationships between major faults with similar average recurrence intervals will emerge. Does the timing of rupture of the Denali fault (Fig. 6), for example, relate in any systematic way to events on the Alaskan-Aleutian megathrust? Could the recurrence intervals of some faults (the Nankai Trough megathrust,

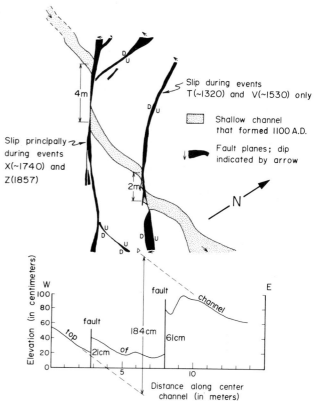

Fig. 26. Plan and cross-sectional views of offset channel at Pallett Creek. Six meters of right-lateral offset has been accomplished during at least four large earthquakes.

Fig. 25. This scarp represents faulting along the main trace of the San Andreas fault during a large earthquake in about 1500 A.D. Lateral slip of over 2 m accounts for the difference between correlative units on opposite sides of the fault. No fault slip has occurred since deposition of the unfaulted layers overlying the fault.

(Fig. 2), for example) vary secularly over many average intervals as a function of time elapsed since the latest great earthquake on a related fault (the Median Tectonic Line, for example)? Slip events along the Garlock and White Wolf faults (Fig. 14) are believed to occur less frequently than events on the nearby south-central segment of the San Andreas fault [Burke, 1979; Stein, 1981]. Are events on the San Andreas fault systematically delayed (or advanced) by occasional large slip events on these related faults?

More complete information on the prehistoric behavior of large faults might reveal that contiguous fault segments have distinctly different average recurrence intervals. Hay et al. [1981] present evidence that the northern reach of the San Andreas fault (Fig. 14) ruptures at least every three centuries, on the average; and it is known that the southern reach of the San Andreas fault has not produced a great earthquake for at least the past 210 yrs -- the period of historical record. Both values are greater than the average recurrence interval on the south-central segment at Pallett Creek. Are these permanent differences in average interval or do the apparent differences simply reflect the incompleteness of our knowledge?

Conclusions

Geological studies play an important role in the characterization of earthquake hazard and in earthquake prediction because they extend understanding of the spatial and temporal behavior of active faults beyond the limits of the historical and instrumental record into the more ancient past.

Several geomorphic, stratigraphic, and structural approaches have met with varying degrees of success in determining the dates of ancient earthquakes and their average and individual recurrence intervals and in characterizing ancient fault ruptures and deformations. Several other promising approaches have not yet been fully implemented.

Current technical limitations, especially in the realm of geochronology, greatly limit the level of detail at which the history of ancient earthquakes can be resolved. However, new methods may soon help remove some of these limitations. Nevertheless, even considering current momentum in this field of investigation, the long-term spatial and temporal history of most major faults, especially in underdeveloped countries, probably will remain obscure for the next several decades. Unfortunately, many of these faults will produce large earthquakes before their recurrence characteristics can be known and utilized in hazard mitigation or earthquake prediction. In such places, post-earthquake geological field investigations of these faults may well be invaluable in determination of their spatial and temporal behavior for future hazard mitigation or prediction.

In areas where vigorous programs of geological investigation are pursued, determination of recurrence intervals and dates of large earthquakes and behavioral characteristics for the major faults or fault segments might reveal systematic migrations, clusterings, or other interactions of large earthquakes that have value in prediction or hazard mitigation.

Acknowledgements. I am indebted to Lynn Sykes and Manik Talwani for the opportunity to attend the Maurice Ewing Symposium on Earthquake Prediction in May of 1980, to David Simpson and Paul Richards for their patience as editors of this volume, and to Shari Asplund who typed the manuscript. L. Seeber and two anonymous reviewers made very helpful suggestions for revising the manuscript first submitted, and many other colleagues offered suggestions that improved earlier drafts. My studies of the San Andreas fault system have been supported principally by the U.S. Geological Survey Earthquake Hazards Reduction Program under Contracts 14-08-0001-15225, -16774 and -18385. Contribution Number 3534, Division of Geological and Planetary Sciences, California Institute of Technology, Pasadena, CA 91125.

References*

Agnew, C. R., and K. E. Sieh, A documentary study of the felt effects of the great California earthquake of 1857, Bull. Seismol. Soc. Amer., 68, 1717-1729, 1978.

Allen, C. R., The tectonic environments of seismically active and inactive areas along the San Andreas fault system, Proceedings of the Conference on Geologic Problems of the San Andreas Fault System, Stanford Univ. Publ. Univ. Ser. Geol. Sci., 11, 70-82, 1968.

Allen, C. R., M. Wyss, J. N. Brune, A. Grantz, and R. E. Wallace, Displacements on the Imperial, Superstition Hills, and San Andreas faults triggered by the Borrego Mountain Earthquake, The Borrego Mountain Earthquake of April 9, 1968, U. S. Geological Survey Prof. Paper 787, 87-104, 1972.

Allen, C. R.. Geological criteria for evaluating seismicity, Geol. Soc. Amer. Bull., 86, 1041-1057, 1975.

Ando, M., Seismotectonics of the 1923 Kanto earthquake, Jour. Phys. of the Earth, 22, 263-277, 1974.

Ando, M., Source mechanisms and tectonic significance of historical earthquakes along the Nankai Trough, Japan, Tectonophysics, 27, 119-140, 1975.

Ando, M., T. Tsukuda and A. Okada, Trenches across the 1943 trace of the Shikano fault in Tottori, Report of the Coordinating Committee for Earthquake Prediction, 23, 160-165, 1980.

Bonilla, M. G., Trench Exposures Across Surface Fault Ruptures Associated with San Fernando Earthquake, San Fernando, California, Earthquake of February 9, 1971, U.S. Dept. of Commerce, 3, 173-182, 1973.

Broecker, W. S., and E. A. Olson, Lamont radiocarbon measurements VIII, Radiocarbon, 3, 176-204, 1961.

Brown, R. D., and R. E. Wallace, Current and historic fault movement along the San Andreas fault between Paicines and Camp Dix, California, Proceeding of the Conference on Geologic Problems of the San Andreas Fault System, Stanford Univ. Publ. Univ. Ser. Geol. Sci., 11, 22-41, 1968.

Burford, R. O., and P. W. Harsh, Slip on the San Andreas fault in central California from alinement array surveys, Bull. Seismol. Soc. Amer., 70, 1233-1261, 1980.

Burke, D. B., Log of a trench in the Garlock fault zone, Fremont Valley, California, U.S. Geological Survey, Miscellaneous Field Studies, Map MF-1028, 1979.

Carlson, R., H. Kanamori, and K. McNally, A survey of microearthquake activity along the San Andreas fault from Carrizo Plains to Lake Hughes, Seismol. Soc. America Bull., 69, 177-186, 1979.

Carter, B., Possible Pliocene inception of lateral displacement on the Garlock fault, California, in *Abstracts with Programs, Cordilleran Section, Geol. Soc. Amer.*, 12, 101, 1980.

Carver, G. A., Quaternary Tectonism and Surface Faulting in the Owens Lake Basin, California, *Technical Report AT-2*, Mackay School of Mines, Univ. of Nevada, Reno, Nev., 1970.

Clark, J. A., W. E. Farrell and W. R. Peltier, Global changes in postglacial sea level: A numerical calculation, *Quat. Res.*, 9, 265-287, 1978.

Clark, M. M., Map showing recently active breaks along the Garlock and associated faults, California, *U.S. Geological Survey, Miscellaneous Geologic Investigations, Map I-741*, 1973.

Clark, M. M., and K. R. Lajoie, Holocene behavior of the Garlock fault, *Abstracts with Programs, Cordilleran Section, Geol. Soc. America*, 6, 156-157, 1974.

Crook, Jr., R., C. R. Allen, B. Kamb, C. M. Payne, and R. J. Proctor, Quaternary Geology and Seismic Hazard of the Sierra Madre and Associated Faults, Western San Gabriel Mountains, California, *Final Technical Report, Contract No. 14-08- 0001- 15258, U. S. Geological Survey Earthquake Hazards Reduction Program*, 1978.

Damon, P. E., A. Long, and E. I. Wallick, Dendrochronologic calibration of the Carbon-14 Time Scale, *Proc. 8th Int'l. Conf. on Radiocarbon Dating, Royal Society of New Zealand, Wellington, New Zealand*, 1, 45-59, 1972.

Davis, T., Late Holocene seismic record, western Big Bend of San Andreas fault, *Abstracts with Programs, Geol. Soc. America, Cordilleran Section*, in press, 1981.

Denton, G. H., and W. Karlén, Holocene Climatic Variations -- Their Pattern and Possible Cause, *Quaternary Research*, 3, 155-205, 1973.

Dutton, C. E., The Charleston earthquake of August 31, 1886, *Ninth Annual Report of the Directorate*, pp. 209-528, U. S. Geol. Survey, Washington, D.C., 1889.

Geological Survey of India, The Bihar-Nepal earthquake of 1934, *Mem. Geol. Surv. India*, 73, 391 pp., 1939.

Hay, E.A., W. R. Cotton, and N. T. Hall, Late Holocene behavior of the San Andreas fault in northern California, *Abstracts with Programs, Geol. Soc. America, Cordilleran Section*, in press, 1981.

Hirano, S., and T. Nakata, A few examples of recurrence periods of active faults, Abstract No. 201 in *Assoc. Japanese Geographers*, 72, 1980.

Irwin, W. P., and I. Barnes, Effect of geologic structure and metamorphic fluids on seismic behavior of the San Andreas fault system in central and northern California, *Geology*, 3, 713-716, 1975.

Ishibashi, K., Comments on the long-term prediction of the Tokai earthquake, *Proc. Earthquake Prediction Research Symp., Seismol. Soc. Japan*, 123-125, 1980.

Kelleher, J. A., Rupture zones of large South American earthquakes and some predictions, *Jour. Geophys. Res.*, 77, 2087-2103, 1972.

Kelleher, J., L. Sykes, and J. Oliver, Possible criteria for predicting earthquake locations and their application to major plate boundaries of the Pacific and the Caribbean, *Jour. Geophys. Res.*, 78, 2547-2585, 1973.

Keller, R. P., C. R. Allen, R. Gilman, N. R. Goulty, and J. A. Hileman, Monitoring slip along major faults in southern California, *Bull. Seismol. Soc. Amer.*, 68, 1187-1190, 1978.

Kieckhefer, R. M. Late Quaternary Tectonic Map of New Zealand 1:50,000, *New Zealand Geological Survey, Dept. of Scientific and Industrial Research*, 1979.

Lahr, J. C., and G. Plafker, Holocene Pacific - North American plate interaction in southern Alaska: Implications for the Yakataga seismic gap, *Geology*, 8, 483-486, 1980.

Lajoie, K. R., J. P. Kern, J. F. Wehmiller, G. L. Kennedy, S. A. Mathieson, A. M. Sarna-Wojcicki, R. F. Yerkes, and P. F. McCrory, Quaternary Marine Shorelines and Crustal Deformation, San Diego to Santa Barbara, California, *Geological Excursions in the Southern California Area*, P.L. Abbott, ed., Dept. Geol Sci., San Diego State Univ., San Diego, Ca., 3-15, 1979.

LaMarche, V. C. Jr., and R. E. Wallace, Evaluation of effects on trees of past movements on the San Andreas fault, northern California, *Geol. Soc. America Bull.*, 83, 2665-2676, 1972.

Lawson, A. C., et al., *The California Earthquake of April 18, 1906, Report of the State Earthquake Investigation Commission*, 2 vols., 641 pp., Carnegie Institute of Washington, Washington, D.C., 1908.

Lensen, G. J., Analysis of Progressive Fault Displacement During Downcutting at the Branch River Terraces, South Island, New Zealand, *Geol. Soc. Amer. Bull.*, 79, 545-555, 1968.

Lensen, G. J., and P. Vella, The Waiohine River Faulted Terrace Sequence, *Recent Crustal Movements, Royal Society of New Zealand, Bull.* 9, 117-119, 1971.

Lomnitz, C., Major earthquakes and tsunamis in Chile during the period 1535 to 1955, *Geologische Rundsch.*, 59, 938-960, 1970.

Lubetkin, L., Late Quaternary activity along the Lone Pine fault, Owens Valley fault zone, California, unpublished Master's Thesis, Stanford University, 85 p., 1980.

Lubetkin, L., and M. Clark, Late Quaternary activity along the Lone Pine fault, eastern California, *EOS*, 61, 1042, 1980.

Matsuda, T., Y. Ota, M. Ando, N. Yonekura, Fault mechanism and recurrence time of major earthquakes in southern Kanto district, Japan, as deduced from coastal terrace data, *Geol. Soc. Amer. Bull.*, 89, 1610-1618, 1978.

Maugh, T. H. II, Radiodating: Direct detection extends range of of technique, *Science, 200*, 635-637, 1978.

McCann, W. R., S. P. Nishenko, L. R. Sykes, and J. Krause, Seismic Gaps and Plate Tectonics: Seismic Potential for Major Boundaries, Pure and Appl. Geophys., 117, 1082-1147, 1979.

Meisling, K. E., and K. E. Sieh, Disturbance of trees by the 1857 Fort Tejon earthquake, California, J. Geophys. Res., 85, 3225-3238, 1980.

Minster, J. B., and T. H. Jordan, Present-day plate motions, Jour. Geophys. Res., 83, 5331-5354, 1978.

Miyashita, K., and M. Matsu'ura, Inversion analysis of static displacement data associated with the Alaska earthquake of 1964, J. Phys. Earth, 26, 333-349, 1978.

Nakata, Geograph. Review of Japan, 15, 118-119.

Nakata, T., M. Koba, W. Jo, T. Imaizumi, H. Matsumoto, and T. Suganuma, Holocene Marine Terraces and Seismic Crustal Movement, 195-204, 1979.

Okada, A., Strike-slip Faulting of Late Quaternary along the Median Dislocation Line in the Surrounding of Awa-Ikeda, Northeastern Shikoku, The Quaternary Research, 7, no. 1, 15-26, 1968. [J-E]

Okada, A., Fault topography and rate of faulting along the Median Tectonic Line in the drainage basin of the river Yoshino, northeastern Shikoku, Japan, Geographical Review of Japan, 43, 1-22, 1970. [J-E]

Okada, A., Quaternary faulting along the Median Tectonic Line in the central part of Shikoku, Geographical Review of Japan, 46, 295-322, 1973. [J-E]

Okada, A., Geomorphic development and fault topography in the Butaipass area along the Atera fault zone, central Japan, Geographical Review of Japan, 48, 72-78, 1975. [J-E]

Okada, A., Quaternary faulting along the median tectonic line of southwest Japan, Memoirs of the Geological Society of Japan, No. 18, 79-108, 1980.

Okada, A., M. Ando, and T. Tsukada, Trenches across the Yamasaki fault in Hyogo Prefecture, Report of the Coordinating Committee for Earthquake Prediction, 24, 190-194, 1980.

Okada, A., and T. Matsuda, A fault outcrop at the Onosawa Pass and recent displacements along the Atera fault central Japan, Geographical Review of Japan, 49, 632-639, 1976. [J-E]

Okada, A., and A. Sangawa, Fault morphology and Quaternary faulting along the Median Tectonic Line in the southern part of the Izumi Range, Geographical Review of Japan, 51, 385-405, 1978. [J-E]

Oldham, R. D., Report on the earthquake of 12th June 1897, Mem. Geol. Surv. India, 29, 379 pp., 1899.

Ota, Y., T. Matsuda, and K. Naganuma, Tilted Marine Terraces of the Ogi Peninsula, Sado Island, Central Japan, Related to the Ogi Earthquake of 1802, Bull. Seismol. Soc. Japan (Zisin) ser. 2, 29, 55-70, 1976.

Ota, Y., and T. Yoshikawa, Regional characteris-tics and their geodynamic implications of late Quaternary tectonic movement deduced from deformed former shorelines in Japan, J. Phys. Earth, 26, Suppl., S 379-S 389, 1978.

Ota, Y., and N. Hori, Late Quaternary Tectonic Movement of the Ryukyu Islands, Japan, The Quaternary Research, 18, no. 4, 221-240, 1980. [J-E]

Penick, J., Jr., The New Madrid Earthquakes of 1811-1812, U. Missouri Press, Columbia, Mo., 181 pp., 1976.

Plafker, G., and M. Rubin, Uplift history and earthquake recurrence as deduced from marine terraces on Middleton Island, Alaska, in Proceedings of Conference VI: Methodology for identifying seismic gaps and soon-to-break gaps, U.S. Geological Survey Open File Report, 78-943, 857-868, 1978.

Plafker, G., T. Hudson, T. Bruns, and M. Rubin, Late Quaternary offsets along the Fairweather fault and crustal plate interactions in southern Alaska, Canadian Jour. Earth Sci., 15, 805-816, 1978.

Richter, C.F., Elementary Seismology, W. H. Freeman and Co., San Francisco, Ca., 1958.

Richter, D. H., and N. A. Matson, Jr., Quaternary Faulting in the Eastern Alaska Range, Geol. Soc. Amer. Bull., 82, 1971.

Sangawa, A., Geomorphic Development of the Izumi and Sanuki Ranges and relating Crustal Movement, Science Reports of the Tohoku University, 7th Series (Geography), 28, no. 2, 313-338, 1978a.

Sangawa, A., Fault activity and geomorphological development in northeastern Shikoku (Northern outskirts of Sanuki Mountains), and the central and western Kii peninsula, MTL, 3, 41-59, 1978b. [J]

Sangawa, A., and A. Okada, Fault outcrops related to recent activity of the Median Tectonic Line in the western Kii peninsula, MTL, 2, 51-60, 1977. [J]

Sarna-Wojcicki, A. M., K. R. Lajoie, S. W. Robinson, and R. F. Yerkes, Recurrent Holocene displacement on the Javon Canyon fault, rates of faulting, and regional uplift, western Transverse Ranges, California, Abstracts with Programs, Cordilleran Section, Geol. Soc. America, 11, 125, 1979.

Scholl, D. W., F. C. Craighead, Sr., and M. Stuiver, Florida Submergence Curve Revised: Its Relation to Coastal Sedimentation Rates, Science, 163, 562-564, 1969.

Schubert, C., and R. S. Sifontes, Boconó Fault, Venezuelan Andes: Evidence of postglacial movement, Science, 170, 66-69, 1970.

Shimazaki, K., and T. Nakata, Time-predictable recurrence model for large earthquakes, Geophys. Res. Lett., 7, 279-282, 1980.

Sieh, K. E., and L. S. Cluff, Studies of recent displacements along the Denali fault, south-central Alaska, Abstracts with Programs, Cordilleran Section, Geol. Soc. Amer., 7, no. 3, 423, 1975.

Sieh, K. E. and R. H. Jahns, Late Holocene slip

rate of the San Andreas fault at Wallace Creek, California, Abstracts with Programs, Cordilleran Section, Geol. Soc. Amer., 12, 152, 1980.

Sieh, K. E., A study of Holocene displacement history along the south-central reach of the San Andreas fault, Ph.D. Dissertation, Stanford University, 219 pp., 1977.

Sieh, K. E., Pre-historic large earthquakes produced by slip on the San Andreas fault at Pallett Creek, California, Jour. Geophys. Res., 83, 3907-3939, 1978a.

Sieh, K. E., Slip along the San Andreas fault associated with the great 1857 earthquake, Bull. Seismol. Soc. Amer., 68, 1421-1448, 1978b.

Sims, J. D., Determining earthquake recurrence intervals from deformational structures in young lacustrine sediments, Tectonophysics, 29, 141-152, 1975.

Stein, R. S., W. Thatcher, and R. O. Castle, Seismic and aseismic deformation associated with the 1952 Kern County, California, earthquake, and relationship to the Quaternary history of the White Wolf fault, submitted to J. Geophys. Res., 1981.

Stout, J. H., J. B. Brady, F. Weber, and R. A. Page, Evidence for Quaternary Movement on the McKinley Strand of the Denali Fault in the Delta River Area, Alaska, Geol. Soc. Amer. Bull., 84, 939-948, 1973.

Stuiver, M., Radiocarbon timescale tested against magnetic and other dating methods, Nature, 273, 271-274, 1978.

Sugimura, A., and T. Matsuda, Atera Fault and Its Displacement Vectors, Geol. Soc. Amer. Bull., 76, 509-522, 1965.

Swan, F. H., III, D. P. Schwartz, and L. S. Cluff, Recurrence of moderate to large magnitude earthquakes produced by surface faulting on the Wastch fault zone, Utah, Bull. Seismol. Soc. Amer., 70, 1431-1462, 1980.

Sykes, L., Aftershock zones of great earthquakes, seisimicity gaps and earthquake prediction for Alaska and the Aleutians, J. Geophys. Res., 76, 8021-8041, 1971.

Sykes, L. R., J. B. Kisslinger, L. House, J. Davies, and K. H. Jacob, Rupture zones and repeat times of great earthquakes along the Alaska-Aleutian arc, 1784-1980, Earthquake Prediction, Maurice Ewing Series, 4, American Geophysical Union, in press, 1981.

Taylor, F. W., B. L. Isacks, C. Jouannic, A. L. Bloom, and J. Dubois, Coseismic and Quaternary Vertical Tectonic Movements, Santo and Malekula Islands, New Hebrides Island Arc, Jour. Geophys. Res., 85, 5367-5381, 1980.

Wallace, R.E., Profiles and ages of young fault scarps, north-central Nevada Geol. Soc. Amer. Bull., 88, 1267-1281, 1977.

Wallace, R. E., Patterns of faulting and seismic gaps in the Great Basin Province, in Proceedings of Conference VI: Methodology for identifying seismic gaps and soon-to-break gaps, U.S. Geological Survey Open File Report, 78-943, 857-868, 1978a.

Wallace, R. E., Size of larger earthquakes, north-central Nevada, in Earthquake Notes, Seismol. Soc. America, 49, 23, 1978b.

Wallace, R. E., Active faults, paleoseismology, and earthquake hazards in the western United States, Earthquake Prediction, Maurice Ewing Series, 4, American Geophysical Union, in press, 1981.

Weber, G. E., W. R. Cotton, and L. Oshiro, Recurrence Intervals for Major Earthquakes and Surface Rupture Along the San Gregorio Fault Zone, Coastal Tectonics and Coastal Geologic Hazards in Santa Cruz and San Mateo Counties, California - Field Trip Guide, Cordilleran Section, Geol. Soc. America, 75th Annual Meeting, 112-119, 1979.

Weldon, R., and K. E. Sieh, Holocene rate of slip along the San Andreas fault and related tilting near Cajon Pass, southern California, Abstracts with Programs, Cordilleran Section, Geol. Soc. America, 12, 159, 1980.

Yang, A., and A. W. Fairhall, Variations of natural radiocarbon during the past eleven millenia and geophysical mechanisms for producing them, Proc. 8th Int'l. Conf. on Radiocarbon Dating, Wellington, New Zealand, 1972.

Yonekura, N., Quaternary tectonic movements in the outer arc on southwest Japan with special reference to seismic crustal deformation, Bull. Dept. Geogr. Univ. Tokyo, 7, 19-71, 1975.

Yonekura, N., T. Matsuda, M. Nogami, and S. Kaizuka, An active fault along the western foot of the Cordillera Blanca, Peru, Jour. Geography Tokyo, 88, 1-19, 1979.

Yonekura, N., and K. Shimazaki, Uplifted marine terraces and seismic crustal deformation in arc-trench systems: a role of imbricated thrust faulting, EOS, 61, 1111, 1980.

* [J] indicates article in Japanese.
  [J-E] indicates article in Japanese with English abstract.

ACTIVE FAULTS, PALEOSEISMOLOGY, AND EARTHQUAKE HAZARDS
IN THE WESTERN UNITED STATES

Robert E. Wallace, U.S. Geological Survey, Menlo Park, CA 94025

Abstract. Active faults are those that may have displacement within a future period of concern to humans. Studies of prehistoric earthquakes --paleoseismology-- show that average recurrence intervals for large displacements and related earthquakes on most active faults in the western United States are generally longer than 1,000 years; for many the average recurrence is greater than 10,000 years. Only on the San Andreas fault and its major branches are average recurrence intervals as short as 10-200 years. In the Great Basin province, a central part may have had no major displacement events (surface faulting events that could accompany a M$\approx$7 or larger earthquake) in late Quaternary time whereas in the western and eastern parts the generalized rate of faulting is in the range of 5×10^{-5} to 5×10^{-6} events per year per 1000 km^2. In some localized areas of western Nevada rates are higher. Probabilistic expressions of the likelihood of future behavior of active faults are needed.

Active Faults

Active faults are those faults that may have displacement within a future period of concern to humans. Most, but not all, are seismogenic. Both the fault displacement and earthquakes generated create hazards.

I suggest that criteria for estimating or predicting the future behavior of faults be kept separate from the definition of an active fault given above. The definition should remain constant, but the technical bases and methods for estimating future behavior are likely to change and improve in the future. Confusion has been created by current so-called "definitions" of active faults which range from "a fault having had displacement within the past 10,000 years" to "a fault having had displacements in late Quaternary time" (see review by Slemmons, 1977). Such criteria incorporate, without so indicating, elements of risk acceptance appropriate to a given project and should not be considered definitions.

Until recently the historic record of seismicity was used almost exclusively for assessing the future behavior of faults and for estimating the seismicity to be expected in a region. Useful as the historic record is, it is inadequate because in many regions large earthquakes occur at intervals greater than the length of the historic record. Even in regions of high seismicity segments of faults that have generated large earthquakes in the past may be currently seismically quiet, and this lack of current earthquake activity can give a falsely low measure of the hazard. Activity can also change with time, and long term changes should be understood and considered.

Paleoseismology (discussed in the next section) is becoming an important basis for estimating the long-term or average behavior of faults and for characterizing the seismic hazard in a province. In addition to estimations of future displacements based on paleoseismology, specific earthquake predictions based, for example, on changes in strain or stress may become feasible, and more accurate prediction of short-term future behavior of faults may become possible.

The rate of past activity can be usefully expressed as: (a) average recurrence intervals for either displacement events or earthquakes of a given size on a single fault, (b) average number (or fractions) of events per year, or (c) number of earthquakes or displacement events on sets of faults or fault segments averaged over unit areas or given lengths of faults for defined periods of time to characterize regions. Average rates of displacement, without an indication of the size of individual displacement events, also carry an implication, albeit incomplete, of the degree of activity, but rate of displacement alone cannot define the size of events to be expected.

Paleoseismology

The study of prehistoric earthquakes --paleoseismology-- is emerging as a critical consideration in evaluating earthquake hazards. Techniques of paleoseismology are principally geologic and include analysis of such features as: (a) microstratigraphic relations along faults as seen in natural or artificial exposures, (b) fault scarps and other fault topography, (c) offset drainage and changes in profiles of channels, (d) regional geologic relations that reveal long-term rates of displacement, (e) seismically induced sedimentary structures, and (f) marine, lacustrine, and river terraces related to uplift and faulting. A variety of techniques that can provide dates on geologically young materials include:[14]C, K/Ar, and other isotopic ratios; dendrochronology, fission track, rates of weathering and soil profile development, amino-acid racemization, and paleomagnetic correlation.

In interpreting prehistoric seismicity from studies of faults, the assumption is generally made that displacement events on faults were sudden and were accompanied by earthquakes. Methods for differentiating in the geologic record between sudden displacement and slow displacement, (i.e., aseismic fault-creep events) are not yet clear, and the problem deserves attention.

Exploration of faults and exposure of them by trenching have become common engineering-geologic practices. In addition to trenching merely to find the location or geometry of a fault, some exploration sites have been selected especially to examine paleoseismic history. The best sites for this purpose have many distinct stratigraphic units that can record displacement events on faults and dateable material to provide a time scale of the events. Examples of sites that have produced significant paleoseismic data are those on the San Andreas fault at Pallett Creek, California, on the Garlock fault at Fremont Valley, California, on the Javon Canyon fault, western Transverse Range, California, and on the Wasatch fault, Utah. For nine events in about 1400 years of record at Pallett Creek, Sieh (1978, 1980) determined an average recurrence interval of about 145 years and a range of 100-230 years. On the Garlock fault, Burke (1979) found evidence for 9 to possibly 17 events in about 15,000 years, or an average recurrence interval in the range of 900 to 1800 years. On the Javon Canyon fault, A. M. Sarna-Wojcicki, K. R. Lajoie and R. F. Yerkes (written commun., 1980) recognized five events in 3500 years. Sites developed on the Wasatch fault have revealed that the average recurrence interval for large-displacement events on the

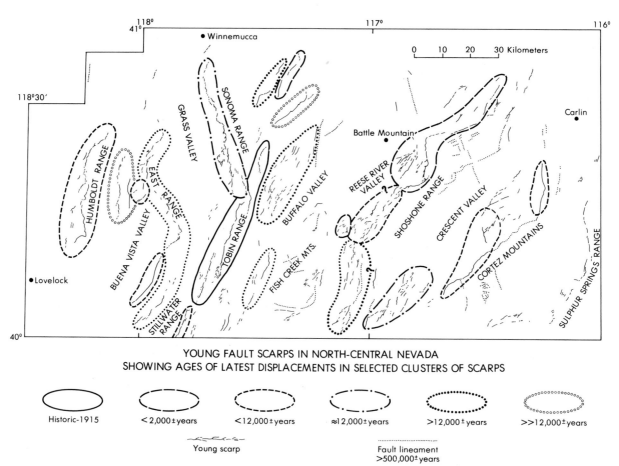

YOUNG FAULT SCARPS IN NORTH-CENTRAL NEVADA
SHOWING AGES OF LATEST DISPLACEMENTS IN SELECTED CLUSTERS OF SCARPS

Historic-1915   <2,000±years   <12,000±years   ≈12,000±years   >12,000±years   >>12,000±years

Young scarp

Fault lineament
>500,000±years

Figure 1. Young fault scarps in north-central Nevada showing ages of latest displacements in selected clusters of scarps.

fault is between 500 and 1,000 years at Kaysville, and between 1500 and 2300 years at Hobble Creek (Swan and others, 1980).

Fault scarps and their characteristics, such as slope angle and sharpness of crest, provide a useful reconnaissance means of dating surface displacement events on faults. Wallace (1977a, 1977b and 1978b) has used this technique to analyze sets of young faults in north-central Nevada (Fig. 1). Bucknam and Anderson (1979) and Bucknam (1980) refined the method by showing that a correlation between scarp slope and height exists for scarps of a given age in

western Utah. Nash (1980) derived a hillslope degradation model that predicts the relations found empirically by Bucknam and Anderson. Grose and Dodge (1979, 1980), in studying the Black Rock fault in western Nevada, developed corrections for slope angles of scarps as a function of different lithologies.

In some regions, especially where displacement has broken the earth's surface, an estimation of the total length and amount of displacement may be possible. From the length of break, a down-dip dimension assumed from tectonic or seismic considerations, and average displace-

Table 1. Average Recurrence Intervals and Rates of Slip on Some Active Faults, and Selected Paleoseismic Data, United States

Fault	Area	Average Recurrence interval, years [1]	Rate of slip mm/ year	Reference	Notes
	CALIFORNIA				
San Andreas	Wallace Creek	<250–330	>30–40	Sieh, 1980	Average recurrence of 10-m slip events
San Andreas	Pallett Creek	145		Sieh, 1978 and 1980	Range of recurrence intervals about 100–230 yrs.
San Andreas	Fort Ross	128–506		La Marche and Wallace, 1972	2-3 tilts of redwood tree between 1400 or 1650 and 1906
Garlock	Fremont Valley	~900–1,800		Burke, 1979	9-17 events in 14,700 ± 130 yrs.
Garlock	Koehn Lake		7	Clark and Lajoie, 1974	80 m offset since late Pleistocene 11,360± 160 yrs.
San Jacinto	NE of Anza		≥8–12	Sharp, 1980a and 1980b	5.7–8.6 km horizontal offset in no more than 0.73 my.
San Jacinto	S. California	<7.4		Sharp, 1980a	Average for entire length of fault
San Jacinto	S. California	100		Sharp, 1980a and 1980b	Average at one locality
Coyote Creek	W. Imperial Valley		3–5	Clark and others, 1972	1.7 m offset of Lake Cahuilla sediments, age 283–478 yrs.
Coyote Creek	W. Imperial Valley	15–200		Clark and others, 1972, and Sharp, pers. commun.	Offset of Holocene Lake Cahuilla beds Original estimate, 160–205 yrs. modified by new data
Coyote Creek	W. Imperial Valley		1–2	Sharp, 1980a and 1980b	10.9 m RL offset of buried stream channel, 5,000–6,800 yrs. old
Javon Canyon	W. Transverse Range	700–1,750	1.1	Sarna-Wojcicki and others, pers. commun., 1980	5 displacements of marine terrace 3,500 yrs. old
Ano Nuevo-	Pt. Ano Nuevo	~16,000		Weber and Cotton, 1980	6 offset events, 0.76–1.37 m/event, in marine terrace 105,000 years old
Frijoles		~6,000		Weber and Cotton, 1979	6 offset soils representing 7 faulting events in 40-50,000 years
Oak Hill	San Fernando	~ 200		Bonilla, 1973	Previous displacement on fault break of 1971
Unnamed	Lone Pine	3-7,000		Clark, 1980	Work by L. Lubetkin. 3 1872-type events in 10-22,000 yrs.
	NEVADA				
Genoa	Genoa	2-3,000		Pease, 1979 and pers. commun., 1979	2-3 displacements in last 6,000 yrs.
Unnamed	Carson City	2-6,000		Pease, 1979 and pers. commun., 1979	At least 2 displacements in 12,000 yrs.

[1] Some calculations may appear to be inconsistent without qualifications fully developed in original texts.

Table 1 (continued)

Fault	Area	Average Recurrence interval, years [1]	Rate of slip mm/ year	Reference	Notes
Pearce	Pleasant Valley	~6,000		Wallace, 1977b, 1978b and 1979	1915 scarp plus one large displacement younger than 12,000 ± years and several older than 12,000 ± yrs.
Pearce	Pleasant Valley	5,000		Bonilla, pers. commun., 1980	
Unnamed	West Flank Humboldt Range	~12,000		Wallace, 1977b, 1978b and 1979	Single-event younger than Lake Lahonatan shoreline
Unnamed	Northwest Flank Stillwater Range	>12,000		Wallace, 1978b, and 1979	Latest scarp is older than Lake Lahontan shoreline
Black Rock Fault	Black Rock Desert	6,250-8,330 6,000-8,000		Grose and Dodge, 1979; Dodge and Grose, 1979	3 or 4 displacements in 25,000 yrs. Latest within last 1400 yrs. 4 or 5 displacements in 24,000 yrs.
	UTAH				
Wasatch	Big Cottonwood Cr.		0.5-1.0	Scott, 1980	Over past 100,000 yrs.
Wasatch	Kaysville	500-1,000		Swan and others, 1980	Minimum of 3 offset events of fan ~6,000 years old. 1.7-3.7 m/event
Wasatch	Hobble Creek	1,500-2,400	1.1	Swan and others, 1980	6-7 offset events in 12,000 yrs., 0.8-2.8 m per event
Wasatch	Wasatch Range	50-400		Swan and others, 1980	Integration of 6-10 segments, each of which behaves like those at Kaysville or Hobble Creek
Unnamed	Fish Springs Range W. Utah	~11,800		Bucknam and Anderson 1979; Bucknam, 1980	Single-event scarp younger than Bonneville shoreline
Unnamed	Drum Mts W. Utah	~11,800		Bucknam and Anderson 1979; Bucknam, 1980	Single-event scarp younger than Bonneville shoreline
	NEW MEXICO				
La Jencia	Rio Grande Valley	5,000-1,000,000		Machette,1980	One event in Middle (?) Holocene to latest Pleistocene
County Dump	Rio Grande Valley	90,000-190,000		Machette, 1978	4 fault events at 20,000, 120,000, 310,000 and 400,000 yrs. ago
	TENNESSEE New Madrid	600-1,000		Russ, 1979	3 large earthquakes in 2,000 years including 1811-12
	ALASKA Middleton Island	500-1,350	10	Plafker, 1978	6 episodes of uplift recorded by marine terraces in 4,300 yrs.
	Otmeloi Island, Yakatat	420		Plafker, pers. commun., 1980	Beach raised 500+ yrs. ago and during 1899 earthquake
	IDAHO Arco	>10,000		Pierce, 1980 and pers. commun.	
	WASHINGTON Puget Sound	23-276		Sims, 1975	14-21 deformed zones within 1,804 varves

[1] Some calculations may appear to be inconsistent without qualifications fully developed in original texts.

ment, the seismic moment of a prehistoric earthquake can be calculated (Kanamori, 1977, Thatcher and Hanks, 1973, Hanks, and others, 1975, Wallace, 1978a) and a correlation to expressions of magnitude are possible.

Sims (1975) has developed techniques to determine recurrence intervals from deformational structures in young lacustrine sediments. The structures observed are intricately deformed thin beds that suggest deformation in units weakened by liquefaction. Sand boils or sand craterlets, which are also liquefaction phenomena, are used by Sieh (1978) as a key to prehistoric earthquakes at Pallett Creek, California.

In summary, geologic techniques can be used to estimate the distribution, size, and average recurrence intervals of large prehistoric dis-

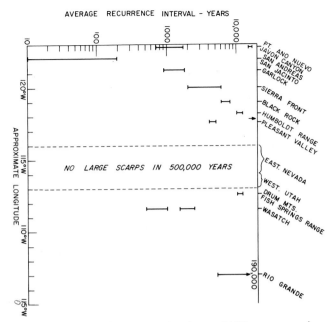

Figure 2. Diagram showing differences in average recurrence intervals of displacements on faults in the western United States from the continental margin eastward to the Rio Grande Rift.

placement events on faults and of related earthquakes.

Earthquake Hazard Evaluation Based on Paleoseismicity

Data in Table 1 demonstrate that most faults that have had repeated displacements in the western United States are characterized by average recurrence intervals at a particular location along a fault of 1000 years or more, or expressed as chances per year, one-thousandth or less; for many the average recurrence is greater than 10,000 years. Only the San Andreas fault system and its major branches, such as the San Jacinto fault, are characterized by recurrence intervals appreciably lower, e.g., in the 10- to 200-year range.

A general increase in average recurrence intervals away from the San Andreas fault system is shown diagrammatically in figure 2, using selected data from table 1 and projecting data points to an east-west line across the western United States. If the displacements on the faults are interpreted as related to interplate motion between the North American and Pacific plates, the San Andreas fault system is accommodating most, perhaps two-thirds, of the motion, and other faults, distributed across at least 18° degrees of longitude, accommodate the remainder in decreasing amounts away from the plate margin. The Wasatch fault zone is an exception and represents a perturbation in the general eastward decrease in rates. Strain is propagated inland from the western continental margin very likely in complex ways, including subduction, oroclinal warping, localized extension, and involving both rheid and elastic response.

For regional evaluations of earthquake hazards derived from paleoseismologic studies of faults, normalization by area is required. In the Great Basin province, a reconnaissance of young fault scarps (Wallace, 1977a, 1978b) indicates that generalized rates of faulting in the range 5×10^{-5} to 5×10^{-6} faulting events per year per 1000 km^2 are characteristic of western Nevada, eastern California, and west-central Utah in Holocene time (12,000+ years) (Fig. 3). A "faulting event" as used here is estimated to have been accompanied by an earthquake of about 7 M or greater, because the scarps preserved are generally a few meters high. By comparison to historic events and by seismic-moment arguments, fault displacements of that size would be accompanied by that size earthquakes. Scarps of only a few centimeters height tend to become obliterated within a few centuries at most. In contrast, an area of

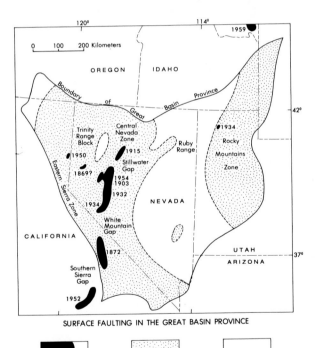

SURFACE FAULTING IN THE GREAT BASIN PROVINCE

HISTORIC FAULTING	HOLOCENE FAULTING	NO LATE QUATERNARY EVENTS
1915		
$3 \times 10^{-3} >$ event $> 1 \times 10^{-4}$ per year per 1000 KM^2	(last 12,000+years) $5 \times 10^{-5} >$ event $> 5 \times 10^{-6}$ per year per 1000 KM^2	(last 5×10^{5}+years)

NOTE: An event is an occurrence of surface faulting large enough to have produced an earthquake $\geqq$ M7. Earthquakes of 1869, 1903, 1934 and 1950 were <M7. Those of 1872, 1915, 1932, 1952 and 1954 were >M7.

Figure 3. Map of the Great Basin province showing zones of various rates of surface faulting based primarily on fault scarp data. See figure 4 for subdivisions derived for western Utah.

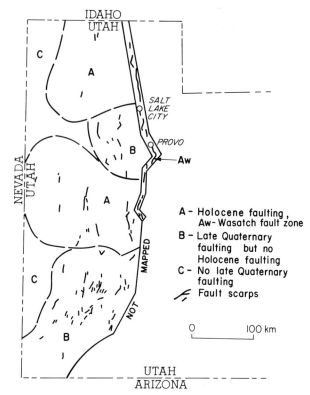

IDAHO
UTAH

C

A

SALT
LAKE
CITY

PROVO

B

Aw

A

A - Holocene faulting,
Aw - Wasatch fault zone
B - Late Quaternary
faulting but no
Holocene faulting
C - No late Quaternary
faulting
Fault scarps

NEVADA
UTAH

NOT MAPPED

C

B

0   100 km

UTAH
ARIZONA

Figure 4. Map of western Utah showing zones of various histories of faulting based on fault scarp data (from Bucknam and others, 1980).

approximately 100,000 km^2 in western Utah and eastern Nevada may have had no 7 M or greater events in late Quaternary time (approximately the past 500,000 years). In western Utah, Bucknam and others (1979, 1980) have recognized subprovinces, some having Holocene faulting, and others no Holocene faulting (Fig. 4). Among the subprovinces, the Wasatch fault zone is characterized by the greatest concentration of prominent scarps along faults active in late Quaternary time.

Larger faults, or sets of faults, are capable of producing large earthquakes on any of a number of segments or units. For example, Swan and others (1980) point out that, inasmuch as each of 6 to 10 segments of the entire Wasatch fault could produce earthquakes at the rate of the segment at Kaysville, the average recurrence interval for the entire fault can be expressed as between 50 and 400 years. Wallace (1970) performed a similar analysis for the San Andreas fault.

Paleoseismology can be helpful in refining the patterns and the values of expected seismic shaking derived from historic records alone. For example, in the study area in Nevada shown in figure 1, paleoseismicity suggests a uniform seismic flux over four counties instead of the wide range of coefficient Aa values (3-7) shown

on the ACT-3 map (Applied Technology Council, 1978). Using paleoseismic data from Salt Lake Valley, Utah, Bucknam and others (1979, 1980) estimate that accelerations with a 10 percent chance of being exceeded in 50 years may be two or three times as large as those estimated on the map by Algermissen and Perkins (1976) which was based principally on historic records.

An important step to be taken in estimating the hazards of faults is the development of probabilistic models of the likelihood of future displacement. Cluff and others (1980) have so analyzed the Wasatch fault zone. They use a model, commonly assumed by geologists and seismologists, in which strain release is constant over long periods and the intervals between large earthquakes can be represented by averages. Given these assumptions, it follows that the longer the time since the last large surface displacement or earthquake, the shorter the time until the next. Uniformity of strain release and of recurrence interval over long periods of time, however, may not be an entirely correct model, because long-term clustering of events in time seems to be real. Given a model in which major events are clustered in time and the clusters are separated by long quiescent periods, the waiting time until the next event cannot be derived by subtracting the elapsed time since the last event from the average recurrence interval. For example, if the present instant falls in a long period of quiescence, a long time since the last event might be followed by a long time until the next event. Conversely, if the present instant falls within a cluster of events, the waiting period until the next event may be short relative to the average recurrence interval (see Kagan and Knopoff, 1979, Mandelbrot, 1977 for discussions related to this problem). The degree and styles of long-term clustering are as yet poorly known.

Paleoseismology is just emerging as a distinct discipline. Despite many unresolved questions, the long-term record available only from pal̲eoseismology already has provided valuable insight into the faulting process and assists in the evaluation of earthquake hazards.

References

Algermissen, S. T. and Perkins, D. M., A probabilistic estimate of maximum acceleration in rock in the contiguous United States: U.S. Geological Survey Open-File Report 76-416, 44 p., incl. 2 maps, 1976.

Applied Technology Council, Tentative provisions for the development of seismic regulations for buildings: National Bureau of Standards 510, 505 p., 1978.

Bonilla, M. G., Trench exposure across surface fault rupture associated with San Fernando earthquake, in San Fernando, California Earthquake of February 9, 1971, Vol. III, Geological and Geophysical

Studies: U.S. Dept. of Commerce, National Oceanic and Atmospheric Administration, 173-182, 1973.

Bucknam, R. C., Characteristics of active faults in the Great Basin, in Summaries of Technical Reports, Vol. IX., National Earthquake Hazards Reduction Program: U.S. Geological Survey Open-File Report 80-6, 94-95, 1980,

Bucknam, R. C., Algermissen, S. T. and Anderson, R. E., Late Quaternary faulting in western Utah and the implication in earthquake hazard evaluation [abs.]: Geological Society of America Cordilleran Section Annual Meeting, Abstracts with Programs, 71-72, 1979.
_____ Patterns of late Quaternary faulting in western Utah and an application in earthquake hazard evaluation, in Proceedings of Conference X, Earthquake hazards along the Wasatch and Sierra-Nevada frontal fault zones: U.S. Geological Survey Open-File Report 80-801, 299-314, 1980.

Bucknam, R. C. and Anderson, R. E., Estimation of fault-scarp ages from a scarp-height-slope-angle relationship: Geology, 7, n. 1, 11-14, 1979.

Burke, D. B., Log of a trench in the Garlock fault zone, Fremont Valley California: U.S. Geological Survey Misc. Field Studies Map MF-1028, 1979.

Clark, M. M., Quaternary faulting in southern California, in Summaries of Technical Reports, Vol. IX, National Earthquake Hazards Reduction Program: U.S. Geological Survey Open-File Report 80-6, 7-8, 1980.

Clark, M. M., Grantz, Arthur and Rubin, Meyer, Holocene activity on the Coyote Creek fault as recorded in sediments of Lake Cahuilla, in The Borrego Mountain earthquake of April 9, 1968: U.S. Geological Survey Professional Paper 787, 112-130, 1972.

Clark, M. M. and Lajoie, K. R., Holocene behavior of the Garlock fault [abs.]: Geological Society of America Cordilleran Section Annual Meeting, Abstracts with Programs, 156-157, 1972.

Cluff, L. S., Patwardhan, A. S., and Coppersmith, K. J., Estimating the probability of occurrence of surface faulting earthquakes on the Wasatch fault zone, Utah: Seismological Society of America Bulletin, 7, n. 5, 1463-1478, 1980.

Dodge, R. L. and Grose, L. T., Seismotectonic and geomorphic evolution of a typical basin and range normal fault, The Holocene Black Rock fault, northwestern Nevada [abs]: Geological Society of America Cordilleran Section Annual Meeting, Abstracts with Programs, 75, 1979.

Dodge, R. L. and Grose, L. T., Tectonic and geomorphic evaluation of the Black Rock fault, northwestern Nevada, in Proceedings of Conference X: U.S. Geological Survey Open-File Report 80-801, 494-508, 1980.

Gross, L. T., and Dodge, R. L. Seismotectonic and geomorphic evolution of the Holocene Black Rock fault, northwestern Nevada: Semi-annual Technical Report for period Aug. 1, 1978 to Jan. 14, 1979 to U.S. Geological Survey, Grant No. 14-08-0001-6-505, 1-17, 1979.

Hanks, T. C., Hileman, J. A., and Thatcher, Wayne, Seismic moments of the larger earthquakes of the southern California region: Geological Society of America, 86, 1131-1139. 1975,

Kagan, Yan and Knopoff, Leon, Statistical study of occurrence of shallow earthquakes: Geophysical Journal of the Royal Astronomical Society, 55, 67-86, 1978.

Kanamori, Hiroo, The energy release in great earthquakes: Journal of Geophysical Research, 82, n. 20, 2981-2987, 1977.

LaMarche, V. C. Jr., and Wallace, R. E., Evaluation of effects on trees of past movements on the San Andreas fault, northern California: Geological Society of America Bulletin, 83, n. 9, 2665-2676, 1972.

Machette, M. N., Dating Quaternary faults in the southwestern United States by using buried calcic paleosols: Journal of Research, 6, n. 3, 369-381, 1978.

Machette, M. N., Seismotectonic analysis Rio Grande Rift, New Mexico, in Summaries of Technical Reports, Vol. IX, National Earthquake Hazards Reduction Program: U.S. Geological Survey Open-File Report 80-6, 56-57, 1980.

Mandelbrot, B. B., Fractals, form, chance and dimensions: W. H. Freeman and Co., San Francisco, 365 p., 1977.

Nash, D. B., Morphologic dating of degraded normal fault scarps: Journal of Geology, 88, n. 3, 353-360, 1980.

Pease, R. C., Fault scarp degradation in alluvium near Carson City, Nevada [abs.]: Geological Society America Cordilleran Section Annual Meeting 1979, Abstracts with Programs, 121, 1979.

Pierce, K. L., Quaternary dating techniques, in Summaries of Technical Reports, Vol. IX, National Earthquake Hazards Reduction Program: U.S. Geological Survey Open-File Report 80-6, 112-114, 1980.

Plafker, George and Rubin, Meyer, Uplift history and earthquake recurrence as deduced from marine terraces on Middleton Island, Alaska, in Isacks, B. L., and Plafker, George, co-organizers, Proceedings of Conference VI, Methodology for identifying seismic gaps and soon-to-break gaps: U.S. Geological Survey Open-File Report No. 78-943, 687-722, 1978.

Russ, D. P., Late Holocene faulting and earthquake recurrence in the Reelfoot Lake area, northwest Tennessee: Geological Society of America, 90, n. 11, 1013-1018, 1979.

Scott, W. E., Quaternary stratigraphy of the Wasatch Front, in Summaries of Technical Reports, Vol. IX, National Earthquake Hazards Reduction Program: U.S. Geological Survey Open-File Report 80-6, 62-63, 1980.

Sieh, Kerry, Prehistoric large earthquakes produced by slip on the San Andreas fault at Pallet Creek, California: Journal of Geophysical Research, 83, no. B8, 3907-3939, 1978.

_____ Late Holocene behavior of the San Andreas fault, in Summaries of Technical Reports, Vol. IX, National Earthquake Hazards Reduction Program: U.S. Geological Survey Open-File Report 80-6, 39-40, 1980.

Sims, J. D., Determining earthquake recurrence intervals from deformational structures in young lacustrine sediments: Tectonophysics, 29, 141-152, 1975.

Sharp, R. V., Variable rates of late Quaternary strike slip on the San Jacinto fault zone, southern California: Journal of Geophysical Research, 1980a, in press.

_____ Salton Trough tectonics, in Summaries of Technical Reports, Vol. IX, National Earthquake Hazards Reduction Program: U.S. Geological Survey Open-File Report 80-6, 37-38, 1980b.

Slemmons, D. B., Definitions of "active fault." U.S. Army Engineer Waterways Experiment Station, Miscellaneous Paper S-77-8, 1-22, 1977.

Swan, F. H. III, Schwartz, D. P., and Cluff, L. S., Recurrence of moderate to large magnitude earthquakes produced by surface faulting on the Wasatch fault zone, Utah: Seismological Society of America Bulletin, 70, n. 5, 1431-1462, 1980.

Thatcher, Wayne, and Hanks, T. C., Source parameters of southern California earthquakes: Journal of Geophysical Research, 78, n. 35, 8547-8576, 1973.

Wallace, R. E., Earthquake recurrence intervals on the San Andreas fault: Geological Society of America Bulletin, 81, n. 10, p. 2875-2890, 1970.

_____ Time-history analysis of fault scarps and fault traces -- a longer view of seismicity: Proceedings of 6th World Conference on Earthquake Engineering, New Delhi, India, 1, 766-769, 1977a.

_____ Profiles and ages of young fault scarps, north-central Nevada: Geological Society of America, 88, no. 9, 1267-1281, 1977b.

_____ Size of larger earthquakes, north central Nevada [abs.]: Earthquake Notes (Seismological Society of America, Eastern Section), 49, n. 1, 23, 1978a.

_____ Patterns of faulting and seismic gaps in the Great Basin province, in Proceedings of Conference VI, Methodology for identifying seismic gaps and soon-to-break gaps, May 25-27, 1978: U.S. Geological Survey Open-File Report 78-943, 858-868, 1978b.

_____ Map of young faults scarps related to earthquakes in north-central Nevada: U.S. Geological Survey Open-File Report 79-1554, 2 sheets, 1979.

Weber, G. E. and Cotton, W. T., Recurrence intervals for surface faulting along the Frijoles fault and the Ano Nuevo thrust fault of the San Gregorio fault zone, San Mateo County, California [abs.]: Geological Society of America Cordilleran Section Annual Meeting, Abstracts with Programs, 134, 1979.

_____ Geologic investigations of recurrence intervals and recency of faulting along the San Gregorio fault zone, San Mateo County, California, in Summaries of Technical Reports, Vol. IX, National Earthquake Hazards Reduction Program: U.S. Geological Survey Open-File Report 80-6, 128-129, 1980.

REPEAT TIMES OF GREAT EARTHQUAKES ALONG SIMPLE PLATE BOUNDARIES

Lynn R. Sykes and Richard C. Quittmeyer

Lamont-Doherty Geological Observatory
and Department of Geological Sciences of Columbia University
Palisades, New York 10964

Abstract. Repeat times of large earthquakes
are obtained for 12 segments of simple plate boundaries of the convergent type and three of the
transform type for which two or more large shocks
are known to have occurred at nearly the same
place. Repeat time varies from as short as
35 years in some areas to more than 150 years in
others and is related not only to the relative
plate velocity but also to the geometry of the
rupture zone. Average displacement in these large
events increases systematically as the dimension
of the rupture zone increases. Static stress drop
increases by a factor of about four for thrust
events of magnitude, M_w, 7.6 to 9.5 and by about
10 for shocks of M_w 6.3 to 8.1 along transform
faults. The static stress drop and average
displacement also vary by a factor of two to four
among events that ruptured nearly the same portion
of a plate boundary, the larger values being
associated with larger dimensions. Our data and
those of other workers support the time-predictable
model wherein the time interval between two large
shocks is proportional to the displacement in the
preceding event, i.e., the shock that starts the
time interval. Values of α, the ratio of seismic
slip to total slip, computed with the time-
predictable model vary from 0.3 to 0.9; whereas
values determined for the slip-predictable model
(average displacement in a large event
proportional to the preceding time interval), show
a much greater variation. Hence, accurate estimates of repeat time depend critically upon the
amount of displacement in the last large shock.
Errors involved in determining displacement and α,
however, can lead to uncertainties as large as a
factor of three in predicted repeat time if knowledge is available for only the last large event
along a given segment of a plate boundary. When
the time interval between two previous large
shocks in an area is known and the ratio of their
displacements can be determined, the time interval
between the last large shock and a future earthquake can be estimated with much greater accuracy.
An application of this calibrating technique to
six regions in which three or more shocks are
known to have occurred yields repeat times within
2% to 50% of those observed. These calculations
also indicate that variations in plate velocity do
not exceed these percentages for time scales of
10's to 100's of years. The northwestern half of
the 1857 rupture zone in southern California,
which experienced large displacement in that
shock, probably will not be the site of a great
earthquake for many decades whereas the southeastern half, which was the locus of smaller
offset in 1857, appears to have a greater potential. The increase of displacement and static
stress drop with the size of the rupture zone
suggests that slip continues longer at a given
place on the rupture surface as size increases.
Hence, large earthquakes cannot be considered to
be merely the sum of moderate-size events placed
end-to-end. The relief of stress is only partial
even in great earthquakes. Stress drop appears to
be governed not by the amount of stress available
but by the presence of barriers, particularly
major transverse boundaries, that control the
dimensions of rupture zones.

Introduction

Most of the world's great earthquakes, i.e.
events with magnitudes, M, greater than 7.7, occur
at shallow depths along plate boundaries of either
the convergent or strike-slip type. These shocks
account for most of the total seismic slip, energy
release, and cumulative seismic moment for the
major plate boundaries of the world. Rupture
during great shocks typically extends 50 to
1000 km along strike with average displacements
from a few meters to as much as 20 m. Most of
these shocks rupture the entire range of depths
over which interplate movement occurs either
partially or totally by sudden seismic slip.

Since the strains that are released in great
earthquakes appear to be built up over decades to
centuries by slow plate movements, great shocks do
not reoccur at the same place in less than about
35 to 150 years [Sykes, 1971; Kelleher et al.,
1973]. Seismic gaps, that is portions of plate
boundaries that have not been the sites of large
shocks for tens to hundreds of years, appear to be

the most likely locations for future large earthquakes [Fedotov, 1965; Mogi, 1968a; Tobin and
Sykes, 1969; Kelleher, 1970; Sykes, 1971; Kelleher
et al., 1973]. In fact, nearly all of the great
earthquakes of the past 15 years have occurred
within what had been seismic gaps for many decades
[McCann et al., 1978, 1979].

Kelleher et al. [1973] and McCann et al. [1978,
1979] use the term seismic potential as a
qualitative measure of the chance that a given gap
will, in fact, be the site of a large earthquake
within the next few decades. McCann et al. assign
the highest potential to areas that are known to
have experienced great earthquakes in the past but
which have not been the sites of such events for
at least 100 years. They arbitrarily picked a
period of 100 years since repeat times of large
shocks are generally poorly known. (We use the
term repeat time to characterize the individual
periods of time that elapsed between large shocks
that rupture nearly the same portion of a plate
boundary.) Obviously, a quantitative understanding of the seismic potential for large
earthquakes must involve a consideration of repeat
time and of how it is related to the rate of plate
movement and to the geometry and thermal
properties of the zone of plate convergence.

In this paper we examine deterministic aspects
of repeat time for large earthquakes along simple
plate boundaries. Kelleher et al. [1974],
Acharya [1979], and others argue qualitatively
that repeat time may be related to one or more of
the following factors (Figure 1): length of

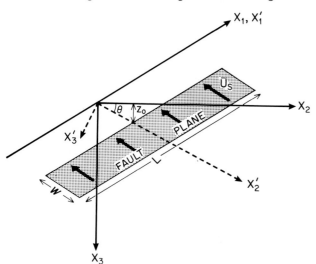

Fig. 1. Schematic model of a pure dip-slip
fault showing the labeling conventions used in
this paper. The fault plane has length, L;
down-dip width, W; and dip, θ. Seismic displacement, $\bar{U}_s$, is parallel to the width of the
fault and is assumed constant over the fault
plane. z_o is the vertical distance from the
free surface to the top of the fault. In the
primed coordinate system the fault plane lies
within the plane defined by $x_3' = 0$.

rupture along strike (L), downdip width (W), or
the dip of the plate boundary (θ), in addition to
the rate of plate convergence (V_c). That factors
other than the rate of plate convergence are
involved is indicated by the variation of repeat
time among regions with relative plate velocities
that are comparable. For example, repeat times in
Mexico average about 35 years [Kelleher et al.,
1973; McNally and Minster, 1980] while those in
parts of the Aleutians are about 60 years [Davies
et al., 1980; Sykes et al., 1980, 1981]. Also,
even for a particular point along a given plate
boundary, repeat times may vary by at least a
factor of 1.5 to 2. This is the case in southwestern Japan where the record of large historic
earthquakes is complete back to at least 1707 and
perhaps as far as 684 A.D. [Ando, 1975].

Our data set, collected on a world-wide basis,
consists of observed repeat times and source parameters for about 15 segments of convergent plate
boundaries for which a history of at least two
large (M > 7) shocks is available. Data were also
collected for about 20 strike-slip earthquakes.
In selecting data we chose events that ruptured
the same or nearly the same segment of a plate
boundary and those that ruptured the entire depth
range over which seismic slip takes place between
two interacting plates. For strike-slip faults
these requirements may be met for events as small
as M = 6, whereas for convergent boundaries the
shocks must generally be larger than magnitude 7
to 7.5. Only events along relatively simple plate
boundaries were examined as strains may not build
up as uniformly along more complicated, multibranched boundaries. Also, earthquakes for which
parameters were estimated indirectly, such as
seismic moment from aftershock area, were
excluded. The results apply only to the
recurrence of large thrust and strike-slip earthquakes and not to those of smaller size, those of
intermediate depth, those involving normal
faulting within trenches, or those occurring near
but not along the main plate boundaries.

We purposely attempt to push the idea of
average properties--i.e. displacements and
stress drops averaged over the entire rupture
surface--as far as possible. While there is no
doubt that many, and perhaps all, large earthquakes consist of multiple events, and that
displacement and stress drop vary considerably
within individual rupture zones, we conclude that
much can be learned from average parameters at
about the factor of 1.5 to 2 level. It is clear
that variations in displacement, stress drop and
energy release along the rupture zone become
important at about that level and must be
considered in future work on repeat times.

We find good support in the globally observed
data on repeat times and displacements for the
proposition that the time interval between large
shocks at a given place is proportional to the
displacement in the preceding earthquake (model a
of Figure 2) rather than to the displacement in
the shock that terminates that interval (model b).

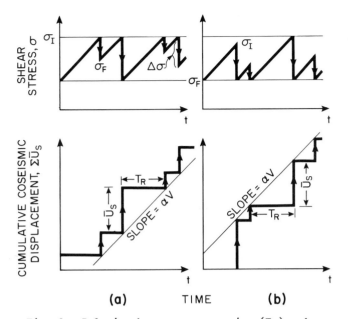

Fig. 2. Relation between repeat time (T_R) and seismic displacement for two models of stress release during earthquakes along nearly same part of a plate boundary [after Shimazaki and Nakata, 1980]. In model (a), known as the time-predictable model, the shear stress at which the earthquake initiates (σ_I) is constant for all shocks but the final value (σ_F), to which the shear stress drops, varies from event to event. The drop in stress during the earthquake must be recovered before the next earthquake will occur. Thus, the repeat time to the next earthquake is proportional to the amount of displacement (or stress drop) that occurred during the previous event. W, the down-dip width is taken to be constant throughout each sequence of shocks. In model (b), known as the slip-predictable model, the shear stress at which an earthquake initiates differs from one event to another, but the final stress is the same for all shocks. In this case, the amount of slip in the next earthquake is proportional to the time since the last earthquake. The constant of proportionality for both cases is αV, where α is the ratio of seismic to total slip and V is the plate velocity. Arrows denote stress drop or slip during earthquakes. Note that in both models the stress drop, $\Delta\sigma$, is not constant among shocks as a function of time.

Bufe et al. [1977] found such a relationship for small shocks along a segment of the Calaveras fault in California while Shimazaki and Nakata [1980] report a similar finding for three sequences of earthquakes and earthquake-related features in Japan.

Our results clearly do not support a model that assumes stress drop is constant among large shocks. The idea of constant stress drop does have a utility when seismic moments are compared that vary by a factor of as much as 10^{13}, but to use the constant stress drop approximation in estimating the repeat times of large shocks can lead to biases or uncertainties of a factor of at least four.

Very little is known about the relative amounts of seismic and aseismic slip along many plate boundaries, especially those of the convergent type where the plate boundary is often water covered. Kanamori [1977a] calculated the average seismic slip, $\bar{U}_s$, from the seismic moment for a number of large earthquakes along subduction zones and compared it with the amount of slip predicted from plate tectonic models for the time interval between it and the last large shock at the same location. He suggested that the seismic slip rate was comparable to the rate of plate motion in Chile and possibly in Alaska while that in the Kuriles and northern Japan was a very small portion, about 1/4, of the total slip. For our global set of data we calculate values of the ratio of seismic slip to total slip, α that range from 0.2 to 1.8 if the average slip that occurred in the last large shock in an area is compared with that calculated from plate tectonic models for the interval between it and the preceding large shock. The computed values of α only vary from 0.3 to 0.9, however, if the calculated plate motion during that interval is compared with the amount of seismic slip in the preceding earthquake, following the ideas of Bufe et al. [1977] and Shimazaki and Nakata [1980].

If the displacements and stress drops are not, in fact, the same in two successive earthquakes at the same location, it is not surprising that the computed values of α reflect mainly these differences rather than those related to differing rheologies. Values of α representative of differences in rheology could, however, be obtained by taking averages over many cycles of large earthquakes or by calculating α for an appropriate model such as the time-predictable model of Shimazaki and Nakata [1980]. We propose that α is close to 1.0 at shallow depths along several subduction zones and becomes small at greater depths as temperature increases along the plate interface.

Our data and those of Bufe et al. [1977] and Shimazaki and Nakata [1980] indicate that the approximate time of occurrence of a future large shock can be estimated from the amount of displacement in the last large event and from one of the following: 1) a calibration interval for which a repeat time and the displacement in the event preceding it are available or 2) an estimate of the ratio, α. If the time interval between two successive shocks is known, but one displacement is not known from observations, it may be possible to estimate that displacement from knowledge of more readily available quantitites such as source dimension, intensity distribution or tsunami height. With this purpose in mind we attempt to determine scaling laws describing the dependence of average displacement upon the source parameters L, W, θ and z_o (Figure 1) for both dip-slip and strike-slip faulting. The use of such a relationship may provide a powerful tool for the estimation of the times of occurrence of future large shocks.

Theoretical and Empirical Expressions
Repeat Time

The repeat time (T_R) as averaged over many earthquake cycles is

$$T_R = \frac{\bar{U}}{V} = \frac{\bar{U}_s}{\alpha V} \qquad (1)$$

where $\bar{U}$ is the average total displacement (seismic plus aseismic), $\bar{U}_s$ is the average displacement that occurs in large earthquakes (coseismic displacement), V is the relative plate velocity and α is the ratio of seismic slip to total slip. The combined contribution of small to moderate size earthquakes, which is typically 0 to 15% of the total slip, is neglected in this analysis. We assume that strains are built up fairly uniformly along each of the simple plate boundaries of the world. We will assume further, that on a time scale of years to centuries, two points well removed from the plate boundary move at the relative rates of plate motion that have been determined on a global basis as averages for the past 1 to 5 million years (m.y.) [e.g. Chase, 1978; Minster and Jordan, 1978]. We show later that our data justify this assumption. Closer to the plate boundary strains are built up for 10's to 100's of years and are then released (or partly released) during large earthquakes.

At a given place along a plate boundary the quantity αV in equation (1) is likely to be a constant over many earthquake cycles if stress is built up fairly uniformly and if α depends on the rheology of the plate boundary and on the distribution of large asperities. If α and V do not, in fact, change appreciably over several earthquake cycles, individual repeat times at a given place would be proportional to $\bar{U}_s$. The time-predictable model of Figure 2a [Shimazaki and Nakata, 1980] indicates that repeat time is proportional to the displacement in the shock that initiates that time interval. In the slip-predictable model of Figure 2b the repeat time is proportional to displacement in the shock that ends the time interval. To the extent that model (a) is correct, the calculation of the times of occurrence of future large shocks could be accomplished through the measurement or estimation of the displacement in shocks that have already occurred. This places a premium on developing accurate scaling laws for displacement and on obtaining historical information that could be used via scaling laws to estimate displacement.

Equation (1) usually cannot be used directly to calculate repeat times, though, since α is unknown and since $\bar{U}_s$ is rarely measured directly. $\bar{U}_s$ can be derived, however, from the observed seismic moment, M_o, as

$$\bar{U}_s = M_o/\mu A \qquad (2)$$

where A = LW is the area ruptured in the shock and μ is the shear modulus. Combining equations (1) and (2)

$$\alpha = \frac{1}{T_R} \cdot \left[\frac{M_o}{\mu A V} \right] = \frac{\bar{U}_s}{V T_R} \qquad (3)$$

If the terms in the brackets are known, α can be calculated from them and from observed values of T_R. Ideally both T_R and the quantity in brackets should be averaged over several cycles of large shocks. In almost all cases, however, observed values of M_o are only available for the shocks of the last 80 years. Hence, some types of scaling laws must be used to infer the displacements for older shocks from the distribution of shaking or from the generating area or heights of seismic sea waves. We derive scaling laws of this type in a later section and use them to estimate $\bar{U}_s$ for older shocks and to estimate α for the time-predictable model.

Chinnery [1969a] and others obtain an expression for the static stress drop, $\Delta\sigma$, on a dislocation surface, A, of length L and width W:

$$\Delta\sigma = C \mu \bar{U}_s/W \qquad (4)$$

where C is a geometrical factor that depends on the ratio of length to width, dip, depth and type of faulting. All but one of the shocks we examined (the Rat Island event of 1965) involved either nearly pure dip-slip or pure strike-slip faulting. In computing the stress drop we use expressions for either pure dip-slip or pure strike-slip faulting. For the 1965 event we used the dip component of displacement in equation (4).

Eliminating $\bar{U}_s$ using equations (1) and (4)

$$T_R = \left[\frac{\Delta\sigma}{\alpha\mu C} \right] \frac{W}{V_c} \qquad (5)$$

for dip-slip faulting, where V_c is the relative plate velocity perpendicular to the plate boundary, i.e. the convergence velocity. A similar expression is obtained for strike-slip faulting by substituting V_p, the plate velocity parallel to the plate boundary, for V_c. If the term in brackets in equation (5) is a constant, B, then equation (5) becomes

$$T_R = B \frac{W}{V_c} \qquad (6)$$

for dip-slip faulting. Under this assumption repeat time is a simple function of rupture width and convergence velocity. Simple fault models suggest that for many data sets C in equation (5) may be taken to be a constant. For an infinite homogeneous medium with a uniform jump in displacement over the entire dislocation surface, C is about 0.9 for dip-slip faulting with $L > W$ [Chinnery, 1969a], an inequality that holds for the rupture dimensions of most large shocks along subduction zones. Similarly, for $L > 2W$, an inequality that holds for almost all large strike-slip events, C is nearly a constant of about 0.65 for strike-slip faults in an infinite medium and 0.32 for vertical strike-slip faults that intersect the surface of a half space. Kanamori and Anderson [1975] and Kanamori [1977b] argue that the stress drop is also nearly a constant for large shocks along various plate boundaries. To a first approximation μ and α may likewise be constant. We examine the validity of the constant stress drop assumption in a later section as well as the accuracy of estimating T_R from equation (6).

Data Analysis

The data used to investigate the above relations and to determine empirical scaling laws for seismic displacement are compiled in Tables 1 and 2. Table 1 lists data for strike-slip earthquakes, while Table 2 contains information on shocks that are predominantly of the thrust type. In general, the earthquakes reported in Table 2 are larger and more recent than those in Table 1.

For strike-slip events the rupture lengths reported in Table 1 are based primarily on the spatial distribution of either aftershocks or offsets of the surface of the earth. For those few cases in which both types of data are available the estimates from aftershocks are sometimes 25 to 50% greater than those based on the the extent of surface faulting; some of the estimates, however, are nearly identical. Widths for more recent strike-slip events are estimated from the depth range over which aftershock activity extends. For earlier events, and for regions where the depth control of aftershocks is inadequate, a width of 10 to 15 km is assumed. Many of the values of average displacement in Table 1 are based on field observations of offset at the surface. Ideally such average values would involve measurements at a large number of representative sample points. In many cases, however, the average is taken over only two or three points. In addition, displacement at the surface may not reflect adequately the average displacement at depth. Values of average displacement determined from geodetic modeling or from seismic moments obtained from long-period surface waves will usually provide a better estimate than those values based solely on offsets at the surface. The stress drops in Table 1 are computed for a model [Chinnery, 1969a] in which constant displacement occurs on a vertical strike-slip fault of rectangular shape that intersects the surface.

For the thrust-type earthquakes in Table 2 the down-dip projection of the epicentral distribution of aftershocks is assumed to be equivalent to the rupture area. The length of the aftershock area usually can be measured in a straightforward manner. The uncertainty of this measurement is related to the accuracy of the epicentral locations which we take to be about 25 km for teleseismically recorded events.

The down-dip width was determined using combinations of the surface projection of W as inferred from the epicenters of aftershocks, the maximum depth of aftershocks, and the fault dip as determined from focal mechanisms. W is usually more uncertain than L. In the few cases for which aftershock data were not available W was estimated by assuming that the down-dip length of the dipping seismic zone from the surface to a depth of about 50 km is equivalent to the rupture width in great shocks. Along the Alaska-Aleutians plate boundary this appears to be the portion of the plate interface that ruptures in great earthquakes [Davies and House, 1979]. In other regions, the interface along which thrust faulting occurs may extend to greater or shallower depths.

The moments listed in Table 2 were determined exclusively from very long-period (> 100 sec) surface waves. No estimates inferred from aftershock area [e.g. Abe, 1975; Kanamori, 1977b] are included. Average displacement was calculated from equation (2) using the rupture area estimated from aftershocks (Table 2) and $\mu = 5 \times 10^{11}$ dyne-cm^{-2}, an average of the shear modulus for the depth range involved in large thrust earthquakes.

Stress drops reported in Table 2 were calculated numerically from expressions for displacement given by Mansinha and Smylie [1971] for a homogeneous half space. The model consists of pure dip-slip movement of constant amount on a rectangular fault of arbitrary dip such that the depth to the top of the fault is 10 km. We report the value of stress drop at the center of the fault. The adequacy of the model used to calculate stress drop and the effect of uncertainties in μ, L and W are described in a later section and in Appendix A. It is important to realize that uncertainties in the model used to estimate $\Delta\sigma$ do not enter into our discussions of scaling laws, estimates of $\bar{U}_s$ and α or the applicability of the time-predictable model.

Relative plate velocities in both Tables 1 and 2 are computed, in most cases, using the rates and poles of rotation given by Minster and Jordan [1978]. (Exceptions are noted in Table 2; no values are given in Table 1 for intraplate events or events in Turkey.) The relative plate velocities are uncertain by less than 10 to 20% [Minster and Jordan, 1978] for nearly all of the plate boundaries we studied. For the area of the 1946 Nankaido earthquake, however, the relative velocity may be more uncertain. For most of the segments of plate boundaries that we examined, the mismatch between the direction of convergence (normal to the trench) and the direction of relative plate motion is less than 20°. Thus,

Table 1. Strike-Slip Earthquakes

Date	Location	Magnitude (M_w)	Moment ($\times 10^{27}$ dyne-cm)	Rupture Zone Length (km)	Rupture Zone Width (km)	Ratio L/W	Average Displacement (cm)	Plate Velocity (cm/y)	Dates of Previous Earthquakes	Average Repeat Time (yr)	Stress Drop: Half-Space Center (bars)	Geometric Factor
Interplate Earthquakes												
10 Jul 1958	Southeastern Alaska	7.7	4.3(D)	350(1,C)	10-15?	28.0	270-380(2,A)	5.81	1847	111?	26	0.32
9 Jan 1857	S. California	7.8	5.3-8.7(D)	360-400(3,H)	10-15(3)	30.4	450-380(3,A)	5.54	545,665,860,965,1190, 1245,1470,1745(4)	164 (S.D. = 76)	36	0.32
18 Apr 1906	Cen. California	7.7	3.5-4.3(D)	420-470(3,H)	10(3)	45.0	280-320(3,A)	5.65	---	---	29	0.32
		7.7	4.9(D)	450(5,H)	5-12(5)	52.9	480-570(5,B)	5.65	---	---	59	0.32
19 May 1940	Imperial Valley, CA	6.8	0.23-0.29(D)	60(6,H)	10(6)	6.0	125-160(6,23,A)	5.41	1901,1922,1934,1966(8)	22 (S.D. = 10)	13	0.34
27 Jun 1966	Parkfield, CA	6.6	0.03(D)	35-40(7,C)	5-10(7)	5.0	30(7,B,C)	5.59	---	---	4	0.36
9 Apr 1968	Borrego Mtn., CA	6.3	0.06-0.11(20,21,F)	30(9,H)-45(9,G)	12(9)	3.1	20-30(10,A)	5.44	---	---	2-3	0.39
15 Oct 1979	Imperial Valley, CA	6.3	0.02-0.03(D)	30(11,H)	10?	3.0	20-35(11,24,A)	5.41	1940	39	3-4	0.42
4 Feb 1976	Guatemala	7.5	2.6(13,E)	240(12,H)	15(13)	16.0	110(12,A)	1.92	---	---	7	0.32
				300(13,G)	15(13)	20.0	190(13,C)	1.92	---	---	12	0.32
16 Oct 1974	Gibbs Fracture Zone	7.0	0.45(14,E)	70-80(14,G)	10-15(14)	6.0	125-215(14,C)	2.25	---	---	14	0.34
26 Dec 1939	Ercincan, Turkey	7.7	4.5(D)	350(22,H)	15?	23.3	~285(22,A)	---	---	---	18	0.32
20 Dec 1942	Erbaa Niksar, Turkey	7.0	0.35(D)	70(22,H)	15?	4.7	~112(22,A)	---	---	---	8	0.36
1 Feb 1944	Gerede-Bolu, Turkey	7.5	2.4(D)	190(22,H)	15?	12.7	~275(22,A)	---	---	---	18	0.32
18 Mar 1953	Gönen-Yenice, Turkey	7.2	0.73(D)	58(22,H)	15?	3.9	~280(22,A)	---	---	---	21	0.38
22 Jul 1967	Mudurnu, Turkey	7.0	0.36(D)-0.88(20,F)	80(22,H)	15?	5.3	~100(22,A)	---	---	---	7	0.35
Intraplate Earthquakes												
4 Dec 1957	Gobi-Altai, Mongolia	8.1	18(15,E)	270-300(15,G)	20-50(15)	8.1	600(15,C)	---	---	---	17	0.33
		7.8	5.9(D)	280(6,G)	20(6)	14.0	350(6,A)	---	---	---	17	0.32
27 Jul 1976	Tangshan, China	8.0	13(16,E)	270(15,G)	50(16)	5.4	320(16)	---	---	---	7	0.35
		7.5	1.8(17,E)	140(17,G)	15(17)	9.3	285(17,C)	---	---	---	19	0.33
7 Mar 1927	Tango, Japan	7.7	4.6(18,I)	35(18,G,I)	13(18)	2.7	250-300(18,A,B)	---	---	---	25	0.44
10 Sep 1943	Tottori, Japan	7.6	3.6(18,F)	33(18,G,I)	13(18)	2.5	200-280(18,B,C)	---	---	---	27	0.46
28 Jun 1948	Fukui, Japan	7.6	3.3(18,F)	30(18,G)	13(18)	2.3	200-250(18,B,C)	---	---	---	25	0.49
31 Aug 1968	Dasht-e Bayāz, Iran	7.1	0.58(D)-0.67(20,F)	80-110(19,H,G)	15(19)	6.3	135(19,A)	---	---	---	9	0.34

Notes:

A Displacement value from surface offset
B Displacement value from modeling of geodetic data
C Displacement value from moment assuming $\mu = 3 \times 10^{11}$ dyne/cm^2
D Moment value from observed displacement and rupture parameters (assuming $\mu = 3 \times 10^{11}$ dyne/cm^2)
E Moment value from modeling of long-period surface waves
F Moment value from modeling of long-period body waves
G Length from aftershocks
H Length from extent of surface offset
I Length from geodetic data

Data Sources:

(1) Sykes [1971]
(2) Plafker et al. [1977]
(3) Sieh [1978a]
(4) Sieh [1978b]
(5) Thatcher [1975]
(6) Brune and Allen [1967]
(7) Scholz et al [1969]
(8) Bakun and McEvilly [1979]
(9) Hamilton [1972]
(10) Clark [1972]
(11) Sharp [1979]
(12) Bucknam et al. [1978]
(13) Kanamori and Stewart [1978]
(14) Kanamori and Stewart [1976]
(15) Okal [1976]
(16) Chen and Molnar [1977]
(17) Butler et al. [1979]
(18) Kanamori [1973]
(19) Ambraseys and Tchalenko [1969]
(20) Hanks and Wyss [1972]
(21) Burdick and Mellman [1976]
(22) Ambraseys [1970]
(23) Clark [1972]
(24) Sieh [personal communication]

Table 2. Large Thrust Earthquakes

Date	Location	Magnitude[§] (M_w)	Moment ($\times10^{27}$ dyne-cm)	Length (km)	Downdip Width (km)	Dip (deg)	Ratio L/W	Avg. Total[++] Displacement (cm)	Avg. Dip-Slip[##] Displacement (cm)	Total Plate[**] Velocity (cm/yr)	Convergent Plate Velocity (cm/yr)	Dates of Previous Earthquakes	Average Repeat Time (yr)	Stress Drop[#] (bars)	Geometric[§§] Factor	σ_A[+]	σ_M[++]
20 Dec 1946	Southern Japan	8.1	---	150(1)	90(1)	30-45(1)	1.67	---	---	4.0	4.0	1605,1707,1854(2)	114 (S.D. = 29)	37	.63	---	---
6 Nov 1958	Kuriles	8.3	44(3)	150(3)	70(3)	30(3)	2.14	840	840 (φ = 0)	10.1	10.1	1780,~1849?,1918(3,6)	61 (S.D. = 13)*	12	.62	1.05	0.78
13 Oct 1963	Kuriles	8.5	67(4)	275(5)	110(5)	22(5)	2.50	445	445 (φ = 0)	9.95	9.4	1667,1763,1856(6)	100 (S.D. = 10)	10	.58	0.29	0.33
16 May 1968	Northern Japan	8.2	22(8)	150(7)	105(7)	20(7)	1.43	355	338 (φ = 18)	10.4	10.3	1893(3)	76	5	.62	0.25	---
11 Aug 1969	Kuriles	8.2	28(7)	230(8)	105(8)	16(8)	2.19	180	179 (φ = 5)	10.2	9.5	1894(3)	79	5	.58	0.18	---
17 Jun 1973	Northern Japan	7.8	6.7(9)	90(9)	105(9)	27(9)	0.86	140	129 (φ = 23)	10.25	9.1				.79		
4 Nov 1952	S. Kamchatka	9.0	350(10)	450(11)	175(11)	30(10)	2.57	890	890 (φ = 0)	9.35	9.3	1737,1841?(6,12)	108? (S.D. = 5)	14	.55	0.86	0.89
1847	Shumagins, Alaska	---	---	---	---	20	---	---	---	7.4	7.2	1788(14)	59				
28 Mar 1964	Southern Alaska	9.2	820(4)	750(15)	120(13)	20(25)	4.17	1215	1215 (φ = 0)	6.55	6.5	1906(14)	59	18	.53	0.96	---
4 Feb 1965	Rat Is., Alaska	8.7	125(4)	650(16,17)	80(17)	18(16)	8.13	480	282 (φ = 54)	8.5	5.0†		32	10	.59		
10 Jan 1973	Colima, Mexico	7.6	3(19)	85(19)	65(19)	30(19)	1.31	110	101 (φ = 23)	5.6	5.6	1941(20)	47-50	5	.69	0.56	1.70
29 Nov 1978	Oaxaca, Mexico	7.6	3(21)	80(22)	70(22)	14(21)	1.14	110	107 (φ = 14)	7.3	6.9	1928-1931(20,22)		5	.69	0.32	---
22 May 1960	S. Chile	9.5	2000(4)	1000(23)	210(23)	10(26)	4.76	1900	1871 (φ = 10)	9.15	8.6	1575,1737,1837(23)	128 (S.D. = 31)	21	.47	1.77	---
17 Oct 1966	Central Peru	8.1	20(25)	80(24)	140(24)	12(24)	0.57	360	360 (φ = 0)	8.75	8.2			12	.92	---	---
1843	N. Lesser Antilles	---	---	260?(20)	70?(13)	30(14)	3.71	---	---	2.0	2.0	1690(20)	153			---	---

Notes:

§ M_w is computed from moment, M_o, using the formula $M_w = 2/3 \log M_o - 10.7$ Kanamori [1977b].

** Plate velocity is computed using the rates and poles of rotation given by Minster and Jordan [1978] with two exceptions. The plate velocity for the Nankaido region is from Seno [1977]; that for the Shumagin region is an average of values from Minster and Jordan [1978] and Chase [1978].

The stress drop computed is for constant dip-slip displacement on a dipping rectangular fault that is embedded in a half-space such that the vertical distance to the top of the fault is 10 km Manshinha and Smylie [1971]. μ is taken as 5×10^{11} dyne-cm^2. The value given is the stress drop at the center of the fault.

+   The convergent plate velocity for the Rat Islands is computed by assuming the plate velocity has the same direction as the average slip vector for the mainshock and aftershocks [Wu and Kanamori, 1973; Stauder, 1968]. If the direction from the results of Minster and Jordan is used, the V = 3.3 cm/yr; W/V = 24.2 x 10^3 yr; and U_s/V = 85.5 yr.

++ $U_s = M_o/Lw\mu$ (μ = 5×10^{11} dyne-cm^2).

φ is the angle in degrees between the slip vector and the normal to the plate boundary.

*   The average repeat time is calculated by assuming 3 earthquake cycles occurred since ~1849, during a gap in the data. i.e. an event is assumed to have occurred in ~1849, during a gap in the data.

§§ The geometric factor is appropriate for use in equation (4) to obtain stress drop at the center of a rectangular fault embedded in a half-space such that Z_o is 10 km.

+   The ratio of seismic slip to total slip for the slip predictable model.

++ The ratio of seismic slip to total slip using U_{sc} for the most recent and T_R averaged over 2 or 3 earthquake cycles.

Date Sources:

(1) Fitch and Scholz [1971] and Mogi [1968]
(2) Ando [1975]
(3) Fukao and Furumoto [1979]
(4) Kanamori and Anderson [1975]
(5) Kanamori [1970a] and Mogi [1968b]
(6) Rikitake [1976]
(7) Kanamori [1971] and Ichikawa [1971]
(8) Abe [1973]
(9) Shimazaki [1974]
(10) Kanamori [1976]
(11) Kelleher et al. [1974] and Fedotov [1965]
(12) Fedotov [1965] and Kondorskaya et al. [1977]
(13) Down-dip width of dipping seismic zone to a depth of ~50 km

(14) Dip estimated from configuration of Benioff zone
(15) Algermissen et al. [1969]
(16) Jordan et al. [1965], Stauder [1968], Mogi [1968b]
(17) Spence [1977]
(18) McCann et al. [1980]
(19) Reyes et al. [1979]
(20) Kelleher et al. [1973]
(21) Stewart et al. [in press]
(22) Singh et al. [1980c]
(23) Kelleher [1972], Plafker [1972], Plafker and Savage [1970]
(24) Abe [1972] and Fedotov [1965]
(25) Kanamori [1970b]
(26) Kanamori and Cipar [1974]

uncertainties in the azimuth of plate velocity do not have a significant affect on the computed velocities of plate convergence except in the western Aleutians where oblique subduction is occurring. For that region we use the magnitude of relative plate velocity from Minster and Jordan [1978], but for the direction we use an average of the slip vectors for the mainshock and aftershocks of the Rat Islands sequence of 1965 [Stauder, 1968].

Reliable estimates of repeat times can be obtained at most for about the last three earthquake cycles. For many of the events listed, however, information is available for only one earthquake cycle, i.e., two large events. Almost all the data on previous earthquakes involve historical accounts of the extent and type of damage produced. We have relied primarily on the interpretations of other investigators to delimit the size and locations of such ruptures. For two events, however, we have made additional assumptions to arrive at repeat times. The only prior earthquakes that are documented in the region of the Kurile earthquake of 1963 occurred in 1780 and 1918. If these shocks form a complete record, repeat times for this region vary from 45 to 138 years. A known gap in the historical record exists, however, during the mid-1800's [Abe, 1977]. We assume an additional earthquake occurred in the area in about 1849; hence, the range of repeat times becomes 45 to about 69 years.

The repeat time of the 1952 Kamchatka earthquake is also hard to assess. Two prior shocks in 1737 and 1841 may have been as large as the 1952 event. The shock of 1841 produced a tsunami whose reported height at Hilo, Hawaii was somewhat larger than that recorded in 1952 [Abe, 1979]. This suggests the displacements during the two events may have been nearly equal. If the rupture areas of the shocks were similar, a repeat time of about 110 years is obtained. Fedotov [1965], however, indicates a rupture zone for the 1841 event that, although poorly known, is only 25 to 30% of that for the 1952 shock. This suggests that the most recent great shock prior to the one in 1952 occurred in 1737 and that the repeat time is therefore 215 years. In Table 2 and in figures involving repeat time, we include the Kamchatka earthquakes assuming the 1841 event is similar in size to those of 1737 and 1952 but mark those data points with a query.

The validity of equation (6), as well as the dependence of repeat time, stress drop, and average seismic displacement on various parameters, is examined using simple linear regression analysis. Each data point was weighted equally. The parameter shown along the horizontal axis in various figures is taken to be an independent variable. When errors are present in the independent variable, the estimated slope can differ from the true or error-free slope [e.g. Guest, 1961]. Since taking such errors into account is complicated and since changes on the

order of $\pm 1/2$ in the slopes we estimate do not significantly affect our results, we choose to ignore errors in the independent variables.

Displacements and Stress Drops for Simple Transform Faults

Many of the observations of surface displacement in earthquakes come from transform faults like the San Andreas of California, the North Anatolian of Turkey, the Motagua of Guatemala, and the Fairweather of southeast Alaska. Since the range of depths of aftershocks for events of magnitude greater than about six is similar for these fault systems and motion occurs along nearly vertical faults, the two geometrical parameters W and θ (Figure 1) are nearly constant. Hence, stress drop and seismic displacement appear to be largely a function of the length of the rupture zone, L. For convergent plate boundaries stress drop and displacement appear to be a more complicated function of W, L and θ. We therefore examine transform boundaries first. Most of the data on repeat times, however, come from convergent plate margins. Repeat times for strike-slip faults are examined after those for convergent plate boundaries.

Scaling Laws for Displacement and Stress Drop

Figure 3 shows the average seismic displacement U_s, for strike-slip faults, as a function of the length of rupture, L. The data points for plate boundaries (solid symbols) indicate a systematic increase in both U_s and in the average stress drop, $\Delta\sigma$, with increasing L. The increase is nearly proportional to L to the first power. Points for earthquakes along the plate boundary in Turkey (T symbols) generally agree with this trend; they are differentiated from the other plate boundary shocks to indicate their poorer quality. Results for strike-slip events within lithospheric plates (open symbols), while also generally consistent with those for plate boundaries, indicate higher average stress drops and larger displacements for a given value of L.

The value of the exponent in a power law expression relating L and $\bar{U}_s$ for earthquakes along strike-slip plate boundaries (solid line in Figure 3) is strongly dependent on the three Californian data points with values of L between 20 and 40 km. Detailed studies of two of these events, the Parkfield earthquake of 1966 [Lindh and Boore, 1981] and the Borrego Mountain earthquake of 1968 [Burdick and Mellman, 1976], indicate that much of the seismic energy was radiated from portions of the fault considerably smaller than the rupture length estimated from the extent of aftershocks or offset at the surface. Consequently, higher stress drops and displacements are obtained for those sub-regions. In fact, if these studies are taken at face value, and the Imperial Valley earthquake in 1979 is assumed to be similar, the replotted data points

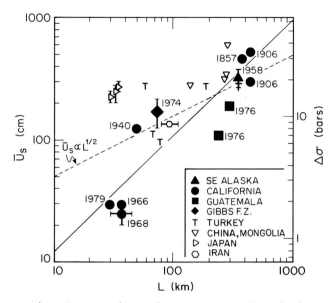

Fig. 3. Strike-slip component of seismic displacement, $\bar{U}_s$, as a function of fault length, L, for moderate and large strike-slip earthquakes. Events that occurred along plate boundaries are plotted as either large filled symbols or "T" symbols. Data for shocks in Turkey are more uncertain than for other events along plate boundaries. Intraplate events are shown as smaller open symbols. The solid line is the best linear fit to the plate boundary earthquakes (log $\bar{U}_s$ = log 2.6 + 0.94 log L; r = .91 where parameter r is the linear correlation coefficient; $\bar{U}_s$ in cm, L in km). Rupture widths (i.e. depth range) for these events can be considered approximately constant (10-15 km). Note that seismic displacement is proportional to length to about the first power for strike-slip earthquakes along plate boundaries. The dashed line has a slope of 1/2 and is visually fit to the complete data set for comparison. The ordinate on the right-hand side is stress drop at the center of the surface trace for a model [Chinnery, 1969a] in which constant displacement occurs on a vertical strike-slip fault, L >> W, μ = 3 x 10^{11} dyne-cm^{-2}, and W = 15 km. For some events more than one estimate of $\bar{U}_s$ and L are shown. Data are summarized in Table 1.

suggest a dependence of $\bar{U}_s$ on L$^{1/2}$ rather than L. On the other hand, perhaps many or all of the other shocks shown on Figure 3 are multiple events wherein the stress drops and displacements of sub-regions are much higher than the average over the entire rupture surface. In this case, all the data points in Figure 3 might be shifted a similar amount. Hence, the overall dependence of $\bar{U}_s$ on L to the first power might still hold. For strike-slip earthquakes along plate boundaries we conclude that $\bar{U}_s$ is dependent on L to a power

between about 0.5 and 1.0, and that a more precise determination of this exponent based on our data set is unwarranted.

Regardless of the precise exponent of L, it is clear from Figure 3 that stress drop cannot be taken as a constant in comparing the slip or repeat time for transform boundaries. Data for strike-slip events as compiled by Chinnery [1969b] also show a general increase of displacement as length increases. In a study of earthquakes from a variety of tectonic environments King and Knopoff [1968] also conclude that stress drop increases with the size of events. The model used to calculate stress drop, which assumes constant displacement over the rupture surface, may lead to a bias in estimates of $\Delta\sigma$. While the absolute values of stress drop in Figure 3 and in Table 1 may be in error, the values of C in equation (4) are likely to be nearly identical for the strike-slip events we examined since L/W is greater than 2.3 and a long fault approximation is appropriate. Hence, biased values of C are unlikely to lead to large errors in estimating the dependence of $\Delta\sigma$ on L, which is of primary interest in this paper.

Displacements, Stress Drops and Repeat Times
for Convergent Plate Boundaries

In this section we examine how repeat time, T_R, is a function of the rate of underthrusting, V_c, the geometry of the rupture surface (Figure 1) and the ratio of seismic slip to total slip (α).

<u>Dependence on Rate of Underthrusting</u>

If T_R were only a weak function of the geometry of the rupture surface and of α, we would expect to observe a strong correlation in the data between T_R and V_c. Figure 4 shows that this is not the case. T_R varies by more than a factor of four among areas in which the rate of underthrusting is about the same. Hence, variations in either geometry, α, or both must be responsible for the large range of repeat times.

<u>Increase in Displacement and Stress Drop with Downdip Width</u>

Average dip-slip components of seismic displacement, $\bar{U}_{sc}$, for a number of large shallow events along subduction zones were calculated using equation (2) and are plotted in Figure 5 as a function of the down-dip width, W. The data show a pronounced and systematic increase of $\bar{U}_{sc}$ with W. The increase is much faster than that predicted ($\bar{U}_s$ proportional to W to the first power) by equation (4) if $\Delta\sigma$, the stress drop, is a constant. Lines of constant stress drop in Figure 5, which were calculated for an infinite homogeneous medium using C = 0.9 in equation (4), indicate a systematic increase in stress drop with increasing W. Stress drop for dip-slip events is shown explicitly in Figure 6 as a function of W for those events from which it could be computed

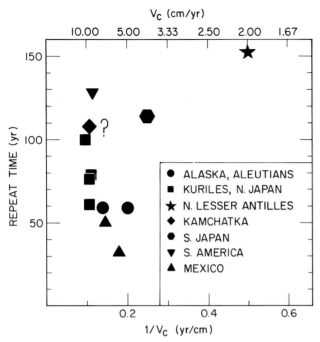

Fig. 4. Repeat times for large thrust-type earthquakes as a function of inverse plate velocity. Velocity in the direction normal to the plate boundary is used. No obvious correlation between repeat time and inverse plate velocity is observed. The repeat time for the 1952 Kamchatka earthquake is questionable.

from the seismic moment using equations (2) and (4). $\Delta\sigma$ was calculated as the drop in shear stress at the center of the rupture surface (Figure 1) for an infinite homogeneous medium where the jump in displacement was held constant over the rupture surface. The values of the geometrical factor, C, in equation (4) were taken from Chinnery [1969a] for the ratio (L/W) appropriate to each data point. In Figure 6, $\Delta\sigma$ increases systematically with W by about a factor of 6.

In Appendix A we show that stress drop increases with W by a factor of four for the more realistic half-space model. We also show in Appendix A that computed values of stress drop are very insensitive to realistic variations in the dip of convergent plate boundaries. The dislocation models that we use, however, are not very realistic. Since the displacement is taken to be constant over the rupture surface, singularities in stress drop occur at the edges of the dislocation. It seems unlikely to us, however, that the use of more realistic distributions of displacement on the rupture surface would result in computed stress drops that differ by as much as a factor of four for the events we studied. Hence, it seems to be inescapable that the average stress drop for large shocks along

subduction zones and along transform faults increases with the size of the rupture zone.

It should be noted in Figure 5 that L is not being held constant in the data set as W increases; W usually increases along with L for most large shocks of the dip-slip type (Figure 7B). Figure 8 shows a great deal of scatter when $\bar{U}_{sc}$ is plotted as a function of L for convergent plate margins. We return later to a more detailed discussion of how $\bar{U}_{sc}$ scales with W and with L.

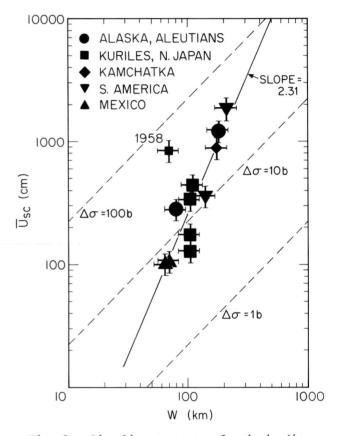

Fig. 5. Dip-slip component of seismic displacement as a function of rupture width in downdip direction, W. Events prior to 1960 are plotted with smaller symbols. The solid line gives the best linear fit to the data (log $\bar{U}_{sc}$ = log (6.1 x 10^{-3}) + 2.31 log W; r = 0.91; $\bar{U}_{sc}$ in cm, W in km). Anomalous Kurile shock of 1958 is not included in the regression analysis. Dashed lines are contours of constant stress drop (in bars) for a model [Chinnery, 1969a] of constant dip-slip motion on a rectangular fault (L = 2W) embedded in an infinte space with $\mu = 5$ x 10^{11} dyne-cm^{-2}. Note that stress drop increases with $\bar{U}_{sc}$ and W. Error bars are taken as $\pm$ 20% uncertainty.

Anomalous Nature of 1958 Shock

The data point for the 1958 earthquake in the Kurile Islands is an anomaly in Figure 5 and in the other data sets we present. In a detailed study of that shock Fukao and Furumoto [1979] point out that it is characterized by an extremely large felt area compared to its relatively small after-shock area and by anomalously large amounts of energy at periods shorter than 20 sec. These observations are consistent with the high stress drop (78 bars) they report and with that computed in this paper.

Several lines of evidence favor a deeper than normal focus for the rupture zone of the 1958 earthquake: the aftershock zone is located closer

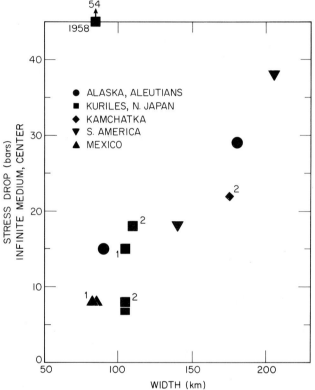

Fig. 6. Stress drop as a function of rupture width. Stress drop is computed for a model in which constant displacement occurs over a rectangular fault embedded in an infinite medium [Chinnery, 1969a]. The stress drop at the center of the fault is used and μ is taken as 5×10^{11} dyne-cm^{-2}. The observed stress drop calculated in this manner varies by a factor of 5 to 6 for the earthquakes studied. The 1958 earthquake is again anomalous. The symbols labelled 1 and those labelled 2 are two subsets of events with similar values for the ratio L/W. Note that within the individual subsets an increase in width is always accompanied by an increase in stress drop.

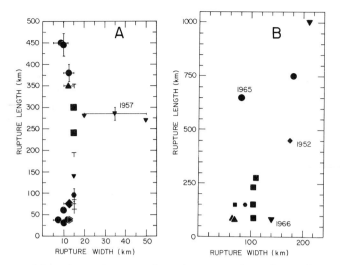

Fig. 7. Rupture length as a function of rupture width. (a) Strike-slip earthquakes; (b) Thrust earthquakes. The width for the thrust earthquakes is measured perpendicular to the plate boundary. For the strike-slip shocks length and width are uncorrelated; for large thrust events these parameters are weakly correlated. The symbols for the strike-slip earthquakes are the same as in Figure 3; those for thrust earthquakes are as in Figure 4.

to the island arc than those of other large recent events; the mainshock is located near the landward (arcward) side of the aftershock zone at a depth of about 70 to 80 km, and aftershock activity is low to non-existent for depths shallower than about 30 km [Fukao and Furumoto, 1979]. Aftershocks of the nearby earthquakes of 1963 and 1969, however, were plentiful at shallow depths, but as in the 1958 event they extend to the unusually great depth of about 90 km. Although the focal mechanism of the 1958 shock, like most large shallow shocks along subduction zones, was characterized by thrust faulting; the dip of the inferred fault plane, 30°, is larger than that commonly found for great shallow shocks. The larger dip also suggests a greater than normal depth for the dislocation surface. Hence, while the mechanism of the 1958 event seems to be indicative of movement between plates, its greater than normal depth and high average stress drop single it out as a very anomalous event compared to the others we studied. Thus, we did not include it in the various regressions we performed.

Repeat Time as Function of W/V_c

Observed values of T_R are compared in Figure 9 with the ratio W/V_c as predicted by equations (5) and (6). The correlation of T_R with W/V_c is, in

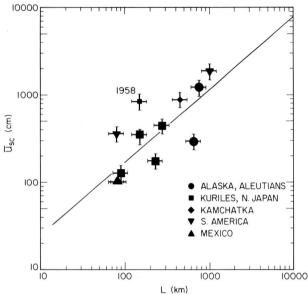

Fig. 8. Dip-slip component of seismic displacement as a function of rupture length, L, for large thrust-type earthquakes. Relative symbol size same as in Figure 5. The solid line shows the best linear fit to the data (log U_{sc} = log 3.77 + 0.83 log L;' r = 0.81; U_{sc} in cm, L in km). Error bars indicate $\pm$ 20% uncertainty.

fact, much better than that found in Figure 4 where T_R was plotted as a function of $1/V_c$ alone. Nevertheless, the data points in Figure 9 scatter by a factor of up to 1.5 about the best fitting line that passes through the origin. This is not unexpected since we have shown that $\Delta\sigma$ is not a constant among our data; equation (6) should only be valid if the ratio $\Delta\sigma/\alpha$ is a constant. As we show later, computed values of α vary by a factor of three. Hence, it is not surprising that observed repeat times vary considerably from those calculated using equation (6). Still, Figure 9 and equation (6) seem to provide the most accurate estimates of repeat time for areas in which a history is lacking of two or more large shocks at nearly the same place.

Values of α for Slip-Predictable Model

We computed the ratio of seismic slip to total slip (α) using equation (3) for those segments of subduction zones for which information was available on both repeat time and seismic moment. Since values of M_o are only available for the most recent event in a sequence, it is relatively easy to compare $\bar{U}_{sc}/V_c$ as determined from that event with the time interval, T_R, that precedes it. This comparison assumes that the slip-predictable model (Figure 2b) is correct. Points shown without a vertical error bar in Figure 10 represent those areas for which only a single time

interval is known. The X's denote the latest interval for regions in which more than two shocks are known to have occurred. The values of α calculated in this way (Figure 10) vary from 0.18 to 1.77 (Table 2), i.e. by a factor of nearly ten. This large range of values and what appears to be a nonphysical result, 1.70 to 1.77, indicate that the α's in Figure 10 are probably more diagnostic of the model used for the computations (i.e. that of Figure 2b) than they are of differences in rheology.

The largest value of α in Figure 10 (1.70 to 1.77) is associated with the Chilean earthquake of 1960. Although the source parameters of that event are poorly determined, allowance for

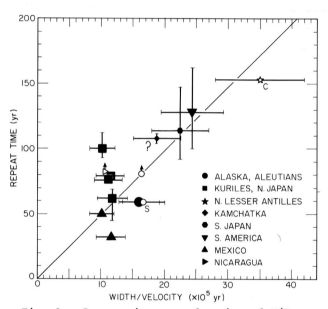

Fig. 9. Repeat time as a function of W/V_c. Relative symbol size is the same as in Figure 5. For open symbols the down-dip width was estimated as the distance along the Benioff zone from 0 to 50 km depth. The solid line, with a slope of 4.83 x 10^{-5} (r = 0.97) represents the best fit of the data to equation (6). Data shown both as filled symbols (except the Kamchatka earthquake) and labeled open symbols (C and S) were included in the regression analysis. The open symbols with attached arrows represent the current (i.e. 1980) status of two seismic gaps - Nicaragua in Central America and Yakataga in southern Alaska. These symbols will move in the direction of increasing T_R until the respective gaps are filled by large earthquakes. They fall within the factor of two scatter of the data points about the regression line. Error bars for W/V_c are plotted as $\pm$ 20%; the actual spread in repeat times is shown when three or more events are known to have occurred at the same place. The repeat time of the 1952 Kamchatka earthquake is questionable.

uncertainties in L, W and μ still leads to values of α greater than one. If Mo for that shock was overestimated, however, the computed value of α from equation (3) would also be overestimated. The energy release and seismic moment of the 1960 earthquake are larger than those of any other shock of the past 80 years [Kanamori, 1977b]. The 1000-km length of the source region of the 1960 shock appears to be about twice that of large events that occurred along that zone in 1575, 1737 and 1837 [Lomnitz, 1970; Kelleher, 1972]. Its great length as well as the large size of its tsunami indicate that the average displacement in the 1960 shock was larger than that of preceding historic events.

In contrast, the smallest value of α in Figure 10 is associated with the Nemuro-Oki earthquake of 1973. Several lines of evidence including the sizes of the tsunami-generating areas, felt areas, numbers of aftershocks and amplitudes of tsunamis, indicate that the 1973 event was considerably smaller than the preceding shock of 1894 [Abe, 1977]. Using the method we employed in Figure 10, large values of α as in the 1960 shock are what we would expect to calculate if the time-predictable model of Figure 2(a) is correct and if the most recent event is much larger than the preceding shock. Conversely, small values of α as in the 1973 shock are also expected when the preceding event was much larger than the most recent one.

Shimazaki and Nakata [1980] find that the time-predictable model (a) of Figure 2 fit the data for

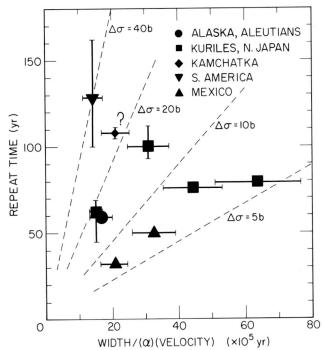

Fig. 11. T_R as a function of $W/\alpha V_c$. Relative symbol size as in Figure 5. The values of α used are computed for the slip-predictable model (b) of Figure 2. Lines of constant stress drop are shown for the same fault model as in Figure 5. Errors shown for $W/\alpha V_c$ are ± 20%; for T_R the observed range is indicated. The lack of correlation between T_R and $W/\alpha V_c$ is interpreted as an indication that either stress drop varies or α is not being computed correctly. The repeat time of the Kamchatka earthquake is not well determined.

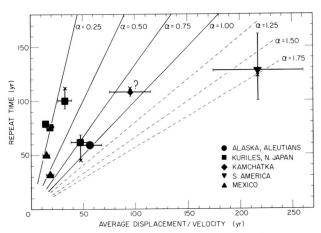

Fig. 10. T_R as a function of $\bar{U}_{sc}/V_c$. Relative symbol size same as in Figure 5. Lines of constant α are plotted for the slip-predictable model (b) of Figure 2. The values of α range from 0.18 to 1.77; however, values greater than one are nonphysical. Error bars for $\bar{U}_{sc}/V_c$ are plotted as ± 20%. The observed range of repeat times is shown with the value for the most recent earthquake cycle indicated by an "x". The repeat time for the 1952 event near Kamchatka is questionable.

three sets of earthquake-produced features in Japan much better than the slip-predictable model (b). In a later section we recompute α according to the time-predictable model by comparing observed repeat times with values of $\bar{U}_s/V$ estimated for the preceding shocks in a sequence. The smaller variation in α (0.3 to 1.0) that we obtain lends considerable support to the time-predictable model.

Another indication of the questionable nature of values of α computed for the slip-predictable model is obtained in Figure 11 by comparing T_R with the ratio $(W/\alpha V)$, where the computed values of α from Figure 10 are used. The scatter in Figure 11 is _much worse_ than that in Figure 9 where the horizontal axis is W/V alone and α was not included. When T_R is compared with $W/\alpha V$ but with α computed for the time-predictable model, the data points show much less scatter (7 to 25 bars) than those of Figure 11. The amount of scatter with that approach is comparable to that in Figure 9.

Scaling Laws for Dip-Slip Displacement

The functional dependence of $\bar{U}_s$ upon the geometry of the dislocation is examined here for those dip-slip events for which $\bar{U}_s$ was computed using (2). Since the free surface has little effect upon the $\bar{U}_s$-$\Delta\sigma$ relationship (Appendix A), $\bar{U}_s$ appears to be largely a function of L and W and not of θ and z_o.

We tried to obtain best fitting solutions of the form

$$\bar{U}_s = bL^m W^n \qquad (7)$$

where b, m, and n are constants. We sought a scaling law that would also fit the data for transform boundaries, i.e. $1/2 \leq m \leq 1$. Values of m larger than 1 and those significantly smaller than 1/2 also give a very poor fit to the data for dip-slip events. For m = 1 the best fitting value of n (Figure 12) is 1.03. A much better fit, however, is obtained (Figure 13) with m = 1/2. In that case n = 1.6. For these two cases m + n is

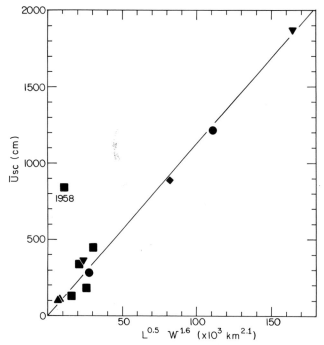

Fig. 13. Seismic slip $\bar{U}_{sc}$ as a function of $L^{.5}W^n$ for n = 1.6. This value of n gives the best least squares fit for the relation $\bar{U}_{sc} = bL^m W^n$ when m = 0.5. Values of b and r are 1.12×10^{-2} and 0.99 respectively. L and W are in km; $\bar{U}_{sc}$ in cm. The 1958 Kurile earthquake was again excluded. Symbol shapes and relative sizes are as in Figure 5.

slightly greater than 2. Hence, the sum of the exponents is fairly well determined. This difficulty in separating out scaling with L from that with W for dip-slip faulting is understandable since increases in L usually accompany those in W (Figure 7). Thus, scaling with L is similar for dip-slip and strike-slip faults, i.e. $1/2 \leq m \leq 1$.

Kanamori and Anderson [1975], Abe [1975] and Kanamori [1977b] obtained an empirical relationship

$$M_o = \eta A^{3/2} \qquad (8)$$

between seismic moment and rupture area A, where η is a constant under the assumption that the ratio L/W is a constant. If L/W and η are constants, $\Delta\sigma$ is also constant. From equations (2) and (8)

$$\bar{U}_s = \eta \frac{A^{1/2}}{\mu} \qquad (9)$$

Hence, equation (9) predicts that displacement will scale as $A^{1/2}$.

If m = 1/2 (Figure 13), a much better fit is obtained for $n \simeq 3/2$ than for n = 1/2, i.e. than

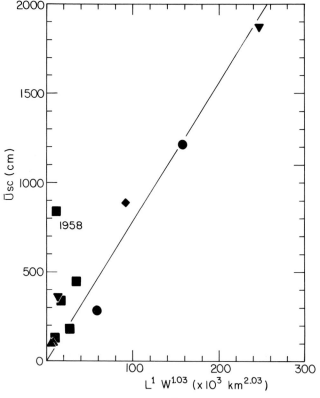

Fig. 12. Seismic slip $\bar{U}_{sc}$ as a function of $L^1 W^n$ for n = 1.03. This value of n gives the best least squares fit for the relation $\bar{U}_{sc} = bL^m W^n$ when m = 1. b equals 7.83×10^{-3} (L, W in km, $\bar{U}_{sl}$ in cm) and r = 0.98. The 1958 Kurile event was excluded from the analysis. Symbol shapes and relative sizes are the same as in Figure 5.

for scaling of $\bar{U}_s$ with $A^{1/2}$ (dashed line in Figure 14). The derivation of the $A^{1/2}$ scaling relationship in equation (9) is strictly empirical and does not result from any theoretical considerations. Thus, the relatively poor fit of our data to this relation implies only that better empirical relations are available. The fit of the data in Figure 14 to a relationship involving $A^{1/2}$ can be improved (solid line) if the regression line is not constrained to pass through the origin. Even for this case, though, empirical relations giving smaller residuals are possible for values of n other than 1/2.

Figure 15 compares the observed values of U_s, i.e. those obtained from the seismic moment using equation (2), with those calculated from the best fitting forms of (7) for m = 1/2 and 1. In these and in other attempts to derive scaling laws, the data set is dominated by the values for the great shocks of 1952, 1960 and 1964. The width, W, is poorly determined for the Chilean shock of 1960. Hence, attempts to derive empirical scaling laws are limited by the small number of very great shocks for which the seismic moment has been determined directly and by uncertainties in determining the moment.

Figure 15 suggests that $\bar{U}_s$ can be estimated from L and W to within 1.5 m. This uncertainty is a large percentage of $\bar{U}_s$ for shocks with $\bar{U}_s < 5$ m. This is understandable since the uncertainties in determining W and L are often about 30 to 50 km,

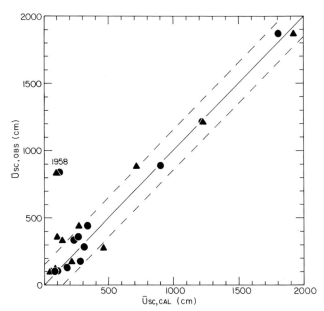

Fig. 15. Comparison of seismic displacement computed from regression relations with observed values. The relations used are those determined in Figures 12 and 13. For $L^m = L^{1/2}$ (circles) all of the calculated values of $\bar{U}_{sc}$ except for the 1958 Kurile event fall within ± 1.5 m (dashed lines) of the observed values, i.e. those obtained from the seismic moment using equation (2). For $L^m = L^1$ (triangles) there is more scatter; still, most calculated values of $\bar{U}_{sc}$ fall within a factor of 1.5 of those observed.

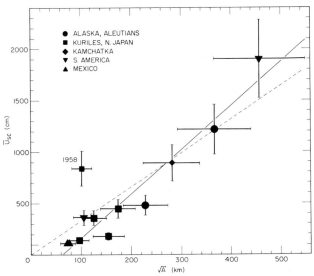

Fig. 14. Dip-slip component of seismic displacement as a function of $A^{1/2}$. Relative symbol size and error bar percentage are the same as in Figure 5. The solid line is the best linear fit to the data ($\bar{U}_{sc} = 273.94 + 4.30\ A^{1/2}$; r = 0.97; $\bar{U}_{sc}$ in cm, $A^{1/2}$ in km). The dashed line is the best fit if the line is constrained to pass through the origin ($\bar{U}_{sc} = 3.28\ A^{1/2}$). The 1958 Kurile earthquake was not included in the regression analysis.

which is a large percentage of the dimensions for shocks with $\bar{U}_s < 5$ m. For very large events it probably is not, in fact, possible to estimate $\bar{U}_s$ to 1.5 m since the best fitting lines are forced to pass close to the data points for the three largest events.

The seismic moment and stress drop can also be estimated from L, W and μ alone as

$$M_o = \mu\ bL^{m+1}W^{n+1} \qquad (10)$$

$$\Delta\sigma = C\ \mu bL^m W^{n-1} \qquad (11)$$

Estimating moment from equation (10) appears to be more precise than estimating it from the area using equation (8). Equations (7), (10) and (11) should be regarded as empirical fits to existing data and should not be extrapolated outside the range $65 \leq W \leq 210$ km or $80 \leq L \leq 1000$ km. The constant $\bar{b}$ in equation (7) must, of course, have a physical dimension if $(m + n) \neq 1$. Equation (11) indicates that stress drop is proportional to source dimension to nearly the first power, i.e. $m + n - 1 \simeq 1$.

Tests of Time-Predictable Model

The wide range of computed values of α in Figure 10 indicates that the ratio of seismic slip to total slip might be better estimated using the time-predictable model of Figure 2(a). For these computations we require estimates of $\bar{U}_s$ for shocks that precede the time intervals, T_R. Such data are presently available for only a few locations along transform faults. To our knowledge there are no direct measurements of $\bar{U}_s$ (and no values of moment from which it could be calculated) for an older event in a sequence of dip-slip shocks along a simple plate boundary. Hence, for convergent margins we must resort to estimating $\bar{U}_s$ in a former shock from that in a recent event using some type of scaling relationship. We do this using the ratio of either 1) the dimensions of the two rupture zones or 2) tsunami heights for a pair of earthquakes at the same observation point. The results of using tsunami heights are contained in Appendix B and are not as definitive as those based on the sizes of rupture zones.

Scaling for Dip-slip Faults Using Lengths of Rupture Zones

Equation (7) could be used to estimate the seismic displacement in a preceding shock, $\bar{U}_{s1}$, from that in a recent event, $\bar{U}_{s2}$, if the lengths and widths of the events are known, i.e.

$$\bar{U}_{s1} = \left[\frac{L_1}{L_2} \right]^m \left[\frac{W_1}{W_2} \right]^{-n} \bar{U}_{s2}. \tag{12}$$

The approximate lengths of rupture zones of older earthquakes often can be estimated from either reports of strong shaking, arrival times of tsunamis or the numbers and distribution of aftershocks [Kelleher, 1972; Ando, 1975; Sykes et al., 1980, 1981]. The width, however, is poorly constrained for most older events.

For events along nearly the same part of a plate boundary the ratio W_1/W_2 probably lies between the following extremes: 1.0 (if the entire down-dip width of the lithosphere is ruptured in each event and if the lithosphere-asthenosphere boundary is sharp) and L_1/L_2 (if L/W remains nearly constant). In either case

$$\frac{\bar{U}_{s1}}{\bar{U}_{s2}} = \left[\frac{L_1}{L_2} \right]^p \tag{13}$$

where for the two cases p equals m or $m + n$, respectively. Best fitting values of m lie between about 0.5 and 1.0 and those for $(m + n)$ are slightly greater than 2.0. Hence, if only the ratio of the lengths is available, the exponent p in equation (13) could be as small as 0.5 or as large as 2.0 for dip-slip faults.

Nankaido Events. Fortunately, the parameter p in equation (13) can be estimated directly from the inferred lengths and vertical displacements (Table 3) associated with the large shocks of 1707, 1854 and 1946 that ruptured similar parts of the plate boundary along the Nankaido region of Japan. Shimazaki and Nakata [1980] show that the time intervals between consecutive shocks are proportional to the respective amounts of uplift in the preceding shock as measured at the port of Murotsu. This proportionality is remarkably accurate since the actual times of occurrence agree with those calculated from the amounts of uplift in the preceeding shock to within 2.5 years i.e. about 2% of the time interval between large earthquakes.

Assuming that the ratio of seismic slip in two events equals the ratio of vertical displacements at Murotsu, values of p of 0.71 and 0.78 are obtained from equation (13) using the paramters listed in Table 3 for the pairs of events 1707-1854 and 1707-1946. The average value of p, 0.75, is consistent with the range of p inferred above. That p is substantially less than $(m + n)$ indicates that W does not change substantially as L varies. The data are consistent, however, with a small increase in W as L increases. This does not seem to be unreasonable, since during events of the greatest size, rupture may propagate somewhat farther down dip into the lower part of the lithosphere where plate motion probably occurs mostly by aseismic processes. Obviously, values of p need to be obtained for other areas before these generalizations can be taken as more than hypotheses.

Geodetic data indicate that a 300 km-long zone ruptured during the Nankaido earthquake of 1946 [Ando, 1975]. Most of the aftershocks that occurred within one day of the mainshock and the generating area of the strongest sea waves, however, were confined to the eastern half of that zone. Hence, slip appears to have occurred more slowly in the western half where a smaller tsunami was also generated [Ando, 1975]. Felt reports of aftershocks and the distribution of sea waves, however, indicate that slip occurred promptly throughout the entire 300-km long zone during the Ansei II event of 1854. Intensity data [Ando, 1975] suggest that displacements associated with

Table 3. Data for Nankaido Earthquakes

Year of Event	Coseismic Uplift* at Murotsu, m	Length[+] of Rupture Zone, (km)
1707	1.8	530
1854	1.2	300
1946	1.15	300

*Data from Shimazaki and Nakata [1980];
[+]Ando [1975].

the great shock of 1707, which ruptured the entire 530 km portion of the plate boundary adjacent to the Nankai trough, occurred slowly in the same 150-km long segment that broke slowly in 1946. The fact that the vertical displacements at Murotsu were nearly the same in 1854 and 1946 suggest that they should be compared with the total length of rupture (prompt plus slow lengths) which was 300 km in each event. The two computed values of p are consistent when L is defined in this way but are quite different if the lengths of only the regions of prompt displacements are used in equation (13).

Values of α for Dip-Slip Events

Figure 16 compares the observed time interval following shocks of both the dip-slip and strike-slip type with estimates of $\overline{U}_s/V$ for those earth-

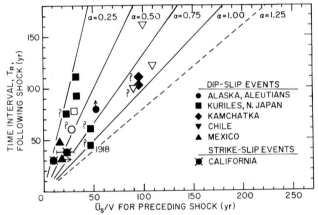

Fig. 16. Repeat time, T_R, as a function of $\overline{U}_s/V$ for the time-predictable model (a) of Figure 2. U_s was estimated for the shock preceding the time interval T_R by scaling the calculated value of $\overline{U}_s$ for the most recent shock in the area by equation (15) with p = 1. L_1 and L_2 are the lengths of the rupture zones of the shock preceding the time interval and the most recent, respectively. Open symbols indicate shocks for which L_1 and L_2 differ considerably; solid symbols denote pairs of events for which the form of the scaling law was not crucial since $L_1 \sim L_2$. Symbol with arrow pointing upward represents average slip computed from seismic moments of two large shocks of 1899 near Yakutat Bay, Alaska. Arrow pointing to right indicates data point for Mexican shock of 1941 which may have been larger than the calibration shock of 1973. Note range of computed values of α, 0.3 to 1.0, i.e. a factor of about 3, is much smaller than the range in Figure 10. This suggests that the time-predictable model (a) of Figure 2 is a better approximation than the slip-predictable model (b).

quakes. Pertinent data are listed in Table 4. For many of the pairs of large events at nearly the same locations, the lengths of the two rupture zones are nearly the same. Those data points are indicated as solid symbols in Figure 16 and are, of course, not sensitive to the scaling relationship used for the older data. The values of α are sensitive to the scaling parameter p for events of quite different lengths that are paired with either the Chilean earthquake of 1960, the Nemuro-Oki event of 1973 or the Aleutian shock of 1965. For p = 1, the values of α in Figure 16 range from 0.29 to 1.05. This is considerably smaller than the range of values computed for the slip-predictable model (Figure 10).

For the time-predictable model the range of α is larger, 0.26 to 1.41, if the scaling parameter p in equation (13) is 0.5; the range is only somewhat larger (0.29 to 1.13) if the value of p obtained for the Nankaido area, 0.75, is used. Values of p larger than 1.0 do not reduce the total range of computed values of α in Figure 16 since the largest and smallest values are then associated with events for which the preceding shock was similar in size to that for the most recent event. Also, values of p greater than about 1.0 are not consistent with the data from the Nankaido region. Thus, the data in Figure 16 and those from Nankaido suggest that p is about 0.75 to 1.0. Values of p in that range lead to smaller scatter in the computed values of α if the time predictable model is assumed in the calculations. The data of Figure 16 along with those of Bufe et al. [1977] and Shimazaki and Nakata [1980] appear to us to provide strong support for the time-predictable model and to cast doubt on the slip-predictable model.

Some of the data points for dip-slip events in Figure 16 have considerable uncertainty associated with them. The zone ruptured by the Oaxaca earthquake of 1978 appears to have broken previously during a sequence of events from 1928 to 1931 [Kelleher, et al., 1973]. The dimensions of the rupture zones of those earlier events are poorly determined. A series of shocks ruptured a long but unspecified portion of the plate boundary in the western and central Aleutians between 1897 and 1907 [Sykes et al., 1980, 1981]. We rather arbitrarily assume that the 650 km long rupture zone of the 1965 shock previously broke in two shocks each 325 km long. The 1918 Kurile event for which a value of α of 1.05 was obtained, only overlapped with the subsequent shock of 1963 over about two-thirds of its length [Fukao and Furumoto, 1979]. The rupture zone of a previous shock in 1780 is shown by Fukao and Furumoto [1979] to be nearly coincident with that of the 1963 event. If we assume the earthquake of 1780 was followed by large shocks of nearly equal size about 1849 and in 1918 and 1963, an average repeat time of 61 years and an α of 0.78 are obtained. If the data point for the earthquake of 1918 is omitted in Figure 16, the range of α is 0.29 to 0.92, a factor of about three.

Table 4. Calculation of α for Dip-Slip Events Using Time-Predictable
Model and Ratio of Source Sizes

Region	Date of Earthquake (Event 2)	Calibrating Earthquake (Event 1, $\bar{U}_{sc}$ known)	T_R* (years)	Length Event 2 / Length Event 1	Calculated** Dip-Slip Displacement (cm)	α
Chile	1837	1960	123	0.50	936	0.88
Chile	1737	1960	100	0.42	786	0.91
Chile	1575	1960	162	(0.46)	861	0.62
Kamchatka	1841	1952	111	1.0	890	0.86
Kamchatka	1737	1952	104	1.0	890	0.92
W. Aleutians	1906	1965	59	(0.5)	141	0.48
N. Japan	1894	1973	79	2.1	275	0.38
N. Japan	1856	1968	112	1.0	338	0.29
N. Japan	1763	1968	93	1.0	338	0.35
Kurile Is.	1918	1963	45	1.0	445	1.05
Kurile Is.	1780	1963	(69)	1.0	445	0.68
Kurile Is.	1893	1969	76	1.2	214	0.30
Mexico	1941	1973	32	1.0	101	0.56
Mexico	1928–31	1978	47–50	1.0	107	0.32
Yakataga Gap, Alaska	1899	post 1980	≥81	---	320[+]	≤0.65[++]

[+] Calculated directly from M_O determined by Thatcher and Plafker [1977].
[++] <0.95 if μ of 3.4×10^{11} dynes/cm^2 appropriate for crust is used instead.
* T_R is time interval from Event 2 to next large earthquake at nearly the same location.
** Calculate from $\bar{U}_2 = \bar{U}_1 (L_2/L_1)^p$ with p = 1.0.

Seismic and Aseismic Slip in Strike-Slip Earthquakes

Values of α are listed in Table 5 for strike-slip earthquakes. Estimates are computed using values of T_R and $\bar{U}_s$ for both the slip-predictable model (Figure 2b) in equation (3) and also for the time-predictable model (Figure 2a). Repeat times in Table 5 are based on instrumental or historical data with one exception. The times of previous earthquakes that broke all or part of the 1857 rupture zone in southern California are based on radiocarbon dating [Sieh, 1978b]. These dates, therefore, have uncertainties associated with them that will affect estimates of α. In particular, the last large earthquake that occurred before the one in 1857 is assigned a date of 1745. An uncertainty of one standard deviation for this determination is +55 years; however, historical constraints limit the latest possible date to 1769 [Sieh, 1978b]. The range of values of α for the 1857 earthquake incorporates this uncertainty.

For the strike-slip earthquakes studied α varies from 0.09 to nearly 1.0 for the slip-predictable model. This range is approximately equal to that for thrust earthquakes except that no nonphysical values are obtained. The value of 0.09 is calculated for the Imperial valley if coseismic slip alone is used. If a value for average slip measured about 10 days after the main shock is used, α is calculated to be 0.17. As in the case for thrust earthquakes, the range of α is reduced when the time-predictable model is used. The low value of 0.09 to 0.17 for the Imperial valley is increased to 0.59 to 0.76. (As discussed in a later section, however, consideration of the distribution of slip during the 1940 earthquake for only that portion of the fault that subsequently reruptured in 1979, leads to a value of 0.33 for α). Since the amount of slip seems to be nearly the same for the events in 1934 and 1966 [Bakun and McEvilly, 1979] in the Parkfield region, the value of α remains relatively low for both the slip- and time-predictable models.

Alternate estimates of the slip rate for portions of the San Andreas fault system in California have been determined from geologic and geodetic data. In central California these data suggest displacements of 3 to 4 cm/yr across the San Andreas fault [Savage and Burford, 1973; Thatcher, 1979a; Sieh, 1977]. If we use a value of 3.5 cm/yr for the slip rate in equation (3), alternate estimates of α for the slip- and time-predictable models in the Parkfield region are 0.27, which are probably more realistic then those obtained using the total rate of plate movement. If the slip in 1966 was 60 cm along a 20 to 25 km segment of the aftershock zone as Lindh and Boore [1980] infer, α becomes 0.54 for a slip rate of 3.5 cm/yr. In the Imperial Valley the accumulated

Table 5. Ratio of Seismic to Total Slip for Strike-Slip Earthquakes

Date	Seismic Displacement (cm)		Repeat Time (yr)	Plate Velocity (cm/yr)	α_A	α_B
	Most Recent Event	Previous Event	Most Recent			
10 Jul 1958	270-380		111?	5.81	0.42-0.59	
9 Jan 1857	450-480		88-167	5.54	0.48-0.95	
27 Jun 1966	30	30?	22	5.59 3.5	0.17 0.27	0.17 0.27
15 Oct 1979	20-35	125-160	39	5.41	0.09-0.17	0.59-0.76

α_A is determined for the slip-predictable model.
α_B is determined for the time-predictable model.

strains probably cannot be explained by slip on a single shallow fault; thus geodetic estimates of relative plate motion are inappropriate for individual faults in that region [Thatcher, 1979b]. Slip on several subparallel strands of the San Andreas may explain the observed strain data [Thatcher, 1979b]. Movement on such sub-parallel faults, as well as aseismic slip (creep), provide a partial explanation of why some of the observed values of α are significantly less than 1.0 regardless of the model chosen. Also, some of the plate motion in the Imperial Valley and Parkfield regions may occur in occasional great events rather than solely in shocks of magnitude less than 7.

Accurate Estimation of Repeat Time

Single Event Method

There are several routes that might be taken in using the results of this paper for the estimation of repeat times of large earthquakes. For regions lacking a history of two or more shocks at nearly the same place but in which one shock is known to have occurred (i.e. the 1964 Alaska zone), equation (1) could be used to estimate T_R if $\bar{U}_s$ can be estimated for the single known event. In this case a value of α must be assumed. In the worst case α could be uncertain by a factor of three (Figure 16), $\bar{U}_s$ by a factor of two, and hence T_R by a factor of six. If a world-wide average for α is used, say 0.65, then T_R would be uncertain by a factor of four. The uncertainty in α might be reduced if that parameter does not vary as much for a given plate boundary and a value from another part of the boundary is used. Also, the data in Figure 9 indicate that equation

(6) can be used to estimate T_R to within a factor of three. Calculations of that kind, however, do not predict explicitly that T_R can vary at a given place as observations indicate since stress drop is taken to be constant. Hence, large uncertainties in T_R are likely to remain for the single event method.

Calibration Using a Pair of Events

The uncertainty in estimating repeat times is likely to be much smaller if time intervals between two or more past events and the ratio of displacements in those shocks can be used to estimate the time interval, T_{R2}, between the last event and the next one in the future, such that

$$T_{R2} = \left(\frac{\bar{U}_{s2}}{\bar{U}_{s1}} \right) T_{R1} \qquad (14)$$

We are, of course, assuming in equation (14) that the time-predictable hypothesis is correct. For most of the sequences of dip-slip events that we have discussed $\bar{U}_{s2}$ was estimated from the seismic moment using equation (2) and $\bar{U}_{s1}$ was obtained from it using a scaling relationship. Systematic errors in determining $\bar{U}_{s2}$ should not effect the ratio (U_{s2}/U_{s1}). Systematic errors in $\bar{U}_{s2}$ (and hence in $\bar{U}_{s1}$), however, will correlate with those in α since

$$U_{s1}/\alpha = V T_{R1}. \qquad (15)$$

Estimates of T_{R2} from equation (14) will depend, of course, on the accuracy with which the ratio $(\bar{U}_{s2}/\bar{U}_{s1})$ can be estimated.

The accuracy of using a pair of events for

calibration in equation (14) is tested in Table 6 for six instances in which three or more events are known to have occurred along nearly the same part of a plate boundary. The percentage errors in estimating the actual time intervals range from 0 to 53%. The uncertainty is obviously much smaller than that involved with the single event method. The largest error in Table 6 is associated with the 1918 Kurile shock, whose rupture zone only coincided with that of the events of 1780 and 1963 over about two-thirds of its area. Also, the error of about 30% for the time interval 1575 to 1737 in Chile could easily be related to an intervening uncataloged event or poor control on the size of the rupture zone of the earlier shock. The two sets of time intervals, T_{R1} and T_{R2} in Table 6 are all associated with preceding shocks of about the same size. Hence, they do not provide an independent check on the accuracy of scaling the ratio of displacement from other types of data.

It is obvious from the relatively small errors in Table 6, however, that the time of occurrence of future earthquakes should be computed whenever possible using a set of large events rather than estimating it from equation (1) using some arc-wide or world-wide average for α. Since the ratio of time intervals in equation (14) should be proportional to the relative velocities of plate motion during those intervals, the latter are unlikely to vary by more than the percentages in Table 6. Uncertainties other than change in velocity or failure of the time-predictable hypothesis, however, are likely to have resulted in the prediction errors of Table 6.

Equation (14) could also be used for a pair of events in which the ratio of displacements could be determined from the relative sizes of the two earthquakes even when the absolute value of one or both displacements is unknown. An example would be the pair of large events in the northern Lesser Antilles in 1690 and 1843 [Kelleher et al., 1973].

If only a single large event is known to have occurred along a given part of a plate boundary, it may still be possible to estimate repeat time by comparing the average displacement in the last shock with the amount of strain built up since that event. Thus, the shock could be predicted to occur at about the time the strain drop in the last shock equaled the amount of strain that had been built up by plate motion. This is essentially the idea put forth by Reid [1910]. That method, however, would be subject to systematic errors in estimating displacement or strain drop and in detecting the cumulative strain built up.

One case in which equation (14) appears to fail by a factor of two should be noted. A great earthquake ruptured a long segment of the plate boundary off Ecuador and Colombia in 1906 [Kelleher, 1972]. Non-overlapping portions of that zone were subsequently re-ruptured in 1942, 1958 and 1979. If we assume displacement was constant throughout the 1906 rupture zone, we obtain repeat times and calculated values of α that differ by a factor of two (73/36 years). Hence, we must conclude that either the displacement in the 1906 shock or α varied by a factor of two along strike or that considerable uncertainties exist in our knowledge of which area in fact broke in 1906. Displacement may well vary

Table 6. Estimates of Repeat Time Using Another Time Interval for Calibration

Area	T_{R1} (Years)	Observed T_{R2} (Years)	$T_{R2} = \left[\dfrac{\bar{U}_{s2}}{\bar{U}_{s1}}\right]^p \cdot T_{R1}$		Precent Error in Prediction	
			p = 1.0	p = 0.75	p = 1.0	p = 0.75
Chile	1960–1837 = 123	1837–1737 = 100	103	100	3%	0%
Chile	1960–1837 = 123	1737–1575 = 162	114	107	30	34
Kamchatka*	1952–1841 = 111	1841–1737 = 104	111	111	7	7
Northern Japan	1968–1856 = 112	1856–1763 = 93	112	112	20	20
Kurile Islands	(1918–1780)/2 = 69[+]	1963–1918 = 45	69	69	53	53
Nankaido area, Southern Japan	1947.0–1855.0 = 92	1855.0–1707.8 = 147.2		138**		6

*Assuming lengths of Kamchatka events are nearly the same.
[+]Two equal sized events assumed during interval.
**p is scaling parameter of equation (13). See Table 4 for other data.
Calculated directly from equation (14) using vertical displacements in Table 3.

considerably along strike for rupture zones, like that of 1906, which extend several hundred kilometers along strike. Thus, although our emphasis on displacement averaged over the rupture surface has been very productive at the factor of 1.5 to 2 level, at that level variations in displacement over a given rupture zone almost certainly assume a critical importance. Data at a number of points along such a rupture zone would probably be needed to estimate variations in displacement and to provide data for the calculation of subsequent repeat times for individual portions of a long rupture zone.

An example of this type of phenomenon can be seen for the Imperial valley earthquakes of 1940 and 1979. Slip during the 1940 earthquake, which averaged about 125-160 cm [Brune and Allen, 1967; Trifunac, 1972], was not evenly distributed along the fault. The average displacement south of the Mexico-United States border ($\sim$ 230 cm) was about 2.5 times the displacement north of the border ($\sim$ 90 cm) [Trifunac, 1972]. Thus, if the time-predictable model is correct and we examine the distribution of slip within the rupture zone rather than only its average value, it is not surprising that the northern portion of the 1940 rupture zone broke again before the southern portion.

Consideration of the actual distribution of slip during earthquakes can also affect estimates of α. In Table 5 the average value of slip during the 1940 shock is used to compute α for the time-predictable model. If instead we use the value of displacement in 1940 averaged over only that portion of the fault that re-broke in 1979 (70 cm), α is reduced to 0.33. This is still over two times larger than the value of α for the slip-predictable model.

Implications for Great Earthquakes in Southern California

One of the most important uses of the time-predictable model might be the estimation of the approximate time of occurrence of a great earthquake in southern California for the 360-400 km long segment of the San Andreas fault that last ruptured in a great event in 1857. Radiocarbon dating and geologic evidence of previous large shocks at one point along that segment (Pallett Creek) yield an average repeat time of about 160 years for the past 1400 years [Sieh, 1978b]. The distribution of events, however, is non-uniform in time (Figure 17), and the "intervals appear to have a bimodal distribution, with clustering of intervals about 100 and 230 years" [Sieh, 1978b]. Most of the events that Sieh concludes are similar in offset at Pallett Creek to that of 1857 are followed by time intervals of about 230 years (L events in Figure 17). Time intervals of about 100 years are associated with what Sieh [1978b] infers to be earthquakes with relative small offset about 1190 A.D., offsets of less certain size about 545 and 860 and large offset about 1745.

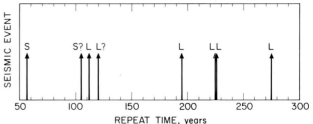

Fig. 17. Time intervals, T_R, (arrows) between large shocks along San Andreas fault at Pallett Creek, California based on C^{14} dating reported by Sieh [1978b]. L and S indicate large and small offsets at Pallett Creek, where large is taken as similar in offset to that in the great shock of 1857. L and S refer to events that initiate the interval T_R (as in the time-predictable model). Note that intervals tend to be distributed bimodally near either 100 or 230 years and that large events tend to be followed by longer time intervals.

These data suggest that the amount of offset may correlate with the succeeding time intervals as in the time-predictable model. Sieh [1978b] presents some evidence that rupture about 1745, which like that in 1857, appears to be associated with several meters of strike-slip offset at Pallett Creek, did not extend as far northwest as it did in 1857. If the length of rupture in 1745 was smaller than that of 1857, the results of this paper suggest that its average displacement should have been smaller than that of 1857. This would be in accord with a relatively small time interval (about 100 years) following the event of 1745 via the time-predictable hypothesis.

If we do not consider variations in displacement along the 1857 rupture zone, the evidence from past sizes and time intervals might seem to favor a 200 to 300 year interval following the 1857 shock. Obviously, whether the interval between great shocks in that zone is about 100 or 200 to 300 years has enormous implications for the assessment of earthquake hazards in southern California. It may be possible to estimate the time of occurrence of future great shocks by the event pair method and equation (14) if the dimensions of the rupture zones of the events of about 1745 and 1470 can be obtained by trenching or other methods and if variations in offset can be measured within their rupture zones.

Right-lateral offsets, associated with the 1857 earthquake, however, vary from about 9.5 m in the Carrizo Plain near the northwestern end of the fault break to between 3 and 6 m in the southeastern half of the rupture zone [Sieh, 1978a]. These marked variations in slip may well affect the timing of future large shocks. For example, another 100 to 200 years may still be required for the level of stress along the northwestern segment of the 1857 zone to build up to the pre-1857 level. Stresses in the southeastern part of the zone, which includes Pallett Creek, however, may be nearing the pre-1857 level since the displacement and stress drop in that segment were not as large in 1857 as they were in the northwestern segment. Hence, if any of the 1857 rupture zone has the potential of being the site of a large to great earthquake during the next 50 years, it seems reasonable that such an event would occur in the southeastern portion of that zone. The amount of displacement in that segment about 1745 could provide a much better estimate of the time of occurrence of a future large shock. The Parkfield area near the northwestern end of the 1857 break, which may have been the site of large foreshocks of the 1857 earthquake [Sieh, 1978c] may well not experience forerunning activity prior to a large shock in the southeastern half of the 1857 zone. Lindh and Boore [1980] suggest that moderate-size shocks near Parkfield may signal the occurrence of a great shock within the 1857 zone. Our proposal that the southeastern part of the 1857 zone should re-rupture sooner than the northwestern segment has an analogy in the fact that the 1979 shock in the Imperial valley, California, broke that portion of the 1940 rupture zone that had experienced the smaller slip in 1940.

Significance and Distribution of
Aseismic Slip

We now examine some of the possible uncertainties in estimating the amounts of seismic and aseismic slip for major plate boundries. We attempt to answer the following questions: Is the factor of three variation in α in Figure 16 indicative of real differences in the mode of slip among and along subduction zones; does α vary with depth, do systematic errors in the data lead to values of α that are biased?

Systematic Errors in α

The factor of three variation in the computed values of α for earthquakes along subduction zones could arise from one or a combination of the following factors: 1) uncertainies in M_o, L, W, or μ; 2) errors in the scaling laws we used to calculate displacements for older shocks; 3) the rupture zones of two consecutive shocks along a portion of a plate boundary may not have been coincident; 4) relative plate motions on a time scale of years to centuries may differ appreciably from the values determined for the last few million years, or 5) the estimated time interval T_R may be too long if major events have been overlooked or missed in historic compilations.

Uncertainties in scaling laws are unlikely to be a major contributor to the scatter in the values of α in Figure 16 since the rupture zones of many of the events that precede a time interval, T_R, are similar in size to those of shocks that terminate the interval. The historic record for Mexico, Japan, and Chile appears to be good enough that it is unlikely that great shocks

have been missed for intervals comparable to the repeat times we infer. It also seems unlikely that a great event has been missed in the area of the 1965 Aleutian earthquake in the last 60 years, our estimate of the most recent repeat time for that region. For the 1952 Kamchatka zone we may have underestimated the repeat time by a factor of two if the preceding great shock in that area actually occurred in 1737 rather than 1841. Hence, if anything, we have underestimated the repeat time for that area which would lead to smaller values of α rather than larger. Such an error for the Kamchatka event would not change the amount of scatter in Figure 16. The contributions of aftershocks and smaller events to the seismic slip would also result in values of $\bar{U}_s$ and α that are only 0 to 20% larger than those in Figure 16.

Very little is known about the relative velocities of plate motion on a time scale of 10's to 1000's of years. An examination of pairs of events in the previous section, however, indicates that variations in V over those intervals cannot be larger than 2 to 50%. Neotectonic and geodetic studies could provide important control on the possibility of small changes in velocity.

Systematic and random errors may be present in the estimates of M_o, L, W, and μ that are used to derive values of $\bar{U}_s$ and α for dip-slip events. We used a single value of the shear modulus, μ, in all of our calculations for subduction zones. If some of the events occurred entirely in the crust, however, while others occurred solely in the mantle, the actual values of α would differ by about a factor of two. For example, Alewine [1974] used geodetic and seismic data to derive a model for the 1964 earthquake in which displacement varies over the downdip width of the dislocation. His determination of the average slip, 18.5 m, differs from ours largely because he uses a μ appropriate to the crust.

Expressions for displacements in the far field that were used in deriving the seismic moment from surface waves are dominated by terms proportional to $\sin(2\theta)$ [Harkrider, 1970] for values of the dip, θ, greater than about 10° to 15°. Systematic errors in θ for faults dipping less than 20° could result in uncertainties in M_o and hence in $\bar{U}_s$ of as much as 50%. Since the dips of many of the thrust events we studied are in that range (Table 2), the calculated values of M_o, $\bar{U}_s$ and α may be uncertain by as much as 50%. Uncertainties in the determination of moment from surface waves or free oscillations become very large for shallow faults of dip less than 10°. This may be an important source of uncertainty for the Chilean earthquake of 1960 (Table 2).

For dip-slip events we calculated $\bar{U}_s$ and α from the seismic moment as determined from waves of periods of about 2 to 50 minutes but the rupture area was determined from aftershocks occurring from days to months after the mainshocks. For most of the events the length of the aftershock zone along strike does not appear to increase much after the first day. The development of the aftershock zone with time in the down-dip direction is not well known. Very little is known about the time history of displacements and about the development of aftershock areas of great shocks on a time scale of minutes to days. Significant amounts of strain release with those time constants could be occurring as so-called slow earthquakes. The failure to include contributions to the seismic moment from signals of that type could lead to systematic underestimates of U_s and α. In our calculations slow release of that type would be assigned to aseismic slip for dip-slip faulting. For many of the strike-slip events, however, such contributions are assigned to seismic slip since $\bar{U}_s$ is usually determined from either surface offset or geodetically measured slip, each of which usually includes contributions from days to months after the main shock.

It is interesting that computed values of α in Figure 16 are close to 1.0 for regions associated with the two largest shocks for which α could be computed, the 1952 Kamchatka and the 1960 Chilean earthquakes. For the area of the great Alaskan earthquake of 1964, the only evidence for repeat time comes from the dating of marine terraces, which yield 500 to 1350 years [Plafker and Rubin, 1978]. Using the value of $\bar{U}_s/V$ for the 1964 earthquake in equation (1), we obtain repeat times of 187 and 623 years for α's of 1.0 and 0.3 respectively. It seems unlikely that $\bar{U}_s$ for the event preceding the 1964 shock would have been much larger than that of 1964. (It is the slip in the preceding event, of course, that should be compared with the time interval preceding the 1964 shock in the time-predictable model.) If we take the 1964 shock to involve mostly seismic release (i.e., $\alpha \simeq 1$) as in the great shocks of 1952 and 1960, the calculated repeat time is considerably smaller than the values reported by Plafker and Rubin [1978]. Nevertheless, their smallest interval, 500 years, is compatible with the smallest value of α in Figure 16. Since they studied displacements associated with imbricate faults within the outermost part of North American plate (Figure 18), however, their results may not pertain to repeat times for the main plate boundary. Individual imbricate faults may not move during every large event along the main plate boundary.

For some of the smaller earthquakes that we studied, such as the Mexican events of 1941 and 1928-31 the dimensions of the rupture zones are quite uncertain. If those events were, in fact, larger than the succeeding shocks of 1973 and 1978, the values of α in Figure 16 should be increased. The magnitude ($M_s = 7.7$) of the 1941 event suggests it may, in fact, have been larger than the shock of 1973 ($M_s = 7.5$). McNally and Minster [1980] summed the seismic moments of Mexican shocks for the period 1898 to 1979 as derived from surface wave magnitudes and conclude that seismic slip accounts for most of the plate motion along that subduction zone if two gaps that have been aseismic for large shocks for 80 years or more [Singh et al., 1980b; McNally and Minster, 1980] are excluded. It is difficult to estimate the uncertainty in the value of α obtained by McNally and Minster [1980] since it is ultimately derived from magnitudes and estimates of felt areas. Their values may not be significantly different from the two we obtained (0.56 and 0.31), especially if the shocks of 1941 and 1928-31 were somewhat larger than those of 1973 and 1978 respectively.

McNally and Minster [1980] obtain a very small average value of α, about 0.2, for Central America by the same method. There is no known systematic error that could increase that estimate to 1.0. Thus, at least some of the variations in in Figure 16 are probably real. A considerable amount of the scatter in Figure 16, however, must be attributed to a combination of systematic errors in M_o, μ, L and W; uncertainties in accounting for slow release of energy; uneven loading rates, and failure to include contributions from smaller events.

Possible Variation of α with Depth.

Except for two Mexican shocks that were discussed already, the remaining low values of α in Figure 16 for dip-slip faults are associated with earthquakes near northern Japan and the Kurile Islands. Several of those events are unusual in that aftershock activity extends to a depth of about 70 to 90 km [Fukao and Furumoto, 1979]. Aftershocks of great thrust earthquakes in Alaska, Mexico and many other regions of plate convergence do not extend deeper than 30 to 50 km. It is possible that this difference may be related to the greater age of the plate that is being subducted in the northwestern Pacific.

A significant part of the total plate motion in the western Pacific for depths from about 40 to 80 km may take place aseismically, perhaps by the occurrence of slow earthquakes or creep. This is not unreasonable to expect for the lower part of the lithosphere where higher temperatures are

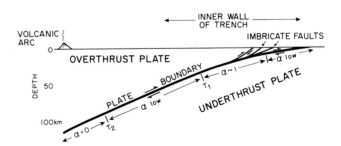

Fig. 18. Schematic vertical cross section of upper 100 km of a subduction zone showing possible variations of α with depth. T_1 and T_2 are isotherms that may move up or down in depth in various arcs. Arrows indicate sense of thrust faulting.

likely to result in aseismic movements. If so, the values of α would be higher than 0.3, (perhaps close to 1.0), within the upper 40 km of the earth (Figure 18). We also suggest in Figure 18 that α is low in the uppermost few kilometers of subduction zones. Displacements in the less competent sedimentary rocks at those depths appear to occur slowly or aseismically [Fukao, 1979].

Transitions in the behavior of α along the plate boundary probably do not occur at the same depths in all subduction zones (Figure 18). The two deeper transitions are likely to be sensitive to the thermal structure and pore pressure. Hence, the length of the zone where α is about one is likely to be a function of both the dip and the depth at which a critical temperature (T_1) is reached such that aseismic slip starts to become appreciable. In regions of steep dip--such as Central America and the Marianas--or higher temperatures at a given depth (perhaps related to back-arc spreading)--such as Tonga and the South Sandwich arc--the length of the zone of $\alpha \simeq 1$ may be very short. This could severly limit the sizes of the largest thrust earthquakes that could occur in those areas. Plate motion could be accommodated by aseismic slip at most depths and by only moderate-size events along the short region where seismic slip is possible.

Discussion and Conclusions

We have found a number of regularities in the observed repeat times of large earthquakes along simple plate boundaries of the convergent and strike-slip type. Average displacements, stress drops, and repeat times are strongly related to the size of the rupture zone. The largest Mexican earthquakes of this century, for example, are characterized by small rupture zones (L and W < 150 km), small displacements ($\simeq$ 1 m), $M_o < 10^{28}$ dyne-cm and short repeat times (30 to 40 years). Earthquakes near Kamchatka and southern Chile, on the other hand, are characterized by long rupture zones (450 to 1000 km), large displacements (8 to 20 m), $M_o > 10^{29}$ dyne-cm and repeat times greater than 100 years. The repeat time of large shocks is not solely a function of the rate of plate movement.

As well as varying from area to area, repeat times are not constant at the same locality but can vary by a factor of two or more. We find support in our data from both thrust and transform plate boundaries and from the data of Bufe et al. [1977] and Shimazaki and Nakata [1980] for the hypothesis that repeat time at a given place can vary and that it is proportional to the amount of displacement in the preceding earthquake. The fact that the 1979 earthquake in southern California ruptured that portion of the 1940 rupture zone that had undergone relatively small displacement in that preceding event also supports the time-predictable model. While the concept of approximately constant stress drop does have a utility for discussing shocks with seismic moments

varying by a factor of 10^{13} or more, we conclude that a more careful examination shows that changes in stress drop of about a factor of four for large dip-slip events and about a factor of 10 to 15 for moderate to large strike-slip earthquakes correlate with the sizes of rupture zones. Variations in displacement have important consequences for the long-term prediction of earthquakes and for the assessment of seismic hazards if the time interval between large shocks is, in fact, proportional to the displacement in the last major shock in a region.

The time-predictable model of Figure 2(a) has both deterministic and random components. The deterministic element, of course, is that T_R is governed by the displacement in the preceding event. A large shock occurs when the stress level reaches (i.e. is recharged to) a value σ_I (Figure 2a) that is the same for all shocks at the same place. This seems reasonable since the stress level at which slip occurs is likely to be governed by the distribution of large asperities and the level of pore pressure, both of which are unlikely to change appreciably over many earthquake cycles. The individual stress drops and displacements in a series of shocks may well be randomly distributed. The rupture time, and hence the average displacement and stress drop, may be very sensitive to stress concentrations, barriers, transverse features, and other inhomogeneities that may cause a propagating rupture to stop at one place in one shock and at another in a different event [Sykes, 1971; Das and Aki, 1977; Das and Scholz, 1980].

A consequence of the variation in stress drop with size for large earthquakes is that the drop in stress must be only a fraction of the total stress, except possibly for the very greatest earthquakes. Large earthquakes appear to be stopped not by the exhaustion of the available stress, but by strong barriers. The rupture zones of even the largest earthquakes are usually delimited by major transverse features, and those features probably serve to arrest motion in these shocks as well. Thus, it seems likely that stress drop is only partial even in the greatest events and that the barriers that limit rupture are regions of high strength. The strongest barriers, i.e. those that are rarely if ever crossed by rupture during a single great event, consist of either large offsets of the entire subduction zone, abrupt changes in the strike of the plate boundary, sudden changes in the dip of adjacent segments of downgoing lithosphere or major features on the downgoing plate, such as aseismic ridges that are entering the subduction zone.

A consequence of the increase in average displacement with size of the rupture zone is that great shocks cannot be regarded simply as the sum of smaller events situated end to end along a fault. Slip appears to continue in an average sense for a longer time the greater the size of the rupture zone. Hence, on the average a given part of a fault zone probably does not heal immediately but continues to slip until stopping phases arrive from major barriers (C. Scholz, personal communication). In a series of experiments simulating rupture along a fault Das [1981] finds that pulses reflected from the free surface tend to promote rather than stop motion at a given point on the fault. Displacement at a given place is finally arrested by stopping phases from the ends of the fault. Hence, the nature of stopping phases and the average rupture time are probably sensitive to the types of boundary conditions and to whether faulting is unilateral or bilateral. Models of that type offer an explanation of why average displacements are greater for larger rupture zones. From these considerations it seems likely that the average displacement will increase with length at a power somewhat less than one, which agrees with our finding that $\bar{U}_s$ scales with L to a power between 1/2 and 1. Scaling of $\bar{U}_s$ with L to the first power would correspond to stopping phases coming in mainly from the opposite end of the fault from that where rupture is initiated. In more realistic situations, stopping phases are likely to arrive from less removed parts of the fault surface as well as from the farthest end. More thought should be given to the boundary condition at the down-dip end of the fault and whether that boundary generates appreciable stopping phases or not.

We find that repeat times can be estimated to within 2% to 50% of observed times if a pair of previous events is available for calibration and if the ratio of the displacements in those events can be estimated either by direct measurement or from scaling relationships. If only one event or no previous shocks are known in a region, repeat times probably can be estimated using either equations (1) or (6) but with an uncertainty of up to a factor of three. Since so little has been known about repeat times of great earthquakes on a world-wide basis, estimates of repeat time with uncertainties of a factor of 0.5 to 3 should still be very useful in making long-term estimates of seismic risk. When the uncertainties can be reduced to the level of 10% or better, estimates of repeat time become important for the long-range prediction of earthquakes.

If the time-predictable model of Figure 2(a) is correct, a second large earthquake should not be expected immediately in a region (like that of the 1973 Nemuro-Oki shock) even when the last shock was smaller than previous ones at that location. Instead, the time interval to the next shock would be expected to be short. Likewise, large portions of the rupture zones of the 1960 Chilean and 1964 Alaskan earthquakes probably will not be sites of great shocks for at least 150 to 200 years. Other portions of those areas, however, may have experienced much smaller slip in those great events and hence could be sites of large earthquakes within somewhat shorter intervals. Variations of a factor of two or three in the amount of slip along the strike of long rupture zones (as observed for the 1857 break in southern California) may be common. A similar gradient in co-seismic displacement for the Colombian earthquake of 1906 may account for the fact that portions of its rupture zone were re-ruptured at time intervals that differ by a factor of two.

We obviously need to recover as much information as possible about previous large shocks so that they can be used to calibrate the times of future earthquakes. Better estimates of repeat time of large shocks will also require greater knowledge of α, of the detailed structure of the upper 50 km of subduction zones, of the gradual release of energy in slow earthquakes and of the detailed distribution of displacement over the rupture zones of previous large shocks. There is obviously a need to go beyond the average quantities used in this paper.

Appendix A: Calculation of Stress Drop for Faulting in a Homogeneous Half-Space

In this appendix we examine how changes in the dip (θ), depth to the top of the rupture zone (z_o) and variations in L/W (Figure 1) affect the computed values of stress drop for a dip-slip fault in a homogeneous half space. We show that stress drop calculated using that model is not a constant when allowance is made for known variations in dip, fault depth and the ratio L/W.

Numerical Calculations. Mansinha and Smylie [1971] give closed analytic expressions for the displacement field for inclined finite dip-slip faults in a homogeneous half-space where U_s is constant over the rupture surface. Using their equations we computed the drop in shear stress, $\Delta\sigma_{3'2'} = 2\mu\,\Delta e_{3'2'}$, close to the fault surface from the change in shear strain $\Delta e_{3'2'}$ by the appropriate numerical differentiations of the displacements in the primed coordinate system shown in Figure 1. The calculations were checked to make certain they were insensitive to the step size used in the numerical differentiation and to the distance from the fault plane. When the dislocation was placed at a depth much larger than that of any of its dimensions, the computed values of C checked with those of Chinnery [1969a] for an infinite space.

Effect of Dip and Depth to Fault Surface. The geometrical factor C was calculated as above from the ratio $W\Delta\sigma/\mu\bar{U}_s$ for various orientations of the rupture surface in a homogeneous half-space, where $\Delta\sigma$ was determined at the center of the dislocation. The results are shown in Figure A1. For L/W > 1 the geometrical factor, and hence $\Delta\sigma$, is very insensitive to variations in dip from 15° to 45°. That range is typical of the dips of the fault planes of large shallow earthquakes along subduction zones. For L/W < 1 such changes in dip can cause variations in C on the order of 20% if the fault lies near the surface of the half-space. C is also insensitive to changes in L/W for L/W > 2. Since L/W ranges from 0.6 to 8.1

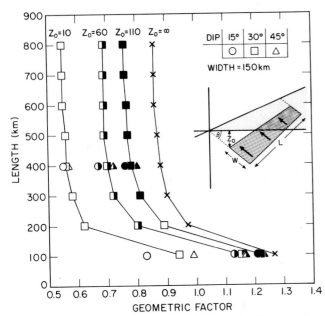

Fig. A1. Variation of the geometric factor, C in equation (4), as a function of rupture length, width, depth, and dip for pure dip-slip movement on a rectangular fault buried in a half-space [Mansinha and Smylie, 1971]. U_s is taken to be constant over rupture surface. For L/W >> 1 the geometric factor, C, is approximately constant for a given z_o and θ. As L/W approaches one or less, C increases rapidly. For a given depth and L/W < 1, varying θ from 15° to 45° causes a change in C of less than 0.05. For shallow depths and L/W > 1, the change in C for such a variation in dip is larger. For a given length, width and dip, increasing the depth to the top of the fault results in an increase in C. The values for $z_o = \infty$ are from Chinnery [1969a]. The values of C shown can be used to determine stress drop at the center of a fault.

(Table 2 and Figure 7) for the dip-slip earthquakes we considered, however, it was necessary to compute the exact value of C for each data point.

The computed stress drop in a half-space can vary by a factor of up to two (Figure A1) depending upon the depth to the top of the fault, z_o. Boore and Dunbar [1977] report a similar effect for vertical strike-slip faults. That quantity is usually poorly known for most large shocks along subduction zones. We set $z_o = 10$ km for all of our calculations for dip-slip faulting in a homogeneous half space. While the absolute level of the stress drops we compute may be uncertain by as much as two, depending upon whether the fault surface can be considered deep or close to the free surface, that uncertainty is not large enough to account for the marked variations in stress drop that we compute for the half-space model.

Average Stress Drop over Rupture Surface. We also investigated the appropriateness of calculating merely the stress drop at the center of the rupture zone rather than an average over the entire surface. For each earthquake we calculated an average stress drop for a grid of 81 points spaced over the rupture surface. In our calculations for a homogeneous half-space the average stress drop is a linear function of the stress drop at the center of the fault. The average stress is, however, larger than the stress at the center of the fault. This effect probably arises from stress singularities at the edge of the dislocation surface for models with U_s constant over the rupture surface. Obviously models involving realistic distributions of stress and displacement are needed in future work as are models in which rigidity varies spatially.

Computed Stress Drops. Figure A2 shows the stress drop for our data set as calculated at the center of the dislocation as a function of W for a homogeneous half-space. The range of values, 5 to 21 bars (a factor of about four), is somewhat smaller than that, 7 to 38 bars, for an infinite medium (Figure 6).

When the slip on each dip-slip fault system is taken to be the same (i.e. 2 m in Figure A3), the computed stress drop for a half-space decreases

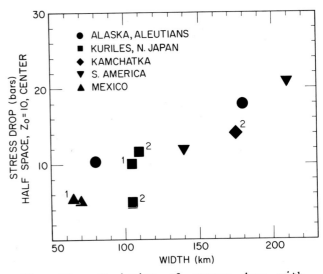

Fig. A2. Variation of stress drop with rupture width for a half-space. Stress drop is calculated numerically for the appropriate fault geometry and average displacement for each data point, using the formulation of Mansinha and Smylie [1971]. The depth to the top of the fault, z_o, is assumed to be 10 km; μ is taken as 5×10^{11} dyne-cm^{-2}. The value of stress drop at the center of the fault is used. As width increases from 65 to 210 km, stress drop increases by a factor of 4, from 5 to 21 bars. The symbols labelled 1 and 2, respectively, denote two subsets of events for which the ratio L/W is approximately constant.

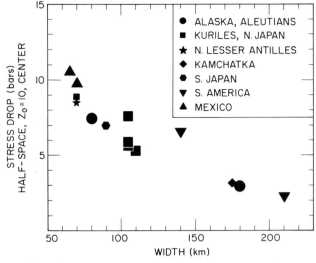

Fig. A3. Stress drop for half-space model as a function of width with displacement held constant at 2 m for all fault geometries. The observed decrease in stress drop of about a factor of 5 results solely from changes in fault geometry (i.e. in the factor C/W of equation (4)).

with increasing W. Hence, the actual displacement increases sufficiently rapidly with the dimensions of the rupture zone to more than offset the effect of geometry alone upon the computed stress drops.

Since the model we used for both half-space and infinite media in which displacement is taken to be constant over the dislocation surface leads to stress singularities, the absolute values of stress drop that we computed may not be correct. In this paper, however, we are mainly concerned with the increase of stress drop as source dimension increases. Data points labelled 1 and those labelled 2 in Figure 6 and Figure A2 each have nearly identical aspect ratios (L/W). Hence, even if our calculated values of C (equation 4) are incorrect, C (whatever its value) must be the same in an infinite medium for a given value of L/W. Hence, if the shear moduli are the same, the ratio of the stress drops from equation (4) for two events of the same L/W must be

$$\frac{\Delta\sigma_2}{\Delta\sigma_1} = \left(\frac{\bar{U}_{s2}}{\bar{U}_{s1}}\right) \cdot \left(\frac{W_2}{W_1}\right) \qquad (A1)$$

(which is the same as the ratio of two values on the vertical axis in Figure 6). For pairs of data points marked 1 and for those labelled 2 in Figures 6 and A2, the ratio in equation (A1) always increases as W becomes larger. Thus, it is clear that $\Delta\sigma$ is not constant among the events we studied, neither for dip-slip nor strike-slip

earthquakes but that it increases with the size of the rupture zone.

Appendix B: Estimates of Source Parameters from Tsunami Data

Some of the pairs of large events that we have examined that occurred at nearly the same place generated tsunamis for which maximum wave heights, H, are tabulated by Abe [1979]. We attempt here to use observations at Hilo, Hawaii for the Kamchatka pair of 1841 and 1952 and for the Chilean pair of 1837 and 1960 to infer the parameters of the earlier events from those of the more recent shocks. The observed periods of the sea waves, about 30 minutes, are much longer than the corner periods associated with the finite size of the generating areas of even the greatest earthquakes such as Chile 1960. Hence, we can take the source to be located at a point and assume that H is proportional to the amount of water displaced:

$$H \sim LW \, (U_s)_z$$

where $(U_s)_z$ is the vertical component of seismic displacement in that portion of the epicentral area that is water covered and W_T is the width of that area in map view. Several factors, however, seem to preclude using older data on tsunami heights to infer more than qualitative aspects of their source properties. Much of the excitation of tsunamis may be associated with vertical movements along secondary imbricate faults of steeper dip than that of the main plate boundary (Figure 18). While we will assume that $(U_s)_z$ is proportional to $\bar{U}_s$, that assumption may not be correct to better than a factor of about two since the vertical displacements on imbricate faults of steeper dip are unlikely to be strictly proportional to movements on the main plate boundary. Also, wave heights for the events of 1837 and 1841 are available from only a single location, Hilo. It is again difficult to ascertain if relative source properties can be determined from single observations to better than a factor of about two. Regions of both uplift and subsidence are typically associated with great earthquakes that involve thrust faulting [Plafker, 1972]. If the downdip width of faulting, W, varies in a pair of shocks, the average water height above the rupture zone is very sensitive to the location of the coastline (since, of course, the area above sea level does not contribute to tsunami generation). For a source area that is completely water covered, H should be proportional to M_o. This is probably the best estimate of source properties that can be made from wave heights unless the contribution of water-covered areas alone is calculated more precisely.

Kamchatka Earthquakes of 1841 and 1952. Maximum heights of sea waves at Hilo were 4.6 and 3.7 m for the events of 1841 and 1952 [Abe, 1979]. Whatever scaling relationship is used, the ratio of heights is close enough to one that the dimensions, moment and displacement for the 1841 shock should be only slightly greater than those for the 1952 event. Fedotov [1965] places the rupture zone of the 1841 event within that of the 1952 earthquake. He indicates a smaller rupture area, however, for the shock of 1841. The descriptions of that earthquake that we have read are so brief that it was not possible to ascertain whether its rupture zone was in fact small or if it extended over a dimension comparable to that of the better described events of 1737 and 1952. While the scaling of the sea waves at Hilo indicates to us that the rupture area of the 1841 shock was probably comparable to that of the 1952 earthquake, we cannot rule out the possibility that the event of 1841 was anomalous, perhaps like the Kurile shock of 1958 or the great tsunami-generating earthquake of 1946 in the Aleutians.

Chilean Earthquakes of 1837 and 1960. Maximum wave heights, H_1 and H_2, at Hilo are reported by Abe [1979] as 6 and 10.5 m for the shocks of 1837 and 1960. Felt reports [Lomnitz, 1970; Kelleher et al., 1973] indicate that the Chilean shocks of 1575, 1737 and 1837 did not rupture as far north as that of 1960. While the southern limit of these older rupture zones is not well defined, it is unlikely that any of these events ruptured farther south than the end of the thrust zone at the Chile Ridge (which formed the southern boundary of the 1960 rupture zone). Hence, the rupture zones of the three older events undoubtedly were smaller than that of 1960. The smaller wave height for the shock of 1837 compared to that for the event of 1960 also indicates that the moment and source dimensions of the former were smaller than those of the 1960 shock. If we take H to be proportional to M_o and relate M_o to L with equations (7), (10) and (13), we obtain lengths of 816 and 830 km for the shock of 1837 if the length of the 1960 zone was 1000 km and p is either 0.75 or 1.0. In view of the various uncertainties in using tsunami heights to infer source properties, however, it seems premature to take these quantitative estimates too seriously. For example, if the ratio of wave heights for the events of 1837 and 1960 is reduced by a factor of two, rupture lengths of 634 and 650 km are obtained for p = 0.75 and 1.0. The known area of strong shaking of the 1837 shock indicates a minimum rupture length of about 500 km.

Acknowledgments. We thank Dr. Chris Scholz for numerous discussions throughout the course of our work and for suggesting that dislocation models may be useful in estimating repeat times. Our work was also stimulated by a seminar presented at Lamont in April 1980 by Dr. K. Mogi in which he proposed that repeat times for the Nankaido region of Japan were related to the size of the rupture zone of preceding events as in the Shimazaki-Nakata model. We thank Drs. Shamita Das, Klaus Jacob, and Paul Richards for critically reading the manuscript and for helpful discussions. Drs. K. McNally, K. Shimazaki and S. K. Singh sent us preprints of relevant work prior to publication. This work was partially supported by grants EAR 78-22770 and EAR 79-26350 from the National Science Foundation. Lamont-Doherty Geological Observatory Contribution Number 3113.

References

Abe, K., Mechanisms and tectonic implications of the 1966 and 1970 Peru earthquakes, _Phys. Earth Planet. Int._, _5_, 367-379, 1972.

Abe, K., Tsunami and mechanism of great earthquakes, _Phys. Earth Planet Int._, _7_, 143-153, 1973.

Abe, K., Reliable estimation of the seismic moment of large earthquakes, _J. Phys. Earth_, _23_, 381-390, 1975.

Abe, K., Some problems in the prediction of the Nemuro-oki earthquake, _J. Phys. Earth_, _25_, suppl., S261-S271, 1977.

Abe, K. Size of great earthquakes of 1837-1974 inferred from tsunami data, _J. Geophys. Res._, _84_, 1561-1568, 1979.

Acharya, H. K., A method to determine the duration of quiescence in a seismic gap, _Geophys. Res. Lett._ _6_, 681-684, 1979.

Alewine, R. W., Application of linear inversion theory toward the estimation of seismic source parameters, Calif. Inst. Techn., Ph.D. Thesis, 1974.

Algermissen, S. T., W. A. Rinehart, R. W. Sherburne, and W. H. Dillinger, Preshocks and aftershocks of the Prince Willian Sound earthquake of March 28, 1964, in: _The Prince Willian Sound, Alaska, Earthquake of 1964 and Aftershocks_, L. E. Leipold, ed., U.S. Dept. of Commerce, Environ. Sci. Services Admin., Coast Geod. Surv., 79-130, 1969.

Ambraseys, N., Some characteristic features of the Anatolian fault zone, _Tectonophysics_, _9_, 143-165, 1970.

Ambraseys, N. N., and J. S. Tchalenko, The Dasht-e Bayaz (Iran) earthquake of August 31, 1968: A field report, _Bull. Seism. Soc. Am._, _59_, 1751-1792, 1969.

Ando, M., Source mechanisms and tectonic significance of historical earthquakes along the Nankai trough, Japan, _Tectonophysics_, _27_, 119-140, 1975.

Bakun, W. H., and T. V. McEvilly, Earthquakes near Parkfield, California: Comparing the 1934 and 1966 sequences, _Science_, _205_, 1375-1377, 1979.

Boore, D. M., and W. S. Dunbar, Effect of the free surface on calculated stress drops, _Bull. Seism. Soc. Am._, _67_, 1661-1664, 1977.

Brune, J. N., and C. R. Allen, A low stress-drop, low-magnitude earthquake with surface faulting: The Imperial, California, earthquake of March 4, 1966, _Bull. Seism. Soc. Am._, _57_, 501-514, 1967.

Bucknam, R. C., G. Plafker, and R. V. Sharp, Fault movement (afterslip) following the Guatemala earthquake of February 4, 1976, Geology, 6, 170-173, 1978.

Bufe, C. G, P. W. Harsh, and R. O. Burford, Steady-state seismic slip - a precise recurrence model, Geophys. Res. Lett., 4, 91-94, 1977.

Burdick, L., and G. R. Mellman, Inversion of the body waves from the Borrego Mountain earthquake to source mechanism, Bull. Seis. Soc. Am., 66, 1485-1499, 1976.

Butler, R., G. S. Stewart, and H. Kanamori, The July 27, 1976 Tangshan, China earthquake - A complex sequence of intraplate events, Bull. Seis. Soc. Am., 69, 207-220, 1979.

Chase, C. G., Plate kinematics: The Americas, East Africa, and the rest of the world, Earth Planet. Sci. Lett., 37, 355-368, 1978.

Chen, W-P., and P. Molnar, Seismic moments of major earthquakes and the average rate of slip in Central Asia, J. Geophys. Res., 82, 2945-2969, 1977.

Chinnery, M. A., Theoretical fault models, Pub. Dominion Observatory, Ottawa, 37, 211-223, 1969a.

Chinnery, M. A., Earthquake magnitude and source parameters, Bull. Seis. Soc. Am., 59, 1969-1982, 1969b.

Clark, M. M., Surface rupture along the Coyote Creek fault, The Borrego Mountain Earthquake of April 9, 1968, Geol. Surv. Prof. Paper 787, 55-86, 1972.

Das, S., and K. Aki, Fault plane with barriers: A versatile earthquake model, J. Geophys. Res., 82, 5658-5670, 1977.

Das, S., and C. H. Scholz, Theory of time-dependent rupture in the Earth, submitted to J. Geophys. Res., 1980.

Das, S., Three-dimensional spontaneous rupture propagation and implications for earthquake source mechanism, submitted to Geophys. J. R. astr. Soc., 1981.

Davies, J., and L. House, Aleutian subduction zone seismicity, volcano-trench separation and their relation to great thrust-type earthquakes, J. Geophys. Res., 84, 4583-4591, 1979.

Davies, J., L. Sykes, L. House, and K. Jacob, Shumagin seismic gap, Alaska peninsula: History of great earthquakes, tectonic setting, and evidence for high seismic potential, submitted to J. Geophys. Res., 1980.

Fedotov, S. A., Regularities of the distribution of strong earthquakes of Kamchatka, the Kurile Islands, and northeastern Japan, Trudy Inst. Fiziki Zemli, Acad. Nauk. SSSR, 36 (203), 66-93, (English translation by S. Ward, 1966), 1965.

Fitch, T. J., and C. H. Scholz, Mechanism of underthrusting in southwest Japan: A model of convergent plate interactions, J. Geophys. Res., 76, 7260-7292, 1971.

Fukao, Y., Tsunami earthquakes and subduction processes near deep-sea trenches, J. Geophys. Res., 84, 2303-2314, 1979.

Fukao, Y., and M. Furumoto, Stress drops, wave spectra and recurrence intervals of great earth-quakes - implications of the Etorofu earthquake of 1958 November 6, Geophys. J. R. astr. Soc., 57, 23-40, 1979.

Guest, P. G., Numerical Methods of Curve Fitting, Cambridge Univ. Press, Cambridge, England, 422 p., 1961.

Hamilton, R. M., Aftershocks of the Borrego Mountain earthquake from April 12 to June 12, 1968, The Borrego Mountain Earthquake of April 9, 1968, Geol. Surv. Prof. Paper 787, 24-30, 1972.

Hanks, T. C., and M. Wyss, The use of body-wave spectra in the determination of seismic-source parameters, Bull. Seism. Soc. Am., 62, 561-589, 1972.

Harkrider, D. G., Surface waves in multilayered elastic media. Part II. Higher mode spectra and spectral ratios from point sources in plane layered earth models, Bull. Seismol. Soc. Amer., 60, 1937-1987, 1970.

Ichikawa, M., Spatial and temporal distributions of aftershocks, in General Report on the Tokachi-oki Earthquake of 1968, Z. Suzuki (ed.), 67-83, 1971.

Isacks, B., J. Oliver, and L. R. Sykes, Seismology and the new global tectonics, J. Geophys. Res., 73, 5855-5899, 1968.

Jordan, J. N., J. F. Lander, and R. A. Black, Aftershocks of the 4 February 1965 Rat Island earthquake, Science, 148, 1323-1325, 1965.

Kanamori, H., Synthesis of long-period surface waves and its application to earthquake source studies - Kurile Islands earthquake of October 13, 1963, J. Geophys. Res., 75, 5011-5025, 1970a.

Kanamori, H., The Alaskan earthquake of 1964: Radiation of long-period surface waves and source mechanisms, J. Geophys. Res., 75, 5029-5040, 1970b.

Kanamori, H., Focal mechanism of the Tokachi-Oki earthquake of May 16, 1968: Contortion of the lithosphere at a junction of two trenches, Tectonophysics, 12, 1-13, 1971.

Kanamori, H., Mode of strain release associated with major earthquakes in Japan, Ann. Rev. Earth Planet. Sciences, 1, 213-239, 1973.

Kanamori, H., Re-examination of the earth's free oscillations excited by the Kamchatka earth-quake of November 4, 1952, Phys. Earth Planet. Int., 11, 216-226, 1976.

Kanamori, H., Seismic and aseismic slip along subduction zones and their tectonic implica-tions, Island Arcs, Deep Sea Trenches, and Back-Arc Basins, Maurice Ewing Series, 1, (American Geophysical Union), 163-174, 1977a.

Kanamori, H., The energy release in great earth-quakes, J. Geophys. Res., 82, 2981-2987, 1977b.

Kanamori, H., and J. J. Cipar, Focal process of the great Chilean earthquake, May 22, 1960, Phys. Earth Planet. Int., 9, 128-136, 1974.

Kanamori, H., and D. L. Anderson, Theoretical basis of some empirical relations in seismology, Bull. Seis. Soc. Am., 65, 1073-1095, 1975.

Kanamori, H., and G. S. Stewart, Mode of strain

release along the Gibb's fracture zone, mid-Atlantic ridge, Phys. Earth Planet. Int., 11, 312-332, 1976.

Kanamori, H., and G. S. Stewart, Seismological aspects of the Guatemala earthquake of February 4, 1976, J. Geophys. Res., 83, 3427-3434, 1978.

Kelleher, J. A., Space-time seismicity of the Alaska-Aleutian seismic zone, J. Geophys. Res., 75, 5745, 1970.

Kelleher, J., Rupture zones of large South American earthquakes and some predictions, J. Geophys. Res., 77, 2087-2103, 1972.

Kelleher, J., L. Sykes, and J. Oliver, Possible criteria for predicting earthquake locations and their application to major plate boundaries of the Pacific and Caribbean, J. Geophys. Res., 78, 2547-2585, 1973.

Kelleher, J., J. Savino, H. Rowlett, and W. McCann, Why and where great thrust earthquakes occur along island arcs, J. Geophys. Res., 79, 4889-4899, 1974.

Kelleher, J., and W. R. McCann, Buoyant zones, great earthquakes, and unstable boundaries of subduction, J. Geophys. Res., 81, 4885-4896, 1976.

King, C-Y., and L. Knopoff, Stress drop in earthquakes, Bull. Seis. Soc. Am., 58, 249-257, 1968.

Kondorskaya, N. V., N. V. Shebalin, and E. A. Khrometskaya, eds., New Catalog of Strong Earthquakes in the Territory of the U.S.S.R. from Ancient Times to 1975, 1977.

Lindh, G., and D. Boore, Control of rupture by fault geometry during the 1966 Parkfield earthquake, Bull. Seismol. Soc. Am., in press, 1980.

Lomnitz, C., Major earthquakes and tsunamis in Chile during the period 1535 to 1955, Geolog. Rundschau, 59, 938-960, 1970.

Mansinha, L., and D. E. Smylie, The displacement fields of inclined faults, Bull. Seis. Soc. Am., 61, 1433-1440, 1971.

McCann, W. R., S. P. Nishenko, L. R. Sykes, and J. Krause, Seismic gaps and plate tectonics: Seismic potential for major plate boundries, Proc. Conf. VI: Methodology for Identifying Seismic Gaps and Soon-to-Break Gaps, U. S. Geol. Surv. Open-File Report 78-943, 441-584, 1978.

McCann, W. R., S. P. Nishenko, L. R. Sykes, and J. Krause, Seismic gaps and plate tectonics: Seismic potential for major boundaries, Pure Appl. Geophys., 117, 1082-1147, 1979.

McNally, K., and J. B. Minster, Non-uniform seismic slip rates along the Middle America trench, submitted to J. Geophys. Res., 1980.

Minster, J. B., and T. H. Jordan, Present-day plate motions, J. Geophys. Res., 83, 5331-5354, 1978.

Mogi, K., Some features of recent seismic activity in and near Japan, I, Bull. Earthquake Res. Inst., Tokyo Univ., 46, 1225-1236, 1968a.

Mogi, K., Development of aftershock areas of great earthquakes, Bull. Earthquake Res. Inst., Tokyo Univ., 46, 175-203, 1968b.

Molnar, P., Earthquake recurrence intervals and plate tectonics, Bull. Seis. Soc. Am., 69, 115-133, 1979.

Okal, E. A., A surface wave investigation of the Gobi-Altai (December 4, 1957) earthquake, Phys. Earth Planet. Int., 12, 319-328, 1976.

Plafker, G., Alaskan earthquake of 1964 and Chilean earthquake of 1960: Implications for arc tectonics, J. Geophys. Res., 77, 901-925, 1972.

Plafker, G., and J. C. Savage, Mechanism of the Chilean earthquakes of May 21 and 22, 1960, Geol. Soc. Am. Bull., 81, 1001-1030, 1970.

Plafker, G., and M. Rubin, Uplift history and earthquake recurrence as deduced from marine terraces on Middelton Island, Alaska, Proc. Conf. VI: Methodology for Identifying Seismic Gaps and Soon-to-Break Gaps, U. S. Geol. Surv. Open-File Report 78-943, 687-722, 1978.

Plafker, G., T. Hudson, T. Bruns, and M. Rubin, Late Quaternary offsets along the Fairweather fault and crustal plate interactions in southern Alaska, Can. J. Earth Sci., 15, 805-816, 1978.

Reid, H. F., The California Earthquake of April 18, 1906, II, The Mechanics of the Earthquake, Carnegie Inst. Washington, Washington, D.C., 31-32, 1910.

Reyes, A., J. N. Brune, and C. Lomnitz, Source mechanism and aftershock study of the Colima, Mexico earthquake of January 10, 1973, Bull. Seis. Soc. Am., 69, 1819-1840, 1979.

Rikitake, T., Recurrence of great earthquakes at subduction zones, Tectonophysics, 35, 335-362, 1976.

Savage, J. C., and R. O. Burford, Geodetic determination of relative plate motion in central California, J. Geophys. Res., 78, 832-845, 1973.

Scholz, C. H., M. Wyss, and S. W. Smith, Seismic and aseismic slip on the San Andreas fault, J. Geophys. Res., 74, 2049-2069, 1969.

Seno, T., The instantaneous rotation vector of the Philippine Sea plate relative to the Eurasian plate, Tectonophysics, 42, 209-226, 1977.

Sharp, R. V., Surface displacement on the Imperial fault and Brawley fault zone associated with the October 15, 1979 Imperial Valley earthquake (abstract), Earthquake Notes, 50, 37, 1979.

Shimazaki, K., Nemuro-oki earthquake of June 17, 1973: A lithospheric rebound at the upper half of the interface, Phys. Earth Planet. Int., 9, 314-327, 1974.

Shimazaki, K., and T. Nakata, Time-predictable recurrence model for large earthquakes, Geophys. Res. Lett., 7, 279-282, 1980.

Sieh, K. E., A study of Holocene displacement history along the south-central reach of the San Andreas fault, Ph.D. thesis, Stanford University, 1977.

Sieh, K. E., Slip along the San Andreas fault associated with the great 1857 earthquake, Bull. Seis. Soc. Am., 68, 1421-1448, 1978a.

Sieh, K., Prehistoric large earthquakes produced by slip on the San Andreas fault at Pallett

Creek, California, J. Geophys. Res., 83, 3907-3948, 1978b.

Sieh, K., Central California foreshocks of the great 1857 earthquake, Bull. Seismol. Soc. Am., 68, 1731-1749, 1978c.

Singh, S. K., L. Astiz, and J. Havskov, Seismic gaps and recurrence periods of large earthquakes along the Mexican subduction zone: A re-examination, submitted to Bull. Seis. Soc. Am., 1980a.

Singh, S. K., J. Yamamoto, J. Havskov, M. Guzman, D. Novelo, and R. Castro, Seismic gap of Michoacan, Mexico, Geophys. Res. Lett., 7, 69-72, 1980b.

Singh, S. K., J. Havskov, K. McNally, L. Ponce, T. Hearn, and M. Vassiliov, The Oaxaca, Mexico, earthquakes of 29 November 1978: A preliminary report on aftershocks, Science, 207, 1211-1213, 1980c.

Spence, W., The Aleutian arc: Tectonic blocks, episodic subduction, strain diffusion, and magma generation, J. Geophys. Res., 82, 213-230, 1977.

Stauder, W., Mechanism of the Rat Island earthquake sequence of February 4, 1965, with relation to island arcs and sea-floor spreading, J. Geophys. Res., 73, 3847-3858, 1968.

Sykes, L. R., Aftershock zones of great earthquakes, seismicity gaps, and earthquake prediction for Alaska and the Aleutians, J. Geophys. Res., 76, 8021-8041, 1971.

Sykes, L. R., J. B. Kisslinger, L. House, J. Davies, and K. H. Jacob, Rupture zones of great earthquakes in the Alaska-Aleutian arc, 1784 to 1980, Science, 210, 1343-1345, 1980.

Sykes, L. R., J. B. Kisslinger, L. House, J. Davies, and K.H. Jacob, Rupture zones and repeat times of great earthquakes along the Alaska-Aleutian arc, 1784-1980, Earthquake Prediction, Maurice Ewing Series, 4, (American Geophysical Union), in press, 1981.

Thatcher, W., Strain accumulation and release mechanism of the 1906 San Francisco earthquake, J. Geophys. Res., 80, 4862-4872, 1975.

Thatcher, W., Systematic inversion of geodetic data in central California, J. Geophys. Res., 84, 2283-2295, 1979a.

Thatcher, W., Horizontal crustal deformtion from historic geodetic measurements in southern California, J. Geophys. Res., 84, 2351-2370, 1979b.

Tobin, D. G., and L. R. Sykes, Seismicity and tectonics of the northeast Pacific Ocean, J. Geophys. Res., 73, 3821-3845, 1968.

Trifunac, M. D., Tectonic stress and the source mechanism of the Imperial valley, California, earthquake of 1940, Bull. Seism. Soc. Am., 62, 1283-1302, 1972.

Utsu, T., Large earthquakes near Hokkaido and the expectancy of the occurrence of a large earthquake off Nemuro, Rept. Coord. Comm. Earthquake Prediction, Geogr. Surv. Inst. Japan, 7, 7-13, (in Japanese), 1972.

Wu, F., and H. Kanamori, Source mechanism of February 4, 1965, Rat Island earthquake, J. Geophys. Res., 78, 6082-6092, 1973.

EARTHQUAKE PREDICTION IN THE INTERMOUNTAIN SEISMIC BELT--
AN INTRAPLATE EXTENSIONAL REGIME

Walter J. Arabasz and Robert B. Smith

Department of Geology and Geophysics, University of Utah, Salt Lake City, Utah 84112

Abstract. The Intermountain seismic belt (ISB)
is characterized by late Quaternary normal faulting, diffuse shallow seismicity, and episodic
scarp-forming earthquakes ($6\frac{1}{2} \leq M \leq 7\frac{1}{2}$) associated
with the complex interaction of subplates within
the western North American plate. Earthquake prediction studies in this area provide a useful complement to similar studies along active plate
margins and in areas of intraplate compression. A
master catalog of earthquakes since 1850 in the
southern half of the ISB has been systematically
examined for foreshock occurrence, precursory
quiescence and clustering, migration of mainshocks, and seismicity gaps. Precursory quie-
science is common and quiescent periods before
the largest historical mainshocks appear to scale
in duration with mainshock size. Well-documented
studies indicate precursory quiescence and clustering before the 1975 Pocatello Valley earthquake (M_L=6.0) and monotonically increasing seismicity before four of its late aftershocks ($M_L \leq$
3.6). Since 1967 there has been a 300 km by
100 km N-S trending seismicity gap ($M_L \geq 3.5$) along
the main axis of the southern ISB that is anomalous with respect to long-term rates of seismic
energy release. This same zone envelops two 70-
km-long microseismicity gaps along the Wasatch
fault zone that coincide with segments that have
ruptured repeatedly in Holocene but not historical
time. Data are inadequate to conclude whether any
of these seismicity gaps are precursory. Extensive quarry-blast monitoring throughout the
Wasatch Front area since 1974 has revealed no
significant velocity variations with time; no
shocks greater than M_L=3.6, however, have occurred
close to any ray path. Problems for earthquake
prediction in the ISB include: (1) a lack of
understanding of the seismic cycle on major normal
faults; (2) relatively low seismic flux, which
makes the detection of precursory seismicity
changes difficult; and (3) uncertainty in the
mechanism of normal faulting, which may be listric
(i.e., flattening with depth) -- causing variations in dynamic source parameters not now
accounted for in models of normal faulting.

Indroduction

The Intermountain seismic belt (ISB) is an
extensive zone of intraplate deformation within
the western United States (Figure 1) that is characterized by late Quaternary normal faulting,
diffuse shallow seismicity, and episodic scarp-
forming earthquakes ($6\frac{1}{2} \leq M \leq 7\frac{1}{2}$) associated with the
complex interaction of subplates within the west-
ern North American plate [e.g., Smith and Sbar,
1974; Smith, 1978]. Because of the intraplate
setting, the predominantly extensional stress
regime, and relatively low strain rates, earthquake prediction studies in this area provide a
useful complement to similar studies along active
plate margins and in areas of intraplate compression. Our objectives here are: (1) to characterize the ISB and assess inherent problems for predicting major earthquakes in such an area, and
(2) to present observations from preliminary
earthquake prediction studies in the Utah region
as probably representative of the ISB.

Characteristics of ISB

Detailed descriptions of the regional seismicity and contemporary tectonics of the ISB are
summarized by Smith [1978] for the ISB as a whole,
by Arabasz et al. [1980] for the Wasatch Front
area of north-central Utah, and in a summary volume edited by Arabasz et al. [1979] for the Utah
region as a whole. Some notable features of the
ISB are: (1) its length (>1300 km) and breadth
(100-200 km); (2) its segmentation into several
sectors with divergent trends; (3) the diffuse-
ness of seismicity, with focal depths almost
exclusively shallower than 15-20 km, and weak correlation with major active faults (based on dense-
network monitoring with both portable and fixed
microearthquake networks); (4) a general predominance of normal faulting reflecting an extensional
stress regime, but with spatially rapid changes in
stress orientation in some areas; (5) the common
occurrence of swarm sequences at various localities throughout the ISB; (6) relatively low rates

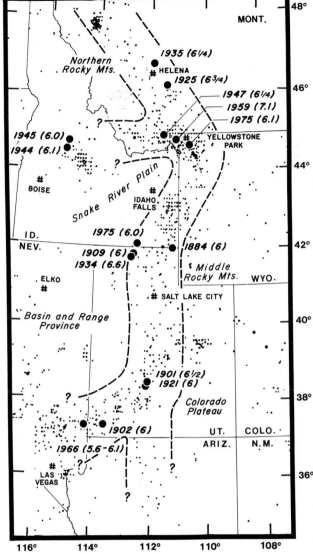

Fig. 1. Index map of Intermountain seismic
belt. Epicenters of historical mainshocks
(M≥6.0) shown as large circles, NOAA epicenters through 1974 as smaller circles.
Schematic dashed outline of seismic belt
based on various seismotectonic studies as
well as depicted seismicity. Magnitudes (in
parentheses) are estimated or measured
values of M_L, except values for 1925-1959
(attributed to Pasadena), which are either
m_b or M_s. Epicenter of 1966 shock (M_L5.6,
m_b6.1) added for reference.

of crustal strain ($\sim 10^{-8}$/yr or less) compared to
those at active plate boundaries; (7) moderate
background seismic flux, which, for comparison, is
lower by a factor of 4-6 than that in the California-Nevada seismic zone [Algermissen, 1969];
(8) apparent stresses that are within the range of

values (0.01-100 bar) computed similarly for
intraplate earthquakes elsewhere [Doser, 1980];
(9) relatively long (>1000 yr) recurrence intervals for surface faulting (excepting parts of the
Wasatch fault) [Swan et al., 1980; Wallace, 1980];
and (10) a historical paucity of large (M>7.0)
surface-faulting earthquakes -- despite abundant
late Quaternary and Holocene fault scarps throughout the ISB [e.g., Bucknam et al., 1980; Swan et
al., 1980; Witkind, 1975].

Along most of its length, the ISB follows the
boundary between thin weak crust and lithosphere
of the Basin and Range Province and thicker more
stable crust and lithosphere of the Rocky Mountains and the Colorado Plateau [e.g., Smith, 1978;
Thompson and Zoback, 1979]. Significantly, the
ISB coincides with a persistent zone of weakness
in the North American continent along which differential crustal movement occurred through the
Paleozoic and early Mesozoic, and which also defined where eastward-moving thrust plates broke to
the surface during late Cretaceous-early Tertiary
time [Stokes, 1976]. We return to this point
later because these preexisting thrust faults
appear to have had a fundamental influence on
Cenozoic normal faulting.

Contemporary Deformation

Contemporary W- to NW-oriented horizontal extension characterizes most of the ISB as well as
the Great Basin itself, apparently reflecting
sources of stress both beneath and at the boundaries of the involved subplates [Smith, 1978;
Zoback and Zoback, 1980]. Recent estimates of
rates of crustal deformation in parts of the ISB
include: (1) uplift rates of the Wasatch Mountains
east of the Wasatch fault from fission-track dating of 0.012 mm/yr, for the period 90 m.y. to 10
m.y., and 0.4 mm/yr for 10 m.y. to present [Naser
et al., 1980]; (2) Holocene rates of dip-slip on
segments of the Wasatch fault of 1-2 mm/yr from
geological trenching studies [Swan et al., 1980];
and (3) estimates of extensional strain rates
(from seismic moment rates) across the southern
ISB between about 36°N and 42°N that range from
2-7 x 10^{-9}/yr [Doser, 1980] to 1-5 x 10^{-8}/yr
[Greensfelder et al., 1980] -- a range compatible
with geological rates of Late Quaternary extension.

With regard to geodetically-measured horizontal
strain, Prescott et al. [1979] have reported data
for 1972 to 1978 that show an "apparently reversed
strain field" north of Salt Lake City. Their data
imply ENE-WSW compression of 0.17±0.05 μstrain/yr
across the Wasatch fault near Ogden, rather than
general E-W extension expected from normal faulting and exhibited by focal mechanisms in the area.
A regional strain tensor computed by Doser [1980]
from 18 earthquakes in the Utah region indicates
a maximum principal strain component oriented
N79°E that is extensional and approximately equal
to the amount of near-vertical compressional
strain.

Seismicity and Recurrence

Since 1850 at least 15 independent earthquakes (i.e., mainshocks or the largest event of a swarm) within the ISB have had an estimated magnitude of 6.0 or greater (Figure 1). There are two documented cases of historical surface fault displacement. Normal fault scarps with maximum surface displacements of 6.7 m and 0.5 m, respectively, were produced by the M=7.1 Hebgen Lake, Montana, earthquake of August 1959 and by the M=6.6 Hansel Valley (Kosmo), Utah, earthquake of March 1934. Regional instrumental monitoring of earthquakes in the ISB dates from about 1950, systematic computerized locations from the early 1960's, and monitoring with telemetered seismic networks since the early to mid-1970's.

As apparent in Figure 1, the ISB comprises several divergent segments, and many of the largest historical shocks have tended to occur where there are pronounced changes in trend of the ISB. North of 42°N, seismicity appears to "wrap around" the Snake River Plain, perhaps reflecting a stress field influenced by the development of a NE-propagating volcanogenic zone along a lithospheric zone of weakness [Smith et al., 1977; Furlong, 1979]. Yellowstone Park, at the apex of the Snake River Plain, marks the apparent convergence of three seismic zones trending E-W, NE-SW, and NW-SE. At about 38°N there is a prominent southward divergence of the ISB. Seismicity chiefly extends southwestward into southern Nevada, but there is a weak continuation southward into northern Arizona.

The epicenter maps in Figure 2 illustrate the problem of correlating accurately-located earthquakes with active faults in the ISB. Figure 2a shows a 5.5-yr sample of earthquakes in the Wasatch Front area located with a 43-station telemetry network [Arabasz et al., 1980]. The most striking feature of Figure 2a is the paucity of small-magnitude earthquakes along most of the 370-km long Wasatch fault, despite the fact that this major normal fault zone has been the most active locus of surface faulting in the eastern Great Basin during Holocene time [Bucknam et al., 1980; Wallace, 1980]. The seismicity pattern shown in Figure 2a, including anomalously low seismicity along the Wasatch fault, diffuse seismicity throughout the Wasatch Front area, and abundant microseismicity a few tens of kilometers east and west of the Wasatch fault has been generally observed since systematic instrumental monitoring began in the area in 1962 [Arabasz et al., 1980]. Another major normal fault shown in Figure 2a is the East Cache fault that forms the eastern boundary of the Cache Valley graben. Abundant seismicity occurs to the east of the fault, beneath the Bear River Range, precluding correlation with the East Cache fault itself. Figure 2c shows the results of extensive microearthquake monitoring by the University of Utah in southeastern Idaho and western Wyoming. Here too the predominance of earthquake locations away from the traces of major active faults such as

the Teton fault and the bounding faults of Swan Valley, Star Valley, and Bear Lake Valley, is striking. Microearthquake monitoring during the 1970's in the vicinity of the Red Canyon and Hebgen faults that ruptured during the 1959 Hebgen Lake earthquake sequence shows a similar result (Figure 2b).

There is accumulating evidence that the repeat time (at a particular location) for large displacement events on faults in the ISB is significantly longer than 1000 years [e.g., Wallace, 1980], with the apparent exception of the Wasatch fault. Swan et al. [1980] demonstrate that the two most recent scarp-forming earthquakes on the Wasatch fault near Kaysville, Utah, occurred within the past 1580±150 years; further, they estimate that the recurrence interval for a magnitude 6½ to 7½ earthquake along the entire Wasatch fault zone may be 50 to 430 years. A maximum magnitude of ∼7½ is generally assumed from the historical experience of Basin and Range faulting in the western United States. Seismic moment rates predict a recurrence interval for an earthquake of magnitude 7.0 to 7.5 somewhere in the Wasatch Front area (Figure 2a) as short as 120 to 280 years [Doser, 1980]. Extrapolating the historical seismicity leads to a recurrence interval of 200 years for the same area and magnitude range [Arabasz et al., 1980].

Earthquake Prediction-Related Observations

Three areas of earthquake prediction research that have been most actively pursued in the Utah region include: (1) long-term monitoring of local quarry blasts for temporal variation in P-wave velocities before local earthquakes, (2) analysis of space-time patterns of Utah seismicity for general forecasting, and (3) rheological modeling of crustal flexure (and predicted stress fields) associated with major normal faulting -- as a means of discerning stages of a seismic cycle on segments of the Wasatch fault zone. We will briefly summarize topic (1) and elaborate on topic (2); topic (3) is discussed by Zandt and Owens [1980].

Temporal Monitoring of Seismic Velocities

The feasibility of detecting temporal changes in P-wave velocities along the Wasatch Front has been investigated by Smith et al. [1980]. Quarry blasts averaging 40,000 lb of explosives at the Kennecott Bingham Canyon copper mine, 25 km SW of Salt Lake City, were used as timed sources. From October 1974 through June 1978, several thousand P-wave arrival measurements were made at 24 selected stations of the Wasatch Front seismic network. Standard deviations of source-corrected travel times averaged ±0.1 s (with that at only one station larger than ±0.14 s). Computed velocities showed remarkable stability for the observation period, and no significant variations in velocity greater than one standard deviation,

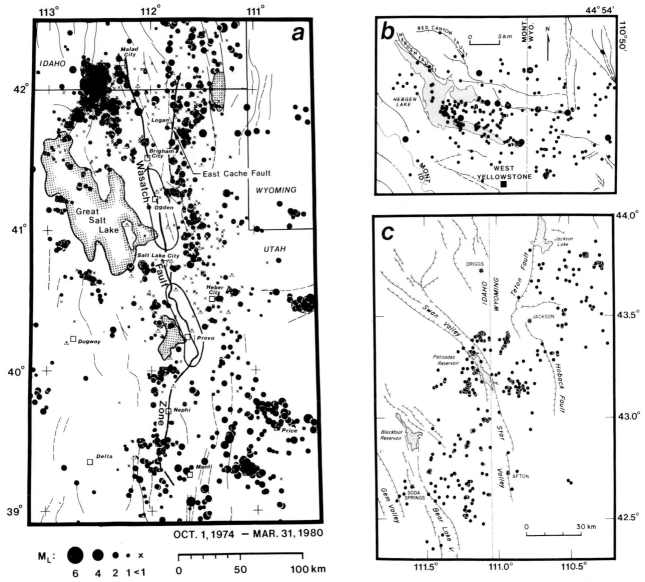

Fig. 2. Diffuse upper-crustal seismicity in the Intermountain seismic belt showing lack of strong correlation with major active faults in (a) the Wasatch Front area [Arabasz et al., 1980], (b) the Hebgen Lake area [Smith et al., 1977], and (c) the Idaho-Wyoming border area [Bones and Arabasz, 1978; Smith et al., 1976]. Earthquakes in (b) and (c) are of $M_L \geq 3.0$. Large and small circles in (b) indicate focal depths greater than and less than 10 km, respectively. Elliptical areas in (a) outline seismicity gaps discussed in text.

which ranged from 1 to 4 percent, were interpreted as meaningful. However, during the observation period no earthquakes larger than $M_L = 3.6$ occurred close to any sampled ray path. Short-term (≤ 60 d) velocity anomalies were identified at 14 stations, but none correlated convincingly with earthquake activity. Measurable velocity variations with time were not found in two cases of special interest: (1) for several sampled ray paths that passed through the hypocentral region of an earthquake swarm ($M_L \leq 3.3$) near Magna, Utah, less

than 20 km from the Bingham Canyon copper mine; and (2) for ray paths through a source region in Pocatello Valley in which seismicity changes occurred before late aftershocks ($M_L \leq 3.6$) of an earthquake of $M_L = 6.0$ (see Figure 4). Monitoring local blasts, even on daily to weekly basis, is not encouraging for the prediction of small earthquakes ($M_L \leq 3.5$) in the Wasatch Front area. However, a meaningful test of this prediction technique for moderate to large earthquakes in the ISB has yet to be made.

Space-Time Seismicity

Major efforts were recently made at the University of Utah to compile a master catalog of all locatable earthquakes in the Utah region (Figure 3) since 1850 [see Arabasz et al., 1979]. The catalog comprises revised historical and instrumental data sets corresponding to periods before and after July 1962, respectively, and now contains more than 5000 earthquakes, including 413 in the historical data set. The catalog has been systematically examined for foreshock occurrence, precursory quiescence and clustering, migration of mainshocks and seismicity gaps [Arabasz and Griscom, 1978; Griscom and Arabasz, 1979; Griscom, 1980]. Figure 3 gives an overview of cumulative seismicity in the defined region; the earthquake record is most complete for the elongate polygonal subarea, designated the sample area, along the main axis of the seismic belt.

Foreshocks. Prominent foreshocks preceded two magnitude 6 earthquakes near Elsinore, Utah, in 1921 [Pack, 1921] and the 1975 Pocatello Valley earthquake of M_L=6.0 near the Idaho-Utah border [Arabasz et al., 1981]. From a systematic study of foreshocks in the Utah region, Griscom [1980] found the following: (1) Defining foreshocks to be antecedent earthquakes of smaller size that occur within 40 days [Jones and Molnar, 1979] and 50 km of a mainshock, identifiable foreshocks have preceded 12% of all historical mainshocks of Modified Mercalli epicentral intensity (I_o) IV or greater, and 43% of all instrumental mainshocks of M_L≥3.5 in the Utah region. (2) Observed foreshock activity is variable and does not simply

characterize any definable geographic region. (3) Examining prior seismicity within 100 days and 50 km of all mainshocks of M_L≥3.5 in the Utah region, the probability of earthquake occurrence is roughly constant for $10<t_p<100$, where t_p is the time preceding the mainshock in days, and increases by about a factor of 2 for $t_p<10$ (peaking during the last 6 hours). (4) The percentage of foreshocks located within 15 km of the mainshock epicenter increases as t_p decreases.

For the best locally recorded foreshock sequence to date, that for the 1975 Pocatello Valley earthquake, an accelerated increase in rate of foreshock occurrence postulated by Jones and Molnar [1979] was not observed [Arabasz et al., 1981]. A foreshock of M_L=4.2 occurred abruptly 22 hours before the mainshock of M_L=6.0, followed by 156 smaller shocks that behaved as aftershocks of the M_L=4.2 event. Systematic increases in frequency of occurrence were observed by Arabasz and Richins [1976] before four late aftershocks (M_L=3.4-3.6) occurring 4.7 to 14.5 months after the Pocatello Valley mainshock. Figure 4 shows their data, which consists of counts of earthquakes (S-P<2.5 s), complete above a magnitude threshold for each case, for overlapping 12-hr time windows with 4 counts per day. The increase in rate of occurrence before each culminating event is roughly linear.

Precursory Quiescence and Clustering. Arabasz and Griscom [1978] observed that before the largest historical earthquakes in the Utah region there were unusually long periods marked by the absence of felt shocks in the epicentral area, based upon records at close towns that had a long earlier history of reporting small shocks. Hundreds of small towns were distributed throughout the active earthquake belt in Utah by the 1870's [see Griscom, 1980]. To test an hypothesis of precursory quiescence, the following simple experiment was devised. For all pre-instrumental mainshocks of I_o≥V in the sample area (Figure 3) during 1900-1959, t_p was measured for the first preceding felt shock located within 75 km of the mainshock. Immediate foreshocks ($t_p<40$ days) were excluded. Also, data points were rejected for which t_p was greater than 2/3 the interval time to the last earthquake of equal or greater size. There are obvious uncertainties in the completeness and accuracy of the earthquake locations; however, the catalog for the selected time period contains 224 small shocks (I_o≥IV) whose locations are fairly well constrained by felt reports. Figure 5 shows the measured values of t_p for cutoff distances of 50 km and 75 km plotted as a function of mainshock size. In terms of models of precursory quiescence and/or clustering [e.g., Wyss et al., 1978], t_p here might reflect the timing either of a precursory burst of seismicity or the onset of a decrease in seismicity.

The results (Figure 5) shows the tendency for t_p to <u>increase</u> with size of mainshock. As a check, the observations were tested by assuming

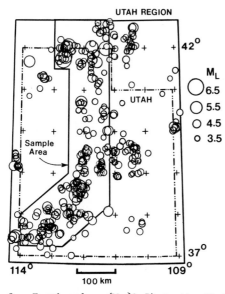

Fig. 3. Earthquakes (M_L≥3.5) in the Utah region, 1850-1979. Pre-instrumental shocks have Modified Mercalli epicentral intensities of IV or greater.

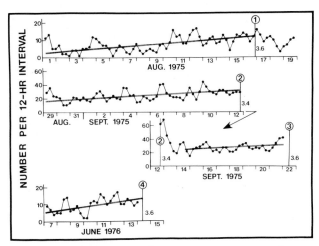

Fig. 4. Frequency of occurrence versus time for time samples preceding four late aftershocks (M_L=3.4-3.6) of the March 28, 1975 Pocatello Valley earthquake (M_L=6.0). Positive linear trends have correlation coefficients of 0.7, 0.7, 0.3, and 0.5 for cases 1 to 4, respectively. [From Arabasz and Richins, 1976].

that the occurrence of felt earthquakes was a Poisson arrival process with mean rate of occurrence λ. From historical data summarized in Arabasz et al. [1979], λ for 210 independent felt earthquakes in the sample area, 1870 to 1959, is 2.6 x 10^{-2} per year per 1000 km^2. (Average recurrence for the Wasatch Front area predicts λ = 2.3 earthquakes ($I_0 \geq IV$) per year per 1000 km^2.) The probability of no felt earthquake in any interval of time or length t is simply $e^{-\lambda t}$, for which values corresponding to circular areas of radius 50 km and 75 km, respectively, are indicated in Figure 5. For the 50-km-cutoff area, for example, it can be seen that all seven data points for earthquakes of $I_0 \geq VII$ have values of t_p corresponding to probabilities less than 0.5, and two have corresponding probabilities less than 0.15. An estimate of the (1-α) 100 percent upper confidence interval on the proportion of interval times greater than t_0 can be made [e.g., Benjamin and Cornell, 1970] as exp $[-\lambda t_0/(1+k_\alpha\sqrt{1/n})]$, where k_α is that value of a standard normal variable that will be exceeded with probability α, and n is the sample size. Noting for the 50-km case that all seven data points for $I_0 \geq VII$ have t_p>1350 days, we can state with 95 percent confidence that the expected proportion should be less than 0.51 for a Poisson process; with identical confidence, the proportion of interval times greater than 750 days for the 75-km case should be less than 0.40, rather than the observed 0.86 for $I_0 \geq VII$. If t_p in Figure 5 does indeed reflect a precursory time interval, and one that scales with mainshock magnitude, it appears that precursory time intervals for seismicity changes in the ISB may be

significantly longer than equivalent times observed elsewhere [Kristy and Simpson, 1980; Wyss et al., 1978], which approximately follow the scale of Scholz et al. [1973] shown in Figure 5 for reference.

For post-July 1962 instrumental recording, seismicity prior to the M_L=6.0 Pocatello Valley earthquake forms the most complete data set in the Utah catalog and shows evident changes (Figures 6 and 7). Figure 6 shows a space-time plot of seismicity within 100 km epicentral distance (Δ) of the March 28, 1975 mainshock. For Δ<50 km, a period of relative quiescence began 6.4 years (t_p= 2327 d) before the mainshock and lasted for 4.4 years until April 14, 1973 (t_p = 712 d) when an earthquake of M_L=4.2 occurred at Δ = 9 km. During the following 200 days, small-magnitude shocks clustered at $\Delta \leq 25$ km. During the 450 days preceding the mainshock there is a weak suggestion of a few small shocks migrating toward the mainshock focus. Recall that a foreshock of M_L= 4.2 (Δ = 1 km) occurred 22 hours before the mainshock. Epicentral patterns corresponding to periods of interest in Figure 6 are shown in Figure 7. A general precursory model involving a significant decrease in background seismicity, interrupted by a burst of earthquakes clustered in the source region, is a reasonable one for the Pocatello Valley earthquake. Let us examine the prior seismicity more critically, again by assuming a Poisson process.

The circular area of 50-km radius centered on the Pocatello Valley earthquake experienced 12 independent earthquakes between July 1, 1962 and the time of the March 28, 1975 mainshock (12.75 years) that were of $M_L \geq 2.5$ -- the catalog's threshold for completeness in the Wasatch Front area.

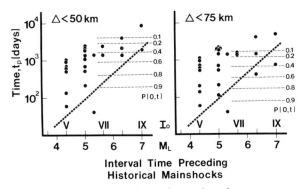

Fig. 5. Interval time of earthquake occurrence, as a function of mainshock magnitude, before all independent mainshocks ($I_0 \geq V$) in the sample area in Figure 3. Time t_p is the interval between a mainshock and the last felt earthquake located within either 50 or 75 km. P[0,t] is the probability of no shock occurring during a time interval corresponding to t_p, assuming a Poisson arrival process (see text). Heavy dashed line indicates precursor time scale of Scholz et al. [1973].

Accordingly, λ = 0.94 events per year. For comparison, the rate predicted from 129 years of seismicity in the Wasatch Front area [Arabasz et al., 1980] is 1.27 events per year. For a Poisson process the probability of the observed 4.4-year period of quiescence would be 0.015, and at the 95 percent confidence level, the proportion of interval times greater than 4.0 years would be less than 0.08. It is noteworthy that the onset of the long quiescence at t_p = 2327 days is consistent with the historical data of Figure 5 for a mainshock of magnitude 6.0, while the onset of the precursory burst of seismicity at t_p = 712 days is close to the time scale of Scholz et al. [1973]. Griscom [1980] has systematically analyzed seismicity prior to other mainshocks in the Utah instrumental catalog and found additional evidence of precursory quiescence and clustering, but inadequate numbers of sampled earthquakes make conclusions tentative.

Seismicity Gaps. Spatial gaps in the distribution of the largest earthquakes in a seismic belt and temporal decreases in seismicity before larger earthquakes now have a widely recognized significance for earthquake prediction. These two cases of "missing" earthquakes have been termed seismic gaps of the first kind and second kind, respectively, by Mogi [1979]. Seismic gaps of the first kind can be recognized in some intraplate areas. Can any be recognized in the ISB? In general, the historical record in the ISB is simply inadequate to define any repetitive cycles of major strain-energy release throughout the belt. There is, however, a 300-km-long segment of the ISB between 38.9°N and 41.5°N that has had a conspicuous absence of historical earthquakes larger than magnitude 5½. This segment, if any, might be considered a gap of the first kind because it encompasses 80 percent of the Wasatch fault zone, which has yet to rupture in historical time. Also, studies of the Utah earthquake catalog have recently called attention to a significant decrease in seismicity along this same segment of the ISB beginning in the late 1960's that might be interpreted as some form of a seismic gap of the second kind [Griscom and Arabasz, 1979; Griscom, 1980]. This is illustrated in Figure 8, a space-time plot of independent mainshocks ($I_o \geq$ IV) within the sample area of Figure 3. The hachured space-time compartment in Figure 8 points out that within a 300 km x 100 km N-S trending zone along the ISB, only one mainshock of $M_L \geq 3.5$ occurred from 1968 through March 1980 -- despite background microseismicity (Figure 2a). And along this same zone, no mainshock larger than intensity VII or about magnitude 5½ has occurred since 1914. In contrast, the sectors of the ISB immediately north and south of this zone have shown more uniform seismic flux since the late 19th century. Moderate-size earthquakes at each end of the zone have occurred roughly every 20-25 years with a tendency to alternate north and south. (Application of the technique of Kasahara [1978] to test

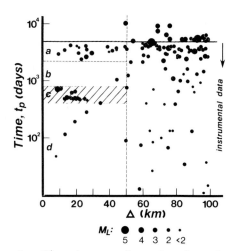

Fig. 6. The time t_p of earthquakes that occurred before the 1975 Pocatello Valley earthquake as a function of epicentral distance Δ from the mainshock. Time periods of interest are keyed to Figure 7.

for migration showed no significant trend for all grouped mainshocks in either the Utah region or the sample area.)

What is the significance of the 300-km-long seismicity gap? From instrumental seismicity recorded since 1962, the mean rate of occurrence of independent mainshocks of $M_L \geq 3.5$ throughout the seismic belt in Utah is ~3 x 10^{-2} events per year per 1000 km^2, which predicts 0.9 shocks per year ($M_L \geq 3.5$) for the hachured space compartment in Figure 8. By any measure, the absence of all but one such event during a 12.25-year period is statistically significant (e.g., the probability assuming a Poisson process is less than 0.01). Figure 8 shows that relative quiescence appears in the same space compartment during 1923-1938, prior to a period of normal seismicity during 1939-1963. The earthquake record is admittedly incomplete prior to 1938, but there probably was a real reduction in seismicity -- based on observations in temporally earlier and spatially adjacent space-time compartments. If true, the 1968-1979 seismicity gap may be one of a series of quiescent periods reflecting progressive stress concentration, and may not itself culminate in a large earthquake [e.g., Ishida and Kanamori, 1980].

The 300-km-long seismicity gap for $M_L \geq 3.5$ envelops two 70-km-long seismicity gaps along the Wasatch Front (see Figure 2a) that have been quiet down to the microearthquake level since at least 1962 [Arabasz et al., 1980]. Evidence of recurrent Holocene surface faulting along these quiet segments of the Wasatch fault argues convincingly for large ($6½ \leq M \leq 7½$) prehistoric earthquakes [Swan et al., 1980] and against long-term aseismicity. Uncertainties discussed by Swan et al. [1980] regarding the timing of the last surface-faulting event and the variation of inter-event times at sites within the microseismicity gaps are such

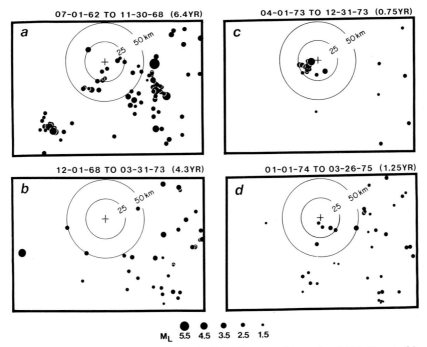

07-01-62 TO 11-30-68 (6.4YR)   04-01-73 TO 12-31-73 (0.75YR)

12-01-68 TO 03-31-73 (4.3YR)   01-01-74 TO 03-26-75 (1.25YR)

M_L 5.5 4.5 3.5 2.5 1.5

Fig. 7. Spatial distribution of seismicity ($M_L \geq 1.5$) preceding the 1975 Pocatello Valley earthquake during consecutive periods of interest outlined in Figure 6.

that one cannot confidently determine whether the corresponding segments of the Wasatch fault are in the initial or later stages of a seismic cycle. Thus the microseismicity gaps are not necessarily gaps of the second kind. Whether or not the anomalous segment of the ISB between 38.9°N and 41.5°N is in a late preparatory stage for a large earthquake remains to be determined. But the space-time seismicity suggests at least that it is one of the most likely candidate areas for future earthquakes of moderate size ($5 \leq M \leq 6\frac{1}{2}$).

Problems for Prediction

No area of the ISB has yet been targeted for any multifaceted earthquake prediction experiment. Consequently, prediction studies to date in the ISB have relied almost exclusively on variations in seismic characteristics, and -- for the foreseeable future -- geodetic data, additional data on ages of faulting, and subsurface structural data from petroleum exploration will likely provide the only supplemental information. With this in mind, let us summarize some significant problems for predicting major earthquakes in the ISB.

Perhaps the foremost problem is a lack of understanding of the seismic cycle on a major normal fault and whether specific segments of active faults are in the initial or later stages of a strain accumulation cycle. This problem is paramount for the Wasatch fault but applies as well to seismically quiet faults with equally spectacular Holocene scarps such as the Teton

fault [e.g., Love and de la Montagne, 1956]. Preliminary data on apparent stress and stress drops in the ISB [Doser, 1980] are consistent with arguments of Raleigh and Evernden [1980] that

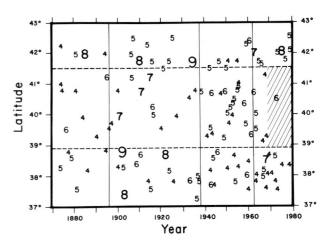

Fig. 8. Space-time plot of earthquakes, in terms of Modified Mercalli epicentral intensity ($I_o \geq 4$), within the sample area of Figure 3 as a function of latitude. Vertical lines are time markers for sample completeness: $I_o \geq 7$ since 1896 and $I_o \geq 6$ since 1938; shocks of $I_o \geq 5$ are complete since 1950, and those of $I_o \geq 4$ or $M_L \geq 3.5$ since 1962 when instrumental monitoring began. Hachured area and dashed lines outline space-time compartments discussed in text.

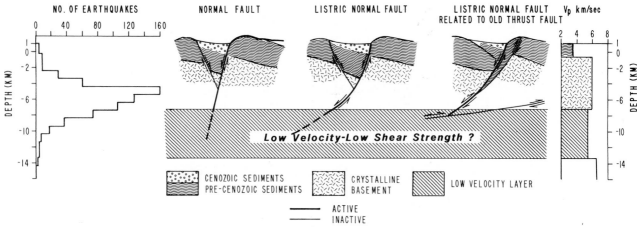

Fig. 9. Idealized models of normal faulting in the Intermountain seismic belt. Histogram of
focal depths from dense-network studies in Pocatello Valley, Idaho-Utah border, and Hansel
Valley, Utah [Arabasz et al., 1980]. Crustal velocity from seismic refraction surveys in central
Utah [Smith et al., 1975].

intraplate earthquakes in the western U.S. should
reflect low mean stress similar to that at the
plate boundary in California. If true, the implications are that precursory changes in rock properties, such as seismic velocity, may be very
localized, and sites of potentially damaging
earthquakes will likely be characterized by low
ambient stress, extensive fracturing, and high
fluid pressure [Raleigh and Evernden, 1980].

We assume that the sites of earthquakes of $M \geq$
7.0 may be reasonably anticipated from the Late
Quaternary geology. It is of course a moot point
whether the sporadic occurrence of large scarp-
forming earthquakes outside the bounds of the main
seismic belt -- such as prehistoric faulting on
the Fish Springs fault in western Utah [Bucknam et
al., 1980] or along the Howe and Arco faults NW of
the Snake River Plain [see Witkind, 1975] -- can
be anticipated unless there is premonitory seismicity. Within the main seismic belt, the pro-
blematical correlation of seismicity with geological structure suggests that candidate sites
for moderate-size shocks as large as M=6½ may not
be targeted. For example, the location and structural complexity of the 1975 Pocatello Valley
earthquake would not have been expected from the
surface geology [Arabasz, et al., 1981].

Another key problem regarding the correlation
of seismicity with geological structure is that of
listric or "sled-runner" normal faults that flat-
ten with depth, rather than maintaining moderate
to steep dips of 45°-70°, as classically interpreted for Basin and Range faulting. New seismic
reflection data from petroleum exploration in the
overthrust belt of Wyoming, Idaho, and Utah show
that many normal faults in the eastern part of the
ISB become listric at depth merging into a single
sole or decollement fault that originally formed
as a low-angle thrust in an earlier period of
horizontal compression [e.g., Royse et al., 1975;

MacDonald, 1976]. The role and extent of listric
faulting as an earthquake generating mechanism in
the ISB is unknown but may be of considerable
importance. For example, does diffuse seismicity
between major faults reflect slip on listric
faults or on steeply-dipping fractures within
basement rocks? Predicted stress levels would
differ greatly.

Three models of normal and listric faulting are
shown in Figure 9, all of which depict normal
faulting with Holocene surface displacement. In
the right-hand model, of particular interest, subsurface structure is controlled by seismic reflection and drilling information and shows late Cretaceous-early Tertiary crustal shortening on
imbricate thrust faults [Royse et al., 1975].
Relaxation of the horizontal compression in Cenozoic time and the predominance of horizontal
extension has led to the development of secondary
normal faults as antithetic slip planes forming a
surficial graben that relieves extensional strain.
The center model in Figure 9 suggests that normal
faults developing under crustal extension may
become curved at depth, either because of variation of the coefficient of internal friction or
increasing effect of anelasticity; the left-hand
model illustrates a classical interpretation of
subsurface structure associated with Basin and
Range faulting.

In the listric fault models, the slip surfaces
eventually merge with a flat zone of decoupling
where horizontal displacement is accommodated. The
speculative boundary between an overlying layer of
brittle deformation and a decoupled extending
plastic layer lies near the top of a distinct
crustal low-velocity zone in the Intermountain
area [Smith et al., 1975], close to the lower
range of earthquakes in the ISB. Whether the low-
velocity layer indeed is a layer of low shear
strength is unknown, but high ambient temperature

in the Basin and Range and the presence of stacked thrust sheets of marine sedimentary rocks as a source of pore fluid make it reasonable. If listric faulting is a plausible seismogenic mechanism, then such parameters as fault zone friction along reactivated zones of weakness, configuration and depth extent of slip surfaces, and variation of stress with slip length will be critical for the predictability of earthquakes in the ISB.

Finally, for statistical analysis of space-time seismicity patterns in the ISB, a serious problem results from the realtively low seismic flux and short historical record. Adequate numbers of earthquakes above a magnitude threshold are typically unavailable for rigorous time-series analysis. Results presented in this paper, however, are encouraging for the detection of significant precursory changes in space-time seismicity patterns -- and the possibility of longer precursor times in the ISB could be distinctly advantageous.

Acknowledgments. This paper benefited from ideas developed in conversations and joint research with George Zandt, Melinda Griscom, and Diane I. Doser. Support from the following sources is gratefully acknowledged: U.S. Department of the Interior, Geological Survey, Contract 14-08-0001-16725; Division of Earth Sciences, National Science Foundation, NSF Grant EAR 77-23706; and the State of Utah.

References

Algermissen, S.T., Seismic risk studies in the United States, Proc. Fourth World Conf. on Earthquake Engineering, 1, 14-27, 1969.

Arabasz, W.J., and M. Griscom, Precursory seismicity patterns in the Utah region: Can regional variations in the precursor time scale be large? (abstract), Eos Trans. AGU, 59, 1126, 1978.

Arabasz, W.J., and W.D. Richins, Late aftershocks -- a search for premonitory seismic changes (abstract), Eos Trans. AGU, 57, 68, 1976.

Arabasz, W.J., W.D. Richins, and C.J. Langer, The Pocatello Valley (Idaho-Utah border) earthquake sequence of March-April 1975, Bull. Seismol. Soc. Amer., 71, in press, 1981.

Arabasz, W.J., R.B. Smith, and W.D. Richins (Eds.), Earthquake Studies in Utah, 1850 to 1978, 552 pp., Special Publication, University of Utah, Salt Lake City, 1979.

Arabasz, W.J., R.B. Smith, and W.D. Richins, Earthquake studies along the Wasatch Front, Utah: network monitoring, seismicity, and seismic hazards, Bull. Seismol. Soc. Amer., 70, 1479-1499, 1980.

Benjamin, J.R., and C.A. Cornell, Probability, Statistics, and Decision for Civil Engineers, 684 pp., McGraw-Hill, New York, 1970.

Bones, D.G., and W.J. Arabasz, Seismicity of the Intermountain seismic belt in southeastern Idaho and western Wyoming, and tectonic implica-tions (abstract), Earthquake Notes, 49(1), 19, 1978.

Bucknam, R.C., S.T. Algermissen, and R.E. Anderson, Patterns of late Quaternary faulting in western Utah and an application in earthquake hazard evaluation, U.S. Geol. Surv. Open File Rep. 80-801, 299-314, 1980.

Doser, D.I., Earthquake recurrence rates from seismic moment rates in Utah, M.S. thesis, Univ. of Utah, Salt Lake City, 1980.

Furlong, K.P., An analytic stress model applied to the Snake River Plain (northern Basin and Range province, U.S.A.), Tectonophysics, 58, T11-T15, 1979.

Greensfelder, R.W., F.C. Kintzer, and M.R. Somerville, Seismotectonic regionalization of the Great Basin, and comparison of moment rates computed from Holocene strain and historic seismicity, U.S. Geol. Surv. Open File Rep. 80-801, 433-493, 1980.

Griscom, M., Space-time seismicity patterns in the Utah region and an evaluation of local magnitude as the basis of a uniform earthquake catalog, M.S. thesis, Univ. of Utah., Salt Lake City, 1980.

Griscom, M., and W.J. Arabasz, Space-time seismicity patterns in the Utah region: A 300-km-long seismicity gap in the Intermountain seismic belt (abstract), Earthquake Notes, 50(4), 69, 1979.

Ishida, M., and H. Kanamori, Temporal variation of seismicity and spectrum of small earthquakes preceding the 1952 Kern County, California earthquake, Bull. Seismol. Soc. Amer., 70, 509-527, 1980.

Jones, L.M., and P. Molnar, Some characteristics of foreshocks, J. Geophys. Res., 84, 3596-3608, 1979.

Kasahara, K., Statistical discrimination of migrating seismic activity, U.S. Geol. Surv. Open File Rep. 78-943, 335-350, 1978.

Kristy, M.J., and D.W. Simpson, Seismicity changes preceding two recent central Asian earthquakes, J. Geophys. Res., 85, 4829-4837, 1980.

Love, J.D., and J. de la Montagne, Pleistocene and recent tilting of Jackson Hole County, Wyoming, in Guidebook, 11th Annual Field Conference, Jackson Hole, edited by R.R. Berg, pp. 169-178, Wyoming Geol. Assoc., 1956.

MacDonald, R.E., Tertiary tectonics and sedimentary rocks along the transition, Basin and Range Province to Plateau and Thrust Belt Province, Utah, in Symposium on Geology of the Cordilleran Hingeline, edited by J. G. Hill, pp. 281-317, Rocky Mt. Assoc. Geol., 1976.

Mogi, K., Two kinds of seismic gaps, Pure Appl. Geophys., 117, 1172-1186, 1979.

Pack, F.J., The Elsinore earthquakes in central Utah, September 29 and October 1, 1921, Bull. Seismol. Soc. Amer., 11, 155-165, 1921.

Precott, W.H., J.C. Savage, and W.T. Kinoshita, Strain accumulation rates in the western United States between 1970 and 1978, J. Geophys. Res., 84, 5423-5435.

Raleigh, C.B., and J.F. Evernden, The Case for low deviatoric stress in the lithosphere, U.S. Geol. Surv. Open File Rep. 80-625, 1, 168-198, 1980.

Royse, F., M.A. Warner, and D.L. Reese, Thrust belt structural geometry and related stratigraphic problems, Wyoming-Idaho-northern Utah, in Symposium on Deep Drilling Frontiers in the Central Rocky Mountains, pp. 41-54, Rocky Mt. Assoc. Geol., 1975.

Scholz, C.H., L.R. Sykes, and Y.P. Aggarwal, Earthquake prediction: A physical basis, Science, 181, 803-810, 1973.

Smith, R.B., Seismicity, crustal structure, and intraplate tectonics of the interior of the western Cordillera, in Cenozoic Tectonics and Regional Geophysics of the Western Cordillera, edited by R.B. Smith and G.P. Eaton, pp. 111-144, Geol. Soc. Amer. Mem., 152, 1978.

Smith, R.B., L. Braile, and G.R. Keller, Crustal low velocity layers: Possible implications of high temperatures at the Basin Range-Colorado Plateau transition, Earth Planet. Sci. Lett., 28, 197-204, 1975.

Smith, R.B., and M.L. Sbar, Contemporary tectonics and seismicity of the western United States with emphasis on the Intermountain seismic belt, Geol. Soc. Amer. Bull., 85, 1205-1218, 1974.

Smith, R.B., J.R. Pelton, and D.L. Love, Seismicity and the possibility of earthquake related landslides in the Teton-Gros Ventre-Jackson Hole area, Wyoming, Wyoming Univ. Contr. Geol., 2, 57-64, 1976.

Smith, R.B., R.T. Shuey, J.R. Pelton, and J.T. Bailey, Yellowstone hot spot: Contemporary tectonics and crustal properties from new earthquake and aeromagnetic data: J. Geophys. Res., 82, 3665-2676, 1977.

Smith, R.B., G. Zandt, and J.E. Gaiser, Temporal monitoring of seismic velocity along the Wasatch Front using quarry blasts, Bull. Seismol. Soc. Amer., 70, 1527-1546, 1980.

Stokes, W.L., What is the Wasatch Line?, in Symposium on Geology of the Cordilleran Hinge-line, edited by J.G. Hill, pp. 11-25, Rocky Mt. Assoc. Geol., 1976.

Swan, F.H., III, D.P. Schwartz, and L.S. Cluff, Recurrence of moderate-to-large magnitude earthquakes produced by surface faulting on the Wasatch fault zone, Bull. Seismol. Soc. Amer., 70, 1431-1462, 1980.

Thompson, G.A., and M.L. Zoback, Regional geophysics of the Colorado Plateau, Tectonophysics, 61, 149-181, 1979.

Wallace, R.E., Active faults, paleoseismology, and earthquake hazards, Proc. Seventh World Conf. on Earthquake Engineering, in press, 1980.

Witkind, I.J., Preliminary map showing known and suspected active faults in Idaho, U.S. Geol. Surv. Open File Rep. 75-278, 71 pp., 1975 [Open File Rep. 75-279 for Wyoming; Open File Rep. 75-285 for Montana.]

Wyss, M., R.E. Habermann, and A.C. Johnston, Long term percursory seismicity fluctuations, U.S. Geol. Surv. Open File Rep. 78-943, 869-894, 1978.

Zandt, G., and T.J. Owens, Crustal flexure associated with normal faulting and implications for seismicity along the Wasatch Front, Utah, Bull. Seism. Soc. Amer., 70, 1501-1530, 1980.

Zoback, M.L., and M. Zoback, State of stress in the conterminous United States, J. Geophys. Res., 85, 6113-6156, 1980.

GREAT DETACHMENT EARTHQUAKES ALONG THE HIMALAYAN ARC
AND LONG-TERM FORECASTING

Leonardo Seeber and John G. Armbruster

Lamont-Doherty Geological Observatory of Columbia University
Palisades, New York   10964

Abstract.   Great ruptures along the Himalayan
arc occur on a shallow dipping detachment and they
conform with the hypothesis of seismic gaps. The
detachment separates the subducting basement from
the overlying sedimentary wedge and extends updip
below the foredeep without reaching the surface.
The detachment appears aseismic in the period
between great earthquakes. Thrust earthquakes of
intermediate and smaller magnitude form a narrow
belt that coincides with the sharpest topographic
break.   These earthquakes mark the down-dip
transition from the detachment to the basement
thrust, along which the more rigid portion of the
overriding slab, the Tethyan slab, is in contact
with the subducting basement. Downdip from this
transition, slip on the basement thrust seems to
occur aseismically.

During the last two centuries great Himalayan
earthquakes have occurred at an average rate of
one about every 30 years. The area of rupture for
the last four of these earthquakes is estimated
primarily from their well documented surface
effects. In general, the severe effects do not
decrease away from a well defined "epicentrum".
Instead, they are scattered throughout a large
meizoseismal area. The variations in intensity in
these areas may primarily reflect near surface
conditions. The boundaries to these areas are
often clearly delineated by a sharp drop in intensity.

The inferred ruptures extend along strike about
300 km in the 1905, 1934, and 1950 events, and
almost twice that amount in the 1897 event. The
1897 and the 1934 ruptures abut along a major
basement structure. A small gap between the 1897
and the 1950 earthquakes may have ruptured shortly
before 1950. Between the 1934 and 1905 ruptures
there is a 600 km long gap. This gap was
partially, or perhaps totally filled by the 1803
and 1833 events.

These results suggest that most of the 2500 km
long Himalayan detachment has ruptured since 1800
except probably portions in Kashmir and Uttar
Pradesh. If the rate of seismic energy release by
the Himalayan detachment during the last 180 years
is representative of the long-term average rate,
we would expect the repeat-time for the typical
Himalayan detachment rupture to be about 180-240
years.   The repeat time for the exceptionally
large 1897 rupture may be longer.

Introduction

The concept of seismic gaps [e.g., Kelleher,
1973; Sykes, 1971] has been applied successfully
to many active plate boundaries.   One of the
important prerequisites has been a good understanding of the tectonics, such as has recently
become available for many of the plate boundaries
where oceanic crust is involved.   Purely
continental plate boundaries are typically more
diffuse and complex and in general still poorly
understood [e.g. Burchfiel, 1980]. Thus, the gap
concept has had very limited application to the
seismicity   along   those   boundaries   [e.g.
Ambraseys, 1970].

The Himalayan arc (Figure 1) stands out along
the Alpine convergence zone as a remarkably long
(2500 km) and uniform feature. Uniformity in
structure   and   stratigraphy   along   this   arc
[Gansser,   1964,   p. 246]   suggests   a   uniform
evolution.   Uniformity in morphology and stratigraphy   suggests   a   uniform   tectonic   regime.
Recently, we incorporated the major features of
the Himalaya into a continental subduction model
characterized by a single tectonic process, rather
than by a set of successive events [Seeber et al.,
1981]. The seismicity, the main constraint to our
model, suggests that the pattern of slip along the
Himalaya and along oceanic subduction boundaries
is similar.   Thus, the concept of seismic gaps
should apply to both types of convergent plate
boundaries.

The purpose of this paper is to examine the
surface effects of great Himalayan earthquakes,
and infer the extent of the associated ruptures.
Some spacial and temporal constraints on the next
great earthquakes can be inferred from the recon-
struction in space and time of the previous
ruptures.

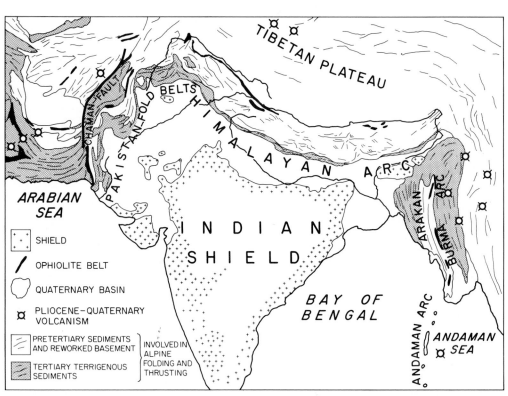

Fig. 1. Regional setting of the Himalayan arc, along which the Indian shield is subducting northward. Tectonic features based on Gansser [1964, plate I]; recent volcanic centers according to Katsui [1971].

LEGEND (in figure):
SHIELD
OPHIOLITE BELT
QUATERNARY BASIN
PLIOCENE-QUATERNARY VOLCANISM
PRETERTIARY SEDIMENTS AND REWORKED BASEMENT
TERTIARY TERRIGENOUS SEDIMENTS
INVOLVED IN ALPINE FOLDING AND THRUSTING

Tectonic Model

Great earthquakes and intermediate to small magnitude earthquakes form two parallel belts of seismicity along the Himalayan arc (Figures 2 and 3). These belts are associated with two distinct seismogenic faults in the Himalayan subduction structure. The subducting Indian shield and the overriding Tethyan slab interact along a poorly defined basement thrust. In our preferred interpretation (Figure 4), this interplate fault is quasi-horizontal, aseismic (at least in the magnitude range of teleseismic data) and terminates updip at the basement thrust front (BTF). The two main seismogenic faults, a detachment and a steeper dipping thrust, merge at the BTF and partially absorb interplate convergence updip from the BTF. These faults bound a relatively weak sedimentary wedge composed of terrigenous sediments as well as low grade metasediments and overthrust sheets of middle to high grade rocks. The sedimentary wedge partially absorbs interplate convergence by pervasive shortening. The detachment separates the sedimentary wedge from the underthrusting Indian shield and slips by great earthquakes. A brittle counterpart of the ductile Main Central thrust (see below) separates the wedge from the Tethyan slab and generates intermediate and smaller magnitude thrust earthquakes. In an earlier version of this model

(Figures 3 and 5) [Seeber et al., 1981] the belt of thrust earthquakes occur on a relatively steep (∼ 30°) portion of the interplate thrust at the BTF. In this interpretation the subducting lithosphere is sharply bent at the BTF.

During the periods between great earthquakes the detachment appears aseismic and the intermediate-magnitude thrust earthquakes are concentrated near the BTF. Downdip (north) of the BTF the basement thrust must extend as an aseismic interplate fault since all the available fault-plane solutions north of the BTF are consistent with intraplate east-west extension and are inconsistent with interplate slip [Ni and York, 1978; Molnar and Tapponnier, 1978]. The main elements in this continental subduction model resemble corresponding elements in oceanic subduction zones, particularly in the ones characterized by a wide sedimentary wedge (accretionary prism) [Figure 5, Seeber et al., 1981].

The narrow belt of thrust earthquakes corresponds closely with the prominent topographic front between the Lesser and the High Himalaya. Topographic front and seismic belt fit remarkably well a small circle over the central 1700 km of the arc (Figure 3). Beyond this central portion the seismic belt and topographic break deviate from the small circle but continue to track each other along the entire arc. Thus, the BTF, which is the structural element

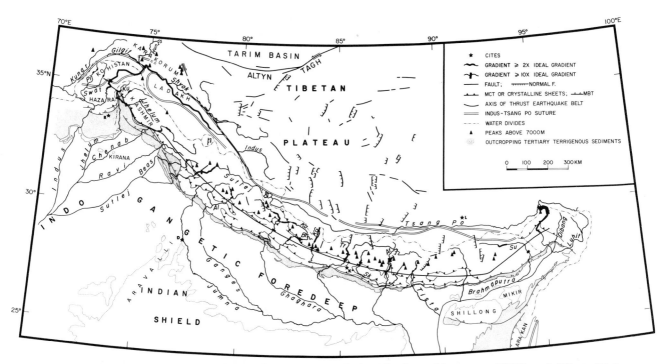

Fig. 2. Main structural and morphologic features of the Himalaya [Gansser, 1977] and Tibet [Molnar and Tapponnier, 1978]. The narrow belt of thrust earthquakes and the topographic front between the High and Lesser Himalaya (BTF, Figure 4) correlate also with the downstream limit of abnormally high river gradients. The MCT generally follows the BTF but deviates locally, as in Hazara, at the northwestern terminus of the arc, and in the Kathmandu Klippe. Pi = Pinjkora, Ku = Kundar, Al = Alaknanda, Ka = Karnali, Bh = Bheri, KG = Kali Gandaki, Tr = Trisuli, SK = Sun Kosi, Ar = Arun, Su = Subansiri, D = Delhi, K = Kathmandu, L = Lhasa. (From Seeber and Gornitz, in preparation).

associated with the seismic belt and the topographic break, marks a fundamental boundary in the Himalayan subduction structure. In our model, the topography south of the BTF, the upper surface of the sedimentary wedge (Lesser Himalaya, Sub-Himalaya, and foredeep), is in dynamic equilibrium with the shear stress at the detachment [Elliott, 1976]. The topography north of the BTF, the upper surface of the Tethyan and Tibetan slabs (High Himalaya, Tethyan Himalaya, and at least part of the Tibetan Plateau), is in static equilibrium with a crust thickened by continental subduction [Powell and Conagan, 1973] and is independent of the shear stress along the basement thrust.

The main boundary thrust (MBT) and the main central thrust (MCT) are the two most prominent surface structures of the Himalayan arc (Figures 2 and 4). In an alternative to the model proposed here, these two faults are similar, but successive intracontinental thrusts in a large scale pattern of deformation that advances progressively toward the Indian foreland, much as the leading edge of a thrust and fold belt does on a smaller scale [Mattauer, 1975; LeFort, 1975; Molnar et al., 1977; Stocklin, 1980]. This alternative model is here named the evolutionary model, in contrast to our preferred, steady-state model.

The MCT is a shear zone separating the Tethyan slab above from the Lesser Himalayan sediments below [Stocklin, 1980]. It dips to the north at a shallow angle, steepening down-dip below the High Himalaya [Gansser, 1964, plate II; Fuchs and Frank, 1970; Pecher, 1978; Hashimoto, 1973]. At places it extends far south covering the Lesser Himalaya with sheets or klippen of crystalline rock [Stocklin, 1980]. The MCT also corresponds with a zone of inverse metamorphism. Mineral assemblages were set at increasing temperature and pressure upward in the Lesser Himalayan sediments, the maximum pressure of about 7-8 kb corresponds with the plane of maximum shearing, while the maximum temperature occurs a few kilometers above this plane in the Tethyan slab. This inverse metamorphic profile is consistent with rapid underthrusting of cold near-surface rocks below a thick slab [LeFort, 1975; Toksoz and Bird, 1977].

The ages of metamorphism and of the anatectic leucogranites, rooted in the zone of highest metamorphic grade above the MCT, cluster between 15 and 20 m.y. [LeFort, 1975; Mehta, 1980]. The crystalline sheets of the Lesser Himalaya are characterized by a thinner and lower grade zone of inverse metamorphism [Stocklin, 1980; Brunel and Andrieux, 1977] and by somewhat older ages of anatectic granitization [Andrieux et al., 1977]. Thermometric analysis of fluid inclusions

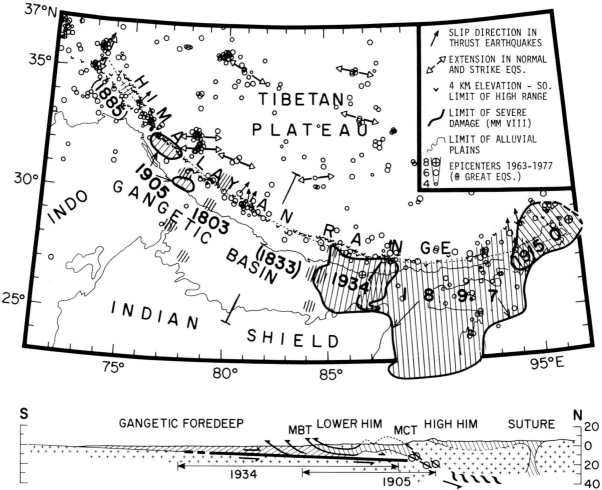

Fig. 3. Epicentral map of the Himalaya (NOAA data, 1963-1977; only epicenters determined with 20 or more P arrival-times; size of circle proportional to magnitude) and intensity VIII contours of possible great Himalayan earthquakes (intensity $\geq$ VIII for each earthquake is shaded with lines perpendicular to the arc, pre 1897 shaded areas may be intensity < VIII). Representative fault plane solutions [Molnar et al., 1977] are indicated by arrows, single arrows = slip direction of thrusting events, pairs of arrows = tension axes of normal faulting events. The sharp topographic front of the High Himalaya (thresholds of 4 km elevation) correlates with the belt of moderate magnitude thrust earthquakes. This morphotectonic feature (BTF - see Figure 4) fits precisely a small circle (dotted line) over the central 2/3 of the Himalayan arc. The cross section (below) is for the central portion of the arc (location is indicated on map) and contains the hypocenters (projected along the arc) and fault dip (northward component only) of the three earthquakes near 81°E and the earthquake near 88°E for which depth and fault plane solution are well-constrained by body-wave modelling. Depth and slip direction of well-constrained earthquakes at the eastern and western terminus of the arc fit the same pattern. The subduction model suggested by this section is somewhat different from the model in Figure 4.

indicates that the equilibrium pressure for this last phase of metamorphism is about half the pressure that characterizes the surrounding metamorphic mineral assemblages [Pecher, 1979].

Thus, if the ambient pressure of metamorphism is close to lithostatic pressure, the outcropping portion of the MCT was active as a ductile shear zone at a depth of 20 to 30 km about 15 m.y. ago. Thereafter, tectonism on this portion of the fault subsided and P-T conditions were altered primarily

by erosion and uplift. About half of the way up to the surface the fluid phases equilibrated and the last metamorphic phase was concluded.

The above interpretation is in contrast with the evolutionary model, according to which the ages of metamorphism and granitization correspond with the climax and then halting of deformation on the entire MCT, followed sometime later ($\sim$10 m.y. ago) by the onset of thrusting on the MBT [LeFort, 1975; Stocklin, 1980]. The equilibration pressure

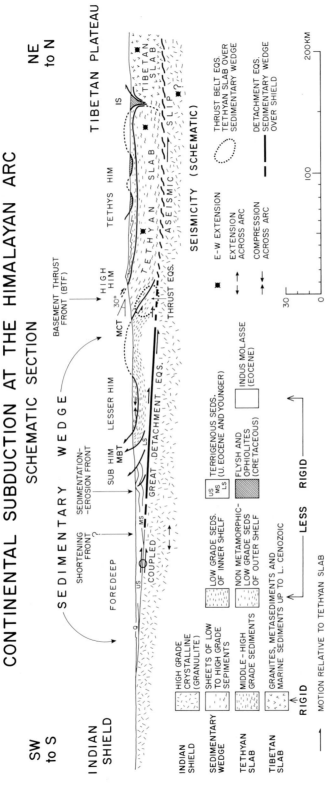

Fig. 4. Model (to scale) of continental subduction in the central portion of the Himalayan arc. A subhorizontal shear zone of interplate convergence below the Tibetan Plateau merges updip into a detachment that slips by great earthquakes and a steeper dipping thrust the "MCT" (see text) which slips by intermediate and smaller magnitude earthquakes. The transition from seismic to aseismic slip occurs at the basement thrust front (BTF) and corresponds with a sharp topographic front along a small circle (Figures 2 and 3). The detachment and the "MCT" bound on either side a sedimentary wedge of terrigenous sediments as well as pre-orogenic metasediments. This wedge is less rigid than the converging slabs and is shortened by pervasive deformation. In an earlier version of this model (Figures 3 and 5) the belt of intermediate magnitude thrust earthquakes is associated with a steeper portion of the basement thrust.

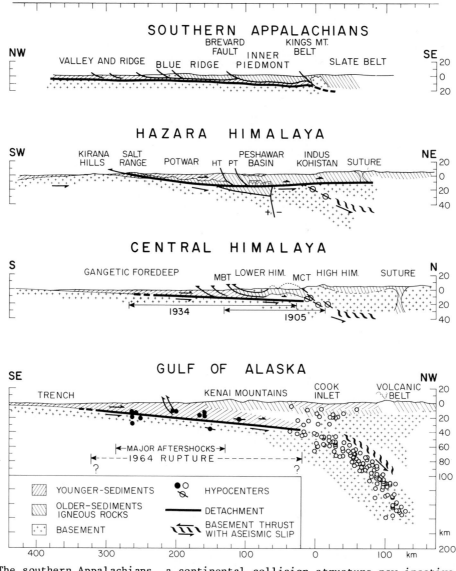

Fig. 5. The southern Appalachians, a continental collision structure now inactive, and the active oceanic subduction structure in the Gulf of Alaska are compared to the Himalayan continental subduction structure in the Central Himalaya, where detachment coupling is probably strong, and in Hazara, the extreme western Himalaya, where the detachment coupling is probably weak. Arrows indicate qualitatively velocity with respect to the overriding slab. Southern Appalachian section is from deep seismic reflection results [COCORP; from Cook et al., 1979]. Recent data [Behrendt et al., 1980] indicate that the detachment extends to the southeast beyond the Kings Mt. belt. In the Alaskan section aftershocks (filled circles) and inferred rupture (slightly modified) of the great 1964 earthquake (M_s = 8.4) are from Plafker [1965]; hypocenters during the interseismic period (April-June, 1972) are from Lahr et al. [1974]. The central Himalaya section is the same as in Figure 3. The section for the Hazara Himalaya is based on data from a local seismic network [Armbruster et al., 1979]. A thick Infracambrian evaporite layer is at the base of the sedimentary wedge in Hazara. This weakly coupled wedge extends above and north of the basement thrust and the detachment may slip aseismically [Seeber et al., 1981]. The detachment/basement-thrust model for continental subduction in the Himalaya may be also applicable to the oceanic subduction structure in Alaska, where the accretionary wedge is very wide.

for fluid inclusions found by Pecher [1979] on two profiles across the MCT raises a problem for this model. In both profiles this pressure increases upsection about 2 kb over a 30 km distance across the MCT. Pecher [1979] suggests that a pronounced topographic gradient across these profiles at the time of equilibration may explain the northward increase in equilibration pressure. This seems very unlikely since the present topographic gradient from the Lesser to the High Himalaya along the profiles is already quite pronounced. The northward increase in equilibration pressure can be better explained by a differential uplift of the High Himalaya relative to the Lesser Himalaya, postdating the fluid inclusions.

We accurately determined the depths of several of the larger thrust earthquakes along the BTF by body-wave synthesis (Figure 3) [Armbruster et al., in preparation]. In all cases these focal depths are close to 15 km, ranging from 10 to 20 km. Thus, the present depth-range of brittle thrust faulting corresponds to the upper limit of the metamorphism and aseismic shearing as deduced from the mineral assemblages.

In the steady state model (Figure 4) the forward edge of the Tethyan slab is continuously being uplifted, partly to maintain isostatic equilibrium and counteract the effect of the high rate of erosion on the south slope of the High Himalaya, and partly because of underthrusting by the sedimentary wedge. As sediments above the northward tilting slab become gravitationally unstable, they slide northward above the Tethyan slab and form discordant recumbent folds verging to the north [Fuchs and Frank, 1970; LeFort, 1975; Stocklin, 1980].

The uplift of the forward edge of the slab also causes the BTF to recede down-dip on the active basement thrust. The inactive portion of the basement thrust at the base of the Tethyan slab is uplifted and reaches the surface as the MCT. Thus, the ages of metamorphism and granitization on the MCT (15-20 m.y.) reflect the time it takes for the lower face of the Tethyan slab to rise from the zone of ductile shearing (20-30 km deep), through the zone of brittle deformation at the BTF (12-20 km deep) to the surface. The 0.5-1.0 mm/yr rate of uplift obtained this way is close to estimates of Himalayan erosion rates determined from the sediment load of rivers (0.7-1.0 mm/yr) [Curray and Moore, 1971] or radiometric dating [Mehta, 1980].

The MCT, defined as the ductile shear zone [Mattauer, 1977] is certainly not active at or near the surface. In the steady state model, however, the Tethyan slab and the sedimentary wedge (Figure 4) are two independent tectonic units which should be separated by an active tectonic boundary. Paleothermometry of fluid inclusions across the MCT in central Nepal (discussed above) suggests that post-metamorphic uplift is faster for the Tethyan slab north of the MCT than for the sedimentary wedge south of the MCT. Similar differential motion is suggested by

the topographic break between these two units and by the profiles of rivers that cross the Himalaya. These profiles show invariably a sharp increase in gradient that closely corresponds to the inferred position of the BTF, even in cases where the trace of the MCT is significantly offset from the BTF (Figure 2) [Seeber and Gornitz, in preparation]. This suggests that the change in river gradient is not caused by differential resistence to erosion, since the MCT is the main lithologic boundary, but rather is the consequence of differential uplift. The thrust faulting associated with the belt of intermediate-magnitude earthquakes can account for differential motion between the Tethyan slab and the sedimentary wedge (Figure 4). At the surface this motion may be taken up by slip on the many bedding plane faults, subparallel and near to the MCT [Gansser, 1964, plate II; Valdiya, 1976; Pecher, 1978, p. 299; Hashimoto, 1973; Fuchs and Frank, 1970].

The MBT is the lithotectonic boundary between the precollisional sediments of the Lesser Himalaya and the terrigenous sediments of the Sub Himalaya and the foredeep. At the surface the MBT is a brittle fault with obvious signs of ongoing activity [e.g., Sinvhal et al., 1973; Krishnaswamy et al., 1970; Stocklin, 1980]. The upper Tertiary Siwalik sediments exposed in the Sub-Himalaya typically become coarser upward in the section, with fan conglomerates occurring only in the Plio-Pleistocene. This pattern of sedimentation is often interpreted as evidence for the progressive rise of the Himalaya, [e.g., Gansser, 1964, p. 261] and, in particular, for the recent (Plio-Pleistocene) onset of tectonism on the MBT [e.g., LeFort, 1975; Stocklin, 1980; Parkash et al., 1980].

Alternatively, the upward coarsening of the terrigenous sediments exposed along the Sub-Himalaya is the result of time transgressive sedimentation along a subduction front. Both coarse and fine sediments are deposited at all times. The coarse sediments are deposited near the subduction front and are subducted first, so that the older exposed sediments are likely to be distal and fine [Johnson et al., 1979]. Moreover, the complete absence of Siwalik terrigenous sediment in the Lesser Himalaya [Gansser, 1964, p. 246] suggests that the MBT is older than these sediments [Parkash et al., 1980]. It is possible, therefore, that the MBT is not a recently developed replica of the MCT, and that instead, it is a distinct feature in a steady-state tectonic regime. In our interpretation the MBT is the sole thrust below the thick sequence of inner-shelf sediments, originally resting on the northern margin of India. It is also one of the imbricate thrusts that contribute to the current shortening of the sedimentary wedge (Figure 4).

Slip on the MBT may occur by earthquake ruptures. In analogy to the slip behavior of imbricate thrusts within the accretionary wedges of oceanic subduction zones, the repeat time for earthquakes on the MBT is expected to be considerably longer than the repeat time of great earthquakes on the underlying detachment [Sykes et al., this volume, Yonekura and Shimazaki, 1980].

Great Himalayan Earthquakes

In the tectonic framework discussed above, the great earthquakes of the Himalaya occur on a quasi-horizontal detachment that extends southward below the foredeep. The consistent absence of primary surface rupturing during the great earthqukes [Oldham, 1899, p. 138; Middlemiss, 1910, p. 329, 349; Dunn et al., 1939, p. 149,153] is a strong evidence for a quaishorizontal causative fault. Further evidence is presented below.

The dimensions of the rupture in a great detachment event are much larger than the depth of the rupture at any point. Thus, the intensity should reflect the close proximity of the rupture in the area above it and should decrease abruptly beyond this area. In general, the intensities of the great Himalayan earthquakes form a pattern of this kind.

The surface effects of the earthquakes discussed here are quantified in either the Rossi-Forel scale (RF) or in the Mercalli I-X scale (ME).

1905 Kangra Earthquake

On the basis of instrumental data the 1905 Kangra earthquake was assigned M = 8.4 by Richter [1958] and M_s = 8 by Kanamori [1977]. The intensity survey of this earthquake was coordinated and compiled by Middlemiss [1910]. This event (Figure 6) is characterized by two distinct areas of high intensity, RF $\geq$ VIII, separated by a 100 km gap where the intensity is RF VII (ME VII-). Both these areas straddle the MBT.

The hypothesis of two separate earthquakes is not considered because neither intensity data [Middlemiss, 1910, p. 332] nor the instrumental data [Richter, 1958, p. 63] suggest two seismic sources separated in time. Thus, the intensity data can be interpreted for either a large detachment rupture, approximately 300 km along strike, and underlaying both regions of RF $\geq$ VIII, or for a smaller rupture underlaying only the more prominent region of high intensity toward the northwest. In the second case there is no straightforward explanation for the high intensity zone toward the southeast. In the first case, both high intensity areas, as well as the associated prominent reentrants [Middlemiss, 1910, p. 336], may be ascribed to corresponding zones of relatively strong coupling on the underlying detachment.

Two independent pieces of evidence suggest that the 1905 rupture extends under both high intensity zones. During the 24 hours before the 1905 main shock at least seven separate foreshocks were felt [Middlemiss, 1910, p. 355]. Five of these were felt only locally so that the locations assigned

according to the felt reports (Figure 6) are probably not grossly incorrect. Thus, the 1905 foreshocks are distributed along a portion of the Himalayan front that approximately corresponds with the full extent of RF intensity $\geq$ VIII. A prominent gap in the teleseismically located seismicity corresponds with this inferred rupture [Menke and Jacob, 1976].

The second indicator for a large rupture is from geodetic data. The line CDE (Figure 6), from the alluvial plains, through the Sub-Himalaya and across the MBT to the Lesser Himalaya, had been surveyed before the main shock (CD in 1862 and DE in 1904). Shortly after the main shock this line was resurveyed. The difference in elevation, presumably caused by the earthquake, is given with reference to point C [Middlemiss, 1910, plate 27]. The Sub-Himalaya had risen with respect to the alluvial plains and the Lesser Himalaya. The amount of measured uplift ($\sim$ 13 cm maximum) should be a minimum value since the reference point C may also have been coseismically uplifted. It is difficult to estimate possible errors in the leveling data since only the results are reported by Middlemiss [1910, p. 348]. The portion DE of the profile is crucial since it crosses the MBT and shows an elevation change contrary to a thrust movement on the MBT. DE was surveyed 11 months before the main shock and then twice 1-month and 20-21 months after the main shock and the measured change in elevation (13 cm) is above random noise but close to conceivable error from poorly calibrated rods [Stein, this volume].

Another line that crosses the MBT about 50 km northwest from the 1905 area of maximum intensity (Figure 6, line AB) was recently relevelled [Chugh, 1974]. This data shows an uplift northeast of the MBT with respect to the area southwest of this fault. This interseismic change in elevation is opposite in sense from the coseismic change in elevations indicated by releveling on line CDE, and suggests elastic reloading of the system.

The most straightforward qualitative interpretation of these levelling data is that (1) the 1905 rupture extends for about 300 km from the highest intensity area through the southeastern zone of high intensity where the coseismic change in elevation was measured, and (2) the rupture was not on the MBT but was on a detachment that extends below the Himalayan front, including the foredeep, with the upper sheet moving southwest [Seeber et al., 1981].

Both relevelling data and the distribution of foreshocks suggest a large 1905 rupture comparable to the other great Himalayan earthquakes (except the 1897 event). In this case, the intensities along the portion of the Himalayan front corresponding with this rupture are comparatively low, reflecting primarily the near absence of liquefaction in the plains and the only sparse landslides in the mountains (Figure 6). It is possible that this rupture was narrow across

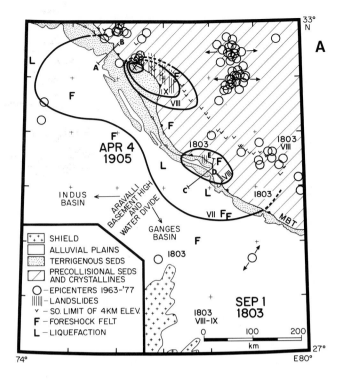

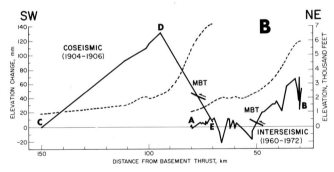

Fig. 6. Rossi-Forel (RF) intensity contours and some specific surface effects of the 1905 Kangra earthquake [Middlemiss, 1910] M = 8.4 [Richter, 1958], M = 8 [Kanamori, 1977]. Note that the two intensity maxima correspond with structural reentrants, suggesting a common cause (see text). Seven foreshocks were felt in the 24 hours prior to main shock [Middlemiss, 1910, p. 356]. Only two were felt at more than one place (only one place plotted). Foreshocks were at least as widely distributed as the area of intensity $\geq$ VIII. The coseismic change in elevation along CDE is relative to point C [Middlemiss, 1910, p. 348]. The interseismic profile AB is relative to point A [Chugh, 1974]. The profiles are aligned according to the distance to the basement thrust front (BTF, see Figure 4) taken as the small circle in Figure 3. The epicentral data (1964-1977) are the same as in Figure 3. Note the aseismic gap in this data along the BTF corresponding

to the full extent along strike of intensity $\geq$ VIII in the 1905 event. "1803" indicates localities with strong effects from that earthquake. Intensity values (RF) are assigned whenever the reports are sufficiently specific.

strike, being limited primarily to the Sub-Himalaya and that the seismic excitation in the foredeep and in the Lesser Himalaya was relatively low. Alternatively, the potential for liquefaction and landslides was unusually low.

The 1905 event happened at the end of the dry season (April 4). At this time of the weather cycle the mountain slopes are at their driest and they are least prone to landsliding. Thus, the timing of the 1905 event, as well as the 1934 event, can account for the relative minor occurrence of coseismic landslides.

The pattern of sedimentation along the Himalayan foredeep is not uniform and differences in the rate of sedimentation along the foredeep may correspond to differences in liquefaction potential. The area of the alluvial plains corresponding to the 1905 rupture is at a higher elevation than any other portion of the plains and is the water divide between the Ganges and the Indus drainages (Figure 6). This topographic high corresponds to a basement high, the northern extension of the Aravalli below the foredeep [Gansser, 1964, page 20]. This portion of the plains is probably subsiding at a low rate and is characterized by an unusually low rate of sedimentation. It is possible that the upper layer of sediments is on the average older, it has experienced more liquefaction events and is now less prone to liquefaction than the corresponding layer elsewhere along the foredeep.

In conclusion, the 1905 event ruptured the detachment for about 300 km along the Himalayan front. This event is characterized by low intensities (RF VII to VIII) above most of this rupture. These low intensities may be ascribed to low liquefaction potential in the plains, perhaps a permanent feature of the 1905 area, and to low landslide potential in the mountain, a characteristic which depends on the season.

1934 Bihar Earthquake

The 1934 Bihar earthquake is the most recent of the great earthquakes to occur in a densely populated and readily accessible area of the Himalayan front, and is, in many ways, the best documented of these events [Dunn et al., 1939]. Gutenberg and Richter [1954] assign $M_s = 8.3$ to the 1934 event, Chen and Molnar [1977] give $M_o = 1.6 \times 10^{28}$ dyne-cm. A well determined epicenter for this event is located near the MBT, about 100 km south of the thrust earthquake belt and 50 km north of the "slump belt", the nearest maximum intensity zone (Figure 7; Seeber et al. [1981]). On this basis alone, the 1934 event must

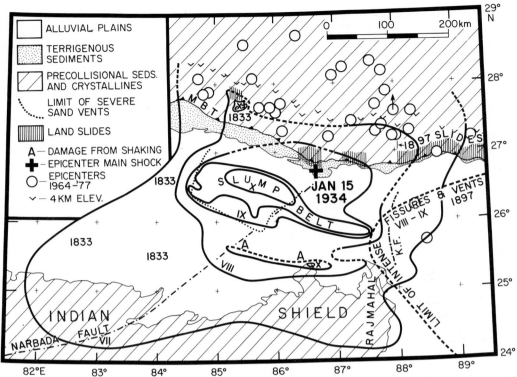

Fig. 7. Intensity contours in the Mercalli I-X scale (ME) and specific effects of the 1934 Bihar earthquake [Dunn et al., 1939; the distribution of sand vents is from Brett, 1935], M_s = 8.3 [Gutenberg and Richter, 1954], M_o = 1.6 x 10^{28} dyne-cm [Chen and Molnar, 1977]. The epicentral data (1964-1977) are the same as in Figure 3. The 1934 epicenter is from Seeber et al. [1981]. The 1897 and 1934 meizoseismal areas (ME $\geq$ VIII) abut near the Kishangang fault (K.F.), a major basement discontinuity that corresponds to an abrupt eastward decrease in the width of the Sub-Himalaya, and to the eastern margin of the outcropping shield. This tectonic boundary may also affect the recent seismicity along the thrust belt. An extension of the Narbada fault under the foredeep, parallel to the structural trends deduced by magnetic data [Gansser, 1964, p. 20] coincides with a number of discontinuities in the intensity distribution, the 1934 epicenter and a marked discontinuity along the Sub-Himalaya. "1833" marks sites with strong effects from that earthquake [Oldham, 1882].

be tectonically differentiated from the thrust earthquake belt. One aftershock was located by the Burma Meteorological Department near the epicenter of the main shock [Dunn et al., 1939, plate 1]. Two aftershocks were reported by the ISS. This suggests that the many felt aftershocks [Dunn et al., 1939, p. 89] were all in the range $M \leq 6$.

Three distinct areas of very high intensity (ME IX-X) are located within the intensity $\geq$VIII perimeter, and are related to different specific effects. The large area of high intensity along the axis of the alluvial plains is associated with ground failure, liquefaction and "slumping". The narrow belt of high intensity near the northern limit of the outcropping shield suffered from high acceleration (high frequency vibration). The destruction in this area was highest on alluvium near the outcrops of crystalline rocks [Dunn, 1939, p. 216-217]. The third area of ME X is in the isolated intramontane Kathmandu basin. Destruction there can be ascribed to a mixture of vibration and ground failure in the unconsolidated

recent lacustrine sediments of the basin. The Kathmandu basin was also heavily damaged by the 1833 earthquake (see below). It is possible that this basin and many other similar basins [e.g. Mexico City basin, S.K. Singh, personal communication] are often characterized by enhanced intensities.

The extrusion of sand and water from fissures and vents was so intense in the 1934 event that damage to agricultural land was feared and a careful survey was carried out [Brett, 1935]. Ejected sand covered an area 150 by 100 km with an average thickness of 5 cm (Figure 7). The amount of water ejected may have been many times the amount of sand, judging from the coseismic rise in ground water level, and subsequent increase in the flow of rivers [Auden and Ghosh, 1935, p. 200]. Liquefaction of this kind is essentially a coseismic dewatering of the sediments. In the narrow and long belt of "slumping" (Figure 7) the surface of the nearly flat alluvial plains was extensively fissured and faulted [Dunn et al., 1939, p. 38-39]. This belt was closely associated with the zone of intense ejection of water and sand, but could still be recognized as a distinct feature. Near surface liquefaction was not apparent in many places that suffered "slumping". There is no evidence that the sediments of the "slump" and liquefaction belts differ from those in the north or south of this belt [Dunn et al., 1939, p. 159] and a non-tectonic cause of these phenomena seems unlikely.

The intense dewatering and the "slumping" can both be the mainfestation of north-south shortening in the sedimentary wedge at the updip end of the detachment rupture. Dewatering causes subsidence but can also account for layer-parellel shortening. On the other hand, "slumping" can be the surface effect of buckling of the sediment layers. Anticlinal buckling causes shortening and also uplift. Thus, the superposition of the subsidence and the uplift expected from a combination of dewatering and buckling may result in a field of small changes in elevation near the updip end of a detachment rupture. Releveling carried out after the 1934 event was rather inconclusive because of the generally poor status of markers in the meizoseismal area, but it suggests no major coseismic changes in elevation [Dunn et al., 1939, p. 40–42]. In agreement with this conclusion, no major coseismic changes in river courses were observed [Dunn et al., 1939, p. 44–45].

The 1934 event is the only well known Himalayan earthquake that produced a "slump belt". The 1905 rupture did not propagate far south of the Sub-Himalaya and it is everywhere deeply buried (Figure 3). The 1950 and 1897 ruptures are deeply buried everywhere because there is no outcropping shield in their foreland (see below). Thus, the updip terminus of the 1934 rupture may be unusually shallow, and the "slump belt" may reflect the large strain near this terminus. The high intensity area south of the "slump belt" near the outcropping shield is not in the near field, and it may reflect a concentration of seismic energy at the thin edge of the sedimentary wedge.

According to the detachment model, the rupture should not extend farther north then the basement thrust front (BTF). The ME VIII contour [Dunn et al., 1939, plate 2] follows the southern flank of the High Himalaya, the surface expression of the BTF, and suggests that the 1934 rupture extended only as far as the BTF.

The narrow "slump belt" (Figure 7) extends for about 300 km in the plains approximately parallel to equal-depth contours of the basement [Gansser, 1964, page 20] about 100 km south of the MBT. In the "slump belt" destruction was nearly total (ME $\geq$ IX). Beyond the two ends of this belt the intensity drops abruptly to relatively low values. Associated with the "slump belt" there is a 300 km long belt of severe destruction along the Lesser Himalayas. Thus, boundaries to the intensity in the Himalayan strike direction, as determined by different phenomena (e.g. liquefaction and vibration) in different tectonic environments

(e.g. the foredeep and the Lesser Himalaya) are similar. These boundaries delimit the meizoseismal area that can be associated with the extent of rupturing in the strike direction.

In our estimate, the 1934 detachment rupture extends 300 km along strike approximately as the ME VIII perimeter, and about 150 km across strike, from the BTF to the "slump belt". To the east this rupture abuts the 1897 rupture, so there is no gap between these two events. This boundary coincides with the Kishangang fault, a major north-south discontinuity in the basement (Figures 7 and 8) [Mathur and Evans, 1964]. This transverse structural boundary also coincides with an eastward drop in recent seismicity along the BTF.

1897 Western Assam

The 1897 earthquake (Figure 8) is the largest well-documented earthquake in India [Oldham, 1899], and probably one of the largest known anywhere. The highest intensity area of this event corresponds approximately with the Shillong Plateau and the northern extension of this tectonic unit below the plains of western Assam. The strongest effects included widespread landslides on the steep south side of the plateau, high acceleration (> 1 g) on the northern side of the plateau, and particularly severe liquefaction followed by floods (subsidence?) in the Assam plains north of the plateau. The area with the strongest effects (ME X) is included in the bell-shaped isoseismal contour drawn by Oldham [1899].

The correlation of the highest intensity with the outcrops of the Shillong crystalline rocks suggests that the ME X contour reflects the surface seismic response of these rocks. The severity of the effects in this area requires that this entire area is in the near field. Thus, the meizoseismal area, as the area above the rupture, may extend beyond the ME X area.

The only other isoseismal contour drawn by Oldham is the "limit to severe damage" (Figure 8), which is drawn around all the known cases of severe damage and of liquefaction, probably the limit of ME VII to VIII. On the basis of Oldham's descriptions we draw another contour around the areas where liquefaction phenomena were pervasive and intense. This contour is probably the limit of ME VIII to IX and extends $\sim$550 km in the E-W direction from the ME VIII contour of the 1934 event to less than 100 km from the ME VIII contour of the 1950 event (Figure 8). The distribution of 1897 landslides along the Himalayan front was not used to draw the ME VIII to IX contour. Nevertheless, the E-W extent of landslides corresponds to this contour which delimits the intense liquefaction in the alluvial plains. In our interpretation, the 1897 rupture extends along strike at least as far as the ME VIII to IX contour.

The western extent of the 1897 landslides and intense liquefaction are sharply defined and abut with the 1934 landslides and "slump belt"

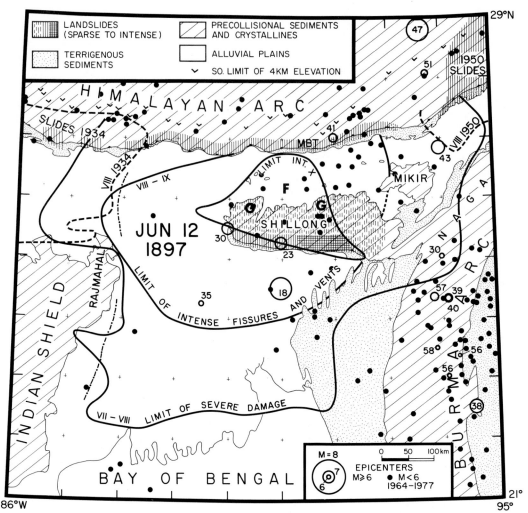

Fig. 8. Surface effects of the Western Assam earthquake of 1897. Intensities are in the Mercalli I-X scale (ME). The "limit of severe damage", probably ME VII to VIII, and the ME X region centered on the Shillong Plateau are from Oldham [1899, map N.1]. The limit of intense fissures and vents, intensity VIII to IX, and the distribution of landslides are from our interpretation of Oldham's compilation. F: zone of severe flooding following the earthquake; G: zones with high acceleration effects. All ISS epicenters are plotted for the period 1964-1977 (no magnitude M > 6 in this period). Only ISS epicenters with M > 6 for the period before 1964 are plotted (numbers indicate last 2 digits of the year in the 20th century). The Kishangang basement fault (K.F.) across the Gangetic basin and the buried scarp at the eastern edge of the Indian shield are indicated by dot-dashed lines. The western boundary of the 1897 rupture may be controlled by these features (see also Figure 7). A probable gap between the 1897 and the 1950 ruptures corresponds to the Mikir Hills. Earthquakes in 1943 and 1947 may have ruptured this gap.

respectively (Figure 7). The 1934-1897 rupture boundary as defined by the intensity coincides with the Kishangang fault, a major transverse basement feature (see section on 1934 event). The eastern extent of rupture is less well defined. The contour delimiting intense liquefaction reaches 93°E. The Mikir Hills, which extends from 93° to 94°E and are morphologically and geologically similar to the Shillong plateau, were effected by considerably lower intensities than the Shillong Plateau. Thus, the 1897 rupture probably did not extend farther than 93°E. This estimate would leave a fairly wide gap (50-100 km) between the 1897 and 1950 ruptures (see below). Two large forerunners of the 1950 event, the 1943, M 7.2 and the 1947, M 7.7 earthquakes, may have ruptured this gap.

Following a general pattern, variations of the 1897 intensity across strike can be associated with systematic changes in the surface seismic response and do not provide strong constraints on the extent of faulting in this direction. There is no available information about the intensity distribution north of the Sub-Himalaya. According to our model, the rupture should not extend farther north of the BTF, approximately the southern limit of 4 km elevation (Figure 8). Toward the south intense liquefaction extended to about 24°N. This corresponds approximately to the southern extent of the Rajmahal hills and with a pronounced bend in the Arakan fold belt (Figure 8).

In conclusion, the 1897 is amongst the greatest known earthquake ruptures, about 550 km long (east-west) and possibly 300 km wide (north-south). No surface rupture of any significance is associated with the 1897 event [Oldham, 1899, p. 138]. As for all the other well-known great Indian earthquakes, the absence of primary surface rupturing is strong evidence for a shallow-dipping, detachment-like fault source. This detachment rupture extends from the presumably continental foreland of the Himalayan arc to the presumably oceanic foreland of the Burma arc. It is possible that these two arcs share the same detachment and that the 1897 and the 1950 events account for subduction at both these structures.

The 1897 detachment source certainly extends under the highest intensity area corresponding with the Shillong Plateau. Thus, the Shillong and related Precambrian outcrops in Assam must be part of a detached crystalling sheet [Mathur and Evans, 1964]. This important conclusion raises the possibility that much of the crust underlaying the area east of the Kishangang fault is oceanic, and that the eastern Himalaya arc is in the process of changing from a continental subduction zone to an oceanic subduction zone [Seeber et al., 1981].

1950 Eastern Assam Earthquake

The 1950 Eastern Assam earthquake ($M_s = 8.4$, Gutenberg and Richter [1954]; $M_w = 8.6$, $M_o^s = 10^{29}$ dyne-cm, Kanamori [1977]) is the most recent of the great Himalayan earthquakes. It is also located in the extreme eastern and most remote portion of the Himalayan front. It has the best instrumental coverage, but not the best set of data on surface effects (Figure 9).

Instrumental data put the epicenter beyond the surface termination of the Himalayan arc, in the Mishmi Mountains that bound the Assam basin towards the east-northeast and trend northwest. Ben-Menahem et al. [1974] fit a strike-slip solution to the first-motion data of the main shock with one of the planes striking northwest. A more satisfactory fit of the data is obtained with the solution shown in Figure 9 [Chen and Molnar, 1977] where one of the planes dips gently northwestward and the slip is in the dip-direction of this plane. This solution suggests that the Himalayan detachment, which according to our model is the locus of all the other known great earthquakes, extends farther east than the surface termination of the Himalaya, and is also the locus of the 1950 Assam event. Seismic network data at

the western terminus of the Himalayan arc shows that basement structures and the detachment also extend well beyond the surface terminus of this arc at the Hazara-Kashmir syntaxis [Seeber and Armbruster, 1979].

The intensity distribution in Figure 9 is based on field observations [Poddar, 1953]. Poddar's ME IX and X contours are not shown in Figure 9 because they seem primarily a function of the type of effect. He assigns intensity X to landsliding in the mountains and intensity VIII-IX to liquefaction in the plains. Each of the effects that characterize the area of intensity ME $\geq$ VIII are considered separately in Figure 9.

The most dramatic feature of the 1950 event was the landslides. Mathur [1953] found by air reconnaissance that landslides covered 15,000 km^2, or about 1/3 of the surface in an area of 46,000 km^2. He concluded that at least 5 x 10^{10} m^3 of material was involved in the sliding. This is about 30 times the average yearly amount of detritus carried by the Brahmaputra river. Since the area effected by the 1950 slides is only a small portion of the drainage of the Brahmaputra, landslides during great earthquakes probably account for a substantial portion of the erosion in the eastern Himalaya.

The degree of water saturation in the weathered layer is a key factor in generating landslides. Thus, the much larger extent of landslides in the 1950 and 1897 events than in the 1934 and 1905 events can be attributed both to the location of the earthquakes, since the average precipitation generally increases eastward along the Himalaya, and to the timing of these events, since the 1950 and 1897 earthquakes occurred in the wet season, whereas the 1934 and 1905 earthquakes occurred in the dry season.

The distribution of landslides in the mountains as mapped by Mathur [1953] corresponds closely to the distribution of extensive liquefaction in the alluvial plain as indicated by the ME VIII contour [Poddar, 1953] or by severe damage to buildings, roads and rails [Dutt, 1953]. Both landslides and liquefaction were widespread and severe over the entire meizoseismal area and these effects drop off sharply at the western border of the meizoseismal area, near 94°E, the better surveyed portion of this boundary [Mathur, 1953]. The Subansiri River near this border was dammed by the landslides for four days [Poddar, 1953] and some of the worst liquefaction damage was reported from the area where this river enters the plains (Figure 9) [Dutt, 1953; Poddar, 1953].

The areas of most severe landsliding as mapped by Mathur [1953] (Figure 9) correspond to the deepest river gorges of the meizoseismal area along the Dihang (Brahmaputra), the Dibang and the Luhit rivers. Thus, these concentrations of landslides may reflect primarily the local relief, and not the strength of the seismic source.

The aftershocks [Chen and Molnar, 1977] are located primarily in the central portion of the

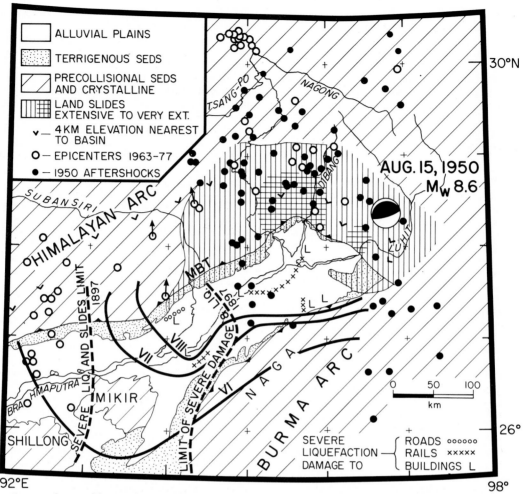

Fig. 9. The surface effects for the East Assam earthquake of 1950; $M_s = 8.4$ [Gutenberg and Richter, 1954]; $M_w = 8.6$ and $M_o = 10^{29}$ dyne-cm [Chen and Molnar, 1977]. Intensity contours (Mercalli) are by Poddar [1953] after a field reconnaissance (contours for intensity higher than VIII are omitted; see text). Data on road and rail damage are from Mathur [1953] and Dutt [1953]. Distribution of land-slides is from an air reconnaissance by Mathur [1953] (limit to landslide area is indicated only where directly verified). Fault plane solution (lower focal hemisphere, compression quadrant is black), epicenters of main shock (indicated by focal sphere) and aftershocks (black dots) are from Chen and Molnar [1977]. Epicenters from 1963 to 1977 (NOAA; open dots) are the same as in Figure 3. Eastern limits of 1897 event are also indicated.

meizoseismal area, and extend considerably beyond the limit of the landslides to the north and to the southeast (Figure 9). According to the tectonic model proposed here (Figure 4), detachment ruptures should not extend beyond the BTF. The topographic and seismic expression of the BTF in the area of the 1950 event correspond approximately to the northern boundary of the landslides. It is possible that aftershocks north of the BTF do not reflect the extent of the main rupture, and are instead caused by stress changes induced by the main rupture north of the BTF, in the area where seismicity reflects primarily east-west extension. However, the tectonic model proposed for the central portion of the Himalayan

arc may not apply to the eastern portion of this arc where oceanic rather than continental crust is perhaps being subducted (see discussion of 1897 event).

Few or no aftershocks are located [Chen and Molnar, 1977] on the western and eastern sides of the meizoseismal area. The location of the main shock confirms that the eastern side of this area corresponds with the rupture. Thus, it is possible that the well defined western side of the meizoseismal area also corresponds with the rupture even if no aftershocks are located there. The extent of the 1950 rupture in this direction is crucial in determining whether a seismic gap exists between the 1950 and the 1897 ruptures

[Khattri and Wyss, 1978] (see discussion on 1897 event).

In conclusion, the 1950 rupture extends about 300 km along the Himalayan strike from the epicenter of the main shock westward to a sharp intensity fall off at approximately 94°E. The intensities fall off rapidly south of the alluvial plain and suggest that the rupture did not extend below the Naga Hills. The eastern and northern extent of the rupture are less well defined.

Discussion

The sizes of the four Himalayan detachment ruptures which we infer from the intensity data are larger than other estimates for the same earthquakes [Chen and Molnar, 1977; Khattri and Wiss, 1978; Singh and Gupta, 1980]. When considering intensity data these and other authors tend to correlate the rupture only with the zone of maximum intensity. For large ruptures (with respect to depth), the zone of highest intensity is generally smaller than the meizoseismal area, the area above the rupture, because the intensity within this area can vary substantially, reflecting variations in source strength and/or surface response and/or other factors.

The size of the rupture can be also estimated from seismic moments obtained from amplitude spectra of recorded surface waves [e.g. Chen and Molnar, 1977, for the 1934 and 1950 earthquakes] or from the spectral ratio of surface waves travelling in opposite direction from the rupture [e.g. Singh and Gupta, 1980, for the 1934 earthquake].

The excitation function for surface waves vanishes when the dip angle and the depth of the rupture approach zero [Mendiguren, 1977]. Consequently, the moments of shallow detachment earthquakes are poorly constrained by surface wave data unless these parameters are well known independently. Intermediate magnitude thrust earthquakes along the Himalaya occur primarily at the downdip end of the inferred detachment ruptures and are generally characterized by steeper dip angles (Figure 3). Thus, 1.6×10^{28} dyne cm [Chen and Molnar, 1977] may be a low estimate of the moment for the 1934 Bihar event since it is based on a fault plane solution assumed to be similar to the solutions of recent smaller thrust earthquakes in the same area. On the other hand, the large moment estimated by the same method for the 1950 Assam earthquake [10^{29} dyne cm; Chen and Molnar, 1977], which is consistent with the 300 km long rupture we infer from the intensity, is based on a fault plane solution obtained from data of that same event (Figure 9). Moreover, the symmetrical dislocation model [Brune, 1970] is a misrepresentation of a detachment where the rigidity in the basement below this fault is much higher than in the sedimentary wedge above. This model may be grossly inadequate for deriving moments from the seismic energy radiated by detachment earthquakes.

The meizoseismal areas that we infer are usually close or somewhat smaller than the ME VIII contours. The 1905 event is exceptional since a large portion of the meizoseismal area corresponds to RF VII (ME VII-) Fortunately, foreshocks and geodetic data provide independent constraints and indicate that the rupture does correspond with this meizoseismal area. Thus, the 1905 data suggest that it is not unreasonable for ME $\geq$ VIII to occur only in the near-field of Himalayan detachment ruptures (at least in the strike direction - see the 1934 event).

Figure 10 is a tentative space-time diagram of Himalayan detachment ruptures after 1800. The data for the great earthquakes after and including 1897 provide one of the best compilations of surface effects from a set of great ruptures along a single structural unit. On the other hand, the data prior to 1897 are scanty [Jones, 1885, for the 1885 Kashmir event; Oldham, 1882, for the other earthquakes; Figures 3, 6 and 7] and the diagram in Figure 10 may be incomplete for this period. The 1803 earthquake was strongly felt (ME $\geq$ VIII) over a large area (Figure 6) and is very likely a great detachment rupture. The association of all the other earthquakes prior to 1897 with the Himalayan detachment is more uncertain.

Since 1897 about 1500 km or 60% of the Himalyan detachment (excluding the Hazara arc) has ruptured. If this rate of activity continues, the entire Himalayan front will rupture in about 130 years and the repeat-time for a typical great rupture (300 km long) would be about 150 years. Great earthquakes were less frequent in the century before 1897 assuming the record is complete. If the activity rate since 1803 is representative, the entire front would rupture in 180-240 years, depending on the extent of rupturing in 1803, 1833 and 1885. Typical great ruptures would occur anywhere along the Himalaya about every 30 years and the repeat time would be about 200-270 years. Ruptures much larger than 300 km, e.g. the 1897 rupture, may have longer repeat times [Sykes and Quittmeyer, this volume]. Since the 1950 event (30 years), there has been no large earthquake (M $\geq$ 7) that could be associated with the Himalayan detachment. A similar lull in seismicity has not occurred at least since before 1869. If the seismicity rate in the last 180 years is representative, the next rupture should be expected soon [Keilis-Borok et al., 1980].

According to the space-time diagram in Figure 10, a number of gaps may not have ruptured since 1800: a possible but unlikely small gap between the 1950 and 1897 ruptures; a possible gap between the 1803 and the 1833 ruptures; a gap between the 1905 and 1885 ruptures in Kashmir; and a gap in the Hazara arc (Figure 2). Great earthquakes that could have ruptured the detachment in the Hazara arc during the last two centuries are missing from the historic record [Quittmeyer and

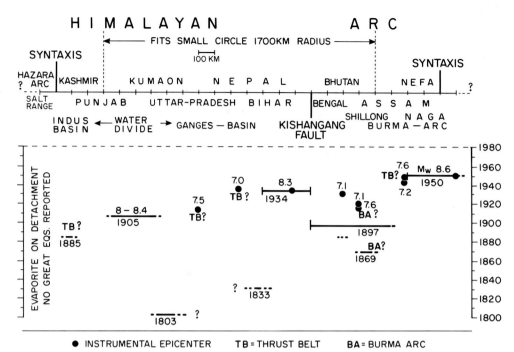

Fig. 10. Space-time diagram of detachment ruptures along the Himalayan arc after 1800. The data are very scanty for the period before 1897 and may be incomplete. The most likely gaps unruptured since 1800 are in Kashmir and in Uttar Pradesh, between the 1833 and 1803 ruptures. The gap between the 1897 and 1950 ruptures was probably filled by the 1943 and 1947 earthquakes. If the 1833 and 1803 ruptures abut, the entire Himalayan detachment, excluding Hazara and Kashmir where aseismic slip may play an important role, has probably ruptured since 1800.

Jacob, 1979]. This portion of the Himalayan detachment coincides with a thick Eocambrian evaporite layer [Seeber and Armbruster, 1979]. This evaporite causes weak coupling on the detachment which is manifested in the Salt Range and other characteristic surface features of the Hazara arc, as well as possible aseismic slip on the detachment. Thus the Hazara gap has probably a low seismic potential.

The Eocambrian evaporite was very widespread along the Tethyan margin of Gondwana [Stocklin, 1977]. Portions of this evaporite formation may underthrust the Himalaya and cause aseismic slip on the detachment east of Hazara, particularly in Kashmir [Seeber et al., 1981]. The active structure of Kashmir may be transitional between the structures of Hazara reflecting weak coupling at the detachment and the structures in the central Himalaya reflecting strong coupling (Figure 5). Thus, it is possible that the 1885 Kashmir earthquake (Figure 10) was a detachment earthquake, and that the gap between this rupture and the 1905 rupture is a likely site for the next great detachment event. However, the active tectonism of Kashmir is poorly constrained and any statement regarding great earthquakes in this region is very speculative.

The wide distribution of high intensities ($\geq$ ME VIII) in 1803 and, to a lesser degree, in 1833 (Figure 3) suggests that these events are great detachment earthquakes. However, the extent of these ruptures are very poorly determined by the available data [Oldham, 1882]. These two probable detachment ruptures abut the 1905 and 1934 ruptures, however, a wide gap may exist between them (Figure 10). Excluding Kashmir, this gap would be the most likely site for the next great event. If the 1803 and the 1833 ruptures are about the same size as the 1905 and 1934 ruptures ($\sim$ 300 km) they would abut. In this case, all the Himalayan detachment, except the portion in Hazara and Kashmir, would have ruptured since 1800 and the most likely next great rupture is either in Kashmir or a repeat of the 1803 event.

The intensity data indicate that the 1897 and the 1950 ruptures did not abut. The gap between them, about 100 km wide, may have been filled by the two large earthquakes in 1943 (M 7.2) and 1947 (M 7.6). Thus, this gap [Khattri and Wyss, 1979] is a possible, but not a likely site for the next great detachment earthquake.

Along some plate boundaries for which more than one rupture cycle is documented, periods of high seismic release rate during which much of the the plate boundary is ruptured in large earthquakes, alternate with periods of quiescence [Sykes et al., 1981]. It is possible that the Himalayan seismicity in the last 180 years represents a complete active cycle during which the entire detachment ruptured, and that the Himalayan front is now entering a relatively long period of quiescence. In this case our estimate of the recurrence rates are too short. On the other hand, if the rate of great Himalayan earthquakes in the last 180 years (a 300 km long rupture on the average every 30 years) is representative, the next great earthquake can be expected during the next decade.

Conclusions

Continental subduction along the Himalayan arc and oceanic subduction along arcs with wide accretionary wedges are structurally similar. Great earthquakes rupture a quasi-horizontal fault, the detachment, in both these structures. The concept of seismic gaps is also applicable to both. The Himalayan detachment extends at least 2600 km along strike, including portions beyond the eastern and western syntaxis. In the dip direction the detachment extends from the basement thrust front (BTF), which corresponds to the boundary between the Lower Himalaya and the High Himalaya, to the foredeep, south of the MBT. This distance may be from as little as 100 km in the Western Himalaya (1905 event) to more than 300 km in the eastern Himalaya (1897 event).

In our interpretation the last four great earthquakes ruptured about 1,400 km of the Himalayan detachment (500 km in 1897; 300 km in 1905; 300 km in 1934; 300 km in 1950). Two large earthquakes prior to the 1950 event (1943 and 1947) probably ruptured a 100 km gap between the 1897 and the 1950 ruptures. 100 km or more of the detachment at the western terminus of the Himalaya coincide with an evaporite layer and may slip aseismically. Excluding this portion, about 3/4 of the Himalayan detachment has ruptured since 1897.

During the century before 1897 there were three possible great detachment earthquakes: 1885 in Kashmir, 1833 in western Bihar and 1803 in Uttar Pradesh. The extent of rupturing for these events is poorly constrained. The most likely gaps unruptured since 1800 are in Uttar Pradesh, between the 1803 and the 1833 ruptures, and in Kashmir, between the 1885 and the 1905 ruptures. The tectonics of the Kashmir Himalaya is poorly understood, and it is not known whether the detachment in this region can generate great earthquakes. On the other hand, a gap in Uttar Pradesh would be expected to generate a great earthquake soon.

If the 1803 and the 1833 ruptures are as large as the more recent great ruptures ($\sim$ 300 km), then they would abut. In this case, excluding Kashmir, the entire Himalayan detachment has ruptured in the last 180 years and the most likely future great earthquake is a repeat of the 1803 event.

If the rate of detachment ruptures since 1803 is representative, and if the detachment slips primarily in 300 km long ruptures (the 1897 event can be considered two such ruptures occurring simultaneously), then the entire Himalayan detachment would rupture in 180 to 240 years and the repeat time of a typical (300 km) rupture would be 200-270 years, depending on the extent of detachment rupturing in the 19th century. This repeat time can be substantially longer if periods of high seismicity during which the entire detachment ruptures alternate with periods of quiescence.

Acknowledgments. We thank Terry Engelder, Klaus Jacob, Richard Quittmeyer, Thomas Fitch, David Simpson, Ross Stein, and Lynn Sykes for critically reading the manuscript and for helpful suggestions. We are also grateful to the Government of Pakistan for assistance in our fieldwork. This work was supported by grants EAR 79-19995 from the National Science Foundation and USGS 19123 from the U.S. Geological Survey. Lamont-Doherty Geological Observatory Contribution Number 3125.

References

Ambraseys, N., and S. Sarma, Liquefaction of soils induced by earthquakes, Bull. Seismol. Soc. Am., 59, 651-664, 1969.

Ambraseys, N. N., Some characteric features of the Anatolian fault zone, Tectonophysics, 9, 143-165, 1970.

Andrieux, J., and M. Brunel, Metamorphism, granitization, and relations with the Main Central Thrust in Central Nepal: 87Rb/87Sr age determinations and discussion, Editions de C.N.R.S., 268, 31-40, 1977.

Armbruster, J., L. Seeber, and K. H. Jacob, The northwestern termination of the Himalayan mountain front: Active tectonics from micro-earthquakes, J. Geophys. Res., 83, no. B1, 269-282, 1978.

Auden, J. B., and A. M. N. Ghosh, Preliminary account of the earthquake of 1934 in Bihar and Nepal, Rec. Geol. Surv. India, 68, 177-239, 1934.

Ben-Menahem, A., E. Aboodi, and R. Schild, The source of the great Assan earthquake - An interplate wedge motion, Phys. Earth Planet. Int., 9, 265-289, 1974.

Behrendt, J. C., R. M. Hamilton, H. D. Ackerman, V. J. Henry, and K. C. Bayer, Cenezoic reactivation of faulting in the vicinity of Charleston, South Carolina earthquake zone, submitted to Geology, 1980.

Bird, P., Initiation of intracontinental subduction in the Himalaya, J. Geophys. Res., 83, 4975-4987, 1978.

Brett, W. B., A report on the Bihar earthquake and on the measures taken in consequence thereof up to the 31st December 1934, Government of Bihar and Orissa, Patna, 101 pp., 1935.

Brune, J. N., Tectonic stress and the spectra of seismic shear waves from earthquakes, J. Geophys. Res., 75, 4997-5009, 1970.

Brunel, M., and J. Andrieux, Deformations

superposees et mecanismes associes au chevauchement central Himalayan "M.C.T.": Nepal oriental, in Proceedings C.N.R.S. Coloquium on the Geology and Ecology of the Himalayas, Paric, 69-84, 1977.

Burchfiel, B. C., Plate tectonics and the continents: A review, in Continental Tectonics, National Academy of Sciences, Washington, D. C., 1980.

Chandra, U., Seismicity, earthquake mechanisms and tectonics along the Himalayan mountain range and vicinity, Phys. Earth Planet. Inter., 16, 109-131, 1978.

Chen, W. P., and P. Molnar, Seismic moments of major earthquakes and the average rate of slip in central Asia, J. Geophys. Res., 82, 2945-2969, 1977.

Chugh, R. S., Study of recent crustal movements in India and future programs, paper presented at the International Symposium on Recent Crustal Movements, Zurich, 1974.

Cook, F. A., D. S. Albaugh, L. D. Brown, S. Kaufman, J. E. Oliver, and R. D. Hatcher, Thin-skinned tectonics in the crystalline southern Appalachians; COCORP seismic-reflection profiling of the Blue Ridge and Piedmont, Geology, 7, 563-567, 1979.

Curray, J. R., and D. G. Moore, Growth of the Bengal deep-sea fan and denudation of the Himalayas, Geol. Soc. Amer. Bull., 82, 563-572, 1971.

Dunn, J. A., J. B. Auden, A. M. N. Ghosh, S. C. Roy, and D. N. Wadia, The Bihar-Nepal earthquake of 1934, Mem. Geol. Surv. India, 73, 1939.

Dutt, G. N., Damage caused by earthquakes of the 15th August, 1950 in Assam, in A Compilation of Papers on the Assam Earthquake of August 15, 1950, Compiled by M. B. Ramachandra Rao, Government of India, 72-75, 1953.

Elliott, D., The motion of thrust sheets, J. Geophys. Res., 81, 949-963, 1976.

Fuchs, G., and W. Frank, The geology of West Nepal between the rivers Kali Gandaki and Thulo, Bheri, Jehrb. Geolog. Bund. Anst., 18, 103 p., 1970.

Gansser, A., Geology of the Himalayas, Inter-Science Publishers, John Wiley and Sons, London, 1964.

Gansser, A., The great suture zone between Himalaya and Tibet a preliminary account, Editions du C.N.R.S., 268, 181-191, 1977.

Gutenberg, B., and C. F. Richter, Seismicity of the Earth and Associated Phenomena, 273 p., Princeton University Press, Princeton, New Jersey, 1954.

Hashimoto, S., Y. Ohta, and Ch. Akiba (editors), Geology of the Nepal Himalaya, Sapporo (Saikon), 292 pp., 1973.

Johnson, G. D., N. M. Johnson, N. D. Opdyke, and R. A. K. Tahirkheli, Magnetic reversal stratigraphy and sedimentary tectonic history of the Upper Siwalik Group, Eastern Salt Range and Southwestern Kashmir, in Geodynamics of Pakistan, edited by A. Farah and K. DeJong,

p. 249, Geol. Surv. Pakistan, Quetta, 1979.

Jones, E. J., Notes on the Kashmir earthquake of 30 May 1885, Records of the Geol. Surv. India, 18, 153-156, 221-226, 1885.

Kanamori, H., The energy release in great earthquakes, J. Geophys. Res., 82, 2981-2987, 1977.

Katsui Y. (editor), List of the World Active Volcanoes, Volcanological Society of Japan, 1971.

Keilis-Borok, V., L. Knopoff, and C. R. Allen, Long-term premonitory seismic patterns in Tibet and the Himalayas, J. Geophys. Res., 85, 813-820, 1980.

Kelleher, J. A., L. R. Sykes, and J. Oliver, Possible criteria for predicting earthquake locations and their applications to major plate boundaries of the Pacific and Caribbean, J. Geophys. Res., 78, 2547-2585, 1973.

Khattri, K., and M. Wyss, Precursory variation of seismicity rate in the Assam area, India, Geology, 6, 685-688, 1978.

Krishnaswamy, V. S., S. P. Jalote, and S. K. Shome, Recent crustal movements in Northwest Himalaya and the Gangetic foredeep and related patterns of seismicity, Proceedings of the 4th Symposium on Earthquake Engineering, University of Roorkee, 19 .

Lahr, J. C., R. A. Page, and J. A. Thomas, Catalog of earthquakes in South Central Alaska, April-June 1972, U.S. Geol. Surv. Open-File Report, 1974.

LeFort, P., Himalayas: The collided range. Present knowledge of the continental arc, Amer. J. Sci., 275A, 1-44, 1975.

Mathur, L. P., Assam earthquake of 15th August 1950 - A short note on factual observa-tions, in A Compilation of Papers on the Assam Earthquake of August 15, 1950, compiled by M. B. Ramachandra Rao, Government of India, 1953.

Mathur, L. P., and P. Evans, Oil in India, Sp. Brochure, Inter. Geol. Congr. 22nd Session, New Delhi, 64-79, 1964.

Mattauer, M., Sur de mechanisme de formation de la schistosite dand l'Himalaya, Earth Planet. Sci. Lett., 28, 144-154, 1975.

Mehta, P. K., Tectonic significance of the young mineral dates and the rates of cooling and uplift in the Himalaya, Tectonophysics, 62, 205-217, 1980.

Mendiguren, J. A., Inversion of surface wave data in source mechanism studies, J. Geophys. Res., 82, 889-894, 1977.

Menke, W., and K. H. Jacob, Seismicity patterns in Pakistan and northwestern India associated with continental collision, Seismol. Soc. Amer. Bull., 66, 1695-1711, 1976.

Middlemiss, C. S., The Kangra earthquake of 4th April 1905, Mem. Geol. Surv. India, 38, 1910.

Molnar, P., W. P. Chen, T. J. Fitch, P. Tapponnier, W. E. K. Warsi, and F-T Wu, Structure and tectonics of the Himalaya: A brief summary of relevant geophysical observations, Editions de C.N.R.S., 268, 269-294, 1977.

Molnar, P., and P. Tapponnier, Active tectonics of Tibet, J. Geophys. Res., 83, 5361-5375, 1978.

Ni, J., and J. E. York, Cenozoic extensional tectonics of the tibetan plateau, J. Geophys. Res., 83, 5377-5384, 1978.

Oldham, T., A catalogue of Indian earthquakes from the earliest time to the end of A.D. 1869, Geol. Surv. India Mem., 19, part 3, 163-215, 1882.

Oldham, R. D., Report on the great earthquake of 12 June 1897, Mem. Geol. Surv. India, 29, 1899.

Parkash, B., R. P. Sharma, and A. K. Roy, The Siwalik group (molasse) - sediments saved by collision of continental plates, Sedimentary Geology, 25, 127-159, 1980.

Pecher, A., Deformations et Metamorphisme associes a une zone de cisaillement, Thesys, University of Grenoble, 354 pp., 1978.

Pecher, A., Les inclusions fluides des quartz d'exsudation de la zone du M.C.T. himalayen au Nepal central: donnees sur la phase fluide dans une grande zone de cisaillement crustal, Bul'. Mineral, 102, 537-554, 1979.

Plafker, G., Tectonic deformation associated with the 1964 Alaska earthquake, Science, 148, 1675, 1965.

Poddar, S. M. C., A short note on the Assam earthquake of August 15, 1950, in A Compilation of Papers on the Assam Earthquake of August 15, 1950, compiled by M. B. Ramachandra Rao, Government of India, 1953.

Powell, C. McA., and P. J. Conaghan, Plate tectonics and the Himalayas, Earth Planet. Sci. Lett., 20, 1-12, 1973.

Powell, C. McA., A. R. Crawford, R. L. Armstrong, R. Prakash, and H. R. Wynne-Edwards, Reconnaissance Rb-Sr dates for the Himalayan central gneiss, northwest India, Indian Jour. Earth Sci., 6, 139-151, 1979.

Quittmeyer, R. C., and K. H. Jacob, Historical and modern seismicity of Pakistan, Afghanistan, Northwestern India and Southeastern Iran, Bull. Seismol. Soc. Am., 69, 773-823, 1979.

Richter, C. F., Elementary Seismology, W. H. Freeman and Company, San Francisco and London, 768 pp., 1958.

Seeber, L., J. G. Armbruster, and R. C. Quittmeyer, Seismicity and continental subduction in the Himalayan arc, Interunion Commission on Geodynamics Working Group 6 Volume, 1981.

Seeber, L., and J. G. Armbruster, Seismicity of the Hazara Arc in northern Pakistan: Decollement vs. basement faulting, in Geodynamics of Pakistan, edited by A. Farah and K. A. DeJong, 131-142, Geol. Surv. Pakistan, 1979.

Sengupta, S., Geological and geophysical studies in western aprt of Bengal Basin, India, Bull. Amer. Assoc. Petrol. Geol., 50, 1001-1012, 1966.

Singh, D. D., and H. K. Gupta, Source dynamics of two great earthquakes of the Indian subcontinent: The Bihar-Nepal earthquake of January 15, 1934 and the Quetta earthquake of May 30, 1935, Bull. Seismol. Soc. Am., 70, 757-773, 1980.

Sinvhal, H., P. N. Agrawal, G. C. P. King, and V. K. Gaur, Interpretation of measured movement at a Himalayan (Nahan) Thrust, Geophys. J. Roy. astr. Soc., 34, 203-210, 1973.

Stein, R. S., Role of slope-dependent errors on the accuracy of geodetic leveling in Southern California, this volume, 1981.

Stocklin, J., Structural correlation of the Alpine ranges between Iran and Central Asia, Mem. h. ser. Soc. Geol. Fr., no. 8, 333-353, 1977.

Stocklin, J., Geology of Nepal and its regional frame, Jour. Geol. Soc. London, 137, 1-34, 1980.

Sykes, L. R., Aftershock zones of great earthquakes, seismicity gaps and earthquake prediction for Alaska and the Aleutians, J. Geophys. Res., 76, 8021-8041, 1971.

Sykes, L. R., J. B. Kisslinger, L. House, J. N. Davies, and K. H. Jacob, Rupture zones and repeat times of great earthquakes along the Alaska-Aleutian arc, 1784-1980, this volume., 1981.

Sykes, L. R., and R. C. Quittmeyer, Repeat times of great earthquakes along simple plate boundaries, this volume, 1981.

Toksoz, M. N., and P. Bird, Modeling of temperatures in continental convergence zones, Tectonophysics, 41, 181-193, 1977.

Valdiya, K. S., Himalayan transverse faults and folds and their parallelism with subsurface structures of north Indian Plains, Tectonophysics, 32, 353-386, 1976.

Yonekura, N., and K. Shimazaki, Uplifted marine terraces and seismic crustal deformation in arc-trench systems: A role of imbricated thrust faulting (abstract), EOS Trans. AGU, 61(46), p. 1111, 1980.

ACTIVE FAULTS AND DAMAGING EARTHQUAKES IN JAPAN

—MACROSEISMIC ZONING AND PRECAUTION FAULT ZONES

Tokihiko Matsuda

Earthquake Research Institute, University of Tokyo, Tokyo 113, Japan

Abstract. The location of active faults and epicenters of damaging earthquakes on land in Japan are compared. Frequency of damaging earthquakes of M≥6.5 is four times higher in areas where active faults reach lenghts ≥10 km as compared to areas characterized by shorter fault segment. About 80 % of recent large damaging earthquakes have occurred along topographically-detectable active fault zones or within five kilometers from fault traces. This suggest that future damaging earthquakes will occur mostly (80 % or more) along presently-known active fault zones. Recent damaging earthquakes occurred successively in four limited areas : Rikuu, Kanto, Nobi and Tango areas. These earthquakes occurred within a short interval of a few tens of years and a few tens of kilometers, indicating their occurrences may be physically coupled to each other. Twelve precaution faults on land in Japan are designated, based on (1) elapse ratio of active faults and (2) existence of a quiet segment in a recently active fault zone.

Introduction

Damaging earthquakes of about magnitude 7 in Japan are commonly associated with surface rupture along recognized active geologic faults. This implies that the location of active Quaternary faults may be a useful key in recognizing future sites of large earthquakes. Recently, data concerning the active Quaternary faults in Japan has been collected and tabulated in a catalogue of sheet maps(Research Group for Active Faults, 1980). 1471 active faults have now been recognized in Japan. In this paper, some attempts are made to evaluate the seismic potential of various regions of Japan using the known distribution of active geologic faults.

Distribution of Active Faults and Damaging Earthquakes

Active fault data used in this paper are taken from Research Group for Active Faults(1980). Fig. 1 shows distribution of well-defined active faults on land(certainty I and II by R.G.A.F., 1980). Earthquake data are from Japanese Meteorological Agency for period of 1926-1979 and from Utsu(1979) for period of 1885-1925. Twenty four damaging earthquakes that occurred on land are shown in Fig. 2. These earthquakes, occurred during the period 1885 to 1978, are of magnitude ≥6.5 located at depth less than 30 km, and are used for comparison with the distribution of active faults.

The Japanese islands were divided into quadrangles on scale of 1:50,000 by the Geographical Institute of Japan(approximately 20 km x 25 km). These may be classified into zones 1, 2 and 3 based on distribution of active faults, as shown in Fig. 3.

Zone 1: area where no active faults are known.

Zone 2: area where an active fault(or faults) shorter than 10 km in length is known.

Zone 3: area where an active fault(or faults) longer than 10 km is present or a part of it is included. Areas having active faults with certainty III(lowest class in certainty) longer than 10km in length are regarded as zone 2.

Future large shallow earthquakes greater than magnitude 7 are expected to occur from zone 3, because all the Japanese historical surface faults over 10 km in length were associated with shallow earthquakes of magnitude 7 or greater (Fig. 4).

Comparison of locations between active faults and epicenters of the shallow damaging earthquakes indicate that there is a close relation between the two. Fig. 5 shows the spatial relation between areas of zones 1, 2, 3 and number of damaging earthquakes from each zone. Most of damaging earthquakes(16 among 24) have occurred in zone 3 areas which occupy only about one third of area of the Japanese Islands. This means that frequency of earthquake occurrence per unit area is four times higher in a region of zone 3 than in zones 1 and 2. Zone 3, therefore, is considered to be an area of higher potential for future shallow damaging earthquakes than the other areas.

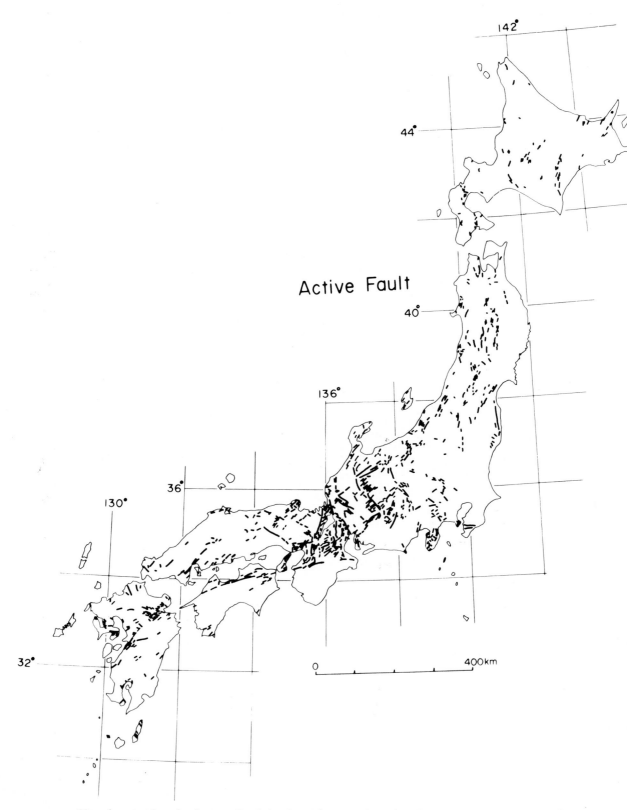

Fig. 1. Active faults on land in Japan(Research Group for Active Faults, 1980).

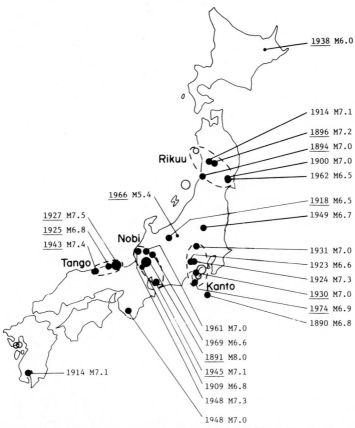

Fig. 2. Very shallow, major damaging earthquakes on-land(solid circles, h≤30 km, M>6.5, period 1885-1978, after J.M.A. and Utsu, 1979). Open circles are those near-off-shore. Underlined are those accompanied surface faulting. Larger circles are M≥7.5. Two earthquakes (M<6.0, smaller solid circles) have been added as they also had associated surface faults.

Seismic Historic Surface Faults and Pre-detectable Active Faults

Among the on-land damaging earthquakes (Fig. 2), twelve earthquakes were accompanied by surface faulting. Some of the surface faults appeared along easily-detectable pre-existing Quaternary fault traces, but some others appeared in a area in which active faults are hardly detectable in the topography before or even after the earthquake.

Degrees of coincidence in location between damaging earthquakes and active faults are classified into following three cases :

Case (1), in which seismic faulting appeared entirely or partially on known Quaternary fault traces which have distinct topographic expressions detectable before the earthquake occurred.

Case (2), in which the seismic surface fault or instrumental epicenter or meisoseismal area of the earthquake was located within 5 kilometers of detectable active faults.

Case (3), in which no corresponding detectable active faults are found in the epicentral area within 5 km of the epicenter.

As shown in Table 1, 19 earthquakes among 24 (about 80%) fall in cases (1) and (2). This implies that most(about 80%) of future, inland, shallow, large damaging earthquakes(M≥6.5) will come from limited areas along presently-known active fault zone. The rest(about 20%) will occur, however, from unexpectable area where no significant active faults are known at present. Such future, unexpectable large earthquakes will occur mostly from C-class faults or less active ones, as will be mentioned in the next section.

Number of Known Active Faults and Frequency of Damaging Earthquakes

Active faults in Japan have been classified into classes A, B, C...., according to their long-term slip rate (S) averaged in late Quaternary time:classes A, B, C are assigned to faults of 1≤S<10 mm/year, 0.1≤S<1 mm/year 0.01≤S<0.1 mm /year, respectively(Matsuda, 1977;Res. Gr. for Active Faults, 1980).

The number of known active faults, on land, belonging to each class is compared, in Fig. 6,

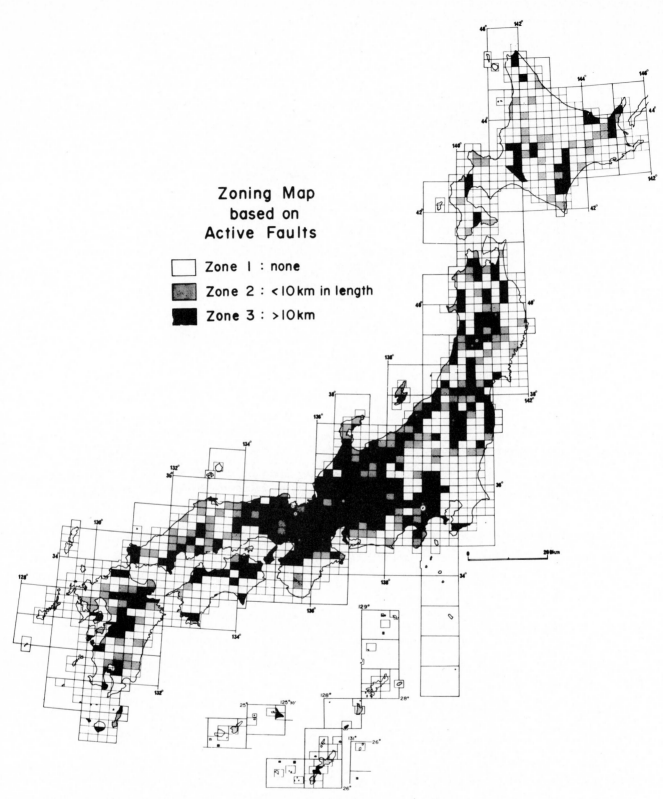

Fig. 3. Zoning map of Japan based on distribution of active faults. Zone 1 : area of no active fault. Zone 2 : area of active faults shorter than 10 km in length. Zone 3 : area of active faults longer than 10 km. (Data after Res. Gr. for Active Faults, 1980).

Length of surface break and earthquake magnitude

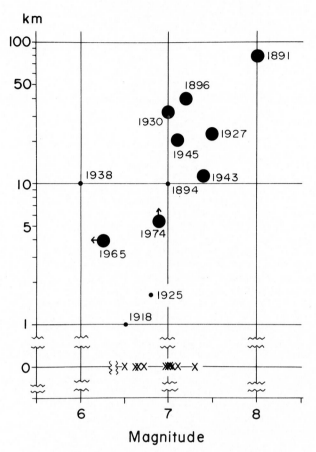

Magnitude

Fig. 4. Relation between surface fault length and earthquake magnitude. Magnitude data for earthquakes in 1885-1925 after Utsu (1979). Small circles indicate inconclusive surface faulting or rupture due to land-slides. Crosses represent very shallow, on-land earthquakes without surface faulting.

with the number of damaging earthquakes that occurred along them respectively. It is interesting to note that two or three large earthquakes had occurred on each of classes A, B and C during the last one hunderd years, in spite of the very large differences in earthquake recurrence intervals among faults of each class. The recurrence intervals of earthquakes become ten times longer successively in the lower class faults than in the higher class faults: recurrence intervals of magnitude 7 earthquakes are calculated to be $0.16-1.6/10^3$ years for class A, $0.16-1.6 \times 10^4$ years for class B and $0.16-1.6 \times 10^5$ years for class C faults, according to equation (1).

The similar number of damaging earthquakes from classes A and B faults can be explained by the fact that number of class B faults so far known is about ten times more than that of class A faults. However, number of C-class faults are far less than ten times of number of B class, even when unspecified or Certainty III faults are regarded as C class. This implies that many more C class faults exist than are yet recognized and that future on-land damaging earthquakes whose locations we may fail to predict will occur largely on such lower class faults that are not described yet in Res. Gr. for Active Faults (1980). C class faults are relatively difficult to find owing to their weak topographic expression.

Calculated and Observed Earthquake Recurrence Intervals

An average recurrence interval of earthquakes from a fault can be calculated from an average rate of displacement in late Quaternary assuming a certain earthquake magnitude or displacement on the fault. For the active faults on land in Japan, the following relation is known (Matsuda 1977).

$$\log R = 0.6 M - (\log S + 1.0) \qquad (1)$$

where R represents an average recurrence interval of earthquakes in years and S represents average rate of displacement in mm per year.

Occurrence of major earthquakes from zones 1, 2, 3

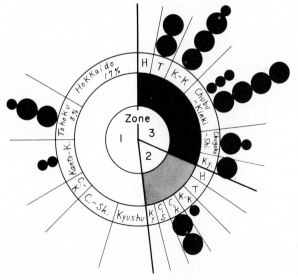

Fig. 5. Diagram showing number of shallow earthquakes from each of zones 1, 2, 3. Earthquakes are shown in Fig. 2 (solid circles). Larger circles : earthquakes of M>7. Smaller circles : earthquakes of 7>M≥6.5.

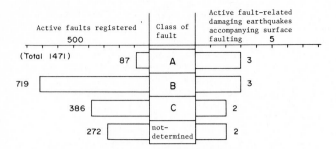

Fig. 6. Number of active faults by classes
registered in R.G.A.F.(1980) and number of
Case 1 (Table 1) damaging earthquakes from
each class of fault.

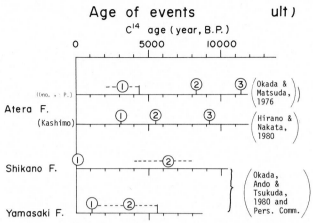

Fig. 7. Pre-historic events along the Atera
fault, Shikano fault and Yamasaki fault(data
from Okada & Matsuda, 1976, Hirano & Nakata,
1980, and Okada, Ando & Tsukuda, 1980 and
personal comm.).

Excavation of the Shikano and the Yamasaki
faults in the western Japan(Okada, Ando & Tsukuda
1980) and the Atera fault in central Japan(Okada
and Matsuda, 1976, Hirano & Nakata, 1980) has
revealed the occurrence of several prehistoric
earthquakes. The results are shown in Fig. 7.

The Atera fault is class A fault that slips (S)
at 3-5 mm/year. The recurrence interval calculated from eqation(1) assuming a magnitude 8
earthquake is about 1300-2100 years. The result
of the excavation shows a few events with intervals of 2000-4000 years. The Shikano fault of B-
C class which means an average recurrence interval of several thousand years, was broken during
the Tottori earthquake of 1943. That was the
first surface faulting event known in historical
time. The excavation shows that the pre-1943
event occurred several thousand years ago. The
Yamasaki fault of A-B class was demonstrated, by
the excavation,to have moved about 1000 years ago
in accordance with historic record:the earthquake

in A.D. 868 was ascribed to this fault. The
second last faulting along the Yamasaki fault was
a few thousand years ago.

The recurrence intervals on these faults, obtained by excavations, are consistent with those
calculated. This supports the earlier estimation
that the recurrence interval of large earthquakes
occurring from the same fault or the same segment
of a fault is longer than several hundred years
even if it is a class A fault(Matsuda, 1977). If
so, one may consider that areas close to a fault
that moved in the last one hundred years or so
are safer in earthquake risk than the other
areas. However, this is not necessarily true
everywhere because of possible increase in

TABLE 1. Coincidence In Location Between Surface Fault Traces Or Epicenters Of
Damaging Earthquakes And Detectable Quaternary Faults.

Case I(good) (10/24)	Case II (moderate) (9/24)	Case III(none) (5/24)
1891 Nobi(A)	1894 Shonai(B)	1890 Miyakejima
1896 Rikuu(B)	1909 Anegawa(B)	1900 Miyagi
1918 Omachi(A-B)	1914 Sakurajima	1914 Senpoku
1927 Tango(C-B)	1923 Yamanashi	1925 Tajima
1930 Kita-Izu(A)	1924 Tanzawa	1962 Miyagi
1943 Tottori	1931 Nishi-Saitama(B)	
1945 Mikawa(C)	1948 Fukui(B)	
1961 Kita-Mino(C)	1948 Tomitagawa	
1969 Gifu-Centr.	1949 Imaichi	
1974 Izu-oki(B)		

Case I-III, See text.
A, B, C in parenthesis represent degree of activity of corresponding Quaternary
fault by Res. Gr. for Active Faults(1980).

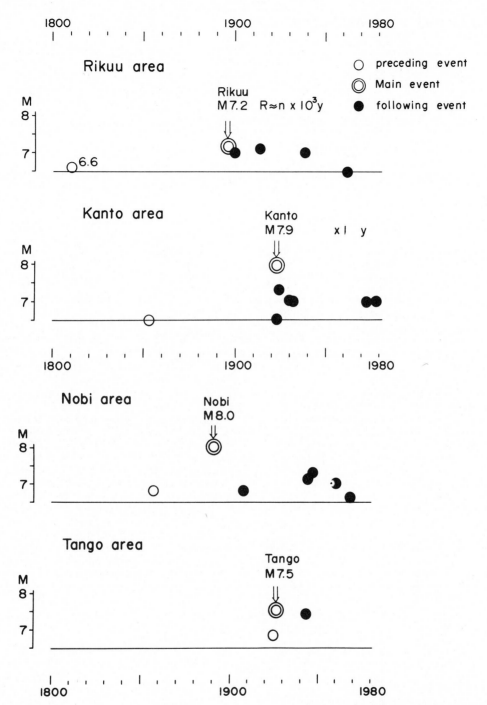

Fig. 8. Successive occurrence of large shallow earthquakes in four areas. For location of areas and earthquakes, see Fig. 2.

seismicity in the neighbouring areas. This problem will be discussed below.

Successive Occurrence of Damaging Earthquakes in a Region

Epicenters of on-land, shallow, damaging earthquakes larger than 6.5 in magnitude which occur- red during the last one hundred years are located mostly in four limited areas, the Rikuu, Kanto, Nobi and Tango areas, as shown in Fig. 2. The time sequences of their occurrence in each area are shown in Fig. 8. It is remarkable in these areas that once a large event had occurred, two or more large damaging earthquakes followed it

TABLE 2. Precaution Faults Based On Elapse Ratio E (E>0.5)

Precaution faults	S mm/y	M	R $\times 10^3$y	t $\times 10^3$y	E t/R
(1) Median Tectonic Line	5-10[1]	8[2]	0.6-1.3	>1	0.8-1.6[2]
(2) Arima-Takatsuki-Rokko fault zone	0.8[3]	7	2	>1	>0.5
(3) Atera fault	3-5	8	1.3-2.1	>1	0.5-0.8[2]
(4) Inadani fault zone	4-7[3]	8	0.9-1.6	>1	0.9-1.6
(5) Itoigawa-Shizuoka Line	0.8[3]	8	7.9	4[4]	0.5
(6) Fujigawa fault zone	3-7[5]	7	0.2-0.5	>1	2-5
(7) Kozu-Matsuda-Kannawa fault zone	1-2	7	0.8-1.6	>1	0.6-1.3[2]

Number on the left correspond to those in Fig. 9.

[1] Okada(1980), [2] Matsuda(1977), [3] R.G.A.F.(1980), [4] Matsushima and Ban(1979),
[5] Yamazaki(1979)

with intervals of several tens of years, which are very short intervals when compared with general recurrence intervals of earthquakes from a fault. Those successive events in each respective area occurred only several tens of kilometers apart from each other. These events are not aftershocks:they were not located in any of the aftershock areas of the preceding events, and the lengths of intervals between events were too long to regard them aftershocks. This successive occurrence of earthquakes in time and space seems to indicate that there are some interactions among different faults in a region and that each respective area had been in a stressed state high enough to produce earthquakes successively.

Precaution Fault Zones

Some faults may be selected for earthquake prediction studied. Based on the following two criteria, certain faults are designated as "precaution fault zones". Criterion 1 is based on the periodicity of earthquake occurrence on the same fault:the longer the time since the last large earthquake, the shorter the time until the next earthquake. Since large earthquakes take place more or less periodically from a fault, the ratio of length of time since the last large earthquake t, to average recurrence interval R (called Elapse ratio, $E = t/R$), represents roughly the potential for near future large earthquakes

TABLE 3. Precaution Fault Zone Based On Recent Historical Seismicity

Location of precaution segment and/or fault zone	Recent historical large Eq.(1800-, M≥7) from nearby segment of the fault zone
(8) Yanagase-Suzuka fault zone	1819(M7.4)
(9) Suruga trough segments of Nankai trough fault zone	1854(8.4), 1944(8.0). 1946(8.1)
(7) Kozu-Matsuda and Kannawa segments of Sagami trough fault zone	1923(7.9)
(10) Nishi-Kanto fault zone	1855(7.0), 1931(7.0)
(11) Shinanogawa fault zone	1847(7.4), 1964(7.5)
(12) Akita-Shonai fault zone	1804(7.1), 1894(7.0)

Numbers on the left correspond to those in Fig. 9.

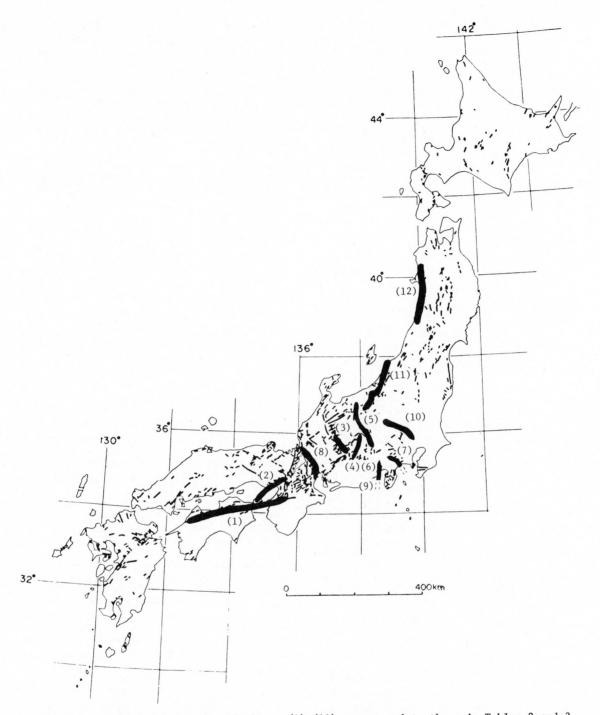

Fig. 9. Location of precaution faults. (1)-(11) correspond to those in Tables 2 and 3.

from that fault(Matsuda, 1977). Most of the active faults in Japan have no record of activity in historical time. This means that t of those faults is usually larger than several hundred years, while R of class A faults is usually 1000 years or so. Therefore, class A faults having no historic record of activity have gener-ally a higher possibility for occurrence of a near future large earthquake than active faults on which a large earthquake occurred in historical time. Thus, active faults with $E>0.5$ are denoted as precaution faults in Table 2. The Median Tectonic Line(Okada, 1980) is an example of precaution fault zones according to criterion

1. The central segment of the Median Tectonic Line has a recurrence interval of 600-1300 years, but has no historical record of damaging earthquakes from that segment during the last 1000 years or so. In this case R = 600-1300 years, t >1000, then the elapse ratio is E>0.6-1.3.

The second criterion is based on the chain reaction of seismicity between segments in a fault zone. In this case, the shorter the time since the last event, the shorter the time until the next event. This is not contradictory to the first criterion. This second criterion is not applicable to a single fault segment, but to a whole fault zone or a certain tectonic province in which various fault segments are included. As shown in the previous section, there must be some trigger action or links in seismicity between various fault segments in a fault zone or in the same province. Therefore, such a fault zone or an area including a segment on which a large event has occurred in historical time, has a higher probability for the next, near future event than other fault zones quiet through the recent historical past. A historically non-seismic fault segment included in such a historically active fault zone or a fault zone including such non-seismic segments can be a precaution fault of criterion 2. They are listed in Table 3.

The Shinanogawa fault zone, for example, produced two great earthquakes in recent history: one is the Zenkoji earthquake(M7.4) of 1847 at the western end and the other is the Niigata earthquake(M7.5) of 1964 at the eastern end. The central segments between the two epicentral areas are regarded as a precaution fault or a kind of seismic gap(seismic gap of the first kind by Mogi 1979). The Kozu-Matsuda and Kannawa segments of the Sagami trough fault zone have no record of activity in recent past(E>0.5) and are situated next to segments from which great earthquakes occurred successively in 1923 and in 1703(Matsuda et al. 1978). Thus, this segment is, according to criterion 2, a seismic gap with E>0.5. Fuji-gawa area(Yamazaki 1979) north of the Suruga Bay is also in a similar situation.

Precaution faults based on either criterion 1 or 2 are shown in Fig. 9.

Precursory Shocks from Active Faults

Since active faults in Japan are usually very low in seismicity, we are especially aware of some cases in which local residents adjacent to an active fault had noticed the successive occurrence of felt earthquakes before a large damaging earthquake. Among seven large damaging earthquake greater than magnitude 7 that were accompanied by surface faulting, shown in Fig. 2, four earthquakes, the 1891 Nobi, 1896 Rikuu, 1930 Kita-Izu and 1945 Mikawa earthquakes were preceded by a series of foreshocks large and frequent enough to frighten residents within a few days before the main event. The other three

earthquakes, the 1894 Shonai, 1927 Tango and 1943 Tottori earthquakes, which were all located on the Japan Sea coast, were not accompanied by such felt foreshocks. From these examples, it is worth to note that an abnormal succession of local felt earthquakes from an area adjacent to faults which have been active in the Quaternary but have been quiet during recent history, may be a possible precursor for a large damaging earthquake.

Summary and Conclusion

Regions designated as zone 3(Fig. 3) are struck by large earthquakes, on the average four times as often as those areas of zone 1 and 2. Most of inland central Japan, including the Chubu and Kinki districts, is characterized by the highest density of active faults and, hence the greatest possibility for large damaging earthquakes. On the other hand, fault-free areas(zone 1) dominate in the northern part of Hokkaido and on the Pacific side of the main islands of Japan. The central and eastern Kanto region is designated as a fault-free area(zone 1) owing to absence of active fault features on the ground surface, although many earthquakes, some damaging, have occurred there. They are of subcrustal origin related to subducting Pacific and Philippine Sea plates. The inland segment of the Sagami trough fault zone, for instance, is designated as a precaution fault zone because of the following two criteria. The first one is high ratio of the geologically-estimated recurrence interval to the length of time elapsed since the last large earthquakes. Moreover, according to the second criterion, this segment is a seismic gap in that fault zones, along which recent large earthquakes have taken place in recent historical time from the neighboring segment of that fault. Based on these two criteria, twelve precaution faults on land in Japan are shown.

Beside the two criteria mentioned in this paper, a third criterion based on the short-term phenomena such as current seismicity and crustal deformation is necessary for prediction of earthquakes. Knowledge about active faults relating to the two criteria derived from the long-term nature of seismicity may serve for evaluation of shorter-term phenomena.

References

Hirano, S. and T. Nakata, Some examples on age of displacement along active faults, Abstract submitted to Geographical Society of Japan, 18, 1980.

Matsuda, T., Estimation of future destructive earthquakes from active faults on land in Japan, J. Phys. Earth, 25, suppl., S251-S260, 1977.

Matsuda, T., Y. Ota, M. Ando and N. Yonekura, Fault mechanism and recurrence time of major earthquakes in southern Kanto district, Japan, as deduced from coastal terrace data, Geol. Soc. Amer. Bull., 89, 1610-1618, 1978.

Matsushima, Y. and N. Ban, The Jomon dwelling site displaced by the Itoigawa-Shizuoka Tectonic Line, southeast of Lake Suwa, central Japan, Daiyonki-Kenkyu(The Quaternary Research), 18, 155-164, 1979.

Mogi, K., Two kind of seismic gaps, Pageoph, 117, 1172-1186, 1979.

Okada, A., Quaternary faulting along the Median Tectonic Line of southwest Japan, Memoir Geol. Soc. Japan, 18, 79-108, 1980.

Okada, A. and T. Matsuda, A fault outcrop at the Onosawa pass and recent displacements along the Atera fault, central Japan, Geogr. Rev. Japan, 49, 632-639, 1976.

Okada, A., M. Ando, and T. Tsukuda, Study of active faults by trenching, Gekkan Chikyu(The Earth Monthly), 1, 608-615, 1979.

Research Group For Active Faults,(ed.), Active faults in Japan:Sheet Maps and Inventories, The University of Tokyo Press, Tokyo, 1-363, 1980.

Utsu, T., Seismicity of Japan from 1885 through 1925 --- A new catalog of earthquakes of M≥6 felt in Japan and smaller earthquakes which caused damage in Japan, Bull. Earthq. Res. Inst., 54, 253-308, 1979.

Yamazaki, H., Active faults along the inland plate boundary, north of Suruga Bay, Japan, Gekkan Chikyu(The Earth Monthly), 1, 570-576, 1979.

GEOLOGICAL ANALYSIS OF SEISMICITY ALONG THE TANCHENG-LUJIANG FAULT ZONE IN EASTERN CHINA

Zhongjing Fang, Menglin Ding, Fengju Ji, and Hongfa Xien

Institute of Geology, State Seismological Bureau
Beijing (Peking), The People's Republic of China

Introduction

Many investigators [Deng et al., 1973; Li and Wang, 1975; Fang et al., 1976; Allen, 1975; York et al., 1976; Tapponnier and Molnar, 1977; Tchalenko et al., 1973; Matsuda, 1976] have pointed out that recent movements along intracontinental strike-slip fault zones and in faulted basins (rift belts) are closely accompanied by strong earthquakes. Questions that are of importance for analysis of risk along active faults include: What is the probability that earthquakes will recur in places along major intraplate fault zones where strong seismic events have previously occurred during historic time? Does the absence of strong seismic activity in a fault zone during historic time signify that the area is one of low seismic risk?

China is an earthquake prone country. Despite the existence of historic records of seismic activity for about the past three thousand years, uncertainty remains in the estimation of seismic risk based only on historic records and instrumental data. Many of the most recent major earthquakes in China (e.g., the Zhaotong, Haicheng, Lunglin, and Tangshan earthquakes) occurred in seismic gaps without previous historic earthquakes. Thus, the search for evidence of motion along active faults during Late Quaternary time is important in the analysis of seismic risk along potentially active fault segments for which there is no record of earthquake activity during historic time. In recent years, we have turned our attention to the seismological study of intraplate seismogenic fault zones in connection with research for immediate-term prediction of seismic risk.

The Tancheng-Lujiang fault zone through eastern China (Figure 1) is one of the principal areas of study. A magnitude 8.5 earthquake, affecting half of China, occurred in the Juxian–Lingyi area on July 25, 1668 along the central segment of the Tancheng-Lujiang fault. In addition, ten earthquakes of magnitude 6 or greater and sixteen events of magnitudes 5.0–5.9 have occurred along the Tangcheng-Lujiang zone since 70 B.C. (Figure 1, Table 1) [The Research Group of History of Seismology, 1956].

Geologic Background of the Earthquakes

The NNE-trending Tancheng-Lujiang fault zone has a long history of movement. The fault zone consists of several deep-seated faults in a horst and graben zone, ranging from several kilometers to 40 km in width. It crosses several different geotectonic units in eastern China, extending from Zhaoxing County on the Heilongjiang River in the north to Guangji County, Hu Bei province in the south (Figure 1). At its southern end, the Tancheng-Lujiang fault zone intersects with the Xiangfan-Guanji fault. The overall length of the fault zone is more than 2400 km.

The Tancheng-Lujiang fault zone can be divided into segments which correspond to the different geotectonic blocks in which the fault is developed [Tectonic Map Compiling Group, 1974; Ma, et al., 1979]. The northern segment lies to the north of Changtu. It is developed in the Hercynian Jiling-Heilongjiang down-warped block and consists of two main northeast-trending faults of Mesozoic to Cenozoic age (the Yilang-Yitong fault system) which form a graben-type fault system. The middle segment lies in the area south of Changtu to the Huaihe River. It constitutes the main body of the Tancheng-Lujiang fault zone, and is developed in the northern China structural block which consists of Archaeen to Early Proterozoic basement. This segment is a complex Mesozoic to Cenozoic horst and graben system 20 to 40 km wide consisting of four major faults. In the Yihe-Shuhe River basin, Shandong Province, the fault zone (called the Yi-Shu fault zone) dates to the Mesozoic and is comprised of two grabens separated by a horst block. These two grabens are filled with Cretaceous volcanic and clastic rocks 4 to 5 km thick. The southern segment lies in the area from the Huaihe River south to Guangji County and is the southwestern continuation of the main fault zone. It is developed in the Yangtze Block which comprises Late Proterozoic basement and the Caledonian to Hercynian northern Huaiyang folded block. From

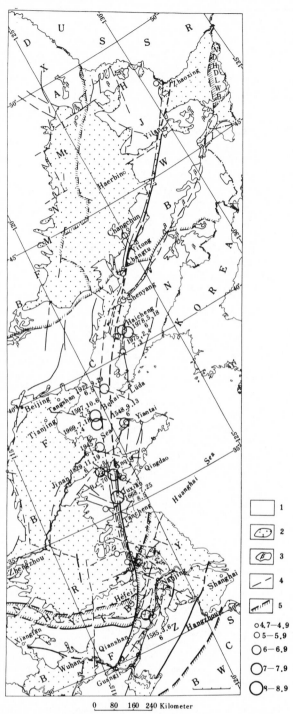

Monggol Folding Block; N.D.H.D.L.W.B. – Nadanhadaling Warping Block; H.J.W.B. – Hei-Ji Warping Block; N.C.F.B.R. – Northern China Faulting Block Region; N.H.Y.F.B. – Northern Huaiyang Folding Block; Y.Z.F.B. – Yangzi Faulting Block; and S.C.W.B. – Southern China Warping Block.

geological, seismological and geophysical data, it is postulated that the Tancheng-Lujiang fault zone is deep-seated, breaking through the crust and into the upper mantle. It has been subjected to polyphase tectonism resulting in complex structures. During the Mesozoic era, periods of intensive compressional tectonism alternated with extensional tectonism with associated magmatism. This resulted in crosscutting of the main fault zone by late WNW and ENE-trending faults [Shandong Seismo-Geologic Group, 1974; Fang et al., 1980] as the orientation of the principal compressive stress of the northern China regional stress system turned from a northwesterly direction in the Mesozoic to a ENE-WSW direction during Cenozoic time. This complex structural zone may be one of potential risk of earthquake generation [Fang et al., 1980; Fang et al., 1979].

In the neotectonic period, the intensity of motion along the various segments of the Tancheng-Lujiang fault zone was distinctly different. In the middle segment, many Late Quaternary faults developed by inheriting the pre-existing NNE and WNW trends of faults, and formed newly faulted uplifts and depressions with distinct fault scarps. The drainage systems across the faults have sharp knickpoints, offset streams, and deeply incised channels on the upthrown fault blocks. In the southern segment, paleotopographic relief is clearly distinguished on the sides of the fault zone. Cut and fill terraces are developed, but few active faults are present. The active faults that are developed are NNE-trending, right-lateral strike-slip faults. However, the Late Pleistocene sediments covering the faults are undisturbed in many areas. Most of the northern segment of the Tancheng-Lujiang fault system is covered by Late Cenozoic sediments with thick plant cover and few outcrops. Further to the north, the fault valley is wider with gentle topography. In places, where less V-shaped gulches are developed in Holocene deposits, and active faults are rare. The middle segment of the fault system may be divided into two distinct parts: a) the northern part in the Low Liaohe-Bohai area represents part of a Cenozoic faulted depression in the North China Plain, that has been subjected to right-lateral wrench faulting and extensional tectonics; b) the southern part of the middle segment is comprised mainly of the right-lateral overthrust in the central Shandong-northern Jiangsu area. For the middle and southern segments of the main fault zone, the duration of neotectonic activity (Figure 2) is different in var-

Fig. 1. Map showing geotectonic setting of the Tancheng-Lujiang fault zone and the distribution of earthquake epicenters. 1- Pre-Cenozoic rocks; 2 - Cenozoic era and basin boundary; 3 - Late Cenozoic basalts; 4 - determined and inferred faults; 5 - boundary of block tectonic element. Letters on map refer to the following blocks: D.X.A.L.Mt.-N.M.F.B. – Daxingangling-Nei

Table 1

Comparison of Seismicity in Various Segments of the Tancheng-Lujiang Fault Zone
(from 70 B.C. to 1978 A.D.)

| Segment | Magnitude | | | | | Total Released Energy |
	4.7-4.9	5-5.9	6-6.9	7-7.9	8-8.9	($Log_{10}E = 11.8 + 1.5 M$)
Northern	1	3				5.4×10^{20}
Middle		11	5	4	1	3.9×10^{24}
Southern	1	2	1			6.8×10^{20}

ious areas. It appears that the motion along the overall fault zone is not synchronous. This may be related to the intersection of the main fault zone by faults of other orientations.

The seismicity along the Tangcheng-Lujiang fault zone is shown in Figure 1 and Table 1. Strong earthquakes with magnitude of 6 or more have occurred in the middle segment as single infrequent but high intensity events. The level of seismicity in the northern and southern segments is lower than that in the middle. During the past two thousand years, the total released energy by earthquakes with magnitude 4.75 or more in the middle segment is about 3200 times that in the southern and northern segments (Table 1). Moreover, the recurrence of strong historic earthquakes at a single locality is very rare, and the interval on recurrence is relatively long. For instance, in the year 70 B.C., the Anqiu earthquake, one of the earliest recorded earthquakes, with magnitude of 7 occurred in this belt. Since then no strong earthquake recurred to date at the same locality, though two earthquakes, one in 1597 and the other in 1969, occurred nearby in the central part of the Bohai Sea. There is no record of recurring earthquakes in those localities. It is of interest that along the middle segment of the Tancheng-Lujiang fault zone there appears to be a correlation between the distribution of earthquake epicenters and points where the main fault is intersected or approached by WNW-trending faults. Therefore, localities with similar structural configuration (i.e., fault intersections), but without historic earthquakes, such as the northern Jiangsu segment, should be closely monitored as areas of potential seismic activity. It is clear that the historic seismicity along the Tancheng-Lujiang fault zone is in close correlation with the subdivision of the fault zone in terms of the tectonic position of various segments and the differences in tectonic movement during Late Quaternary time. The northern segment, developed in the Jiling-Heilongjiang warping block with relatively young basement, exhibits intensive magmatism, low rigidity and low Late Quaternary tectonic activity, so it is referred to

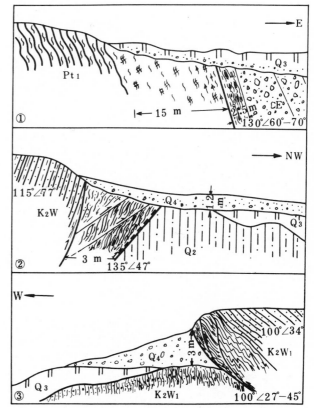

Fig. 2. Map showing the time differences for recent movement along various segments of Tancheng-Lujiang fault zone. (1) Geological profile of southern segment, in the Jiangjia-Huawu area, Qianshan County, Anhui province, showing that no recent movement occurred since middle Pleistocene (Q_2). (2) Geological profile of middle segment, in Qiaotougaunzhuang area, Tancheng County, Shandong province, showing no recent movement occurred in Holocene (Q_4). (3) Profile of the middle segment in the area of Sangzhuang, Tancheng County, Shandong province, showing recent movement occurred in Holocene (Q_4).

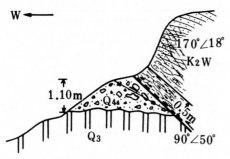

Fig. 3. Profile of Hezhuang active fault in Xingyi County.

Fig. 5. Outcrop of overthrust fault in Late Quaternary in Sangzhuang, Tancheng County.

as a segment of low seismic activity. The middle segment, developed in the northern China region with old, highly rigid basement, shows intensive faulting and intensive neotectonic differential movement, so it is referred to as a segment of strong seismic activity. The southern segment is developed in the northern Huaiyang folded block and Yangtze faulted block. Though its basement was deformed and consolidated relatively early, its rigidity is low and the recent movement is weak, so it is referred to as a segment of intermediate activity. Therefore, the Tancheng-Lujiang fault zone may not be referred to, over-all, as a zone of strong seismic activity.

Stick-Slip Motion Along the Active Faults and Recurrence of Earthquakes

There are many modes of displacement along active faults, but two main types may be distinguished depending on the coefficient of friction on the fault plane; one is creep and the other is stick-slip. The latter often can easily accumulate the strain energy necessary for strong earth-

quakes. For instance, in some recent and ancient seismic areas such as in Linfeng, Xichang, Haiyuan, Changma, Luho, Haichen, Longlin and Tangshan areas, the traces of stick-slip motion along active faults or paleoseismogenic faults were found in meizoseismal areas and their vicinities. In general, an active stick-slip fault can be considered as a geological indicator for estimation of seismic risk in an area. So investigation of fault traces for evidence of stick-slip motion along the active faults can be conducted to find evidence for prehistoric and ancient earthquakes, extending backward prolonging the historic seismic record and knowledge about intervals of earthquake recurrence.

In the southern portion of the central segment, the southern Shandong-northern Jiangsu area, many outcrops showing a stick-slip characteristic of faulting crop out from Juxian County southward thorugh Qijishan Mt. and Malingshan Mt. in Tancheng County, Malingshan Mt. in Xingyi County, Fengshan Mt. in Squien County, and Chonggangshan Mt. and Fengshan Mt. in Sihong County, reaching to Tunghe at the mouth of Huaihe River. Along the fracture zone, forming the eastern graben border, a Late Quaternary fault is developed discontinuously for a distance of 200 or more kilometers. Cretaceous volcanic and clastic rocks and Mesozoic granites are thrust over Late Pleistocene to Holocene eluvial, alluvial, and diluvial sediments as a result of wrench faulting in response to east-west compression [Fang, et al., 1976; Fang et al., 1979] (Figures 5-8). A large number of active faults are developed in this region. The intense surface deformation reflects deep-seated strain along the fracture zone, as shown by the results of triangulation surveys that indicate east-west crustal shortening and north-south extension, and of reversals on short-line level surveys indicating an uplift of the graben block. We have found outcrops with traces of intermittent stick-slip motion along active faults. Two examples follow:

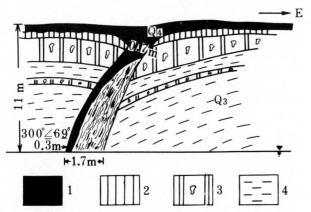

Fig. 4. Profile of active fault on left bank of Xingyi River east of Zhangshan sluice in Xingyi County. 1- black soil; 2 - grey-black sandy soil; 3 - brown-yellow sandy clay with numerous calcareous nodules; 4 - yellow-brown sand clay with iron-manganese nodules.

Fig. 6. Outcrop of Late Quaternary overthrust fault 1 km west of Chengdinzi, Donghai County.

Fig. 8. Outcrop of Late Quaternary overthrust fault on eastern side of Fengshan Mt., Suqian County.

The first example is an outcrop on the Hezhuang active fault in Xingyi County (Figures 3 and 9). It is located along the western border fault in the eastern graben on the western slope of Malingshan Mt., the Malingshan block uplift. The fault dips to the east. Oblique slickensides, raking 40° north, are well preserved on the fault plane. The outcrop shows the right-lateral overthrust of Late Cretaceous sandstone on a Late Pleistocene loess soil unit (Figure 3, Q_3) and colluvial sediments (Q_4). The fault can be traced for a distance of several kilometers. The fault profile clearly records at least two period of reverse motion along the active fault. Initially, the Late Pleistocene unit was overthrust by the Cretaceous sandstone along the reverse fault forming a scarp. Then colluvial sediments accumulated on the footwall by the gravitational collapse of the scarp. After that, the fault moved again and the colluvial sediments were overthrust to form a new scarp along the old fracture zone. Presum-

ably, this intermittent motion along the active fault was accompanied by strong earthquake activity.

The second example is an outcrop 35 km to the south of Hezhuang village, on the left bank of the Xingyi River, 1 km to the east of Zhangshan sluice. This outcrop of the main fault of Late Pleistocene age shows a compressional fault zone 1.7 m wide (Figures 4 and 10). After the formation of a Holocene black soil layer (^{14}C dating yields an age of 8230 ± 120 years), the active fault again moved in a right-lateral sense and formed en echelon extensional crack that then filled with superficial black soil to a depth of 11 or more meters. It is not easy to determine whether these cracks are associated with the earthquake of magnitude 8.5 that occurred in 1668. However, the multiple motion along the active fault and the possibility of earthquake recurrence are clearly reflected at this locality.

The above examples show the nature of intermittent motion along active faults in the southern

Fig. 7. Outcrop of Late Quaternary overthrust fault in Xilinien, Donghai County.

Fig. 9. Showing intermittent motion along Late Quaternary active fault in Hezhuang, Xingyi County.

Fig. 10. Profile of paleoseismic crack development on the Late Quaternary active fault, left bank of Xingyi River, 1 km east of Zhangshan sluice in Xingyi County. Black colour shows black bog soil filling cracks.

portion of the middle segment. Each abrupt dislocation may have been accompanied by an earthquake. However, this segment has remained as a seismic gap during the past two thousand years or more. These traces of intermittent motion along an active fault gives us an important geological indicator for evaluating the future seismic risk in northern Jiangsu province.

Summary and Conclusion

From the above discussion we can conclude:
1. The historic seismic activity in the Tancheng-Lujiang fault zone has obvious subdivisions that correlate well with differences in fault movement since Late Quaternary and tectonic blocks in which various segments of faults have formed.
2. The identification of intermittent motion along active faults is a geological indicator for the recurrence of ancient earthquakes. A complex history of multiple motion at a single locality indicates the possibility for recurrence of earthquakes at a single epicenter along a fault. However, the recurrence interval depends on earthquake magnitude, tectonic media and its mechanical properties, and evolutionary stages of the fault. This is the important seismotectonic problem to be studied for more accurate determination of long-term seismic risk in an area.
3. Parts of fault systems that are seismic gaps (no magnitude 6 or greater earthquakes in historic times), but along which exist seismotectonic conditions necessary for fault movement and evidence for Late Pleistocene or Holocene fault activity, should be regarded as segments only temporarily quiet. Along these fault segments seismic risk may be greater than along other segments where historic earthquakes have occurred,

relieving accumulated stress. An example is the northern Jiangsu area at the southern end of the middle segment of the Tancheng-Lujiang fault zone. Such an area must be monitored for long term earthquake prediction.

Acknowledgments. The authors wish to thank De-fu Shen, who translated the manuscript into English. Especially we would like to thank S. Kirby, M. Carr and Yongnian He for reviewing the manuscript. All photographs were taken by Dongyin Shi

References

Allen, C.R., Geological criteria for evaluating seismicity, Geol. Soc. Amer. Bull., 86, 1975.

Deng, C. T., et al., On the tendency of seismicity and their geological set-up of the seismic belt of the Shansi graben, Sci. Geol. Sin., 1, 1973.

Fang, Z. J. et al., The characteristics of Quaternary movements along the middle segment of the old Tancheng-Lujiang fracture-zone and their seismogeologic conditions, Sci. Geol. Sin., 4, 1976.

Fang, Z. J. et al., Preliminary study of Late Cenozoic tectonic stress field in Jiangsu-Shandong-Anhui region, Seismology and Geology, 1, 1979.

Fang, Z. J., et al., A preliminary study on the block-faulting characteristic of southern part of the North China fault block region and its surroundings: Formation and development of the North China fault block region, Science Press, 1980.

Li, P., and L. M. Wang, Exploration of the seismogeological features of the Yunnan-West Sichuan region, Sci. Geol. Sin., 4, 1975.

Ma, X. Y. et al., Tectonics of the North China platform basement, Acta Geol. Sin., 53, 1979.

Matsuda, T., Active faults and preestimation of earthquakes, Symposium on the Studies of Earthquake Prediction, 1976.

Research Group of History of Seismology, The Committee of Seismological Work, Academia Sinica, Chronological table of seismic records in China (Volume I), Science Press, 1956.

Shandong Seismo-Geologic Group, Institute of Geology, Academia Sinica, Characteristics of the block-faulting tectonics of Shandong region and a preliminary division of its earthquake belt, Sci. Geol. Sin., 4, 1974.

Tapponnier, P., and P. Molnar, Active faulting and tectonics in China, J. Geophys. Res., 82, 1977.

Tchalenko, J.S., B. Berberian, and H. Behzadi, Geomorphoric and seismic evidence for recent activity in the Dorunch fault Iran, Tectonophysics, 19, 1973.

Tectonic Map Compiling Group, Institute of Geology, Academia Sinica, A preliminary note on the basic tectonic features and their developments in China, Sci. Geol. Sin., 1, 1974.

York, J. E., R. Cardwell, and J. Ni, Seismicity and Quaternary faulting in China, Bull. Seismol. Soc. Amer., 66, 1976.

SPECIFICATION OF A SOON-TO-OCCUR SEISMIC FAULTING IN THE TOKAI DISTRICT, CENTRAL JAPAN, BASED UPON SEISMOTECTONICS

Katsuhiko Ishibashi

International Institute of Seismology and Earthquake Engineering
1 Tatehara, Oho-Machi, Tsukuba-Gun, Ibaraki Pref., 305 Japan

Abstract. The 'Tokai earthquake' expected in the Tokai district, the Philippine Sea coast of central Japan, is the most important target now in the Japanese earthquake prediction program. Its long-term and short-term prediction efforts are among the most advanced cases in the world. It is also a big social problem in Japan because of its preestimated catastrophic disaster. This paper reviews in detail the long-term prediction research of this earthquake and presents the most probable fault model. The expected event is considered an interplate earthquake due to the underthrusting of the Philippine Sea plate beneath southwest Japan at the Nankai - Suruga trough. Based on the analyses of source mechanisms of the 1854 Ansei Tokai (M = 8.4) and the 1944 Tonankai (M = 8.0) earthquakes, the most probable rupture zone of the coming earthquake is the 'Suruga trough thrust' in Suruga Bay and its southwestern extension. The first-order approximation fault model is as follows: dip direction, N72°W; dip angle, 34°; length, 115 km; width, 70 km; reverse dip slip, 3.8 m; left-lateral strike-slip, 1.3 m; seismic moment, 1.6×10^{28} dyne·cm; stress drop, 50 bar; magnitude, 8.3. Crustal movements in the Suruga Bay region during the last 80 years or so are well interpreted as the preparatory strain accumulation for this faulting. Geomorphological features and historical seismic activities in this region are generally in good harmony with this prediction. The seismic gap for 126 years on the Suruga trough thrust since 1854 and the considerable amount of strain accumulation in this region estimated from geodetic survey data suggest a fairly high probability of a near-future occurrence of this faulting. And, if the event is a little smaller-scale, it may occur earlier. The seismogenic tectonism in the Suruga Bay region is rather complicated due to the collision of the Izu Peninsula with the Japanese Islands, and the subduction of the Philippine Sea plate at the Suruga trough has not been established yet by geophysical evidences. For further refinement of the long-term prediction of the Tokai earthquake, it is necessary to obtain more clear view on the real dynamical and physical process of plate convergence in the Izu collision zone as a total system.

1. Introduction

Along the Nankai trough off southwest Japan (Figure 1), where the Philippine Sea plate is underthrusting northwestward beneath the Eurasian plate [Fitch and Scholz, 1971; Kanamori, 1972; Sugimura, 1972; Ando, 1975b], there have occurred a series of great interplate earthquakes of magnitude around 8 with remarkable temporal and spatial regularities. Twelve such events are known at present for the last 1,300 year time interval, all of which have caused widespread serious damage to the Nankai and the Tokai districts (Figure 1), the Pacific coast of southwest and central Japan. The latest events are the 1944 Tonankai (M ≃ 8.0, more than 1,200 dead) and the 1946 Nankai (M ≃ 8.1, more than 1,400 dead) earthquakes.

And now, the seismic gap in the eastern half of the Tokai district (Shizuoka Prefecture and its vicinity, Figure 1), which was delineated by the 1944 Tonankai earthquake, has been regarded as the most dangerous one in Japan and become the most important target in the Japanese earthquake prediction program. The anticipated earthquake in this gap, already named 'the Tokai earthquake', is considered one of such interplate great earthquakes as above, along the Suruga trough, the northeastern extension of the Nankai trough (Figure 1). Because of the condition that the subduction boundary is very near to land in this region, and owing to rather abundant data of high quality on crustal movements (both recent and Quaternary), seismic activities (both recent and historical), submarine topography, and so on, the most probable fault model of the Tokai earthquake at this moment has been set up through investigations on seismogenic tectonism in this region. And, as the possibility of its near future occurrence is considered rather high, a continuous watch system for various kinds of short-term and immediate precursors has been consolidated and the Earthquake Assessment Committee to evaluate anomalous data has been established.

The Tokai earthquake is also a big social problem now in Japan, since the long-term prediction on the earthquake image (place, size, faulting type) and on its possible imminency and the estimation of damage caused by it are fairly con-

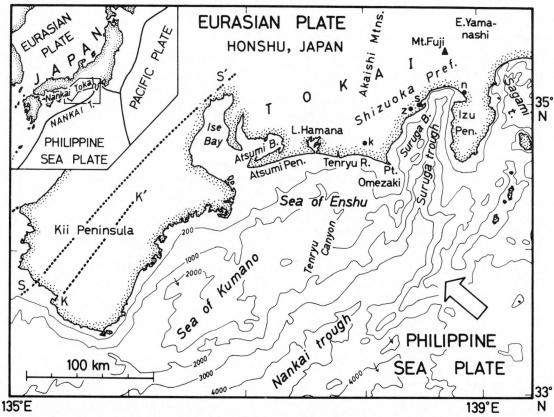

Figure 1. Index map of the Tokai district and its vicinity. The white arrow shows the moving direction of the Philippine Sea plate relative to the Eurasian plate after Seno [1977a]. Isobaths (in meters) are after Geological Survey of Japan [1978]. SS', leading edge of the Philippine Sea plate inferred by Shiono [1974]; KK', that inferred by Kanamori [1972]. k, Kakegawa; n, Numazu; s, Shimizu; z, Shizuoka.

vincing. As the most likely source region is almost right beneath the densely populated area, once it occurs, the disaster will become catastrophic and Japan's industries and economy may suffer serious damage. So, a special law called 'Large-Scale Earthquake Countermeasures Act' has been enacted, and under this law the whole nation is taking various countermeasures of disaster prevention for this earthquake based on the hypothetical fault model.

The case of the Tokai earthquake prediction is a good example of the two-stage earthquake prediction strategy proposed by Ishibashi [1978c] as an alternative to the prevailing step-by-step one. According to this strategy, the practical procedure for prediction of a tectonically significant large earthquake consists of two stages; the first one is to specify an expected faulting and to make sure its near future possibility through elaborate investigations on seismogenic tectonism and fault properties in the questioned region (i.e., to set up a target), and the second one is to predict its occurrence time by detecting short-term and immediate precursors by concentrated observations (i.e., earthquake predic-

tion itself in a narrow sense): though an actual procedure may go on, more or less, in the form of a mixture of them. The Tokai earthquake prediction operation has been proceeding basically along this line and now entered the second stage of short-term vigilance. However, the target specification of the first stage, of course, has not been accomplished yet. There are considerable difficulties in drawing a convincing picture of the coming seismic faulting and there still remains much obscurity concerning its true nature, particularly due to the complicated mode of plate convergence in this region, which is the northernmost part of the Philippine Sea - Eurasian plate boundary where the Izu Peninsula is colliding with Honshu Island (Figure 1). And the improvement of the expected fault model is directly connected with the basic problems concerning the actual process and mechanism of plate convergence at subduction boundaries.

This paper reviews in detail the first stage work of the Tokai earthquake prediction, describing how the fault model has been set up, and showing what are important problems awaiting future investigations. At first we will look back rather in detail upon the progress of prediction research of this earthquake for more than a decade, for it is worth knowing in order to understand the present state of earthquake prediction work in Japan. As it also tells us eloquently how the target setting based on basic studies is important for stimulating the public to efforts to prevent earthquake disaster, effects of researches to the public at each step are also referred to.

2. History of the Prediction Research of the Tokai Earthquake

When the Coordinating Committee for Earthquake Prediction, Japan (CCEP, president, Dr. T. Hagiwara) started functioning in April 1969, the Tokai district had been already regarded vaguely as one of the most likely place in Japan to be hit by an offshore large earthquake in the near future. The main reason was that the region was considered what we now call a seismic gap since the 1854 Ansei Tokai earthquake (M = 8.4 after Kawasumi [e.g., Usami, 1975]), although some seismologists thought that the 1944 Tonankai earthquake (M ≃ 8.0 to 8.3 after Kanamori [1972]) had filled in the gap. The 'Red Print' published in June 1969 [Commission on Earthquake Prediction, 1969] mentions this district as an example of 'the area of special observation' for that reason. In October 1969 at the monthly meeting of Earthquake Research Institute of the University of Tokyo (ERI) Mogi pointed out that, according to the first-order triangulation analysis by Harada and Isawa [1969], the Tokai district, from the Izu Peninsula through Suruga Bay to the coastal area of Enshu-nada (the Sea of Enshu) (Figure 1), had been considerably compressed northwestward from the Pacific side for about 70 years. He interpreted this deformation by a model on crustal movements related to occurrence of offshore large earthquakes, which is based on the sea-floor spreading hypothesis and the elastic rebound theory, and suggested the possibility of a large earthquake to release this compressive stress [Mogi, 1970a,b]. Following this suggestion CCEP examined observed data in the Tokai district at the 5th meeting held in November 1969, and then designated the district as 'the area of special observation' officially, taking into account steady subsidence of the west coast of Suruga Bay since 1889 [Geographical Survey Institute (GSI), 1970a], frequent occurrences of great earthquakes in this region, and absence of them since 1854. At that time, however, the image of the target was fairly vague because the tectonic framework of this region was still obscure and the understanding of the 1854 and the 1944 earthquakes was rather poor. So, though the designation and Mogi's research were reported sensationally by some newspapers and weekly journals, they did not hold public attention so much.

By 1971, the subduction process of the Philippine Sea plate beneath southwest Japan along the Nankai trough had been made clear through investigations of the 1944 Tonankai and the 1946 Nankai earthquakes [Fitch and Scholz, 1971; Kanamori, 1972]. In 1972 Sugimura [1972] pointed out that the Philippine Sea - Eurasian plate boundary to the northeast of the Nankai trough would run along the Suruga trough and lead to the Sagami trough traversing inland to the north of the Izu Peninsula (Figure 1), and suggested the existence of a plate boundary megathrust along the southern half of the Suruga trough. In May 1973 at the spring meeting of the Geodetic Society of Japan Fujii [1973] discussed the future Tokai earthquake from the viewpoint of crustal movement with these studies as background. He showed that the crustal movement in the Tokai district was characterized by, (1) subsidence around Point Omaezaki (Figure 1) and uplift in its back mountain area during interseismic periods, and (2) uplift around Point Omaezaki and subsidence in the Lake Hamana - Atsumi Bay region (Figure 1) at the time of great earthquakes. And he attributed these crustal movements to the subduction of the Philippine Sea plate beneath the Tokai district at the mouth of Suruga Bay and the Sea of Enshu. He estimated the recurrence time of the Tokai earthquake at 150 ∿ 200 years on the basis of inferred coseismic tilt of $3 \sim 4 \times 10^{-5}$ at the time of the 1707 and the 1854 great earthquakes and secular tilting rate of 2×10^{-7}/year since 1889, both around Point Omaezaki, and stated that the next great earthquake would occur within this century in sooner case.

On June 17, 1973 the Nemuro-oki earthquake of magnitude 7.4 occurred off eastern Hokkaido which had been designated as 'the area of special observation' by CCEP [e.g., Abe, 1977; Utsu, 1979]. This earthquake had been expected by Utsu [1970, 1972] and other Japanese seismologists for almost the same reason as that for the Tokai earthquake. And accidentally shortly before the occurrence of this earthquake Rikitake testified in the Diet and wrote on a newspaper that the most likely places in Japan where interplate great earthquakes would occur were off eastern Hokkaido and off the Tokai district. He had calculated cumulative probabilities of earthquake occurrence in these two areas as more than 0.8 ∿ 0.9 by crustal strain analysis [Rikitake, 1974]. Under these circumstances the public had been impressed strongly that the next great earthquake in Japan would hit the Tokai district in the near future. However, from those days the epicentral region of the future Tokai earthquake had come to be said as the Sea of Enshu without any specific bases, the word 'the Sea-of-Enshu earthquake' had come to be used frequently, and little attention had come to be paid to Suruga Bay.

In November 1973 at the fall meeting of the Seismological Society of Japan (SSJ) Ando proposed an expected fault model for the future Tokai earthquake based on the investigation on source mechanisms of recent and historical great earthquakes along the Nankai trough and presented estimations of coseismic crustal deformations, tsunamis, and seismic intensities [Ando, 1975a, b]. According to him, the source region of the 1854 Ansei Tokai earthquake had been the Sea of Enshu and its western adjacency (Kumano-nada, or the Sea of Kumano, Figure 1), and that of the 1944 Tonankai earthquake, the Sea of Kumano; the Sea of Enshu, therefore, had been a seismic gap for 119 years, whereas the Sea of Kumano had repeated an earthquake in 90 years. His fault model for the future event was a right-lateral reverse faulting of 4 meter dislocation along a fault plane of 100×70 km^2 dipping toward N30°W by 25 degrees occupying the Sea of Enshu almost entirely. Besides the long-term seismic quiescence and the remarkable crustal deformations as described before, Ando [1975a] paid particular attention to the uplift near Kakegawa (Figure 1) at the time of the 1944 Tonankai earthquake [Sato, 1970; Sato and Inouchi, 1975; Inouchi and Sato, 1975]. He interpreted this uplift as due to a creep-like slip along the eastern bottom of the above-mentioned fault plane, which he regarded as a kind of pre-slip of the coming faulting. Thus, he considered that the risk of a large earthquake was very high. He suggested that along the Suruga trough the Philippine Sea plate was colliding with the continental plate and therefore large earthquakes resulting from the underthrusting mechanism were unable to occur in Suruga Bay. Meanwhile, at the 23rd meeting of CCEP held in the same month, Japan Meteorological Agency (JMA) reported a large seismicity gap corresponding to a magnitude 7.6 ∿ 7.7 earthquake in the Sea of Enshu [Sekiya and Tokunaga, 1974a, 1975], and GSI [1974b] reported increase of strain velocity of Omaezaki rhombus base-line net. Ando's research and these observed data were reported extensively by newspapers etc. and impressed the public more deeply that the Tokai earthquake was the main target of the Japanese earthquake prediction program. On the other hand, however, the image of 'the Sea-of-Enshu earthquake' had been almost fixed, which made a vague impression on the public that the seismic center would be somewhat far from the land. GSI [1974a] considered the remarkable crustal subsidence along the west coast of Suruga Bay during the period 1889 - 1967 to be nonseismic but structural. At the 24th meeting held in February 1974 CCEP upgraded the Tokai district to 'the area of intensified observation' based on the examination at the previous meeting and the social importance of the district.

In October 1974 at an unofficial symposium on the Tokai earthquake [Ando and Fukao, 1975] the following difficulties in expecting the Sea-of-Enshu earthquake were pointed out: (1) The west half of the expected source region might have ruptured at the time of the 1944 Tonankai earthquake judging from its aftershock distribution, coseismic crustal deformation, and tsunami source area [Sekiya, 1975; Fujii, 1975; Hatori, 1975a; Harada, 1975; Mikumo, 1975]; (2) The secular crustal movement along the west coast of Suruga Bay was hard to understand by plate motion relating to the Sea-of-Enshu earthquake [Fujii, 1975; Mikumo, 1975]; (3) The frequency of great interplate earthquakes in the Tokai district might be about a quarter of that in the Nankai district judging from the uplift rate of marine terraces at Point Omaezaki [Matsuda, 1975]. Mikumo [1975] suggested that the source region of the future Tokai earthquake might be the eastern half of the Sea of Enshu - south off Suruga Bay - the southern tip of the Izu Peninsula, and Matsuda [1975] suggested the potentiality of 'the Suruga Bay earthquake' from the viewpoint of geology and geomorphology. From those days not a few seismologists had come to doubt the near future possibility of the Tokai earthquake itself due to these difficulties and ambiguities. However, as the general feeling that the Tokai district was seismically dangerous had not vanished among the persons concerned, various kinds of observations had been planned and carried out aiming vaguely at the Sea-of-Enshu earthquake [e.g., Hagiwara, 1975; Utsu, 1975a, b]. On the other hand the public had gradually come to pay less attention to the Tokai earthquake, and its disaster prevention program had scarcely progressed.

In May 1976 at the 33rd meeting of CCEP Ishibashi presented a report which asserted that the most probable rupture zone of the future Tokai earthquake would not be the Sea of Enshu but Suruga Bay and its southwestern extension and that this region should be upgraded to 'the area of concentrated observation', the highest rank, because the earthquake might be imminent [Ishibashi, 1977b]. The newly assigned source region was a much worse place than the previous one from the viewpoint of earthquake disaster. In order to clarify the difference from the former Sea-of-Enshu earthquake Ishibashi called this coming earthquake 'the Suruga Bay earthquake'. (In October 1976 at the ad hoc committee of the Geodetic Council of Japan the official calling of 'the Tokai earthquake' instead of 'the Suruga Bay earthquake' was decided. In this paper, however, the word 'Suruga Bay earthquake' is sometimes used as the meaning of the Tokai earthquake taking place in the Suruga Bay region just for the simplicity of writing.) This new idea was an outcome of the investigation on seismogenic tectonism in the northernmost margin of the Philippine Sea plate. Ishibashi hypothesized that Suruga Bay is not a collision zone but a subduction zone along the Suruga trough up to its deepest part for the following reasons: (1) Around Suruga Bay there is no remarkable seismic activity reflecting collision whereas in the eastern Yamanashi area to the northeast of Suruga Bay (Figure 1) a remarkable earthquake swarm suggests the plate collision [Ishibashi, 1976a]; (2) An elastic rebound of the west coast of Suruga Bay must have taken place along the entire Suruga trough at the time of the 1854 Ansei Tokai earthquake judging from its seismic intensity distribution compiled by Hagiwara [1970]; (3) The secular crustal movement around Suruga Bay as mentioned before can be interpreted most reasonably as the result of the northwestward underthrusting of the Philippine Sea plate at the Suruga trough; (4) As pointed out by Sugimura [1972] and Matsuda [1975] the geomorphic features on the west side of the Suruga trough suggest active seismic crustal movements during the late Quaternary. And he concluded that Suruga Bay and its southwestern extension remains a huge seismic gap since 1854. Ishibashi searched for the historical documents on the 1854 earthquake which would support this hypothesis, and found the facts that the west coast of northern Suruga Bay uplifted remarkably at the time of the earthquake and that tsunami hit the head of the bay immediately after the shock. Almost coincidently Hatori [1976] presented old documents on this earthquake which had been accidentally discovered. They contained abundant evidence of the 1854 coseismic crustal uplift of the entire west coast of Suruga Bay. Consequently, 'the Suruga Bay earthquake hypothesis' has been generally accepted among the persons concerned in a short time, and reexamination of all existing data in the Tokai district has been started immediately by the CCEP's subcommittee for the Tokai district (chairman, Dr. T. Asada). At the 34th meeting of CCEP held in August 1976 Ishibashi proposed an expected fault model of the Suruga Bay earthquake as described later based on reexaminations of source mechanisms of the 1854 and the 1944 earthquakes and other seismological, geodetic, and geomorphological data in the framework of plate tectonics [Ishibashi, 1977b]. As for the occurrence time he considered that it might be imminent because the seismic gap had lasted for 122 years and the accumulated strain around Suruga Bay was inferred to be close to an ultimate value [Ishibashi, 1977b].

In October 1976 at the fall meeting of SSJ Ishibashi [1976c] presented a paper on the Suruga Bay earthquake. Besides scientific argument, he emphasized that the expected earthquake would bring unprecedented catastrophe in the worst case and the existing system of earthquake prediction was quite insufficient to reduce its extraordinary damage, and proposed that a powerful and comprehensive organization unifying short-term/immediate prediction and disaster prevention for this specific earthquake should be created as speedily as possible. The Suruga Bay earthquake hypothesis shocked inhabitants and local governments in the Tokai district. The problem of the Tokai earthquake has become a matter of nationwide concern, and the strong desire for early realization of practical earthquake prediction has arisen. Thus, in the same month the Headquarters for Earthquake Prediction Promotion was organized in the Ministry. In November at the 35th meeting CCEP announced its official view on the Tokai earthquake that its rupture zone would be Suruga Bay - south off Point Omaezaki but no long-term precursor indicating its occurrence time had been observed by that time. In December the Geodetic

Council of Japan recommended to the Government the intensification of various kinds of observations and surveys in the Tokai district for long-term prediction and establishment of a continuous watch system and a special committee to evaluate anomalous phenomena for short-term prediction.

Under these circumstances, the data from various observation networks deployed in the Tokai district by several institutions have been concentrated at JMA on the real-time basis for continuous watching of short-term and immediate precursors of the Tokai earthquake. And the Earthquake Assessment Committee for the Tokai district (chairman, Dr. T. Hagiwara) was organized in April 1977. In 1978, the Large-Scale Earthquake Countermeasures Act was introduced by the Government to the Diet in April, promulgated in June, and put in force in December [e.g., Kino, 1979]. Such a conspicuous progress in the field of prediction and disaster prevention of earthquakes had been also promoted by the anomalous crustal activities in the Izu Peninsula around 1976[e.g., Tsumura, 1977] and the Izu-Oshima earthquake of January 14, 1978 (M = 7.0) which killed 25 persons. Furthermore as a background we must enumerate the success of predictions and disaster preventions of the 1975 Haicheng and other earthquakes in China. In the meantime, various kinds of observations and surveys have been intensified in and around Suruga Bay since 1976 to make clearer the mode of plate convergence and the true nature of the coming earthquake, and a considerable amount of new findings have been brought out during the last four years. And lively discussions including several objections to the Suruga Bay earthquake hypothesis have been continuing up to now. These works after 1976 will be referred to in the following chapters on occasion.

3. Geomorphological Feature in the Tokai District

The Nankai trough is a juvenile trench as deep as about 5,000 m running along the Philippine Sea coast of southwest Japan in the SWW-NEE or SW-NE direction (Figure 1) [e.g., Yonekura, 1975]. The submarine geomorphic features and seismic reflection profiles in the Seas of Enshu and Kumano including the northern margin of the Shikoku Basin, south off the Tokai district [e.g., Yonekura, 1975; Maritime Safety Agency (MSA), 1976a, b, c, 1977; Mogi, A., 1977], as well as those off the Nankai district, are consistent with the northwestward underthrusting of the Philippine Sea plate beneath southwest Japan at the Nankai trough [Fitch and Scholz, 1971; Kanamori, 1972; Sugimura, 1972; Ando, 1975b]. In particular, a topographically remarkable manifestation is the arrangement of deep sea plains at the foot of the upper continental slope and 'the ridge and trough zone' on the lower continental slope (Figure 2 (a)). The ridge and trough zone in this region, in general, consists of a row of outer ridges of deep sea plains (inside), another row of ridges

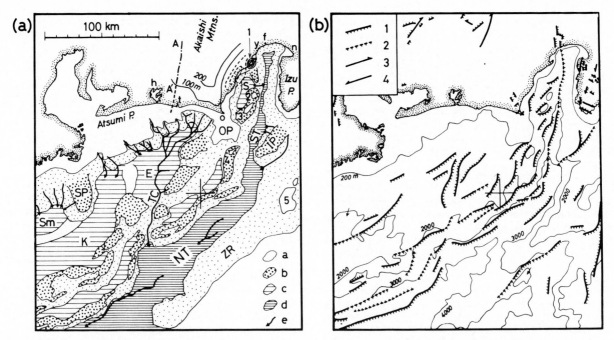

Figure 2. (a) Geomorphological map in the Tokai district. Submarine geomorphic features are after Mogi, A. [1977]. a, Continental shelf; b, Spur, ridge, bank; c, Deep sea plain, basin, depression; d, Trough; e, Canyon. Contours on the west coast of Suruga Bay represent deformation of the Maki-nohara coastal plain (about 10^5 years of age) after Tsuchi [1968b]; broken curves show the subsiding area. The cross near the center indicates the position of (34°N, 138°E). A-A', Akaishi Tectonic Line; E, Enshu basin; IP, Izu spur; K, Kumano basin; NT, Nankai trough; OP, Omaezaki spur; S, Seno-umi basin; Sm, Shima basin; SP, Shima spur; ST, Suruga trough; TC, Tenryu canyon; ZR, Zenisu ridge; f, Fuji River; h, Lake Hamana; n, Numazu; o, Omaezaki; t, Tenryu River; y, Yui; 1, Udo Hill; 2, Seno-umi north bank; 3, Seno-umi south bank; 4, Kanesuno-se bank; 5, Zenisu. (b) Distribution of active faults after The Research Group for Active Faults [1980]. On land, only active faults of certainty I are shown, and in the sea, only those of dip-slip type with scarp higher than 200 m are shown. 1, Dip-slip fault; 2, Inferred dip-slip fault; 3, Right-lateral strike-slip fault; 4, Left-lateral strike-slip fault. The scale of the map is the same as (a).

on the inner margin of the Nankai trough (outside), and small-scale depressions between them. Though these ridges, depressions, and deep sea plains are separated into blocks and running in echelon, the general trend of arrangement is parallel to the Nankai trough as a whole. On the continental slope, especially on the lower one, many submarine active faults are recognized (Figure 2(b)) [e.g., MSA, 1976a, c; The Research Group for Active Faults (RGAF), 1980]. Most of them are north-side-up reverse faults trending parallel to the Nankai trough. It should be noted that a remarkable submarine canyon called Tenryu Canyon is deeply carving Enshu basin and its outer ridges from north to south and reaching the Nankai trough at a depth of 4,000 m [e.g., MSA, 1976b; Mogi, A., 1977]. As it is situated on the southern extension of the Akaishi Tectonic Line, a north-south trending left-lateral fault to the west of the Akaishi Mountains, it is suspected that Tenryu Canyon was formed along a tectonic line. The chain of outer ridges and the inner margin of the Nankai trough appear to have

been left-laterally offset at this canyon. The steep inner walls of the canyon as high as 200 m are regarded as of fault origin based on seismic reflection profiles.

At the eastern end of the Nankai trough, south off Point Omaezaki, the trough axis sharply turns to the north by more than 30 degrees and almost straight enters into Suruga Bay (Figures 1, 2). This north-south running trough is called the Suruga trough. According to MSA [1977], around the turning point the uppermost sediment layer in the trough remains undeformed, and the trough appears tectonically inactive here compared with its other parts.

The submarine topography in Suruga Bay shows the following conspicuous geomorphic features (Figure 2) [Misawa and Yoshiwara, 1968; Nasu et al., 1968; Mogi, A., 1977; MSA, 1978a, c, d, 1980a, b, c]. First of all, the deep and narrow Suruga trough, 2,500 m deep at the mouth of the bay and 1,300 ~ 1,500 m deep at its northern end, stretches nearly to the head of the bay. And the topography is quite different on its east and west sides. The western edge of the trough forms a very steep submarine cliff as high as more than 1,500 m. On the west side of this cliff there is Seno-umi bank which is 32 m deep at its shallowest part. This bank continues to Udo Hill on land to the north, and to Kanesuno-se bank south off Point Omaezaki, forming a series of uplifting belt. To the west of Seno-umi bank Seno-umi basin forms a depression with a depth of several hundred meters. The above-mentioned uplifting belt is apparently the extension of the outside row of ridges in 'the ridge and trough zone' in the Sea of Enshu, and Seno-umi basin, that of the row of depressions inside it. In general, the submarine topography and geological structure on the west side of the trough are much deformed and full of undulations. On the contrary, the eastern edge of the trough continues to the west coast of the Izu Peninsula fairly smoothly through continental slope and continental shelf, and the geological structure on this slope is scarcely deformed. Based on these geomorphic features and on the deformation of the sediment layer at the bottom of the trough, the western edge of the Suruga trough is regarded as a westerly-dipping active reverse fault [Sugimura, 1972; Matsuda, 1975; MSA, 1978b, 1980a; RGAF, 1980]. There are some other submarine active faults recognized on the west side of the trough [RGAF, 1980]. Very recently Mogi and Sakurai [1980] reported that the shelf break on the east coast of Suruga Bay (mostly 200 ∿ 250 m deep) is deeper than that on its west coast (100 ∿ 150 m deep) and that the inclination of the unconformity plane formed in the maximum Würm glacial age is larger on the east coast (2 ∿ 3°) than on the west coast (1 ∿ 2°). And on the east coast, according to them, the shelf break and 'the 300 ∿ 500 m deep break' deepen from north to south along the west coast of the Izu Peninsula. They interpreted these facts as evidence of the recent subduction of the Izu block at the Suruga trough beneath its west side.

According to Tsuchi [1967], the wide flat-top of Seno-umi bank with a depth of 75 m, which is inferred to have been formed by wave-cut in the end of the Würm glacial stage, is tilting westward by about 7 degrees (According to Mogi and Sakurai [1980], however, the shelf break depth of Seno-umi bank, 60 ∿ 110 m, shows southeastward tilt). On the west coast of the bay Udo Hill also shows a remarkable landward tilt of 7 degrees for the last 10^5 years, and the long-term uplift rate during the same period attains to $3 \text{ m}/10^3\text{y}$, which is comparable to those at the southern Boso Peninsula and Oiso Hill facing the Sagami trough of high seismic activity [Tsuchi, 1967, 1968a, b].

At Point Omaezaki, where considerable coseismic crustal uplifts took place at the time of the 1707 and the 1854 great earthquakes, an uplifted marine terrace as high as 50 m is developing. Matsuda [1975] adopted the age of this terrace of about 10^5 years following Tsuchi [1968b] and es-

timated its uplift rate at $0.4 \text{ m}/10^3\text{y}$, which is about a quarter of that at Muroto Point, the southeastern tip of Shikoku Island in the Nankai district. Recently, as reviewed by Kakimi [1977], the age of this terrace has been estimated at younger, $6 \sim 8 \times 10^4$ years B.P. But, still, the uplift rate is $0.8 \text{ m}/10^3\text{y}$, about a half of that at Muroto, even if the youngest age is adopted. The tilting rate of this terrace, on the contrary, is about $1.1 \%/10^5\text{y}$ (assuming the age of 8×10^4 years B.P.), 3 times or more of those at other typical landward tilting terraces facing the Nankai and the Sagami troughs [Kakimi, 1977]. And the tilting direction, southwest, is noticeable since it is apparently inconsistent with the coseismic tilt due to an earthquake along the Nankai trough proper. Rather, it seems to show an effect originated in Suruga Bay. However, as this Omaezaki terrace is very narrow, it is somewhat questionable whether it represents the general tendency in this area or not [Kakimi, 1977]. As for the movement of the Holocene marine terrace at Omaezaki, there remains much more ambiguity at present. There are two interpretations on the height of the former shoreline, around 5 m and $10 \sim 13$ m [Kakimi, 1977]. The former implies that interseismic recoveries of coseismic uplifts are almost 100 % and that the mode of the seismic crustal movement at Omaezaki may have changed in the Holocene. The latter gives, however, an uplift rate for the last 6,000 years almost comparable to that at Muroto. Kakimi [1977] and Sakamoto et al. [1978] suspect that the latter is the case. Anyway, further intensive investigations are necessary.

Hatano et al. [1979] investigated the height distribution of the highest Holocene shoreline of about 6,000 years B.P. on the northern coast of Suruga Bay. According to them, at the west side of the mouth of the Fuji River (Figure 2(a)) the height of the former shoreline is estimated at 25 m, which gives an uplift rate of nearly $4 \text{ m}/10^3\text{y}$, and it descends gradually to the west; at Yui (Figure 2(a)) about 17 m, and at Shimizu (Figure 1) 9 m +. On the contrary, on the east side of the river the former shoreline is inferred to be buried in the ground below sea level; near the riverside, at a depth of about 20 m, which gives a subsidence rate of more than $3 \text{ m}/10^3\text{y}$, becoming shallower to the east, and at Numazu (Figure 2(a)) almost zero. Hatano et al. [1979] considered these results to be consistent with the large slip rate of the Omiya and the Iriyamase faults ($5 \sim 10 \text{ m}/10^3\text{y}$ for the last $2 \sim 3 \times 10^4$ years, west side up) which are roughly trending north-south along the lower course of the Fuji River [GSI, 1977b]. GSI [1977b] and Hatano et al. [1979] regarded these faults as the northern extension of the Suruga trough. Yamazaki [1979] investigated several active faults to the north of the Suruga trough including the Omiya and the Iriyamase faults. According to him, though the vertical slip rate in this region is almost the highest in Japan (west side up), the

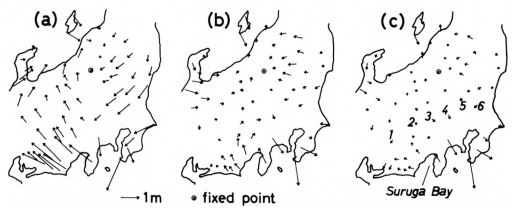

Figure 3. (a) Horizontal displacement vectors of the first-order triangulation stations in the Tokai district and its adjacent regions for the period 1882/1899 ~ 1950/1961 deduced by Harada and Isawa [1969]. (b) Result of the correction in which the positions of stations for the second survey are rotated around the fixed point by 0.8" counterclockwise. (c) Result of the second correction in which the corrected positions in (b) are further made to diverge by 1×10^{-6} from the fixed point. (All these figures have been reproduced from Fujita [1973]) 1, Ena-san; 2, Kai-Komagatake; 3, Kokushi-dake; 4, Dodaira-yama; 5, Teruishi-yama; 6, Tsukuba-san.

mode of faultings is not so simple; normal and reverse faults are mixed and running in echelon suggesting larger-scale left-lateral shear. Tsuneishi and Shiosaka [1978, 1979] and Tsuneishi et al. [1980] emphasized a north-south trending left-lateral strike-slip fault (west side up) as the northern extension of the Suruga trough and called it the Fujikawa fault.

Tsuchi [1968b] divided the Quaternary crustal movements on the west side of the Suruga trough into two types based on the deformation of dissected fans such as the Makinohara and the Ogasa surfaces; one is undulations with axes of NE-SW direction running in echelon, and the other is an upwarping with an axis of NW-SE direction forming the projection of the Omaezaki Peninsula (Figure 2(a)). Yonekura [1975], however, showed that the deformation of the Makinohara dissected fan in the Omaezaki Peninsula was an upwarping with an axis of NE-SW direction, by assuming original fan. Concerning the tectonism on the west side of the Suruga trough, Kakimi [1977] emphasized undulations with axes of NE-SW direction running in echelon, rather than landward tiltings, following Tsuchi [1968b]. GSI [1977b] proposed a schematic representation of morphotectonic division in the Suruga Bay region, in which the region characterized by undulations was called 'the west Suruga Bay mobile belt'.

Kakimi [1977] pointed out that the late Quaternary crustal movement in the Tokai district appears somewhat different in the region facing the Sea of Enshu and in the region facing Suruga Bay. Concerning the former, he suggested that the landward tilting rate at the Atsumi Peninsula (Figures 1, 2) for the last 10^5 years may be comparable to that at Muroto Point and therefore the frequency of Tokai earthquakes in the Sea of Enshu may be the same as that of Nankai earthquakes. He also suggested, based on recent studies, that

the commencement of repeated occurrences of Tokai earthquakes causing seismic crustal movements in the Atsumi Peninsula may be later than 10^5 years B.P.

As a remarkable phenomenon characterizing the Quaternary crustal movement in the Tokai district, we should pay a special attention to 'the Akaishi uplift' of the Akaishi Mountains with an area of about 40×100 km^2 about 50 km west of the Suruga Bay coast (Figures 1, 2). The total amount of uplift is roughly 10^3 m for the last 10^6 years, which gives an uplift rate of 1 m/10^3y [Kasahara, 1975]. This uplift appears to continue to the above-mentioned upwarping in the Omaezaki Peninsula with an axis of NW-SE direction.

4. Recent Crustal Movements in the Tokai district

4-1. Horizontal Displacements

Harada and Isawa [1969] deduced horizontal displacement vectors of the first-order triangulation stations in Japan except for Hokkaido for the period between two nationwide surveys, 1882-1904 and 1948-1964, under the assumption that five stations chosen in rather stable regions had been stationary during this period. In their result all stations in the Tokai district represented remarkable northwestward displacements of around 2 m for the period of about 70 years as is shown in Figure 3(a). Mogi [1970a, b] interpreted this result in terms of his elastic rebound model for offshore thrust earthquakes as representing considerable interseismic strain accumulation as described in Chapter 2. However, Fujita [1973] suggested that these displacement vectors might contain some systematic errors in rotation and scale. He showed that these vectors almost vanish if the positions of stations in the second survey are rotated around one fixed point counterclockwise by 0.8" and made to diverge from the fixed point by 1×10^{-6}, as shown in Figures 3(b) and 3(c). On the contrary, Sato [1977] estimated that the systematic error in scale might be negligible because the selection of fixed stations along the axis of the Honshu arc is tectonically reasonable, and that the displacement vectors shown in Figure 3(c) had been overcorrected. He concluded that the crust of the Tokai district had been probably displaced toward inland by more than 1 m relative to the middle part of central Honshu judging from the horizontal crustal strains in this district.

Very recently Fujii and Nakane [1980] calculated horizontal displacements of the first-order triangulation stations in the Tokai and the Kanto districts during the period 1882/1891 ∿ 1955/1957 by fixing six inland stations, Ena-san, Kai-Komagatake, Kokushi-dake, Dodaira-yama, Teruishi-yama, and Tsukuba-san (see Figure 3(c)), and obtained rather smaller magnitudes, less than 1 m toward the northwest, on the west coast of Suruga Bay. However, their assumption that Ena-san had been stationary during this period is questionable because the G.D.P traverse survey in the Tokai and its adjacent regions revealed remarkable horizontal contractions throughout the survey route up to the Japan Sea side [GSI, 1974c]. So it is very probable that the northwestward displacements on the west coast of Suruga Bay deduced by them are somewhat underestimated.

After all, reliable horizontal displacement vectors in the Tokai district have not been obtained yet. But, thus reviewing, Sato's [1977] estimation as above that the crust on the west coast of Suruga Bay has probably been displaced by more than 1 m during the last nearly 100 years seems reasonable.

4-2. Horizontal Crustal Strain in and around Suruga Bay

The first-order triangulation net covering Suruga Bay including supplementary stations (Figure 4) has been surveyed three times; in 1884, 1931 and 1973 (At the time of the second nationwide first-order survey of 1948/1967 supplementary stations were not surveyed). The former two surveys were triangulations and the last one was mainly trilateration by Geodimeter-8.

For the period 1931 ∿ 1973, principal strains and maximum shear strains in the net were estimated by GSI [1974d]. The result represented remarkable crustal contraction across Suruga Bay in the NWW-SEE direction with the mean velocity of the maximum shear strain of 4.3×10^{-7}/year, which was often referred to by Ishibashi [1976c], Rikitake [1977], and others. However, this calculation contained a deficiency that the net adjustment for the 1931 survey was not unified [GSI, 1974d, 1977c; Fujii and Nakane, 1979]. Then, GSI [1977c] made a careful recalculation for the same period, by which the mean value of the maximum

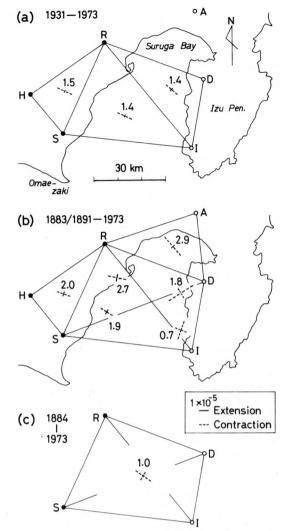

Figure 4. Strain changes in the first-order triangulation net covering Suruga Bay. Solid and open circles indicate the first-order stations and supplementary ones respectively: A, Ashitaka-yama; D, Daruma-yama; H, Hakko-san; I, Iwashina-mura; R, Ryuso-zan; S, Sakabe-mura. Horizontal principal strains in each triangle are shown by solid and broken bars. Numerals represent maximum shear strains in 10^{-5}. (a) For the period 1931 ∿ 1973 after GSI [1977c]. (b) For the period 1883/1891 ∿ 1973 after Fujii and Nakane [1979, 1980]. (c) Strains in the quadrilateral net for the period 1884 ∿ 1973 after Dambara [1980a, b].

shear strain velocity over the bay was reduced to 3.3×10^{-7}/year (Figure 4(a)). But this was still larger than the ordinary value of 1 ∿ 2 × 10^{-7}/year in whole Japan [GSI, 1972b; Nakane, 1973], and fairly uniform contraction of the whole region in the NWW-SEE direction was shown again. So, Aoki [1977], Utsu [1977b], and others paid much attention to it. GSI [1977c] did not use the side of Sakabe-mura - Daruma-yama in calculation because this direction had not been directly measured in 1931.

As for the period 1884 ∿ 1973, horizontal crustal strains over the bay was calculated by GSI [1977a]. Very recently, however, this calculation was corrected considerably in Rep. Coord. Comm. Earthq. Predict., Vol.23, p.177, February 1980, in consequence that a significant error during data processing relating to Iwashina-mura station in the first survey of 1884 was found out [Fujii and Nakane, 1979; GSI, 1980d]. The revised result showed 2.0×10^{-7}/year mean value of the maximum shear strain velocity over four triangles covering central Suruga Bay, which is almost the same as the above-mentioned ordinary value and inconsistent with the result for the period 1931 ∿ 1973.

On the other hand, Fujii and Nakane [1979, 1980] estimated horizontal crustal strains over Suruga Bay for three periods, 1883/1891 ∿ 1931, 1931 ∿ 1973, and 1883/1891 ∿ 1973, systematically. They obtained the result that, though Suruga Bay had been contracted in the NW-SE direction as a whole, the maximum shear strain velocity averaged over four triangles covering central Suruga Bay had changed from 4.8×10^{-7}/year during the period 1883/1891 ∿ 1931 to 2.0×10^{-7}/year during the period 1931 ∿ 1973. Although their result for the total period of 1883/1891 ∿ 1973, with 2.1×10^{-7}/year maximum shear strain velocity as a simple mean (Figure 4(b)), is almost the same as GSI's revised result, that for the period 1931 ∿ 1973 is considerably different from that by GSI [1977c]. The apparent disorder of strain accumulation in the triangles containing the Daruma-yama - Iwashina-mura side as is seen from Figure 4(b) may suggest the effect of a presumable large crustal deformation in the Izu Peninsula associated with the 1923 Kanto and the 1930 North-Izu earthquakes and relating crustal activities.

In the 1931 triangulation survey only measurements of horizontal angles were carried out, and so a proper estimation of the size of the net in this survey is rather difficult [Fujii and Nakane, 1980]. Fujii and Nakane [1980], after trying two kinds of net adjustments for the 1931 survey based on two different assumptions, concluded that the 1931 survey should not be used in order to see dimension-depending strain factors such as magnitudes of principal strains and dilatations. They considered that the comparison of the 1883/ 1891 and the 1973 surveys gave the proper estimation of horizontal crustal strains in the Suruga Bay region. However, there is another opinion that a careful comparison of the 1931 and the 1973 surveys gives the most reliable information because the accuracy of measurements in 1931 is much better than that in 1884 [GSI, 1977c; Sato, 1977; Sato H., personal communication, 1980].

Very recently Dambara [1980a, b] calculated horizontal crustal strain of the quadrilateral

net covering Suruga Bay for the period 1884 ∿ 1973 using GSI's unpublished data of side-lengths of six sides as is shown in Figure 4(c).

In February/March 1977 another distance-measurement survey over Suruga Bay was carried out, and GSI [1977d] estimated horizontal crustal strains over the bay for the period 1884 ∿ 1977. (This result was also corrected in Rep. Coord. Comm. Earthq. Predict., Vol.23, p.177, February 1980, due to the above-mentioned Iwashina-mura trouble.) The strain during this period was affected remarkably by the 1974 Off-Izu-Peninsula earthquake (M = 6.9) which occurred at the southern tip of the Izu Peninsula. But, in the triangle RSD (see Figure 4) the NWW-SEE trending crustal contraction is considered to have been continuing also during the period 1973 ∿ 1977 [Dambara, 1980a].

After all, it is certain that the crust in the Suruga Bay region has been contracted in the NWW-SEE direction rather steadily during the last about 100 years, apart from the Izu Peninsula. But, the magnitude of contraction, as well as di-

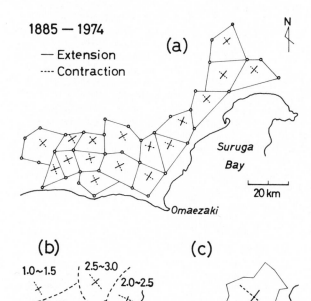

Figure 5. Horizontal crustal strains in the Tokai district accumulated during the period 1885 ∿ 1974 as deduced from the second-order triangulation data. (a) Distribution of principal strain axes as deduced by GSI [1976]. (b) Schematic map showing the distribution of horizontal crustal strains drawn by Sato [1977]. Numerals represent maximum shear strains in 10^{-5}. (c) Principal strain axes as average in the illustrated polygon obtained by Dambara [1980b] based on GSI's [1976] result.

latation, is still uncertain due to the difficulty in estimating properly the scale of the triangulation net. On the other hand, the maximum shear strain can be estimated more easily and has been the main subject of discussion. However, it does not seem so important for the expected Suruga Bay earthquake. On this point we will discuss later in connection with the time of occurrence of the coming earthquake.

The second-order triangulation net in the Tokai district was first surveyed in 1885, and resurveyed in 1974 as the survey of primary control points mainly by distance measurements. GSI [1976] deduced horizontal crustal strains from these two surveys. One of the results are reproduced in Figure 5(a). Based on these results Sato [1977] showed the pattern of the horizontal crustal strains in this district schematically as is reproduced in Figure 5(b). The NWW-SEE or NW-SE contraction is predominant except near Lake Hamana, which is in good harmony with the results of the first-order triangulation analyses. The NNE-SSW extension near Lake Hamana is inferred as the effect of the 1944 Tonankai earthquake [GSI, 1976; Sato, 1977]. Dambara [1980b] obtained averaged horizontal crustal strain in the domain shown in Figure 5(c), on the southwest coast of Suruga Bay, based on GSI's [1976] result. According to him, the maximum compressional strain is about 2×10^{-5} in the NW-SE direction, and the maximum shear strain and dilatation, 3.2×10^{-5} and -1.2×10^{-5}, respectively. In 1977, another survey of primary control points was carried out on the west coast of Suruga Bay and in the further inland mountain region [GSI, 1978b]. It revealed that the strain change in the Akaishi Mountains during the period 1885 ∿ 1977 is fairly small compared with that in the coastal region, and that the strain accumulation on the west coast of Suruga Bay during the period 1974 ∿ 1977 is negligible [GSI, 1978b].

4-3. Changes of Side Lengths in the Suruga Bay Region

In order to see the linear strain across Suruga Bay all available results at present concerning changes of side lengths in this region are summarized in Figure 6. Although Fujii and Nakane [1979, 1980] calculated changes of some side lengths in this region for the periods 1883/1891 ∿ 1931 and 1931 ∿ 1973, and GSI [1977c], for the period 1931 ∿ 1973, we better avoid using the 1931 survey to estimate the distance change for the reason as described in 4-2.

According to Figure 6(a) four sides across Suruga Bay and D-I side in the Izu Peninsula were considerably contracted during the period 1883/ 1891 ∿ 1973. Their linear strains and linear strain velocities are summarized in Table 1. During the period between 1973 and 1977 February/ March, sides of R-I, S-I, and D-I were remarkably extended as seen in Figure 6(b) and in Table 1, which is interpreted as mostly due to the consid-

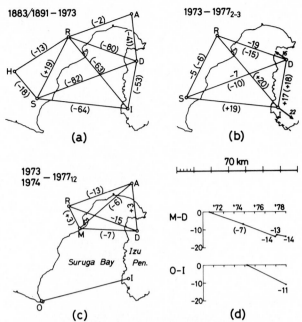

Figure 6. Changes of side lengths in the Suruga Bay region. Numerals without parentheses represent differences between two distance-measurements. Numerals with parentheses show that calculated side lengths have been used. Unit is cm. A, D, H, I, R, S, the same as in Figure 4; M, Muramatsu-mura; O, Omaezaki. (a) Result of comparison of two net adjustments of the 1883/1891 and the 1973 triangulations [Fujii and Nakane, 1979, 1980]. (b) Comparison between the 1973 triangulation/trilateration and the 1977 Feb/Mar distance-measurement [GSI, 1977d]. Horizontal displacement vectors for D and I stations, whose magnitudes are shown by italic numerals in cm, have been deduced by R and S stations being fixed [GSI, 1977d]. A broken line at the southern tip of the Izu Peninsula represents the right-lateral strike-slip fault associated with the 1974 Off-Izu-Peninsula earthquake (M = 6.9). (c) Comparison of the 1977 Dec distance-measurement with the 1973 Nov one (numerals without parentheses), and with the net adjustment of 1974 Nov primary-control-point survey (numerals with parentheses) [GSI, 1978b]. (d) Changes of M-D and O-I side lengths [GSI, 1979b].

erable southeastward displacement of Iwashina-mura station caused by the NW-SE trending right-lateral strike-slip faulting at the time of the Off-Izu-Peninsula earthquake of May 9 (JST), 1974 (M = 6.9) [GSI, 1977d]. On the contrary, according to the same Figure and Table the linear strain velocities of contractions of R-D and S-D sides were increased during this period.

Apparent acceleration of contraction across Suruga Bay in recent years can be seen also in

TABLE 1. Linear strains (in 10^{-6}) and linear strain velocities (in parentheses, in 10^{-7}/y) of four sides across Suruga Bay and D-I side in the Izu Peninsula during a few survey intervals based on Figure 6

Side (km)	I∿II (89y)	II∿III (3.3y)	III∿IV (0.8y)	II∿IV (4.1y)
R-D (42.5)	-19 (-2.1)	-4.5 (-14)	+0.9 (+12)	-3.5 (-8.6)
R-I (54.0)	-12 (-1.3)	+3.7 (+11)		
S-I (53.5)	-12 (-1.3)	+3.6 (+11)		
S-D (61.7)	-13 (-1.5)	-1.1 (- 3.4)		
D-I (27.3)	-19 (-2.2)	+6.2 (+19)		

Survey epoch: I, 1884; II, 1973 Nov; III, 1977 Feb/Mar; IV, 1977 Dec.
+, Extension; -, Contraction.

Figure 6(d). Five measurements of M-D side (33.8 km), in 1971 August, 1974 November (not directly measured), 1977 December, 1978 February, and 1978 December, show a linear strain velocity of about -6×10^{-7}/year. And measurements of O-I side (51,9 km) in 1975 March and in 1978 December show a linear strain velocity of -5.6×10^{-7}/year. The contraction rate of A-R side (39.2 km) was also increased, from -5.9×10^{-9}/year during the period 1883/1891 ∿ 1973 (Figure 6(a)) to -1.1×10^{-6}/year during the period 1974 ∿ 1977 (Figure 6(c)). However, concerning these apparent accelerations of contraction across Suruga Bay we should bear in mind that changes of side lengths during the period 1883/1891 ∿ 1973 have not been deduced by direct measurements but by comparing results of net adjustments of old and new triangulations. And the estimation of the net size of the old survey may contain considerable error as mentioned in 4-2.

It is noticeable that the R-D side was extended by 4 cm between February/March and December in 1977 as is seen in Figures 6(b) and (c). This movement may have some relation to the remarkable crustal activity in the Izu Peninsula around this time. An apparent interruption of contraction of the M-D side after 1977 December may have some relation to the occurrence of the Izu-Oshima earthquake of January 14, 1978 (M = 7.0), though it may be just an error.

The Omaezaki rhombus base-line net (Figure 7 (a)) which was set up in November 1970 has been measured three times; in 1970 November, 1971 November, and 1973 October [GSI, 1971, 1972a, 1974b]. The result illustrated in the upper right in Figure 7(b) shows no significant crustal deformation during these three years. As Sakabe-mura and Takatenjin are the first-order triangulation stations, the S-Tt side (18.4 km) was measured indirectly twice before 1970, in 1890 and 1956. According to the result shown in Figure 7(b), this side was contracted by a rate of 0.7×10^{-7}/year between 1890 and 1956, but the rate was increased remarkably to about 6×10^{-7}/year after 1956, although the accuracy of data for 1890 and 1956 is open to question.

The Omaezaki radial base-line net (Figure 7

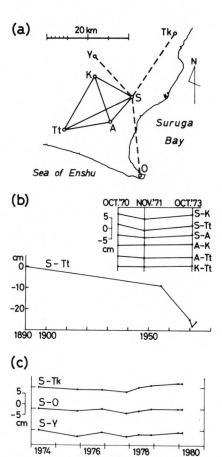

Figure 7. (a) Locations of the Omaezaki rhombus base-line net (solid lines) [GSI, 1972a, 1974b] and the Omaezaki radial base-line net (broken lines) [GSI, 1974e]. A, Akadohara; K, Kikugawa-mura; O, Omaezaki; S, Sakabe-mura; Tk, Takakusa-yama; Tt, Takatenjin; Y, Yoko-oka. (b) Secular changes of side lengths of the Omaezaki rhombus base-line net [GSI, 1974b]. (c) Secular changes of side lengths of the Omaezaki radial base-line net [GSI, 1980b].

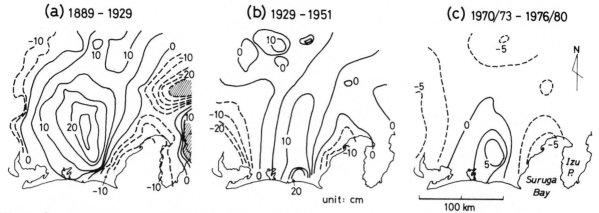

(a) 1889 – 1929   **(b)** 1929 – 1951   **(c)** 1970/73 – 1976/80

unit: cm

100 km

Figure 8. Vertical crustal movements in the Tokai district during the periods 1889 ∿ 1929 (a) and 1929 ∿ 1951 (b) deduced by Dambara [1968], and 1970 Sep/1973 Jul ∿ 1976 Jan/1980 Feb contoured by Dambara [1980b] based on GSI's [1980c] result. Contour intervals are 5 cm in (a) and (b), and 2.5 cm in (c).

(a)) which was set up in April 1974 has been measured eight times; in 1974 April, 1975 November, 1976 November, 1977 November, 1978 May, 1978 November, 1979 December, and 1980 February [GSI, 1974e, 1977a, 1978b, 1979a, b, 1980b]. The result is illustrated in Figure 7(c), which shows no significant crustal movement. Dambara [1980b] pointed out, however, a change of tendency of movement around the beginning of 1978.

4-4. Vertical Crustal Movement

Dambara [1968] arranged precise leveling surveys in the Tokai district to three epochs of 1888.8, 1929.2, and 1951.1, and deduced vertical crustal movements in the Tokai district and its adjacent regions for the periods 1889 ∿ 1929 and 1929 ∿ 1951 as shown in Figures 8(a) and (b). Figure 8(c) is the contour map of the vertical crustal movement in this district during the period 1970 Sep./1973 Jul. ∿ 1976 Jan./1980 Feb. drawn by Dambara [1980b] based on GSI's [1980c] data. These figures represent clearly the basic pattern of the recent vertical crustal movement in this district; the subsidence of the west coast of Suruga Bay and the uplift of the western inland area, a manifestation in the geodetic domain of the Akaishi uplift as mentioned in Chapter 3.

As the attention to the Tokai district increased, leveling surveys in this district including the second-order routes have been intensified [GSI, 1970b, 1974a, 1977a, d, 1979a, 1980a, b]. Figure 9 shows contour maps of the vertical crustal movement on the west coast of Suruga Bay for the last about 80 years as deduced through these surveys by GSI [1978a, 1979a].

Figures 10(a) and (b) show secular vertical movements of five representative bench marks which are distributed extensively on the west coast of Suruga Bay relative to stable bench marks. In Figure 10(a) the 6 ∿ 7 cm coseismic

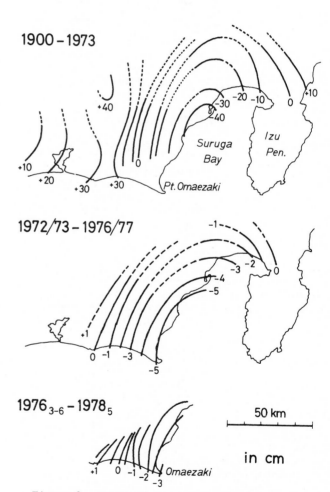

1900 – 1973

1972/73 – 1976/77

1976₃₋₆ – 1978₅

50 km

in cm

Figure 9. Vertical crustal movements on the west coast of Suruga Bay during the last about 80 years. The top and the middle have been reproduced from GSI [1978a], and the bottom, from GSI [1979a].

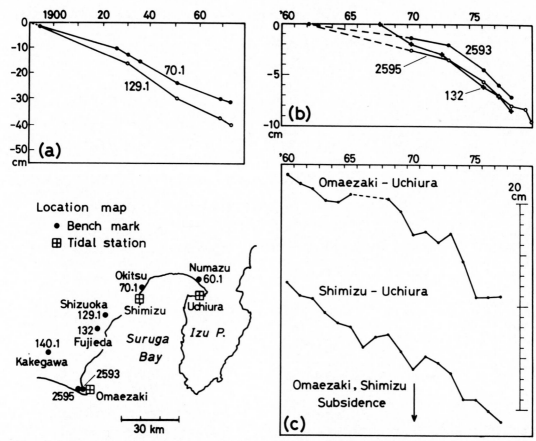

Figure 10. (a) Secular vertical movements of BM 70.1 (at Okitsu) and BM 129.1 (at Shizuoka) relative to BM 60.1 (at Numazu) after Sato [1977]. (b) Secular vertical movements of BM 132 (at Fujieda), BM 2593 and BM 2595 (both near Omaezaki) relative to BM 140.1 (at Kakegawa) [GSI, 1979a, 1980b]. (c) Differences of anual mean sea level among tidal stations in Suruga Bay [GSI, 1979a].

subsidence of BM 60.1 due to the 1923 Kanto earthquake has been corrected. As BM 60.1 has been stationary except this subsidence, this figure represents approximately movements relative to the mean sea level [Sato, 1977]. As for Figure 10(b), the datum point BM 140.1 may have a slight tendency to uplift, but the amount is very small [GSI, 1979a; Dambara, 1980b]. Figure 10(c) represents differences of anual mean sea level among tidal stations in Suruga Bay. Since Uchiura tidal station, near BM 60.1, is almost stationary, this figure shows approximately crustal subsidence at Omaezaki and Shimizu relative to the mean sea level. It is remarkable that Shimizu has been subsiding for the last nearly 20 years at almost a constant rate of about 8 mm/year [Sato, 1977; GSI, 1977a, 1979a]. Apparently unsteady subsidence at Omaezaki may be partly because Omaezaki tidal station faces the open sea [GSI, 1977a].

Figures 9 and 10 show that the west coast of Suruga Bay as a whole has been steadily subsiding since around 1900 parallel to the Suruga trough at almost a constant rate of about 5 mm/year, and that this rate has increased in recent years to 10 mm/year or more especially in the southern part. In particular, the considerable acceleration of subsidence at Omaezaki since 1973 ∿ 1974 as seen in Figures 10(b) and (c) has been much noted [GSI, 1977d, 1979a, 1980b; Mogi, K., 1977; Dambara, 1980b]. According to Dambara [1980b] the commencement of this acceleration occurred at the beginning of 1974. Mogi, K. [1977] and GSI [1977d] suggested its possible relation to the occurrence of the 1974 Off-Izu-Peninsula earthquake. GSI [1979a], on the other hand, pointed out that the subsiding rate around Omaezaki might have overtaken that at Fujieda as suggested from Figure 10(b). The subsidence rate of 5 mm/year is the same as that at Kushimoto, the southern tip of the Kii Peninsula, preceding the 1946 Nankai earthquake (M ≃ 8.1) [Sato, 1977]. Incidentally, the subsidence rate at Aburatsubo, the southern tip of the Miura Peninsula, preceding the 1923 Kanto earthquake (M ≃ 8.0) was 6 mm/year [Sato, 1977].

After all, the total amount of subsidence on the west coast of Suruga Bay at present since around 1900 is estimated at nearly 50 cm. On the other hand, the amount of the Akaishi uplift probably exceeds 40 cm during the last 80 years. Dambara [1980b] pointed out that this uplift also has been accelerated in recent years as seen in Figure 8.

5. 1854 Ansei Tokai Earthquake and the Suruga trough thrust

The Ansei Tokai earthquake of December 23, 1854 (M = 8.4 after Kawasumi [e.g., Usami, 1975]), together with the Ansei Nankai earthquake (M = 8.4 after Kawasumi [e.g., Usami, 1975]) which followed the former in only 32 hours, formed an entire rupture of the boundary between the Philippine Sea plate and southwest Japan along the Nankai trough [Ando, 1975b]. Ando [1975b] presented a fault model of this earthquake as a right-lateral reverse faulting of 4 m dislocation along a fault plane of 230 × 70 km² dipping toward N30°W by 25 degrees and lying beneath the Seas of Kumano and Enshu. He did not extend the fault into Suruga Bay because at that time there was no information on coseismic vertical crustal movements inside Suruga Bay except the 80 ~ 100 cm uplift of Point Omaezaki and because he regarded the Suruga trough as an aseismic collision boundary. However, this fault model apparently conflicted with the violent shaking on the entire west coast of Suruga Bay as high as 7 in the JMA seismic intensity scale (corresponding to 10 and more in the MM scale) (Figure 11(a)). Although Ando [1975a, b] attributed this strong shaking to the effect of the eastward rupture propagation and to the thick alluvium covering the coastal area, generally, the area of seismic intensity 6 and more in average should be considered to be inside of the source region as pointed out by Muramatsu [1972].

This problem has been clearly solved in 1976 by an accidental finding of an old document. It was an official report presented by the Governor of Shizuoka Prefecture in 1893 at the request of the president of the University of Tokyo concerning tsunamis and ground deformations at coastal areas of the Izu Peninsula, Suruga Bay and the Sea of Enshu mainly due to the Ansei Tokai earthquake [Hatori, 1976; Earthquake Research Institute, 1977]. By virtue of this report it has been disclosed that many places on the west coast of Suruga Bay, up to its deepest part, had uplifted remarkably at the time of this earthquake, whereas on its east coast vertical displacements had been scarcely noticed. Based on the descriptions, it is certain that these uplifts are not surface phenomena due to tsunami, etc., but crustal movements. And the report gave a vivid description of tsunami generation in Suruga Bay witnessed from a mountain on the west coast of the Izu Peninsula as follows; simultaneously with the strong shock the sea surface of the center of the bay was upheaved furiously with a thunderous detonation, instantly the enormous swelling collapsed into a huge water ring, advanced against the coast in every direction, then the center

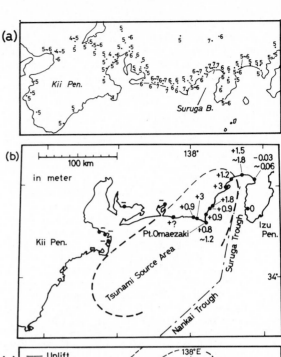

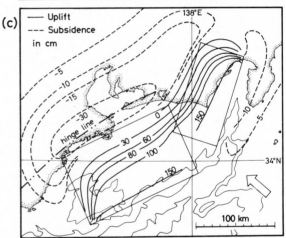

Figure 11. (a) Seismic intensity distribution (in the JMA scale) of the 1854 Ansei Tokai earthquake compiled by Hagiwara [1970]. (b) Vertical surface displacements observed along the coast at the time of the 1854 Ansei Tokai earthquake compiled by Ishibashi [1977b] and the tsunami source area inferred by Hatori [1976]. (c) Vertical surface displacement field associated with the fault model of the 1854 Ansei Tokai earthquake [Ishibashi, 1976c, 1977b]. Two rectangles represent horizontal projections of the SW and the NE fault planes. The white arrow shows the moving direction of the Philippine Sea plate relative to the Eurasian plate used in this calculation (N50°W). The hinge line of the late Quaternary crustal movement after Ando [1975b] and isobaths of 2,000, 3,000 and 4,000 m are also shown.

became hollow, and the sea surface repeated several up-and-down motions. According to another newly discovered document, Numazu, a town at the head of Suruga Bay (Figures 1, 2), was struck by tsunami immediately after the shock [Ishibashi, 1976c]. Vertical crustal movements associated with this earthquake are summarized in Figure 11(b) together with the tsunami source area inferred by Hatori [1976].

The remarkable crustal movements in the Suruga Bay region during this century consisting of the subsidence and northwestward displacement of the west coast, and the NWW-SEE contraction across the bay as reviewed in Chapter 4, the conspicuous coseismic uplift and the violent shaking in the whole region on the west coast of the bay at the time of the 1854 earthquake as described above, and the outstanding geomorphic features in the region along the Suruga trough as seen in Chapter 3 lead us to the following conclusion naturally; along the Suruga trough the Philippine Sea plate is underthrusting northwestward beneath the Tokai district as well as along the Nankai trough, and at the time of the 1854 Ansei Tokai earthquake a resultant elastic rebound of the continental lithosphere took place not only along the Nankai trough but also along the Suruga trough. Thus, what was suggested by Sugimura [1972] as mentioned in Chapter 2 has been established. Although Sugimura [1972] considered that along the northern half of the Suruga trough the two plates were colliding, the underthrusting of the oceanic plate and the resultant rebound of the continental plate have proved to be occurring along the entire extension of the trough. Following Sawamura's [1953, 1954] naming of 'the Nankai thrust' which was emphasized by Sugimura [1972], we will call the plate boundary thrust along the Suruga trough 'the Suruga trough thrust' hereafter in this paper, mainly for convenience of description.

From the viewpoint of the above-mentioned tectonics Ishibashi [1976c, 1977b] derived a revised static fault model of this earthquake as follows. The actual faulting of this earthquake is inferred to have taken place along some deformed area as is illustrated in Figure 12, because the plate boundary along the Nankai-Suruga trough is curved markedly probably due to the northward collision of the Izu-Bonin arc with central Honshu [e.g., Matsuda, 1978]. For calculation easiness, however, the faulting has been approximated by a two-fault model, one along the Nankai trough (SW fault) and the other on the Suruga trough thrust (NE fault). The strike or the dip direction, the dip angle, the slip angle, and the position of the upper edge of the SW fault plane have been assumed as the same as those of the 1944 Tonankai's fault as will be explained in the next chapter. As for the geometry of the NE fault plane, or the geometry of the Suruga trough thrust itself, at first a plate boundary with a strike parallel to the trough (N13°E) and dipping westward by 40 degrees has been tentatively assumed. Here, the dip angle is assumed following

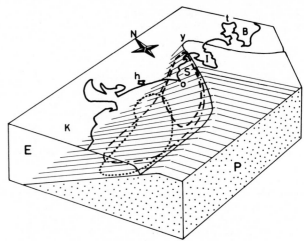

Figure 12. Sketch map of inferred rupture areas of the 1854 Ansei Tokai earthquake (chain line), the 1944 Tonankai earthquake (dotted line), and the anticipated Tokai earthquake (broken line) on the plate boundary thrust in the Tokai district. B, Boso Peninsula; E, Eurasian plate; h, Lake Hamana; I, Izu Peninsula; K, Kii Peninsula; o, Point Omaezaki; P, Philippine Sea plate; S, Suruga Bay; t, Tokyo; y, eastern Yamanashi area.

Ishibashi's [1976a] suggestion that the dip angle of the plate boundary at its northern tip (in the eastern Yamanashi area, Figures 1, 12) is around 45 degrees. Then, by a condition that the plate boundary should continue smoothly to that along the Nankai trough, the dip direction and the dip angle became N72°W and 34 degrees respectively, both in average, beneath the Suruga Bay area.

In order to evaluate the slip angle of the NE faulting, or the ratio of a dip-slip component to a strike-slip component, we must have an idea on the relative moving direction of the Philippine Sea plate to the Eurasian plate along the Suruga trough. As for this direction along the Nankai trough in the Sea of Kumano and its western adjacency, Kanamori [1972] obtained the estimation of N50 ∿ 54°W, and Ando [1975a, b] used this value. Seno [1977a] also obtained the estimation of N54°W around the Sea of Enshu. However, it seems possible that along the Suruga trough the direction may be somewhat different from this value. For example, Kaizuka [1972, 1974, 1975] proposed a hypothesis that the Izu-Bonin inner arc ('Izu Inner Bar') was shifting in a direction of N10°W along the Suruga trough and its southern extension ('the Nishi-shichito Fault Zone'), differentially from the Philippine Sea plate accompanying the northward-moving Izu-Bonin outer arc ('Izu Outer Bar'), being caused by the oblique convergence of the Pacific plate at the Izu-Bonin trench. Ando et al. [1973] also presented almost the same idea. Kakimi [1977] supported Kaizuka'z hypothesis emphasizing a sinistral echelon topography on the west side of the Suruga trough, and considered the fault along the Suruga trough to be a high-angle reverse fault with a left-lateral strike-slip being predominant. Moreover, in Sagami Bay on the east of the Izu Peninsula, the moving direction of the Philippine Sea plate of N22 ∿ 35°W seems preferable through analyses of the 1923 Kanto earthquake (M ≃ 8) [Ando, 1971, 1974; Ishibashi, 1976a; Matsu'ura et al., 1980], although estimation around N50°W have been also obtained concerning the same earthquake [Kanamori, 1971]. And, Fujii and Nakane [1980] obtained the NNW direction of displacements of the first-order triangulation stations on the west side of Suruga Bay for the period 1883/1891 ∿ 1955/1957, though the assumption employed in their calculation seems somewhat questionable as described in 4-1. On the other hand, Seno [1980] discussed the reality of 'Izu Bar' proposed by Kaizuka [1974, 1975] in detail and concluded that at present there is no firm evidence for it. He also suggested that the relative motion direction of the Philippine Sea plate to the Eurasian plate off the southern Kanto district has been about N50°W since the 1923 Kanto earthquake based on the investigation on horizontal crustal movements by means of a finite element method. In general, the actual situation of plate interaction in such a region of complex tectonism as now under consideration is probably that the relative motion direction between two plates is rather fluctuating in a shorter time range and changes from one event to another or from one interseismic period to another, though after all resulting in almost a constant direction as a very long-term average. In conclusion we have no reason to adopt a slip direction along the Suruga trough different from that along the Nankai trough. Thus, the direction of N50°W is tentatively assumed here. But it should be added that the estimation of the present slip direction at the Suruga trough by means of precise analyses of geodetic and seismological data is very important if the Suruga trough should be a subduction boundary at all.

In this geometrical framework other fault parameters have been determined so that the model is the best fit to the coseismic vertical crustal displacement field (Figure 11(b)) and also to the hinge line of the late Quaternry crustal movement in the Kii Peninsula - Atsumi Bay region (Figure 11(c)) and is in harmony with the inferred tsunami source area (Figure 11(b)) and the general tendency of the seismic intensity distribution (Figure 11(a)). The NE fault plane has been placed so that its upper extension breaks the surface on the western steep cliff of the Suruga trough. The obtained fault model is as follows (in this paper the notation of fault parameters as is illustrated in Figure 13 is used): The SW fault, right-lateral reverse; ϕ = N25°W (θ = N65°E), δ = 24°, L = 150 km, W = 100 km, H_0 = 3 km, λ = 113°, D = 4.0 m (D_d = 3.7 m, D_r = 1.6 m), M_0 = 3.0 × 10^{28} dyne·cm (rigidity is assumed as 5 × 10^{11} dyne·cm^{-2}), α = N50°W; the NE fault,

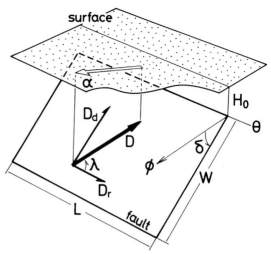

Figure 13. Notation of static parameters of a reverse fault buried in a half space used in this paper: ϕ, dip direction; θ, strike; δ, dip angle; L, length; W, width; H_0, depth of the upper edge; λ, slip angle (if λ<90°, left-lateral; if λ>90°, right-lateral); D, dislocation amount; D_d, dip-slip component; D_r, right-lateral strike-slip component (in this case, negative); α, motion direction of the foot wall side.

left-lateral reverse; ϕ = N72°W (θ = N18°E), δ = 34°, L = 115 km, W = 70 km, H_0 = 2 km, λ = 71°, D = 4.0 m (D_d = 3.8 m, D_r = -1.3 m), M_0 = 1.6 × 10^{28} dyne·cm, α = N50°W. Figure 11(c) shows the vertical surface displacement field associated with this fault model. The calculation has been made for a semi-infinite medium of V_P = 6.0 km/sec and V_S = 3.4 km/sec, by using the computer program composed by Sato and Matsu'ura [1974].

In order to see the adequacy of the assumed slip direction of the Philippine Sea plate along the Suruga trough thrust, in regard to the NE fault only, vertical and horizontal surface displacement fields associated with five different fault models in which δ and α are changed as follows have been additionally calculated; δ = 34° (α = N30°W, N72°W), δ = 60° (α = N30°W, N50°W, N72°W). The result is shown in Figure 14. All cases of δ = 60°, only one of which is shown in Figure 14, conflict with a report that the 1854 coseismic vertical crustal displacements were scarcely noticed on the west coast of the Izu Peninsula. And their horizontal surface displacement fields in which the west coast of Suruga Bay are not displaced southeastward at all whereas the Izu Peninsula is remarkably displaced westward seem to be unacceptable. As for cases of δ = 34°, the model with α = N50°W, which is our final result as before and is illustrated in Figure 18 also, appears to be better than the other two, compared with the pattern of the recent horizontal crustal movement as described in Chap.4.

Recently Kobayashi and Midorikawa [1979] cal-

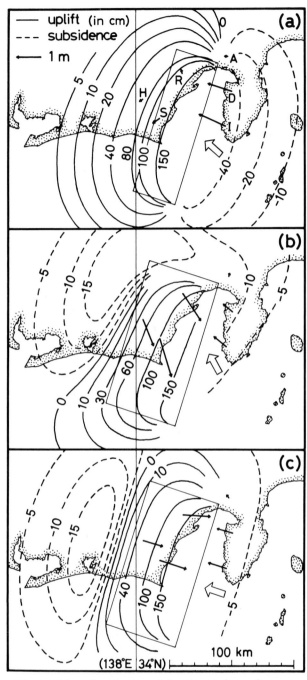

Figure 14. Vertical and horizontal surface displacement fields associated with hypothetical fault models on the Suruga trough thrust in which δ and α have been changed from those of the NE fault in Figure 11(c). Rectangles are horizontal projections of fault planes. White arrows represent α. A, D, H, I, R, and S are the same as in Figure 4. (a) δ = 60° and α = N50°W. (b) δ = 34° and α = N30°W. (c) δ = 34° and α = N72°W.

culated the distribution of ground-surface peak accelerations based on the above-mentioned our final model taking site geological conditions into account, and showed that the obtained distribution was in good harmony with the seismic intensity distribution compiled by Usami [1977], which is basically the same as, but more detailed than, that in Figure 11(a'). Meanwhile, Aida [1979b] made numerical experiments on the tsunami due to this earthquake, and found that our final model as above gave an extremely good result concerning the distribution of maximum inundation heights from the Boso to the Kii Peninsulas including inside of Suruga Bay. The maximum inundation heights of this tsunami are, according to the minute field and documentary investigations by Hatori [1977], about 5 m above mean sea level at most places on the Suruga Bay coast and about 6 m along the Enshu coast. Hatori [1977] pointed out that the rather lower maximum inundation height at the head of Suruga Bay and an asymmetric pattern of height distribution on the east and west coasts in the northern part of the bay support the inferred tsunami source area extending up to the head of the bay as in Figure 11(b).

Recently, Tsuneishi and Sugiyama [1978], Tsuneishi and Shiosaka [1978] and Tsuneishi [1980] expressed the opinion that the fault along the Suruga trough was not a plate boundary reverse fault but a vertical left-lateral strike-slip fault, which they called 'Suruga Bay fault'. And Tsuneishi and Shiosaka [1978, 1979] asserted that the 1854 Ansei Tokai earthquake had been due to the slip of this fault and its inland extension to the north, which they called 'Fujikawa fault'. However, the remarkable coseismic crustal uplift and the prominently strong shaking on the west coast of Suruga Bay at the time of this earthquake are very difficult to explain by this hypothesis, though Tsuneishi and Shiosaka [1978, 1979] made no comment on this point. Aida [1979b] showed by numerical experiments that the distribution of maximum inundation heights of the Ansei tsunami can not be explained by this model. Left-lateral displacements emphasized by them along the Fujikawa fault do not necessarily conflict with the idea of the Suruga trough thrust because the latter also contains a considerable amount of left-lateral strike-slip motion. Rather, west-side-up surface displacements described by them along the Fujikawa fault seem to support our interpretation. Thus, this new hypothesis is still much open to question.

But, Tsuneishi and Shiosaka's [1978, 1979] suggestion that the 1854 seismic faulting extended far inland to the north of Suruga Bay is important. Kanbara and Matsuoka 'Jishin-yama' (earthquake hills), about 3 m uplifted ground bodies with areas of roughly 1,000 (NS) × 50∿100 (EW) m² each, which appeared near the mouth of the Fuji River at the time of this earthquake [GSI, 1977b; Tsuneishi and Shiosaka, 1978, 1979], the north-south stretching zone to the north of Suruga Bay with seismic intensities of 6 ∿ 7 in

TABLE 2. Fault parameters of various models of the 1944 Tonankai earthquake

			ϕ	θ	δ	L (km)	W (km)	H_0 (km)	λ	D (m)	(D_d, D_r) (m)	M_0 *) $(10^{28}$dyne·cm$)$	α
Kanamori[1]			N54°W	N36°E	10°	120	80	16	90°	3.1	(3.1, 0.0)	1.5	N54°W
Ando[2]			N30°W	N60°E	25°	130	70	–	108°	4.0	(3.8, 1.3)	1.8	N50°W
Inouchi	SW	fault	N46°W	N44°E	30°	154	67	–	72°	4.7	(4.5,-1.5)	2.5	N25°W
and Sato[3]	NE	fault	N46°W	N44°E	30°	84	78	–	72°	3.2	(3,0,-1.0)	1.0	N25°W
Ishibashi[4]	SW	fault	N25°W	N65°E	24°	110	70	3	113°	4.0	(3.7, 1.6)	1.5	N50°W
	NE	fault	N25°W	N65°E	24°	80	80	20	113°	4.0	(3.7, 1.6)	1.3	N50°W
Fujii[5]	SW	fault	N25°W	N65°E	24°	110	70	3	113°	5.0	(4.6, 2.0)	1.9	N50°W
	Mid	fault	N25°W	N65°E	24°	80	80	20	113°	6.0	(5.5, 2.3)	1.9	N50°W
	NE	fault	N45°W	N45°E	24°	100	80	20	93°	0.2	(0.2, .01)	0.08	N48°W

1) Kanamori [1972], 2) Ando [1975b], 3) Inouchi and Sato [1975], 4) Ishibashi [1976c, 1977b], 5) Fujii [1980a, b]. *) Seismic moment, the rigidity of 5×10^{11} dyne·cm^{-2} is assumed. Some parameters which are not explicitly shown in original papers have been calculated by the author. For notation, see Figure 13.

the JMA scale [Usami, 1977, 1979], and north-south trending west-side-up reverse active faults in this region as described in Chapter 3 seem to suggest that the NE fault plane in our final model better be extended further inland. However, the location of the northern edge of this fault plane remains open to future investigations closely in connection with the problem where the northernmost part of the plate boundary runs through.

6. 1944 Tonankai Earthquake

Only 90 years after the 1854 Ansei Tokai earthquake, the Tonankai earthquake of December 7, 1944 (M ≃ 8.0 to 8.3 after Kanamori [1972]) occurred again along the Nankai trough, hit the western part of the Tokai district severely, and killed more than 1,200 people [Iida, 1977]. Kanamori [1972] revealed this earthquake as an interplate low-angle reverse faulting caused by the underthrusting of the Philippine Sea plate beneath southwest Japan at the Nankai trough. This event and the successive 1946 Nankai earthquake (M ≃ 8.1 to 8.4 after Kanamori [1972], more than 1,400 victims) ruptured almost the entire boundary between the Philippine Sea and the Eurasian plates from the Ryukyu to Izu-Bonin arcs [Fitch and Scholz, 1971; Kanamori, 1972; Ando, 1975b]. However, the rupture zone of the 1944 earthquake was rather small, which has left to us the problem of the future Tokai earthquake. Kanamori [1972] derived a fault model of this earthquake as is shown in Table 2 and Figure 15 on the basis of the relocated hypocenter of the main shock, its fault-plane solution, rough estimate of seismic moment, and aftershock distribution. Referring to this model and furthermore taking into account coseismic crustal deforma-

tions and tsunamis Ando [1975b] proposed another static fault model, which is also shown in Table 2 and Figure 15. Meanwhile, tsunami source area, aftershock distribution, and coseismic vertical crustal movement associated with this earthquake have been reexamined by Hatori [1974], Sekiya and Tokunaga [1974a, 1975], and Sato and Inouchi [1975], respectively, despite the difficulties that the observations had been poor because the earthquake had occurred near the end of the World War II. Their results are compiled in Figure 16. In this figure the coseismic crustal subsidence of 20 cm at Kushimoto, the southern tip of the Kii Peninsula [Tsumura, personal communication, 1976; Ando, 1975b], is also shown. As pointed out by Mikumo [1975], Kanamori's and Ando's fault planes look too narrow compared with these newly analyzed data. Inouchi and Sato [1975] composed another static fault model as is shown in Table 2 and Figure 15 based on the vertical crustal movement, and Ishibashi [1976c, 1977b] proposed a revised static fault model by examining preceding works. The bases of Ishibashi's model are as follows.

First of all, the fault plane, which is assumed to be a part of the plate boundary thrust, has been assumed to be parallel to the axis of the Nankai trough (strike ≃ N60°E) as was the case in Ando's [1975b] modeling, and its upper extension has been assumed to break the surface on the steep inner wall of the trough. And Kanamori's [1972] hypocenter (depth = 30 km) has been adopted. Then the dip angle of the fault plane becomes about 20 degrees. This geometry of the plate boundary is in harmony with the leading edge of the Philippine Sea plate inferred by Kanamori [1972] and Shiono [1974] beneath the Kii Peninsula (Figure 1). Although the strike of the fault plane or its dip direction assumed here is

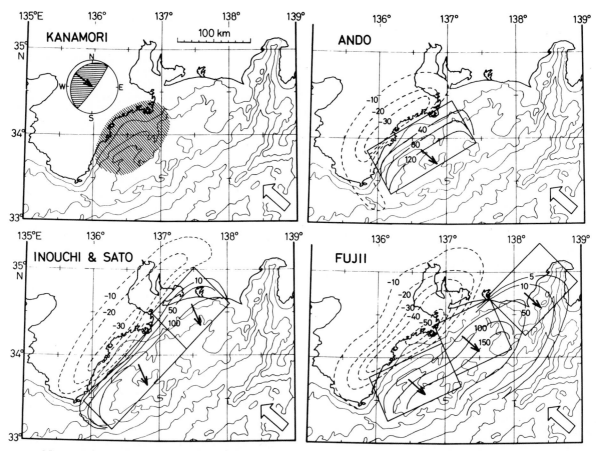

Figure 15. Fault models of the 1944 Tonankai earthquake proposed by Kanamori [1972], Ando [1975b], Inouchi and Sato [1975], and Fujii [1980a, b]. Motion directions of hanging wall sides of faults are shown by solid arrows. White arrows represent the moving direction of the Philippine Sea plate relative to the Eurasian plate obtained by Kanamori [1972] and Seno [1977a] (N50 ～ 54°W). Concerning Kanamori's model the assumed fault surface and the fault-plane solution (Wulff projection of the lower focal hemisphere, compression field is hatched) are shown. Concerning the other three horizontal projections of fault planes and calculated vertical surface displacement fields (in cm) are shown.

somewhat different from Kanamori's [1972] solution, the latter was rather technically indeterminate as he mentioned. Next the moving direction of the Philippine Sea plate, or the reverse of the slip direction of the hanging wall side of the fault, has been fixed as N50°W following Kanamori [1972] and Ando [1975b], by which the slip angle on the fault plane becomes a known parameter. In this framework remaining static fault parameters have been found by trial-and-error method so that an obtained model is the best fit to the coseismic vertical crustal displacement field shown in Figure 16. In this crustal deformation special attention should be paid to the uplift in the Enshu coastal region to the northeast of the inferred tsunami source area (Kakegawa uplift) [Sato, 1970, 1977; Sato and Inouchi, 1975; Inouchi and Sato, 1975; GSI, 1977d]. Ando [1975a] interpreted this uplift as due to a separated motion from the main rupture, a creep-like

slow slip on a small area at the eastern bottom edge of the fault plane which he assumed for the future Sea-of-Enshu earthquake. But, as Mikumo [1975] has pointed out, it seems more reasonable to interpret it as due to the northeasternmost slip of the main rupture. As the uplift is not so large-scale, especially around Point Omaezaki almost zero, and as the tsunami source area which represents the area of large-scale vertical surface displacement at the sea-bottom is narrower than the area of crustal uplift on land, the faulting seems to have occurred in some deeper part beneath the Sea of Enshu. Although the actual rupture zone is inferred to have stretched as is illustrated in Figure 12, a model comprising two rectangle fault planes has been adopted for calculation simplicity. After all a fault model of the first-order approximation as is shown in Figure 17 and Table 2 has been obtained. Sato [1977] considered that the surface trace

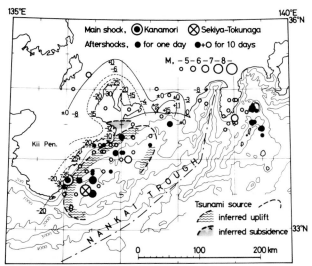

Figure 16. Compilation of various data on the 1944 Tonankai earthquake. Epicenters of the main shock are after Kanamori [1972] and Sekiya and Tokunaga [1974a, 1975]. The aftershock distribution is after Sekiya and Tokunaga [1974a, 1975]. The tsunami source area is after Hatori [1974]. Contours and numerals on land represent coseismic vertical crustal movement (in cm) after Sato and Inouchi [1975]. The 20 cm subsidence at the southern tip of the Kii Peninsula (Kushimoto) is after Ando [1975b] and Tsumura [personal communication, 1976]. Isobaths are in meters.

of the 1944 faulting had been located on the west side of Point Omaezaki since this place had not been uplifted unlike the 1854 case. He suggested a possible important role of north-south running Tenryu Canyon in the Sea of Enshu which was described in Chapter 3 (Figures 1, 2) in relation to this point. Aoki [1977] also attached much importance to Tenryu Canyon, which he considered a major tectonic boundary of the block structure in this region and had bounded the 1944 faulting on the east. In these respects Sato [1977] and Aoki [1977] supported Inouch and Sato's fault model of the 1944 Tonankai earthquake. Meanwhile, Aida [1979a] examined the above-mentioned four fault models from the viewpoint of tsunami generation and found that Inouchi and Sato's model with its dislocation reduced by a factor of 0.45 was the best. This modified model could reproduce observed tsunami records at 6 tide-gage stations fairly well by numerical experiments. However, the large difference between the motion direction of the foot wall side in Inouchi and Sato's model (N25°W) and the moving direction of the Philippine Sea plate obtained by Kanamori [1972] and Seno [1977a] (N50 ∿ 54°W) (Figure 15) seems unacceptable. And this model produces obviously too much uplift along the coast of the Sea of Enshu. 0.45 times reduced dislocation still brings larger uplift there and, on the con-

trary, much smaller subsidence along the coast of the Sea of Kumano. In regard to Tenryu Canyon Yonekura [1979] pointed out that the north-south trending structure along the canyon appears to have been cut by the structure parallel to the Nankai trough, and suggested that the former may be older than the latter. As for tsunami generation Ishibashi's model seems to be also satisfactory if it is slightly revised and if dynamic rupture process is taken into account.

Very recently Fujii [1980a, b] proposed another fault model of this earthquake in which a faulting beneath the Suruga Bay region was added to slightly modified Ishibashi's model, based on the inference that at the time of this earthquake Okitsu (see Figure 10), the west coast of deep Suruga Bay, had been possibly uplifted by 5 or 10 cm [Fujii, 1975, 1980a, b] and across Suruga Bay the east-west expansion might have taken place [Fujii and Nakane, 1979; Fujii, 1980a, b]. This model is also shown in Table 2 and Figure 15. The surface trace of the newly added northeastern fault plane was assumed to coincide with the East-off-Izu Tectonic Line proposed by Ishibashi [1976b], in order to produce some expansion across Suruga Bay. Aftershocks in and around Suruga Bay which occurred within ten day period from the main shock as seen in Figure 16, considerable seismic damage of housings around Okitsu [e.g., Iida, 1977], and rather early arrival of tsunami at Uchiura on the head of Suruga Bay (see

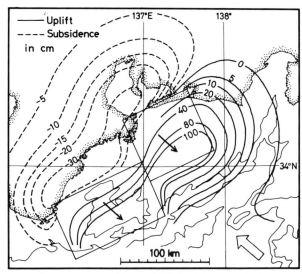

Figure 17. Vertical surface displacement field associated with Ishibashi's [1976c, 1977b] fault model of the 1944 Tonankai earthquake described in Table 2. Two rectangles are horizontal projections of fault planes. Solid arrows represent motion directions of the hanging wall sides of the faults. The white arrow is the same as in Figure 15. Isobaths of 2,000, 3,000 and 4,000 m are also shown.

Figure 10) presumed by Sekiya [1975, 1977] appear to support Fujii's model. However, it is certain that no coseismic crustal uplift was detected by a precise leveling at Point Cmaezaki [Inouchi and Sato, 1975; Sato, 1977; Fujii, 1980b]. And the coseismic uplift at Okitsu seems still open to question. On the other hand, the expansion across Suruga Bay has become doubtful by a re-analysis of old triangulation data [Fujii and Nakane, 1980] as was described in 4-2. Anyway, the coseismic vertical crustal movement on the west coast of Suruga Bay at the time of this earthquake is evidently quite different from the remarkable coseismic crustal uplift there at the time of the 1854 Ansei Tokai earthquake as described in the previous chapter. Therefore, after all, it seems almost certain that the main rupture of this earthquake did not enter into Suruga Bay. But, it seems also true that the 1944 rupture had a potentiality to propagate further eastward. So, it is very important for the prediction of the Tokai earthquake to clarify where, how, and why the main rupture of the 1944 Tonankai earthquake terminated. Concerning this point, Ishibashi [1977b] suggested an effect of the activity of the East-off-Izu Tectonic Line since 1923 [Ishibashi, 1976b], and Mogi [1980] suggested an effect of the 1891 Nobi earthquake (M ≈ 8), an exceptionally great intraplate rupture accompanied with a NW-SE trending left-lateral strike-slip faulting. But both of them seem to be still open to further discussions.

Tada et al. [1980] interpreted that the Kakegawa uplift had been caused by a high-angle reverse sub-faulting in the continental lithosphere which had been triggered by the faulting on the main thrust. And they inferred that the main thrust had ruptured up to Suruga Bay to some extent. In general, it seems very important to take sub-faults or splay faults into account for further interpretation of great earthquakes in the Tokai district.

7. Coming Tokai Earthquake

As a result of discussions in the preceding two chapters it is naturally concluded that the most probable rupture zone of the next large earthquake in the Tokai district, if it should occur at all in some near future, is mainly the Suruga trough thrust beneath Suruga Bay and the east part of the Sea of Enshu [Ishibashi, 1976c, 1977b]. This part of the plate boundary thrust has remained unruptured for 126 years, whereas that beneath the Seas of Kumano and Enshu repeated rupturing in 90 years. Thus, as the first-order approximation of the static fault model of the future Tokai earthquake, we may well adopt the NE fault in the fault model of the 1854 Ansei Tokai earthquake. The parameters are repeated as follows: dip direction (ϕ), N72°W; dip angle (δ), 34°; length (L), 115 km; width (W), 70 km; slip angle (λ), 71°; dislocation amount (D), 4.0 m; dip-slip component (D_d), 3.8 m; left-lateral strike-slip component ($-D_r$), 1.3 m. These parameters give seismic moment of 1.6×10^{28} dyne·cm, stress drop of 49 bar, and magnitude about 8.3. Of course the actual faulting will be more complicated on such an distorted surface as is illustrated in Figure 12 due to the complex tectonic setting; the dip angle may be higher in the northern part and lower in the southern part, the width may be narrower in the northern part, and it may propagate further inland as discussed in Chapter 5. And the ratio of the dip-slip component to the strike-slip component, which depends on the dip direction, the dip angle, and the interplate slip direction, remains open to question to some extent also as discussed in Chapter 5. Figure 18(a) shows the vertical surface displacement field due to the present hypothetical fault model. Figure 18(b) illustrates horizontal displacements of the first-order triangulation stations and horizontal strain changes in some triangles in the Suruga Bay region expected from the same model. This calculation has been made by a computer program based on Sato and Matsu'ura's [1974] program. These expected coseismic crustal deformations can be well interpreted as a release of crustal strains accumulated through the recent crustal movements as reviewed in Chapter 4, which suggests that our hypothetical fault model is reasonable.

Due to this faulting the Tokai district and surrounding central Japan will suffer seismic shakings as violent as those in 1854 as shown in Figure 11(a). And coastal regions from the Boso to the Kii Peninsulas, especially the Suruga Bay and Enshu coasts, will be struck by severe tsunami. So, if this 'Suruga Bay earthquake' actually occurs, the disaster will become catastrophic in the worst case. As the rupture takes place almost right beneath the land, Tokaido megalopolis and the main arteries such as Shinkansen and Tomei Highway connecting Tokyo and Kyoto will be severely damaged. Shizuoka Prefectural Government estimated the loss of life in the prefecture unofficially at about 97,500 in July, 1980.

Long-period dynamic displacements at several places due to the present hypothetical fault model have been calculated by Ishibashi and Sato [1977], and long-period ground velocities due to the same model have been calculated by Muramatsu and Mikumo [1979]. Recently Suzuki and Sato [1979] tried to obtain the distribution of maximum horizontal accelerations due to a slightly modified fault model, empirically considering higher frequency. And Midorikawa and Kobayashi [1980] calculated the distribution of ground-surface peak accelerations based on the present fault model taking site geological conditions into account. Based on this result, Central Disaster Prevention Council, Japan, declared in August 1979 'the area under intensified measures against earthquake disaster' containing about 5.7 million inhabitants where the seismic intensity is expected to be 6 or more in the JMA scale. As for tsunami due to this hypothetical earthquake, Sai-

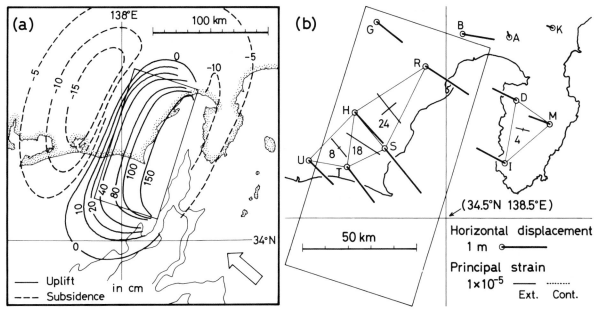

Figure 18. (a) Vertical surface displacement field associated with the hypothetical fault model of the future Tokai earthquake [Ishibashi, 1976c, 1977b]. The rectangle is the horizontal projection of the fault plane. The white arrow shows the moving direction of the Philippine Sea plate relative to the Eurasian plate used in this calculation (N50°W). Isobaths of 2,000 and 3,000 m are also shown. (b) Horizontal displacements of the first-order triangulation stations in the Suruga Bay region calculated for the same model as in (a): A (Ashitaka-yama), 13 cm; B (Habuna-mura), 72 cm; D (Daruma-yama), 59 cm; G (Daimuken-zan), 71 cm; H (Hakko-san), 91 cm; I (Iwashina-mura), 71 cm; K (Kamuriga-dake), 14 cm; M (Banjiro-dake), 50 cm; R (Ryuso-zan), 113 cm; S (Sakabe-mura), 124 cm; T (Takatenjin), 96 cm; U (Uenomi-shinden), 74 cm. Released horizontal strains in some triangles expected from this hypothetical faulting are also shown. Numerals represent dilatations in 10^{-6}. The rectangle is the same as in (a).

to and Kosuge [1979], for example, has made some numerical experiments.

The evaluation of the occurrence time of the Suruga Bay earthquake is extremely difficult at the present long-term prediction stage. If, simply, a 4 meter seismic slip is divided by the 3.3 cm/year convergence rate of the Philippine Sea plate in this region estimated by Seno [1977a] (Figure 21), the mean return period of great earthquakes with 4 meter dislocation becomes about 121 years. Actually, the 1854 Ansei Tokai earthquake occurred in 147 years after the 1707 Hoei earthquake along the Suruga trough, and in the Seas of Enshu and Kumano the 1944 Tonankai earthquake occurred in only 90 years after the 1854 earthquake. Thus, the seismic gap for 126 years on the Suruga trough thrust since 1854 seems significant. Rikitake [1977] estimated probabilities of occurrence of the Suruga Bay earthquake on the basis of the statistical analysis of recurrence tendency and crustal strain accumulation, and obtained rather high cumulative probability of around 65 % and hazard rate of 30 % for 1976.

If we assume that the west coast of Suruga Bay has been subsiding since after the 1854 Ansei Tokai earthquake at a constant rate of several

mm/year, which is the rate for the last 80 years or so as described in Chapter 4, then the total amount of subsidence since 1854 exceeds 60 cm. Furthermore, if a rapid recovery of the 1854 coseismic uplift during the post-seismic stage being taken into account, as was the case with Point Muroto for the 1946 Nankai earthquake, the subsidence may attain to 70 or 80 cm. This is about 50 ∿ 60 % of the expected uplift due to the hypothetical fault model for the Suruga Bay earthquake, which is 130 ∿ 150 cm as is seen in Figure 18(a). The ratio of the interseismic subsidence to the coseismic uplift is said to be 40 % in the southern Kanto district and 80 % in Shikoku Island. Thus, the inferred interseismic subsidence of 50 ∿ 60 % of the expected coseismic uplift suggests that the occurrence of the Suruga Bay earthquake is not in the distant future. Seno [1977b] estimated the above-mentioned ratio and the recurrence time of large earthquakes to be 0.63 ± 0.07 and 152 ∿ 275 years respectively concerning Shimizu, and 0.86 ± 0.08 and 150 ∿ 300 years respectively concerning Omaezaki, by means of what he called d-T diagram.

Ishibashi [1976c, 1977b], Rikitake [1977], Aoki [1977], Utsu [1977b], and many others attached much importance to a large amount of maximum

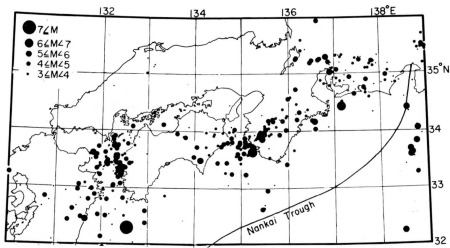

Figure 19. Distribution of subcrustal earthquakes with focal depths of 30 ∿ 100 km in southwest Japan for the period January 1961 ∿ April 1976 located by JMA (reproduced from Mizoue [1977]).

shear strain inferred to have accumulated in the Suruga Bay region since 1854 based upon GSI [1974d, 1977a, c]. Fujii and Nakane [1979], on the contrary, emphasized that this amount had proved to be considerably smaller. However, from the viewpoint of the elastic rebound mechanism of earthquake occurrence, such horizontal maximum shear strains with, say, NWW-SEE and NNE-SSW principal strain axes as have been considered so far should basically relate to a strike-slip earthquake along a NNW-SSE or NEE-SWW trending fault. On the other hand, the Suruga Bay earthquake now under consideration is basically related to a shear strain along the westerly dipping fault plane which is considered to be produced by a maximum compressive stress, say, in the NWW-SEE direction and a minimum compressive stress presumed almost vertical. In this case the maximum horizontal strain axis, say, in the NNE-SSW direction corresponds to the intermediate principal stress under the ground. Accordingly, for the purpose of evaluating the near future possibility of the Suruga Bay earthquake based on the observed horizontal crustal strains, it is inadequate to see the horizontal maximum shear strain over Suruga Bay. Rather, we should compare magnitudes of maximum compressive strains and dilatations accumulated on the west coast of Suruga Bay, the hanging wall side of the assumed fault plane, with an expected strain release due to the hypothetical fault model, which is partly shown in Figure 18(b). However, at present reliable estimations of these dimension-depending strain factors have not been obtained yet due to some difficulties in examining old survey data and choosing reasonable assumptions. Therefore, a more detailed discussion on the evaluation of the occurrence time of the Suruga Bay earthquake on the basis of recent horizontal crustal strains, of course as a long-term prediction, remains a future problem.

8. Recent Seismic Activity in the Tokai district

Along the Nankai trough an inclined seismic plane has not yet been developed well due to the recent commencement of the subduction of the Philippine Sea plate [Kanamori, 1972; Fitch and Scholz, 1971]. Mizoue [1977] showed, however, that the distribution of subcrustal earthquakes is manifesting rather clearly the subduction process of the Philippine Sea plate as in Figure 19. In this figure it should be noted that subcrustal earthquakes are lacking on the northwest side of Suruga Bay. Very recently Yamazaki et al. [1980] showed based on the data of high quality obtained by the microearthquake observation network of Nagoya University during the period from 1978 April through 1979 that the distribution of micro- and small earthquakes with focal depths deeper than 20 km inclined toward the NNW direction suggests low-angle subduction of the Philippine Sea plate rather clearly in the region around Lake Hamana. On the contrary, Yamazaki et al. [1980] and Aoki [1980] pointed out that in the Suruga Bay region the hypocentral distribution is somewhat complicated and such distribution as indicates the plate subduction from the Suruga trough is difficult to recognize. According to them, in the northern half of Suruga Bay earthquakes with focal depths deeper than 25 km are lacking. Kasahara [1980], on the other hand, showed that a seismic plane inclined toward the NWW direction from Suruga Bay is recognized based on the data obtained mainly by the microearthquake observation network of the National Research Center for Disaster Prevention (NRCDP) during the period 1978 ∿ 1979. Concerning this problem, however, data is yet insufficient at present, and further accumulation of observation data is necessary. It should be noted, in addition, that an inclined seismic plane may be obscure in the Suruga Bay

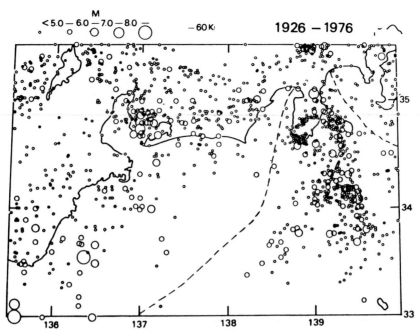

Figure 20. Distribution of shallow earthquakes with focal depths less than 60 km in the Tokai district and its vicinity for the period 1926 ∿ 1976 [JMA, 1977].

area compared with the western region if the area is now at the preseismic quiescent stage.

Figure 20 represents the distribution of shallow earthquakes with focal depths less than 60 km during the recent 50 years after JMA [1977]. The quiescent area prominent in the central part of this figure had been much noted since 1973 [Sekiya and Tokunaga, 1974a, b, 1975; Hagiwara, 1975] as mentioned in Chapter 2. But this area is stretching across the Nankai trough, and seems different from usual quiescent regions relating to interplate thrust earthquakes. In this figure the Suruga Bay region does not appear quiescent, but Utsu [1977b] and Asada [1979] suggested that we can observe a seismicity gap in the Suruga Bay - Omaezaki region if earthquakes of magnitude less than 5 are neglected. Yamazaki and Ooida [1979], who investigated the microseismicity in and around Suruga Bay mainly based on the data by Nagoya University during the period 1975 July ∿ 1977 December, pointed out that the seismicity in the northern half of Suruga Bay was very low throughout the observation period, whereas the seismicity in the southern half had been low until September 1976 and then turned out high. They suggested the existence of a NWW-SEE trending tectonic line dividing these seismic and aseismic areas.

The intraplate seismicity in southwest Japan before and after great interplate earthquakes along the Nankai trough has been discussed by Mogi [1969], Utsu [1974], and Seno [1979]. According to Mogi [1969] the seismic activity in areas surrounding the focal regions of the 1944 Tonankai and the 1946 Nankai earthquakes increased remark-

ably during about 20 years prior to these events. Utsu [1974] pointed out that the rate of occurrence of destructive earthquakes in southwest Japan during 50 years before and 10 years after the great earthquakes had been about four times as high as the rate for other periods. Seno [1979] showed that intraplate earthquakes in southwest Japan tended to cluster in the area adjacent to the expected rupture zone of the Suruga Bay earthquake during the period from 1957 to 1976. According to him, this clustering of intraplate events was significant within a 96 % confidence level, and the level of seismicity in that area was 18 times higher than the normal level.

In the Suruga Bay region there have occurred two strong earthquakes since 1926 when the calculation of epicenters and focal depths was started by JMA. They are the Shizuoka earthquake of July 11, 1935 (M = 6.3) and the Shizuoka earthquake of April 20, 1965 (M = 6.1) in the lower crust beneath the middle part of the west coast of Suruga Bay, both of which caused slight damages including several victims. The former was a NEE-SWW trending left-lateral strike-slip faulting with a principal compressional axis in the N27°E direction after Takeo et al. [1979]. The latter seems a similar faulting according to its fault-plane solution by Maki [1974]. Yamazaki and Ooida [1979] showed, on the other hand, that a small earthquake of October 26, 1976 (M ≤ 4) which had occurred near Shizuoka city had been a NWW-SEE trending left-lateral strike-slip faulting with a principal compressional axis in the east-west direction. These results may suggest a change in the direction of principal compressional axes on

the west side of the Suruga trough in recent
years. However, detailed investigation on stress
distribution within continental and oceanic lithospheres in the Suruga Bay region based on focal
mechanism analyses remains a future problem. The
seismic activity in the Izu Peninsula has become
fairly high since the 1974 Off-Izu-Peninsula
earthquake (M = 6.9) [e.g., Tsumura, 1977; Mogi,
1979], which may represent the high stress level
in the Suruga Bay region.

9. Historical Great Earthquakes in the Tokai District

Hypocenters (epicenters), magnitudes and
rather detailed descriptions for more than 600
destructive earthquakes in and around Japan since
416 A.D. up to the present have been published in
book form by Usami [1975] based on abundant historical documents collected by former investigators [e.g., Usami, 1978]. Mainly based on this
book or its antecedents the space-time distribution of great interplate earthquakes in the Tokai
district has been discussed by Utsu [1974, 1977a,
b], Ando [1975b], Rikitake [1976], Aoki [1977],
and others together with that in the Nankai district. However, careful and critical reading of
original historical documents from the viewpoints
of both seismology and historical science sometimes leads us to a different interpretation of
each earthquake from the prevailing one. Moreover, since around 1976 a considerable amount of
historical documents on old earthquakes in the
Tokai district which had been unknown to seismologists have come out, because local people have
come to pay considerable attention to old earthquake records, and also because vigorous work of
collection of these documents have been continued
by Professor T. Usami, Dr. Y. Tsuji, and others.
And as a result some new interpretations which
are significant for the Tokai earthquake prediction have been brought about. In the following
discussion the conventional way to divide rupture
zones along the Nankai-Suruga trough into five,
A, B, C, D, and E, is adopted as is shown in Figure 21.

9-1. 1707 Hoei Earthquake

The Hoei earthquake of October 28, 1707 (M =
8.4 after Kawasumi [e.g., Usami, 1975]) is considered in general a complete rupture of the entire plate boundary along the Nankai trough
[Ando, 1975b]. It is well known that around
Point Omaezaki crustal uplifts of 1 ∿ 2 m took
place during this earthquake. The point of particular interest in connection with the prediction of the Tokai earthquake is whether the rupture zone of this event extended deep into Suruga
Bay or not. This problem is a key to know the
recurrence time of large earthquakes in Suruga
Bay and to evaluate whether the considerable
crustal movement in and around Suruga Bay during
the last nearly 100 years is elastic or not.

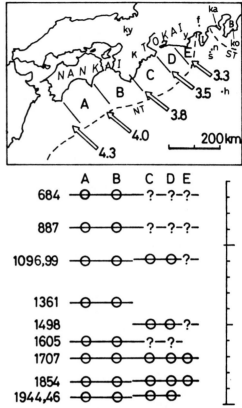

Figure 21. Space-time distribution of great
earthquakes along the Nankai-Suruga trough.
White arrows with numerals (in cm/year) in
the upper figure represent slip vectors of
relative motion between the Philippine Sea
and the Eurasian plates calculated by Seno
[1977a]. Bars with circles show ranges of
rupture zones of great earthquakes. NT, Nankai trough; ST, Sagami trough; B, Boso Peninsula; I, Izu Peninsula; K, Kii Peninsula; f,
Mt. Fuji; h, Hachijo Island; ka, Kamakura; ko,
Kominato; ky, Kyoto; n, Niijima Island; s,
Shikine Island; y, Yaizu.

Mogi, K. [1977] judged that the rupture zone had
occupied only the southern half of the bay based
on the seismic intensity distribution compiled by
Hagiwara [1974]. He made this interpretation as
a support of his hypothesis that a NW-SE trending
active tectonic line divided the bay into two
blocks as is mentioned later. Hatori [1977]
pointed out 1.5 ∿ 2 m subsidence at Shimizu (Figure 1) and its vicinity at the northern part of
the west coast of Suruga Bay and suggested that
the rupture might have stopped near Point Omaezaki. Tsuji [1979] also thought much of the subsidence around Shimizu. However, Ishibashi
[1977e] reexamined historical documents and concluded that the places having subsided were parts
of sand bar and a breakwater made by soil and
stone, and that the apparent subsidence had been

the effect of severe tsunami, etc., and at Shimizu no crustal subsidence had taken place. As for the seismic intensity distribution, Usami [1977] revised it adding newly found records. According to his result the difference in seismic intensity between the southern and the northern halves along the west coast of Suruga Bay is unperceptible; the whole region suffered very strong shaking of 5 ∿ 7 in the JMA seismic intensity scale. Ishibashi [1977e] compared seismic intensities due to the 1707 and the 1854 events at sixteen places along the entire west coast of the bay by original historical descriptions, and concluded that the difference of the distribution between two events were unrecognized though the 1707 earthquake seemed to have caused somewhat weeker shaking as a whole in the questioned region than the 1854 earthquake. As for the tsunami due to the 1707 Hoei earthquake, Hatori [1977] showed the distribution of maximum inundation heights inside Suruga Bay almost the same as that for the 1854 Ansei Tokai tsunami, though the data are not sufficient.

Usami [1977] suggested the occurrence of a remarkable aftershock on the next day of the main shock in Yamanashi Prefecture to the north of Suruga Bay. On the other hand, following the 1703 Genroku Kanto earthquake (M ≃ 8.2) [e.g., Matsuda et al., 1978], whose rupture zone presumably extended far inland northwestward from the Sagami trough [Ishibashi, 1977f], many aftershocks including destructive ones occurred successively in the southeastern Yamanashi until the occurrence of the 1707 Hoei earthquake. And one and a half months after the Hoei earthquake, Mt. Fuji (Figures 1,21) erupted violently. Thus surveyed, the 1703 Genroku Kanto earthquake, the 1707 Hoei earthquake, and the 1707 eruption of Mt. Fuji appear to form an entire activity along the northernmost part of the Philippine Sea - Eurasian plate boundary, and the rupture zone of the Hoei earthquake is inferred to enter deep into Suruga Bay. However, positive proofs that some northern part of the west coast of Suruga Bay coseismically uplifted in 1707 have not been found so far. And, as mentioned before, the shaking on the west coast of the bay in 1707 seems to have been somewhat weeker than that in 1854 as a whole. These matters suggest that in the northern part of E region the faulting of the Hoei earthquake may have been buried in some deeper part of the crust.

9-2. 1605 Keicho Earthquake

The Keicho earthquake of February 3, 1605 (M = 7.9 after Kawasumi [e.g., Usami, 1975]) has been generally considered to consist of two events which occurred simultaneously off the Nankai district and off the Boso Peninsula [e.g., Usami, 1975; Hatori, 1975b]. However, Utsu [1977a, b] suggested the possibility that a slow slip occurred off the Tokai district based on the strong tsunami along the Tokai coast. Ishibashi [1976b, 1977a] and Utsu [1977b] suggested that the uplift

inferred by Fukutomi [1934, 1938] at the southern tip of the Izu Peninsula and Niijima and Shikine Islands (Figure 21) might correspond to this and the 1498 earthquakes. Ishibashi [1978b] examined historical records critically and stated that the source having been assumed off the Boso Peninsula was much doubtful and that this event might have been a tsunami earthquake along the entire Nankai trough. Tsuji [1979], who has searched for unknown historical documents extensively in Shizuoka Prefecture, reported that no record on this earthquake had been found except a few records on tsunami. Thus the true nature of this earthquake is still much obscure.

9-3. 1498 Meio Earthquake

The Meio earthquake of September 20, 1498 (M = 8.6 after Kawasumi [e.g., Usami, 1975]) is said to have shaken a broad area from the Boso Peninsula to the east down to the Kii Peninsula to the west severely. It is certain that in Kyoto (Figure 21), more than 100 km north of the coast of the Sea of Kumano, the shaking was fairly strong and many aftershocks were felt every day for at least half a month. It is almost certain as well that coastal areas from the Boso to Kii Peninsulas were hit by tsunami. Hatori [1975b] inferred the tsunami source area of this earthquake as locating in the Sea of Enshu and extending to south off the Izu Peninsula by attaching importance to the severe tsunami at Kamakura (Figure 21) on the northeastern coast of Sagami Bay, Kominato (Figure 21) on the southeastern coast of the Boso Peninsula, and Hachijo Island (Figure 21). Utsu [1977a, b] followed this inference and suggested that this earthquake might not be an interplate earthquake along the Nankai-Suruga trough. Aoki [1977] and Utsu [1977b] refer to this event as an important basis for their estimation of the source region of the future Tokai earthquake as mentioned later. Hatori [1975b], Ishibashi [1976b, 1977a], and Utsu [1977b] suggested that the uplift inferred by Fukutomi [1934, 1938] at the southern tip of the Izu Peninsula and Niijima and Shikine Islands might correspond to this earthquake.

However, very recently Ishibashi [1980b] pointed out that it is quite uncertain due to unreliability of historical documents that Kamakura was hit by severe tsunami at the time of this earthquake. As for Kominato and Hachijo Island he also pointed out that we can not judge whether or not the tsunami was so strong as to require a source region extending to south off the Izu Peninsula because historical records are very poor. Rather, according to newly found reliable historical documents, the shaking was surely so strong as to cause severe damage at Shimizu (Figure 1), northern part of the west coast of Suruga Bay, and near Kakegawa (Figure 1), to the northwest of Point Omaezaki [Tsuji, 1979; Ishibashi, 1980b]. And it is also certain that inside Suruga Bay tsunami was very severe, which has been already known. Considering all these things Ishibashi [1980b] suggested the strong probability that the rupture zone of this earthquake had entered into Suruga Bay. Tsuji [1979] reported a subsidence near Yaizu (Figure 21) on the west coast of Suruga Bay, but whether this was a crustal movement or merely a surface phenomenon due to strong tsunami is open to question. After all, the 1498 Meio earthquake seems basically of the same type as the 1854 Ansei Tokai earthquake.

9-4. Other Older Earthquakes

Prior to the 1498 Meio earthquake five great earthquakes as shown in Figure 21 are known to have occurred off southwest Japan along the Nankai trough through their typical features of macroseismic effects, land deformations, and tsunamis [e.g., Usami, 1975, 1978; Utsu, 1977a, b]. However, except the 1096 earthquake, it is very obscure whether or not their rupture zones extended to C, D, and E regions, because in these early times of Japanese history the Tokai district was not so civilized as western districts and no historical record concerning seismic effects in this district has been known so far. As for the 1096 earthquake it is very probable that the faulting occurred off the Tokai district because strong tsunami in its coastal area has been recorded. But it is uncertain whether the rupture zone entered into Suruga Bay or not.

Utsu [1977a, b] paid attention to five earthquakes which are said to have occurred off the Kii Peninsula, in C or C+D regions, and whose magnitudes are given as 7.0 so far. They are the 922, 1360, 1403, 1408, and 1520 earthquakes. Utsu [1977a, b] suggested that their magnitudes might be 7.5 or larger and pointed out that they had occurred within 50 years from the great earthquakes. Although this suggestion is interesting, historical records on these events are extremely poor both in quantity and in quality at present. So their locations, sizes, and even occurrences themselves are open to future investigations. On the other hand, we should keep in mind a probability that some great earthquakes in older times are missing from historical records.

10. Discussion and Conclusion

The so far described prediction of the Tokai earthquake on its spatial factors and near future possibility is rather simple and clear. However, there have been proposed a few different ideas on the future Tokai earthquake from the above-mentioned Suruga Bay earthquake hypothesis.

The first one, which was not offered by a specific investigator but came to minds of several persons concerned, is as follows [e.g., Asada, 1979]: As there is no historical case in which Suruga Bay (E region in Figure 21) alone ruptured, in the next time as well Suruga Bay will rupture together with the Seas of Kumano and Enshu (C and D regions in Figure 21); And as the time for C

and D to rupture is expected to be about 100 years after 1944, the Tokai earthquake will not occur until the middle of the next century. A similar logic has been offered by Usami [1978] as follows: Four historical Tokai earthquakes among five in all were followed by Nankai earthquakes within two years; So, if a great earthquake occurs in the Tokai district in the near future, another one will occur in succession in the Nankai district within two years with an 80 % confidence level; But this conflicts the empirical recurrence time of Nankai earthquakes of about 100 years because the last event occurred in 1946.

Concerning these opinions, first of all, we must keep in mind imperfection of historical records and shortness of historic times compared with time ranges of crustal activities. And we must examine whether there are some tectonic reasons why E region always breaks together with C and D regions. As the convergence rate of the Philippine Sea plate is roughly the same all along the Nankai trough, simultaneous or successive breakage of neighboring regions will be easily brought about. However, according to Seno's [1977a] calculation, the variation of the convergence rate along the Nankai trough is not negligible as is seen in Figure 21. Moreover due to a complex subduction process around the Izu Peninsula, which is mentioned later, the stress accumulation rate in E region is considered to be somewhat smaller in average. So, it may sometimes happen that the breakage of E region can not follow D. In this case, however, if stress is built up sufficiently, E region may be able to break independently of D region. The vivid description of the 1854 Ansei tsunami generation in Suruga Bay as mentioned in Chapter 5 seems to suggest that at least one initial break, even if this earthquake was multiple, took place somewhere beneath Seno-umi bank (Figure 2(a)) in the center of E region. This means that E region can rupture for itself without following D's rupture.

Aoki [1977] and Utsu [1977b] discussed the analogy between the earlier half (684 ∿ 1361) and the later half (1605 ∿ 1946) of the 1,300 year time series of twelve great earthquakes shown in Figure 21 including great earthquakes in Sagami Bay and large eruptions of Mt. Fuji, and suggested a possibility that the next Tokai earthquake might be of the 1498 Meio type. As they followed the prevailing interpretation of the Meio earthquake as described in 9-3, they considered that, in this case, the rupture zone of the future Tokai earthquake would not enter into Suruga Bay but extend to south off the Izu Peninsula. Aoki [1977] estimated the occurrence time of this type of Tokai earthquake at a year of 2030 ± 28 based on the relation of earthquake occurrence between the two series. Comments by the present author on this opinion are as follows: (1) The simple grouping of old earthquakes based on incomplete historical records seems unreasonable; (2) So far prevailing interpretation of the 1498 Meio earthquake is considerably questionable as discussed in 9-3, rather this event is supposed to be of the 1854 Ansei Tokai earthquake type; (3) Remarkable crustal deformatin along the west coast of Suruga Bay, which is regarded as mainly elastic strain accumulation, can be hardly released by a faulting south off Suruga Bay; (4) If a great earthquake is to occur south off the Izu Peninsula, its southern tip is expected to be subsiding and south-/southeastward tilting secularly at least to some extent, but such a crustal movement has not been observed since around 1900 [e.g., Fujii, 1969; Dambara and Tsuchi, 1975].

Mogi, K. [1977] proposed the existence of an active tectonic line running in the NW-SE direction through the Irozaki fault on the southern tip of the Izu Peninsula, passing across the Suruga trough, and extending to the west coast of Suruga Bay, on the basis of the submarine topography of Suruga Bay including the right-lateral displacement of the trough axis and high seismic activity along that line. He suggested that the recurrence time of rupture along the Suruga trough was different between the northern and the southern sides of this tectonic line, and that the rupture zone of the next large earthquake in the Tokai district would not extend to the northern half of Suruga Bay. Mogi, K. [1977] cited the 1707 Hoei earthquake as an example that the northern side of the hypothetical tectonic line did not rupture. But probably that was not the case as discussed in 9-1. And the general feature of the recent crustal movement in the Suruga Bay region shows no significant difference between the northern and the southern parts. Therefore, if the recent crustal movement should represent the preparation to a future seismic slip at all, the northern half of Suruga Bay is considered to have the same potential of an earthquake risk as the southern half.

Fujii [1980b] suggested a possibility that the large earthquake in Suruga Bay next to the 1854 event had been over, even if somewhat incomplete, at the time of the 1944 Tonankai earthquake and therefore the next Suruga Bay earthquake would be in some distant future, on the basis of his fault model of the 1944 earthquake as described in Chapter 6. However, the faulting beneath the Suruga Bay region during the 1944 Tonankai earthquake inferred by Fujii [1980a, b] is still questionable as discussed in the same chapter. Even if it actually occurred, it must have been small-scale and was not on the Suruga trough thrust. Moreover, the 1944 coseismic strain release on the continental side of the Suruga trough thrust is estimated to be inconsiderable as a whole based on the geodetic survey data as reviewed in Chapter 4. Thus, the possibility suggested by Fujii [1980b] seems very little.

It should be added here that, as pointed out by Ishibashi [1977b] and Aoki [1977], the main seismic rupture of the 1854 Ansei Tokai earthquake may have occurred in D + E region in Figure 21 and that of the 1944 Tonankai earthquake, in C region. If this is the case, one cycle of rup-

ture of C + D + E region was completed by a set of the 1854 and the 1944 earthquakes, and E region may not be an imminent seismic gap.

On the other hand, it should be much noted that, even if the strain accumulation on the hanging wall side of the Suruga trough thrust is insufficient yet for the Suruga Bay earthquake of 4 m dislocation as expected in Chapter 7, there is a possibility of earlier occurrence of a little smaller earthquake, as were actually the cases with the 1973 Nemuro-oki (M = 7.4) [e.g., Abe, 1977; Utsu, 1979] and the 1978 Miyagi-oki (M = 7.4) [e.g., Seno et al., 1980] earthquakes. The accumulated strain in this region since 1854 is inferred to be fully sufficient to bring about such an interplate earthquake with, say, seismic moment of 0.5×10^{28} dyne·cm in order, dislocation of around 2 m, and magnitude of about 7.5. And an earthquake of this size can be destructive enough to the Tokai district and surrounding regions.

The subduction of the Philippine Sea plate at the Suruga trough has not yet been geophysically established. As already mentioned in Chapter 8, hypocentral distribution of micro- and small earthquakes in the Suruga Bay region does not necessarily suggest plate subduction. The multichannel seismic reflection profiling survey, which was carried out by MSA in January 1978 on three east-west trending track lines across the Suruga trough in Suruga Bay, did not manifest plate subduction, unlike in other typical subduction zones such as off northeast Japan [MSA, 1978b; Sakurai and Mogi, 1980]. Hirahara [1980], who investigated three-dimensional P-wave velocity structure beneath southwest Japan associated with the subducting Philippine Sea plate by means of an inversion technique, found that the crust and the upper mantle beneath the Izu Peninsula - Suruga Bay region are of low-velocity, whereas to the west of Point Omaezaki distinct high-velocity zone suggests plate subduction. These geophysical investigations concerning plate subduction at the Suruga trough, however, have just started, and further intensive works are necessary. In spite of the above-mentioned ambiguities, Nakamura [1981] showed that the recent remarkable crustal deformation in the Izu Peninsula - Suruga Bay region can be interpreted well as due to a typical subduction process at the Suruga trough accompanied with a collision to the north. Nakamura [1979, 1980, 1981] argued that the stress field on the east side of the Suruga trough including the Izu Peninsula is ruled by flexure of the lithospheric plate associated with its subduction at the Suruga trough.

However, the less frequent occurrence of Suruga Bay earthquakes compared with Nankai earthquakes which is suspected by historical records and by the uplift rate of the marine terrace at Point Omaezaki, and a presumably incomplete rebound at the time of the 1707 Hoei earthquake seem to suggest some more complex seismotectonic process than a simple subduction along the Suruga trough thrust. Concerning this point, we should discuss by all means the process of plate convergence at the northern extremity of the Philippine Sea plate including the Izu Peninsula and Sagami Bay as a total system, because the complex seismotectonics in the Suruga Bay region is considered to be ruled fundamentally by the collision of the Izu Peninsula with Honshu Island.

Ishibashi [1976b] proposed a working hypothesis that a nascent subduction process had been developed east off the Izu Peninsula in Sagami Bay in recent times, where a part of the northwestward movement of the Philippine Sea plate was being consumed, in consequence of the long-term collision of the Izu Peninsula. Ishibashi [1977d, 1978a, 1980a] developed this hypothesis furthermore and suggested that three elements were basically important for consuming northwestward motion of the Philippine Sea plate; the Suruga trough thrust in the west, 'West Sagami Bay thrust' in the northeast, and 'Izu Transform Belt' between them which occupied most part of the Izu Peninsula and transformed the latter subduction to the former. He considered that this dual subduction process brought about an apparent irregularity of ruptures of the Suruga trough thrust. And this model explained well the inter- and coseismic crustal movements of the 1923 Kanto earthquake on the northwest coast of Sagami Bay, the 1930's and 1970's 'Izu uplift', rather frequent occurrence of destructive earthquakes in the northwest part of Sagami Bay, one order of magnitude smaller slip rates of so far believed interplate inland faults than the rate of plate motion, NW-SE trending right-lateral strike-slip active faults and earthquakes in the Izu Peninsula, and so on [Ishibashi, 1977a, c, d, 1978a, 1980a]. On the other hand, Mogi, K. [1977], as described before, emphasized a single tectonic line or a very narrow belt instead of a broad transform belt in Ishibashi's model. And he did not consider the West Sagami Bay thrust. Somerville [1978] proposed a somewhat similar model as Ishibashi's, but neither considered the West Sagami Bay thrust. Aoki [1977] proposed a block structure containing multiple thrusts in the Sea of Enshu, Suruga Bay, south off the Izu Peninsula. Aoki [1977, 1980] considered that in Suruga Bay, especially in its northern part, two continental lithospheres were colliding each other and large earthquakes occurring in this bay were intraplate ones. It is true that the crust - mantle structure beneath the Izu Peninsula is not oceanic, but it seems unconvincing to deny subduction in Suruga Bay judging from what we have seen. Anyway, all these models and hypotheses are awaiting further investigations to verify them.

Ishibashi [1977d] examined historical strong earthquakes in the northwest part of Sagami Bay during the last 4 centuries, and attributed the following to the faulting of the West Sagami Bay thrust on the basis of their macroseismic effects and tsunamis; the 1633 Kan'ei Odawara earthquake (M = 7.1), the 1782 Tenmei earthquake (M = 7.3),

and the 1853 Ka'ei Odawara earthquake (M = 6.5). He also inferred that at the time of the 1703 Genroku Kanto earthquake (M = 8.2) the West Sagami Bay thrust ruptured simultaneously with the major fault mainly based on the most severe seismic shaking at Odawara on the hanging wall side of this thrust. Then, adding the 1923 faulting during the Kanto earthquake [Ishibashi, 1977c], the time series of ruptures of the West Sagami Bay thrust has become remarkably regular with a constant recurrence time of about 70 years. Furthermore it should be much noted that there seems to be a remarkable space-time correlation between the earthquake occurrence on the Suruga trough thrust and on the West Sagami Bay thrust during the last 4 centuries as follows. In both cases of the 1707 Hoei and the 1854 Ansei Tokai earthquakes on the Suruga trough thrust, they followed ruptures on the West Sagami Bay thrust within a few years. And in the 1853 - 1854 series the rupture of the West Sagami Bay thrust was rather small, probably restricted within some deeper small area, and instead the rupture of the Suruga trough thrust was so large-scale as to extend to its northern extremity and to the shallowest part ('Ka'ei - Ansei type'). In the 1703 - 1707 series, on the contrary, the rupture of the West Sagami Bay thrust is inferred to have been large-scale based on tsunami and seismic shaking in the Izu Peninsula, and the rupture of the Suruga trough thrust is suspected to have been incomplete ('Genroku - Hoei type'). If we admit the before-mentioned dual subduction model in this region, this kind of correlation can be rather easily understood at least qualitatively, and it may help us in predicting future. It is supposed that the selection of a combination of slip modes on the West Sagami Bay and the Suruga trough thrusts is made rather accidentally during post- or inter- seismic periods, but probably we can tell which type has been selected on the basis of the interseismic crustal movements in the region. At present, it seems that the 'Ka'ei - Ansei type' has been selected, because the secular subsidence of the west coast of Suruga Bay is remarkable up to its deepest part as has been often mentioned and the secular subsidence of the northwest coast of Sagami Bay is almost zero, which took place preceding the 1923 faulting of the West Sagami Bay thrust. If a large earthquake on the West Sagami Bay thrust actually occurs before the Tokai earthquake in the near future, say, during the 1990's, following the above-mentioned temporal regularity, then the prediction of the latter both in space and time is expected to become much more accurate.

In conclusion, although there remains considerable obscurity concerning the true nature of seismogenic tectonism in the Suruga Bay region, it is an appropriate estimation at present to assume that the most probable faulting of the future Tokai earthquake is due to a rebound of the continental lithosphere along the Suruga trough thrust. And we can well adopt the hypothetical fault model presented in Chapter 7 as the first-order approximation. The preparatory strain accumulation for this seismic faulting is inferred to have reached a considerable amount, and we have no convincing reason for judging that its occurrence is in some distant future. Therefore, the programs of short-term earthquake prediction and disaster prevention of the Tokai earthquake now carried out nationwidely aiming at this Suruga Bay earthquake is considered also basically appropriate. For further refinement of the specification of the coming Tokai earthquake, it is indispensable to obtain much more clear view on the real dynamical and physical process of plate convergence in the Izu collision zone through further intensive and extensive field investigations.

Acknowledgments. I wish to express my gratitude to Professor T. Asada for his continuous encouragement throughout my research on the Tokai earthquake prediction. I also wish to acknowledge Professor Y. Fujii for invaluable suggestions, informations, and encouragement. I thank Professor H. Sato for helpful discussions and informations, and Dr. M. Matsu'ura for giving me permission to use a computer program for calculation of static deformation. I benefited from discussions with Drs. K. Yamashina and T. Seno.

References

Abe, K., Some problems in the prediction of the Nemuro-oki earthquake, J. Phys. Earth, 25, Suppl., S261-S271, 1977.

Aida, I., A source model of the tsunami accompanying the Tonankai earthquake of 1944 (in Japanese), Bull. Earthq. Res. Inst., Univ. Tokyo, 54, 329-341, 1979a.

Aida, I., A source model of the tsunami accompanying the Ansei Tokai earthquake of 1854 (in Japanese), Abstracts, Seismol. Soc. Japan, 1979 No.2, 177, 1979b.

Ando, M., A fault-origin model of the great Kanto earthquake of 1923 as deduced from geodetic data, Bull. Earthq. Res. Inst., Univ. Tokyo, 49, 19-32, 1971.

Ando, M., Seismo-tectonics of the 1923 Kanto earthquake, J. Phys. Earth, 22, 263-277, 1974.

Ando, M., Possibility of a major earthquake in the Tokai district, Japan and its pre-estimated seismotectonic effects, Tectonophysics, 25, 69-85, 1975a.

Ando, M., Source mechanisms and tectonic significance of historical earthquakes along the Nankai trough, Japan, Tectonophysics, 27, 119-140, 1975b.

Ando, M., and Y. Fukao (Eds.), Off-Tokai Earthquake (in Japanese), 61 pp., 1975.

Ando, M., T. Matsuda, and K. Abe, Stress field of the upper crust in the Japanese Islands (in Japanese), Abstracts, Seismol. Soc. Japan, 1973 No.1, 66, 1973.

Aoki, H., Possibility of a great earthquake in Tokai district (in Japanese), in Proc. Symp. on Earthq. Predict. (1976), 56-68, 1977.

Aoki, H., Deep seismic zone to the west of Suruga Bay, central Japan (in Japanese), in Proc. Earthq. Predict. Res. Symp. (1980), 97-102, 1980.

Asada, T., Problems on the Tokai earthquake (in Japanese), in Progress of the Coordinating Committee for Earthquake Prediction, Japan, during the First Decade, CCEP (Ed.), Geogr. Surv. Inst., 110-120, 1979.

Commission on Earthquake Prediction, National Committee of Geodesy and Geophysics, Science Council of Japan, A Long-Term Yearly Plan of Earthquake Prediction Research (for 1970 Fiscal Year) (in Japanese), 60 pp., 1969.

Dambara, T., Vertical movements of Japan during the past 60 years; IV. Chubu district (in Japanese), J. Geod. Soc. Japan, 13, 66-74, 1968.

Dambara, T., Horizontal strain of the Suruga Bay base-line net (in Japanese), Rep. Coord. Comm. Earthq. Predict., 24, 159-161, 1980a.

Dambara, T., Some problems on prediction of the Tokai earthquake in future (in Japanese), in Proc. Earthq. Predict. Res. Symp. (1980), 85-95, 1980b.

Dambara, T., and R. Tsuchi, Crustal movements in the southern Izu Peninsula (in Japanese), in Rep. Earthq. off the Izu Peninsula, 1974, and the Disaster, 103-106, 1975.

Earthquake Research Institute, Univ. Tokyo, Reports from Shizuoka Prefecture on the Investigations of the Tokai Earthquake of December 23, 1854 (in Japanese), 20 pp., 1977.

Fitch, T. J., and C. H. Scholz, Mechanism of underthrusting in southwest Japan: A model of convergent plate interactions, J. Geophys. Res., 76, 7260-7292, 1971.

Fujii, Y., Crustal movement in Izu Peninsula (in Japanese), J. Geod. Soc. Japan, 14, 62-71, 1969.

Fujii, Y., Seismic crustal movements in the Tokai district (in Japanese), Abstracts, Geod. Soc. Japan, No.39, 72, 1973.

Fujii, Y., Crustal movements in the Tokai district (in Japanese), in Off-Tokai Earthquake, M. Ando, and Y. Fukao (Eds.), 18-27, 1975.

Fujii, Y., On the source region of the 1944 Tonankai earthquake (in Japanese), Abstracts, Seismol. Soc. Japan, 1980 No.1, 43, 1980a.

Fujii, Y., Crustal movements in the Kanto - Tokai districts and the source region of the 1944 Tonankai earthquake (in Japanese), in Earthquakes: Dialogue between Seismologists and Geologists, R. Sugiyama et al. (Eds.), Tokai Univ. Press, 41-63, 1980b.

Fujii, Y., and K. Nakane, Earth's horizontal strain across the Suruga Bay, Honshu, Japan (1) (in Japanese), J. Geod. Soc. Japan, 25, 289-301, 1979.

Fujii, Y., and K. Nakane, Earth's horizontal strain across the Suruga Bay, Honshu, Japan (2) (in Japanese), J. Geod. Soc. Japan, 26, 104-112, 1980.

Fujita, N., Horizontal displacement vectors in Kanto and Chubu districts (in Japanese), Rep.

Coord. Comm. Earthq. Predict., 10, 64-67, 1973.

Fukutomi, T., Evidences of past land upheaval on the coast of the southern Izu Peninsula (preliminary) (in Japanese), Zisin (J. Seismol. Soc. Japan), Ser.1, 6, 351-355, 1934.

Fukutomi, T., Evidences of past land upheaval near Niijima and Shikine Islands (in Japanese), Zisin (J. Seismol. Soc. Japan), Ser.1, 10, 1-4, 1938.

Geographical Survey Institute, Vertical crustal movements in the Tokai district (in Japanese), Rep. Coord. Comm. Earthq. Predict., 2, 49-53, 1970a.

Geographical Survey Institute, Recent vertical crustal movement in the Omaezaki region (in Japanese), Rep. Coord. Comm. Earthq. Predict., 4, 41-43, 1970b.

Geographical Survey Institute, Deformation of Omaezaki rhombus (in Japanese), Rep. Coord. Comm. Earthq. Predict., 5, 44-45, 1971.

Geographical Survey Institute, Deformation of Omaezaki rhombus (2) (in Japanese), Rep. Coord. Comm. Earthq. Predict., 7, 36-37, 1972a.

Geographical Survey Institute, Horizontal strain in Japan (in Japanese), Rep. Coord. Comm. Earthq. Predict., 8, 99-105, 1972b.

Geographical Survey Institute, Vertical movements in Tokai district (2) (in Japanese), Rep. Coord. Comm. Earthq. Predict., 11, 102-104, 1974a.

Geographical Survey Institute, Deformation of Omaezaki rhombus (3) (in Japanese), Rep. Coord. Comm. Earthq. Predict., 11, 105-106, 1974b.

Geographical Survey Institute, G.D.P. traverse survey of high precision in Chubu and Tokai districts (in Japanese), Rep. Coord. Comm. Earthq. Predict., 11, 107-108, 1974c.

Geographical Survey Institute, First order triangulation in Tokai district (in Japanese), Rep. Coord. Comm. Earthq. Predict., 12, 131, 1974d.

Geographical Survey Institute, Results of side-length observations for concentrated base-line network in Tokai district (in Japanese), Rep. Coord. Comm. Earthq. Predict., 12, 132-134,1974e.

Geographical Survey Institute, Horizontal strains in Tokai district (in Japanese), Rep. Coord. Comm. Earthq. Predict., 15, 103-105, 1976.

Geographical Survey Institute, Crustal deformation in the Tohkai district (in Japanese), Rep. Coord. Comm. Earthq. Predict., 17, 113-115, 1977a.

Geographical Survey Institute, Morphotectonic investigation in Tokai region (1) (in Japanese), Rep. Coord. Comm. Earthq. Predict., 17, 116-125, 1977b.

Geographical Survey Institute, Horizontal earth's strain Suruga Bay (in Japanese), Rep. Coord. Comm. Earthq. Predict., 17, 146-148, 1977c.

Geographical Survey Institute, Crustal movements in the Tokai district (in Japanese), Rep. Coord. Comm. Earthq. Predict., 18, 75-80, 1977d.

Geographical Survey Institute, Crustal movements in the Tokai district (in Japanese), Rep. Coord. Comm. Earthq. Predict., 19, 96-98, 1978a.

Geographical Survey Institute, Horizontal strains in the Tokai district (in Japanese), Rep. Coord. Comm. Earthq. Predict., 20, 166-171, 1978b.

Geographical Survey Institute, Crustal movements in the Tokai district (in Japanese), Rep. Coord. Comm. Earthq. Predict., 21, 122-129, 1979a.

Geographical Survey Institute, Horizontal crustal movement in the Tokai district (in Japanese), Rep. Coord. Comm. Earthq. Predict., 22, 159-162, 1979b.

Geographical Survey Institute, Crustal movement in the Tokai district (in Japanese), Rep. Coord. Comm. Earthq. Predict., 23, 88-92, 1980a.

Geographical Survey Institute, Crustal deformation in the Tokai district (in Japanese), Rep. Coord. Comm. Earthq. Predict., 24, 152-158, 1980b.

Geographical Survey Institute, Vertical movements in Chubu and Kinki district (in Japanese), Rep. Coord. Comm. Earthq. Predict., 24, 201-214, 1980c.

Geographical Survey Institute, Revision of horizontal strain in and around Suruga Bay (in Japanese), Rep. Coord. Comm. Earthq. Predict., 24, 286, 1980d.

Geological Survey of Japan, 1:2,000,000 map series, 18, Active faults in Japan (main islands), 1978.

Hagiwara, T., On the seismic intensity distribution of the Tokai earthquake of 1854 (in Japanese), Rep. Coord. Comm. Earthq. Predict., 3, 51-52, 1970.

Hagiwara, T., Distribution of seismic intensity of the great earthquakes in 1854 and 1707 (in Japanese), Rep. Coord. Comm. Earthq. Predict., 12, 143-145, 1974.

Hagiwara, T., Earthquake Prediction Research in Japan, Coord. Comm. Earthq. Predict. Japan, 30 pp., 1975.

Harada, T., On the Tokai earthquake (in Japanese), in Off-Tokai Earthquake, M. Ando, and Y. Fukao (Eds.), 37-38, 1975.

Harada, T., and N. Isawa, Horizontal deformation of the crust in Japan: Result obtained by multiple fixed stations (in Japanese), J. Geod. Soc. Japan, 14, 101-105, 1969.

Hatano, S., M. Tsuzawa, and Y. Matsushima, Vertical movement in Holocene age and in the geodetic time along the northern coast of Suruga Bay (in Japanese), Rep. Coord. Comm. Earthq. Predict., 21, 101-106, 1979.

Hatori, T., Sources of large tsunamis in southwest Japan (in Japanese), Zisin (J. Seismol. Soc. Japan), Ser. 2, 27, 10-24, 1974.

Hatori, T., Sources of the Hoei and the Ansei tsunamis and comparison of them with the 1944 Tonankai tsunami (in Japanese), in Off-Tokai Earthquake, M. Ando, and Y. Fukao (Eds.), 27-34, 1975a.

Hatori, T., Sources of large tsunamis generated in the Boso, Tokai and Nankai regions in 1498 and 1605 (in Japanese), Bull. Earthq. Res. Inst., Univ. Tokyo, 50, 171-185, 1975b.

Hatori, T., Documents of tsunami and crustal deformation in Tokai district associated with the Ansei earthquake of Dec. 23, 1854 (in Japanese), Bull. Earthq. Res. Inst., Univ. Tokyo, 51, 13-28, 1976.

Hatori, T., Field investigation of the Tokai tsunamis in 1707 and 1854 along the Shizuoka coast (in Japanese), Bull. Earthq. Res. Inst., Univ. Tokyo, 52, 407-439, 1977.

Hirahara, K., Three-dimensional P-wave velocity structure beneath southwest Japan: Subducting Philippine Sea plate (in Japanese), Abstracts, Seismol. Soc. Japan, 1980 No.2, 160, 1980.

Iida, K., Disaster and Seismic Intensity Distribution due to the Tonankai Earthquake of Dec. 7, 1944 (in Japanese), Disaster Prevention Council of Aichi Prefecture, 120 pp., 1977.

Inouchi, N., and H. Sato, Vertical crustal deformation accompanied with the Tonankai earthquake of 1944, Bull. Geogr. Surv. Inst. Japan, 21, 10-18, 1975.

Ishibashi, K., Seismotectonics at the northern extremity of the Philippine Sea plate: The eastern Yamanashi earthquake swarm and the 1923 Kanto earthquake (Part 2) (in Japanese), Abstracts, Seismol. Soc. Japan, 1976 No.1, 37, 1976a.

Ishibashi, K., 'East-off-Izu Tectonic Line' and 'West-Sagami-Bay fault' as an origin of the Izu Peninsula uplift: Double structure of the northernmost boundary of the Philippine Sea plate (in Japanese), Abstracts, Seismol. Soc. Japan, 1976 No.2, 29, 1976b.

Ishibashi, K., Re-examination of a great earthquake expected in the Tokai district: Possibility of the 'Suruga Bay earthquake' (in Japanese), Abstracts, Seismol. Soc. Japan, 1976 No. 2, 30-34, 1976c.

Ishibashi, K., Creep dislocation model of the Izu Peninsula uplift: Significance of the East-off-Izu Tectonic Line (in Japanese), Rep. Coord. Comm. Earthq. Predict., 17, 65-67, 1977a.

Ishibashi, K., Re-examination of a great earthquake expected in the Tokai district, central Japan: Possibility of the 'Suruga Bay earthquake' (in Japanese), Rep. Coord. Comm. Earthq. Predict., 17, 126-132, 1977b.

Ishibashi, K., Re-evaluation of the 1923 Kanto earthquake (2): Significance of 'the West-Sagami-Bay fault' (in Japanese), Abstracts, Seismol. Soc. Japan, 1977 No.1, 129, 1977c.

Ishibashi, K., A possibility of 'the West-Sagami-Bay earthquake' and its relation to the Tokai earthquake: Seismotectonics in the Sagami Bay - Izu Peninsula - Suruga Bay region (in Japanese), in Rep. Subcomm. Tokai Distr., Coord. Comm. Earthq. Predict., Geogr. Surv. Inst., 53-68, 1977d.

Ishibashi, K., Did the rupture zone of the 1707 Hoei earthquake not extend to deep Suruga Bay? (in Japanese), in Rep. Subcomm. Tokai Distr., Coord. Comm. Earthq. Predict., Geogr. Surv. Inst., 69-78, 1977e.

Ishibashi, K., Source region of the 1703 Genroku Kanto earthquake and recurrence time of major earthquakes in Sagami Bay, Japan (1) (in Japanese), Zisin (J. Seismol. Soc. Japan), Ser. 2, 30, 369-374, 1977f.

Ishibashi, K., Plate convergence around the Izu collision zone, central Japan: Development of a new subduction boundary with a temporary transform belt, in Abstracts of Papers, Intern. Geodynamics Conf. 'Western Pacific' and 'Magma Genesis', Tokyo, March 13-17, 66-67, 1978a.

Ishibashi, K., On the source region of the 1605 Keicho earthquake: Doubt on the prevailing view (in Japanese), Abstracts, Seismol. Soc. Japan, 1978 No.1, 164, 1978b.

Ishibashi, K., Practical strategy for earthquake prediction (in Japanese), in Earthquake Prediction Techniques, T. Asada (Ed.), Univ. Tokyo Press, 193-209, 1978c, English edition to be published in Sept., 1981.

Ishibashi, K., Modern tectonics around the Izu Peninsula (in Japanese), The Earth Monthly, Kaiyo Shuppan Co., Ltd., Tokyo, 2, 110-119, 1980a.

Ishibashi, K., Comments on the long-term prediction of the Tokai earthquake (in Japanese), in Proc. Earthq. Predict. Res. Symp. (1980), 123-125, 1980b.

Ishibashi, K., and R. Sato, Long-period dynamic displacements due to a hypothetical Tokai earthquake (in Japanese), Abstracts, Seismol. Soc. Japan, 1977 No.1, 32, 1977.

Japan Meteorological Agency, Seismic Activity in the Tokai District (in Japanese), 169 pp., 1977.

Kaizuka, S., Macro-topography of the island arc system and plate tectonics (in Japanese), Kagaku, Iwanami Shoten, Tokyo, 42, 573-581, 1972.

Kaizuka, S., Position of the Kanto district in the island arc system and its Quaternary crustal movement (in Japanese), in Earthquakes and Crustal Deformations in the Kanto District, T. Kakimi, and Y. Suzuki (Eds.), Lattice Pub. Co., Tokyo, 99-118, 1974.

Kaizuka, S., A tectonic model for the morphology of arc-trench systems, especially for the echelon ridges and mid-arc faults, Jap. J. Geol. Geogr., 45, 9-28, 1975.

Kakimi, T., Late Quaternary crustal movement in the Tokai district (in Japanese), Rep. Subcomm., Coord. Comm. Earthq. Predict., 1, 28-34, 1977.

Kanamori, H., Faulting of the great Kanto earthquake of 1923 as revealed by seismological data, Bull. Earthq. Res. Inst., Univ. Tokyo, 49, 13-18, 1971.

Kanamori, H., Tectonic implications of the 1944 Tonankai and the 1946 Nankaido earthquakes, Phys. Earth Planet. Interiors, 5, 129-139, 1972.

Kasahara, K., Akaishi uplift (in Japanese), in Off-Tokai Earthquake, M. Ando, and Y. Fukao (Eds.), 44-45, 1975.

Kasahara, K., On the pattern of earthquake occurrence in the Kanto district (in Japanese), Abstracts, Seismol. Soc. Japan, 1980 No.2, 66, 1980.

Kino, Y., The "Large-Scale Earthquake Countermeasures Act" in Japan, in Abstracts of Contributed Papers, Intern. Symp. Earthq. Predict., Unesco Headquarters, Paris, 90, 1979.

Kobayashi, H., and S. Midorikawa, On seismic faults and seismic intensity distributions of the Ansei Tokai and the Nobi earthquakes (in

Japanese), in Abstracts, 4th Symp. Earthq. Engineering, Appl. Geosci., 39-40, 1979.

Maki, T., On the earthquake mechanism of the Izu-Hanto-oki earthquake of 1974 (in Japanese), Spec. Bull. Earthq. Res. Inst., Univ. Tokyo, 14, 23-36, 1974.

Maritime Safety Agency, Submarine topography, geological structure and magnetic anomaly of Enshu-nada (in Japanese), Rep. Coord. Comm. Earthq. Predict., 15, 109-114, 1976a.

Maritime Safety Agency, Ocean environmental chart of the adjacent seas of Nippon, Bathymetric chart of Iro-saki to Muroto-saki (1:500,000), 1976b.

Maritime Safety Agency, Ocean environmental chart of the adjacent seas of Nippon, Submarine structural chart of Iro-saki to Muroto-saki including geomagnetic total intensity anomaly (1:500,000), 1976c.

Maritime Safety Agency, Deformation of the uppermost sediment layer in the Suruga and Nankai troughs (in Japanese), Rep. Coord. Comm. Earthq. Predict., 17, 109-112, 1977.

Maritime Safety Agency, Submarine topography and geological structure, northward of the Suruga Bay (in Japanese), Rep. Coord. Comm. Earthq. Predict., 20, 133-134, 1978a.

Maritime Safety Agency, Structure of the Suruga trough from multichannel seismic reflection profiling (preliminary report) (in Japanese), Rep. Coord. Comm. Earthq. Predict., 20, 135-137, 1978b.

Maritime Safety Agency, Bathymetric chart, No. 6362-5, Northern part of Suruga-wan (1:50,000), 1978c.

Maritime Safety Agency, Submarine structural chart, No. 6362-5-S, Northern part of Suruga-wan (1:50,000), 1978d.

Maritime Safety Agency, Submarine topography and geological structure, southward of the Suruga Bay (in Japanese), Rep. Coord. Comm. Earthq. Predict., 23, 77-79, 1980a.

Maritime Safety Agency, Basic map of the sea in coastal waters (1:50,000), Southeastern part of Suruga-wan, 1980b.

Maritime Safety Agency, Basic map of the sea in coastal waters (1:50,000), Southwestern part of Suruga-wan, 1980c.

Matsuda, T., Geomorphological and geological data on the Tokai earthquake (in Japanese), in Off-Tokai Earthquake, M. Ando, and Y. Fukao (Eds.), 15-17, 1975.

Matsuda, T., Collision of the Izu-Bonin arc with central Honshu: Cenozoic tectonics of the Fossa Magna, Japan, J. Phys. Earth, 26, Suppl., S409-S421, 1978.

Matsuda, T., Y. Ota, M. Ando, and N. Yonekura, Fault mechanism and recurrence time of major earthquakes in southern Kanto district, Japan, as deduced from coastal terrace data, Geol. Soc. Amer. Bull., 89, 1610-1618, 1978.

Matsu'ura, M., T. Iwasaki, Y. Suzuki, and R. Sato, Statical and dynamical study on faulting mechanism of the 1923 Kanto earthquake, J. Phys. Earth, 28, 119-143, 1980.

Midorikawa, S., and H. Kobayashi, Isoseismal map in near-field with regard to fault rupture and site geological conditions (in Japanese), Trans. Archit. Inst. Japan, No. 290, 83-94, 1980.

Mikumo, T., On the so-called Off-Tokai earthquake (in Japanese), in Off-Tokai Earthquake, M. Ando, and Y. Fukao (Eds.), 39-43, 1975.

Misawa, Y., and T. Yoshiwara, Submarine topography in Suruga Bay (in Japanese), in Fossa Magna, Y. Fujita et al. (Eds.), 196-200, 1968.

Mizoue, M., Some remarks on the characteristics of subcrustal earthquake activities (in Japanese), in Proc. Symp. on Earthq. Predict. (1976), 97-105, 1977.

Mogi, A., An Atlas of the Sea Floor around Japan: Aspects of Submarine Geomorphology (in Japanese), Univ. Tokyo Press, 90 pp., 1977, English edition published in 1979.

Mogi, A., and M. Sakurai, A study on the deep shelf break of the west coast of Izu Peninsula: Suggesting Suruga trough to be a subduction zone (in Japanese), in Proc. Earthq. Predict. Res. Symp. (1980), 117-121, 1980.

Mogi, K., Some features of recent seismic activity in and near Japan (2): Activity before and after great earthquakes, Bull. Earthq. Res. Inst., Univ. Tokyo, 47, 395-417, 1969.

Mogi, K., Recent horizontal deformation of the earth's crust and tectonic activity in Japan (1), Bull. Earthq. Res. Inst., Univ. Tokyo, 48, 413-430, 1970a.

Mogi, K., An interpretation of horizontal crustal deformation (in Japanese), Rep. Coord. Comm. Earthq. Predict., 2, 85-87, 1970b.

Mogi, K., An interpretation of the recent tectonic activity in the Izu-Tokai district (in Japanese), Bull. Earthq. Res. Inst., Univ. Tokyo, 52, 315-331, 1977.

Mogi, K., Izu: The recent crustal activity (in Japanese), in Progress of the Coordinating Committee for Earthquake Prediction, Japan, during the First Decade, CCEP (Ed.), Geogr. Surv. Inst., 121-140, 1979.

Mogi, K., Relationship between the future Tokai earthquake and the 1891 Nobi earthquake (in Japanese), Rep. Coord. Comm. Earthq. Predict., 24, 162-163, 1980.

Muramatsu, I., Relationship between disastrous shaking areas and aftershock areas (in Japanese), Abstracts, Seismol. Soc. Japan, 1972 No. 2, 115, 1972.

Muramatsu, I., and T. Mikumo, Estimation of long-period ground-velocities in the Tokai district due to fault models of large earthquakes (in Japanese), Abstracts, Seismol. Soc. Japan, 1979 No.1, 93, 1979.

Nakamura, K., σH_{max} trajectories east of Suruga trough, Japan: An effect of flexure of lithospheric plate (in Japanese), Zisin (J. Seismol. Soc. Japan), Ser. 2, 32, 370-372, 1979.

Nakamura, K., Tectonics of Izu and flexure of

lithospheric plate (in Japanese), The Earth Monthly, Kaiyo Shuppan Co., Ltd., Tokyo, 2, 94-102, 1980.

Nakamura, K., An interpretation of current crustal deformations around the Suruga Bay, Japan (in Japanese), Zisin (J. Seismol. Soc. Japan), Ser. 2, 34, 1981, in press.

Nakane, K., Horizontal tectonic strain in Japan (I) (II) (in Japanese), J. Geod. Soc. Japan, 19, 190-199, 200-208, 1973.

Nasu, N., R. Tsuchi, and E. Honza, Submarine geological structure of the west part of Suruga Bay (in Japanese), in Fossa Magna, Y. Fujita et al. (Eds.), 191-195, 1968.

Rikitake, T., Probability of earthquake occurrence as estimated from crustal strain, Tectonophysics, 23, 299-312, 1974.

Rikitake, T., Recurrence of great earthquakes at subduction zones, Tectonophysics, 35, 335-362, 1976.

Rikitake, T., Probability of a great earthquake to recur off the Pacific coast of central Japan, Tectonophysics, 42, T43-T51, 1977.

Saito, A., and S. Kosuge, Study of tsunami-responses in Suruga Bay by numerical modeling (in Japanese), J. Fac. Marine Sci. Tech. Tokai Univ., No. 12, 71-85, 1979.

Sakamoto, T., H. Yamazaki, I. Isobe, K. Ito, and S. Goto, On the Holocene marine terraces near Omaezaki, Shizuoka Pref., central Japan (in Japanese), Bull. Geol. Surv. Japan, 29, 133-135, 1978.

Sakurai, M., and A. Mogi, Multichannel seismic reflection profiling in the Suruga trough (in Japanese), Rep. Hydrogr. Res., Mar. Safety Agcy., No. 15, 1-21, 1980.

Sato, H., Crustal movements associated with the 1944 Tonankai earthquake (in Japanese), J. Geod. Soc. Japan, 15, 177-180, 1970.

Sato, H., Crustal movements in the Tokai district as revealed by geodetic surveys (in Japanese), Rep. Subcomm., Coord. Comm. Earthq. Predict., 1, 19-27, 1977.

Sato, H., and N. Inouchi, Vertical land movement in Tokai district associated with the Tonankai earthquake of 1944 (in Japanese), Zisin (J. Seismol. Soc. Japan), Ser. 2, 28, 489-491, 1975.

Sato, R., and M. Matsu'ura, Strains and tilts on the surface of a semi-infinite medium, J. Phys. Earth, 22, 213-221, 1974.

Sawamura, T., Relation between the activities of the outer earthquake zone in southwestern Japan and the geologic structure and crustal movements of Shikoku and its vicinity (in Japanese), Res. Rep. Kochi Univ., 2, 1-46, 1953.

Sawamura, T., On the Nankai thrust and the distribution of initial motions of seismic waves by the Nankai earthquake in 1946 (in Japanese), Res. Rep. Kochi Univ., 3, 1-6, 1954.

Sekiya, H., Seismic activities around the Sea of Enshu (in Japanese), in Off-Tokai Earthquake, M. Ando, and Y. Fukao (Eds.), 6-14, 1975.

Sekiya, H., On the seismicity gap and seismic ac-

tivities in the Sea of Enshu (in Japanese), Rep. Subcomm., Coord. Comm. Earthq. Predict., 1, 9-18, 1977.

Sekiya, H., and K. Tokunaga, On the seismicity near the Sea of Enshu (in Japanese), Rep. Coord. Comm. Earthq. Predict., 11, 96-101, 1974a.

Sekiya, H., and K. Tokunaga, On the seismicity near the Sea of Enshu (2) (in Japanese), Rep. Coord. Comm. Earthq. Predict., 12, 114-119, 1974b.

Sekiya, H., and K. Tokunaga, On the seismicity gap near Enshunada (in Japanese), Quart. J. Seismol., 39, 83-88, 1975.

Seno, T., The instantaneous rotation vector of the Philippine Sea plate relative to the Eurasian plate, Tectonophysics, 42, 209-226, 1977a.

Seno, T., Recurrence times of great earthquakes in the seismotectonic areas along the Philippine Sea side coast of southwest Japan and south Kanto district (in Japanese), Zisin (J. Seismol. Soc. Japan), Ser. 2, 30, 25-42, 1977b.

Seno, T., Pattern of intraplate seismicity in southwest Japan before and after great interplate earthquakes, Tectonophysics, 57, 267-283, 1979.

Seno, T., Review of tectonics of the Izu Peninsula (in Japanese), The Earth Monthly, Kaiyo Shuppan Co., Ltd., Tokyo, 2, 81-86, 1980.

Seno, T., K. Shimazaki, P. Somerville, K. Sudo, and T. Eguchi, Rupture process of the Miyagi-Oki, Japan, earthquake of June 12, 1978, Phys. Earth Planet. Interiors, 23, 39-61, 1980.

Shiono, K., Travel time analysis of relatively deep earthquakes in southwest Japan with special reference to the underthrusting of the Philippine Sea plate, J. Geosci. Osaka City Univ., 18, 37-59, 1974.

Somerville, P., The accomodation of plate collision by deformation in the Izu block, Japan, Bull. Earthq. Res. Inst., Univ. Tokyo, 53, 629-648, 1978.

Sugimura, A., Plate boundaries in and near Japan (in Japanese), Kagaku, Iwanami Shoten, Tokyo, 42, 192-202, 1972.

Suzuki, Y., and R. Sato, Estimation of short-period acceleration, velocity, and displacement due to a fault model (2): Hypothetical Suruga Bay earthquake (in Japanese), Abstracts, Seismol. Soc. Japan, 1979 No.2, 201, 1979.

Tada, T., M. Kaizu, M. Tsuzawa, and Y. Nakahori, Kakegawa uplift and the source region of the Tonankai earthquake (in Japanese), Abstracts, Seismol. Soc. Japan, 1980 No.1, 188, 1980.

Takeo, M., K. Abe, and H. Tsuji, Mechanism of the Shizuoka earthquake of July 11, 1935 (in Japanese), Zisin (J. Seismol. Soc. Japan), Ser. 2, 32, 423-434, 1979.

The Research Group for Active Faults, Active Faults in Japan: Sheet Maps and Inventories (in Japanese), Univ. Tokyo Press, 363 pp., 1980.

Tsuchi, R., Terraces on land and submarine topography (in Japanese), Umi, 5, 80-84, 1967.

Tsuchi, R., Quaternary crustal movement and Neogene structure in the southern Fossa Magna (in Japanese), in Fossa Magna, Y. Fujita et al. (Eds.), 72-82, 1968a.

Tsuchi, R., Crustal movements deduced from the deformation of dissected fans, with reference to those in the Tokai region (in Japanese), The Quat. Res., 7, 225-234, 1968b.

Tsuji, Y., Earthquakes and tsunamis off Tokai region as investigated by historical documents (in Japanese), Marine Sciences Monthly, Kaiyo Shuppan Co., Ltd., Tokyo, 11, 32-44, 1979.

Tsumura, K., Anomalous crustal activity in the Izu Peninsula, central Honshu, J. Phys. Earth, 25, Suppl., S51-S68, 1977.

Tsuneishi, Y., On the Fujikawa fault and tectonics of the Izu Peninsula (in Japanese), Abstracts, Seismol. Soc. Japan, 1980 No.2, 108, 1980.

Tsuneishi, Y., Kanto Regional Construction Bureau, and K. Shiosaka, Boring investigation of the Fujikawa fault (in Japanese), Abstracts, Seismol. Soc. Japan, 1980 No.2, 107, 1980.

Tsuneishi, Y., and K. Shiosaka, Fault of the Ansei-Tokai earthquake (1854) (in Japanese), Rep. Coord. Comm. Earthq. Predict., 20, 158-161, 1978.

Tsuneishi, Y., and K. Shiosaka, Additional data on the Fujikawa fault (1) (in Japanese), Rep. Coord. Comm. Earthq. Predict., 22, 149-154, 1979.

Tsuneishi, Y., and Y. Sugiyama, Sunzu fault across the Suruga Bay (in Japanese), Rep. Coord. Comm. Earthq. Predict., 20, 138-141, 1978.

Usami, T., Descriptions of All Disastrous Earthquakes in Japan (in Japanese), Univ. Tokyo Press, 327 pp., 1975.

Usami, T., Intensity distribution of the Hoei (1707) and the Ansei (1854) earthquakes (in Japanese), Rep. Coord. Comm. Earthq. Predict., 17, 84-88, 1977.

Usami, T., Great earthquakes of the past (in Japanese), in Earthquake Prediction Techniques, T. Asada (Ed.), Univ. Tokyo Press, 12-28, 1978, English edition to be published in Sept., 1981.

Usami, T., Intensity distribution of the Ansei (1854) earthquake (in Japanese), Rep. Coord. Comm. Earthq. Predict., 22, 216, 1979.

Utsu, T., Recent seismic activity and seismological observation in Hokkaido (in Japanese), Rep. Coord. Comm. Earthq. Predict., 2, 1-2, 1970.

Utsu, T., Large earthquakes near Hokkaido and the expectancy of the occurrence of a large earthquake off Nemuro (in Japanese), Rep. Coord. Comm. Earthq. Predict., 7, 7-13, 1972.

Utsu, T., Space-time pattern of large earthquakes occurring off the Pacific coast of the Japanese Islands, J. Phys. Earth, 22, 325-342, 1974.

Utsu, T., On the possibility of the Sea-of-Enshu earthquake (in Japanese), in Off-Tokai Earthquake, M. Ando, and Y. Fukao (Eds.), 1-5, 1975a.

Utsu, T., Microseismicity in and around the Sea of Enshu (in Japanese), in Abstracts, 12th Symp. on Natural Disasters, 375-376, 1975b.

Utsu, T., Historical large earthquakes off Tokai region (in Japanese), Rep. Subcomm., Coord. Comm. Earthq. Predict., 1, 1-8, 1977a.

Utsu, T., Possibility of a great earthquake in the Tokai district, central Japan, J. Phys. Earth, 25, Suppl., S219-S230, 1977b.

Utsu, T., Nemuro-Hanto-Oki earthquake (in Japanese), in Progress of the Coordinating Committee for Earthquake Prediction, Japan, during the First Decade, CCEP (Ed.), Geogr. Surv. Inst., 79-87, 1979.

Yamazaki, H., Active faults at a plate boundary (in Japanese), The Earth Monthly, Kaiyo Shuppan Co., Ltd., Tokyo, 1, 570-576, 1979.

Yamazaki, F., and T. Ooida, The seismicity in and near Suruga Bay (in Japanese), Zisin (J. Seismol. Soc. Japan), Ser. 2, 32, 451-462, 1979.

Yamazaki, F., T. Ooida, and H. Aoki, Subduction of the Philippine Sea plate in the Tokai district (in Japanese), Abstracts, Seismol. Soc. Japan, 1980 No.2, 161, 1980.

Yonekura, N., Quaternary tectonic movements in the outer arc of southwest Japan with special reference to seismic crustal deformations, Bull. Dept. Geogr., Univ. Tokyo, No. 7, 19-71, 1975.

Yonekura, N., Submarine active faults off Tokai region (in Japanese), The Earth Monthly, Kaiyo Shuppan Co., Ltd., Tokyo, 1, 577-582, 1979.

LONG- AND INTERMEDIATE-TERM SEISMIC PRECURSORS TO EARTHQUAKES — STATE OF THE ART

Martin Reyners

Lamont-Doherty Geological Observatory of Columbia University

Palisades, New York 10964

Abstract.   Research into long- and intermediate-term seismic precursors to earthquakes in the last few years has resulted in both the consolidation and refinement of established methods and experimentation with new ones. Precursory changes in seismic velocities, focal mechanism and b-value continue to be useful in the prediction of some earthquakes. Additional seismic parameters for which precursory variations have recently been reported include seismic spectrum, source parameters, the P-wave amplitude of teleseisms and the amplitude and direction of approach of microseisms. Searches for precursory seismicity patterns have also attracted much attention. Diverse patterns have been reported, including preseismic activation, quiescence, clustering, earthquake migration and combinations of these. At the same time, however, other careful studies to identify seismic precursors to earthquakes have met with negative results.
Seismic precursors can be interpreted in different ways. They may reflect changes in rock properties intimately connected with the earthquake preparatory process, and the mixed results obtained in the identification of these precursors may arise because of differences in experimental method and regional differences in anomaly sign, size, and shape. Alternatively, such precursors may result from systematic spatio-temporal seismicity fluctuations in areas of spatially varying rock properties. In light of the mounting evidence that the characteristics of seismic precursors differ with tectonic environment, there is a need to continue baseline measurements in regions of interest, both to determine the reliability of such precursors in terms of successes, failures and false alarms, and to isolate background parameter variations not connected with the earthquake preparatory process.

Introduction

Since the early 1960's, when reports of temporal changes in seismic velocities prior to earthquakes first suggested that earthquake prediction on a scientific basis might be a realizeable goal, there has been great interest in earthquake precursors. Diverse precursors, both seismic and non-seismic, have been proposed, with precursor times ranging from tens of years to minutes. Here we will consider long- and intermediate-term seismic precursors to earthquakes, with precursor times ranging from years to weeks. A wide variety of such precursors have been reported, yet enthusiasm as to their general usefulness for prediction has been tempered by a significant number of studies in which these precursors have not been observed.

The identification, verfication and application of long- and intermediate-term precursors to earthquakes is no simple matter. Firstly, it must be established that an observed anomaly is real, and not simply a product of noise in the data. Secondly, the uniqueness of the anomaly has to be ascertained, by extensive sampling of the normal background variations of the parameter measured. And thirdly, for the anomaly to be precursory, it should be shown or argued that it bears some physical relationship to the main shock. To date many studies of seismic precursors have been retrospective and as such are limited, since the occurrence of the main shock has reduced the number of degrees of freedom of these experiments. Progress in the real-time identification of seismic precursors has also suffered owing to our low level of understanding of the physical nature of the earthquake preparatory process and its regional variability.

Rikitake [1975, 1979] has made a thorough classification of earthquake precursors, both seismic and non-seismic, and the reader is referred to these papers for documentation of long- and intermediate-term seismic precursors reported prior to 1977. The present work will concentrate on research carried out since 1977, with special reference to false alarms and negative results. (A compendium of U.S. work on seismic precursors during this period is included in Ward [1979]). The seismic precursors have been subdivided into groups simply for convenience; as discussed later, various apparent precursors may in fact be different manifestations of some single underlying precursory change.

Precursory Seismicity Patterns

The last few years have seen an upsurge in the number of reports of spatio-temporal patterns of seismicity preceding larger earthquakes (e.g., Brady, 1977; Evison, 1977a, b, c; Ishida and Kanamori, 1977, 1980; McNally, 1977; Ohtake et al., 1977; Sekiya, 1977; Talwani, 1979; Wyss and Habermann, 1979; Keilis-Borok et al., 1980a, b, c]. This is not surprising, considering the growth in recent years of earthquake catalogues suitable for the identification of such patterns. Diverse patterns have been reported, including preseismic activation, quiescence, clustering, earthquake migration and combinations of these. Schematic representations of some of these patterns are shown in Figure 1.

The seismicity patterns reported fall into two general groups: those that are not necessarily tied to the immediate area of an impending earthquake, and those that are. Notable among authors who have not required precursory seismicity patterns to be spatially closely connected to an impending earthquake have been V. I. Keilis-Borok and his colleagues [Keilis-Borok et al., 1980a, b, c]. These authors identify three precursory seismicity patterns, each of which seems to represent a different facet of "bursts of seismicity" — that is, abnormal clustering of earthquakes in the space-time-energy domain. In almost all cases, these patterns occur at some distance from subsequent strong earthquakes, suggesting the possibility that small but significant stress redistribution due to medium-sized earthquakes at moderate distances can act as a triggering agent. Although the mechanical interaction of large areas implied by these patterns may at first seem physically unappealing, examples of such triggering (or alternatively, regional strain episodes) are not uncommon. For instance, Wang [1979] finds that the frequency of aftershocks of the 1966 Xingtai earthquake increased before a number of subsequent strong earthquakes occurring throughout north China, while Wu et al. [1979b] find a good correlation between earthquake occurrence in northeastern China and Japan.

The possibility that remote earthquakes provide precursory information begs the question of what areas should be analyzed together in a search for precursory seismicity patterns. As pointed out by Keilis-Borok et al. [1980a] there is as yet no unique answer to this question. These authors investigate the dependence of their results on area of surveillance by varying the boundaries of the area and repeating their analyses. Also, a precursory "burst of seismicity", once identified, gives no indication of the exact location of an impending earthquake. Nevertheless, it may be of value since it provides an experimental long-term enhancement of the probability that a large earthquake will occur, and thus suggests the

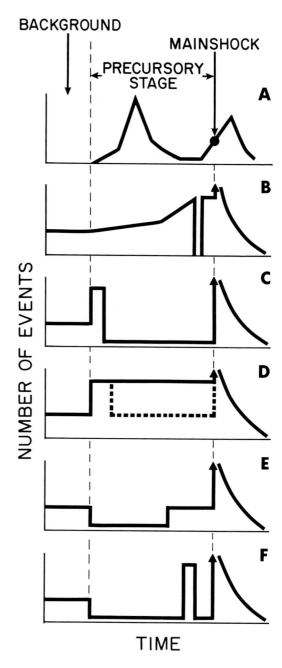

Fig. 1. Schematic representations of reported precursory seismicity fluctuations within or near earthquake rupture zones. Legend: A - Brady [1977] for rock bursts in mines in Idaho, USA; B - Talwani [1979] for reservoir-induced microearthquakes; C - Evison [1977a, b, c] for earthquakes in New Zealand and California; D - Aggarwal and Perez [1980] for 11 events with M_s or M_w > 7; E - Ohtake et al. [1978] for 15 large earthquakes in Mexico, Central America and California; and F - Wyss et al. [1978] for various earthquakes worldwide.

need to search for more localized precursors. Often conclusions based on such bursts of seismicity complement those based on more localized seismicity patterns, thus enhancing both conclusions. For example, for the great 1950 Assam-Tibet earthquake (M = 8.6), Keilis-Borok et al. [1980a] identify a precursory burst of seismicity in a large region encompassing parts of China, Nepal, Bhutan, Bangladesh and Burma, while Khattri and Wyss [1978] identify a precursory quiescence lasting about 30 years in the general vicinity of the earthquake. Similarly, Keilis-Borok et al. [1980c] find a burst of seismicity in a region covering most of New Zealand before the 1976 Milford Sound earthquake (M = 7.0), whereas Evison [1977a] finds a precursory swarm followed by a precursory gap in the immediate vicinity of the earthquake. Indeed, the influences of both remote and nearby earthquakes have been combined in the phenomenon of "induced foreshocks" described by Fedotov et al. [1977]. These authors suggest that a large remote earthquake is closely followed by small earthquakes ("induced foreshocks") in the area of preparation of a future large earthquake.

The bulk of reported precursory seismicity patterns fall into the second group — those that bear a close spatial relationship to an impending earthquake. Of course this may simply be a consequence of the fact that most of these studies have been retrospective. The immediate vicinity of an earthquake is the most obvious place in which to conduct a retrospective search for precursory seismicity patterns. Many of the patterns reported are statistically significant, and cannot be considered random fluctuations. For example, McNally [1977] determines patterns in a 25-year record of seismicity in central and southern California after identifying and removing a random component of the seismicity. The patterns

also appear related to earthquake preparatory processes, since precursor times derived from them often scale with magnitude in a manner similar to other precursors. This is illustrated in Figure 2, in which precursor time-magnitude relationships derived from seismicity patterns are compared with the relationship of Rikitake [1979] determined using 15 precursors, both seismic and non-seismic.

How the area occupied by the precursory seismicity patterns relates to the magnitude of the ensuing event is at present not well established. Clearly this area will usually be relatively poorly determined. For example, the apparent dimensions of an area of relative seismic quiescence will depend on both the level of background seismicity and the accuracy of hypocentre location. On the basis of data from 19 strong earthquakes exhibiting precursory relative seismic quiescence Wu et al. [1976] derive the following empirical relations:

$$M = 3.00 \log L - 0.07 \quad (\pm 0.33)$$

and

$$M = 1.55 \log Q + 0.31 \quad (\pm 0.34)$$

where M is the magnitude of the large earthquake in the quiescent zone, L is the length of major axis of the zone in kilometres and Q is its area in km^2. On the other hand, in a study of patterns of seismicity preceding 15 large earthquakes that have occurred since 1964 in Mexico, Central America and California, Ohtake et al. [1978] find that there is little or no correlation between the dimensions of the zones of anomalous seismicity (quiescence followed by a resumption of activity) and the magnitude of the main shock. Similarly, there appears to be no clear-cut relationship between areas of anomalous seismicity and the aftershock zones of the ensuing earthquakes. Many of the areas of precursory seismicity (earthquake swarm followed by quiescence) identified by Evison [1977a, b, c] for earthquakes in New Zealand and California are several times larger than the aftershock zones of the events, while the areas of precursory seismic quiescence proposed by Wu et al. [1976] are extremely large compared with the aftershock zones of the events. In contrast, Wyss et al. [1979] suggest a 50% decrease in seismicity during the 3.5 years preceding the 1975 Kalapana, Hawaii earthquake (M = 7.2) involved a crustal volume with dimensions only about half those of the earthquake source.

If precursory seismicity patterns of this second group are real and related to earthquake preparatory processes, what is the significance of the diversity of the patterns reported? Rock fracturing experiments [eg., Mogi, 1962; Scholz, 1968a] indicate that the details of precursory seismicity patterns depend on the strength and heterogeneity of a rock specimen, as well as on the strain rate. Thus different patterns may reflect differences in tectonic environment. Alternatively, the diversity of the patterns may

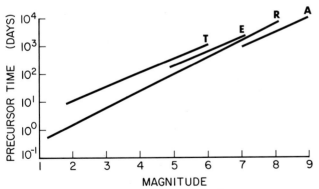

Fig. 2. A comparison of precursor time-magnitude relationships derived from seismicity patterns with the relationship of Rikitake [1979] determined using 15 precursors, both seismic and non-seismic. Legend: A - Aggarwal and Perez [1980]; E - Evison [1977b]; R - Rikitake [1979]; T - Talwani [1979].

be at least partly due to differences in the experimental design adopted by various authors. Experimental design is important considering the large number of degrees of freedom in the determination of a spatio-temporal seismicity pattern. This has been recognized by Keilis-Borok et al [1980b], who contend that "there are so many adjustable parameters that we run the risk of self-deception... hence it may be difficult to perform the job of data fitting honestly". For example, details of precursory seismicity patterns may differ with magnitude range. Evison [1977c] defines a precursory swarm followed by a precursory gap for the 1968 Inangahua, New Zealand, earthquake (M = 7.1) by studying seismicity of magnitude 4.5 and greater. However, when he takes into account events with magnitudes down to nearly five units less than that of the main shock, evidence for the swarm is strengthened while the gap is obscured [Evison, 1978]. Similarly, Tsai et al. [1979] find precursory increases in the monthly frequency of occurrence of microearthquakes in the 2.5-3.5 magnitude range before several moderate earthquakes of magnitude 5.7-6.8 in eastern Taiwan, yet no precursory increase in the frequency of occurrence of larger earthquakes is apparent. It should also be kept in mind that recipes for precursory swarms, activation, quiescence and earthquake migration adopted by different authors are generally arbitrary and certainly non-unique. The parameters of these recipes, such as space, time and magnitude, have usually been chosen in such a way as to optimize the score of a given predictor for a particular data set.

Given the likelihood that seismicity patterns will vary with tectonic environment and method of experiment, it is clear that we are at an early stage in the development of such patterns as useful predictors. Systematic baseline measurements in specific areas need to be made to determine the reliability of the patterns in terms of successes, failures and false alarms. The usefulness of a long period of baseline measurements in the identification of precursory seismicity patterns has been demonstrated by Mizoue et al. [1978]. By studying thirteen years of microearthquake data in the vicinity of Wakayama City, Japan, these authors find that a systematic variation in the mode of seismicity accompanies earthquakes of magnitude 4.7 to 5.2. A seismicity gap appears 2-3 years before a moderate earthquake, and the occurrence of a precursory shock of magnitude as large as 4.3 at the southern border of this gap serves as a reliable indicator of a moderate earthquake 6-7 months later on a pre-existing fault across the gap. Similarly, thirteen years of continuous observation of microearthquakes near the Yamasaki fault in southwest Japan has resulted in the identification of variations of seismic activity with a four-year period, and a migration of activity along the fault [Oike, 1979].

Efforts to improve hypocentre location accuracy

should accompany baseline measurements of seismicity. Up till now, many studies of seismicity patterns have largely been concerned with the epicentral distribution of earthquakes, and the depth dimension of the pattern has not been addressed, usually because of poor depth control. There is no reason to suspect that seismicity patterns will be most evident in the distribution of epicentres. Indeed, such patterns may be more obvious in the depth distribution of earthquakes, especially in areas of pronounced vertical heterogeneity, such as subduction zones. Depth variation of seismic activity may in fact obscure seismicity patterns when these are determined using epicentres alone. For example, Dewey and Spence [1979] find that shallow earthquakes in the subduction zone in coastal Peru occur principally in two distinct zones, the "coastal-plate interior" zone being 50 km inland and 30 km deeper than the "interface thrust" zone. Accurate location of shallow microearthquakes in the North Island of New Zealand has similarly revealed the existence of two distinct zones of seismicity, with activity at the top of the subducted plate being underlain by significant activity in the interior of the plate [Reyners, 1980]. As pointed out by Dewey and Spence [1979], the presence of such adjacent seismic zones, each corresponding to different stress regimes, may greatly hinder the identification of seismic gaps or anomalous precursory activity in either zone if the hypocentres of earthquakes in the two zones cannot be distinguished by accurate location methods or, possibly, by focal mechanism studies. Reports of precursory variations in the depth distribution of seismicity are not uncommon. In a depth-time plot of intermediate depth seismicity before the intermediate depth Vrancea, Romania earthquake of 1977 (M = 7.2), Mârza [1979] identifies a precursory pattern which is very similar to epicentral patterns found by Evison [1977a] before shallow earthquakes (see Figure 3). In addition, Bufe et al. [1974] and Gupta [1975a] have reported vertical migration of seismic activity before shallow events. The three-dimensional nature of seismicity patterns clearly warrants further study, not only to elucidate the predictive value of the patterns themselves, but also because hypocentral migration patterns may be responsible for other observed seismic precursors, as discussed later.

The usefulness of a seismicity pattern for prediction would certainly be enhanced if the properties of the radiated waves of earthquakes in such a pattern differ from those of the background seismicity. Recent results suggest this may be the case, at least for some earthquakes. A precursory cluster of seismicity before the 1971 Cape Kamchatka earthquake (M_s = 7.8) is made up of events recognized by Fedotov et al. [1977] as possible precursors on the basis of their exceedingly low ratio of P- to S-wave energy [Wyss et al., 1978]. Also, Ishida and Kanamori [1978] have reported that the 1971 San Fernando, California, earthquake (M_L = 6.4) was preceded by a

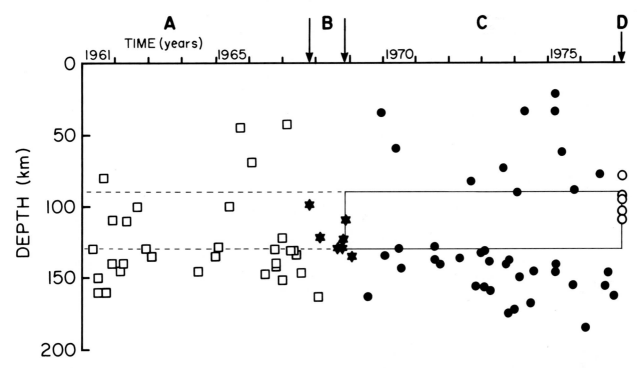

Fig. 3. Precursory pattern evident in a depth-time plot of seismicity prior to the intermediate depth Vrancea, Romania earthquake of 1977 (M_L = 7.2). The seismic episodes indicated are as follows: A - normal seismicity (open squares); B - precursory "swarm" (stars); C - precursory "gap" (filled circles; the gap area is outlined); D - major multiple event (open circles). After Mârza [1979].

precursory cluster consisting of earthquakes having different signal characteristics from background events. These authors also find that the frequency of the spectral peak of tightly clustered events that occurred near the epicentral area during the 1 1/2 years immediately prior to the 1952 Kern County, California, earthquake (M_s = 7.7) was systematically higher than for events before 1949 [Ishida and Kanamori, 1980]. Further research in this field may provide valuable insights into how seismicity patterns are related to the earthquake preparatory process.

Precursory Changes in Seismic Velocities

Early evidence that temporal changes in seismic velocities may precede earthquakes was presented by Kondratenko and Nersesov [1962]. In a study of shallow earthquakes of magnitude greater than about five in the Garm region of Tadjikistan, USSR, these authors found that values of the ratio of compressional to shear velocity (V_p/V_s), and in particular of V_p itself, were anomalously low for a period before the events, and that the duration of the anomalous period increased with the earthquake magnitude. These results sparked off searches for similar precursory velocity changes in other areas, and many such changes have been reported. A recent summary of precursory velocity changes and the physical models that have been postulated to explain them is included in Lukk and Nersesov [1978]. In this section we will discuss both positive and negative results that have been reported in the last few years in searches for precursory velocity changes, some of the problems and difficulties in detecting these, alternative explanations as to their origin, and the reliability and generality of the changes as precursors.

The method used in the early determinations of precursory velocity changes in Tadjikistan, that of the Wadati diagram, continues to be popular in the identification of such changes. In a Wadati diagram, t_s-t_p is plotted against t_p for the data of a single earthquake, where t_p and t_s are the arrival times of the P- and S-waves, respectively. If one assumes that Poisson's ratio is the same in all layers traversed by waves of the earthquake (i.e., the P- and S-waves travel the same path), the Wadati diagram is a straight line with a slope of V_p/V_s-1. The usefulness of the Wadati diagram arises from the fact that it yields important information about V_p/V_s without requiring the development of travel-time curves or the location of the source event. In a recent series of papers, Feng et al. [1976, 1977, 1978] have summarized V_p/V_s anomalies determined with the Wadati diagram for 16 earthquakes with magnitudes between 5.1 and 7.9 in western China. The general shape of the V_p/V_s variation prior to an earth-

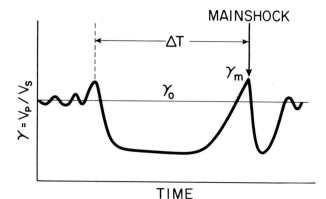

MAINSHOCK

TIME

Fig. 4. General shape of the V_p/V_s variation prior to an earthquake found by Feng et al. [1976]. ΔT denotes the precursor time, γ_0 denotes the average background value of the velocity ratio, and γ_m denotes the maximum observed value of the velocity ratio following the precursory bay in the ratio.

quake found by these authors is shown in Figure 4. This shape is similar to that found earlier in the Garm region of the USSR (see Figure 6), with a notable refinement being the high values of V_p/V_s immediately before and after the precursory bay in the ratio. It has been found that earthquakes with high values of V_p/V_s (~ 2.0) following a bay in V_p/V_s can be useful for short-term prediction because, as well as predicting the location of the epicentre of the main shock, the maximum value of V_p/V_s observed (γ_m in Figure 4) is directly proportional to the magnitude of the main shock. Other characteristics of precursory V_p/V_s variations reported by these authors are similar to those found previously. The amplitude of the bay in V_p/V_s is independent of magnitude, while the duration of the precursory period of anomalous V_p/V_s increases with magnitude. Also, the area of the region of anomalous V_p/V_s increases with magnitude, according to the relation

$$M_s = -0.7 + 1.7 \log S \pm 0.7$$

where M_s is the main shock magnitude and S is the area in km^2 [Feng et al., 1977]. It is clear that P- and S-waves from an earthquake occurring in such an area of anomalous velocity ratio recorded at a station outside the area will not travel the same path because of lateral variation in the V_p/V_s ratio, and the Wadati diagram will no longer be linear. Feng et al. [1978] find that Wadati diagrams always appear curved during periods of anomalous V_p/V_s before earthquakes, and thus the curvature of the diagram may be used for observation of the velocity anomaly process before an earthquake.

Apart from China, precursory variations in V_p/V_s determined by using the Wadati diagram have continued to be reported from other areas in recent years, including Iran [Hedayati et al., 1978], New Zealand [Rynn and Scholz, 1978] and the USA [Talwani, 1979]. Precursory changes in V_p/V_s have also been identified using other methods. For example, Evison [1975] finds that the ratio of S- to P-wave travel times from a local earthquake at a single station gives a good estimate of the average V_p/V_s for the wave path, providing the origin time of the event has been moderately well determined. He goes on to show that for three New Zealand earthquakes with $M_L \geq 6.0$, precursory changes in the velocity ratio can be identified and mapped from standard observatory data. A similar approach in the measurement of V_p/V_s at a single station has been adopted by Fedotov et al. [1977]. These authors find that large V_p/V_s variations of up to $\pm 12\%$ tend to precede earthquakes in Kamchatka with $m_b \geq 5.0$ near the same station as was used to measure V_p/V_s. The deviations begin 10-20 days before an earthquake and last for some time after it. In contrast, Feng [1977] uses the records of two stations in the identification of a V_p/V_s anomaly prior to the M = 6.1 Xinfengjiang, China, earthquake.

It has been reported that precursory changes in V_p/V_s in the focal zones of earthquakes are largely due to changes in V_p [e.g. Whitcomb et al., 1973]. Consequently, searches for precursory variations in P-wave travel times from explosions and P-wave residuals of teleseisms have been undertaken, as such variations may reflect local variation of V_p. A refinement of the P-wave residual technique has been the determination of all residuals relative to a reference station within or near the area of interest, thus isolating near station effects from those due to source and path. Recent reports of precursory variations in P-wave residuals include those of Cramer et al. [1977], Johnston [1978], Engdahl and Kisslinger [1979] and Gupta and Singh [1979].

Despite the numerous positive results discussed above, temporal variation in seismic velocities is currently not the highly touted precursor it once was. The reason for this is that many negative results have been reported in searches for such variations, some recent examples being those of Bolt [1977], Peake et al. [1977], Steppe et al. [1977], Chou and Crosson [1978] and Murdock [1978]. Why are temporal changes in seismic velocities apparent before some earthquakes and not others? The answer to this question is not straightforward, and can be approached in two rather different ways. On the one hand, we can take the view that such changes are a real phenomenon, intimately connected with the earthquake preparatory process, and that mixed results in detecting them arise from differences in experimental method, differences in the structure of experimental areas, and differences in anomaly sign, size and shape peculiar to an impending event. On the other hand, we can take the view that temporal changes in seismic velocities are not related to temporal changes in physical rock properties, and that these changes, when observed, merely reflect systematic spatio-temporal seismicity fluctuations in areas of spatially varying rock properties.

Evidence that velocity anomalies observed in the field may reflect physical changes in rock properties is provided by laboratory experiments in rock fracturing. These have shown that the velocities of seismic waves in a rock sample vary with applied stress [e.g. Sobolev et al., 1978; Spetzler, 1978]. However, translation of these results into the field requires a knowledge of whether the typical 10 to 100 bar stress drop during an earthquake occurs in a high ambient stress field of the order of a kilobar required for the observation of dilatancy-induced effects, or in a low ambient stress field of tens to hundreds of bars. Such knowledge of the ambient stress field is currently not available. Stress differences have been used by Johnston [1978] to explain the fact that the velocity anomaly observed prior to the M_s = 7.2 Kalapana, Hawaii, earthquake of 1975 encompassed only a small fraction of the aftershock zone. He contends that the small anomalous region represents a very restricted zone of intense stress build-up where rupture initiated, whereas the rest of the much larger rupture surface remained at a lower stress state throughout the observed precursory period.

Rupture mechanism will have a bearing on the level of ambient stress at which an earthquake occurs. Dung et al. [1979] invoke this to explain the conflicting findings reported in searches for precursory velocity variations. They maintain that re-slippage on pre-existing faults (stick-slip) could be quite different from the fresh fracturing of an intact rock medium (shear rupture). In the former case, the stress level is relatively low, causing little or no dilatancy and velocity variations, while in the latter case the stress level may be much higher, and considerable dilatancy and velocity variations could occur. Supporting evidence for these ideas is provided by Wang [1975], who finds in laboratory experiments that velocities of both P- and S-waves decrease significantly prior to the occurrence of shear rupture, while no change is observed during stick-slip motion. Herein lies an explanation of the negative results obtained in searches for velocity variations prior to earthquakes in the San Andreas fault zone of California. As this zone represents a well-established plate boundary, earthquakes there can reasonably be considered re-slippage on pre-existing faults, and thus are likely to occur at a low ambient stress level. In contrast, many earthquakes showing precursory velocity anomalies, such as the Yanyuan-Ninglang and other Chinese events discussed by Dung et al. [1979], occur in regions far from simple plate boundaries, where only small-scale faults have been found. Dung et al. [1979] postulate that these events may represent fresh fracturing.

Both laboratory and theoretical studies indicate that changes in seismic velocity with stress

in dilatant rock should be substantially anisotropic [eg., Anderson et al., 1974; Hadley, 1975]. Thus if dilatancy-induced seismic velocity variations do occur in the field, then anisotropy should be a prominent feature of the phenomenon. Possible premonitory variations in the anisotropy of S-wave velocities before earthquakes in Nevada, USA, have been reported by Gupta [1973]. Anisotropy is also suggested by the results of Feng [1975], who finds that a correlation seems to exist between the orientation of a velocity anomaly area and the source mechanism of the main shock. The fact that Steppe et al. [1977] find no temporal variations in P-wave velocity anisotropy before three earthquakes with magnitudes between 4 and 5 that occurred on the San Andreas fault in central California may indicate that significant dilatancy did not occur before these events. Even in cases when dilatancy does occur, a velocity anomaly may be disguised by dilatancy anisotropy. For example, Murdock [1978] uses such anisotropy to explain the lack of a velocity anomaly prior to the M_L = 6.0 Yellowstone, Wyoming, earthquake of 1975. He contends that if dilatancy occurred as a prelude to this normal faulting event, the cracks formed were probably mainly vertical, and therefore the waves studied (which travelled upward through the crust) might show no measureable anomaly.

The possibility of anisotropy in dilatancy-induced velocity anomalies highlights the need for intelligent experimental design in the search for such anomalies. Wadati diagrams may not be too limited by such anisotropy, as seismic waves from earthquakes in a dilatant region to stations of an array often sample a wide range of ray paths. In contrast, seismic waves from explosion sources may be affected by anisotropy, as these travel largely horizontal paths. In addition, these waves may be refracted around a lens of dilatant, lower velocity material. Thus it would be prudent to complement seismic refraction studies with seismic reflection and P-wave residual studies, so that a wide range of ray paths is sampled. A further complication that should be considered in designing experiments to detect precursory variations in seismic velocities is the fact that the signature of an anomaly may differ with seismic frequency. This has been pointed out by Raikes [1978], who notes that the size of dilatant cracks may have differing effects on teleseismic and local waves; small cracks will have a larger effect on the velocity of the higher frequency P-waves from local earthquakes.

An alternative approach one can take to precursory changes in seismic velocities is that they do not reflect physical changes in rock properties, but rather are an artefact of sampling. A detailed re-examination by Lindh et al [1978b] of the data for anomalies in travel-time residuals of about 0.2 sec reported to precede two magnitude 5 earthquakes along the San Andreas fault in central California shows that the anomalies were more likely caused by differences in the depth and magnitude of the source earthquakes during the "anomalous" periods, and were unrelated to any premonitory material property changes. Additional data chosen to mimimize such problems show that travel times before the two earthquakes were in fact stable to within a few hundredths of a second for rays that passed within a few kilometres of the hypocentres. These authors stress that earthquake arrival-time residuals can be a function of location, magnitude and fault-plane solution, and that as these parameters cannot be assumed to be drawn from a stationary population, velocity changes inferred from such data must be explicitly shown not to be due to such causes before they constitute evidence for an in situ material property change. Similarly, Lockner and Byerlee [1978] report that a detailed analysis of apparent velocity anomalies observed in the laboratory prior to both failure of intact samples and violent slip in samples containing saw cuts reveals that these anomalies are related to sampling errors resulting from arrivals of small magnitude events being picked late. Sampling problems in the analysis of velocity anomalies have also been discussed by Jackson and Ergas [1979], from a statistical standpoint. These authors illustrate the problems that are caused by an incomplete knowledge of local time-independent structure and the statistical properties of the data.

Time-independent spatial variations of V_p/V_s and travel-time residuals are not uncommon. In early studies of the Garm region of the USSR, Semenov [1969] showed that in addition to temporal V_p/V_s anomalies being present prior to large earthquakes, the region exhibits both lateral and depth variations of V_p/V_s. Lateral and depth variations of V_p/V_s have also been reported by Rynn and Scholz [1978] for the Alpine fault zone and Fiordland region of the South Island of New Zealand, while Healy and Peake [1975], in a study of aftershocks of two moderate earthquakes along

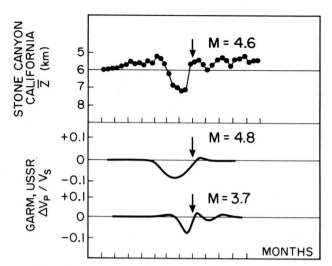

Fig. 6. A comparison of the variation in mean apparent focal depth ($\bar{Z}$) of microearthquakes preceding the 1972 Stone Canyon, California, earthquake with V_p/V_s anomalies at Garm, USSR, preceding moderate earthquakes [after Bufe et al., 1974].

the San Andreas fault in central California, find that V_p/V_s is considerably different on each side of the fault, being 1.79 for stations on the east side and 1.71 for stations on the west side. In addition, marked azimuthal variation of P-wave residuals related to spatial heterogeneity in crustal structure has been reported in California, for both local source events [Steppe et al., 1977] and teleseisms [Raikes, 1978]. Not only is spatial zonation in V_p/V_s and travel-time residuals common, but there is evidence that a close relationship exists between such zones and earthquake zones. For example, Asimov et al. [1979] report that not only can the Frunze region of Kirgizia, USSR, be zoned according to V_p/V_s, but also strong shocks in the region gravitate towards zones with lower values of the ratio (see Figure 5). Similarly, in a study of V_p/V_s zones in the crust and descending lithospheric slab in the northeastern Japan arc, Horiuchi et al. [1977] find that earthquakes are occurring preferentially in regions where the ratio is relatively small.

Given the above observations of spatial variations in V_p/V_s and travel-time residuals, and their relationship to earthquakes, it is clear that precursory velocity anomalies unrelated to temporal changes in rock properties may be observed if larger earthquakes are preceded by some systematic spatio-temporal seismicity pattern. As discussed earlier, evidence for such patterns is mounting. The premonitory vertical migration of microearthquakes before the magnitude 4.6 Stone Canyon, California, event of 1972 reported by Bufe et al. [1974] provides an example of such a pattern. As can be seen in Figure 6, the shape of the average depth anomaly observed for

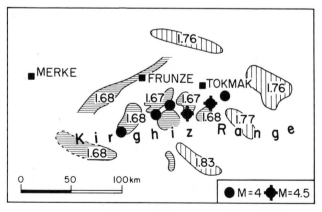

Fig. 5. V_p/V_s zonation and larger earthquakes in the Frunze region of Kirgizia, USSR [after Asimov et al., 1979].

the microearthquakes is very similar to the shapes of V_p/V_s anomalies preceding moderate earthquakes at Garm, USSR, and it is clear that the V_p/V_s variation observed at Garm could result from a precursory variation in microearthquake depths if V_p/V_s varies with depth. Of course, this interpretation is not unique — the observed average depth anomaly can also be attributed to dilatancy biasing of the microearthquake hypocentres.

The mixed results regarding precursory changes in seismic velocities, and the diverse possible explanations of such results, suggest that we are still at an early stage in understanding such phenomena. However, there is no need for discouragement. Even if not related to physical changes in rock properties in some instances, precursory velocity anomalies can still be useful in themselves as earthquake predictors. As the characteristics of a velocity anomaly may be different in different regions, we have little alternative but to continue baseline measurements in regions of interest to determine the reliability of such anomalies as precursors. An example of such a baseline study is that of Calhaem [1979], who has measured teleseismic P-wave residuals at New Zealand seismograph stations for the period 1965-1977. He finds a complex situation, with certain events being well correlated with changes in residuals, others showing no correlation at all, and some changes of residual occurring at times of quiet seismic activity. Similarly, a search for velocity anomalies in the Tsengwen reservoir area of Taiwan by Wu et al. [1979a] reveals anomalies, but these are not precursory to large earthquakes, as the duration of the anomalies would imply. In studying the reliability of velocity anomalies for prediction, we must also isolate background velocity variations not connected with the earthquake preparatory process. Theoretical calculations suggest that earth tides can produce velocity variations of 10^{-5}, but Bungum et al. [1977] report observed variations as large as 10^{-3}. In addition, Clymer and McEvilly [1978] have shown that seasonal variations of up to 1% in travel times in central California measured using a controlled source are a response to changes in water table depth, with a sensitivity of about one millisecond of travel-time change per metre of water table change.

Precursory Changes in Focal Mechanism
and the Amplitude Ratio of P- and S-Waves

One of the many changes in the stress field prior to earthquakes of magnitude 4.5 to 5.7 in the Garm region of the USSR observed by Nersesov et al. [1973] was a rotation of compressional stress axes in the microearthquake background. These axes shifted suddenly by about 100° from three to five months before an earthquake. This result stimulated searches for similar precursory changes in focal mechanism in other areas. Recent reports of such changes include those of Ishida

and Kanamori [1978], Engdahl and Kisslinger [1979], Gir Subash et al. [1979] and Wyss et al. [1979]. At the same time, other searches for precursory focal mechanism changes have produced negative results [e.g. Warren, 1979], while observed changes in the focal mechanisms of small earthquakes have not been followed by a larger event [McNally et al., 1978].

As the amplitude ratio of P- and S-waves is dependent on focal mechanism, measurement of this ratio may also give information on any precursory stress reorientation. However, care must be exercised in the interpretation of results of this approach, as the variation of the amplitude ratio with distance from an event depends not only on focal mechanism, but also on source depth, crustal structure and properties of the crustal medium. Lindh et al. [1978a] prefer to explain changes in the amplitude ratio accompanying three recent California earthquakes in terms of small systematic changes in stress or fault configuration in the source region rather than attenuation changes, because the amplitude ratio changes were accompanied by simultaneous first motion changes at one or more stations, and changes were also noted in the ratio of P-wave amplitudes recorded at two different stations. In contrast, Feng [1974] prefers to explain changes in the amplitude ratio accompanying large earthquakes in China primarily in terms of changes in the attenuation characteristics of the crust. A variation in the use of the amplitude ratio of P- and S-waves for earthquake prediction has been that of Fedotov et al. [1977]. These authors have studied changes in the parameter $\log(E_s/E_p)$, where E_s and E_p are the radiated energies of the short period S- and P-waves. They find that areas for which this parameter for small earthquakes is low may be the sites of future large earthquakes.

As with velocity anomalies, precursory changes in focal mechanism and in the amplitude ratio of P- and S-waves can be explained from two standpoints, one requiring temporal changes in physical rock properties, the other a spatio-temporal seismicity pattern. It should be noted that the dilatancy-diffusion model of earthquake precursors does not per se offer an explanation for precursory changes in focal mechanism. Such changes are suggested, however, by the so-called dilatancy-instability model of earthquake precursors developed by Russian seismologists [e.g. Mjachkin et al., 1975]. Slip on a fault can produce a local rotation of principal stress axes, as the theoretical calculations of Rudnicki [1979] demonstrate. Gupta [1975b] has shown that a possible explanation of precursory changes in focal mechanism can be offered by assuming systematic vertical migration of foreshock activity within a region of thrust tectonics. Furthermore, depth variation of focal mechanism for crustal earthquakes has been reported [e.g. Okano and Kimura, 1979; Reyners, 1980]. Clearly such spatial variation in focal mechanism will produce an apparent precursory variation in focal mechanism if systematic spatio-temporal seismicity patterns precede larger earthquakes, especially if such larger earthquakes occur near the boundaries of regions of differing focal mechanism, as suggested by Grin et al. [1977].

Precursory Changes in b-Value

Statistics on the number of earthquakes as a function of magnitude are often summarized by the well-known frequency-magnitude relation $\log N = a - bM$, where N is the number of earthquakes of magnitude M or greater, and a and b are constants. The parameter b turns out to be useful in describing seismicity, as it measures the relative numbers of smaller and larger events in a given period. Furthermore, there is good evidence from both laboratory studies of rock deformation [e.g. Scholz, 1968b; Weeks et al., 1978] and studies of aftershock sequences [e.g. Bufe, 1970; Gibowicz, 1973] that changes in b-value are inversely related to changes in stress. Temporal variations in b-value may thus be useful in monitoring the earthquake preparatory process. Indeed, many precursory decreases in b-value have been reported, some recent examples being those of Li et al. [1978], Cagnetti and Pasquale [1979] and Guha [1979]. A pattern of b-value variation found by Li et al. [1978] to precede the 1976 Tangshan, China, earthquake (M_s = 7.8) is illustrated in Figure 7. However, not all temporal decreases in b-value are followed by a significant earthquake [e.g. Wyss and Lee, 1973; Robinson, 1978].

As b-value is a measure of seismicity, many of the problems inherent in the identification of precursory seismicity patterns also beset experiments designed to detect precursory b-value variations. Notably, the pattern of b-value variation depends strongly on area of surveillance. For example, Ma [1978], in a study of variations in b-value before several large earthquakes in north China, has found that for smaller areas around the earthquake epicentres b-values vary with time from higher to lower values as the earthquake approaches, while for larger areas peak values of b appear immediately before the earthquakes. The time duration of these peak values seems to be related to the magnitude of the impending event. In addition, Ma finds a spatial variation of b-values, with areas of low b-value in the vicinity of the earthquake epicentres being sur-

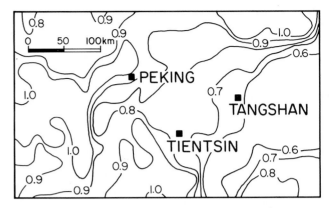

Fig. 8. Spatial variation of b-value in the Peking-Tientsin-Tangshan region for the period 1974.1.1-1975.12.31 [after Li et al., 1978].

rounded by areas of high b-value. An example of spatial variation of b-value is shown in Figure 8. Given the likelihood of spatial variability in b-value, it is possible that some reported temporal variations in b-value are in fact caused by seismicity migrating from one region to another, and thus reflect a precursory spatio-temporal seismicity pattern rather than precursory changes in rock properties [e.g. Udias, 1977].

Other Seismic Precursors

The advent of broadband, digital seismic recording has stimulated searches for precursory variations in seismic spectrum and source parameters in recent years. However, results to date have been mixed. The literature contains reports of increases in the relative high-to-low frequency content of seismic waves recorded prior to larger earthquakes [Sadovsky and Nersesov, 1974; Ishida and Kanamori, 1980], as well as decreases [Fedotov et al., 1972; Gusev et al., 1979]. Other authors have found no appreciable precursory variation in seismic spectrum. For example, in a study of high-quality data for two of the largest earthquakes to occur in central California in recent years, Bakun and McEvilly [1979] find that foreshock radiation is neither universally higher nor lower frequency than that of comparable aftershocks or normal background earthquakes. Similarly, Tsujiura [1978], studying the Izu-Oshima-kinkai, Japan, earthquake of 1978 (M = 7.0), finds no systematic difference in source spectra between foreshocks, aftershocks and normal shocks. He interprets an observed scatter in source dimensions for these events in terms of local spatial variation in source spectrum reflecting the inherent tectonic nature of different areas, rather than the influence of the occurrence of the main shock.

In an analysis of over 1000 local earthquakes in the Garm region, USSR, Rautian et al. [1978] find

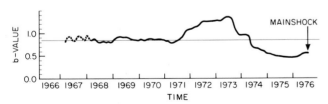

Fig. 7. Variation of b-value with time preceding the 1976 Tangshan, China, earthquake (M_s = 7.8) [after Li et al., 1978].

that a limited amount of data are consistent with a change in spectral content and stress drop of earthquakes in the vicinity of and before stronger earthquakes, but a clear, consistent pattern is not yet evident. In contrast, recent results from China [Zheng et al., 1977; Zhu et al., 1977b, c] indicate that for both the 1975 Haicheng earthquake (M = 7.3) and the 1976 Tangshan earthquke (M = 7.8), the stress drop of foreshocks was larger than that of aftershocks. A similar result is obtained by Sacks et al. [1979] for foreshocks and aftershocks of a magnitude 6.4 transform fault earthquake in Iceland.

Another seismic parameter for which precursory variations have been sought is the P-wave amplitude of teleseisms. Laboratory studies of P-wave propagation in rocks stressed to failure indicate large changes in amplitude which can be correlated with increasing stress [Gowd, 1970]. While Murdock [1978] fails to find an anomaly in P-wave amplitudes preceding the 1975 Yellowstone, Wyoming, earthquake (M_L = 6.0) at two stations within 20 km of the epicentre, Nersesov and Nikolayev [1978] report that sharp temporal variations in the amplitude of teleseismic P-waves precede strong earthquakes in the Garm region of the USSR. The possibility that microseisms might provide precursory information has also been investigated. Preliminary studies by Zhu et al. [1977a] and Niu and Zhu [1978] suggest variations in the amplitude and direction of approach of microseisms may precede larger earthquakes in China.

Discussion

So where do we stand at present in predicting earthquakes using long- and intermediate-term seismic precursors? All we can say with certainty is that many variations in seismic parameters have been reported to precede strong earthquakes, and past experience in recognizing such changes has resulted in the successful real-time prediction of some events. Much less is known about the nature of such changes, and as yet there has been little consensus on the physical basis of earthquake prediction, although there has been no lack of hypotheses.

Clearly earthquake prediction using seismic precursors is still at an early stage. It is thus highly desirable to adopt a broad and open-minded approach in the search for such precursors, for fear of throwing out the baby with the bathwater. The reader may have been surprised by the large number of Chinese and Russian studies discussed above. One reason for this is the philosophy with which earthquake prediction is approached in these countries. Seismologists there do not believe it is essential to understand the physical basis of a precursor before it can be useful for prediction. The overriding recommendation of a precursor is that it works — why it works is a question which can be more fully addressed when more and varied data have been accumulated. Such an observational

approach — in essence "fishing" for anomalies — is perfectly legitimate, given earthquake prediction is still in its wildcat days. Physical models of seismic precursors that have been proposed to date, such as the dilatancy-diffusion model and the dilatancy-instability model, have yet to be shown to be valid or universally applicable, and thus there is a danger in limiting earthquake prediction studies to tests of hypotheses based solely on such models.

Nature does not give up its secrets easily — the only way we will learn these is by continued, systematic observation. Thus studies of time-independent crustal structure and spatio-temporal seismicity patterns should be continued, together with laboratory studies of rock deformation, to increase our understanding of the nature of seismic precursors. In so doing we may come closer to determining the most useful precursor in a given region. In addition, baseline measurements must be continued in areas of interest, to determine the reliability of a given precursor in terms of successes, failures and false alarms. Estimates of precursor reliability are necessary input for any earthquake forecasting scheme which is to be generally useful to public agencies. In determining such reliability, efforts should also be made to isolate background parameter variations not connected with the earthquake preparatory process.

Finally, since large earthquakes cannot (as yet) be made to order in areas of dense seismographic coverage, we must accept that our knowledge of seismic precursors will necessarily be patchy for some time to come.

Acknowledgments. The author appreciates critical reviews of this manuscript by Drs. Bill Ellsworth, Klaus Jacob and David Simpson. Thanks are also due to Linda Murphy for typing the manuscript and Kazuko Nagao for drafting the figures. This work was supported by contract EAR 79-26350 of the National Science Foundation. Lamont-Doherty Geological Observatory Contribution Number 3091.

REFERENCES

Aggarwal, Y. P., and O. Perez, Precursory changes in seismicity, I: Observations (abstract), EOS, Trans. Amer. Geophys. Union, 61, 108, 1980.
Anderson, D. L., B. Minster, and D. Cole, The effect of oriented cracks on seismic velocities, J. Geophys. Res., 79, 4011-4015, 1974.
Asimov, M. S., Zh. S. Yerzhanov, K. Ye. Kalmurzaev, M. K. Kurbanov, G. A. Mavlyanov, S. Kh. Negmatullaev, I. L. Nersesov, and V. I. Ulomov, The state of earthquake prediction research in the Soviet republics of Central Asia, Internat. Symp. on Earthquake Prediction, UNESCO, Paris, 2-6 April 1979, Rept. III-12, 1979.
Bakun, W. H., and T. V. McEvilly, Are foreshocks

distinctive? Evidence from the 1966 Parkfield and the 1975 Oroville, California sequences, Bull. Seismol. Soc. Amer., 69, 1027-1038, 1979.

Bolt, B. A., Constancy of P travel times from Nevada explosions to Oroville Dam station 1970-1976, Bull. Seismol. Soc. Amer., 67, 27-32, 1977.

Brady, B. T., Anomalous seismicity prior to rock bursts: Implications for earthquake prediction, Pure Appl. Geophys., 115, 357-374, 1977.

Bufe, C. G., Frequency-magnitude variations during the 1970 Danville earthquake swarm, Earthquake Notes, 41(3), 3-6, 1970.

Bufe, C. G., J. H. Pfluke, and R. L. Wesson, Premonitory vertical migration of microearthquakes in central California: Evidence of dilatancy biasing, Geophys. Res. Lett., 1, 221-224, 1974.

Bungum, H., Risbo, T., and E. Hjortenberg, Precise continuous monitoring of seismic velocity variations and their possible connection to solid earth tides, J. Geophys. Res., 82, 5365-5373, 1977.

Cagnetti, V., and V. Pasquale, The earthquake sequence in Friuli, Italy, 1976, Bull. Seismol. Soc. Amer., 69, 1797-1818, 1979.

Calhaem, I. M., Long term changes in teleseismic P residuals at N. Z. stations 1965-1977 (abstract), Internat. Union Geodesy Geophys. XVII General Assembly, Canberra, Australia, ICG Abstract 06/12, 1979.

Chou, C. W., and R. S. Crosson, Search for time-dependent seismic P travel times from mining explosions near Centralia, Washington, Geophys. Res. Lett., 5, 97-100, 1978.

Clymer, R., and T. V. McEvilly, Travel-time monitoring in central California with vibroseis — an update (abstract), Earthquake Notes, 49(4), 38, 1978.

Cramer, C. H., C. G. Bufe, and P. W. Morrison, P-wave travel-time variations before the August 1, 1975, Oroville, California earthquake, Bull. Seismol. Soc. Amer., 67, 9-26, 1977.

Dewey, J. W., and W. Spence, Seismic gaps and source zones of recent large earthquakes in coastal Peru, Pure Appl. Geophys., 117, 1148-1171, 1979.

Dung, S. S., H. C. Ge, Y. L. Lo, C. Y. Hsu, and F. C. Wang, Earthquake prediction on the basis of V_p/V_s variations — a case history, Phys. Earth Planet. Inter., 18, 309-318, 1979.

Engdahl, E. R., and C. Kisslinger, Seismological precursors of earthquakes in an island arc (abstract), Phys. Earth Planet. Inter., 18, 349-350, 1979.

Evison, F. F., Determination of precursory velocity anomalies in New Zealand from observatory data on local earthquakes, Geophys. J. Roy. astron. Soc., 43, 957-972, 1975.

Evison, F. F., Fluctuations of seismicity before major earthquakes, Nature, 266, 710-712, 1977a.

Evison, F. F., The precursory earthquake swarm, Phys. Earth Planet. Inter., 15, P19-P23, 1977b.

Evison, F. F., Precursory seismic sequences in New Zealand, N. Z. J. Geol. Geophys., 20, 129-141, 1977c.

Evison, F. F., Long-term seismic precursor to the 1968 Inangahua earthquake, New Zealand, N. Z. J. Geol. Geophy., 21, 531-534, 1978.

Fedotov, S. A., A. A. Gusev, and S. A. Boldyrev, Progress of earthquake prediction in Kamchatka, Tectonophysics, 14, 279-286, 1972.

Fedotov, S. A., G. A. Sobolev, S. A. Boldyrev, A. A. Gusev, A. M. Kondratenko, O. V. Potapova, L. B. Slavina, V. D. Theophylaktov, A. A. Khramov, and V. A. Shirokov, Long- and short-term earthquake prediction in Kamchatka, Tectonophysics, 37, 305-321, 1977.

Feng, D. Y., Anomalies of the amplitude ratio of S and P waves from near earthquakes and earthquake prediction, Acta Geophys. Sinica, 17, 140-154, 1974 (English translation in Chinese Geophysics, 1, 1-15).

Feng, D. Y., Amomalies of seismic velocity ratio before the Yongshan-Daguan earthquake (M = 7.1) on May 11, 1974, Acta Geophys. Sinica, 18, 235-239, 1975 (English translation in Chinese Geophysics, 1, 47-53).

Feng, D. Y., S. H. Zheng, G. Y. Sheng, Z. X. Fu, S. L.Gao, R. M. Luo, and B. C. Li, Preliminary study of the velocity anomalies of seismic waves before and after some strong and moderate earthquakes in western China (1) - the velocity ratio anomalies, Acta Geophys. Sinica, 19, 196-205, 1976.

Feng, D. Y., Z. Y. Wang, J. P. Gu, G. Y. Sheng, and S. L. Gao, Preliminary study of the velocity anomalies of seismic waves before and after some strong and moderate earthquakes in western China (II) — the anomalous regions and their characteristics, Acta Geophys. Sinica, 20, 115-124, 1977.

Feng, D. Y., J. P. Gu, Z. Y. Wang, G. Y. Sheng, R. M. Luo, and K. L. Li, Preliminary study of the velocity anomalies of seismic waves before and after some strong and moderate earthquakes in western China (III) — variations of curvature of the Wadati diagrams, Acta Geophys. Sinica, 21, 292-309, 1978.

Feng, R., On the variations of the velocity ratio before and after the Xinfengjiang reservoir impounding earthquake of magnitude 6.1, Acta Geophys. Sinica, 20, 211-221, 1977 (English translation in Chinese Geophysics, 1, 111-123).

Gibowicz, S. J., Variation of the frequency-magnitude relation during earthquake sequences in New Zealand, Bull. Seismol. Soc. Amer., 63, 517-528, 1973.

Gir Subhash, S. M., Y. Aggarwal, and R. Gir, Precursory changes in seismic activity and stress reorientation prior to earthquakes at Blue Mountain Lake, New York (abstract), EOS, Trans. Amer. Geophys Union, 60, 589, 1979.

Gowd, T. N., Changes in absorption of ultrasonic energy travelling through rock specimens stressed to fracture, Phys. Earth Planet. Inter., 4, 43-48, 1970.

Grin, V. P., Z. A. Medzhitova, and T. Ya. Serebryanskaya, Spatial distribution of the focal parameters of weak earthquakes in the territory of the Chu Basin and their relationship with strong earthquakes, Izv. Acad. Sci. USSR, Earth Physics, 13, 166-173, 1977.

Guha, S. K., Premonitory crustal deformations, strains and seismotectonic features (b values) preceding Koyna earthquakes, Tectonophysics, 52, 549-559, 1979.

Gupta, I. N., Premonitory variations in S-wave velocity anisotropy before earthquakes in Nevada, Science, 182, 1129-1132, 1973.

Gupta, I. N., Premonitory seismic-wave phenomena before earthquakes near Fair View Peak, Nevada, Bull. Seismol. Soc. Amer., 65, 425-437, 1975a.

Gupta, I. N., Precursory reorientation of stress axes due to vertical migration of seismic activity?, J. Geophys. Res., 80, 272-273, 1975b.

Gupta, H. K., and V. P. Singh, Is Shillong region, northeast India, undergoing a dilatancy stage precursory to a larger earthquake? (abstract), Internat. Union Geodesy Geophys. XVII General Assembly, Canberra, Australia, ICG abstract 6/3, 1979.

Gusev, A. A., A. N. Semenov, and L. G. Sinelnikova, The earthquake spectral anomaly estimate by the M_{LH} to m_B relation and its possible application to earthquake prediction, Phys. Earth Planet. Inter., 18, 326-329, 1979.

Hadley, K., Azimuthal variation of dilatancy, J. Geophys. Res., 80, 4845-4850, 1975.

Healy, J. H., and L. G. Peake, Seismic velocity structure along a section of the San Andreas fault near Bear Valley, California, Bull. Seismol. Soc. Amer., 65, 1177-1197, 1975.

Hedayati, A., J. L. Brander, and R. G. Mason, Instances of premonitory crustal velocity ratio changes in Iran, Tectonophysics, 44, T1-T6, 1978.

Horiuchi, S., T. Sato, S. Hori, A. Yamamoto, T. Kono, K. Hashimoto, and E. Murakami, V_p/V_s structure in the crust and upper mantle beneath the Tohoku District, Abstr. Annual Meeting Seismol. Soc. Japan, May 1977, p. 8, 1977.

Ishida, M., and H. Kanamori, The spatio-temporal variation of seismicity before the 1971 San Fernando earthquake, California, Geophys. Res. Lett., 4, 345-346, 1977.

Ishida, M., and H. Kanamori, The foreshock activity of the 1971 San Fernando earthquake, California, Bull. Seismol. Soc. Amer., 68, 1265-1279, 1978.

Ishida, M., and H. Kanamori, Temporal variation of seismicity and spectrum of small earthquakes preceding the 1952 Kern County, California, earthquake, Bull. Seismol. Soc. Amer., 70, 509-527, 1980.

Jackson, D. D., and R. A. Ergas, Statistical analysis of travel-time anomalies, Phys. Earth Planet. Inter., 18, 303-308, 1979.

Johnston, A. C., Localized compressional velocity decrease precursory to the Kalapana, Hawaii, earthquake, Science, 199, 882-885, 1978.

Keilis-Borok, V., L. Knopoff, and C. R. Allen, Long-term premonitory seismicity patterns in Tibet and the Himalayas, J. Geophys. Res., 85, 813-820, 1980a.

Keilis-Borok, V. I., L. Knopoff, and I. M. Rotvain, Bursts of aftershocks, long-term precursors of strong earthquakes, Nature, 283, 259-263, 1980b.

Keilis-Borok, V. I., L. Knopoff, I. M. Rotvain, and T. M. Sidorenko, Bursts of seismicity as long-term precursors of strong earthquakes, J. Geophys. Res., 85, 803-811, 1980c.

Khattri, K., and M. Wyss, Precursory variation of seismicity rate in the Assam area, India, Geology, 6, 685-688, 1978.

Kondratenko, A. M., and I. L. Nersesov, Some results of the study of change in the velocity of longitudinal waves and the relation between the velocities of longitudinal and transverse waves in a focal zone, Tr. Inst. Fiz. Zemli Akad. Nauk. SSSR, 25, 130-150, 1962.

Li, Q. L., J. B. Chen, L. Yu, and B. L. Hao, Time and space scanning of the b-value: A method for monitoring the development of catastrophic earthquakes, Acta Geophys. Sinica, 21, 101-125, 1978.

Lindh, A., G. Fuis, and C. Mantis, Seismic amplitude measurements suggest foreshocks have different focal mechanisms than aftershocks, Science, 201, 56-59, 1978a.

Lindh, A. G., D. A. Lockner, and W. H. K. Lee, Velocity anomalies: An alternative explanation, Bull. Seismol. Soc. Amer., 68, 721-734, 1978b.

Lockner, D. A., and J. D. Byerlee, Velocity anomalies: An alternative explanation based on data from laboratory experiments, Pure Appl. Geophys., 116, 765-772, 1978.

Lukk, A. A., and I. L. Nersesov, Character of temporal changes in the velocities of elastic waves in the earth's crust of the Garm region, Izv. Acad. Sci., USSR, Earth Physics, 14, 387-396, 1978.

Ma, H. C., Variations of the b-values before several large earthquakes occurred in north China, Acta. Geophys. Sinica, 21, 126-141, 1978.

Mârza, V. I., The March 4, 1977 Vrancea earthquake seismic gap, Bull. Seismol. Soc. Amer., 69, 289-291, 1979.

McNally, K., Patterns of earthquake clustering preceding moderate earthquakes, central and southern California (abstract), EOS, Trans. Amer. Geophys. Union, 58, 1195, 1977.

McNally, K. C., H. Kanamori, J. C. Pechmann, and G. Fuis, Earthquake swarm along the San Andreas fault near Palmdale, southern California, 1976 to 1977, Science, 201, 814-817, 1978.

Mizoue, M., M. Nakamura, Y. Ishiketa, and N. Seto, Earthquake prediction from microearthquake observation in the vicinity of Wakayama City, north western part of Kii Peninsula, central Japan, J. Phys. Earth, 26, 397-416, 1978.

Mjachkin, V. I., W. F. Brace, G. A. Sobolev, and J. H. Dieterich, Two models for earthquake forerunners, Pure Appl. Geophys., 113, 169-181, 1975.

Mogi, K., Study of elastic shocks caused by the fracture of heterogeneous materials and its relations to earthquake phenomena, Bull. Earthquake Res. Inst, Tokyo Univ., 40, 125-173, 1962.

Murdock, J. N., Travel time and spectrum stability in the two years preceding the Yellowstone, Wyoming, earthquake of June 30, 1975, J. Geophys. Res., 83, 1713-1717, 1978.

Nersesov, I. L., and A. V. Nikolayev, Time variation of fluctuations of amplitude of teleseismic P-waves measured at the Garm earthquake — forecasting and test area, Doklady Acad. Nauk. SSSR, Earth Sciences, 232, 21-24, 1978.

Nersesov, I. L., A. A. Lukk, V. S. Ponomarev, T. G. Rautian, B. G. Rulev, A. N. Semenov, and I. G. Simbireva, Possibilities of earthquake prediction, exemplified by the Garm rgion of the Tadzhik SSR, in Earthquake Precursors, ed. M. A. Sadovsky, I. L. Nersesov, and L. A. Latynina, Acad. Sci. USSR, Moscow, 216 pp., 1973.

Niu, Z. R., and C. Z. Zhu, Preliminary study of the microseismis in relation to earthquakes (II), Acta Geophys. Sinica, 21, 325-331, 1978.

Ohtake, M., T. Matumoto, and G. V. Latham, Seismicity gap near Oaxaca, southern Mexico as a probable precursor to a large earthquake, Pure Appl. Geophys., 115, 375-385, 1977.

Ohtake, M., T. Matumoto, and G. V. Latham, Patterns of seismicity preceding earthquakes in Central America, Mexico and California, in Methodology for Identifying Seismic Gaps and Soon-to-Break Gaps, U.S. Geol. Surv. Open-File Rept. 78-943, 585-610, 1978.

Oike, K., Seismic activity and crustal movement on the Yamasaki fault in southwest Japan, Phys. Earth Planet. Inter., 18, 341-344, 1979.

Okano, K., and S. Kimura, Seismicity characteristics in Shikoku in relation to the great Nankaido earthquakes, J. Phys. Earth., 27, 373-381, 1979.

Peake, L. G., J. H. Healy, and J. C. Roller, Time variance of seismic velocity from multiple explosive sources southeast of Hollister, California, Bull. Seismol. Soc. Amer., 67, 1339-1354, 1977.

Raikes, S. A., The temporal variation of teleseismic P-residuals for stations in southern California, Bull. Seismol. Soc. Amer., 68, 711-720, 1978.

Rautian, T. G., V. I. Khalturin, V. G. Martynov, and P. Molnar, Preliminary analysis of the spectral content of P and S waves from local earthquakes in the Garm, Tadjikistan region, Bull. Seismol. Soc. Amer., 68, 949-971, 1978.

Reyners, M., A microearthquake study of the plate boundary, North Island, New Zealand, Geophys. J. Roy. astron. Soc., 63, 1-22, 1980.

Rikitake, T., Earthquake precursors, Bull. Seismol. Soc. Amer., 65, 1133-1162, 1975.

Rikitake, T., Classification of earthquake precursors, Tectonophysics, 54, 293-309, 1979.

Robinson, R., Seismicity within a zone of plate convergence — the Wellington region, New Zealand, Geophys. J. Roy. astron. Soc., 55, 693-702, 1978.

Rudnicki, J. W., Rotation of principal stress axes caused by faulting, Geophys. Res. Lett., 6, 135-138, 1979.

Rynn, J. M. W., and C. H. Scholz, Study of the seismic velocity ratio for several regions of the South Island, New Zealand: Evaluation of regional t_s/t_p and near source V_p/V_s values, Geophys. J. Roy. astron. Soc., 53, 87-112, 1978.

Sacks, I. S., A. T. Linde, and R. Stefansson, Stress field changes during a tectonic episode in northern Iceland (abstract), EOS, Trans. Amer. Geophys. Union, 60, 738, 1979.

Sadovsky, M. A., and I. L. Nersesov, Forecasts of earthquakes on the basis of complex geophysical features, Tectonophysics, 23, 247-255, 1974.

Scholz, C. H., Experimental study of the fracturing process in brittle rock, J. Geophys. Res., 73, 1447-1454, 1968a.

Scholz, C. H., The frequency-magnitude relation of microfracturing in rock and its relation to earthquakes, Bull. Seismol. Soc. Amer., 58, 399-415, 1968b.

Sekiya, H., Anomalous seismic activity and earthquake prediction, J. Phys. Earth, 25S, 85-93, 1977.

Semenov, A. M., Variations in the travel-time of transverse and longitudinal waves before violent earthquakes, Izv. Earth Phys., 4, 245-248, 1969 (English translation).

Sobolev, G. H., H. Spetzler, and B. Salov, Precursors to failure in rocks while undergoing anelastic deformations, J. Geophys. Res., 83, 1775-1784, 1978.

Spetzler, H., Seismic velocity changes during fracture and frictional sliding, Pure Appl. Geophys., 116, 732-742, 1978.

Steppe, J. A., W. H. Bakun, and C. G. Bufe, Temporal stability of P-velocity anisotropy before earthquakes in central California, Bull. Seismol. Soc. Amer., 67, 1075-1090, 1977.

Talwani, P., An empirical earthquake prediction model, Phys. Earth Planet. Inter., 18, 288-302, 1979.

Tsai, Y. B., T. Q. Lee, and Z. S. Liaw, A study of microearthquake activity preceding some moderate earthquakes in eastern Taiwan (abstract), EOS, Trans. Amer. Geophys. Union, 60, 884, 1979.

Tsujiura, M., Spectral analysis of seismic waves for a sequence of foreshocks, main shock and aftershocks: The Izu-Oshima-kinkai earthquake of 1978, Bull. Earthquake Res. Inst. Tokyo Univ., 53, 741-759, 1978.

Udias, A., Time and magnitude relations for three microaftershock series near Hollister, California, Bull. Seismol. Soc. Amer., 67, 173-185, 1977.

Wang, C. Y., Variations of V_p and V_s in granite

permonitory to shear rupture and stick-slip sliding: Application to earthquake prediction, Geophys. Res. Lett., 2, 309-311, 1975.

Wang, Z. G., Frequency of aftershocks of the 1966 Xingtai earthquake and the subsequent strong earthquakes of north China, Acta Seismol. Sinica, 1, 150-153, 1979.

Ward, P. L., Earthquake prediction, Rev. Geophys. Space Phys., 17, 343-353, 1979.

Warren, D. H., Fault-plane solutions for microearthquakes preceding the Thanksgiving Day, 1974, earthquake at Hollister, California, Geophys. Res. Lett., 6, 633-636, 1979.

Weeks, J., D. Lockner, and J. Byerlee, Changes in b-values during movement on cut surfaces in granite, Bull. Seismol. Soc. Amer., 68, 333-341, 1978.

Whitcomb, J. H., J.D. Garmany, and D. L. Anderson, Earthquake prediction: Variation of seismic velocities before the San Francisco earthquake, Science, 180, 632-635, 1973.

Wu, F. T., Y. H. Yeh, and Y. B. Tsai, Seismicity in the Tsengwen reservoir area, Taiwan, Bull. Seismol. Soc. Amer., 69, 1783-1796, 1979a.

Wu, J. Y., S. J. Yu, and S. Y. He, The correlation of earthquake occurrence between northeastern China and Japan, Acta Geophys. Sinica, 22, 415-438, 1979b.

Wu, K. T., M. S. Yue, H. Y. Wu, S. L. Chao, H. T. Chen, W. Q. Huang, K. Y. Tien, and S. D. Lu, Certain characteristics of the Haicheng earthquake (M = 7.3) sequence, Acta Geophys. Sinica, 19, 95-109, 1976 (English translation in Chinese Geophysics, 1, 289-308).

Wyss, M., and R. E. Habermann, Seismic quiescence precursory to a past and a future Kurile Island earthquake, Pure Appl. Geophys., 117, 1195-1211, 1979.

Wyss, M., and W. H. K. Lee, Time variations of the average earthquake magnitude in central California, Stanford Univ. Publ. Geol. Sci., 13, 24-42, 1973.

Wyss, M., R. E. Habermann, and A. C. Johnston, Long term precursory seismicity fluctuations, in Methodology for Identifying Seismic Gaps and Soon-to-Break Gaps, U.S. Geol. Surv. Open-File Rept. 78-943, 869-894, 1978.

Wyss, M., A. C. Johnston, and S. M. Ihnen, Precursors to the 1975 Hawaii earthquake (abstract), EOS, Trans. Amer. Geophys. Union, 60, 589, 1979.

Zheng, Z. Z. , Z. C. Hu, Y. P. Guo, and R. J. Wang, Spectral changes of fore- and after-shocks of the Haicheng earthquake, Acta Geophys. Sinica, 20, 125-130, 1977.

Zhu, C. Z., M. S. Fang, Z. W. An, and S. L. Luo, Preliminary study of the microseisms in relation to earthquakes, Acta Geophys. Sinica, 20, 20-32, 1977a.

Zhu, C. Z., C. H. Fu, and S. L. Luo, Source parameters for small earthquakes before and after the M = 7.8 Tangshan earthquake, Acta Geophys. Sinica, 20, 264-269, 1977b (English translation in Chinese Geophysics, 1, 353-359).

Zhu, C. Z., C. H. Fu, Z. K. Jung, and S. L. Luo, Source parameters for small earthquakes and the Q of the medium before and after the Haicheng earthquake, Acta Geophys. Sinica, 20, 222-231, 1977c (English translaton in Chinese Geophysics, 1, 125-137.

CHANGES IN THE SEISMICITY AND FOCAL MECHANISM OF SMALL EARTHQUAKES
PRIOR TO AN M_S 6.7 EARTHQUAKE IN THE CENTRAL ALEUTIAN ISLAND ARC

Selena Billington*, E. R. Engdahl**, and Stephanie Price***

*Cooperative Institute for Research in Environmental Sciences (CIRES),
Box 449, University of Colorado/NOAA, Boulder CO 80309
**U. S. Geological Survey, Mail Stop 967, Box 25046, Federal Center,
Denver CO 80225
***ENSCO, Inc., P. O. Box 2578, Indian Harbour Beach, FL 32937

Abstract. On November 4 1977, a magnitude M_S 6.7 (m_b 5.7) shallow-focus thrust earthquake occurred in the vicinity of the Adak seismographic network in the central Aleutian island arc. The earthquake and its aftershock sequence occurred in an area that had not experienced a similar sequence since at least 1964. About 13 1/2 months before the main shock, the rate of occurrence of very small magnitude earthquakes increased abruptly in the immediate vicinity of the impending main shock. To search for possible variations in the focal mechanism of small events preceding the main shock, a method was developed that objectively combines first-motion data to generate composite focal-mechanism information about events occurring within a small source region. The method could not be successfully applied to the whole study area, but the results show that starting about 10 1/2 months before the November 1977 earthquake, there was a change in the mechanism of small- to moderate-sized earthquakes in the immediate vicinity of the hypocenter and possibly in other parts of the eventual aftershock zone, but not in the surrounding regions.

Introduction

On November 4, 1977 an M_S 6.7 (m_b 5.7) earthquake occurred near Adak Island in the central Aleutian island arc (U. S. Geological Survey Preliminary Determination of Epicenters parameters: origin time 09h:52m:55.7s; 51.659°N.; 175.952°W.; 33 km depth). This paper reports the results of a successful search through data from a local seismographic network for possible seismological precursors to this earthquake.

The Adak network (Figure 1) consists of thirteen short-period two-component stations in addition to the six-component station ADK, which

has been operating since 1965. Eight of the network stations were installed in July 1974; three more stations were installed in 1975 and the remaining two were installed in 1976 and 1977. Each station is instrumented with vertical-component and east-west horizontal-component seismometers. Peak magnifications of over 500K are obtained at a frequency of 10 to 12 Hz.

The Aleutian island arc marks the site of the subduction of the Pacific lithospheric plate under the North American lithospheric plate. The seismicity in the vicinity of Adak Island (Figure 1; see also Engdahl, 1977) results from the convergence of these plates. The epicenters in Figure 1 were determined from data taken from the local network. These events are of about m_b 2.7 or greater and occurred between August 1974 and December 1978. While seismicity of all depths is shown in Figure 1, the shallow seismicity of the thrust zone between the two plates is seen as the band of seismicity south of and parallel to the island arc. The dense cluster of shallow seismicity just east of 176°W. longitude marks the most active part of the aftershock sequence to the November 1977 earthquake.

The results of this study of precursory phenomena to a moderate-sized event are presented in the next two sections. The first section discusses in detail the distribution of seismicity in space and time. This includes teleseismically detected historical earthquakes in the entire Adak region, smaller earthquakes in the area of network coverage to the southeast, and finally very small events in the eventual aftershock zone. The second section discusses variations over time in the focal mechanisms of earthquakes which occurred in that part of a 7600 km^2 region that includes the main shock and aftershock zone (approximately 525 km^2) and where network coverage is sufficient to resolve differences in the mechanism.

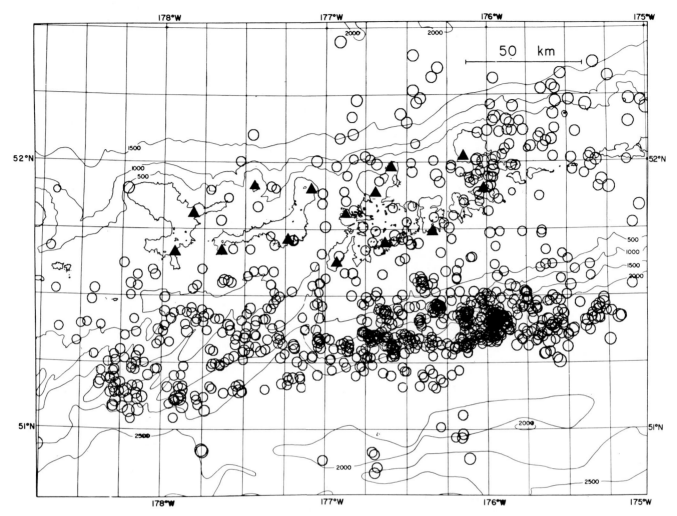

Figure 1. Map view of seismicity from August 1974 through December 1978 in which only free depth solutions for events with duration magnitude greater than or equal or 2.2 (about m_b 2.7) are plotted. Bathymetric contours are at 500 fathom intervals (1 fathom = 1.83 meters). Two-component seismographic stations of the Adak local network are indicated by filled triangles.

Seismic Activity in Space and Time

Teleseismically detected earthquakes in the area of local network coverage area

A study of the distribution of teleseismically detected earthquakes in the immediate vicinity of the Adak area [Engdahl, 1978; Dewey et al., 1979] was updated using two types of hypocenter data. First, for the time period from 1964 until the July 1974 installation of the Adak seismographic network, shallow-depth teleseismically-detected earthquakes were relocated by the Master Event method. This method assumes that within a source region limited in lateral extent and depth the travel-time bias to a teleseismic receiving station, caused primarily by the higher seismic velocity within the downgoing lithospheric slab, will be the same for each event occurring in that source region. Thus, if residuals for a well-located (master) event are applied as station corrections to a poorly located event in the same region, the source biases will be reduced and the location of this second event will be improved considerably relative to the master event. The master events for this study were the same two as were used in the relocation of a sequence of events which occurred in 1971 in the Adak Canyon region southwest of Adak Island [LaForge and Engdahl, 1979]: one an m_b 5.0 event which occurred February 22, 1976 at a depth of 25 km; and the other an m_b 5.4 event which occurred March 12, 1975 at a depth of 19 km. All of the events selected for relocation were m_b 4.5 (USGS) or greater.

Second, for the time period from August 1974 through 1979, the USGS PDE data were searched for shallow-focus thrust-zone events occurring within

2 degrees of the Adak network. From these events, all of those which had been routinely located by the local network were selected. The magnitudes of the selected events are all m_b 4.0 (USGS) or greater. The network hypocenters for these events comprise the second set of data used in this study.

The two sets of hypocenter data for shallow-focus teleseismically-detected earthquakes which have occurred since 1964 near Adak are shown in Figure 2. This figure is a plot of the distribution of events in time as a function of distance along the island arc. This figure shows that in the Adak region, the largest earthquakes often occur in sequences. These sequences seem to occur in adjoining but not overlapping regions, and also in regions that have not previously experienced similar earthquake sequences since at least 1964. In particular, the November 4, 1977

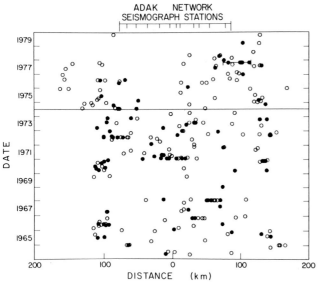

Figure 2. Plot of the occurrence of earthquakes from the Adak region over time as a function of distance along the arc. Distance is arc length with respect to a center of curvature defined by the trench and volcanic arc, referenced at a radius approximately at the band of shallow seismicity. Data are shallow-depth teleseismically-detected earthquakes. Before August 1974, the locations are from master event relocations of earthquakes with m_b greater or equal to 4.5. After August 1974, the locations are from routine network determinations and the magnitudes are all above m_b 4.0. Filled points are for events of m_b greater than or equal to 5.0. The aftershock area for the November 1977 earthquake is shown as a horizontal line segment. The projection of the location of the Adak seismographic network is shown above the seismicity plot.

earthquake and its largest immediate aftershocks (m_b 5.2, 5.4, and 5.3) fill a large portion of one such region along the thrust zone. These observations may extend the concept of seismic gaps developed for large earthquakes [eg: McCann et al., 1978] to the case of seismic gaps of smaller dimension that seem to precede moderate-sized earthquakes.

Earthquakes in the area southeast of Adak Island

The M_S 6.7 earthquake of November 4, 1977 occurred in the southeast portion of the area monitored by the Adak seismographic network (Figure 1; near 176.0°W. longitude and 51.4°N. latitude). The shallow seismicity in the southeast area alone is seen in Figure 3. Epicenters on this map are for events which occurred from August 1974 through June 1978, had magnitudes of about m_b 2.7 or greater, and had depths less than 50 km. The magnitude threshold was imposed on the data to ensure a hypocenter data set which is complete at least above the threshold despite annual variations in background noise (due to weather) and despite a time period in which the effective network was halved due to wind damage.

Earthquakes throughout the Adak area tend to occur in spatial clusters (Figures 1 and 3), at least for the time period studied. Using these natural clusters to describe the seismicity, we defined small regions that encompass individual clusters. The small regions, which we used to classify the seismicity in the southeastern portion of the network coverage area, are shown in Figure 3. The epicenter of the November 1977 main shock is shown on the map as a solid square; it is located on the boundary between the small source regions which we term SE2 and SE4. Aftershocks of this earthquake occurred only in the regions we term SE2, SE4, and SE6. These three regions might, therefore, provide likely sites for the possible occurrence of seismological precursors. However, the level of activity in both the SE4 and SE6 regions was relatively low compared to that in the SE2 region during the three years preceding the main shock, so that SE2 proved to be the best region in which to search for precursors.

The filled triangles in Figure 3 represent epicenters of earthquakes with m_b of 4.5 or greater that occurred before the main shock of November 1977. The one such earthquake in the SE2 region occurred in September 1976 only 4 km from the hypocenter of the eventual main shock. The heavy open triangles in Figure 3 represent epicenters of earthquakes with m_b 4.5 or greater which occurred after the main shock. Of particular interest are the three moderate-sized aftershocks in the SE6 region. These three events all had magnitudes of 5.0 or greater and all occurred within 29 hours of the main shock; the first occurred only 10 minutes after the main shock. Indeed, all of the aftershock activity in the east half of the SE6 region was over by 36

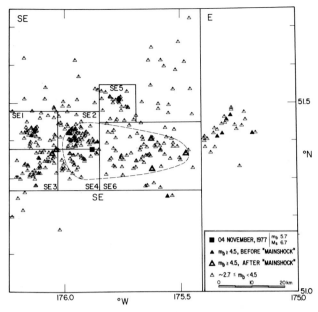

Figure 3. Map view of seismicity in the southeastern portion of the network coverage area for the time period from August 1974 through June 1978. Triangles mark events with m_b greater than or equal to about 2.7. The filled square is the epicenter of the November 4, 1977 main shock, and the dashed line shows the aftershock zone. Epicenters of events with m_b greater than or equal to 4.5 which occurred before the main shock are marked by filled triangles; those which occurred after the main shock are marked by heavy open triangles. One m_b 5.0 earthquake, which probably occurred in the map area in August 1976, was not locatable and is not shown; it occurred when one of the two Develocorders was off and the galvonometer traces on the other were being interchanged. The rectangles define small regions used to classify small spatial clusters of seismicity.

hours after the main shock. This same area had also been inactive above a threshold magnitude of about m_b 2.7 since August 1974 and was inactive following the aftershock activity through June 1978.

Small earthquakes in the epicentral region

The region that we term SE2 was the only one which had aftershocks and that had also been moderately active in the years before the main shock. Therefore one of the aspects of the search for seismological precursors to the earthquake of November 4, 1977 was the determined eking out of hypocenters for every possible event in the SE2 region, including events which are normally considered too small to locate and that have

magnitudes well below the annual average magnitude threshold [Price, Master's Thesis, in preparation, 1981].

One result of the study of the detailed seismicity in the SE2 region is seen in Figure 4. This plot of the cumulative number of earthquakes as a function of time shows that starting about 13 1/2 months before the November 4, 1977 earthquake, the rate of occurrence of small magnitude events increased abruptly in the immediate vicinity of the impending main shock. Unfortunately, in late October 1977, half of the seismological network was downed by an unusually severe storm, so that the data base for this study could not be extended beyond the November 1977 main shock, although it could be kept complete above a threshold magnitude of about m_b 2.7. Without the data for the rate of small magnitude earthquakes that occurred after the November 1977 earthquake, it is not possible to resolve whether the rate increase in September 1976 was associated with the beginning of an abnormally high rate before the main shock or with the end of an abnormally low rate which may have started in mid-1975. The difference in rate between the time periods of August 1974 through August 1976 and

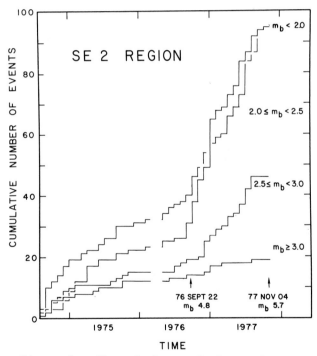

Figure 4. Plot of the cumulative number of events (by half-magnitude ranges) as a function of time for the SE2 region. The gap in the data during the spring of 1976 was caused by a lightning strike which temporarily downed the network. The m_b 4.8 earthquake of September 22, 1976 occurred only 4 km from the hypocenter of the main shock on November 4, 1977.

September 1976 through October 1977 is significant, according to the t-test, at the 99% confidence level. The increased rate started at about the same time that an m_b 4.8 earthquake occurred (September 1976), which, as noted above, was located 4 km from the hypocenter of the main shock. To ensure that this change was not artifactual, a special effort was made to confirm that the network detection or location capability did not increase in late 1976. Also, because the increased rate of activity occurred over an entire year, the variation in rate cannot be attributed to noise due to annual weather cycles. Therefore it seems likely that the observed change of rate is real and may be precursory to the November 1977 earthquake.

Seismicity rate changes have been reported before many earthquakes, including an m_b 5.0 earthquake which occurred in 1976 in the region due south of Adak Island [Engdahl and Kisslinger, 1977]. Engdahl and Kisslinger observed a period of decreased seismicity, beginning about 4 1/2 months prior to this earthquake. If the period of decreased seismicity beginning about 2 1/2 years before the November, 1977 earthquake was real, then this period may be analogous to the rate decrease reported for the smaller earthquake.

Focal Mechanisms

Precursory changes in focal mechanism have been reported before several moderate-sized earthquakes [eg: Nersesov et al., 1973; Engdahl and Kisslinger, 1977; see also Lindh et al., 1978; Warren, 1979]. Of all of the events shown in Figure 3, only for the magnitude M_s 6.7 main shock of November 4, 1977 can a teleseismic focal mechanism solution be reliably determined. The solution [R. LaForge and J. Pohlman, personal communication, 1980] is a thrust mechanism, similar to solutions for other shallow-focus earthquakes in the central Aleutian island arc [eg: Stauder, 1968, 1972; LaForge and Engdahl, 1979]. In order to search for possible precursory variations in the mechanism of the earthquakes mapped in Figure 3, it is necessary to rely solely upon local network data. For this study, the P-wave first motions for all of the events mapped in Figure 3 were carefully re-read from Develocorder records of thirteen vertical-component seismometers of the Adak network, and the polarities of the stations verified independently from teleseismic data. Unfortunately, an average of less than six clear P-wave first motions are available for each event, and in addition, only a small part of the focal sphere is sampled by the local stations. Therefore, for any one event, there is not enough information to uniquely determine a focal-mechanism solution based on local P-wave first-motion data alone. There are also no additional data, such as S-wave first-motions, and it so happens that the ratio of P-wave to SV-wave amplitudes [Kisslinger, 1980], at the one station for which we routinely record

these values, does not clearly differentiate between the particular thrust focal mechanisms interpreted for events in this region.

One approach to overcome these difficulties is to generate composite focal-mechanism solutions. A method was developed for this study that objectively combines first-motion data to generate composite focal mechanism information [Billington, in preparation, 1981]. The method is based on a comparison of the set of observed first-motion data to each of the 2^{13} (= 8192) mathematically possible patterns of first motions for the thirteen stations of the Adak network. It turns out that, for each of the five smallest regions shown in Figure 3, only a few of the 2^{13} mathematically possible patterns are well matched by the observed data. Moreover, these special patterns are compatible with physically plausible focal-mechanism solutions, and different sets of events are associated with each of the few special patterns. Generally, the data for each of these few special patterns is insufficient to uniquely determine a solution, but it can clearly be shown that the different special patterns of first motions represent different mechanisms.

Since different sets of events are associated with each of the few special patterns, an immediate product of this method is obviously a sorting of events according to the focal mechanism with which they are associated. Sorting events within small regions by mechanism allows a search for possible variations in mechanism as a function of time. Time-space plots of events with different mechanisms in the SE2 region are shown in Figure 5. The left column shows the distribution in time of all of the events in the SE2 region that have magnitudes of about m_b 2.7 or greater (Figure 3). The other four columns in Figure 5 each show a subset of these events. The three central columns represent the temporal distribution of events associated with the three special patterns of first motions determined for the SE2 region, and the column on the right includes all remaining events in SE2 that could not be associated with any particular pattern or focal mechanism.

The first distinct pattern of interest is called the dilatation pattern because only dilatational first motions are recorded at the local network for the events associated with this pattern. The pattern is compatible with the thrust mechanism expected from the direction of convergence of the Pacific and North American lithospheric plates and with the thrust mechanism solutions determined teleseismically for larger earthquakes in the central Aleutians. About half of the events in Figure 3 are associated with the this pattern of first motions, and as seen in Figure 5, events which fit this pattern occurred nearly continuously over the time period studied.

For two reasons, events which fit the dilatation pattern were treated separately from the rest of the events in each region. First, the main interest in this study was in changes in

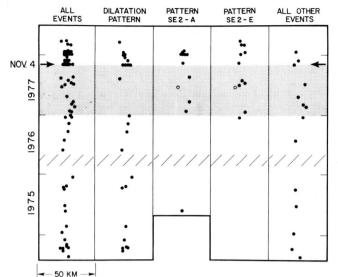

| ALL EVENTS | DILATATION PATTERN | PATTERN SE 2 - A | PATTERN SE 2 - E | ALL OTHER EVENTS |

|— 50 KM —|

Figure 5. Five time-space plots for events located in the SE2 region. The vertical axis is time from August 1974 through June 1978. Each of the horizontal axes is distance from west on the left to east on the right. The hatched stripe in early 1976 represents a time period in which no data were collected due to a lightning strike. The time-space plot in the left column shows all events in SE2 with m_b of about 2.7 or greater. The three central columns show subsets of events in SE2, corresponding to the first-motion patterns with which the events are associated. The open circles show events which might be associated with either pattern SE2-A or SE2-E. The column on the right shows events which fit none of the three special first-motion patterns. The stippled region shows the time period in which the mechanism of most events in the SE2 region changed from the typical background mechanism (dilatation pattern) to different mechanisms (patterns A and E). Stations which are critical for determining pattern SE2-A were not installed until the summer of 1975.

mechanism from the regular background one. Second, the relatively large number of these events with a simple first-motion pattern nearly obscured the other (more interesting) patterns. The events shown in the three columns on the right in Figure 5 were sorted by the above described method after the removal of the dilatational-type events from the data set. Because of missing data at critical (possibly compressional) stations, some of the removed dilational-type events might be associated with the other special patterns instead. The only dilatation-type event occurring in the SE2 region in the 10 1/2 months prior to the main shock is one such event.

In contrast to the events which have been associated with the dilatation pattern, events

associated with the two other special patterns found in the SE2 region are seen to occur mostly after the beginning of 1977. During this same time period, there is also an unusual lack of events which have the background dilatation pattern. These comparisons show that starting about 10 1/2 months before the November 1977 earthquake, the mechanism of most small to moderate-sized events in the immediate vicinity of the epicenter changed.

Similar space-time plots are shown in Figure 6 for events occurring in regions SE1, SE3, SE4, and SE5. An examination of these plots suggests that the change in focal mechanism before the November 4, 1977 main shock was localized in space. For example, there is no apparent evidence of a change in mechanism in the SE5 region (Figure 6d) during the time period in which a change was observed in the SE2 region; there is also no conclusive change during that time period in the SE3 region (Figure 6b). However, during the last two-thirds of the time period of interest, it appears that earthquakes in the SE4 region (Figure 6c) occurred with special patterns of first motion rather than with the background dilatation one. The only other place and time period in which a change in the mechanism of earthquakes is observed is in the SE1 region (Figure 6a) during the first six months of 1977. This period, in which only unusual mechanisms occurred, was terminated by an earthquake of m_b 5.0 in the SE1 region (filled triangle in Figure 3).

To summarize the variations seen in the space-time plots, no change in mechanism associated with the November 4, 1977 M_s 6.7 earthquake is seen in regions SE1, SE3, or SE5. None of these regions was later part of the aftershock zone. A clear change in the mechanism of most small to moderate-sized earthquakes is seen in the SE2 region before the November 1977 main shock, and also perhaps for a shorter time period in the SE4 region. Aftershocks to the M_s 6.7 main shock occurred only in the SE2, SE4, and SE6 regions. This suggests that the zone over which the mechanism of earthquakes changed before the November 1977 earthquake might be the same area as the eventual aftershock zone. Unfortunately, the method to objectively composite first-motion data for focal mechanism information was not successful in the SE6 region because the region is relatively large, because there were very few earthquakes occurring in this region before November 1977, and because of the greater distance from this region to the seismographic network.

If there is much variation in the depths of earthquakes considered together in the analysis, then the resulting differences in take-off angles could result in differences in the observed first-motion patterns. However, given the hypocentral depth resolution of the network and the small range of focal depths in the thrust zone, a clear correlation between the different mechanisms and depth of earthquakes could not be distinguished.

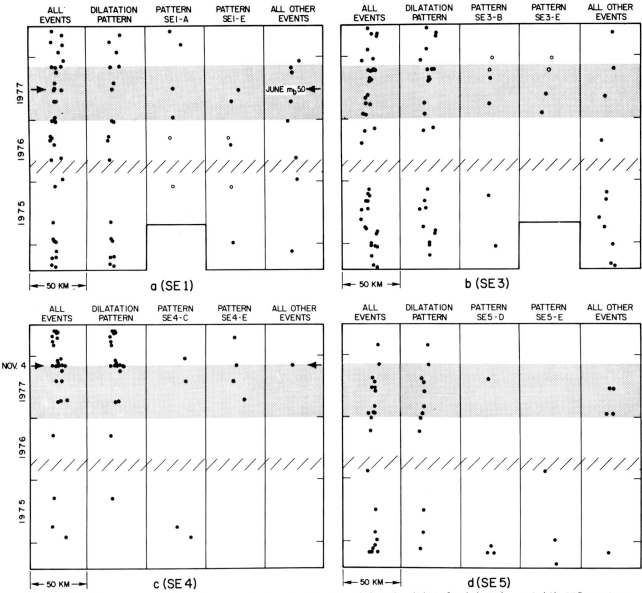

Figure 6. Time-space plots for events located in the (a) SE1, (b) SE3, (c) SE4, and (d) SE5 regions. The caption to Figure 5 describes the plots. In each of these plots, the stippled portion marks the time period in which the change in mechanism is seen in the SE2 region. Stations which are critical for determining patterns SE1-A and SE3-E were not installed until the summer of 1975.

We conclude that a real change in mechanism occurred, independent of focal depth.

The nature of the change in mechanism is seen by looking at the first-motion data for each of the special patterns for regions SE1 - SE5 on lower hemisphere projections of the focal sphere (Figure 7). For any one of the special patterns of first motions, there is insufficient data to uniquely determine a focal mechanism solution. However, the dilatation pattern for each of the regions apparently represents the same mechanism, which is probably a thrust mechanism similar to

the composite focal-mechanism solution determined by LaForge and Engdahl (1979) for earthquakes in the Adak Canyon area to the west of the region in this study. There is also a satisfying correlation between the special patterns in each region which have been labeled pattern E. Apparently the same focal mechanism is shared by all of the events associated with the various patterns labeled E. The E patterns themselves differ from region to region, of course, because the azimuth and take-off angles to the stations of the network differ from region to region. This

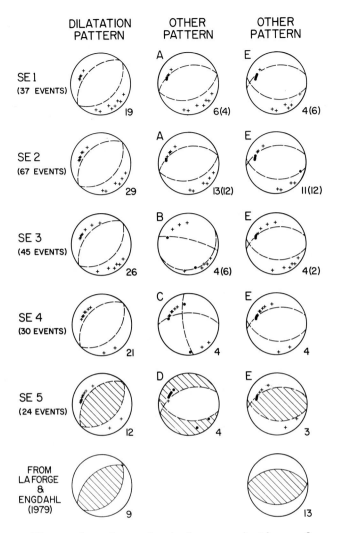

DILATATION PATTERN | OTHER PATTERN | OTHER PATTERN

SE 1 (37 EVENTS) 19 A $6(4)$ E $4(6)$

SE 2 (67 EVENTS) 29 A $13(12)$ E $11(12)$

SE 3 (45 EVENTS) 26 B $4(6)$ E $4(2)$

SE 4 (30 EVENTS) 21 C 4 E 4

SE 5 (24 EVENTS) 12 D 4 E 3

FROM LAFORGE & ENGDAHL (1979) 9 13

Figure 7. Lower hemisphere projections of the focal sphere for hypothetical hypocenters located in the middle of regions SE1 - SE5. Data are P-wave first motions of the special first-motion patterns determined by the method of objectively compositing data for focal mechanisms. The special patterns are labelled by letters corresponding to pattern names in Figures 5 and 6. The subscripts indicate the number of events associated with each pattern. For the few cases in which an event could be associated with either of the special patterns A (or B) or E, subscripts in parentheses indicate different possible interpretations of the number of events associated with each pattern. Plus-symbols are dilatational P-wave first motions at stations of the local seismographic network; black dots are compressional first motions. To aid in the comparison of these data with composite focal mechanisms for other earthquakes in the Adak area (last row; from LaForge and Engdahl, 1979), the compressional quadrant has been hatched for a few of the

focal spheres. The nodal planes drawn as dashed lines represent only one of the possible focal mechanism solutions allowed by the first motion data.

___

correlation was not apparent until the special patterns had been chosen for each of the regions independently.

One possible interpretation of each of the first-motion patterns is shown in Figure 7 by the dashed nodal planes. The particular interpretations shown represent a change in mechanism from a background thrust mechanism (dilatation pattern) to a thrust mechanism (pattern E) in which the strike of the thrust planes is rotated roughly 45° clockwise. The T-axis remains fixed (nearly vertical). This interpretation of the first-motion data of pattern E compares well to another composite focal-mechanism solution by LaForge and Engdahl (1979). Moreover, this interpretation of the first-motion data is similar both to the (counterclockwise) 65° rotation of the strike of the thrust planes observed for small earthquakes before an m_b 5.0 earthquake [Engdahl and Kisslinger, 1977] in the Adak region and to the rotation of nodal planes between foreshocks and aftershocks observed by Lindh et al. (1978) in California.

An alternative interpretation involves a change from the background thrust mechanism (dilatation pattern) to a pure strike-slip mechanism (nodal planes striking NNE-SSW and ESE-WNW), in which the P-axis remains nearly fixed. There is no obvious interpretation of the first-motion data in which the change in mechanism occurs such that the northwest-dipping thrust plane and B-axis remain fixed, while the slip vector rotates. This type of change might be expected if pre-existing faults were providing the rupture planes of these small earthquakes.

Summary and Discussion

The November 1977 earthquake and its aftershock sequence occurred in a region which had not experienced a similar earthquake sequence since before 1964. This is typical of the larger earthquakes which occur in the vicinity of Adak Island and suggests that the phenomenon of small seismic gaps may precede moderate-sized earthquakes. Several interesting variations in the seismicity and focal mechanism of small earthquakes occurred prior to the M_s 6.7 shallow-focus earthquake of November 4, 1977 in the central Aleutian island arc. First, about 13 1/2 months before the November 1977 earthquake, the rate of occurrence of very small magnitude events increased abruptly in the immediate vicinity of the impending main shock. Second, about 10 1/2 months before the November 1977 earthquake, the mechanism of most small- to moderate-sized events in the immediate vicinity of the hypocenter changed. For the particular source regions in which different focal mechanisms could be distinguished, only in those regions that were a part of the eventual aftershock zone was there a change in the mechanism of earthquakes before the main shock.

It is certainly possible that any one of the above phenomena is coincidental and not a seismological precursor to the M_S 6.7 earthquake. It is also possible that these phenomena are genuine precursors to this earthquake but are unique either to this earthquake or to moderate-sized shallow-focus earthquakes in the central Aleutian island arc. However, if any of the observed variations in seismicity or focal mechanism are indeed repeatable precursors, then the variations are important in two ways:

First, similar variations should be recognizable in future data from the Adak local seismographic network and perhaps elsewhere. For instance, the routine monitoring of both P-wave first-motion patterns and P-wave to SV-wave amplitude ratios [Kisslinger, 1980] is being implemented with incoming Adak network data. Second, while there is no unique interpretation of the first-motion data presented in this paper, there are still valuable constraints on possible interpretations of the precursory focal mechanisms compared to the background focal mechanisms. The asperity model [Brady, 1974; Stewart, 1979; Wyss et al., 1981; Kanamori, this volume, 1981] predicts simply a re-orientation of the stress field as weaker asperities break during the preparation time. Perhaps evidence such as we have presented for the range of possible re-orientations of the stress field will be useful in the development of more detailed modelling of the source preparation process for moderate-sized or larger earthquakes.

Acknowledgements. The authors thank G. Choy, J. W. Dewey, S. Ihnen, C. Kisslinger, A. Lindh, J. Whitcomb, M. Wyss and an anonymous reviewer for critically reviewing this paper, and thank R. E. Habermann for helpful discussions. Figures were drafted by R. McDonald. The research was supported under U.S.G.S Contract No. 14-08-0001-16716.

References

Brady, B. T., Theory of earthquakes, Part I: A scale independent theory of rock failure, Pure Appl. Geophys., 112, 701-725, 1974.

Dewey, J. W., S. Billington, E. R. Engdahl and W. Spence, Teleseismic search for seismicity patterns precursory to large earthquakes in Peru and near Adak, Alaska, paper presented at UNESCO International Symposium on Earthquake Prediction, Paris, April 1979.

Engdahl, E. R., Seismicity and plate subduction in the central Aleutians, in Island Arcs, Deep Sea Trenches and Back-Arc Basins, American Geophysical Union, Washington D. C., 1977.

Engdahl, E. R., Seismicity patterns precursory to larger earthquakes in the Adak region, in Proceedings of Conference VI: Methodology for Identifying Seismic Gaps and Soon-to-Break Gaps, U. S. Geological Survey Open-File Report 78-943, 163-171, 1978.

Engdahl, E. R. and C. Kisslinger, Seismological precursors to a magnitude 5 earthquake in the central Aleutian islands, J. Phys. Earth, 25, S243-S250, 1977.

Ishida, M. and H. Kanamori, The foreshock activity of the 1971 San Fernando earthquake, California, Bull. Seism. Soc. Amer., 68, 1265-1279, 1978.

Kisslinger, C., Evaluation of S- to P-amplitude ratios for determining focal mechanisms from regional network observations, Bull. Seism. Soc. Amer., 70, 999-1014, 1980.

LaForge, R. and E. R. Engdahl, Tectonic implications of seismicity in the Adak Canyon region, central Aleutians, Bull. Seism. Soc. Amer., 69, 1515-1532, 1979.

Lindh, A., G. Fuis and C. Mantis, Seismic amplitude measurements suggest foreshocks have different focal mechanisms from aftershocks, Science, 201, 56-59, 1978.

McCann, W. R., S. P. Nishenko, L. R. Sykes and J. Krause, Seismic gaps and plate tectonics: Seismic potentialfor major plate boundaries, in Proceedings of Conference VI: Methodology for Identifying Seismic Gaps and Soon-to-Break Gaps, U. S. Geological Survey Open-File Report 78-943, 441-584, 1978.

Nersesov, I. L., A. A. Lukk, V. S. Ponomarev, T. G. Rautian, B. G. Rulev, A. N. Semenov, and I. G. Simbireva, Possibilities of earthquake prediction, exemplified by the Garm area of the Tadzhik S. S. R., in Earthquake Precursors, Academy of Sciences of the U. S. S. R., Moscow 1973.

Stauder, W., Mechanism of the Rat Island sequence of February 4, 1965, with relation to island arcs and sea-floor spreading, J. Geophys. Res., 73, 3847-3858, 1968.

Stauder, W., Fault motion and spatially bounded character of earthquakes in Amchitka Pass and the Delarof Islands, J. Geophys. Res., 77, 2072-2080, 1972.

Stewart, W. D., Aging and strain softening model for episodic faulting, Tectonophysics, 52, 613-626, 1979.

Warren, D. H., Fault-plane solutions for microearthquakes preceding the Thanksgiving Day, 1974, earthquake at Hollister, California, Geophys. Res. Lett., 6, 633-636, 1979.

Wyss, M., A. C. Johnston, and F. W. Klein, Earthquake prediction: A multiple asperity model, in press, Nature, 1981.

TEMPORAL AND SPATIAL CHARACTERISTICS OF THE VARIATIONS OF SEISMIC VELOCITY RATIO ANOMALIES

De-Yi Feng and Jin-Ping Gu

The Seismological Institute of Lanchow, State Seismological Bureau
Lanchow, The People's Republic of China

Abstract. This paper summarizes preliminary results on the temporal and spatial characteristics of seismic velocity ratio anomalies before and after strong and moderate-sized earthquakes that occurred in China during the period 1962-1979. This paper also studies their relationship to types of earthquake sequences. The data for V_p/V_s anomalies before and after 36 earthquakes of M > 5.0 are presented here, including recently obtained data on the V_p/V_s anomaly before the M = 7.8 Tangshan earthquake.

First, we have investigated and discussed some general temporal and spatial characteristics of V_p/V_s anomalies; namely, the time-dependent characteristics of the V_p/V_s variations, the spatial characteristics of the regions with anomalous V_p/V_s, and their development process with time. Emphasis is given to some of the important characteristics of the V_p/V_s anomalies.

Secondly, the relationship between the characteristics of V_p/V_s anomalies and types of earthquake sequences are investigated.

Finally, we present a brief discussion of questions recently raised concerning the study of velocity anomalies, their application to earthquake prediction, and their theoretical interpretation.

Introduction

For approximately the last 10 years Chinese seismologists have been studying the seismic velocity anomalies before and after earthquakes, with emphasis on studies of V_p/V_s anomalies before and after major earthquakes [Feng, D. et al., 1974; Feng, D., 1975; Feng, D. et al., 1977; Feng, D., et al., 1978; Feng, D. et al., 1978; Feng, D. et al., 1980; Feng, R., 1977, Feng, R. et al., 1976; Dung, et al., 1979]. Some of the V_p/V_s data were obtained and identified as anomalous before the occurrence of the earthquakes, and these data were immediately used in the earthquake prediction. Since many seismologists hold different viewpoints on using premonitory seismic velocity variations as earthquake precursors, we must continuously make detailed studies of the different characteristics of the seismic velocity anomalies, especially their temporal and spatial characteristics.

In this paper, we have collected data for the V_p/V_s anomalies before and after 36 earthquakes of $M \geq 5.0$ which occurred in China between 1962-1979 including data recently obtained for the V_p/V_s anomaly before the Tangshan earthquake (M = 7.8). The corresponding characteristics of the V_p/V_s anomalies are shown in Table 1.

In this paper, using the data of the V_p/V_s anomalies, we first summarize and investigate some general temporal and spatial characteristics of seismic velocity ratio anomalies; namely, the time-dependent characteristics of the V_p/V_s variations, the spatial characteristics of the V_p/V_s anomalous regions, and their development process with time. Secondly, the relationships between the characteristics of the V_p/V_s anomalies and types of earthquake sequences are investigated, and some differences in the characteristics of V_p/V_s anomalies between multiple main-shock sequences and single main-shock sequences are given. We also present here a special account of what happened when moderate and strong events occurred in succession in the seismic zone during a relatively short time period. In these cases, the form of the curves of the V_p/V_s variation with time is often abnormal and the superposition and migration of anomalous regions might also appear. Finally, on the basis of the observational results described in this paper, we shall briefly discuss existing questions about the study of velocity anomalies and their application to earthquake prediction as well as their theoretical explanation.

General Temporal and Spatial Characteristics of Seismic Velocity Ratio Anomalies

Variations of the Seismic Velocity Ratio with Respect to Time

Before the major earthquakes the average decrease in the velocity ratio during the anomalous period always fluctuates between 4% and 7% (see Table 1). From the beginning of the V_p/V_s anomaly to the occurrence of the main-shock, the duration of ΔT is the precursory time.

Table 1

Seismic Velocity Ratio Anomalies Before and After Earthquakes (M $\geq$ 5) in China

No.	Date of Event	Area of Epicenter	Ms	Total Anomaly ΔT (Mo.)	ΔTo (Mo.)	Average Anomaly Change	Notes*
1	62/3/19	Xinfengjing Kwangtung	6.1	11	1.5	5%	1,F
2	64/9/23	Xinfengjing, Kwangtung	5.3	4.5			1,F
3	66/9/28	Chungtien, Yunnan	6.4	(20)		5%	2
4	70/1/05	Tunghai, Yunnan	7.7	60		6%	2
5	70/2/24	Tai, Szechuan	6.2	27.5	1.2	5%	2
6	70/7/31	Leipo, Szechuan	5.4	5.5			2
7	70/12/3	Xiji, Ninghsia	5.1	10.5	0.1	4%	3
8	71/6/28	Wuchung, Ninghsia	5.1	10			2
9	71/8/16-8/18	Mabien, Szechuan	5.9	10	1.5	4%	2
10	72/1/23	Shinbin, Yunnan	5.6	6.7		6%	2
11	72/4/8	Kangting, Szechuan	5.2	4.3			2
12	72/9/27-9/30	Kanting, Szechuan	5.7	10	1.0	4%	2
13	73/2/6	Luhuo, Szechuan	7.9		65		2
14	73/6/3	Jinhe, Sinkiang	6.0	16	2.0	4%	4; 5
15	73/6/29-6/30	Mabien, Szechuan	5.4	8	1.0	5%	2
16	73/8/11	Nanpin, Szechuan	6.5	23.2	3.2	5%	2
17	74/5/11	Yongshan, Yunnan	7.1	101	12.8	7%	9
18	74/9/24	Aheji, Sinkiang	5.5	(15)		4%	4
19	74/9/30	Dakancheng, Sinkiang	5.0	8		6%	4,A
20	74/11/17	Nanpin, Szechuan	5.7	8.1	1.4	5%	2,B
21	75/1/15	Jiulong, Szechuan	6.2	21	1.3		2
22	75/2/4	Haicheng, Liaoning	7.3	>48	10		10
23	75/5/27	Aheji, Sinkiang	5.0	5	0.1	4%	4
24	75/9/10	Bachu, Sinkiang	5.3	8		4%	4
25	76/1/10	Kuche, Sinkiang	5.8	13		5%	4,C; 6
26	76/5/29	Longling, Yunnan	7.4	48			4,C; 6
27	76/8/3	Keshi, Sinkiang	5.2	(9)		5%	4

Table 1. (continued)

No.	Date of Event	Area of Epicenter	Ms	Total Anomaly ΔT (Mo.)	ΔTo (Mo.)	Average Anomaly Change	Notes*
28	76/7/28	Tangshan,	7.8	>80	54.7		11
29	76/8/16	Songpan Szechuan	7.2	51.5	6.1	5%	12,D
30	76/11/7	Yanyuan, Szechuan	6.7	86		4%	13
31	77/7/23	Kuche, Singkiang	5.5	15		5%	4
32	77/12/19	Xiker, Sinkiang	6.1	(25)	2.4	5%	4
33	78/4/22	Kuerle, Sinkiang	5.8	(24)		5%	4,B; 7,D
34	78/7/13	Heisui, Szechuan	5.4	7			4,B; 7,D
35	79/3/29	Kuche, Sinkiang	6.0	36	1.8	5%	4,C; 8,E
36	79/7/9	Liyang, Kiangsu	6.0	15		5%	4,C; 8,E

* Data taken from: 1 - Feng, R. [1977]; 2 - Feng, D. et al. [1976]; 3 - Feng, D. et al. [1974]; 4 - Wang [personal communication]; 5 - Seismological Bureau of Sinkiang [personal communication]; 6 - Seismological Bureau of Yunnan [personal communication]; 7 - Seismological Bureau of Szechuan [personal communication]; 8 - Seismological Bureau of Kaingsu [personal communication]; 9 - Feng, D. [1975]; 10 - Feng, R. [1976]; 11 - Feng, D. [personal communication]; 12 - Feng, D. [1980]; 13 - Dong [1979].

Notes: A - the V_p/V_s anomaly was discovered before the earthquake; B - the earthquakes was predicted on the basis of V_p/V_s anomaly; C - a medium-term prediction was made on the basis of V_p/V_s anomaly before the earthquake; D - the Seismological Bureau of Szechuan discovered V_p/V_s anomaly and made a medium-term prediction; E - Liyang Seismological Station, Kiangsu Province discovered a V_p/V_s anomaly before the earthquake; F - reservoir impounding earthquakes.

Figure 1 shows the relationship between the variation of ΔT and magnitude M using the data from Table 1. From this figure we can see that, except for the fact that ΔT for multiple mainshock sequences and reservoir impounding earthquakes may be shorter, other points (lgΔT, M) related to major earthquakes show clustering about a linear increase in lgΔT with magnitude. The maximum magnitude variation, ΔM, for fixed ΔT is approximately 0.8 (Figure 1). For a basis of comparison, the V_p/V_s precursor times of ΔT for five events in Japan [Rikitake, 1979] were included in Figure 1. This figure shows that, except for the somewhat indeterminate V_p/V_s anomaly data for the Matsushiro earthquake, the values of ΔT for the four other earthquakes in Japan were well within the limits of the relation ΔT(M) obtained from events which occurred in China.

In the later stages of the anomalous period, the velocity ratio, V_p/V_s increases to recover its normal value, but its recovery in a great majority of cases is not completed at one time. In other words, V_p/V_s may decrease again or may obviously fluctuate after its increase and recovery. Therefore, we can determine the duration, ΔTo, of the V_p/V_s recovery as the interval between when it first returns to its normal value at the end of an anomalous period and the occurrence of the main shock [Feng et al., 1977]. Based on the data in Table 1, we also show the mutually dependent relationship between ΔTo and M in the lower part of Figure 1. From this figure we can see that within some range of fluctuation, lgΔTo also increases statistically with M in a linear fashion.

In order to once again verify the fluctuation characteristics of V_p/V_s recovery, we will briefly describe the results of the velocity ratio anomaly which were recently obtained prior to the strong (M = 7.8) earthquake which occurred at Tangshan. Using the data of arrival times of $\bar{P}$ and $\bar{S}$ waves from events which have occurred since 1970 in Tangshan and adjacent areas (recorded by more than 50 stations of the networks of Hopei, Peking, Tientsin, Shantung and the telemetry network of the Institute of Geophysics of the State Seismological Bureau) the variation of the average velo-

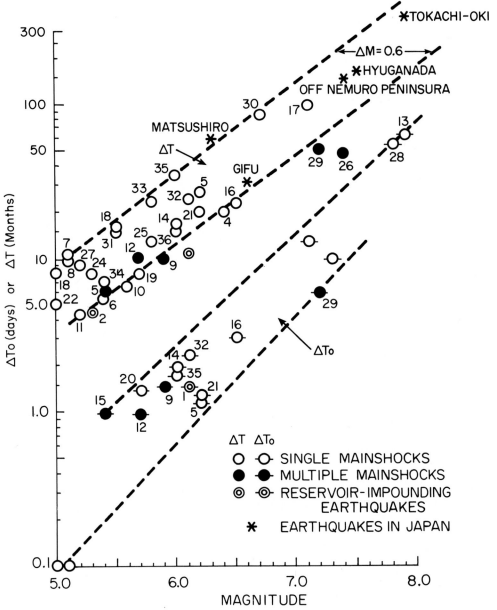

Fig. 1. Relationship between magnitude (M) and duration of the velocity ratio anomaly (△T) and its recovery time (△To).

city with respect to time is obtained. The preliminary results are shown in Figure 2. Because of the paucity of stations and absence of sufficient data, it is very difficult to study the V_p/V_s anomaly before 1970.

The V_p/V_s data shown in Figure 2 are generally average monthly values. If in a given month only 1 to 2 earthquakes occurred, we obtained an average by including events which occurred in months that were closest to the month lacking sufficient events to obtain an average. The horizontal bars in Figure 2 denote the intervals used to obtain averages, and the vertical bars represent standard errors.

Because of the lack of sufficient data before 1970, it is difficult to determine the beginning of the V_p/V_s anomaly and normal V_p/V_s values before the appearance of the V_p/V_s anomaly. For comparison and analysis, we take the average V_p/V_s value of 1.76 for the surrounding regions as a reference normal value.

Figure 2 indicates that a V_p/V_s anomaly existed before 1970 and in January 1972 it increased to above the reference normal value while at the same

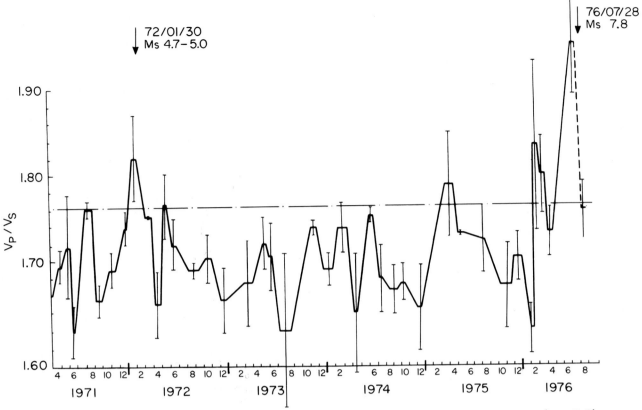

Fig. 2. Variation of average velocity ratio before the 1976 Tangshan earthquake (M = 7.8).

time an event of magnitude 5 occurred. After the earthquake, the V_p/V_s decreased back to its anomalous state and then again it increased sharply after February 1976 until it exceeded 1.95 immediately before the main shock. Since January 1972 when the velocity ratio returned to above the normal value, ΔT_o is calculated to be 4 1/2 years. This is in good agreement with the relation $(\Delta T_o, M)$ obtained from other earthquakes (see Figure 1). If we calculate earthquake magnitude from the empirical formula

$$M = 1.10 \lg \Delta T_o(\text{days}) + 4.3$$

[Feng, D. et al., 1980], we find that M = 7.8.

Spatial Characteristics of the Regions of Anomalous Velocity Ratio

Detailed preliminary results have been published [Feng, D. et al., 1977] describing the methods used for determing regions of V_p/V_s anomalies outlining some of the basic characteristics of these regions (i.e., form, size and orientation). In this section we will emphasize the spatial characteristics of the regions of anomalous seismic velocity ratio.

An important characteristic in the spatial distribution of the regions of anomalous velocity ratio is that the area in which the V_p/V_s anomaly appears is often located at the side of the epicenter of the main shock. In other words, the main shock always occurs at a peripheral portion of the anomalous region instead of at its center. Moreover, the epicenter of the main shock is often located outside the area of the strongest V_p/V_s anomaly and often at a considerable distance from the most anomalous area. Several examples of this are shown in Figure 3 in which the smaller areas approximately outline the regions of strongest V_p/V_s anomaly within the V_p/V_s anomalous regions. The epicenters of the main shocks are all located outside of the most anomalous areas.

The results shown in Figure 3 indicate that V_p/V_s anomalous regions are extremely asymmetrical in relation to the epicenters of main shocks. From some observational data, we see that, prior to earthquakes, other premonitory phenomena also frequently demonstrate a similar asymmetric spatial distribution. For instance, before the Songpan earthquake (M = 7.2) in August 1972, many precursors were noted in a large anomalous area to the south as far as Wengchuan and Guanxian which were over 200 km from the epicenter. In the Wuda region, however, located 100 km north of the epicenter of the main shock, the majority of premonitory anomalies were not extraordinary except for a few unique cases. Therefore, it can be noted

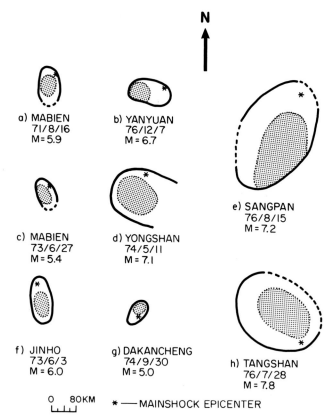

N

a) MABIEN
71/8/16
M=5.9

b) YANYUAN
76/12/7
M=6.7

c) MABIEN
73/6/27
M=5.4

d) YONGSHAN
74/5/11
M=7.1

e) SANGPAN
76/8/15
M=7.2

f) JINHO
73/6/3
M=6.0

g) DAKANCHENG
74/9/30
M=5.0

h) TANGSHAN
76/7/28
M=7.8

0   80KM   * —MAINSHOCK EPICENTER

Fig. 3. Approximate extent of the anomalous
V_p/V_s region and the interior anomalous areas
before several earthquakes in China.

that the region in which other precursor anomalies
appeared was approximately consistent with the
region of anomalous velocity ratio (compare Figure 3 and Feng et al. [1980]).

Because of the asymmetrical spatial distribution of V_p/V_s anomalies and because V_p/V_s variations may also depend on the direction of stress
and the orientation of micro-cracks, the seismic
velocity anomalies which are observed on the
Earth's surface may have a preferred orientation
(i.e., at the same distance from the epicenter of
a main shock the observed results in different
direction may be different).

Again using the Songpan earthquake (M = 7.2) as
an example, we designate the stations Songpan,
Wudu and Chengxian, which lie in an approximate
northeast-easterly direction, as group A; and
designated stations Wengxian, Wudu and Mingxian,
which lie in an approximate east-west direction,
as group B. The epicenteral distances of stations
Songpan and Wengxian and Chengxian and Mingxian
are approximately the same. In finding average
V_p/V_s values for group A and group B separately
for the period from late 1974 to mid-1976, it can
be seen that for group A the velocity ratio anomaly ratio is quite obvious, its minimum value is
1.52, and the average anomalous decrease is about
14%. For the same period it can be seen that for

group B the velocity ratio anomaly is not as obvious as that in group A. Its minimum value is
only 1.66, and the average anomalous decrease is
only 7%.

Development of the Region of Anomalous Velocity
Ratio with Time

Determination of the approximate extent of the
velocity ratio anomalous region at various intervals enables us to study the development process
of the V_p/V_s anomalous region with time. Figure 4
illustrates the temporal variation of the maximum
linear dimensions of the V_p/V_s anomalous region
before the Yanyuan-Ninglang earthquake (M = 6.7).
This variation was determined by using the method
of combining data obtained at two separate stations into groups. From Figure 4, we can see that
during the initial stage of the period of anomalous V_p/V_s (September 1969 to early 1970) the
extent of the anomalous region increased rapidly
and reached its extreme bounds by April 1970. It
then fluctuated slightly, and during the final
stage of the period of anomalous V_p/V_s (March -
May 1975) it increased somewhat at first and then
decreased rapidly. The extent of the anomalous
region again increased in October 1976 and then
decreased on November 4. Finally, the main shock
(M = 6.7) occurred on November 7.

The variations of the region of anomalous velocity ratio mentioned above indicate the following
two important characteristics: 1) in the initial
stage it rapidly increased and reached its extreme
bounds; and 2) in the final stage it followed a
decrease-increase-decrease pattern. Correspondingly, the average seismic velocity ratio
(determined from the Wadati diagram for data obtained at several stations) rapidly reduced to the
lower value in the beginning stage while in the
later stage it might have decreased and fluctuated
again after the V_p/V_s recovery.

Relationship Between Temporal and Spatial
Characteristics of Seismic Velocity Ratio Anomalies
and Types of Earthquake Sequences

Difference in the Characteristics of Velocity
Ratio Anomalies Between Two Basic Main Shock Types
of Earthquakes

Preliminary results indicate that there are some
differences in the characteristics of V_p/V_s anomalies between multiple main shock sequences and
single main shock sequences. These differences
are clearly delineated by the following two
aspects.

First, from Table 1 and Figure 1, we can see
that the total anomalous times, ΔT, before multiple main shock sequences are often shorter than
the total anomalous times, ΔT, before single main
shock sequences with the same magnitudes. This
characteristic is especially striking in the Songpan and Longling earthquakes which both have
magnitudes >7.

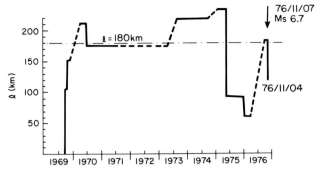

Fig. 4. Variation of the maximum linear dimension of the anomalous V_p/V_s region before the Yanyuan-Linglang earthquake (M = 6.7) of November 7, 1976.

Secondly, the extent of the anomalous region before multiple main shock sequences may be larger than that before single main shock sequences with the same magnitude. This characteristic is also especially striking in the Songpan earthquakes (M = 7.2) (Figure 5).

Also, differences between the two main shock types of earthquake sequences can be seen by comparing V_p/V_s anomaly characteristics with seismic activity before a main shock. If, in the initial and final stages of the V_p/V_s anomalous period prior to the main shock, there are two "signal" earthquakes or earthquake sequences which occurred separately showing a relatively maximum in magnitude for the whole anomalous period, the forthcoming main earthquake often falls into the single main shock sequence category of earthquakes. On the other hand, if there is only one "signal" earthquake or earthquake sequence with maximum magnitude during the anomalous period between the initial and final stages, the ensuing main event often belongs to the multiple main shock sequence category of earthquakes. Some examples of signal earthquakes have already been published [Feng et al., 1976]. As additional examples we present four large earthquakes (M > 6.5) which occurred in China in 1976. Their signal earthquakes are shown in Table 2. From Table 2 we can see that there is only one signal earthquake before the Songpan or Longling multiple main shock sequence and there are two signal earthquakes before the Tangshan and Yanyuan-Ninglang single main shock sequences.

Temporal and Spatial Characteristics of Velocity Ratio Anomalies Before and After Earthquakes Which Occurred in Succession

When moderate to strong earthquakes occur in the same seismic zone during a relatively short period, the temporal and spatial characteristics of V_p/V_s anomalies may be complicated. First of all, the form of the curves of the V_p/V_s variation with time becomes abnormal. Its main characteristic is that V_p/V_s does not recover after the "post-shock anomalous period" but continues to remain anomalous. The V_p/V_s anomaly may even be more evident than it was before. Secondly, the extent of the V_p/V_s anomalous region in these cases may be quite large (Figure 5). However, the most important characteristic of the V_p/V_s anomalies is that superposition and migration of anomalous regions may also appear.

Figure 6 shows a schematic representation of the superposition and migration of V_p/V_s anomalous regions before and after some earthquakes which occurred in succession in western China. In Figure 6, three examples from Sinkiang earthquakes (Figure 6b, c, d) are based on the research results obtained and provided by Wang, G. L. et al. of the Seismological Bureau of the Sinkiang Vighur Nationality Autonomous Region.

The examples of the three groups of earthquakes shown in Figure 6a, b, d are basically similar. For instance, region A plus region B constitutes the total V_p/V_s anomalous region prior to the first major earthquake. After this earthquake, V_p/V_s soon recovered only in region A while in region B the V_p/V_s anomaly continued. Consequently, a new anomaly consisting of regions B and C was formed, and finally the second major earthquake occurred. The formation time of region C was either: 1) prior to the first major earthquake (the Songpan earthquake, Figure 6a); 2) at the same time as the first major earthquake; or 3) soon after the occurrence of the first major earthquake (the Kuerle and Xikel earthquakes, Figures 6b, d).

In Figure 6, the character of the anomalous region of the Kuche event (M = 6.0) is that region A is the region of anomalous V_p/V_s for the Kuerle event (M = 5.8), and that after its occurrence the V_p/V_s anomaly not only continued in this region but also developed into region B. Therefore, region A added to region B became the V_p/V_s anomalous region. The corresponding characteristics of the velocity ratio anomalies in relation

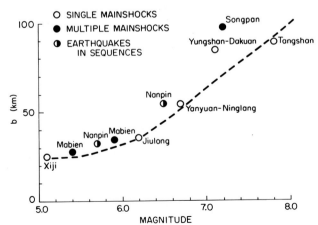

Fig. 5. Relationship between the minor half-axis of the anomalous V_p/V_s region (b) and magnitude (M).

Table 2

The Signal Shocks Before Four Earthquakes (M $\geq$ 6.5) in China During 1976

MAIN EARTHQUAKE					FIRST SIGNAL SHOCK				SECOND SIGNAL SHOCK			
Date	District	N	E	M	Date	N	E	M	Date	N	E	M
76/8/16	Songpan*	32.8°	104.3°	7.2	73/8/11	32.9°	104.0°	6.5				
76/5/29	Longling*	24.5°	99.0°	7.4	73/6/1-2	25.0°	98.7°	5.0				
76/7/28	Tangshan	39.4°	118.0°	7.8	45/9/23	39.7°	118.7°	6¼	70/5/25	39.5°	118.0°	4.8
									72/1/30	39.1°	118.8°	5.0
76/11/7	Yanyuan-Ninglang	27.5°	101.1°	6.7	69/6/22	27.9°	99.8°	4.3	75/12/1	27.3°	100.8°	4.9

The symbol (*) indicates a multiple main shock sequence.

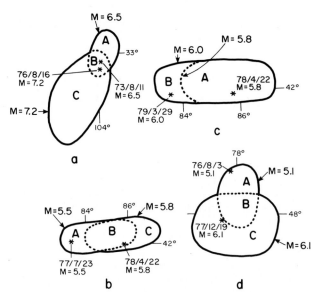

Fig. 6. Schematic diagram showing the migration and partial superposition of anomalous V_p/V_s regions before and after some earthquakes which occurred in sequences.

to the earthquakes mentioned above can be seen in Table 1. Obviously, region B (Figure 6a, b, d) and region A (Figure 6c) may be regarded as regions of the superposition of V_p/V_s anomalous regions which are related to both the first and second major earthquakes.

Discussion

Some Existing Questions Concerning the Study of Velocity Anomalies and Their Application to Earthquake Prediction

Although in recent years more abundant data for premonitory velocity anomalies (including V_p/V_s anomalies) have been obtained and some attempts to use V_p/V_s anomalies in earthquake prediction have been made, there are still a few questions which need to be further studied.

First, it is necessary to enhance the observational precision and increase the density of networks in the seismic zones in China. For the examples of V_p/V_s anomalies which occurred prior to earthquakes that have been discussed in this paper, the majority of the original data and records were carefully re-examined and corrected and the necessary statistical methods for their selection and investigation were applied. Therefore, from a statistical viewpoint, the results obtained are reliable. In addition, the development of man-made vibro-source methods will have a significant bearing on the study of velocity anomalies.

Secondly, we must take into account the methods of observation for velocity anomalies. From the preliminary results we described in this paper, it can be understood that anomalous velocity regions are often asymmetric relative to the positions of the epicenters of main shocks and moreover the velocity anomalies in themselves may be preferentially oriented. Consequently, the seismic velocity anomalies may be observed in some regions or some directions, but not observed in other regions or in other directions. For example, prior to the Songpan earthquake (M = 7.2) in 1976, the identified V_p/V_s anomalies had been observed at the network located south of the epicenter of the main shock, but this observation was difficult to see at the Kansu network (north of the main shock) because the V_p/V_s anomalous region was limited to the south [Feng et al., 1980]. If the observations are limited to only a single direction, even though the distance to the epicenter of the main shock is small enough and the precision of observation in high enough, it is still not certain if velocity anomalies will be observed. Therefore, it is possible that we may obtain negative results.

Moreover, we must continue to study intensively the temporal and spatial characteristics of velocity anomalies, especially their spatial characteristics. Preliminary results demonstrate that the phenomena of velocity anomalies may sometimes be complicated. As an example, we use the January 14, 1975 Miaulgou earthquake (M = 4.8) which occurred in Xinkiang (43.6°N, 86.6°E). According to results obtained by Wang G. L. et al. of the Seismological Bureau of the Sinkiang Vighur Nationality Autonomous Region the V_p/V_s anomaly before this earthquake lasted for about 16 months. Differences in types of velocity anomalies are probably related to earthquake types as well as to specific seismic regions or other factors. If at the same time we study the spatial distribution of the seismic velocity or velocity ratio for these earthquakes, we shall discover that although the anomalous period is rather long, the anomalous region is rather small. Therefore, the magnitudes of the main shock may be correspondingly rather small. Inversely, for the strong Songpan and Longling events, the duration of V_p/V_s anomalies was short but the extent of anomalous regions was rather large and the main shocks were strong earthquakes. In general, if we simultaneously take into account the temporal and spatial characteristics of velocity anomalies, we can obtain more precursory information on earthquakes.

On the Theoretical Explanation of Velocity Ratio Anomalies

There are some difficulties in completely explaining the previously discussed preliminary results on the temporal and spatial characteristics of seismic velocity ratio anomalies by either of the two models (i.e., the dilatancy-diffusion (DD) model [Nur, 1972] and the Institute of Physics of the Earth (IPE) model [Miachkin, 1978]). For instance: the main shock always occurs at the peripheral portion of an anomalous region instead of at its center; the most anomalous area is often located some distance from the epicenter of the main shock; at a later time V_p/V_s may decrease again or fluctuate after its increase and recovery; and the anomalous region may also change in the decrease-increase-decrease pattern illustrated in Figure 4.

It is rather difficult for the two models given to explain the above facts. For example, the DD model suggests a rather symmetric anomalous region about the epicenter; a gradual decrease of the anomalous region away from the epicenter; and its rates of change is not supposed to be higher than the average velocity of the diffusion of water into newly created pores. This model does not agree well with the rather quick decrease-increase-decrease fluctuation in a region of anomalous velocity ratio. The IPE model also fails to give a satisfactory explanation for the extreme asymmetry (as well as some other characteristics) of the velocity ratio anomalous region.

We believe that stick-slip may play an important role in the late stage of V_p/V_s anomalies. It is possible that stick-slip results in a relaxation of stress; the cracks close and V_p/V_s is restored. However, the stopping of stick-slip may cause stress to concentrate again and cause the velocity ratio anomaly to reappear. From Figure 4 we can ascertain that, before the Yanyuan-Ninglang earthquake (M = 6.7), stick-slip had probably occurred twice in the source region. The first frictional sliding was rather slow and it probably occurred between mid-1975 and the beginning of 1976. The second stick-slip occurred immediately prior to the main shock (early in November 1976). Its speed was rather fast, and this slip most likely corresponded with the short-term or imminent anomalies.

In the near future, we plan to study further the details of the process of the formation of an earthquake source as well as a more reasonable theoretical explanation for the temporal and spatial characteristics of seismic velocity ratio anomalies.

References

Dung, S. et al., Earthquake prediction on the basis of V_p/V_s variations - a case history, Physics Earth Planet. Inter., 18, 1979.

Feng, D., Anomalous variations of seismic velocity ratio before the Yongshan-Daguan earthquake (M = 7.1) on May 11, 1974, Acta Geophysica Sinica, 18, 1975; Chinese Geophysics, 1(1), 1978.

Feng, D. et al., Velocity anomalies of seismic waves from near earthquakes and earthquake prediction, Acta Geophysica Sinica, 17, 1974.

Feng, D. et al., Preliminary study of the velocity anomalies of seismic waves before and after some strong and moderate earthquakes in western China (I) - The velocity ratio anomalies, Acta Geophysica Sinica, 19, 1976.

Feng, D. et al., Preliminary study of the velocity anomalies of seismic waves before and after some strong and moderate earthquakes in western China (II) - The anomalous regions and their characteristics, Acta Geophysica Sinica, 20, 1977.

Feng, D. et al., Preliminary study of the velocity anomalies of seismic waves before and after some strong and moderate earthquakes in western China (III) - Variations of curvature of the Wadati diagrams, Acta Geophysica Sinica, 21, 1978.

Feng, D. et al., Anomalous variations of seismic velocity ratio before the Songpan-Pinwu earthquakes (M = 7.2) of 1978, Acta Seismologica Sinica, 2(I), 1980.

Feng, R., On the variations of velocity ratio before and after the Xinfengjiang impounding-reservoir earthquake of M 6.1, Acta Geophysica Sinica, 20, 1977.

Feng, R. et al., Variations of V_p/V_s before and after the Haicheng earthquake of 1975, Acta Geophysica Sinica, 19, 1976.

Mjachkin, V. I., The processes of preparing earthquakes, Moscow, 1978.

Nur, A., Dilatancy, pore fluids, and premonitory variations of t_s/t_p travel times, Bull. Seismol. Soc. Amer., 62, 1972.

Rikitake, T., Classification of earthquake precursors, Tectonophysics, 54, 1979.

VELOCITY CHANGES IN THE CHARLEVOIX REGION, QUEBEC

G.G.R. Buchbinder

Division of Seismology and Geothermal Studies, Earth Physics Branch
Department of Energy, Mines and Resources, Ottawa, Ont. K1A 0Y3

Abstract. In the seismically active region of Charlevoix, Quebec, seismic travel time experiments have been undertaken from 2 shot points that were recorded at up to 13 stations. Over a period of 1870 days, 12 shots were set off on the northshore and 8 shots on the southshore of the St. Lawrence River. Epicentral distances vary from 12 to 68 km. Significant long term changes in travel time residuals of up to 46 $\pm$ 11 ms have been observed. The maximum difference from one shot point to two of the stations is 76 $\pm$ 16 ms.

A maximum in residual was observed SW of the center of the Charlevoix impact structure in 1976, this maximum spread towards the NE in 1978 and the SW in 1978 and 1979. The relative average shot residuals from the north shore increased from 1974 to 1977 then decreased in 1978 and remained steady through 1979. For the south shore a maximum was observed in 1976 after which it gradually decreased. Two small earthquakes on 31 October 1976 and 5 July 1979, are not likely to have resulted in short term velocity changes, but a magnitude 5.1 event certainly did not. The hypothesis that the vertical component of the earth tide produced changes in travel-time was tested experimentally and rejected. It is concluded that the observed negative changes in velocity, can be explained in terms of the dilatancy - diffusion model. Positive changes in velocity may be due to the closing of cracks.

Introduction

Seismic travel time monitoring experiments have been conducted since 1974 in a seismically active region in Eastern Canada. The experiments up to 1977 have been described in detail by Buchbinder and Keith (1979). The region of concern is the La Malbaie - Baie-St-Paul region on the St. Lawrence River, northeast of Quebec City. Six magnitude 5-7 earthquakes were felt in the region between 1663 and 1870 (Smith, 1962). The first large instrumentally located events occurred in 1924 and 1925 (Hodgson, 1950). However, these estimates of magnitude did not take the high Q into account. Microearthquakes have been recorded during short field surveys in 1970 (Leblanc et al., 1973) and in 1974 (Leblanc and Buchbinder, 1977). A telemetered array has been operating in the region since the fall of 1977 (Anglin and Buchbinder, 1980). All these data indicate that the earthquakes are confined in a region of about 90 by 30 km centered on the river.

Two recent events have occurred at the northeastern end of the zone; 19 Aug. 1979, $m_b(Lg)$ = 5.0, and 23 Oct. 1975, $m_b(Lg)$ = 4.2 (Anglin and Buchbinder, 1980). Stevens (1980) has shown that older instrumentally recorded events occurred at either end but predominantly at the northeastern one. Because of the seismic history of this zone it is highly probable that another large event will occur there. In an effort to answer the question as to when it may occur, the Earth Physics Branch has undertaken a multidisciplinary research project in the region, which involves not only seismic but also magnetic, gravimetric, geodetic, water level tilt and strain studies.

This report deals with the interpretation of calibration explosions to monitor changes in seismic velocities. In the past few years there have been a large number of studies of changes in travel time before earthquakes and the attendent interpretation in terms of the dilatancy hypothesis. These studies used earthquakes or explosions to study travel time changes. There have been very few studies, however, (Ward, 1979) that have reported changes of travel time before an earthquake and that have led to the prediction of an earthquake.

In the euphoria that existed in the early to mid seventies about the dilatancy hypothesis and its use in earthquake prediction, it was assumed by the author that if dilatancy occurred in the Charlevoix region, producing changes in velocity of a few percent, this would result in changes of travel time of the order of 100 milliseconds. It was on this basis that the calibration experiment was designed. When no such large changes in travel time were observed the experimental procedure and analysis were revised to monitor travel time anomalies to permit an insight into the question of the accuracy of our absolute travel time measurements and the precision from one shot to the next. This experiment is continuing, although the limit of a few milliseconds error has been reached and there will be no further substantial decreases in these errors.

Recently Winkler and Nur (1978) have shown that dilatancy is not a guarantee for an observable velocity anomaly. If dilatancy exists it may be highly depth dependent and confined to the upper 5 km of the crust and that the change in V_P/V_S can be very small.

The experiments up to 1977 have been described by Buchbinder and Keith (1979) and it was reported that significant changes in travel time had occurred over a period of 38 months.

Geology and Structure

The region for this experiment lies 100 km N.E. of Quebec City straddling the St. Lawrence River. The rocks of the north shore belong to the Precambrian Grenville Province, and dip under the river at about 20° degrees to the south-east where they are covered by Paleozoic sediments. The contact between the two rock types is generally referred to as Logan's line. The Precambrian rocks are faulted by very high angle faults, dipping to the south-east (Roy, 1978). Leblanc and Buchbinder (1977) have shown that the microearthquakes are confined to the Precambrian rocks and Anglin and Buchbinder (1980) have confirmed this and show that the events are limited to the north-west by the steeply dipping faults.

Another major tectonic feature in the region is the Charlevoix impact structure formed some 350 Ma BP (Rondot, 1968; Robertson, 1968), the outline of the present eroded rim valley is shown by a dashed line in Figure 1. The association of this structure with the microearthquakes has been suggested but is tenuous (Leblanc et al., 1973), since only about one half the crater was underlain by seismic activity and this activity extends beyond the crater to the north-east.

Experiments

The shot and station geometry is such that, from the relation for the size of a dilatant zone (L dil.) as given by Anderson and Whitcomb (1975) Log (L dil.) = .26M + .46, which gives 32 km for M = 4 and 57 km for M = 5, most widely felt earthquakes and certainly all damaging earthquakes should be detected if they are preceded by velocity changes of a percent or more. The stations and shot points are indicated in Figure 1.

The north shore shot point (NSSP) is the St.-Jerome Mine about 10 km NW of

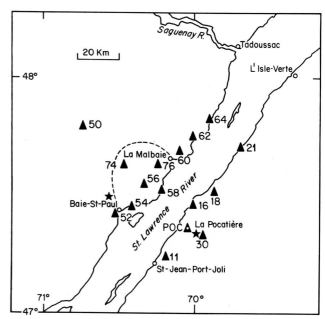

Fig. 1. Seismometer stations are triangles, A are array stations, stars are shot points.

Baie-St.-Paul. This is a water filled quarry. The explosive charge (230-2300 kg) is distributed in 4 to 5 equal parts in the deepest part of the quarry. The centre of the charge does not vary by more than ± 3 m, which translates to ± .5 ms in arrival time at the receiving stations. Between June 1974 and September 1979, shots 3-14 were set off in the quarry.

At the south shore shot point (SSSP) near La Pocatière, holes were drilled into an outcrop of near-vertically dipping Paleozoic sediments. Shots numbered 101 - 108 were detonated between October 1975 and September 1979. For most shots one hole containing 450 kg of explosives was used. Until 1975 Hydromex was used, after that Powerfrac was used. In order to account for shot point migration, hole No. 1 was used as origin and all travel times from the other holes were corrected to make it appear that they came from this hole by computing the changes in distances and using a phase velocity of 6.00 km/s, which keeps the uncertainty of the correction to less than 1 ms; the maximum correction is 6 ms.

In 1974 the shot in the mine was detonated by an ordinary commercial blaster box. The P arrival from the shot was recorded through a nearby geophone on a highspeed strip chart recorder together with a radio time signal. The shot instant was computed from the known shot-geophone distance and an assumed velocity; the estimated error was about 7 ms.

From 1975 to 1978 the shots were triggered by a chronometer controlled relay. The chronometer was a digital clock that can be synchronized to a radio time signal or a master clock if no radio time signal is available. The shot instant is delayed by the relays in the shooter box that detonate the charge. This delay, here called the relay delay, is estimated to be 26 $\pm$ 5 ms for the 1975 to 1977 shots. In 1978 a similar box with a delay of 35 $\pm$ 5 ms was used. These delays were applied as corrections to all travel times. In 1979 a new shooter box was obtained that had delays of a fraction of a millisecond. Finally, the shot instant is also affected by the error in the shooter clock correction of $\pm$ 2 ms.

Instruments

The instruments that were used have been described in detail by Buchbinder and Keith (1979), only the salient points and changes will be given here.

The array, referred to in the following by A, is a 7 - element vertical component telemetered analogue magnetic tape recording array. Four of these stations are on the north shore and three on the south shore (fig. 1). Timing is through a temperature compensated chronometer that puts a time code on the tape. In addition, before and after each shot a radio time signal is also placed on the tape. The band pass filter applied to the data is 1.0 to 25 Hz. The analogue tape is subsequently digitized at 120 samples /s.

The portable backpack instruments referred to here as BP are microprocessor controlled digital recording magnetic tape systems. Two horizontal components and one vertical component were usually deployed. The band pass of these instruments is 1.0 to 20 Hz. The internal temperature compensated chronometer places time at the start of each block of 80 samples. The early version of the microprocessor only permitted recovery of the start of the digitized samples to $\pm$ 8 ms and the sampling frequency was 60 samples /s. Starting in 1978 this uncertainty was removed, and the sampling frequency was increased to 120 samples /s.

The temperature compensated clocks in the shooting system and recorders were initially compared to radio time signals on highspeed strip charts with an uncertainty of $\pm$2 ms. As of 1977 the clocks were compared to radio time signals on an oscilloscope, with an uncertainty of $\pm$ 1 ms.

The time base was CHU from Ottawa, for which no correction was applied; when WWV signals from Colorado were used they were corrected by 7 ms to allow for the time difference in propagation time for the two time signals to the region of the experiment.

Data Analysis

Buchbinder and Keith (1979) had considerable problems in the analysis of travel times recorded by different recording systems (e.g. visual and magnetic tape). Therefore, here only data will be used that were recorded on magnetic tape, since the aim of the analysis is to determine changes in travel time by cross-correlating wave forms.

The cross-correlation program determines the cross-correlation coefficient for two waveforms. The maximum of the coefficient is determined by interpolation to 1/10 the cross-correlation lag. Analysis of a number of cross-correlation functions had shown this to be reasonable. The cross-correlation lag is 1/120 s for all combinations of instruments except that it was 1/60 s for BP-BP and BP-A in 1977 and before, since the digitization rate was 1/60 s for the BP.

Buchbinder and Keith (1979) showed that the difference in shot size of 230 kg and 2300 kg had no significant effect on the wave forms. This was the range of charges that were used. The difference in the band pass frequency filters of the 2 recording systems is only apparent when the waveform contains 30 Hz energy. This is only observed at station 56, when cross-correlating BP and A wave forms, however, the results were still very stable. To allow for the difference in phase response of the different systems 5 ms are added to the errors when cross-correlating BP and A wave forms.

After considerable experimentation with the length of the waveforms to be cross-correlated it was determined that 0.8 seconds gave the most satisfactory results. The difference was not in the cross-correlation lag but rather in the maximum of the coefficient that tended to be higher for the shorter sample length. For waves that travel in the crust beneath the St. Lawrence river the first arrival is usually not the maximum amplitude so that a shorter sample length could not be used.

All the necessary corrections were applied to the arrival times and all possible combinations of cross-correlating wave forms for each station were undertaken. It has already been pointed out by Buchbinder and Keith (1979) that depending on the combination in going from one shot to some other one (for example from 14 to 12 and from 12 to 10 compared to from 14 to 10 directly) the difference in cross-correlation lag is very small. In Table 1 are shown the time lags for the St. Jerome mine shots in milliseconds for each station. The lags are with respect to shot No. 14 since it was recorded at most stations. Stations 18 and 30 did not record shot 14 and are therefore given with respect to the first shot recorded at the station. In all, 98 cross-correlation lags were obtained for the NSSP. Table 2 shows similar data for the La Pocatiere shot point on the south shore. Here shot 108 was usually taken as reference since it is the shot common to most stations. Similarly as for the NSSP stations 18

TABLE 1   ST. JEROME MINE, CROSS-CORRELATIONS IN MILLISECONDS

SHOT No.	3	4	5	6	7	8	9	10	11	12	13	14
Day No.	204	267	350	243	151	152	236	306	306	178	179	249
Cons.	1	429	877	1136	1408	1409	1493	1563	1563	1800	1801	1871
Day M.	23.7	24.9	15.12	31.8	31.5	1.6	24.8	2.11	2.11	27.6	28.6	6.9
Y	74	75	76	77	78	78	78	78	78	79	79	79
Stn.												
52 12.21Km					BP 2		BP 0	BP 2	BP 3	BP -2	BP 10	BP 0
54 13.77	A -29	A -15	BP -6	A 2	A -12	A -1	A -6	A 5	A 2	A 2	A 13	A 0
56 17.34			BP 4	BP 15	BP 0		A -2	A -3	A -4	A -1	A 16	BP A 0
58 25.87	A -15	A 7		BP 19				BP 1		BP -4	BP 7	BP 0
60 39.02	A -13	A 9	BP 14	A 33	A -2	A BP -4 2	A -2	A -1	A -2	A -2	A 13	A 0
62 47.75					BP 1		BP 1	BP 1	BP 2	BP -2		BP 0
64 59.66				A 3	A -9	A -1	A -3	A 5	A 5	A 3	A 13	A 0
74 13.11			BP 8	BP 19	BP -2		BP -2	BP -2				BP 0
76 26.63				BP 20			BP -1	BP -2	BP -1	BP -5	BP 5	BP 0
10/11 42.24				A 8	A 1	A 8	A -3	A 9	A 7	A 6	A 17	A 0
16 41.82					A -11	A 1	A -2	A 6	A 4	A 4	A 15	A 0
18 52.18	A (0)	A (15)		BP (26)								
20/21 67.84				A 3	A -14		A -6	A 6	A 2	A 2	A 10	A 0
30 51.39	A (0)	A (15)	BP (27)	BP (6)		BP (6)				BP (4)		
50 31.30									BP 0	BP 1		

and 30 start with the earliest shots. These stations will not enter into the interpretation. Station 54 is given with respect to shot 107, since shot 108 was not recorded, on the assumption that there was not much change between the two shots. Here a total of 54 cross-correlation lags were obtained.

The data from Tables 1 and 2 are shown graphically in Figure 2. For each station the horizontal line is the base line or zero lag

TABLE 2   LAPOCATIERE,   CROSS-CORRELATIONS IN MILLISECONDS

SHOT	101	102	103	104	105	106	107	108
Day No.	280	302	351	242	150	235	177	248
Cons.	1	389	438	694	967	1052	1359	1430
Day M.Y.	7.10.75	28.10.76	16.12.76	30.8.77	30.5.78	23.8.78	26.6.79	5.9.79
52 39.39Km					BP -1	BP 6		BP 0
54 32.60	A (-30)		BP (-13)	A (-9)	A (-17)		A (0)	
56 32.61		BP 40	BP 37	BP 11	BP 1	A 3	A -6	BP 0
58 24.72	A -3	BP 20	BP 8	BP 11			BP -3	BP 0
60 38.55	A 4	BP 25	BP 20	A 34	A -1	A 6	A -2	A 0
62 44.73				BP 9		BP 9	BP -4	BP 0
64 53.84				A 5	A -9	A 7	A 5	A 0
74 46.28		BP 12		BP 19		BP 3		BP 0
76 36.92				BP 7	BP -2	BP 7	BP -3	BP 0
10/11 18.42				A 5	A -5	A 5	A 0	A 0
16 13.13			BP -14		A -12	A 6	A -2	A 0
18 21.84	A (0)			BP (5)				
20/21 46.23				A 5	A -11	A 5	A 1	A 0
30 5.53	A (0)				BP (1)	BP (7)	BP (-6)	

line. Chronologic time is from left to right. The dot on the right is the last shot, the preceeding shots are represented by a dot for zero lag change and a vertical line for a finite lag change, the amplitude of which is the lag in milliseconds with respect to the reference shot. Lines above the base line indicate travel time increases, lines below the base line are travel time decreases. Shot 13 has been omitted in Figure 2, this will be explained later.

ERRORS

The RMS errors with respect to shots 14 and 108, the latest shots, are given for all the earlier shots and all possible instrument combinations, in the last three lines of Table 3. In the following paragraphs some pertinent details are given.

The number and size of the errors in the data have been decreasing over the course of the experiment, this is shown in Table 3. At the

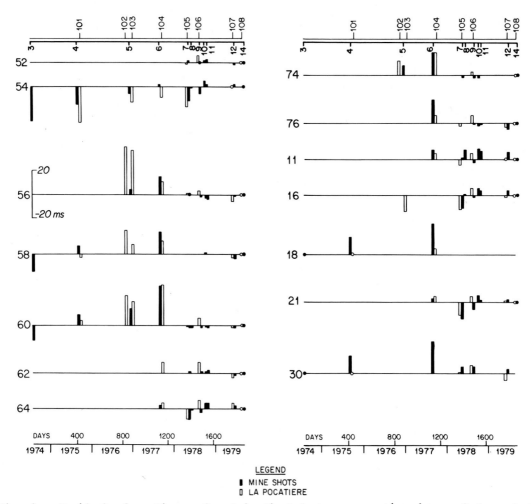

Fig. 2. Residuals from Figures 2 and 3 adjusted to common time base of last shot. Shot 13 is not included.

top of the table the shot numbers on the north and south shores are given and below that the years. Below this are seven lines that give various errors in milliseconds for the shots to which they apply. The first line applies to array (A) data only and is the error in digitizing the analogue array data. The next four lines refer to both array (A) and backpack (BP) data, in descending order are given the following errors: blaster box, shooter clock, recorder clock and commercial blaster box error. The next error applies only to the BP's and is due to the uncertainty in the start of the digitizing of the BP's. The last line of errors is the amount of uncertainty that accounts for the different instrumental responses when cross-correlating between BP and A data. The last three lines give the RMS error for the different combinations of instruments in the cross-correlation process. The RMS errors given were obtained by combining the errors of shot 14 with each of the earlier shots and similarly for shot 108. As an example for the

BP-BP combination shots 14 and 6 the RMS error is $(2(2^2) + 3(2)^2)^{1/2}$ = 5 ms or for the A-BP combination of shots 14 and 5 $(4^2+2(2^2) + 3(2^2) + 8^2 + 5^2)^{1/2}$ = 11 ms.

Thus the RMS errors for BP - BP combinations range from ±4 to ± 9 ms, those for A - A combinations from ±7 to ±10 ms and those for the mixed instruments from ±8 to ±11 ms. For the BP's the errors have thus decreased by a factor of two during the course of the experiments.

The errors given above are the instrumental or timing errors only. There is also the error due to the cross-correlation process, which may amount to 1-4 ms but is difficult to estimate. At any rate the error given for the new BP-BP combinations is most likely the lower limit that may be achieved for the overall process using our systems and because of the frequency content of the wave forms.

Interpretation

In order to detect any possible long term trends in the data for the two shot points

TABLE 3

SUMMARY OF ERRORS

SHOT	NUMBERS	YEAR	Array digit.	Blaster box	Shooter clock	Recorder clock	Commerc blaster	BP Micropr.	BP-A	BPxBP	A x A	BPXA
				ARRAY				BP		RMS ERRORS		
3		1974	4			2	7		5		10	
4	101	1975	4	2	2	2			5		7	
5	102 103	1976	4	2	2	2		8	5	9	7	11
6	104	1977	4	2	2	2		8	5	9	7	11
7			4	2	2	2			5	5	7	8
8	105	1978	4	2	2	2			5	5	7	8
9	106		4	2	2	2			5	5	7	8
10			4	2	2	2			5	5	7	8
11			4	2	2	2			5	5	7	8
12	107		4		2	2			5	4	7	8
13		1979	4		2	2			5	4	7	8
14	108		4		2	2			5			

consider Figure 2. Shot 13 has been eliminated from Figure 2, later it will be shown to have been systematically late by 11 ms. Here we are interested in the long term drift of the residuals and not short term variations.

Referring to Figure 2 it is evident that the residuals are essentially stationary and zero at station 52 in 1978 and 1979. Stations 62, 64, 11 and 21 show the same since 1977. Station 60 typifies the change in residual as seen from both shot points: an increase from 1974 to 1977, a sudden drop in 1978 and no change in 1979. The same pattern is seen, although with fewer data points, at stations 74, 76, 18 and 30. Stations 56 and 58 are similar to station 60 but only as seen from the NSSP. Seen from the SSSP the peak occurs one year earlier, in 1976. Station 54 shows similar changes in residual as seen from both shot points. Disregarding two early residuals in 1977 it is not clear whether the maximum has occurred in 1978 or 1979 or whether it has in fact been reached yet.

Short term negative travel time residuals have been observed from shots 105 and 7 at least at stations 64, 16 and 21. They may be present at other stations also but are not readily separated from the long term drift.

There are some general considerations that should be noted in the interpretation of the data, based on the observed micro-seismicity and the dilatancy-diffusion model. a) There are no earthquakes in the Paleozoic sediments under the south shore stations, thus no dilatancy related changes in velocity are expected in SSSP path to south shore stations for paths less than about 25 km. b) Dilatancy and diffusion explains only positive anomalies, whether long or short term. c) Negative anomalies may be due to crack closing that increases velocity.

Thus the long term positive changes in residual can be explained in terms of a) and b). The short terms negative residuals at stations 64, 16 and 21 can be explained in terms of c).

The sketch in Figure 3 is shown as an aide for the interpretation given above. The 5 stations that show no or few anomalies are indicated by double circles around the station location. The anomaly labelled 1976 and seen by stations 56 and 58 has to be between these stations and the SSSP. The anomaly 1977 has to extend beyond the paths from NSSP to stations 74 and 30 and beyond the path to station 60 to the SSSP, however, it does not necessarily have to include the stations 60 and 74 as shown. The anomaly 1978/79 should not extend over the path NSSP to 11.

Thus the interpretation for the long term velocity changes is that a maximum was observed near to the centre of the zone, the maximum migrated in 1977 and 1978 towards the southwest and in 1977 also towards the north east. The maximum collapsed for station 56 starting in

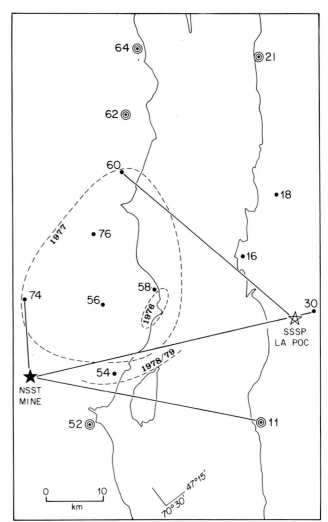

Fig. 3. Sketch of change in the maximum of travel time residuals. Boundaries are arbitrary except for minimum extent as given in the text.

1977 and for the other stations in and around the impact crater in 1978. The contour labelled 1978/79 indicates that the maximum may have propagated to this region, the problem is that the maximum at station 54 is not pronounced and may not yet have been reached.

Discussion

From the cross-correlation lags of the NSSP in Table 1 the average shot residuals for each shot were determined. Since they are relative to shot 14 they are referred to as the relative average shot residual (RASR). They are shown in Figure 4. RASR were also determined for the SSSP data from Table 2 and they are also shown in Figure 4.

If the RASR is zero, it not only means that

there was no long or short term average change between shot 14 and the comparison shot, but there was also no significant error in the shot time, unless fortuitously one cancels the other. If the RASR is not zero, at least three interpretations are possible: (1) it is due to transient velocity changes near the shot points, (2) it is due to timing errors of the shot instant, (3) it is due to the long term trend. Unfortunately any combination of the above is possible with the present data set which poses problems in the interpretation.

The RASR for the NSSP increases uniformly from -19 ms to +14 ms between shots 3 and 6 respectively. After that they are essentially zero except for shot 13.

Shot 13 was set off 24 hours after shot 12 and the same instruments occupied each site. The difference in residual, shot 13 minus shot 12 has a mean and 95% confidence limits of 11.2+1.7 ms for 9 degrees of freedom. The difference is therefore significantly different from zero, shot 13 being 11.2 ms later on the average than shot 12. Because of the small scatter from the mean of shot 13, +1.7 ms, the shot may have an error in its origin time. Alternatively there occurred a change in velocity near to the shot point and after shot 12. This interpretation is less likely since the nearest larger event had magnitude 2.9 and occurred 20 km SSW from the shot point 8.5 days after shot 12.

Shots 10 and 11 were planned to test the hypothesis that the vertical component of the solid earth tide may cause changes in velocity. On November 2, 1978 relatively large tidal differences of 220 μ gals were expected and the shots were set off at 05 and 22 hours. The observed difference in residual, shot 11 minus 10, ranged from -4 to +1 ms with an average and 95% confidence limits of -1.0 + 1.3 ms. The difference is thus not significant and the hypothesis of tidal effects is rejected. This is also evidence that the accuracy of our experiments is well within the errors that were foreseen. Therefore for the NSSP there was a slow long term increase in RASR from 1974 to 1977 with no significant long term change after that.

For the SSSP there was a peak in long term drift in 1976. While it is correct to estimate the significance of changes in RASR for the NSSP shots 10 to 13, this cannot be done for the remainder of the shots because the station occupation was very variable, fewer stations were occupied for each shot, which results in poor statistics so that the RASR cannot be taken too seriously but may only be representative of a trend.

Shot 102 was followed 2.5 days later by a M_L 2.3 event. Thus this event and the one mentioned previously can only be responsible for velocity changes if one admits a threefold increase in the precursor time as given by Whitcomb et al. (1973).

The simplest interpretation for the RASR is that the north shore shot point shows a maximum in 1977, whereas the south shore shot point shows it one year earlier. Unfortunately, because of the unstable station configuration it is difficult to assess the significance of this.

The positive travel time changes, long and short term, can be interpreted by dilatancy-diffusion as can the long term decreases as observed at station 56 from the south shore point. The short term negative travel time changes can be interpreted in terms of crack closing.

Buchbinder (1980) has also determined the S-wave travel time changes by the same method that was used here for the determination of the P-wave residuals. The travel time changes for P and S are about equal, unless they are due to timing errors, which leads to the relation $d\alpha = 3 d\beta$, where α and β are the P- and S-wave velocities. This relation between the velocities can be interpreted by the theory of O'Connell and Budiansky (1977) for viscoelastic properties of fluid-saturated cracked solids. This theory predicts the crack density parameter increases and the partial saturation parameter decreases for negative values of both $d\alpha$ and $d\beta$. Our observations are therefore compatible with dilatancy-diffusion.

All the earthquakes of magnitude up to 3 that could be connected to travel time anomalies have been mentioned above. In addition to these, only 2 larger events occurred during the course of the experiments. A magnitude 4.2 event occurred on 23 October 1976 off St. Simeon near to station 64. With an empirical precursor time of a month this event could not have been detected. A magnitude 5.0 event occurred on 19 August 1979 off St. Fidele near to station 62. With an empirical precursor time of 5 months it should have produced an anomaly for shots 107, 12 and 13. Shots 108 and 14 were set off 17 days after this event. No significant changes in travel time were observed, with respect to the 3 earlier shots.

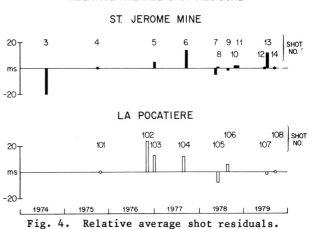

Fig. 4. Relative average shot residuals.

CONCLUSION

Significant changes in P travel times from two shot points on either side of the St. Lawrence River, Quebec have been observed at a number of stations. The maximum change at a station is 46 $\pm$ 11 ms. The maximum difference from one shot point to two stations is 76 $\pm$ 16 ms. The simplest and uniform interpretation of the data is that a minimum in velocity was observed in 1976 near the centre of the Charlevoix impact structure. By 1977 the minimum had migrated outward to cover some 30 - 40 km parallel to the river. By 1978 the minimum, had moved further towards the SSW but the extension towards the NE is not known. These conclusions are in agreement with dilatancy-diffusion. Stations at the periphery have not shown changes in travel time. Negative residuals or increases in velocity can be explained by crack closing. Two small magnitude 2.5-3 earthquakes are not likely to have produced short term changes in travel time residuals. A magnitude 5.0 event with a thrust type focal mechanism solution at 10 km depth certainly did not produce significant changes in residuals.

The experiment is continuing to shed more light on the speculative interpretation given here and to resolve the inconsistancies. Other geophysical measurements such as magnetic, gravimetric, geodetic, water level, tilt and strain in the region are also continuing so that a synthesis of all the data may be produced.

Acknowledgements. Drs. A. Green and C. Wright critically read the manuscript, their and the annonymous reviewers comments lead to its improvement. Mr. A.J. Wickens help in the data analysis is greatfully acknowledged.
Contribution from the Earth Physics Branch No. 897.

References

Anderson, D.L. and H. Whitcomb (1975). Time dependent seismology, J. Geophys. Res. 80, 1497-1503.

Anglin, F.M. and G.G.R. Buchbinder (1980). Microseismicity associated with structural features in the Charlevoix region, Quebec, Can. Jour. Earth Sciences (submitted to).

Buchbinder, G.G.R. and C.M. Keith (1979).

Stability of P travel times in the region of La Malbaie, Quebec, Bull. Seism. Am. 69, 463-481.

Buchbinder, G.G.R. (1980). Precise P- and S-wave velocity variations, crack density and saturation changes, submitted to Jour. Geophys. Res., 1980.

Hodgson, E.A. (1950). The Saint Lawrence Earthquake March 1, 1925, Dominion Observatory Publications VII, 10, 365-436.

Leblanc, G. and G. Buchbinder (1977). Second microearthquake survey of the St. Lawrence Valley near La Malbaie, Quebec, Can. J. Earth Sci. 10, 2778-2789.

Leblanc, G., A.E. Stevens, R.J. Wetmiller, and R. Duberger (1973). A micro earthquake survey of the St. Lawrence Valley near La Malbaie, Quebec, Can. J. Earth Sci. 10, 42-53.

O'Connell, R.J. and B. Budiansky (1977). Viscoelastic properties of fluid-saturated cracked solids, Jour. Geophys. R., 82, 5719-5735.

Robertson, P.B. 1968. La Malbaie structure, Quebec - a Paleozoic meteorite impact site. Meteoritics, 4, pp. 89-112.

Rondot, J. 1968. Nouvel impact meteoritique fossile. La structure semi-circulaire de Charlevoix. Canadian Journal of Earth Sciences 5, pp. 1305-1317.

Roy, D.W. (1978). Origin and evolution of the Charlevoix cryptoexplosion structure, Ph. D. Thesis, Princeton University.

Smith, W.E.T. (1962). Earthquakes of eastern Canada and adjacent areas 1534-1927, Dominion Observatory Publications 26, 269-301.

Stevens, A.E. (1980). Re-examination of some larger La Malbaie, Quebec, earthquakes (1924-1978), Bull. Seism. Soc. Am. 70, 529-557.

Ward, P.L. (1979). Earthquake prediction, Reviews of Geophys. and Space Physics, 17, 343-535.

Whitcomb, J.H., J.D. Garmany, and D.L. Anderson (1973). Earthquake prediction: Variation of seismic velocities before the San Fernando Earthquake, Science 180, 632-635.

Winkler, K.W. and A. Nur (1978). Depths constraints on dilatancy induced velocity anomalies, in Earthquake Precursors, C. Kisslinger and Z. Suzuki, Editors, Center for Academic Publications Japan, Japan Societies Press, pp. 231-241.

A CONCENTRATION CRITERION FOR SEISMICALLY ACTIVE FAULTS

G. A. Sobolev and A. D. Zavialov

Institute of Physics of the Earth, Moscow, USSR

According to the kinetic theory of strength of materials [Zhurkov, 1953] the failure of brittle solids is a long-term process which depends upon material properties, stress and temperature. Utlimate failure is preceeded by formation of cracks and their accumulation in some volume of the material. The critical crack density which preceeds macrofracture formation may be expressed as follows

$$K = N_*^{-1/3}/L \qquad (1)$$

where K is the concentration criterion, N_* is the number of randomly distributed cracks occurring in a unit of volume prior to crack coalescence and macrofracture propagation, and L is the average crack length.

A mean concentration criterion of K = 3 to 5 has been determined in laboratory studies [Zhurkov et al., 1977] for different materials (including both minerals and rocks) greatly varying in both crack length and number.

Theoretical analysis has shown [Panasuk, 1968] that two co-linear cracks may be treated as isolated if the distance between them exceeds 3 times the crack length (i.e., $a/l \geq 3$; where a = the distance between cracks of length 1). The same result has been obtained from theoretical analysis of the coalescence of randomly distributed cracks [Petrov, 1979].

The theory of scale independence in the failure process [Miachkin et al., 1975; Sobolev et al., 1979] led us to evaluate the concentration criterion for seismic faults (K_{sf}). A catalog of local earthquakes which occurred in the Kamchatka region during the period 1963-1978 has been used. This catalog lists 28,000 earthquakes of energetic class $K^{F68} \geq 8.5$ [Fedotov, 1968] and with depths ≤ 100 km. The seismically active zone of Kamchatka has been divided into overlapping elementary volumes. Values of K_{sf} have been computed for each volume over a time interval, T, beginning with T_p. For calculations involving in situ faults, the variables N_* and L of equation (1) must be redefined as follows

$$N_* = N_\Sigma/V \qquad (2)$$

where N_Σ is the total number of earthquakes in volume V for the time interval T;

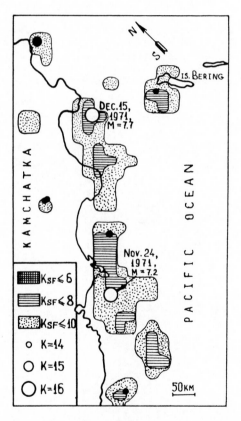

Fig. 1. Map of K_{sf} values in the Kamchatka seismoactive area from the period January 1, 1963 to November 1, 1971. The range of K_{sf} values are: 1. 4.1-6.0; 2. 6.1-8.0; 3. 8.1-10.0. Earthquakes denoted by black circles occurred between January 1, 1963 and November 1, 1971. White circles denote future earthquakes of November 24, 1971 (M = 7.2) and December 15, 1971 (M = 7.7). The energetic class of earthquakes is given as K^{F68}.

$$L = \frac{\sum\limits_{i=1}^{N_{\Sigma}} l_i}{N} \qquad (3)$$

where l_i is the fault length of a seismic event of energetic class K_i and can be evaluated [Risnichenko, 1976] using the empirical equation

$$\lg l_i = 0.244 K_i - 2.666.$$

The values of K_{sf} calculated in this manner are shown in the maps of the Kamchatka region (Figures 1-3). The dimensions of the elementary volumes used were $(100 \text{ km})^3$, $(50 \text{ km})^3$ and 200 km x 200 km x 100 km. The time interval $T - T_o$, over which K_{sf} is calculated, continually increases. T_o remains constant and K_{sf} is recalculated as T^o increases by 0.5 year intervals. Analysis of these data show low values of K_{sf} occur in future source areas 4.5 +3 years before a large event.

The concentration criterion is a cumulative parameter, thus the choice of T_o is critical to calculations. Fedotov [1968] studied the duration of seismic cycles in Kamchatka and found the period of premonitory seismic activity prior to large

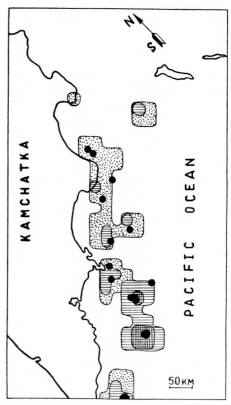

Fig. 3. Map of K_{sf} values in the Kamchatka seismoactive area for the period January 1, 1973 to December 31, 1978. Symbols and values are the same as those in Figure 1.

earthquakes to be about 10 years. Since our data span a 15 year interval we feel justified in our analyses.

The earliest seismic events used in the study cannot pre-date those in the comprehensive catalog of Kamchatka earthquakes (i.e., 1 January 1963). This value for T_o is satisfactory for analysis of seismicity preceding the two strongest Kamchatka earthquakes, which occurred on November 24, 1971 and December 15, 1971. Up until November 1, 1971 K_{sf} values in the epicentral region of these and two additional earthquakes were between 10 and 8 for elementary volumes of 200 km x 200 km x 100 km; 8 and 6 for volumes of $(100 \text{ km})^3$; and approximately 5.5 for volumes of $(50 \text{ km})^3$. The maps in Figures 1-3 show K_{sf} values for elementary volumes of 50 km x 50 km x 100 km. One can see from Figure 1 that the zones of small K_{sf} values do not overlap.

Most zones of small K_{sf} values which are void of earthquakes in Figure 1 later experienced earthquakes of magnitude $\geq$ 5.5 (Figure 2). The seismicity surrounding Bering Island is the only exception to this correlation. Here, K_{sf} values as low as 6.4-6.7 occur near the island during the time interval January 1, 1973 to December 31, 1978, yet no major earthquakes occur. The same situation

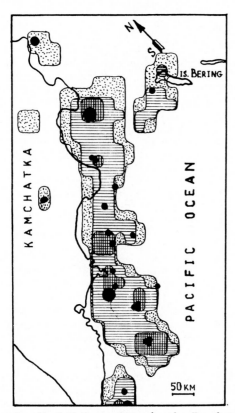

Fig. 2. Map of K_{sf} values in the Kamchatka seismoactive area for the period January 1, 1963 to December 31, 1978. Symbols and values are the same as in Figure 1.

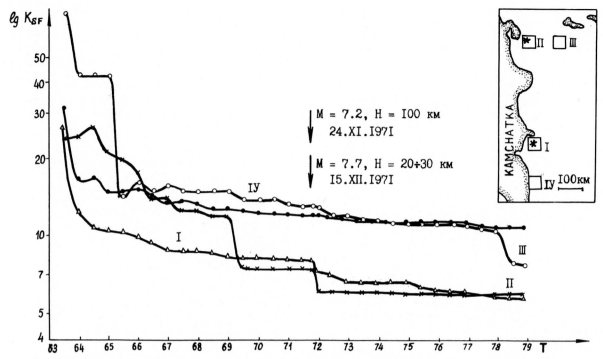

Fig. 4. Time variations of K_{sf} values in the (I) epicentral zone of the November 24, 1971; M = 7.2 earthquake; (II) the epicentral zone of the December 15, 1971; M = 7.7 earthquake; (III) near Bering Island and (IV) in an "empty" region of quiet seismicity. These are regions I, II, III and IV on the inset.

occurs near Bering Island for the period 1973 to 1978 (Figure 3). However, the influence of the aftershocks of the two large earthquake of 1971 previously mentioned is now eliminated. (The sharp drops in K_{sf} values in the Ust-Kamchatka and Petropavlovsk regions during 1972 are due to the aftershock sequences of these two earthquakes.) Comparison of Figures 2 and 3 shows that the choice of T_o (the lower time limit) does not affect the major trends exhibited by the data.

We will now discuss the use of K_{sf} values as a function of time for four small regions. These are the epicentral areas of earthquakes which occurred in the Petropavlovsk region (November 24, 1971; M = 7.2), the Ust-Kamchatsk region (December 15, 1971; M = 7.7), the area adjacent to Bering Island, and the "empty" zone in southern Kamchatka (Figure 4).

The rapid decrease of all K_{sf} values from 1963 to 1966 resulted from the initiation of the count. Prior to the magnitude 7.7 earthquake, K_{sf} values reached below 7 in the Ust-Kamchatsk region. The previous (1969) drop in K_{sf} in this region was due to an earthquake swarm. K_{sf} ceases to drop in the epicentral region after the earthquake and its aftershocks of December 15, 1971, and even increases slightly through 1979. This break in the slope, occurring in 1972, signifies that the zone has entered a period of quiescence and provides a logical time to restart the summation of the K_{sf} values.

In the Petropavlovsk region values of K_{sf} decreased to less than 8 prior to the deep earthquake of magnitude 7.2. In contrast to the magnitude 7.7 earthquake discussed above, K_{sf} values continue to decrease after this earthquake and they may indicate a forthcoming large event. Three earthquakes of magnitude ≥ 5.5 have occurred near this epicentral zone between 1973 and 1978.

In the Bering Island region the values of K_{sf} dropped sharply during the 1978 swarm and continue to decrease up to the present time.

In some volumes of the Kamchatka seismoactive zone the values of K_{sf} remain above 10. In these "empty" volumes no earthquakes with magnitude ≥ 5.5 occurred between 1963 and 1978. In contrast, K_{sf} values were less than 8 prior to 13 of the 15 earthquakes which occurred between 1973 and 1978 with magnitudes ≥ 5.5 (Figure 3). K_{sf} decreased to less than 10 in the region of the remaining two earthquakes prior to the events.

K_{sf} magnitudes are probably greater than analogous values of K, which are calculated from laboratory test specimens, for two reasons. First, lower magnitude earthquakes which may occur during a preparatory phase prior to T_o are not included in the calculations. Second and more importantly, it is reasonable to assume that some percentage of faults in a seismically active region develop slowly, releasing stress via a series of small events. The seismic energy radiated by each small stress drop may not be sufficient for detection by the network of seismological stations.

From our analysis of the data from the Kamchatka catalog we make the following two conclusions:

1. The narrow range of K_{sf} (the concentration criterion for seismic faults) ($5 \leq K_{sf} \leq 8$) falls in close proximity to analogous K values calculated and experimentally determined for rocks and other materials ($3 \leq K \leq 5$). This implies that the same process of crack accumulation governs the failure process both on the laboratory and in situ scales.

2. The seismic fault concentration criterion (K_{sf}) may be a valuable tool for earthquake prediction. More detailed studies are required in order to estimate the behavior of K_{sf} prior to large earthquakes in different seismically active regions.

Acknowledgments. We wish to gratefully thank S. A. Fedotov for letting us use some of his seismological material.

References

Fedotov, S. A., On the seismic cycle, possibility of quantitative seismic zoning and long-term seismic prediction (in Russian), Seismic Zoning in the USSR, pp. 121-150, Nauka, Moscow, 1968.

Miachkin, V. I., B. V. Kostrov, G. A. Sobolev, and O. G. Shamina, Principles of source physics and earthquake precursors, Physics of Earthquake Source, pp. 6-29, Nauka, Moscow, 1975.

Panasuk, V. V., Equilibrium of brittle solids with fractures, Naukova Dumka, Kiev, 1968.

Petrov, V. A., About mechanism and kinetic of failure, Physics of Solids, 21, 3681-3686, 1979.

Risnichenko, U. V., Crustal earthquake source size and seismic moment, Physics of Earthquake Study, Nauka, Moscow, 1976.

Sobolev, G. A. et al., Laboratory Study of Failure Precursors, Gerlands Beitr. Geophys., vol. 88, pp. 140-153, 1979.

Zhurkov, S. N., and B. N. Narsulaev, The temporal dependence of solids strength, Journal of Technical Physics, 23, 1953.

Zhurkov, S. N. et al., On the prediction of rock failure, Izvestia of Academy of Science USSR, Physics of the Earth, 6, 11-18, 1977.

EARTHQUAKE PREDICTION STUDIES IN SOUTH CAROLINA

Pradeep Talwani

Department of Geology, University of South Carolina, Columbia, S. C. 29208

Abstract. Continuous reservoir induced seismicity has been monitored in South Carolina at Lake Jocassee since October 1975 and at Monticello Reservoir since December 1977. In addition, at Lake Jocassee, the radon concentration in groundwater has been monitored for over 3 years, and the water level in an observation well for one year. These form part of a multifaceted approach to seek out precursory changes associated with the larger, $2.0 \leq M_L \leq 4.0$ events. Sequential earthquake source regions inferred from seismicity, changed in brittle state from heterogeneous (January 1976) to moderately heterogeneous (February 1977), to relatively uniform (August 1979). The seismic parameters studied include the spatial and temporal behavior of seismicity, changes in t_s/t_p ratios and P/SV amplitude ratios. The observed seismicity was used to test some statistical algorithms. A fair degree of success was achieved with these methods. Results so far include the formulation of an empirical earthquake prediction model and successful prediction of two small events. Long-period ($\sim$ 44 weeks) and short-period ($\sim$ 1 day) changes were noted in the radon concentration in an observation spring. The water level in an observation well was also found to be associated with short term ($\sim$ 1 day) changes preceding the larger events. The amplitude of these anomalies was found to depend on the distance of the well from the hypocenter and on the magnitude of the earthquake. These observations suggest that a multifaceted approach is a useful one, and that the conclusions and models arrived at after monitoring small earthquakes in a small region can perhaps be extended to larger tectonic ones.

Introduction

Reservoir induced seismicity (RIS) has been observed at Lake Jocassee since October 1975 and at Monticello Reservoir since December 1977. Seismicity at both these sites (Figure 1) located in the Piedmont region of South Carolina, has been continuous, shallow (< 5 km), low level ($M_L < 4.0$) and in each case concentrated in a small area ($\sim$ 100 km^2). Continuous monitoring of this seismicity revealed that it was accompanied by changes in several parameters. Consequently each of these locations serves as a large scale laboratory, where seismicity and other parameters can be observed in space and time. Observation of these makes for a better understanding of the physics of the earthquake process, which in turn leads to an evaluation of various proposed parameters as earthquake precursors.

Most of the discussion in this paper is centered on Lake Jocassee where we have several years' data, and where we have employed a multifaceted approach. In addition to a study of various seismological factors, we have also monitored various geochemical (principally radon concentration in groundwater) and hydromechanical (water level in an observation well) parameters.

This paper is divided into four parts. First, we discuss the nature of our interpretation of reservoir induced seismicity at Lake Jocassee. Second, there is a brief description of various parameters that are being monitored and the results thereof. Third, we present an example of an earthquake that was accompanied by four precursors, and finally, we present some data from Monticello Reservoir which allow us to estimate the size of the anomalous zone associated with $M_L \sim 2.5$ earthquake.

Nature and Cause of RIS at Lake Jocassee

Seismological Background

The Jocassee development project, consisting of a 177-m-high dam and four reversible pump turbines, is a pumped-storage hydroelectric facility located in Oconee and Pickens counties, South Carolina. Construction of the dam was completed in 1972, and the lake, with a capacity of 1.43 km^3 was filled in April 1974.

This region had been regarded as aseismic until an MM intensity III-IV earthquake occurred near the dam in October 1975. This was followed by several felt events, including an M_L 3.2 event on November 25, 1975. The seismic activity at Lake Jocassee has been monitored continuously

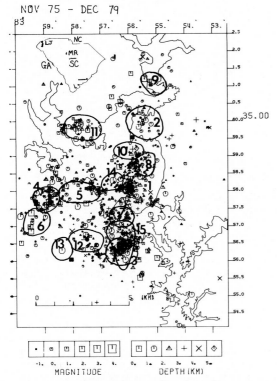

Fig. 1. Lake Jocassee earthquakes from November 1975 to December 1979. Inset shows the location of Lake Jocassee (LJ) and Monticello Reservoir (MR) in South Carolina. The seismicity was monitored on 4-7 portable seismographs (through 1978), which were deployed at 22 sites around the lake. Solid squares show the location of 3 permanent stations deployed since October 1978. These are supplemented by portable seismographs whenever there is an increase in seismicity. The sequential location of larger events and their aftershock zones are indicated by numbered solid rings.

since then. Between 1975 and 1978 the seismicity was monitored on 4-7 portable seismographs. Since January 1979 it has been monitored on 3 permanent stations, which are supplemented by 3 to 6 portable seismographs following any large event or when there is a marked increase in the seismicity. Although there was a decrease in the number of seismographs in 1979, their locations (Figure 1) were such that there was no decrease in the detection threshold ($\sim M_L$ -0.8).

Figure 1 shows the locations of about 800 well-located events that occurred in the period 11/75 to 12/79. The threshold for locatable events is about M_L = -0.6. Most of the seismicity is concentrated in a 100 km^2 area in the vicinity of the lake, and is shallower than 5 km.

Association of Seismicity with Reservoir Level

The seismic activity at Lake Jocassee is concentrated in the heavily fractured Henderson augen gneiss unit. In situ stress measurements have been made at Bad Creek, about 10 km NW of Jocassee dam. These include measurements made by using the hydraulic fracturing technique in a borehole (Haimson, 1975), and by overcoring in a pilot tunnel (Schaeffer et al., 1979). Both these measurements indicate the existence of large stresses in the top 300 m. The largest principal stress is horizontal, about 250 bars (at a depth of 236 m), and oriented about N60°E. Composite fault plane solutions indicate that strike-slip faulting is the predominant mechanism in the region (Talwani et al., 1979). This mechanism is associated with earthquakes to a depth of about 2 km.

We have also monitored hourly lake levels. An analysis of 10-day average lake levels and changes, and comparison with seismicity, suggests that periods of large energy release and/or larger earthquakes follow periods of rapid sustained lake level increase (Talwani et al., 1979). This observation together with an analysis of the stress data, focal mechanisms and detailed mapping of the surface fractures, lead us to conclude that the observed seismicity is triggered (induced) by pore-pressure changes in a highly pre-stressed rock. These pore-pressure changes are caused by lake level fluctuations, and the seismicity is related to an existing network of fractures, rather than to breaking of new rock (Talwani et al., 1979; and in preparation).

Change in the Nature of Seismicity with Time

Without any change in the detection threshold, we observed that the frequency of seismic activity declined from about 5-6 events a day in early 1976, to about 2 events a day in 1977. By the end of 1979 it had dropped farther to one every 2 or 3 days. Along with a decrease in the earthquake frequency, there was also a change in the nature of seismicity. This is illustrated in Figures 2a-2c, which show the daily number of events for 20-day periods before and after three large events ($2.0 \leq M_L \leq 4.0$). The events were chosen to cover 3 distinct periods—periods of intense activity, lesser activity and markedly decreased activity, including events on January 14, 1976, February 23, 1977 and August 26, 1979 respectively. The seismicity during each period is compared with frequency of shocks in laboratory experiments by Mogi (1963, 1967). In these experiments it is assumed that earthquakes are caused by brittle fracturing of the stressed earth's crust, and the fracturing can be simulated in the laboratory. Then the frequency curves of the elastic shocks have a close relation with structural states of material and space distribution of applied stresses. The observed behavior in January 1976 (Figure 2a) when there was a marked increase in the daily number of events is comparable to that of heterogeneous brittle rocks. Note that the largest number of events per day occur <u>before</u> the main shock. The observed behavior in February 1977 (Figure 2b)--foreshock, main shock and aftershocks--has been classified as Mogi Type II (Mogi, 1963). In this case, the largest number of events per day occur <u>after</u> the main shock. It is observed in regions of moderately heterogeneous structure and/or non uniform stress conditions. There was a marked decrease in seismicity before the 1979 main shocks (note change in scale in Figure 2c), and the seismicity pattern resembles a Mogi Type I pattern (Mogi, 1963) which is characteristic of homogeneous structure under the application of uniform stress.

It is important to note that there was no change in the number of instruments deployed after the events on January 14, 1976 and February 23, 1977. The number of instruments was increased within a day after the event on August 26, 1979. Hence the patterns illustrated in Figures 2a-c reflect actual changes in seismicity and are not a result of instrument changes.

Thus in the seismically active regions in the vicinity of Lake Jocassee there was a change in the inferred brittle state of the respective source regions. The inferred brittle source regions changed from heterogeneous (January 1976) to moderately heterogeneous (February 1977) to relatively uniform (August 1979). The nature of the source region, as will be shown later, has a bearing on whether or not a particular precursor is seen. Clearly, when the source region is homogeneous and there is little if any foreshock activity, we would not expect to observe seismological precursors.

In Figure 3 changes in b-values are compared with the earthquake frequency and energy release. The frequency and seismic energy release are again computed for 10-day periods. The b-value was obtained by least squares from a plot of cumulative number of events plotted as a function of their duration. The b-values on the top row were calculated for foreshock-aftershock sequences (arrows indicate time of events with $M_L \geq 2.0$) and for any period chosen, at least 100 events were considered. In the second row, the histogram was obtained for events occurring in 3-month periods. We note that the b-values

Fig. 2. Number of events per day for 20-day periods before and after 3 large events, compared to laboratory data on rock fracture (see text). a. For M_L 2.5 event on 01/14/76, b. M_L 2.3 event on 02/23/77 and c. M_{bLg} 3.7 event on 08/26/79. The number of events are for 24 hour days keyed to "time zero", the time of the main shock. A check next to Vp/Vs etc. indicates that precursory changes were observed in that parameter.

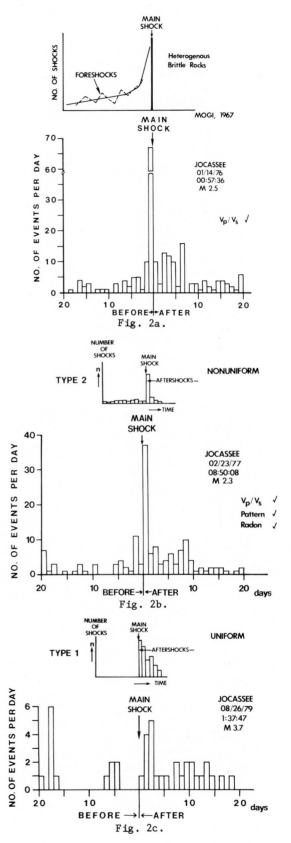

Fig. 2a.

Fig. 2b.

Fig. 2c.

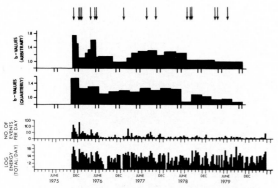

Fig. 3. b-values for the 5 year period
compared with the frequency and seismic
energy release. The b-values were obtained
for foreshock-aftershock sequences (top row)
and quarterly (second row). The arrows
indicate the larger ($\geq M_L$ 2.0) events.

decrease with time, as do the number of events,
whereas the seismic energy release does not.
The implication of this observation is discussed
later.

Epicentral Growth and the Location of Large Events

Figure 4 shows the growth of the seismically
active region observed in the period between
November 1975 (when seismicity began) and May
1976. The areal growth was governed by the local
geology, and it occurred in the first six months
of seismicity. Subsequent seismicity (through
1979) was contained, for the most part, within
the epicentral envelope defined by May-June 1976.

Figure 1 shows the sequential location of all
the larger ($M_L \geq 2.0$) events and their aftershock
zones. These occurred in November 1975 (event
number 1) through August 1979 (event number 15).
We note that the events occur in an area where
no previous large event has occurred. The larger
events and their aftershocks appear to fill a
void in prior seismicity. This observation was
utilized to define an additional criterion in
predicting the location of a future earthquake
(Sauber and Talwani, 1980). Event number 15
was deeper than earlier events in the area (no.
3 and 7) and its aftershock zone underlay those
of an earlier event.

Summary

Figure 5 is a cartoon summarizing the observed
reservoir induced seismicity at Lake Jocassee.
It represents a NW-SE cross-section across the
lake. Knowledge of local geology, seismic
reflection profiling at Rosman, N.C. (12 km to
the NE) by Clark et al. (1978) and the hypocentral locations of earthquakes were all
incorporated in compiling the cartoon, which
represents our interpretation.

The early seismicity that followed the
impoundment of Lake Jocassee was shallow, and
spread laterally. We infer that the shallow
seismicity was associated with an abundance of
fractures in the crystalline Henderson gneiss
rocks. The seismic behavior was similar to that
observed in heterogeneous crystalline rocks under
laboratory conditions. Large b-values were
observed. The larger events were associated with
the larger of these fractures (thick lines).
However once the 'locked' portion of these
fractures had been 'unlocked', i.e. once a large
event occurred, there was no further seismicity
associated with that fracture system.

Gradually, the nature of seismicity changed
to a Mogi Type II pattern, that of moderately
heterogeneous conditions. There was some
deepening in the seismicity, as well as a
decrease in the frequency of events. In the
following years, the seismicity deepened further,
and there being a marked decrease in the number
of fractures available, there was a marked
decrease in activity. The source regions now
behave as more homogeneous ones (Mogi Type I).
Further downward growth of seismicity is constrained by the presence of an impervious layer
of Brevard phyllites. Thus the seismicity is
contained in a 'bowl' of fractured Henderson
augen gneiss.

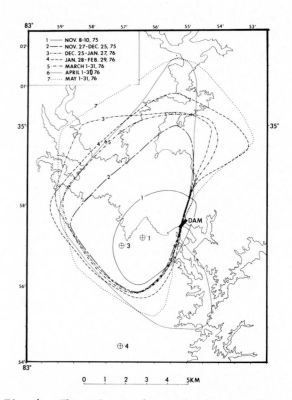

Fig. 4. The epicentral growth of seismicity
at Lake Jocassee in the first six months
after its inception in November 1975.

Earthquake Prediction Program at Lake Jocassee

In view of the continuous nature of the observed reservoir induced seismicity at Lake Jocassee, a multifaceted earthquake prediction program was started in 1976. The various elements of the program are summarized in Table 1. Several elements of this program have been discussed elsewhere and they will be dealt with briefly.

Seismic Parameters

Spatial and Temporal Variation of Seismicity or the Empirical Earthquake Prediction Model. The seismic activity at Lake Jocassee is concentrated in a small area (Figure 1). Our ability to accurately locate small events ($M_L \sim -0.6$) has enabled us to follow the spatial and temporal variation of seismicity. This model and its development have been described earlier (Talwani, 1979) and will be summarized here.

1. The Model. The model is shown schematically in Figure 6. The precursor sequence begins with a period of slow (or no) increase in seismicity, called the α-phase, which is followed by a period of rapid increase in seismicity, termed the β-phase. Within the β-phase, there is a period of quiescence, when the increased seismicity suddenly stops. The duration of the maximum period of quiescence within the β-phase is termed the γ-phase.

The onset of the precursor activity, (or the start of α-phase) was recognized by observing one or more of the following: an increase in the number of events, an anomalous change in the concentration of radon in groundwater, or by a decrease in the t_s/t_p ratio values.

The location of a main event was predicted by observing the locations of earthquakes in the α and β phases. In the α-phase, the period of slow (or no) increase in seismicity, the location of events are scattered in the vicinity of the 'target' area, although there is an absence of activity in the 'target' area itself. In the β-phase, the period of increased seismicity, the activity is concentrated in a small cluster,

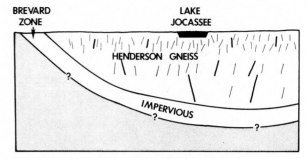

Fig. 5. A cartoon along a NW-SE cross section across Lake Jocassee to depict our understanding of the nature of seismicity.

which defines the 'target' area--the location of the main shock.

The time of the main shock is predicted by identifying the γ-phase or the period of quiescence within the β-phase. The main shock occurs within 2 days of the resumption of activity.

The magnitude (M_L) of the main shock is predicted from the empirical relation
$$M_L = 2 \log D - 0.07$$
where D is the duration, in days, of the fore-shock sequence, i.e. α + β.

2. Results. For the period 1976-78, ten large ($M_L \geq 2.0$) events occurred in the area. The α-phase was identified for 7 of them, the β-phase for 8 and the quiescence or γ-phase was observed for all. Of the 8 cases where clustering was observed (β-phase), the main shock occurred in the target area (often within a circle of 1 km diameter for M_L 2.0-2.5 events) 5 times; for the others, it was located from 2 to 7 km away from the cluster (Tables 2A and 2B).

t_s/t_p Ratio Changes. The travel-time ratio of shear to compressional waves (t_s/t_p) was determined for each located event by using Wadati plots. Since December 1975 we have obtained over 600 such values. (A detailed paper on these observations is under preparation). The mean value is 1.69 with a standard deviation of 0.09, so we define a t_s/t_p anomaly when we observe 3 or more values less than 1.60 occurring in one week. When we observe 2 or 3 such anomalous values in a week, the anomaly is classified as 'maybe'. Three examples of events associated with precursory t_s/t_p anomalies are shown in Figure 7. These events occurred on January 14, 1976 (M_L 2.5), April 23, 1976 (M_L 2.1) and November 25, 1977 (M_L 2.2). The events in January 1976 and November 1977 were successfully predicted, based on the t_s/t_p ratio anomalies (Stevenson et al., 1976, 1978; Talwani et al., 1978). Some details pertaining to the November 1977 event are given below. A magnitude 2.3 event on February 23, 1977 was associated with a t_s/t_p ratio as well as other anomalies and is discussed in a later section.

TABLE 1. Earthquake Prediction Program at Lake Jocassee

I. Seismic Parameters
   a. Spatial and Temporal Pattern of Seismicity
   b. t_s/t_p Ratio Changes
   c. P/SV Amplitude Ratio Changes
II. Geochemical Parameters
   a. Radon Concentration in Groundwater
   b. Radon Concentration in Soil
   c. Other Geochemical Parameters
III. Hydromechanical Parameters
   a. Water Level in an Observation Well

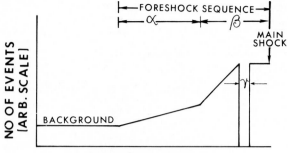

FORESHOCK SEQUENCE

MAIN SHOCK

NO OF EVENTS (ARB. SCALE)

BACKGROUND

TIME (ARB. SCALE)

Fig. 6. Schematic earthquake prediction model. The foreshock sequence consists of slow (or no) increase in seismic activity (α-phase) followed by a rapid increase (β-phase). The earthquake occurs following a period of quiescence (γ-phase).

1. Prediction of the November 25, 1977 M_L 2.2 Event. The t_s/t_p ratio values that led to this prediction are shown in Figure 7. The ratio decreased from a normal background value of about 1.69 on November 10 to 1.41 by November 20 and returned to the background value on November 23. This implied a 10-13 day 'bay', and based on the precursor time-magnitude relationship for Lake Jocassee (Stevenson, 1977) we predicted an earthquake of magnitude 2-2.5 to occur in the next few days. Figure 8 shows the sequential location of events associated with the t_s/t_p 'bay'. Based on their clustering (events 1, 3, 4, 8) and their identification with β-phase, we predicted the location of the main shock to be within the cluster, located about 2 km NW of BL2

(Figure 8). This prediction was telephoned in to Dr. Wesson of the U.S. Geological Survey of November 23, 1977.

A magnitude M_L 2.2 event occurred on November 25, 2977 (22:22 h UTC) as predicted. However its location (Figure 8, solid circle) and that of its aftershocks (events 10 and 11) was to the north of the lake, about 7 km to the northeast of the predicted location.

2. Results. In the three year period (1976-78) we noted ten t_s/t_p anomalies. Large ($M_L \geq$ 2.0) earthquakes occurred after 6 of them and there were 4 false alarms. For 10 events with $M_L \geq 2.0$ that occurred in this period, 6 were associated with t_s/t_p anomalies, 3 were associated with 'maybe' anomalies and one event was not preceded by a low value (Tables 2A and 2B). This particular event in September 1977, was over 2.5 km deep. When we studied all the t_s/t_p ratios, we noted that low values were associated with the shallower (Z < 2.5 km) earthquakes only (Talwani and Stevenson, in preparation).

3. Travel-time Residuals from Liberty Quarry. We monitored the relative travel-time differences between stations BG3, BL2 and KTS (Figure 8) for blasts at Liberty Quarry, located about 30 km southeast of Lake Jocassee. For the period December 1975 - February 1978 the relative travel-time residual changes were not distinguishable from the noise.

P/SV Amplitude Ratio Changes. We studied the ratio of P and S wave amplitudes (P/SV) and polarity of P-wave arrivals on vertical component seismographs for the foreshocks and aftershocks

TABLE 2A. Summary of Results: Precursors of Earthquakes with $M_L \geq 2.0$ (Aftershocks removed)

No.	Date Y M D	M	Z(km)	α-phase	β-phase	γ-phase	Target	ts/tp	P/SV (no)[1]	P.R. (No.)[3]	Algorithm KB[4]	KM[5]	Radon	Well	Nature Mogi Type	Remarks Event No. in Fig. 1
1.	75 11 25	3.2														1.
②.	76 01 14	2.5	0.5	Yes	Yes	Yes	Yes	Yes	Yes (1)	No	No	Yes	No		III	2.
3.	76 04 23	2.1	0.1	Yes	Yes	Yes	No (2)	Yes	Yes (1)	Yes (1)	Yes	Yes	Yes (S)			5.
4.	76 06 02	2.0	0.9	Yes	Yes	Yes	Yes	Yes	No	No			Yes (S)			6.
5.	76 06 11	2.2	0.5	Yes	Yes	Yes	No (3)	M[7]	No	No	]YES	]NO	No			6.
6.	77 02 23	2.3	1.6	Yes	Yes	Yes	Yes	Yes	Yes (2)	Yes (1)	Yes (2)	No	Yes (L)	Yes	II	7.
7.	77 09 07	2.5	2.7	No	Yes	Yes	Yes	No	Yes (2)	Yes (1)	Yes (2)	Yes (2)[8]	Yes (L)	Yes		8.
⑧.	77 11 25	2.2	0.7	Yes	Yes	Yes	No (7)	Yes		No	No	Yes	No			9.
9.	78 08 21	2.3	1.9	No	No	Yes	No	M		Yes (1)			No			10.
10.	78 09 15	2.3	1.7	Yes	Yes	Yes	Yes	Yes		No			No			11.
11.	78 10 05	2.3	1.5	No	No	Yes	No	M		No			No		I	12.
12.	79 05 01	2.1	2-3*													13.
13.	79 05 28	2.5	2-3*				TOO FEW FORESHOCKS										14.
14.	79 08 26	3.7	2-3*													15.

Notes.
1. β-phase cluster in target area. If not distance to target area (km).
2. No. of stations showing P/SV anomaly.
3. No. of stations showing Polarity reversal.
4. Modified Keilis-Borok algorithm.
5. Modified K. McNally algorithm
6. Short period (S) or long period (L) anomaly.
7. "Maybe" ts/tp anomaly (see text)
*Depths based on aftershock depths.
8. Two swarms.
○ Predicted Event, precursor used underlined.

A.   Empirical earthquake prediction model (1976-1978)

No. of Earthquakes:

Used	Phase Identified α	β	γ	Located in Cluster Defining β-phase
10	7	8	10	5

B.   ts/tp Ratio Changes (1976-78)

No. of Alarms	Earthquakes Predicted	False Alarms
10	6	4

No. of Earthquakes	Anomalies Observed	"Maybe" Anomaly	No Anomaly
10	6	3	1 (depth 2.7 km)

Anomaly period   15 days for $\sim M_L$ 2.3 events
Anomalous zone   4 km (radius)

C.   Earthquake Prediction Algorithms

	Alarms	Earthquake Predicted	False Alarms	Earthquake Missed
Modified Keilis Borok	8	5	3	2
Modified McNally	6	4	2	2

D.   Radon Anomaly

Both increases and decreases observed. Long period anomaly when sensor in preparation zone.

Short period anomaly (co-or post-seismic) when sensor outside preparation zone.

E.   Well Level Change

Earthquakes $M_L > 2.0$	Anomaly Observed for	No Data for
3	2	1 (Nov 77)

Anomaly threshold is a function of M_L and hypocentral distance.

Short period anomaly observed $\sim$ 1 day (coseismic/preseismic).

of the larger events. The sequences studied were located in small clusters (1 km X 1.5 km). The individual sequences occurred in a period of a few days.

The amplitude ratio was found to depend (among other factors) on the azimuth and distance of a station with respect to the hypocenter. Changes in the ratio and in P-wave polarity would be observed at some stations and not at others. The observed systematic changes in the ratio and polarity of P-waves at some stations are possibly due to precursory changes in stress patterns, fault plane orientation or in the strength of materials (See Talwani et al., 1978 and Talwani and Rastogi, 1980 for a detailed discussion).

1. Results. The results of a search for precursory changes in P/SV ratios, and in the polarities of the P-wave first arrivals are summarized in Table 2A. Four events were associated with possible changes in P/SV ratio, two of them at two stations. Polarity reversals were noted for 4 events (at one station each) and not for 6 events.

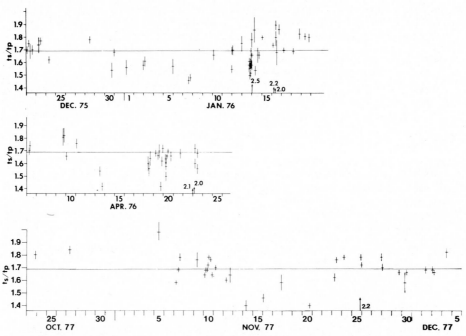

Fig. 7. t_s/t_p ratio values of some events preceding main shocks
in January 1976, April 1976 and November 1977 indicated by arrows.
The mean value for t_s/t_p ratio is 1.69 ± 0.09 (1 s.d.). The
height of the cross is inversely proportional to the quality of
the t_s/t_p ratio value.

Pattern Recognition Techniques. Different
pattern recognition techniques have been suggested
to identify a significant increase in foreshock
activity in the future rupture zone, (identified
as the β-phase in the empirical prediction model).
Among these are the Keilis-Borok (Keilis-Borok
et al., 1977; Caputo et al., 1979) and McNally
(1978) algorithms. In the Keilis-Borok algorithm,
developed for large Italian earthquakes, earthquake 'swarms' are assumed to precede large
earthquakes, and testing criteria were developed
to recognize them. A 'swarm' was identified when
a group of earthquakes, clustered in space and
time, occurred at a time when the regional
seismicity exceeded the normal background
seismicity.

In the McNally algorithm, developed for moderate
central California earthquakes, a series of
earthquake occurrences was broken into non-random
clustered components and a random Poisson-distri-
buted component. The non-random component which
clusters in time is determined by trying different
time interval thresholds, while assuming background
seismicity is randomly distributed in time.

Sauber and Talwani (1980) applied a modified
form of these algorithms to Lake Jocassee data.
Events ranging in magnitude from -0.6 to 2.0 were
used to 'predict' the events of magnitude between
2.0 and 2.6. In the modified Keilis-Borok
algorithm a time window of 2 days was chosen. A
'swarm' was recognized when at least half the

number of events in a given time interval
occurred within an area 1 km in radius. If this
swarm occurred in an area outside the aftershock
zone of previous large events, the 'swarm' was
recognized as a precursor of a large shock. An
earthquake of magnitude ∿ 2.0 is predicted to
occur within 10 days after the end of the 'swarm',
and located in the vicinity of that the swarm
which occurred in an area of quiescence. Of the
8 alarms given by this algorithm, 5 were successful predictions, and 3 were false alarms; 2 events
were note predicted (Table 2A).

In the modified McNally method, precursory
activity was recognized by trying different trail
times, in order to exclude related events until
the remaining data approximated a Poisson random
process. The events that occurred in succession
within this trail time (180 minutes for Lake
Jocassee data) are identified as clusters. In
this method when the cluster size (number of
events/cluster) is larger than the mean cluster
size (three), a large earthquake is predicted.
Of the six alarms given by this algorithm, 4 were
successful predictions, and 2 were false alarms;
2 events were not predicted (Table 2A).

Geochemical Parameters

Radon Concentration in Groundwater. Along
with the monitoring of seismicity in the Lake
Jocassee area, detection of radon concentration

in some wells and a spring began in February 1976. The object of these investigations was to seek changes in the radon concentration associated with the larger $(2.0 < M_L < 4.0)$ earthquakes. A description of the techniques used, and of the results after about three years of simultaneous monitoring of radon and seismicity has recently been published (Talwani et al., 1980) and is summarized below.

1. There are long-term fluctuations (in the form of 50-100% decrease over an 8 to 9-week period) in the radon concentrations in the spring with a period of about 44 weeks. Whether these are related to soil temperatures, water table depth, and/or other environmental factors or to periods of decreased seismicity is not clear. The radon concentration in the spring is diluted by rainwater runoff, although it is not affected by lake level fluctuations.

2. There are some positive and negative radon anomalies possibly associated with the larger seismic events.

3. When the monitoring site is in the epicentral area, for example, February 1977 (or on the same fracture system?), stress changes responsible for the earthquake may cause detectable long-period (over 10 days for $M_L \sim 2.0$ events) precursory radon anomalies.

4. When the monitoring site is distant from the epicentral area, short-period (~ 1 day) changes are observed both before and after the event. These are liable to be missed if samples are not collected within 12 hours of an event. Discrete sampling does not allow us to preclude the possibility of the changes being coseismic rather than precursory. Thus some of the negative results (Table 2A) can be because the discrete sampling rate of about 3 times per week is inadequate to 'catch' the short period changes in the radon concentration.

5. The source of the anomalous radon lies in the immediate vicinity of the sensor rather than in the epicentral area.

6. The exact time of onset of the radon anomaly is probably controlled both by the distance to the source and by the geological conditions.

Radon Concentration in Soil. We also experimented with track etch cups to measure soil radon. We deployed arrays of 25 cups. These were similar to the cups described by King (1978) and were buried at depths of 0.7 m for periods of 14-50 days. Our conclusion after about a 1-year period of their use is that the soil radon method is useful in determining areas of high and low radon concentrations and possibly in detecting long-period anomalies (Talwani et al., 1980).

Other Geochemical Parameters. Starting in February 1978 we began to measure the conductivity and chlorinity of water samples collected from the spring for radon analysis. The object of these measurements was to get an idea of the

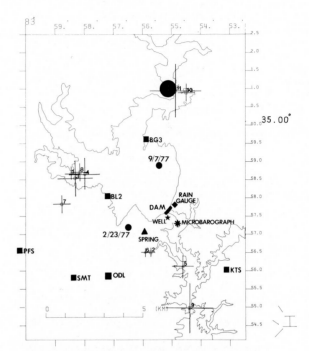

Fig. 8. Sequential location of events (numbered crosses) with precursory low values associated with the November 25, 1977 event (solid circle). Solid squares show location of portable seismographs. The locations of the events on 2/23/77 and 9/7/77, Jocassee dam, Lake Jocassee and Keowee, the observation well and spring, microbarograph and the rain gauge are also shown.

sources of the spring water. The conductivity of the water reflects the amount of total dissolved solids. The chlorinity reflects the amount of Chlorine ions and is generally greater for connate water as compared with surface or rainwater. No significant anomaly in these features was observed in a period of about two years.

Water Level Changes in an Observation Well

The water level in a 100 m deep observation well located in the epicentral area (Figure 8) was monitored for about one year (December 29, 1976-December 29, 1977) in a search for precursory changes associated with the larger earthquakes (Van Nieuwenhuise, 1980). The effect of ocean tides, earth tides, and wind on the response of the well aquifer system were taken to be negligible. The water level in the well was found to be independent of that in Lake Jocassee. Corrections were made for recharge from Lake Keowee, and for changes in barometric pressure. Smoothed, corrected well level curve for the study period is shown in Figure 9, wherein the occurrence time of all earthquakes with magnitudes greater than 1 are also indicated.

During the study period, 35 earthquakes (1.0 < M_L < 2.3) occurred at hypocentral distances between 1.3 and 7.2 km from the well. Of these only 8 had possible associated changes in the well water level. The two large events ($M_L \geq$ 2.3) that occurred on February 23, 1977 and September 7, 1977 were found to be associated with the larger related changes in the well level. Van Nieuwenhuise (1980) determined these changes to be ∿ 10 cm and 9 cm respectively at hypocentral distances of ∿ 3.8 and 5 km respectively. He also found six other events (1.7 < M_L < 2.7) which were possibly associated with water level changes. It is reasonable to assume that the amplitude of the well level changes due to the occurrence of these events is a function of both their magnitude and then hypocentral distance from the well. This amplitude is also likely to be dependent on the rock and aquifer characteristics around the observation well. In Figure 10 the magnitude of events (M_L > 1.0) occurring during the period for which we have well data, were plotted against the hypocentral distance (D km) from the well. The circled data represent those events for which possible associated well level changes were observed. An amplitude (A cm) value of ∿ 5 cm represents the threshold of a detectable anomaly. This is represented by a pair of slant dashed lines. Earthquakes falling to the right (lower magnitude and/or greater hyposentral distance) did not produce detectable anomalies; those of the left did. The amplitude of the anomaly (A cm) was found to fit an empirical relation $A = 8 M_L - 2D$, or $M_L = 0.25D + 0.12A$ where D is the hypocentral distance in km.

Although these data are only for one year, and were obtained for small earthquakes it is interesting to note that the observation that the threshold water level anomaly depends on the

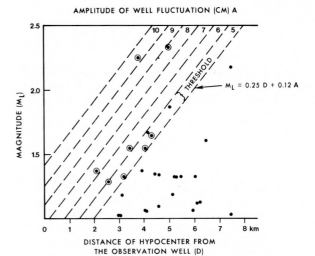

AMPLITUDE OF WELL FLUCTUATION (CM) A

$M_L = 0.25 D + 0.12 A$

MAGNITUDE (M_L)

DISTANCE OF HYPOCENTER FROM THE OBSERVATION WELL (D)

Fig. 10. Magnitude vs. hypocentral distance from the observation well of events (M_L > 1.0) recorded during the study period. The events associated with water level changes are circled. A threshold change of 5 cm in the well level is indicated.

magnitude of, and hypocentral distance to the earthquake was also noted by Sadovskly et al. (1979) for magnitudes 3 to 5.5 South Kurile earthquakes. However, for both the larger events at Lake Jocassee the duration of the anomaly was small (∿ 1 day).

The M_L 2.3 Earthquake of February 23, 1977

The M_L 2.3 event that occurred on February 23, 1977 was located close to the observation well and the observation spring (whose radon concentration is monitored). Their location is shown in Figure 8. This event was accompanied by 4

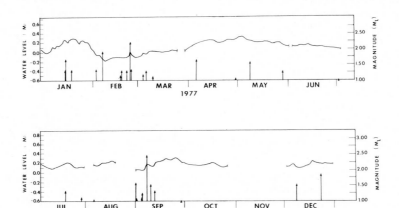

Fig. 9. Smoothed, corrected well level data and the seismic events occurring during the study period. The height of the arrows indicating the events are proportional to their magnitudes.

precursors, which are shown in Figure 11. From top to bottom are the corrected water level in the well; radon concentration in the spring (continuous data using an ionization chamber, and by discrete sampling), daily rainfall, t_s/t_p ratios for located events and water level in Lake Jocassee. The figure shows data for the period January 1 to February 28, 1977; the time of the M_L 2.3 event is indicated by a vertical line on February 23.

We note that the event was preceded by a rapid increase in the corrected well level, which occurred about 1 day before the event. The decline in radon emanation began about 2 weeks before the event. (Also see Talwani et al., 1980 for a more detailed discussion of this radon anomaly). The event was also accompanied by a precursory decrease in the t_s/t_p ratio values, and the spatial and temporal distribution of its foreshocks were in agreement with the empirical earthquake prediction model (Talwani, 1979). Interestingly this event occurred at a time characterized by Mogi Type II seismicity. Implications of these observations are discussed in a later section.

Size of the Anomalous Zone

(This section summarized the results of some observations at Monticello Reservoir (Rastogi and Talwani, 1980). It has been included because the conclusion arrived at is pertinent to earthquake prediction studies at Lake Jocassee).

In the period July–September 1978 induced seismicity at Monticello Reservoir (Figure 1) was recorded on 3-6 portable seismographs together with 10 telemetered stations. Precursory low t_s/t_p ratio values were noted for the foreshocks of two events (M_L 2.7 and 2.3). Low t_s/t_p values were observed only for stations where the observed S-P values were within 1.0 sec. For those stations where the observed S-P time was greater than 1 sec, normal t_s/t_p values were obtained. The hypocentral distance to a station with an S-P time of 1 sec is about 8 km. This observation suggests that the dilatant or otherwise affected area associated with the low anomalous ratios preceding $M_L \sim 2.5$ earthquakes is less than 8 km in extent.

Results and Conclusions

After monitoring seismicity at Lake Jocassee for over 4 years, radon concentration in ground water for over 3 years, and water level in an observation well for one year, we can evaluate the results of this multifaceted approach. The results of various methods used are summarized in Tables 2A and 2B. Other results and conclusions of this study are:

a. Using various parameters, an empirical earthquake prediction model was formulated. It

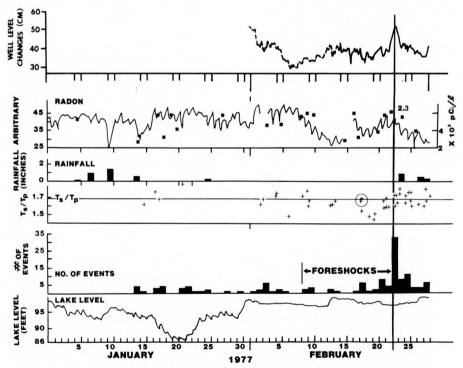

Fig. 11. Various parameters for January and February 1977, and their relationship to a magnitude 2.3 event on February 23, 1977 (see text for details).

was used to successfully predict an M_L 2.2
earthquake on November 25, 1977. The M_L 2.5
January 14, 1976 earthquake was predicted based on
precursory changes in the t_s/t_p ratios.

b. The February 23, 1977, M_L 2.3 event was
found to be associated with 4 precursors. These
results justified our belief in using the reservoir
induced seismicity as a large scale laboratory to
predict earthquakes and understand the physics
of their occurrence.

c. While using seismic parameters, the best
change of success in predicting an earthquake is
when the seismicity is characterized by Mogi Type
II activity.

d. When seismicity is characterized by Mogi
Type I behavior, other parameters, e.g., geochemical and hydrodynamic, should be carefully
monitored.

e. t_s/t_p anomalies were found to be associated
with events shallower than about 2.5 km, and the
earthquake preparation (or the observed anomalous)
zone was about 4 km in radius.

f. The anomalous relative travel-time
residuals from quarry blasts were not distinguishable from the noise in those data.

g. Several factors controlled the observed
P/SV ratios. Extremely good station coverage
(spatially) as well as knowledge of velocity
structure are needed if precursory changes in
P/SV can be used to predict earthquakes.

h. The source of anomalous radon lies in
the immediate vicinity of the sensor rather than
in the epicentral area.

i. When the radon monitoring site is in the
earthquake preparation zone, detectable long-
period (over 10 days for M_L > 2.0 events)
precursory radon anomalies are observed. When
it is outside it, or distant, short-period
($\sim$ 1 day) changes are observed both before and
after the event.

j. For radon monitoring to be successful in
predicting earthquakes, a network of continuous
recording monitors is needed.

k. The anomalous changes in water level were
found to depend on both the magnitude of and
hypocentral distance to the earthquake from the
observation well. Long term (10's of days)
changes were not noted, even when the well was
located in the preparation zone.

l. A network of observation wells might be
able to determine the size and nature of the
earthquake preparation zone.

Acknowledgements. This study was partially
funded by U. S. Geological Survey contract No.
14-08-0001-17670 and NSF Grant EAR76-22722. I
am grateful to former and present students at
South Carolina who made this study a success.
These include David Amick, Jeanne Sauber, Robert
Van Nieuwenhuise and Donald Stevenson.
Discussions with B. K. Rastogi were helpful. I
am grateful to Bill Ellsworth, John Lahr, Gary
Mavko and B. K. Rastogi and the anonymous reviewer for reviewing the manuscript and suggesting
improvements.

References

Caputo, M., P. Gasperini, V. Keilis-Borok, L.
Marcelli, E. Ranzman, and I. Rotwain, Earthquake swarms as precursors to strong earthquakes in Italy, Ann. Geofis. Rome, in press,
1979.

Clark, H., J. Costain, and L. Gover, Structure
and seismic reflection studies on the Brevard
ductile deformation zone, near Rosman, North
Carolina; Am. J. Sci. 1788, 419–441, 1978.

Haimson, B. D., Hydrofracturing stress measurements, Bad Creek pumped storage project,
Report for Duke Power Company, 19 pp., 1975.

Keilis-Borok, V., et. al., Long term prediction
of Earthquakes, Program of Int. Assoc. Seis.
Phys. Earth's Interior, Durham Assembly,
(Abstract), p. 124, August 1977.

King, C. Y., Radon emanation on San Andreas
Fault, Nature, 271, 516–519, 1978.

McNally, K. C., Patterns of earthquake clustering
preceding moderate earthquakes, central
California, (preprint), 1978.

Mogi, K., Some discussions on aftershocks, foreshocks and earthquake swarms--the fracture of
a semi-infinite body caused by an inner stress
origin and its relation to earthquake phenomena, 3, Bull. Earthquake Res. Inst., Tokyo
Univ., 41, 615–658, 1963.

Mogi, K., Earthquakes and Fractures, Tectonophysics, 5, 35–55, 1967.

Rastogi, B. K., and P. Talwani, Spatial and
temporal variations in t_s/t_p at Monticello
Reservoir, South Carolina, Geophy. Res. Lett.,
7, 781–784, 1980.

Sadovskly, M. A., F. I. Monakhov, and A. N.
Semenov, Hydrodynamic Precursors of South
Kurile earthquakes, Doklady, Earth Sci. Sect.,
(Engl. Transl.) 236, 3–6, 1979.

Sauber, J., and P. Talwani, Application of Keilis-
Borok and McNally prediction algorithms to
earthquakes in the Lake Jocassee area, South
Carolina, Phys. Earth Planet. Interiors, 21,
267–281, 1980.

Schaeffer, M. F., R. E. Steffans and R. D. Hatcher,
In-situ stress and its relationship to joint
formation in the Toxaway gneiss, Northwestern
South Carolina, Southeastern Geology, 20, 129–
143, 1979.

Stevenson, D. A., Vp/Vs anomalies associated
with Lake Jocassee, South Carolina earthquakes,
M. S. Thesis, Univ. of South Carolina, Columbia
S. C., 77 pp., 1977.

Stevenson, D. A., P. Talwani and D. C. Amick,
Recent seismic activity near Lake Jocassee,
Oconee County, South Carolina, preliminary
results and a successful earthquake prediction,
EOS Trans. AGU, 57(4), 190, 1976.

Stevenson, D. A., D. Amick and P. Talwani, t_s/t_p
anomalies and an earthquake prediction at
Lake Jocassee, South Carolina, EOS Trans. AGU,
59, (4), 328, 1978.

Talwani, P., An empirical earthquake prediction
model, Phys. Earth Planet. Interiors, 18, 288–
302, 1979.

Talwani, P., D. Stevenson, J. Sauber, B. K. Rastogi, A. Drew, J. Chiang, and D. Amick, Seismicity studies at Lake Jocassee, Lake Keowee and Monticello Reservoir, South Carolina (October 1977–March 1978), Seventh Technical Report, contract 14-08-0001-14553, p. 190, U.S. Geological Survey, Washington, D.C., 1978.

Talwani, P., W. S. Moore, and J. Chiang, Radon anomalies and microearthquakes at Lake Jocassee, South Carolina, J. Geophys. Res., 85, 3079–3088, 1980.

Talwani, P., and B. K. Rastogi, Search for precursory changes in amplitude ratio of P and S waves for Jocassee, South Carolina earthquakes, Final Technical Report for NSF contract EAR7818126, 119 pp., 1980.

Van Nieuwenhuise, R. E., Well level fluctuations relation to the occurrence of earthquakes at Lake Jocassee, South Carolina, M. S. Thesis, Univ. of South Carolina, Columbia, 119 pp., 1980.

CRUSTAL DEFORMATION STUDIES AND EARTHQUAKE PREDICTION RESEARCH

Wayne Thatcher

U.S. Geological Survey
Menlo Park, California 94025
U.S.A.

Abstract

Repeated geodetic surveys provide data relevant to earthquake hazard assessment and precursor monitoring and supply constraints on the mechanisms of stress build up and relief in the crust and upper mantle. The detection of steady deformation indicates a seismic hazard, the degree of which depends on the rate of deformation and the time elapsed since a previous large earthquake. If the coseismic strain drop is known, the measured deformation rate can be used to estimate an earthquake recurrence interval. Practically, monitoring for precursory deformation can be done on a regional scale with resurvey intervals of years to decades. On a local scale, observations can be almost continuous. Significant experience with the first type of monitoring and limited experience with the second has uncovered several possible presursors but no unambiguous ones. Reported anomalies have amplitudes of $\sim 10^{-6}$ in strain or tilt. Using careful observing procedures, precision of $\sim 2 \times 10^{-7}$ is attainable. Resolving ambiguities is thus impeded not primarily by measurement imprecision, but rather by unexplained episodic movements (such as aseismic uplift in southern California) that are not obvious precursors. The plate tectonic setting and Quaternary deformation history provide clues to the origins of such aseismic movement. Deformation modeling is useful in assessing the significance of suspected precursory deformation, in deciphering observed movement patterns, and in constraining the mechanical properties of the lower lithosphere and uppermost asthenosphere, where precursory movements may originate.

Introduction

Since their inception in the late 19th century, repeated geodetic surveys have provided data sufficiently precise to detect the growth of mountains, the relative motions of lithospheric plates across major boundaries, and the elastic strain accumulation taking place in seismically active regions. The basic observations are relatively simple, being measurements of angles, lengths, or elevation differences typically repeated at intervals that range from years to decades. The length scale of the survey networks is usually ~ 10 km or greater, comparable to the depths at which rupture initiates in shallow focus crustal earthquakes. Consequently, the surface deformation being measured is frequently sensitive enough to reflect strain changes occurring both regionally and around the actual earthquake focus itself. Although the surface strain rates are usually rather low, typically a few parts in 10^7 per year, these faint deformation signals can commonly be extracted from even the earliest survey data.

In the absence of major earthquakes these crustal movements, in the long term at least, are uniform and occur at a relatively constant rate. The sense of movement, its spatial distribution, and the rate of straining define the regional deformation pattern, aid in assessing earthquake risk, and provide a natural complement to studies of seismicity, focal mechanisms, regional tectonics, seismic gaps, and earthquake recurrence. Superimposed on this long-term, steady deformation pattern, episodic aseismic movements are also observed. Best understood are the postseismic transients that follow major crustal earthquakes and persist for periods of years to tens of years. Apart from the fairly superficial surface fault creep that is sometimes observed, these movements reflect deformation that occurs predominantly in the lower lithosphere or uppermost asthenosphere at depths of ~ 20–100 km. Postearthquake surveys thus provide constraints on the mechanical properties of regions on or adjacent to the failure plane of major earthquakes, and form a link both with laboratory studies of aseismic fault slip and mantle rheology and with precursor models. Other episodes of aseismic movement have been documented, but none of these are yet well understood; some may have been

TABLE 1. Recurrence intervals from crustal strain rates.

Region	Strain Rate (μstrain/yr)	Estimated Recurrence Interval (years)	Earthquakes Last	Earthquakes Next
E. Hokkaido, Japan (Shimazaki, 1974)	0.40	125	1894	~2019 (Occurred in 1974)
N. San Andreas Fault, California (Thatcher, 1975b)	0.76	225	1906	~2130
Tokai District, Japan (G.S.I., 1976)	0.35	140	1854	~1994
Owens Valley, California (Savage et al., 1975)	0.12	420	1872	~2302
Fairview Peak, Nevada (Prescott et al., 1979)	0.09	560	1954	~2514

Recurrence intervals estimated using observed secular strain rates and estimates of coseismic strain drop. For the northern San Andreas fault a strain drop of 170 μstrain is obtained from the geodetic measurements of coseismic movements in 1906 (Hayford and Baldwin, 1908). In all other regions a coseismic strain drop of 50 μstrain has been assumed.

earthquake precursors, others are not obviously related to earthquake occurrence.

In summary, geodetic measurements have diverse applications in studies related to earthquake hazard mitigation and prediction. In this review I shall consider three areas where such measurements are particularly useful: first, in the assessment of longterm earthquake hazard; second, in monitoring for possible precursory crustal deformation; and third, in constraining the mechanism of stress build up and release in the brittle crust.

Hazard Evaluation

The detection of steady secular deformation by itself indicates some degree of earthquake risk, and in general high deformation rates and long quiescent intervals should be taken as indicators of a heightened earthquake potential. Subject to several caveats to be discussed later, the average recurrence interval between large earthquakes may be estimated if the secular rate and average coseismic strain drop are both known.

The coseismic strain or stress drops of large crustal earthquakes can be obtained both from geodetic measurements and from analysis of seismic waves. The former indicate an average strain drop of 50-100 μstrain (e.g., Rikitake, 1974) and the latter suggest stress drops of 10-100 bars for earthquakes above M=7 (Kanamori and Anderson, 1975). (If the average crustal shear modulus is 3×10^{11} dynes/cm2, these two

independent estimates are quite consistent with each other.) In several well-studied regions, both the secular deformation rate and the coseismic strain drop in a previous large earthquake are known. Estimated recurrence intervals for several fault zones are listed in Table 1.

For several reasons, recurrence estimates obtained in this way cannot be used for purposes more refined than generalized risk forecasting. Even at major plate boundaries, recurrent earthquakes on the same fault are neither equally spaced in time nor of precisely the same size (e.g., Ando, 1975). Furthermore, deformation rates are in general unlikely to be uniform during the entire strain build up period between large earthquakes. For example, the rate of shear straining has decreased by about a factor of 4 since the 1906 San Francisco earthquake (Figure 1), and subsidence rates declined by comparable amounts in the 70 years since the 1896 Riku-u earthquake (see Figure 17a). In addition, abrupt changes in long-term movement patterns have been documented in several regions (see Figures 12a, 15b).

Finally, there is no compelling reason to assume that all the observed secular deformation represents elastically stored strain energy. Steady fault creep is observed on the San Andreas system in California (e.g., Burford and Harsh, 1980) and on the North Anatolian fault in Turkey (Aytun, 1981), and upper crustal rocks are not uncommonly intensely folded as well as faulted (e.g., Matsuda, 1962; Yeats, 1977).

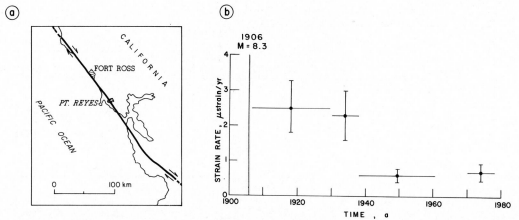

Fig. 1. Postseismic shear strain rates (shearing parallel to San Andreas fault) since 1906 San Francisco earthquake. (a) Location map, with 1906 surface faulting (heavy solid line) and location of two triangulation nets (hatched areas). (b) Strain rates, with horizontal line indicating interval between surveys and vertical line showing one standard deviation error bars for each determination. Data for 1906-30 from Fort Ross network (Thatcher, 1975a), other data from Pt. Reyes (Thatcher, 1975b; Prescott et al., 1979).

Figure 3 illustrates the close association between active late Quaternary folding and substantial rates of secular straining during the past 100 years in the South Fossa Magna of central Japan. In this region, historical seismicity levels are relatively low, and no earthquakes larger than M=7 have been documented (see Figure 2), certainly consistent with the notion that significant permanent deformation is taking place in the upper crust. Nonetheless,

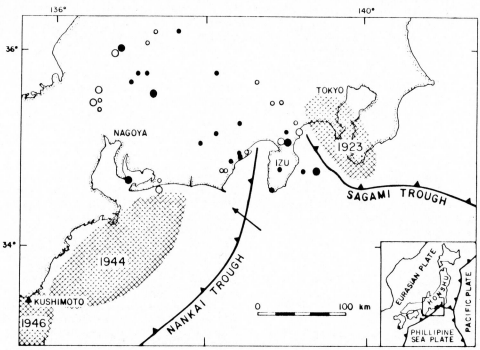

Fig. 2. Location map of central Honshu, Japan, showing major plate boundaries (underthrust zones, teeth on upthrown plate), great earthquakes in this century (hatched, with dates), and historic earthquakes with M > 6 since 599 A.D. (open dots, before 1900; solid dots, since 1900; smaller dots, M=6-7; larger dots, M > 7). Inset shows plate tectonic setting. Modified from Thatcher and Matsuda (1981).

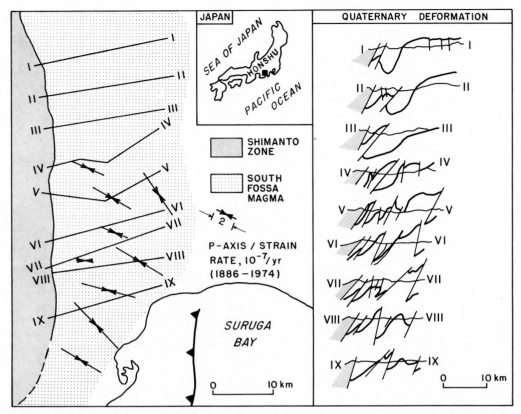

Fig. 3. Quaternary deformation and contemporary horizontal straining near the western edge of the South Fossa Magna, central Japan. The area of this figure can be seen on the regional map of Fig. 2. Cross-sections show Quaternary faulting and folding of Miocene and younger rock units (from Matsuda, 1962) and geodetically-determined compressive strain axes and rates are from G.S.I. (1977).

such evidence is seldom available, particularly in regions of strike-slip faulting, and it is generally difficult either to demonstrate the importance of these effects or to estimate the relative proportions of elastic and inelastic deformation.

Precursor Monitoring

Resolving whether or not clearly anomalous crustal movements precede damaging earthquakes is undoubtedly one of the highest priorities in prediction research. Geodetic surveys can play a major role in accomplishing this task, although it is important to recognize the limitations imposed by the ultimately achievable precision of the measurements as well as by necessary compromises between temporal resolution and areal coverage. Perhaps more importantly, unresolved problems remain in defining preseismic anomalies and in clearly distinguishing them from other episodic movements that are not precursors. In what follows, I shall use survey results from California and Japan to illustrate both the

advantages and shortcomings of these measurements and to emphasize some of the difficulties involved in identifying unambiguous earthquake precursors.

Methods, Precision, and Practical Limitations

Carefully made, geodetic measurements can detect changes in horizontal strain and ground tilt to a precision of a few parts in 10^7 over distances of about 10 km or greater. Prior to ~1960, most horizontal control surveys were done using triangulation, repeated measurements of the angular separation of distant survey monuments, and these angles were as accurate as 1 second of arc, or about 5 parts in 10^6. Since then, however, laser ranging methods have become widespread, and lines as long as ~50 km can routinely be measured to within 1 part in 10^6. Further increases in precision can be attained if airborne measurements of temperature and humidity are made to correct for the refractivity of the air along the laser path at the time of ranging. Figure 4 shows the repeatable precision obtained by the U.S.

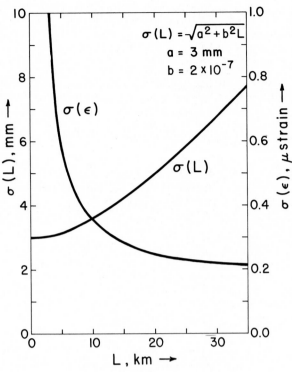

$$\sigma(L) = \sqrt{a^2 + b^2 L}$$
$$a = 3 \text{ mm}$$
$$b = 2 \times 10^{-7}$$

Fig. 4. Precision of line length determinations by laser ranging methods used by U.S. Geological Survey (from Savage and Prescott, 1973).

Geological Survey (USGS) using such procedures.

Currently the USGS measures 800 lines annually in the western United States and southern Alaska at an average cost of $400 per line. Except for a few areas singled out for special studies, it is practical only to resurvey each net once a year, and not all regions of interest can be covered at present. For example, Figure 5 shows the USGS nets in southern California. Despite the incomplete coverage and temporal sampling, coherent changes can be seen from net to net in the largest strain component, e_{22}, and significant temporal rate fluctuations are evident as well (Figure 6).

Recent development of a multi-wavelength distance-measuring laser geodimeter at the University of Washington (Huggett et al., 1977) has made it feasible to carry out more frequent length measurements. Using two laser light sources of differing wavelength the refractivity along the path can be accounted for, and with line lengths up to ~10 km, precision is two to three times better than that obtained by the USGS using a single-wavelength geodimeter and aircraft-based refractivity corrections. Daily observations have been made since late-1975 on a network of 10 lines located at Hollister, California, near the junction of the creeping portions of the Calaveras and San Andreas faults. The network configuration and some typical results are shown in Figure 7. The maximum range of this instrument is at present too short to survey most existing horizontal

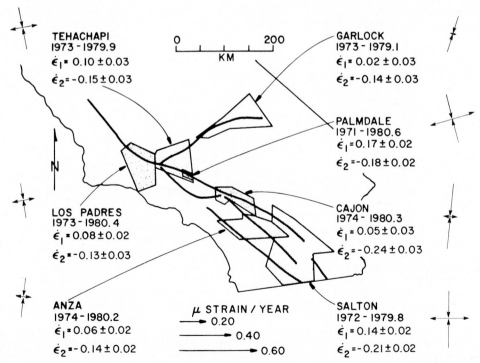

Fig. 5. Geodolite networks in southern California, with main active faults and principal strain rates averaged for 1973-80 for each net (from Savage et al., 1981).

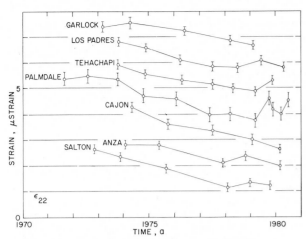

Fig. 6. Annual changes in north-south contractional strain component e_{22} for each of the geodolite networks shown in Fig. 5. Error bars are one standard deviation values determined from the least squares fit of a uniform strain field to the line length changes for each network.

strain networks. However, its evident stability and capability for recording strain changes nearly continuously is of potentially great use in detailed monitoring of areas of intensified study in the United States and elsewhere.

Although time consuming (~2 km/day) and comparatively expensive (~$300/km), first-order geodetic leveling is intrinsically the simplest and still one of the most precise surveying methods in use today. Measurement procedures have changed relatively little over the past century, and even the earliest leveling measurements are only about half as precise as current survey results. Measurements are made in linear traverses or closed loops, usually along roads or railway lines.

Random errors due to observational uncertainties in individual level rod readings are well understood and relatively small. The statistical uncertainty in level difference increases monotonically with the length of the route, and a one-standard-deviation estimate of this quantity is given by

$$\sigma = \pm \, \alpha \sqrt{L},$$

where σ and α are in mm, and L, the route length, is in km. The empirical constant α is usually determined from the cumulative elevation changes around many closed leveling loops, since for perfect leveling these changes should sum to zero. Its value varies between about 1 and 3 for first-order leveling (Bomford, 1971, p. 244). Error bars plotted for leveling data in this paper will in each case refer to this cumulative random error.

Elevation-dependent systematic errors due to unequal refraction or rod miscalibration are potentially much larger but are usually very difficult to detect or adequately correct for. The occasional visually obvious correlations between local topographic roughness and elevation change indicate these errors can be at least locally important. Whether or not they are common contaminants of leveling data and accumulate systematically over long routes have long been matters of debate (see Bomford, 1971, pp. 239-243; Savage and Church, 1974, pp. 694-697), and these issues remain contentious at the present time (see Jackson, 1981; Reilinger and Brown, 1981; Stein, 1981; this volume, and Mark et al., 1981; Strange, 1981). I consider these questions further below in a discussion of recent uplift in southern California.

Preearthquake Anomalies

Retrospective studies in several regions have shown that unusual crustal movements detected geodetically have preceded at least some earthquakes of M 6-1/2 or greater. Although problems of interpretation do exist, it seems likely that the examples discussed below represent actual tectonic deformation and are not due to measurement error or other non tectonic effects. However, in no case is it certain that these movements are indeed causally related to the earthquakes that follow them—equally unusual deformation episodes occur unaccompanied by major earthquakes.

In the 8 years prior to the 1964 Niigata, Japan, earthquake, level changes occurred that departed notably from the long-term trend established by earlier data (Figure 8a). During this time tilts as large as 2×10^{-6} developed, changes considerably larger than the expected cumulative random errors of first-order leveling. Although these data could be interpreted as uplift of the coast adjacent to the eventual rupture plane, the tidal gage records (Figure 8b) are not entirely consistent with the leveling results and further suggest an alternative interpretation. The differenced tidal gage data do indicate about 5 cm of relative uplift at Nezugeseki, but this is over twice that shown by the leveling data (see Benchmark 1, Figure 8a). Furthermore, the record for Kashiwazaki, although noisy, suggests that most of the relative movement is due to subsidence at this station, over 60 km from the 1964 fault plane, rather than uplift at Nezugeseki, immediately adjacent to it. Based on these data alone, it is not clear which interpretation, if either, is the correct one.

Tilts of comparable magnitude have been documented in the region immediately to the north of the epicenter of the 1971 M=6.4 San Fernando, California, earthquake, also a thrust event (Figure 9). Results also indicate a reversal in tilt direction between 1965-68 and 1968-69 (Figure 9b,c). Although this

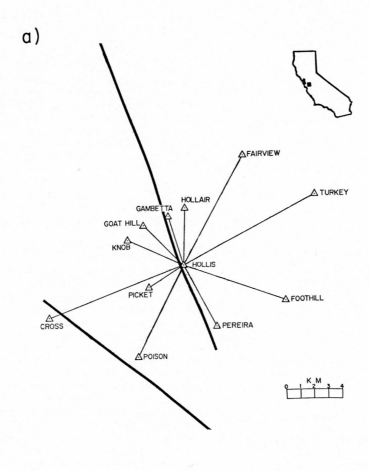

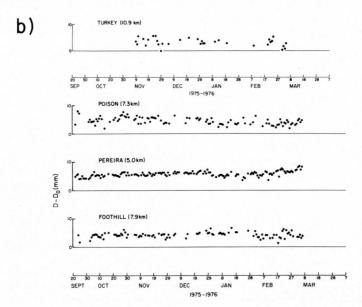

Fig. 7. (a) Network of lines measured daily by two-wavelength laser geodimeter located at Hollister, California. (b) Line length changes versus time (From Huggett et al., 1977).

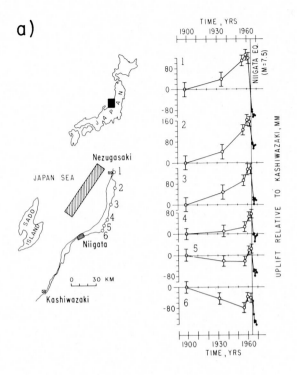

a)

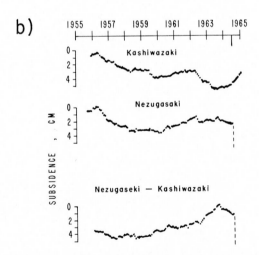

b)

Fig. 8. Level changes preceding the 1964 M=7.5 Niigata earthquake. (a) Location map, with successive movements of six selected benchmarks relative to Kashiwazaki, the southern-most point on the leveling route. Postearthquake points are shown as solid dots without error bars. The hatched rectangle is the surface projection of the coseismic fault plane. Figure is taken from Savage (1977). (b) Uncorrected tidal gage records, and their difference for two stations located in Fig. 8a, above (from Tsubokawa et al., 1968).

observation depends critically on data from the 1968 survey, the change is independently supported by 1968 and 1969 surveys on the adjacent leveling route BD (located in Figure 9a). These data indicate about 60 mm of relative uplift of benchmark B, in this case with respect to a point 30 km farther to the east (see Castle et al., 1974, Figure 5). The level changes plotted in Figure 9 can be accounted for by a model in which aseismic slip propagates updip on a north-dipping thrust fault towards the eventual 1971 earthquake hypocenter (Thatcher, 1976; Stuart, 1979).

Leveling fortuitously carried out in the several days before and after the 1944 Tonankai earthquake (M=8.0) provides an unusual glimpse of immediate pre- and post-seismic tilting that occurred near the focal region of a great underthrust earthquake. Figure 2 locates this earthquake with respect to major plate boundaries and Figure 10a shows the survey route relative to the 1944 fault plane. The relevant measurements were made between adjacent benchmarks (5260 and 5259) located about 30 km NE of the 1944 fault plane, and this 2-km-long segment was effectively a long baseline tiltmeter that was read 4 times in the 3 days surrounding the earthquake. Figure 10b shows the coseismic and long-term interseismic (1944-1933) level changes between these two benchmark relative to those on adjacent sections of leveling route. Figure 10c plots the relative displacements between 5260 and 5259 as a function of time from the 1944 earthquake. The two immediate pre- and post-earthquake points are differences of forward and backward runs between the benchmarks made on successive days, the normal procedure for first-order double-run leveling. These changes are significantly larger than the expected observational errors and seem to represent true tectonic tilts: Sato (1977; unpublished data, 1978) states that these differences usually agree within ± 2 mm. He further showed that out of 20 measurements made between BM's 5268 and 141 during the 12 days that span the earthquake occurrence, the two largest discrepancies are those shown in Figure 10c for BM's 5260 and 5269, where the coseismic level changes were also largest (Figure 10b). The immediate pre- and post-seismic tilts agree in sense with the earthquake deformation but are opposite in direction to the interseismic trend and the large amplitude postseismic tilt of 1944-67. Although the unexpectedly large tilts of 06-07 and 08-09 December may be evidence of a protracted failure process, the data are fragmentary and their precise relation to the Tonankai earthquake is uncertain. Note that the actual 1944 earthquake epicenter occurred at the SW end of the coseismic fault plane, over 200 km from BM's 5259 and 5260.

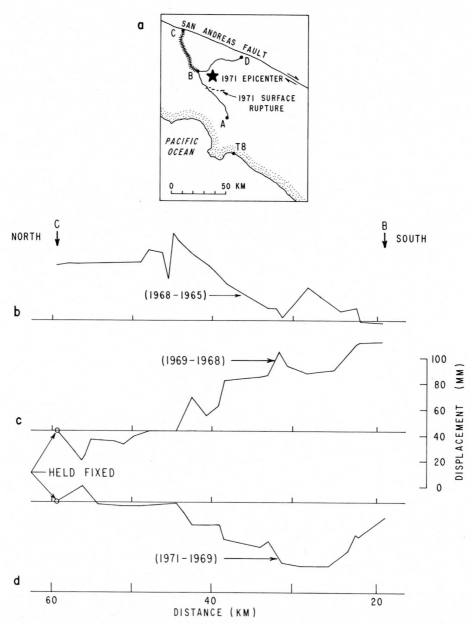

Fig. 9. Level changes preceding the 1971 M=6.4 San Fernando, California earthquake (modified from Castle et al., 1974). (a) Location of level lines, earthquake epicenter, and surface faulting that accompanied the 1971 shock. (b,c,d) Successive level changes on line segment CB, to the north of the epicenter, shown in Fig. 9a above. The 1968–1965 changes are relative to tidal benchmark T8, other changes are relative to point C. The 1971 survey was carried out after the earthquake.

Aseismic Uplift and Episodic Movements

Aseismic deformation episodes also take place that are neither followed by nor obviously related to significant earthquakes. Although the southern California uplift is the most notable episode of this kind, similar movements have been documented elsewhere. In this section

I first briefly review the evidence for repeated uplift in southern California, then discuss it in relation to the Quaternary deformation and plate tectonic setting of the region, and finally compare it with aseismic uplift observed in two regions of central Japan.

The movements in southern California (Figure 11) span a large region, and land uplifts during

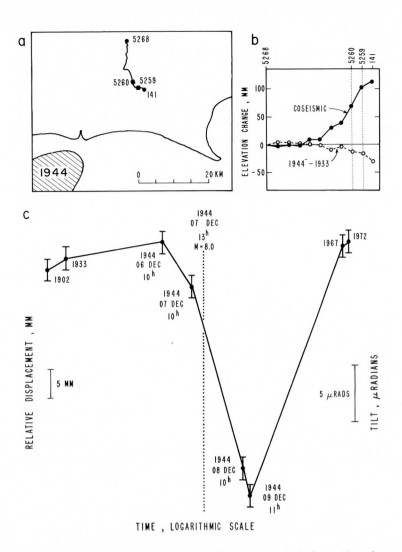

Fig. 10. (a) Location of leveling route relative to 1944 Tonankai earthquake rupture. (b) Level changes between BM's 5268 and 141, with locations of 5260 and 5259 noted. (c) Relative displacements and tilts between BM's 5260 and 5259 plotted on a logarithmic time scale relative to the 1944 earthquake.

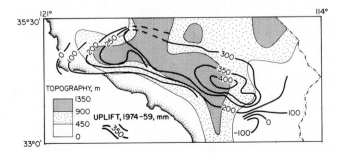

Fig. 11.   The southern California uplift for the interval 1959-74 superimposed on smoothed topography (Stein, 1980; 1981).   Uplift contours are from Castle (1978) and are relative to tidal benchmark T8 shown in Figure 12b.

1959-74 have exceeded 100 mm in an area at least 600 X 100 km2. Available data further indicate that two episodes of uplift followed by partial collapse have occurred since 1900 (Figure 12a). For the most part, the uplifted area is confined to the vicinity of the Transverse Ranges, a young mountain chain that has risen as much as 2 km during Quaternary time (Jahns, 1954). The uplift also coincides rather well with the big bend of the San Andreas fault, a 300-km-long segment that is deflected 20° – 30° from its regional trend through California. Since this trend closely matches the direction of relative motion between the Pacific and North American plates, the big bend introduces a significant collisional component into the motion occurring in southern California. Thus,

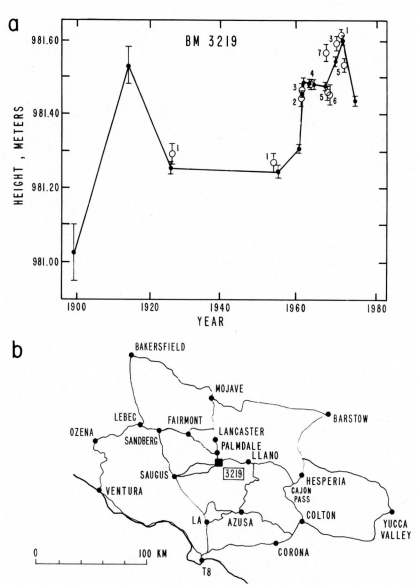

Fig. 12. (a) Uplift history of leveling benchmark 3219, located about 10 km south of Palmdale. Elevation changes are relative to T8 (see Fig. 12b), reached by the most direct route (small solid dots) and by various other routes (numbered open dots). 1 = Saugus-Sandberg-Lebec-Bakersfield-Mojave-Palmdale. 2 = Los Angeles-Azusa-Colton-Hesperia-Barstow-Mojave-Palmdale. 3 = Los Angeles-Azusa-San Gabriel Mountains-Llano-Palmdale. 4 = Saugus-Sandberg-Fairmont-Palmdale 5 = Corona-Colton-Cajon Pass-Palmdale. 6 = Corona-Colton-Yucca Valley-Hesperia-Palmdale. 7 = Ventura-Ozena-Lebec-Sandberg-Lancaster-Palmdale. (From Mark et al., 1981). (b) Leveling routes and locations referred to in (a) above.

in a general way the high relief, active mountain building, range-front thrust faulting, and modern geodetic uplift are all consequences of the oblique convergence of continental lithospheric blocks in southern California.

Because of the general correlation of uplift with topography, it has recently been argued that the apparent movements are in fact an artifact of height-dependent systematic leveling errors (Jackson et al., 1980; Strange, 1981).

The proposed errors are of two kinds, those due to miscalibration of leveling rods and those caused by unequal refraction due to a stable but non-linear thermal gradient near the ground surface. The important issues with respect to the credibility of the 1959-1974 uplift are whether these errors are very common rather than extremely rare, and whether they accumulate systematically or tend to randomize over long leveling routes.

TOPOGRAPHY

contours in meters

QUATERNARY UPLIFT

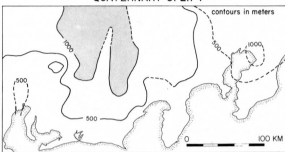

contours in meters

0   100 KM

GEODETIC UPLIFT ~1900-1970

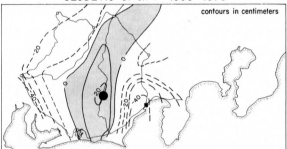

contours in centimeters

Fig. 13. Comparison of topography, Quaternary uplift and recently-measured geodetic uplift in Tokai district, central Japan (from Thatcher and Matsuda, 1981). Dot in bottom frame locates BM 5303, whose movement history is plotted in Fig. 14.

In my view, the existing leveling data themselves suggest the proposed errors are insufficient to negate the existence of the uplift. Even if these errors were large and accumulated, both are expected to vary considerably in sign and magnitude over the more than 5000 km of leveling routes used to define the uplift: hundreds of rod pairs are involved, and temperature, wind conditions, and topographic slope (all of which affect unequal refraction) vary widely. If such errors were common, a consistent picture of the movement history would be difficult, if not impossible, to obtain. Contrary to these expectations the considerable body of data defining the uplift tends to show a consistent and orderly picture.

One example is shown in Figure 12a, where elevation differences between T8 and BM 3219, near Palmdale, are determined by traversing 8 different routes at various times. Some routes overlap in part, but the lines shown represent considerable diversity in measuring conditions, terrain coverage, and topography, and yet the agreement between the different estimates remains impressive.

Undoubtedly differences of interpretation will persist among investigators examining particular level lines and small portions of the uplifted region. However, substantial objections to the existence of the uplift must be based on examination of large collections of consistent data like that shown in Figure 12. To date this task has not been attempted.

The continued uplift of Quaternary mountain ranges in central Japan has also been detected by repeated leveling surveys. Figure 13 compares topography, Quaternary uplift, and geodetic uplift in central Japan. The region of modern uplift forms a narrow belt 30 km wide and 250 km long (Dambara, 1970; Figure 1), and only the southern half is shown in Figure 13c. The aseismic displacement history of BM 5303 (Figure 14) is rather typical, and movements have been roughly uniform since leveling was begun in 1902 (see also Thatcher and Matsuda, 1981, Figure 5). Although this uniformity contrasts markedly with the episodic uplift observed in southern California, maximum uplift rates, about 3 mm/a, are comparable to time-averaged rates observed in California (e.g., 5 mm/a at BM 3219, Figure 12a).

In common with the growth of the Transverse Ranges, the Quaternary uplift of central Honshu is an effect of continental collision. In this case the convergence is between the Izu Peninsula, on the Philippine Sea plate, and the island of Honshu, on the Eurasian plate (see Figure 2).

Although geodetic measurements of modern movements are not available, the late Cenozoic uplift of the Southern Alps in New Zealand is also rather similar. There, oblique convergence between the Australian and Pacific plates has resulted in uplift and crustal shortening as

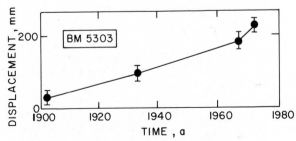

Fig. 14. Movement history of BM 5303 (see Fig. 13, bottom) in the Akaishi Mountains, central Honshu, Japan.

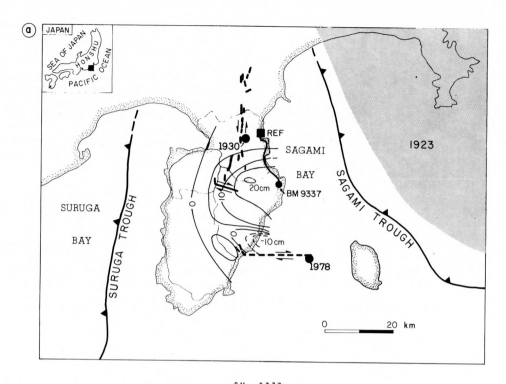

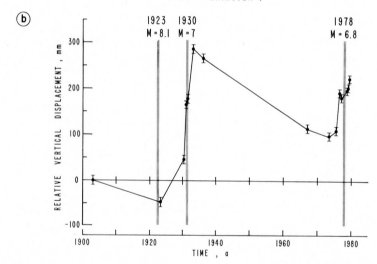

Fig. 15. (a) Tectonic setting of Izu peninsula, central Japan. Contours show uplift measured during 1967-79 on leveling routes shown by faint lines (from Geographical Survey Institute, 1980). Surface faulting in 1930 and 1978 earthquakes is shown with sense of strike-slip movement. Position of the source region of the great 1923 Kanto earthquake is shaded. (b) Movement history of BM 9337 relative to reference benchmark (REF), both of which are located in (a) above.

well as large strike-slip offsets across the Alpine fault (Walcott, 1978).

In sharp contrast with the regularity of movements just 100 km to the west (Figure 14), localized uplift on the Izu Peninsula has been

notably irregular since 1900 (Figure 15). Two uplift episodes have taken place, the first followed by partial collapse, the second apparently still in progress. Both have occurred in about the same region, a roughly

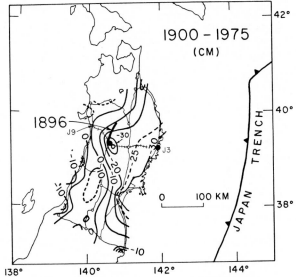

1900 - 1975
(CM)

JAPAN TRENCH

1896

Fig. 16. Level changes, 1900–1975, in
Northern Honshu, Japan. Faint lines with
open circles show leveling routes used to
construct contour map. Solid dots denote
benchmarks with movement histories plotted
in Fig. 17. Railroad track symbol shows
level line for which elevation changes are
plotted in Fig. 18. Region of surface
faulting in the 1896 M=7.5 Riku-u earthquake, on the Senya fault, is also shown.
Note the localized postseismic subsidence
near this fault. Figure modified from Kato
(1979).

circular area 30 km in diameter. Repeated
gravity surveys on the leveling routes shown in
Figure 15a have provided independent
confirmation of the movement pattern outlined by
the leveling results (Earthquake Research
Institute, 1980).

The uplifted region includes several
Quaternary volcanoes, and most of the peninsula
is broken up by a dense array of active
conjugate strike-slip faults. Two of these
fault systems ruptured in earthquakes of about
M=7, in 1930 and in 1978. Although movements
were centered about 20 km from the faults
themselves, each event was preceded by episodes
of substantial aseismic uplift (Figure 15b).
While a direct causal relation between uplift
and earthquake is unclear, the repetitive
behavior does suggest a more than coincidental
relationship, and these movements are perhaps
more appropriately classified with the
preearthquake anomalies discussed in the
previous section.

Deformation Mechanisms

Geodetic measurements are also of considerable
use in constraining the physical mechanisms of

stress build up and relief in active seismic
zones. It may be that this information is not
needed, and strictly empirical monitoring may be
sufficient to predict earthquakes. However,
experience to date has uncovered no unambiguous
precursors, and the existence of episodic
movements that are not obviously related to
significant earthquakes suggests it may be
worthwhile to understand the basic processes a
little better. Here I use several examples to
illustrate how deformation modeling is useful in
deciphering observed movement patterns and in
constraining the mechanical properties of the
lithosphere and uppermost asthenosphere.

The often detailed complexity of observed
movements is illustrated by Figure 16, which
shows recent level changes in northern Honshu,
Japan. There, at least three deformation fields
are contributing to the measured displacements.
The dominant effect is trenchward tilting and
coastal subsidence related to the westward
subduction of the Pacific plate at the Japan
Trench. Superimposed on this pattern are two
more local effects, uplift of Quaternary
mountains to the west of the subsidence region,

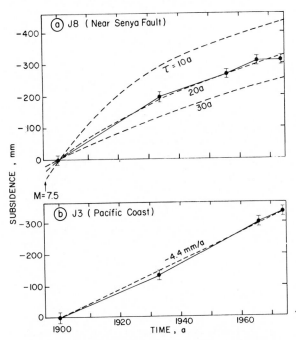

Fig. 17. Subsidence of benchmarks J8 and
J3 (see Fig. 16) plotted versus time.
Solid lines connect observed values, with
error bars showing cumulative random error
from benchmark J9. Dashed lines in (a)
show computed time histories for models of
postearthquake asthenospheric relaxation
discussed in the text. Dashed line in (b)
indicates straight line fit to the data
(Note: a is abbreviation for year). From
Thatcher et al. (1980).

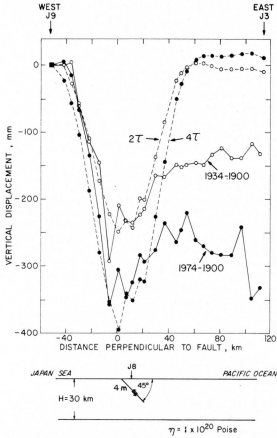

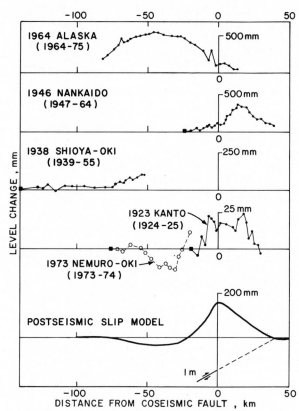

Fig. 18. Comparison of observed (solid lines) and computed (dashed) cumulative post-1896 level changes on the east-west route shown in Fig. 16. Model is shown to scale below the profile, which is projected perpendicular to the coseismic fault. Asthenospheric relaxation time, $\tau = 2\,\eta/\mu$, where η = asthenospheric viscosity, μ = shear modulus. Model fit requires $\tau = 20$ a. Modeling makes no attempt to match observed changes beyond $+40$ km, which are presumed to be effects of subduction of the Pacific plate beneath the Japan trench. From Thatcher et al. (1980).

following thrust faulting in the overlying lithosphere (Thatcher and Rundle, 1979). Although the same process occurs in subduction zones, the absence of a descending slab and other plate boundary complications makes the modeling rather more straightforward for intraplate events like the 1896 earthquake. In fact, once the fault geometry is fixed independently, both the effective elastic thickness of the lithosphere and the viscosity of the uppermost asthenosphere can be determined by matching observed and model-predicted movements, as is illustrated in Figures 17a and

Fig. 19. Observations of postseismic level changes for large underthrust earthquakes, compared with model of postseismic slip. In each case the origin for the distance scale (absissa) is taken as the surface projection of the bottom edge of the coseismic fault obtained from published determinations of the faulting parameters. The model calculation uses a two dimensional dislocation model. Corresponding coseismic fault plane for the model is indicated by a dashed line. Solid square shows assumed fixed points on observed profiles. Data from Brown et al. (1977), Thatcher and Rundle (1979), Abe (1977, Fig. 10, with 5 subsiding benchmarks deleted near Joban coalmines), Kasahara (1975), and Y. Fujii (written communication, 1978).

and marked local downwarping near the Senya fault, site of the 1896 Riku-u earthquake.

The difference in movement histories between the Pacific coast and the vicinity of the Senya fault (Figure 17) clearly distinguishes the two modes of deformation: while the subsidence of J3 occurs at a nearly constant rate, movements at J8 have decreased monotonically since 1900. This decrease, as well as the localization of movements near the Senya fault, suggests an association with the 1896 earthquake, an M=7.5 thrust event. Furthermore, the deformation pattern is just that expected from the stress relaxation of a viscoelastic asthenosphere

18. The decrease in subsidence rate at J8 determines the relaxation time τ and thus the asthenospheric viscosity η (Figure 17a). Once the fault geometry is fixed, the spatial extent of postearthquake deformation is directly proportional to the lithospheric thickness H. As Figure 18 shows, the match between the observed and predicted profiles near the fault is quite good. Changing H by more than + 5 km would significantly degrade this fit. Farther than 40 km east of the fault, observed movements are not related to the 1896 earthquake, and the match breaks down; Thatcher et al (1980, Figure 8) show that the movements here are due to the subduction of the Pacific plate at the Japan Trench.

Because the asthenospheric viscosity is relatively high, deformation related to this earthquake has continued for nearly a century. For great plate boundary shocks, discernible motions may persist for many hundreds of years.

Postseismic transients of much shorter duration are also commonly observed and are particularly well documented for great underthrust earthquakes. In contrast with the longer term movements, these displacements largely involve uplift. Although detectable uplift may continue for several decades, the major movements occur in the 1 to 5 years following the earthquake. As Figure 19 demonstrates, the observations are mimicked rather well by a simple model in which aseismic slip occurs on the downdip extension of the coseismic fault. Although the point of maximum uplift is displaced somewhat from the model-predicted location, this disagreement is of no great concern, since the downdip decrease in coseismic slip is undoubtedly gradual and its termination point uncertain by at least a few tens of kilometers.

Flow due to power law creep in the lower lithosphere (Kirby, 1977) may also account for the postseismic movements shown in Figure 19. The mechanism in fact differs little from slip on a discrete fault, since in either case deformation will tend to localize along the downdip extension of the coseismic fault. Surface displacements will be rather similar and it seems unlikely that the data can distinguish between the two mechanisms.

Acknowledgment. Careful reviews were provided by J.C. Savage and A.G. Lindh.

References

Abe, K., Tectonic implications of the large Shioya-oki earthquakes of 1938, Tectonophysics, 41, 269-289, 1977.

Ando, M., Source mechanisms and tectonic significance of historical earthquakes along the Nankai trough, Japan, Tectonophysics, 27, 119-140, 1975.

Aytun, A., The creep measurements of the North Anatolian fault zone in the Ismet Pasa region, Proceedings of Interdisciplinary Conference on Earthquake Prediction Research in the North Anatolian Fault Zone, Istanbul, March 31-April 5, 1980, in press, 1981.

Bomford, G., Geodesy, third edition, Oxford University Press, London, 731 pp, 1971.

Burford, R.O., and P.W. Harsh, Slip on the San Andreas fault in central California from alinement array surveys, Bull. Seismol. Soc. Amer., 70, 1223-1261, 1980.

Brown, L.D., R.E. Reilinger, S.R. Holdahl, and E.I. Balazs, Postseismic crustal uplift near Anchorage, Alaska, J. Geophys. Res., 82, 5349-5359, 1977.

Castle, R.O., J.N. Alt., J.C. Savage and E.I. Balazs, Elevation changes preceeding the San Fernando earthquake of February 9, 1971, Geology, 61, 61-66, 1974.

Castle, R.O., Leveling surveys and the southern California uplift, U.S. Geological Survey Earthquake Information Bulletin, 10, 88-92, 1978

Coastal Movements Data Center, Tables and Graphs of Annual Mean Sea Level Along the Japanese Coast, 37 pp., Tokyo, 1976.

Dambara, T., Synthetic vertical crustal movements in Japan during the recent 70 years (in Japanese), J. Geod. Soc. Japan, 17, 100-108, 1970

Earthquake Research Institute, Gravity change rate in Izu Peninsula, Rep. Cord. Comm. Earthq. Pred., 24, 113-116, 1980

Geographical Survey Institute, Horizontal earth's strain in Suruga Bay, Rep. Coord. Comm. Earthq. Predict., 16, 105-107, 1976 (in Japanese).

Geographical Survey Institute, Horizontal strains in the Tokai district, Rep. Coord. Comm. Earthq. Predict., 20, 166-171, 1978 (in Japanese).

Geographical Survey Institute, Crustal deformation in the eastern Izu district, Rep. Coord. Comm. Earthq. Pred., 23, 48-52, 1980 (in Japanese).

Hayford, J.F., and A.L. Baldwin, The earth movements in the California earthquake of 1906, in The California Earthquake of April 18, 1906, Report of the State Earthquake Investigation Commission, vol.1, pp.114-145, Carnegie Institution of Washington, Washington, D.C., 1908.

Huggett, G.R., L.E. Slater, and J. Langbein, Fault slip episodes near Hollister, California: initial results using a multiwavelength distance measuring instrument, J. Geophys. Res, 82, 3361-3368, 1977.

Jackson, D.D., Analysis of leveling data to determine tectonic uplift, Proc. Fourth Maurice Ewing Symposium, Earthquake Prediction, this volume, 1981.

Jackson, D.D., W.B. Lee and C.C. Liu, Aseismic uplift in southern California, an alternative interpretation, Science, 210, 534-536, 1980.

Jahns, R.H., editor, Geology of southern California, Calif. Division of Mines Bull. 170, San Francisco, 1954.

Kanamori, H. and D.L. Anderson, Theoretical basis of some empirical relations in seismology, Bull. Seismol. Soc. Amer., 65, 1073-1096, 1975.

Kasahara, K., Aseismic faulting following the 1973 Nemura-oki earthquake,. Hokkaido, Japan (a possibility), Pure Appl. Geophys., 113, 127-139, 1975.

Kirby, S.H., State of stress in the lithosphere: Inferences from the flow laws of olivine, PAGEOPH, 115, 245-258, 1977.

Kato, T., Crustal movements in the Tohoku district, Japan, during the period 1900-1975, and their tectonic implications, Tectonophysics, 60, 141-167, 1979.

Matsuda, T., Crustal deformation and igneous activity in the South Fossa Magna, Japan, Geophys. Monogr. Am. Geophys. Union, 6, 140-150, 1962.

Mark, R.K., J.C. Tinsley III, E.B. Newman, T.D. Gilmore, and R.O. Castle, An assessment of the accuracy of the geodetic measurements that define the southern California uplift, J. Geophys. Res., 86, in press, 1981.

Prescott, W.H. Savage, J.C. and Kinoshita, W.T., 1978. Strain accumulation rates in the Western United States between 1970 and 1976, J. Geophys. Res., 84, 5423-5435, 1979.

Reilinger, R.E., and L.D. Brown, Geodetic leveling, crustal movement, and earthquakes in the U.S., Proc. Fourth Maurice Ewing Symposium, Earthquake Prediction, this volume, 1981.

Rikitake, T., Probablility of earthquake occurrence as estimated from crustal strain, Tectonophysics, 23, 299-312, 1974.

Sato, H., Some precursors prior to recent great earthquakes along the Nankai trough, J. Phys. Earth, 25, Suppl., S 115-S 121, 1977.

Savage, J.C., Earth deformation associated with earthquakes, Proceedings of AIAA Guidance and Control Conference, San Diego, Aug. 16-18, 1976, American Institute of Aeronautics and Astronautics, 11, 405-412, 1976.

Savage, J.C., and J.P. Church, Evidence for postearthquake slip in the Fairview Peak, Dixie Valley, and Rainbow Mountain fault area of Nevada, Bull. Seismol. Soc. Amer., 64, 687-698, 1974.

Savage, J.C., and W.H. Prescott, Precision of geodolite distance measurements for determining fault movements, J. Geophys. Res., 78, 6001-6008, 1973.

Savage, J.C., W.H. Prescott, M. Lisowski and N. King, Strain accumulation in southern California 1973-1980, J. Geophys. Res., 86, in press, 1981.

Savage, J.C., J.P. Church, and W.H. Prescott, Geodetic measurement of deformation in Owens Valley, California, Bull. Seismol. Soc. Amer., 65, 865-874, 1975.

Shimazaki, K., Pre-seismic crustal deformation caused by an underthrusting oceanic plate, in eastern Hokkaido, Japan, Phys. Earth Planet. Interiors, 8, 148-157, 1974.

Stein, R.S., Contemporary and Quaternary deformation in the Transverse Ranges of southern California, Ph.D. thesis, Stanford University, Stanford, California, 121 p., 1980.

Stein, R.S., Role of elevation-dependent errors on the accuracy of geodetic leveling in the southern California uplift, Proc. Fourth Maurice Ewing Symposium, Earthquake Prediction, this volume, 1981.

Strange, W.E., The impact of refraction correction on leveling interpretations in southern California, J. Geophys. Res., 86, in press, 1981.

Stuart, W.D., Strain softening instability model for the San Fernando earthquake, Science, 203, 907-910, 1979.

Thatcher, W., Strain accumulation and release mechanism of the 1906 San Francisco earthquake, J. Geophys. Res., 80, 4862-4872, 1975a.

Thatcher, W., Strain accumulation on the northern San Andreas fault zone since 1906, J. Geophys. Res., 80, 4873-4881, 1975b.

Thatcher, W., Episodic strain accumulation in southern California, Science, 194, 691-695, 1976.

Thatcher, W., T. Matsuda, T. Kato and J.B. Rundle, Lithospheric loading by the 1896 Riku-u earthquake, northern Japan: Implications for plate flexure and asthenospheric rheology, J. Geophys. Res., 85, 6429-6435, 1980.

Thatcher, W., and T. Matsuda, Quaternary and modern crustal movements in the Tokai district, central Honshu, Japan, J. Geophys. Res., 86, in press, 1981.

Thatcher, W. and J.B. Rundle, A model for the earthquake cycle in underthrust zones, J. Geophys. Res., 84, 5540-5556, 1979.

Tsubokawa, I., T. Dambara, and A. Okada, Crustal movements before and after the Niigata earthquake, in General Report on the Niigata Earthquake, edited by H. Kawasumi, Tokyo Electrical Engineering College Press, Tokyo, 129-139, 1968.

Walcott, R.I., Present tectonics and late Cenozoic evolution of New Zealand, Geophys. J. R. Astr. Soc., 52, 137-164, 1978.

Yeats, R.S., High rates of vertical crustal movements near Ventura, California, Science, 196, 295-298, 1977.

DELAYS IN THE ONSET TIMES OF NEAR-SURFACE STRAIN AND TILT PRECURSORS TO EARTHQUAKES

Roger Bilham

Lamont-Doherty Geological Observatory of Columbia University
Palisades, New York 10964

Abstract. An examination of reported anomaly size, epicentral distance and earthquake magnitude for 40 events shows that more than 70% of strain and tilt anomalies are observed by instruments located between 20 km and 200 km from a magnitude 6 to 7 earthquake and that the magnitude of the anomaly lies between 10^{-7} and 10^{-5} strain. A probable reason for the shortage of data for other distances and event magnitudes is that a relatively small number of events greater than M = 7 have occurred close to tilt and strain measurement arrays and that the spacing of instruments is inadequate to monitor the small radius of deformation associated with events with magnitudes less than M = 6. The noise level in strain and tilt measurements accounts for the approximate 10^{-7} detection threshold indicated by the data.

The lead time for strain and tilt anomalies preceding earthquakes is usually less than that associated with other types of precursive phenomena, and for a given magnitude earthquake, the observed lead times for strain and tilt anomalies show a scatter of up to 3.5 orders of magnitude. This result renders strain and tilt measurements of limited use for earthquake prediction studies unless it can be shown that the scatter in the data is due to systematic errors in measurement.

Two possible explanations for the low lead times observed for strain and tilt precursors are examined. One explanation is the possibility of a bias in the reporting of anomalies in noisy data. The other is that observed strain anomalies are associated with local changes in strain at the instrument location that occur at the surface only at an advanced stage in the precursory process.

A suggested mechanism involves the influence on local strain rates of near-surface, fluid-filled joints whose stiffness is dependent on the strain rate and stress in the surrounding medium.

Introduction

The acquisition of strain and tilt data in the epicentral region has long been considered important to constrain models of earthquake mechanism. As a monitor of preseismic processes, strain measurements are of fundamental importance in seismic gap studies where the approximate magnitude and location of the future event has been forecast. A currently held view, however, is that strain and tilt data are inadequate to provide a clear understanding of the precursory processes that are suggested by other forms of anomalous epicentral behavior. Inadequacies in strain and tilt data include a poor signal-to-noise ratio and a general absence of spatial coherence where the same strainfield has been observed simultaneously at several locations. Strain and tilt precursors to earthquakes constitute approximately 25% of all reported precursors yet show less amplitude and duration consistency than do all other forms of precursor.

The single most important feature of reported physical precursors to earthquakes is the apparent relationship between the magnitude (M) of the earthquake and the duration (T) of the observed precursor. The values assigned by various authors to the relation $\log_{10} T = aM - b$ vary according to the data available, the average values being $a = 0.7 \pm 0.1$ and $b = 1.5 \pm 0.5$ [Scholz et al., 1973; Whitcomb et al., 1973; Tsubokawa, 1973; Rikitake, 1976, 1979]. Using 391 reported cases of precursory changes in land deformation, tidal strain amplitude, foreshock behavior, tilt and strain, seismicity, seismic wave velocity, geomagnetism, earth currents and resistivity, radon emission and variations in underground water level and aquifer pressure, Rikitake [1979] obtained the relationships shown in Figure 1.

Three distributions are distinguished in Figure 1 which are classified by Rikitake as precursors of the first, second, and third kind. The first kind of precursor obeys the relationship $\log T = 0.6M - 1.01$, the second kind of precursor is unrelated to earthquake magnitude but occurs a few hours before the event, and the third type of precursor occurs at any time between these two extremes. Rikitake notes that water-tube tiltmeter data and some borehole tiltmeters can be grouped with the first kind of precursor but that most tilt and strain precursors are precursors of the third kind. In this article tilt and strain precursors to earthquakes greater than magnitude 3 will be examined. For brevity, all con-

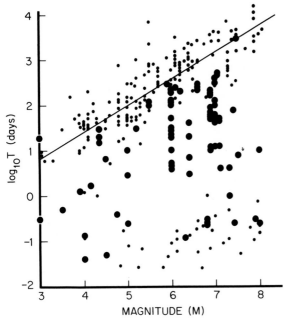

Figure 1. Precursor duration versus earthquake magnitude. Small dots represent precursors excluding foreshock data documented by Rikitake [1976, 1979]. Large dots represent strain and tilt precursors listed in Table 1. The line represents $\log T = 0.6M - 1.01$ determined for 'precursors of the first kind' by Rikitake [1979]. The cluster of small dots between 0 and -2 (mostly resistivity anomalies) are Rikitake's 'precursors of the second kind'.

tinuous measurements of tilt and strain will be discussed as 'strain'. There are two features to the data that are considered important, the almost universal occurrence of strain precursors later than that predicted by the log T versus magnitude relationship and the wide scatter in the onset time of strain precursors. The late arrival of strain precursors will be referred to as a delay. The delay is important since it suggests that strain measurements may provide indications of seismic imminence, the scatter is important since unless its cause can be understood, strain or tilt measurement will be of little value in the precise indication of the time of future earthquakes.

Data Quality

The strain data in Figure 1 are based on values listed in Table 1 and are primarily those enumerated by Tanaka [1964], Rikitake [1976, 1979], and Dobrovolsky et al. [1979] with a few additions. Where possible, the original sources are named including references to later discussions of the data where these exist. It will be noted that although 40 earthquakes are reported to have been associated with strain precursors, there exist

fewer than 10 events in which two or more locations have simultaneously experienced a precursor and only four events in which more than three observatories have monitored precursory strain anomalies. Perhaps the most convincing precursory strain anomaly precedes the magnitude 6 Odaigahara earthquake of 1961 which was observed by at least ten instruments at five locations within 100 km of the epicenter [Nishimura and Tanaka, 1963].

No treatment of strain or tilt data is complete without a discussion of the noise level associated with the measurements. Table 1 lists the depth of each observation. Although the data acquisition depth varies from 800 m to less than 10 m, half of the data were acquired less than 20 m from the surface and more than 60% of the data were acquired less than 50 m from the surface. Measurements within 20 m of the Earth's surface are often contaminated by atmospherically induced noise. The published data rarely provide sufficient information to enable an independent assessment of the noise level at different frequencies although recent results are significantly better documented than are earlier accounts. In most cases it is necessary to accept the often careful, but occasionally optimistic, assessment of the authors that a particular span of data is anomalous. An apparent contradiction in some of the data is that different onset times are reported for precursors observed by different instruments at the same observatory. The coincidental alignment of instruments with zero strain or tilt directions can explain some but not all of these discrepancies. A possible contributory factor to some instruments at a particular site detecting signals before others is that if all the instruments are operated at the same sensitivity, the amplification or attenuation of tectonic signals by underground cavity geometry may dominate the data [King and Bilham, 1976]. Some of the early Japanese data obtained from horizontal pendulum tiltmeters are particularly contaminated by systematic errors of this form. In such cases the direction and magnitude of the computed tilt vector or strain tensor axes will be incorrect, although the timing of anomalous changes will be unaffected.

A general feature of the data in Table 1 is that for a given epicentral distance, observed strain and tilt precursors are unexpectedly large. It has been argued that since the strain at failure in most rocks is of the order of 10^{-4}, coseismic epicentral strains should not exceed 10^{-4}, and it is unlikely that preseismic anomalies exceed 1% of this value [Sacks, 1978]. An inverse power law diminution of strain with epicentral distance implies that at distances of the order of several fault dimensions strain amplitudes are significantly smaller. The data are illustrated graphically in Figure 2. There is no consistent pattern suggestive of either a relationship between anomaly size versus distance or earthquake magnitude versus distance. It is possible to

TABLE 1. Precursors to earthquakes equal to or greater than magnitude 3 observed by strainmeters and tiltmeters. Data compiled from Tanaka [1964], Rikitake [1976, 1979], Dobrovolsky et al. [1979] and original sources.

Earthquake		Precursor			Observatory			
Magnitude	Date (depth)	amplitude in microstrain or microradians	duration in days	delay, $\log_{10}$ day	Location	distance km	depth, m	Source
Kanto 7.9	1923	15 (P)	0.3	4.2	Tokyo	80	–	14,47
Kii 5	Jul 04, 1929	1.3 (P)	13	1.5	Tanabe	20	–	15
Tottori 7.4	Sep 10, 1943	0.5 (P)	0.25	3.9	Ikuno	60	237	33,37,50
Tonankai 8.0	Dec 07, 1944	0.2 (P)	0.24	4.3	Kamigamo	160	9	33,50
Fukui 5	1948	1 (W)	0.008	4.1	Bandojima	20	–	
Nanki 6.7	Apr 26, 1950	0.75 (P)	0.29	3.5	Tamamizu	80	35	3,33,50
		0.2 (P)	0.28	3.6	Kamigamo	120	9	
		0.1 (P)	0.23	3.7	Kochi	200	40	
Daishoji-oki 6.8	Mar 07, 1952 (20 km)	300 (P)	90	1.2	Ogoya	40	300	33,35,34,37
		100 (P)	10	2.0	Ogoya	11	11	
Yoshino 7.0	Jul 18, 1952 (70 km)	0.3 (P)	400 (15)	0.6	Kamigamo	80	10	13,33,35,41, 50,57
		10 (P)	150 (15)	1.0	Kishu	60	60	
		10 (P)	15	2.0	Yura	80	30	49
		2.5 (S)	300	0.7	Osakayama	50	150	
		1.5 (S)	120	1.1	Ide	35	35	
		3 (S)	60	1.4	Ide	35	35	
		1 (S)	11	2.1	Ide	35	35	
Odaigahara 6.0	Dec 26, 1960 (60 km)	3 (P)	200	0.2	Kishu	40	60	33,34,35,36,37
		5 (P)	100	0.5	Kishu	40	60	
		3 (P)	10	1.5	Kishu	40	60	
		15 (P)	120	0.5	Shionomisaki	90	5	
		2 (P)	20	1.2	Shionomisaki	90	5	
		33 (P)	200	0.2	Yura	90	30	
		4 (P)	110	0.5	Yura	90	30	
		3 (P)	20	1.2	Yura	90	30	
		15 (P)	30	1.1	Oura	90	10	
		3 (P)	5	1.9	Oura	90	10	
		6 (P)	30	1.1	Kamigamo	100	9	
		1 (P)	5	1.9	Kamigamo	100	9	33,35,58
Hyuganada 7.1	Feb 27, 1961 (20 km)	0.5 (P)	12	2.1	Makamine	120	25	
		0.5 (P)	4	2.6	Makamine	120	25	
Hyogo swarm 5.9	May 07, 1961 (40 km)	6 (P)	300	0	Ikuno	30	237	59
		(P)						
Kitamino 7.0	Aug 19, 1961 (3 km)	25 (P)	50	0.5	Ogoya	40	150	33,34,35,37
		5 (P)	15	1.5	Ogoya	40	150	59
		1 (P)	50	0.5	Kamioka	60	484	
		0.5 (P)	15	1.5	Kamioka	60	800	
Shirahama-oki 6.4	Jan 04, 1962 (40 km)	1 (P)	20	1.5	Yura	35	30	36
		0.5 (P)	7	1.9	Yura	35	30	
		1 (P)	40	1.2	Kishu	65	60	
		10 (P)	3	2.3	Kishu	65	60	

TABLE 1. (cont.)

Earthquake		Precursor			Observatory			Source
Magnitude	Date (depth)	amplitude in microstrain or microradians	duration in days	delay, $\log_{10}$ day	Location	distance km	depth, m	
Echizenmisaki-oki		18 (P)	180	0.8	Kamigamo	80	10	36
	6.9 Mar 27, 1963	5 (P)	60	1.3	Kamigamo	80	10	
	(03 km)	3 (P)	15	1.9	Kamigamo	80	10	
		5 (P)	180	0.8	Ogoya	90	150	
		7 (P)	70	1.3	Ogoya	90	150	
		4 (P)	15	1.9	Ogoya	90	150	
		2 (P)	180	0.8	Ikuno	110	237	
		2 (P)	60	1.3	Ikuno	110	237	
		1 (P)	10	2.1	Ikuno	110	237	
Niigata	7.5 1964	15 (W)	3000	0	Maze	80	10	21
Matsushiro	~4.5 1966	0.6 (W)	0.05	3.0	Matsushiro	5	60	63
Gifu	6.6 1969	0.4 (W)	250	1.5	Inuyama	48	–	54
		0.4 (S)	250	1.5	Inuyama	48	–	
		0.4 (W)	250	1.5	Kamitakara	60	40	
Atsumi	6.1 1971	0.4 (W)	250	0.2	Inuyama	90	–	54
Nemuro	7.4 1973	0.6 (P)	76	1.5	Enino	250	40	20
Ashkhabad	4 Apr 25, 1957	0.5 (P)	0.04	2.8	Ashkhabad	25	10	39
U.S.S.R.	4 Oct 13, 1958	0.3 (P)	0.12	2.2	Alma Ata	250	15	39,40
Afghanistan	5 Mar 02, 1959	(P)	0.25	2.6	Kondara	245	50	39,40
	3.5 Mar 16, 1959	(P)	0.5	1.4	Kondara	300		
U.S.S.R.	6.3 1963	0.05 (P)	0.12	3.6		2800	10	42
		0.5 (P)	0.12	3.6	Kondara	600		42
Dushambe	4.5 Jan 14, 1965	0.07 (S)	6	0.9	Kondara	100	40	25
Hindu Kush	7.5 Mar 14, 1965	0.5 (S)	8	2.5	Kondara	300	50	25
		1 (S)	8	2.5	Kondara	300	10	
Tien Shien	6 May 04, 1965	0.09 (S)	15	0.9	Talgar	250	40	23,25
		0.05 (S)	4	2.0	Talgar			26,27
U.S.S.R.	5.5 Oct 03, 1967	3.5 (S)	120	0.2	Kondara	20	50	27
Djungarskoie	5 Aug 20, 1967	0.05 (S)	10	1.0	Talgar	320	40	25
Pamirs	3 Jan 09, 1969	0.02 (S)	2	1.8	Garm	50	50	25
Kurile	8 Mar 22, 1978	0.1 (S)	10	2.7	Shikotan	200	15	29
Danville	4.3 1970	0.1 (M)	30	0.1	Berkeley	30	10	62
San Fernando	4.7 Feb 21, 1971	500	0.4	2.2	Sylmar	20	0	56
Mina swarm	3 Jun 1972	0.2 (S)	20	-0.5	Mina	15	100	43
Hollister	4.3 Jan 10, 1974 (7.7 km)	1 (B)	15	0.4	Nutting	17	3	18

TABLE 1. (cont.)

| Earthquake | | Precursor | | | Observatory | | | |
Magnitude	Date (depth)	amplitude in microstrain or microradians	duration in days	delay, log$_{10}$ day	Location	distance km	depth, m	Source
Hollister 5.2	Nov 28, 1974	7 (B)	30	0.6	San Juan	11	3	30
Briones 4.3	Jan 08, 1977	2 (B)	20	0.3	Berkeley	6	10	19
San Jose 4.2	Jul 29, 1977	10 (B)	1.7	1.2	Hamilton	6.7	3	31
3.9	(9 km)	10	1.8			9	3	
Haicheng 7.3	Feb 04, 1975	0.3 (P)	1	2.7	Shihpengyu	20	0	44
		2 (P)	4	3.3	Shenyay			
Koyna 7	Dec 10, 1967	70 (P)	300 (100)	0.6	Koyna	5	10	10
New Zealand 5.5	Jan , 1976	2 (VP)	100	0.4	Khandalla	50	15	8
6.2	Jan , 1977	3 (VP)	200	0.3	Khandalla	60	15	

Note 1: The letters in parenthesis in column 3 refer to instrument type. P = horizontal pendulum, VP = vertical pendulum, M = mercury tube tiltmeter, W = water tube tiltmeter, B = borehole tiltmeter, and S = strainmeter.

Note 2: Precursor duration in column 6 refers to the time between the onset of the reported precursor and the time of the earthquake or the largest event in the case of a swarm of earthquakes.

Note 3: Numbers in parenthesis in column 4 are later precursors also observed in the data but not plotted in Figure 1.

Note 4: The numbers in column 5 are the observed error in the precursor lead time compared to the least squares fit for "precursors of the first kind" [Rikitake, 1979]. An apparent delay in the onset of a precursor is positive in this column.

Note 5: Observatory depths from Hosoyama [1956], Bilham [1973], and Okada [1977].

Note 6: Source numbers appear at the end of each reference.

conclude from Figure 2 that if an inverse power law applies to the fall-off in strain with distance, then many of the data are an order of magnitude larger than might be expected from a precursory slip model or dilatancy model with strains of the order of 10^{-6} in the epicentral zone. For events larger than magnitude 7 predicted precursor magnitudes for a soft-inclusion model do not contradict the observed data [Dobrovolsky et al., 1979]. Most reported anomalies have amplitudes between 10^{-7} and 5×10^{-6} and are monitored at distances between 80 and 120 km from the epicenter. The low values in Figure 2 are presumably near the threshold of detectability given noisy data. It should be noted also that the limited range of early recording methods influenced the lower limit of detectability for short period signals more than for long period signals which could be maintained on scale by manual adjustment.

Precursor strain rate and precursor anomaly size as a function of the duration of the precursor are plotted in Figures 3 and 4. These figures are of value in recognizing certain deficiencies in the data. No precursors are identified with strain rates less than 10^{-9}/day reflecting uncertainties in the long-term stability of strain and tilt data. A probable reason for the shortage of data for other distances and event magnitudes is that a relatively small number of events greater than M = 7 have occurred close to tilt and strain measurement arrays and that the spacing of instruments is inadequate to monitor the small radius of deformation associated with events with magnitude less than M = 6. Few precursors with a duration of approximately a day have been reported, presumably due to the high level of thermal and atmospherically induced noise at or near diurnal periods. It is possible that this 'hole' in the data gives rise to the apparent bimodal distribution in strain and tilt precursors reported by Rikitake [1979].

There has been no attempt to relate precursor data to the type of earthquake or its tectonic setting. That such considerations are important is clear from the significant differences between the properties of observed precursors reported from Central Asia, Japan, and the U. S. (Table 2).

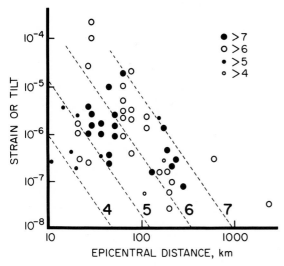

Figure 2. Strain or tilt precursor amplitude as a function of distance from epicenter. The dotted lines are the predicted strain amplitudes at various distances for given magnitude events obtained from a 'soft-inclusion' model of an earthquake source region [Dobrovolsky et al., 1979].

Discussion

It is clear in Figure 1 that the log T versus earthquake magnitude relationship that applies to a wide variety of precursory anomalies does not apply in any simple way to strain and tilt data. Approximately 70% of strain and tilt data are located below the $\pm$ 1 order-of-magnitude scatter about the least squares fit observed for radon emission, land deformation resistivity, magnetotelluric and water level precursors. The total scatter in strain and tilt data is more than three orders of magnitude. This scatter is interpreted to signify that a delay of variable duration may occur between the onset of most strain and tilt precursors and the earlier appearance of other forms of anomaly. This conclusion is tenable if it can be demonstrated that all anomalies are related to the same physical process and that there are no alternative explanations. Strain variations in the epicentral zone are at the heart of all theoretical models invoked to explain precursive phenomena either directly in the form of precursive slip or indirectly due to changes in elastic moduli. It is thus unreasonable to require fundamentally different physical processes for different types of precursive phenomena. A real possibility is that the delay is an artifact in the reporting of anomalies. For example the determination of the initiation of a strain precursor is frequently a subjective process. Errors are likely to be biased in a conservative direction, that is, toward a later time at which a definite change in strain or tilt gradient is clearly evident. The poor signal-to-

noise ratio in the data makes the identification of low rates of change difficult, particularly if there is no abrupt onset to a precursor. Moreover, instruments in certain azimuths may not be subjected to strain and tilt variations initially but may monitor subsequent complex strain changes arising from rotation, displacement or adjustment of tectonic stresses [Kasahara, 1973; Scholz, 1979].

Can the apparent delay in strain precursor onset time be explained by a fundamental difference between strain and tilt measurements and other forms of precursory manifestation? Whereas measurements of land deformation, radon emission, magnetotelluric currents, seismic velocity and water level variation integrate physical variations in the upper kilometer or more of the Earth's crust, strain and tilt measurements are significantly affected by conditions local to the Earth's surface. In a region of elastic inhomogeneity strainfields can be distorted, resulting in major variations in observed amplitude but only small variations in phase. If the region is partly viscoelastic, it is possible also to introduce errors in phase [Bilham and Beavan, 1979]. A possible source of stress-dependent elasticity is the inferred presence of surface fissures of large aspect ratio [Reasenberg and Aki, 1974; Herbst, 1976; Zschau, 1977].

Figure 5 illustrates a possible mechanism for generating apparent delays between the onset of regionally applied stress and the observation of local strain in fissured terrain. 1, 2, and 3

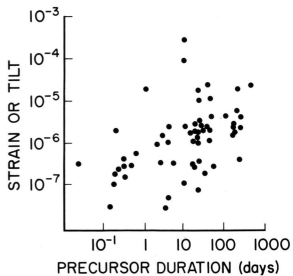

Figure 3. Observed strain or tilt precursor amplitude as a function of precursor duration. Strains larger than 10^{-4} (tilts larger than 10^{-4} radian) are close to the failure strain of most rocks. Note the near absence of data points near 1 day common to Figures 3 and 4 which may be attributable to diurnal atmospherically-induced noise.

416   BILHAM

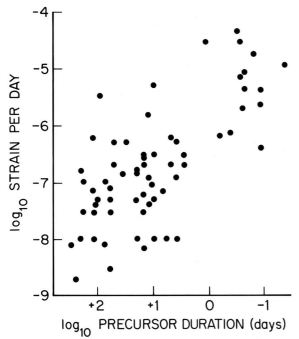

Figure 4. Strain rate (defined as maximum strain amplitude divided by precursor duration) as a function of precursor duration. The noise level of most tilt and strain instruments near the surface is approximately inversely proportional to the frequency at which it is measured. Hence the noise level for a given instrument is approximately parallel to the horizontal axis of Figure 4. For 'quiet' instruments (laser strainmeters) the average noise level corresponds to less than 10^{-9}/day.

formly increasing strain, strainmeter 1 monitors a non-linearly decreasing strain rate and strainmeter 2 monitors initially little strain but subsequently a strain rate approaching that of the strain rate observed by strainmeter 3. The model can explain the apparent delay between the onset of deep seated anomalies and their detection by surface strainmeters if there exists sufficient noise in the strain data to mask the low initial strain rates associated with the onset of the precursor (line 2 or 2' in Figure 5). The values in Figure 5 are based on the laboratory results of Goodman [1975], Siegfried and Simmons [1978] and Walsh and Grosenbaugh [1979]. Horizontal stresses encountered near the Earth's surface are commonly larger than those predicted from hydrostatic loading and vary from a few bars to more than a hundred bars, lower values being observed in tectonically active areas [Dahlgren et al., 1980]. Commonly reported values for stress in the uppermost 20 m of the Earth's surface place the stress/strain behavior of most observing sites in the non-linear region of Figure 5 [Jamison and Cook, 1980].

One of the implications of the model is that applied strainfields at all frequencies will be subject to some degree of attenuation, phase lag or non-linearity depending on their amplitude. This is weakly supported by data from 17 strain observatories in Europe where body tide amplitudes are found to be 20% too small. However, these data have not been corrected for topographic and geologic inhomogeneities and for locations where such corrections have been applied, observed tidal amplitudes are shown to be in close agreement with theory [Bilham and Beavan, 1979]. Site-corrected tidal data from central Asia are as much as a factor of 3 too large across active faults and 20% too low in other locations [Latynina et al., 1979], but these data can be explained by known geologic zones of weakness. An improved test would be to examine strain and tilt tidal data from only those instruments for which delayed precursors are reported but these data are generally not available. Earth tidal phases are generally in good agreement with theory and non-linearity greater than 1% has not been detected [Berger and Beaumont, 1970]. A negative result does not necessarily exclude the applicability of the model, since if fluids are present in near-

represent strainmeters near the surface of a dry inhomogeneous medium. The inhomogeneities take the form of fissures in contact at asperities. Strainmeter 2 is in a location commonly preferred by investigators, away from obvious signs of fissuring or weakness. Strainmeter 3 is a long baseline instrument placed in homogeneous material and strainmeter 2 is an instrument placed across a known fissure. Under uniformly increasing tectonic stress, strainmeter 3 monitors uni-

TABLE 2. Characteristics of Central Asian, Japanese, and North American Data Sets

	USSR	Japan	U.S.
Range in precursor epicentral distance (km)	20 - 2800	5 - 250	3 - 30
Range in event magnitude	3.5 - 7.5	4.5 - 8.0	3 - 5.2
Mean logarithmic precursor delay	2.2	1.6	0.6

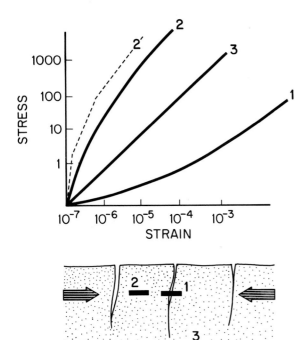

Figure 5. Model invoked to explain the formation of apparent delays in the observation of applied strainfields at the Earth's surface. The arrows indicate a uniformly increasing stress applied to a region in which surface fissures exist. In their presence, strainmeter (1) will monitor a rapidly decreasing strain and strainmeter (2) a rapidly increasing strain which may emerge above instrumental noise levels only at a much later stage in the precursory process. Strain curve (2') indicates the rapid changes in strain rate expected if several different populations of fissures exist.

surface fissures, it is possible to invoke not only stress-dependent behavior but also strain-rate dependent behavior. In a fluid-filled model, joint stiffness can be large for seismic or diurnal periodic strain but can be significantly smaller for periodic strains of longer period. Hence, the effect may be weak at diurnal periods but significant for tectonic strains exceeding several days. In support of this hypothesis it is noted that precursors to earthquakes with magnitudes less than 3 (duration < 10 days) are not associated with strain delays [Rikitake, 1979].

The closure pressure for different sets of fissures in a rock varies with the aspect ratio and orientation of the fissures relative to the applied strain. If different populations of fissures with widely different properties exist, the observation of a steadily changing stress field will be characterized by rapid changes in strain rate. If the strainmeter baseline includes ele-

ments of one population of fissures and not another, it may detect either a decrease or increase in strain rate under monotonically changing applied stress. Changes in strain rate or inflections in tilt vectors are common in strain precursor data. If the measurement baseline is long compared to the spacing of fissures, the model predicts that strain data should agree with other observations of precursive phenomena. Four out of six precursors observed by long-baseline water-tube tiltmeter data agree with the log T/magnitude relationship, which implies that if the model is correct, the spacing of fissures must be of the order of tens of meters or less. In situ studies of surface fractures [Engelder and Sbar, 1977] and near-surface fractures [Zoback et al., 1978] indicate that fracture spacings of this order play a significant role in determining the bulk properties of near-surface rocks.

The closure of one population of fractures when their closure pressure is exceeded by the ambient stress level effectively transfers further increases of stress to other populations of fractures if these exist. This has important consequences in earthquake prediction studies. The transfer can be caused either by a change in applied-stress orientation or by a change in stress amplitude. When it occurs, a hitherto dormant or relatively inactive set of fissures may generate anomalous physical behavior. Stress changes late in the precursory process appear to occur as indicated by azimuthal variations in tilt vectors and the axes of strain tensors. Observed resistivity and water-level changes hours before earthquakes (precursors of the second kind [Rikitake, 1979]) may be attributable to these effects.

Conclusions

A conservative view of the reported log T versus earthquake-magnitude relationship [Rikitake, 1979] based on all reported physical precursors is that the relationship describes an earliest time before which anomalous behavior related to the rupture process is not observable. In order to improve the numerical constants in the relationship, Rikitake [1979] found it necessary to exclude data that do not agree with a simple scaling law. These data include all strain and tilt measurements and foreshock observations. A systematic delay of variable degree in strain and tilt precursor data is possible. The data are of variable quality and the delay may be an artifact of observation. That is, the data are frequently contaminated by signals of non-tectonic origin and the identification of the onset of a precursor will be biased toward a later time when it is clearly manifest above the apparent noise level. Although this argument applies to all forms of precursory observation, it appears that surface strain measurements are either less sensitive to small variations in strain at depth or that these variations are masked by the relatively high noise level associated with near surface strain observations. However, the frequent occurrence of clear onsets to strain anomalies or sudden changes in strain rate indicates that noise alone cannot account for the observations and that if all precursors are attributable to the same physical process, a non-linearly increasing strainfield or a delay mechanism must be responsible.

A possible mechanism to account for both non-linear surface strain-rates and apparent delays in strain-precursors is proposed in which near surface fissures absorb initial increases in stress. Strainmeters or tiltmeters located between such fissures will at first witness little or no strain change. Subsequent stiffening of populations of fissures at increasingly large stresses results in the appearance of changes in strain rate. Instruments with baselines long compared to the spacing of fissures will experience no delay in the monitoring of applied strains. Precursor data from water-tube tiltmeters with baselines of the order of 50 m record precursors that are not delayed, which implies that the spacing of fissures contributing to the effect must be of the order of meters, a result that is not contradicted by in-situ studies of fracture spacing near the Earth's surface.

If surface fractures affect strain precursors, the mechanism should be possible to test independently using signals of known chronology. The earth tide signal provides evidence that the effect is small at least at diurnal periods; however, the ubiquitous presence of fluid-filled fissures will render the mechanism strain-rate dependent. Since the duration of most delayed precursors is of the order of 10-100 days, a more telling test would be to examine the attenuation, linearity, and phase properties of known long-period tides or reservoir impounding.

One of the consequences of the mechanism is that other local physical changes should be observable near the strainmeter or tiltmeter at the time of the onset of delayed precursors. Changes of body tide amplitude may be apparent but so also should local variations in resistivity, seismic velocity, water flow, etc. It is noted that some late-stage anomalies (notably resistivity) may reflect such changes in local conditions but that there appears to be no simple one-to-one relationship between strain precursors and other physical precursors. It may be that this highlights a basic deficiency in the data in that there are relatively few earthquakes for which a complete spectrum of physical precursors has been monitored. It is hoped that future studies will provide these data. A further consequence of the model is that measurements made across known fissures (major joints, faults or block boundaries) provide the earliest warning of precursory deformation. Simultaneous measurements in the surrounding medium potentially can provide an insight into the state of stress in the region.

In the absence of a conclusive explanation for the observed scatter in strain precursor onset time, it appears that strain and tilt data are of limited use for predicting the time and magnitude of an impending quake. This is perhaps only surprising in view of the log T versus magnitude relationship that describes approximately 50% of known precursors. In principle, strain and tilt measurements contain far more information than the time of the start of a precursory process leading to seismic rupture. Since many of the systematic errors that contaminated earlier strain and tilt data are now understood, it is perhaps time to reinterpret existing data and to direct future measurements guided by the lessons of past inadequacies.

Acknowledgments. I am indebted to John Beavan, Terry Engelder, Egill Hauksson, Richard Plumb, and Chris Scholz with whom I have discussed aspects of the data and to an anonymous reviewer who made several useful suggestions. Thanks also to Pat Gibson and Gianine Lupo who researched some of the original references. The work was supported by U. S. Geological Survey contracts 14-08-0001-17644 and 14-08-0001-18272. Lamont-Doherty Geological Observatory Contribution No. 3107.

References

Numbers in parenthesis after each reference refer to sources in Table 1.

Berger, J., and C. Beaumont, An analysis of tidal strain observations from the United States of America, II. The inhomogeneous tide, Bull. Seismol. Soc. Amer., 66, 1821-1846, 1976. (1)

Bilham, R. G., The location of earth strain instrumentation, Phil. Trans. Roy. Soc. Lond., A, 274, 429-433, 1973. (2)

Bilham, R. G., and R. J. Beavan, Tilts and strains on crustal blocks, Tectonophysics, 52, 121-138, 1979. (3)

Dahlgren, J. P., R. M. Richardson, M. L. Sbar, and T. Engelder, The influence of thermally induced stress upon near-surface stress data near the San Andreas fault, Palmdale, California, J. Geophys. Res., in press, 1980. (4)

Dobrovolsky, I. P., S. I. Zubkov, and V. I. Miachkin, Estimation of the size of earthquake preparation zones, Pure Appl. Geophys., 117, 1025-1044, 1979. (5)

Engelder, T., and M. L. Sbar, The relationship between in situ strain relaxation and outcrop fractures in the Potsdam sandstone, Alexandria Bay, New York, Pure Appl. Geophys., 115, 41-55, 1977. (6)

Fujii, Y., Relation between duration period of the precursory crustal movement and magnitude of the earthquake, J. Geod. Soc. Japan, 27, 197-214, as reported in T. Rikitaki, Tectonophysics, 54, 293-309, 1979. (7)

Gerard, V. B., Mid-New Zealand Earth strain and tilt observations and precursors of earthquakes, in Terrestrial and Space Techniques in Earthquake Prediction Research, pp. 501-510, edited by Vogel, Vieweg, 1979, 712 p. (8)

Goodman, R. E., Methods of Geological Engineering in Discontinuous Rocks, pp. 472, West Publishing, 1976. (9)

Guha, S. K., Premonitory crustal deformations, strains and seismotectonic features (b-values) preceding Koyna earthquakes, Tectonophysics, 52, 549-559, 1979. (10)

Hagiwara, T., T. Rikitaki, K. Kasahara, and J. Yamada, Observation of ground tilting and strain at Hokuga Village, Fukui Prefecture, Rept. Spec. Comm. Investigation of Fukui Earthquake, pp. 61-64, Science Council Japan, Tokyo, 1948. (11)

Herbst, K., Interpretation von Neigungsmessungen im Periodenbereich oberhalb der Gezeiten, Ph.D. dissertation, Technischen Universitat Clausthal, 1976. (12)

Hosoyama, On secular observation of tilting motion of the ground, Bull. Dis. Prev. Res. Inst., (Memorial Issue), 31, 20-27, 1956. (13)

Imamura, A., On the tiltings of the Earth preceding the Kwanto earthquake of 1923, Proc. Imper. Acad. Japan, 4, 4, 1928.(14)

Imamura, A., On crustal deformations preceding earthquakes, Jap. J. Astr. Geophys., 10, 82-92, 1933. (15)

Imamura, A., and T. Kodaira, On the preseismic Earth-tilting and mechanism of occurrence of the Kii earthquake of July 4, 1929, Proc. Imper. Acad. Japan, 5, 460-462, 1929. (16)

Jamison, D. B., and N. G. W. Cook, Note on measured values for the state of stress in the Earth's crust, J. Geophys. Res., 85, 1833-1838, 1980. (17)

Johnston, M. J. S., and C. E. Mortensen, Tilt precursors before earthquakes on the San Andreas fault, California, Science, 186, 1031-1034, 1974. (18)

Johnston, M. J. S., A. C. Jones, W. Daul, and C. E. Mortensen, Tilt near an earthquake (M_L = 4.3), Briones Hills, California, Bull. Seismol. Soc. Amer., 68, 169-173, 1974. (19)

Kasahara, K., Tiltmeter observation in complement with precise levellings, J. Geod. Soc. Japan, 19, 93-99, 1973 (in Japanese). (20)

Kasahara, K., Premonitory crustal movement observed at Enino before the earthquake of the Enimo Peninsula on June 17, 1973, Proc. Symp. Earthquake Prediction, Seism. Soc. Japan, 3-14, 1977 (in Japanese). (21)

King, G. C. P., and R. G. Bilham, Tidal tilt measurement in Europe, Nature, 243, 74-75, 1973. (22)

Latynina, L. A., On horizontal deformations at faults recorded by extensometers, Tectonophysics, 29, 421-427, 1975. (23)

Latynina, L. A., and R. M. Karmaleyeva, On certain anomalies in the variations of crustal strains before strong earthquakes, Tectonophysics, 9, 239-247, 1970. (24)

Latynina, L. A., and R. M. Karmaleyeva, Measurement of slow movements in the Earth's crust as a method of seeking forewarnings of earthquakes, in Physical Bases of Seeking Methods of Pre-dicting Earthquakes, edited by M. A. Sadoski, Acad. Sci. USSR, Moscow, 1970, (English translation by D. B. Vitaliano). (25)

Latynina, L. A., and R. M. Karmaleyeva, Extenso-meter measurements: The measurement of crustal strain, Nauka, Moscow, pp. 152, 1978. (26)

Latynina, L. A., R. M. Karmaleiva, S. D. Rizaeva, E. Ya. Starkova, and B. Mordonov, Deformation of the Earth's surface at Kondava before the earthquake of 3-x-1967, in The Search for Earthquake Forerunners at Prediction Test Fields, Nauka, Moscow, 1974 (in Russian). (27)

Latynina, L. A., R. M. Karmaleieva, A. V. Tikhomiov, and L. E. Khasilev, Secular and tidal deformations recorded in the fault zone of the N.E. Tien Shan, in Terrestrial and Space Techniques in Earthquake Prediction Research, edited by Vogel, Vieweg, 1979, pp. 712. (28)

Monakhov, F. I., L. I. Bozhkova, A. M. Khantaev, and E. v. Khaidurova, and Yu. A. Shluyev, Short range precursors and the prediction of the Kurile Islands earthquakes of March 22-28, 1978, Izv. Akad. Nauk. SSR Fizika Zemli, 7, 78-80, 1979. (29)

Mortensen, C. E., and M. J. S. Johnston, Anomalous tilt preceding the Hollister earthquake of November 28, 1974, J. Geophys. Res., 81, 3561-3566, 1974. (30)

Mortensen, C. E., and E. Y. Iwatsubo, Short term tilt anomalies preceding local earthquakes near San Jose, California, Bull. Seismol. Soc. Amer., 70, 2221-2228, 1980. (31)

Nersesov, I. L., L. A. Latynina, and R. M. Karmaleyeva, On the relationship of crustal strain to earthquakes on the basis of data of the Talgar station, in Earthquake Precursors, edited by M. A. Sadovskiy, Acad. Sci. USSR, Moscow, 1973 (English translation by Vitaliano, USGS). (32)

Nishimura, E., Anomalous tilting movement of the ground observed before destructive earthquakes, First Int. Symp. on Recent Crustal Movements, Leipzig, Dis. Prev. Res. Inst. reprint 1-18, 1962. (33)

Nishimura, E., and K. Hosoyama, On tilting motion of ground before and after the occurrence of an earthquake, EOS, Trans Amer. Geophys. Un., 34, 597-599, 19 . (34)

Nishimura, E., and Y. Tanaka, On peculiar mode of secular ground tilting connected with a sequence of earthquakes in some restricted areas, Geophysical Papers Dedicated to Prof. Kenzo Sassa, Kyoto University, 365-378, 1963. (35)

Nishimura, E., and Y. Tanaka, On anomalous crustal deformation observed before some recent earthquakes, Part II, Bull. Dis. Prev. Inst., Kyoto Univ., 7, 66-75, 1964. (36)

Nishimura, E., Y. Tanaka, and T. Tanaka, On anomalous crustal deformation observed before some recent earthquakes, Part I., Bull. Dis. Prev. Res. Inst., Kyoto Univ., 5A, 28-43, 1962. (37)

Okada, Y., Results of continuous observations of crustal movements in Japan, Collected Papers of the Dis. Prev. Res. Inst., 1977. (38)

Ostrovskiy, A. E., On changes in tilts of the Earth's surface before strong near earthquakes, in Physical Bases of Seeking Methods of Predicting Earthquakes, 152 pp., edited by M. A. Sadovsky, Acad. Sci. USSR, Moscow, 1970, (English translation by Vitaliano, USGS). (39)

Ostrovskiy, A. E., Tilts and earthquakes, in Earthquake Precursors , 216 pp., edited by M. A. Sadovsky, I. L. Nersesov, and L. A. Latynina, Acad. Sci. USSR, Moscow, 1973, (English translation by Vitaliano, USGS). (40)

Ozawa, I., On the observation of crustal deformation at Osakayama, Bull. Dis. Prev. Inst., Kyoto Univ., 5th Anniv. Memorial Issue, November 1956, 31, 14-19, 1956. (41)

Panasenko, G. D., Observation of a tilt storm as a precursor of a strong remote earthquake, Izv. USSR Acad. Sci. Ser. Geophysics, 10, 19 . (42)

Priestley, K. F., Possible premonitory strain changes associated with an earthquake swarm near Mina, Nevada, Pure Appl. Geophys., 113, 250-256, 1975. (43)

Raleigh, B., G. Bennet, H. Craig, T. Hanks, P. Molnar, A. Nur, J. Savage, C. Scholz, R. Turner, and F. Wu, Prediction of the Haicheng earthquake, EOS Trans. Am. Geophys. Un., 58, 236-272, 1977. (44)

Reasenberg, P., and K. Aki, A precise, continuous measurement of seismic velocity measurement for monitoring in situ stress, J. Geophys. Res., 79, 399-406, 1974. (45)

Rikitake, T., Earthquake precursors, Bull. Seismol. Soc. Amer., 65, 1133-1162, 1975. (46)

Rikitake, T., Earthquake Prediction: Developments in Solid Earth Geophysics, 9, pp. 357, Elsevier, 1976. (47)

Rikitake, T., Classification of earthquake precursors, Tectonophysics, 54, 293-309, 1979. (48)

Sacks, I. S., Borehole strainmeters, U.S. Geol. Surv. Conf. Measurement of Ground Strain Phenomena Related to Earthquake Prediction, Carmel, September 7-9, 1978. (49)

Sassa, K., and I. Nishimura, On phenomena forerunning earthquakes, EOS Trans. Am. Geophys. Union, 32, 1-6, 1951. (50)

Sassa, K., and I. Nishimura, On phenomena forerunning earthquakes, Pub. Bur. Centrale Seism., A-19, 277-285, 1954. (51)

Scholz, C. H., A physical interpretation of the Haicheng earthquake prediction, Nature, 267, 121-124, 1977. (52)

Scholz, C. H., L. R. Sykes, and Y. P. Aggarwal, Earthquake prediction -a physical basis, Science, 181, 803-810, 1973. (53)

Shichi, R., Continuous observation of crustal movement and its possible improvement, Proc. Symp. Earthquake Prediction, Dec. 1972, pp. 26-34, Seism. Soc. Japan, Tokyo, 1973 (in Japanese). (54)

Siegfried, R. S., and G. Simmons, Characterization of oriented cracks with differential strain analysis, J. Geophys. Res., 83, 1269-1278, 1978. (55)

Sylvester, A. G., and D. D. Pollard, Observation of crustal tilt preceding an aftershock at San Fernando, California, Bull. Seismol. Soc. Amer., 62, 927-931, 1972. (56)

Takada, M., On the crustal strain accompanied by a great earthquake, Bull. Dis. Prev. Res. Inst., 27, 29-46, 1959. (57)

Tanaka, Y., Relation between crustal and subcrustal earthquakes inferred from the mode of crustal movements, Spec. Contrib., Geophys. Inst., Kyoto Univ., 4, 19-28, 1964. (58)

Tanaka, Y., On the stages of anomalous crustal movements accompanied with earthquake, Spec. Contrib., Geophys. Inst. Kyoto Univ., 8, 1-17, 1965. (59)

Tsubokawa, T., On relation between duration of precursory geophysical phenomena and duration of crustal movement before earthquake, J. Geod. Soc. Japan, 19, 116-119, 1973 (in Japanese). (60)

Whitcomb, J. H., J. D. Garmany, and D. L. Anderson, Earthquake prediction: variation of seismic velocities before the San Fernando earthquake, Science, 180, 632-635, 1973. (61)

Walsh, J. B., and M. A. Grosenbaugh, A new model for analysing the effect of fractures on compressibility, J. Geophys. Res., 84, 3532-3536, 1979. (62)

Wood, M. D., and R. V. Allen, Anomalous microtilt preceding a local earthquake, Bull. Seismol. Soc. Amer., 61, 1801-1809, 1971. (63)

Yamada, J., A water-tube tiltmeter and its application to crustal movement studies, Rept. Earthquake Res. Inst., 10, 1-147, 1973 (in Japanese). (64)

Zoback, M. D., J. H. Healy, J. C. Roller, G. S. Gohn, and B. B. Higgins, Normal faulting and in-situ stress in S. Carolina coastal plain near Charleston, Geology, 6, 147-152, 1978. (65)

Zschau, J., Air pressure induced tilt in porous media, in Proc. 8th Int. Symp. Earth Tides, Bonn, 1977. (66)

NEOTECTONIC DEFORMATION, NEAR-SURFACE MOVEMENTS AND SYSTEMATIC ERRORS IN U.S. RELEVELING MEASUREMENTS: IMPLICATIONS FOR EARTHQUAKE PREDICTION

Robert Reilinger and Larry Brown
Department of Geological Sciences
Cornell University
Ithaca, NY 14853

Abstract. Analyses of U.S. releveling measurements indicate that derivative crustal movement estimates may reflect tectonic deformation, near-surface movements, and/or systematic errors. Discriminating the contributions of these factors is especially crucial for unambiguous geodetic detection of possible precursory seismic deformations. While reliable leveling measurements of co-seismic and post-seismic movements are well documented for some of the larger (M > 6) dip-slip earthquakes, leveling evidence for pre-seismic motion is generally sparse and often ambiguous. Subtle earthquake-related motions may be masked by both aseismic movements and systematic errors. Deep magma injection and surficial groundwater withdrawal are two mechanisms which are shown to cause surface movements which, under some circumstances, could be misidentified as seismic-related. Of more concern, perhaps, are systematic measurement errors. Topography dependent errors are an exceptionally troublesome type, perhaps affecting as much as 20% of U.S. leveling. However, other varieties of systematic error also contribute to the uncertainty. Discrepancies between leveling and tide gauge data and within nets of leveling alone suggest large, long baseline accumulations of error. In many cases, aseismic and erroneous contributions can not be unequivocally determined ex post facto. However, a comprehensive examination of the NGS crustal movement data base, representing a large sampling of the entire U.S. Level Net, provides perspective and criteria needed to begin to recognize movement directly related to earthquake activity.

Perhaps the most extensive set of measurements relevant to earthquake activity exists in southern California, where much attention has recently been focused. Reevaluation of some of these leveling observations indicates that while some appear to reflect tectonic deformation, others are suspect because of indications of systematic errors and near-surface, non-tectonic movements. Specifically, possible preseismic movements reported for the 1971 San Fernando earthquake in the vicinity of the earthquake fault as well as approximately 30 km northwest of the epicenter may be due to systematic errors. Movements near the San Gabriel fault, initially ascribed to the Palmdale Bulge and more recently to preseismic effects of the 1971 San Fernando earthquake apparently reflect near-surface sediment compaction due to water table fluctuations. Similarly, there is strong evidence of contamination by rod calibration errors in some of the releveling observations used to define the southern portion of the Palmdale Bulge (Llano to Azusa, California). The reality of the Palmdale Bulge itself must be questioned in view of this reevaluation. In contrast, possible tilting southwest of Palmdale between 1961 and 1964 is not easily related to systematic errors or near-surface movements and thus may represent tectonic deformation. Whether this tilt anomaly was due to preseismic effects of the San Fernando earthquake or a mechanically separate tectonic event is presently unknown.

Introduction

Vertical movements of the earth's crust are commonly expected to accompany the various phases of strain buildup and release associated with major earthquakes. Observations of vertical co-seismic and post-seismic movements using precise leveling are well-documented. However, reports of preseismic movement in the U.S. are rare and, as will be argued, questionable. Recognition of true pre-seismic motion is complicated by

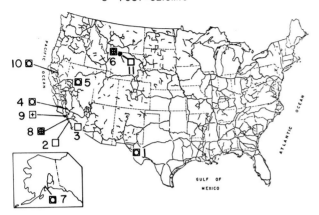

Figure 1. Leveling routes for which crustal movement information has been obtained in the U.S. Locations of U.S. earthquakes for which releveling evidence of crustal movement has been reported are also shown. Numbers refer to Table 2.

systematic leveling errors, near-surface non-tectonic processes (e.g., fluid withdrawal), the general lack of sufficiently redundant and extensive surveys, and the fact that significant changes in elevation have been identified which are unrelated to earthquakes. Such "noise" could easily hide a pre-seismic signal. Considerable uncertainty exists as to the extent and magnitude of these obscuring "movements". Direct determination of their effects is often extremely difficult. However, some perspective on these problems can be obtained from empirical analyses of existing National Geodetic Survey (NGS) releveling (Figure 1). Such an analysis forms the basis of this report.

This paper reviews some factors that must be considered when attempting to extract tectonic information, especially those relevant to earthquake prediction, from historic releveling observations. Evidence on the extent and nature of systematic errors, non-tectonic movements, and tectonic deformation (both earthquake related deformation and tectonic movements unassociated with earthquakes) from U.S. releveling measurements is presented. Specific criteria to help recognize suspect movements are developed and illustrated by application to a reevaluation of certain southern California leveling results of particular interest in earthquake prediction.

Systematic Errors in Leveling

At the root of much of the current debate regarding leveling-derived estimates of crustal motion is the prevailing uncertainty as to the role of systematic measurement errors. In particular, systematic errors which accumulate with relief have become a central issue in crustal movement research . While errors of this type have been known to geodesists for some time (e.g., Bomford, 1971), their influence has been considered too small to be of concern in most geodetic applications. However, new field experiments carried out by the NGS (Whalen, 1980) and empirical analyses (Brown et al., 1980) confirm earlier suspicions (e.g., Savage and Church, 1974; Brown and Oliver, 1976; Citron and Brown, 1979; Jackson et al., 1980; Chi et al., 1980) that topography-induced systematic errors are larger and more common than heretofore established and consequently that such errors can be and probably have been misinterpreted as tectonic motions of the crust.

Topography-correlated errors can arise from improperly calibrated leveling rods and from unequal atmospheric refraction of the foresight and backsight readings. The effects of these two sources of error should differ in a number of respects and thus in principle may be distinguished. For example, ficticious movements resulting from rod calibration errors should correlate rather closely with detailed topography, and change magnitude only where rod pairs are changed. In contrast, atmospheric refraction will be independent of the rods used in the survey. Furthermore, refraction errors can be expected to accumulate in a more complex manner because refraction depends on a variety of parameters which may vary significantly during the course of a given survey (e.g., near surface temperature gradients, individual sight lengths, wind, etc.). Because of procedural changes (a tendency towards shorter sight lengths in newer surveys), atmospheric refraction should more often than not result in ficticious movements which show a positive correlation with topography (i.e., high areas will appear to be rising) while errors due to rod miscalibration should have no preference for positive or negative correlations. In practice, these distinctions are not always easy to draw. However, a preliminary survey of NGS releveling estimates of elevation change which correlate with topography indicates that about 75% display positive correlations. While rod calibration errors appear to effect some leveling observations (e.g. Jackson et al., 1980; this paper) this preliminary result suggests that refraction may be the more pervasive source of elevation correlated error.

The expected magnitude for refraction error is a point of considerable uncertainty. According to one approximation (i.e., Kukkamaki, 1938), this error is proportional to the height difference between benchmarks, the temperature difference between 0.5 m and 2.5 m above the ground, and the square of the sight length used in the observation. Figure 2 shows representative values for refraction error using constants given by Holdahl (1980). For reasonable temperature differences (1-2 degrees C: Whalen, 1980) and sight lengths (25-75 m), errors as large as 30-40 mm or more can easily accumulate over height differences of 100 m (300-400 ppm). Since refraction error will usually have the same sign, its effect should tend to cancel when differencing surveys to compute movement. This rationale has often been used as an argument for ignoring the effect. However, if surveys are conducted using different sight lengths (Strange, 1981) and/or under different micrometeorological conditions, the refraction effect will result in what appear to be movements that roughly correlate with relief.

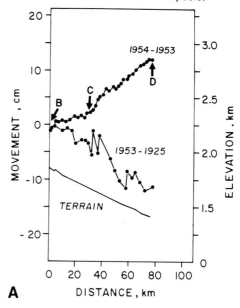

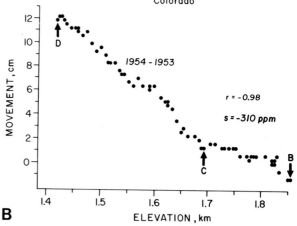

Figure 3. a) Profiles of apparent elevation change and topography from Colorado Springs to Leadville, Colorado. Reversal of apparent tilt and correlation with topography strongly suggest elevation correlated error.
b) Apparent elevation change versus elevation difference for 1954-1953 profile in Figure 3a. Correlation coefficient (r) and regression slope (magnitude of error) are also shown.

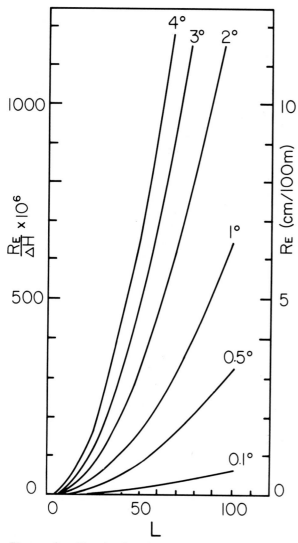

Figure 2. Magnitude of refraction error (R_E) normalized by height difference (ΔH) versus sight length (L) for various temperature differences. Based on relationship and theoretically determined constant given by Holdahl (1980).

Examples of apparent movement correlating with elevation are numerous (Brown et al., 1980). Approximately 20% of U.S. releveling observations show visual correlations between apparent movement and topography. The magnitude of the effect often reaches 30-40 mm per 100 m change in

Table 1. Magnitude of systematic discrepancies between repeated levelings (apparently due to leveling errors) in areas of subdued relief. Discrepancies are given in terms of apparent tilt (mm/km) and apparent tilt rate ($10^{-8}yr^{-1}$). Distance over which discrepancies accumulate monotonically and average relief along route are shown.

AREA	MAGNITUDE (MM/KM)	MAGNITUDE (10^{-8} YR^{-1})	DISTANCE (KM)	AVERAGE RELIEF (M)
East Coast U.S.	0.5	1.6	650	<5
West Coast U.S.	0.3	3.3	380	<10
Midcontinent U.S.	1.0	2.6	140	<100

Quoted Limit for Systematic Error in a Single Survey <0.2 mm/km (Bomford, 1971)

elevation. Although, correlation with topography alone is insufficient to warrant rejection of a tectonic interpretation (e.g., see Reilinger et al., 1977), it is clearly grounds for suspicion particularly when correlations persist at short wavelengths (e.g., Figures 18 and 19) or when multiple surveys give inconsistent results. For example, Figure 3a shows apparent vertical movements and terrain along the route from Colorado Springs to Leadville, Colorado based on surveys conducted in 1925, 1953, and 1954. The reversal of the 1953-1925 apparent tilt during the period 1954-1953 strongly suggests systematic error, and the correlation with terrain (Figure 3b) points toward an elevation correlated error. Since elevation-dependent errors may contaminate a significant portion of the NGS releveling data base, the possibility of such errors must always be considered prior to invoking tectonic explanations for apparent movements which correlate with relief.

Topography dependent error is not the only type of systematic error affecting U.S. leveling measurements. Table 1 lists a number of areas where comparison of repeated leveling measurements show large systematic discrepancies which may be due to errors in the observations. These examples occur in areas of generally subdued relief, thus ruling out elevation-correlated errors. If leveling errors are solely responsible for these discrepancies, whatever their cause, they range in magnitude from .3 mm/km to 1 mm/km and remain systematic (i.e., accumulate monotonically) for distances ranging from about 100 km to over 600 km.

Figure 4 shows three different estimates of elevation change (assuming constant rates of movement over the time period between levelings) along the east coast of the U.S. from Maine to Florida: 1) from unadjusted leveling measurements; 2) from tide gauge records

(squares); and 3) from the same leveling observations adjusted with standard least squares procedures for consistency with other repeated leveling lines which form circuits extending inland from the coast (tide gauge data were not used in the adjustment, Jurkowski et al., 1979). The leveling measurements span an approximately 30 yr. time interval. The fact that the relative movement between Maine and Florida is substantially reduced through the adjustment and the fact that the adjusted leveling profile is

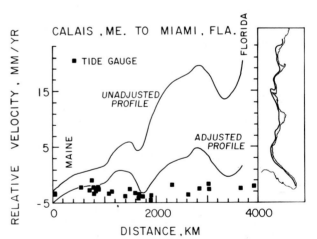

Figure 4. Elevation change profiles along east coast of U.S. (map at right). Unadjusted profile based on observed elevations from leveling assuming constant velocity movement (modified from Brown, 1978). Adjusted profile is same data adjusted by other levelings lying inland from coast (tide gauge data not used in adjustment). Squares show similar profile derived from tide gauges.

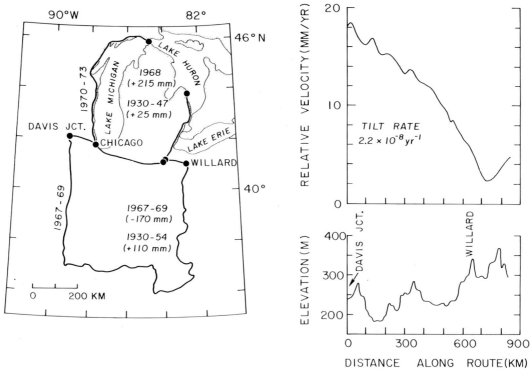

Figure 5. Leveling loops used to investigate apparent tilting between Davis Junction, Illinois and Willard, Ohio. Elevation change profile and topography along Davis Junction to Willard route shown at right. Misclosures for each loop, given in brackets, suggest systematic error in 1967–1969 leveling from Davis Jct. to Willard, Ohio.

more consistent with tide gauge data (although serious discrepancies still remain) indicate that the regional north-south tilt results from systematic errors in the leveling observations. The error remains more or less systematic over distances of 1000's of kilometers, and on some sections (e.g., 1800-2600 km) reaches .5 mm/km.

Comparison of leveling and tide gauge estimates of crustal movement along the west coast of the U.S. between Astoria, Oregon, and Crescent City, California show similar discrepancies (Brown et al., 1980). Unlike the east coast profile, apparent crustal movements along the west coast were derived from only two surveys and were thus not subject to possible temporal bias due to stringing together segments covering different time intervals. For the west coast profile, the north-south error reaches .3 mm/km and remains systematic over a distance of 380 km.

The cause of the apparent errors in these coastal surveys is presently unknown. The predominantly north-south orientation of the coastal profiles may suggest unequal lighting or other factors which are believed to preferentially accumulate on north-south lines (Bomford, 1971). However, the substantial reduction of the apparent tilt indicated by the east coast profile when adjusted with inland data suggests that the error may be related to the proximity of the leveling route to the coast (i.e., the error did not effect, or had less of an effect, on profiles further inland).

Suspect movements are not restricted to coastal and north-south profiles. For example, consider the large apparent tilt across the U.S. midcontinent identified by Brown and Oliver (1976) from releveling between Davis Junction, Illinois and Willard, Ohio (Figure 5). This tilt is perhaps the largest apparent movement defined by leveling in the eastern U.S. The tilt anomaly shows no clear relationship to geologic structure and is inconsistent with movements inferred from comparisons of water level gauges in the great lakes (Brown and Oliver, 1976). Figure 5 shows the results of a loop closure analysis (see Chi et al., 1980 for discussion of method) for circuits including the Davis Junction to Willard route. The fact that misclosures are considerably larger when the circuits are closed with the 1967-1969 surveys between Davis Junction and Willard than with the 1930-1947 surveys, even though the remainder of the loop was surveyed more closely in time to the 1967-1969 interval, suggests that the large apparent tilt of the interior plains is due to systematic error and not real ground motion. Such an error would have to reach 1 mm/km and remain systematic for distances of well over 500 km.

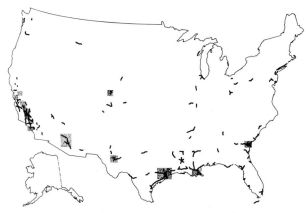

Figure 6. Location of elevation change profiles which indicate subsidence possibly due to groundwater effects. Shaded areas represent previously published cases of near surface subsidence.

In both the coastal and interior examples cited above, the discrepancies are characterized by consistent accumulation over large distances. The effects of these long baseline discrepancies are often reduced below the level of concern for most geodetic applications by network adjustments. However, when examined in the context of evaluating crustal movements such discrepancies become quite significant. Although the net apparent movement over a profile can be large, the tilt rates resulting from these errors are rather low, especially when compared with those exhibited by unequivocal examples of real movement. Tilt may therefore be the more diagnostic parameter in evaluating reliability of crustal movement estimates.

Near-Surface Movements

In addition to systematic errors, releveling measurements are influenced by near-surface movements which can mask, or be mistaken for deep-seated tectonic motion. Near-surface effects include: a) Benchmark instability (frost heave, soil moisture and temperature changes, human disturbance), b) Surface failure (land slides, mine and cavern collapse), c) Loading (reservoir impoundment, building settlement), and d) Fluid withdrawal (water, oil, gas). Benchmark instability and surface failure are often easily identified or are of such local extent that they are not a serious problem for regional tectonic studies. Such effects can, however, complicate local investigations - for example, of movements near earthquake faults. Near-surface soil or sediment compaction due to earthquake ground-shaking may be responsible for the predominance of subsidence over uplift near many earthquake faults (Savage and Hastie, 1966). Subsidence due

to surface loading and fluid withdrawal is, in general, easily related to human activity. In fact, leveling has proven quite effective at monitoring such movements, with important engineering applications (e.g., Poland and Davis, 1969). However, near-surface movements, and in particular movements due to variations of water levels in aquifers, appear to be more widespread than previously reported. In addition, such effects can be subtle and subsequently misidentified as tectonic deformation.

Figure 6 shows NGS releveling profiles in the U.S. which indicate subsidence relative to surrounding areas and which overlie aquifer systems which have experienced variations in water levels due either to pumping or natural causes. These movements may therefore represent sediment compaction associated with these water level variations, and are consequently suspect as indicators of tectonic motion. It is interesting to note from Figure 6 that these apparently near-surface movements are quite common in southern California, affecting much of the area peripheral to the Palmdale Bulge. Arguments will be presented later suggesting that in at least one case such groundwater subsidence in southern California has been misidentified as tectonic uplift of the adjacent areas.

A particular example, not previously reported, which illustrates criteria which can be used to recognize near-surface sediment compaction is the relative subsidence of the Los Angeles basin. Figure 7 gives a map of the L.A. Basin showing

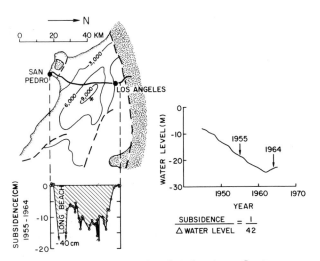

Figure 7. Map of Los Angeles basin. Contours indicate depth to basement (M), heavy dashed lines are faults, stippled areas are bedrock outcrops, heavy solid line (San Pedro to Los Angeles) is leveling route - crustal movements for period 1955-1964, indicating sediment compaction, shown below map. Asterisk shows location of observation well for which water level history is shown at right.

contours of basement depth and the location of a leveling route traversing the basin. Also shown are elevation changes along the leveling route and the history of water level decline measured in an observation well near the center of the basin. Subsidence near the center of the basin reaches 15 cm relative to the periphery and extends over a distance of 40 km. The observed subsidence correlates spatially with aquifer geometry and temporally with the history of water level decline. In addition, the magnitude of the effect (i.e., the ratio of subsidence to water level decline) is comparable to observations in other areas (Poland and Davis, 1969). Had the relationship between aquifer geometry and subsidence not been noticed, it is possible that these measurements could have been misinterpreted as tectonic motion.

Vertical Movements and Earthquakes

In spite of the substantial difficulties associated with releveling estimates of crustal movement, some of which have been described in previous sections, the capability of the leveling technique for monitoring tectonic earth movements is well established. In a number of cases, relatively subtle earth movements (i.e., tilts few x 10^{-6} rad and tilt rates few x 10^{-8} rad/yr) have been identified. In this section we briefly review releveling evidence for earthquake related deformation in the U.S. and use specific examples to illustrate some of the criteria employed to identify real tectonic movements.

The best examples of tectonic deformations measured by leveling in the U.S. are those in the vicinity of major earthquakes. Figure 1 and Table 2 review those U.S. earthquakes for which vertical movements have been reported. All of these earthquakes are associated with faults that have a significant component of dip-slip movement (with the possible exception of the 1940 Imperial Valley earthquake). Up to the present, there is no clear evidence from U.S. releveling measurements for permanent vertical deformation associated with purely strike-slip faulting although present observations are not sufficient to rule such out.

The most obvious vertical movements are those accompanying the earthquake (coseismic). Coseismic deformation has been well-documented for several of the larger (M > 6) normal and thrust earthquakes which have occurred in areas of preexisting geodetic control (Table 2). Observed movements range in magnitude from a few cm to a few m depending on the size of the earthquake and the proximity of the leveling measurements to the epicentral area. In general, coseismic movements are well explained by elastic dislocation theory (Savage and Hastie, 1966) although complications can arise from such factors as near-surface soil or sediment compaction due to ground shaking.

Post-seismic vertical movements have also been observed by releveling for some of the larger dip-slip earthquakes (see Table 2). These movements are usually smaller than associated coseismic movements; however like coseismic movements they can often be identified by their close spatial and temporal association with earthquakes, and in some cases surface faulting. Where sufficient observations exist, post-seismic deformation rates appear to decrease exponentially from the time of the earthquake. For example, movements near Anchorage following the 1964 Alaska earthquake are shown in Figure 8 (Brown et al., 1977). The Alaska earthquake, one of the largest events ever recorded, occurred where the oceanic Pacific plate is being thrust under the continental North American plate at a rate of over 5 cm/yr (Plafker, 1972). Savage and Hastie (1966) and Hastie and Savage (1970), using a dislocation model of thrust faulting, showed that the coseismic displacements were consistent with low-angle thrusting. Post-seismic movements near Anchorage (Figure 8) amounted to as much as 0.55 m of land uplift at an exponentially decreasing rate during the decade following the earthquake. Additional evidence for deformation following the Alaska earthquake was reported by Prescott and Lisowski (1977) from analysis of detailed leveling arrays on Middleton Island in the Gulf of Alaska. Tilts associated with the Alaska post-seismic movements were on the order of 10^{-5} to 10^{-6} rad. There is still considerable debate as to the mechanism responsible for post-seismic movements, but at least some of the observations appear consistent with after-slip on the fault, or an extension of the fault that ruptured during the earthquake, although other explanations have been proposed (e.g., Nur and Mavko, 1974; Wahr and Wyss, 1980).

While co-seismic and post-seismic movements are well established for at least some earthquakes, clear evidence for preseismic deformation from U.S. releveling measurements is quite rare. This may be due to a lack of appropriate measurements as opposed to the absence of such movements since it is unusual to have multiple levelings of sufficient proximity prior to an earthquake. Precursory vertical movements have been suggested from leveling measurements for only three U.S. earthquakes: the 1959 magnitude 7.1 Hebgen Lake, Montana, the 1971 magnitude 6.4 San Fernando, California and the 1973 magnitude 6.0 Point Mugu, California earthquakes (see Table 2). The evidence for preseismic movement near Point Mugu is marginal, both because the proposed movements are barely significant relative to random error estimates and because the area was subject to surficial subsidence due to groundwater withdrawal during the period of interest (Castle et al., 1977). The 1971 San Fernando earthquake is exceptional in that significant releveling was available for the epicentral area prior to the earthquake. These observations were analyzed after the earthquake

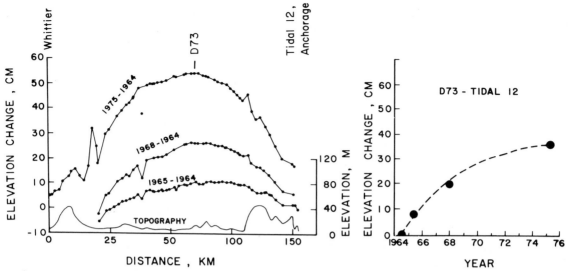

Figure 8. Elevation changes and topography between Anchorage and Whittier following 1964 Alaska earthquake (modified from Brown et al., 1977). Profiles are tied to sea level at Anchorage. Elevation change versus time for benchmark near center of uplift is shown at right. Note exponentially decreasing uplift.

and were interpreted to indicate precursory movements (Castle et al., 1974). However, reevaluation of the relevant leveling observations, described in a later section of this paper, cast some doubts on the reliability of these measurements and hence on their tectonic significance. Reilinger et al. (1977) found evidence for possible precursory uplift throughout a broad region surrounding the area of major co-seismic movement of the 1959 Hebgen Lake earthquake which apparently accumulated at a rate of 3-5 mm/yr (Figure 9). The zone of uplift is defined by five independent elevation change profiles derived from 12 independent surveys. Although three of the five movement profiles show positive correlation with topography (i.e. high areas going up), one shows a negative correlation and one shows no correlation; yet all indicate a consistent sense of movement. This consistency argues strongly against elevation correlated errors as the cause of the observed uplift. The doming stands out distinctly in relation to movements in surrounding areas, and shows a close spatial correlation with the zone of major co-seismic deformation and aftershock activity for the 1959 earthquake (Figure 9). In addition, the geodetically measured deformation is consistent in sign with Cenozoic deformation deduced from geologic structure (Reilinger et al., 1977). Tilts associated with this uplift range from 3 - 7 x 10^{-6} rad with associated tilt rates between 1 - 3 x 10^{-7} rad/yr. Although Reilinger et al. (1977) suggest that doming began prior to the earthquake, because of the limited number of pre-earthquake leveling measurements, it is impossible to prove that the activity was precursory (i.e., doming may have accompanied,

and/or immediately followed the earthquake). In sum, therefore, leveling evidence for vertical movements preceding any U.S. earthquake is weak in both quantity and quality.

Other Tectonic Deformation

Recognizing real tectonic deformation from releveling, although necessary, is not sufficient grounds to infer that they are directly relevant to the earthquake prediction problem. Earthquake

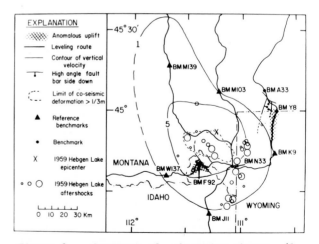

Figure 9. Contours of elevation change (1 mm/yr and 5 mm/yr contours shown) for doming of Hebgen Lake region. Movements may reflect preseismic movements for 1959 Hebgen Lake earthquake (Reilinger et al., 1977).

Table 2. Earthquakes for Which Vertical Movements Derived from Leveling have been Reported. Maximum Observed Deformation and Typical Dimension of Effected Area are also Given. Numbers refer to Figure 1.

	Earthquake	Movement(cm)	Dimensions(km)	Comments	Reference
1	1931 Valentine, Texas M$\sim$6.1 coseismic	10	30	Observed movements could include possible preseismic and/or postseismic deformation	Ni et al., 1980
2	1933 Long Beach, California; M$\sim$6.3 coseismic	20	15	Complicated by effects of fluid withdrawal	Parkin, 1948
3	1940 Imperial Valley, California; M$\sim$7.1 coseismic	20	25	Complicated by possible near-surface compaction	Miller et al., 1970
4	1952 Kern County, California; M$\sim$7.7 coseismic	60	30		Lofgren, 1966
5	1954 Western Nevada series of earthquakes; M$\sim$6.6 to M$\sim$7.1 coseismic postseismic	100 7	35 8		Whitten, 1957 Savage and Church, 1974
6	1959 Hebgen Lake, Montana; M$\sim$7.1 preseismic coseismic	20 700	100 35	Movements interpreted as preseismic could include coseismic and postseismic effects	Reilinger et al., 1977 Myers and Hamilton, 1964

Table 2 (cont'd.)

Earthquake	Movement (cm)	Dimensions (km)	Comments	Reference
7 1964 Alaska; M∿8.4				
coseismic	600	400		Plafker, 1969
postseismic	55	100		Brown et al., 1977 Prescott and Liskowski, 1977
8 1971 San Fernando, California; M∿6.4			Preseismic movements questionable – see text	
preseismic	20	30		Castle et al., 1974
coseismic	230	30		Burford et al., 1971
postseismic	6	10		Savage and Church, 1975
9 1973 Point Mugu, California; M∿6.0			Complicated by sub- sidence due to groundwater effects	
preseismic	4	50		Castle et al., 1977
coseismic	3	50		
10 1975 Oroville, California; M∿5.7				
coseismic	18	20		Savage et al., 1977
postseismic	3	20		
11 1975 Yellowstone, Wyoming; M∿6.0			Complicated by pos- sible postseismic movements of 1959 Hebgen Lake earth- quake	
coseismic	6	8		Pitt et al., 1979

related movements must be separated from movements due to other deep seated processes, such as inelastic deformation (fault creep, folding; e.g. Thatcher, 1981), isostatic adjustments and magmatic activity. These mechanisms are believed, on the basis of observational evidence, to result in contemporary vertical movements which are sufficiently rapid to be detected by releveling measurements.

Movements due to subsurface magmatic activity, for example, are not restricted to volcanically active regions (e.g., Hawaii, Iceland, Japan), having been reported in Yellowstone National Park (Reilinger et al., 1977; Pelton and Smith, 1979), and the Rio Grande rift (Reilinger et al., 1980) as well. Crustal uplift in the Central Rio Grande rift illustrates tectonic deformation which appears to be unrelated to major earthquake activity. The existence of an active magma body beneath the central Rio Grande rift was inferred primarily on the basis of geophysical, and some geological information (Sanford et al., 1977). The magma body is believed to consist of a thin sill at a depth of about 20 km (Figure 10). Elevation change profiles along the routes shown in Figure 10 are given in Figure 11. All three profiles indicate uplift of the area overlying the magma body. The observed uplift is believed to be due to tectonic deformation and not measurement errors or near-surface movements because: 1) uplift is defined by three independent elevation change profiles; 2) while the two east-west profiles show a rough negative correlation with topography near the area of uplift, the north-south profile shows no correlation, thus ruling out elevation-dependent errors as the primary cause of the observed movements; 3) the Belen to Amarillo profile demonstrates that the uplift of the rift is anomalous relative to points to the east; 4) geomorphic evidence for post-Pliocene deformation (Bachman and Mehnart, 1978) is consistent in sign with the geodetic observations; 5) anomalous uplift occurs directly above the magma body; and 6) modeling studies indicate that uplift could result from activity within the magma body. If uplift is accumulating more or less continuously as suggested (Reilinger et al., 1980), it is characterized by an average rate of 4 mm/yr with corresponding tilt rates of 5 to 10 x 10^{-8} rad/yr. In spite of the rather compelling independent evidence for an active magma body beneath the area of uplift, it has been suggested that these movements may be due to an impending earthquake (Koseluk and Bischke, 1981).

Reliability Criteria

The selected cases described above demonstrate both the utility and limitations of geodetic leveling to detect tilts of a few x 10^{-6} rad and

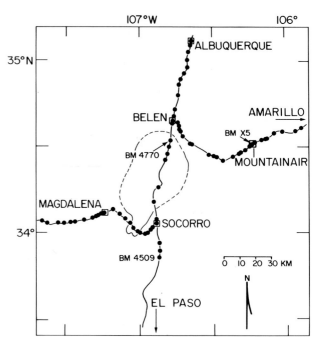

Figure 10. Locations of leveling routes and benchmarks (dots) in Socorro-Albuquerque, New Mexico area. Outline of mid-crustal magma body is also shown (from Rinehart et al., 1979).

tilt rates of a few x 10^{-8} rad/yr. Thus while non-tectonic influences (e.g. systematic error) can obscure real earth movement, the technique has clearly proven effective at monitoring relatively subtle tectonic deformation. It is essential, however, that individual releveling observations be examined in detail for possible contamination by systematic errors and near surface movements prior to invoking tectonic explanations. Particularly effective quantitative techniques include comparison of forward and reverse levelings (e.g., Savage and Church, 1974) and loop closure analysis (e.g., Chi et al., 1980). In addition, the following qualitative criteria, some of which were illustrated by the previous examples, have proven useful for evaluating the reliability of particular data sets (Brown et al., 1980): 1) magnitude of apparent movements relative to possible errors (since many errors remain poorly understood this is equivalent to determining whether the movements in question stand out in relation to "background noise"); 2) consistent temporal behavior when multiple levelings are available (e.g., Alaska); 3) relations with independent geophysical or geologic estimates of recent movement (e.g., tide guage, lake levels, tilt meters, horizontal movements, geomorphic evidence, etc.); 4) consistent movements when multiple leveling lines cross a given feature (e.g., Hebgen Lake, Rio Grande rift); 5) correlation with geologic structure and tectonic

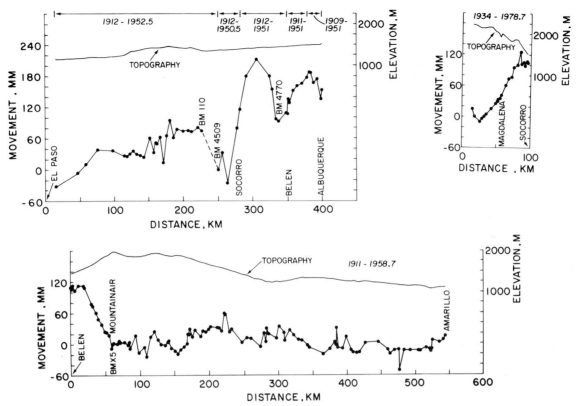

Figure 11. Profiles of elevation change and topography used to infer uplift above Socorro magma body. Dates of leveling are indicated at the top of each plot.

activity (e.g., Hebgen Lake); 6) lack of correlation with topography ruling out possible elevation correlated errors; 7) lack of relationship to possible near-surface processes (e.g., fluid withdrawal, reservoir impoundment, etc.); 8) lack of relationship between apparent movements and procedural changes (changes in sight lengths, rod or instrument changes); and 9) consistency of inferred mechanism with tectonic setting (e.g., Alaska).

Although no single criteria is sufficient to warrant acceptance or rejection of a set of observations, when multiple criteria give consistent results they provide strong evidence for the reliability of the measurements in question.

A Case Study: Southern California Releveling Measurements

Much attention has recently been focused on leveling in southern California, where there is both considerable concern about future earthquakes and an abundance of leveling observations. Using the above reliability criteria, developed through analysis of the much broader data base of U.S. releveling, we have reevaluated some of the observations used to deduce preseismic movements for the 1971 San Fernando earthquake as well as the Palmdale Bulge. Our reevaluation, representing a different perspective, suggests that many of the southern California measurements are significantly affected by both topography-dependent errors and near-surface movements. On the other hand, at least some of the observations may still reflect tectonic deformation. Thus, the configuration of the Palmdale Bulge will, at the very least, require revision in light of improved understanding of those factors which can influence releveling measurements.

In our analysis of southern California releveling observations, data have been displayed in terms of relative movements for sequential time intervals along the pertinent segments of the leveling routes. This contrasts with previous attempts to tie the observations to a tide gauge in order to relate movements to sea level. Analyzing relative movements minimizes the effects of systematic errors, which can accumulate to rather substantial amounts over the 100-200 km distance to the tide gauge, and as will be demonstrated, greatly simplifies interpretation of the observations.

Figure 12 shows those leveling routes in Southern California for which crustal movement information has been investigated for this study.

This information was used to define both the configuration of the Palmdale Bulge and preseismic movements for the 1971 San Fernando earthquake. Possible preseismic movements were reported by Castle et al. (1974) in the vicinity of the earthquake fault (segment I) and 30 km northwest of the epicenter (segment II). Strange (1981) subsequently reported evidence of preseismic movements just north and east of Saugus (along segment II and III). These observations were previously used to define the Palmdale Bulge (Castle et al., 1976; Castle, 1978). Apparent movements along segment III and along the Llano to Azusa route (Figure 12) were also instrumental in determining the configuration of the Palmdale Bulge (Castle et al., 1976; Castle, 1978). Each of these features will be discussed separately below.

The sequence of movements along segment I crossing near the area of surface faulting are shown in Figure 13. Coseismic movement consisting of relative subsidence south of the San Fernando Fault and relative uplift north of the fault are clearly indicated for the 1969-1971 interval. These movements are roughly consistent with elastic rebound accompanying thrust faulting (Savage et al., 1975). Possible preseismic tilting up to the north is indicated by the profiles for the time intervals 1955-1961, 1961-1964, and 1964-1965. Apparently no tilt accumulated along this section between 1965-1969. Figure 14a shows relative movements between points near the ends of this profile segment

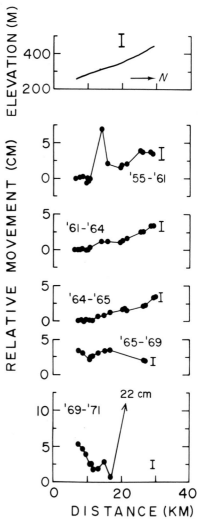

Figure 13. Relative movements for sequential time intervals and topography for route I in Figure 12.

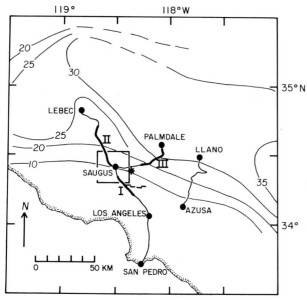

Figure 12. Leveling routes in Southern California for which crustal movements have been investigated in this study. 1971 San Fernando earthquake epicenter (*) and surface fault are shown along with contours of Palmdale Bulge as reported by Castle (1978).

plotted as a function of time. The temporal consistency of these movements is, in itself, normally evidence that the measurements reflect real movements (criteria 2). However, there are two reasons to suspect systematic error, and in particular refraction errors, rather than true ground motion.

Examination of the profiles in Figure 13 indicates that the observed tilting correlates with topography (criteria 6). This correlation, although suggestive, is not sufficient to confirm systematic error because real movements can in some cases correlate with relief (e.g., Reilinger et al., 1977). However, the sequence of apparent tilts between 1955 and 1969 show a systematic relationship to the sight lengths used for different surveys (Figure 14b); a relationship that is consistent with that expected from refraction errors (Holdahl, 1980) (criteria 8).

RELATIVE MOVEMENT VS. Δ(SIGHT-LENGTH²)

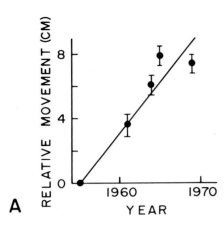

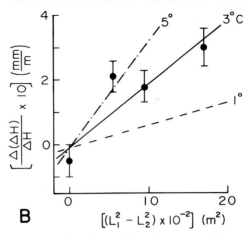

A

B

Figure 14. a) Relative movement between benchmarks near ends of profile shown in Figure 13 plotted versus time.
b) Relative movement [Δ(ΔH)] indicated by profile in Figure 13 normalized by height difference between these points (ΔH) for different time intervals plotted versus difference in square of sight length for corresponding leveling surveys. The straight lines represent the expected relationship from refraction error for a range of temperature differences (from relationship given by Holdahl, 1980).

The $3^\circ C$ temperature difference that results in a good fit to the observations, although somewhat higher than daily average temperature differences observed in other areas ($1-2^\circ C$, Whalen, 1980), may not be unreasonable for the spring and summer months in southern California (Holdahl, 1980 reports that temperature differences "may frequently attain values up to $4^\circ C$ shortly after noon during the summer"). Although Strange (1981) contends that significant tilting persists along this segment of the profile in spite of refraction corrections, the effectiveness of corrections made without on site temperature measurements is not well established. In view of the possibility of refraction errors, the tectonic significance of the sequence of apparent tilts shown in Figure 13 remains ambiguous.

Figure 15 shows profiles of relative elevation change for sequential time intervals crossing the two areas of reported preseismic deformation northwest of the epicenter (Segment II; Figure 12). During the 1953-1964 interval the main deformation consisted of subsidence in the vicinity of Saugus relative to points farther north (ruled area of plot). This movement was originally attributed to the Palmdale Bulge (Castle et al., 1976; Castle, 1978) and more recently to preseismic effects of the San Fernando earthquake (Strange, 1981). However, analysis of releveling measurements throughout the Saugus Basin indicates that relative subsidence shows a close correlation with the geometry of the Saugus aquifer and the history of water level decline (Reilinger, 1980). The spatial correlation between the zone of subsidence and the Saugus aquifer is shown in

Figure 16. In addition to the spatial and temporal correlation, the degree of subsidence of individual benchmarks within the aquifer is roughly proportional to the product of aquifer thickness and water level decline (Figure 17), the expected relationship for near surface compaction (e.g. Terzaghi and Peck, 1967). These observations strongly suggest that subsidence above the Saugus aquifer results from near-surface sediment compaction due to fluctuations of the water level and not from tectonic deformation (criteria 7). This result is particularly important to the current controversy surrounding the Palmdale Bulge (e.g. Stein, 1981) since, unlike many of the measurements defining the Bulge, those in the Saugus area do not correlate with relief.

The other large possible movements shown in Figure 15 occurred between 1965 and 1968 and between 1968 and 1969. These observations were the primary evidence used to infer pre-earthquake slip at depth on the earthquake fault (Thatcher, 1976). The 1965-1968 movements consisted of uplift of the north section relative to the south by about 6 cm. The 1968-1969 movements were, in essence, a reversal of the 1965-1968 movements (criteria 2). The important point is that both sets of apparent movements were dependent upon the 1968 survey. This is illustrated by the bottom-most plot in Figure 15, which shows the general absence of movement for the 1965-1969 interval. Therefore, either we were fortunate enough to catch preseismic deformation at a time of significant deflection (1968), and again when the movements had exactly reversed themselves (1969), or the 1968 survey was in error. While oscillatory movements may have occurred, the possibility of errors in the 1968 survey is at least as likely, particularly in light of the now suspect results south of the epicenter, and similarly suspect trends identified in leveling observations in other parts of the country.

The possibility that refraction errors contaminate leveling measurements south of the epicenter naturally raises the question as to whether this same effect is responsible for the apparent error in the 1968 survey northwest of the earthquake. The steep tilts indicated by the 1965-1968 and the 1968-1969 movement profiles occur where there is a corresponding steep slope in topography (25 to 40 km). However, the topographic slope is so steep (>.04 rad) that only short sight-lengths could be used, making it unlikely that atmospheric refraction coupled to sight-lengths was a significant effect. Unusual

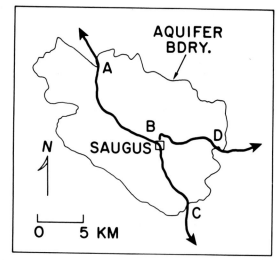

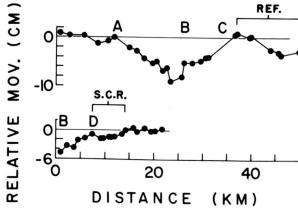

Figure 16. (Top) Leveling routes traversing the Saugus aquifer.
(Bottom) Relative movements along routes shown at top. Note correlation between subsidences and aquifer location. Section labelled S.C.R. occurs within the alluvial aquifer of the Santa Clara river and may be subject to subsidence due to compaction of this aquifer (Mark et al., 1981). Section labelled REF may reflect refraction error (Figure 14b and text).

near-surface temperature differences at the time of the survey, or other elevation-correlated errors, such as miscalibrated leveling rods (Jackson et al., 1980) may have affected these observations (see also Stein, 1981).

Apparent movements along the route from Llano to Azusa, California were used to define the configuration of the Palmdale Bulge (Fig. 12). Elevation change profiles along this route derived from surveys in 1934, 1962, and 1971 are shown in Figure 18. The 1962-1971 profile indicates about 10 cm uplift at Llano relative to Azusa. However, the following observations strongly suggest that this apparent movement is due to systematic error, and in particular rod

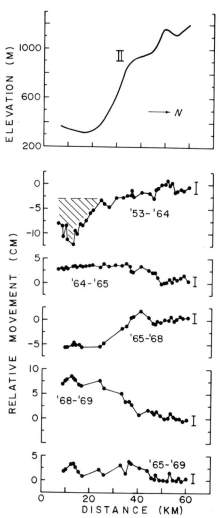

Figure 15. Relative movement for sequential time intervals and topography for route II in Figure 12.

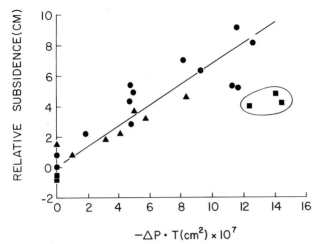

Figure 17. Plot of subsidence versus change in potentiometric surface (i.e., water level: ΔP) times aquifer thickness (T) for benchmarks in and immediately adjacent to aquifer (from Reilinger, 1980). Different symbols refer to different releveled segments: circles–Saugus to North (1953-1964); squares–Saugus to South (1955-1964); triangles–Saugus to East (1955-1961). The three circled points lie in the southeastern part of the aquifer and may reflect either different sediment characteristics or some effect other than sediment compaction in this area.

calibration error in the 1962 survey (note that these are different data than those reported by Jackson et al., 1980): 1) elevation changes along the segment of the 1962-1971 profile where most of the relative movement occurs (between A and B, Fig. 18) show a detailed correlation with topography along the leveling route. The relationship between apparent movement and terrain, which has a correlation coefficient of .98 is illustrated in Figure 19. Such a correlation with detailed topographic features is suggestive of systematic errors (criteria 6); 2) point A in Figure 18 where the correlation between movement and terrain ends rather abruptly is precisely the point where the leveling instrument and rods were changed in the 1962 survey (criteria 8); 3) the elevation change profile derived from the 1934 and 1962 surveys along this route indicates movements in the opposite sense and of the same magnitude as the 1962-1971 profile. This is clearly illustrated by the 1934-1971 profile (Fig. 18) which shows no significant movement. This reversal in the sense of apparent movement casts serious doubts on the 1962 survey (criteria 2). These three points strongly suggest that apparent movements between Llano and Azusa, previously used to define the Palmdale Bulge, actually result from rod calibration errors in the 1962 survey.

Movements along leveling route III in Figure 12

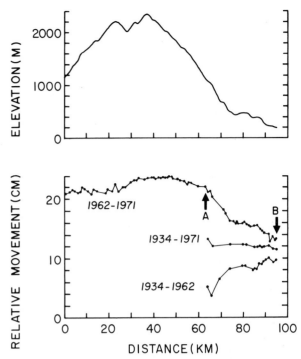

Figure 18. Profiles of apparent elevation change from Llano to Azusa, California used to define the Palmdale Bulge (see Figure 12 for location).

were also important for determining the geometry of the Palmdale Bulge (Castle et al., 1976; Castle, 1978). In contrast to the previous examples, these apparent movements do not seem to be due to either systematic errors or near-surface effects and thus may represent tectonic deformation.

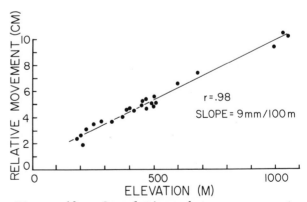

Figure 19. Correlation between apparent movement and terrain for segment (A to B) of 1962-1971 profile shown in Figure 18. Correlation coefficient (.98) and slope of regression line are also given. Correlation and reversal of apparent movement strongly suggest elevation-dependent systematic error.

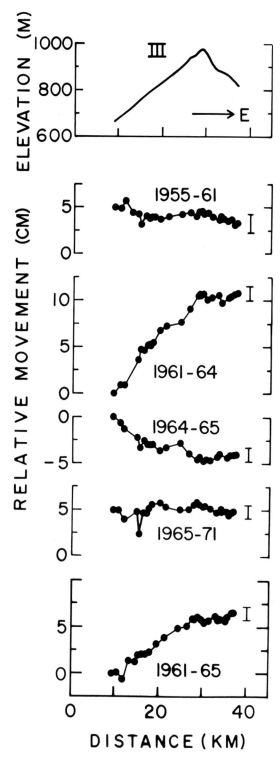

Figure 20. Relative movement for sequential time intervals and topography for route III in Figure 12.

Figure 20 shows the sequence of relative movements along the survey route south of Palmdale. The major tilt event occurred between 1961 and 1964 and amounted to more than 10 cm of relative movement over a distance of 20 km. This corresponds to a tilt of 5×10^{-6} rad. The general absence of movements for the 1955-1961 interval, the 1964-1965 interval and the 1965-1971 interval attests to the reliability of all of these surveys (i.e., comparison of the 1955 or 1961 surveys with any of the later surveys will give roughly the same result). This implies that the 1961 to 1964 tilt event is in fact defined by five independent surveys. This is illustrated by the bottom most profile in Fig. 20 which shows significant tilting for the period 1961-1965 which, although reduced in amplitude, is quite similar to that shown by the 1961-1964 profile (criteria 2). In addition, this tilt anomaly does not show a strong correlation with topography (i.e., the direction of tilting does not reverse where the topographic slope reverses) (criteria 6). Furthermore, the sequence of relative movements show no relationship either to changes in leveling rods or to changes in sight lengths (criteria 8). This evidence suggests that the apparent tilting south of Palmdale reflects real crustal movements. Whether the tilt anomaly was a precursor to the San Fernando earthquake or represented a mechanically separate event is presently unknown.

Discussion and Conclusions

In spite of laudable progress in developing sophisticated new geodetic methods (e.g., VLBI, Laser Ranging, GPS) releveling measurements continue to be the most accurate (over appropriate distances) and widespread source of information on contemporary vertical movements of the continental crust. As such they constitute an important input to the earthquake prediction problem. Previous investigations, a few of which have been described here, clearly demonstrate the potential of the technique for monitoring subtle earth movements. However, it is equally clear that releveling estimates of crustal movement are influenced by near-surface movements and as yet poorly understood systematic errors which can obscure or be mistaken for tectonic deformation. Thus, uncritical interpretation of releveling observations can lead to erroneous tectonic conclusions, which in the case of earthquake prediction could entail serious social ramifications. The checking techniques (e.g., circuit closure analysis) and reliability criteria illustrated in this study, represent an attempt to quantify specific procedures for evaluating the tectonic significance of particular leveling data sets. Although not foolproof, these procedures have proven effective in a number of cases at discriminating tectonic movements from suspect effects. However, even when spurious effects can be eliminated, relating observed deformation to preseismic mechanisms may be quite difficult because of our limited understanding of precursory phenomena and our general inability to distinguish them from vertical movements due to other causes (e.g., magmatic activity, isostatic movements, aseismic orogenic deformation, etc.). Furthermore, the sparse distribution of leveling surveys in both space and time, even in areas like southern California, makes it highly unlikely that precursory movements for all but the largest earthquakes will ever be detected. In order for leveling to become more than an accidental contributor to earthquake prediction, a systematic leveling program designed for geodynamic rather than geodetic objectives is needed to develop the observational background required to recognize possible preseismic movement.

Acknowledgements. We thank the National Geodetic Survey for supplying the leveling data used for this study, and its personnel for many informative discussions. Greg Jurkowski, Christie Chi, Michael Bevis, and Dave Miesen provided technical assistance. We are grateful to Jack Oliver, Michael Bevis and Mauwia Barazangi for helpful comments on earlier versions of this paper. This research was supported by U.S. Geological Survey Grant 14-08-0001-17625, NASA Grant NAG5-40 and USNRC contract AT(49624-0367). Contribution No. 684 of the Department of Geological Sciences, Cornell University.

REFERENCES

Bachman, G.E., and H.H. Mehnert, New K-Ar data and the late Pliocene to Holocene geomorphic history of the central Rio Grande region, New Mexico, Geol. Soc. Am. Bull., 89, 283-292, 1978.

Bomford, G., Geodesy, third ed., England, Clarendon, Oxford, 226 p., 1971.

Brown, L.D., Recent vertical crustal movements along the East Coast of the United States, Tectonophysics, 44, 205-231, 1978.

Brown, L.D., and J.E. Oliver, Vertical crustal movements from leveling data and their relation to geologic structure in the eastern United States, Rev. Geophys. and Space Physics, 14, 13-35, 1976.

Brown, L.D., R.E. Reilinger, S.R. Holdahl, and E.I. Balazs, Post-seismic crustal uplift near Anchorage, Alaska, J. Geophys. Res., 82, 3369-3378, 1977.

Brown, L.D., R.E. Reilinger, and G.P. Citron, Recent vertical crustal movements in the U.S.: Evidence from precise leveling, in Earth Rheology, Isostasy and Eustasy, ed. N.A. Morner, John Wiley and Sons, 389-405, 1980.

Brown, L.D., D.L. Miesen, R.E. Reilinger, and G.A. Jurkowski, Geodetic leveling and crustal movement in the U.S.: Part I, Topography and vertical motion, (abstract), EOS, Trans. Am. Geophys. Union, 1980 Spring Meeting Program, Changes and Corrections, 4, 1980.

Burford, R.O., R.O. Castle, J.P. Church, W.T. Kinoshita, S.H. Kirby, R.T. Ruthven, and J.C. Savage, Preliminary measurements of tectonic movement in the San Fernando, California Earthquake of February 9, 1971, U.S. Geol. Surv. Prof. Pap., 733, 80-85, 1971.

Castle, R.O., J.N. Alt, J.C. Savage, and E.I. Balazs, Elevation changes preceding the San Fernando earthquake of February 9, 1971, Geology, 2, 61-66, 1974.

Castle, R.O., J.P. Church, and M.R. Elliott, Aseismic uplift in Southern California, Science, 192, 251-253, 1976.

Castle, R.O., J.P. Church, M.R. Elliott, and J.C. Savage, Preseismic and coseismic elevation changes in the epicentral region of the Point Mugu earthquake of February 21, 1973, Bull. Seism. Soc. Am., 67, 219-231, 1977.

Castle, R.O., Leveling surveys and the southern California uplift, U.S. Geol. Surv. Earthquake Inf. Bull., 10, 88-92, 1978.

Chi, S.C., R.E. Reilinger, L.D. Brown, and J.E. Oliver, Leveling circuits and crustal movements, J. Geophys. Res., 85, 1469-1474, 1980.

Citron, G.P., and L.D. Brown, Recent vertical crustal movements from precise leveling surveys in the Blue Ridge and Piedmont provinces, North Carolina and Georgia, Tectonophysics, 52, 223-236, 1979.

Hastie, L.M., and J.C. Savage, A dislocation model for the Alaska earthquake, Bull. Seism. Soc. Am., 60, 1389-1392, 1970.

Holdahl, S.R., 1980, An assessment of refraction error and development of methods to remove its influence from geodetic leveling, Final Technical Report Contract 14-08-0001-17733, USGS, 18 p., 1980.

Jackson, D.D., W.B. Lee, and C. Liu, Aseismic uplift in southern California: An alternative interpretation, Science, 210, 534-536, 1980.

Jurkowski, G., L.D. Brown, S.R. Holdahl, and J.E. Oliver, Map of apparent vertical crustal movements for the eastern United States (abstract), EOS, Trans. Am. Geophys. Union, 60, 315, 1979.

Koseluk, R.A., and R.E. Bischke, An elastic rebound model for normal fault earthquakes, J. Geophys. Res., 86, 1081-1090, 1981.

Kukkamaki, T.J., Uber die nivellitsche refraktian, Publication of the Finnish Geodetic Institute, 25, 48 p., 1938.

Lofgren, B.E., Tectonic movement in the Grapevine Area, Kern County, California, U.S. Geol. Surv. Prof. Pap., 550-B, p. B6-B11, 1966.

Mark, R.K., J.C. Tinsley, E.B. Newman, T.D. Gilmore, and R.O. Castle, An assessment of the accuracy of the geodetic measurements that define the Southern California uplift, J. Geophys. Res., in press, 1981.

Miller, R.W., A.J. Pope, H.S. Stettner, and J.L. Davis, Crustal movement investigations, Operational data report, U.S. Department of Commerce, Coast and Geodetic Survey, DR-10, 1970.

Myers, W.F., and W. Hamilton, Deformation accompanying the Hebgen Lake earthquake of August 17, 1959, U.S. Geol. Surv. Prof. Paper 435, 55-98, 1964.

Ni, J.F., R.E. Reilinger, and L.D. Brown, Vertical crustal movements in the vicinity of the 1931 Valentine, Texas, earthquake, Seism. Soc. Am. Bull., in press, 1981.

Nur, A., and G. Mavko, Post-seismic viscoelastic rebound, Science, 181, 204-206, 1974.

Parkin, E.J., Vertical movement in the Los Angeles region, 1906-1946, Trans. Am. Geophys. Union, 29, 17-26, 1948.

Pelton, J.R., and R.B. Smith, Recent crustal uplift in Yellowstone National Park, Science, 206, 1179-1182, 1979.

Pitt, A.M., C.S. Weaver, W. Spence, The Yellowstone Park earthquake of June 30, 1975, Bull. Seism. Soc. Am., 69, 187-205, 1979.

Plafker, G., Tectonics of the March 27, 1964 Alaska earthquake, U.S. Geol. Surv. Prof. Pap. 543-I, 1-74, 1969.

Plafker, G., Alaskan earthquake of 1964 and Chilean earthquake of 1960: Implications for arc tectonics, J. Geophys. Res., 77, 901-925, 1972.

Poland, J.F., and G.H. Davis, Land subsidence due to withdrawal of fluids, in Reviews of Engineering Geology II, 187-269, 1969.

Prescott, W.H., and M. Lisowski, Deformation at Middleton Island, Alaska, during the decade after the Alaskan earthquake of 1964, Bull. Seism. Soc. Am., 67, 579-586, 1977.

Reilinger, R.E., Elevation changes near the San Gabriel Fault, Southern California, Geophys. Res. Lett., 7, 1017-1019, 1980.

Reilinger, R.E., G.P. Citron, and L.D. Brown, Recent vertical crustal movements from leveling data in southwestern Montana, western Yellowstone National Park, and the Snake River Plain, J. Geophys. Res., 82, 5349-5359, 1977.

Reilinger, R.E., J.E. Oliver, L.D. Brown, A.R. Sanford, and E.I. Balazs, New measurements of crustal doming over the Socorro magma body, New Mexico, Geology, 8, 291-295, 1980.

Rinehart, E.J., A.R. Sanford, and R.M. Ward, Geographic extent and shape of an extensive magma body at mid-crustal depths in the Rio Grande rift near Socorro, New Mexico, in Riecker, R.E., ed., Rio Grande Rift: Tectonics

and Magmatism, American Geophysical Union, Washington, D.C., 237-251, 1979.

Sanford, A.R., and others, Geophysical evidence for a magma body in the crust in the vicinity of Socorro, New Mexico, in Heacock, J.E., ed., The Earth's Crust, AGU Monograph 20, 385-403, 1977.

Savage, J.C., and L.M. Hastie, Surface deformation associated with dip slip faulting, J. Geophys. Res., 71, 4897-4904, 1966.

Savage, J.C., and J.P. Church, Evidence for postearthquake slip in the Fairview Peak, Dixie Valley, and Rainbow Mountain fault areas of Nevada, Bull. Seism. Soc. Am., 64, 687-698, 1974.

Savage, J.C., R.O. Burford, and W.T. Kinoshita, Earth movements from geodetic measurements, Calif. Div. Mines and Geol. Bull., 196, 175-186, 1975.

Savage, J.C., and J.P. Church, Evidence for afterslip on the San Fernando fault, Bull. Seism. Soc. Am., 65, 829-834, 1975.

Savage, J.C., M. Lisowski, W.H. Prescott, and J.P. Church, Geodetic measurements of deformation associated with the Oroville, California earthquake, J. Geophys. Res., 82, 1667-1671, 1977.

Stein, R.S., Role of elevation-dependent errors on the accuracy of geodetic leveling in the southern California uplift, this volume, 1981.

Strange, W.E., The impact of refraction corrections on leveling interpretations in southern California, J. Geophys. Res., in press, 1981.

Terzaghi, K., and R.B. Peck, Soil Mechanics in Engineeering Practice, John Wiley and Sons, New York, 729, p., 1967.

Thatcher, W., Episodic strain accumulation in Southern California, Science, 194, 691-695, 1976.

Thatcher, W., Crustal deformation studies and earthquake prediction research, this volume, 1981.

Wahr, J., and M. Wyss, Interpretation of postseismic deformation with a viscoelastic relaxation model, J. Geophys. Res., 85, 6471-6477, 1980.

Whalen, C.T., Refraction errors in leveling - NGS test results, Proceedings Second International Symposium on Problems Related to the Redefinition of North American Vertical Gedeotic Networks, ed. G. Lachapelle, Canadian Institute of Surveying, Ottawa, Canada, 757-782, 1980.

Whitten, C.A., The Dixie Valley-Fairview Peak, Nevada, earthquakes of December 16, 1954: Geodetic measurements, Bull. Seism. Soc. Am., 47, 321-325, 1957.

DISCRIMINATION OF TECTONIC DISPLACEMENT FROM SLOPE-DEPENDENT ERRORS
IN GEODETIC LEVELING FROM SOUTHERN CALIFORNIA, 1953-1979

Ross S. Stein

Lamont-Doherty Geological Observatory
of Columbia University, Palisades, New York 10964

Abstract. Precise geodetic leveling from southern California carried out between 1953 and 1979 contains linear slope-dependent correlations with a weighted mean value of $(0.3 \pm 2.3) \times 10^{-5}$ times the topographic height difference. This is equivalent to an error of 3 ± 46 mm at the 95% confidence interval over the roughly 1000 m relief of the Transverse Range leveling routes. Linear regression of geodetic tilt onto topographic slope for 1100 km of leveling surveys that are not subject to significant atmospheric refraction error demonstrates that neither the sign nor the magnitude of the correlations changes significantly with time, despite alterations in rod calibration and field procedures during the 1960's. The dominant cause of correlation is errors in the applied rod correction. The errors do not accumulate over several relevels of a route or over distances greater than about 80 km on an individual route; they can be treated as a source of random noise. The rod-corrected uplift from 1953 to 1968 at Bench Mark G54 near Grapevine, a characteristic point on the southern California uplift, north of the San Andreas fault, is 149 ± 18 mm with respect to Saugus (165 ± 9 mm, observed) on a route without significant differential optical refraction. Episodic uplift and collapse along the 100 km ridge route that includes G54 cannot be ascribed to rod or refraction errors, and no more than 48 mm can be caused by ground water withdrawl from the alluvial aquifer beneath Saugus.

Introduction

The search for tools for earthquake prediction has created a need for reliable measurements of displacement and deformation at the earth's surface. Since the rate of deformation appears to be slow and broadly distributed between earthquakes, the detection of aseismic crustal movements demands the highest precision, the longest period of observation, and the greatest areal coverage that can be provided. For measurement of vertical displacements, conventional geodetic leveling best fulfills these requirements. It is a simple, optical, and highly redundant procedure for measuring changes in elevation (Figure 1). It has been carried out in essentially the same manner for almost one hundred years in many places with frequent resurveys. On flat terrain, precise leveling can currently reproduce elevations to within 10 mm over a distance of 100 km [Heiskanen and Moritz, 1967; Bomford, p. 226-280, 1971], while the standard random error for leveling from the first part of this century is only twice as large [Vaniček et al., 1980].

But how accurately can elevation changes be measured using successive leveling surveys in regions of great topographic relief? Recently the accuracy of leveling carried out prior to currently established standards set by the Federal Geodetic Control Committee [1974, 1975] has been challenged by Jackson et al. [1980]. Jackson et al. contend that leveling before 1964 suffers from the accumulation of elevation-dependent errors in excess of one part in ten thousand times the topographic height difference (dH), or greater than 100 mm over 1000 m of relief. Jackson et al. attribute the errors predominantly to mismeasurement of the leveling rods by as much as 1 mm over the 3 m rod length, non-uniform graduation of the rods, and changes in procedures for rod calibration by the National Bureau of Standards.

Castle et al. [1976] assembled 10,000 km of southern California resurveys using the observed elevation differences supplied by the National Geodetic Survey (NGS). These are the measured elevations corrected for the thermal coefficient of expansion for invar and measured linear elongation or contraction of the rods relative to a standard, the rod excess. Castle [1978] concluded that a 70,000 km^2 region within the Transverse Ranges and lying athwart the San Andreas fault's Big Bend underwent 200-400 mm of uplift during 1959-74 (Figure 2a), followed by a partial collapse. If leveling carried out before the uplift commenced were consistently contaminated by elevation-dependent errors as large as a few parts in ten thousand, then the observed change in elevation could be merely an artifact of the

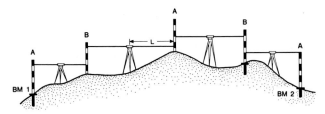

Figure 1. Leveling Procedure: elevation differences between benchmarks (BM's) are measured by sighting adjacent rods with a horizontal telescope, the level instrument. A backsight is made to the A rod, then a foresight is made to the B rod. This is repeated on two scales on each rod. The rods are alternated at every observation; the length of each foresight and backsight pair maintained equal.

error. This would provide one explanation for the similarity between contours of uplift and smoothed topography shown in Figure 2a. If, on the other hand, slope-dependent errors are significantly smaller than 10^{-4} times the height difference, dH, and do not accumulate with time, the observed uplift cannot be ascribed to a linear measurement error. If real, uplift and subsidence of the Transverse Ranges would then demonstrate that these youthful mountains are actively growing and deforming, both during and between earthquakes.

Using a number of independent tests, the contention of Jackson et al. [1980] will be probed on 1100 km of levels carried out between 1953 and 1979 within the uplifted region. Because Strange [1981] presents evidence that accumulating differential refraction, a non-linear elevation-dependent error, can significantly modify observed elevations, this study will be confined to leveling routes where differential refraction must be minimal.

Strategy

The intent of this statistical analysis is to determine the maximum magnitude of accumulating leveling errors, and to remove these errors from a representative and critical resurveyed leveling route. It is assumed in the analysis that elevations measured from any given leveling survey, n, contain: (1) a linear, rod-related error, (2) a non-linear atmospheric refraction error, and (3) real earth movement, all of the same order of magnitude. The rod error can be approximated by

$$dH_n = (1+e_n)dH, \qquad n = 1,2 \qquad (1)$$

where e is the rod excess; e = measured minus true rod length. Differencing the two surveys to obtain elevation change, dh,

$$dh = dH_{n+1} - dH_n = (e_{n+1} - e_n)dH \qquad (2)$$

$$\text{or} \quad dh = e_{net} \, dH \qquad (3)$$

where $e_{net} = e_{n+1} - e_n$.

The simplest expression that is commonly invoked for the atmospheric refraction error, from Kukkamäki [1938] for resurveyed elevations is

$$dh = \gamma^*[(L_{n+1}^2 \Delta T_{n+1} - L_n^2 \Delta T_n)]dH, \qquad (4)$$

where L is the sight length between rod and instrument (Figure 1), ΔT is a linear approximation of the vertical temperature gradient along the line of sight, and γ^* is a physical constant. Since observations of the thermal gradient and hence ΔT do not exist for historical leveling, the assumption is made here that $\Delta T_1 \approx \Delta T_2$, as the squared sight length term will in any event dominate. Then (4) becomes

$$dh = \gamma(L_{n+1}^2 - L_n^2)dH \qquad (5)$$

where $\gamma = \gamma^*\Delta T$.

Because any profile of elevation, or elevation change from resurvey, may exhibit a trend - a mean slope not equal to zero - bench marks spaced far apart along the leveling route will be more likely to show both a larger dh and dH than those closely spaced. To remove this bias, tilt measured by resurvey (dh/dx) can be compared with topographic slope or grade (dH/dx), where dx is the distance

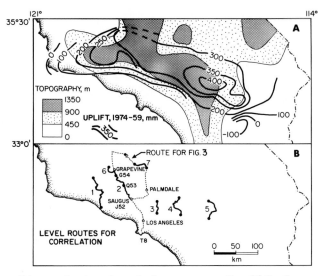

Figure 2. A. Smoothed contours of uplift from 1959 to 1974 from Castle [1978] superposed on smoothed Southern California topography shaded. Note the striking similarity between uplift and topography, except in the south (the Salton Sea and Peninsular Range), and in the north (Sierra Nevada Range). B. Level routes. The dashed circuit through Saugus, Bakersfield, and Palmdale is used for calculation of misclosures. Route 2 is called Ridge in the text.

between adjacent BM's; tilt and slope are of course independent of BM-spacing. Dividing eqns. (3) and (5) by dx and adding to these expressions uncorrelated residual tilt, ε, results in a combined simplified expression,

$$\frac{dh}{dx} = [e_{net} + \gamma(L_{n+1}^2 - L_n^2)]\frac{dH}{dx} + \varepsilon, \quad n=1,2 \quad (6)$$

Eqn. (6) must be solved under conditions mutually satisfactory for e_{net}, γ, and ε. Resolution of the rod term, e_{net}, demands that (6) be evaluated only for resurveys where rods have not been changed. If in survey n or n+1, more than one pair of rods is used, the correlation associated with each rod pair difference can be obscured in the regression. Also, since e_{net} is linear, successful correlation requires a large variance in dH/dx (slope) to overcome sources of random error, soil-, and BM-instability. This requires rough topography, with large excursions from the mean slope. Discrimination of the refraction term, on the other hand, requires evaluation over constant atmospheric and ground conditions to ensure that γ is constant; this is probably optimized over a uniform slope. Also, the variance of $(L^2 - L^2)$ must be large, which requires gentle slopes (≤ 0.02 or 2%). For grades less than 2%, L will not be constrained by the useable rod height (2.5 m) to be the same for both surveys n and n+1.

Thus, despite gross simplifications made to obtain eqn. (6), the conditions necessary for multiple linear regression to find e_{net} and γ are incompatible, because to solve for e_{net} requires a large variance in dH/dx, the slope, while to find γ requires uniform and gentle dH/dx. By pursuing limiting cases of (6), two strategies are possible. One that will be employed in this paper is to consider only routes where both $\Delta T_{n+1} \approx \Delta T_n$ and $L_{n+1} \approx L_n$, so that γ, and hence refraction error, becomes negligible; this requires rough, steep topography. The alternative is to take cases where dH/dx is constant, in which case e_{net} and ε become indistinguishable. This approach suffers because real earth movement cannot be segregated from rod error. Also, the assumption of nearly constant ΔT becomes critical to the resolution of γ. If $L_n = L_{n+1}$, (6) simplifies to

$$\frac{dh}{dx} = e_{net}\frac{dH}{dx} + \varepsilon, \quad n=1,2 \quad (7)$$

A linear least-squares regression of tilt onto topographic slope is performed for each resurveyed segment. No segment contains any rod changes that mask the correlation (e.g., containing both positively and negatively correlated segments). Because random leveling errors grow with the square root of distance, tilt or elevation change can be more accurately measured from benchmarks farther apart [Bomford, 1971]. To give more weight to the better data, weights, w, are assigned, where w = dx, the distance between

benchmarks. Both the weighted and unweighted coefficients of correlation, r, are calculated. To favor the identification of a correlation, the highest of the weighted and unweighted absolute value of r is chosen. The significance of r is checked by an equal-tails test of the null (or no correlation) hypothesis. Refer to the Appendix and Stein [1980] for a more detailed discussion of the weighted linear regression, and the basis for assignment of weights. Because the least-squares regression is not robust, outliers, or sections with extreme values of tilt or slope, must be removed. If this is not done, a significant line can be fit by connecting a cluster of points with one extreme point. However, no significant line will be found if this one point is removed. From 10 to 20% of the total benchmark population is removed; this includes marks that differ by 5 mm from adjacent marks, outliers, and in some cases additional marks whose removal will strengthen the correlation. If r is significant at the 95% level of confidence, the correlation is accepted as an error, e_{net}. If the values of e_{net} for adjacent segments do not differ by more than one standard deviation, the releveled segments are correlated together in the same regression, and retained as one segment if the significance of the correlation increases. In this manner the rod pair responsible for the correlation can be isolated from rod changes that have no effect.

To estimate the topographic roughness required for significant correlation, a 30 km segment from a railroad grade in the Transverse Ranges has been isolated because its slope is almost uniform (Figure 3). If the dh profile were the product of elevation-dependent error, it should display a constant tilt. However, the error cannot be found simply by dividing the mean tilt by the (constant) slope because that is the long wavelength (30-km) signal; it is the mark-to-mark correlation that is sought. To reduce the noise in this as in any elevation change profile, some benchmarks must be eliminated. If the mark rejection criterion of Jackson et al. [1980] is stiffened to favor correlation, such that any mark that differs 5 mm from both its neighbors is eliminated (they use a 10 mm difference), then the standard deviation about the mean tilt of the remaining 80% of the benchmarks shrinks from 2 to 1 μrad (10^{-6} radian). To find an elevation-dependent error, e_{net}, of 10^{-4}, a 2% grade (0.02 slope) range is required to overcome the bench mark scatter; a 10% range in grade of the topography is needed to resolve a 5×10^{-5} error. Therefore, only leveling routes with grades that vary by about 10% will be selected for regression, and only segments that contain this range can be isolated from the remainder of the survey route. This criterion effectively eliminates all railroad grades from consideration because their slopes usually vary by no more than 2%, and hence, lack the topographic variation necessary for regression. So while railroad routes can contain useful leveling data, they cannot be used to

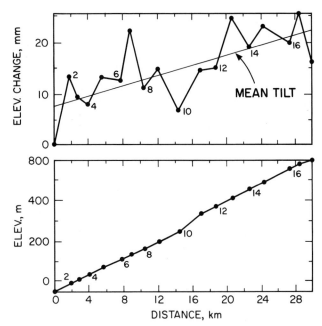

Figure 3. Variation of tilt over constant slope - this 30 km railroad segment is shown in Figure 2b. Since the topographic slope is almost uniform (lower profile), there is no signal to correlate with tilt (upper profile). The roughness or variation in slope in the lower profile must exceed that of the upper profile for significant correlation.

detect significant elevation-dependent systematic error.

Results

1. The Ridge Route

In a number of respects the 100 km leveling route from Saugus to Grapevine is ideally suited to test for real tilt and rod errors in eqn. (7); it will be referred to as Ridge in this report (Figure 2b). Its full length has been resurveyed six times from 1953 to 1974, spanning the years during which observed uplift and a change in rod calibration and leveling procedures took place that Jackson et al. [1980] contend is significant. Elevation changes can be tied to the long-term tide gauge station at San Pedro through benchmark Tidal 8 (T8 in Figure 2b) as Castle et al. [1976] have done. The route transects the southern California uplift identified by Castle et al. [1976]. Ridge develops a 1000 m gain and loss in elevation, with the majority of section slopes distributed between 7 and -7%. The sight lengths for Ridge have been short and roughly constant for all surveys, (the mean sight length, L, varies from 20-25 m; see Figure 1), and each survey was performed in the Spring. Both of these factors reduce the potential error caused by differential

optical refraction to below 10 mm [Strange, 1981]. Most of the benchmarks are emplaced in 10 m.y. old consolidated sediments and 100-200 m.y. old weathered granites [Jenkins, 1975]. North of Grapevine lies a well-documented region of major pumping-induced subsidence that cannot be subjected to analysis [Lofgren, 1975].

Profiles of topography, observed elevation change, and adjusted elevation change are displayed for successive survey intervals in Figures 4a through 8a. Saugus is at 0 km, and Grapevine is located at 95 km. Except for the first (1953) and last (1974) surveys, each survey is used twice, as the latest and earliest survey of the paired differences (e.g., 1965-64, 1968-65). The plot of tilt as a function of slope for each survey difference follows as Figures 4b through 8b. Rod changes are demarked at vertical lines, and since the rods used in either the early or late survey may change, the year of the change is indicated beneath the upper profile.

During the first and longest resurvey interval, 1964-53, tilt is not correlated with slope in rod segment A (Figure 4a). 80 mm of uplift takes place over essentially flat terrain, whereas the elevation increases only 10 mm over a region of 600 m relief. Since segment A is uncorrelated, it is left unadjusted in the top profile of Figure 4a. Both segments B and C are positively correlated, as can be seen by inspection, and yield a very large combined slope-dependent error, e_{net}, of $(13.2 \pm 1.1) \times 10^{-5}$ (Figure 4b). The correlation coefficient for the weighted regression of tilt onto slope, r, is 0.84, equivalent to 99% confidence that the tilt and slope are correlated. In other words, 84% of the observed tilts measured between benchmarks are equal to $(13.2 \pm 1.1) \times 10^{-5}$ times the topographic slope between those marks.

To remove the correlated error, e_{net}, from the observed elevation change profile, the correlated component of tilt, which is equal to e_{net} times the slope, is subtracted from the observed tilt. This operation is performed for each section in the segment, and the resulting profile is plotted as the adjusted elevation change. This adjustment need not remove the entire observed tilt, since some portion of the tilt may be uncorrelated with topography. Rather, the adjusted elevation change has no dependence on slope; a regression of adjusted tilt onto slope would produce both e_{net} and r equal to zero. For this 11 year interval (1964-53), the observed and adjusted elevation changes display nearly the same net uplift with respect to Saugus, despite the large error in segments B and C. This is because most of the elevation changes take place in the absence of topographic relief, while the correlated segment rises only an additional 300 m to the peak elevation. In contrast to the southern end of the route, the elevation of Grapevine (at the 95-km position, Figure 4a) is 90 mm higher in the adjusted profile relative to the observed profile.

During the survey interval, 1965-64, the same

segments that displayed a large positive correlation in 1964-53 show an almost equally large negative correlation (36-98 km distance, Figure 5). Once again, segment A (0-36 km) remains uncorrelated, with nearly uniform tilt over a great range of slopes. The most straightforward explanation for the reversal in sign of e_{net} for segments at 36-98 km is that the 1964 tilts are positively correlated, and the 1953 and 1965 tilts are relatively free of correlations.

Rod segments A and B have nearly the same correlation in the 1968-65 survey interval (Figure 6). If segments A and B contain elevation-dependent errors, the value of the error does not differ between them by more than 1.5×10^{-5}, which is roughly the limit of resolution. The high amplitude displacements at 25-35 km distance (Figure 6a, elevation change profile) cannot be correlated with elevation; the excursion in tilt over these sections is much larger than for the remainder of the segments. However, if those five benchmarks are removed from the regression as part of the 20% mark deletion, a correlation with 99% confidence can be obtained. In addition to a real elevation-dependent correlation that persists for 75 km, there is a confined anomaly. After correction of the entire segment including the anomaly, the form of the anomalous region is essentially preserved in the adjusted elevation change profile. The adjusted uplift is about 20 mm less than the observed uplift during the 1968-65 resurvey interval.

The 1971 survey was run soon after the M6.4 San Fernando earthquake. Because the first 20 km of Ridge lies within the aftershock zone, this portion of the route for the interval 1971-68 underwent seismic displacement that overwhelms slope-dependent correlation [Castle et al., 1975]. So while the southernmost portion of segment A (0-7 km, Figure 7a) cannot be correlated, the remaining two portions of segment A (16-41 km and 46-77 km) yield a correlation of $(-6.8 \pm 0.8) \times 10^{-5}$ (Figure 7b). This value of e_{net} has been removed from all portions of segment A in the adjusted profile. Segments A and B alternate in the profiles (Figure 7a) because two rod sets were periodically exchanged during the 1971 survey. The adjusted elevation change during this interval shows 35 mm greater uplift than the apparent elevation change.

In the final resurvey interval, 1974-71, rod segment A is again correlated, but this time positively (Figure 8). The negative 1971-68 and positive 1974-71 correlations indicate that the 1971 survey within segment A is negatively correlated. Its value for e_{net}, for comparison, is about one-third of the 1964 error.

Net uplift from 1953 to 1974. The observed and adjusted elevation changes for the entire 21-year interval can now be compared by summing the changes at a few benchmarks common to all resurveys (Figure 9). The total observed elevation change near Grapevine, benchmark G54 at the 82-km position, with respect to Saugus, J52 at 0 km, is

121 $\pm$ 9 mm. Here the standard deviations represent the random error. The total adjusted elevation change for the same interval is 128 $\pm$ 24 mm at Grapevine with respect to Saugus, where the larger standard deviation contains both the random error and the uncertainty of adjustment. The adjustment error is calculated from the standard deviation of e_{net} for successive resurvey intervals. The observed and adjusted values of uplift agree closely despite the fact that 65% of the 470 km of releveled segments in Ridge produce elevation-dependent correlations with 99% confidence. How can this happen? The good agreement arises because the mean value of e_{net} averaged over successive intervals is nearly zero, and because many of the larger tilts are uncorrelated.

2. Compilation of correlations for all routes

1700 km of leveling surveys were investigated for correlation. 1100 km of the levels proved to contain both the topographic variation and the absence of regions of pumping-induced subsidence, necessary for regression. The route locations are shown in Figure 2b and a summary of these correlations is presented in Figure 10. The resurvey intervals for rod segments are grouped roughly into chronological order. Since each correlation derives from two differenced surveys and the resurvey interval varies, the temporal sequence is not exact. Note that e_{net} of eqn. (7) is the combined error from two rod pairs, one from each survey. Both significant (r or $r_w \geq 95\%$) and insignificant correlations are plotted and used in computations. The variance of correlation coefficients by maximum likelihood is used to calculate the weighted mean, μ, and variance, σ_o^2, of the population. This yields $\mu = 0.34 \times 10^{-5}$, and $\sigma_o^2 = 5.2 \times 10^{-5}$, which means that the mean true correlation is $0.3 \pm 4.6 \times 10^{-5} \times dH$, at the 95% confidence interval.

Discussion

The near-zero mean error and the general lack of a significant change in sign or magnitude of the error with time form a crucial finding of this work. The errors are equally abundant and nearly equal in magnitude before, during, and following the period of observed uplift in southern California, 1959 to 1968. Had rods tended to shorten with time, and this length change gone undetected by calibration, the mean correlation would have been positive, rather than close to zero. This is because every value of e_{net} derives from a difference of a later rod pair from an earlier pair. The same circumstances would result if the rods maintained stability but the length measured from calibration erroneously increased with time. If during a particular epoch, rods shortened or calibrations increased, the mean value of e_{net} for that period would be positive. Neither effect emerges in Figure 10.

The mean error and population standard devi-

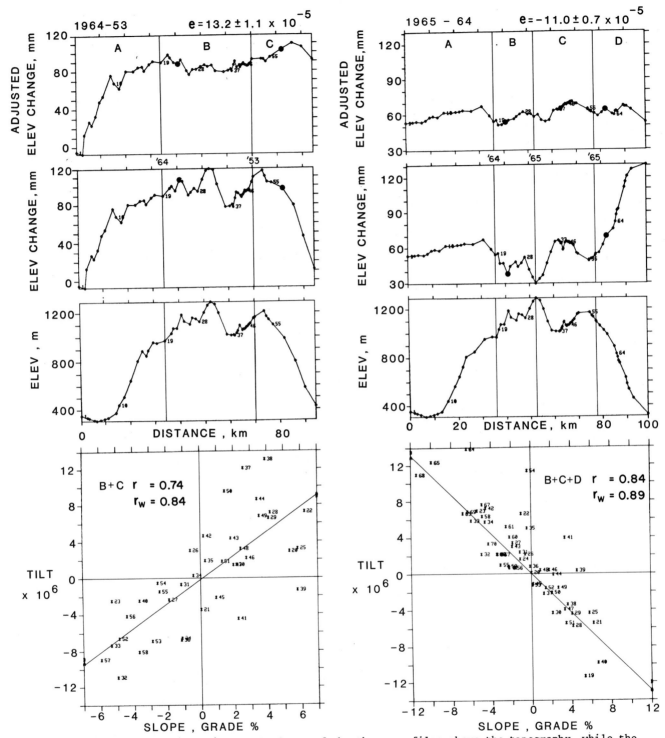

Figures 4 (left) and 5 (right). <u>A</u>: The lower of the three profiles shows the topography, while the elevation change from resurvey is shown in the middle profile. The rod changes are demarked by vertical lines, with the year of each change indicated. If a segment is uncorrelated, it is reproduced in the adjusted top profile, whereas if it is correlated at r or $r_w \geq 95\%$ interval of confidence, the correlation is removed in the upper profile. The slope of the regression, e_{net}, is shown in the upper right. Large dots are BM's shown in Figure 9. <u>B</u>: Plot of tilt as a function of slope for the correlated segments.

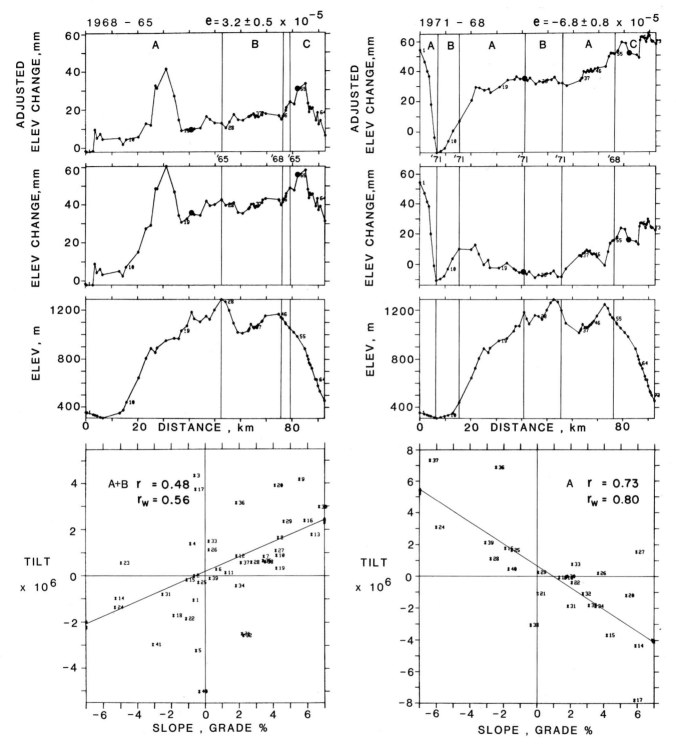

Figure 6.   Same explanation as for Figures 4 and 5.

Figure 7.   Same explanation as for Figures 4 and 5.

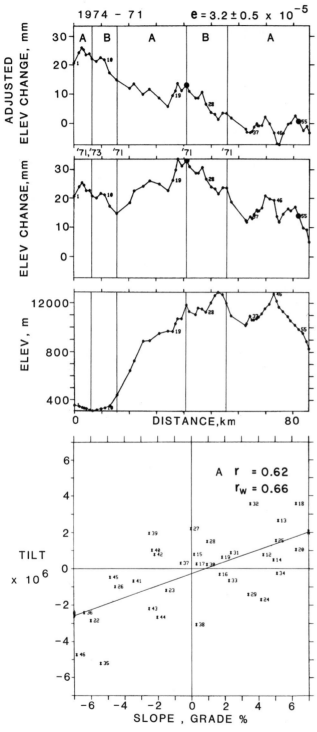

Figure 8. Same explanation as for Figures 4 and 5.

ation is $(0.3 \pm 2.3) \times 10^{-5} \times dH$. Because the variance of $\pm 5.2 \times 10^{-5}$ produces no more than 50 mm of artificial uplift over the maximum 1000 m relief, over three of these errors must accumulate with time to equal the 149 mm of uplift at G54 in 1968. In other words, there is greater than 99% confidence that the uplift shown in Figure 9 is unrelated to the systematic error tested for in eqn. (7). Since positive and negative errors are almost equally distributed and persist for distances less than 80 km, these errors do not accumulate. The observed elevation changes in steep terrain should therefore be accurate to 50 mm (2σ) in most cases, and to 80 mm (3σ) in almost all cases.

The two unusually large errors visible in Figure 10 have the same rods in common (rods 312-268, -274), used in some of the 1964 leveling. Jackson et al. [1980] cite rod 268 as typical, displaying non-uniform elongation that is not removed by the linear rod correction used to reduce the measured elevation changes. When compared to the one hundred other National Geodetic Survey rods calibrated by the National Bureau of Standards after 1964, it is clear that rod 268 is exceptional: the standard error of a linear fit to the rod differs by 3σ from the population mean [see Mark et al., 1981, Figure 5]. For both the Ridge and the Saugus to Palmdale routes there are surveys before and after 1964. The large positive (1964-53) and negative (1965-64) correlations almost cancel, leaving almost no cumulative effect on the observed elevation change. Specifically, the Ridge error for 1964-53 is $(13.2 \pm 1.1) \times 10^{-5}$, while that for the succeeding interval, 1965-64, is $(-10.8 \pm 0.7) \times 10^{-5}$. Adding the two intervals gives the net correlated error for 1965-53, $(2.4 \pm 1.3) \times 10^{-5}$, within the mean standard error for all resurveys.

Significance of residual tilt

The mean residual tilt, ε, of eqn. (7) is rarely found to be statistically significant for a regression regardless of its correlation coefficient, r. This can be seen by inspection of Figures 4 though 8; the y-intercept does not differ significantly from zero. Physically, this means that over a leveling segment that can be subjected to regression, the tilt is most often not uniform. Only rarely does tilt between any three successive bench marks exceed 2×10^{-6} or 2 μradians. The standard error for ε of the regressions is in all cases greater than $\pm 1 \times 10^{-6}$, or ± 1 μradian. For the 30 km average segment length in Figure 10, a 60 mm uplift of one end with respect to the other would be required to achieve 95% confidence as a real tilt. For Ridge, it can be seen from Figure 9 that only during the first time interval does such a large elevation change take place. Thus while the residual displacements shown in Figure 9 are essentially free of both rod and refraction errors, and while the resultant uplift is significantly larger than both expected random errors and the error of adjustment, the mean tilt is not significant.

Tests for rod-related errors

1. If the correlations are related to leveling rods, the error, e_{net}, should change significantly when rods are changed. For the Ridge resurveys, this proved to be the case for each of the five resurvey intervals, although the error did not differ significantly for every rod change. Five other resurvey intervals display significant changes in e_{net} with rod change, along routes 1, 2, 4, 5, and 7 (Figure 2b). These relevels span the years 1953 to 1979. Note that even if the entire rod population contained significant errors, the correlations would not always change detectably with change of rods. This is because the errors of some adjacent rods may differ by less than the correlation resolution from a perfect rod standard.

2. The error contribution of a specific rod set, e_n, can be distinguished from e_{net} under special circumstances. For route 5 (Figure 2b), correlated errors with 99% confidence are obtained for the resurvey intervals 1979-78 and 1978-76 over 23-32 sections with a grade range of 8%. The rods used for these surveys overlap two to five benchmarks leveled with different rods. Four groups of resurveyed sections exist, called lap sections, where a number of benchmarks were releveled with the changed rods rather than the more common procedure where each rod pair shares only one common benchmark. The elevation-dependent error for each lap section can be

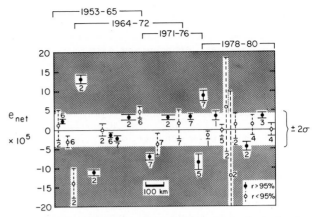

Figure 10. Summary plot of the slope-dependent correlations from 1100 km of relevels, arranged chronologically. Routes without significant correlations are dashed. The width of the brackets indicates the distance over which the correlation is maintained. Numbers correspond to the routes in Figure 2b.

reduced to five approximate linear equations for the errors of five rod pairs (standard deviations of about $\pm 2 \times 10^{-5}$):

$$A-B = -10.0 \times 10^{-5}$$
$$A-D = -8.8 \times 10^{-5}$$
$$A-E = -8.3 \times 10^{-5}$$
$$D-B = 0.6 \times 10^{-5}$$
$$D-C = 0.4 \times 10^{-5}$$

where each letter corresponds to a separate rod pair. Solving the simultaneous equations yields

$$A = -9.0 \pm 2.0 \times 10^{-5}$$

$$B = C = D = E = 0 \pm 2.0 \times 10^{-5}$$

A large negative correlation is found each time rod pair A is used with D and B, whereas B and D exhibit negligible correlations when compared to each other or two other rods. The rod pair A (316-132180, -87849) error, together with the 1964 and 1971 rod pair errors from Ridge that could also be isolated from e_{net}, demonstrate that these errors are neither time- nor location-dependent, but can be assigned to specific rods.

3. A third test for rod-related errors can be designed by considering the complete 354 km level circuit that contains Ridge, Saugus and Palmdale (Figure 2b, dashed). This circuit has been leveled within two years on four occasions (1953/55, 1964/65, 1972/74, and 1978), both before and after the period of observed uplift. Rods were changed from 9 to 24 times during each circuit. If no period contains elevation-dependent errors significantly larger than the mean error, the misclosures, or differences between initial and final elevations of the Saugus terminus of the

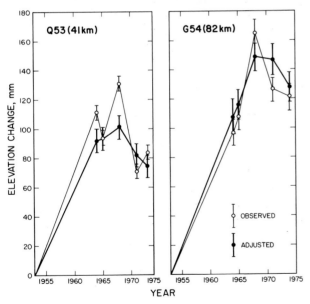

Figure 9. Uplift histories for two representative BM's along the Ridge route. The BM's are indicated by large dots in Figures 4-8. The observed and adjusted values are usually within one standard deviation.

route, should be similar in magnitude. Further, if the errors are fixed to the rods, the errors should randomize and the misclosure should be of the same order as the mean rod error. The mean misclosure is -10 ± 36 mm, consistent with the observed elevation-dependent error of 3 ± 45 mm over the 1000 m elevation difference (Table 1). This test holds regardless of the impact of differential refraction, since surveys with dissimilar sight lengths are not differenced, and because temperatures do not vary significantly during the survey. (Strange [1981] applied about a 10-30 mm refraction correction to the misclosures.)

4. Correlations that change significantly without an associated rod change constitute a strong test for errors that are independent of rods. Because a segment must be about 15 km long to achieve a correlation with 95% confidence, this test can only be performed in special cases where rod changes were infrequent. No cases have been found where a 15 km segment adjacent to a correlated rod segment displays the same value of e_{net} within one standard deviation at 95% confidence.

5. Another test for errors independent of rods can be performed by isolating segments where the same rod pairs were used in both surveys of a route, eliminating the effect of rods regardless of the rod error, under the assumption that no rod strain and no real earth movement took place between surveys. Only two such cases have been located in the 1700 km searched, both 6-10 km segments leveled in the early 1970's (Figure 11). In neither case does the elevation change for the resurveys differ by more than 2 mm over the length of the segment. The assumptions therefore appear justified unless rod strain balanced rod movement, suggesting that no errors emerge in the absence of rod differences.

Sources of rod-related errors

A number of factors contribute to discrepancies in the measurement of and correction for rod excess, or the difference between actual and measured rod lengths.

Improperly encoded calibrations. The National Geodetic Survey produces a computer encoded list

TABLE 1. Circuit Misclosures

Years	Rod Changes	Orthometrically corrected misclosure, mm
1953/55	9	+24
1964/65	13	-70
1972/74	9	+11
1978	24	-5
Mean	14	-10 ± 36

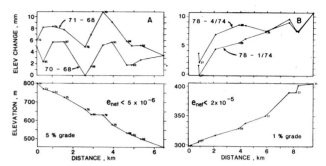

Figure 11. Segments surveyed twice with the same rod pair show no slope-dependent correlations $> 2 \times 10^{-5}$.

of rod calibration records, the RIF (Rod and Instrument File) that is used to correct field elevations for measured rod excess. There are internal inconsistencies in this list that result in improper corrections. For example, the encoded rod excess for Los Angeles County rods 315-95, -96, calibrated in 1977, is -3.1×10^{-5}, inconsistent with the listed calibration values that indicate an excess of -0.3×10^{-5}. The error deduced from statistical analysis for leveling with the RIF calibration is $(-5.1 \pm 0.9) \times 10^{-5}$ (Figure 12), only slightly larger than the encoding error.

Improperly calibrated rods. The points of observation for calibration have never been standardized for first order leveling, and at least three agencies have performed the calibrations: the National Bureau of Standards, Navy Gauge and Standards Center, U. S. Geological Survey [Kumar and Poetzsche, 1980]. Some rods are observed for calibration at 200 mm from the footplate, despite the fact that the first order leveling procedure precludes sighting the rod below 500 mm [Federal Geodetic Control Committee, 1974, 1975]. This leads to a probable 3.4×10^{-5} error in calculated rod excess for NGS rod 316-87849, calibrated in 1977 (Figure 13).

Damage to rod in use. Undoubtedly some rods are damaged during leveling so that the pre-survey calibration is no longer appropriate for correction of field elevations. Rod 312-268, used in the 1964 survey of Ridge, shows both a large and non-uniform excess of $(8 \pm 7) \times 10^{-5}$ in its post-survey 1965 calibration. However, the statistically measured excess for this rod and its mate is $(-12.4 \pm 0.7) \times 10^{-5}$ (Figures 5 and 6). Since both the calibration and statistically calculated errors are large but different, damage appears likely. Unfortunately, the rod serial number is attached only to the rod frame; the invar tape is periodically changed and its strain or damage cannot be traced.

Thermal coefficient of expansion for invar. The values of TCE for invar range from $(-2.5$ to $+3.0) \times 10^{-6}/°C$, and standard field temperatures range from about 10-30°C [Kumar and Poetzschke,

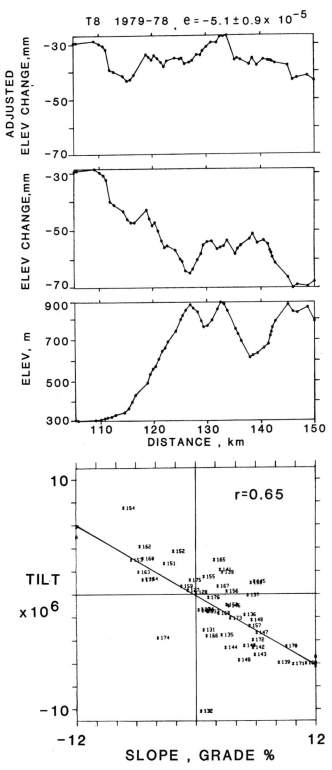

T8   1979-78 $e = -5.1 \pm 0.9 \times 10^{-5}$

ADJUSTED ELEV CHANGE, mm

ELEV CHANGE, mm

ELEV, m

DISTANCE, km

r=0.65

TILT $\times 10^6$

SLOPE, GRADE %

Figure 12. 50 km segment of 200 km route from T8 to G54 (route shown in Figure 2b). The encoding error in the RIF is 3.1×10^{-5}.

1980]. The NGS assigns a value of $0.8 \times 10^{-6}/°C$ for the same commercial rod (Kern) that the USGS applies $2.3 \times 10^{-6}/°C$, for temperature corrections. This can lead to discrepancies of 2.5×10^{-5} under the temperature range expected during leveling in southern California.

Thus at least some of the errors shown in Figure 10 are related to encoding, computation, and assumptions about thermal response, rather than miscalibration. What is important in considering these factors is that they are not systematic in time: none cause most older rods to be longer than their nominal length, or cause current rods to be shorter than their nominal length. Neither the statistical evidence compiled for Ridge (Figures 4-8), the 1100 km of relevels from 1953-1979 (summarized in Figure 10), nor an examination of potential causes of the elevation-dependent correlations, indicates that the errors are systematic in time, or confined to a particular location. The errors can therefore be treated as a source of random noise.

Comparison with other studies

Leveling rod errors. Strange [1980] produced a maximum estimate of leveling rod errors by comparing 64 resurveys over 17 leveling routes with varying topography. The elevation, dH_n, of the endpoint of each resurveyed route was subtracted from the initial endpoint elevation, dH_1. This method measures real earth movement as part of the error. Nevertheless, the 95% confidence error reported was about $6 \times 10^{-5} \times dH$, which is only slightly larger than that obtained in this analysis, where rod errors and earth movement are separated.

The central contention of Jackson and Lee [1979] and Jackson et al. [1980] is that the temporal change in elevation measured from resurveys across the southern California uplift correlates with elevation at the kilometer scale; that is, the apparent elevation change (dh) between surveys mimics the elevation difference (dH) from one benchmark to the next. Jackson et al. [1980, their Figure 1] demonstrate this correlation with a plot of the spatial variation in incremental elevation change, d(dh)/dx, with the incremental change in topography, d(dH)/dx. The technique of Jackson et al. has been applied to the survey years 1965-64 over the segment Saugus to Palmdale where 160 mm observed elevation change is concentrated (route location, Figure 2b; plot, Figure 14a). In Figure 14b, the same technique is employed to correlate a straight line - a uniform tilt with endpoints as in the observed elevation change, and with benchmarks spaced along the route as in reality - with the true topography. Because the tilt of the straight line is uniform, there can be no benchmark-to-benchmark correlation with topography in Figure 14b. Despite this, both Figures 14a and 14b display what appear to be impressive correlations of similar magnitude. Why

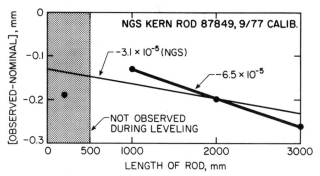

Figure 13. Calibration over the useable length of rod differs by 3.4×10^{-5} from the NGS RIF applied rod excess.

does the technique of Jackson et al. [1980] fail to discriminate between topography correlated to actual leveling surveys, and topography correlated to a uniform tilt devoid of any short wavelength (mark-to-mark) signal? The problem lies in the fact that the topography, the real elevation change, and the uniform tilt have precisely the same variation in benchmark spacing, as Stein [1980] and Mark et al. [1981] have pointed out, a property common to all leveling data. Because the elevation, elevation change, and straight line all have a trend (a mean slope not equal to 0), benchmarks spaced farther apart will tend to show both larger dh and dH than those more closely spaced, though their tilt (dh/dx) and slope (dH/dx) are of course independent of spacing. This can be seen in Figure 14c, which shows the two profiles that are correlated in Figure 14b. The benchmark spacing for a leveling route will always vary more than an order of magnitude, in this case from 0.02-4.27 km, so that unless the values of dh and dH are normalized to the distance between marks, the technique of Jackson et al. [1980] correlates benchmark spacing rather than elevation-dependence.

Strange [1981, his Figure 5] employs another technique to argue for slope-dependence that also suffers from the influence of a trend. His plot of cumulative elevation change as a function of cumulative elevation will always produce a highly significant correlation as long as the trend of neither elevation change (dh) nor topography (dH) curve changes sign. Any tilting elevation change profile and sloping topographic profile will correlate and this can be misinterpreted as a mark-to-mark correlation.

Jackson et al. [1980] do not derive their values for the slope of the regression of elevation change onto elevation, or elevation-dependent error, from the plots such as shown in Figure 14a. Instead they remove the trend from both dh and dH curves by separately fitting each to a 3rd- or 4th-order polynomial. They correlate the residuals to find e_{net}. Implicit in this technique is the assumption that the long wavelength or large scale contribution to the eleva-

tion change profile represents tectonic signal, and that the residual high frequency or mark-to-mark signal can be isolated for elevation-dependent correlation. This may not be valid: Abundant evidence from both the field and laboratory suggests that faulting takes place on all spatial scales. Deformation may occur on the scale of the benchmark spacing or over the entire level route. There can be no ideal order poly-

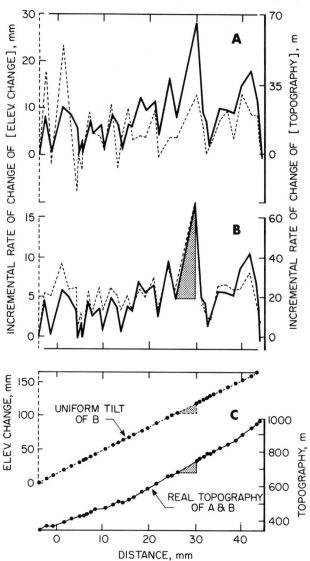

Figure 14. Correlation plot after Jackson et al. [1980] of the 1965-64 surveys of the route from Saugus to Palmdale (shown in Figure 2b). <u>A</u>: correlation of observed elevation change from resurvey, with topography. <u>B</u>: correlation of a uniform tilt with same endpoints and BM-spacing as observed elevation change, with topography. <u>C</u>: the segments correlated in B are shown. Note that when BM's are spaced far apart, a large signal is produced (stippled) since both curves have a positive trend.

nomial. Even if such an a priori segregation of long and short wavelength signals can be justified, residual fitting still suffers from other problems. Since benchmark spacing is never uniform, a sampling bias will persist in the minimization of residuals when fitting the curve. Regions of closely spaced marks will be fit better than those regions where marks are spaced farther apart, and a misfit curve introduces a trend. If the curve fit to the elevation change has too low an order, the trend will not be removed over segments of the profile, and benchmark spacing effects can then dominate. As the order of the polynomial increases, the magnitude of the residuals must drop, and the signal to be correlated consequently diminishes.

Subsidence caused by groundwater withdrawal. Reilinger [1980] argues that the relative subsidence of Saugus with respect to bench marks to the north and south during 1953-1964 (Figure 4a, 0-15 km) is best explained by compaction of the Saugus aquifer, which has a maximum saturated or effective thickness of 600 m. From 1945 through 1967, groundwater withdrawal greatly exceeded recharge. However, records for the two deep wells that tap the Saugus aquifer do not display the long-term head decline that occurred in the overlying alluvial aquifer. Rather, from 1953 to 1966, 90% of the water was pumped from the unconfined alluvial aquifer, which has a maximum saturated thickness of 60 m [Robson, 1972, page 39 and Table 7]. The 0- to 15-km section of the Ridge route (Figures 4a-8a) traverses the alluvial aquifer. It is about thirty times more permeable than the Saugus aquifer, because it is composed of poorly bedded unconsolidated gravel, sand, and silt. To estimate the maximum subsidence within the alluvial aquifer caused by the increase in effective stress, or the stress borne by the aquifer matrix, during the period of pumping, the aquifer compressibility must be measured or modeled. From the hysterisis loops of in situ vertical well extensometers, compressibilities of eight San Joaquin Valley, California, confined aquifers have been measured by Poland et al. [1975], yielding a mean value of $(5 \pm 2.0) \times 10^{-2}$ Nm^{-2}. Because the alluvial aquifer is unconfined and has a coarse grained skeleton, its matrix bears a greater load than the aquifers measured by Poland et al.; pore pressure and seepage stress are lower than in a fine-grained aquifer, resulting in higher permeability and transmissibility, and lower compressibility. Thus these values provide an extreme upper limit. Using the maximum aquifer thickness (60 m) and the portion of the roughly linear 1945-1964 water table decline that occurred between the 1953 and 1964 surveys at Saugus (14 m), differential subsidence of 9 ± 3 mm results. A compressibility of an order of magnitude higher would be required to account for the observed 90 mm differential elevation change between Saugus and bench marks 15 km north or south that Reilinger [1980] ascribes to water withdrawal.

Consider an alternate and independent approach to estimate subsidence related to groundwater withdrawal: the 1953/55-1926 survey includes eight to ten years of water table decline, or about 10 m. No more than 21 ± 7 mm of differential elevation change is evident between Saugus and BM's 15 km to the south, and 12 ± 6 mm with respect to BM's 10-20 km to the north during this period, although 40% of the water was pumped from 1945-1953/55 [Robson, 1972, his Figure 12]. Although pumping continued to 1968 at reduced rates, no more than 10 ± 7 mm of differential subsidence at Saugus can be seen during the 1965-64 and 1968-65 resurveys (Figures 5a and 6a). A maximum estimate of subsidence of 48 ± 20 mm can be obtained by extrapolating the 1945-63 23-m head decline [Robson, 1972, plates 5 and 7] to 30 m through 1968.

Conclusions

Significant correlations between elevation change and elevation have been identified for 65% of the leveling route segments in steep terrain that are suitable for regression. The mean correlation of $(0.3 \pm 2.3) \times 10^{-5} \times dH$ does not change with time; the correlations are as abundant in the 1950's as in the 1970's despite changes in leveling procedure and calibration during this period. The correlations can be removed by a straightforward adjustment. An 100 km long segment through the southern California Transverse Ranges that has been releveled six times shows 165 ± 9 mm of observed uplift at BM G54 near Grapevine with respect to Saugus (J52), and yields 149 ± 18 mm after removal of level rod-related slope-dependent errors between 1953 and 1968. Because sight lengths on this route must be short and were almost the same for each resurvey, differential refraction should have no significant effect on these values of elevation change.

Subsidence caused by groundwater withdrawal from an alluvial aquifer beneath Saugus from 1945-1968 can be approximated by the product of maximum values of the aquifer thickness (60 m) head decline (30 m), and compressibility (5 ± 2 $\times 10^{-2}$ Nm^{-2}). This predicts an upper limit of 48 mm of non-tectonic subsidence, leaving a minimum of 100 ± 18 mm of tectonic displacement at G54 during 1953-68.

The 1100 km of resurveys subjected to the test for slope-dependent correlation comprise about 15% of the total population of levels that define the southern California uplift of Castle [1978]. If this sample is representative, the observed elevation change for southern California surveys should be accurate to within $\pm 4.6 \times 10^{-5} \times$ the topographic relief, or about 50 mm for most surveys, and 90 mm for 95% of the surveys. The magnitude of the errors is considerably smaller than that of the observed uplift, and about one-fifth the magnitude claimed by Jackson et al. [1980]. Therefore uplift in southern California cannot be the product of any linear slope-dependent error.

A number of tests confirm that the linear slope-dependent errors are related to leveling rods. Fifteen correlations differ significantly where rods are changed, although the spatial resolution of the correlation is precise only to within about 5 km of the rod change. Errors can be assigned to specific rod pairs where multiple lap sections are available, which precludes location- or time-dependent correlations for those cases. Significant changes in correlation without change in rod have not been located, although in some instances these may be masked by frequent rod changes. No elevation-dependent correlations can be found where the same rod was used in both surveys of a route. Finally, circuit misclosures are within the expected mean rod error when many rod changes take place, further substantiating that the errors do not accumulate over distances as large as 300 km.

Several factors contribute to the rod errors, although their relative importance is difficult to assess. Redesign of calibration and encoding practices can reduce current discrepancies, and rod errors larger than about 2×10^{-5} can be identified and removed from some but not all of the historic leveling data.

Appendix: Weighted Linear Regression

The regression equation predicts values of y for given values of x to within an assessable random fluctuation. We assume that $Y' = a + x$, where Y = tilt, Y' = the predicted Y, x = slope, and b, the slope of the regression, is synonomous with e_{net} of eqn. (3), so that

$$b = e_{net} = \frac{n\Sigma xy - \Sigma x \Sigma y}{n\Sigma x^2 - (\Sigma x)^2} \quad , \qquad (8)$$

$$a = \frac{y \Sigma x^2 - \Sigma x \Sigma xy}{n\Sigma x^2 - (\Sigma x)^2} \qquad (9)$$

The y intercept, a, represents the mean uncorrelated tilt, or in other words, the residual tilt after removal of e_{net}. The population regression line is estimated by the method of least squares; the sum of the squares of the deviations in Y are minimized by the regression line. The standard deviation of the slope (here, the standard deviation of e_{net}), S_b, is found by differentiating the equation for slope with respect to Y.

$$S_b = \left(\frac{(Y - bx - a)^2}{n\Sigma x^2 - (\Sigma x)^2} \right)^{1/2} \qquad (10)$$

The sample correlation coefficient [from Crow et al., 1960] is

$$r = b\frac{\sigma_x}{\sigma_y} = \frac{n\Sigma xy - \Sigma x \Sigma y}{([n\Sigma x^2 - (\Sigma x)^2] [n\Sigma y^2 - (\Sigma y)^2])^{\frac{1}{2}}} \qquad (11)$$

The significance of r is checked by an equal tails test of the null hypothesis - that no correlation exists. The test requires a greater value of r as the population decreases in size. No correction is applied for auto-correlation. This favors correlation because values of e_{net} are not strictly independent; each survey is used twice.

The weighted regression. In this analysis it is assumed that Y varies from point to point and that X is known exactly. In fact, from one survey to another, Y varies by a factor of 1-10, whereas X varies by no more than 10^{-4}. First the insertion of weights, w, into the equations, and then the rationale for assigning those weights are presented. To adjust the regression line to give more weight to better data, each X_i and Y_i are associated with a W_i inside the summation, and w replaces n. Thus, $Y = a + bx$ becomes $w_i y_i = aw_i + bw_i x_i$, leading to

$$b' = \frac{\Sigma w \Sigma wxy - \Sigma wx \Sigma wy}{\Sigma w \Sigma wx^2 - (\Sigma wx)^2}, \quad a' = \frac{\Sigma wy \Sigma wx^2 - \Sigma wx \Sigma wxy}{\Sigma w \Sigma wx^2 - (\Sigma wx)^2} \quad (12)$$

$$\text{and } S_b' = \left(\frac{\Sigma [w(y - b'x - a')^2]}{\Sigma w \Sigma wx^2 - (\Sigma wx)^2} \right)^{\frac{1}{2}} \quad . \qquad (13)$$

Similarly, the weighted regression coefficient of the sample becomes,

$$r_w = \frac{(\Sigma w \Sigma wxy - \Sigma wx \Sigma wy)}{\{ [\Sigma w \Sigma wx^2 - (\Sigma wx)^2] [\Sigma w \Sigma wy^2 - (\Sigma wy)^2] \}^{\frac{1}{2}}} \cdot \qquad (14)$$

To estimate the mean, μ, and the weighted variance, σ_o^2, of the true correlation, for the compilation of all data shown in Figure 10, the variance of correlation coefficients by maximum likelihood is used [Anderson and Bancroft, 1952]. Because the variance of the true correlation is not necessarily equal to that of the sample, we assume that errors in measuring the correlations, y_i, are Gaussian with a known variance, σ_i^2, determined from the linear regression of each resurvey. The true correlations are also assumed to be taken from a Gaussian population. By maximizing the probability with respect to the mean value, we solve

$$\mu = \frac{\Sigma [y_i / (\sigma_i^2 + \sigma_o^2)]}{\Sigma [1/(\sigma_i^2 + \sigma_o^2)]} \quad . \qquad (15)$$

The probability of obtaining the observed sample is maximum when

$$\Sigma \left\{ \frac{[(y_i - \mu)^2 - \sigma_i^2 - \sigma_o^2]}{(\sigma_i^2 + \sigma_o^2)^2} \right\} = 0 \qquad (16)$$

is satisfied. Note that eqns. (15) and (16) weight the observations by the reciprocal of their variances. σ_o^2 is first set to a value obtained from the population standard deviation of the weighted mean [which yields $(0.9 \pm 4.6) \times 10^{-5}$] and iterated through (15) and ($\overline{16}$) until the equality of (16) holds.

   Assignment of weights. We assume a Gaussian distribution with known variance, and adopt the relationship between w and the variance [Bacon, 1953], where

$$w_i = 1/\sigma_i^2$$

letting $w_i \sigma_{y_i}^2 = w_2 \sigma_2^2 = \ldots w_n \sigma_n^2 = \sigma^2$,

where the w's are the ratios of the variance of each point to some convenience variance taken as the standard. Therefore an observation made where the variance is σ/w is worth w observations in a region where the variance is σ^2. This weighting scheme is employed for both calculation of the population variance in Figure 10 [eqns. (15) and (16)] and for consideration of random errors in the regression [eqns. (12) and (14)].

   Because tilts established over longer distances can be more accurately measured, they become a more important tool for testing the dependence of tilt on slope. In geodetic leveling, there is a standard error associated with each measurement of height (at time, t_o, and t_1), where if $y = dH$,

$$\text{tilt, } Y = \frac{y_{t_1} - y_{t_0}}{dx},$$

$$\text{and } S_y = [(\delta y/\delta t_1)^2 + (\delta y/\delta t_o)^2 S_{t_0}^2]^{\frac{1}{2}} \quad (17)$$

Taking the partial derivatives with respect to time,

$$\delta y/\delta t_1 = 1/dx \text{ and } \delta y/\delta t_o = -1/dx, \quad (18)$$

$$\text{so } S_y = (2/dx)^{\frac{1}{2}}$$

Random leveling errors lead to accuracies of $(\text{distance})^{\frac{1}{2}}$; this formula arises from consideration of the sums of squares [Bomford, 1971]. S_{t_0} and S_{t_1} are set equal to each other, since a constant factor, k, will not affect w. Thus,

$$S_y = [(k/dx)^2 dx + (k/dx)^2 dx]^{\frac{1}{2}} = k (1/dx)^{\frac{1}{2}}.$$

Since $w = 1/S_y^2$, $\underline{w = dx}$, the distance between benchmarks.

   Acknowledgments. For their help, consultation, and insight, I am indebted to R. O. Castle, Wayne Thatcher, Nancy King, and Bernard Hallet. Reviews by Robert Reilinger, J. C. Savage and Art McGarr improved an earlier version of this manuscript. David D. Jackson suggested the use of maximum likelihood techniques. 'Magic fingers' Stan Silverman made music on the computer, and Mary Anne Avins graciously typed and retyped this manuscript. I appreciate the generous support from Stanford University while I was a student there, and from the U. S. Geological Survey's Office of Earthquake Studies. This work was completed under Lamont-Doherty Post-Doctoral Fellowship 6-90288C. Lamont-Doherty Geological Observatory Contribution No. 3133.

REFERENCES

Anderson, R. L., and T. A. Bancroft, _Statistical Theory in Research_, 399 pp., McGraw-Hill, New York, 1952.

Bacon, R. H., The best straight line among the points, _Am. Jour. Physics_, _21_, 24, 1953.

Bomford, G., _Geodesy_, 816 pp., Oxford Univ. Press, 1971.

Castle, R. O., Leveling surveys and the southern California uplift, _U. S. Geol. Surv. Earthquake Inf. Bull._, _10_, 88-92, 1978.

Castle, R. O., J. P. Church, and M. R. Elliot, Aseismic uplift in Southern California, _Science_, _192_, 251-253, 1976.

Castle, R. O., J. P. Church, M. R. Elliot, and N. L. Morrison, Vertical crustal movements preceding and accompanying the San Fernando earthquake of Feb. 9, 1971: A summary, _Tectonophysics_, _29_, 127-140, 1975.

Crow, E. L., F. A. Davis, and M. W. Maxfield, _Statistics Manual_, 288 pp., Dover Publications, New York, 1960.

Federal Geodetic Control Committee, _Classification, standards of accuracy, and general specifications of geodetic control surveys_, N.O.A.A., Rockville, Maryland, 1974.

Federal Geodetic Control Committee, _Specifications to support classification, standards of accuracy, and general specifications of geodetic control surveys_, N.O.A.A., Rockville, Maryland, 1975.

Heiskanen, W. A., and H. Moritz, _Physical Geodesy_, W. H. Freeman, San Francisco, 1967.

Jackson, D. D., and W. B. Lee, The Palmdale bulge - an alternate interpretation (abstract), _EOS, Trans. AGU_, _60_, 810, 1979.

Jackson, D. D., W. B. Lee, and C. Liu, Aseismic uplift in Southern California: An alternate interpretation, _Science_, _210_, 534-536, 1980.

Jenkins, O. P., Geologic map of California: Los Angeles Sheet, _Cal. Div. Mines and Geol._, 1975.

Kukkamäki, T. J., Über die nivellitische refraktion, _Veroff Finn. Geod. Inst._, _25_, 1938.

Kumar, M., and H. Poetzschke, Level rod calibration procedures at the National Geodetic Survey, _NOAA Tech. Memo._, Rockville, Maryland, 1980.

Lofgren, B. E., Land subsidence due to groundwater withdrawal, Arvin-Maricopa area, California, U.S. Geol. Surv. Prof. Paper 437-O, 1975.

Mark, R. K., J. C. Tinsley III, E. B. Newman, T. D. Gilmore, and R. O. Castle, An assessment of the accuracy of the geodetic measurements that led to the recognition of the Southern California uplift, J. Geophys. Res., 86, in press, 1981.

Poland, J. F., B. E. Lofgren, R. L. Ireland, and R. G. Pugh, Land subsidence in the San Joaquin Valley, California, as of 1972, U. S. Geol. Surv. Prof. Paper 437-H, 1975.

Robson, S. G., Water resources investigations using analog model techniques in the Saugus-Newhall area, Los Angeles County, California, U.S. Geol. Surv. Open-File Rept. 5021-04, 58, 1972.

Reilinger, R., Elevation changes near the San Gabriel fault, Southern California, Geophys. Res. Lett., 7, 1017-1019, 1980.

Stein, R. S., Accuracy of geodetic leveling for the detection of rapid crustal deformation, in Contemporary and Quaternary Deformation in the Transverse Ranges of Southern California, unpub. Ph.D. disser., Stanford Univ., June 1980.

Strange, W. E., The effect of systematic errors in geodynamic analysis, 2nd Int. Symp. Probs. Redefinition No. Amer. Vertical Geodetic Networks, Ottawa, 705-727, 1980.

Strange, W. E., The impact of refraction correction on leveling interpretations in southern California, J. Geophys. Res., 86, in press, 1981.

Vaniček, P., E. I. Balazs, and R. O. Castle, Geodetic leveling and its applications, Rev. Geophys. Space Phys., 18, 505-524, 1980.

HEIGHT DEPENDENT ERRORS IN SOUTHERN CALIFORNIA LEVELING

David D. Jackson, Wook B. Lee, Chi-Ching Liu

Department of Earth and Space Sciences, University of California
Los Angeles, California, 90024

Abstract. For many profiles in southern California, the tilts inferred from repeat leveling
data are strongly correlated with slopes. The
ratio of tilt to slope at short wavelengths is
nearly equal to that for long wavelengths, suggesting that the correlation results from height
dependent systematic errors. Such errors are well
known in geodesy, and they may result from rod
miscalibration, or atmospheric refractions, among
other causes. However, the magnitude of error
that we suggest (spurious tilt on the order of
10^{-4} times the slope) is larger than previously
appreciated. There is evidence for rod miscalibration, as well as for height dependent errors
from other sources, presumably atmospheric refraction.

Subsidence due to fluid extraction could cause
elevation dependent changes, because sediments
at low elevations would subside more than harder
materials at higher elevations. This effect may
be widespread in southern California, and could
in some cases be mistaken for tectonic tilting or
elevation dependent errors. However, elevation
dependent errors can be clearly identified in
areas where subsidence is negligible.

A careful analysis of leveling data for southern
California suggests to us that the reported
aseismic uplift (Castle et al., 1976, 1977) can
be more plausibly explained as the result of
nontectonic effects including systematic leveling
errors.

Introduction

Precise leveling data have been heavily used in
tectonic studies and some surprising consequences
have been inferred from these data. For example,
Castle et al. (1976, 1977) and Castle (1978), have
reported that a large region of southern California
was uplifted by as much as 450 mm during the period 1961-1974, only to be downdropped a similar
amount after 1974. The uplift apparently took
place in rapid episodes, with Palmdale rising 150
mm with respect to Long Beach in the period 1961-
1962. This would imply an uplift rate of 100 mm/
year, or a tilt rate of $2 \times 10^{-6} \mathrm{yr}^{-1}$, since the
apparent uplift was concentrated in a zone approximately 50 km wide. These rates are quite high by
geological standards. The geologically determined
uplift rate of 10 mm/yr averaged over 600,000
years across the Oak Ridge fault north of Ventura
is considered an exceptionally high rate (Yeats,
1977) and it has probably been accomplished
primarily through large earthquakes. The episodic
and reversible nature of the apparent southern
California uplift is enigmatic.

Reports of tectonic uplifts and downdrops
inferred from leveling data abound for other parts
of the U.S. and the world as well. Table 1 lists
some of these observations. Not all are as
geologically distinctive or as ominous as the
southern California uplift, but certainly all have
great tectonic significance if they are real.
Because of the importance of these observations,
we take a critical look at the leveling data. As
you know from the abstract, we have found evidence
for systematic errors in leveling data, at least
for southern California. We have not studied the
data for the other areas listed in Table 1, but
our results would suggest that great caution is
required in interpreting any leveling data to
infer tectonic motions.

The primary evidence for systematic errors is a
strong correlation between apparent uplift and
topography, or equivalently, between apparent tilt
and topographic slope. Such correlations are well
known to occur as the result of certain types of
systematic error, especially leveling rod miscalibration and atmospheric refraction. (Bomford,
1973, p. 240). However, correlations between
uplift and topography might also result from tectonic deformation, from subsidence in sedimentary
valleys, or simply by chance.

Strong correlations between uplift and topography
in U.S. leveling data were first pointed out by
Brown and Oliver (1976), who examined data for the
Appalachians and the Atlantic coastal plain.
Because the uplifts were also strongly correlated
with geological boundaries, and because systematic
errors were expected to be too small to explain
the uplifts, the authors preferred a tectonic
explanation. Citron and Brown (1979) examined
these data in more detail, and concluded that
either systematic errors were larger than expect-

TABLE 1. Reported aseismic tilting from leveling data

location	time interval	vertical motion	elevation contrast	profile length	tilt/slope × 10⁶	reference
So. Calif.	1960–1974	450 mm	1000 n	200 km	450	1
Atlantic Coast	1935–1968	300 mm	250 m	300 km	400	2
Appalachian Highlands	1934–1969	200 mm	450 m	100 km	420	2, 3
Interior Plains	1947–1969	280 mm	150 m	500 km	1200	2
U.S. Basin & Range	1911–1955	400 mm	1000 m	900 km	400	4, 5
Hungary	1955–1957	50 mm	?	84 km	?	6
Israel	1958–1966	80 mm	1000 m	60 km	80	7

References for Table 1.

1. Castle, 1977

2. Brown & Oliver, 1976

3. Citron & Brown, 1979

4. Brown Reilinger & Citron, 1980

5. Reilinger, 1978

6. Bendefy, 1965

7. Karcz and Kafri, 1971

ed or the apparent movement was real and correlated with topography. Chi et al. (1980) showed evidence that some leveling data for the Sierra Nevada, which showed apparent uplifts highly correlated with topography, were also involved in significant circuit misclosures. They concluded that systematic errors in the data were likely.

Jackson et al., (1979, 1980) first presented evidence that tilt/slope correlations were significant in data for southern California, and that the apparent aseismic uplift in southern California might be the result of systematic errors. The latter paper also presented evidence that rod calibration errors were responsible for some of the correlations. Recently Strange (1980, 1981) has also argued that the apparent uplift resulted from systematic errors, but that refraction errors are the dominant problem. Stein (1980, 1981) argued that tilt/slope correlations probably result from rod miscalibrations, but that substantial residual tilt, presumably of tectonic origin, still remains after correcting for these errors. Mark et al. (1981) contended that tilt/slope correlations are not pervasive, and that many of the data that exhibit these correlations can be explained by surficial subsidence rather than systematic errors. Reilinger (1980) argued that much of the apparent uplift could be explained by subsidence due to groundwater pumping in the Saugus basin. Stein (1980, 1981) and Mark et al. (1981) both questioned the methodology used in our earlier papers (Jackson et al., 1979; 1980). In this paper we consider some of the arguments raised

in these recent papers. We discuss the methodology in detail, and we present some new data that reinforce the arguments made in our 1980 paper.

Evidence for Systematic Errors:
Correlation of Apparent Tilt With Slope

We present below some evidence for systematic errors that result in a strong correlation between apparent tilt and topographic slope. While it is possible that tectonic tilts could also be correlated with slope, there is strong circumstantial evidence that the observed correlation is caused largely by systematic errors. Before presenting the evidence, a discussion of the data processing is appropriate.

Leveling data come in the form of height differences between successive benchmarks,

$$\Delta h_i = h_i - h_{i-1} \qquad (1)$$

where i is a benchmark index, and the heights are measured with respect to some reference surface, usually an equipotential surface. Because of changes in the benchmark heights, the reference surface, or both, measured elevation differences will change with time. The temporal change in elevation difference is

$$\Delta^2 h_i = \Delta h_i - \Delta h'_i \qquad (2)$$

where $\Delta h'_i$ is the height difference observed on a previous survey called the reference survey.

We compute relative elevation differences according to equation 2, and fit them with a model that accounts for both regional tilting (presumably tectonic) and elevation dependent effects (including systematic errors). This model is:

$$\Delta^2 h_i = \alpha \Delta h_i + \beta \Delta x_i + e_i \qquad (3)$$

where Δx_i is the horizontal distance between two adjacent benchmarks, α and β are parameters to be determined from the data, and e_i is an experimental error. The significance of (3) may be more clearly realized if we divide through by Δx_i: then

$$t_i = \alpha s_i + \beta + e_i' \qquad (4)$$

where $t_i = \Delta^2 h_i / \Delta x_i$ is apparent tilt, $s_i = \Delta h_i / \Delta x_i$ is a topographic slope, and $e_i' = e_i / \Delta x_i$ is the error in measuring tilt. The parameter β is interpreted as uniform tilting, presumably of tectonic origin, occurring between the time of the reference survey and the test survey. The parameter α is the tilt/slope ratio.

Equations (3) and (4) are both examples of systems of weighted equations

$$\frac{\Delta^2 h_i}{\sigma_i} = \alpha \frac{\Delta h_i}{\sigma_i} + \beta \frac{\Delta h_i}{\sigma_i} + \frac{e_i}{\sigma_i} \qquad (5)$$

Such weighting is appropriate if the standard deviations of the observations vary along the profile, perhaps as the result of uneven benchmark spacing. Equation (3) corresponds to $\sigma = \sigma_0$, a constant, and (4) corresponds to $\sigma_i = a\Delta x_i$, where a is a constant. We tried several different formulas relating σ_i to Δx_i, and found that the estimated values of α and β were not terribly sensitive to the choice of σ_i. For ease of interpretation, we present here the results for $\sigma_i = a\Delta x_i$, as assumed in equation (4).

Depending on the causes of elevation correlated effects and regional tilts, both α and β could be expected to vary along the profile. We have attempted to subdivide the profiles into sections such that α and β are relatively constant. This requires that the direction of the survey should be relatively constant, because the actual regional tilt will be a vector in two dimensions, and the apparent tilt observed in the leveling data will be the projection of this vector onto the direction of the profile. We believe that height dependent effects result partly from rod miscalibrations, so we shall select subsets of data for which a single pair of rods is employed in the test survey, and another single pair is used in the reference survey.

In the discussion to follow, we shall be testing the hypotheses that either or both of α and β vanish. The hypothesis that α is zero corresponds to the case that elevation correlated errors are negligible, as assumed in most literature on this subject to date. The hypothesis that $\beta = 0$ corresponds to the case that there is no regional

tilting, so that the observed tilts result from height dependent effects and random errors only. The hypothesis that $\alpha = 0$ and $\beta = 0$ corresponds to the case that the observations are due entirely to random error. To test these hypotheses we employ the Fisher F-test (discussed in Draper and Smith, 1966, p. 67). In this test, two alternate hypotheses are tested against each other, by comparing the difference in the sum of squared residuals to that expected for a set of random data. We tested four hypotheses in pairs. The four hypotheses are:

(a) $\alpha = 0 \quad \beta = 0$
(b) $\alpha = 0 \quad \beta \neq 0$
(c) $\alpha \neq 0 \quad \beta = 0$
(d) $\alpha \neq 0 \quad \beta \neq 0$

In the discussion to follow, we shall make use of the mean tilt and mean slope.

$$\bar{t} = \frac{1}{n} \sum_{i=1}^{n} t_i, \qquad \bar{s} = \frac{1}{n} \sum_{i=1}^{n} s_i \qquad (6)$$

where n is the number of segments in the profile. We define residuals as:

$$r_i = t_i - \hat{\alpha} s_i - \hat{\beta} \qquad (7)$$

where $\hat{\alpha}$ and $\hat{\beta}$ are estimated by least squares, constrained as required. Under hypothesis (a), $\hat{\alpha} = \hat{\beta} = 0$, and the residuals are simply the observed apparent tilts. Under hypothesis (b), we have

$$\hat{\alpha} = 0 \qquad \hat{\beta} = \bar{t} \qquad (8)$$

so that the residuals are the observed tilts, reduced by the mean tilt. Under hypothesis (c), we have

$$\hat{\alpha} = \Sigma s_i t_i / \Sigma s_i^2, \qquad \hat{\beta} = 0 \qquad (9)$$

Under hypothesis (d), α and β must be estimated simultaneously. The result is:

$$\hat{\alpha} = \Sigma (s_i - \bar{s})(t_i - \bar{t}) / \Sigma (s_i - \bar{s})^2, \qquad \hat{\beta} = \bar{t} - \hat{\alpha} \bar{s} \qquad (10)$$

In quoting the significance of a correlation between tilt and slope, we refer to the significance of hypothesis (d) compared with (b). That is, the regional tilt in not assumed to vanish. In quoting the significance of regional tilting, we compare hypothesis (c) and (d); the correlation between tilt and slope is not neglected.

Some results will be presented as scatter plots of tilt vs. slope. Under hypothesis (a), the tilts should be scattered randomly about the value zero, regardless of topographic slope. Under (b), the tilts should lie about a horizontal line, whose intercept is the regional tilt. Under (c), the tilts should lie near a straight line through the origin. The slope of this line is α. Under (d), the data should again lie on a straight line, whose slope is α and whose intercept is β (the predicted tilt for zero slope).

There are some leveling surveys for which expected changes in α and β do not occur at the same places in either the test survey, the reference survey, or both. For instance, rod changes will be made at some locations, and changes in the direction of surveying will occur at other locations. In these cases we treated the data two separate ways. First, we divided the profile into subsets, such that both rod usage and survey direction were uniform within each subset for both the test survey and the reference survey. This often left too few data within each subset to make any significant conclusions. Second, we built more complicated models, of the form

$$t_i = \alpha_j s_i + \beta_k + e_i \qquad (11)$$

where j indexes subsets of the profile over which rod usage was uniform, and k indexes subsets over which the survey direction was uniform. In this way we could combine data into larger subsets to improve the resolution of the unknown parameters. The statistical testing then becomes a bit more complicated, and the number of hypotheses that may be tested increases substantially. For simplicity, in these cases we generally tested the following hypotheses in pairs:

(a) all $\alpha_j = 0$, all $\beta_k = 0$
(b) all $\alpha_j = 0$, all $\beta_k \neq 0$
(c) all $\alpha_j \neq 0$, all $\beta_k = 0$
(d) all $\alpha_j \neq 0$, all $\beta_k \neq 0$

Results

Locations of the leveling profiles are shown in Figure 1. Also shown in Figure 1 are regions covered by Quaternary sediment, major faults crossed by the leveling profiles, and some landmarks referred to in the text.

Figure 2 presents apparent tilt and slope as a function of distance along the profile from Burbank to Saugus to Palmdale for the test year 1965, using the three reference years 1955, 1961, and 1964. Slope is measured positive for elevation increasing to the NE and tilting is measured positive if the NE end moves up with increasing time. Unstable benchmarks (those that move either up or down by more than 10 mm with respect to stable benchmarks on <u>both</u> sides) were edited out. We also edited some benchmarks simply because they were too close (less than 1 km) to their neighbors. Data plotted in this way are independent of any reference benchmark. Furthermore, tilt errors do not accumulate along the profile (elevation errors do) so that the reader can mentally isolate a particular section of the profile and consider it independent of the other sections.

Letters in Figure 2 indicate the specific rod

pairs used in the surveying. The first letter gives the rod pair used on the test survey, and the second letter gives that for the earlier reference survey. A single rod pair (H) was used for the 1965 test survey. The parts of the profile over competent rock and over recent sediments are indicated at the bottom. For each year, the apparent tilt from that year to the reference year 1965 is given on top, and the local slope is given at the bottom. The local slope appears to differ from year to year only because of differences in benchmark sampling. Both the benchmark and the rod notations are explained in Table 2.

Note that the slope is positive everywhere except just south of Saugus, and between Vincent and Palmdale. The topmost diagram in Figure 2 shows the apparent tilt between 1955 and 1965. Like the slope, the apparent tilt is generally positive (NE end up) except for the 10 km region south of Saugus, and for the region between Vincent and Palmdale. The generally positive correlation is also evident for small scale features, such as the narrow region of positive tilt just north of Saugus, and the local maximum in tilt at about 80 km.

The middle diagrams in Figure 2 show tilt and slope for the interval 1961-1965. Again, the tilt is generally positive except for the regions south of Saugus and north of Vincent. Three different pairs of rods were used in 1961, as indicated by the letters B, C, and D. If there were uniform tectonic tilting during the time interval 1961-1965 one would expect to see a difference in mean tilt north and south of Saugus, because the direction taken by the leveling survey changes there (see Figure 1). While there is a slight suggestion of higher average tilting north of Saugus, this could be the result of a tilt/slope correlation, because the average slope is also greater north of Saugus. Certainly the variability of the tilt is greater to the south of Saugus. This could be caused by benchmark instability or by cultural noise in the more populous San Fernando Valley.

Apparent tilt, and slope, for the interval 1964-1965 are shown in the lower diagrams in Figure 2. Note that two rod pairs, F and G, were used in 1964, with a change occurring at Saugus. The correlation between tilt and slope is positive for the portion Burbank-Saugus, and negative for the portion Saugus-Palmdale. The specific location of the change in correlation is not clear from the diagram. It occurs near Saugus, but any correlation between tilt and slope in the Saugus basin itself is unclear. Apparent tilts are always highly variable in this region, possibly because of benchmark instabilities or cultural noise.

Scatter plots of tilt vs. slope for the profile Los Angeles-Saugus and Saugus-Palmdale are shown in Figure 3. The correlation between tilt and slope is obvious in each case. It is also clear that the average tilt in each section (the predicted tilt for zero slope) is relatively small compared to the slope correlated tilts. For the

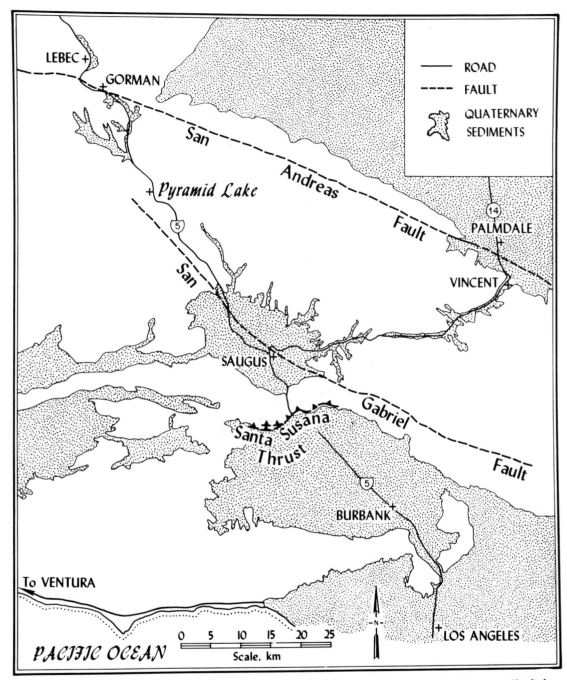

Figure 1. Map of southern California showing leveling profiles studied here. Shaded areas are covered by recent alluvium.

1955 data, the tilt/slope ratio is clearly different for the two subsets of data, even though the same rods were used in both subsets. This implies that either severe atmospheric refraction, or elevation correlated subsidence, has contributed to the apparent tilts between Burbank and Saugus. The poorest linear fit between tilt and slope is

for the case 1961a, between Saugus and Palmdale. There were two different rod pairs employed over this profile in 1961, which may explain some of the large residuals. The reversal of the sign of the tilt slope ratio at Saugus is easily seen in the comparison of cases 1964a and 1964b.

We estimated "tilt/slope" ratios and mean tilts

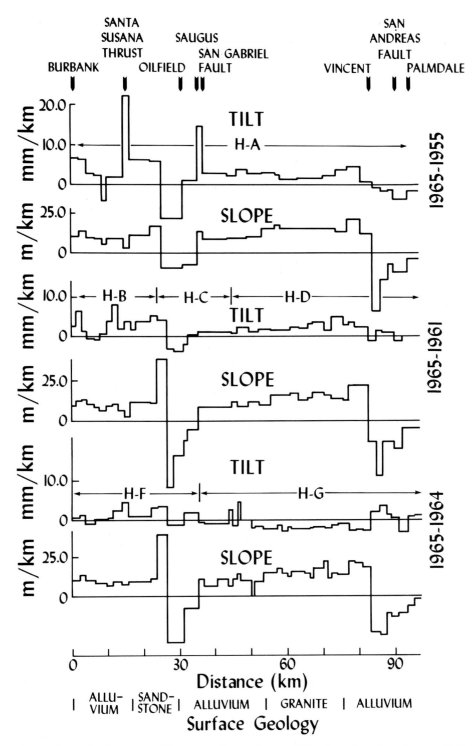

Figure 2. Tilt and slope vs. distance along the profile Los Angeles-Saugus-Palmdale. Leveling rods used are shown as upper case letters. Leveling rod codes, and specific benchmarks corresponding to place names, are shown in Table 2.

TABLE 2. Notation

Benchmarks

'Los Angeles'	=	L141		'Lang'	=	N486
'Burbank'	=	H43 RESET		'Acton'	=	A487
'Saugus'	=	J52		'Vincent'	=	3219
'Pyramid Lake'	=	G992		'Palmdale'	=	PALMDALE2
'Lebec'	=	F54				

'San Gabriel Fault' = V370 (on Saugus-Lebec profile)

'San Gabriel Fault' = Y898 (on Saugus-Palmdale profile)

'San Andreas Fault' = A54 (on Saugus-Lebec profile)

'San Andreas Fault' = J811 (on Saugus-Lebec profile)

Rods

Our notation	rod code	serial numbers	Our notation	rod code	serial numbers
A	312	368, 387	F	312	248, 254
B	312	391, 459	G	312	268, 274
C	312	301, 304	H	317	163, 263
D	312	308, 322	I	316	87815, 87859

simultaneously over the several segments of the profile using equation 11. The results are given in Table 3. For each year, we estimated two separate values of β_k, corresponding to regional tilts south and east of Saugus, respectively. We also estimated a separate value of α_j for each rod pair used in the test survey. Thus for the 1965-1955 comparison we used only one α; for 1965-1961, we used three; and for 1965-1964 we used two. For 1965-1955 and 1965-1961, all α_j were tested for significance as a group, and similarly all β_k. For 1965-1964, both the α_j and β_k were divided at Saugus, so the data sets were partitioned there as well and the parameters were tested separately. As shown in Table 3, the tilt/slope ratios are significantly different from zero at 97% confidence or greater in each case. The regional tilts are significant only for 1965-1961. As discussed above, the estimated regional tilts are highly dependent on a few observations at the margin of the Antelope Valley. If these are disregarded, the regional tilt is not significant at 90% confidence.

Apparent tilt between 1964 and 1965, and slope, are shown for the profile Los Angeles-Lebec in Figure 4. The only editing performed on these data was to delete some benchmarks that were closer than 1 km to their nearest neighbors. The apparent tilt undergoes some rapid sign reversals, such as that at 28 km, that may be caused by unstable benchmarks. Other sign reversals may be the result of correlation with the slope, which also exhibits many sign reversals, especially north of

Castaic Junction. The horizontal dashed lines show average tilt and slope for two parts of the profile. The boundary between the two parts is at Castaic Junction, north of Saugus. Rod changes took place near Pyramid Lake in 1964, and 20 km north of Pyramid Lake in 1965.

Scatter plots of tilt and slope for the sections Castaic Junction-Pyramid Lake, and Pyramid Lake-Lebec are shown in Figure 5. The estimated tilt/slope ratios are 0.5×10^{-4} and -1.3×10^{-4} respectively. As in previous examples, the correlation between tilt and slope is significant at the 95% confidence level. The regional tilt is nominally significant for the northern section, but differs substantially from north to south.

The leveling data shown in Figures 4 and 5 provide, in concise form, examples of many of the problems in leveling data for southern California. Taken at face value, these data would suggest that Lebec was uplifted about 70 mm with respect to Los Angeles, with most of the uplift occurring between Los Angeles and Saugus. However, there is no reason to expect tectonic motion on such a scale during the seismically quiet year between the two surveys. South of the San Gabriel Fault, there is evidence of height correlated errors, although this evidence is partly obscured by benchmark instabilities and the uniformity of topographic slope. It is also conceivable that local subsidence contributes to the apparent tilt. North of the San Gabriel fault, the benchmarks are primarily on bedrock, and subsidence is not likely to be important. Evidence for height related

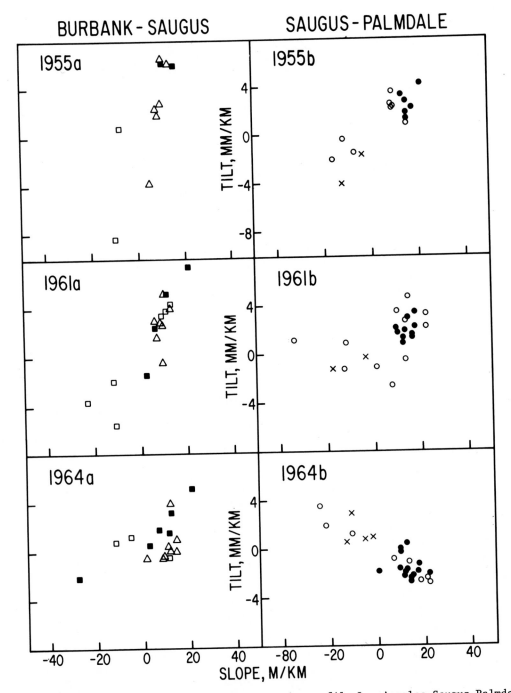

BURBANK - SAUGUS   SAUGUS - PALMDALE

Figure 3. Scatter plots of tilt and slope for the profile Los Angeles-Saugus-Palmdale: Filled symbols indicate segments for which both benchmarks are on bedrock. Symbols show locations of segments: Δ = south of Santa Susana Thrust; □ = Santa Susana Thrust to San Gabriel Fault; O = San Gabriel Fault to San Andreas Fault; × = north of San Andreas Fault, test year is 1965, and reference year is shown.

464   JACKSON ET AL.

TABLE 3.   Regression Results

Tilt/Slope Ratio

Years	Rods	Profile	Alpha, ppm	R^2	Significance, %
1965-1955	H-A	Burbank-Palmdale	170 ± 21	0.40	99
1965-1961	H-B	Burbank-Saugus	250 ± 85	.74	99
1965-1961	H-C	Burbank-Palmdale	233 ± 41	*	*
1965-1961	H-D	Saugus-Palmdale	105 ± 29	*	*
1965-1964	H-F	Burbank-Saugus	106 ± 35	0.41	97
1965-1964	H-G	Saugus-Palmdale	-109 ± 10	0.63	99
1965-1964	H-F	Saugus-Pyramid	61 ± 15	0.21	97
1965-1964	H-G, I-G	Pyramid-Lebec	-133 ± 07	0.83	99

Regional Tilt

Years	Profile	Beta, microradians	Significance, %
1965-1955	Burbank-Saugus	0.91 ± 0.44	60
1965-1955	Saugus-Palmdale	0.25 ± 0.34	45
1965-1961	Burbank-Saugus	1.25 ± 0.65	94
1965-1961	Saugus-Palmdale	0.57 ± 0.39	*
1965-1964	Burbank-Saugus	0.76 ± 0.44	75
1965-1964	Saugus-Palmdale	-0.06 ± 0.47	30
1965-1964	Saugus-Palmdale	-0.06 ± 0.47	30
1965-1964	Castaic Pyramid	-1.16 ± 0.49	95
1965-1964	Pyramid-Lebec	-0.11 ± 0.22	50

*For 1965-1961, three values of alpha were estimated and tested simultaneously, and all share the same significance. Similarly, two values of beta were estimated and tested simultaneously. R^2 is the correlation coefficient; for the 1965-1961 data, R^2 is the multiple correlation coefficient, reflecting the effect of fitting three separate coefficients.

errors in this section is overwhelming. Furthermore, as discussed below, refraction errors should not be significant on such steep slopes. The reversal in tilt/slope ratio at Pyramid, where rods were changed, would seem to implicate rod errors, at least on part of the profile. Because the rod pairs G and H were also involved in the leveling from Saugus to Palmdale in 1964 and 1965, with about the same tilt/slope ratio, it is highly likely that one or both of these rod pairs were miscalibrated.

We have examined other data with mixed results. The profile from Saugus-Palmdale was leveled in 1971 (after the San Fernando earthquake) and in 1978. We have compared both of these surveys to 1965. The 1971 results are stongly dominated by the San Fernando earthquake, and it is therefore difficult to assess the presence or absence of height dependent errors. The 1978 survey was done as part of the cooperative Southern California Releveling Project (SCRP), and many different rod pairs were used. We found no statistically significant correlation between slope and tilt over the whole profile Saugus-Palmdale, and dividing the profile according to rod usage left too few data in each subset. An alternate (northern) route

between Saugus and Palmdale was surveyed in 1968 and 1973. The occurrence of the San Fernando earthquake between these two surveys makes it difficult to confirm or reject any tilt/slope correlation.

We have examined data for the Saugus-Lebec profile for 1953, 1968, 1974 and 1978. We see no significant regional tilting in the time periods 1953-1964, 1964-1965, 1964-1968, or 1964-1974, for the section of the profile north of the San Gabriel Fault. Apparent tilts are large and apparently random between Saugus and the San Gabriel Fault, possibly because of benchmark instabilities or the effects of fluid extraction in the Saugus basin. The Saugus-Lebec data are also discussed by Stein (1980, 1981), whose conclusions differ somewhat from ours. Stein applies a correction to remove the tilt/slope correlation in those sections where it is statistically significant. After applying this correction, there are generally residual tilts, primarily concentrated in the sections where no correction was applied. The largest residual tilt is in the region south of the San Gabriel Fault in the interval 1953-1964. We believe that the residual tilting is due to subsidence in the Saugus basin, as proposed by

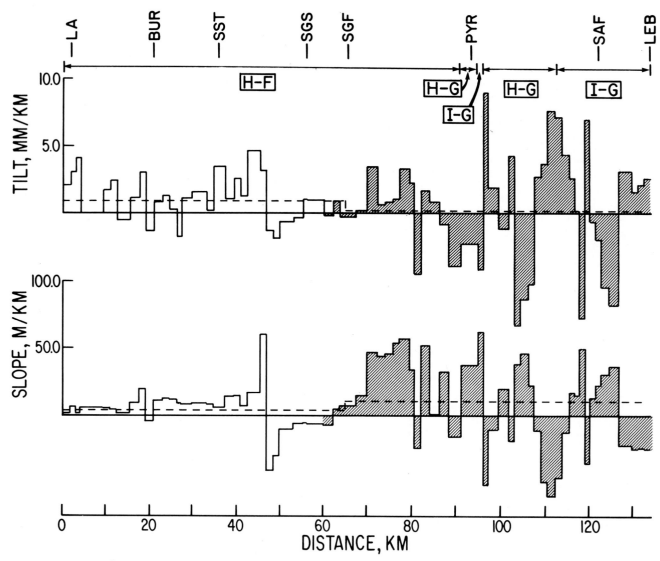

Figure 4. Tilt and slope vs. distance along the profile Los Angeles-Saugus-Lebec.
Surface geology is bedrock between Castaic Junction and Gorman. LA = Los Angeles,
BUR = Burbank, SST = Santa Susana Thrust, SGS = Saugus, SGF = San Gabriel Fault, PYR =
Pyramid Lake, SAF = San Andreas Fault, LEB = Lebec.

Reilinger (1980), rather than tectonic uplift of
Lebec.

We examined data for the coastal profile from
Santa Monica to a point north of Ventura. This
profile is subject to subsidence (and uplift?)
from fluid extraction and injection near Ventura.
Apparent tilts were large and highly variable;
the apparent regional tilting (up to the north)
was within the uncertainties implied by the large
scatter in the data.

Interpretation

The smooth contours of aseismic uplift reported
for southern California (e.g. Castle et al., 1976,

fig. 2) would imply nearly uniform tilting over
profiles of 50-100 km length as considered here.
We have found that the apparent uplift is not well
represented by uniform tilting, but is rather
strongly correlated with topographic slope. In
this section we discuss possible causes for the
observed tilt/slope correlation.

It is certainly possible that tectonic tilting
would be correlated with slope. On the basis of
geological information, it is known that the
Transverse Ranges are currently being uplifted
(e.g., Yeats et al., 1977), and it is reasonable
to propose that the higher peaks are being uplifted faster. This would imply a correlation
between tilt and slope. However, tilting from

tectonic causes should be relatively smooth on a scale comparable to the minimum depth at which tectonic activity occurs. In the case of aseismic displacement on a fault surface, this depth would be that of the top of the displaced part of the fault. The data of figures 2 and 4 show that significant variations in apparent tilt take place with wavelengths of only a few km, and that these variations are relatively coherent with the slope over distances of 30-50 km. Thus a tectonic model of these apparent tilts would require some sort of stress variation that is correlated with topography, occurs at shallow depths, is spatially consistent over 30-50 km intervals, yet which can reverse its sign in time periods of a few years or less. One frequently suggested model is that local tilt variations occur near faults or other crustal inhomogeneities in response to variations in a regional stress field. This model fails to account for the large magnitude of the apparent tilts, or for the sudden reversals of the tilt/slope ratio between 1964 and 1965 observed on two different profiles.

Another possibility is that the local, short wavelength correlation between tilt and slope is caused by the effects of fluid extraction (or injection). As fluids are extracted, sediment compaction would be expected. This would serve to increase the height difference between higher benchmarks on hard rock and lower benchmarks on sediments, resulting in tilting that is correlated with slope. Ideally we would like to use only benchmarks on solid rock, far removed from exploited aquifers and oilfields. In fact, many of the important leveling routes follow roads or railroad tracks along canyons and river valleys, and it is difficult to dismiss artificial subsidence anywhere that there are recent sediments.

Subsidence within a sedimentary basin between two successive surveys should result in negative tilts on the southern margin of the basin, and positive tilts on the northern margin, for the sign conventions used in Figures 2 and 4. It is very likely that fluid extraction does contribute significant tilting in the Los Angeles and Saugus basins, and in the San Fernando and Antelope valleys. It is also possible that fluid extraction may cause some of the correlation between tilt and slope, especially in the data of Figure 2. Subsidence could explain the large negative tilts between south of Saugus and Palmdale for the intervals 1955-1965 and 1961-1965. However, subsidence does not easily explain the close agreement between the tilt/slope ratio obtained from the linear regression, and that obtained from the regional averages of tilt and slope. (The fact that the "regional tilt", β, was usually not significantly different from zero indicates that the short wavelength correlation explained the average tilting over the profile as well.) The subsidence model does not explain the reversal in sign of the correlation between slope and tilt, both at Saugus and at Pyramid, between 1964 and 1965. Furthermore, subsidence does not explain

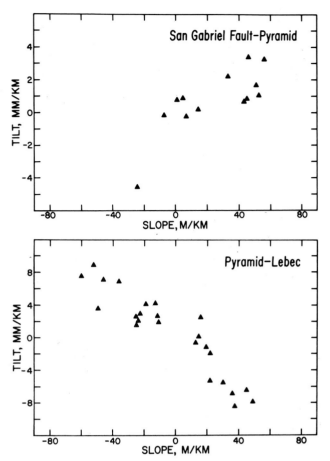

Figure 5. Scatter plots of tilt vs. slope for: (a) Castaic Junction-Pyramid, 1965-1964, and (b) Pyramid-Lebec, 1965-1964.

the excellent correlation between tilt and slope in the region from the San Gabriel Fault to Lebec (Figure 4), where the leveling route is primarily on bedrock and very little fluid extraction takes place. Thus the effects of fluid extraction may contribute to the tilt/slope correlation, but they in no way provide a complete explanation for the observed correlation. The major cause of the effect must be sought elsewhere.

In studying other possible sources for the tilt/slope correlation, it would of course be desirable to isolate the possible effects of subsidence. One could do this either by attempting to correct for subsidence using observed fluid extraction rates or by avoiding altogether any region where subsidence could be a problem. Neither approach is totally satisfactory. Adequate modeling of the subsidence due to fluid withdrawal would require better geological models and more complete fluid production data than are now available. Total avoidance of areas subject to fluid extraction leaves very few data to deal with satisfactorily. We have thus adopted a modified version of the total avoidance porcedure. Where adequate data exist in regions safe from fluid extraction, we avoid oil fields and aquifers altogether. This situation applied to the Saugus-Lebec profile (Figure 4) where the Castaic Junction-Lebec portion is relatively safe. In other areas, we neglect subsidence, estimate tilt/slope ratio (α) and regional tilt (β) using all available data, and then examine residual tilts for evidence of fluid extraction. If fluid extraction were dominating the tilt/slope correlation, one would expect residual tilts to be undercorrected in regions of subsidence and overcorrected elsewhere. We found this behavior in the 1955-1965 data for Burbank-Palmdale, but not in other data.

The other explanations for the tilt/slope correlation that come to mind involve systematic errors in the leveling. One of the most widely feared sources of systematic error is atmospheric refraction. The velocity of light in air depends primarily on the temperature, and to a smaller extent on pressure and humidity. Light rays will be curved upward if the temperature decreases with elevation, as it usually does.

Several mathematical models of the effects of refraction are described and tested against recent experimental data by Whalen (1980). One of the more successful models, and one which has guided much of the recent thinking about refraction errors, is based on the pioneering work of Kukkamaki (1939). According to this model, a temperature gradient in the atmosphere causes curvature of the line of sight, which is proportional to the temperature gradient. Thus for a forward sighting, for example, the difference between the elevation of the leveling instrument, and the position sighted on the rod, is

$$e_f = A \left(\overline{\frac{dT}{dz}}\right)_f L_f^2$$

where A is a constant, $\left(\overline{dT/dz}\right)_f$ is the average vertical temperature gradient along the forward line of sight, and L_f is the length of the forward sighting. The backward sighting would give a similar formula, but with subscripts "b". The error in elevation caused by differencing the backward and forward sighting is then

$$e = A \left[\left(\overline{\frac{dT}{dz}}\right)_b L_b^2 - \left(\overline{\frac{dT}{dz}}\right)_f L_f^2 \right]$$

It is common practice to make foresights and backsights as nearly equal in length as possible, so that

$$e = AL^2 \left[\left(\overline{\frac{dT}{dz}}\right)_b - \left(\frac{dT}{dz}\right)_f \right]$$

where $L = L_f = L_b$. Assuming that the temperature gradient varies linearly with elevation above the ground, then $(dT/dz)_b - (dT/dz)_f = sL \, d^2T/dz$ where d^2T/dz^2 is assumed constant and s is the topographic slope. If it is further assumed that $d^2T/dz^2 \approx (dT/dz)_{z_0}$, where z_0 is a standard height such as 1 m, then the elevation error

due to unequal refraction between the foresight and backsight becomes

$$e = AL^3 s \, (dT/dz)_{z_0} \qquad (12)$$

The error in determining the slope is then $e/L = AL^2 s \, (dT/dz)_{z_0}$ and the relative error is

$$\alpha = AL^2 \left(\frac{dT}{dz}\right)_{z_0} \qquad (13)$$

One of the most important features of (13) is the strong dependence on sight length. While many overly simple assumptions were made in deriving this equation (see Whalen, 1980), the dependence on L^2 is fundamental. This is because the light ray is constrained to be horizontal at the eyepiece of the leveling instrument, so the path of a curved ray must be represented by a quadratic or higher order function in distance from the instrument. The average sight length for surveys in the U.S. is known to have shortened with time, especially in the early 1960's. This change took place partly because of recognition of the refraction problem, and partly because it was required by changes in instrumentation. This change in sight length has been invoked to explain the systematic discrepancies between younger and older leveling data (Reilinger, personal communication, 1980; Strange, 1980).

For very steep slopes, the sight length is limited by the length of the leveling rod. If the forward and backward rods are to be simultaneously visible, we must have $\Delta h = 2L \cdot s \le \ell$, where ℓ is the rod length. When the slope is greater than 5%, the sight length is limited to 25 m for rods with 2.5 m useable length. At this sight length, refraction effects are expected to be negligible (that is, $A \, dT/dz \, L^2 < 10^{-5}$) for typical values of A and dT/dz. Thus on very steep slopes refraction is thought to be no problem because of the short sight length required. The linear dependence of apparent tilt on slope would then hold only for small slopes (say up to 2% or 3%).

Based on the known changes in sight lengths with time, and estimated values of A and dT/dz (Strange, 1980; Holdahl, 1980; Whalen, 1980) we expect that refraction errors of at least $3 \times 10^{-5} \times$ slope must exist in the southern California data, and they could be somewhat larger. Such errors could cause some of the observed correlation between tilt and slope. However, we do not find convincing direct evidence of refraction errors, nor can refraction explain some of the important features of the data. We do not see direct evidence for "saturation" of the tilt/slope correlation for high slopes, as predicted by the refraction model. In fact, some of the most striking evidence for tilt/slope correlation occurs between Saugus and Lebec (see Figure 4), where the terrain is rugged and the slopes are quite high (greater than 5% in some places). We do find suggestive evidence that differential refraction contributes some part of the tilt/slope correlation. In many of the scatter diagrams of Figure 3, the ratio of apparent tilt to slope is smaller in magnitude for negative slopes than for positive slopes. By definition, negative slopes are facing northward or northeastward, so that the flux of sunlight and the thermal gradient in the air will be less than for areas of positive slope. Thus the effects of refraction should be smaller in the areas of negative slope. There are, however, other explanations for the observed variation in tilt/slope ratio. With few exceptions, the negative slopes occur at the southern margin of the Saugus basin, and within a few km of the San Andreas fault. The southern margin of the Saugus basin is the site of a producing oilfield, and the rocks near the San Andreas fault are notoriously unstable, as discussed above.

Refraction errors do not provide an acceptable explanation of the reversals in tilt/slope ratio at Saugus and at Pyramid between 1964 and 1965. Because refraction errors depend both on the sighting length and the temperature gradient in equation (13), it is possible in principle for refraction errors to cause a reversal in tilt/slope ratio if either the sight lengths or the temperature gradient reversed in one survey, relative to the other. Data on actual sight lengths are not readily available, but the average sight lengths were nearly equal for the 1964 and 1965 surveys. Furthermore, the sight lengths will be limited by the topography over most of the Saugus-Lebec profile, so a significant reversal in relative sight lengths is unlikely there. We don't have dT/dz data for these surveys, but we do have data on average temperatures at about 1 m elevation on the rods. There were changes in the average temperatures as a function of distance along the profile, but these changes appear quite random, and they do not mimic the change in tilt/slope ratio. The fact that the reversals in tilt/slope ratio occurred at points where equipment changes were made seems to implicate the equipment itself.

As far as we know, the only parts of the equipment likely to produce elevation correlated errors are the leveling rods. If these are not exactly the specified length, and if proper calibration corrections are not made, then all measured elevation differences will be in error. Calibration errors must account for some of the observed tilt/slope correlation. The magnitude of the tilt/slope correlation implies that the calibration error must be on the order of one part in 10^4 for the profiles we have studied.

Examination of rod calibration data reveals that there are many differences between the procedures recommended in Bomford (1971) and actual practice (Kumar and Poetzschke, 1980). One problem is that many of the rods were calibrated infrequently, and there are in many cases significant changes (greater than 100 ppm) between calibrations. A much more serious problem is that for Fischer rods the calibration measurements were made between scribed markings on the invar strip, which were different from the painted markings read by the observer in the field. The scribed markings were at the nominal distances of 0.2, 1.2, 2.2, and 3.2 m along the rod. The painted markings were sprayed on using a 1 m long invar mask that was aligned with the inscribed marks using a 'special comparator'. The agreement between the painted marks and the inscribed marks is presumed to have been generally within 100 microns, but apparently no test of the presumption was recorded in practice. Even if the mask were in error by less than 100 μm, this error would tend to be concentrated near the scribe marks and could cause substantial error accumulation in the leveling (possibly 100 μm on each sighting). The Fischer rods were used almost exclusively by the National Geodetic Survey from 1927 through 1964, when NGS began to phase them out in favor of Kern rods. Thus the rod calibrations are effectively unknown for NGS data taken prior to 1965, and for some data taken after 1965. For rods acquired after 1965, the calibration procedures improved substantially. However, the calibration procedure in use at the present time is still inadequate because it assumes that any distortions are uniformly distributed throughout the rod. Even if the calibration measurements are error free, nonuniform rod graduation can lead to serious systematic errors in the leveling data.

To correct for rod distortions, observed elevation differences are generally multiplied by a factor $(1 + \varepsilon)$, where ε is a constant derived from calibrating the rod. The factor ε is commonly known as the "rod excess", or "excess length". In practice the rod is calibrated over several different intervals, and a weighted average of the relative length deviations is used as the value of ε. Possible calibration errors could arise in at least two ways. First, the calibration measurements themselves could be in error, or the length of the rod could change after it has been calibrated, so that the "excess length" may not represent the true deviations. Second, the rod errors may not be uniformly distributed along the length of the rod, so that the weighted average used in the data reduction may not be appropriate. On very steep topography, sightings tend to be on the top and bottom of the rod, so that this second type of rod error should not be excessive. However, on gentle topography, sightings tend to be concentrated near the center of the rod, where the relative rod error may be quite different from the average. The first type of rod error can only be directly detected by frequent recalibration of rods, using independent standards. The second type may be assessed, at least approximately, from the calibration data themselves. We outline below a method for estimating the relative errors caused by nonlinear distortions. While the errors are substantial (on the order of 50 ppm) they do not explain the magnitude of the observed tilt/slope ratio (sometimes greater than 100 ppm). We conclude that either substantial biases exist in the calibration data themselves, or that the rods are changing in length after they are calibrated.

Many of the rods are calibrated over three different intervals, and variations in the relative errors over the different intervals are sometimes as large as 100 ppm. For instance, rod 312-274, one of the rods in pair "G", was calibrated in 1965. The length deviations were 20 ppm in the lower portion of the rod, 50 ppm in the middle of the rod, and -110 ppm in the top portion of the rod. The calculated excess length was only 3.7 ppm. Consider "rod strain" $e(x)$ to be the derivative of the length deviation $y(x) - x$, where x is a nominal position on the rod and $y(x)$ is the true position as measured in a calibration test. Then the excess length, and relative deviations for the portions of the rod, are each weighted averages of the rod strain. In general, we should expect the magnitude of the average to decrease, as the "width" of the weighting function increases, and this is generally observed in the calibration data. The excess length (a 3 m average) is smaller in magnitude than the 1 m averages. The relative height error due to rod distortion may also be looked at as a weighted average of the rod strain $e(x)$. We again expect that the relative errors will decrease as more of the rod is used, so that the weighting function has greater width.

Quantitative estimates of the weighting function $g(x)$ used in estimating the excess length may be calculated from an analysis of the rod calibration procedure used by the National Geodetic Survey in reducing the data. Quantitative estimates of the appropriate weighting function $f(x)$ for relative height errors can be obtained from an analysis of the actual sightings on the rods in a field survey. The weighting function $f(x)$ will naturally depend on the terrain; $f(x)$ will have greater width for steep topography, where sightings will be at the top and bottom of the rod. Plots of the weighting function $f(x)$ derived from actual field notes, are shown in Figure 6 for several types of terrain. The data come from a survey on California Highway 39 north of Azusa. Nearly level means that the topographic slope was less than 20 m/km, and steep mean greater than 40 m/km.

The residual error, after correcting for average rod distortion, is again a weighted average of the rod strain. Here the appropriate weighting function is the difference $f(x) - g(x)$. As seen in Figure 6, this difference may be very large for nearly level surveys, but becomes smaller for steep surveys. We can make a rough estimate of the expected residual error by assuming a given function $e(x)$, and computing the indicated average. This is equivalent to making a computer simulation of a leveling survey, using a hypothetical set of rods whose strain functions, averaged for the two rods, are the specified function $e(x)$. We derived the hypothetical function $e(x)$ from the rod calibration data, assuming either that the measured rod deviation was uniformly distributed between the calibration points, or that it was concentrated in the upper 10 mm of each interval between cal-

ibration points. The latter assumption would be appropriate for Fischer rods, if the mask were perfect, and if it were registered perfectly at the lower scribe mark in each interval. Results showed that the residual rod errors would be most significant for nearly level slopes, as indicated by Figure 6. The residual error could easily exceed 100 ppm for nearly level slopes, but would only rarely exceed 50 ppm for steep slopes. The infamous rod 312-268, cited by Stein (1981) as a worst case, yielded relative errors of 332 ppm for nearly level slopes, but only -52 ppm for a moderate slope and 20 ppm for a steep slope, assuming that rod strain was concentrated at the tops of the calibrated intervals. Assumptions about the distribution of strain within the calibrated intervals were significant for individual rods, especially for nearly level surveys, but did not change the general conclusions about the magnitudes of expected errors.

The changes in tilt/slope ratios at Saugus and Pyramid in the interval 1964-1965 provide nearly direct evidence that rod errors as large as 100 ppm exist, even on steep topography. Less dramatic but statistically significant changes, such as that between rod paris C and D in Figure 2 and Table 3, also suggest rod errors. The similarity of values of the tilt/slope ratio for separate comparisons of rod combinations H-G and H-F in Table 3 lends further support. Refraction, as currently understood, cannot explain the reversals in tilt/slope ratio, and neither subsidence nor refraction should be significant on the mountainous section of the Saugus-Lebec profile. While one of the rods of pair G is known from the calibration data to be highly nonuniform, a direct accounting of the nonuniformity does not explain the magnitude of the observed tilt/slope ratio. Furthermore, the expected decrease in the magnitude of the tilt/slope ratio for steep topography is not observed. Thus, while nonuniformity may contribute significant errors, it is apparently not the whole story. The true cause of the tilt/slope correlation is unknown, although a bias in the calibration data themselves, physical changes in the rods after they are calibrated, or large errors in the template used to paint the Fischer rods, may contribute.

It has been argued (Stein 1980; Strange 1980) that one rod with extremely nonuniform distortion (rod 312-268) has introduced some systematic errors, but that this rod is unique and the rod calibration problem is not widespread. In fact, rod pair G (NGS rods 312-268 and 312-274) were involved in the 1964 leveling from Saugus-Palmdale and from Pyramid to Lebec. These rods have also been used in some levelings that show large loop misclosures. They have apparently been raising hell everywhere that they have gone. The rod calibration data show that these rods were highly nonuniform, although other rods used in southern California were nearly as bad. However, there was apparently something wrong with rod pair G that is not explained by the rod calibration data,

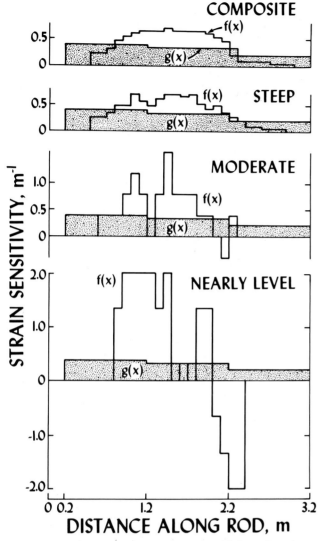

COMPOSITE

STEEP

MODERATE

NEARLY LEVEL

STRAIN SENSITIVITY, m⁻¹

DISTANCE ALONG ROD, m

Figure 6. Rod usage function $f(x)$ and rod calibration function $g(x)$ for three parts of a leveling profile in the San Gabriel Mountains of southern California. Nearly level means that the slope was less than 0.02; steep means that the average slope was greater than 0.04.

and other rods may be expected to have the same problem, whatever it is.

We conclude that the method used to correct for rod distortion is inadequate, and that uncorrected errors on the order of 10^{-4} times the true height difference remain in the published leveling data for southern California. The same calibration procedures are used for other U.S. data as well. Similar methods are used throughout the world except for a few special studies (e.g. Mälzer and Schlemmer, 1975) where rods are calibrated at each graduation (usually each 10 mm). We suggest that rod miscalibration may cause significant errors

in much of the existing leveling data, and that interpretation of these data in terms of tectonic motion should be made with great care. Plots of apparent tilt and slope, as we have shown here, are helpful in recognizing the presence of such errors, but the tilt/slope correlation may often be hidden in the presence of local subsidence, benchmark instabilities, large random errors, or true tectonic motion.

Conclusions

Leveling data for southern California show apparent tilts that are strongly correlated with topographic slope. The correlation occurs both at long wavelengths (greater than 50 km) and at short wavelengths (as short as a few km). The ratio of apparent tilt to slope is on the order of 10^{-4}, about the same as the ratio of apparent uplift to height in the previously reported aseismic uplift. Possible causes for the observed correlation include tectonic motion, subsidence from fluid extraction, refraction errors, and rod calibration errors, or some combination of these effects. Abrupt changes in the tilt/slope ratio at locations where rods were changed strongly suggests that rod calibration errors are responsible for the correlation. Additional evidence comes from the strong tilt/slope correlation observed between Castaic Junction and Lebec, where subsidence is improbable and refraction should be minimized by short sight lengths. Examination of rod calibration data show that nonuniformities in rod length deviations may contribute a substantial error, but that this error does not explain the observed tilt/slope correlations.

We have examined data for Los Angeles-Saugus, Saugus-Palmdale, Saugus-Lebec, and Santa Monica-Ventura. In no case do we find statistically significant evidence for regional tectonic tilting in the interval 1960-1970. There are profiles over which tectonic tilting cannot be excluded. In these cases there is insufficient evidence to distinguish tectonic tilting from systematic errors.

Acknowledgments. We thank R.S. Stein, W.E. Strange, and W.R. Thatcher for conceptual contributions. We thank M. Kumar of the National Geodetic Survey, and R.J. Mitchell of the County of Los Angeles, for providing the data used here. This work was performed under USGS contract USDI-14-08-0001-17687, and NSF Grant EAR-80-08288.

References

Bendefy, G., Studies on recent crustal movements, Second Symposium on Recent Crustal Movements, Aulanko, Finland, 5-15, 1969.
Bomford, G., Geodesy, Oxford University Press, London, 1971.

Brown, L.D., and Oliver, J.E., Vertical crustal movements from leveling data and their relation to geologic structure in the eastern United States, Rev. Geophys. Space Phys., 14, 13-35, 1976.

Brown, L.D., Reilinger, R.E., and Citron, G.P., Recent vertical crustal movements in the U.S.: evidence from precise leveling, in Earth Rheology and Late Cenozoic Isostatic Movements, John Wiley, New York, in press.

Castle, R.O., Church, J.P., Elliott, M.R., Aseismic uplift in southern California, Science, 192, 251-254, 1976.

Castle, R.O., Elliott, M.R., and Wood, S.H., The southern California uplift (Abstract), EOS, Trans. Amer. Geophys. Union, 58, 495, 1977.

Castle, R.O., Leveling surveys and the southern California uplift, Earthquake Information Bull., 10, 88-92, 1978.

Citron, G.P., and Brown, L.D., Recent vertical crustal movements from precise leveling surveys in the Blue Ridge and Piedmont Provinces, North Carolina and Georgia, Tectonophysics, 52, 223-236, 1979.

Draper, N.R., and Smith, H., Applied Regression Analysis, John Wiley and Sons, New York, 1966.

Holdahl, S.R., A model of temperature stratification, in Proceedings, Second International Symposium on Problems Related to Redefinition of North American Vertical Geodetic Networks, Ottawa, Canada, 647-676, 1980.

Jackson, D.D., and Lee, W.B., The Palmdale Bulge - An alternate interpretation, EOS, Trans. Amer. Geophys. Un., 60, 810, 1979.

Jackson, D.D., Lee, W.B., and Liu, C.C., Aseismic uplift in southern California: an alternate interpretation, Science, 210, 534-536, 1980.

Karcz, E., and Kafri, U., Geodetic evidence of recent crustal movements in the Negev, southern Israel, J. Geophys. Res., 76, 8056-8065, 1971.

Kukkamäki, T.J., Formulas and tables for calculation of leveling refraction, Finnish Geodetic Institute, Report no.27, Helsinki, Finland, 1939.

Kumar, M., and Poetzschke, H., Leveling rod calibration procedures at the National Geodetic Survey Draft, NOAA Technical Memorandum National Geodetic Survey, Rockville, Md., 20852, 1980.

Mälzer, H., and Schlemmer, H., Geodetic measurements and recent crustal movements in the southern upper Rhinegraben, Tectonophysics, 29, 275-282, 1975.

Reilinger, R., Vertical crustal movements from repeated leveling data in the Great Basin of Nevada and Western Utah (Abstract), EOS, Trans. Amer. Geophys. Un., 58, 1238, 1978.

Reilinger, R., Elevation changes near the San Gabriel Fault, southern California, Geophys. Res. Lett., 11, 1017-1019, 1980.

Stein, R.S., Contemporary and quaternary deformation in the Transverse Ranges of southern California, Ph.D. Thesis, Department of Geology, Stanford University, Stanford, California, 1980.

Stein, R., Role of elevation-dependent errors on the accuracy of geodetic leveling in the southern California uplift, J. Geophys. Res., this volume.

Strange, W.E., The effect of systematic errors on geodynamic analysis, in Proceedings, Second International Symposium on Problems Related to the Redefinition of North American Vertical Geodetic Networks, Ottawa, Canada, 705-727, 1980.

Strange, W.E., The impact of refraction correction on leveling interpretations in southern California, J. Geophys. Res., in press.

Thatcher, W.R., Crustal deformation studies and prediction research, J. Geophys. Res., this volume, 1980.

Whalen, C.T., Refraction errors in leveling - NGS test results, in Proceedings, Second International Symposium on Problems Related to the Redefinition of North American Vertical Geodetic Networks, Ottawa, Canada, 757-782, 1980.

Yeats, R.S., High rates of vertical crustal movement near Ventura, California, Science, 196, 295-298, 1977.

A SEARCH FOR LONG-TERM EARTHQUAKE PRECURSORS IN GRAVITY DATA

IN THE CHARLEVOIX REGION, QUEBEC

A. Lambert and J. O. Liard

Gravity and Geodynamics Division, Earth Physics Branch
Department of Energy, Mines and Resources
Ottawa, Canada K1A 0Y3

Abstract. A precise gravity network comprising fifteen stations located in the seismically active region of Charlevoix, Quebec was surveyed eight times from 1976 to 1979 to search for evidence of crustal deformation. A LaCoste and Romberg model D gravimeter used on all surveys over the four year period of the experiment provided data on temporal variations in gravity to an estimated accuracy of 4 µGal with respect to the mean gravity of the network. The surveys revealed temporal variations in station values of the order of 10 µGal. A simple uniform-absorption model for the gravity effect of ground water helped to identify attraction effects at one station but did not explain suspected groundwater effects at another station. A gravity anomaly that could not be explained by groundwater effects was tentatively identified in the data of June 1977. This gravity anomaly together with available data on seismic travel-time variations, geodetic leveling and microseismicity is consistent with the occurrence of a zone of dilatancy beneath the north-shore of the St. Lawrence river between Baie-St-Paul and Cap-aux-Oies from late 1976 to mid 1977 prior to unusually high levels of microseismicity. However, there was no correspondence between gravity variations and individual earthquakes in the period 1976 to 1979 that was consistent with current earthquake prediction models.

Introduction

The study described in this paper is part of a research project of the Earth Physics Branch that involves the monitoring of temporal variations in gravity, seismic velocity, electromagnetic impedance, tilt and strain at Charlevoix, Quebec. The project aims at a better understanding of the earthquake generating processes at the fundamental level.

In 1976 a fifteen station gravity network was established along the north shore of the St. Lawrence river for the purpose of monitoring possible crustal deformation associated with the seismicity of the Charlevoix seismic zone. Seismic array studies (Leblanc et al., 1973, Leblanc and Buchbinder, 1977, and Anglin and Buchbinder, 1981) have delineated a well defined zone of microearthquakes (Figure 1) restricted to the Precambrian rocks at depths from 5 to 25 km. Since the year 1650 five earthquakes of magnitude 6.0 or greater have been associated with this zone: the latest event in 1925 having a magnitude of 6.9 (Basham et al., 1979). A study of the epicenters since 1925 (Stevens, 1980) shows that larger events tend to occur at the ends of the zone. The geological and structural setting of the region are summarized by Buchbinder (1981, this volume).

In recent years repeated gravity surveys have been exploited in studies of earthquake processes in highly active areas (e.g. Kisslinger, 1975; Hagiwara et al., 1977; Chen et al., 1979). At Charlevoix, although activity is confined to a relatively small area, only earthquakes of magnitude less than 4.0 can be expected to occur regularly every year. Consequently, high accuracy was emphasized from the beginning through careful procedures and strong network structure. Unfortunately, to achieve the desired accuracy (3-4 µGal, or 30-40 nm/s^2) with available resources, sampling frequency was limited to two surveys per year. This is the major weakness of the experiment.

In this paper gravity data are analyzed with a view to distinguishing variations due to crustal processes from those due to superficial effects. Possible relationships between gravity variations and seismicity are investigated through comparison with other deformation data that provide similar spatial coverage: seismic velocity data (Buchbinder, 1981, this volume) and leveling data. The present paper is not intended to be a comprehensive analysis of all available data from Charlevoix.

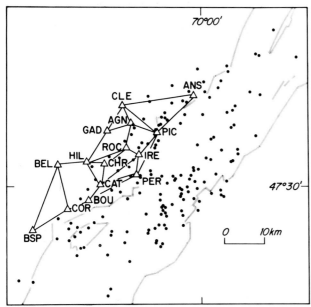

Fig. 1. Location and structure of the
Charlevoix precise gravity network.
Triangles denote all stations except
auxiliary station AER which is located just
north of IRE. Solid lines between stations
represent direct connections comprising an
average of eight consecutive gravity ties on
each survey. Small dots denote epicenters
of microearthquakes located by seismic array
studies in 1974 (Leblanc and Buchbinder,
1977) and since late 1977 (Anglin and
Buchbinder, 1981).

Experimental method and treatment of data

Network surveys

The Charlevoix precise gravity network (Figure
1) was surveyed at roughly six month intervals
from spring 1976 to autumn 1979 using LaCoste
and Romberg (LCR) model D gravimeters.

All stations in the gravity network were
established on what was judged to be bedrock;
later investigations have shown that two
stations (BOU and CAT) are probably sited on
large buried bolders. Each station in this
network consists of two instrument pads set
about 0.5 m apart to allow reoccupation of the
station to within 1 mm by two instruments
simultaneously. Each pad consists of three
brass pins cemented into the surface of the
rock. On each resurvey a particular pad at each
station was always reoccupied by gravimeter LCR
D6. The other pad was occupied by one of LCR
D13, D27 or D28. Thus, a two-instrument
resurvey of the network is equivalent to the
resurvey of two independent but colocated
networks. In October 1977 these two networks
were tied together by measuring ten times with

LCR D6 the small (~ 10 µGal) gravity difference
between the two pads at each station.

The network was surveyed using the "reset"
technique where the dial readings of a
particular instrument are reset to within about
150 µGal of the values in the initial survey by
means of a reset screw provided on the LCR model
D gravimeter. With this technique the unknown
non-linearities in gravimeter calibration
represent constant systematic errors that can be
neglected in an analysis of temporal variations
as long as the same instrument is used from
survey to survey.

Further details of field procedures are given
in Dragert et al., 1981.

Instrument Calibration

In general there is a nonlinear relationship
between gravity (g) and the dial reading (R) on
an LCR model D gravimeter. This relationship is
expressed in terms of a scale factor and an
interval factor curve which describes the
variation $\Delta g/\Delta R$ across the range of the
gravimeter. Interval factor curves were
determined for all four LCR model D gravimeters
used in the Charlevoix experiment by means of a
laboratory technique called Cloudcroft Junior
devised by the manufacturer. The accuracy of
the laboratory interval factor curves for the
four D meters can be tested by solving for
relative interval factor curves in a least
squares adjustment involving many thousands of
observations from different networks over a
period of five years. Preliminary results show
that significant relative variations among
results for different gravimeters still exist.
This suggests that the laboratory method of
adding a small calibrated weight to the
gravimeter beam does not exactly simulate real
changes in gravity. Thus, the true interval
factor curves for our D meters remain unknown at
the µGal level and only relative interval factor
curves can be determined currently.

In the present paper only the observations
made with gravimeter D6 are used. Since the
results from this instrument are available
throughout the four year period of the
experiment, unknown calibration nonlinearities
which are assumed to remain constant can be
neglected in an analysis for temporal
variations. The same principle cannot be
applied to the second gravimeter used in each
survey, since due to malfunctions, the same
second instrument was used in consecutive
surveys only twice. Computations of relative
interval factor curves for LCR gravimeters D13,
D27 and D28 which would enable a second
gravimeter to be included in each survey in a
combined solution with D6 are not yet complete.
Until the combined solutions become available
the results for gravimeter D6 alone must be
considered preliminary.

In the analysis of D6 data it is assumed that

no temporal changes in the unknown interval factor curve of D6 occurred over the period 1976 to 1979. The stability of the calibration of D6 can be verified by observing that there is no tendency for changes in gravity difference in networks observed with D6 to be proportional to the gravity differences themselves. Over the four year period of the experiment the variation in the weighted mean gravity difference between all directly connected pairs of stations was found to be less than one part in ten thousand. Thus, variations in station values due to scale factor changes should be less than 2 μGal at Charlevoix. Two-point calibration ranges on Mount Seymour, British Columbia and Mont. Ste Marie, Quebec which were originally intended to provide independent scale factor control for D-meters have proved to be unsuitable and have now been replaced by a multi-point range.

Calculation of gravity variations

For each of eight surveys of the network of Figure 1 a least-squares adjustment was carried out involving about two hundred fifty gravity observations obtained over a period of two to three weeks. Fifteen station values and an average of five linear drift terms were calculated in each adjustment. In order to obtain gravity values and error estimates independent of spatial origin, the network was solved with respect to the mean gravity of all the stations. This scheme recognizes the mean gravity as the most objective datum for the observations in the absence of an absolute reference point or repeated gravity ties to an external network. Furthermore, the error estimates for station values with respect to the spatial mean gravity are relatively uncorrelated compared to estimates with respect to a single reference station. The significance of spatial and temporal changes of gravity is, therefore, more easily determined. The precision of the least squares network solution for each survey can be evaluated from the distribution of residuals formed by subtracting the solution gravity differences from the observed gravity differences. The standard deviation of unit weight of the residuals for the eight surveys varied from 4.1 μGal to 7.9 μGal with a mean of 6.4 μGal. Standard errors on station values estimated from the least squares fitting procedure range from 2 to 3 μGal. Due to the network structure, gravity values for stations around the outer periphery of the network are less well determined than are the stations in the interior.

In spite of the small magnitude of the error on station values quoted above for a single-instrument adjustment, comparisons of adjusted station values for different instruments show larger disagreement than would be expected. Preliminary results of network adjustments involving two different gravimeters

employed simultaneously at Charlevoix show that 4 μGal is a more realistic estimate of the standard error on station values, although later use of suitable relative calibration curves is expected to improve this. Similarly, comparisons of temporal variations for two different D meters on Vancouver Island (Dragert et al., 1981) suggest that standard errors of 3.5 μGal are appropriate. In recognition of the presence of systematic instrumental effects that do not apparently affect closure in networks, the standard errors on station values for gravimeter D6 have been arbitrarily raised to 4 μGal for the purpose of the present preliminary interpretation. In our judgement 4 μGal is a conservative estimate of the accuracy of the station values.

A four year temporal mean gravity value was calculated for each station from the station gravity values determined with respect to the spatial mean for each survey. Temporal variations with respect to the four year mean were then calculated for each station and are displayed in Figure 2 together with the variations at an auxiliary station (AER). The survey of November 1976 was a partial survey involving only seven stations. Station values for this partial survey are related to the values for the other seven surveys through a comparison of the fifteen station network mean with the seven station network mean over the four year interval of the experiment. The calculated difference between the two datums has a standard error of less than 1 μGal.

Effect of known mass changes

Before the gravity variations of Figure 2 can be interpreted in terms of crustal deformation, the effect of known, non-tectonic mass movements must be considered.

Groundwater

The most obvious mass effect is associated with rainwater and meltwater entering the ground at varying rates through the year. At Charlevoix the average annual input of moisture from rainfall and melting snow is 64 cm and 17 cm, respectively. In general the rate of input falls to a minimum in December and January and reaches maximum at the time of the spring thaw in March and April and again in the summer and early autumn (J. Aubin, personal communication, 1980). The year-to-year variation in precipitation at Charlevoix has been unusually large over the period from 1976 to 1979 (Figure 3).

A 10 cm thick layer of water causes a vertical attraction of 4.3 μGal. Therefore, gravity variations of the same order as those observed (Figure 2) could be produced by temporarily concentrating groundwater beneath some stations or by efficiently draining moisture from the

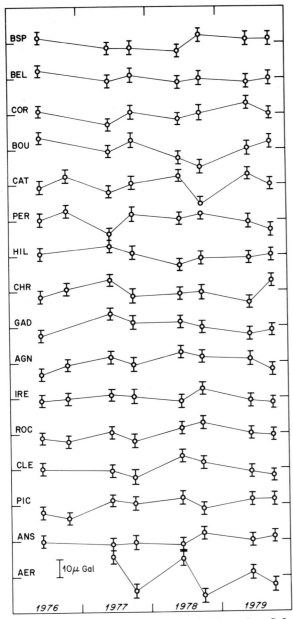

Fig. 2. Temporal gravity variations in µGal
for stations of the Charlevoix gravity
network. Gravity values are computed with
respect to the mean gravity of the network.
All vertical bars are ± 4 µGal and denote
standard error. (1µGal = 10nm/s^2).

areas, and (4) the time-constants of groundwater
flow into and out of the local material. Since
continuous monitoring of groundwater is carried
out at only a single location, no theoretical
models are available for the local hydrology
beneath the gravity stations at Charlevoix.
Therefore, possible variations in groundwater
attraction were estimated using a simple
exponential model for the gravitational response
to a moisture impulse in a manner similar to
Whitcomb et al., 1980.

The gravity effect of groundwater at each
station was calculated by convolving the
moisture input curve of Figure 3, augmented with
data for two previous years, with the product of
0.43 µGal/cm, the infinite slab value for water,
and an exponential function of time. A range of
possible groundwater effects was calculated by
considering a range of decay constants (time to
drop-off to 1/e) from one month to six months
(Figure 4). Six months was taken as a
reasonable upper limit for the decay constant,
since most stations are on fractured bedrock and
all stations are on or near sloping terrain.

The observable ground moisture effect at a
particular station is given by the difference
between the predicted absolute variations for
the station and the predicted absolute
variations of the reference datum. The
appropriate response function for calculating
the temporal variations of the reference datum
will be a weighted sum of exponential functions
for a range of decay constants. The predicted
absolute gravity variation for the reference
datum at Charlevoix has been calculated assuming
all response conditions are equally represented
in the network (middle curve, Figure 4). The
predicted gravity variation for slowly draining
sites with respect to the reference datum is
given by the difference between the top and
middle curves of Figure 4. This difference

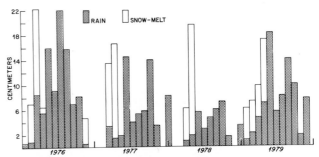

Fig. 3. Histogram of water-equivalent
moisture input to highland areas of
Charlevoix, Quebec in 30 day intervals.
Input from melting snow is postulated to
occur only when there is a rapid decrease in
recorded snow depth when temperature rises
to near 0°C or higher (assume old snow
yields 20% equivalent water by depth, new
snow yields 10% equivalent water).

vicinity of others. The groundwater response in
the vicinity of a station to the regional
moisture input of Figure 3 will be a complicated
function of time depending on: (1) the
efficiency of direct recharge of moisture into
the ground around the station, (2) the specific
storage capacity of the local material, (3) the
existence of hydraulic connections to adjacent

increases smoothly to a maximum of 11-12 μGal in the second half of 1976, diminishes to a low of 5 μGal at the end of 1978, and increases again to 9-10 μGal at the end of the four year period. The predicted change for rapidly draining sites has the opposite sign and approximately the same magnitude. It is given by the difference between the bottom and middle curves of Figure 4. Thus, an increase or a decrease of 6 μGal relative to the reference datum could be expected from Autumn 1976 to Autumn 1978. If the top or bottom curves of Figure 4 are more representative of the average response in the network, ground moisture induced variations at stations could double in magnitude but would follow the same temporal pattern.

As predicted by the simple model described here the observed gravity at stations BOU, ROC and CLE all change smoothly to maxima or minima in 1978 (Figure 2). Except possibly for station BOU, where a minimum of the order of 10 μGal is seen, the accuracy of the observations is not sufficient, however, to allow verification of the general validity of the model. Station BOU

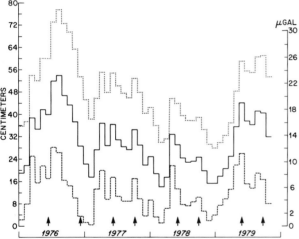

Fig. 4. Thirty-day estimates of retained ground moisture and corresponding gravity change for three different response functions for the period 1976 to 1979. The top curve represents predicted absolute gravity change for a slowly draining site calculated using an exponential response function with a decay constant of six months. The middle curve represents the predicted absolute gravity change for the reference datum assuming that all response conditions are equally represented in the network. The bottom curve represents predicted absolute gravity change for a rapidly draining site calculated using an exponential response function with a decay constant of one month. The vertical arrows identify the 30 day interval corresponding to the beginning of each gravity survey.

is located on a large buried boulder on unconsolidated material near the summit of a hill and, therefore, could well satisfy the assumption of the model that moisture entering from the surface be distributed uniformly in the material underlying the station.

Where groundwater is concentrated beneath a station by horizontal flow more abrupt changes in gravity could be expected than are predicted by the uniform absorption model. The rapid drop from spring 1978 to autumn 1978 at station CAT which is on unconsolidated material on the flanks of a valley could be due to a more complex response to moisture input. For comparison, a highly nonlinear response to available moisture is probably manifested in the large variations (bottom, Figure 2) at an auxiliary station AER which is situated on a concrete slab at Ste Irénée airport (north of IRE) in an area underlain by a thick wedge of glacial till. This station was established as a temporary point for the purpose of tying by air to external reference stations and is not considered to be stable enough to be included in a search for precursory effects.

St. Lawrence river effect

Gravitational attraction by tidal and seasonal variations in water level in the St. Lawrence river could also influence gravity values at coastal stations. Theoretical calculations show that the attraction effect of the river varies from 1.3 to 2.3 μGal/m at stations COR, BOU, PER, IRE and ANS. At station PIC which is close to the shore the effect is known to be as high as 5 μGal/m from examination of survey data (Lambert and Bower, 1977). Since the amplitude of the mainly semi-diurnal tide in the river is 2.2m, a single gravity reading could be biased by as much as 10 μGal at station PIC or 3 to 5 μGal at the other coastal stations. Fortunately, multiple connections between stations in the network are made at random times with respect to the tides and each connection consists of at least eight ties made over a three hour period. The effect of water level variations on the adjusted gravity values is, therefore, appreciably reduced to less than 4 μGal at station PIC and less than 3 μGal at other coastal stations. The tidal attraction effect on the six coastal stations probably contributes internal inconsistencies into the Charlevoix network which result in increased uncertainties in station values derived from the least squares network solutions.

Corrections for attraction effects on individual gravity observations at Charlevoix have not yet been implemented on a routine basis pending experimental verification of the theoretically derived amplitudes at all coastal stations. In the meantime, the gravity variations at station PIC must be interpreted with caution.

Seasonal variations in river level along the north shore are normally less than ± 15 cm (A. Bolduc, personal communication, 1980) and, therefore, should not cause significant attraction effects. It is not known whether seasonal variations could exceed these limits close to the mouths of tributaries such as the La Malbaie river.

Interpretation of gravity variations

Significant gravity changes

Neglecting auxiliary station AER Figure 2 shows that several stations depart significantly from the zero mean at the 90% confidence level (≥ 5 µGal). Among these stations BOU and CAT show changes in October 1978 that are significant from zero at the 99% confidence level (≥ 10 µGal). To examine further the spatial prominence of the variations which are for the most part marginally significant when taken individually, the result for each of the eight surveys was plotted on a separate diagram of the gravity network (Figure 5). Comparison with computer generated values normally distributed about zero with a standard deviation of 4 µGal revealed two anomalous groups of values that are significant at the 95% confidence level. With reference to Figure 5 the anomalies are identified as follows.

1. In June 1977 gravity values for adjacent stations COR, BOU, CAT and PER taken as a group are significantly negative with respect to neighbouring stations BEL, HIL, CHR, ROC, IRE, PIC, AGN, GAD. The mean of the first group is −4.5 µGal with a standard deviation of 1.3 µGal contrasted with the mean of the second group of +3.2 µGal with a standard deviation of 1.0 µGal. For all other surveys except that of October 1978 these paired groups of stations yield a contrast that tends toward the opposite sign (i.e. positive values for the group COR, BOU, CAT and PER). The contrast for the other surveys is not, however, generally significant at the 95% confidence level.

2. In October 1978 two anomalously negative values occur at adjacent stations BOU and CAT. The probability of two adjacent stations attaining these values by chance is less than one part in one thousand.

Changes in gravity can be caused by both vertical movements and by movements of mass. To assess the possible relevance of the identified anomalies to a study of tectonic processes a comparison is made with other evidence of aseismic deformation of the crust in the Charlevoix region.

Comparison with leveling

First-order leveling was carried out by the Geodetic Survey of Canada along the railway between Baie-St-Paul and La Malbaie in 1926,

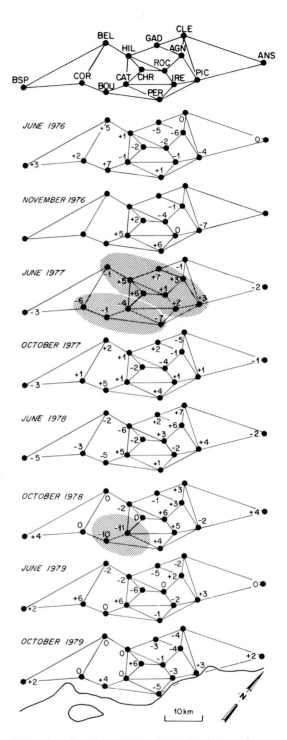

Fig. 5. Spatio-temporal variations of gravity from 1976 to 1979 at Charlevoix, Quebec. Gravity values are computed with respect to the mean gravity of the network and are given in µGal (1 µGal = 10 nm/s²). Estimated standard error on all values is ± 4 µGal. Hatched areas denote locations of identified anomalies.

1963, 1977 and as far as Cap-aux-Oies in 1978 (Figure 6). This line is ideal for monitoring crustal movements, since the elevation change of only 5 m along the 40 km route should ensure that refraction effects are minimal. Elevation changes in the three intervals 1926 to 1963, 1963 to 1977 and 1977 to 1978 are shown in Figure 6.

1. Neglecting the change at the first bench mark at Baie-St-Paul (evidence from another line suggests an isolated displacement here) there is an uplift of about 1 cm between Cap-aux-Oies and Pointe-au-Pic (PIC) from 1926 to 1963; the changes in elevation on both sides of the maximum occur over a distance of about 6 km and are, therefore, significant with respect to the estimated error propagation of 0.5cm at 85% probability.

2. From 1963 to 1977 there is a relative decrease in elevation of about 2 cm at the La Malbaie end of the line.

3. From 1977 to 1978 leveling from Baie-St-Paul to Cap-aux-Oies indicated no significant change.

The 1977 and 1978 relevelings were carried out during the June 1977 and October 1978 gravity surveys and, therefore, direct comparisons can be made with the gravity results. The relative change in gravity and elevation along the coast from Baie-St-Paul to La Malbaie in June 1977 is consistent with a volume expansion of the crust south-west of Cap-aux-Oies (Figure 6), provided the 1963 levels represent "unperturbed" elevations at Charlevoix. The leveling results between Baie-St-Paul and Cap-aux-Oies from June 1977 to October, 1978, unfortunately, did not extend far enough toward La Malbaie to reveal whether the possible relative uplift of 1977 has disappeared or not.

As in June 1977 the releveling of October 1978 also coincides with a tentatively identified negative gravity anomaly (Figure 5). A comparison of the leveling and gravity changes from June 1977 to October 1978 shows that there is no detailed agreement. The two ends of the leveling segment would be expected to be one or two centimeters lower in October 1978 with respect to June 1977, if both gravity and leveling are responding to volumetric expansion and contraction.

A 40m - diameter leveling array near station CHR (Figure 1) has been releveled to a precision of 1 µ radian at one to two month intervals since September 1977 (Gagnon et al., 1980). Results from this array are consistent with uplift of several centimeters in the vicinity of the gravity low of October 1978. Unfortunately, local movements of the water table cannot yet be ruled out as a cause of the observed signal. The low precipitation during the summer and autumn of 1978 (Figure 3) caused unusually low water-table levels beneath the leveling array (D.R. Bower, personal communication, 1980) that could have induced a large, local, north-east

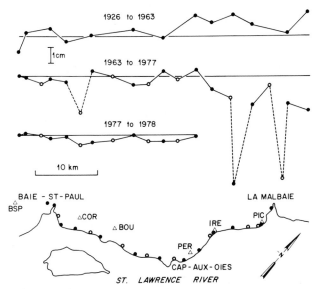

Fig. 6. Elevation changes over time intervals 1926 to 1963, 1963 to 1977 and 1977 to 1978 along the north-shore of the St. Lawrence River at Charlevoix, Quebec. Solid dots denote 1926 bench marks. Open dots denote 1963 bench marks. Triangles denote stations of the precise gravity network. Dashed lines indicate changes in elevation that are suspected to be local.

tilt. The interpretation of the leveling array data is continuing.

Comparison with seismic travel-time changes

The monitoring of seismic travel-time at Charlevoix provides some preliminary evidence that the gravity variations may be related to a real deep-seated phenomenon operating in the region. Since 1974 seismic travel-times have been recorded about twice a year in the Charlevoix region from two shot points to a network of up to thirteen seismic stations (Buchbinder, 1981, this volume). The Precambrian crust beneath the gravity network was sampled by shots at points west and southeast of the network that were recorded by seismic stations on the north-shore of the St. Lawrence river. Results showed that travel-times which had been increasing throughout the region since 1974 reached a maximum in late 1976 in a localized area that roughly coincides with the negative part of the gravity anomaly of June 1977 (Buchbinder, 1981, Figure 3, this volume). Between December, 1976 and August, 1977 this local travel-time maximum collapsed by 25msec as travel times at other stations on the north-shore continued to increase. In early 1978 travel-times at virtually all stations showed a marked decrease. Most stations have subsequently remained at the lower travel-time values, except for three or four stations, including one near gravity station BOU, where a return to higher travel-times occurred. This minor increase was complete by November 1978 shortly after the gravity low of October 1978. Unlike the travel-time maximum of 1977 the second, less pervasive maximum has not been followed by a drop in travel-times up to the end of 1979.

Relationship to seismicity

During the four-year period from 1976 to 1979 there were twenty-one events with magnitudes greater than 2.0 in the Charlevoix seismic zone (Figure 7). Only two of these events were greater than magnitude 4.0; an M=4.2 event occurred on October 23, 1976 and an M=5.1 event on August 19, 1979, both located toward the north-east end of the seismic zone. The temporal and spatial relationship between the gravity and these individual events, if any, is not yet clear. For example, although an M=2.6 event occurred on July 18, 1977 within 5 km of the gravity low of June 1977, other seismic

events, such as that of November 3, 1977, were not associated with a gravity low. Nor were there any events greater than M=2.0 in the vicinity of the more localized gravity low of October 1978. The only obvious relationship between the M=5.1 event and the gravity data is that the tentatively identified gravity anomalies occurred within a three year interval before the event, although at a distance of 30 km from the epicenter.

A histogram of the number of events per month recorded by a seismometer (LMQ) near gravity station CHR shows a peak in activity during the second half of 1977 followed by a number of smaller peaks separated by roughly six month intervals (Figure 8). Again there is no obvious spatial correlation between enhanced microseismicity and gravity anomalies, except that the negative gravity anomalies are located in the area closest to the highest density of microearthquakes beneath the St. Lawrence river (between 10 km and 30 km from seismic station LMQ, Buchbinder, personal communication, 1980). As far as temporal correlations are concerned, it might be significant that the gravity anomaly of June 1977 just precedes the large peak of microseismicity of the latter part of 1977. The later anomaly of October 1978 occurred between the peaks of July - September, 1978 and February - March, 1979. The rather infrequent sampling interval for gravity hampers any attempt to test for a definitive correlation between short-term variations in gravity and seismic activity.

Discussion

Precise gravimetry is a relatively new technique whose usefulness in earthquake prediction has not yet been fully demonstrated. In this paper it is shown that at the accuracy level of a few microgals there are difficulties in identifying signals of tectonic origin among those of a more superficial nature. There is evidence for highly non-linear response to moisture input at station CAT and at auxiliary station AER that is not predicted by a uniform absorption model. These effects are difficult to model theoretically and could only be modelled empirically with much more frequent sampling of gravity. Considering the fact that the two-station anomaly of October 1978 occurs at a time of unusually low moisture input and that both stations are located on unconsolidated material, no tectonic significance should be attached to it. There is also no support for corresponding vertical movement from leveling results. On the other hand, the negative anomaly of June 1977 occurs in the same area after a period of average moisture input. Thus, it is not possible to explain all the observed gravity variations by groundwater alone. Tidal attraction from the St. Lawrence river is also estimated to be too small to account for the anomalies.

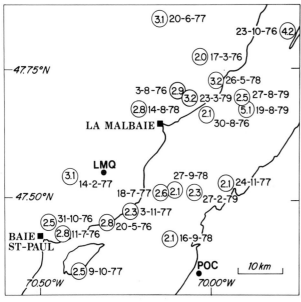

Fig. 7. Epicenters of Charlevoix earthquakes of magnitude 2.0 or greater during the period January 1, 1976 to December 31, 1979. Each event is labelled with a magnitude (M_N of M_L) and a date. The aftershocks of the October 23, 1976 and August 19, 1979 events are omitted. LMQ and POC are seismograph stations of the regional network of the Canadian Seismological Service. The data were obtained from Wetmiller and Horner (1978), Horner et al. (1979a, 1979b), and Wetmiller (personal communication, 1980).

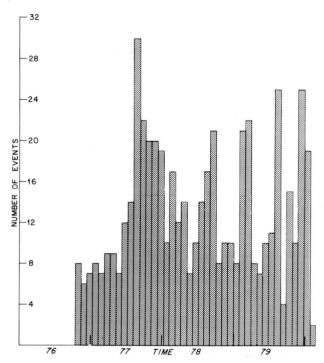

Fig. 8. Histogram of number of seismic events recorded in thirty-day intervals at seismograph station LMQ from 1976 to 1979 (adapted from Anglin and Buchbinder, 1981).

On the basis of the data presented here no one-to-one relationship between gravity variations and individual seismic events is revealed. Nevertheless, an intriguing coincidence has been noted between an unexplained temporal gravity anomaly and other observations: a coincidence that could reflect a process related indirectly to seismicity. In addition to the arguable evidence for uplift from leveling, there is a coincidence of the gravity anomaly of June 1977 with the onset of a peak in microseismic activity and a localized, transient high in seismic travel-time. All three anomalies are unique features of their respective data sets over the four-year period from 1976 - 1979. The magnitude and sense of the changes would be consistent with accelerated increase in crack volume accompanied by surface uplift beneath the coastal area between Baie-St-Paul and Cap-aux-Oies from late 1976 until mid 1977. Closure of cracks in this area would then have occurred between June and late August 1977. The onset of microseismic activity throughout the region could have signaled a regional stress redistribution associated with the broadening and collapse of the dilatant volume. A similar relationship between gravity, uplift and microearthquakes was observed on the Izu peninsula, Japan but no travel-time anomaly greater than $\pm$ 10 msec was identified in the

available data (results summarized by Tsumura, 1977).

Considering the preliminary nature of the gravity results reported here it is premature to dwell further upon the possible tectonic significance of the observed gravity changes. It is worth noting, however, that precursory effects detectable at the $\pm$ 4 μGal accuracy level can be reasonably expected on the basis of earthquake prediction models and empirical data on deformation preceeding some large earthquakes. According to the empirical expression $\log L = 0.26 M + 0.46$ (Anderson and Whitcomb, 1975) the diameters (L) of the zones of preparation for the $M \geqslant 2.0$ earthquakes in Figure 7 are expected to range from 10 km to 60 km. Thus, anomalous zones associated with several of the events in Figure 7 should extend beneath the Charlevoix gravity network. Evidence from the M = 7.5, 1964, Niigata, Japan earthquake (Scholz et al., 1973) and from the M = 6.4, 1971, San Fernando earthquake (Hanks, 1974; Anderson and Whitcomb, 1975) suggests that 10^{-5} can be taken as a conservative estimate of the volume strain preceeding these earthquakes. Similar volume strains might be expected at Charlevoix, although the shape and depth of the zone of preparation could be highly variable depending on the mechanical structure and stress state (Mogi, 1978). Earthquake source mechanism studies at Charlevoix show that the most common mode of failure has been oblique thrust (Hasegawa and Wetmiller, personal communication, 1980). It is reasonable, therefore, to assume that the direction of minimum stress is vertical and that most of the expected volume increase prior to an earthquake could be manifested in vertical expansion. For a volume strain of 10^{-5} the gravity change associated with the vertical expansion of a right circular cylinder of height 10 km (zone associated with M = 2.0 earthquake) is 12 μGal. Larger vertical dimensions for the anomalous zone corresponding to larger magnitude events give rise to proportionately larger gravity effects.

The lack of evidence for frequent gravity changes at Charlevoix prior to the many smaller events may be explained by the expected short duration of the anomalies. One estimate of the precursor-earthquake time separation is given by the empirical expression $\log T = 0.685 M - 1.57$ (Scholz et al., 1973) where T is the time in days and M is the earthquake magnitude. The expression predicts a precursor-earthquake time interval of a few days for events of magnitude 2.0 to 3.0. The expected gravity anomalies for these small events would clearly not be observable in a survey that took two or three weeks to complete. On the other hand, the above empirical expression predicts a precursor-earthquake time interval of about three months for the M = 5.1 event of August 19, 1979 at Charlevoix. A gravity anomaly might have been expected in the survey of June 1979. Although not very likely, one possible explanation for the lack of a gravity expression before this earthquake could be the insensitivity of the present gravity network to long-wavelength effects. This insensitivity is due to the lack of a stable external or absolute reference. External air ties made through station AER (Ste-Irenee airport) in 1976 to airport stations at Quebec City (130 km) and Bagotville (100 km) have not yet been repeated.

The gravity changes that have been identified at Charlevoix are preliminary and must be confirmed by a joint analysis of data from two different gravimeters. Although some interesting correlations with other data are seen, the tentatively identified anomaly of June 1977 cannot be related to local earthquakes by current earthquake prediction models. Moreover, there is insufficient evidence from aseismic regions to enable the small gravity variations and travel time changes at Charlevoix to be classified as unique features of seismically active regions.

Conclusions

Temporal variations of the order of 10 µGal were observed in the Charlevoix region, Quebec over the period 1976 to 1979. A spatially coherent anomaly that could not be explained by groundwater effects was identified in one of eight semi-annual surveys. When compared with seismic travel-time variations, geodetic leveling and microseismicity, this gravity anomaly is consistent with dilatancy accompanied by uplift but cannot be related to individual seismic events by current earthquake prediction models. Before any attempt is made to model the observed effects the gravity anomaly must be confirmed by other available gravity data. No other evidence for precursory gravity changes is seen at Charlevoix over the period 1976 to 1979.

Acknowledgements. We are grateful to H. Dragert for assistance in development of field procedures, R.K. McConnell and colleagues in the Gravity Data Centre for assistance with the management and adjustment of the gravity data and to the Geodetic Survey of Canada for leveling data. Gravity observations were carried out by Les Consultants BMJ Inc., Lavoie/Gaucher et Ass. and Géomines Ltée under contract to the Earth Physics Branch, EMR, Ottawa. The paper was improved by suggestions from anonymous reviewers and the editors. Contribution from the Earth Physics Branch No. 904.

References

Anderson, D. L. and J. H. Whitcomb, Time-dependent seismology, J. Geophys. Res., 80, 1497-1503, 1975.

Anglin, F.M. and G.G.R. Buchbinder, Microseismicity in the mid St. Lawrence Valley Charlevoix Zone Quebec, submitted to Bull. Seismol. Soc. Amer., 1981.

Basham, P.W., D.H. Weichert, and M.J. Berry, Regional assessment of seismic risk in eastern Canada, Bull. Seismol. Soc. Amer., 69, 1567-1602, 1979.

Buchbinder, G.G.R., Velocity changes in the Charlevoix region, Quebec, in Earthquake Prediction, edited by D.W. Simpson and P.G. Richards, Maurice Ewing Ser., Vol. 4, AGU, Washington, D.C., 1981.

Chen Yun-tai, Gu Hae-ding, and Lu Zao-xun, Variations of gravity before and after the Haicheng earthquake, 1975, and the Tangshan earthquake, 1976, Phys. Earth Planet. Inter., 18, 330-338, 1979.

Dragert, H., A. Lambert and J. Liard, Repeated precise gravity measurements on Vancouver Island, British Columbia, submitted to J. Geophys. Res., 1981.

Gagnon, P., J. Jobin, R. Sanchez and Y. Van Chestein, Une méthode spéciale de nivellement géométrique pour l'étude des déformations locales de la croute terrestre, Proc. Second International Symposium on Problems Related to the Redefinition of North American Vertical Geodetic Networks, edited by G. Lachapelle, pp. 353-371, Canadian Institute of Surveying, Ottawa, 1981.

Hagiwara, Y., H. Tajima, S. Izutnya and H. Hanada, Gravity changes associated with earthquake swarm activities in the eastern part of Izu peninsula (in Japanese), Bull. Earthquake Res. Inst. Tokyo Univ., 52, 141-150, 1977.

Hanks, T.C., Constraints on the dilatancy-diffusion model of the earthquake mechanism, J. Geophys. Res., 79, 3023-3025, 1974.

Horner, R.B., A.E. Stevens and R.J. Wetmiller, Canadian Earthquakes - 1977, Seismological Series, Number 81, Earth Phys. Br., Dept. of Energy, Mines and Resources, Ottawa, 1979a.

Horner, R.B., A.E. Stevens and R.J., Wetmiller, Canadian Earthquakes 1978, Seismological Series, Number 83, Earth Phys. Br., Dept. of Energy, Mines and Resources, Ottawa, 1979b.

Kisslinger, C., Processes during the Matsushiro, Japan, earthquake swarm as revealed by leveling, gravity, and spring-flow observations, Geology, 3, 57-62, 1975.

Lambert, A. and D.R. Bower, Gravity tide effects on precise gravity surveys, Proc. 8th Int. Symp. Earth Tides, edited by M. Bonatz and P. Melchior, pp. 536-539, Bonn, 1977.

Leblanc, G., A.E. Stevens, R.J. Wetmiller and R. Duberger, A microearthquake survey of the St. Lawrence Valley near LaMalbaie, Quebec, Can. J. Earth Sci., 10, 42-53, 1973.

Leblanc, G., and G. Buchbinder, Second microearthquake survey of the St. Lawrence Valley near LaMalbaie, Quebec, Can. J. Earth Sci., 10, 2778-2789, 1977.

Mogi, K., Dilatancy of rocks under general triaxial stress states with special reference to earthquake precursors, in Earthquake Precursors, edited by C. Kisslinger and Z. Suzuki, Advances in Earth and Planetary Sciences 2, pp. 203-217, Japan Scientific Societies Press, Tokyo, 1978.

Scholz, C.H., L.R. Sykes, and Y.P. Aggarwal, Earthquake prediction: a physical basis, Science, 181, 803-809, 1973.

Stevens, A.E. Re-examination of some larger LaMalbaie, Quebec, earthquakes (1924-1978), Bull. Seismol. Soc. Amer., 70, 529-557, 1980.

Tsumura, K., Anomalous crustal activity in the Izu peninsula, central Honshu, in Earthquake Precursors, edited by C. Kisslinger and Z. Suzuki, Advances in Earth and Planetary Sciences 2, pp. 137-146, Japan Scientific Societies Press, Tokyo, 1978.

Wetmiller, R.J. and R.B. Horner, Canadian Earthquakes - 1976, Seismological Series, Number 79, Earth Phys. Br., Dept. of Energy, Mines and Resources, Ottawa, 1978.

Whitcomb, J.H., W.O. Franzen, J.W. Given, J.C. Pechmann, and L.J. Ruff, Time-dependent gravity in southern California, May 1974 to April 1979, J. Geophys. Res., 85, 4363-4373, 1980.

GRAVITY PROFILES IN SOUTHERN CALIFORNIA

Jack F. Evernden

U. S. Geological Survey, Menlo Park, California 94025

Abstract. Relative gravity data obtained by John Fett using Lacoste-Romberg Model D gravimeters along several profiles crossing the San Andreas Fault in southern California are analyzed as regards reproducibility of gravity values and explanations of observed changes in gravity. Gravity changes of several tens of microgals were associated with water extraction and recharge in several alluvial basins. There appears to have been correlated gravity (approximately 30 microgals) and elevation (10 centimeters) along one profile during 1979. No gravity changes associated with earthquakes were detected.

Introduction

Under auspices of the National Earthquake Hazard Reduction Program, Earth Science and Engineering, Inc. (John Fett, principal investigator) is conducting a program of relative gravity measurements along eight profiles in southern California (see Figure 1). The same two instruments, Lacoste-Romberg gravimeters D3 and D19, have been used at all times. Beginning in November 1976, the repetition rate was once every two months on most profiles with one loop (base station-profile-base station) on each reoccupation. Since November 1977, most profiles have been reoccupied on a monthly basis with double loops (base station-profile-base station-profile-base station) using two instruments. This type of profiling yields four estimates of relative gravity between the profile base station and any other station of the profile. Where pertinent, the details of the schedule of observations at an individual station will be given in the text or in figures. Table 1 presents brief descriptions of pertinent features of the geological environment of each station. All profiles are along roads that have been surveyed repeatedly by first order leveling.

The profile base stations, A1, C1, F1, G6, and H6, are tied by monthly observations (single loop with two instruments) to one or both of the primary base stations at Hemet and Lytle Creek. Profile B is observed relative to A1 and profiles D and E are observed relative to Lytle Base. Lytle Base is very near the location of a UCSD cryogenic gravimeter. This juxtaposition and the direct

observation of many stations relative to Lytle Base allow evaluation of relative performance of the Lacoste-Romberg gravimeters and the cryogenic gravimeter. The evaluation will be presented subsequent to demonstration of the quality of the Lacoste-Romberg data.

Earth Science and Engineering, Inc. is responsible only for making the observations and reducing these meter readings to relative gravity values via corrections for linear drift of the instruments and solid earth tides. All stations (except station H1 of H profile) are at sufficient distance from the coast that relative gravity values along the profiles are not significantly affected by tidal loading. Standard formulas are used when correcting for solid earth tides. These reduced data are submitted to the USGS in semiannual reports or when requested. I have taken the responsibility for analysing these data as to adequacy of present field procedures, resultant size of detectable gravity changes and correlation of detected gravity changes with tectonic phenomena and with changes in water level in alluvial basins and reservoirs. Fortuitously, the period of observations includes the last year of the recent drought in California so that large changes in water level occurred in some basins where gravity stations were located. These changes in water level are evident in the gravity data.

Given that the data may display a very complicated behavior of gravity with time with occasional rapid changes in gravity, analysis was conducted via use of a program based on a least squares cubic spline (Thompson, 1973). Where data extend to the left or right of limiting partition points, these data are fit by a linear curve while fulfilling the normal continuity conditions at partition points. Partition points are placed where deemed appropriate to evaluate "real" gravity changes and to ignore "noise". The patterns of partition points chosen for the several profiles are presented either in the text or in figures. F statistic values are given on each figure, the comparison being between the calculated spline curve and the hypothesis of no gravity change with time. The legend of each figure indicates the associated probability that the spline

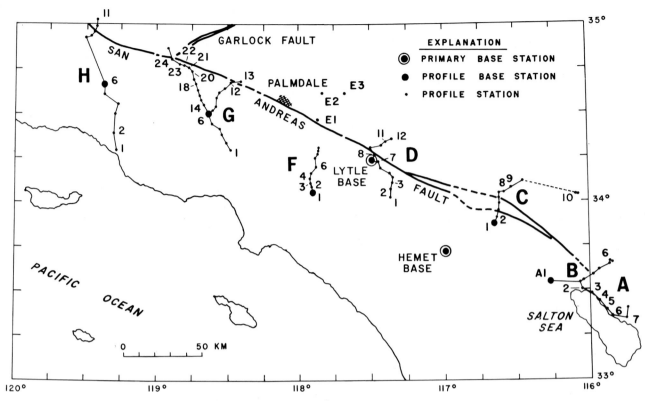

FIG. 1 Locations of gravity stations.

curve is a better fit to the data than is the hypothesis of no gravity change with time.

Quality of Data

Reproducibility of Gravity Values. The first step in analysis is to evaluate the reproducibility of the pairs of reported relative gravity values obtained with each instrument when doing double loops. Such reproducibility puts a lower limit on detectable gravity change. Table 2 gives such data for stations of profile C for the first year of double looping (November 1977 to December 1978) and for the last year of double looping (January 1979 to February 1980). During both periods, there was a pronounced difference between the performances of the two instruments (D3 and D19), D19 being far better (ideally, both parameters of Table 2 would be zero). If two instruments of D19 quality were used while obtaining data of the quality of the C profile , one would have high confidence of detecting gravity changes of about 20 microgals given a year of data of monthly double loops.

Note that the x (mean difference of the two gravity values made each month with each instrument at each station) and s_x (RMSD of such paired observations re mean difference x) values for the stations of profile C are quite similar for the early and late periods of observation. The data

of this profile suggest a correlation of x value and t_{obs} (time interval between observation at profile base station and any other station, measured in minutes), x values increasing as t_{obs} increases.

For the sake of brevity, similar details on the data of other profiles will not be given. However, certain observations which are discordant with conclusions drawn from the data of profile C will be mentioned:

(a) Although the D19 x values are always smaller than the D3 values for the data of a complete profile (except as noted below under (c)), the large differences observed at some stations of profile C are not observed on other profiles.

(b) The correlation of t_{obs} and x values suggested in the data of profile C is not as clear or is not present in the data of other profiles. In this regard, it is interesting to note the x values for the few stations for which observations are made with only one instrument (D19). Six stations, H1, H2, H11, A4, A5, and A6 are so observed with t_{obs} values going as high as 120-140 minutes (H2). None of the x values for these several stations are greater than 21 microgals, the average being 18.1. Although the details are certainly not clear, it seems that the routine and discipline of the observations must have a significant influence on the quality of the observations. The data suggest that improvements in

486   EVERNDEN

Table 1
Geologic Ground Conditions

Station A1. On a boulder atop Recent alluvium at the edge of Coachella   Valley.
Station A2. Recent fan deposit.
Station A3. Quaternary lake deposit.
Station A4. Quaternary lake deposit.
Station A5. Quaternary lake deposit.
Station A6. Quaternary lake deposit.
Station A7. Quaternary lake deposit.
Station A8. Tertiary volcanic rock.

Station B1. Recent fan deposit.
Station B2. PreCretaceous metasedimentary rock atop sediments in Box Canyon Wash.
Station B3. PreCretaceous metasedimentary rock atop sediments in Box Canyon Wash.
Station B4. Weathered granitic rock inselberg.
Station B5. Recent fan deposit.
Station B6. Granitic rock.

Station C1. Pre-Cretaceous metamorphic rock. Borders a large water basin to the north. Water level is rising in the basin due to recharge project subsequent to the drought.
Station C2. Unconsolidated Recent sediments.
Station C3. Unconsolidated Recent sediments.
Station C4. Pre-Cambrian metamorphic rock.
Station C5. Recent sediments.
Station C6. Slightly fractured crystalline rock.
Station C7. Thick Recent sediments in valley.
Station C8. Pleistocene nonmarine sediments atop bedrock.
Station C9. Granitic rock.
Station C10. Granitic rock.

Station D1. Granitic boulder.
Station D2. Recent alluvium.
Station D3. Recent alluvium.
Station D4. Weathered Pre-Cretaceous metamorphic rock.
Station D5. Weathered Pre-Cretaceous metamorphic rock.
Station D6. Recent alluvium close to granitic bedrock.
Station D7. Recent alluvium in canyon.
Station D8. Impermeable sandstone.
Station D9. Recent sediments in canyon.
Station D10. Pliocene nonmarine sediments.
Station D11. Pliocene nonmarine sediments.
Station D12. Recent sediments.
Station E1. Highly weathered sediments (Paleozoic limestone?).
Station E2. Highly weathered granitic butte surrounded by Recent sediments.
Station E3. Moderately weathered granitic rock. Alluvium basin to the south.

Station F1. Compacted fill atop Upper Miocene marine sediments.

Station F2. Recent sediments. Ground water basin beneath -- pumping during drought, more pronounced near station F4.
Station F3. Recent sediments. Ground water basin beneath -- pumping during drought, more pronounced near station F4.
Station F4. Recent sediments. Ground water basin beneath -- pumping during drought, more pronounced near this station.
Station F5. Recent sediments at edge of canyon.
Station F6. Pre-Cambrian igneous or metamorphic rock. Close to and somewhat higher than Morris Reservoir.
Station F7. Residual granitic boulder.
Station F8. Granitic boulder in highly weathered area.
Station F9. Highly weathered granitic rock.
Station F10. Slide area in granitic rock.

Station G1. Plio-Pleistocene marine sediments.
Station G2. Pliocene marine sediments.
Station G3. Plio-Pleistocene nonmarine sediments.
Station G4. Quaternary sediments.
Station G5. Plio-Pleistocene nonmarine sediments.
Station G6. Upper Miocene marine sediments.
Station G7. Atop sediments near and somewhat higher than a reservoir.
Station G8. Upper Miocene marine sediments.
Station G9. Pre-Cambrian metamorphic rock.
Station G10. Pre-Cambrian metamorphic rock.
Station G11. Pre-Cambrian metamorphic rock.
Station G12. Pre-Cambrian metamorphic rock.
Station G13. Granitic rock.
Station G14. Upper Miocene marine sediments.
Station G15. Pliocene nonmarine sediments.
Station G16. Pliocene nonmarine sediments.
Station G17. Pliocene nonmarine sediments.
Station G18. Pliocene nonmarine sediments.
Station G19. Pliocene nonmarine sediments.
Station G20. Pliocene nonmarine sediments.
Station G21. Pliocene nonmarine sediments.
Station G22. Pliocene nonmarine sediments.
Station G23. Possible slide.
Station G24. Recent alluvium in canyon.
Station G25. Recent alluvium in canyon.
Station H1. Pleistocene marine deposit near coast.
Station H2. Miocene marine sediments.
Station H3. Eocene marine sediments.
Station H4. Eocene marine sediments.
Station H5. Eocene marine sediments.
Station H6. Eocene marine sediments.
Station H7. Pliocene nonmarine sediments.
Station H8. Plio-Pleistocene nonmarine sediments.
Station H9. Plio-Pleistocene nonmarine sediments.
Station H10. Tertiary sediments.
Station H11. Tertiary sediments.

Table 2
Mean Difference of Paired Observations

Station Pair (t_{obs} (min))	Inst.	Nov77-Dec78 x	s_x	Jan79-Feb80 x	s_x
C2 re C1	D3	30.6	19.8	25.5	24.9
(15-30)	D19	9.8	9.4	24.7	25.9
C3 re C1	D3	34.4	18.9	44.9	31.3
(30-50)	D19	11.5	12.7	12.3	11.6
C5 re C1	D3	42.4	17.6	26.7	20.1
(50-80)	D19	15.6	12.1	19.2	12.0
C7 re C1	D3	39.7	28.1	36.3	15.0
(75-140)	D19	17.0	12.3	14.5	10.9
C9 re C1	D3	26.5	21.5	33.6	14.3
(100-200)	D19	24.9	22.7	28.1	21.8
C10 re C1	D3	27.6	28.5	24.3	18.5
(70-160)	D19	19.2	16.1	24.0	19.2
All	D3	33.5	22.8	31.9	21.5
	D19	16.3	14.8	20.5	17.9

x = average difference between paired observations (microgals)

s_x = RMSD of differences from average (microgals)

---

data quality could be achieved by careful analysis of present procedures and resultant improvements in field operations.

(c) On the D profile, the data for D19 show a profound degradation with time. The D19 x and s_x

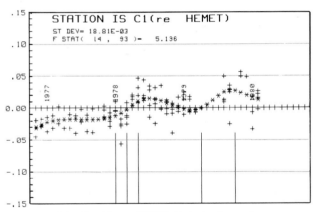

FIG. 2 <u>C1 re HEMET</u> Vertical scale: unit is ten microgals. Horizontal scale:unit is one month with the year designation placed over January. Symbol (+): observations. Symbol (*): calculated least squares cubic spline (Thompson, 1973). Vertical line segments: partition points (one at zero except where noted). Probability value P given in each legend is probability that calculated spline is a better fit to the observations than is a horizontal line (i.e., no gravity change with time). P>.99.

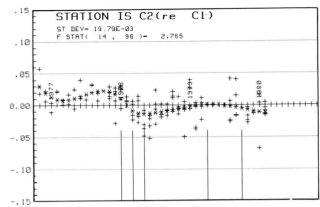

FIG. 3 <u>C2 re C1</u> See legend of Figure 2 for description of figure. P>.99.

values for Late 1977--Early 1978 are 11.3 and 10.4, using data of all 12 stations. For the time period Late 1979-- Early 1980, the D19 x and s_x values for the data of the entire profile are 31.2 and 25.0. The D3 values showed no significant change (26.9 and 20.3 for the early data, 22.1 and 16.9 for the late data). Why such a drastic degradation of D19 data occurred on profile D while such degradation did not occur in the D19 data of other profiles is unknown.

The only conclusion that one can draw is that field procedures are greatly influencing measurements, x values varying from 11 to greater than 30 for no apparent reason. If one could routinely obtain x values of 10 to 15 microgals, elevation changes of 5 centimeters could be detected with high confidence.

<u>Identification of Systematic Changes.</u> The second step in analysis is to investigate the level of noise in the data when attempting to fit the observations to a pattern of gravity change, i. e.,

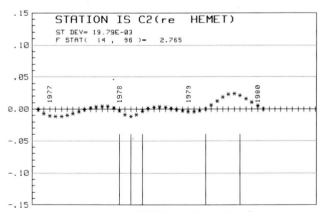

FIG. 4 <u>C2 re HEMET</u> See legend of Figure 2 for description of figure. Calculated curve is (<u>C2 re C1</u>) + (<u>C1 re HEMET</u>). St. dev. and F stat are for (<u>C2 re C1</u>).

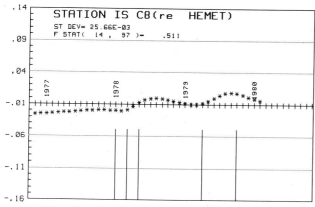

FIG. 5 <u>C8 re HEMET</u> See legend of
Figure 2 for description of figure.
Calculated curve is (<u>C8 re C1</u>) +
(<u>C1 re HEMET</u>). St. dev. and F stat
are for (<u>C8 re C1</u>). P<.90.

we now investigate the reproducibility of mean
values from month to month when allowing for systematic change. Again, profile C was chosen for
detailed illustration.First, the data of the local base (C1) show a pronounced change in gravity
relative to Hemet Base due to fluid extraction
and replacement. The "change" at all other stations relative to C1 is reversed and of the same
amplitude, thus establishing the reality of the
change and its localization at C1. Second, the
spread in t_{obs} values is greater for the C profile than for most of the other profiles,thus
providing a good opportunity for evaluating the
degradation of data quality as the time between
observation at the base station and at another
station on the profile increases. Third, the
schedule of monthly observations and double looping has been rigidly observed along this profile.
Finally, there have been two earthquakes on or
within 20 kilometers of this profile (M=4 and
M=5.5).
   The data of this profile have a surprising
character. Although all calculated values are
supposedly relative to a fixed value of gravity

Table 3

Root Mean Square Deviation of Residuals
(Mean of D3 values = Mean of D19 values)

A-All Stations re C1			B-Seq. Pairs of Stations		
Sta.Pair	RMSD	t_{obs}(m)	Sta.Pair	RMSD	t_{obs}(m)
C2 re C1	17.9	15 - 30	C2 re C1	17.9	15 - 30
C3 re C1	20.3	30 - 50	C3 re C2	15.9	15 - 20
C4 re C1	22.1	40 - 65	C4 re C3	20.5	10 - 15
C5 re C1	22.8	50 - 80	C5 re C4	19.4	10 - 15
C6 re C1	26.2	60 - 110	C6 re C5	21.7	10 - 30
C7 re C1	24.5	75 - 140	C7 re C6	20.3	15 - 30
C8 re C1	25.0	90 - 160	C8 re C7	16.3	15 - 20
C9 re C1	25.1	100 - 200	C9 re C8	21.2	24 - 40
C10 re C1	22.6	70 - 160	C10 re C9	20.6	45 - 55

Table 4
95 % Confidence Limits -- Simulated Random Data

(Hypothesis:No gravity change--S.D.=25 microgals)

Schedule	Insts	Loops	Slope per Yr.	End Values
Annually	2	2	±37 μgals	±43 μgals
Annually	2	20	10	10
Semian'lly	2	2	34	39
Monthly	2	2	26	17
Biweekly	2	2	17	11
Weekly	2	2	14	8

Slope per Year = 95% confidence limit on gravity
                 change during year
End Values = 95% confidence limit on estimated
             values of relative gravity at be-
             ginning and end of year. If the data
             are adequate in number for a given
             RMSD, the number under "End Values"
             will be one-half the number under
             "Slope per Year"(i.e., the condition
             where st. dev. of the mean is very
             small).

at C1, and though the two instruments are carefully calibrated, the mean reported relative gravity values (relative to C1) by each instrument
are greatly different at some stations. This difference increases with increasing time interval
between observation at C1 and at the station,
reaching a value of 75 microgals at C9. This phenomenon is not present in data of several of the
profiles, is present only in less exaggerated
form on other profiles, and is totally inexplica-
ble. To eliminate this effect, the means of all
relative gravity values for each instrument at
each station relative to its local base station
are made equal. Since both instruments are supposedly detecting and reporting the same relative
gravity values, this seems to be a legitimate
step.

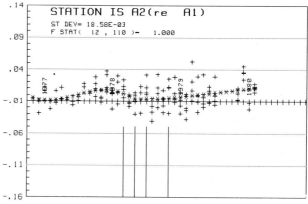

FIG. 6 <u>A2 re A1</u> See legend of Figure
2 for description of figure. P<.90.

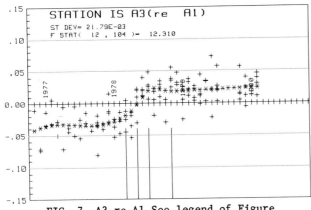

FIG. 7 <u>A3 re A1</u> See legend of Figure 2 for description of figure. P>.99.

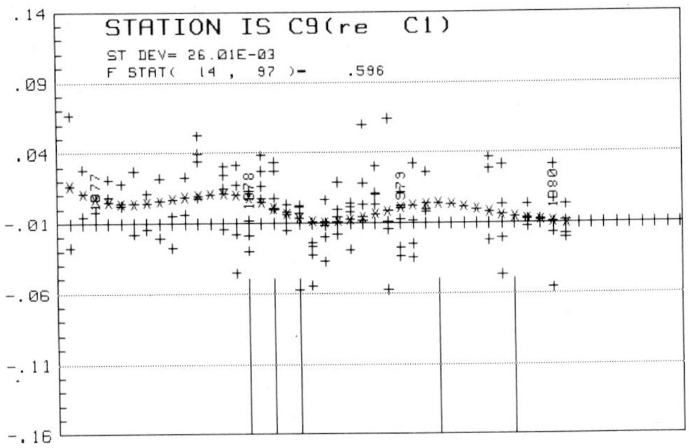

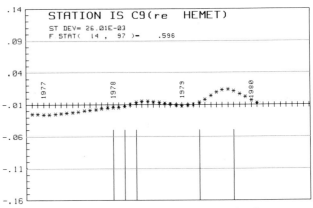

FIG. 9 <u>C9 re C1</u> See legend of Figure 2 for description of figure. P<.90.

Given the use of partition points as in Figures 2, 3, and 4 (C1 re Hemet Base, C2 re C1, and C2 re Hemet Base), this placement of partition points being selected to achieve fit to the pattern of gravity change occurring at C1, Table 3A presents the root mean square deviation (RMSD) of residuals from the least squares cubic spline fit for stations C2 through C10 re C1. Comparison of predicted shapes of gravity changes at stations C2 and C8 (Figures 4 and 5) illustrates that the increase in RMSD at C8 is not related to a failure of the selected pattern of partition points to fit a behavior of gravity values at C8 which was distinctly different from that at C1 and C2. Table 3A includes a listing of the normal spread in time interval between observations at C1 and at each station. There is apparent correlation between length of this time interval and the RMSD values. To illustrate further that this increase in RMSD may be a time-dependent effect, Table 3B gives the RMSD values for the data of successive station pairs when using data acquired since double looping has been conducted. These values are generally smaller than those of Table 3A. It appears that there is a random or nonlinear com-

ponent of drift in these instruments that leads to increasing RMSD with increase of time interval between observation at two stations.

The conclusion of the analysis above is that RMSD values of lower than 25 microgals will require carefully adjusted instruments and improved field procedures. Routine RMSD values as low as 10-15 microgals at all stations of a profile relative to the profile base station may be attainable by such procedures, this estimate of attainable accuracy being based on Lambert's results (Lambert, this volume).

<u>Analysis of Artificial Data.</u> As the result of procedures presently used in southern California, one must contend with RMSD values so large as to make changes in elevation of 5 to 10 centimeters (gravity change of 15 to 30 microgals) undetectable at high confidence with programs of monthly observation. Of course, individual sets of data may demonstrate such gravity changes at high confidence. The point stressed is that there is the real possibility of randomness of the data ob-

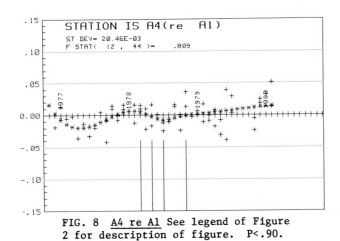

FIG. 8 <u>A4 re A1</u> See legend of Figure 2 for description of figure. P<.90.

FIG. 10 <u>C9 re HEMET</u> See legend of Figure 2 for description of figure. Calculated curve is (<u>C9 re C1</u>) + (<u>C1 re HEMET</u>). St. dev. and F stat are for (<u>C9 re C1</u>).

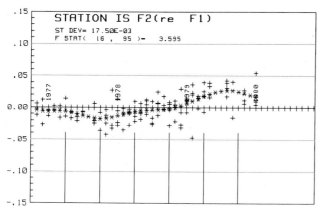

FIG. 11 <u>F2 re F1</u> See legend of Figure 2 for description of figure. P>.99.

scuring such changes. To illustrate this point, several analyses of totally random data were conducted with the results being presented in Table 4. Several schedules of observation were hypothesized for a fixed time period, zero change in gravity was hypothesized (any imagined linear change would yield the same estimates of confidence limits), and a standard deviation of 25 microgals was assumed for the artificial gravity data. Sets of artificial data were generated using random numbers of zero mean and specified normal standard deviation. These artificial data were then analyzed via a linear least squares program for calculated gravity change, RMSD of residuals, and 95 per cent confidence limits on gravity change. Table 4 presents the result of 100 runs on such simulated data for the cases of 2, 3, 12, 26, and 52 repeat observations. Case 1 is intended to simulate annual observations while Case 5 simulates weekly observations. It is clear that one cannot have high confidence of detecting gravity changes equal to the RMSD of the data by use of data taken on a monthly basis. Weekly occupation of the profile would allow such a gravity change to be detected with high confidence. Data observed on an annual schedule (2 instruments and double loops) can lead to such inaccurate interpretations as to render such data worse than useless (one seldom can accept just how bad one's data are). However, the second simulation (many loops conducted in a very short time) indicates the expectable improvement in estimates of gravity change by such procedures. A program of 20 loops once a year is superior to doing the same number of loops throughout the year if looking only for annual change.

If the gravity change is greater than 1½ times the RMSD of the data, there is high confidence of detecting a change in gravity along the profile when using monthly data.

<u>Comparison of Lacoste-Romberg and UCSD Cryogenic Gravimeters</u>. It is concluded that gravity changes of less than 25 microgals may be undetectable at high confidence by routine schedules of monthly

observation with present field procedures, but that gravity changes of 40 microgals or greater are detectable at very high confidence. In this regard, it is pertinent to note the comparative performances of the Lacoste-Romberg instruments and the UCSD cryogenic gravimeter at Lytle Base. As mentioned above, many stations in a variety of geologic and tectonic environments are observed directly relative to Lytle Base, including Hemet Base, D1, F1, G6, H6, E1, E2, and E3. The data of all of these stations show no evidence of gravity change at these stations relative to Lytle Base. The logical conclusion is that none of these stations, including Lytle Base, has experienced detectable gravity change during 1976-1979. The fact that these several stations show nearly flat gravity profiles relative to Lytle Base plus the fact that all stations of the D profile other than D2 and D3 (Lytle Base being along this profile) appear to have experienced little or no gravity change during this time period suggest that very little change in gravity has occurred at Lytle Base during this time period.

In order to achieve a better estimate of gravity change at Lytle Base during 1978 and 1979, eight stations directly observed relative to Lytle Base were used as a "base station" for estimating change at Lytle Base, the assumption being that the average change in gravity at these several stations was near zero. The average calculated gravity change at Lytle Base relative to the mean behavior of these eight stations (D1, D2, D11, D12, E2, E3, F1, H6) is considered an estimate of the potential gravity change at Lytle Base. The maximum gravity change calculated for Lytle Base between January 1978 and January 1980 is 4.7 microgals. The calculated change in gravity at Lytle Base between November 1978 and February 1979 is less than 0.1 microgal.

These conclusions are stated explicitly because it has been suggested (Goodkind, 1980) that permanent changes in gravity at Lytle Creek in excess of 100 microgals may have occurred within one or two days in December 1978. Further "in-

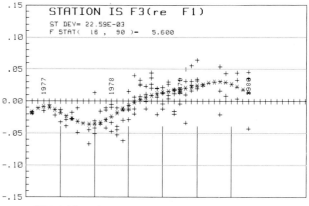

FIG. 12 <u>F3 re F1</u> See legend of Figure 2 for description of figure. P>.99.

creases" in the following weeks and months resulted in an apparently persistent change of approximately 200 microgals occurred by late January 1979. Goodkind (op. cit.) has suggested that the step-like changes indicated by his instrument are probably fallacious but that the apparent ramp-like change of a few tens of microgals during 1978 is real. The Lacoste-Romberg data deny this interpretation of the data of the cryogenic gravimeter, suggesting rather that the ramp-like change in the cryogenic data is also an instrumental artifact. Data to be presented below will confirm that present procedures, when using Lacoste-Romberg instruments, can and have detected gravity changes of 40 to 70 microgals at high confidence, and have shown that such gravity changes occur on timescales appropriate for hydrologic and tectonic processes (no earthquakes having occurred which could have caused nearly instantaneous changes in gravity).

Detected Gravity Changes and Their Interpretation

Profile A. Discussion of detected gravity changes and their correlation with tectonic or hydrologic processes will begin with the southernmost profile A and proceed northward to H. Figures 6, 7, and 8 show the gravity values and calculated best fit curves (*) for A2, A3, and A4 re A1, using a pattern of partition points designed to fit A3 (one partition point is at zero). Note that neither A2 nor A4 shows any gravity change re A1, while A3 shows a very pronounced change with this calculated change having a very high confidence $(F(12,104)=12.3)$ of being a better fit to the data than no gravity change $(F_{.99}(12,104)=3.48)$. Table 1 shows station A3 to be located within an alluvial basin, thus allowing for hydrologic influences on the gravity value. The fact that stations A3 to A7 are located within the same alluvial basin but only A3 shows a detectable gravity change is probably due to local conditions of extraction and recharge. Clarification of this point is being attempted.

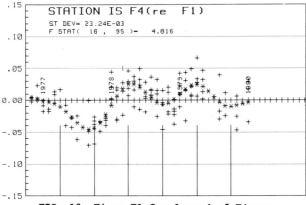

FIG. 13 **F4 re F1** See legend of Figure 2 for description of figure. P>.99.

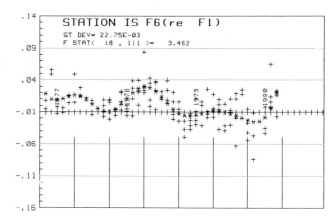

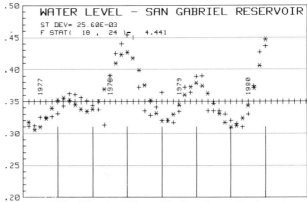

FIG. 14 **A. F6 re F1** See legend of Figure 2 for description of figure. P>.99.
B. Water level – San Gabriel Reservoir See legend of Figure 2 for description of figure. Vertical scale: (kilofeet–1). P>.99.

Since the Imperial Valley earthquake of October 15, 1979 (M=6.5) caused surface breakage within 45 kilometers of station A7 (1-1.5 source lengths from the end of observed displacement), a change of gravity at A7 relative to A1 during the year prior to the earthquake was sought. Placement of the last partition point at January 1979 gave an estimated change in gravity through January 1980 at A7 re A1 of 15 microgals with an F(8,103) value of 0.63. Similar estimates of gravity change at each station were made and these numbers were then treated as a pattern of change along the A profile. An estimated gravity increase of 14 ±9 microgals was calculated from A1 to A7. There may have been a small increase in gravity along the profile but the quality of available data make it impossible to demonstrate such an increase at high confidence.

Profile B. No clearly detectable gravity changes at any station occurred along profile B. The maximum calculated departures from zero were 10 microgals when using three partition points at No-

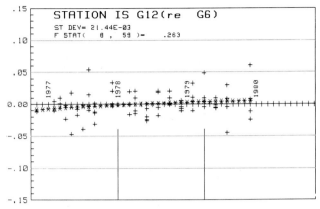

FIG. 15 <u>G12 re G6</u> See legend of Figure 2 for description of figure. P .90.

vember 76, January 78, and April 79 with associated $F(8,117)$ values of 0.51. When fitting all data of each station to a linear curve, the maximum calculated change in gravity re A1 was 14 microgals at B6 with an $F(2,123)$ value of 1.4. Fitting of these calculated changes at each station to a change vs. distance profile gave a calculated change of 8 ±6 microgals; i.e., no evidence of change.

Profile C. Regarding profile C, the behavior of C1 has already been noted. Table 1 indicates that the observed gravity changes at C1 were probably the result of changes of water level in the alluvial basin on which station C1 is located. It is noted in Table 1 that there was a deliberate effort to recharge this basin after the drought as a hedge against a future drought. An earthquake of approximately M 4 occurred on July 4, 1978 very near C1. However, the fact that the rise in gravity at C1 began in the last quarter of 1978 and the fact that there has been no return of gravity values to the mid-1978 values (see Figure 2) suggest strongly that the observed changes in gravity are related to recharge and not to this small earthquake. If there was a small effect on gravity values at C1 associated with the earthquake, this effect is undetectable in the presence of a 35 to 40 microgal effect associated with recharge.

An earthquake of M 5.5 (Goat Mountain earthquake) occurred at 34.31°N 116.44°W on March 15, 1979; i.e., 15-20 kilometers from station C9. Figures 9 and 10 (C9 re C1 and C9 re Hemet Base) indicate no detected changes in gravity at C9 prior to this 5.5 earthquake.

Profile D. The stations of profile D are on a variety of rock types and located from 15 kilometers south of the San Andreas Fault to 15 kilometers north of it. Stations D2 and D3 are located on an alluvial groundwater basin. A few other stations are on Recent alluvium in canyons. No station on shallow alluvium in canyons on any

profile has shown detectable gravity change as a response to the drought or subsequent heavy rainfall. The data of station D2 suggest an increase of 30 microgals re Lytle Base between mid-1977 and mid-1978 while the data of D3 suggest a maximum of 15 microgal increase during the same time period. These calculated increases are so small that one must conclude that there is no strong evidence for gravity change at these stations since late 1976. No other station of the D profile gives any better evidence of gravity change re Lytle Base since late 1976. Maximum calculated changes are 15 to 20 microgals with very low confidence in their reality.

Profile F. Profile F has three stations (F2, F3, F4) located in a heavily pumped alluvial basin, as well as one station (F6) located near the San Gabriel and Morris Reservoirs. Figures 11, 12, and 13 present observed gravity values (re F1) at F2, F3, and F4 from November 1976 to date, as well as the least square cubic spline (*) associated with the indicated partition points (one at zero). Table 1 indicates that these three stations are located within the same alluvial basin and that there was excessive pumping related to the recent drought with greatest pumping near F4. The drought was extreme for two years, followed in the winter of 1977-1978 by abnormally heavy rainfall. The observed pattern of gravity values seems to reflect the excessive extraction and subsequent recharge.

Station F6 is located on Pre-Cambrian metamorphic rocks (Table 1) near San Gabriel and Morris Reservoirs. Figure 14A indicates observed relative gravity values at F6 re F1 as well as the cubic spline associated with the indicated distribution of partition points (one being at zero). Figure 14B indicates the level of water (expressed in (kilofeet-1)) in San Gabriel Reservoir from late 1976 to date. A cubic spline of the same characteristics as that used for Figure 14A is fit to these data . The general correlation of the shapes of the two curves of Figure 14

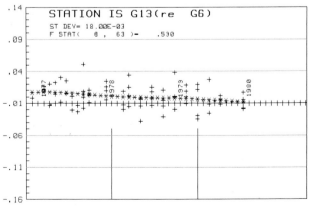

FIG. 16 <u>G13 re G6</u> See legend of Figure 2 for description of figure. P<.90.

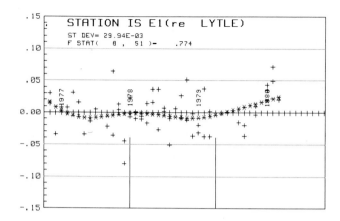

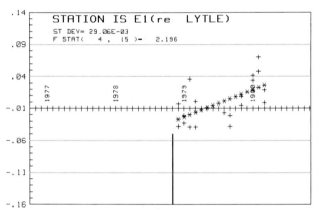

FIG. 17 See legend of Figure 2 for description of figures. A. All data. P<.90. B. Data from Nov. 1979 to Mar. 1980. Linear fit. One partition point at Oct. 1979. P<.90.

indicates that the gravity changes observed at F6 are related to changes of water level in the reservoir.

Other stations located on hard rock along all profiles show no statistically significant changes in gravity.

Profile G. Profile G is of particular interest since it follows the only first order level line that reportedly showed pronounced changes in elevation between January 1978 and January 1979 (W. Thatcher, personal communication). The reported change in elevation is ten centimeters (down towards the north, G1 to G24). Unfortunately, profile G was not observed as intensively as profiles A through D. The USGS and Fett agreed on the need for this decreased effort when faced with inadequate funds to do all profiles properly. Therefore, the conclusions to be given are only marginally significant but very interesting. Fortunately, there were no pronounced hydrologic effects for any stations of this profile.

To investigate the possibility for detecting

the measured elevation changes along this profile between January 1978 and January 1979, the change in gravity at each station during this period was first estimated. The data for station G18 are so limited and confusing that they were ignored. The estimated gravity changes at each station were fit to a linear least squares model of gravity change along the profile. A gravity change (increase to the northeast) of 26 microgals with a 95% confidence limit of ±11 microgals was obtained, a gravity change essentially equal to that predicted by the elevation change.

Profile H. No detectable gravity changes were found for any stations of the H profile.

Gravity Observations and Reported Stress Changes during 1979. The last item of interest is to investigate the possibility of gravity changes since March 1979 in the Palmdale area in order to ascertain whether detectable gravity changes occurred contemporaneously with the stress relaxation detected by J. Savage(C. B. Raleigh, this volume). Inspection of Figure 1 indicates that very few gravity stations are located so as to be able to detect such an effect if it occurred.

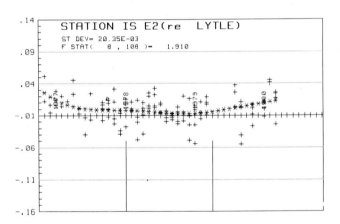

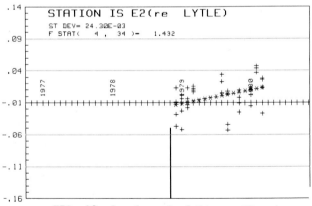

FIG. 18 See legend of Figure 17 for description of figures. A. P<.90. B. P<.90.

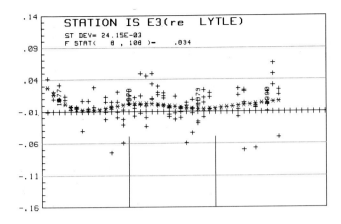

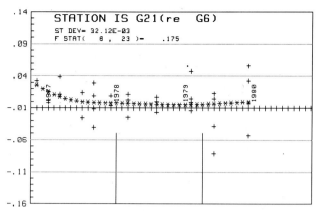

FIG. 21 G21 re G6 See legend of
Figure 2 for description of figure.
P<.90.

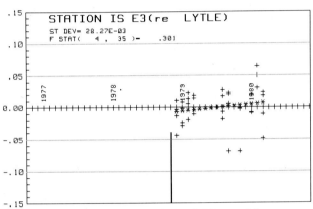

FIG. 19 See legend of Figure 17
for description of figures.
A. P<.90. B. P<.90.

and G13 show no detectable changes in gravity
during late 1979. Actually, there have been no
detectable changes at these stations since the
beginning of observations in 1976. None of the
stations along the D profile except possibly D2
and D3 (see above) have shown detectable changes
in gravity, particularly D8 through D12. At station E1, the station closest to Palmdale, observed gravity values suggest a change in behavior during 1979 (Figure 17). The F value given
on the figure is a function of both A and B in
$Y = A + B*X$. The 95 per cent confidence limit on the
slope term (B) leads to a 95 per cent confidence
limit on the calculated change in gravity indicated in Figure 17B of ± 29 microgals, significantly less than the calculated change of 52 microgals. Station E2, located well north of the
San Andreas Fault but equally near Palmdale,
shows a behavior similar to but less pronounced
than that of E1 (see Figure 18). The 95% confidence limit on the change in gravity is ±22 microgals while the calculated change is 27 microgals. Station E3 shows no apparent change in gravity during 1979 (see Figure 19).

Stations G20, G21, G22, G23, G12, G13, E1, D7,
and D8 are located just south of, on, or just
north of the San Andreas Fault from Lebec to Cajon Pass with only E1, E2, G12, and G13 being
near Palmdale. Figures 15 and 16 for stations G12

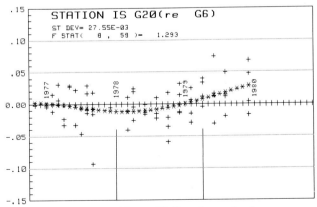

FIG. 20 G20 re G6 See legend of
Figure 2 for description of figure.
P<.90.

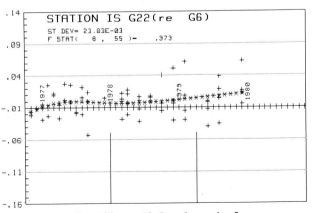

FIG. 22 G22 re G6 See legend of
Figure 2 for description of figure.
P<.90.

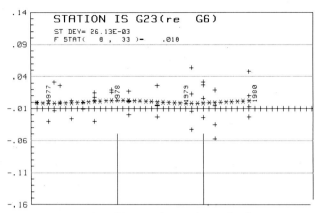

FIG. 23 G23 re G6 See legend of
Figure 2 for description of figure.
P<.90.

Stations G20 to G23 display gravity values suggesting changes during 1979. However, the simplest explanation of these changes is continuation of the gravity increases (i.e., elevation decreases) observed during calendar 1978 and discussed above (Figures 20 through 23). Therefore, the only stations that would seem to possibly show gravity changes linked directly to the relaxation reported by Savage are E1 and E2.

Conclusions

(A). When using Lacoste-Romberg Model D gravimeters in a profiling mode, presently employed field procedures yield relative gravity values with an RMSD of 17 to 26 microgals.

(B). One cannot have high confidence of detecting gravity changes of less than the RMSD of the data without biweekly or weekly observational schedules.

(C). One can have high confidence of detecting gravity changes of 1½ times the RMSD of the data by a program of monthly observations for one year (assuming double looping and use of two instruments).

(D). Annual gravity observations, when double looping and using two instruments, cannot detect gravity changes of 1.5 to 2.5 times the RMSD of the data at high confidence, depending upon mode of interpretation.

(E). The capability attainable with present field procedures is more than adequate to detect elevation changes of 20 centimeters (50 to 60 microgals) by realizable schedules of observation. The only way to have high confidence of detecting elevation changes of 5 centimeters is by a program of many reoccupations if field procedures cannot be improved markedly.

(F). At present, the Lacoste-Romberg Model D gravimeter is the best available for measuring relative gravity values in a mobile mode. Therefore, efforts to detect gravity changes of less than the RMSD of the data should not be undertaken unless very laborious schedules of observation are planned. Such investigations will require at least biweekly observations continued for a year or more.

(G). The data obtained at Lytle Base and surrounding stations indicate that the large changes in gravity reported by the UCSD cryogenic gravimeter at Lytle Base actually did not take place.

(H). All profiles other than profile G provide no evidence of gravity changes during calendar year 1978 other than those due to extraction of water from alluvial basins and/or recharge. The only profile suggesting a real change in gravity during this time period is profile G, the profile along which a ten centimeter change (northeast end down) in elevation was reportedly observed by first-order leveling during this time period. Somewhat fortuitously, the calculated increase in gravity along this profile agrees nicely with the calculated changes in elevation.

(I). The gravity stations were poorly located to detect any gravity changes near Palmdale during 1979. The data of stations E1 and E2 are suggestive while the data of all other stations in the area give no suggestion of gravity change during this period that could be correlated with the stress relaxation reported by J. Savage.

References

Goodkind, John M., 1980, "Continuous gravity measurements in the region of the Palmdale Uplift", U.S.G.S. Open File Report 806 (Summaries of Technical Reports, Vol. IX), pp. 313-316.

Thompson, Richard F., 1973, "Least Squares Splines", NASA X-692-73-321, Goddard Space Flight Center, October 1973.

SOME FEATURES OF MEDIUM- AND SHORT-TERM ANOMALIES BEFORE GREAT EARTHQUAKES

Guomin Zhang and Zhengxiang Fu

The Center of Seismic Analysis and Earthquake Prediction, State Seismological Bureau
Beijing (Peking), The People's Republic of China

Abstract. In this paper, we analyze the temporal and spatial distribution of medium- and short-term anomalies prior to four large earthquakes (Haicheng, M_s = 7.3; Longling, M_s = 7.4; Tangshan, M_s = 7.8; Songpan, M_s = 7.2) which occurred in northern and southwestern China in 1975 and 1976. The mechanism of these anomalies and their significance to earthquakes prediction are discussed.

Introduction

During the eleven year period 1966 to 1976, fifteen large earthquakes with a magnitude of 7 or greater took place in China (Figure 1 and Table 1). Nine of these events occurred in northern and southwestern China. Based upon observational data, Ding et al. [1979] proposed the classification of precursory anomalies into four types: long-, medium- and short-term, and imminent precursory anomalies. In this paper, short-term and imminent anomalies are discussed together as one type, namely, short-imminent anomalies.

In this study, we will give an account of certain features of temporal and spatial variations of medium- and short-term anomalies prior to four great earthquakes which occurred in China in 1975 and 1976. These events are the Haicheng, Songpan, Longling and Tangshan earthquakes. Because earth-

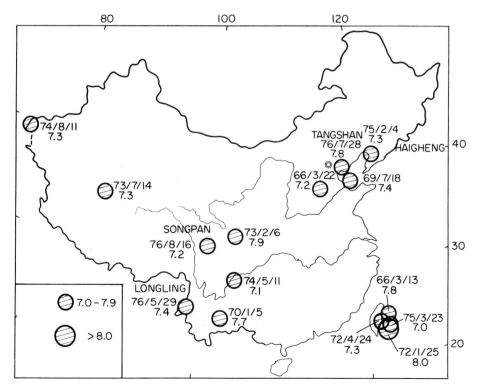

Fig. 1. Epicentral distribution map of China ($M_s \geq 7.0$, 1966-1976).

Table 1. Catalogue of Strong Earthquakes in China (Ms ≥ 7.0; 1966-1976)

No.	Date	Epicenter E	Epicenter N	Ms	h (km)	Location
1	66/3/13	122.6°	24.1°	7.8	63	Hualian, Taiwan
2	66/3/22	115°03'	37°32'	7.2	9	Ningjin, Hebei
3	69/7/18	119.4°	38.2°	7.4	35	Bo Hai
4	70/1/5	102.7°	24.0°	7.7	13	Tonghai, Yunnan*
5	72/1/25	122.4°	22.6°	8		Taidong, Taiwan
6	72/4/24	121.6°	23.6°	7.3		Dagangkou, Taiwan
7	73/2/6	100.4°	31.5°	7.9	17	Luhuo, Sichuan*
8	73/7/14	86.5°	35.3°	7.3	33	Mani, Xisang
9	74/5/11	103.9°	28.2°	7.1	14	Daguan, Yunnan*
10	74/8/11	73.8°	39.4°	7.3		Wuoia, Xinjian
11	75/2/4	122°48'	40°39'	7.3	12	Haicheng, Liaoning
12	75/3/23	122°36'	22°58'	7.0	33	Xiagang, Taiwan
13	76/5/29	98°45'	24°33'	7.4	20	Longling, Yunnan
14	76/7/28	118.2°	39.6°	7.8	16	Tangshan, Hebei*
15	76/8/16	104°06'	32°42'	7.2	15	Songpan, Sichuan

* macroseismic epicenter

quake prediction in our country is still in its infancy and the observational data are not extensive, the results of our studies are preliminary.

Temporal and Spatial Features of Medium-Term Anomalies

The temporal character of medium-term anomalies (seismicity, fault activity, ground tilt, resistivity, radon content, level of underground water, etc.) before the four great earthquakes are shown in Figures 2 through 5.

Seismicity

Curves a, b, c and d in Figure 2 respectively show the changes in earthquake magnitude ($M_L \geq 2.0$), frequency ($M \geq 2.0$) in Laoning Province, b-value [Li et al., 1978], and velocity ratio [Feng et al., 1976] before the Haicheng earthquake. These curves indicate that, in the region around Haicheng, the intensity and frequency of seismic activity increased remarkably from early 1973 to the occurrence of the main Haicheng earthquake. The total number of earthquakes recorded reached about 580 in 1974, but only about 70 were recorded in 1971 and 1972. The magnitude 4.8 Benxi earthquake occurred in December 1974. The velocity ratio had a low value which lasted from 1972 to 1973; it then returned to its normal value in 1974. During the same time period b-values showed an anomalous increase and then decrease.

The features of seismicity anomalies (earthquake frequency and b-value) before the three other great earthquakes are similar to those before the Haicheng earthquake (see Figures 3a, b; 4a, b and 5a, b). In Figures 3, 4 and 5, the dashed lines represent the level of normal variation or range.

Regional Migration of Seismicity and Seismic Gaps

In Figures 6 through 8 we show that the distribution of regional seismicity at several hundred kilometers from the future epicenter migrated towards the epicentral area of the future main shock for several years prior to the Haicheng, Tangshan and Longling earthquakes.

Figure 9 shows a map of small earthquake epicenters before the Haicheng event. This map shows that in 1972 a gap in the occurrence of small shocks occurred in the northern part of the Tancheng-Lujiang fault (Figure 9a). After late 1972, small shocks started to gradually occur in the gap (Figure 9b). After the occurrence of these small shocks the great Haicheng earthquake occurred within the gap. It should be noted that during this period the seismic activity remained relatively calm in the source area (the aftershock area of the main earthquake). Only nine small shocks (maximum magnitude 1.3) were recorded from 1971 to January 1975.

Figure 10 shows that a seismic gap formed in 1973 before the Songpan earthquake; the main shock of which occurred within the gap in August 1976

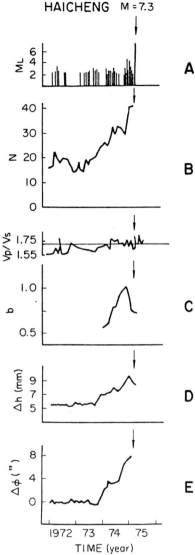

Fig. 2. Changes in the medium-term anomalies before Haicheng earthquake. A) change in magnitude with time in Liaoning Province; B) small earthquake frequency in south Liaoning Province; C) ratio of wave velocity; D) fault displacement at Jin XIan short levelling station (Δ = 190 km); F) ground tilt at Yingkou station (Δ = 18 km).

[Sichuan Province Seismological Bureau, 1979]. In Figures 9 and 10, the gaps are not coincident with the main shock's aftershock zone. Chen et al. [1979] also reported the occurrence of a narrow band of anomalous seismicity near the future main shock before the four great earthquakes.

Fault Displacement

Anomalous fault displacements were detected at station Jin Xian (Δ = 190 km) before the Hiacheng

earthquake [The Geodetic Survey Brigade for Earthquake Research, State Seismological Bureau, 1977]; at Xia Guan (Δ = 170 km) before the Longling earthquake [Chen et al., 1979]; at Tahueichang (Δ = 180 km) and Niukouyu (Δ = 200 km) before the Tangshan earthquake [Zhang et al., 1979]; and at station Songpan (Δ = 50 km) before the Songpan earthquake [Sichuan Province Seismological Bureau, 1979] (Figures 2e, 3c, 4g, h and 5e respectively). These data indicate that obvious fault displace-

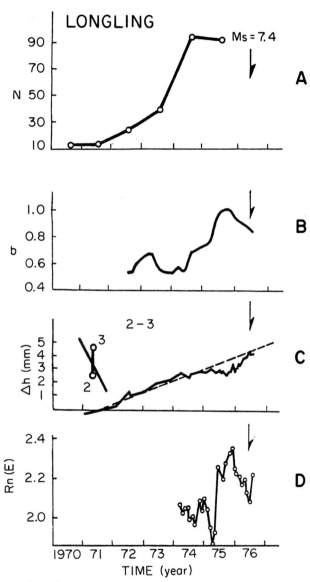

Fig. 3. Changes in medium-term anomalies before the Longling earthquake. A) earthquake frequency in Longling area and its surroundings; B) b-value; C) fault displacement at Xiaquan short levelling station (Δ = 170 km); D) radon content at Xiaquan station (Δ = 170 km).

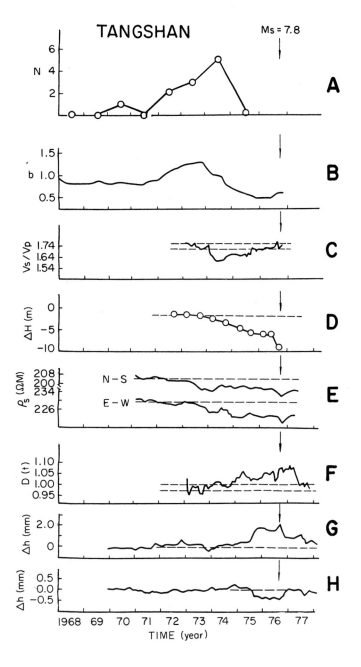

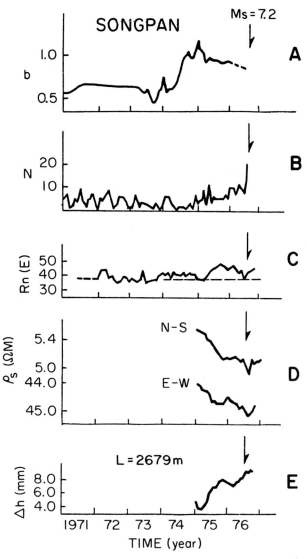

Fig. 4. Changes in medium-term anomalies
before the Tangshan earthquake. A) earthquake frequency (M ≥ 4.0) in the Tianjin-
Tangshan region; B) b-value in Tangshan
region; C) ratio of wave velocity in the
Tianjin-Tangshan region; D) Level of underground water at Tangshan well; E) resistivity at Changli station (Δ = 70 km); F)
average change in radon content of underground water in Beijing-Tianjin-Tangshan
region; G) fault movement at Tahuishag station (Δ = 200 km); H) fault movement at
Niukouyu station (Δ = 180 km).

Fig. 5. Changes in medium-term anomalies
before the Songpan earthquake. A) b-value;
B) small earthquake frequency in Longmen
Shan area; C) radon content at Songpan station (Δ = 50 km); D) resistivity at Wudu
station (Δ = 100 km); E) short levelling at
Songpan station (Δ = 50 km).

ment was recorded during a period from 1 to 2
years before the main shocks in the region surrounding the epicentral area. The amplitude of
the fault displacement reached one to several
millimeters.

As shown in Figure 2f, tilt in an E-W direction
at Yingkou station (Δ = 18 km) deviated from the
normal trend of yearly variation from the beginning of 1974 until the occurrence of the Haicheng
event; the deviation, Δφ, reached 8". This anomalous change is about 20 times greater than the
root-mean-square error (0.4").

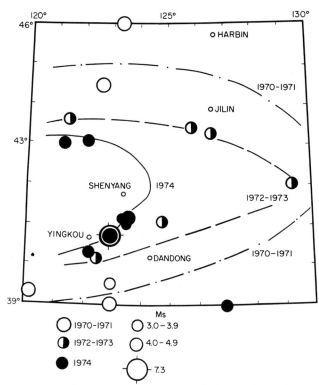

Fig. 6. Regional seismicity migration ($M_s > 3.0$) before the Haicheng earthquake (1970-1974).

Radon Content

The anomalous change in radon content of underground water before these earthquakes is shown in Figures 3d, 4f and 5c. From these curves, we can see that increases in radon content were observed from 1 to 2 years before the main shocks [Chen et al., 1979; Zhang and Qiuo., 1979; Sichuan Province Seismological Bureau, 1979]. In Figure 4f [from Zhang and Qiuo, 1979] D(t), the average index value of the variation of radon content in eight

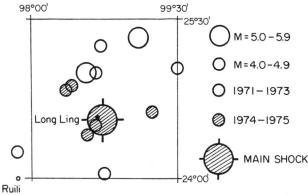

Fig. 7. Epicentral distribution ($M_L \geq 4.0$) in the Longling area (1971-1975).

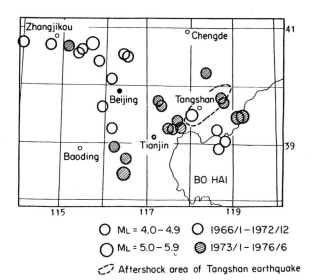

Fig. 8. Epicentral distribution ($M_L \geq 4.0$) before the Tangshan earthquake (January 1966 - June 1976).

wells in Bejing-Tianjin-Tangshan region, is defined as

$$D(t) = \frac{1}{8} \sum_{i=1}^{8} \frac{Rni(t)}{\overline{Rni}}$$

where Rni(t) denotes the monthly average value of radon content in well i and $\overline{Rni}$ is its average monthly value during the period 1972-1975.

Water Level and Resistivity

The water level at Tangshan deep well dropped more than five meters from the first half of 1973 to the occurrence of the Tangshan main shock [Wang, 1979]. At the same time, the electric resistivity at Changli station ($\Delta = 84$ km) decreased by three percent or more before the Tangshan earthquake (Figure 4d, e) [Zhao et al., 1978].

After analyzing the spatial and temporal development of medium-term anomalies before four great earthquakes with M 7, we summarize some of the common features as follows:
1. Medium-term anomalies within 200 to 300 km from the epicenter may be observed three to four years prior to a large (M > 7) earthquake.
2. Seismicity anomalies such as earthquake frequency, b-value, earthquake migration, etc. appear first. After an increase in seismicity other anomalies such as changes in water level and radon content of underground water, electric resistivity, fault displacement, etc. appear. They do not appear simultaneously.
3. With regard to the distribution of anomalies, seismicity in the future source area (after-

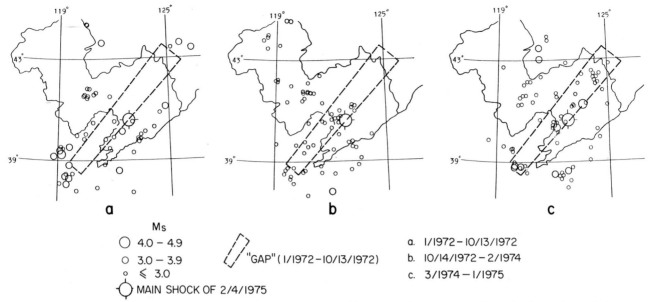

a

b

c

Ms

○ 4.0 – 4.9

○ 3.0 – 3.9

∘ ≤ 3.0

⊖ MAIN SHOCK OF 2/4/1975

⬭ "GAP" (1/1972–10/13/1972)

a. 1/1972–10/13/1972

b. 10/14/1972–2/1974

c. 3/1974–1/1975

Fig. 9. Epicentral distribution map in Liaoning Province and its surroundings before the Haicheng earthquake.

shock area of the main shock) remained relatively quiet before the occurrence of the great earthquakes (Figures 6, 8, 10). The area which showed the strongest anomalous variations sometimes coincides with the future source area of the large earthquake. For example, Figure 11 shows this characteristic before the Tangshan earthquake.

4. There are certain similarities in the anomalous patterns before the four earthquakes. For example, all b-values first increase and then decrease; resistivity usually decreases; and the radon content of underground water increases.

In order to understand the physical implication of medium-term anomalies on the preparatory pro-

cess of large a earthquake, we compare the effects during anomalous time with the results of experiments in rock failure [Zhang et al., 1979].

Figure 12 shows the experimental curves for rock rupture under high pressure. Included are microfracture, apparent resistivity, radon content, and pore pressure in saturated or partially saturated rock samples. The abscissa in the figure

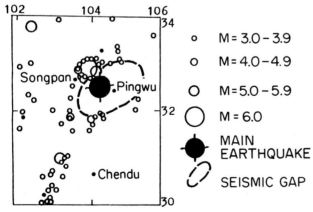

Fig. 10. Epicentral distribution ($M_L \geq 3.0$) in the Songpan-Longmen Shan area before the Songpan earthquake.

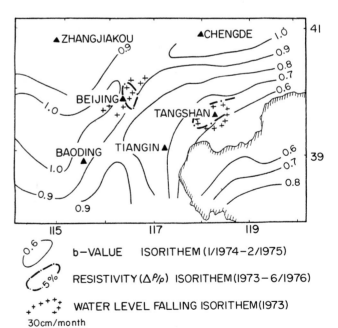

b–VALUE ISORITHEM (1/1974–2/1975)

RESISTIVITY (Δρ/ρ) ISORITHEM (1973–6/1976)

WATER LEVEL FALLING ISORITHEM (1973)

30cm/month

Fig. 11. Map of b-value, resistivity and variation in water level before the Tangshan earthquake.

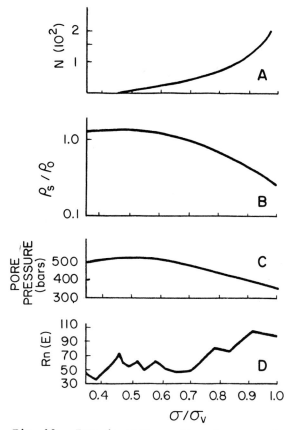

Fig. 12. Experimental results of some rock samples under high pressure. A) micro-fracture; B) apparent resistivity; C) pore pressure; D) radon content.

denotes the precentage stress, σ/σ_v. When the loading stress, σ, on the rock sample exceeds 0.5 σ_v (fracture strength), the frequency of micro-fracture increases, the resistivity and the pore pressure decrease, and radon content increases. The characteristics of these curves are similar to the anomalous changes in the frequency of small earthquakes, electrical resistivity, and the level of and radon content of underground water prior to some large earthquakes. The above comparison may explain why the medium-term anomalies which occur before a large earthquake may relate to micro-fracture development and interaction, fluid movement, etc. during the process of stress accumulation in rock within the crust. Therefore, the medium-term anomalies correspond to the second and third stages of the IPE (Institute of Physics of the Earth) model and the DD (dilatancy-diffusion) model for earthquake forerunners [Myachkin et al., 1975].

There are differences between the medium-term characteristics of the four earthquakes and other research results. First of all, Rikitake [1976] collected statistical data and worked out relations between the anomalous duration and magni-

tude. These relations show that the medium-term anomaly lasts for 6 to 7 years prior to an earthquake with magnitude greater than 7. Based on the data of the above four great earthquakes, the duration of most of the medium-term anomalies is less than three or four years. The anomalous time for the velocity ratio lasted for only three years prior to the Tangshan earthquake (M = 7.8) [Jin et al., 1979], whereas, according to the DD model, the duration would be expected to be 10 years. In addition, it is well known that the two models (DD and IPE) of earthquake forerunners indicate that various anomalies should be observed at the same time. However, our results show that the beginning and the development of various precursors in the medium-term anomalies are not synchronous; normally, the seismicity anomalies appear one or two years earlier than the others.

Secondly, with respect to the anomalous pattern, there are remarkable differences in b-value curves. The DD and IPE models suggest that the b-value should gradually decrease before a great earthquake, but the variation of the b-value in our results demonstrates a pattern which first increases and then decreases before the four great earthquakes.

Finally, some phenomena, such as a narrow band of seismicity, a seismic gap or migration of seismicity are not emphasized in the second stage (medium-term) of the DD or IPE models of earthquake forerunners.

Temporal and Spatial Features of Short-Imminent Anomalies

In the later stage of medium-term anomalies, obvious changes in the mean rate of some anomalous curves are found from about 1 to 3 months prior to the four great earthquakes. These changes may be considered to mark the start of the short-term anomalous stage; for example, changes in the level of underground water and resistivity before the Tangshan earthquake (Figure 4d, e) and the b-value and fault displacement prior to the Haicheng event (Figure 2d, e). During the short-imminent anomaly stage sudden variations with large amplitudes are sometimes recorded. The characteristics of short-imminent anomalies are summarized in the following sections.

Anomaly Patterns

We will discuss first the impulsive patterns. Prior to the Songpan earthquake there was an impulsive increase in radon content (Figure 16c) [Sichuan Province Seismological Bureau, 1979]. There had been no obvious change in earth current at the Tengcheng middle school (Δ = 40 km) since 1976. However, a sudden impulse change (amplitude reaching 100 μA) was observed within two months prior to the Longling earthquake (Figure 13b). The spontaneous eruption of an abandoned oil well on the Cangzhou fault at Qing Xian occurred sometime within two months prior to the Tangshan earthquake (Figure 15d). However, since the oil well had been drilled and closed in 1973, this eruption went unobserved.

Secondly, anomalous curves go up and down (step changes) lasting until the occurrence of an earthquake. For example, an anomalous change in the ground tilt ($\Delta\phi$) at Shenyang station ($\Delta = 150$ km) was detected (Figure 13a, $\Delta\phi = d\phi_{Ni} - Dd\phi_{Ei}$; $D = $ const. ϕ_N and ϕ_E denote tilt components in the N-S and E-W direction, respectively). The value decreases with larger amplitude three days prior to the Haicheng event. The amplitude of the anomalous decrease was seven times greater than the root-mean-square error in the normal period (Figure 13a). The earth current at Yingkou station (Figure 13b) and the radon content at the Liuhetang Hot Spring (Figure 13f) showed a step variation three days before the Haicheng earthquake. In addition, a step change in the water temperature (amplitude = 10°C) at the Palazhang Hot Spring ($\Delta = 10$ km) was observed one and a half months prior to the Longling earthquake (Figure 14c).

Finally, we note gradual changes of large amplitude. For example, the level of underground water at the Pagouzhuang well, the radon content at Langfang station; and the electric resistivity at Chanci station prior to the Tangshan earthquake are considered to fall into this category (Figure 15a, b, c). It is noted that the appearance of impulsive and step variations are superimposed on the background of the continuous changes.

Spatial Distribution of Anomalies

Macroseismic anomalies, such as underground water, unusual animal behavior, etc., are sometimes observed in the future epicentral area or along and in the vicinity of seismogenic faults. For instance, the distribution of macroseismic anomalies before the Haicheng earthquake is consistent with the NW strike of the seismic fault which generated the Haicheng event (Figure 17).

At a depth of 80 cm, the ground temperature rose 1.5 degrees in the vicinity of the epicentral area prior to the Tangshan main shock [Chen et al., 1979] (Figure 18).

Synchronization of the Time of Anomalies

The variation in foreshocks, ground tilt, earth current, radon content of underground water, animal behavior, etc. appeared almost synchronously three days prior to the Haicheng earthquake. These variations showed a step increase. In addition to these variations, unusual weather, such as a mysterious light and ground mist, were observed during this period.

It therefore seems that short-imminent anomalies mainly last for a few months or several days prior to a great earthquake. According to [Monakhov, 1979], the duration of short-term and imminent precursors is no more than 10 days. The correlation between the duration of these anom-

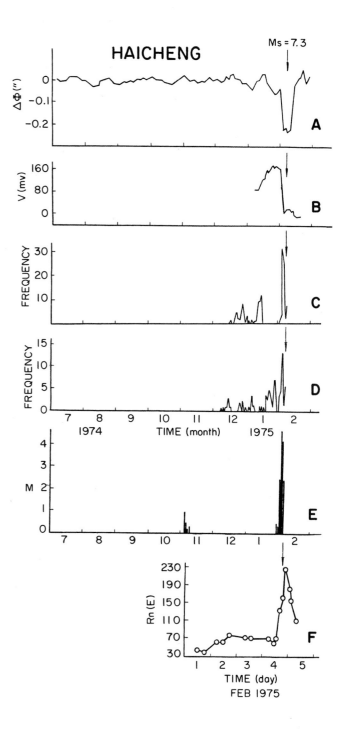

Fig. 13. Changes in short-term anomalies before the Haicheng earthquake. A) ground tilt at Shenyang station ($\Delta = 150$ km); B) electrical potential difference at Yangkou station ($\Delta = 15$ km); C) number of unusual animal behavior in the Dandong area ($\Delta = 150$ km); D) number of unusual underground water in Dandong area ($\Delta = 150$ km); E) foreshocks; F) radon content at the Liuhetang Hot Spring ($\Delta = 72$ km).

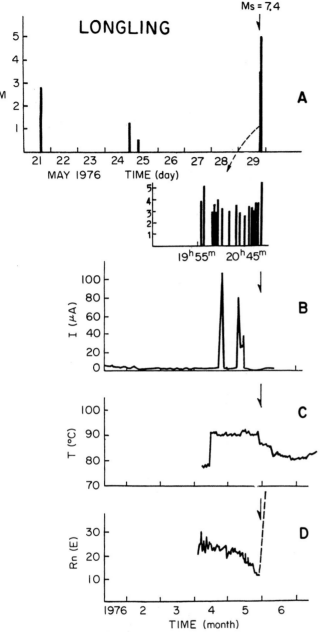

Fig. 14. Changes in short-term anomalies before the Longling earthquake. A) fore-shocks; B) sudden changes in earth current at Tenchong Middle School ($\Delta = 40$ km); C) water temperature at Palazhang Hot Spring ($\Delta = 10$ km); D) radon content at Palazhang Hot Spring ($\Delta = 10$ km).

alies and magnitude of the main shock may be uncertain.

Difference in Anomalies

Observational data show that prior to the four earthquakes the kinds of anomalies were not the same. For instance, a few days before the Hai-cheng and Longling earthquakes there was obvious foreshock activity, however this was not the case prior to the Tangshan and Songpan events.

It is very important to analyze the features of short-imminent anomalies for their role in the prediction of future great earthquakes. At this time, however, our observations of short-imminent precursors of great earthquakes are not sufficient and systematic enough.

Discussion

Recently, Myachkin et al. [1975] described two models (DD and IPE), based on the dilatation of

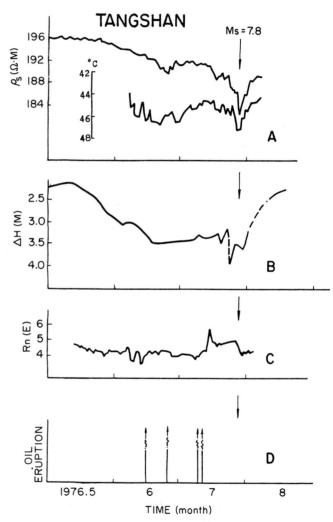

Fig. 15. Changes in short-medium anomalies before the Tangshan earthquake. A) resistivity and temperature of well water at Changli station ($\Delta = 70$ km); B) water level of Paguozhuang well ($\Delta = 40$ km); C) radon content at Lanfang station ($\Delta = 130$ km); D) oil eruption at abandoned well at Qing Xian ($\Delta = 150$ km).

rock and the development and interaction of cracks, which can explain some features of medium- and short-term anomalies. However, they did not discuss in detail the sudden changes with high amplitude during the short-imminent stage. The IPE model states that the short-term changes relate to the formation of small cracks before the main shock [Myachkin et al., 1976]. Rikitake [1976] hypothesized that short-range precursors

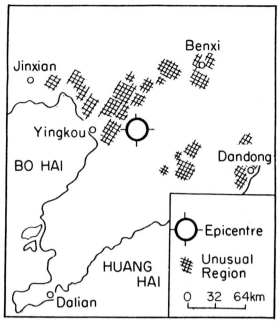

Fig. 17. Regions where unusual animal behavior took place before the Haicheng earthquake.

are related to seismic creep in the focal zone.

In this paper, we attempt to explain some of the characteristics of medium- and short-term anomalies by analyzing the "creep-stress concentration" along crustal faults.

Some results of laboratory experiments show that a number of factors (e.g., the presence of weak alteration minerals, high temperature, high pore pressure, low effective confining pressure, large thicknesses of fault gouge, etc.) could be responsible for promoting fault creep [Byerlee and Summers, 1975].

Depending on differences in confining pressure, rock strength, temperature, etc., at different depth in the crust (Table 2), it is possible that some factors such as low confining pressure, high porosity, lower rock strength, more water, etc., could be responsible for promoting fault creep in the upper crust. In the lower crust fault creep is promoted by higher temperature [Brace and Byerlee, 1970], and the crustal conditions between the upper and lower crust may be conducive to rapid fault rupture. In fact, the distribution of the depth of hypocenters is principally confined to a certain depth within the crust. Thus, at the depth between the upper and lower crust, there are some favorable conditions for the accumulation of elastic energy and the release of energy by rapid fault dislocation. We call this part of the crust the "energy accumulation sphere". We may therefore roughly divide the crust into three parts: the upper crust, the energy accumulation sphere and the lower crust. Here, in a qualitative way, as shown in Figure 19, we present a "creep-stress

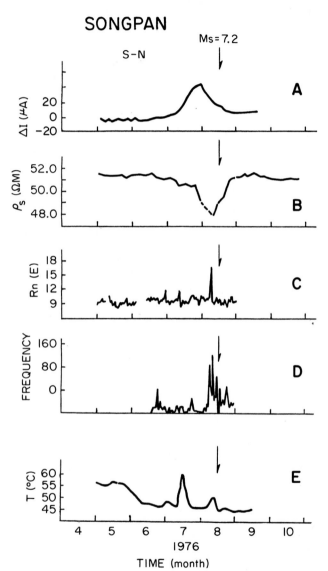

Fig. 16. Changes in short-term anomalies before the Songpan earthquake. A) earth current at Kangding Middle School (Δ = 350 km); B) resistivity at Wudu station (Δ = 120 km); C) radon content of underground water at Guza station (Δ = 350 km); D) number of unusual animal behavior and underground water in Songpan-Longmen Shan area; E) water temperature at Litang Hot Spring (Δ = 350 km).

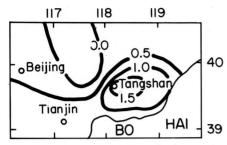

Fig. 18. Isogeotherm map (°C) at 80 cm prior to the Tangshan earthquake (76/7/23-7/25).

concentration" in the seismic fault and the area surrounding it and discuss the relationship between "creep concentration" and the process of medium-term and short-term anomalies.

Under a regional dynamic force, shear creep may take place along the fault surface in the lower crust since it has a higher temperature. Because the "energy accumulation sphere" is stronger, the fault is locked to the fault creep in the lower crust. Thus, it is possible that stress accumulation and concentration begins and develops in the "energy accumulation sphere". When the level of stress accumulation rises to about 50% of the rock strength, the process of microcrack formations starts and "stick-slip" begins to occur around the fault in the energy accumulation sphere. At the same time, we record anomalous change of seismic activity (such as the increase of small earthquakes, the appearance of narrow seismic bands, etc.) which directly reflect the process of stress accumulation in the sphere. In conjunction with the increase of stress accumulation in the sphere, the stress level rises in the upper crust, so that the formation of microcracks begins, develops, and at the same time interacts with "stable sliding". Observational methods (e.g., electrical resistivity, level of radon content of ground water, land deformation, etc.) will record medium-term anomalies which reflect the stress-strain process in the upper crust. These variations will be observed at a

later time than the anomalous seismicity in the energy accumulation sphere (Figure 19).

Stress-strain greatly increases and larger stable sliding can take place along some parts of the seismic fault in the upper crust as the stress level reaches a critical level prior to rapid earthquake dislocation in the sphere. Thus, a series of processes, i.e., the development and interaction of microcracks, expansion and closing, fluid movement, etc. become more complicated and intense, and consequently the average rates of some medium-term observational curves change remarkably showing a sudden and unusual variation. Therefore, short-term anomalies may be related to the fault creep process in the upper crust. Of course, fault creep in the upper crust will also affect the stress state in the energy accumulation sphere.

Finally, the whole preparatory cycle for an earthquake will culiminate in a rapid rupture at a fault surface in the energy accumulation sphere.

Using inversion of geodetic data, Chen et al. [1979] and Zhang [1979] concluded that creep occurred from 15 to 34 km in the lower crust and from 1 to 8 km in the upper crust before the Tangshan earthquake. The results and inferences from

Table 2. Confining Pressure, Rock Strength and Temperature in the Crust

Depth (km)	Confining Pressure (kb)	Temperature (C°)	Granite Strength (kb)
0	0	0	1.5
5	1.3	200	5.95
10	2.7	360	7.50
20	5.5	550	7.35
30	8.5	700	6.6
40	11.5	850	4.8

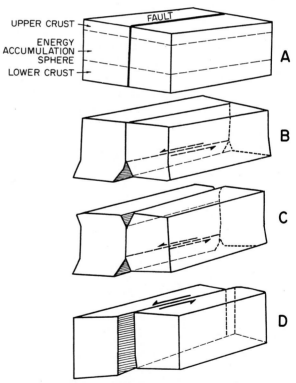

Fig. 19. "Creep-stress concentration" figure. A) crust; B) creep in the lower crust, stress concentration in the energy accumulation sphere; C) creep in the upper crust; D) dislocation along the fault surface in the energy accumulation sphere.

these data are major foundation towards the establishment of a pattern of creep-stress concentration.

As mentioned above, we can understand some of the characteristics of medium- and short-term anomalies based on "creep-stress concentration". For instance, why the starting time of various medium-term anomalies are not synchronous; why large anomalous amplitudes are observed in and near the epicentral area of the main shock; and the reason for the existence and sudden variation in short-term anomalies. However, this pattern cannot explain such phenomena as the migration of regional seismicity over an area of several hundred kilometers. This points out that our model lacks the ability to completely explain the preparatory process of an earthquake and all of the anomalous phenomena which occur before a great earthquake.

Conclusion

In this paper, we have given a preliminary analysis of some of the temporal and spatial features of medium- and short-term anomalies before great earthquakes with magnitudes seven or larger which occurred in China. The results indicate that certain medium-term anomalies (including changes in seismicity, land deformation, level of and content of radon in underground water, electrical resistivity, etc.) were recorded within a radius of 200 to 800 km from the future main epicenter several years prior to a great earthquake. Seismicity anomalies were observed at an earlier time than the other anomalous variations. Several types of anomalous changes with larger amplitudes were sometimes distributed in and near the future epicentral area. In the latter part of the medium-term stage, obvious changes in the mean rate of some anomalous processes were found approximately a few months prior to a great earthquake. This may mark the start of the short-term stage where sudden variations with large amplitudes were recorded. In addition, such remarkable phenomena as underground water, animal behavior, mysterious light, ground mist, etc. were observed in the epicenter area and near the seismic fault.

We consider that medium- and short-term anomalies may be related to a series of factors and processes such as development and interaction of microfractures, stable-sliding and stick-slip of cracks, fluid movement, and seismic fault creep in crustal rock before a great earthquake. We also tried to use a pattern of "creep-stress concentration" along faults to understand some of the characteristics of medium- and short-term anomalies.

For earthquake prediction, it is very important to study the characteristics of medium-, short-term and imminent anomalies before great earthquakes. However, anomalous variations are not necessarily precursors of earthquakes, and they can occur without the subsequent occurrence of an earthquake. In other words, the anomalous vari-

ations discussed in this paper are not only related to the processes of the development of a large earthquake but also to other aseismic factors, and they are also affected by climatic and astronomical factors. In conclusion, all indications point to the fact that a great deal of future research is necessary in the field of earthquake prediction.

References

Brace, W. F., and J. D. Byerlee, California earthquake: why only shallow focus, Science, 168, 1573-1575, 1970.

Byerlee, J. D., and R. Summers, Stable sliding preceding stick-slip on fault surfaces in granite at high pressure, Pure Appl. Geophys., 113, 63-68, 1975.

Chen, F. et al., Tangshan Earthquakes, Seismology Press, 1979.

Chen, L. et al., Longling Earthquakes in 1976, Seismology Press, 1979.

Chen, Y. et al., A dislocation model of the Tangshan earthquake of 1976 from the inversion of geodetic data, Acta. Geophysica Sinica, 22, 202-217, 1979.

Chen, Z. et al., Characterics of regional seismicity before major earthquakes, International Symposium of Earthquake Prediction, contributed paper, Paris, 1979.

Ding, G., S. Mei, and Z. Ma, Method of earthquake prediction, International Symposium on Earthquake Predictin, contributed paper, Paris, 1979.

Feng, R., Q. Pang, and Z. Fu, Variation of V_p/V_s before the Haicheng earthquake of 1975, Acta Geophysica Sinica, 19, 295-305, 1976.

Geodetic Survey Brigade for Earthquake Research, State Seismological Bureau, Ground surface deformationof the Haicheng earthquake of magnitude 7.3, Acta Geophysica Sinica, 20, 251-263, 1977.

Guo, Z., and Q. Bauyan, Physics of Seismic Source, Seismology Press, 1979.

Jin, A. et al., The study on the changes of the velocity of the seismic waves propagating through the medium underneath the area surrounding Peking, The First Conference of the Seismological Society of China, 1979.

Li, Q. et al., Time and space scanning of the b-value, Acta Geophysica Sinica, 21, 101-124, 1978.

Monakhov, F. I., The duration of short-term earthquake precursors, Fezika Zemli, 1, 91-94, 1979.

Myachkin, V. I. et al., Two models for earthquake forerunners, Pure Appl. Geophys., 113, 169-181, 1975.

Myachkin, V. I., B. V. Kostrov, and G. A. Sobolev, Physics of the focus and earthquake precursors, FAN, Tashkent, Uzbek SSR, 121-131, 1976.

Rikitake, T., Earthquake prediction, in Development in Solid Earth Geophysics, 9, Elsevier Science Publishing Co., 1976.

Sichuan Province Seismological Bureau, Songpan

Earthquakes in 1976, Seismology Press, 1979.

Wang, C., The characters of the variation of water-level in deep wells before and after the 1976 Tangshan earthquake, International Symposium on Earthquake Prediction, contributed paper, Paris, 1979.

Zhang, G., and J. Qiuo, An analysis of the process of preparation and the medium-term precursors of the 1976 Tangshan earthquake (M_s = 7.8), International Symposium on Earthquake Prediction,, contributed paper, Paris, 1979.

Zhang, Y., The study of creep along the fualt near Tangshan before Tangshan earthquake, in Research in Earthquake Sciences, 1, Seismology Press, pp. 51-52, 1979.

Zhao, Y., and F. Qian, Electrical resistivity anomaly observed in and around the epicentral area prior to the Tangshan earthquake of 1976, Acta Geophysica Sinica, 21, 181-190, 1978.

ON PRECURSORY PHENOMENA OBSERVED AT THE YAMASAKI FAULT, SOUTHWEST JAPAN,
AS A TEST-FIELD FOR EARTHQUAKE PREDICTION

Yoshimichi Kishimoto

Disaster Prevention Research Institute, Kyoto University, Kyoto, Japan

Abstract. The Yamasaki fault is a typical left-lateral strike-slip fault in the northern Kinki District, Southwest Japan. In 1977 the Yamasaki fault was chosen as a test-field for earthquake prediction, and a composite observation was commenced. During three years since 1977, four earthquakes of M3.5 or more occurred near the fault. The first one of M3.7 in 1977 was the first example of a rather small earthquake which accompanied various precursors of short- and long-term. Some short-term precursors and a spatio-temporal pattern are explained. These four earthquakes revealed a common anomalous change of chlorine ion content in underground water. This seems to indicate that the underground water is very sensitive to earthquake occurrence at the Yamasaki fault area, and that the chlorine ion content might possibly become an effective precursor. It is important to investigate the mechanism of any precursor in order to apply the result at the test-field to a great earthquake.

SEISMOTECTONIC FEATURES OF THE YAMASAKI FAULT

The Yamasaki fault is a left-lateral strike-slip fault, situated about 100 km to the west of Kyoto City. It extends 80 km long in a direction nearly NW to SE, as indicated in Fig.1. According to geological and geomorphological evidence, this fault has been very active for the last several hundreds-of-thousands of years.

Microseismicity in the vicinity of the Yamasaki fault had been elucidated by the end of the 1960's by the Tottori Microearthquake Observatory, Disaster Prevention Research Institute, Kyoto University. Fig.1 shows distributions of epicenter and active fault in the northern Kinki and eastern Chugoku Districts, at the center of which the Yamasaki fault exists. Various close correlations between the seismic activity and active fault are discernible in this figure. Along the Yamasaki fault, in particular, a linear arrangement of epicenters is recognized. In other words, two aseismic areas are adjacent to each other bounded by a belt of high seismicity along the Yamasaki fault. This means that the Yamasaki

fault is a boundary between two geological blocks and stresses are released primarily at that boundary.

Investigations of earthquake mechanism in the northern Kinki, including the Yamasaki fault area, revealed that most earthquakes are of the strike-slip type with vertical null vector, and that their maximum compressive stresses are in the EW direction [Nishida, 1973; Kishimoto and Nishida, 1973; Huzita et al., 1973]. Taking into consideration both this seismological result and various geological evidence, it may be said that a tectonic EW compression has acted upon this area throughout recent geological times, including the present.

Seismic activity along the Yamasaki fault is pretty high not only for microearthquakes but also for earthquakes of moderate magnitudes. In 1961, an earthquake of M5.9 occurred at the cen-

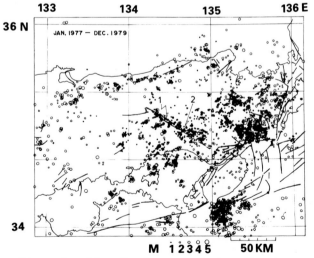

Figure 1. Distributions of microearthquake and active fault in the northern Kinki and eastern Chugoku Districts, Southwest Japan. Lines extending from NW to SE denoted by 1 is the Yamasaki fault system, and a triangle denoted by 2 represents the position of the observation tunnel shown in Fig.2.

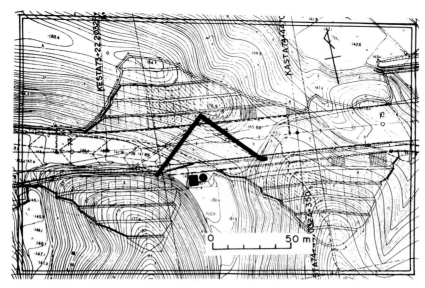

Figure 2. The observation tunnel of L-shape (thick line) and the former topography in the vicinity. A belt in EW direction indicates the Chugoku Highway which corresponds approximately to the fault. The tunnel is constructed 8 m below the road surface. It is recognized that ridges are bent left-laterally at the fault. Square and circle denote the telemetering room and 20 m-well, respectively.

tral part of the fault, which has been the largest one in the Yamasaki fault area for the past several decades. The largest earthquake which has occurred since 1965, the period of microearthquake observation, was M5.1 on Sept. 21, 1973.

A Gutenberg-Richter's relation, derived from microearthquake observation for more than 15 years, shows that in the Yamasaki fault area earthquakes of magnitudes 4, 5 and 6 are expected once in 2, 8 and 40 years, respectively. Also, a very large earthquake of M7 or more can be expected to occur once in several hundreds of years. This result is not so unreasonable, because we have record of a large historical earthquake of M7.1 in 868, and moreover, reality of this earthquake has been confirmed by a trench test carried out at this fault very recently [Okada et al., 1980].

TEST-FIELD PROJECT

Test-field project is one item in the Japanese Earthquake Prediction Program. It aims at experimental investigations of prediction, by the use of the moderate earthquake, which occurs rather frequently but brings about no damage, as well as many other kinds of observation as possible.

The Yamasaki fault is the only one test-field in Japan at present. Composite observations of many kinds are being carried out by the "Research Group for the Yamasaki Fault" consisting of about 20 members from several universities. An L-shaped observation tunnel has been constructed across the fault and 8 m below the ground sur-

face, as shown in Figs. 1 and 2. Inside or in the vicinity of the tunnel, observations of ground strain, radon emission, discharge and level of underground water, geoelectric potential, geomagnetic force, electric conductivity of rock and others, are in progress. Geochemical observations of underground water are done mainly at a hot spring 10 km to the east of the tunnel. Most of these observations are telemetered to the Disaster Prevention Research Institute in Kyoto.

In the following section, we shall first discuss a small earthquake of M3.7 which occurred near the tunnel on Sept.30, 1977 and was accompanied by various precursors. Next, we shall mention some short-term precursors which seem particularly significant at the Yamasaki fault.

SOME PRECURSORY PHENOMENA OBSERVED
AT THE YAMASAKI FAULT

An earthquake of M3.7 on Sept.30, 1977

This earthquake was located nearly on the Yamasaki fault, as shown in Fig.6. It was fortunate that various observations had been commenced 6 months before the earthquake and so much data was available. Detailed examination of the data verified that several precursors, of long- and short-term, accompanied this earthquake. Fig.3 indicates some observational results over several months including the earthquake. Precursory changes of chlorine ion content and strain rate are remarkable.

Chlorine ion content is observed once a day by chemical analysis of the water from a self-

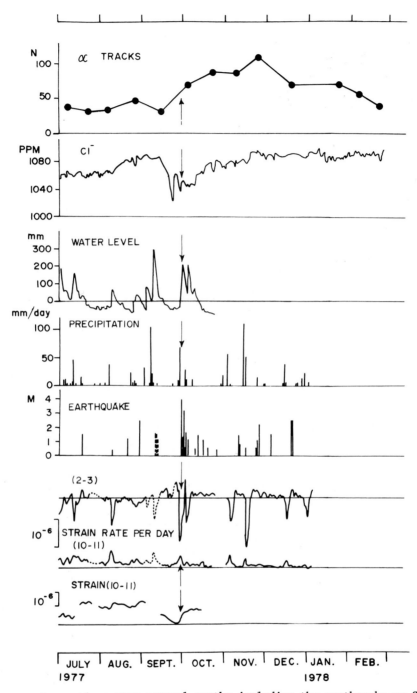

Figure 3. Various observations over several months including the earthquake on Sept.30, 1977 represented by arrows. Two pairs of numerals, (2-3) or (10-11), in the graphs of strain or strain rate denote two pairs of measuring points between which strain was observed.

gushing, 300 m-deep well at a hot spring mentioned earlier. In Fig.3, the chlorine ion content showed a sudden decrease about 14 days before the earthquake, and then returned to the previous level gradually. This phenomenon will be discussed in the next subsection in some detail.

Large changes of strain rate just before the earthquake have already been discussed by Oike [1977] and Oike and Kishimoto[1977]. This change corresponds to right-lateral slip of the fault caused perhaps by a sudden rainfall after a relatively long and dry period. This slip is op-

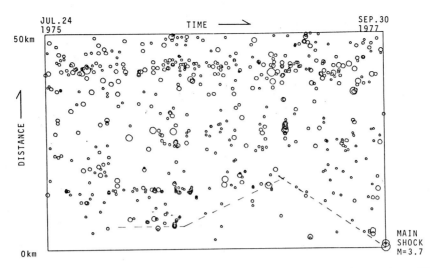

Figure 4. A spatio-temporal pattern in the case of the earthquake on Sept.30, 1977.

posite to the normal left-lateral slip of this fault. It is said that such rainfall triggers an anomalous movement of the fault and then the earthquake occurs shortly after that movement. Therefore, such anomalous change of strain would be of use as an immediate precursor.

As a long-term phenomenon, the preseismic quiescence was observed even for a small earthquake. Fig.4 shows an expression of this phenomenon. The right end of the abscissa corresponds to the earthquake in the problem. The abscissa and ordinate denote time and epicentral distance of another earthquake from that in problem.

Seismic activity begins to subside away from the vicinity of the epicenter of earthquake to come some hundreds of days before it occurs, and after some time again reactivates back toward that epicenter. If, accordingly, we can determine this turning point, this phenomenon would be another useful long-term precursor.

This phenomenon was also observed in other earthquakes near the fault. Fig.5 shows a superposed graph of three earthquakes, designated by an inserted figure. This graph points out that preseismic quiescence appears in the same manner for the earthquakes along or near this fault. It seems important to examine whether this phenome-

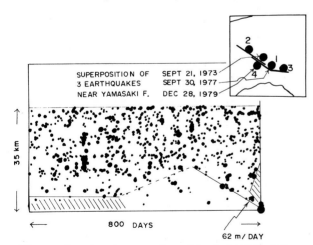

Figure 5. A superposed spatio-temporal pattern of three earthquakes along the Yamasaki fault. A shaded portion on the left side denotes a small aseismic area which has lasted pretty long and has been filled up by the earthquakes concerned. That on the right side suggests generation of a small aseismic portion near the epicenter just before its occurrence.

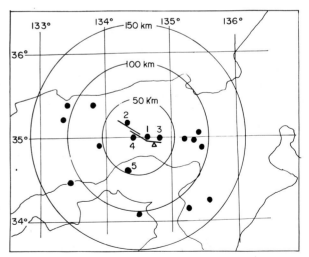

Figure 6. Epicenters of earthquakes of M3.5 or more for three years from March, 1977. Nos. 1 to 5 are especially near earthquakes to the test site, and Nos. 1 to 4 are referred to the text.

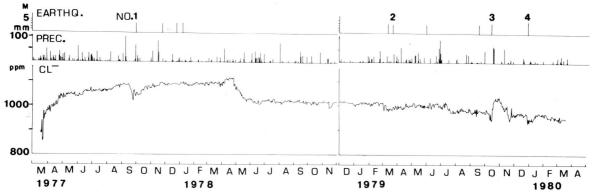

Figure 7. Time variation of chlorine ion content for three years together with main earthquakes and precipitation. Epicenters of earthquakes Nos. 1 to 4 are referred to Fig.6.

non is observable or not for small earthquakes which occur in other places or have different mechanisms.

Chlorine ion content and other observations

For about three years from 1977 up to the present, four earthquakes of magnitude of 3.5 or more have occurred near the Yamasaki fault, as shown in Fig.6. M3.5 was assigned as a standard of magnitude for which some precursors may be expected, referring to earthquake, No.1 below, discussed in the previous subsection. These earthquakes are as follows:

    No.1 Sept.30, 1977, M3.7
    No.2 March 22, 1979, M3.5
    No.3 Oct.13, 1979, M4.3
    No.4 Dec.28, 1979, M4.9

The most remarkable fact was that changes of chlorine ion content were observed for three earthquakes, Nos.2 to 4, in the same manner as in earthquake No.1, as shown in Fig.7. In the case of No.3, the chlorine ion content showed a sudden decrease about 10 days before the earthquake, but the change in No.4 was coseismic. In the case of No.2 the change of chlorine ion content is not so clear as those in other three earthquakes, perhaps owing to its distance. It may be concluded, however, that these changes of chlorine ion content are really related to earthquake occurrence, since the postseismic behavior of it going back to the previous level gradually is the same for all four earthquakes. These results regarding the chlorine ion content suggest that the same precursor has been observed when the earthquake occurred under the same conditions. And, this fact is considered to indicate one of the most important characteristics needed for precursors.

Some new observations were commenced just after No.3 earthquake and were in operation at the time of No.4. Some of these are shown in Fig.8.

Electric conductivity measured in the same underground water showed similar behavior to chlorine ion content, as seen in the figure.

This may be reasonable, if chlorine is abundant in the water and largely controls electric conductivity. While the chlorine ion content is observed by chemical analysis once a day, electric conductivity is measured by continuous observation, and therefore, the latter is suitable to determine the fine process of change of conductivity. Fig.9 shows a time variation of electric conductivity, in which the hourly value is plotted. It is clearly recognized that the change of conductivity was just coseismic, and moreover, it did not move stepwise but gradually over several days. It is also observed in the figure that plotted points show pretty large fluctuation over several days before the earthquake, but they become stable thereafter. This may indicate mixing of some different kinds of underground water, although its mechanism is not clear yet. If this fluctuation of electric conductivity is always observable, it would be also of use as a precursor. In Fig.8, we also show that discharge of the underground water behaved similarly to the electric conductivity during the postseismic period.

Thus, referring to observational results mentioned in this subsection, it is said that, particularly for earthquakes in the Yamasaki fault area, the state of underground water is closely related to their occurrence and its changes have some possibility to become a sensitive precursor.

The lowest two figures in Fig.8 indicate observations of geoelectric potential difference. One of them is set inside the tunnel and the other outside. The former and the latter are 30 and 500 m long, respectively, and both are perpendicular to the fault. Large preseismic changes before earthquake No.4 noted in both observations may be interpreted as a precursory change, because these are too large to attribute only to rainfall, and moreover the geoelectric potential is considered to have changed as well in the case of earthquake No.1. However, these observations had not sufficiently long period before the earthquakes, and so more investigations are necessary.

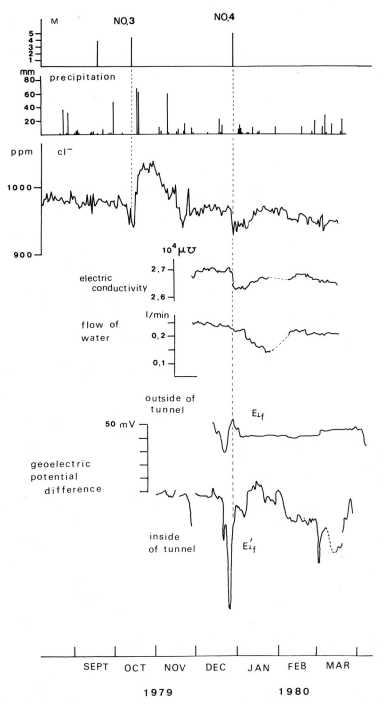

Figure 8. Observations of some new kinds as well as chlorine ion content, and their relations to earthquake No.4.

CONCLUDING REMARKS

We have observed four earthquakes of M3.5 or more which occurred near our test site for three years since the beginning of composite observation, and some clear precursors were recognized especially for three earthquakes of these four.

For the first of them, the earthquake on Sept. 30, 1977, precursors of various time durations were detected. This example demonstrated that even such a relatively small earthquake reveals some precursors and the test-field project is promising. Two earthquakes, Nos.3 and 4 mentioned earlier, were notably accompanied by

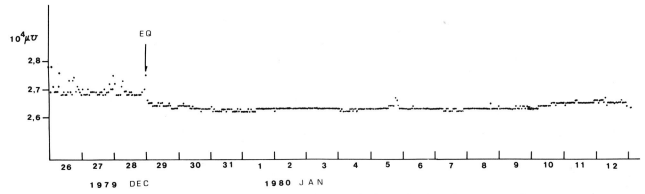

Figure 9. Fine process of change of electric conductivity at the time of Dec.28, 1979 earthquake, in which hourly value is plotted.

changes of chlorine ion content in underground water which was also noted in the first earthquake. This repetition of the same phenomenon is considered to be a very important result concerning precursors.

The preseismic quiescence was observed over a few hundreds of days before the earthquake occurrence in several earthquakes along the Yamasaki fault.

Thus, we may expect successful results of the test-field project in future. However, the results obtained at the present test-field may not necessarily be applied directly to the earthquakes in other regions or of other mechanisms. Furthermore, it seems that there exist characteristic precursors of respective regions in accordance with their seismotectonic conditions. Therefore, investigations of mechanism of the precursor are also important, in order to utilize the results in the test-field.

REFERENCES

Huzita, K., Y. Kishimoto and K. Shiono, Neotec-
tonics and seismicity in the Kinki area, Southwest Japan, Journ. Geoscience, Osaka City Univ., 16, 93-124, 1973

Kishimoto, Y. and R. Nishida, Mechanism of microearthquakes and their relation to geological structures, Bull. Disas. Prev. Res. Inst., Kyoto Univ., 23, 1-25, 1973

Nishida, R., Earthquake generating stress in eastern Chugoku and northern Kinki Districts, Southwest Japan, Bull. Disas. Prev. Res. Inst., Kyoto Univ., 22, 197-233, 1973

Oike, K., On the relation between rainfall and the occurrence of earthquakes, Ann. Disas. Prev. Res. Inst., Kyoto Univ., 20B-1, 35-45, 1977 (in Japanese)

Oike, K. and Y. Kishimoto, The Yamasaki fault as a test-field for the earthquake prediction, Symposium on Investigation of Earthquake Prediction(1976), 83-90, 1977 (in Japanese)

Okada, A., M. Ando and T. Tsukuda, Trenches across the Yamasaki fault in Hyogo Prefecture, Report of Coord. Committee for Earthq. Prediction, 24, 190-194, 1980 (in Japanese)

SOME EMPIRICAL RULES ON FORESHOCKS AND EARTHQUAKE PREDICTION

Ken'ichiro Yamashina

Earthquake Research Institute, University of Tokyo, Tokyo 113, Japan

Abstract. The identification of foreshocks and their relationship to the magnitude and time of the main shock is an important problem in earthquake prediction. A study of earthquake sequences in the ISC catalogue shows that the magnitude and time of the largest earthquake in a sequence(main shock) can be statistically related to the magnitudes and temporal distributions of earlier events in the sequence(foreshocks). Earthquakes are grouped in a sequence if they occur within specified limits of time and space. Within a sequence, M_i is defined as the magnitude of the i-th largest event before time t. F_1 and A_1 are defined as the magnitudes of the largest events before and after M_1. The probability is high that a larger event will occur after time t if, for example, the magnitude difference M_1-F_1 or M_1-A_1 is small. Based on the results of the present study and a previous study of Japanese earthquakes, nine empirical rules are presented for identifying foreshocks and estimating the averaged size and time of the largest event in a sequence.

Introduction

Usually foreshocks cannot be identified as "foreshocks" until the main shock occurs. Therefore discrimination of foreshocks prior to the occurrence of the main shock is important for earthquake prediction. Keilis-Borok et al.(1978, 1980a,b) and McNally(1980a,b) proposed prediction algorithms using essentially the number of earthquakes occurring in a certain time window and the distribution of time intervals of earthquakes, respectively. Sauber and Talwani(1980) applied them to earthquakes at Lake Jocassee, South Carolina. Temporary changes in b values(e.g., Suyehiro et al., 1964, 1972; Suyehiro, 1966; Papazachos et al., 1967; Papazachos, 1975) and time-space patterns of seismicity (e.g., Talwani, 1979) may also be useful to identify earthquake sequences as foreshocks. Recent progress in spectrum study and analysis of waveforms (e.g., Tsujiura, 1977, 1978, 1979; Bakun and McEvilly, 1979; Ishida and Kanamori, 1980) may also be important. Yamashina(1980) proposed a method of probabili-

ty prediction for Japanese earthquakes based on the earthquake catalogue compiled by the Japan Meteorological Agency (JMA). When at least two earthquakes occur in the same area, this method can assess the likelihood that they are foreshocks of a forthcoming larger event. In an earthquake sequence, M_1 is defined temporarily as the magnitude of the largest event which has already occurred before time t (Figure 1, Table 1). F_1 and A_1 are the largest foreshock and aftershock with respect to M_1. Studying statistically the magnitude differences M_1-F_1 and M_1-A_1, empirical rules were proposed as follows: [1] When seismic activity occurs with $M_1-F_1 \leq 0.4$ within about a week, it may be a foreshock sequence followed by a larger event. The probability is 25-30 % for $M_1-F_1 \leq 0.2$, and 20 % for $0.3 \leq M_1-F_1 \leq 0.4$. [2] A similar rule is also applicable for $M_1-A_1 \leq 0.2$. The probability is 20 %. [3] If $0.5 \leq M_1-F_1$ and $0.3 \leq M_1-A_1$, the probability is 5-10 % or less. In these cases, M_1 is expected with high probability to be a main shock. [4] The magnitude of the expected event is larger than M_1 by about 0.5 on the average. [5] About 40-45 % and 80 % of the expected events occur within a day and a week, respectively, after criterion 1 or 2 is satisfied. Time intervals $F_1 \backsim M_1$ and $M_1 \backsim A_1$ may also help to estimate the time of the expected event. [6] If no larger events actually occur within a day or a week after criterion 1 or 2 is satisfied, the probability that a larger event is still expected drops to 3/5 or 1/5 of the initial value, respectively.

The present paper is concerned with whether criteria 1-6 proposed by Yamashina(1980) for Japanese earthquakes can be applied to other regions in the world. The result is that criteria 1-6 are generally valid for ISC data with only slight revision. The present paper also proposes additional criteria using other parameters for magnitude distribution. Application of these criteria to a couple of earthquake sequences will be shown as examples.

Data

The present statistical study is based on the earthquake catalogue for 1971-1975 compiled by

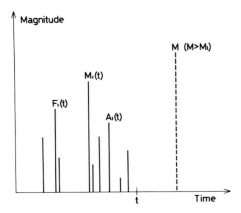

Fig. 1. A schematic example of an earthquake sequence and the present notation for magnitudes; F_1, M_1, A_1 and M.

the International Seismological Centre(ISC). The body wave magnitude(mb) determined by ISC is used in the following discussions.

Earthquakes are classified as part of the same sequence, if they occur within a certain time interval and within a certain distance from each other(Yamashina, 1978). This is the only rule to define earthquake sequences, and is similar to that proposed by Savage and Ellsworth(1979). For the JMA earthquake catalogue Yamashina(1980) used the values of 10 days in time interval and 20 km (+ errors for location) in epicentral distance. These values should, however, be revised for other data sets according to the detectability, accuracy of hypocentral determination, background seismicity, and so on. In the present paper, for the ISC catalogue, 10 days and 50 km (+ errors for location) are assumed for the time interval and epicentral distance. For example, when two earthquakes, A and B, occur with a 5-day interval and a 60-km distance with epicentral errors of 8 km and 10 km, respectively, they are clustered with each other, because 5 days < 10 days and 60 km < (50 +8 +10)km. If another event C, which occurs 8 days after B, is clustered with B, A, B and C are all recognized to belong to the same sequence. In each sequence, the largest event is called the "main shock".

The 10-day interval is based on the fact that many foreshocks precede the main shock within 10 days(e.g., Rikitake, 1975; Papazachos, 1975; Jones and Molnar, 1976, 1979; Utsu, 1978). In some cases, however, foreshock-like activity, which sometimes forms a doughnut-shaped pattern, precedes the main shock by more than several months or years (e.g., Mogi, 1968, 1969, 1979; Kelleher and Savino, 1975; Brady, 1976; Ohtake, 1976; Sekiya, 1976, 1977; Engdahl and Kisslinger, 1977; Evison, 1977a, b, c; Caputo et al., 1977; Ishida and Kanamori, 1978, 1980; Keilis-Borok et al., 1978, 1980a, b; Fuis and Lindh, 1979; Yamashina and Inoue, 1979). Such intermediate-term or long-term activity is left for further investigation, and the present study

deals only with short-term phenomena of several days' duration.

The distance of 50 km is assumed considering that the ISC catalogue compiles earthquakes with minimum magnitude larger than that in the JMA catalogue. If we assumed 20 km, the same as for the JMA catalogue(Yamashina, 1980), earthquake sequences would be unreasonably divided into many subgroups. On the other hand, 100 km would be too large to identify appropriate sequences, because it would result in many interconnections between independent activities. The limit of the distance, 50 km, is raised where errors in location are large. In this way, we can treat earthquakes throughout the world with different detectability and accuracy in hypocentral location. The reason why we use epicentral instead of hypocentral distance is that large errors in depth are inevitable in hypocentral determination. However, we exclude cases in which the difference of the depths exceeds 100 km.

About 49,200 earthquakes are included in the ISC catalogue for 1971-1975. Among them, the magnitudes of about 19,500 earthquakes were determined by ISC. The remaining events were not large enough to determine their magnitudes. According to the present definition for sequences, there were about 12,300 sequences with a main shock whose magnitude was determined. Most of these sequences(about 10,200) did not include any foreshocks and aftershocks for which magnitudes were determined. In the following sections, therefore, we can discuss only the remaining 2,100 sequences. This ratio should be raised when the detectability and the minimum magnitude in the catalogue is improved in the future.

The homogeneity of the data set used here may be criticized. The present analysis, however, does not necessarily require the completeness for all data. It merely assumes that the lower cut-off level of magnitudes is constant in each earthquake sequence, that is to say, in each area and in each short time interval. In most cases, this condition is likely to be satisfied.

M_1-F_1 and M_1-A_1

Probability for M_1-F_1 or M_1-A_1

Figure 2 shows how the probability that M_1 is followed by a larger earthquake (M) is related to

TABLE 1. Notation of Magnitude

M_0	main shock
F_0	the largest foreshock
M_1	the largest event before time t
F_1	the largest event before M_1
A_1	the largest event between M_1 and t
M	the event larger than M_1 after t
M_2	the second largest event before t
M_3	the third largest event before t
M_4	the fourth largest event before t

the value of the magnitude differences M_1-F_1(left) and M_1-A_1(right). For example, when M_1-F_1 was 0.3 (185 cases in total), about 24 % (45 cases) were followed by a larger event. In other words, when earthquakes occur with M_1-F_1= 0.3, we can say that the activity may be a foreshock sequence and a larger event will occur with a probability of about 24 %.

In Figure 2, the probability obtained for the JMA catalogue for 1961-1979 (Yamashina, 1980) is also shown. Shaded bars represent the average probability of the two neighboring values. When the largest two events are equal to each other in magnitude, we cannot conclude which is M_1. Hence, both M_1-F_1 and M_1-A_1 are defined to be 0.0 in this case.

Distributions of the probability for the ISC and JMA data are essentially similar. Two remarkable features can be seen in common in Figure 2: (1) The probability increases distinctly with decreasing M_1-F_1 or M_1-A_1, and (2)the distribution

is not symmetrical. The probability for M_1-F_1 is, as a whole, systematically larger than that for M_1-A_1. In the case of the ISC data, the probability is 20-25 % for $M_1-F_1 \leq 0.3$, about 15 % for $0.4 \leq M_1-F_1 \leq 0.6$, about 10 % for $0.7 \leq M_1-F_1 \leq 1.1$, and 0 % for $1.2 \leq M_1-F_1 \leq 2.1$. Similarly, it is also 20-25 % for $M_1-A_1 \leq 0.2$, about 10 % for $0.3 \leq M_1-A_1 \leq 0.5$, about 8 % for $0.6 \leq M_1-A_1 \leq 0.9$, and less than 5 % for $1.0 \leq M_1-A_1 \leq 2.0$.

Irrespective of F_1 and A_1, M_1 is followed by a larger event with a probability about 3-13 %. This ratio, however, depends on the magnitude M_1, and a smaller event is more liable to be followed by a larger event than a larger event is. For example, it is 13 %, 8 % and 6 % for $M_1 = 4.0$, 5.0 and 6.0, respectively. It is left for further study whether this correlation suggests that there is some difference in physical property between larger and smaller events, or whether it is only a result of the present method of clustering. In the data set used here, the number of cases for

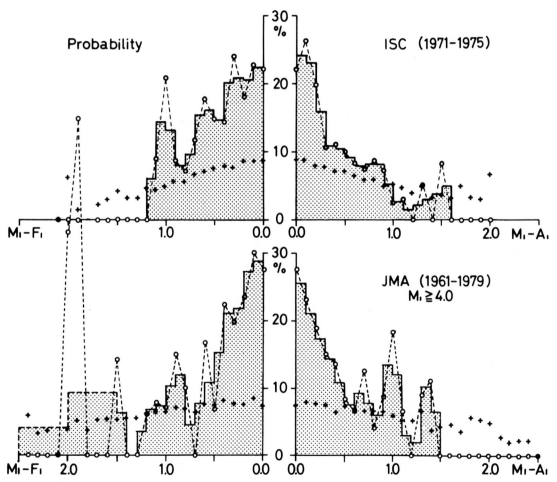

Fig. 2. Probability that a forthcoming larger event is expected with respect to the magnitude difference M_1-F_1(left) and M_1-A_1(right). Upper and lower figures represent the results obtained from the data compiled by ISC (the present study) and by JMA (Yamashina, 1980), respectively. Crosses are the hypothetical probability obtained from a model that M_1 is followed by a larger event without any physical correlation with M_1-F_1 or M_1-A_1.

which $M_1 = F_1 = 4.0$ was 7. If there were no correlation between M_1-F_1 and the probability that M_1 would be followed by a larger event, 0.91 (= 7 x 0.13) out of 7 cases would be expected to be followed by a larger event. Similar calculation can be done for all other cases for which $M_1 = F_1$(i.e., $M_1 = F_1 = 4.1$, 4.2 etc.). When all these cases are summed, the expected probability is 9 % for M_1-F_1 = 0.0. The actual probability, as shown in Figure 2, is 22 %. Because of the existence of the lower cut-off level of magnitudes, such calculations for all M_1-F_1 and M_1-A_1 result in a false weak correlation between M_1-F_1 or M_1-A_1 and the probability as shown in Figure 2 by crosses. The actual probability for small magnitude differences, however, exceeds this level sufficiently, and supports the conclusion in the previous paragraph.

The numbers of M_1-F_1 and M_1-A_1 with and without subsequent larger events are shown in Figure 3. Fluctuations in the probability for large M_1-F_1 and M_1-A_1 (for example, the high probability for $M_1-F_1 = 1.0$, as shown in Figure 2) must be due to lack of sufficient data.

Magnitude of the Expected Event

Figure 4 shows the magnitude difference between M and M_1. Here, M is the magnitude of the event which follows M_1 with magnitude $M_1 < M$. This figure suggests a method for estimating the magnitude of a forthcoming larger event, M, when M_1-F_1 or M_1-A_1 is small. Based on both the ISC and JMA data shown in Figure 2, we will pay attention to the case of $M_1-F_1 \leq 0.4$ or $M_1-A_1 \leq 0.2$, in which the probability exceeds about 20 %. In both cases (solid symbols in Figure 4), the average of $M-M_1$ for ISC data is about 0.35. The distribution of $M-M_1$, however, does not apparently correlate with respective values of M_1-F_1 and M_1-A_1. The averages for all M_1-F_1 and M_1-A_1 are 0.32 and 0.35, respectively.

The present average, about 0.35, does not necessarily suggest the magnitude of the main shock, because M may be succeeded by a still larger event. In order to know the magnitude of the main shock, the distribution of M_0-F_0 may be more useful. Here, M_0 and F_0 are the magnitudes of the main shock and the largest foreshock, respectively. The distribution of M_0-F_0 is equal to that of M_1-F_1 without subsequent larger event (open circles in Figure 3). The average of M_0-F_0 is 0.39, slightly larger than that of $M-M_1$.

In the present paper, we discuss all the cases of $0.1 \leq M-M_1$. Thus we make no distinction between foreshock-main-shock sequences and so-called earthquake swarms. If we want to know only the cases where M is larger than M_1 by a certain value(e.g., Utsu, 1978), the probability in Figure 2 should be reduced using the result shown in Figure 4. For example, the probability for $0.3 \leq M-M_1$ and $0.7 \leq M-M_1$ can be obtained by reducing the vertical scale in Figure 2 by about 1/2 and 1/10, respectively.

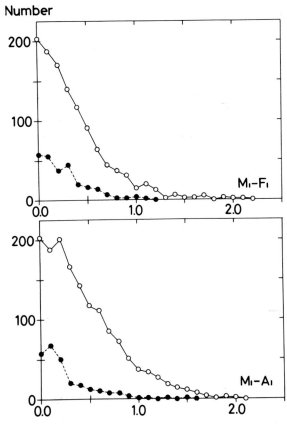

Fig. 3. The number of M_1-F_1(above) and M_1-A_1 (below). Solid and open circles represent the cases with and without a subsequent larger event. The open circles thus represent the cases in which M_1 is the main shock M_0.

Time of the Expected Event

Figure 5 represents the time interval between M and M_1(when M_1-F_1 is determined) or A_1(when M_1-A_1 is determined). About 40 % occurred within a day. Similarly about 55 % and 75 % occurred within 3 days and a week, respectively. It suggests that, when M_1-F_1 or M_1-A_1 is small, special attention to a forthcoming larger event should be paid especially for the first few days. Figure 5 is also useful for finding the probability that a larger event will occur in a given time period. For example, when we want to know the probability that a larger event will occur within a day and within a week, the vertical scale in Figure 2 should be reduced by about 2/5 and 3/4, respectively.

In Figure 5, the number abruptly decreases for intervals longer than 10 days, because of the present definition of an earthquake sequence. Therefore the ratio of the cases longer than 10 days, about 10-15 %, may be underestimated. However it is clear that such cases must be far fewer than the rest, as has been shown by previous

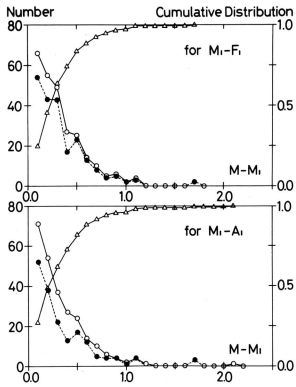

Fig. 4. The number of M –M₁ and its cumulative distribution in the cases that M_1-F_1 (above) and M_1-A_1 (below) are defined. Solid circles show the number for $M_1-F_1 \leq 0.4$(above) and for $M_1-A_1 \leq 0.2$(below).

studies which include various definitions of foreshocks(e.g.,Rikitake, 1975; Papazachos, 1975; Jones and Molnar, 1976, 1979; Utsu, 1978).

Figure 6 shows poor correlations between time intervals of $F_1 \sim M_1$ and $M_1 \sim M$, and of $M_1 \sim A_1$ and $A_1 \sim M$. However, the points between the dashed lines in Figure 6 show that 2/3 of $M_1 \sim M$ and $A_1 \sim M$ lie between 1/10 and 10 times as long as $F_1 \sim M_1$ and $M_1 \sim A_1$, respectively. Thus $F_1 \sim M_1$ and $M_1 \sim A_1$ may also help to estimate the time of the expected event.

Correlation between M_1-F_1 and M_1-A_1

In the previous sections, M_1-F_1 and M_1-A_1 are discussed independently. Figure 7 represents the probability with respect to both M_1-F_1 and M_1-A_1. Although the result shows large fluctuations due to the lack of sufficient data, high values for small M_1-F_1 and M_1-A_1 are note-worthy. When $0.1 \leq M_1-F_1 \leq 0.4$ and $0.1 \leq M_1-A_1 \leq 0.2$, the probability attains as much as about 40 % on the average.

M_1-M_2, M_1-M_3 and M_1-M_4

Next we will examine how the probability varies with respect to M_1-M_2, M_1-M_3 and M_1-M_4. Here, Mi (i = 1, 2, 3 and 4) is the magnitude of the i-th largest event before time t. Figure 8a is the probability for M_1-M_2. It is, generally speaking, similar to but systematically smaller than that for M_1-F_1. Accordingly the parameter M_1-M_2 is less useful than M_1-F_1 for earthquake prediction. Such a result has already been suggested by the difference between M_1-F_1 and M_1-A_1 in Figure 2, because M_2 is equal to F_1 or A_1.

When the parameter M_1-M_3 can be defined, it may be more useful than M_1-M_2, since it gives a high probability for small M_1-M_3 as shown in Figure 8b. When $M_1-M_3 \leq 0.2$, the probability is about 35 %. It is nearly 20 % for $0.3 \leq M_1-M_3 \leq 0.4$, about 15 % for $0.5 \leq M_1-M_3 \leq 0.8$, about 10 % for $0.9 \leq M_1-M_3 \leq 1.1$, and 0 % for $1.2 \leq M_1-M_3 \leq 2.0$.

M_1-M_4 is also useful as shown in Figure 8c. The probability attains about 40 % for $M_1-M_4 \leq 0.1$, about 30 % for $0.2 \leq M_1-M_4 \leq 0.3$, about 20 % for $0.4 \leq M_1-M_4 \leq 0.6$, about 10 % for $0.7 \leq M_1-M_4 \leq 1.1$ and less than 5 % for $1.2 \leq M_1-M_4 \leq 2.4$.

Empirical Rules on Probability Prediction

The present results support the criteria 1-6 proposed for Japanese earthquakes (Yamashina,

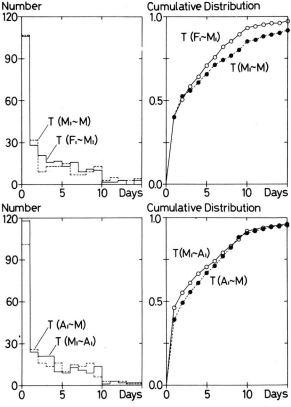

Fig. 5. The number(left) and its cumulative distribution(right) of time intervals between F_1 and M_1, M_1 and M(above), M_1 and A_1, and A_1 and M(below).

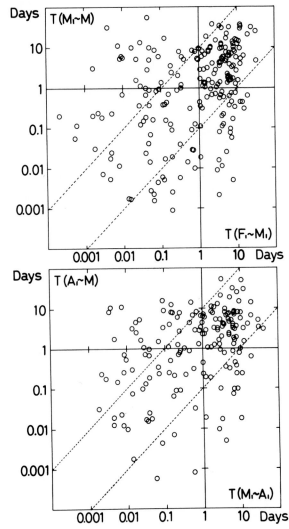

Fig. 6. Correlation between time intervals $F_1 \sim M_1$ and $M_1 \sim M$ (above), and between $M_1 \sim A_1$ and $A_1 \sim M$ (below).

1980). In order to apply the criteria both to the JMA and ISC data, they are slightly revised as follows. In addition, the present results suggest three further criteria 7-9.

[1] When seismic activity occurs with $M_1 - F_1 \leq 0.4$ within about a week, it may be a foreshock sequence followed by a larger event. The probability is about 20-30 %.

[2] A similar rule is also applicable for $M_1 - A_1 \leq 0.2$. The probability is about 20-25 %.

[3] If $0.5 \leq M_1 - F_1 \leq 0.6$ or $0.3 \leq M_1 - A_1 \leq 0.4$, the probability is about 10-15 %. It is only about 5-10 % or less for $0.7 \leq M_1 - F_1$ and $0.5 \leq M_1 - A_1$. In these cases, M_1 can be expected to be a main shock with high probability.

[4] The magnitude of the expected event is larger than M_1 by about 0.4-0.5 on the average.

[5] About 40-45 % and 75-80 % of the expected

events occur within a day and a week, respectively, after criterion 1 or 2 is satisfied. Time intervals $F_1 \sim M_1$ and $M_1 \sim A_1$ may also help to estimate the time of the expected event.

[6] If no larger events actually occur within a day or a week after criterion 1 or 2 is satisfied, the probability that a larger event is still expected drops to 3/5 or 1/5 of the initial value, respectively.

[7] When $0.1 \leq M_1 - F_1 \leq 0.4$ and $0.1 \leq M_1 - A_1 \leq 0.2$, the probability that a larger event is expected attains about 40 %.

[8] We should also pay attention to $M_1 - M_3 \leq 0.4$. The probability is about 35 % for $M_1 - M_3 \leq 0.2$, and nearly 20 % for $0.3 \leq M_1 - M_3 \leq 0.4$.

[9] The cases of $M_1 - M_4 \leq 0.6$ are also worthy of attention. The probability is about 40 % for $M_1 - M_4 \leq 0.1$, about 30 % for $0.2 \leq M_1 - M_4 \leq 0.3$, and nearly 20 % for $0.4 \leq M_1 - M_4 \leq 0.6$.

Examples of Probability Prediction

Two examples of the application of the present criteria for probability prediction of earthquakes are shown in Table 2. In this table, only the events which contributed to $M_1 - F_1$ and/or $M_1 - A_1$ are listed. The 1971 New Ireland earthquakes (Ms = 7.9 on July 14 and 26) were preceded by remarkable activity. $M_1 - F_1$ was defined as 0.3 by the mb = 5.6 event on July 10. According to criterion 1, a forthcoming larger event was expected with a probability of 20-30 %. Such a larger event actually occurred on July 14 as expected. $M_1 - F_1$ was again determined as 0.4, which suggested a still larger event, also with a probability of 20-30 %. On July 18, criterion 7 was satisfied by the mb = 5.8 event. The probability increased to

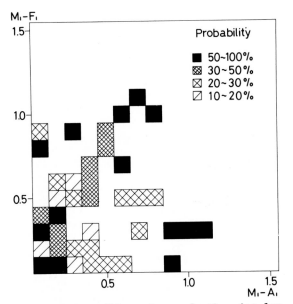

Fig. 7. Probability that a forthcoming larger event is expected with respect to both $M_1 - F_1$ and $M_1 - A_1$.

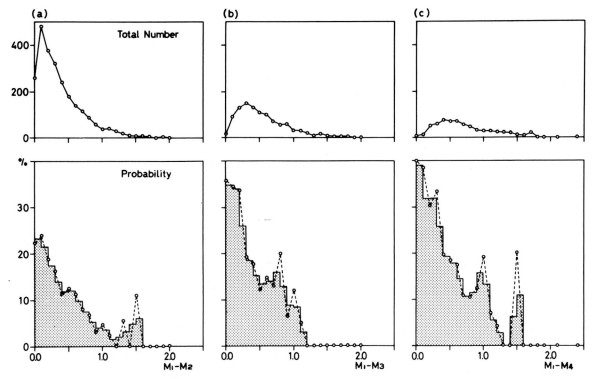

Fig. 8. Probability that a forthcoming larger event is expected versus (a) M_1-M_2, (b) M_1-M_3, and (c) M_1-M_4. Their total numbers both with and without a susequent larger event are also shown.

40 %. This was realized by the mb = 6.6 event on July 26. At this point, this event was expected to be the main shock with a probability of 85-90 % by criterion 3. Hereafter M_1-A_1 did not become less than 0.5, and a still larger event did not occur. The present discussion should, however, be re-examined using a more appropriate magnitude scale which would enable us to compare all earthquakes in the sequence uniformly, since mb magnitude can not represent the overall size of large earthquakes.

An alarm for a larger event with a probability of 20-30 % also preceded the 1975 northeastern China (Haicheng) earthquake, as long as the ISC magnitudes(mb) were concerned. M_1-F_1 was determined as 0.3 by the events which occurred about 9 and 12 hours before the main shock. After the main shock, M_1-F_1 was 1.7, and M_1-A_1 was no less than 1.0. Consequently, it was correctly expected to be the main shock with a probability more than 90 % (criterion 3).

Discussion and Conclusions

The present paper proposes 9 empirical rules for foreshocks as described in a previous section. According to these, when at least two earthquakes occur in the same area within about a week, we can statistically predict whether (1) they may be foreshocks of an impending larger earthquake, or (2) the largest event which has already occurred will be the main shock and no larger events will follow.

Moreover, the present method may also be useful even when only one earthquake is observed. If we merely know the detectability (i.e., the observable minimum magnitude) in the area in question, we may be able to say that the observed sole event will be the largest shock. For example, when the observed event exceeds the level of detectability by 0.5 or 1.0 in magnitude, criterion 3 states that it will be the largest shock with a probability of 85-90 % or 90-95 %, respectively.

The present method has the merit that it can be applied to teleseismic data. Thus it is not necessary to set up an instrument within the exact region in which a forthcoming larger event is expected, if a sufficiently high detectability is available.

At present, a method of deterministic prediction of earthquakes in the short-term has not been established. Accordingly, the probability prediction method is proposed as one of the useful approaches to earthquake prediction. We can not continually pay full attention to minor activity in every region. Taking advantage of the present method, the time and space to which we must pay special attention can be determined automatically.

Criteria 1-2 give 20-30 % probability that a larger event is expected. Even the highest probability in criteria 7-9 is no more than 40 %.

TABLE 2

Examples of the Present Probability Prediction

Date	h m	mb ISC	M_1-F_1	M_1-A_1	Prediction
(a) New Ireland					
1971Jul 9	0949	5.3	-	-	
1971Jul10	0523	5.6	0.3*	-	Alarm(20-30%)
1971Jul14	0611	6.0	0.4*	-	Alarm(20-30%)
1971Jul14	0737	(5.6)	0.4*	0.4	↑Ms=7.9
1971Jul14	0741	(5.7)	0.4*	0.3	
1971Jul18	1431	(5.8)	0.4*	0.2*	Alarm(40%)
1971Jul19	0014	(5.9)	0.4*	0.1*	
1971Jul26	0123	6.6	0.6	-	Main shock?(85-90%)
1971Jul26	0153	(5.7)	0.6	0.9	↑Ms=7.9
1971Jul26	0202	(5.9)	0.6	0.7	
1971Jul26	0225	(6.1)	0.6	0.5	
(b) Northeastern China (Haicheng)					
1975Feb 3	2350	4.1	-	-	
1975Feb 4	0235	4.4	0.3*	-	Alarm(20-30%)
1975Feb 4	1136	6.1	1.7	-	Main shock?(90% <)
1975Feb 4	1148	(5.1)	1.7	1.0	↑Ms=7.4

Stars represent the alarm for a larger event. Percents in parentheses show the probability of the present prediction. Ms magnitudes are after the Preliminary Determination of Epicenters, U. S. Geol. Surv.(monthly).

Consequently, if we are afraid of false predictions and we only want to predict with higher confidence, we had better always say that M_1 will be the main shock and it is not necessary to expect a larger event. However, a probability of only 20 or 30 % may be scientifically useful. If we can re-examine various other data with special attention or we can start new temporary observations in the expected epicentral region, we may improve the probability (Utsu, 1977) or we may be able to predict diterministically in some cases.

Although criteria 7-9 give probabilities higher than 1-2, the former are not always more useful than the latter. As Figure 3 suggests, the numbers of events which could be predicted by criteria 1-2 are 217 and 175, respectively. Against this, the numbers predicted by criteria 7-9 are only 42, 132 and 83, respectively. In the present earthquake catalogue, about 13,500 earthquakes whose magnitudes were determined were not preceded by events with larger or equal magnitudes. This is the number of earthquakes excluding aftershocks and aftershock-like activities. Accordingly, the percentages of earthquakes predicted by criteria 1-2 and 7-9 are 1.6, 1.3, 0.3, 1.0 and 0.6 %, respectively. These small ratios, however, should increase when the detectability is improved. In fact, when we discuss only the sequences in which magnitudes of at least two events were determined, including aftershocks, these percentages are raised to 6.7, 5.3, 1.3, 4.0 and 2.5 %, respectively.

Although the present criteria of probability prediction are written in terms of earthquake magnitudes, they may be more usefully rewritten by moments, amplitudes, signal durations and so on as follows.

A magnitude is a convenient parameter to express earthquake size. However, a conventional definition of magnitude can not necessarily express the overall size of earthquakes (Kanamori, 1977, 1978). In particular, the mb magnitude analyzed here is not suitable for large and great earthquakes. Physically, seismic moment Mo or the new magnitude scale Mw determined using Mo (Kanamori, 1977, 1978) may be a better parameter for comparing the size of earthquakes. If the seismic network enables us to easily obtain the moment of current earthquakes, criteria 1-3 and 7-9 should be replaced by Mw or rewritten using moment assuming an empirical rule between moment Mo and magnitude M: $\log_{10}Mo = \alpha M + \beta$ (e.g., Wyss and Brune, 1968; Aki, 1969; Thatcher and Hanks, 1973; Johnson and McEvilly, 1974; Kanamori and Anderson, 1975; Kanamori, 1977, 1978; Hanks and Kanamori, 1979; Singh and Havskov, 1980). If we assume $\alpha = 1.5$, $M_1-F_1 = 0.2$ and 0.4 correspond to $Mo(M_1)/Mo(F_1) = 2.0$ and 4.0, respectively. Here, $Mo(M_1)$ and $Mo(F_1)$ represent the moments of M_1 and F_1, respectively, and α and β are constants.

Local magnitude ML is given by $ML = \log_{10}A - \log_{10}A_0$ using maximum trace amplitude A (e.g., Richter, 1935, 1958; Tsuboi, 1954). Here, A_0 is a function of epicentral distance. Since we are concerned only with each sequence in a limited area, $M_1-F_1 = 0.2$ and 0.4 approximately correspond to $A(M_1)/A(F_1) = 1.6$ and 2.5, respectively, at an arbitrary station. Here, $A(M_1)$ and $A(F_1)$ are maximum trace amplitudes of events with M_1 and F_1, respectively. However, wave forms of earthquakes in the same sequence at the same station are usually similar to each other. Consequently, $A(M_1)$ and $A(F_1)$ may be replaced by amplitudes of an arbitrary corresponding phase of M_1 and F_1, respectively.

The present criteria can also be rewritten using signal duration τ. Duration magnitude is given by $M\tau = C_0 + C_1\log_{10}\tau$ when the distance is not large (e.g., Kawasumi, 1954; Bisztricsany, 1958; Solov'ev, 1965; Tsumura, 1967; Crosson, 1972; Lee et al., 1972; Real and Teng, 1973; Watanabe, 1973; Herrmann, 1975; Lee and Wetmiller, 1976; Bakun and Lindh, 1977). Here C_0 and C_1 are constants. If we assume $C_1 = 2$ or 3, $M_1-F_1 = 0.2$ and 0.4 can be rewritten as $\tau(M_1)/\tau(F_1) = 1.3$ and $1.6(C_1 = 2)$ or 1.2 and $1.4(C_1 = 3)$, respectively.

In the present paper, we have treated world-wide earthquake data as if they were homogeneous. Even the Japanese data discussed by Yamashina(1980) include both inter- and intra-plate earthquakes, and both shallow and deep earthquakes. The criteria for earthquake prediction may be different for earthquakes in different tectonic regions. The most suitable criteria should be obtained by further study of various regions and of data collected by different organizations.

Acknowledgements. The author would like to express his gratitude to Drs. P. Somerville and D.

W. Simpson, and Professors K. Nakamura and K. Shimazaki for their many valuable suggestions. He is also grateful to Professors K. Tsumura, D. Vere-Jones, V. I. Keilis-Borok, W. J. Arabasz and F. F. Evison, Drs. W. L. Ellsworth, W. U. Savage, D. J. Andrews and T. Kato, Mr. I. Kayano, Ms. L. M. Jones, and the reviewers for their helpful and /or critical comments.

References

Aki, K., Analysis of the seismic coda of local earthquakes as scattered waves, J. Geophys. Res., 74, 615-631, 1969.

Bakun, W. H., and A. G. Lindh, Local magnitudes, seismic moments, and coda durations for earthquakes near Oroville, California, Bull. Seismol. Soc. Amer., 67, 615-629, 1977.

Bakun, W. H., and T. V. McEvilly, Are foreshocks distinctive? Evidence from the 1966 Parkfield and the 1975 Oroville, California sequences, Bull. Seismol. Soc. Amer., 69, 1027-1038, 1979.

Bisztricsany, E. A., A new method for the determination of the magnitude of earthquakes, Geofiz. Kozlemen, 7, 69-96, 1958 (in Russian).

Brady, B. T., Theory of earthquakes, IV, Pure Appl. Geophys., 114, 1031-1082, 1976.

Caputo, M., P. Gasperini, V. Keilis-Borok, L. Marcelli, and I. Rotwain, Earthquake's swarms as forerunners of strong earthquakes in Italy, Ann. Geofis., 15, 269-283, 1977.

Crosson, R. S., Small earthquakes, structure, and tectonics of the Puget Sound region, Bull. Seismol. Soc. Amer., 62, 1133-1171, 1972.

Engdahl, E. R., and C. Kisslinger, Seismological precursors to a magnitude 5 earthquake in the central Aleutian Islands, J. Phys. Earth, 25, S243-S250, 1977.

Evison, F. F., The precursory earthquake swarm, Phys. Earth Planet. Inter., 15, P19-P23, 1977a.

Evison, F. F., Fluctuations of seismicity before major earthquakes, Nature, 266, 710-712, 1977b.

Evison, F. F., Precursory seismic sequences in New Zealand, New Zealand J. Geol. Geophys., 20, 129-141, 1977c.

Fuis, G. S., and A. G. Lindh, A change in fault-plane orientation between foreshocks and aftershocks of the Galway Lake earthquake, M_L = 5.2, 1975, Mojave Desert, California (abstract), Tectonophysics, 52, 601-602, 1979.

Hanks, T. C., and H. Kanamori, A moment magnitude scale, J. Geophys. Res., 84, 2348-2350, 1979.

Herrmann, R. B., The use of duration as a measure of seismic moment and magnitude, Bull. Seismol. Soc. Amer., 65, 899-913, 1975.

Ishida, M., and H. Kanamori, The foreshock activity of the 1971 San Fernando earthquake, California, Bull. Seismol. Soc. Amer., 68, 1265-1279, 1978.

Ishida, M., and H. Kanamori, Temporal variation of seismicity and spectrum of small earthquakes preceding the 1952 Kern County, California, earthquake, Bull. Seismol. Soc. Amer., 70, 509-527, 1980.

Johnson, L. R., and T. V. McEvilly, Near-field observations and source parameters of central California earthquakes, Bull. Seismol. Soc. Amer., 64, 1855-1886, 1974.

Jones, L., and P. Molnar, Frequency of foreshocks, Nature, 262, 677-679, 1976.

Jones, L. M., and P. Molnar, Some characteristics of foreshocks and their possible relationship to earthquake prediction and premonitory slip on faults, J. Geophys. Res., 84, 3596-3608, 1979.

Kanamori, H., The energy release in great earthquakes, J. Geophys. Res., 82, 2981-2987, 1977.

Kanamori, H., Quantification of earthquakes, Nature, 271, 411-414, 1978.

Kanamori, H., and D. L. Anderson, Theoretical basis of some empirical relations in seismology, Bull. Seismol. Soc. Amer., 65, 1073-1095, 1975.

Kawasumi, H., Intensity and magnitude of shallow earthquakes, Pub. Bureau Central Seismol. Intern. Ser. A, Travaux Scientifiques, 19, 1954, 99-114.

Keilis-Borok, V. I., I. M. Rotvain, and T. S. Sidorenko, Increased stream of aftershocks as a forerunner of strong earthquake, Dokl. Akad. Nauk SSSR, 242, 567-569, 1978 (in Russian); to be translated into English in Dokl. Earth Sci. Sections, 242, 1980 (in preparation).

Keilis-Borok, V. I., L. Knopoff, and I. M. Rotvain, Bursts of aftershocks, long-term precursors of strong earthquakes, Nature, 283, 259-263, 1980a.

Keilis-Borok, V. I., L. Knopoff, I. M. Rotvain, and T. M. Sidorenko, Bursts of seismicity as long-term precursors of strong earthquakes, J. Geophys. Res., 85, 803-811, 1980b.

Kelleher, J., and J. Savino, Distribution of seismicity before large strike slip and thrust-type earthquakes, J. Geophys. Res., 80, 260-271, 1975.

Lee, W. H. K., R. E. Bennett, and K. L. Meagher, A method of estimating magnitude of local earthquakes from signal duration, U. S. Geol. Surv. Open-File Report, 18 p., 1972.

Lee, W. H. K., and R. J. Wetmiller, A survey of practice in determining magnitude of near earthquakes: Summary report for networks in North, Central and South America, U. S. Geol. Surv. Open-File Report, 76-677, 1976.

Mogi, K., Source locations of elastic shocks in the fracturing process in rocks (1), Bull. Earthquake Res. Inst., Univ. Tokyo, 46, 1103-1125, 1968.

Mogi, K., Some features of recent seismic activity in and near Japan (2), Bull. Earthquake Res. Inst., Univ. Tokyo, 47, 395-417, 1969.

Mogi, K., Two kinds of seismic gaps, Pure Appl. Geophys., 117, 1172-1186, 1979.

McNally, K. C., Patterns of earthquake clustering preceding moderate earthquakes, central California, submitted to Bull. Seismol. Soc. Amer., 1980a.

McNally, K. C., A statistical analysis of earthquake clustering, San Andreas fault, central California, submitted to Bull. Seismol. Soc. Amer., 1980b.

Ohtake, M., Search for precursors of the 1974 Izu-Hanto-oki earthquake, Japan, Pure Appl. Geophys., 114, 1083-1093, 1976.

Papazachos, B. C., N. Delibasis, N. Liapis, G.

Moumoulidis, and G. Purcaru, Aftershock sequences of some large earthquakes in the region of Greece, Ann. Geofis., 20, 1-93, 1967.

Papazachos, B. C., Foreshocks and earthquake prediction, Tectonophysics, 28, 213-226, 1975.

Real, C. R., and T. L. Teng, Local Richter magnitude and total signal duration in southern California, Bull. Seismol. Soc. Amer., 63, 1809-1827, 1973.

Richter, C. F., An instrumental earthquake magnitude scale, Bull. Seismol. Soc. Amer., 25, 1-32, 1935.

Richter, C. F., Elementary Seismology, W. H. Freeman and Company, San Francisco, p. 338-345, 1958.

Rikitake, T., Earthquake Precursors, Bull. Seismol. Soc. Amer., 65, 1133-1162, 1975.

Sauber, J., and P. Talwani, Application of Keilis-Borok and McNally prediction algorithms to earthquakes in the Lake Jocassee area, South Carolina, Phys. Earth Planet. Inter., 21, 267-281, 1980.

Savage, W. U., and W. L. Ellsworth, Microearthquake clustering preceding the Coyote Lake earthquake (abstract), EOS, Trans. Amer. Geophys. Un., 60, 891, 1979.

Sekiya, H., The seismicity preceding earthquakes and its significance to earthquake prediction, J. Seismol. Soc. Japan, 29, 299-311, 1976 (in Japanese with English abstract).

Sekiya, H., Anomalous seismic activity and earthquake prediction, J. Phys. Earth, 25, Suppl., S85-S93, 1977.

Singh, S. K., and J. Havskov, On moment-magnitude scale, Bull. Seismol. Soc. Amer., 70, 379-383, 1980.

Solov'ev, S. L., Seismicity of Sakhalin, Bull. Earthquake Res. Inst., Univ. Tokyo, 43, 95-102, 1965.

Suyehiro, S., T. Asada, and M. Ohtake, Foreshocks and aftershocks accompanying a perceptible earthquake in central Japan, Papers Meteorol. Geophys., 15, 71-88, 1964.

Suyehiro, S., Difference between aftershocks and foreshocks in the relationship of magnitude to frequency of occurrence for the great Chilean earthquake of 1960, Bull. Seismol. Soc. Amer., 56, 185-200, 1966.

Suyehiro, S., and H. Sekiya, Foreshocks and earthquake prediction, Tectonophysics, 14, 219-225, 1972.

Talwani, P., An empirical earthquake prediction model, Phys. Earth Planet. Inter., 18, 288-302, 1979.

Thatcher, W., and T. C. Hanks, Source parameters of southern California earthquakes, J. Geophys. Res., 78, 8547-8576, 1973.

Tsuboi, C., Determination of the Gutenberg Richter's magnitude of earthquakes occurring in and near Japan, J. Seismol. Soc. Japan, 7, 185-193, 1954 (in Japanese with English abstract).

Tsujiura, M., Spectral features of foreshocks, Bull. Earthquake Res. Inst., Univ. Tokyo, 52, 357-371, 1977.

Tsujiura, M., Spectral analysis of seismic waves for a sequence of foreshocks, main shock and aftershocks: the Izu-Oshima-kinkai earthquake of 1978, Bull. Earthquake Res. Inst., Univ. Tokyo, 53, 741-759, 1978 (in Japanese with English abstract).

Tsujiura, M., The difference between foreshocks and earthquake swarms, as inferred from the similarity of seismic waveform (preliminary report), Bull. Earthquake Res. Inst., Univ. Tokyo, 54, 309-315, 1979 (in Japanese with English abstract).

Tsumura, K., Determination of earthquake magnitude from total duration of oscillation, Bull. Earthquake Res. Inst., Univ. Tokyo, 45, 7-18, 1967.

Utsu, T., Probabilities in earthquake prediction, J. Seismol. Soc. Japan, 30, 179-185, 1977 (in Japanese with English abstract).

Utsu, T., An investigation into the discrimination of foreshock sequences from earthquake swarms, J. Seismol. Soc. Japan, 31, 129-135, 1978 (in Japanese with English abstract).

Watanabe, H., Determination of earthquake magnitude at regional distance in and near Japan (second report), J. Seismol. Soc. Japan, 26, 160-170, 1973 (in Japanese with English abstract).

Wyss, M., and J. N. Brune, Seismic moment, stress, and source dimensions for earthquakes in the California-Nevada region, J. Geophys. Res., 73, 4681-4694, 1968.

Yamashina, K., A definition of foreshock, main shock and aftershock, Programme Abst. Seismol. Soc. Japan, 1978-1, 85, 1978 (in Japanese).

Yamashina, K., Magnitude difference in foreshock sequences and earthquake prediction, Programme Abst. Seismol. Soc. Japan, 1980-1, 145, 1980 (in Japanese).

Yamashina, K., and Y. Inoue, A doughnut-shaped pattern of seismic activity preceding the Shimane earthquake of 1978, Nature, 278, 48-50, 1979.

PRECURSORY CHANGES IN GROUNDWATER PRIOR TO THE 1978 IZU-OSHIMA-KINKAI EARTHQUAKE

Hiroshi Wakita

Laboratory for Earthquake Chemistry, Faculty of Science,
University of Tokyo, Bunkyo-ku, Tokyo, Japan

Abstract. Precursory changes in groundwater prior to the 1978 Izu-Oshima-kinkai earthquake are described in this paper. Changes in the radon concentration, temperature, water level and flow rate of groundwater were observed at five stations located in the area within 90 km from the epicenter. The patterns and occurrences of these changes are quite similar to each other. The mid-term anomalies started from the end of October, 1977 and significant short-term anomalies were also observed a few days before the main shock. Occurrences of these variations related to the movement of groundwater coincide with those of other geophysical phenomena.

Introduction

The Izu Peninsula, located in central Honshu, is one of the most tectonically active areas in Japan. Crustal activities in the Izu Peninsula have been significant since the occurrence of the Izu-Hanto-oki earthquake with a magnitude of 6.9 on May 9, 1974.

Earthquake swarms observed around Mt. Togasayama became frequent in August, 1975 and an anomalous crustal upheaval with the center at Nakaizu was subsequently found at the end of the year. Soon after the occurrences of earthquake swarms at Hokkawa, on the east coast of the peninsula in February, 1976, very intensive studies including survey activities were organized under the direction of the Coordinating Committee for Earthquake Prediction. A few months later, two moderate earthquakes with M5.4 and M4.5 occurred at Kawazu, the southernmost edge of the uplift area, in August, 1976. Earthquake swarm activities on the east coast occurred again both in the spring and the fall of 1977. Then the Izu-Oshima-kinkai earthquake (M7.0) occurred on January 14, 1978 in the area between the Izu Peninsula and Izu-Oshima (Fig. 1).

As crustal activities in the Izu Peninsula are closely related to the possibility of an impending Great Tokai earthquake, the observed crustal upheaval were carefully examined by means of geophysical, geological and geochemical studies connected with the earthquake prediction proj-ect. Geochemical observations have been made focusing upon the area around Nakaizu, the center region of the uplift, since the spring of 1976.

With observation experiences gained over two years in the Izu Peninsula, observations made by scientists of several universities and the Geological Survey of Japan became more reliable by the time of the earthquake. The Izu-Oshima-kinkai earthquake took place after the occurrences of several moderate earthquakes and earthquake swarms in the peninsula, so that the monitoring system was almost completely functioning at that time.

Apparent precursory changes in groundwater were observed at five observation stations in the area within 90 km from the epicenter (Fig. 1). These include changes in the radon concentration, water level and temperature of the groundwater.

Changes in the radon concentration of groundwater

It is well known that changes in the radon concentration of groundwater are often observed prior to an earthquake. Geochemical studies for the purpose of earthquake prediction have been organized in Japan at the end of 1973 (Wakita, 1979a). At the present time, continuous measurements of the radon concentration in groundwater are being carried out at more than 15 stations in Japan. The Izu-Oshima-kinkai earthquake was the first large earthquake with a magnitude of 7 which took place near the observation sites.

Changes in the radon concentration of the groundwater obtained from an artesian well (SKE-1) with a depth of 350 m is shown in Fig. 2. A continuous radon monitoring equipment, Model NW-101 (Aloka Co. Ltd.), was used. Though the principle of the measurement was essentially the same as that described by Noguchi and Wakita (1977), many improvements were made. Fig. 2 shows the 9-point running average of the data from the bi-hourly readings on the original record which covers the period between May, 1977 and February, 1978.

As seen in Fig. 2, seasonal variations follow-

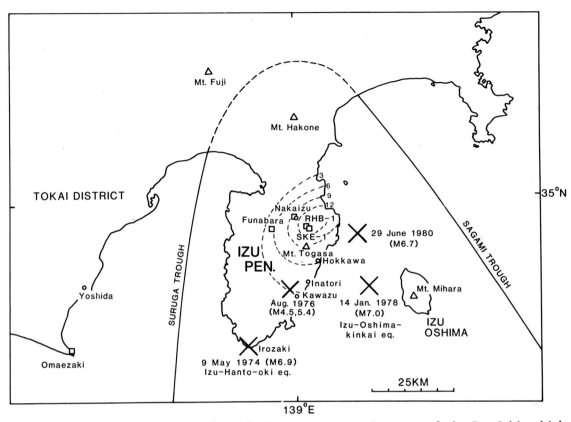

Fig. 1. Locations of observation wells (□) where precursory phenomena of the Izu-Oshima-kinkai earthquake were observed and the epicenters (X) of the Izu-Hanto-oki earthquake (M6.9), the Kawazu earthquake (M4.5, 5.4), the Izu-Oshima-kinkai earthquake (M7.0) and the East off Izu-Hanto earthquake (M6.7) are shown. The contour lines (in centimeters) of uplift during the period between 1967-1969 (Crustal Dynamics Division, GSI, 1976) are also shown.

ing a smooth curve are drastically disturbed in the period prior to and after the earthquake. The radon concentration began to decrease in the middle of October, 1977 and fluctuated rapidly thereafter. By the end of December, the radon concentration returned to normal at the lower concentration level. On January 8, six days before the earthquake, the radon concentration suddenly dropped and remained at the minimum level for 8 hours on January 9 (Wakita et al., 1980a). And then the radon concentration began to increase rapidly and reached the higher level. After the earthquake, a significant increase in the radon concentration was observed and the recovery to the normal level followed. It is inferred that variations starting from the middle of October and the rapid change which occurred on January 9 are the mid-term and short-term anomalies related to the occurrence of the Izu-Oshima-kinkai earthquake, respectively.

The cause of the absence of data in time segments during the period between October 24 and December 15, 1977 shown in Fig. 2 was attributed to an imperfect connection of the signal cable to the recorder (Wakita et al., 1980a). Unprocessed

data taken every 2 hours for the period between October, 1977 and February, 1978 are plotted in Fig. 3 and 5-1. Mid-term and short-term anomalies and after effects of the earthquake are clearly seen.

Similar changes in the radon concentration in the groundwater have been observed in the discrete measurements performed by the Geological Survey of Japan. Ikeda et al., (1979) reported the results obtained from weekly measurements. Among them, the changes in the I-09, an artesian well located at Nakaizu, are quite similar to those in the short-term anomaly of SKE-1.

It is noted that such changes observed prior to the Izu-Oshima-kinkai earthquake were not observed during an observation period over three years at SKE-1, even though during that period moderate earthquakes which occurred near SKE-1, i.e. two Kawazu earthquakes of 18 and 26 August, 1976 (M5.4 and 4.5, Δ=20 km). Also, no significant changes were found during the aftershocks of the Izu-Oshima-kinkai earthquake of a M5.0 (November 23, 1978, Δ=18 km), and M5.4 (December 3, 1978, Δ=17 km). Even the close-by East off Izu-Hanto earthquake (M6.7, June 29, 1980, Δ=15 km)

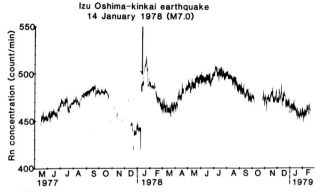

Fig. 2. Precursory radon concentration changes in groundwater observed at Nakaizu prior to the 1978 Izu-Oshima-kinkai earthquake. Plotted are nine-point running averages of the data from each two-hour period. The epicentral distance in about 25 km.

did not produce a radon anomaly.

Precursory radon concentration changes of the Izu-Oshima-kinkai earthquake observed in this study are compared with those reported in China and USSR. In general, mid-term anomalies in the radon concentration of groundwater are observed 2 to 3 months before an earthquake (Wakita, 1979 b). The concentration increases in one case and decreases in the other. The patterns of anomalies vary significantly with sampling sites and they are not easily recognizable.

On the contrary, a step-like or a spike-like change in the radon concentration is considered as one of the possible short-term anomalies. These changes are quite distinct and easy to define the occurrence time. Fig. 4 shows a compilation of the occurrence times of short-term anomalies of radon concentration, the magnitudes of the subsequent earthquakes and their epicentral distances. Characteristics of the short-term radon anomalies can be summarized as follows:
1) Precursory time varies between a few days to ten days prior to an earthquake.
2) In most of the cases, anomalies are observed for earthquakes greater magnitude 7.
3) No systematic regularity exists on the epicentral distance.

Here we like to offer an explanation why any meaningful change was not observed at RHB-1, but was observed at SKE-1 before the earthquake, even though both wells were located in the central part of the uplift area. SKE-1 was situated at the bottom of a narrow valley with an altitude of 270 m. The location of the well was isolated and was at least 2 km away from the nearest village. The well was an artesian with a depth of 350 m. On the other hand, RHB-1 belongs to a rehabilitation hospital, with an altitude of 440 m. The depth of the well was only 150 m. A large amount of water was pumped and consumed for the hospital

use. The groundwater intermittently pumped was used for continuous radon measurement, so that the radon concentration fluctuated greatly. This made it difficult to identify possible changes attributable to the earthquake.

Several explanations can also be offered for the radon concentration changes prior to the Izu-Oshima-kinkai earthquake observed at SKE-1. A plausible model is the deformation of the artesian layer in the vicinity of the well. Changes in the confined aquifer system and in the velocity of water flow reflecting changes in the stress of the region may also be a possibility.

Changes in water level and flow rate of groundwater

The patterns and precursory time of anomalous radon concentration changes mentioned above are quite similar to the changes in the water level in wells, water temperature and flow rate of hot springs.
1) Nagai et al. (1979) reported precursory changes in water temperature of an artesian well with a depth of 500 m at Nakaizu. The distance from the epicenter to the well was about 30 km. The water temperature began to fall on December 10, 1977 and fluctuated excessively, thereafter. These changes had been continued until the occurrence of the earthquake (Fig. 5-2).
2) Asada (1978) reported changes in water temperature of an artesian well with a depth of 700 m at Nakaizu. A significant drop in water temperature occurred 7 days before the earthquake and fluctuation followed.

Temperature changes of groundwater mentioned in 1) and 2) may be attributed to a secondary effect caused by a decrease in the flow rate of artesian wells.

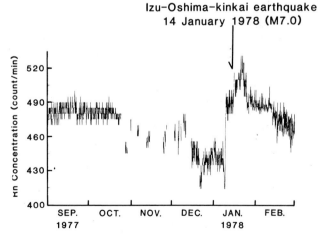

Fig. 3. Precursory radon concentration changes of the 1978 Izu-Oshima-kinkai earthquake. Data for every two-hour segment are plotted for the period between October 24 and December 15, 1977.

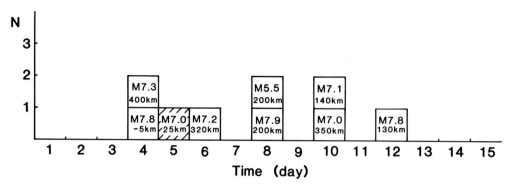

Fig. 4. A compilation of short-term anomaly of radon concentration in groundwater. Data are obtained from the following earthquakes: the Luhuo earthquake (Feb. 6, 1973, M7.9, Δ=200 km), the Mapien earthquake (June 29, 1973, M5.5, Δ=200 km), the Yongshan-Dagnan earthquake (May 11, 1974, M7.1, Δ=140 km), the Gasli earthquake (May 17, 1976, M7.3, Δ=400 km, Sultankhodjaev et al., 1976), the Tangshan earthquake (July 28, 1976, M7.8, Δ=5 and 130 km), the Sungpan-Pingwu earthquake (Aug. 16, 1976, M7.2, Δ=320 km), the Izu-Oshima-kinkai earthquake (Jan. 14, 1978, M7.0, Δ= 25 km, Wakita et al., 1980a) and the Alai earthquake (Nov. 2, 1978, M7.0, Δ=350 km, Sultankhodjaev et al., 1979). The references to five Chinese earthquakes are taken from a compilation by Wakita (1979b).

3) The water level of a 500 m well (I-6) at Nakaizu began to fall in early December. This trend reversed to rise in early January. The earthquake occurred at the time when the water level almost recovered to the previous level (Kishi, 1979). A marked fall in water level was observed after the earthquake (Fig. 5-3).
4) Yamaguchi and Odaka (1978) reported the changes in the water level of a 600 m well at Funabara. The epicentral distance was about 35 km. The water level of the well began to fall in the middle of December, 1977 and the trend accelerated at the end of December. A coseismic fall up to 7 m was observed. Fig. 5-4 shows the water level changes after correcting for the barometric pressure effect.
5) According to Wakita et al. (1980b), significant changes in water level were observed in a 500 m deep well located at Omaezaki, about 90 km from the epicenter. The well was drilled for the purpose of earthquake prediction studies in 1977. The water level of the well is least affected by meteorological and cultural factors, such as precipitations, tide, atmospheric pressure and human activities. The Izu-Oshima-kinkai earthquake occurred during a period that the water level of the well turned rapidly to the equilibrium water level from the disequilibrium stage caused by drilling of the well (Fig. 5-5).
The water level began to fall suddenly on December 28, 1977 and dropped by 25 cm. Then the level began to rise on January 5, 1978. A total 30 cm of coseismic fall was observed for the main shock and the largest after shock occurred on January 15.
6) Anomalous changes in groundwater were observed by members of Namazu-no-kai (Catfish Club), a volunteer group for earthquake prediction (Hot Spring Res. Inst., 1978). These in-

cluded changes in water level and temperature of about 30 shallow wells located in the area within 150 km from the epicenter. The occurrence times of these changes were mostly in the period between 7th and 9th of January, a few days before the earthquake.

Measurements of chemical composition of groundwater and gases

Chemical analyses of groundwater, pH and conductivity measurements included, are periodically performed at 32 sites in the peninsula by the Geological Survey of Japan. No significant changes have been reported (Nagai et al., 1979).
A paper by Sugisaki (1978) described anomalous changes in ratios of He/Ar and N_2/Ar in fault airs prior to the Izu-Oshima-kinkai earthquake. The observation was made at Inuyama, Nagoya with the epicentral distance of about 220 km.

Conclusion

Anomalous changes in groundwater observed prior to the Izu-Oshima-kinkai earthquake are quite similar to each other (Fig. 5). The patterns and precursory time of these changes are essentially the same. Furthermore, the occurrences of these changes in groundwater coincide with those of other geophysical precursors. The occurrence of earthquake swarms at the end of October and early November in the hypocentral region may be regarded as the mid-term anomaly in terms of foreshocks (Fig. 5-6).
Foreshocks including two with M4.9 which occurred on January 13 (Tsumura et al., 1978) may be considered as short-term anomalies. Rapid changes in stress obtained from a borehole strainmeter installed at Irozaki (Yamagishi et

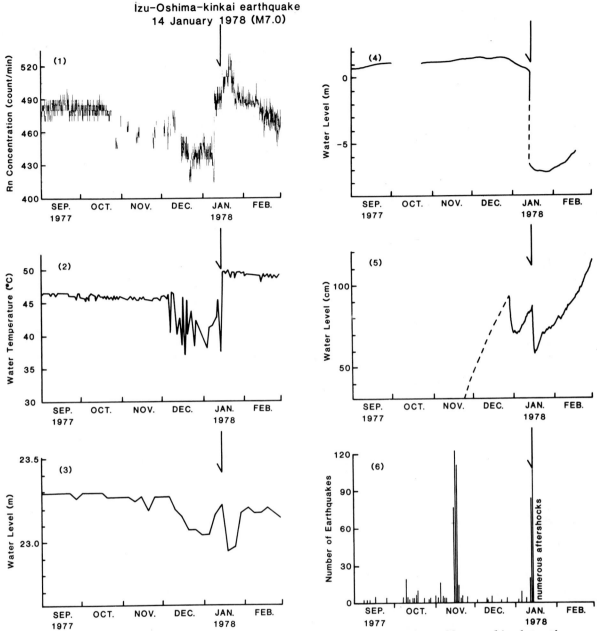

Izu-Oshima-kinkai earthquake
14 January 1978 (M7.0)

Fig. 5. Precursory changes of the 1978 Izu-Oshima-kinkai earthquake, all normalized to the same time scale.
1) Radon concentration changes of a 350 m well at Nakaizu (Δ=25 km).
2) Water level changes of a 500 m well at Nakaizu (Δ=30 km). This figure is based on Fig. 2 in a paper by Kishi (1979).
3) Water temperature changes of a 500 m well at Nakaizu (Δ=30 km). Based on Fig. 22 in a paper by Nagai et al. (1979).
4) Water level changes after correcting for barometric pressure effect of a 600 m well at Funabara (Δ=35 km). Based of Fig. 1 in a paper by Yamaguchi and Odaka (1978).
5) Water level changes of a 500 m well at Omaezaki (Δ=90 km). Based on Fig. 1 in a paper by Wakita et al. (1980b).
6) Seismic activities of the eastern part of the Izu Peninsula. The number of earthquakes per day observed at Kamata in Ito by Japan Meteorological Agency (LMA). Numerous aftershocks following the main shock were not plotted. Data are obtained from the Seismological Division, JMA.

al., 1978) may also be a short-term anomaly.

A model or mechanism to yield such changes has not been constructed at the present stage and remains to be introduced in the future.

In the case of the Izu-Oshima-kinkai earthquake, if various kinds of observation data were collected at a certain data gathering center by a telemetering system, and the evaluation and assessment of data were made in real time, recognition of precursory anomalous changes might not be too difficult.

Acknowledgments. The author wishes to express his appreciation to Dr. T. Asada for helpful comments, to Drs. K. Notsu and Y. Nakamura for assistance in the field work and valuable discussions and to Miss N. Tsushimi for drawing the illustrations for this paper. Work supported in part by the grant of the Ministry of Education, Science and Culture, Japan.

References

Asada, T., The relation between the occurrence of regional earthquakes and the temperature variations of self-spouting hot spring at Nakaizu (2), (in Japanese), Rept. Coord. Comm. Earthq. Predct., 20, 86-89, 1978.

Crustal Dynamics Division, Geograph. Sur. Inst., Crustal deformation in the central part of Izu Peninsula, (in Japanese), Rept. Coord. Comm. Earthq. Predict., 16, 82-87, 1976.

Hot Spring Res. Inst., Water-level anomalies prior to the west off Izu-Oshima Earthquake, Jan. 14. 1978, (in Japanese), Hot Spring Res. Inst., 1-49, 1978.

Ikeda, K., S. Nagai, H. Goto, K. Abe, S. Nagata and N. Oba, Variation of the Izu Peninsula, with special reference to the Izu-Oshima-Kinkai Earthquake, 1978, (in Japanese), Sp. Rept. No. 7, Geol. Surv. Jpn., 129-149, 1979.

Kishi, K., Change of water level and flow rate of wells in the eastern part of Izu Peninsula at the time of Izu-Oshima Kinkai earthquake, 1978, (in Japanese), Sp. Rept. No.7, Geol. Surv. Jpn., 71-85, 1979.

Nagai, S., K. Ikeda, H. Goto, K. Abe, S. Nagata,

and N. Oba, Change of ground water temperature and quality before and after the Izu-Oshima-Kinkai Earthquake, 1978, (in Japanese), Sp. Rept. No.7, Geol. Surv. Jpn., 87-118, 1979.

Noguchi, M., and H. Wakita, A method for continuous measurement of radon in groundwater for earthquake prediction, J. Geophys. Res., 82, 1353-1357, 1977.

Sugisaki, R., Changing He/Ar and N_2/Ar ratios of fault air may be earthquake precursors, Nature, 257, 209-211, 1978.

Sultankhodjaev, A.N., I.G. Chernov and T. Zakirov, Hydrogeoseismological forerunners of the Gasli earthquake, Doklady Akad. Nauk, UzSSR, No.7, 51-53, 1976.

Sultankhodjaev, A.N., S.U. Latipov and T. Zakirov, Hydrogeoseismological effects of the Alai earthquake, November 2, 1978, Doklady Akad. Nauk, UzSSR, No.3, 56-58, 1979.

Tsumura, K., I. Karakama, I. Ogino, and M. Takahashi, Seismic activities before and after the Izu-Oshima-kinkai earthquake of 1978, (in Japanese), Bull. Earthq. Res. Inst., 53, 841-854, 1978.

Wakita, H., Earthquake prediction by geochemical techniques, Recent Progress of Natural Sciences in Japan, 4, 69-75, 1979a.

Wakita, H., Earthquake Chemistry, (in Japanese), Kagaku no Ryoiki, 33, 92-103, 1979b.

Wakita, H., Y. Nakamura, K. Notsu, M. Noguchi, and T. Asada, Radon anomaly: A possible precursor of the 1978 Izu-Oshima-kinkai earthquake, Science, 207, 882-883, 1980a.

Wakita, H., Y. Nakamura, and T. Asada, Water level variations observed prior to the 1978 Izu-Oshima-kinkai earthquake and the 1978 Miyagi-ken-oki earthquake, (in Japanese), Rept. Coord. Comm. Earthq. Predict., 23, 60-62, 1980b.

Yamagishi, Y., H. Sekiya, Y. Suzuki, and K. Sato, Programme Abstr. Seismol. Soc. Jpn., No.1, p. A20, 1978.

Yamaguchi, R., and T. Odaka, Precursory changes in water level at Funabara and Kakigi before the Izu-Oshima-kinkai earthquake of 1978, (in Japanese), Bull. Earthq. Res. Inst., 53, 841-854, 1978.

THE DETECTION OF NANOEARTHQUAKES

Ta-liang Teng and Thomas L. Henyey

Department of Geological Sciences, University of Southern California
Los Angeles, California 90007

Abstract. High-frequency, wideband (20 Hz to 16 kHz) recording instruments for the detection of minute seismic emissions in borehole environments have been designed, developed, and deployed in three deep wells within seismically active regions. Two of these well sites are within a few km of the San Andreas fault near Palmdale, California; the third site is at the Monticello Reservoir, South Carolina. Seismicity near the Monticello Reservoir is induced by the recent impounding of water. The sensor is a highly-sensitive hydrophone emplaced at the bottom of the fluid-filled well. The surface recording package is an analog event recorder complete with event-detecting logic and digital delay circuit. At all three sites, numerous minute seismic emissions were detected with dominant spectral energy in the band 0.5 to 5 kHz and a peak at about 2 kHz. These events have durations of the order of 10 to 100 milliseconds and waveforms similar to a near-field earthquake greatly scaled down in size. Risking downward extrapolation of the duration time vs. magnitude relationship, these events may be assigned magnitudes in the range of -1 to -5; as such we call them nanoearthquakes. With their high-frequency content and in view of the strong attenuation in the upper crust, these nanoearthquakes are probably occurring at distances less than one km from the sensor. If laboratory results are applicable to field situations, the frequency of occurrence of nanoearthquakes may reflect the state of ambient stress, and their rate of occurrence may be useful in identifying the approximate time when a large earthquake is imminent.

Introduction

Minute elastic radiation is known to occur in a stressed rock. Repeated observations in the laboratory and in the field (mainly coalfields) show that these emissions begin to occur as the applied stress reaches a certain threshold characteristic of the rock medium. The frequency of occurrence increases rapidly as the stress increases. As time progresses, the sources of these emissions tend to converge toward a location where the final rupture usually would take place. This pre-rupture elastic radiation process would be very attractive in light of possible application to the prediction of natural earthquakes.

Much has been reported on pre-earthquake anomalous animal behavior; some of these animals are noted for their hearing sensitivity. Reports of audible sounds before the occurrence of large earthquakes are also common. Although many of these reports lack scientific credibility, especially after-the-fact narrations, some of them appear to be valid observations and cannot be easily dismissed. Davison (1938) has compiled descriptions of sounds accompanying or preceding earthquakes; he concludes that 58 to 72 percent of the sounds occurred before earthquakes. Lawson (1908) also reports sounds before the 1906 San Francisco earthquake and its aftershocks. The report of earthquake sounds before shocks in China dates back to 474 A.D. A partial compilation of these reports is given in Table 1. Reports after 1968 are generally obtained through group observations that cannot be explained by individual illusions. The descriptions of earthquake sounds in this table are very similar; they are also similar to those given by Davison (1938) and others (see, for example, Wallace and Teng, 1980). Of particular interest in these reports is a probable rough correlation of earthquake magnitude with precursory time. For large earthquakes (M > 6 1/2), precursory times from hours to days are reported. Such phenomena, if shown to be real, may furnish a very desirable precursory signal for imminent earthquake predictions.

Hill et al. (1976) conclude that under favorable conditions acoustic waves can be excited by seismic P and SV waves impinging at the earth's surface. However, this seismic-to-acoustic excitation is not an efficient process and earthquake sounds are likely to be perceptible only in the immediate neighborhood of seismic disturbances. Calculations by Hill et al. (1976) suggest that it is much more efficient to study minute seismic emissions by placing sensors in the earth rather than to attempt microphone pickups of seismically-induced earthquake sounds or acoustic emissions. This paper discusses our work on instrument de-

TABLE 1a.   Earthquake Sound Reports from Historical Documents

Epicentral Area	Date	M	Precursory Time	Description	Notes
Shanxi	7/474	4-5	Minutes	Sounded like thunder a dozen times, earthquake followed.	A, p. 184
Liaoning	10/24/1594	4-5	Minutes	In the morning, a rumbling of distant thunder approached from NW. As it came close, windows rattled. Shortly after, buildings started shaking.	A, p. 322
Shandon	7/25/1668	8.5	One day	The day before the main shock, the rumbling sound of a rushing river was heard.	A, p. 636
Liaoning	12/11/1855	5-6	Minutes to Hours	Before the quake, a roll of thunder was heard that alarmed people to go outdoors.	A, p. 339
Sichuan	9/12/1850	7	One hour	Shortly after lunch, a burst of sound suddenly came from the NW, about an hour later river overflow and strong ground motions followed.	A, p. 1210

Notes: A.   The historical earthquake data of China, two volumes (in chinese, compiled by Academia Sinica, 1973).

sign, and the detection and analysis of seismic excitations recorded by hydrophones in deep water-filled boreholes.

Experimental Background

There is ample experimental evidence leading to the present field study. In the laboratory, deformation experiments of brittle rocks are accompanied by emissions of elastic energy which Mogi (1962 a, b; 1968) called elastic shocks, and Scholz (1968 a, b, c) referred to as microfractures. Studies of these events have led to an improved physical understanding of the failure mechanism that is the most probable underlying cause of earthquake occurrence (Lockner and Byerlee, 1977; Weeks, Lockner and Byerlee, 1978; Dieterich, 1978; and Kranz, 1980). There is a close similarity between laboratory deformations of brittle rocks and deformation of the crust. Both generate events that follow a similar frequency-magnitude distribution. One can define a continuous spectrum of elastic excitations from 10^{-3} Hz to almost 1 MHz that would accommodate the wave phenomena of large earthquakes (10^{-3} - 10^0 Hz), microearthquakes (10^0 - 10^2 Hz), landslides ($\sim 10^2$ Hz, Cadman and Goodman, 1967), rock bursts (10^2 - 10^3 Hz, Antsyferov, 1966), and laboratory rock fracturing experiments of varying scales (10^5 - 10^6 Hz). Assuming a typical velocity of 4 km/sec for crustal rocks, we find that laboratory fracturing of brittle rocks produces elastic waves with wavelengths in the range of 4 cm to 4 mm, corresponding to microfracturing across grain boundaries, or boundaries between small heterogeneities. For rock bursts with wavelengths in the range 40 m to 4 m, one cannot invoke a grain boundary dislocation mechanism. Crustal irregularities of larger dimensions (large cracks, joints, etc.) are necessary to explain the generation of these excitations.

Rock bursts have been intensively studied (see volumes edited by Antsyferov, 1966; and Hardy and Leighton, 1977). Using a recording system with flat response from 10 to 500 Hz, McGarr and Green (1978) recorded numerous micro-tremors during aftershock sequences to several M=1 to M=1.5 events in a deep mine. Application of a high-frequency seismological method to geomechanics problems, particularly associated with mining operations in coalfields, has led to successful prediction of rock bursts, which bear close resemblance to scaled-down natural earthquakes. It is regularly observed that, prior to rock bursts,

Epicentral Area	Date	M.	Precursory Time	Description	Notes
Shandong	4/02/68	5.2	One Minute	Sounded like a peal of thunder, or the passing of a tractor.	B
Quandong	7/26/69	6.4	Two Days	Divers reported sound heard below water.	C
Yun-nan	1/05/70	7 3/4	One Day	A rumbling sound came from the mountains.	D
Shandong	8/10/70	5	Minutes to Hours	Before the quake, it sounded like thunder, strong wind, a train going over a bridge, or a jet airplane.	E
Yun-nan	5/29/76	7.6	Minutes to half an hour	Rumbling thunder, passing tanks or tractors heard up to a distance of 80 km, over extensive surrounding areas. Sound came from consistent direction when heard from large distance, appeared non-directional when heard very near the epicenter.	F
Sichuan	8/16/76	7.2	An hour to half a month	Sounds of an airplane, single bursts of frog croaking, and long rolls of a thunder were heard at different times and locations.	G

Notes:
B. Shandong earthquake brigade, 1969, a brief report on the Quan Hsien, Shandon and Da Ming, Hebei earthquake of April 2, 1968 (in Chinese). PRC State Seismological Bureau Report.
C. Yangjiang earthquake investigation team for Macroscopic Phenomena, 1969, investigation report, PRC State Seismological Bureau Report.
D. State Seismological Bureau of People's Republic of China, 1970, a macroscopic investigation on the seismo-geological phenomena of the Tunghai, Yunnan earthquake of January 5, 1970, PRC State Seismological Bureau Report.
E. Shandon Earthquake Research Center, 1970, a survey report on Qu-pu earthquake, Appendix I, PRC State Seismological Bureau Report.
F. Chen, L.D., and Chao, W.C., 1979, Lungling earthquake of 1976, a compilation of observations, Earthquake Publishing Company, Beijing, China.
G. Sichuan Provincial Seismological Bureau, 1979, Sungpan earthquake of 1976, a compilation of observations, Earthquake Publishing Company, Beijing, China.

an anomalous increase in the number of minute seismic emissions first takes place. However, an anomalous occurrence of this seismic radiation is not found to be a sufficient condition for a subsequent rock burst; such an occurrence may correspond to a natural earthquake swarm as a scaled-up counterpart.

With these background experimental results, we speculate that in a stressed rock body of $10 - 100$ km in dimensions, there may exist a continuous radiation of minute seismic events, the activity level of which serves as one manifestation of the ambient stress state. To differentiate these field events from those observed in the laboratory, and from microearthquakes which have much lower frequency spectra and much higher energy content, we call them nanoearthquakes. Nanoearthquakes have magnitudes of less than -1, and are the natural downward continuation of microearthquakes. Nanoearthquakes may have a more-or-less homogeneous distribution over the entire volume of the stressed rock body, that is, a volume with dimensions substantially larger than those of the eventual rupture surface. This postulated homogeneous occurrence of nanoearthquakes may correspond to Mogi's (1968) stage B in a laboratory experiment. As the ambient stress increases and it enters Mogi's stage C, two conditions could exist: (1) If the field observation sites were in close proximity to the eventual rupture surface, one might observe a continuing increase in the level of occurrence of these nanoearthquakes. This increase would become more rapid as the rupture is approached; and (2) If the field observation sites were at some distance ($\sim$ 10 - 100 km) from the eventual rupture surface, Mogi's stage C would be accompanied by a gradual decrease in the activity level before the eventual rupture. Unless coverage of an earthquake prone area was extensive, case (2) would be more likely. For such observations to be of practical use, we would first need to know if these nanoearthquakes are numerous enough to be analyzed statistically, and second what the time duration of the field equivalent of Mogi's stage C is likely to be. Considering the data in Table 1, this duration may be of the order of hours to days for large (M > 6 1/2) earthquakes, which may be sufficiently long for purposes of earthquake prediction. To test these hypotheses, we have designed and constructed appropriate field equipment, and have deployed the equipment in deep boreholes drilled into rock bodies known to have been under ambient stress loading. A discussion of our experimental work follows.

Instrumentation

A high-frequency seismic event recording system suitable for field deployment was designed and constructed to detect and study the possible occurrence of nanoearthquakes in stressed rocks. The entire system consists of a downhole sensor, a downhole preamplifier, and a surface event recorder with a time code generator and regulated power supply. A cable up to several thousand feet long connects the downhole instruments with the surface recorder. This cable carries power to supply the downhole preamplifier and transmits the signal from the downhole sensor to the surface recording package. The sensor is a broad-band (1 Hz to 100 kHz) hydrophone with a sensitivity of 1 volt/µbar. One stage of preamplifier with 40 db gain is applied to the signal next to the sensor and used as a line driver for the long cable.

In situ recording is necessary since the expected signals contain frequencies too high for telephone telemetry. However, in the future, rather than relying on waveform analysis and recording the entire waveform, the recording system can be greatly simplified to merely an event counter. Thus its output becomes a low sample rate signal (in terms of number of events per hour) which can be brought back to the laboratory through a variety of telemetry devices.

Figure 1 illustrates a block diagram of the recording system, which is a broad-band analog event recorder normally operated in a standby mode. A 60 db gain amplifier is placed at the front end. When an input signal amplitude exceeds a preset level, an event detector circuit triggers and initiates a fast-start instrument-quality analog tape recorder which runs at a speed of 3 - 3/4 ips and delivers a frequency response up to 16 kHz. A signal delay system has been introduced to avoid the loss of signal during the initial 400 ms start-up of the tape recorder. It consists of an A/D converter with 8 bits parallel output. Each parallel bit is followed by an array of 20 shift register chips with 1024 bits per chip. With a 40 kHz clock drive, a total signal delay of 512 ms is realized. The recombined delayed signal is passed through a D/A converter and a low-pass filter to remove any noise higher than 16 kHz. The signal is then recorded on one track of the tape recorder. Since the low-frequency cutoff of the tape recorder is about 20 Hz, the overall system has a bandwidth of 20 Hz to 16 kHz.

Parallel to the input signal is the timing signal generated by a crystal clock. Since the input signal is expected to be short in duration (a fraction of a second), the after-event duration has been set to be 10 sec. to conserve the recording tape. A special circuit has been constructed to generate a 10 sec. time frame. This consists of a digital clock, calendar circuit, and its accompanying display and the time code generator circuit. In each ten-second period, time of day is displayed once every second for eight seconds, and the month and date once every four seconds. A Ni-Cd rechargeable battery is used to power the clock to prevent loss of clock information due to power failure. The entire recording system is packaged in a 26" x 20" x 16" fiberglass case for field deployment (Figure 2).

Field Studies

Field sites were prepared parallel to instrument development. Downhole deployment is necessary to avoid surface noise (Bacon, 1975). Deep boreholes of sufficiently large diameter are used for emplacement of the sensor-preamplifier package. Three field sites have been used to date. In southern California, two abandoned wildcat oil wells near the San Andreas fault in the Palmdale area were selected. The Skelton well (34°30'N, 118°14'W) was drilled entirely in granitic rock 3 km NE of the San Andreas fault. This well has 10 3/4" surface casing for the first 65 m and is

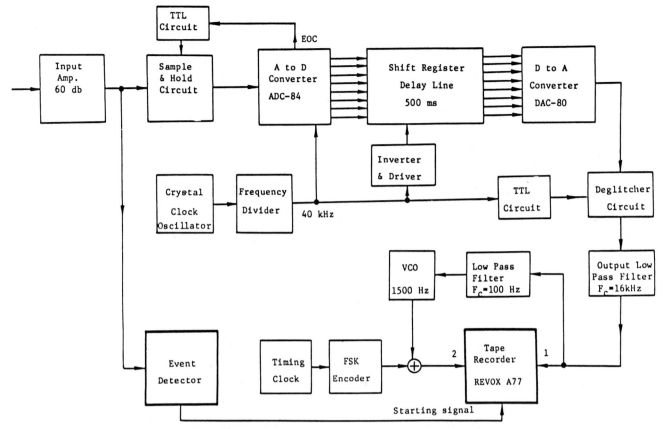

Figure 1. A schematic diagram of the high-frequency event recorder.

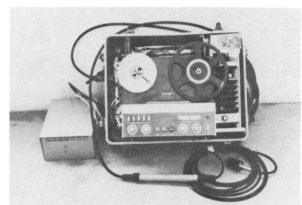

Figure 2. The complete high-frequency event recorder system (minus downhole cable). In the center is an instrument quality analog tape recorder with fast start (400 ms). To its right are the detection and digital delay circuits, plus a crystal clock. The box to its left is a regulated power supply. The cylinder at the bottom is the downhole preamplifier; the sensor hydrophone is in the middle of the small coil of cable.

uncased 8 3/4" hole to an accessible depth of 300 m. The Del Sur well ($34^0 39'$N, $118^0 14'$W) was drilled in Tertiary/Quaternary sandstones and shales 6 km NE of the San Andreas fault. The well is cased with 6 5/8" casing to a total depth of 700 m. A third field site, the "Monticello #2" well ($34^0 18'$N, $81^0 19'$W) was drilled by the U.S. Geological Survey for experiments related to microearthquake activity induced by the impounding of water in Monticello Reservoir. Precise hypocenter locations show the seismicity to be in clusters at depths up to about one km (Talwani, 1979). The well was drilled to a depth of 1 km into granite/gneiss and is uncased except near the surface.

In each case, the hydrophone-preamplifier package was lowered by winch to the well bottom. Sufficient cable was fed into the well to allow for total relaxation of the cable. AC power was connected to the regulated power supply of the surface recording system through an isolation transformer to prevent spurious power line noises from being introduced into the recording circuit. After the sensor was properly emplaced, a few hours were allowed for the sensor to equilibrate with the ambient environment. The recorder gain was adjusted for the maximum sensitivity permit-

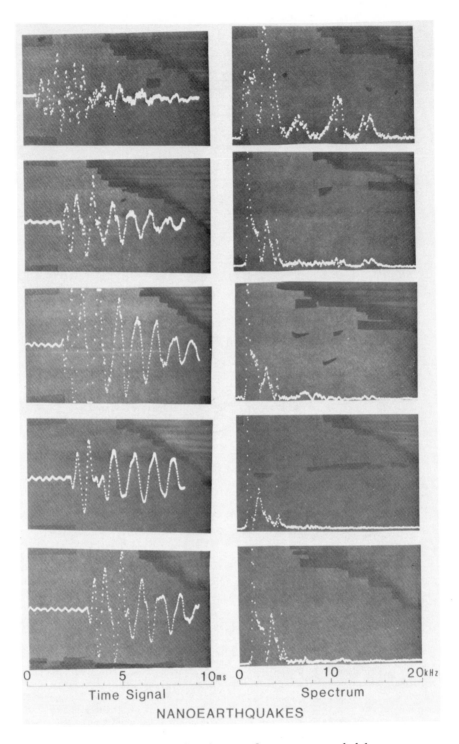

0   5   10ms   0   10   20kHz

Time Signal   Spectrum

NANOEARTHQUAKES

Figure 3. Some typical nanoearthquake waveforms as recorded by a pressure
transducer (left) and their corresponding spectrums (right) recorded at the 300 m deep bottom of the Skelton well near Palmdale,
California. The well is 3 km from the San Andreas fault.

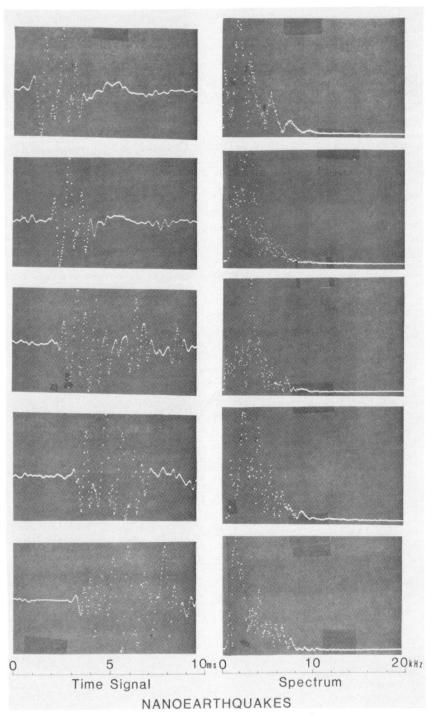

| 0 | 5 | 10 ms | 0 | 10 | 20 kHz |

Time Signal   Spectrum

NANOEARTHQUAKES

Figure 4.   Some typical nanoearthquake waveforms as recorded by a pressure
transducer (left) and their corresponding spectrums (right) recorded
at the 1 km deep bottom of the Monticello #2 well in the immediate
vicinity of the Monticello dam in South Carolina.   The impounding
of the reservoir has induced numerous seismic swarm activities.
These swarms have focal depths up to about 1 km, and into the swarm
hypocenter clusters are the Monticello wells drilled by the USGS.

ted by the background noise. Data gathered from
the three field deployments are discussed below.

Data Analysis and Discussion

A large number of nanoearthquakes were recorded
during the three separate field deployments described above. For illustration, a number of these
events are reproduced in Figures 3 and 4. These
are typical of events from the Skelton and the
Monticello wells.

Recordings from the Skelton and Monticello wells
were subjected to waveform analysis. At this
stage of our experimental work, we seek to answer
two questions: 1. Is high-frequency seismic
radiation generated in stressed rocks in the
field? And if so, 2. What is the character of its
waveforms? The existence of numerous events such
as those in Figures 3 and 4 indicates the answer
to the first question to be affirmative. To study
the waveforms, the analog recording tapes were
played back through a fast sample-and-hold device
and in turn the digitized data were fed through a
minicomputer which stored the digitized events on
disk. At a sampling rate of 40 kHz, the playback
system was consistent with the bandwidth of the
recording system. Signals on the disks were
dumped one file at a time back through the minicomputer and displayed on an oscilloscope. Signals were carefully examined, and windowed for
spectral analysis. Figures 3 and 4 give, respectively, some typical unclipped nanoearthquake
waveforms and their corresponding spectra. Since
photographs were taken from the oscilloscope,
some of the signal tails were omitted although
the accompanying spectrums were obtained for the
entire waveforms. Durations of the nanoearthquakes are primarily in the range of 5 to 100 ms.
The unclipped amplitude is of the order of 10
μbar. Risking the downward extrapolation of the
total duration vs. magnitude scale (Real and Teng,
1973), the corresponding magnitudes for these
events are -1.5 to -4 (using an equation by
McGarr and Green, 1978, the magnitude range is -2
to -5). From the spectra, one finds the dominant
energy peak to be about 2 kHz. This peak drops
off sharply at the low-frequency end. The data
from the Skelton well show little or no energy
below 1.5 kHz, whereas the Monticello data have
energy down to 1 kHz and below. The high-
frequency end does not drop off as sharply;
occasional secondary peaks are present. Little
energy exists beyond 6 kHz. It is interesting
to note that the nanoearthquakes have very
similar waveforms and spectrums whether from the
Skelton well in California or from the Monticello
well in South Carolina. Since the sensor is a
hydrophone emersed in the borehole fluid, no shear
waves were recorded; the converted compressional
waves from incident shear waves at the rock-fluid
interface were probably small, and no such phase
was identifiable. Assuming the S-P interval
velocity to be ∿ 10 km/sec (or 10 m/ms) and that
shear-converted compressional waves are not negli-

gible in amplitude, the maximum epicentral distance of these nanoearthquake sources should not
exceed 1 km. For a signal duration of 10 ms, the
source distance probably is less than 100 m. The
playback data also show that these nanoearthquakes
occur rather sporadically in groups.

During the recording period, at least 30 events
per day were recorded at the Skelton well and
about 500 at the Monticello well. For these two
periods of recording, instruments were set at comparable gain. If all other ambient conditions are
the same, a high event rate would suggest a high
stress state (Mogi, 1968; Lockner and Byerlee,
1977). Recent results from hydro-fracturing
stress measurements at sites near Palmdale and at
the Monticello well site also suggest that the
Monticello well is subject to a higher ambient
stress (Zoback, 1980). For a given monitoring
site, if the number of nanoearthquakes detectable
is large enough to make the time series statistically meaningful, then the rate of nanoearthquake
occurrence may be used as a reference for the
ambient stress state, and thus as an earthquake
precursory parameter. The short detection distance (<< 1 km) is not a severe drawback, so long
as during stress buildup (e.g. Mogi's stage B),
the occurrence of nanoearthquakes is more-or-less
homogeneous over the entire source region. This
region may have dimensions on the order of 10-100
km, depending on the magnitude of the impending
earthquake. Thus, for a given monitoring site
away from the eventual rupture surface, first a
gradual increase of the event rate of nanoearthquakes may be observed, followed by a short

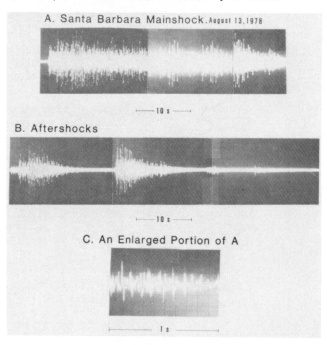

A. Santa Barbara Mainshock. August 13, 1978

——— 10 s ———

B. Aftershocks

——— 10 s ———

C. An Enlarged Portion of A

——— 1 s ———

Figure 5.   Recording of the Santa Barbara earthquake of August 13, 1978 and its aftershocks.

quiescent period before the final rupture.

At the Del Sur well, recorded signals had much higher frequency content. The waveforms also show a characteristic pulse shape that is not expected from a typical disturbance of seismic origin. We suspect that these signals may be related to the casing-wall rock interaction since the Del Sur well is cased over its entire length. The Del Sur recorder also captured the M=5.3 Santa Barbara earthquake of August 13, 1978, and many of its aftershocks (Figure 5). The epicenter of the Santa Barbara earthquake was ∿ 120 km W of the recording site. Much of the high-frequency (>> 20 Hz) content of the seismic waves was lost along the relatively long travel path, and frequencies lower than 20 Hz were not well recorded by the band-limited tape recorder. Nevertheless, the events were still large enough to be recorded due to the high gain setting, and the onsets of signal arrivals were preserved.

Our findings are quite preliminary. Further confirmation on the seismic origin of these detected signals must be carefully established. This work suggests that future nanoearthquake detection or monitoring should be performed in uncased wells. Better identification of the source distance would be desirable. A waveform recorder as described in this paper is important for the first phase of the project, during which characterization of nanoearthquakes and their sources are of prime importance, especially to establish whether they originate from rock failures. If so, for the next phase of practical application, the complex recorder can be replaced by an event counter that produces low sampling rate data of perhaps one datum point (in terms of number of events) per minute or per hour as compared to a 16 kHz sampling rate for the current instrument. The simplification of recording instruments would make widespread applications practical and the low data rate would make on line monitoring through telemetry methods possible.

Acknowledgments

We would like to thank Liang-Fang Sun for his instrumentation and field deployment work. John McRaney and Derek Manov assisted in various phases of the field work. Advice and cooperation from Barry Raleigh, Mark Zoback, and Pradeep Talwani were instrumental in the results from the Monticello well of South Carolina. This research was funded by U.S. Geological Survey contracts 14-08-0001-16613 and -16745.

References

Antsyferov, M.S., Principals of the application of seismo-acoustics to coal seams subject to rock bursts, in Seismo-Acoustic Methods in Mining, ed. by M.S. Antsyferov, Consultants Bureau, New York, pp. 1-8, 1966.

Armstrong, B.H., Acoustic emission prior to rockbursts and earthquakes, Bull. Seismol. Soc. Am., 59, no. 3, pp. 1259-1279, 1969.

Bacon, C.F., Acoustic emission along the San Andreas fault in southern central California, Calif. Geol., 28, pp. 147-154, 1975.

Byerlee, J.D., and P. Lockner, Acoustic emission during fluid injection into rock, proceedings of the first conference on acoustic emission in geologic structures and materials, Trans. Tech Publications, Claustal, W. Germany, 1977.

Cadman, J.D. and R.E. Goodman, Landslide noise, Science, 158, pp. 1182-1184, 1967.

Davison, C., Earthquake sounds, Bull. Seismol. Soc. Am., 28, pp. 147-161, 1938.

Dieterich, J.H., Preseismic fault slip and earthquake prediction, J. Geophys. Res., 83, no. B8, pp. 3940-3949, 1978.

Hardy, H.R. Jr., and F.W. Leighton, ed: Proceedings first conference on acoustic emissions/microseismic activity in geologic structures and materials, Trans. Tech Publications, Clausthal, Germany.

Hill, D.P., F.G. Fischer, K.M. Lahr, and J.M. Coakley, Earthquake sounds generated by bodywave ground motion, Bull. Seismol. Soc. Am., 66, no. 4, pp. 1159-1172, 1976.

Kranz, R.L., The effects of confining pressure and stress difference on static fatigue of granite, J. Geophys. Res., 85, no. B4, pp. 1854-1866, 1980.

Lee, T.Y., and X.K. Hu, A preliminary study on the relationship between earthquake and earthquake sound, Acta Geophysica Sinica, 23, no. 1, pp. 94-101, 1980.

Lockner, L., and J. Byerlee, Acoustic emission and creep in rock at high confining pressure and differential stress, Bull. Seismol. Soc. Am., 67, pp. 247-258, 1977.

Lockner, D., A. Lindh, and J. Byerlee, Apparent velocity anomalies and their dependence on amplitude (abstract), E.O.S. Trans. Am. Geophys. Union, 58, p. 433, 1977.

Lawson, C.A., The California earthquake of April 18, 1906, Report of State Earthquake Investigation Commission, VI, Carnegie Institution of Washington, 1980.

McGarr, A., and R.W.E. Green, Microtremor sequences and tilting in a deep mine, Bull. Seismol. Soc. Am., 68, no. 6, pp. 1679-1697, 1978.

Mogi, K., Study of elastic shocks caused by the fracture of heterogeneous materials and its relation to earthquake phenomena, Bull. Earthquake Res. Inst., 40, pp. 125-173, 1962a.

Mogi, K. Magnitude-frequency relation for elastic shocks accompanying fractures of various materials and some related problems in earthquakes, 2, Bull. Earthquake Res. Inst., 40, 831-853, 1962b.

Mogi, K. Source locations of elastic shocks in the fracturing process in rocks, Bull. Earthquake Res. Inst., 46, pp. 1103-1125, 1968.

Pao, Y.H., Theory of acoustic emission, elastic waves and non-destructive testing of materials, AMD - 29, ed. Y.H. Pao. The American Society of Mechanical Engineers, New York, pp. 107-128, 1978.

Real, C.R., and T.L. Teng, Local Richter magnitude and total signal duration in southern California, *Bull. Seismol. Soc. Am.*, 63, No. 5, pp. 1809-1827, 1973.

Scholz, C.H., Microfracturing and the inelastic deformation of rock in compression, *Geophys. Res.*, 73, no. 4, pp. 1417-1432, 1968a.

Scholz, C.H., Experimental study of the fracturing process in brittle rock, *Geophys. Res.*, 13, no. 4, pp. 1447-1454, 1968b.

Scholz, C.H., The frequency-magnitude relation of microfracturing in rock and its relation to earthquakes, *Bull. Seismol. Soc. Am.*, 58, no. 1, pp. 399-415, 1968c.

Talwani, P., Induced seismicity, earthquake prediction and crustal structure studies in South Carolina, *Summaries of Technical Reports, VIII*, National Earthquake Hazard Reduction Program, U.S.G.S., 1979.

Weeks, J., D. Lockner, and J. Byerlee, Changes in b - values during movement on cut surfaces in granite, *Bull. Seismol. Soc. Am.*, 68, pp. 333-341, 1978.

Wallace, R.E., and T.L. Teng, Prediction of the Sungpan-Pingwu earthquakes of August 16, 1976, *Bull. Seismol. Soc. Am.*, 70, no. 4, pp. 1199-1224, 1980.

Zoback, M., Personal Communication, 1980.

A PRELIMINARY ANALYSIS OF REPORTED CHANGES IN GROUND WATER AND ANOMALOUS

ANIMAL BEHAVIOR BEFORE THE 4 FEBRUARY 1975 HAICHENG EARTHQUAKE

Deng Qidong and Jiang Pu

Institute of Geology, State Seismology Bureau, Beijing, People's Republic of China

Lucile M. Jones and Peter Molnar

Department of Earth and Planetary Sciences, Massachusetts Institute of Technology,

Cambridge, Mass. 02139

Abstract. We have examined the spatial and temporal distributions of some 570 reports of changes in ground water and 670 reports of anomalous animal behavior in the three months before the Haicheng earthquake (4 February 1975, M=7.3). These changes and anomalies were reported from a very large area, extending more than 150 kilometers in nearly all directions from the epicenter with no concentration near it. There are suggestions (1) of correlations in time and space of the two types of anomalies with the ground water changes preceding the aberrant animal behavior by a day or two, (2) of a greater concentration of reports near major active faults than far away from them, and (3) of migrations in time of the area in which there were frequent observations. The number of reported changes in ground water increased abruptly on 1 February 1975, the day of the first recorded foreshock. The number of reported observations of anomalous animal behavior increased dramatically on 3 February, the day on which the first foreshock that was large enough to be felt occurred. These data suggest that some animals may have responded to a shaking of the ground due to foreshocks and others may have sensed changes in the ground water (level, composition, or other properties).

Introduction

The 4 February 1975 Haicheng earthquake was the first major earthquake to be predicted. It was preceded by precursors of many different types (Haicheng Earthquake Work Brigade, 1975; Raleigh et al., 1977; Zhu Fungming, 1976), but among these precursors, foreshocks and macroscopic anomalies were the primary basis of the short term and imminent predictions. (Macroscopic anomalies are anomalies recorded by local observers rather than by instruments. For Haicheng, they included changes in ground water which were usually changes in the level or color of the well water, aberrant animal behavior, the appearance of a low ground fog, and others.) In 1970, Liaoning province was recognized by the State Seismology Bureau as a region susceptible to large earthquakes. In June 1974, from an increase in the rate of surface deformation revealed by short leveling lines and long term magnetic anomalies, it was singled out as a likely place for a large earthquake in the following year or two. An increase in anomalous animal behavior in December 1974, and early January 1975 prompted a short term prediction to be made in mid-January (effective for the first half of 1975). Then on 1 February, foreshock activity began and was accompanied by a dramatic increase in ground water changes. On 3 February, one day before the event, there was a large increase in the number of observations of aberrant animal behavior. An imminent prediction (effective for the next day or two) was issued by the Liaoning Provincial Seismological Bureau at 1000 of the day of the earthquake, which occurred at 1936 local time. Some communes had, in fact, already made predictions of their own the day before (3 February, 1975). More detailed accounts of the prediction and summaries of the data on which it was based are given by Raleigh et al. (1977) and Zhu Fungming (1976). A map of the structure and geology of the epicentral area is shown in Figure 1.

In the present paper, we carried out a more detailed analysis of the distribution in time and space of observations of changes in ground water and of anomalous animal behavior than those given in the references noted above. We had three primary objectives in this study. We sought correlations between the two types of anomalies that might suggest a causal mechanism for the animal behavior. We sought spatial relationships between the observations and the geologic struc-

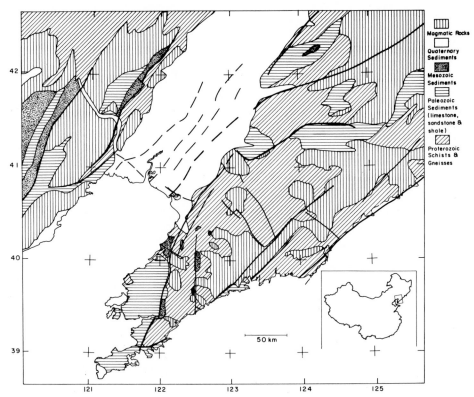

Fig. 1. A geologic map of Liaoning Province.

ture that might enlighten us about the causes of the ground water changes. Finally, we sought variations in time in both the spatial distribution and in the types of reported changes or anomalies which might suggest physical mechanisms and causes of them. For each case, correlations appear to exist, but none are overwhelmingly clear. The most apparent features in the spatial distribution of observations are the vast area covered by them and the absence of a concentration of them near the epicenter, at least until the day before the earthquake.

Data

We used preliminary listings both of changes in ground water observed in wells, underground vegetable storage areas, springs, streams and ponds and of anomalous animal behavior. The vast majority of the ground water reports were from wells, and for most of them, the observations were of changes in the level or a muddying of the water. Other changes reported include other changes in color, taste or smell, oiliness, and less often, a bubbling or swirling of the water. Sometimes, more than one well, storage area, spring, etc. in the same locality showed a change. In only about half of the observations of changes in water level was a figure given for the amount of the change. Usually very little information was given beyond the type of change, the number of wells, vegetable

storage areas, springs, etc. showing the changes, the location and the date. Similarly, observations of aberrant animal behavior were brief, often only noting the type of animal, the location and the date. For a given location and date, often more than one animal or more than one species was noted.

Clearly the data are very crude. Only a tiny fraction were obtained by technically trained people. Usually only a date was listed with no hour or time of day. Moreover, a large majority of the observations were obtained after the earthquake when people (including one of us, D.Q.) from the Liaoning Provincial Seismology Bureau, the Institute of Geology of the State Seismology Bureau (SSB), and other work units of the SSB asked local people about possible precursors. This might have introduced a bias into the data particularly for the period immediately preceding the earthquake. Nevertheless, lacking any criteria for accepting or rejecting specific reported observations, we have used all of the data on the lists. At present, the Liaoning Provincial Seismology Bureau is re-examining each observation and is compiling more complete and accurate lists than are available now. All of the complete written reports of animal anomalies that were available are listed in Table I and mapped in figure 2.

In our analysis, we treated the ground water and animal behavior data separately. For each, we separated the data into different types. For ground water, we distinguished observations of

Table 1. **Examples of Unusual Animal Behavior**

---

(1) 9 January 1974. A snake was found at the entrance of its hole with the upper half of its body frozen stiff and the lower half still moving.
123°10'
41°12'

(2) 24 January 1974, approximately 12:00. More than 40 mice who acted very dazed were found on the floor of a room. Someone slaped at them with a board 3 times and killed 9 of them. One slap alone killed 5 of them. (They did not run away even though they were being attacked.)
124°20'
40°22'

(3) 2 February 1975. Twenty piglets, which were born in October 1974 were found in the pigsty crying wildly. More than half had had their tails bitten off and eaten. One of the stumps originally 17 cm now 2.5 to 2.6 cm long, was bleeding profusely. The manager of the commune's earthquake office investigated this before the earthquake and, considering other anomalies as well, made a prediction to the commune. The commune leaders reported this to the appropriate agency and at 0800, 4 February they called an emergency meeting. They took measures to prepare for an earthquake.
121°58'
40°49'

(4) 2 February 1975, afternoon. A small dog barked wildly and refused to eat. During the night his barking kept people awake. Afterwards he was dragged out and tied to a stake, where he continued to bark wildly until the time of the earthquake.
122°30'
40°38'

(5) 2 February 1975, 1600. Fish in a commercial fish pond began to swim near the surface of the water. 1830. The fish began jumping out of the water and jumped onto the banks.
4 February 1975,
121°24'
41°13'

(6) 2-3 February 1975. A four-year old bull repeatedly bellowed and ran about wildly so that people were not able to get close to him. On the
122°31'
41°24'

afternoon of 3 February, he was wilder than before.

(7) 3 February 1975. Homing pigeons flew away and did not return until after the earthquake.
41°07'
121°55'

(8) 3 February 1975. Among 6 cows in a pen, 4 fought with one another with their horns. The other two dug at the ground with their hoofs. The people who saw this thought that it might be a precursor and issued a warning to their commune.
40°24'
122°53'

(9) 3 February 1975, 1500. Two geese flew away from a hillside and flew at an elevation of 50 m for a distance of more than 1 li (1/2 km).
39°56'
122°23'

(10) 3 February 1975, 1900. Nine horses became very agitated and jumped around.
40°33'
122°43'

(11) 3 February-4 February (morning) 1975. On a deer farm, there were 213 deer kept in 9 corrals arranged in rows of 3. At 1050 on 4 February a deer in the middle corral of the first row started running about wildly and tried to get out. Deer in the first corral began to show the same behavior. They broke down the gate of the corral and ran off in many directions. One, a 3-year old that weighed 90 kg, had its left hind and right front legs broken. The deer farm reported this to the Anshan seismology office.
41°02'
123°08'

(12) 3-4 February 1975. In the Anshan city park aviary, there were over 100 birds. Many of them picked up their eggs and flew out of their nests. While they were flying their eggs fell and smashed.
41°06'
123°00'

(13) 4 February 1975. Many rats who were not afraid of people were found in the hallway of a hostel.
40°10'
122°08'

(14) 4 February 1975. Sheep cried most of the day.
41°02, 121°47'

(15) 4 February 1975, daytime. Three bulls cried
41°08', 123°36'   strangely for 3 or 4 hours.

(16) 4 February 1975, p.m. A production brigade
41°28'   discovered that a frog had
122°32   left its hole where it had
been hibernating.

(17) 4 February 1975, p.m. A mother pig picked up
40°58'   her young piglets in her
122°16'   mouth and carried them
outside.

(18) 4 February 1975, p.m. Four or five mice were
40°43'   seen running up the side of
122°33'   a wall.

(19) 4 February 1975, 1300-1700. At 1300 the
40°11'   keeper noticed 17 normally
124°21'   quiet pigs climbing the
wall of their pen and
smashing against the gate.
Four had escaped. Seven-
teen other (male) pigs,
who were also usually
peaceful, were acting up.
At 1600 the keeper told the
earthquake office of the
pig farm. At 1700 the
leader of the pig farm
investigated and found the
pigs still boisterous. The
pig farm's earthquake
office increased the number
of people on duty, made
preparations for an impending earthquake, and
notified surrounding
families.

(20) 4 February 1975, 1500. Deer in their stable
41°16'   were very frightened and
121°30'   were not quiet.

(21) 4 February 1975, approximately 1600. An old
40°07'   roan horse and a mule
124°21'   refused to enter their
corral. When they were
forced in, they broke
away and ran back outside.

(22) 4 February 1975, 1600. A dog barked at the
39°24'   air as if it had discover-
122°21'   ed something and scratched
at the ground digging three
shallow holes.

(23) 4 February 1975, 1600. Three rabbits, kept
41°07'   by a family, refused to
121°08'   enter their cages and both
ears stood up as though
they were listening to
something.

(24) 4 February 1975, 1700. A pig belonging to a
40°32'   family, cried wildly
122°08'   and tried to break down its
pen.

(25) 4 February 1975, 1730. A goose belonging to a
41°32'   family flew up onto the
122°12'   wall and cried in an odd
way continuously for
several minutes.

(26) 4 February 1975, 1800. A goose belonging to
41°07'   a family refused to enter
its pen and cried at a high
pitch.

(27) 4 February 1975, 1800. Seven horses broke
40°55'   down the fence around their
123°35'   enclosure and ran away.

(28) 4 February 1975, 1830. A sheep started crying
40°32'   mournfully and continued
122°25'   to do so for 30 minutes.
The owner made a report to
the county Seismology
brigade.

(29) 4 February 1975, 1830. A horse began to bite
41°16'   and paw at the ground,
121°30'   continuing for a half hour.

(30) 4 February 1975, 1830. A sheep refused to
41°00', 122°16'   enter its pen.

(31) 4 February 1975, 1900. Twenty-two deer on a
39°51'   deer farm ran around
122°41'   wildly bumping into each
other crying wildly. The
next day it was discovered
that one small deer had
been crushed.

(32) 4 February 1975, 1900. A small bull started
40°17'   running about wildly and
123°24'   calling wildly within the
village. The leader of the
brigade thought about the
fact that a big earthquake
was supposed to be coming
and he called all of the
people out of their houses.

(33) 4 February 1975, 1900. The hair on the body
40°49'   of a small cat stood on
121°03'   end, its tail was straight
up in the air, and it ran
away.

(34) 4 February 1975, 1900+. A girl discovered
40°50'   many mice (or rats) running
122°10'   about wildly in her house.
She deduced that there
could be an imminent earthquake and made proper

preparations. At the time of the earthquake she picked up a small child and ran out of the building.

(35) 4 February 1975, 1900+. A dog had given birth
40°24'
122°21'
to puppies earlier in the day. After 1700, the mother picked up the puppies in her mouth and ran outside with them. She had to hit against the door twice to get it to open. The owner thought that this was strange and immediately remembered the announcement of the earthquake prediction. He went outside and the earthquake occurred 4 or 5 minutes later.

(36) 4 February 1975, 1920. A sparrow kept by a
39°09'
121°38'
family began flying wildly in its cage and hitting against the sides. Afterwards, it was discovered that it had killed itself by hitting the cage too hard.

changes in surface waters (rivers, ponds, and reservoirs), which were few in number, from those that seemed to be more certainly due to changes in ground water (wells, springs, underground storage areas and water coming out of the ground). We then grouped the observations into four categories: rising level, falling level, changes in composition (muddying, changes in taste, color, or smell, or oiliness), and motion of the water (bubbling or swirling). Only a few of the last category were reported and later we combined it with the third category.

For the animals we distinguished reports of snakes, mice and rats, fish, frogs, dogs, cats, farm birds (chicken, geese and ducks), pigs, herbivorous domesticated animals (horses, mules, cows, sheep, and a few domesticated rabbits and deer), wild birds, wild mammals, caged birds, and insects (plus one observation of 400-500 spiders). The last four categories contributed only a very small fraction of the reports, mostly on the day of the earthquake, so we combined them into one category, other animals.

We then constructed maps for each day beginning on 25 December 1974 with separate maps for ground water and animal behavior. For the different groupings in each category we used different symbols. We began with 25 December 1974 because of the sparcity of observations before then. The

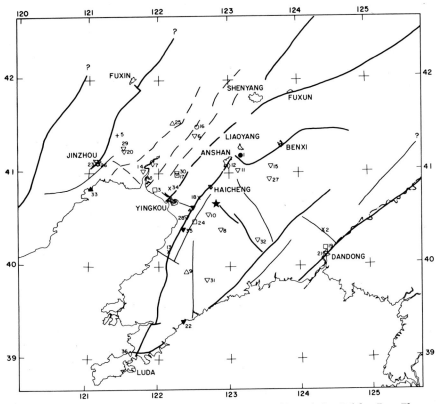

Fig. 2. Map of the reports of anomalous animal behavior listed in Table I. The symbols are the same as in Figure 3.

maps were made at a scale of 1:1,000,000. The locations of the large work brigades where the observations were made were listed in degrees and minutes of arc. Therefore, the uncertainty in the locations is about 2 kilometers. In order to examine larger quantities of data, we made new maps covering longer periods: all of November plus 1-24 Dec. 1974, 25-31 Dec. 1974, 1-7 Jan. 1975, 8-14 Jan. 1975, 15-21 Jan. 1975, 22-26 Jan. 1975, 27-31 Jan. 1975, and 1, 2, 3, and 4 Feb. 1975 (Figure 3). We also constructed maps for the entire period exclusive of the last four days from November through January (Figure 4), and we made day-by-day histograms both for the total number of observations and for different subsets of the observations (Figure 5).

Before calling attention to some of the features to be seen on the maps, it is worth noting some of the sources of contamination or noise in the data. Clearly the population distribution will influence the probability that changes or anomalies will be observed. Moreover, the existence of cities or farms will affect whether or not particular animals or ground water changes can be observed (e.g. big cities have running water so people do not visit the wells). In some areas, particularly in the lowlands near Yingkou, there are few shallow wells. Moreover, the coastline shapes the region in which observations can be made, and the border with Korea limits this area to the east. Finally, the psychological effects of first feeling foreshocks begin on the evening of 3 February and then being asked about precursors subsequent to the mainshock may have influenced the people's perception of possible precursors. In our discussion below, we try to keep these considerations in mind, but regardless, we hope that the data have been presented in such a form that the reader can evaluate these for himself.

RESULTS

General Observations

From the maps (Figures 3 and 4), it is apparent that observations of changes in ground water and of anomalous animal behavior were made at a great distance from the epicenter. In fact, there does not seem to be a clustering near the epicenter except on the day before the mainshock occurred. Although one could construct non-physical, non-geological explanations for this fact, the implication of it is that an area much larger than that defined by the aftershock zone or by that in which foreshocks were felt was affected by changes in stress or material properties preceding and preparatory to the mainshock.

One might expect that slip on faults or a re-distribution of stress might precede large earthquakes and manifest itself in some measurable way. We had hoped that a simple spatial pattern in the changes in water level might appear, such as a quadrantal distribution of rising and falling

levels in wells, but no such pattern is obvious. In many localities, neighboring wells showed water level changes in opposite directions. Perhaps a more complete study, which took into account the depths of the wells, could reveal a simpler pattern. Alternatively, the absence of a simple pattern might be a reflection of very non-uniform precursory deformation, as might be expected to occur if the crust was deforming by the movement of blocks with respect to one another along a network of faults, fissures and cracks of various dimensions.

Among the ground water changes, the number of rising levels far exceeds that of falling levels. This is almost certainly not due to precipitation, because the air temperature during most of this time was below freezing and precipitation was slight.

Fig. 3. Maps of locations and dates of changes in ground water and reports of anomalous animal behavior for different periods preceding the Haicheng earthquake. Dates are given by the day of the month when the month is obvious and by day/month when the month is not obvious. The location of the foreshock activity is shown by a large star on all maps. For animal behavior, other seismicity is shown by small stars. The date of occurrence is also given except for the Benxi swarm when there were events on almost every day. Important recently active faults are shown by solid lines if exposed and dashed if buried. Thicker lines indicate more important faults. A question mark at the ends of faults signifies some doubt whether the faults are presently active.

Symbols for changes in ground water are the following:

● - rising ground water (see text)
▲ - rising surface water
○ - falling ground water
△ - falling surface water
+ - other ground water changes (composition or water)

Animal behavior anomalies are usually shown by symbols listed below, but occasionally when there where too many observations at one location, the letters listed below were used.

●, S - snakes
○, T - toads and frogs
+, R - rats and mice
✖, F - fish
△, B - farm birds
▽, H - domesticated herbivores
□, P - pigs
▲, C - cats
▼, D - dogs
◇, O - other

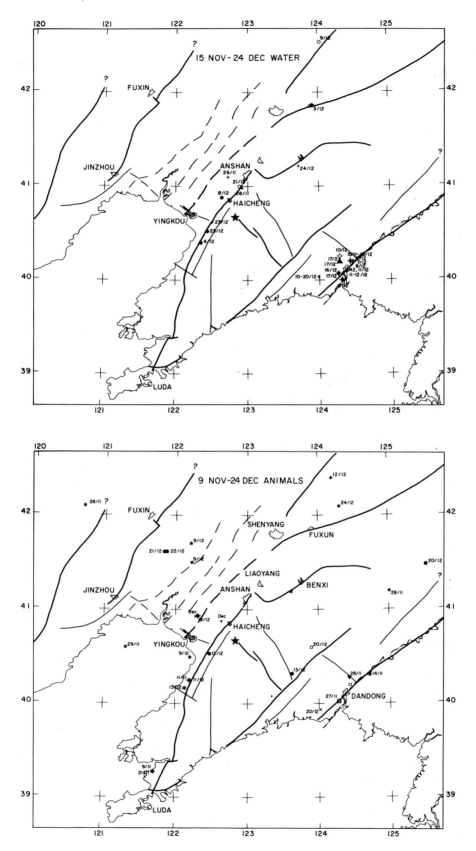

15 NOV-24 DEC WATER

9 NOV-24 DEC ANIMALS

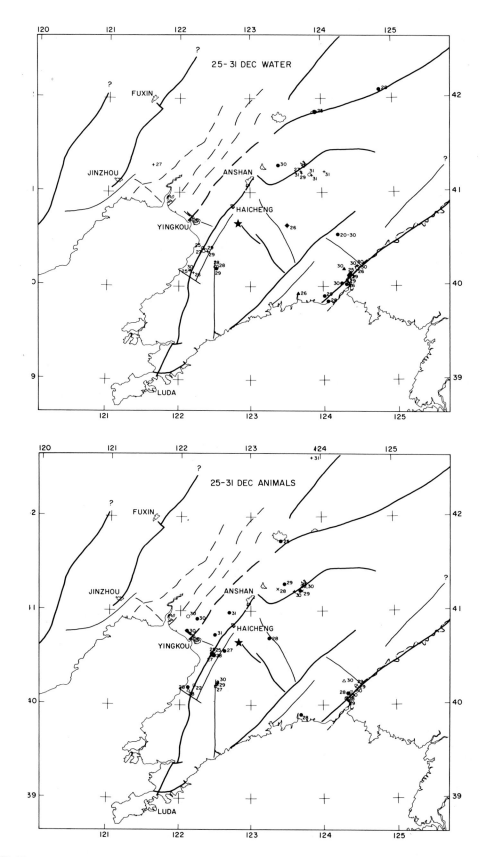

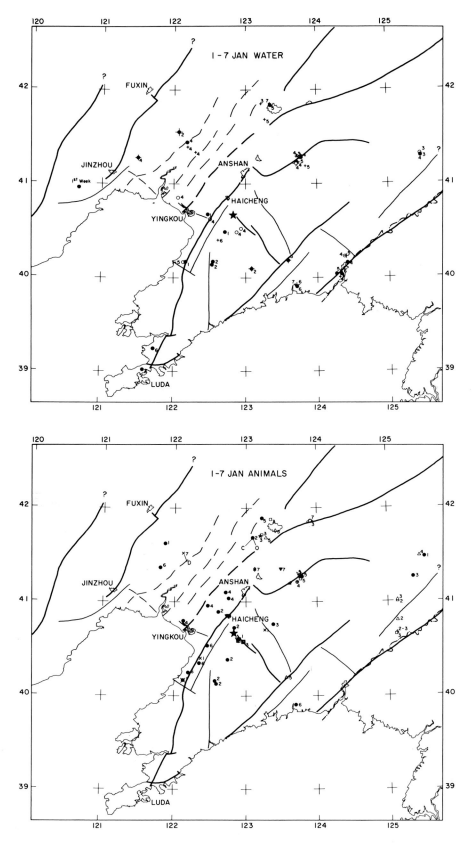

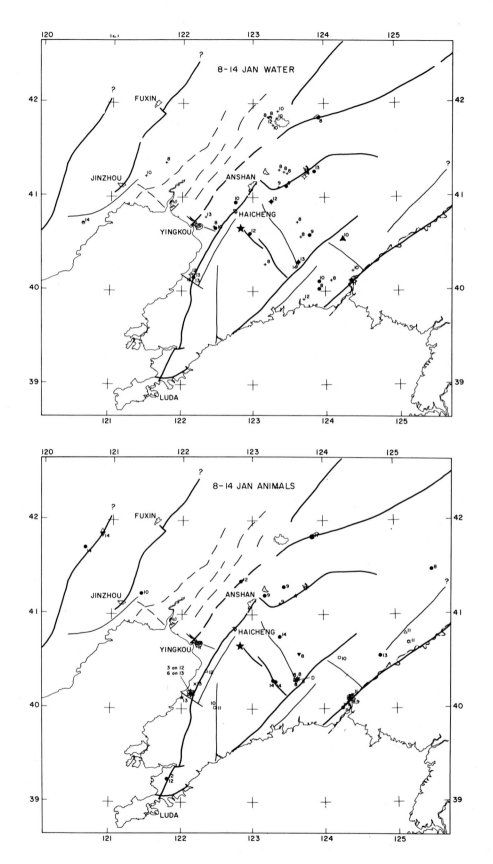

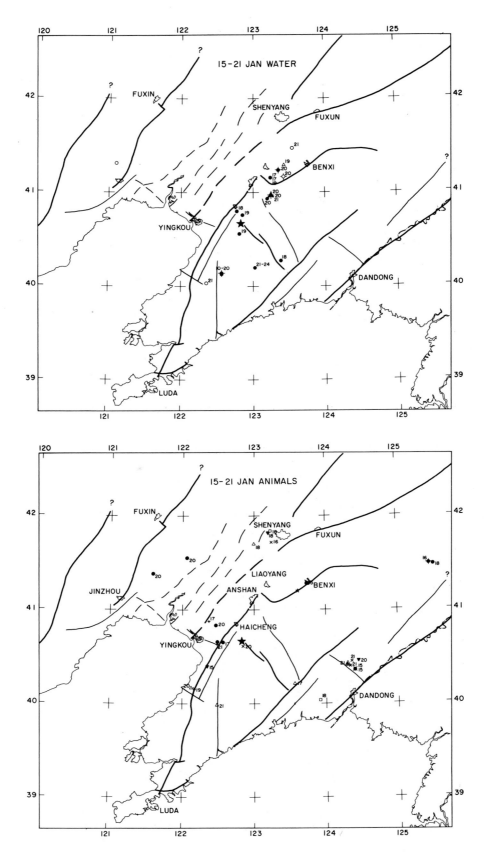

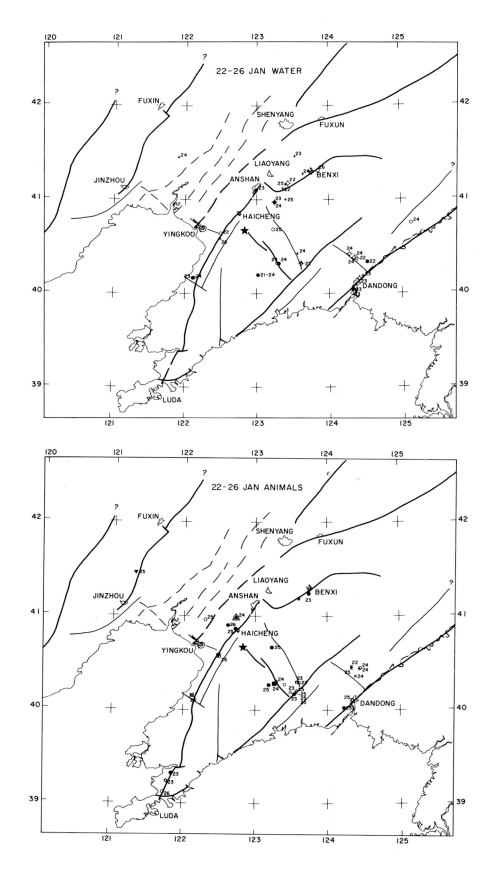

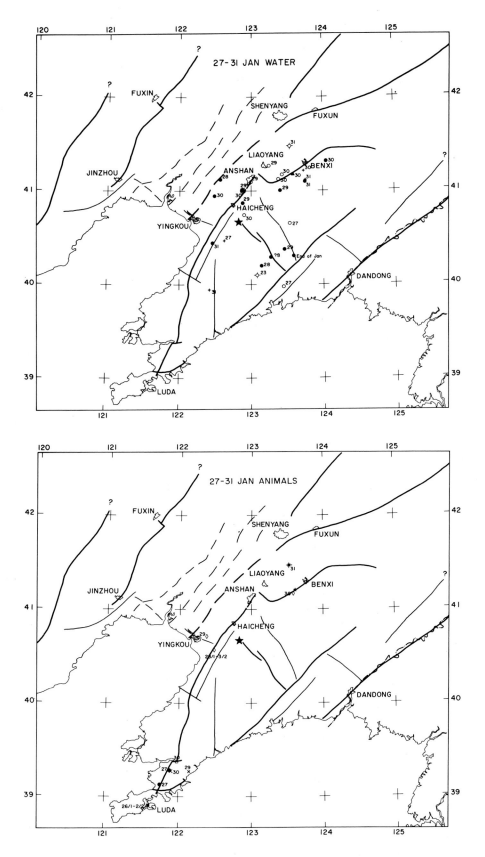

27-31 JAN WATER

27-31 JAN ANIMALS

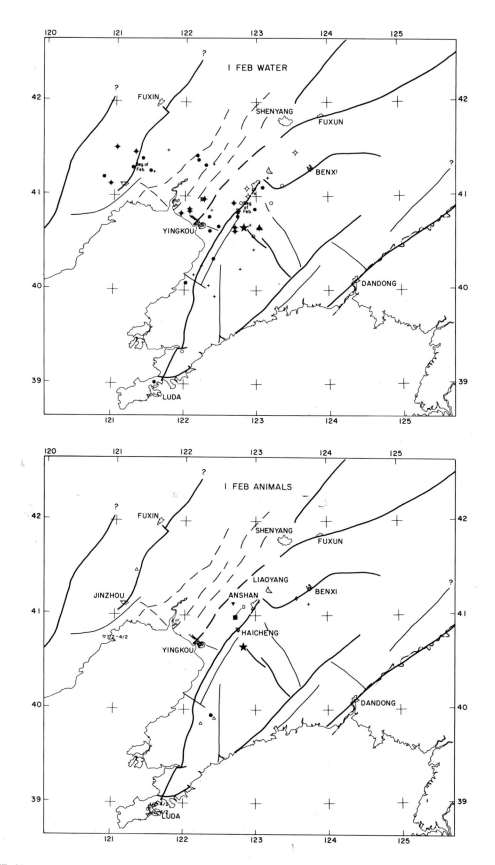

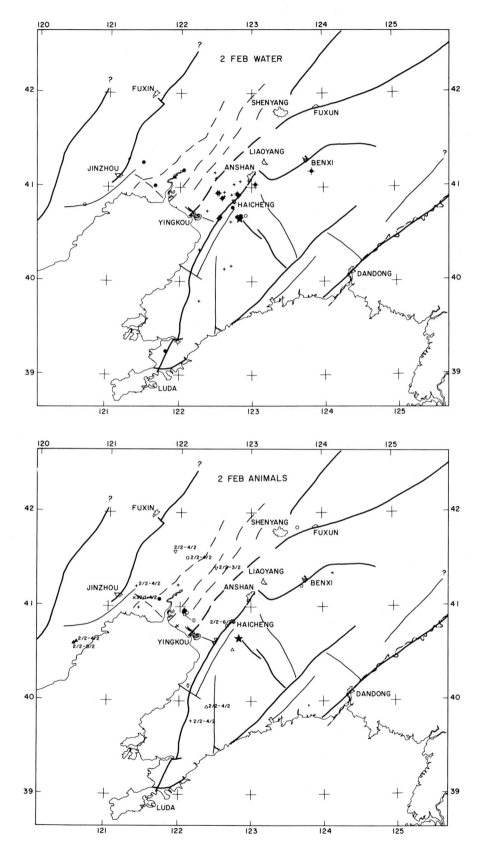

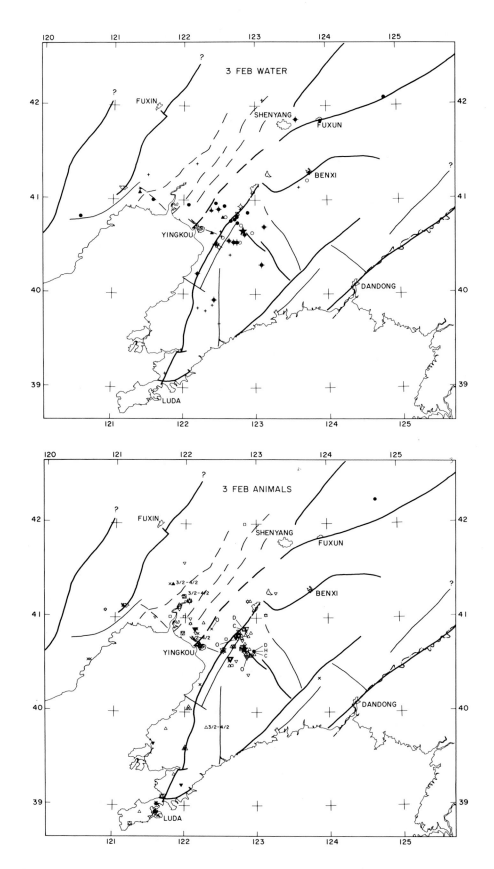

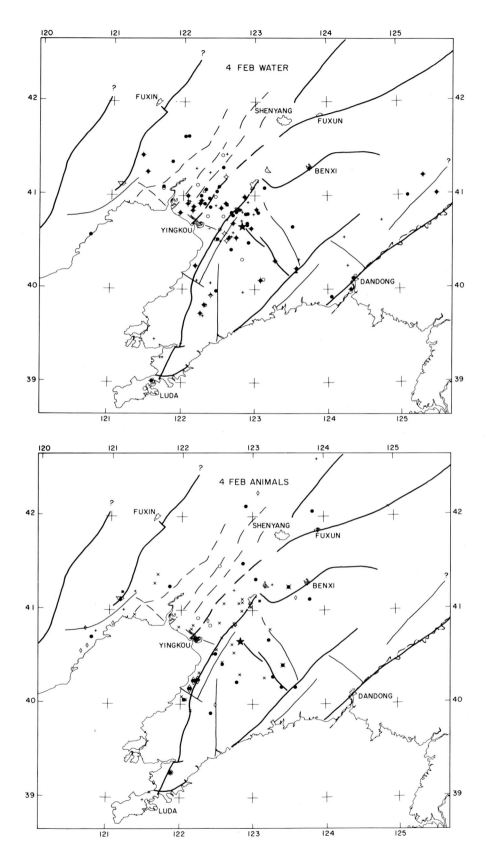

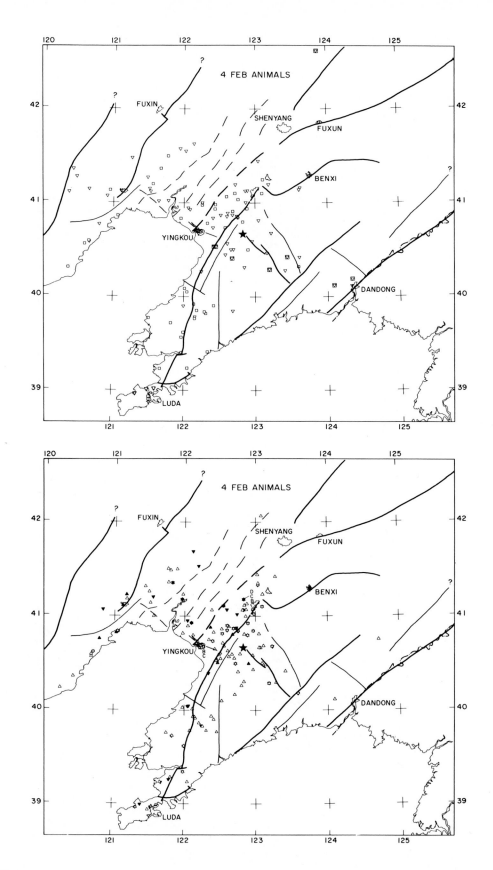

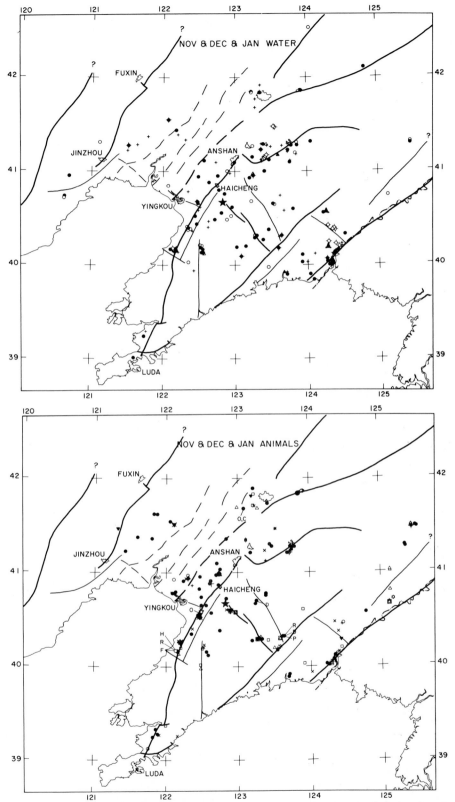

Fig. 4. Maps of locations of changes in ground water and anomalous animal behavior for November and December, 1974 and January, 1975. Symbols are the same as Figure 3.

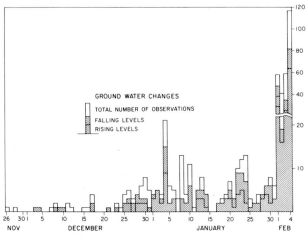

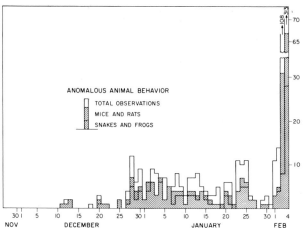

Fig. 5. Histograms of daily numbers of reports of
changes in ground water and of anomalous animal
behavior. A single report can include observations of more than one type of water change or
species of animal. Also shown are histograms for
observations of rising levels, of falling levels,
of mice and rats, and of snakes and frogs (cold
blooded hibernating animals). Again one
observation from a given locality can include
several wells or several individuals of the given
species.

Finally, the most obvious and exceptional anomaly
in the animal behavior was the number of observations of snakes in December and January. Most of
the snakes seen were frozen and some had come only
part of the way out of their holes when they froze.
An observation of snakes in winter in this area is
generally a very rare occurrence. Although the
reported mean temperatures in Liaoning in January
1975 were a few degrees above their usual -10 to
-15°C (Jiang Jinchang, 1980; Haicheng Earthquake
Work Brigade, 1975), experts on animal behavior in
China do not think that such a slight increase in
temperature would drive snakes out of their holes,
even if it awakened them (Jiang Jinchang, personal communication, 1980). It is worth noting,

however, that the temperature on 3 January was
unusually high, as much as 12°C above the typical
daily average for the epicentral area (Haicheng
Earthquake Work Brigade, 1975).

Correlations Between Ground Water and Aberrant Animal Behavior

From a comparison of the maps of ground water
changes and aberrant animal behavior, there is a
suggestion of a correlation in some locations amid
a wide scattering of both sets of data. In
particular, in the Dandong region, in regions
south and southwest of Yingkou (40°10'N, 122°08'E;
and 40°17'N, 123°15'E), and near Benxi (41°05'N,
123°30'E) both types of observations are reported
within a day or two of each other, particularly in
December, in early January, and between 22 and 26
January (Figure 3). In other areas, the correlation is less good. For instance, at the southern
end of the Liaoning peninsula (near Luda) there are
numerous reports of anomalous animal behavior in
late January and no reported changes in ground
water (Figure 3). A closer look at the areas where
both types of anomalies are observed suggests that
the changes in ground water precede the anomalous
animal behavior by from zero to two days (Figure
3). This lag of animals behind the water is also
reflected in the histograms in Figure 5, and is
perhaps clearer in the 2-day averages in Figure 6.
The increase in observations in late December, the
minimum in mid-January, the peak in observations
in late January (22-24 for water and 23-25 for
animals), the minimum at the end of the month, and
the escalation of reports in February all show
this lag.

These correlations call attention to the possibility that changes in ground water are responsible for the anomalous behavior of the animals -
possibly by flooding of holes, by changes in the
chemical composition of the water, by forcing a
degassing of unsaturated cracks (Madden and
Williams, 1977) that drives off charged aerosols
(Tributsch, 1978) or other mechanisms. Unfortunately the data are inadequate to prefer any of
these, or even to show a conclusive correlation
between the two types of observations. In
particular, the correlation between rising water
levels and the appearance of snakes is not
adequate to conclude that they have been driven
out by flooding of their holes.

Correlation of Observations With Faults

Elsewhere we have noted that precursors often are
more common in or near prominent fault zones
(Jones et al., 1980). Such a correlation accords
with the idea that deformation occurs partly by
movement of blocks along narrow zones of concentrated deformation. The data shown in Figures 3
and 4 are consistent with this correlation, but
are inadequate to prove it. When the spatial
distribution of observations is viewed as a whole
there is a north-northeast trend, parallel to the

main faults that control the structure and topography of the region (Figure 4). Although part of this trend is due to the shape of the Liaoning peninsula, we think that the trend is real. There is also a clustering of observations along the Yalu River fault (along the China-Korea boundary in Figures 3 and 4). Perhaps, if there were data from Korea, the linear trend of the observations would be less apparent, but the spatial correlation nevertheless exists. There is also a concentration of observations along the Jinshangling fault (extending south-southeast from Anshan to Luda), particularly in November, December and early January, and near, if not on, the Taizihe fault (extending east from Anshan past Benxi in Figure 4). This latter area experienced a swarm of seismic activity throughout December and January, with a peak in activity on 22 December 1974 (shown by a star without a date in all of the maps for December and January). The correlation between locations of observations and faults is also somewhat better for animal behavior than for ground water. However, the distributions of

observations do not define narrow bands that coincide with the important active faults, and therefore we doubt that the data presented here are adequate enough to prove that such a correlation exists.

Patterns in Space and Time

The sequence of maps in Figure 3 show some changes in time that are worth noting. Particularly for the ground water, many observations were made far from the epicenter in November, December and January. Except for the pulse in activity near Dandong in the fourth week of January, there is a tendency for the region with observations to contract with time during January. Then in February there is an abrupt change. On 1 February there is a large increase in the number of reported ground water changes. Whereas until this time there had been very few reports in the area northwest of the Jinshangling fault, on 1 February, several observations were made there. Together with the preceding data, this gives the impression of a westward or northwestward migration of the region with observations. Then on 4 February, the observations spread over a large area again.

Most of these features can be seen with the data for animal behavior, but they are less clear than with the ground water. Unlike the ground water, the escalation in reports of anomalous animal behavior begins on 3 February. This escalation is apparent with all types of animals, but is more pronounced for domestic animals than wild animals. Observations of snakes and of mice and rats were the most common in December and January, but on 3 and 4 February, the numbers of observations of farm birds, of herbivorous farm animals, of pigs and of dogs each exceeded those of snakes or of mice and rats. The frequency of observations of mice and rats also increased dramatically on 3 February as can be seen in Figures 5 and 6. The observations on 3 and 4 February crudely surround the epicenter, but with the peculiarity that on 3 February there were observations close to the epicenter, but not on 4 February (Figure 3). Observations of snakes and of mice and rats do not cluster around the epicenter at all.

We do not think that the longer term variations are sufficiently clear or simple to merit speculation on their possible causes. The increase in February, however, correlates in a simple way with the foreshock activity (Figure 7). The first foreshock occurred on 1 February coincident with the increase in observations of ground water changes. All of the foreshocks occurred only within a few kilometers of the epicenter of the mainshock but the first might have reflected the beginning of more widespread accelerating deformation. This accelerating deformation might then cause changes in ground water.

The first foreshock large enough to be felt occurred in the evening of 3 February. Unfortunately times were not reported for most observations of anomalous animal behavior, but the majority of

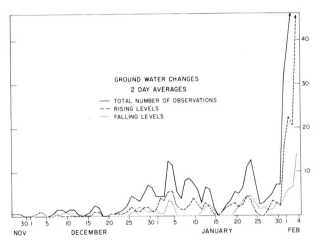

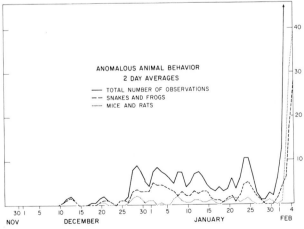

Fig. 6. Two-day averages made daily of the data in Fig. 5.

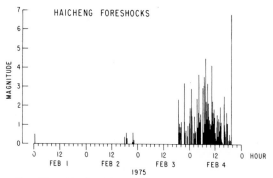

Fig. 7. Magnitude-time plot of the larger events
of the Haicheng foreshock sequence [after
Raleigh et al., 1977].

those that did have times reported were during the
evening. This suggests that some of the anomalous
animal behavior might have been in response to
shaking of the ground from foreshocks. On the
other hand, this explanation is not likely to be
applicable to all of these observations because
many of them were made in areas far from the
epicenters, where the shaking was small, at least
too small for humans to have felt it. Moreover,
the largest foreshock was slightly smaller than
the largest event in the Benxi swarm, on 22
December 1974, and anomalous animal behavior cannot
be associated with it, or with its epicentral area
until after the swarm began to subside. The fact
that there was a notable lack of clustering of
observations around the epicentral area of the
mainshock (where all of the foreshocks occurred)
as noted above also suggests that the foreshocks
are not the sole cause of the anomalies. In
addition, among the reported anomalous animal
behavior, the response of some animals was different from that of others. In general, farm animals
and most domestic animals became nervous and
agitated, but mice and rats acted dazed or temporarily stunned. These observations suggest that
foreshocks are not the only cause of anomalous
animal behavior and that more than one cause might
be important.

As noted above, the increase in foreshock
activity might indicate further accelerating
deformation in the epicentral areas that caused
not only a readjustment of the water table but
other changes that might affect the animals as
well. Two notable features of the Haicheng earthquake were the numerous reports both of a low
ground fog and of an unusual smell, often sulfur-
ous before the earthquake (Deng et al., 1976;
Haicheng Earthquake Work Brigade, 1975). These
observations are consistent with the animals
responding to chemical changes in the air or water
or to charged aerosols in the atmosphere (Deng et
al., 1976; Tributsch, 1978). Finally, although we
consider it unlikely, we cannot eliminate the
possibility that the increase in observations of
anomalous behavior on 3 February was due to the
onset of the foreshock activity heightening the

awareness of the people and the subsequent
occurrence of the mainshock forcing them to reflect
back on what they had seen. This explanation is
even less likely for the changes in ground water
because they began two days before the foreshocks
were felt.

Summary

The data used here are very crude and the results
obtained above are preliminary. Probably, several
other similar studies of much higher quality data
will be needed before we will be able to find
definite patterns that can be used to infer
unequivocally the physical causes of the observations or processes taking place before earthquakes.
Nevertheless, it is worth noting some tentative
conclusions.

The most obvious observation is that reports both
of changes in ground water and of anomalous animal
behavior came from a very large area, extending
more than 150 kilometers from the epicenter of the
mainshock.

Among the observations of changes in the level of
ground water, there is no obvious geographic
distribution of rising and falling levels. Sometimes neighboring wells changed levels in opposite
directions. There is a suggestion, however, that
both anomalous animal behavior and changes in
ground water are more prevalent near major active
fault zones than far from them.

There is a suggestion of a correlation in space
and time between changes in ground water and
anomalous animal behavior, with the latter
occurring zero to two days after the former. There
is also a suggestion of a contraction of the area
in which observations were made as well as a hint
of a westward migration of the area with frequent
observations.

There is a clear increase in the number of
observations of changes in ground water on 1
February 1975, the day on which the first foreshock
occurred, with many observations northwest of the
epicenter where previously there has been very few.
There is also a very large increase in the number
of reports of anomalous animal behavior on 3
February 1975, the first day on which foreshocks
were felt.

Finally, we think that some of the animals might
have been responding to shaking caused by foreshocks and others might have been affected by
changes in ground water. We think that the changes
in ground water might have been due to a redistribution of stress caused by slip on faults or movement of relatively underformed blocks with respect
to each other along narrow zones of deformation.
We realize, however, that the data do not require
such explanations and that other possibilities are
reasonable.

Acknowledgments: We thank the Liaoning Provincial Seismology Bureau for making their data
available and, in particular, the workers who
prepared lists of observations especially for us.

Two of us (Lucile M. Jones and Peter Molnar)
thank the State Seismology Bureau for the courtesy
and hospitality extended during their visit to
Beijing. One of us (L.M.J.) acknowledges support
from the U.S. Geological Survey under contract No.
14-08-001-17759.

References

Deng Qidong, Wang Tingmei, Li Jianguo, Xiang Hongfa
and Cheng Shaoping, A discussion on source model
of Haicheng earthquake, Scientia Geologica
Sinica, 195-204, (in Chinese), July, 1976.

Haicheng Earthquake Work Brigade, A Preliminary
Study of the Haicheng Earthquake, Vol. III,
Earthquake Precursors, Internal Report of the
State Seismology Bureau, (in Chinese), 1975.

Jiang Jinchang, ed., The Strange Instinct - A
Simple Discussion of Animals Forecasting Earthquakes, Seismology Publishing House, Beijing,
97 pp., 1980.

Jiang Pu and Deng Qidong, The development of
precursory field and the tectonomechanical
condition in the Haicheng - Tangshan earthquake
series, Seismology and Geology, 2(2), 31-42,
(in Chinese), 1980.

Jones, L.M., Deng Qidong, and Jiang Pu, The
implication of conjugate faulting in the earthquake brewing and originating process, Seismology
and Geology, 2(1), 19-26, (in Chinese), 1980.

Madden, T. and E. Williams, Possible mechanism for
stress associated electrostatic effects, in
Conference I, Abnormal Animal Behavior Prior to
Earthquakes, U.S. Geological Survey, Office of
Earthquake Studies, Menlo Park, California,
427-429, 1976.

Raleigh, C.B., G. Bennet, H. Craig, T. Hanks, P.
Molnar, A. Nur, J. Savage, C. Scholz, R. Turner,
and F. Wu, Prediction of the Haicheng Earthquake,
EOS Trans. Amer. Geophys. Un., 58, 236-272, 1977.

Tributsch, H., Do aerosol anomalies precede earthquakes?, Nature, 276, 606-607, 1978.

Zhu Fungming, An outline of prediction and forecast
of Haicheng earthquake of M=7.3, J. Seismol. Soc.
Jap., 15-26, 1976, (Available in English in
Proceedings of the Lectures by the People's
Republic of China, Spec. Publ. 43-32, ed. by
J.P. Muller, pp. 11-19, Jet Propulsion Labora-
tory, California Inst. of Tech., Pasadena, 1976).

A PROBABILISTIC SYNTHESIS OF PRECURSORY PHENOMENA

Keiiti Aki

Department of Earth and Planetary Sciences,
Massachusetts Institute of Technology, Cambridge, Massachusetts   02139

Abstract. The concept of probability gain associated with a cursor may be useful for unifying various areas of earthquake prediction research. Judging from the success of predicting the Haicheng earthquake of 1975, the probability gain at each stage of long-term, intermediate-term, short-term and imminent prediction in this case is estimated as a factor of about 30. For many independent precursors, the Bayesian theorem shows that the total probability gain is approximately the product of individual gains. The probability gain for an individual precursor may be calculated as its success rate divided by the precursor time (Utsu, 1979). The success rate can only be determined from the accumulation of experiences with actual earthquakes. The precursor time, on the other hand, may be studied experimentally and theoretically. A review of these studies leads to a suggestion that the loading rate may be faster for smaller earthquakes. The existence of so called "sensitive spots" where precursory strain, radon or other geochemical anomalies show up even for distant earthquakes suggests that some sites may have stress amplification (concentration) effect which may also account for higher loading rate for a small earthquake. The concept of fractals (a family of irregular or fragmented shapes) developed by Mandelbrot (1977) is applied to the fault plane to gain some insight into its geometry. If we use the idea of a barrier model in which smaller earthquakes are generated by the segmentation of a large earthquake, the fractal dimension of the fault plane becomes equal to $3b/c$ where b is the b value of magnitude-frequency relation, and c is the log-moment vs magnitude slope ($c \simeq 1.5$). For $1 < b < 1.5$, which is usually observed, the fractal dimension varies from 2 (filling up plane) to 3 (filling up volume). For $0.5 < b < 1.0$, which is sometimes observed for foreshocks, the model corresponds to fault lines trying to fill up a plane. The Goishi model of Otsuka (1972) and branching model of Vere-Jones (1976) have such geometry. Assuming that the total length of branches is branches is proportional to earthquake energy, the b value for these models becomes 0.75, corresponding to the fractal dimension of 1.5. The probability gain for the tectonic stress increase

by $\Delta\sigma$ can be expressed as exp $(\beta\Delta\sigma)$. The coefficient β has been obtained in laboratories and in the field using various methods. The value of β varies wildly, but tends to show a higher value when the stress is applied in a large scale. This may also be explained by a stress amplification due to the fractal nature of fault plane. Deterministic studies of inhomogeneities, irregularity and fragmentation oi fault zone will be important for understanding precursor phenomena.

Introduction

The most impressive accomplishment in seismology during the last decade was the success of our Chinese colleagues in predicting several major earthquakes. Let us take the Haicheng earthquake of 1975 and consider the probability of its occurrence before the earthquake. When the warning of earthquake occurrence was issued and people were kept outdoors in cold winter temperatures, the hazard rate, that is, the probability of earthquake occurrence per unit time, must have been on the order of 1 per several hours. The area is normally aseismic and historic records indicate the hazard rate to be on the order of 1 per thousand years. In other words, the information gathered by Chinese colleagues was able to raise the probability by a factor of about 10^6. This remarkable accomplishment was made in four stages, namely, long-term, intermediate-term, short-term and imminent prediction. Figure 1 illustrates schematically how the unconditional hazard rate estimated from historic data was raised by each stage of prediction. Assuming an equal gain for all the stages, we find that each stage contributed to the probability gain of a factor of about 30.

In order to achieve this amount of probability gain, many, many specialists and nonspecialists were engaged in collecting information on various precursory phenomena. Some of the key precursors at each stage are indicated in Figure 1.

The purpose of the present paper is to unify various areas of earthquake prediction research by the concept of probability gain. The probability gain for a particular precursor may be studied

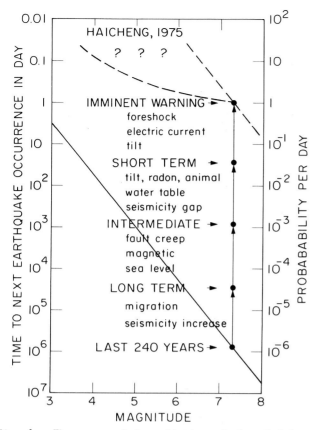

Fig. 1. The successful prediction of the Haicheng earthquake implies that the hazard rate (probability of occurrence of earthquake per unit time) had been increased from about 1 per 1000 years to 1 per several hours through acquisition of precursory information. The probability gain at each of four stages of prediction is about a factor of 30.

empirically using past experiences with actual earthquakes. It may be studied in a laboratory scale model under controlled conditions. More fundamentally, the probability gain may be determined by the increase of tectonic stress, which can be estimated from geodetic data. If these studies can develop a means to determine probability gain as a function of given precursors, the results can be translated into objective quantitative measure for the grade of concern about an earthquake occurrence, which will be helpful to the public officials in charge of public safety.

Let us start with a few definitions.

Definitions

First we specify the area in which an earthquake is predicted to occur. Then, we can define the average frequency of occurrence of earthquakes with a certain magnitude range in that area. For example, if the number $N(M)$ of earthquakes with a magnitude greater than M is recorded during

the total time period T, the average rate of occurrence p_0 per unit time is given by

$$P_0 = \frac{N(M)}{T} \qquad (1)$$

For a short time interval τ, then, the unconditional probability $P(M)$ of occurrence of an earthquake with magnitude greater than M in that area is given by

$$P(M) = p_0 \tau \qquad (2)$$

We shall divide the time axis into consecutive segments with the constant interval τ as shown in Figure 2. The crosses indicate the occurrence of an earthquake with magnitude greater than M. The interval τ is taken short enough so that each segment contains, at most, one earthquake. We shall write the total number of segments as

$$N_0 = \frac{T}{\tau} \qquad (3)$$

Let us now introduce precursors, and designate them as A, B, C, ... For example, A may be a swarm of small earthquakes which may be characterized by the duration, the maximum magnitude and the b value. B may be a ground upheaval characterized by the duration, extent and amount of uplift. C may be a radon anomaly observed in the area characterized by the duration and amplitude. Suppose that, in the total period of observation, the precursor A showed up for time intervals shown in Figure 3.

Consider those segments during which the precursor A existed. Of those segments, let the number of segments containing an earthquake be n_A, and the number of segments containing no earthquake be $\tilde{n}_A$. Then, we can define the conditional probability $P(M|A)$ of occurrence of an earthquake within a time interval τ under the condition that the precursor A is existing as

$$P(M|A) = \frac{n_A}{n_A + \tilde{n}_A} \qquad (4)$$

Since, for small τ, $P(M|A)$ is proportional to τ, we can write

$$P(M|A) = P_A \tau \qquad (5)$$

P_A is the probability of an earthquake per unit time under the condition that the precursor A is existing.

We define similarly n_B, $\tilde{n}_B$ and p_B for the precursor B, and so on. Here, we simplified our problem by neglecting the details of each precursor phenomenon, which may be included by the explicit use of multiple parameters as done by Rhoades and Evison (1979).

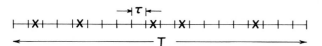

Fig. 2. The cross indicates an earthquake. The time interval τ is taken short enough so that each interval contains, at most, one earthquake.

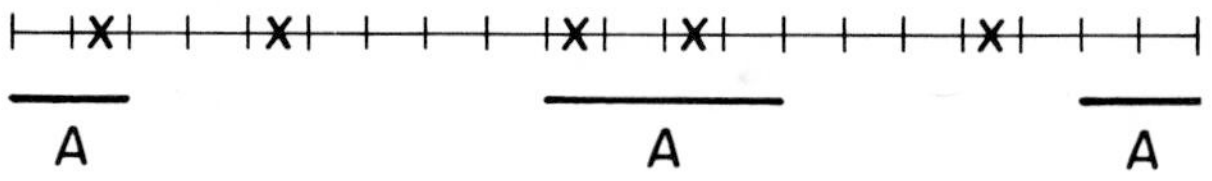

Conditional Probability for Multiple Independent Precursors

Let us find the probability $P(M A,B,C,...)$ of occurrence of an earthquake (with a magnitude greater than M in a specified area) under the condition that n independent precursors A, B, C, ... appeared simultaneously. According to the Bayes' theorem,

$$P(M A,B,C) = \frac{P(A,B,C|M)\ P(M)}{P(A,B,C)} \qquad (6)$$

$$= \frac{P(A,B,C|M)\ P(M)}{P(A,B,C|M)\ P(M) + P(A,B,C|\tilde{M})\ P(\tilde{M})}$$

where $\tilde{M}$ means the non-occurrence of an earthquake.

Since we assume the statistical independence among the precursors,

$$P(A,B,C|M) = P(A|M)\ P(B|M)\ P(C|M) \qquad (7)$$

and

$$P(A,B,C|\tilde{M}) = P(A|\tilde{M})\ P(B|\tilde{M})\ P(C|\tilde{M}) \qquad (8)$$

On the other hand, from the definitions given in the preceding section,
and

$$P(A|M) = \frac{n_A}{N} \qquad (9)$$

$$P(M) = \frac{N}{N_o} \qquad (10)$$

$$P(A|\tilde{M}) = \frac{\tilde{n}_A}{N_o - N} \qquad (11)$$

and

$$P(\tilde{M}) = \frac{N_o - N}{N_o} \qquad (12)$$

Putting the equations (7) through (12) into (6), we obtain

$$P(M|A,B,C) = \frac{1}{1+(\frac{\tilde{n}_A}{n_A}\frac{\tilde{n}_B}{n_B}\frac{\tilde{n}_C}{n_C}...)(\frac{N}{N_o - N})^{n-1}} \qquad (13)$$

Using equation (4), the above relation can be rewritten as

Fig. 3. The precursor A occurs in time intervals marked A.

$$P(M A,B,C) = \frac{1}{1 + (\frac{1}{P(M A)} -1)(\frac{1}{P(M B)} -1)(\frac{1}{P(M C)} -1).../{(\frac{1}{P(M)} -1)^{n-1}}} \qquad (14)$$

The above equation was obtained by Utsu (1979) without the use of the Bayes' theorem. For a small τ the above probability is proportional to τ. We can, then write

$$P(M|A,B,C) \simeq p\tau,$$

where

$$p = p_o \cdot \frac{p_A}{p_o}\frac{p_B}{p_o}\frac{p_C}{p_o} \qquad (15)$$

The above extremely simple relation shows that, for multiple independent precursors, the conditional rate of earthquake occurrence can be obtained by multiplying the unconditional rate p_o by the ratios,

$$\frac{\text{conditional probability}}{\text{unconditional probability}}$$

for all the precursors. We shall call the above ratio the probability gain of a precursor.

A quantitative measure of the grade of concern for an earthquake occurrence is illustrated in Figure 4 which shows the probability of occurrence per day of an earthquake with magnitude greater than a specified value. The unconditional probability is determined from the precursor data by equation (1). The precursors A, B, and C increase this probability approximately by a factor of $(p_A/p_o \cdot p_B/p_o \cdot p_C/p_o)$ as shown in equation (15). The example shown in Figure 4 corresponds to the earthquake with M > 6 1/2 in the Izu-Oshima area in Japan just before the earthquake of January 14, 1978. Precursors A, B and C are uplift (including gravity change), forechange) respectively. The probabilily gain for each precursor was assigned by Utsu (1979) in a manner described later. The conditional probability for the three precursors almost reached the highest grade of concern VI. Although the evaluation of conditional probability was not made in real time, the Japan Meterological Agency nevertheless issued an earthquake information at 10h50m, on January 14 stating that there was a possiblity of occurrence of an earthquake

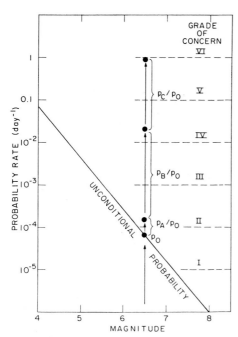

Fig. 4. The probability gains calculated by Utsu (1979) for precursors A (uplift), B (foreshock) and C (radon and water table) for the Izu-Oshima earthquake of 1978.

causing minor damage. An earthquake with M = 7 took place at 12h24m on the same day.

How to Assign the Probability Gain for a Particular Precursor

The evaluation of an earthquake prediction is very simple using the procedure described in the preceding section, if we know the probability gain pA/po, pB/po, ... for given precursors.

For theo Izu-Oshima earthquake of 1978, Utsu (1979) estimated the probability gain in the following manner. The precursor A is the uplift in the Izu Peninsula which was confirmed also by gravity change. The diameter of uplifted area is about 2S km, which may correspond to the source size of an earthquake with M = 6 1/2. According to a summary by Sato and Iuchi quoted in Utsu (1979), only 17 % of anomalous uplifts were connected directly to earthquake occurrence. However, since the uplift in this case is so conspicuous, Utsu assigned the probability of 1/3 instead of 17%. He also assigned the life-time uplift (or precursor time) to be five years. This gives the conditional probability rate $p_A = 1/(3 \times 5 \times 365)$ per day.

The precursor B is the earthquake swarm taking place in the area. According to statistics, one out of twenty swarms may be followed by a major earthquake. However, the Izu area is known for relatively frequent foreshocks. Utsu, thus, assigns the probability of 1/10 instead of 1/20

for the chance of a swarm to be followed by a major earthquake in this area. Utsu made a study of foreshocks for 26 major earthquakes, and found that the mainshock occurred within three days of the biggest foreshock for 19 cases out of 26, and that the difference between the magnitude of the main shock and that of the biggest foreshock was greater than 1.6 for 10 cases out of 26. Since the biggest foreshock in the present case was M 4.9, the difference between the magnitude of the presumed main shock (M 6.5) and that of the biggest foreshock is 1.6. Thus, he assigned the conditional probability rate p_B to be 1/10 19/26 10/26 1/3 $\simeq 10^{-2}$ per day.

Finally, the precursor C for the Izu-Oshima earthquake is the composite of radon anomaly, anomalous water table change, and volumetric strain anomaly. He considers that these three precursors may be closely related, and treats them as one precursor. He just assigns the precursor time of one month, and the probabililty that the precursor is followed by a major earthquake to be 1/10. This gives $p_C = 1/300$ per day.

With the above estimates of conditional probability rates p_A, p_B, p_C and the unconditional probability rate $\bar{p}_O$ based on the past seismicity in the area, he calculates the probability gains as shown in Figure 4. The precursor A (uplift gives only the probability gain of about 2, while the precursors B (radon) and C (foreshock) gives the gain of about 100. The former, however, was important for assigning the magnitude of predicted earthquake. The main reason for the high gain of Latter precursors is their short lifetime.

The probability gain is a function of magnitude of earthquake to be predicted. For a given earthquake swarm the probability that the swarm will be accompanied by a major earthquake will decrease sharply with the magnitude of the latter. The longer lifetime of precursor for greater earthquakes as proposed by various people also tend to diminish the probability gain for precursors of greater earthquakes.

On the other hand, a tremendous gain is possible, if a particular short-term precursor is expected with a high degree of certainty. For example, Sieh's (1978) suggestion that one out of several Parkfield earthquakes may become a foreshock of the next 1857 great California earthquake will give a very high conditional probability rate, say, 10^{-1} per day. This means the probability gain of more than 10^4. With several additional precursors of moderate or small gains, it may be possible to issue a high-grade concern before the next 1857 earthquake.

Precursor Time for Various Models of Rock Failure

As described in the preceding section, Utsu (1979) estimates the probability gain for a given precursor to be equal to its success rate divided by the precursor time. The success rate can only be determined from the accumulation of experiences with actual earthquakes. The precursor time, on the other hand, may be studied using rock samples in the laboratory under various conditions, or by analyzing models of rock failure theoretically. Here we shall make a review of proposed models with regard to the question "what determines the precursor time?".

Theoretical studies of the above problem were made by Rice (1979) and Rice and Rudnicki (1979) for a fluid-filled porous medium, and they concluded that not only the fault length of the impending earthquake but also the loading rate and the constitutive relation may play important parts. Their result showed that the precursor time is closer to proportional to L (fault length) rather than to L^2 as suggested in various empirical formulas (Tsubokawa 1969, Scholz, et al. 1973, Rikitake 1976). Dieterich (1979) also suggests that the precursor time may be proportional to L on the basis of the precursory creep observed in his laboratory experiment and reproduced theoretically by a frictional model of slip-weakening instability. His precursor time is the travel time of precursory creep over the fault length, thus proportional to the latter.

Brady (1974) claims that the L^2 dependence of precursory time on fault length can be derived for a dry-dilatancy model without fluid diffusion. A close look at the derivation of his equation (19) reveals an unacceptable assumption made on the average strain ε within a volume element dV_{fr} due to closure of an average sized microcrack. The average strain can be written as

$$\varepsilon = \frac{\int_{dV_{fr}} \varepsilon(\underset{\sim}{x}) dV}{dV_{fr}}$$

Since $\varepsilon(x)$ is a decreasing function with distance from the microcrack, beyond a certain size of dV_{fr}, the numerator will reach a constant asymptotically. Since, by definition, dV_{fr} should be large enough to include many cracks, ε will be inversely proportional to dV_{fr} instead of a constant as assumed by Brady. If we correct for this, the precursor time becomes independent of L.

This conclusion is expected from a simple consideration that the successive stages of dilatancy model (Miachkin et al., 1975, Sobolev, et al., 1978) are primarily determined by the stress relative to the failure stress (independent of L), and the precursory time is mainly determined by the loading rate. There is no obvious reason why the precursory time should depend on fault length for a dry dilatancy model.

If the loading rate determines the precursor time, the empirical relation between precursor time and fault length means that the loading rate is higher for smaller earthquakes.

This magnitude dependence of loading rate is

somewhat difficult to conceive for a homogenous continuum model of the earth and a common source model of stress for all earthquakes, that is, the plate motion.

One disturbing thing about the empirical relation between the precursor time and earthquake magnitude is the fact pointed out by Tsubokawa (1969) and Anderson and Whitcomb (1975) that the slope of log precursor time vs. magnitude is identical to that of log recurrence time vs. magnitude. This means that the observed precursor time is roughly a fixed fraction of the recurrence time. Since no one tries to pick up a precursor for an earthquake before the time of occurrence of the preceding one, it might be suspected that those precursors may be just noises. On the other hand, these precursors may be real signals as demonstrated by the successful prediction of the Haicheng and other earthquakes.

If we accept the reality of precursors and the magnitude dependence of precursor time, we may have also to accept that the loading rate may be higher for smaller earthquakes.

A higher loading rate for a localized region relative to the surrounding is possible if the stress in the region is somehow amplified. Inhomogeneities such as joints and inclusions can cause such an amplification through stress concentration. The existence of so called "sensitive spots", where precursory strain, radon or other geochemical anomalies tend to show up even for distant earthquakes, as well as the fact that even in the near-field of an earthquake, some precursory phenomena (such as the anomalous water table change) occur only at a small number of sites (wells) also suggest that some sites may be more sensitive because of the greater stress amplification.

Barrier Model and Fractal Dimensions of Fault Planes

The inhomogeneities of the fault zone, sometimes called "patches", "barriers" and "asperities", also introduce stress concentration. These inhomogeneities appear to exist in all scales. Microscopic pictures of the sections of rock sample after failure show that the zone of failure is not a continuous plane but fragmented. Similar fragmentation of fault has been observed at the site of rock burst in a deep mine (e.g., Spottiswoode and McGarr, 1975) and in the epicentral areas of major earthquakes (e.g., Tchalenko and Berberian, 1975). Das and Aki (1977) made a numerical experiment on rupture propagation over a fault plane with distributed barriers, and showed that some of the barriers may remain unbroken after the rupture propagation, offering a mechanism to account for fragmented fault.

The stress concentration around unbroken barriers may become the source of aftershocks (Otsuka, 1976). Aki (1978) summarized the relation between the barrier interval and the maximum slip obtained by various methods, and found that the barrier interval increases with the slip even for the same fault zone. This is consistent with the observed high Griffith fracture energy for greater earthquakes (Aki, 1979), because greater earthquakes break stronger fracture energy barriers with resultant longer barrier intervals.

Andrews (1978) pointed out, from a consideration of energetics, that the stationary occurrence of a large number of small earthquakes cannot be explained by the load of smoothly varying tectonic stress alone, but requires a generation of short wavelength self stress by a large earthquake, unless fault creep, varying in amplitude of all length scales, prepares the fault for small earthquakes. The barrier model offers a physical mechanism for such a roughening of self stress in the fault zone after a major earthquake.

The above line of reasoning suggests a generic process of the whole ensemble of earthquakes, in which an earthquake prepares the stress field for the smaller earthquakes. This is similar to the phenomena of turbulence, in which a large eddy splits into smaller ones, generating a hierachy of eddies linked by a cascade.

The concept of "fractals" developed in a book by Mandelbrot (1977) may be useful for describing the geometry of the assemblage of fault planes. A fractal is a family of irregular or fragmented shapes. An example is the length of the coast of Britain, which increases indefinitely as the scale of map is made finer. Topologically, a coast is a line with dimension 1. To describe the departure of the coast line from a simple line with finite length, he introduces the fractal dimension. For example, the trace of Brownian motion of a particle has the fractal dimension of 2 because it fills up the plane, and the fractal dimension of the west coast of Britain is determined to be about 1.25. One way to obtain the fractal dimension D is, given a segment, to find the number N of subsegments which has a linear dimension r times the segment. Then, D is given by $\log N/\log(1/r)$. For example, if a straight line is divided into N segments, $r = 1/N$ and therefore $D = 1$. If a square is divided into N squares, $r = 1/\sqrt{N}$ and therefore $D = 2$.

Assuming a process in which smaller earthquakes are generated by a large earthquake in the manner in which aftershocks are generated by the barrier model, we can interpret the magnitude-frequency relation and find the fractal dimension of fault plane (we allow an overlap of fault planes, in accordance witH the barrier model). Let the number of earthquakes (in a given time-space range) with fault length greater than L be $N(L)$. The slope of $\log N(L)$ against $\log L$ is $3b/c$, where b is the slope of log frequency-magnitude relation and c is the slope of log moment-magnitude relation, and we assume that the seismic moment is proportional to L^3 (self-similarity). To be compatible with the process for generating segments, we shall restrict the possible fault length to be $L_n = r^n L_o$, where n is the integer.

From the n to $(n + 1)$ step, the length is reduced by a factor r. In this process, the number of earthquakes is multiplied by $(1/r)$, because $\Delta L = (L_{n+1} - L_n)$ is proportional to L and $\Delta L\, dN/dL$ is proportional to $L^{3b/c}$ Thus, the fractal dimension of fault plane is $D = 3b/c$. We shall consider the usual case of $c = 1.5$ (Hanks and Kanamori, 1979). Then, if $b = 1$, the fractal dimension of fault plane is 2, same as its topological dimension. If $b = 1.5$, on the other hand, the fractal dimension becomes 3, which corresponds to filling up volume with fault planes. In most cases, the observed b value falls between the above two extremes. The value slightly greater than 1 observed for the world, imply that the assemblage of fault planes, "the plate boundary" in the context of plate tectonics, is a little more than a plane.

For $0.5 < b < 1.0$, the fractal dimension becomes between 1 and 2. For them, one can no longer consider the fault as a plane, because the fractal dimension must be greater than the topological dimension. It is possible, however, to imagine fault lines trying to fill up a plane. As a matter of fact, the Goishi model of Otsuka (1972) (see also Saito et al., 1973 and Maruyama, 1978) and the branching model of Vere-Jones (1976) have such geometry. In fact, the corresponding log frequency-magnitude relation becomes linear (in the critical case) with the b value equal to 0.75, assuming that the total length of branches is proportional to earthquake energy. This suggests that the fractal dimension of the branching model is 1.5.

Otsuka's model has been shown to be essentially the same as the model used in the study of percolation process. Otsuka proposed this model to describe the growth of an earthquake fault. His model is not based on the elasto-dynamics of rupture propagation along a fault plane, but is based on a probabilistic growth of a tree-like shape. Seismological observations clearly show that an earthquake involves a propagation of rupture with the speed comparable to elastic wave velocities and is not a percolation process. On the other hand, his model may be useful for studying the stage of preparation of a large earthqua... model used for generating a hierachy of earthquakes is adequate for aftershocks and probably for the normal earthquake, but not for foreshocks which will precede a larger earthquake. If there is a basic generic difference between foreshocks and normal earthquakes, there remains a hope to discriminate them. Smaller b values observed for some foreshocks than for aftershocks certainly agree with the idea that the percolation model applies to the former and the barrier model to the latter. Needless to say, the precursor time will be longer for larger earthquakes if the percolation model is applicable to the preparation stage. The cross section of branches at the earth's surface will be points, and may be very difficult to be detected without numerous observations. This may explain the reason for the successful predictions of the Chinese.

Probability Gain by an Increase in Tectonic Stress

So far, we have mainly considered the geometry of the assemblage of fault planes, somewhat phenomenologically, without paying much attention to the detailed state of stress and dynamic process going on over the fault plane. More explicit discussions of these subjects have been given by Hanks (1979) and Andrews (1980). Here we shall stay out of the stress distribution in the fault zone and consider only its response to applied external stress, namely, the problem of estimating the probability gain of an earthquake occurrence by an increase in tectonic stress.

How the probability of earthquake occurrence depends on the tectonic stress is an extremely complex problem. According to Mogi (1962), the probability of occurrence of fracture in a rock sample increases exponentially with the applied stress. When a constant stress σ is applied at $t = o$, he found that the probability of occurrence of fracture between t and $t + dt$ is independent of t and given by

$$\mu(t)dt = \mu_o \exp(\beta\sigma) \qquad (16)$$

where $\mu(t)$ is what we called "hazard rate" earlier, and β is determined as 0.37 bar^{-1} for the tensile fracture of granite samples in the atmosphere. The experiment by Scholz (1972) on static fatigue of quartz under uniaxial compression shows the same functional dependence on stress, but with β about one hundred times smaller than Mogi's value. The probability gain due to stress increase by $\Delta\sigma$ is simply $\exp(\beta\Delta\sigma)$, which will be wildly different whether we use Mogi's value or Scholz's value for β.

The coefficient β can be estimated from the recurrence time of major earthquakes in a fault zone and the associated stress drop. For the Hokkaido-Kuril region, Utsu (1972) constructed a statistical model based on equation (16) and determined the model parameter by fitting the data on recurrence time. Combining Utsu's result with the average stress accumulation rate (0.3 bar per year) inferred from the stress drops in major earthquakes in this zone (Fukao and Furumoto, 1979), we find that β is 1.1 bar^{-1}. Earlier, Hagiwara (1974) obtained β of about 0.3 bar^{-1} by applying equation (16) to the statistical distribution of ultimate strain obtained by geodetic measurements.

From time to time, we find reports on an apparent sensitivity of earthquake occurrence to small stress changes such as due to earth tide and atmospheric pressure. For example, Conrad (1932) showed impressive evidence for the increase of local earthquake frequency by 30% due to the atmospheric pressure gradient across the Alps by 5^{mm} Hg (6.7 x 10^{-3} bars) or greater. This gives β to be about 40 bar^{-1}.

In this conference, Barry Raleigh reported about the increase of seismicity in southern California by a factor of 2 which was associated with a strain change of 10^{-6}. This corresponds to the value of β about 2 bar^{-1}, which is much greater than laboratory values but comparable to the Hokkaido-Kuril result.

The value of β estimated by various methods, thus, ranges from 0.004 to 40 bar^{-1} over four decades. There is some suggestion of increasing β with increasing scale length. In other words, the probability of earthquake occurrence is more sensitive to stress change, when stress is applied in a larger scale.

Earlier, we have discussed the scale dependence of loading rate and suggested that stress concentration or amplification may be occurring in a cascade from a larger scale to successively smaller scales. The scale dependence of β may also be attributed to such stress amplification.

Discussion

I feel that fractal models of fault planes would be useful for understanding the precursory phenomena especially with regard to their sensitivity to tectonic stress. The currently most reliable data for estimating tectonic stress are geodetic data supplied from levelling and geodimeter survey. One promising approach toward understanding precursory phenomena is to determine the stress change at depth from geodetic data, and then correlate the estimated stress change with the stress-sensitive phenomena such as changes in seismicity and magnetic field. A preliminary result from the Palmdale area obtained by Ikeda (1980) is encouraging. He inverted the geodetic data into 3-D stress distribution and found that the state of incremental stress during the downwarp period was in agreement with the fault plane solutions for the swarm of earthquakes during the same period given by McNally et al. (1978). From the magnitude of incremental stress (estimated at about 10 bars) and increase in the frequency of small earthquakes, the β-value was estimated to be about 0.3 bar. This magnitud of stress increase was consistent with that estimated by Johnston et al. (1979) from the observed magnetic anomaly.

In the present paper, we have discussed the fractal aspect of fault planes only from the statistical point. For the purpose of earthquake prediction, however, it may be necessary to study them deterministically. For example, we need to know where the sensitive spots are in order to observe precursors. Recent work by Bakun et al. (1980) on the relation between detailed seismicity and geometry of fault fragmentation along a part of the San Andreas fault demonstrated that such a deterministic approach may be feasible and promising.

Acknowledgements. I thank Joe Andrews for directing me to the most interesting book by Mandelbrot. This work was supported by the U.S. Geological Survey under contract 14-08-0001-19150.

References

Aki, K., A quantitative model of stress in a seismic region as a basis for earthquake prediction Proc. Conf. III, Fault Mechanics and its relation to earthquake prediction, U.S. Geological Survey Open file Report 78-380, 7-30, 1978.

Aki, K., Characterization of Barriers on an Earthquake Fault, J Geophys. Res., 84, 6140-6148, 1979.

Anderson, D. L. and J. H. Whitcomb, Time-dependent seismology, & J. Geophys. Res., 84, 718-732, 175.

Andrews, D. J., Coupling of energy between tectonic processes and earthquakes, J. Geophys. Res., 83 2259-2264, 1978.

Andrews, D. J., Stochastic Fault model - I Static Case, J. Geophys. Res., in press, 1980.

Bakun, W. H., R. M. Stewart, C. G. Bufe, and S. M. Marks, Implication of seismicity for failure of a section of the San Andreas fault, Bull. Seis. Soc. Am., 70, 185-201, 1980.

Brady, B. T., Theory of earthquakes, I. A scale independent theory of rock failure, Pageoph 112, 701-725, 1974.

Conrad, V., Die Zietlich Folge der Erdbeben und Bebenaulosande Ursachem, Handbuch der Geophysik, Bd. IV, 1007-1185, 1932.

Das, S. and K. Aki, Fault plane with barriers: A versatile earthquake model, J. Geophys. Res., 82, 5658-5670, 1977.

Dieterich, J. H., Modeling of Rock Friction. 1. Experimental results and constitutive equations. 2. Simulation of preseismic slip, J. Geophys., Res., 84, 2161-2175 1979.

Fukao, Y., and M. Furumoto, Stress drop, wave spectrum and recurrence interval of great earthquakes - implication from the Etorofu earthquake of Nov. 6, 1958, Geophys. J. Roy. Astr. Soc., 57 23-40, 1979.

Hagiwara, Y., Probability of earthquake occurrence estimated from results of rock fracture experiments, Tectonophysics, 23, 99-103, 1974.

Hanks, T. C., b-values and $\omega^{-\gamma}$ seismic source models: Implications for tectonic stress variations along active crustal fault zones and the estimation of high-frequency strong ground motion, J. Geophys. Res., 84 2235-2242, 1972.

Hanks, T. C. and H. Kanamori, A moment magnitude scale, J. Geophys. Res., 84, 2348-2350, 1979.

Ikeda, K., 3 - dimensional goodetic inversion method for stress modeling in the lithosphere, MS thesis, M.I.T., Jan. 1980.

Johnston, M. J. S., F. J. Williams, J. McWhister and B. E. Williams, Tectonomagnetic anomaly during the southern California downwarp, J. Geophys. Reviews, 84, 6026-6030, 1979.

Mandelbrot, B. B., Fractals, W. H. Freeman and Co., San Francisco, pp. 365, 1977.

Maruyama, T., Frequency distribution of the sizes of fractures generated in the Branching Process --Elementary Analysis, Bull Earth. Res. Inst. 53 407-421, 1978.

Miachkin, V. I., W. F. Brace, G. A. Sobolev, J. J. H. Dieterich, Two models for earthquake forerunners, Pure Appl., Geophys. 113, 169-181, 1975.

McNally, K. C., H. Kanamori, J., Pechmann and G. Enis, Seismicity increase along the San Andreas fault, southern California, Science, 201, 814, 1978.

Mogi, K., Study of elastic shocks caused by the fracture of heterogeneous material and its relation to earthquake phenomena, Bull. Earth. Res. Inst., Univ. Tokyo, 40, 125-173, 1962.

Otsuka, M. A simulation of earthquake occurrence, 5, an interpretation of aftershock phenomena, Zisin, Ser. II, 29, 137-146, 1976.

Otsuka, M., A chain-reaction type source model as a tool to interpret the magnitude-frequency relation of earthquakes, J. Phys. Earth., 20, 35-45, 1972.

Rhoades, D. A., and F. F. Evison, Long-range earthquake forecasting based on a single predictor, Geoph, J. R., Astro. Soc. 59, 43-56, 1979.

Rice, J.R., and J. W. Rudnicki, Earthquake precursor effects due to pore fluid stabilization of weakening fault, J. Geophys. Res., 84, 2177-2194, 1979.

Rice, J. R., Theory of precursory processes in the inception of earthquake rupture, Gert. Beitr. Geoph. Leipzig 88, 91-127, 1979.

Rikitake, T., Earthquake prediction. Elsevier, Amsterdam, 1976.

Saito, M., M. Kikuchi and K. Kudo, An analytic solution for Otsuka's Goishi model, Zisin, II 26, 19-25, 1973.

Scholz, C. H., L. R. Sykes and Y. P. Aggarwal, Earthquake predicton: A physical basis, Science, 181, 803-810, 1973.

Scholz, C. H., Static fatigue of quartz, J. Geophys. Res., 77, 2104-2114, 1972.

Sieh, K. E., 1978, Central California foreshocks of the great 1857 earthquakes, Bull. Seis. Soc. Am., 68, 1731-1750, 1978.

Sobolvev, G., H. Spetzler and B. Salov, Precursors to failure in rocks while undergoing anelastic deformations, J. Geophys. Res., 83, 1775.

Spottiswoode, S. M. and A. McGarr, Source parameters of tremors in a deep-level gold mine, Bull Seis. Soc. Am., 65, 93-112, 1975.

Tchalenko, J. S. and M. Berberian, Dasht-e Bayas Fault, Iran: Earthquake and earlier related Structures in bedrock, Geol. Soc. Am. Bull., 86, 703-709, 1975.

Tsubokawa, I., 1969, On relation between duration of crustal movement and magnitude of earthquake expected, J. Geod. Soc. Japan, 15, 75-88, 1969.

Utsu, T., Large earthquakes near Hokkaido and the expectancy of the occurrence of a large earthquake off Nemuro, Rep. Coord. Comm. for Earthq. Prediction, Geographic Survey Inst. Ministry of Construction, Japan, 7,Construction, Japan, 7, 7-13, 1972.

Utsu, T., Calculation of the probability of success of an earthquake prediction (in the case of Izu-Oshima-Kinkai earthquake of 1978), Rep.Coord. Comm. for Earthquake Prediction, 164-166, 1979.

Vere-Jones, D., A branching model for crack propagation, Pageoph, 114, 711-725, 1976.

A FRAMEWORK FOR EARTHQUAKE PREDICTION

Cinna Lomnitz and Jorge Lomnitz-Adler

Universidad Nacional Autónoma de México, México 20, D.F.

Abstract. Earthquake precursors of potential use in prediction include changes in the local or regional strain patterns, or fluctuations in groundwater flow or in the electromagnetic field which may be attributed to such changes. The observed effects, including P-wave anisotropy, v_P/v_S fluctuations, and S-wave splitting, may be described as second-order anelastic effects. For a second order viscoelastic solid, a finite pre-strain leads to anisotropic wave propagation. Expressions for the seismic velocities parallel and perpendicular to the pre-strain direction are derived in terms of the second-order Lamé constants.

Introduction

The feasibility of earthquake prediction hinges on large earthquakes being preceded by strain fluctuations on a scale of cubic kilometers. If they exist, such fluctuations must be superimposed on the regional strain field: they could be detected either directly or indirectly, i.e. by means of geodetic and other strain measurements, seismicity changes, seismic wave velocity anomalies, S-wave splitting, electromagnetic or geochemical anomalies, changes in ground water flow and so on.

A model based on continuum mechanics would seem to be appropriate for discussing these effects, because the wavelength of seismic signals and of other precursors is several orders of magnitude greater than the mechanisms, such as fluid-filled cracks, which may be postulated as physical explanations of these effects.

Consider an arbitrary volume of the earth's lithosphere which obeys the general constitutive equation

$$\sigma_{ij} = f_{ij}[\varepsilon_{kl}, t] \qquad (1)$$

where σ_{ij} and ε_{ij} are the stress and strain tensors, t is the time, and f_{ij} is a functional which depends on the entire history of the strain ε_{ij}. Expanding about $\varepsilon_{ij} = 0$ and setting $f[0,t] = 0$ we obtain (cf. Green and Rivlin, 1957, 1960;

Green, Rivlin and Spencer, 1959):

$$\sigma_{ij}(t) = \int_{-\infty}^{t} G_{ijkl}(t-t_1)\dot{\varepsilon}_{kl}(t_1)dt_1 +$$

$$\int_{-\infty}^{t} dt_1 \int_{-\infty}^{t} Q_{ijklrs}(t-t_1, t-t_2)\dot{\varepsilon}_{kl}(t_1)\dot{\varepsilon}_{rs}(t_2)dt_2 + \ldots (2)$$

where G_{ijkl} is the linear viscoelastic modulus tensor and Q_{ijklrs} is the second-order viscoelastic modulus tensor, and so on.

For a linear isotropic material the form of the viscoelastic modulus tensor is simply

$$G_{ijkl}(t) = \lambda(t)\delta_{ij}\delta_{kl} + \mu(t)\{\delta_{ik}\delta_{jl} + \delta_{il}\delta_{jk}\} \qquad (3)$$

where $\lambda(t)$ and $\mu(t)$ are the Lamé functions. However, since these Lamé functions depend only on $(t-t_1)$ in the convolution (2) the effects of successive strain changes are superimposed on each other: this is known as the "superposition principle" (Boltzmann, 1876).

A seismic signal propagates at a velocity which is governed by $\lambda(0)$ and $\mu(0)$, commonly known as the Lamé constants. In linear viscoelastic materials this velocity is independent of strain fluctuations; hence such materals cannot exhibit precursory effects. It follows that any precursory effects must be connected with higher-order terms in strain. In the following we develop a second-order stress-strain theory based on eq(2) and we show that an initially isotropic material becomes anisotropic following application of a long-term axial strain.

Theory

The second-order modulus in eq(2) may be written in terms of four second-order Lamé functions Λ_i:

$$Q_{ijklrs}(t_1, t_2) = \Lambda_1(t_1, t_2)\delta_{ij}\delta_{kl}\delta_{rs} +$$

$$\Lambda_2(t_1, t_2)\delta_{ij}(\delta_{kr}\delta_{ls} + \delta_{ks}\delta_{lr}) + \Lambda_3(t_1, t_2)\{\delta_{kl}(\delta_{ir}\delta_{js} + \delta_{is}\delta_{jr}) + \delta_{rs}(\delta_{ik}\delta_{jl} + \delta_{il}\delta_{jk})\} + \Lambda_4(t_1, t_2)\{\delta_{ik}(\delta_{lr}\delta_{js}$$

$+\delta_{1s}\delta_{jr})+\delta_{il}(\delta_{jr}\delta_{ks}+\delta_{js}\delta_{kr})+\delta_{ir}(\delta_{j1}\delta_{ks}+\delta_{jk}\delta_{1s})+$
$\delta_{is}(\delta_{jk}\delta_{1r}+\delta_{j1}\delta_{rk})\}$ (4)

which is analogous to (3), except that four functions Λ_1, Λ_2, Λ_3, Λ_4 govern the second-order behavior of the material in addition to the linear Lamé functions $\lambda(t)$ and $\mu(t)$.

Let us assume that the material is initially isotropic at time $t=-T$, and let a long-term constant strain $A(t)$ be applied in the 1-direction. After a suitably long time interval $T\gg t$ the velocity of propagation of a given seismic signal $E_{ij}(t)$ will no longer be the same in every direction. Indeed, the strain as a function of time may be written

$$\varepsilon_{ij}(t) = A(t)H(T+t)\delta_{i1}\delta_{j1}+E_{ij}(t)H(t) \qquad (5)$$

where $H(t)$ is the Heaviside function which is zero for argument $t<0$ and 1 for $t\geq0$. Differentiating and replacing in (2) we obtain

$$\sigma_{ij}(t) = \Sigma_1+\Sigma_2+\Sigma_3 \qquad (6)$$

where Σ_1 stands for the contribution of the uniaxial strain $A(t)$ over the time T:

$$\Sigma_1 = \int_{-T}^{t}G_{ij11}(t-t_1)\dot{A}(t_1)dt +$$
$$\int_{-T}^{t}\int_{-T}^{t}Q_{ij1111}(t-t_1,t-t_2)\dot{A}(t_1)\dot{A}(t_2)dt_1dt_2 \qquad (7)$$

which is approximately a constant if $t\ll T$. Let us designate this constant as $\sigma_{ij}(A,T)$. The term Σ_2 is quadratic in the seismic signal and may therefore be neglected:

$$\Sigma_2 = \int_{0}^{t}\int_{0}^{t}Q_{ijklrs}(t-t_1,t-t_2)\dot{E}_{kl}(t_1)\dot{E}_{rs}(t_2)dt_1dt_2 \qquad (8)$$

while the term Σ_3 contains the contributions which are linear in the seismic signal E_{ij}:

$$\Sigma_3 = \int_{0}^{t}G_{ijkl}(t-t_1)\dot{E}_{kl}(t_1)dt +$$
$$2\int_{0}^{t}\int_{-T}^{t}Q_{ijkl11}(t-t_1,t-t_2)\dot{E}_{kl}(t_1)\dot{A}(t_2)dt_1dt_2 . \qquad (9)$$

By replacing into (6) we finally obtain
$$\sigma_{ij}(t) = \sigma_{ij}(A,T) + \int_{0}^{t}G_{ijkl}(t-t_1)\dot{E}_{kl}(t_1)dt_1 +$$
$$2\int_{0}^{t}\int_{-T}^{t}Q_{ijkl11}(t-t_1,t-t_2)\dot{E}_{kl}(t_1)\dot{A}(t_2)dt_1dt_2 . \qquad (10)$$

Equation (10) represents the propagation of a seismic signal $E(t)$ in an anisotropic linear viscoelastic material with an equivalent modulus tensor

$$\tilde{G}_{ijkl}(t) = G_{ijkl}(t)+2\int_{0}^{T}Q_{ijkl11}(t,t_1)\dot{A}(t-t_1)dt_1, \quad (11)$$

a fact which may easily be verified by replacing (11) into (10) and comparing the result with (2). In conclusion, the initially isotropic material of modulus G has become an axially anisotropic material of modulus $\tilde{G}$ as a result of applying a uniaxial strain A over a long period of time.

Earthquake precursors

Let now 1 be the direction of the major axis of the tectonic strain tensor (assumed horizontal), and let 2 be the other horizontal axis. The velocities v_P and v_S of the seismic body waves can be derived from the solutions of the equation

$$\det\left|\Gamma_{ik}-\rho v^2\delta_{ik}\right| = 0 \qquad (12)$$

where $\Gamma_{ik}=\tilde{G}_{ijkl}(0)n_jn_1$, and the n_m are the direction cosines of propagation of the signal. Solving for the velocities and introducing our Lamé functions we obtain

$$\rho v^2_{(1)} = \begin{cases} \lambda_0+2\mu_0+\alpha_1+2\alpha_2+4\alpha_3+8\alpha_4 \\ \\ \mu_0+\alpha_3+2\alpha_4 \end{cases} \qquad (13)$$

where the constants are defined by $\lambda_0=\lambda(0)$, $\mu_0 = \mu(0)$, and

$$\alpha_n = 2\int_{0}^{T}\Lambda_n(0,T-t)\dot{A}(t)dt . \qquad (14)$$

Similarly, in the 2-direction

$$\rho v^2_{(2)} = \begin{cases} \lambda_0+2\mu_0+\alpha_1+2\alpha_3 \\ \\ \mu_0+\alpha_3+2\alpha_4 \\ \\ \mu_0+\alpha_3 \end{cases} \qquad (15)$$

A general discussion of this result may be found in Achenbach(1973). It is interesting to point out the existence of three wavefronts which propagate at different speeds. The first and fastest wave in (13) and (15) corresponds to what would ordinarily be identified as a P-wave on a seismogram while the second and third arrivals would seem to be analogous to SV and SH. However, for any arbitrary direction of propagation the particle motion is complicated and will not correspond, in general, to any of these three wave types.

(a)Dilatancy. The effect known as "dilatancy" (cf. Brace, Paulding and Scholz,1966) amounts to the finding that, under certain test conditions, the transverse strain in uniaxial compression increases faster than the axial strain. The effect is time-dependent(see, e.g., Scholz and Cranz,1974), and the form of the time dependence is not well understood. Nor is the possible dependence on sample geometry.

The net volume change under increasing stress is always negative: no net volume increase under compression has actually been observed. The distinction between elastic and "dilatant" deformation corresponds roughly to the contributions of first-order and higher-order terms in the stress-strain equation (1). The dilatant behavior of rocks is therefore well represented by the introduction of nonlinearity in the constitutive equations of the material.

(b)the v_P/v_S effect. Comparing the P-wave velocities in the 1-and 2-directions we may write (cf. eqs.13 and 15):

$$v_{P_1}^2 - v_{P_2}^2 = (2\alpha_2 + 2\alpha_3 + 8\alpha_4)/\rho \quad . \tag{16}$$

Observations by several authors(Raitt,1963; Hess, 1964;Raitt et al.,1969;Morris et al.,1969;Shor et al.,1973;Keen and Tramontini,1970) give values of v_{P_1}/v_{P_2} in the oceanic lithosphere, near the base of the crust, in the range of 1.05 to 1.08 when the 1-and 2-directions are taken parallel and normal to the direction of spreading. This provides a constraint on the values of the constants in eq (16).

The v_P/v_S effect does not necessarily indicate anisotropy in the focal region itself. Available observations (Aggarwal et al.,1975) seem to indicate, however, that v_P/v_S is larger in the 1-direction than in the 2-direction in regions of compressional tectonics. From (13) we find

$$v_{P_1}/v_{S_1} = \sqrt{(\lambda_0 + 2\mu_0 + \alpha_1 + 2\alpha_2 + 4\alpha_3 + 8\alpha_4)/(\mu_0 + \alpha_3 + 2\alpha_4)} \tag{17}$$

in the 1-direction, and, from (15),

$$v_{P_2}/v_{S_2} = \sqrt{(\lambda_0 + 2\mu_0 + \alpha_1 + 2\alpha_3)/(\mu_0 + \alpha_3 + 2\alpha_4)} \tag{18}$$

in the 2-direction. The relative magnitude of v_P/v_S in different direction thus provides another constraint on the material constants, which might thus be determined in different ways in the field or in the laboratory.

Application of the v_P/v_S anomaly to earthquake prediction was first proposed by Soviet investigators(Kondratenko and Nersesov,1962), and has been applied with varying success elsewhere(e.g., Nur, 1972;Ohtake,1973;Aggarwal et al.,1975). Previously suggested explanations for the v_P/v_S effect, as fluid flow in pores at depth, are not required in the present treatment.

(c)S-wave splitting. The experimental field evidence for S-wave splitting has been difficult to establish(Ryall and Savage,1974). However, it is predicted from the above considerations. Let S_I and S_{II} be the two S-type phases derived in eq (15); then we may write for the 2-direction, where the effect is maximum,

$$v_{S_1}^2 - v_{S_{11}}^2 = 2\alpha_4/\rho \quad . \tag{19}$$

Obviously, there is no S-wave splitting in the 1-direction, as eq(13) shows. It is interesting to note that the S-wave splitting effect depends only on a single second-order material constant.

S-wave splitting in actual seismograms was first described and proposed as a earthquake precursor by Gupta(1973). Because of practical difficulties in recognizing successive S-phases the method has not been extensively tested, though detection may be improved by using changes in the direction of particle motion.

Discussion

Observations of earthquake precursors must be interpreted by comparing the observed effects with the fluctuations predicted from a theoretical model. Such a model must be sufficiently general to embrace all effects, in order to relate one observation to another.

Changes in seismicity, including foreshocks and seismic gaps, can be used as precursors provided that cogent statistical criteria can be developed in order to discriminate between random or spurious fluctuations and true anomalies. Doubts have been expressed about the feasibility of achieving significant detections on this basis, at least for short-range prediction(Garza and Lomnitz,1979).

The future of earthquake prediction thus seems to hinge on the possibility of systematically monitoring and detecting strain-related parameters such as tilt, seismic velocity, electrical conductivity, magnetic reluctance, ground-water flow and other effects which may be caused by variations in local and regional strain patterns. Above all, a coherent theoretical framework must be provided for the analysis and interpretation of such potential earthquake precursors. No such framework appears to be available at present, at least not in a complete form. In this paper we propose to use a continuum-mechanics formulism involving second-order material properties in a general viscoelastic material. This approach is well-known in other geophysical applications; but the second-order viscoelastic parameters of rocks have not yet been determined in the field or in the laboratory.

Early experimental work in viscoelasticity of rocks led to a linear constitutive equation of the form

$$\sigma_{ij} = \int_{-\infty}^{t} G_{ijkl}(t-\tau)\dot{\varepsilon}_{kl}(\tau)d\tau \tag{20}$$

where $G_{ijkl}(t) = q_{ijkl} \log(1+at)$ (Lomnitz,1957). At the present time, this form still seems to provide for adequate agreement with the available experimental data at ordinary seismic frequencies. In this paper, eq(20) has been extended and generalized to include second-order effects, such as are to be expected at very long periods, or for loads beyond the elastic range. Our derivation follows closely the general theory developed by Green, Rivlin and their co-workers. In this theoretical framework the stress-strain behavior depends on the entire strain history; but the effect of a transient of short duration can be approximated by superposition on the long-term effects of prior strains.

Second-order anelasticity involves six different modulus functions, all of which depend on time. For most purposes, including an understanding of the more relevant earthquake precursors, it is sufficient to know a set of six material constants which are related to the values of the modulus functions at zero time. The experimental determination of these constants should be helpful in advancing our present understanding of earthquake precursors.

Acknowledgments. The assistance of K. Walton, both as reviewer and as correspondent, is gratefully acknowledged. This work was partially supported by a fellowship grant from the Consejo Nacional de Ciencia y Tecnología(Mexico).

References

Achenbach, J.D., Wave Propagation in Elastic Solids, North Holland, Amsterdam, 425 pp., 1973.

Adams, R.D., The Heicheng, China earthquake of 4 February 1975: The first successfully predicted earthquake, Bull.N.Z.Nat.Soc.Earthq.Eng., 9, 32-42, 1976.

Aggarwal, Y.P., L.R.Sykes, D.W.Simpson, and P.G. Richards, Spatial and temporal variations in t_S/t_P and in P-wave residuals al Blue Mountain Lake, New York: application to earthquake prediction, J.Geophys.Res., 80, 718-732, 1975.

Brace, W.F., B.W.Paulding, and C.Scholz, Dilatancy in the fracture of crystalline rocks, J.Geophys.Res., 71, 3939-3953, 1966.

Chu Fung-Wing, An outline of prediction and forecast of Haicheng earthquake of M=7.3, Proc. Lectures Seism.Delegation of the People's Republic of China, NASA, Jet Prop. Lab. Publ. SP 42-32, 11-19,1976.

Garza, T., and C. Lomnitz, The Oaxaca Gap: a case history, P.Appl.Geophys., 117, 1187-1194,1979.

Green, A.E., and R.S.Rivlin, The mechanics of non-linear materials with memory, Arch.Rat.Mech. Anal., 1, 1-21, 1957.

Green, A.E., and R.S.Rivlin, The mechanics of nonlinear materials with memory, II, Arch. Rat. Mech. Anal., 3, 82-90, 1959.

Green, A.E., and R.S.Rivlin, The mechanics of nonlinear materials with memory, III, Arch. Rat. Mech. Anal., 4, 387-404, 1960.

Gupta, I.N., Dilatancy and premonitory variations of P, S, travel times, Bull.Seis.Soc.Am., 63, 1157-1161, 1973.

Hess, H.N., Seismic anisotropy of the uppermost mantle under oceans, Nature, 203, 629-631, 1964.

Keen, C., and C. Tramontini, A seismic refraction survey on the Mid-Atlantic Ridge, Geophys. J. Roy. Astr. Soc., 20, 473-491,1970.

Kondratenko, A.M., and I.L.Nersesov, Some results on the study of change in the velocity of longitudinal wave and relations between the velocities of longitudinal and transverse waves in a focal zone, Trudy Inst.Fiz.Zemli, 25,130-150, 1962.

Lomnitz, C., Global Tectonics and Earthquake Risk, Elsevier, Amsterdam, 320 pp., 1974.

Lomnitz, C., Linear dissipation in solids, J.Appl. Phys., 28, 201-205, 1957.

Morris, G.B., R.W.Raitt, and G.G.Shor, Velocity anisotropy and delay-time maps of the mantle near Hawaii, J.Geophys.Res., 74, 4300-4316,1969.

Nur, A., Dilatancy, pore fluids, and premonitory variations of t_S/t_P travel times, Bull. Seis. Soc. Am., 62, 1217-1222, 1972.

Ohtake, M., Change in the v_P/v_S ratio related with occurrence of some shallow earthquakes in Japan, J.Phys.Earth, 21, 173-184, 1973.

Raitt, R.W., The crustal rocks, in The Sea, vol. 3, The Earth Beneath the Oceans, Interscience, New York, 85-102, 1963.

Raitt, R.W., G.G.Shor, T.J.G.Francis, and G.B. Morris, Anisotropy of the Pacific upper mantle, J. Geophys. Res., 74, 3095-3109, 1969.

Rikitake, T., Probability of earthquake occurrence as estimated from crustal strain, Tectonophysics 23, 299,312, 1974.

Ryall, A., and W.V.Savage, S-wave splitting: key to earthquake prediction? Bull.Seis.Soc.Am., 64, 1943-1951, 1974.

Scholz, C.H, and R.Cranz, Notes on dilatancy recovery, J.Geophys.Res., 79, 2132-2135, 1974.

Schlue, J.W., and L.Knopoff, Shear wave anisotropy in the upper mantle of the Pacific basin, Geophys.Res.Letters, 3, 359-362, 1976.

Shor, GG., R.W.Raitt, M.Henry, L.R.Bentley, and G.H.Sutton, Anisotropy and crustal structure of the Cocos Plate, Geof.Intern.,13,337-362, 1973.

AN ASPERITY MODEL OF LARGE EARTHQUAKE SEQUENCES

Thorne Lay and Hiroo Kanamori

Seismological Laboratory, California Institute of Technology, Pasadena, California 91125

Abstract. The variation in maximum rupture extent of large shallow earthquakes in circum-Pacific subduction zones is interpreted in the context of the asperity model of stress distribution on the fault plane. Comparison of the historic record of large earthquakes in different zones indicates that four fundamental categories of behavior are observed. These are: (1) the Chile-type regular occurrence of great ruptures spanning more than 500 km; (2) the Aleutians-type variation in rupture extent with occasional ruptures up to 500 km long, and temporal clustering of large events; (3) the Kurile-type repeated failure over a limited zone of 100-300 km length in isolated events; and (4) the Marianas-type absence of large earthquakes. Southern Chile, Alaska, Southern Kamchatka, and possibly the Central Aleutians are grouped in the first category. The Rat Island portion of the Aleutians, Colombia, Southwest Japan, and the Solomon Islands zones demonstrate the temporal variation of rupture length and multiple earthquake sequences that characterize category 2. The New Hebrides and Middle America have earthquake clustering on a more moderate scale, and are intermediate between categories 2 and 3. Category 3 includes the Kurile Islands, Northeast Japan, Peru and Central Chile. Zones lacking large earthquakes (category 4) include the Marianas, Izu-Bonin, and large portions of Tonga-Kermadec. By loosely grouping each subduction zone into these categories and comparing the general range in behavior with a simple fault model, which is used in a numerical simulation, the parameters governing large earthquake development are clarified. Interpretation of the four categories in terms of asperity distribution and interaction permits some inferences of the nature of stress distribution in particular zones. Two factors appear to dominate in the development of large earthquake failure zones; the nature and degree of coupling on the fault contact, and the extent of lateral segmentation of the subduction zone by transverse stress barriers. Strong coupling and uniform stress distribution on the fault plane produces larger events, whereas more heterogeneous stress distributions produce smaller ruptures and temporal variation in rupture length. Segmentation of the subduction zone that may result in stress barriers affecting rupture length is produced by subduction of transverse structures such as aseismic ridges, and is reflected by submarine canyons and geometric variations in trench configuration.

Introduction

The variations in rupture length of large shallow earthquakes around the Pacific reflect variations in mechanical interaction between subducting and overriding lithospheric plates. Numerous studies have shown that these rupture zones usually occur over discrete, non-overlapping segments of the convergent zone [Fedotov, 1965; Mogi, 1968a, 1969a; Kelleher, 1970, 1972; Sykes, 1971; Kelleher et al., 1973; Utsu, 1974; McCann et al., 1980]. Individual segments tend to fail over the same region in successive cycles of large earthquakes [Imamura, 1928; Kelleher, 1972; Ando, 1975; Fukao and Furumoto, 1979], though there are temporal variations in which a section of trench will fail in a single great event during one sequence, and in several smaller events during another [Ando, 1975; Sykes et al., 1981].

The width of the lithospheric interface has been correlated with maximum length of rupture zones by Isacks et al. [1968] and Kelleher et al. [1974], with regions of broad interface contact having the longest rupture zones. The presence of topographic features on the subducted sea floor [Kelleher and McCann, 1976, 1977] and the presence of transverse features such as ridges, submarine canyons, or changes in the strike of the trench that indicate segmentation of the subduction zone [Mogi, 1969b; Vogt, 1973; Carr et al., 1974; Kelleher and Savino, 1975; Vogt et al., 1976; Spence, 1977; Chung and Kanamori, 1978a,b] modify the relationship between contact area and rupture length. Carr et al. [1974] and Nishenko and McCann [1979] discuss variations in trench structure, volcanic activity, earthquake mechanisms and seismicity distribution that may reflect slab segmentation or variations in the nature of subduction along convergent boundaries.

Blocklike behavior of the overthrust plate may also be important [Ando, 1975; Spence, 1977]. The presence of back-arc spreading has been shown to correlate with an absence of large earthquake ruptures [Uyeda and Kanamori, 1979; Ruff and Kanamori, 1980], and is thus associated with low coupling and aseismic subduction.

Little is actually known about the stress distribution and failure process of large earthquakes. Some general inferences about the state of stress in subduction regimes have been made based on maximum rupture lengths in each zone. Kanamori [1971b, 1977] proposed that the degree of coupling on the fault plane is reflected in maximum earthquake dimensions, with strong coupling being manifested by very large rupture zones, and decoupling associated with the absence of large earthquakes. Kanamori [1971b] proposed a model of gradual thinning and weakening of the ocean-continent lithospheric boundary to account for the differences in coupling and maximum rupture area.

Few studies have been performed yielding any resolution of the detailed nature of stress distribution in subduction zones. Lay and Kanamori [1980] studied body waves and surface waves of recent large events in the Solomon Islands in an attempt to resolve the stress distribution on the thrust plane. They suggested that body wave complexity can be used to infer the degree of heterogeneity of stress, particularly in conjunction with foreshock and aftershock analysis. Other efforts to elaborate on the nature of failure in large earthquakes have demonstrated the complexity of the rupture process. Relative timing of body wave arrivals has indicated the multiple rupture nature of some events as manifested in body wave complexity [Imamura, 1937; Miyamura et al., 1965; Wyss and Brune, 1967; Trifunac and Brune, 1970; Nagamune, 1971; Fukao and Furumoto, 1975; Wu and Kanamori, 1973], while synthetic seismograms have been used in more detailed analysis of other complex events [Fukao, 1972; Chung and Kanamori, 1976, 1978a; Abe, 1977; Kanamori and Stewart, 1978; Stewart and Kanamori, 1978; Rial, 1978; Fukao and Furumoto, 1979; Stewart et al., 1980; Ebel, 1980; Boatwright, 1980]. The more detailed combined analysis of body waves and surface waves includes events in the Solomon Islands, New Hebrides, Middle America, Japan Trench, and Kurile Islands. In general it has been found that the body wave moment and inferred body wave source area are smaller than the surface wave determinations, particularly for the multiple rupture events [e.g., Kanamori and Stewart, 1978; Boatwright, 1980].

Kanamori [1978a] appealed to the asperity model to explain the complexity of large events. This model, an outgrowth of laboratory experiments on rock friction, was first proposed by Byerlee [1970] and further developed by Scholz and Engelder [1976]. They suggested that the two sides of a fault are held together by areas of high strength called asperities. The stress on the asperities is high relative to the average stress on the entire fault plane. The nature of the stress concentrations may be variations in geometric orientation or heterogeneities of the frictional strength of the contact zone. There should be a random distribution of stress concentrations of various scale lengths existing on any given fault. Localized slip occurs when the shear stress on the fault surface exceeds the yield stress of the asperities, and this slip is accompanied by increase of stress on stronger asperities. Various aspects of foreshock activity [Wesson and Ellsworth, 1973; Jones and Molnar, 1979; and Ishida and Kanamori, 1978, 1980] and preseismic quiescence [Kanamori, 1981] have been modeled utilizing these ideas. Interactions between adjacent zones of large asperities can induce triggering [Lay and Kanamori, 1980], and the development of large earthquake ruptures or multiple events. In this model we distinguish between multiple events (isolated triggered events) and multiple rupture events (complex body wave events) because the character of the failure process reflects the degree of asperity interaction. The interaction between asperities will clearly be a function of the nature of the stress distribution of a given region, as well as of the presence of stress barriers segmenting the downgoing slab along a subduction zone.

Kelleher et al. [1974] suggested that the size and frequency of large earthquakes would not vary substantially over geologically short time periods, since the interface geometry and regional parameters of the subduction process would not change over short periods. This suggests the validity of characterizing the short-term past and future seismic behavior by examining historic rupture characteristics. In the following sections the nature of large earthquake occurrence in each subduction zone shown in Figure 1 is categorized, and a simple fault model is analyzed to provide a format within which to discuss the regional behavior.

Regional Characteristics

An extensive review of published source studies, seismicity patterns, and earthquake catalogues of circum-Pacific subduction zones has been undertaken [Lay and Kanamori, 1981]. The motivation for this project has been to quantify the common and individual characteristics of large earthquake occurrence in the subduction zones. Particular attention has been paid to studies of body wave complexity and foreshock-aftershock sequences, which most directly reflect the stress distribution on the fault plane. A qualitative compilation of the observations is given in Table 1, where the comments and numerical values are gross features of a particular zone, with substantial deviation from these values possible within that zone. For

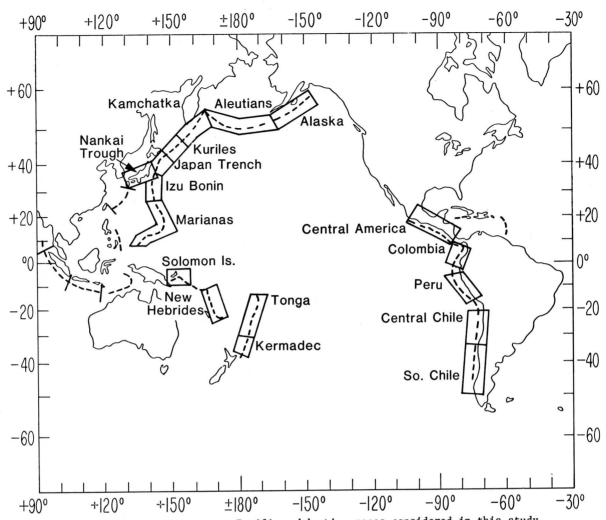

Figure 1. Map showing the circum-Pacific subduction zones considered in this study.

the purposes of discussion, we define 'great' earthquakes as having rupture lengths greater than 500 km long, and 'large' earthquakes as having rupture lengths of 200-500 km.

On the basis of rupture length it is possible to divide the subduction zones into four basic groups, as shown in Table 2. Category 1 is exemplified by Southern Chile (Figure 2a), in which great earthquakes tend to recur regularly in time over approximately the same rupture zone. The 1964 Alaska and 1952 Kamchatka earthquakes occurred in regions which appear to have this mode of rupture. The historic record for the Central Aleutians, which failed in the 1957 event, is unclear [Sykes et al., 1981], but this area may behave similarly. These regions have a large percentage of seismic slip, though this is uncertain for Alaska (Table 1), and are inferred to be strongly coupled. The margins of these large rupture zones tend to abut large transverse features such as the Chile Rise and the narrowing of the plate interface south of Kamchatka.

Category 2 has smaller rupture dimensions as typified by the portion of the Aleutians which failed in the 1965 Rat Island event (Figure 2b). This region, as well as the subduction zones in Colombia, northern Kamchatka, and the Nankai Trough have demonstrated temporal variation in rupture mode, with occasional very large ruptures spanning segments of the trench that fail individually at other times. The Solomon Islands zone is grouped with these because of the frequent occurrence of large doublet events that are temporally and spatially linked.

Category 3 includes the Kurile Islands mode of failure (Figure 2c), in which large earthquakes repeatedly rupture the same portion of the subduction zone, but without coalescing to generate larger events. Peru, Central Chile, and the northern segment of the Japan Trench demonstrate Kurile-type behavior, although the length scale of the subzones ranges from 100-300 km. The New Hebrides and Middle America zones have similarly small rupture dimensions, but a

Table 1. Characteristics of Circum-Pacific Subduction Zones Inferred from Large Earthquakes*

Zone	Rupture Lengths (km)	Repeat Times (yrs)	Seismic Slip %	Sea Floor Topography	Transverse Structures	Anomalous Foreshock-Aftershock Behavior	Large Normal Fault Trench and Tsunami Events	Contact Width(km) Dip (°)	Body Wave Complexity[1]	Back-Arc Spreading
Southern Chile	~1000	~100	75-100%	Smooth	Bounded by Chile Rise, Some canyons	Extensive foreshock activity just before the main shock	Slow events on Chile Rise	200 km 10-20°	No record	No
Alaska	100-200 along penin- sula, 800 in east	Uncertain 50-75 in west, >100 in east	Uncertain 30-100%	Gulf of Alaska Seamounts	Geologic provinces in overthrust slab	Increased preseismic activity near focus	No record	100-300km ~9°	Complex multiple ruptures	No
Kamchatka	500 in south 150 in north	~108[2]	~60%	Smooth	Little evidence	Increased preseismic activity near focus	No record	200 km Uncertain	No record	No
Aleutians	Temporal variation 100-1000	Uncertain ~60 in west	Uncertain May be 100%	Smooth	Changes in strike, canyons ridges	Some foreshocks	Large tsunami & normal fault events	150 km 15-20°	Complex multiple ruptures	No
Colombia	Temporal variation 150-600	Variable 36->73	Uncertain 30-55%	Smooth	Little evidence	Little foreshock activity for 1979 event	No record	150 km Uncertain	Complex for 1979 event	No
Nankai Trough	Temporal variation 150-300	170±70	Uncertain 70%[3]	Smooth	Canyons, terraces 75-100km	No record	No record	100 km 10°	Uncertain	No
Solomon Islands	Doublets 100-300	25-40	50%	Smooth	Woodlark Ridge and New Britain Trench	Few foreshocks, aftershocks	No record	<100 km 30-40°	Simple events, Doublets	No

Region										
New Hebrides	100–200	25–40	Uncertain 50%	Islands Ridges	D'Entrecasteaux and East Rennel Ridges	Foreshocks for some events	Slow events on D'Entrecasteaux Ridge	<100 km 35–40°	Sequences of simple events	Yes
Middle America	100–200	~35	Variable 10–100%	Fracture zones & Ridges	Large ridges & changes in strike	Moderate foreshock activity	No record	<100 km	Sequences of simple events	No
Kurile Islands	200–300	Variable 79–>140	25%	Smooth	Some canyons, 200 km	Many foreshocks	Frequent tsunami events	150 km 20°	Complex multiple ruptures	No
Japan Trench	150 decreasing southward	100 (north) 800 (south)	40% (north) 5% (south)	Seamounts in south	Canyons, terraces 100–200 km	Few foreshocks	Large tsunami & normal fault events	~150 km 10°	Sequences & multiple source events	No
Central Chile	400 (north) 200 (south)	100	Uncertain >50%	Juan Fernandez Ridge	Ridges, changes in strike & dip of Benioff zone	No record	No record	100–150km 10–15°	No record	No
Peru	150	Variable 100	Uncertain	Nazca Ridge	Nazca Ridge	Split aftershock zones	Intermediate depth normal fault events	150 km 12°	Complex multiple events	No
Izu–Bonin Marianas	No events $M_s > 7.4$	––	Uncertain probably very small	Many ridges, seamounts	Many ridges, seamounts	No record	No record	<100 km Uncertain	No record	Yes
Tonga Kermadec	<150	Highly variable	Uncertain 0–90%	Smooth along Tonga, Louisville Ridge	Louisville Ridge	Overlapping aftershock zones	Large normal fault events	100–150km	Uncertain	Yes

*Extracted from Lay and Kanamori (1981).

[1] As observed on WWSSN Long Period Seismograms.

[2] Based on tsunami records.

[3] Based on convergence rate of Seno (1977).

Table 2. Subduction Zone Categories

Category	Zones	Characteristics
1	Southern Chile Southern Kamchatka Alaska Central Aleutians	Regular occurrence of great ruptures (> 500 km long). Large amounts of seismic slip.
2	Western Aleutians Colombia Nankai Trough Solomon Islands	Variations in rupture extent, with occasional ruptures 500 km long. Close clustering of large events and doublets.
2-3	New Hebrides Middle America	Intermediate size and small events with no great earthquakes, but clustering of activity.
3	Kurile Islands Northeast Japan Trench Peru Central Chile	Repeated ruptures over limited zones. No great events. Large component of aseismic slip, or subducting ridges.
4	Marianas Izu-Bonin Southeast Japan Trench Tonga-Kermadec	Large earthquakes are infrequent or absent. Back-arc spreading and large amounts of aseismic slip are inferred.

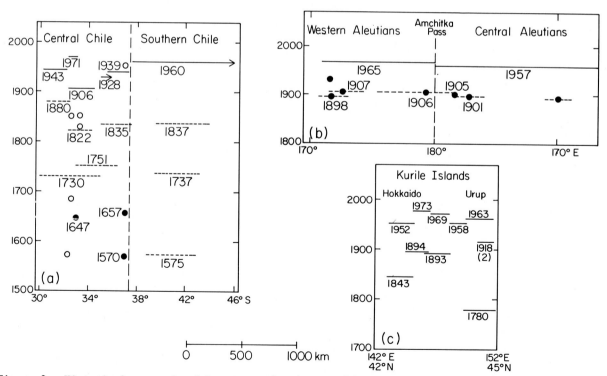

Figure 2. Historical records of large earthquakes in (a) Chile (after Stauder, 1972), (b) the Aleutians (after Sykes et al.,1981), and (c) the Kurile Islands. Inferred rupture extents are shown for events with adequate data.

stronger tendency for events to cluster in time and space, indicating that they are intermediate between groups 2 and 3.

Category 4 is characterized by the Marianas-type absence of great earthquakes. The Izu-Bonin, southern Japan Trench and large portions of the Tonga-Kermadec zones are placed in this category. These particular zones are distinctive because of the presence of active back-arc spreading, and probably of a large component of aseismic slip.

Asperity Model

The regional characteristics discussed in the previous section can be modeled in terms of distribution and interaction of asperities. Asperity size and distribution govern the degree of loading of adjacent asperities when a large asperity fails. The basis of this model is the analysis of earthquake doublets in the Solomon Islands [Lay and Kanamori, 1980]. As shown in Figure 3a, the individual rupture zones in the Solomon Islands are represented by a distribution of asperities. Failure of one of the asperities would cause an increase in stress on the adjacent asperities. In the Solomon Islands region, where the asperities are large and of similar size, these incremental stresses are large enough to trigger failure of an adjacent asperity, producing two or more distinct, but similar events in a sequence. For zones with more complex stress distributions (Figure 3b), the

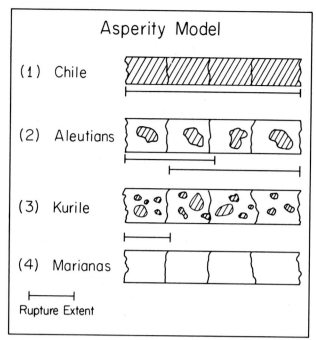

Figure 4. An asperity model indicating the different nature of stress distribution in each subduction zone category. The hatched areas indicate the zones of strong coupling.

variation in asperity size inhibits efficient loading of adjacent large asperities, as much of the load is alleviated by failure of smaller asperities. When the smaller asperities fail, the magnitude of the stress increment induced on adjacent areas is small, and stress builds up on the larger asperities gradually. Thus, a heterogeneous stress distribution is not likely to generate very large ruptures or multiple events. The strength of coupling on the fault is governed by the product of the area of contact and the average stress on the contact zone, with large strength resulting from large asperity area and high friction coefficient or asperity breaking stress. In the presence of aseismic slip or weak plate coupling the asperity contact area is small.

In the framework of this model, a possible interpretation of the nature of stress distribution in each subduction zone category summarized in Table 2 may be schematically given as in Figure 4. For the Chile-type behavior, the lithospheric plates are strongly coupled, and the asperity distribution is essentially uniform over the entire contact zone. Rupture occurs throughout the region in great events due to this uniformity, with the observed complexity of these ruptures stemming from their large size and presence of some lateral segmentation which slightly delays the stress release. Such complexity in large events rupturing across blocklike source regions has been proposed by Nagamune [1978]. The Chile-type zones have

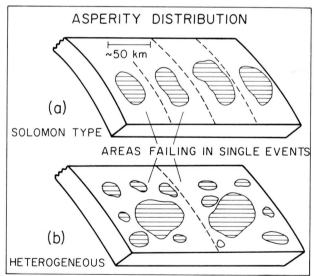

Figure 3. Asperity model representation of coupling on the plate interface in subduction zones. The Solomon Islands region has a uniform distribution of comparable size zones of coupling. Other regions such as the Japan Trench and Kurile Islands have a more heterogeneous stress distribution like that shown in (b). (After Lay and Kanamori, 1980).

relatively simple, geometrically uniform subduction regimes, a condition which is probably favorable for developing uniform stress levels and effective asperity interaction.

Slightly smaller, but still relatively homogeneous asperity distributions are indicated for Aleutians-type zones. Here, the failure of the larger asperities effectively loads adjacent zones causing some large ruptures. Frequently, conditions are not favorable for immediate triggering of adjacent subzones, either due to the stress distribution being heterogeneous due to previous events in adjacent regions or due to stress barriers segmenting the trench, and smaller ruptures occur, possibly as doublets. At other times stress conditions are relatively uniform, with adjacent large asperities, which may fail together producing atypically large ruptures. The subduction regimes are relatively uniform for these regions as well as for the Chile-type zones, though the influence of transverse segmentation appears to be stronger in this case.

For the Kurile-type rupture mode there are numerous small asperities within a given subzone. Failure of the larger of these leads to failure throughout the subzone, but because of the relatively small size of the asperities, the stress increments communicated to adjacent zones are inadequate to cause larger rupture propagation. The heterogeneous stress distribution produces complicated ruptures and foreshock-aftershock activity as numerous small and intermediate size asperities fail along with larger ones. Though the transverse boundaries in this case may be fairly strong, delimiting the rupture zones, the heterogeneity of the stress distribution inhibits large rupture development as well. For the zones in Central Chile, Peru, the New Hebrides, and Middle America, the subduction of large transverse structures may partially account for the irregular stress distribution. For the Japan Trench and Kurile Islands regions, the large amount of aseismic slip may reflect the lack of large, uniform asperities.

The final category is the Marianas-type behavior for which there are no large asperities and hence no large earthquakes. There is relatively weak coupling on the fault plane and extensive aseismic slip.

In this model, an important factor is physical segmentation of the trench caused by transverse structures on the subducting or overriding plates and geometric irregularities in the subduction zone. Such boundaries commonly delimit large failure zones, apparently functioning as barriers to lateral rupture propagation. The other dominant factor is the nature of coupling on the fault plane, which is influenced by the mechanical properties of the plates, the breadth and geometry of the contact zone, age of the subducting slab, and previous rupture history in the zone.

Numerical Simulation

In order to simulate the types of rupture sequences observed along subduction zones, we have adopted a simple fault model based on asperity distributions and interactions, as illustrated in Figure 5. The fault zone consists of a number of subfaults of uniform size, with relative strengths selected randomly from a normal distribution characterized by a mean strength $\bar{s}$ and a standard deviation Σ. A uniform regional loading stress on the system as well as incremental stresses induced stepwise when adjacent subfault failures occur are applied to each subfault. The regional loading stress has the form

$$\sigma_o = \sigma_{oo} + \alpha t. \qquad (1)$$

The subfault stresses are then given by

$$\sigma_i = \sigma_o + \Delta\sigma_i \qquad (2)$$

where $\Delta\sigma_i$ is the stress increment due to failure of adjacent subfaults. We select a simple form of subfault interaction governed by an interaction parameter c:

$$\Delta\sigma_i = c \cdot S_j / d_{ij} \qquad (3)$$

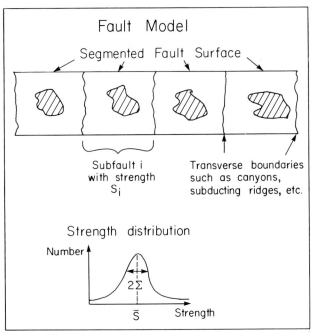

Figure 5. The fault model adopted for this study. The fault is segmented into subfaults, each with its own strength of coupling. The subfault strengths are selected randomly from a Gaussian distribution with mean strength $\bar{s}$ and standard deviation Σ.

where S_j = the stress drop of the failed block, and d_{ij} is the distance to the failed region. We assume that the stress drop is equal to the strength assigned to the subzone. The question of whether the subfault interaction is a static or a dynamic effect is not addressed in this simple model. When the subfault stress exceeds the assigned strength the subfault fails, and the stress increment induced on adjacent zones is computed according to (3) in the same time step. Stress increments due to any additional failures are then computed and applied throughout the zone until all subfault stresses are below their strengths, at which time the failed zones are reassigned their previous strengths and the next time step is performed. This model is very similar to one analyzed by Ito [1980], though we adopt a different form of subfault interaction. Deterministic experimental and numerical fault models have frequently been adopted in simulations of seismicity, as reviewed by Cohen [1979], yet this very simple model reproduces many of the general characteristics observed in more sophisticated models. Time dependent effects are not included in this model, though they clearly exist in reality and may influence long term behavior of a region.

The parameters that govern the extent of rupture in this fault model are the range in strengths along the zone controlled by Σ, and the interaction parameter c. These parameters do not in general affect the resulting seismicity patterns in an identical manner, though cases can be found where their effects are similar. Increasing the interaction produces sequences of large ruptures with regular recurrence, quickly suppressing the effects of variable strength. The physical interpretation of this model is as follows. Subdivision of the fault zone into subfaults of somewhat independent behavior is suggested by the observed physical segmentation of subduction zones, and allows us to introduce stress barriers between subfaults by varying the asperity interaction along the fault. The variability in strength of the subfaults corresponds to variations in amount of seismic slip, subducted topography, and convergence rate along a given zone. The interaction parameter incorporates the efficiency with which failure of a particular zone induces stress increments on adjacent regions, corresponding to the size distribution of asperities. Thus, low c reflects weak triggering interaction which is associated with heterogeneous asperity size distribution.

Figure 6 shows some representative earthquake sequences generated by this model in which the interaction parameter is varied, but the initial random strength distribution is not. For each case c is constant throughout the trench. The vertical axis in each figure represents time and the horizontal axis represents position along the fault, thus each column shows the failure history of a given subfault, with circles marking the time of failure. In Figure 6a there is no

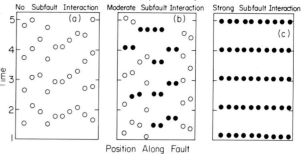

Figure 6. Earthquake sequences generated with the fault model for a fixed value of Σ, and variable subfault interaction parameter, c. In (a) c = 0.0, and the subfaults are completely independent. In (b) c = 0.05, and frequent clustering and occasional long ruptures occur. In (c) c = 0.15 and long ruptures repeatedly rupture the entire fault. Solid circles indicate subzone failures that cluster in time.

interaction between subfaults, the random distribution of strengths governs the failure history and variation in recurrence interval along the trench. This pattern was produced with only a small range in strengths (small Σ), yet clustering of events seldom occurs. The independent behavior of individual subfaults is similar to behavior of the Kurile Islands category. Thus, the subzone scale here is on the order of 100 to 300 km.

Even a small interaction between subfaults produces a markedly greater tendency for subzone failures to cluster in time as shown in Figure 6b, though well-separated subfaults show little obvious interaction. Closely related events are indicated by solid circles to clarify this clustering. Notice that a given length of fault sometimes experiences ruptures spanning only one or two subzones, but occasionally a very long rupture occurs. This temporal variation is similar to the Aleutians-type category. Unusual uniformity in stress conditions along the trench produces the very long rupture, but such events are not necessarily the norm for the regional behavior. Very strong subfault interaction produces the Chile-type pattern shown in Figure 6c. There is regular recurrence of great earthquakes rupturing a large portion or the entire length of the fault.

Individual subduction zones tend to show distinct behavior with greater complexity than appears in the sequences shown in Figure 6. Consider, for example, the Nankai Trough, with the extensive historic record indicated in Figure 7a. The subdivision of this zone into four segments was suggested by Ando [1975] on the basis of transverse geologic and bathymetric structures on the underthrusting and overriding plates. The margins of the zone are bounded by

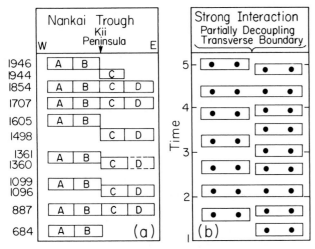

Figure 7. (a) The historical earthquake sequence in the Nankai Trough (after Ando, 1975), and (b) a model earthquake sequence generated by allowing subfault interaction to vary along the fault. A weak stress barrier separates two zones of strong subfault interaction in (b).

the Kyushu-Palau ridge to the west and the Sagami trough to the east. An interesting observation is that large earthquakes have commonly broached the inferred transverse boundaries between zones A and B and between C and D, and occasionally the entire trench has ruptured together. The boundary between B and C lies along the southern extension of the Kii Peninsula, and although this feature is sometimes crossed by ruptures such as in 1707, it appears to exert a stronger segmenting influence than the features separating C and D or A and B. In a simple attempt to incorporate this variation in subfault interaction along the trench we allow the interaction parameter c to vary spatially. In particular, a weak stress barrier in the middle of the fault (between B and C) is introduced, separating two regions of strong subfault interaction. The resulting pattern is shown in Figure 7b, and provides a reasonable approximation to the observed behavior. The weakly segmenting boundary serves to delimit the rupture zones of the doublet type events, as well as to occasionally permit total trench ruptures by allowing some triggering interaction between the adjacent regions. Additional flexibility can be introduced into the model by allowing the stress drop upon failure to vary, thus producing an irregular recurrence interval for a given subfault.

By varying the degree of subfault interaction the entire spectrum of observed great earthquake behavior can be reproduced. Subfaults interact to produce larger ruptures by triggering when the interaction parameter is large. Two basic factors influence the degree of subfault

interaction; the nature of the rupture process within individual subfaults, and the presence of transverse segmentation between subfaults. The influence of these factors cannot be easily segregated. Consider the Chile-type model in Figure 6c. It is possible that the rupture occurs throughout the entire fault because there are no stress barriers segmenting the fault. Alternately, it is possible that the nature of coupling and stress release are such that triggering efficiency of adjacent zones is enhanced, overcoming whatever segmentation exists. Either explanation suggests uniformity throughout the region (with greater heterogeneity indicated for the other rupture modes). In the light of the asperity model, it is the nature of the stress distribution which principally determines the subfault interaction, however, the presence of transverse stress barriers probably modifies the loading of adjacent asperities as well.

Discussion

The fault model presented above is not intended to reproduce the detailed seismic record of any particular zone, rather it highlights the dominant parameters producing particular rupture patterns. The regional variations in these parameters inferred from source studies and other lines of evidence are generally consistent with the asperity model results. This indicates that the asperity model provides a useful intuitive framework within which to analyze large earthquake ocurrence. As regional stress characteristics become better defined by additional detailed studies, the model may be refined, permitting a better understanding of the subduction process as well as a physical basis upon which to assess regional seismic potential.

Many aspects of the source complexity of large events can also be interpreted in the context of the barrier model proposed by Das and Aki [1977] and Aki [1979]. The quantitative aspects distinguishing the barrier and asperity models discussed by Husseini et al. [1975] and Rudnicki and Kanamori [1981], do not have adequate resolution in the seismic record to discriminate which model is more appropriate for the rupture process in great earthquakes. Lacking detailed resolution of the state of stress on the fault plane and physics of the rupture process, we have implicitly incorporated aspects of both models into the fault model presented here, including the regional asperity type stress concentrations which dominate the behavior of fault subzones, and the effects of lateral variations in asperity interaction by which we can introduce stress barriers modifying the rupture process. Both factors contribute to the occurrence of large ruptures, and merging various aspects of each model may yield a better understanding of the subduction process in the future, particularly of the triggering phenomena. Brune [1978] discussed the difficulties in assessing seismic hazard due to the potential for triggering interaction between adjacent zones. This aspect of subfault interaction is clearly reflected in the temporal variation of rupture extent observed in some regions.

The nature of coupling on the fault plane appears to be influenced by the history of subduction. Figure 8 reproduces the evolutionary subduction model proposed by Kanamori [1977]. Shallow dipping, broad, strongly coupled zones such as in Southern Chile and Alaska produce extensive ruptures. The thrust zone may be weakened and partially decoupled by repeated fracturing, yielding smaller rupture lengths as in the Kurile Islands and Japan Trench. Large normal events such as the 1933 Sanriku earthquake in the Japan Trench may represent a transition to

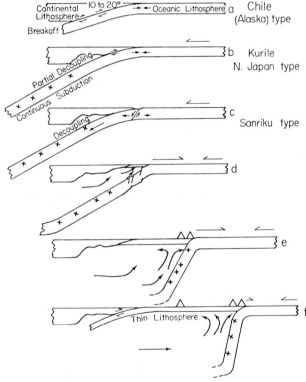

Figure 8. The evolutionary subduction model proposed by Kanamori, 1977. (a) Strong coupling between oceanic and continental lithospheres results in great earthquakes and break off of the subducting lithosphere at shallow depths. (b) Partial decoupling results in smqller earthquakes and continuous subduction. (c) Further decoupling results in aseismic events and intraplate tensional events. (d) Sinking plate results in retreating subduction and formation of a new thin lithosphere. (e) Episodic retreat and formation of ridges. (f) Decelerated retreat and commencement of new subduction.

tensional stress in the slab and complete decoupling of the plate interface which may result in the development of back-arc basins by trench retreat. The variations in structure of the sedimentary wedges, upper slope basins, and terraces in the trench, which correlate with maximum rupture lengths, may reflect the transition between zones of strong and weak coupling on the fault plane.

Ruff and Kanamori [1980] tested correlations between coupling on the fault, parameterized by regional maximum earthquake magnitudes, with other parameters of the subduction zone. They found that convergence rate and lithospheric age correlate with coupling. It was also determined that penetration depth correlates with lithospheric age and contact width with convergence rate, results compatible with work done by Isacks et al. [1968], Vlaar and Wortel [1976], and Wortel and Vlaar [1978]. These correlations suggest that large earthquake generation correlates with interface geometry or with age and rate in some other manner. Ruff and Kanamori [1980] suggest a qualitative model in which convergence rate and lithospheric age determine the horizontal and sinking rates of the slab, thus controlling the dip of the Benioff zone and normal stress distribution on the fault plane. The area of coupling may change due to reducing cross section with increasing dip or degradation of the interface. This model indicates that the downgoing plate determines the stress state in the slab and on the interface, as suggested in the evolutionary subduction model of Kanamori [1977].

Conclusions

Regional variations in large shallow earthquake fault lengths between circum-Pacific subduction zones are used to infer the basic stress distribution in each zone. The degree of coupling and segmentation of the stress regime by transverse boundaries influence the development of large rupture dimensions. Very strong coupling on the fault plane overcomes the effect of lateral segmentation to produce regularly recurring great earthquakes in Southern Chile, Alaska, Southern Kamchatka, and possibly in the Central Aleutians. These zones have predominantly seismic slip and generate large multiple ruptures by sequential failure of adjacent subzones of relatively uniform stress distribution. The variability of rupture length in the Western Aleutians, Colombia, and Nankai Trough regions reflects interaction between coupling and segmentation, and indicates the need to understand these factors when assessing regional seismic potential. The Solomon Islands and Nankai Trough have distinctive large earthquake clustering, perhaps reflecting uniformity in the stress regime along these zones, though transverse boundaries modify the development of large ruptures. The seismic record indicates that zones with substantial aseismic slip and heterogeneous stress distribution generate moderate size, but complex ruptures, as in the Japan Trench, Central Chile, Peru, and the Kurile Islands. The large amount of aseismic slip and weak coupling in these zones may result from progressive weakening of the fault contact as subduction progresses. The New Hebrides and Middle America appear to have small scale but fairly uniform stress distributions, producing occassional clustering of events, but no great earthquakes. A simplified fault model incorporating aspects of the asperity and barrier models has been shown to support the inferences of the state of stress in each region.

Acknowledgments. We appreciate helpful suggestions made by T. Wallace and J. Ebel and the anonymous reviewers. Research supported by the Earth Sciences Section National Science Foundation Grant No. (EAR 78-14786) and United States Geological Survey Contract No. (14-08-0001-18321). One of the authors, T. L., was supported by a National Science Foundation Graduate Fellowship. Contribution No. 3492, Geological and Planetary Sciences, Seismological Laboratory, California Institute of Technology, Pasadena, California 91125.

References

Abe, K., Tectonic implications of the large Shioya-Oki earthquakes of 1938, Tectonophysics, 41, 269-289, 1977.

Aki, K., Characterization of barriers on an earthquake fault, J. Geophys. Res., 84, 6140-6148, 1979.

Ando, M., Source mechanisms and tectonic significance of historical earthquakes along the Nankai Trough, Japan, Tectonophysics, 27, 119-140, 1975.

Boatwright, J., Preliminary body-wave analysis of the St. Elias, Alaska earthquake of February 28, 1979, Bull. Seismol. Soc. Amer., 70, 419-436, 1980.

Brune, J. N., Implications of earthquake triggering and rupture propagation for earthquake prediction based on premonitory phenomena, Proceedings of conference IV: Fault mechanics and its relation to earthquake prediction, USGS Open File Report #78-380, 71-82, 1978.

Byerlee, J. D., Static and kinetic friction of granite under high stress, Int. J. Rock Mech. Min. Sci., 7, 577-582, 1970.

Carr, M. J., R. E. Stoiber and C. L. Drake, The Segmented nature of some continental margins, in C. A. Burk and C. L. Drake, (eds.), The Geology of Continental Margins, Springer-Velag, 105-114, 1974.

Chung, Wai-Ying and H. Kanamori, Source process and tectonic implications of the Spanish deep focus earthquake of March 29, 1954, Phys. Earth Planet. Interiors, 13, 85-96, 1976.

Chung, Wai-Ying and H. Kanamori, Subduction process of a fracture-zone and aseismic ridges - the focal mechanism and source characteristics of the New Hebrides earthquake of January 19, 1969 and some related events, Geophys. J. R. Astr. Soc., 54, 221-240, 1978a.

Chung, Wai-Ying and H. Kanamori, A mechanical model for plate deformation associated with aseismic ridge subduction in the New Hebrides arc, Tectonophysics, 50, 29-40, 1978b.

Cohen, S. C., Numerical and laboratory simulation of fault motion and earthquake occurrence, Reviews Geophy. Space Phys., 17, 61-72, 1979.

Das, S. and K. Aki, Fault planes with barriers; A versatile earthquake model, J. Geophys. Res., 82, 5658-5670, 1977.

Ebel, J. E., Source processes of the 1965 New Hebrides Islands earthquakes inferred from teleseismic waveforms, Geophys. J. R. astr. Soc., 63, 381-404, 1980.

Fedotov, S. A., Regularities of the distribution of strong earthquakes of Kamchatka, the Kurile Islands, and northeastern Japan, Trudy Inst. Fiz. Zemli Akad. Nauk. SSSR, 36, 66-93, 1965 (in Russian).

Fukao, Y., Source process of a large deep-focus earthquake and its tectonic implications - The western Brazil earthquake of 1963, Phys. Earth Planet. Interiors, 5, 61-76, 1972.

Fukao, Y. and M. Furumoto, Foreshocks and multiple shocks of large earthquakes, Phys. Earth Planet. Interiors, 10, 355-368, 1975.

Fukao, Y. and M. Furumoto, Stress drops wave spectra and recurrence intervals of great earthquakes - Implications of the Etorofu earthquake of 1958 November 6, Geophys. J. R. astr. Soc., 57, 23-40, 1979.

Husseini, M. I., D. B. Jovanovich, M. J. Randall and L. B. Freund, The fracture energy of earthquakes, Geophys. J. R. astr. Soc., 43, 367-385, 1975.

Imamura, A., On the seismic activity of central Japan, Jap. J. Astron. Geophys., 6, 119-137, 1928.

Imamura, A., Theoretical and Applied Seismology, Marozen, Tokyo, 358pp., 1937.

Isacks, B., J. Oliver and L. R. Sykes, Seismicity and the new global tectonics, J. Geophys. Res., 73, 5855-5899, 1968.

Ishida, M. and H. Kanamori, The foreshock activity of the 1971 San Fernando, California earthquake, Bull. Seismol. Soc. Amer., 68, 1265-1279, 1978.

Ishida, M. and H. Kanamori, Temporal variation of seismicity and spectrum of small earthquakes preceding the 1952 Kern County, California earthquake, Bull. Seismol. Soc. Amer., 70, 509-527, 1980.

Ito, K., Periodicity and chaos in great earthquake occurrence, J. Geophys. Res., 85, 1399-1408, 1980.

Jones, L. M. and P. Molnar, Some

characteristics of foreshocks and their possible relationship to earthquake prediction and premonitory slip on faults, J. Geophys. Res., 84, 3596-3608, 1979.

Kanamori, H., Seismological evidence for a lithospheric normal faulting – The Sanriku earthquake of 1933, Phys. Earth Planet. Interiors, 4, 289-300, 1971a.

Kanamori, H., Great earthquakes at island arcs and the lithosphere, Tectonophysics, 12, 187-198, 1971b.

Kanamori, H., Seismic and aseismic slip along subduction zones and their tectonic implications, in M. Talwani and W. C. Pitman III, (eds.), Island Arcs, Deep Sea Trenches and Back-Arc Basins, Maurice Ewing Series, I, 163-174, AGU, Washington, D. C., 1977.

Kanamori, H., Use of seismic radiation to infer source parameters, Proceedings of conference IV: Fault mechanics and its relation to earthquake prediction, USGS Open File Report 78-380, 283-318, 1978a.

Kanamori, H., Nature of seismic gaps and foreshocks, Proceedings of conference VI: Methodology for identifying seismic gaps and soon to break gaps, USGS Open File Report 78-943, 319-334, 1978b.

Kanamori, H., The nature of seismicity patterns before major earthquakes, 1981 (this volume).

Kanamori, H. and G. S. Stewart, Seismological aspects of the Guatemala earthquake of February 4, 1976, J. Geophys. Res., 83, 3427-3434, 1978.

Kelleher, J. A., Space-time seismicity of the Alaska-Aleutian seismic zone, J. Geophys. Res., 75, 5745-5756, 1970.

Kelleher, J. A., Rupture zones of large South America earthquakes and some predictions, J. Geophys. Res., 77, 2087-2103, 1972.

Kelleher, J. and W. McCann, Bathymetric highs and the development of convergent plate boundaries, in M. Talwani and W. C. Pitman III, (eds.), Island Arcs, Deep Sea Trenches and Back-Arc Basins, Maurice Ewing Series, I, 115-122, AGU, Washington, D.C., 1977.

Kelleher, J. and W. McCann, Buoyant zones, great earthquakes, and unstable boundaries of subduction, J. Geophys. Res., 81, 4885-4896, 1976.

Kelleher, J. and J. Savino, Distribution of seismicity before large strike slip and thrust-type earthquakes, J. Geophys. Res., 80, 260-271, 1975.

Kelleher, J., J. Savino, H. Rowlett and W. McCann, Why and where great thrust earthquakes occur along island arcs, J. Geophys. Res., 79, 4889-4899, 1974.

Kelleher, J., L. Sykes and J. Oliver, Possible criteria for predicting earthquake locations and their application to major plate boundaries of the Pacific and the Caribbean, J. Geophys. Res., 78, 2547-2585, 1973.

Lay, T. and H. Kanamori, Earthquake doublets in the Solomon Islands, Phys. Earth Planet. Int., 21, 283-304, 1980.

Lay, T. and H. Kanamori, The asperity model and the nature of great earthquake occurrence, in preparation, 1981.

McCann, W. R., S. P. Nishenko, L. R. Sykes and J. Kraus, Seismic gaps and plate tectonics: Seismic potential for major boundaries, Pageoph., 117, 1087-1147, 1980.

Miyamura, S., S. Omote, R. Teisseyre and E. Vesanen, Multiple shocks and earthquake series pattern, Int. Inst. Seismol. Earthquake Eng. Bull., 2, 71-92, 1965.

Mogi, K., Development of aftershock areas of great earthquakes, Bull. Earthquake Res. Inst., 46, 175-203, 1968a.

Mogi, K., Sequential occurrences of recent great earthquakes, J. Phys. Earth, 16, 30-36, 1968b.

Mogi, K., Some features of recent seismic activity in and near Japan (2) Activity before and after great earthquakes, Bull. Earthquake Res. Inst., 47, 395-417, 1969a.

Mogi, K., Relationship between the occurrence of great earthquakes and tectonic structures, Bull. Earthquake Res. Inst., 47, 429-451, 1969b.

Nagamune, T., Source regions of great earthquakes, Geophys. Mag., 35, 333-399, 1971.

Nagamune, T., Tectonic structures and multiple earthquakes, Zisin: J. Seismol. Soc. Japan, 31, 457-468, 1978 (in Japanese).

Nishenko, S. and W. McCann, Large thrust earthquakes and tsunamis: Implications for the development of fore arc basins, J. Geophys. Res., 84, 573-584, 1979.

Rial, J. A., The Caracas, Venezuela earthquake of July 1967: A multiple source event, J. Geophys. Res., 83, 5405-5414, 1978.

Rudnicki, J. W. and H. Kanamori, Effects of fault interaction on moment stress drop and strain energy release, in press, 1981.

Ruff, L. and H. Kanamori, Seismicity and the subduction process, Phys. Earth Planet. Int., 23, 240-252, 1980.

Scholz, C. H. and J. T. Engelder, The role of asperity indentation and ploughing in rock friction, I. Asperity creep and stick slip, Int. J. Rock Mech. Min. Sci., 13, 149-154, 1976.

Seno, T., The instantaneous rotation vector of the Philippine Sea plate relative to the Eurasian plate, Tectonophysics, 42, 209-226, 1977.

Spence, W., The Aleutian arc: tectonic blocks, episodic subduction, strain diffusions and magma generation, J. Geophys. Res., 82, 213-230, 1977.

Stewart, G. S., E. C. Chael and K. C. McNally, The 1978 November 29 Oaxaca, Mexico earthquake – a large single event, Geophy. J. R. Astron. Soc., in press, 1980.

Stewart, G. S. and H. Kanamori, Complexity of rupture propagation in large earthquakes (abstract), Trans. AGU, 59, 1127, 1978.

Sykes, L. R., Aftershock zones of great

earthquakes, seismicity gaps, and earthquake prediction for Alaska and the Aleutians, <u>J. Geophys. Res.</u>, <u>76</u>, 8021-8041, 1971.

Sykes, L. R., J. B. Kisslinger, L. House, J. N. Davies and K. H. Jacob, Rupture zones of great earthquakes, Alaska-Aleutian arc, 1784-1980, <u>Science,</u> in press, 1981.

Trifunac, M. D. and J. N. Brune, Complexity of energy release during the Imperial Valley, California earthquake of 1940, <u>Bull. Seismol. Soc. Amer.</u>, <u>60</u>, 137-160, 1970.

Utsu, T., Space-time pattern of large earthquakes occurring off the Pacific coast of the Japanese Islands, <u>J. Phys. Earth</u>, <u>22</u>, 325-342, 1974.

Uyeda, S. and H. Kanamori, Back-arc opening and the mode of subduction, <u>J. Geophys. Res.</u>, <u>84</u>, 1049-1061, 1979.

Vlaar, N. and M. Wortel, Lithospheric aging, instability, and subduction, <u>Tectonophysics,</u> <u>32</u>, 331-351, 1976.

Vogt, P. R., Subduction and aseismic ridges, <u>Nature,</u> <u>241</u>, 189-191, 1973.

Vogt, P. R., A. Lowrie, D. Bracey and R. Hey, Subduction of aseismic oceanic ridges: Effects on shape, seismicity and other characteristics of consuming plate boundaries, <u>Geol. Soc. Amer. Spec. Paper,</u> <u>172</u>, 59p., 1976.

Wesson, R. L. and W. L. Ellsworth, Seismicity preceding moderate earthquakes in California, <u>J. Geophys. Res.</u>, <u>78</u>, 8527-8546, 1973.

Wortel, M. and N. Vlaar, Age-dependent subduction of oceanic lithosphere beneath western South America, <u>Phys. Earth Planet. Interiors,</u> <u>17</u>, 201-208, 1978.

Wu, F. T. and H. Kanamori, Source mechanism of February 4, 1965, Rat Island earthquake, <u>J. Geophys. Res.</u>, <u>78</u>, 6082-6092, 1973.

Wyss, M. and J. Brune, The Alaska earthquake of 28 March 1964: A complex multiple rupture, <u>Bull. Seismol. Soc. Amer.</u>, <u>57</u>, 1017-1023, 1967.

592

FLUID FLOW ACCOMPANYING FAULTING: FIELD EVIDENCE AND MODELS

Richard H. Sibson

Department of Geology, Imperial College, London SW7 2BP, England

Abstract. Direct evidence that channel flow
of aqueous fluids accompanies shallow crustal
faulting, in some instances at least, comes from
the observation of transitory surface effusions
following some moderate to large earthquakes in
consolidated rocks, and the textural characteristics of the hydrothermal vein systems often
found associated with ancient, exhumed faults.
These phenomena are examined in relation to two
alternative models. In the first, the transitory post-seismic flow results from the collapse
of pre-failure dilatant fractures in accordance
with the dilatancy/fluid-diffusion hypothesis, so
that the fault system can be regarded as a 'pump'.
In the second model, the fault/fracture system
functions as a 'valve' on a fluid reservoir.
Rising fluid pressure induces fault slip,
creating a temporary fracture permeability which
allows partial draining of the reservoir. As
the fluid pressure drops, self-sealing of the
fracture system occurs by deposition of hydrothermal minerals, and the whole cycle can then be
repeated. In some instances there is convincing
evidence that extensive fracture dilatancy has
developed close in to faults (usually of normal
dip-slip character), and that dilatancy pumping
has occurred as a result of post-slip collapse
of these fracture systems. Dilatant extension
fractures can only form or re-open by hydraulic
fracturing adjacent to faults provided they retain some cohesive strength, but an important
upper limit is thereby placed on the permissible
level of differential stress. The more frequent
and intense development of extension vein arrays
adjacent to normal faults in comparison with
thrusts is in accord with theory, which suggests
that hydraulic fracturing should occur at significantly lower fluid pressures around the former
than the latter.

Introduction

Following the seminal work of Frank (1965),
examining the potential role of cyclic dilatancy
and fluid flow accompanying faulting, much
attention has been paid to the development of the
dilatancy/fluid-diffusion model for explaining
observed precursory effects to shallow crustal
earthquakes (Nur, 1972; Whitcomb et al., 1973;
Scholz et al., 1973). Yet as Nur (1975) points
out, mechanisms contributing to a bulk dilatant
strain may include microcrack dilatancy, sand
pile dilatancy and macroscopic fracture dilatancy.
Both the form of the constitutive law and the
spatial extent of the dilatant zone will depend
on which dilatancy process is dominant. Alternative models which do not invoke fluid flow to
account for the observed precursors have also
been developed (e.g. Stuart, 1974; Mjachkin et
al., 1975). Reservoir and fluid-injection induced seismicity (e.g. Raleigh et al., 1972) and
the changes in groundwater level (Kovach et al.,
1975) and oil/gas well pressures (Arieh and
Merzer, 1974; Wu, 1975) sometimes observed to
accompany shallow earthquakes, demonstrate that
fluid flow both affects and is affected by
seismic activity. However, the extent to which
it forms an integral part of the earthquake
failure process remains unresolved and a subject
for debate (Ward, 1979).

Here, this important question is approached
using two direct lines of evidence which to date
have received surprisingly little attention. The
distribution, orientation and textural characteristics of hydrothermal vein systems often found
associated with ancient exhumed faults, and the
transitory post-seismic effusions of groundwater
which are sometimes observed, both provide useful
information on the nature and extent of active
fracture systems and their role in fluid migration accompanying faulting.

Post Seismic Groundwater Effusions

In arid terrain particularly, there are
reports of changes in well water level, spring-
flow and occasional dramatic effusions of groundwater immediately following moderate to large
shallow earthquakes (e.g. Tchalenko, 1973;
Ambraseys, 1974). However, inadequate documentation usually prevents quantification of such
phenomena and ambiguities in interpretation also
arise. Thus, the outpourings of warm groundwater
and allied phenomena accompanying the Matsushiro
earthquake swarm, which Nur (1974) interpreted
in terms of the dilatancy/fluid-diffusion model,

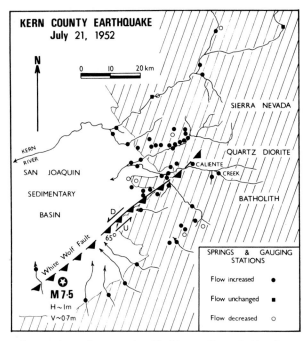

Figure 1. Changes in fluid outflow following the Kern County earthquake. Data from Briggs and Troxell (1955), with geology generalised from Dibblee (1955).

have also been attributed to the inflation of a shallow magma reservoir (Stuart and Johnson, 1975). Again, in the case of the 1930 M7.3 Salmas Earthquake in N.W. Iran (Tchalenko and Berberian, 1974), there were widespread post-seismic changes in spring activity, water was expelled along the main fault trace (dextral-reverse? slip, locally > 5m) and a broad area of the alluvial plain on the downthrown side became fissured and water-logged. This latter behaviour is usually attributed to the lique-faction of unconsolidated alluvium as a result of shaking. Note though, that sand volcanoes produced by earthquakes in similar terrain (e.g. Tangshan, 1976; Imperial Valley, 1979) often lie along linear traces, indicating probable near-surface control of effusion by planar fractures (Leivas et al., 1980).

Kern County Earthquake, 1952

This large event (M_s7.5), involving left-lateral reverse slip, produced a complex surface rupture zone extending for some 35 km along the White Wolf Fault (Buwalda and St. Amand, 1955) (Figure 1). Regular monitoring of stream flow in the region, coupled with the earthquake's occurrence in high summer with no significant precipitation in the following two months, made it possible to quantify post-seismic groundwater pertubations to an unusual extent. Noteworthy changes in springflow and well water level

occurred up to 130 km from the epicentre (Briggs and Troxell, 1955) while casing pressures in oil wells some 12 km south of the fault rose dramati-cally in the days after the event and then de-clined over a period of about a fortnight (Johnston, 1955).

Of particular interest, however, are the out-pourings of groundwater in the region around the rupture (Figure 1) where almost all springs and gauging stations showed a marked increase in flow immediately after the earthquake. In the few cases where flow decreased, it was frequently noted that new springs had broken out in the neighbourhood (Briggs and Troxell, 1955). The following points are of particular importance.

(i) It is clear that this post-seismic effusion cannot be attributed to the consoli-dation of surficial sediments, as most of the springs occur in the crystalline rocks of the Sierra Nevada Batholith.

(ii) Spring flow generally increased very rapidly after the earthquake (certainly within a few hours in some instances); this rapid response suggests that channel flow was dominant. In this regard, it is worth noting that the crudely orthogonal drainage pattern probably follows a network of major fractures in the batholith.

(iii) Excess discharge could be recognised for at least two months after the earthquake (and in some cases for over a year) until in-creasing precipitation made discrimination difficult. The measured outflow attributable to the earthquake totalled c. 6×10^9 litres over the two-month period; a figure of at least 10^{10}

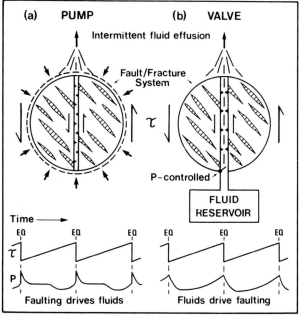

Figure 2. 'Pump' and 'Valve' models for fluid outflows following shallow crustal earthquakes (EQ).

litres seems reasonable for the total excess discharge over the same interval.

(iv) Significantly, more than half the measured excess discharge was from Caliente Creek, near the end of the surface fault break where aftershock activity was also at its most intense (Richter, 1955). This is in line with theoretical analyses concerning the redistribution of pore fluids as a consequence of inhomogeneities in the stress field at the end of a rupture (Nur and Booker, 1972; Weertman, 1974).

(v) The area of intensive outflow is about 1000 km^2. Accepting a total discharge of 10^{10} litres, the dilatant strain relieved $\Delta V/V = 10^{-5}z^{-1}$, where z is the depth of the dilatant zone in kilometres. For z = 1 to 10 km, $\Delta V/V$ ranges between 10^{-5} and 10^{-6}, values consistent with other estimates of the dilatant strains associated with shallow earthquakes (Scholz et al., 1973).

Interpretative Models

The effusion following the Kern County Earthquake can be readily interpreted in terms of

Figure 3. Vein-fill in cross-fault, Gower Peninsula.

Figure 4. Steeply dipping normal faults associated with vertical extension veins infilled with fibrous gypsum, Konarak.

the dilatancy/fluid diffusion hypothesis, with the fluid expulsion resulting from the collapse of a broad network of dilatant fractures enclosing rigid crustal blocks, in response to the decrease in deviatoric stress accompanying slip. Essentially the fault/fracture system is acting as a 'pump' with aqueous fluids being slowly drawn into the dilatant fracture system as elastic strain accumulates, to be rapidly expelled in a pulse every time slip occurs on the main fault (Sibson et al., 1975) (Figure 2a).

From a strictly empirical viewpoint, other interpretations for post-seismic effusions are possible and, in some circumstances, even probable. A diametrically opposed model is based on the premise that a fault/fracture system may act as a self-sealing 'valve' on a fluid reservoir (Figure 2b). In such a model, it is the slow build-up of fluid pressure that periodically induces fault slip, rather than the other way around. Once failure has occurred (in accordance with equations (1) and (2) below) freshly created fracture permeability on the fault and allied fractures allows fluids to escape. Pressure in the reservoir drops, progressive deposition of hydrothermal minerals re-seals the fracture system, and the whole cycle may be repeated. Self-sealing of fracture systems by hydrothermal mineralisation is well known in geothermal steam fields (Facca and Tonani, 1967) where the search for permeable production zones often has to be concentrated on the most recently active faults (Grindley and Browne, 1976). However, the process is not restricted to the upper levels of the crust; Ramsay (1980) has described high aspect ratio extension veins showing multiple increments of opening and sealing (>500 in one case) from rocks deformed under upper greenschist facies conditions, corresponding to depths probably well in excess of 10 km. Environments where the pressurised fluid reservoirs needed for this 'valve' mechanism are likely to occur include regions of active high-level plutonism and areas where prograde regional metamorphism with progressive dewatering is going on at depth (Fyfe et al., 1978).

Vein Systems associated with Faults

Hydrothermal vein systems are common in the upper, frictional regimes of fault zones occupying perhaps the top 10-15 km of continental crust (Sibson, 1977). Because they are sometimes of economic importance, detailed descriptions of vein systems, their textures and filling history are widespread in the mining literature (e.g. Bateman, 1950; Moore, 1975; Park and McDiarmid, 1975). A general inference from such vein textures is that hydrothermal mineralisation is usually episodic, with pulses of fluid associated in time with increments of shear and extensional displacement on fracture systems (Newhouse, 1942; Sibson et al., 1975). Examples which are of particular interest, because their tectonic setting is analogous to the Kern County rupture, are the Mother Lode and Grass Valley gold-quartz deposits (Knopf, 1929; Johnston, 1940). However, major fault zones often have a lengthy and complex movement history which makes interpretation of associated veining difficult. As a consequence, we restrict ourselves here to consideration of vein systems accompanying faults with fairly low finite displacements (tens of metres at most), and their implications for fluid flow in relation to the failure process.

Vein-fills in Cross-faults, S. Wales

In the Gower Peninsula of South Wales a number of steep cross-faults cut at high angles across upright, open, E-W trending Hercynian folds in an Upper Palaeozoic sequence dominated by Carboniferous limestones. George (1940) has demonstrated that the shear displacements on the faults were contemporaneous with fold growth. Rotational movements have occurred, so that slip vectors vary over the fault surfaces, though strike-slip usually predominates. Displacements range up to several hundred metres but are generally an order of magnitude less. Stratigraphic and other considerations suggest that the folding and faulting

took place at depths of 3 to 7 km (Garton, 1977).

Vein-fills of calcite with subordinate haematite occupy many of the faults; their textures record a complex movement history and the passage of large volumes of hydrothermal fluid. Comparatively little veining occurs off the faults. In the example illustrated (Figure 3) which occurs just west of Limeslade Bay (Grid reference: SS625871), an early history of banded crustification was followed by episodic brecciation and recementation of the vein-fill. The initial banded crustification with inwardly terminated crystals is characteristic of intermittent crystal growth from the walls of a progressively opening fissure (Park and McDiarmid, 1975). Thus it seems probable that this fault originated as a hydraulic extension fracture (Secor, 1965), opening incrementally in the direction of the least principal compressive stress which was periodically exceeded by the fluid pressure (see equation (5) below). At least 9 episodes of extensional opening can be recognised. Later, a change in stress field orientation led to a component of shear stress acting along the fracture. Episodic shearing, brecciation and recementation of the vein-fill then occurred, the total slip probably being of the order of 10m.

Thus in this case the 'valve' analogy seems to hold. Periodic increases in fluid pressure (perhaps arising from deformation and dehydration reactions proceeding at depth) have induced increments first of extension, and then of shear, across the same fracture. Each incremental movement was followed by the passage of hydrothermal fluids until self-sealing took place by mineral deposition.

Veining associated with Normal Faults, S.E. Iran

At Konarak, near Chah Bahar on the Makran coast of S.E. Iran, a number of E-W striking normal faults, mainly downthrown to the north though sometimes occurring in conjugate sets, disrupt a flat-lying sequence of Neogene shales and sandstones exposed in cliffs beneath a flight of late Quaternary raised beach deposits (Figure 4). The throw on individual faults, which dip 50-70° north, ranges up to a few metres and fault surface striations indicate almost pure dip-slip. One of the faults, at least, displaces the raised beaches (Vita-Finzi, 1979).

For about a hundred metres or so each side of the faults, the sediments are cut by innumerable vertical, parallel-striking extension fractures infilled with fibrous gypsum. The orientation of these tension gashes indicates that they developed in the same stress field as the normal faults with the maximum compressive stress (σ_1) vertical and the least compressive stress (σ_3) horizontal at right angles to their strike, but nowhere do they cross the faults. Growth of gypsum fibres appears to have occurred from the walls of the veins and a median parting is usually evident (Figure 5). Sometimes there is

Figure 5. Detail of extension vein, Konarak.

evidence for more than one episode of progressive extension. This extensional veining is most intensely developed in the more shaly units (Figure 6), where the lateral extension (and thus the volumetric strain) may locally exceed 10^{-2}, but diminishes going away from the faults.

Gypsum is also locally present on some of the faults, though the fibres are then invariably bent, broken and smeared-out along the slip direction. Elsewhere, however, clastic sandstone dykes are developed initially running sub-parallel to (and cutting across) the gypsum tension gashes before feeding onto the faults (Figure 7).

From the field evidence, this fault/vein system can be readily interpreted in terms of the 'pump' model. The array of gypsum-infilled extension fractures is inferred to have developed in a phase of pre-slip fracture dilatancy, presumably by hydraulic fracturing. When slip occurred, the sudden collapse of dilatant fractures led to expulsion of water and the mobilisation of sand into clastic dykes. Note, though, that because of their fibrous infilling, only partial closure of the extension fractures could occur so that relief of the dilatant strain was incomplete. From the presence of gypsum (implying temperatures less than c.50°C - Heard and Rubey, 1956) and the observation that at least one of the faults affects the capping sequence of raised beach deposits, it is apparent

that this fault/vein system developed under near-surface conditions. It is interesting to speculate whether the collapse of similar near-surface arrays of dilatant fractures may in some cases be responsible for the aligned sand blows accompanying earthquake faulting, rather than pervasive liquefaction.

Veining Around a Reactivated Fault, Anglesey

At Ogof Gynfor on the north coast of Anglesey in N. Wales (Grid reference: SH 378948), finely bedded black shales overlying quartzite breccia-conglomerate, all of Ordovician age, rest unconformably on melange rocks of the Mona Complex (Barber and Max, 1979). Caledonian deformation has induced a sub-vertical, E-W striking cleavage and the whole succession is disrupted by a series of faults which also strike roughly E-W (Bates, 1974).

The particular fault under consideration dips north at a moderate angle (Figure 8). Its detailed structure is fairly complex (Callow, 1975) but aspects relevant to this discussion are summarised schematically in Figure 9. Finite reverse slip across the fault is in excess of 20 m, but there is evidence suggesting that the reverse movements were preceded by a phase of normal dip-slip. First, a tongue of black shale has been smeared down the footwall for well over a metre below the subhorizontal shale/conglomerate boundary, indicating normal shear. Secondly, two dominant sets of extension fractures infilled with fibrous quartz occur in association with the fault, and can be related to the two postulated movement phases. Sparsely developed subhorizontal extension veins, formed in a stress regime compatible with the reverse movements (i.e. σ_3 vertical), cut across vertical extension veins whose orientation, striking sub-parallel to the fault, is consistent with σ_3 horizontal and normal slip (Figure 10). This first-formed vertical veining is comparatively strongly

Figure 6. Intense array of extension veins, Konarak.

Figure 7. Clastic dyke cutting across gypsum extension veins and injecting onto fault, Konarak.

developed with clusters of extension veins, or individual veins showing multiple increments of opening (Figure 11), occurring every few metres across strike. The subsequent reverse shear has deformed these veins close to the fault, with quartz-fibre slicken-sides developing along the fault itself.

In both sets of veins the textures are again characteristic of fissure filling, with syntaxial growth of fibrous quartz from the walls of the extension fractures (cf. Durney and Ramsay, 1973). Although in this case there is no specific evidence that collapse of dilatant extension fractures accompanied discrete slip episodes, it seems clear nonetheless that repeated hydraulic fracturing has accompanied both periods of fault movement, the process being most intense during the earlier phase of normal slip. Controls on the deformation environment are poor. However, the presence of a pressure solution cleavage which intensifies close in to the fault and incipient plastic deformation of quartz fibres within the veins suggests temperatures > 250°C (Sibson, 1977) and an overburden, perhaps arising from tectonic thickening, of at least 7 km assuming 'normal' geothermal gradients.

Controls on Hydraulic Fracturing Associated With Faults

In the last two examples, the orientation, textural characteristics and high aspect ratio of the extension veins have led to the supposition that the vein arrays developed by repeated hydraulic fracturing in the same stress regime as the faults. Where such extensional veins abut onto faults, they are known as 'feather veins', descriptions of such features being widespread in the geological literature (e.g. Dennis, 1967). It is necessary, therefore, to examine the conditions under which cyclic 'hydro-fracture' dilatancy can occur adjacent to an intermittently slipping fault. The case of an existing fault and extension fracture array in otherwise homogeneous, isotropic rock is considered (Figure 12a). We assume first that the simple law of effective stress (Hubbert and Rubey, 1959) holds, so that effective normal stresses are given by

$$\sigma'_n = \sigma_n - P \qquad (1)$$

Figure 8. Reactivated fault at Ogof Gynfor.

where P is the pore fluid pressure. The effective principal compressive stresses are thus $\sigma_1' > \sigma_2' > \sigma_3'$ and failure criteria can be represented on a modified Mohr plot of shear stress, τ, versus σ_n' (Figure 12b). In the compressional field, shear failure along the fault is taken to be governed by a linear frictional criterion of Coulomb form,

$$\tau = C + \mu\sigma_n' \qquad (2)$$

where C is the cohesive strength of the fault and the coefficient of friction $\mu = \tan\phi$, ϕ being the angle of friction. The fault is also assumed to be oriented at the optimum angle to σ_1 for frictional sliding, so that $\theta_f = 45° - \phi/2 = \frac{1}{2}\tan^{-1} 1/\mu$. Equation (2) may be rewritten in terms of the effective principal stresses at failure giving,

$$\sigma_1' = 2\sqrt{K}C + K\sigma_3' \qquad (3)$$

where $K = (\sqrt{1+\mu^2} + \mu)^2$ (Sibson, 1977). A reasonable value of $\mu = 0.75$ gives $\phi = 37°$, $\theta_f = 27°$ and $K = 4$. The extension fractures are assumed to have a 'healed' tensile strength (t) which is less than the tensile strength of the intact rock (T) and is also small in comparison with C (strictly $t < \frac{1}{2}C$ for $\mu = 0.75$), so that shear

Figure 10. Subhorizontal extension veins cutting vertical extension veins, Ogof Gynfor. Both sets are infilled with fibrous quartz.

failure always occurs in the compressional field. Thus the condition for re-opening of the extension fracture is

$$\sigma_3' = -t \qquad (4)$$

or

$$P = \sigma_3 + t \qquad (5)$$

The first essential point to be made is that hydraulic fracturing requiring $\sigma_3' < 0$, can only occur when the shear failure envelope makes a positive intercept with the τ-axis (i.e. C>0). Thus, given the above conditions, hydraulic extension fracturing should not develop adjacent to cohesion-less faults. However, natural faults are likely to possess a small cohesive strength through a combination of surface roughness and hydrothermal cementation. Its value is unlikely to exceed 10 bars (Hoek and Bray, 1977) unless cementation is marked, in which case it may approach the cohesive strength of intact rock, perhaps 100 bars or so.

Differential Stress Levels

An important consequence, if the opening of hydraulic fractures accompanies faulting, is the constraint thereby placed on the prevailing level of differential stress. From equations (3) and (4), the critical differential stress for simultaneous shear failure and hydraulic extension fracturing, as represented in Figure 12b, is

$$(\sigma_1 - \sigma_3) = 2(C - \mu t)(\sqrt{1+\mu^2} + \mu) \qquad (6)$$

and when $\mu = 0.75$

$$(\sigma_1 - \sigma_3) = 4C - 3t \qquad (7)$$

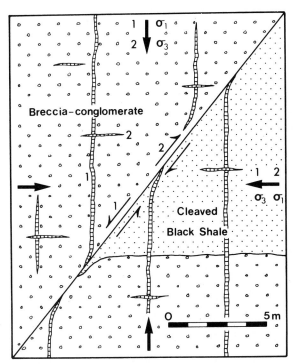

Figure 9. Schematic diagram of fault and associated veining at Ogof Gynfor (width of veins exaggerated). (1) Period of normal fault slip; (2) Period of reverse slip.

Figure 11. Vertical extension vein of quartz, showing multiple increments of opening, Ogof Gynfor.

Above this critical stress hydraulic fracturing is not possible and only shear failure can occur. Thus, accepting C = 10 bars and taking the extreme case when t → 0, the maximum differential stress allowing hydraulic fracturing is just 40 bars. For intact rocks or a hydrothermally cemented fault, where C may be of the order of a hundred bars or so, the maximum stress difference permitting hydraulic fracturing is correspondingly greater.

Hydraulic Fracturing in Different Stress Regimes

Using the boundary condition imposed by the earth's surface, which generally constrains one of the three principal stresses to be vertical, Anderson (1951) recognised three basic stress regimes with associated faulting modes. Thrust faults occur when the vertical stress $\sigma_V = \sigma_3$, strike-slip faults when $\sigma_V = \sigma_2$, and normal faults when $\sigma_V = \sigma_1$. Here we examine the values of the pore-fluid factor ($\lambda_V = P/\sigma_V = P/\rho gz$, ρ being the average rock density, g the gravitational acceleration, and z the depth) required to allow hydraulic fracturing in association with thrust, strike-slip and normal faulting at different depths in the crust. Hubbert and Willis (1957) and Secor (1965) pointed out that λ_V values should be respectively greater and less than unity for hydraulic fractures formed in association with thrust and normal faults, but their analysis did not take account of the limiting differential stress condition expressed in equation (6).

For thrust faults, since $\sigma_V = \rho gz = \sigma_3$, the condition for opening of hydraulic fractures (5) can be written

$$\lambda_V \, \sigma_V = \sigma_V + t \qquad (8)$$

giving

$$\lambda_V = 1 + t/\rho gz \qquad (9)$$

so that $\lambda_V \geqslant 1$, decreasing with depth if t is non-zero.

In the case of hydro-fracturing associated with normal faulting, $\sigma_V' = \sigma_1'$ and $\sigma_3' = -t$, so from equation (3),

$$\sigma_V' = \rho gz(1-\lambda_V) = 2\sqrt{K}C - Kt \qquad (10)$$

With strike-slip faulting, we specify the particular case where $\sigma_V' = \sigma_2' = \frac{1}{2}(\sigma_1' + \sigma_3')$. Again, $\sigma_3' = -t$, so from equation (3) we obtain

$$\sigma_V' = \rho gz(1-\lambda_V) = \sqrt{K}C - \frac{1}{2}t(K+1) \qquad (11)$$

Equations (9), (10) and (11) can be used to separate potential fields of hydraulic fracturing and shear failure for the three faulting modes on plots of λ_V versus depth. In the example shown (Figure 13), likely extreme values for the failure parameters (C = 100 bars and t = 10 bars) have been chosen and the delimiting curves drawn for the top 10 km of a crust with average density $\rho = 2.55$ g/cm^3 and $\mu = 0.75$ (i.e. K = 4). The barbed side of each curve represents the field of potential hydraulic fracturing, except for the thrust regime where the curve marks conditions under which hydraulic fracturing must occur.

Fluid pressures at least equivalent to the hydrostatic head ($\lambda_V = 0.39$ in this case) may

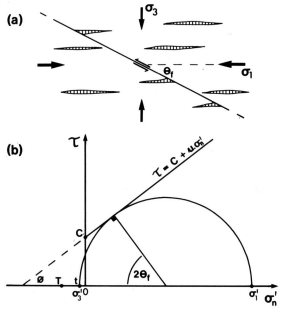

Figure 12. Hydraulic fracturing in association with faults: (a) field relationships; (b) critical stress condition for simultaneous shear failure and re-opening of extension fractures.

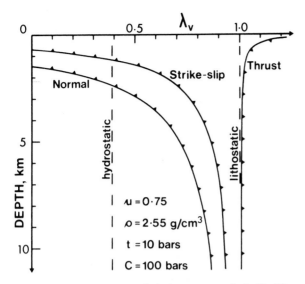

Figure 13. Curves delimiting potential fields of shear failure and hydraulic fracture (barbed side) for different faulting modes.

generally be expected in established fault zones, so that in the uppermost regions of both normal and strike-slip faults, hydraulic fractures should inevitably accompany faulting provided the faults retain some cohesive strength. This may account for the extensive fissuring often observed in association with normal faults at the surface (Abdallah et al. 1979). For hydraulic fracturing at greater depths, fluid pressures must exceed the hydrostatic head for all faulting modes. Such abnormal fluid pressures may arise through a variety of mechanisms but generally require highly impermeable 'cap' layers of some kind (Fyfe et al. 1978). However, throughout the frictional regime the required λ_v values remain significantly lower for normal faults than for thrusts, with a strike-slip regime having intermediate values. At any given depth, bearing in mind the limitations imposed by our assumption of homogeneity, the formation or re-opening of vertical extension fractures in normal and strike-slip fault terrains should also act as a regulatory mechanism, preventing λ_v from rising above the critical values once they exceed the hydrostatic head.

Discussion

Post-seismic effusions and the hydrothermal vein systems often found in exhumed fault zones provide strong evidence that, in at least some instances, intermittent channel flow of fluids accompanies faulting throughout the frictional regime. In some cases the patterns of veining and their textural characteristics accord best with the 'valve' model (e.g. Gower); in others they can be more readily interpreted in terms of

the 'pump' model, involving the development of pre-slip fracture dilatancy and its subsequent collapse at failure (e.g. Konarak, and possibly Ogof Gynfor). Note, though, that where the extension fractures have been infilled with hydrothermal minerals, relief of dilatancy can only have been partial and substantial permanent strains (locally > 10^{-2} close in to the faults) have resulted (cf. Ramsay, 1980). However, in some circumstances extension fractures may open and close without mineral deposition, leaving little or no evidence that cyclic dilatancy has occurred.

From their habit and textures, the arrays of extension veins adjacent to the faults at Konarak and Ogof Gynfor appear to be the product of repeated hydraulic fracturing. Theory suggests that such hydro-fracture dilatancy is essentially a low differential stress phenomenon which should be best developed in association with normal faults, especially at shallow depths. This seems to be borne out by the field evidence. The process is therefore in direct contrast to models of high-stress microcrack dilatancy and the views of Nur et al. (1973), who argue that pre-slip dilatancy should be most pronounced around thrust faults.

The pattern of fluid flow and pressure build-up during the earthquake stress cycle may be complex and probably varies with faulting mode. What seems clear is that for pre-slip hydro-fracture dilatancy to occur at other than shallow depths around normal and strike-slip faults, fluid pressures must be in excess of the hydrostatic head during the phase of stress accumulation, so that λ_v exceeds the critical value for the particular depth, faulting mode and failure parameters (Figure 13). Once hydro-fracturing has occurred, fluid pressures should drop towards the hydrostatic, resulting in an increase in effective normal stress and the strengthening of the fault. Shear failure then becomes possible once differential stress has risen sufficiently to overcome this dilatancy hardening effect.

Acknowledgements Part of the fieldwork leading to this paper was accomplished during the 1976 Royal Geographical Society - Imperial College expedition to the Iranian Makran. The work forms part of a project sponsored by the U.S. Geological Survey under the National Earthquake Hazards Reduction Program, Contracts No. 14-08-0001-G-377, 14-08-0001-G-466 and 14-08-0001-17662.

References

Abdallah, A., V. Courtillot, M. Kasser, A-Y. Le Dain, J-C. Lepine, B. Robineau, J-C. Ruegg, P. Tapponier, and A. Tarantola, Relevance of Afar seismicity and volcanism to the mechanics of accreting plate boundaries, Nature, 282, 17-23, 1979.

Ambraseys, N.N., Historical seismicity of North-Central Iran, Geol. Surv. Iran Report, 29, 47-95, 1974.

Anderson, E.M., The Dynamics of Faulting, Oliver and Boyd, Edinburgh, 1951.

Arieh, E., and A.M. Merzer, Fluctuations in oil flow before and after earthquakes, Nature, 247, 534-535, 1974.

Barber, A.J., and M.D. Max, A new look at the Mona Complex (Anglesey, North Wales), J. Geol. Soc. London, 136, 407-432, 1979.

Bateman, A.M., Economic Mineral Deposits, John Wiley, New York, 1950.

Bates, D.E.B., The structure of the Lower Palaeozoic rocks of Anglesey, with special reference to faulting, Geol. J., 9, 39-60, 1974.

Briggs, R.C., and H.C. Troxell, Effect of Arvin-Techapi earthquake on spring and stream flow, Cal. Div. Mines Bull., 171, 81-98, 1955.

Buwalda, J.P., and P. St. Amand, Geological effects of the Arvin-Techapi earthquake, Cal. Div. Mines Bull., 171, 41-56, 1955.

Callow, M.J.W., A field and analytical investigation of a thrust fault complex, M.Sc. dissertation, Imperial College, Univ. of London England, 1975.

Dennis, J.G., International Tectonic Dictionary, Am. Assoc. Petrol. Geol. Mem., 7, 95-96, 1967.

Dibblee, T.W., Geology of the southeastern margin of the San Joaquin Valley, California, Cal. Div. Mines Bull., 171, 23-34, 1955.

Durney, D.W., and J.G. Ramsay, Incremental strains measured by syntectonic crystal growths, in Gravity and Tectonics, eds. K.A. DeJong and R. Scholten, John Wiley, New York, 67-96, 1973.

Facca, G. and F. Tonani, The self-sealing geothermal field, Bull. Volcanologique, 30, 271, 1967.

Frank, F.C., On dilatancy in relation to seismic sources, Rev. Geophys., 3, 485-503, 1965.

Fyfe, W.S., N.J. Price and A.B. Thompson, Fluids in the Earth's Crust, Elsevier, Amsterdam, 1978.

Garton, M.R., The Oxwich Fault, M.Sc. dissertation, Imperial College, Univ. of London, England, 1977.

George, T.N., The structure of Gower, Q.J. Geol. Soc. London, 96, 131-198, 1940.

Grindley, G.W., and P.R.L. Browne, Structural and hydrological factors controlling the permeabilities of some hot-water geothermal fields, in Proceedings 2nd U.N. Symposium on Development and Use of Geothermal Resources, New York, 377-386, 1976.

Heard, H.C., and W.W. Rubey, Tectonic implications of gypsum dehydration, Bull. Geol. Soc. Am., 77, 741-760, 1966.

Hoek, E., and J.W. Bray, Rock Slope Engineering, Institution of Mining and Metallurgy, London, 1977.

Hubbert, M.K., and W.W. Rubey, Role of fluid pressure in the mechanics of overthrust faulting, Bull. Geol. Soc. Am., 70, 115-166, 1959.

Hubbert, M.K. and D.G. Willis, Mechanics of hydraulic fracturing, Trans. Am. Inst. Min. Engrs., 210, 153-168, 1957.

Johnston, R.L., Earthquake damage to oil fields and to the Paloma cycling plant in the San Joaquin Valley, Cal. Div. Mines Bull., 171, 221-226, 1955.

Johnston, W.D., The gold-quartz veins of Grass Valley, California, U.S. Geol. Surv. Prof. Paper, 194, 1940.

Knopf, A.D., The Mother Lode system of California U.S. Geol. Surv. Prof. Paper, 157, 1929.

Kovach, R.L., A. Nur, R.L. Wesson, and R. Robinson, Water-level fluctuations and earthquakes on the San Andreas Fault zone, Geology, 3, 437-440, 1975.

Leivas, E., E.W. Hart, R.D. McJunkin, and C.R. Real, Geological setting, historical seismicity and surface effects of the Imperial Valley Earthquake, October 15th, 1979, Imperial County, California, in Reconnaissance Report Imperial County, California Earthquake October 15th, 1979, Earthquake Eng. Res. Inst., 1980.

Mjachkin, V.I., W.F. Brace, G.A. Sobolev, and J.H. Dieterich, Two models for earthquake forerunners, Pure Appl. Geophys., 113, 169-181, 1975.

Moore, J. McM., Fault tectonics at Tynagh Mine, Trans. Inst. Min. Metall., 84, B141-B145,

Newhouse, W.H., Ore Deposits as Related to Structural Features, Princeton University Press, Princeton, 1942.

Nur, A., Dilatancy, pore fluids, and premonitory variations of t_s/t_p travel times, Bull. Seismol. Soc. Am., 62, 1217-1222, 1972.

Nur, A., Matsushiro, Japan, earthquake swarm: confirmation of the dilatancy - fluid diffusion model, Geology, 2, 217-221, 1974.

Nur, A., A note on the constitutive law for dilatancy, Pure Appl. Geophys., 133, 197-206, 1975.

Nur, A., and J.R. Booker, Aftershocks caused by pore-fluid flow? Science, 175, 885-887, 1972.

Nur, A., M.L. Bell, and P. Talwani, Fluid flow and faulting, 1: a detailed study of the dilatancy mechanism and premonitory velocity changes, in Proceedings of the Conference on Tectonic Problems of the San Andreas Fault System, Stanford Univ. Publ. Geol. Sci., 13, 391-404, 1973.

Park, C.F., and R.A. McDiarmid, Ore Deposits, Freemans, San Francisco, 1975.

Raleigh, C.B., J.H. Healy, and J.D. Bredehoeft, Faulting and crustal stress at Rangely, Colorado, Am. Geophys. Union Monogr., 16, 275-284, 1972.

Ramsay, J.G., The crack-seal mechanism of rock deformation, Nature, 284, 135-139, 1980.

Richter, C.F., Foreshocks and Aftershocks, Cal. Div. Mines Bull., 171, 177-198, 1955.

Scholz, C.H., L.R. Sykes, and Y.P. Aggarwal,

Earthquake prediction: a physical basis, Science, 181, 803-810, 1973.

Secor, D.T., Role of fluid pressure in jointing, Am. J. Sci., 263, 633-646, 1965.

Sibson, R.H., Fault rocks and fault mechanisms, J. Geol. Soc. London, 133, 191-213, 1977.

Sibson, R.H., J. McM. Moore, and A.H. Rankin, Seismic pumping - a hydrothermal fluid transport mechanism, J. Geol. Soc. London, 131, 653-659, 1975.

Stuart, W.D., Diffusionless dilatancy model for earthquake precursors, Geophys. Res. Lett., 1, 261-264, 1974.

Stuart, W.D., and M.S. Johnston, Intrusive origin of the Matsushiro earthquake swarm, Geology, 3, 63-67, 1975.

Tchalenko, J.S., The Kashmar (Turshiz)1903 and Torbat-e-Heidariyeh (South) 1923 earthquakes in Central Khorassan (Iran), Ann. di Geofis, 26, 29-40, 1973.

Tchalenko, J.S. and M. Berberian, The Salmas (Iran) earthquake of May 6th, 1930, Ann. di Geofis, 27, 151-212, 1974.

Vita-Finzi, C., Contributions to the Quaternary geology of Southern Iran, Geol. Min. Surv. Iran Report, 47, 1979.

Ward, P.L., Earthquake prediction, Rev. Geophys. Space Phys., 17, 343-353, 1979.

Weertman, J., Water flow paths around a dislocation on an earthquake fault, J. Geophys. Res. 79, 3291-3293, 1974.

Whitcomb, J.H., J.D. Garmany, and D.L. Anderson, Earthquake prediction: variation of seismic velocities before the San Fernando earthquake, Science, 180, 632-635, 1973.

Wu, F.T., Gas well pressure fluctuations and earthquakes, Nature, 257, 661-663, 1975.

ACOUSTIC EMISSION DURING STRESS CORROSION CRACKING IN ROCKS

Barry Kean Atkinson and Rees D. Rawlings

Department of Geology, and Department of Metallurgy and Materials Science,
Imperial College of Science and Technology, London SW7 2BP, Great Britain.

Abstract Stress corrosion of Black gabbro and
Westerley granite in air and liquid water at 20°C
has been studied using single cracks in double
torsion specimens. The acoustic response (event
rates, ring-down count rates and amplitude distributions) were continuously monitored during the
experiments. Both these materials give significant acoustic emission, probably due to extension
of the macrocrack, at even the slowest crack
velocity (10^{-9}m.s^{-1}) studied. The rate of
emission can be used as an indirect measure of
crack velocity. The acoustic response, especially the amplitude distribution of events, is a
sensitive indicator of the micro-mechanism of
crack extension, which in turn depends upon the
stress intensity factor, crack velocity and
degree of "humidity" at the crack tip. The
critical stress intensity factors for Black gabbro
and Westerley granite were respectively
2.88 MN.m$^{-3/2}$ and 1.74 MN.m$^{-3/2}$. The stress
corrosion index for Black gabbro was between 32
and 36 for tests in air or in liquid water at
high stress intensity factors, but fell to 29 for
tests in liquid water at low stress intensity
factors. The stress corrosion index for Westerley
granite was between 35 and 39 for tests in air
and liquid water. Monitoring acoustic emission/
microseismic activity in seismically active zones
could potentially be used as a means of detecting
stress corrosion in rocks and the nucleation of
earthquakes.

Introduction

The purpose of this study was to investigate
stress corrosion crack growth in two common
crustal rocks, granite and gabbro, and to establish if acoustic emission could be used to monitor
this deformation process.

At the outset of this work there were a number
of unanswered questions relating to acoustic
emission and stress corrosion cracking in rocks.
Although some preliminary studies had been made
of acoustic emission during slow crack growth in
quartz (Scholz, 1972) and glass (Byerlee and
Peselnick, 1970), Anderson and Grew (1977) were
of the opinion that the question of whether during
geological deformation rocks undergo slow crack
growth without acoustic emission had not yet been
satisfactorily answered by experiments. There is
ample evidence, however, that ceramics undergo
acoustic emission during stress corrosion cracking
(Evans and Linzer, 1973; Evans et al., 1974).

Assuming that stress corrosion of rocks is
accompanied by acoustic emission then what are
the rates and characteristics of this emission?
In particular, can the rate of emissions be
correlated with the crack velocity and the stress
intensity factor? Preliminary work by Scholz
(1972) has shown that, at least for single
crystals of quartz, the number of microseisms
per unit time is proportional to the crack
velocity during stress corrosion. Further
questions of interest are whether acoustic emission from rocks undergoing stress corrosion
ceases at very low crack velocities, and how is
acoustic emission influenced by rock type? As
will be seen later in this paper, answers to
these and other questions have been found.

Abstract reports of this work have been presented by Atkinson and Rawlings (1979a, b). See also
the abstract of Swanson (1980a) for related work.
Recently, considerable interest in acoustic
emission, crack growth and rock fracture has been
shown by a group of scientists at CIRES in
Boulder. Granryd et al. (1980) noted an enhancement of high frequency spectral components of
waveforms from acoustic events immediately prior
to rock failure. Kurita et al. (1980) showed
that the greater the initial crack density in a
rock the larger is the volumetric strain at
failure, but that there is no significant change
in fracture stress. Even in the closure of cracks
some kind of microfracturing is involved which can
generate acoustic emission (Mizutani et al.,
1980). Under hydrostatic compression some portion
of the cracks in a rock close irreversibly and at
that time local fracture occurs of asperities on
crack surfaces.

Studies of stress corrosion are of great importance in modelling earthquake sources as can be
gauged from the large number of seismic phenomena
that are believed to be controlled by this process (Scholz, 1972; Martin, 1972; Bonafede et al.,

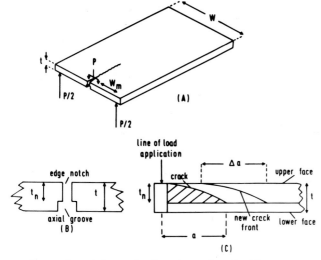

Figure 1. Schematic drawing of a double torsion specimen (A) General view. (B) Axial cross section. (C) Longitudinal cross section.

1976). Recently, Das and Scholz have developed a comprehensive theory of time-dependent rupture in the earth based upon a fracture mechanics description of stress corrosion crack growth (Das and Scholz, 1980 a,b; Scholz and Das, 1980). This theory has the encouraging feature from the standpoint of earthquake prediction that there must be a nucleation stage prior to an earthquake and also predicts its form. Laboratory studies of stress corrosion crack growth can serve to constrain our estimates of the form and duration of this nucleation phase.

Experimental Methods

The experiments described here were conducted on double torsion plates of Westerley granite and Black gabbro in air of 30% relative humidity and liquid water at 20°C. The double torsion plates were nominally sized 6cm in width by 12cm in length and were 0.4cm thick. Their bottom surfaces were grooved along a longitudinal axis to a depth of approximately 1/3 of the nominal thickness to facilitate directional control of the fracture and one end was notched along this axis by a diamond saw. A schematic drawing of the test specimen is shown in Figure 1. Specimens were deformed by four point bending at the notched end and a single mode I (tensile) fracture propagated along the specimen length. The loading frame used for this study was a commercial one made by the Instron Corporation.

The stress intensity factor for mode I cracks (the driving force acting to propagate the crack) is given by

$$K_I = \sigma_a Y c^{\frac{1}{2}} \qquad (1)$$

where K_I is the mode I stress intensity factor,

σ_a is the remote applied stress, Y is a geometrical constant and c is a characteristic crack length. For the double torsion specimen, K_I is independent of crack length and can be determined from a knowledge of the specimen dimensions, Poisson's ratio (ν) and the applied load (P). Thus

$$K_I = PW_m(3 (1 + \nu)/Wt^3t_n)^{\frac{1}{2}} \qquad (2)$$

where the previously undefined variables are as given in Figure 1. By rapidly loading a test specimen to a load close to P_c (that load associated with K_{Ic}, the critical stress intensity factor for catastrophic crack propagation), then holding the cross head of the deformation machine steady and allowing the crack to grow under a steadily diminishing applied load, a complete K_I-crack velocity (v) diagram can be constructed from the load relaxation data. Only one measure of crack length is needed. Thus,

$$v = -\phi a_f P_f (1/P^2)(dP/dt) \qquad (3)$$

where a is the crack length, f denotes measurements made at the end of a test and t is the time. ϕ is a geometrical constant. Velocity measure-

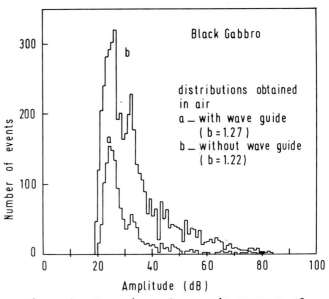

Figure 2. Comparison of acoustic response of Black gabbro obtained using the same transducer both with and without a wave guide in air. Number of acoustic emission events are plotted against amplitude in dB. The amplitude distribution coefficient, b, (see equation 4) is given for each case. Distributions were obtained by loading double torsion specimens at a cross-head speed of 0.01 cm/min. until catastrophic failure occurred.

ments are accurate to 10% over the whole of the K_I-v curves.

Specimens were precracked by loading slowly until crack extension was noted by a deviation from linearity in the load/time curve. For further details of the double torsion testing technique see Williams and Evans (1973), Evans (1972), and Atkinson (1979 a,b).

Acoustic emission was monitored using standard, commercially available Dunegan/Endevco equipment which allowed event rates, ring down count rates and amplitude distributions to be monitored simultaneously. Amplitude distributions were obtained by sorting events into 100 bins each 1 dB in width. For tests in air an acoustic emission transducer of resonant frequency 250 kHz was attached with a constant force spring to the end of the specimen furthest away from the loading points. A smear of vacuum grease was placed between the specimen and the transducer to ensure good acoustic contact. For tests in liquid water a stainless steel waveguide was used to conduct the acoustic signals out of the test chamber and into a dry atmosphere where the transducer could be used. Both the specimen and the transducer were bonded to the waveguide with quick-setting epoxy cement to ensure good acoustic contact.

Some loss of sensitivity occurred when a wave guide was used, but the form of the amplitude distribution was virtually unaltered (Figure 2). The b value in this diagram is the exponent

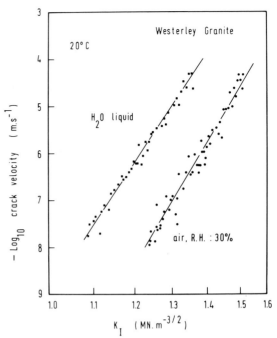

Figure 4. Plot of $\log_{10}$ crack velocity against $\log_{10}$ stress intensity factor for mode I crack propagation (K_I) in Westerley granite. Data are for material cracked in air of 30% R.H. and liquid water at 20°C.

characterising the amplitude distribution given by

$$n(a) = (a/a_o)^{-b} \qquad (4)$$

where n(a) is the fraction of the emission population whose peak amplitude exceeds amplitude a and a_o is the lowest detectable amplitude.

Test Materials

The Westerley granite used in this study was very similar to that described by Brace (1965). Modal analysis of the granite gave the following composition by volume: quartz, 27%; microcline, 36%; plagioclase, 30%; phyllosilicates (micas, clays), 6%. Total porosity was approximately 1%. Mean grain size was 0.75mm.

The Black gabbro consisted of the following constituents in decreasing proportions: plagioclase, 50%; monoclinic pyroxene, 38%; magnetite, 8%; a feldspar/quartz eutectic, 3%. Total porosity was also approximately 1%. Grain size varied from 0.4 to 1 mm.

Fracture Mechanics Results

Critical Stress Intensity Factors

Precracked double torsion specimens were loaded to failure at a fast cross head speed of 2cm.min^{-1}

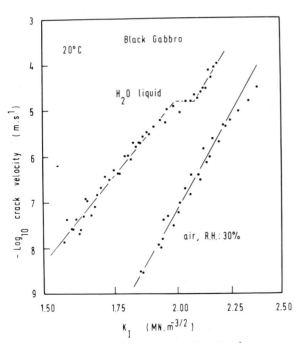

Figure 3. Plot of $\log_{10}$ crack velocity against $\log_{10}$ stress intensity factor for mode I crack propagation (K_I) in Black gabbro. Data are for material cracked in air of 30% R.H. and liquid water at 20°C.

to establish the critical load for catastrophic crack propagation. The critical stress intensity factor, K_{Ic}, was calculated from this load using equation (2). Previous studies have shown that for other rock types and ceramics increasing the rate of deformation has virtually no effect on K_{Ic} at this rate of cross head motion.

The K_{Ic} values obtained for Westerley granite and Black gabbro were respectively 1.74 ± 0.05 $MN.m^{-3/2}$ and $2.88\pm0.05MN.m^{-3/2}$. These figures are the mean and standard deviation of six tests on each material.

The result for Westerley granite agrees well with the independent estimate of K_{Ic} made by Swanson (1980b) of $1.8\pm0.1MN.m^{-3/2}$. It also agrees well with the K_{Ic} calculated by the author from Swain and Lawn's (1976) fracture surface energy data of 1.79 $MN.m^{-3/2}$. It does not agree with Schmidt and Lutz's (1980) value of $2.5MN.m^{-3/2}$ nor Zoback's (1978) estimate of $0.9MN.m^{-3/2}$.

The result for gabbro is much greater than the $0.84MN.m^{-3/2}$ found by Zoback (1978) for K_{Ic} of Academy Black gabbro. Our result is in line, however, with other determinations of K_{Ic} we have made for basic rocks (unpublished data) and confirms that they have the highest room temperature-room pressure K_{Ic} values we have so far measured in geological materials.

While allowing for some variability due to different specimens, the most likely cause of the discrepancies in the above results lies in the different experimental techniques involved. This point will be developed further in a forthcoming publication (Atkinson, in preparation).

Stress Corrosion Data

As described in an earlier section, stress corrosion data for Westerley granite and Black gabbro were obtained from load relaxation curves of double torsion specimens in liquid water and air of 30% R.H. at 20°C. The results are shown in Figures 3 and 4.

The data for gabbro show the characteristic trimodal form of similar curves obtained for glasses and polycrystalline ceramics (see for example Wiederhorn, 1974; 1978). The data for Westerley granite do not show this trimodal form in the range of our experimental observations. The data of Swanson (Swanson and Spetzler, 1979; Swanson, 1980b) suggest that bending of crack velocity/stress intensity factor curves for Westerley granite in toluene (effective R.H. = 11.3%) only occurs at crack velocities of the order of $10^{-3}m.s^{-1}$. In air of 30% R.H. and liquid water a change in slope of the K_I-v curve is expected to occur at even higher crack velocities (see Figure 13 and Discussion). These velocities are outside the range of the present experiments.

Note that no stress corrosion limit was found even at crack velocities as low as ca.$10^{-9}m.s^{-1}$. The lowest crack velocities measured in this study correspond to 0.54 K_{Ic} for Black gabbro and 0.63 K_{Ic} for Westerley granite.

The crack velocity/K_I data can be described by the following equation (Evans, 1972; Atkinson, 1979b; 1980)

$$v = \alpha K_I^n \qquad (5)$$

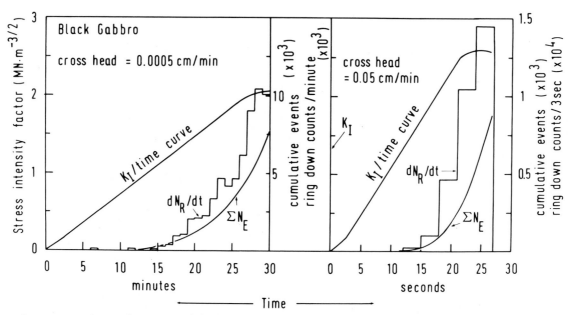

Figure 5. Stress intensity factor/time curves for double torsion specimens of Black gabbro cracked in air of 30% R.H. at two cross-head speeds: 0.05cm/min. and 0.0005cm/min. The acoustic response of the gabbro during loading is shown on the same plots: the ring-down count rate (dN_R/dt) and the cumulative number of acoustic emission events (ΣN_E).

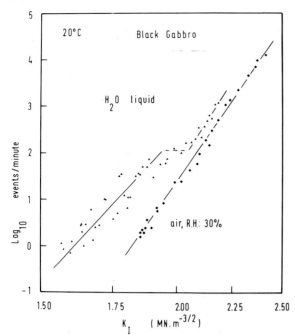

Figure 6. Log_{10} acoustic emission event rate (dN_E/dt) plotted against log_{10} stress intensity factor for mode I crack propagation in Black gabbro in air of 30% R.H. and liquid water at 20°C.

where α and n are constants. The values of n, the stress corrosion index, found in this study are given in the next section and are compared with the slopes of acoustic emission/K_I curves.

Acoustic Emission Results

Acoustic emissions in the frequency range 100 kHz to greater than 1 MHz were detected for both Westerley granite and Black gabbro. The data described below were obtained using a band-pass filter that eliminated all emissions except those that fell in the range 100 to 300 kHz.

Emissions Obtained at Different Loading Rates

In preliminary studies of acoustic emission during crack propagation in Westerley granite and Black gabbro, double torsion specimens were loaded at different cross head speeds and acoustic emissions continuously recorded. This was done both to determine the likely rate of emission during stress corrosion tests and to relate emissions to K_I/time curves. Typical results for Black gabbro in air are shown in Figure 5.

It can be seen that acoustic emission begins before significant bending of the K_I/time curves occurs, and hence before rapid crack propagation begins. Similar behaviour was noted for Westerley granite. This emission is associated with sub-critical crack propagation. For gabbro it begins at 0.35 K_{Ic} and 0.49 K_{Ic} at cross-head speeds of

respectively 0.0005 cm/min and 0.05 cm/min. Note also that the cumulative number of emissions is greater in the slower test than in the faster test. Westerley granite showed the same sort of behaviour; acoustic emission beginning at 0.39 K_{Ic} and 0.50 K_{Ic} at cross-head speeds of respectively 0.0005 cm/min and 0.05 cm/min for double torsion specimens cracked in air.

Event and Ring-Down Count Rates

Acoustic emission event rates and ring-down count rates were determined for both test materials as a function of K_I at the same time as crack velocity measurements were made during load relaxation tests. The results are shown in Figures 6 and 7 and Figures 8 and 9. These figures should be compared with Figures 3 and 4, the K_I-crack velocity diagrams.

Significant acoustic emission occurs at even the lowest K_I (lowest crack velocity) studied. Furthermore, the form of the emission rate/K_I curves for a given material is similar to the form of its crack velocity/K_I curve.

Assuming that the acoustic emission rate/K_I data can be described by similar equations to (5) the slopes can be compared with those for crack velocity/K_I curves. The acoustic emission event rate (dN_E/dt) is given by

$$dN_E/dt = \beta \, K_I^{n_E} \qquad (6)$$

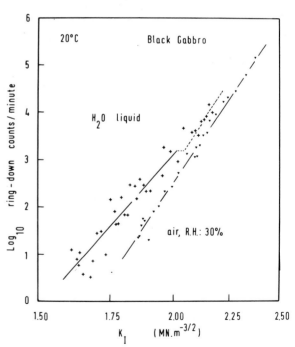

Figure 7. Log_{10} ring-down count rate (dN_R/dt) plotted against log_{10} stress intensity factor for mode I crack propagation in Black gabbro in air of 30% R.H. and liquid water at 20°C.

and the ring-down count rate (dN_R/dt) is given by

$$dN_R/dt = \gamma \, K_I^{\,n}R \qquad (7)$$

where β and n_E and γ and n_R are constants. Values of n, n_E and n_R are rather similar when determined for the same material and conditions. They are presented in Tables 1 and 2.

Because of the close similarity in the values of n, n_E and n_R the acoustic emission rates can be used as an indirect measure of crack velocities.

Amplitude Distributions

Plots of typical amplitude distributions are shown in Figures 10 and 11. Both data sets show a decrease in the exponent b in the well-known amplitude distribution power law (equation 4) with increasing K_I. Physically, this means that at high stress intensity factors the acoustic emission population is characterised by a greater number of large events than at lower stress intensity factors. Scholz (1968), Weeks et al. (1978) and Stesky (1975) have noted similar changes in acoustic emission amplitudes with increasing stress in fracture and friction experiments.

These results are presented in different form in Figure 12. It will be seen from this figure that for both gabbro and granite the b values for the wet tests are higher and show a more marked

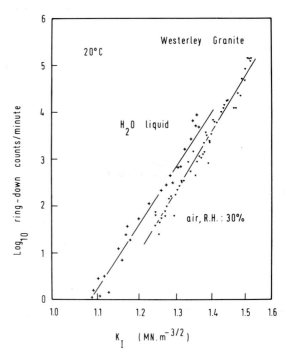

Figure 9. $\log_{10}$ ring-down count rate (dN_R/dt) plotted against $\log_{10}$ stress intensity factor for mode I crack propagation in Westerley granite in air of 30% R.H. and liquid water at 20°C.

dependence on K_I than the tests on these materials in air. The trends in the two data sets approach each other at high K_I values, however, where the influence of environment on crack propagation is expected to be less marked.

These observations are consistent with the idea that at low K_I values different crack propagation mechanisms operate in wet and dry gabbro and granite, but at high fractions of K_{Ic} the crack propagation mechanism that is active in dry rocks also controls the behaviour of wet ones.

In this context it is noteworthy that K_I values at which the b values for wet and "dry" gabbro are similar cover a range over which the slopes of K_I-v diagrams for wet and "dry" gabbro are also similar. The b values diverge over a range of K_I where the K_I-v curves have quite different slope.

The K_I-v data for Westerley granite fall in the range of K_I values for which b is relatively constant. The slopes of these K_I-v diagrams are also rather similar.

Deformation Mechanisms

For both Westerley granite and Black gabbro a decrease in stress intensity factor was accompanied by an increase in the proportion of intergranular as opposed to intragranular fractures. A similar observation was made by Swanson (1980b)

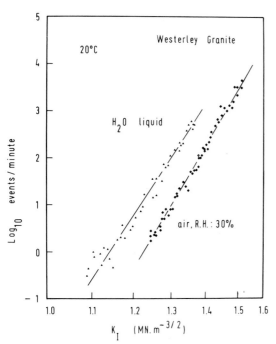

Figure 8. $\log_{10}$ acoustic emission event rate (dN_E/dt) plotted against $\log_{10}$ stress intensity factor for mode I crack propagation in Westerley granite in air of 30% R.H. and liquid water at 20°C.

Table 1. Values of n, n_E and n_R for Black gabbro

| | Air (30% R.H.) | Water | |
		Stage 3 (High K_I)	Stage 1 (Low K_I)
n	32.1 (0.989)	36.1 (0.928)	28.6 (0.977)
n_E	34.3 (0.996)	36.7 (0.940)	26.8 (0.941)
n_R	35.0 (0.992)		26.3 (0.965)

*Figures in brackets are correlation coefficients.

in his study of Westerley granite. He found that the greatest changes took place in the crack path when quartz grains were encountered. Our study confirms this observation. Following Swanson (1980b) we can explain this behaviour by noting that the grain boundaries will be the main access routes for moisture in the test specimens. At low K_I values they will stress corrode before even the more highly stressed intragranular cracks. Thus increasing the probability of the microfracture following the grain boundary.

Scanning electron microscopy of fracture surfaces of Black gabbro showed that there is a noticeable difference between the fracture of plagioclase feldspar under wet and dry conditions at slow crack velocities. The feldspar as well as being heavily twinned contains alteration bands of sericite particles. In water the transgranular fracture of the feldspar is located preferentially down these bands. Transgranular fracture of feldspar in air, however, takes place on cleavage planes inclined to these bands.

Discussion

Sources of Acoustic Emission

There are potentially four main sources of acoustic emission in these experiments. They are: (1) increments in the primary crack; (2) subsidiary microcracking in the process zone around the main crack tip; (3) dislocation motion; and (4) twinning. Sources (3) and (4) can probably be neglected for experiments at room temperature, leaving sources (1) and (2) to consider.

Acoustic emission from the primary crack is likely to depend on the crack growth rate

Table 2. Values of n, n_E and n_R for Westerley granite

	Air (30% R.H.)	Water
n	39.1 (0.968)	34.8 (0.987)
n_E	39.6 (0.991)	31.1 (0.938)
n_R	38.8 (0.972)	35.8 (0.969)

*Figures in brackets are correlation coefficients

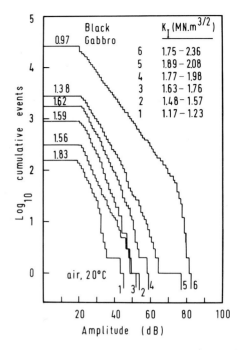

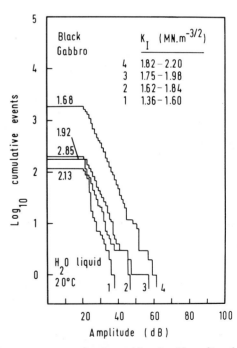

Figure 10. Typical amplitude distributions for double torsion specimens of Black gabbro cracked (A) in air of 30% R.H., and (B) in liquid water at 20°C. The stress intensity factor is given in the column on the right of the figure and the b-value (equation 4) associated with each curve is indicated.

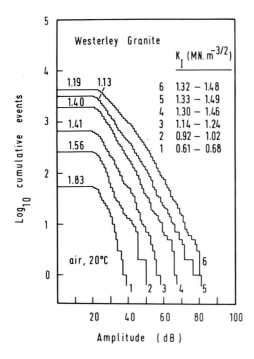

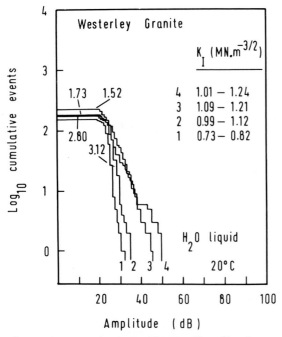

Figure 11. Typical amplitude distributions for double torsion specimens of Westerley granite cracked (A) in air of 30% R.H., and (B) in liquid water at 20°C. The stress intensity factor is given in the column on the right of the figure and the b-value (equation 4) associated with each curve is indicated.

(Evans et al., 1974). Thus,

$$dN/dt \propto \{(da/dt)/\Delta a\} \ln \{\psi K_I (\ell \Delta a)^{\frac{1}{2}}\} \quad (8)$$

where Δa is the crack length increment producing the emission, ℓ is the length of the crack front, ψ is a constant depending upon the threshold voltage of the transducer and the proportion of strain energy converted into elastic energy, da/dt is the crack growth rate and dN/dt is an event or ring-down count rate. When the background emission rate is exceeded, Evans et al. (1974) find that

$$dN/da \propto \ln K_I \quad (9)$$

From an analysis of acoustic emission from subsidiary cracks forming at the perimeter of the process zone at a macrocrack tip in ceramics Evans et al. (1974) find that in this case

$$dN/da \propto q K_I^2 \quad (10)$$

where q is the density of microcrack sources. Although Evans et al. (1974) neglected the time dependence of microcrack formation stress and the statistical nature of grain size and stress at the process zone perimeter, both of which should influence the result, even if these terms were included they are unlikely to modify the strong K_I^2 dependence of dN/da. Subsidiary microcracking may be a more important contributor to acoustic emission than macrocrack extension if there is a high density of microcrack sources, q, in the process zone.

If extension of the primary crack is the main source of acoustic emission in these experiments then the exponents n, n_E and n_R should be virtually the same. This is what we find for both Black gabbro and Westerley granite. It strongly suggests that acoustic emission from the process zone in these experiments is small compared to that deriving from macrocrack extension, probably because the process zone is itself small.

A summary schematic drawing of the relation between crack velocity, acoustic emission, stress intensity factor and "humidity" is shown in Figure 13.

Evans and Linzer (1973) have noted that acoustic emission during subcritical cracking in porcelain in water at 20°C can also be accounted for mainly by macrocrack extension. Emission from Lucalox alumina under similar deformation conditions, however, seems to be dominated by microcracking in a process zone (Evans et al., 1974). By locating the sources of emission in Westerley granite Swanson (1980a,b) was able to establish that some acoustic events, probably associated with microcracking, occurred ahead of the macrocrack tip when crack velocity exceeded $10^{-5} m.s^{-1}$. These events were not detected at lower crack velocities.

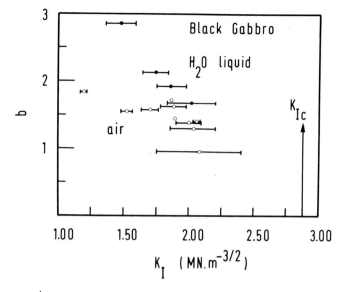

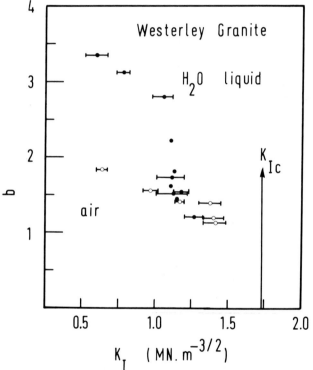

Figure 12. Synoptic diagrams showing the variation in the amplitude distribution parameter b (equation 4) with stress intensity factor for tests in liquid water and air of 30% R.H. at 20°C. (A) Black gabbro, (B) Westerley granite.

Attenuation at Water/Rock Boundary

It is of considerable interest to determine whether the differences we have noted in the acoustic response of gabbro and granite during

stress corrosion are due to changes in crack propagation mechanism or to attenuation at the boundary between the specimen and its medium.

According to Pollock (1973 a,b) the number of ring-down counts, N_R associated with N_E events with amplitudes in excess of V_o is given by

$$N_R = (N_E f\tau/b)(V_o/V_t)^b \qquad (11)$$

for $b > 0$, where f is the resonant frequency of the transducer, τ is the decay time of the envelope (determined by attenuation in the structure and the transducer), V_t is the threshold amplitude. If we wish to compare the acoustic response of a material at two gains then we can write

$$N_{R(1)}/N_{R(2)} = \{(N_E f\tau/b)(g_{(1)})^b\}_{(1)}/ \\ \{(N_E f\tau/b)(g_{(2)})^b\}_{(2)} \qquad (12)$$

where $g_{(1)}$ and $g_{(2)}$ are gains. If the only parameter changed is the gain and b = 1 then this equation reduces to

$$N_{R(1)}/N_{R(2)} = g_{(1)}/g_{(2)} \qquad (13)$$

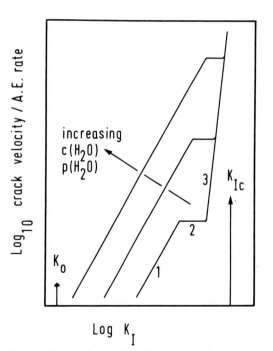

Figure 13. Schematic diagram depicting the stress corrosion and acoustic emission behaviour expected for granite and gabbro as a function of "humidity" at the crack tip in the light of the present experiments. Figure shows $\log_{10}$ crack velocity/acoustic emission rate plotted against $\log_{10}$ mode I stress intensity factor.

Square wave pulses were generated with a pulse generator and transmitted to specimens via a wave guide. Acoustic emissions were detected at the other end of a specimen via a transducer and wave guide. Pulses were generated with different frequencies from 300 Hz to 2kHz but with a fixed pulse width of 50 microseconds. Ring-down counts and events were counted for fixed time intervals both with the specimen in air and then in liquid water.

Typical results are shown in Figure 14. It can be seen that the presence of water at the specimen boundary rather than air results in an effective reduction in gain of 10dB. In other experiments the reduction in gain was as low as 7dB. Although the total number of ring-down counts are very different for the two environments, the total number of events are rather similar. Similar results were obtained for both gabbro and granite.

If we apply equation (13) to these results then for a reduction in gain of 7 - 10dB the ratio $N_{R(air)}/N_{R(water)}$ should fall in the range 2.2 to 3.2. Which is close to the ratio observed in the pulsing experiments.

For our acoustic emission studies of crack propagation in rock if the only difference between the results for air and water was an effective difference in gain of 7 - 10dB then: (a) the amplitude distribution for water would be displaced by 7 - 10dB to lower levels compared to the result for air; (b) the b-values would remain the same; (c) there would be less events observed for tests in water; and (d) fewer ring-down counts would be observed in water.

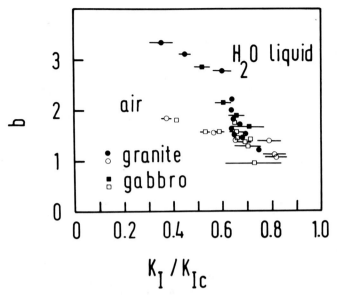

Figure 15. Plot of the amplitude distribution parameter b, (equation 4) versus stress intensity factor normalised with respect to K_{Ic}, the critical stress intensity factor for catastrophic crack propagation for Black gabbro and Westerley granite. Data are for experiments in air of 30% R.H. and liquid water at 20°C.

Our observations on Black gabbro show the following, however: (a) amplitude distributions for water are displaced to lower levels of dB than predicted from pulsing experiments (up to 20dB); (b) b-values for tests in air and water are different (except at high fractions of K_{Ic}); (c) for a given K_I more events were noted in water than in air; (d) for a given K_I more ring-down counts were recorded in water than in air. Moreover, if the difference in effective gain due to attenuation at the rock/medium interface is taken into account then the difference between the number of ring-down counts in air and water would be even more marked.

For Westerley granite we note the following: (a) amplitude distributions for water are displaced to levels of dB that are broadly predicted from pulsing experiments; (b) b-values for tests in air and water are similar over the range of K_I-v data (but not at low fractions of K_{Ic}); (c) for a given K_I more events were noted in water than in air; (d) for a given K_I more ring-down counts were recorded in water than in air.

These results strongly support our contention that different crack propagation mechanisms operate in water and in air in gabbro, except at high values of K_I. At high K_I in gabbro and over the range of K_I - crack velocity data obtained for granite probably the same crack propagation mechanism occurs for a given material in water and in air. However, this mechanism may well occur to

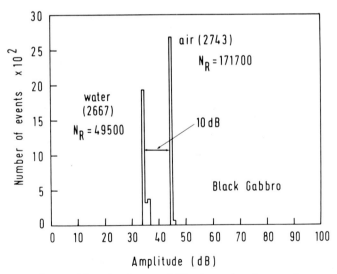

Figure 14. Apparent shift in gain observed during pulsing experiments with Black gabbro in air of 30% R.H. and in liquid water. Reduction in amplitude arises from attenuation at the specimen surface. Shown in brackets are the number of acoustic emission events recorded as well as N_R, the number of ring-down counts measured.

a greater extent in water as more ring-down counts and events were observed.

The attentuation of acoustic emission signals by the presence of water at the surface of specimens we attribute to the damping of surface waves.

Amplitude Distributions

It will be apparent from the foregoing that we believe the mechanisms of crack propagation during stress corrosion are dependent on the crack tip environment, and that the acoustic response can be used to monitor the mechanisms of crack propagation. In particular, the amplitude distribution is a sensitive indicator of crack propagation mechanism. The amplitude distribution parameter, b, should not be influenced by attenuation but only by mechanism.

In Figure 15 we show a plot of b versus K_I normalised with respect to the K_{Ic} of the material in question. There is a surprisingly good correlation between the trends in these data for the different materials. For a given material, the trends in $b/(K_I/K_{Ic})$ data depend upon crack propagation mechanism and hence on (K_I/K_{Ic}), crack velocity and the degree of "humidity" at the crack tip. Surprisingly, however, the trends in $b/(K_I/K_{Ic})$ data are also similar for different materials, presumably undergoing different crack propagation mechanisms.

To date it has not been possible for us to investigate crack growth in Westerley granite at $(K_I/K_{Ic}) < 0.6$ as this involves extremely slow crack velocities of $< 10^{-9}$m.s^{-1}. Although this area is currently under study, we cannot at this stage explain the apparent onset of different crack growth mechanisms in water and in air at these low values of K_I as implied by a change in the trends of $b/(K_I/K_{Ic})$ data (Figure 15).

Conclusions and Applications to Earthquakes

Two common crustal rocks, granite and gabbro, have been shown to undergo stress corrosion crack growth at 20°C in air and water. The K_{Ic} of Black gabbro (2.88 MN.m$^{-3/2}$) is much greater than that of Westerley granite (1.74 MN.m$^{-3/2}$).

A stress corrosion limit has not been detected in these experiments even at crack velocities as slow as ca. 10^{-9}m.s^{-1}. Moreover, by inductive reasoning based on the data of Wilkins (1980) we suggest that in Westerley granite a corrosion limit may not be encountered until crack velocities have slowed to at least 10^{-12}m.s^{-1}.

Significant acoustic emission accompanies macro-crack propagation at even the slowest crack velocities observed in this study. The mechanism of crack propagation depends upon K_I and crack tip environment. Acoustic emission rates and, especially, amplitude distributions are a sensitive indicator of crack propagation mechanism.

It follows, therefore, that if acoustic emission can be used as an analogue of seismicity in the earth then changes in amplitude distributions and seismicity rates could in principle be related to changes in crack growth mechanism, stress, crack velocities and the degree of "humidity" in the environment of crack tips.

A major problem here is the large number of seismic events needed to establish statistically meaningful changes in seismicity patterns, and the potentially short nucleation stage of some earthquakes (see Das and Scholz, 1980b). A possible way round at least the first of these obstacles is to monitor acoustic emissions in the field. Acoustic emissions with frequencies in the range 0.5 to 5 kHz have been detected with deeply buried geophones in seismically active areas (Teng and Henyey, 1981). See also Weeks et al. (1978). These emissions occur on a time-scale measured in seconds or minutes, rather than the much longer time-scale associated with low frequency seismic events.

It is possible, therefore, that stress corrosion in the earth could be detected by appropriate arrays of deeply buried acoustic emission transducers. Such arrays have been used to monitor propagating fractures in petroleum reservoir rocks (Shuck and Keech, 1977).

Finally, Fracture Mechanics data for gabbro and granite rocks, especially the stress corrosion index and K_{Ic} and the upper boundary to the stress corrosion limit, should be incorporated into analyses of earthquake source processes, such as those of Das and Scholz (1980b).

Acknowledgements This study was supported by funds from the US Geological Survey as part of the Earthquake Hazards Reduction Program, Contract Numbers: 14-08-0001-17662 and 14-08-0001-18325. Additional funds were provided by the British Natural Environment Research Council with grant GR3/3716.

References

Anderson, O.L., and P.C. Grew, Stress corrosion theory of crack propagation with application to geophysics, Rev.Geophys. Space Phys. 15, 77-104, 1977.

Atkinson, B.K., Fracture toughness of Tennessee Sandstone and Carrara Marble using the double torsion testing method, Int. J. Rock Mech. Min. Sci. and Geomech. Abstr., 16, 49-53, 1979a.

Atkinson, B.K., A fracture mechanics study of subcritical tensile cracking of quartz in wet environments, Pure Appl. Geophys., 117, 1011-1024, 1979b.

Atkinson, B.K., Stress corrosion and the rate-dependent tensile failure of a fine-grained quartz rock, Tectonophysics, 65, 281-290, 1980.

Atkinson, B.K., and R.D. Rawlings, Acoustic emission characteristics of subcritical and fast tensile cracking in rock, in Proceedings International Conference on Acoustic Emission and Materials Evaluation, Chelsea College, London, Proceedings Institute of Acoustics, 1979a (in press).

Atkinson, B.K., and R.D. Rawlings, Acoustic emission during subcritical and fast tensile cracking of Westerley granite and a gabbro, Eos. Trans. AGU, 60, 740, 1979b.

Bonafede, M., D. Fazio, F. Mulargia, and E. Boschi, Stress corrosion theory of crack propagation applied to the earthquake mechanism. A possible basis to explain the occurrence of two or more large seismic shocks in a geologically short time interval, Bolletino di Geofisica, teorica ed applicata, XIX, 377-403, 1976.

Brace, W.F., Some new measurements of linear compressibility of rocks, J. Geophys. Res., 70, 391-398, 1965.

Byerlee, J.D., and L. Peselnick, Elastic shocks and earthquakes, Naturwissenschaften, 57, 82-85, 1970.

Das, S., and C.H. Scholz, Subcritical rupture in the earth, II: Theoretical calculations, Eos Trans. AGU, 61, 305, 1980a.

Das, S., and C.H. Scholz, Theory of time-dependent rupture in the earth, J. Geophys. Res., 1980b (in press).

Evans, A.G., A method for evaluating the time-dependent failure characteristics of brittle materials - and its application to polycrystalline alumina, J. Materials Sci., 7, 1137-1146, 1972.

Evans, A.G., and M. Linzer, Failure prediction in structural ceramics using acoustic emission, J. Amer. Ceram. Soc., 56, 575-581, 1973.

Evans, A.G., M. Linzer, and L.R. Russell, Acoustic emission and crack propagation in polycrystalline alumina, Mat. Sci. and Eng., 15, 253-261, 1974.

Granryd, L.A., C.H. Sondergeld, and L.H. Estey, Precursory changes in acoustic emissions prior to the uniaxial failure of rock, Eos. Trans. AGU, 61, 372, 1980.

Kurita, K., I.C. Getting, and H.A. Spetzler, The effect of thermal cycling on volumetric strain, Eos. Trans. AGU, 61, 372, 1980.

Martin, R.J., III, Time-dependent crack growth in quartz and its application to the creep of rocks J. Geophys. Res., 77, 1406-1419, 1972.

Mizutani, H., K. Kurita, and T. Waza, Acoustic emission in hydrostatic compression, Eos. Trans. AGU, 61, 371, 1980.

Pollock, A.A., Acoustic emission amplitudes, Nondestructive Testing, 6, 264-269, 1973a.

Pollock, A.A., Acoustic emission, A review of recent progress and technical aspects, in Acoustics and Vibration Progress, vol. I. edited by R.W.B. Stephens and H.G. Leventhall, Chapman and Hall, London, 50-84, 1973b.

Schmidt, R.A., and T.J. Lutz, K_{Ic} and J_{Ic} of Westerley Granite: Effects of thickness and in-plane dimensions, ASTM STP (1980) in press.

Scholz, C.H., The frequency magnitude relation of micro-fracturing in rock and its relation to earthquakes, Bull. Seism. Soc. Am., 58, 399-415, 1968.

Scholz, C.H., Static fatigue of quartz, J.Geophys. Res., 77, 2104-2114, 1972.

Scholz, C.H., and S. Das, Subcritical rupture in the earth, I: Physics and observations, Eos. Trans. AGU, 61, 304, 1980.

Shuck, L.Z., and T.W. Keach, Monitoring acoustic emission from propagating fractures in petroleum reservoir rocks, in Proceedings First Conference on Acoustic Emission/Microseismic Activity in Geological Structures and Materials, edited by H.R. Hardy, Jr., and F.W. Leighton, Trans. Tech. Publications, Clausthal, Germany, 309-338, 1977.

Stesky, R.M., Acoustic emission during high-temperature frictional sliding, Pure Appl. Geophys., 113, 31-43, 1975.

Swain, M.V., and B.R. Lawn, Indentation fracture of brittle rocks and glasses, Int. J. Rock Mech. Min. Sci. and Geomech. Abstr., 13, 311-319, 1976.

Swanson, P.L., Observations of the fracture process in Westerley granite double torsion specimens, Eos. Trans. AGU, 61, 372, 1980a.

Swanson, P.L., Stress corrosion cracking in Westerley granite: An examination by the double torsion technique, M.Sc. thesis, University of Colorado, 1980b.

Swanson, P.L., and H. Spetzler, Stress corrosion of single cracks in flat plates of rock, Eos. Trans. AGU, 60, 380, 1979.

Teng, T., and T.L. Henyey, The detection of nano-earthquakes, this volume, 1981.

Weeks, J., D. Lockner, and J. Byerlee, Change in b-values during movement on cut surfaces in granite, Bull. Seism. Soc. Am., 68, 333-341, 1978.

Wiederhorn, S.M., Subcritical crack growth in ceramics, in Fracture Mechanics of Ceramics, vol. 2, edited by R.C. Bradt, D.P.H. Hasselman, and F.F. Lange, Plenum Press, New York, 613-646, 1974.

Wiederhorn, S.M., Mechanisms of subcritical crack growth in glass, in Fracture Mechanics of Ceramics, vol. 4, edited by R.C. Bradt, D.P.H. Hasselman and F.F. Lange, Plenum Press, New York, 549-580, 1978.

Wilkins, B.J.S., Slow crack growth and delayed failure of granite, Int. J. Rock Mech. Min. Sci. and Geomech. Abstr., 17, 365-369, 1980.

Williams, D.P., and Evans, A.G., A simple method for studying slow crack growth, J. Testing and Evaluation, 1, 264-270, 1973.

Zoback, M.D., A simple hydraulic fracturing technique for determining fracture toughness, Proceedings 19th U.S. Rock Mechanics Symposium, Stateline, Nevada, 83-85, 1978.

A SLOW EARTHQUAKE SEQUENCE FOLLOWING THE IZU-OSHIMA EARTHQUAKE OF 1978

I. Selwyn Sacks and Alan T. Linde

Dept. of Terrestrial Magnetism, Carnegie Inst. of Washington, Wash., D.C. 20015

J. Arthur Snoke

Department of Geological Sciences, VPI&SU, Blacksburg, Virginia 24061

Shigeji Suyehiro

Japan Meteorological Agency, Tokyo, Japan

Abstract. Slow earthquakes provide a mechanism for temporal nonlinearity in stress redistribution. The Izu-Oshima earthquake of 1978 was followed by a series of slowquakes which were recorded on three Sacks-Evertson borehole strainmeters. These records can be matched by modelling which invokes a sequence of slowquakes whose geometries and dimensions are constrained by the aftershock activity. Motion for up to one hour on two quasiorthogonal faults with a combined moment of 4.45×10^{25} dyne-cm account for the borehole strain data as well as levelling and triangulation data. This moment is about 0.4 that of the submarine Izu-Oshima earthquake. The largest aftershocks (m = 5.1, 5.8) occur on one of these faults after the major stress change due to the slow events.

Introduction

The development of the plate tectonic framework has enabled an increased understanding of the broad scale processes involved in the occurrence of earthquakes. The earth, however, behaves in a non-linear fashion and thus far we have a poor understanding of the details of many seismic phenomena. Of particular interest to us here is the apparent nonlinearity in stress buildup with time. Rikitake (1976) cataloged observations in which there is evidence for rapid changes (over hours or days) in the local stress field immediately preceding an earthquake; it is well known that the time scale for events in a foreshock-aftershock sequence is much longer than that necessary for elastic redistribution of stress. It is the thesis of this paper that at least some of these stress redistributions take place in the form of slow earthquakes. These slowquakes resemble normal earthquakes but have longer time scales; i.e. lower rupture velocities and/or longer rise times.

Studies of particular earthquakes (e.g., Kanamori and Stewart 1979) have led to the inference of coseismic processes with time scales longer than those normally associated with earthquakes. Because such processes are infraseismic (in the sense that their spectra are dominated by periods which lie below the passband of conventional seismograph systems), the conclusions were derived from such characteristics as anomalous long period spectral character or unusually long surface wave coda.

We have previously reported on direct observations of strain changes which occur as slow earthquakes (Sacks et al 1978, 1980). As pointed out below, these strain changes are of very long period and theoretically decrease with the third power of distance from the source. Thus recording of such data requires installation in the near source region of a highly sensitive long period instrument which is linear in its response. The data discussed in this paper (and in our earlier work) were recorded by a network of Sacks-Evertson borehole strainmeters installed by the Japanese Meteorological Agency along the seismically active Pacific coast of Honshu south of Tokyo. These instruments have large dynamic range and high sensitivity from zero frequency to several hertz (Sacks et al., 1971). Strains smaller than 10^{-11} can be observed. Vibration tests using explosions have demonstrated that the instrument behaves linearly even when subjected to high accelerations. Because these highly sensitive borehole strainmeters are installed in sites close to earthquake activity, they record reliably not only the radiated waves but also the strain steps produced by nearby earthquakes.

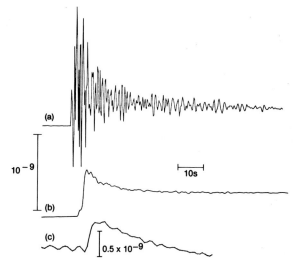

Fig. 1. Records of normal and slow earthquakes. (a) m=4.0 normal earthquake on 19 Sep 1973 at a distance of 15 km from MAT, the recording station. (b) result of low-pass filtering #(a); frequencies above 1 Hz are highly attenuated. (c) a small slowquake recorded at Irako on 25 Aug 1976. Note the similarity between (b) and (c).

In Sacks et al (1978,1980) we described a number of single station observations which we interpreted in terms of slowquakes. In one case the strainmeter recorded a sequence of foreslowquakes, main slowquake and afterslowquakes which exhibited properties similar to those for normal earthquakes (see Fig. 4 in Sacks et al 1980). Because of the rapid decrease of strain change with distance, instruments need to be close together if any but large-moment slowquakes are to be recorded by more than one instrument. Here we review the basis for our modelling and analyse data from three strainmeters which all recorded a sequence of slowquakes following the Izu-Oshima earthquake of 1978.

Normal and slow earthquake straingrams: observed and synthetic

Figure 1 illustrates the similarity between the straingram for a slow earthquake and that for a normal earthquake (after low pass filtering). We are not suggesting that the focal processes are similar; they may be related to different physical properties of the medium. Dieterich (1978), from laboratory model studies, and Das and Scholz (1980), from mathematical modelling, have suggested that slow events may be an integral part of the earthquake faulting process. The general characteristics of the resulting waveforms differ only in their time scales - typically a factor of at least 10 to 100. We can therefore use the same

synthetics programs to model both slowquakes and normal earthquakes. Our synthetics are obtained by calculating the response of a dilatometer installed at the free surface of a uniform, homogeneous, elastic halfspace to a double-couple point source (with a variable time history) located within the halfspace. (The program for calculating the dilatation is patterned after displacement-field programs developed by Johnson (1974) and Anderson (1976).)

Figure 2 shows synthetic straingrams for one event time history at two different distances both in the near field and so the strain step is not negligible compared with the radiated pulse. In the extreme near field (Fig. 2a) only the strain step is observed. As distance increases (Fig. 2b) the radiated pulse becomes more dominant because it drops off less rapidly with distance than does the strain step.

In Fig. 3 waveforms for both normal and slow earthquakes are compared with synthetics. The known location and focal mechanism were used for the normal earthquake, and the time history used was a parabolic ramp with a duration of 0.5 seconds. Because the slow earthquake was detected on only one instrument, its location, depth and radiation pattern were not known. An adequate fit is obtained by assuming a

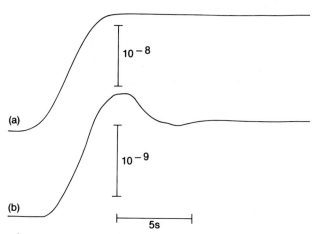

Fig. 2. Theoretical waveforms of dilatant strain vs. time for two different distances. (b) is for a distance 4 times that for (a). The source time function is a parabolic ramp with a duration of 6 seconds. The difference between the final and initial strain levels is the strain step produced by the earthquake. The magnitude of the step is inversely proportional to the cube of the distance between source and observation. The radiated pulse consists of the strain changes which take place during the arrival of seismic waves and which decrease less rapidly with distance than does the strain step. At greater distances, the radiated pulse dominates and in the far field the strain step is negligible.

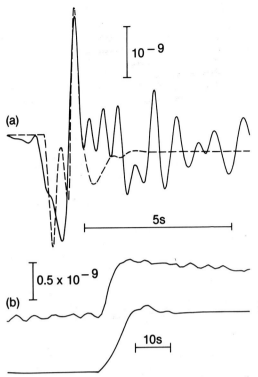

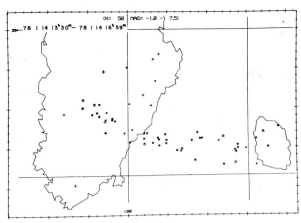

Fig. 4. Epicentral distribution of aftershocks in the period 4:30 - 7:59, January 14, 1978. Circles are earthquakes, crosses are stations. Note the relatively quiet zone inland near the coast. The pattern of onland activity suggested the F3 geometry. From Tsumura et al (1978).

Fig. 3. Comparison of data and model waveforms for (a) a normal earthquake and (b) a slowquake. The first cycle of the earthquake (MAT, depth 3 km, 0.05 degrees, m=3.4, 9 Jun 1973) is matched by a rise time of 0.24 seconds. The slowquake, upper trace in (b), (IRO, 25 Aug 1976) is modelled with a rise time of 8 seconds. Both events are modelled as vertical strike slip events.

strike-slip event with a 16 second duration. As discussed more fully by Sacks et al (1978), for such observations on a single instrument there is an unresolvable ambiguity between an event being near, slow and of a certain moment or further, slower and of a larger moment.

Izu-Oshima quake of 1978 and subsequent activity

On January 14, 1978, a large earthquake (JMA magnitude 7.0) occurred under the sea between Oshima Island and Izu Peninsula. Fault parameters for this earthquake determined from seismograph recordings of the radiated field, show that the earthquake was confined to the area between the island and the coast of Izu (Shimazaki and Somerville, 1978). However aftershock activity covered a much larger area and developed into the interior of the Izu peninsula. Figures 4 and 5 taken from Tsumura et al (1978) show the seismicity as a function of time. Note that the region near the coast remains relatively aseismic until some time after the main shock. There is also a definite

migration of the activity inland with time. In addition, crustal deformation observations require motion on a fault extending well inland (Inouchi and Sato, 1979). The interpretation of these observations was that the length and dislocation for the on-land fault are about half those for the submarine fault. Because of the small amount of vibration damage to nearby houses, Umeda and Murakami (1978) suggest that the on-land faulting motion was accompanied by only low accelerations.

The three nearest borehole strainmeters (20-60 km distance) recorded static field changes

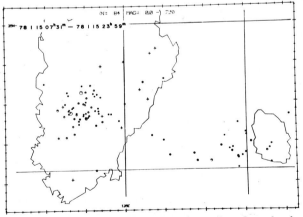

Fig. 5. Epicentral distribution of aftershocks in the period 22:31, January 14 to 14:59, January 15, 1978. The secondary aftershock region extended southwestwards (F4) following the largest aftershock at 22:31 (January 14) with m=5.8, which caused some property damage in the western part of the Izu Peninsula. From Tsumura et al (1978).

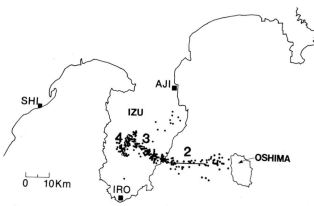

Fig. 6. Map of southern-central Honshu. Borehole strainmeters IRO, AJI, and SHI recorded strain changes due to the normal earthquake F2 (m=7.0, 14 Jan 78) and due to F3 and F4, a slowquake sequence. Epicenters of normal aftershocks of these events are also shown.

resulting from motion on the faults. Fig. 6 shows the location of these three instruments and the aftershock region of the earthquake. (Also shown are the positions of the faults as determined from seismic radiation and from crustal deformation.) Each strainmeter has two channels of output; one (DT) with a flat response from a few hertz to zero frequency, the other has higher gain and is high-pass filtered at a period of 25 minutes. These data are recorded digitally at a remote location and there is also an on-site potentiometric chart recorder. The digital data from the DT channel are used for most of the analysis presented here. Before digitizing, the data are low-pass filtered at approximately 80 sec period with a 12 db/octave fall-off, even though the sampling rate is once every 5 seconds. The station at IRO suffered power failure due to the earthquake and although local recording resumed after a break of about 4 minutes, some hours of digital information was lost. Data which we use for IRO were hand-digitized from the analogue recording and so do not cover the interval immediately following the main shock; differences in filtering between these data and the electronically digitized data are not significant for the time scales considered (see Fig. 15). The strain field changes coincident with and following the earthquake are shown in Fig. 7.

For approximately an hour following the earthquake, slow strain changes were recorded on these widely spaced (~60 km) instruments. The magnitudes of these changes are comparable to the strain steps due to the earthquake. Thus the source of this strain redistribution must have dimensions similar to those of the earthquake. These data, together with the crustal deformation observations (Inouchi and Sato, 1979) lead us to conclude that the source

was a slowquake sequence with a moment of the same order as that of the normal earthquake. All the data (slow strain changes, crustal deformation, aftershocks) provide strong evidence that these slowquakes occurred on the Izu Peninsula. Not until these slowquakes redistributed the stress did the first large (normal) aftershock (m=5.1, 77 minutes after the main normal earthquake) occur in central Izu.

Coseismic strain changes

In this paper we are primarily concerned with understanding the postseismic strain changes. Analysis of the coseismic variations is complicated by several effects, and a full discussion of these will be presented elsewhere. Here we merely indicate the nature of the problems and show that the recorded data are consistent with previously determined values for the earthquake parameters.

The low-pass filtering of the digital data is such that the signal due to the coseismic strain change does not reach full value before the recording of deformation due to the subsequent activity. In addition to the slow changes discussed below, Tsuneishi et al. (1978) have proposed, on the basis of surface traces of faulting, that there was motion on a fault northwest of Inatori. They estimated that slip of 1.28 m occurred over a fault length of 3 km

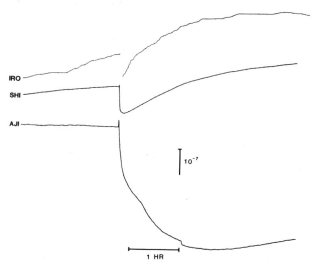

Fig. 7. Records from the 3 borehole strainmeters at locations shown in Fig. 6. Compression is down. The strain steps (the rapid upward excursions) are due to the earthquake F2. As a result of low-pass filtering and following strain changes, the recorded strain steps do not reflect the full magnitude of the coseismic strain changes. Slow strain changes of comparable magnitude occur over the following hour. We associate these slow changes with motion on the faults F3 and F4.

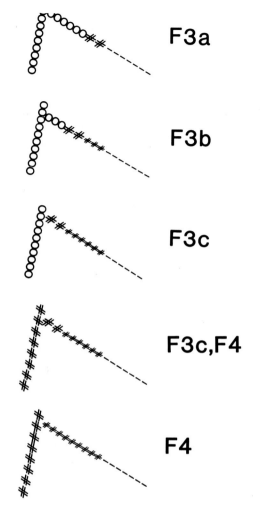

F3a

F3b

F3c

F3c,F4

F4

Fig. 8. The proposed faulting sequence on F3 and F4. The dashed part of F3 has been excluded from our analysis (see text). Circles indicate fault sections which have not yet begun to move. Large # signs represent the current activity and small # signs, the regions which have previously ruptured. The time scale for this process is indicated in Fig. 9. Note that our fault subdivision scheme is for purposes of calculational simplicity; the faulting appears to be continuous.

with a strike N40W. Since there was vibration damage in this region, it is possible that this faulting was fairly rapid and immediately followed F2. If we assume that the fault had a 5 km width, its seismic moment was about 6% that of the main quake.

The calculated strain changes at AJI due to the main shock (F2) fall in the range 1.8 to 5.2 x 10^{-7} depending on the value for the strike for F2. (AJI is near a nodal plane and the difference of 5° between the strike of Okada's (1978) model and that of the Shimazaki and

Somerville (1978) model results in the large change in the calculated strain. Further westward propagation of F2 will reduce the overall strain at AJI.) The Inatori fault adds about 1.3×10^{-7} to these values. Preliminary analysis (in which we allow for the filter and the duration of faulting) of the first 10 seconds of the AJI digital record gives a value for coseismic strain change of about 2.4×10^{-7}. At SHI, the calculated strain due to F2 is 1.8×10^{-7}; including the Inatori fault results in a total strain change of -0.2×10^{-7}. Because of the low pass filtering of the digital data the actual recorded strain would depend on the relative timing of the two faults since the strains are of opposite sense. In addition SHI is sufficiently distant that synthetic modeling of the straingram is necessary. At the time of writing we are able to say only that the coseismic strain change recorded at SHI is between -0.3 to $+0.4 \times 10^{-7}$.

Thus the initial strain changes at AJI and SHI are consistent with preliminary calculations for seismically induced strain steps. In a separate report, we will present a detailed study of the effects mentioned here and extend the study to include the strain data recorded by a number of more distant borehole strainmeters.

Slowquake modelling

In our attempts to match the strain data, we have constrained the geometries of our models for slowquakes so that they are consistent with the aftershock activity and also with parameters for onland faulting as determined by Okada (1978) and Inouchi and Sato (1978). In particular, the aftershock activity indicates an inland migration with time (Figs. 4 and 5) and this has influenced our thoughts on the sequence of events.

The technique we have used in this modelling is a pseudo-static one. We approximate a propagating fault by calculating the strain fields at different times using as sources fault sections which are at different locations. The faulting process appears to be continuous; we arbitrarily select a few times at which to calculate the corresponding strain values. Because the instrument has a flat response at long periods, the recorded data at any time is the integral of the strain changes which have occurred. We calculate incremental strains and so, as faulting proceeds, these must be added to the accumulated values computed previously. We have calculated the strain fields due to strike-slip dislocations using a modified version of Sato and Matsu'ura's (1974) routine for a double-couple source embedded in an elastic half-space. The strain values calculated at the various times are then used for comparison with the corresponding data. We concentrate on an analysis of the strain changes following the main earthquake and so we have

Table 1. Fault parameters.

Fault	Strike E of N	Dip	Slip	L km	W km	D m	Moment 10^{25} dyne-cm	Reference
F3a	302	80	8	3	7	1.0	0.74	this study
F3b	302	80	8	3	7	1.0	0.74	this study
F3c	302	80	8	3	7	1.0	0.74	this study
+ F3c	302	80	8	3	7	0.5	0.37	this study
F4	189	90	0	10	7.5	0.07	0.18	this study
F4	189	90	0	10	7.5	0.63	1.67	this study
F4(tot)	189	90	0	10	7.5	0.7	1.85	this study
F4	220			10	7.5			Okada
F3(tot)	302	80	8	9	7		2.6	this study
F3	300			16	7.5	0.5	2.1[*]	Okada
F3	300	80	8	12	8	0.71	2.4[*]	Inouchi and Sato
F3	303	75	2.4	6	6.5	1.2	1.6	Shimazaki and Somerville
F2							11.0	Shimazaki and Somerville

+   concurrent faulting
*   calculated from their parameters

restricted our analysis to the period starting 20 seconds after the earthquake.

Choice of fault parameters

All the evidence indicates that there was motion on more than one fault and that the onland faulting was slow compared with a normal earthquake. Shimazaki and Somerville (1978) analysed only the far field data which required a fault totally submarine but noted that faulting must have continued onland. Inouchi and Sato (1978), Tsuneishi et al (1978), Okada (1978) and Shimazaki and Somerville (1979) all showed that faulting continued to well within the Izu peninsula. Aftershock distributions, levelling data and distance measurements were used in their determinations of the fault parameters. We use these determinations to constrain our modelling and will use the fault labelling system employed by Okada. F1 and F2 were submarine earthquakes (according to Shimazaki and Somerville (1978), F1 was a precursor to the main shock, F2, which had a magnitude of 7.0). F3 and F4 were onland slowquakes and we match the strain data by analysing these in a piece-wise fashion (Fig.6). We use Inouchi and Sato's (1978) values for the slip and dip directions and vary the strike and moment values to obtain a good fit, but require overall consistency with the aftershock activity. The choice of the F3-F4 faulting system is supported not only by the aftershock activity but also by previous seismicity on the Izu Peninsula. The 1976 Kawazu earthquake aftershocks were essentially on the east-west F3 fault; north-south faulting also occurs as evidenced by a 1930 left lateral earthquake on the Tanna fault, although this is to the north of our F4.

Figure 8 shows schematically the development sequence for the slowquakes on faults F3 and F4. The locations are set primarily by the seismicity and we start our analysis at a time after the normal earthquake such that the strain changes are dominated by the slowquakes. Table 1 gives the parameters for these faults. The following discussion describes our analysis technique. For calculational convenience, we have artificially divided the faulting process into a number of stages (see Table 1); the last stage of motion on F3 overlaps the beginning of activity on F4. Fig. 9 shows a plot of our estimate of cumulative seismic moment for the F3-F4 faulting system and the points at which we have sampled for our analysis.

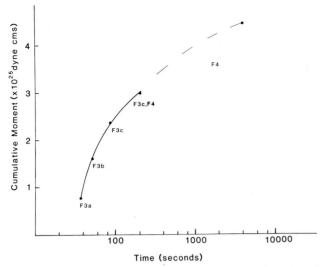

Fig. 9. Cumulative moment vs. time. The points correspond to the fault stages shown in Fig. 8. The smooth curve joining the points indicates that the process was continuous but became slower with time. The final section is dashed because we did not calculate intermediate stages for F4.

F3a Stage

The starting time for our analysis is set at 20 seconds after the normal earthquake and the starting location for faulting is at the eastern section of the onland aftershock region as developed within the first hour. This is about 8 km along strike from the coast. The dip and slip values for all sections of F3 are taken from Inouchi and Sato (1978). The fault parameters for F3a are a length of 3 km, width 7 km, slip of 1 m, strike N302E, dip 80N and a slip angle of 8 degrees. The depth to the center of the fault is 5 km. Our preferred strike direction differs from that given by Inouchi and Sato by only 2 degrees. The strain field produced by this dislocation is shown in Fig 10. AJI is very close to a nodal plane while SHI is well within the compressional quadrant. As the faulting propagates, AJI will accumulate little strain because it is near the nodal region while the strain at SHI will increase rapidly. The time domain record (Fig. 11) shows this variation.

F3b Stage

At this time, 20 seconds after F3a, we take the motion to be occuring on F3b (with parameters as for F3a). This produces a strain field similar to that from F3a and the cumulative strain is as shown in Fig. 11. The nodal plane has moved away from AJI and the station is now experiencing compressional strain.

F3c Stage

The faulting has now migrated a further 3 km. The strain pattern is shown in Fig. 12. Now AJI is well within a compressional quadrant, as is SHI, and this is consistent with the recorded data as shown in Fig. 11. The fault parameters for F3a, b and c are all identical except for location.

F3c-F4 Stage

We interpret the flattening of the SHI curve (see Fig. 11) as due to motion commencing on F4 while F3c is still active (because F3 and F4 produce strains of opposite polarities at SHI). We use the same fault orientation for F3c as above but with a slip of 0.5 m. The F4 fault is not well constrained by the aftershock activity and we have no local crustal deformation data to help in determining slip angles. The dimensions of F4 are those suggested by Okada (1978) on the basis of the seismicity. We have taken a vertical pure strike-slip left lateral fault with a strike N189E, length 10 km, width 7.5 km and, during this initial stage, a slip of 0.07 m. The depth to the center of the fault is 5 km. This, together with a concurrent slip of 0.5 m on F3c, results in a small strain at SHI while AJI continues to experience strong compression. The strain field is shown in Fig. 13 and the corresponding accumulation to the strain is given in Fig. 11.

F4 Stage

The final stage of matching to the data involves motion only on the F4 fault. It

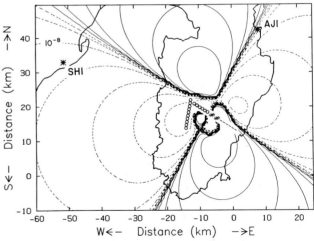

Fig. 10. Strain field produced by the F3a stage of faulting superposed on a map of the Izu Peninsula. Solid lines represent dilatation. There are two contours per decade. The fault coding scheme is as given in Fig. 8. AJI is near the nodal region while SHI is well within a compressional quadrant.

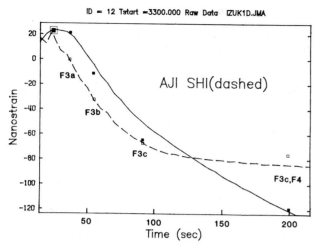

Fig. 11. Observed strain (curves) at AJI and SHI vs. time after the main earthquake. Compression is negative (down). The squares (closed for AJI, open for SHI) are the cumulative strains calculated from the model. The large symbols at t=20 s represent initial reference values. The strain levels shown in Figs. 10, 12-14 are the differences in strain between the symbols for the corresponding time and the preceding symbol.

appears that this continues for over an hour. The hand-digitized data from IRO are also used in constraining the model. The orientations for F4 are as given above, but with a slip of 0.5 m. The strain field is as shown in Fig. 14. We have compared the model with the data by using the strain field to determine the relative amplitudes at the three stations. These values

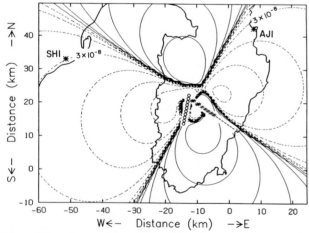

Fig. 12. As for Fig. 10. Faulting has progressed to the west end of F3 (F3c). Both stations are in compressional quadrants and this is consistent with the rapid compressional trend of the recorded data (see Fig. 11) at this time.

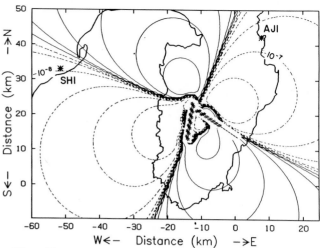

Fig. 13. As for Fig. 10. Both F4 and the western section of F3 are active. AJI is strongly compressional but, but because the separate faults produce opposite polarities at SHI, its strain rate is considerably reduced, although still compressional. This agrees well with the recorded data (Fig. 11).

are used to rescale the data such that, if the model was a perfect fit, the scaled data from the three would be identical. The original data are shown in Fig. 7 and Fig. 15 shows the rescaled data. We find the agreement to be very good. Since it is possible that more than one

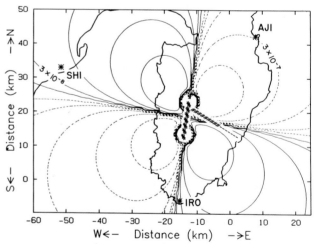

Fig. 14. As for Fig. 10. This activity (F4 alone) starts at about 200 seconds after the main quake, lasts for about 1 hour, and is continuous with the previous stage. Note that the strain at SHI has changed sign and is now dilatational. IRO is also dilatational and with somewhat larger value than SHI. AJI remains compressional and with significantly larger amplitude than the other two stations. This variation can be seen in the recorded data shown in Fig. 7.

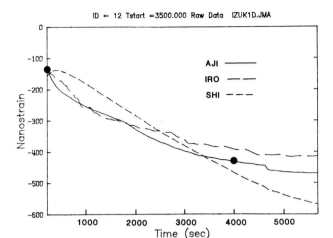

Fig. 15. Normalized strain data for the duration of the F4 activity. The recorded data are normalized to the AJI location by means of factors calculated as in Fig. 14. The calculated strain change at AJI is indicated by solid circles. The first hundred or so seconds of the plot show the competing effects of F3c and F4 on SHI. Note that, by use of simple scaling factors, we have been able to transform the data shown in Fig. 7 to the similar curves shown here.

fault may be active in the F4 region, the strike direction we determine for an assumed single fault may not be the same as any of the actual strike directions. Further, the seismicity suggests that another fault orthogonal to F4 may have developed at the southern end of F4. Incorporating this into our model has little effect on strain values calculated at the positions of the stations, although the strike of the F4 fault would be altered somewhat in order to obtain the best fit.

The faulting process: an overview

The sequence of activity began with a precursory earthquake (F1) followed 6 seconds later by the main earthquake (F2) with magnitude 7.0. The rupture for this shock lasted about 6 seconds (Shimazaki and Somerville 1979). It appears that the rupture then began to slow down. Although we have chosen to begin our analysis 20 seconds after initiation of F2, there was almost certainly relatively slow rupturing occuring in the region of F3 shown with dashes in our figures (e.g. see Shimazaki and Somerville). The rupture then propagated inland along the F3 fault with the total duration of motion being about 100 seconds on that fault. Before the F3 faulting was complete, the F4 fault was activated and continued moving for about one hour.

From the strain data, we calculate that the seismic moments of the slowquakes were 2.6×10^{25}

dyne-cm for F3 and 1.85×10^{25} for F4. Okada's (1978) parameters give a moment of 2.1×10^{25} dyne-cm for F3 and those determined by Inouchi and Sato (1978) correspond to a moment of 2.4×10^{25} dyne-cm. For comparison, Shimazaki and Somerville (1979) gave a value 11×10^{25} dyne-cm for F2, the main earthquake. Thus the total moments for the slowquakes was about 0.4 that of the normal earthquake.

In the analysis of the strain data, we have used as a starting model fault parameters based on those determined by previous workers. These parameters are not equally well constrained by our data. The geometry is such that the

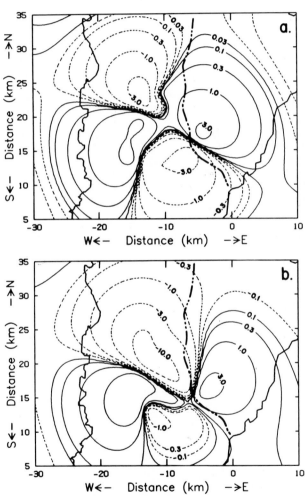

Fig. 16. Vertical displacement amplitudes due to the composite of the F3-F4 faulting system. (a) is for vertical pure strike-slip faults; in (b) F3 has 70° dip and 20° slip. Note that these changes move the nodal planes such as to reverse the sign of the vertical displacements on the levelling route. Similar effects result from extending F3 eastwards (see Okada 1978). Thus these levelling data provide only a moderate constraint on the details of the model.

location, fault length and strike of the fault are well determined, while the dip and slip angles are poorly constrained. The effects due to relatively large variations in dip and slip angles can be cancelled by small changes in strike and moment. Models for F3 by Shimazaki and Somerville (1979), and Inouchi and Sato (1978), based on crustal deformation, leave the western end of F3 terminated about 10 km or less along strike from the coast compared with our distance of 18 km. These will not satisfy the strain data; for example, the strain at AJI has the wrong polarity. Okada's (1978) model for F3 has a longer fault length (extending to 15 km from the coast) and would be in better agreement with the strain data.

Comparison with geodetic data

Further checks on our model can be made by comparing calculated displacements with the results of levelling surveys and triangulation measurements (Crustal Dynamics Department, GSI, 1976, 1978a, 1978b). Postseismic levelling surveys were conducted along most of the routes in the Izu Peninsula. Significant vertical movements were observed in three areas. In the north east, the uplift appears to be part of a long term (over 10 years) trend of vertical movements. A line along the coast from Atami to Kawazu shows changes which are due mainly to the submarine fault F2 while the levels on a more central line, from Shuzenji, at its northern end, to Kawazu, are affected primarily by motion on faults F3 and F4. The eastern section of this line at Kawazu is affected somewhat by the submarine activity (see Fig 11 in Shimazaki and

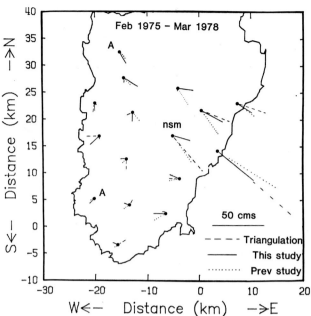

Fig. 18. Map showing stations in triangulation survey. Dashed lines represent horizontal changes determined from the surveys by the Geographical Survey Institute, who assume that stations marked A did not move. Solid lines are horizontal displacements calculated from our model (including the F2 fault of Shimazaki and Somerville, 1978) and dotted lines show the displacements resulting from the model of Shimazaki and Somerville (1979). Station NSM is close to a nodal plane and small changes in fault location (in particular depth of burial) will result in large changes in the calculated value for NSM displacement. Our model gives a satisfactory fit to the data including the significant displacements near the west coast.

Somerville, 1978). The faulting is primarily strike-slip and so vertical displacements are second order effects which are sensitive to the slip angle and, to a lesser extent, the dip angle. Vertical displacements due to a vertical pure strike-slip fault are shown in Fig. 16(a) and in 16(b) those due to the same a fault with changes of $20°$ in the dip and slip angles. The component of dip-slip moves one nodal plane along the strike direction while changes in the dip angle have a smaller effect on the other nodal plane. The results from the model in Fig. 16(b) are compared with the data in Fig. 17.

There appear to be at least three families of faults which can match the levelling data. Short faults east of the levelling route (Shimazaki and Somerville, 1979; Inouchi and Sato, 1978); long faults crossing the levelling route and contiguous with F2 (Okada, 1978) and short faults crossing the levelling route but with a greater component of dip slip (this

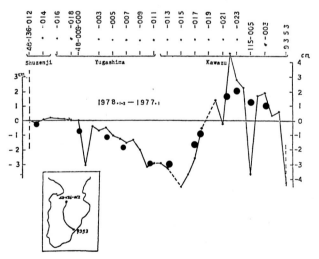

Fig. 17. After Inouchi and Sato (1978). Vertical displacements along the levelling route shown in Fig. 16 compared with calculated values for the model in Fig. 16(b) (filled circles - our data). The left hand side of the figure corresponds to the northern end of the line.

study). The value for the slip angle used in Fig. 16(b) and Fig. 17 is certainly an upper limit; motion on the dashed (near coast) section of F3 (see, e.g., Fig. 10) will considerably reduce the dip slip component required to match the levelling data (see Okada 1978). For the other levelling lines, which showed no significant changes, our model gives vertical changes less than or equal to the ambient changes determined by repeated postseismic surveys.

Triangulation data, also from the Geographical Survey Institute, allow calculation of horizontal movements for the time interval February 1975 to March 1978. This is a long time interval and the relative displacements will be affected by all tectonic movements, both seismic and aseismic, in the time interval (e.g. the Kawazu earthquake of August 1976). In Fig. 18 we show the comparison between the values calculated from measurements and the results of our model. Since horizontal displacements are little affected by the vertical component of faulting, there is little difference between the values shown and those for a similar model with increased dip and slip angles. Also shown are values calculated by Shimazaki and Somerville (1979). Their model included, in addition to F2, a 6 km. long F3 with moment 1.6×10^{25} dyne-cm. The length for their fault was constrained to fit the vertical levelling data. Despite the differences between the models, they fit the data more or less equally well; our model appears to do a little better in the west and south.

Discussion

Previous reports on slow earthquakes have relied on indirect observations or on single station recordings of strain events. We have presented here an analysis of a sequence of slowquakes which were recorded on three widely spaced Sacks-Evertson borehole strainmeters. The basic philosophy used in the analysis was to match our strain data by using fault models based on previous determinations by other workers using completely independent data.

There are other strong indications that slow faulting occurred. There was little damage due to high accelerations on land but the crustal deformation results require onland faulting. About half of the aftershock region is onland but far-field seismic data are satisfied by a wholly submarine earthquake.

In our analysis, we have attempted to keep to the simplest fault geometries consistent with other results. Improved agreement between our calculations and observations could be obtained by allowing small variations in the geometry. In particular, allowing the fault traces to curve rather than form straight lines could produce excellent agreement.

We find that modelling the initial activity on the Izu Peninsula as slowquakes gives good agreement with the strain data. The model is also consistent with the horizontal displacement data and, with minor variations in some of the fault parameters, with the levelling data.

Acknowledgements

We thank Miss K. Sato for help with the data preparation. Suggestions made by A. McGarr and an anomynous reviewer have helped us improve and clarify the manuscript. We are particularly grateful to the reviewer for pointing out an error in our original vertical displacement calculations. This work was partially supported by a grant from the Sloan Foundation and by National Science Foundation Grant EAR 78-01786.

References

Anderson J. G., Motions near a shallow rupturing fault: evaluation of effects due to the free surface, Geophys. J. R. Astr. Soc., 46, 575-593, 1976.

Crustal Dynamics Department, Geographical Survey Institute, Crustal deformation in the central part of the Izu Peninsula (in Japanese), Rep. Coord. Comm. Earthquake Predict., Geographical Survey Institute, Tokyo, 16, 82-87, 1976.

Crustal Dynamics Department, Geographical Survey Institute, Crustal deformation in the central part of the Izu Peninsula (4), (in Japanese), Rep. Coord. Comm. Earthquake Predict., Geographical Survey Institute, Tokyo, 19, 71-75, 1978a.

Crustal Dynamics Department, Geographical Survey Institute, Crustal deformation in the Izu Peninsula (in Japanese), Rep. Coord. Comm. Earthquake Predict., Geographical Survey Institute, Tokyo, 20, 92-99, 1978b.

Das S., and C. H. Scholz, Subcritical rupture in the earth, II: Theoretical calculations, Trans. Am. Geophys. Union, 61, 305, 1980.

Dieterich J. H., Time-dependent friction and the mechanics of stick-slip, Pure Appl. Geophys., 116, 790-806, 1978.

Inouchi N., and H. Sato, Crustal deformation related to the Izu-Oshima Kinkai earthquake of 1978, Bull. Geographical Survey Institute, 23, 14-24, 1979.

Johnson L. R., Green's function for Lamb's problem, Geophy. J. R. Astr. Soc., 37, 99-131, 1974.

Kanamori H., and G. S. Stewart, A slow earthquake, Phys. Earth and Planet. Interiors, 18, 167-175, 1979.

Kitagawa T., and K. Yamamoto, An experiment on the rupture velocity of a tensile crack, Zisin: J. Seismol. Soc. Japan, 28, 207-214, 1975.

Okada Y., Fault mechanism of the Izu-Oshima-Kinkai earthquake of 1978, as inferred from the crustal movement data (in Japanese), Bull. Earthquake Res. Inst., 53, 823-840, 1978.

Rikitake T., Developments in solid earth geophysics, in Earthquake Prediction, 9, pp112-119, Rikitake, T., ed., Elsevier, New York, 1976.

Sacks I. S., Suyehiro, D. W. Evertson, and Y. Yamagishi, Sacks-Evertson strainmeter, its installation in Japan and some preliminary results concerning strainsteps, Pap. Meteorol. Geophys., 22, 195-208, 1971.

Sacks I. S., S. Suyehiro, A. T. Linde and J. A. Snoke, Slow earthquakes and stress redistribution, Nature, 275, 599-602, 1978.

Sacks I. S., S. Suyehiro, A. T. Linde, and J. A. Snoke, Stress redistribution and slow earthquakes, Tectonophysics, in press, 1980.

Sato R., and M. Matsu'ura, Strains and tilts on the surface of a semiinfinite medium, J. Phys. Earth, 22, 213-221, 1974.

Shimazaki, K., and P. Somerville, Summary of the static and dynamic parameters of the Izu-Oshima-Kinkai earthquake of January 14, 1978, Bull. Earthquake Res. Inst., University of Tokyo, 53, 613-628, 1978.

Shimazaki, K., and P. Somerville, Static and dynamic parameters of the Izu-Oshima, Japan earthquake of January 14, 1978, Bull. Seismol. Soc. Am., 69, 1343-1378, 1979.

Tsumura K., I. Karakama, I. Ogino and M. Takahasi, Seismic activities before and after the Izu-Oshima-kinkai earthquake of 1978 (in Japanese), Bull. Earthquake Res. Inst., 53, 675-706, 1978.

Tsuneishi Y., T. Ito and K. Kano, Surface faulting associated with the 1978 Izu-Oshima-kinkai earthquake, Bull. Earthquake Res. Inst., 53, 649-674, 1978.

Umeda, Y., and H. Murakami, Lineament of cracking of roads due to Izu-Oshima earthquake and the area damaged by its largest aftershock (in Japanese), Programme and Abstracts, Seis. Soc. of Japan, 1, 22, 1978.

RUPTURE MECHANICS OF PLATE BOUNDARIES

Amos Nur

Department of Geophysics, Stanford University
Stanford, California 94305

Abstract. A mechanical instability--drasti-
cally different from stick-slip type models may
be responsible for every large earthquakes,
M > 7 or so. In the model presented here, the
beginning of the rupture cycle consists of
aseismic yielding at the bottom part of the
lithosphere. The yield zone propagates upwards,
initially very slowly, with accelerating speed.
It breaks out at sonic speeds, with surface
rupture length which is determined by the
thickness of the lithosphere. Predicted
values for surface rupture range from 100 to
500 km or so for strike-slip events, in
agreement with observed values.

Introduction

One of the most important questions in the
design of long range and long duration experiments to deterime the deformation of the
earth's surface near plate boundaries is the
expected time-space pattern of this deformation.
Positioning of ground stations, and their
density, repeat rates of measurements, and
expected spatial coherency can all be greatly
improved with some preknowledge of the character
of the field to be measured. In this paper we
develop a conceptual model for inter-, pre-,
and co-seismic surface deformation at and around
active plate boundaries, in order to provide
the range of surface deformation patterns which
we can anticipate from available physical models
of rock yielding.

It is generally recognized that large earthquakes (M > 7.0 or so) are responsible for
most of the seismic slip at plate boundaries.
In spite of the dramatic model of instability
of these slip events--in the form of destructive
earthquakes with rupture lengths extending
hundreds of kilometers, most of the deformation
at plate boundaries is not seismic. In large
thrust earthquakes, seismic rupture occurs to
depth of only about 1/2 the plate thickness
(e.g., the Alaska 1964 earthquake) leaving
the lower part unruptured. In large strike-
slip events this depth is typically even
smaller--10 or 15 km as compared with 70 to

100 km for the thickness of the lithosphere.
Consequently, a major portion of the plate
boundary--below the seismogenic zone--must
yield aseismically.

Savage (1975) proposed a simple model shown
in Figure 1, invoking steady creep in a narrow
zone, extending underneath the seismogenic
trace of the plate boundary. Turcotte and
Spence (1974) suggested a model in which the
stress remains constant with time in the
aseismic zone, with slip rate which varies
with depth. Prescott and Nur (1980) suggested,
on the basis of laboratory observed creep
behavior of hot rock, and field observations of
interseismic crustal deformation, that the
aseismic zone is more likely to be of some
width in which shear strain rate is anomalously
high.

None of these current models treat the problems of seismic yielding as a truly mechanical
one--in which motion is more or less rigorously
related to the acting forces or the stresses.
Such motion must include the time history not
only of particles in the boundary region, but
also the possible motion of edges of the zone
in which plate boundary yielding does actually
occur.

Furthermore, the correct mechanism of the process must also link the aseismic yielding to
the fact that seismic gaps for large earthquake
(M > 7 or so) seem to be filled in a systematic
and regular fashion--unlike smaller earthquakes (M < 6 or so) which fail to show any
significant regularity. The mechanical model
must also be able to describe the development
of yielding and rupture in time and space and
provide understanding of the factors which
control the length of rupture of very large
earthquakes. Typical ruptures are hundreds of
kilometers long, several times the thickness
of the lithosphere in which they occur.

In this paper we have attempted to extend the
Savage model for the rupture at plate boundaries,
to include an aseismic as well as the seismic
instability. In the next section we outline
the basic concepts of the model and show preliminary results. The mechanics considered

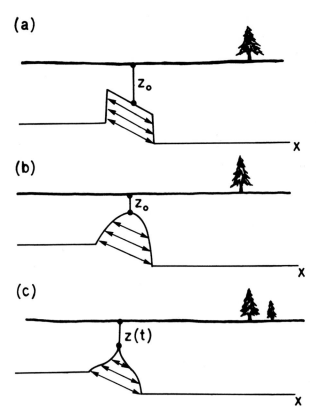

(a)

Z_o

X

(b)

Z_o

X

(c)

$z(t)$

X

Fig. 1. Models for rupture of the lithosphere: (a) The Savage Model with fixed dislocation at depth Z_o; (b) The Turcotte model with fixed crack at Z_o; (c) Proposed model with moving dislocation/crack or dislocation flux. The depth to tip $Z(t)$ decreases with time.

provide very specific predictions of the time history of surface deformations associated with plate boundary yielding to distances of hundreds of kilometers away from the boundary and along the boundary. The conceptual model as given here is exceedingly simple-minded, and should be developed in the future in much more detail, with the mechanics involved to be refined in order to provide a basis for the design of long and large scale (10-300 km) repeated geodetic deformation measurements which might be necessary to reveal the actual crustal process leading to, during, and following very large events. The development of this model by incorporating more detailed rheology may thus provide working hypotheses for intermediate range prediction of very large earthquakes.

An Instability Model for Large Earthquakes

We begin by considering the simple screw dislocation model of Savage and Burford (1970), which represents an infinitely long trans-

current fault in a flat earth. In the model a dislocation, with Burger's vector b_o, is situated at depth Z_o below the free surface as shown in Figure 1.

In the Savage model the stress at the neighborhood of the tip of the slip zone becomes very high--with the consequence that the tip must become unstable, with a tendency to move. Turcotte's model (1974) attempts to overcome part of this problem, by imposing a constant stress condition on the slip zone for $Z > Z_o$, as shown in Figure 1. However, the stress acting on the tip is again very large, suggesting again that this configuration is not stable.

Since models with fixed slip zones lead invariably to very high tip stresses, we may consider relaxing the assumption that the tip depth is fixed. Instead we assume that the tip can move in response to the forces acting on it. The simplest model is obtained from classical dislocation theory, where motion of a dislocation is related to the force acting on it through some simple law of motion (Head, 1972). For the configuration on Figure 1, the force acting on the dislocation is simply the attraction to the free surface (e.g., Cottrell, 1964)

$$F = \frac{\mu b^2}{4\pi} \cdot \frac{1}{Z} \qquad (1)$$

where F is force per unit length of dislocation line, b is the Burger's vector, and μ is the shear modulus.

A similar expression holds for the attractive glide force acting on an edge dislocation towards the free surface

$$F = \frac{\mu b^2}{4\pi(1 - \nu)} \cdot \frac{1}{Z}. \qquad (2)$$

A noteworthy feature of equations (1) and (2) is that the force acting on the dislocation is inversely proportional to its distance from the free surface to which it is attracted.

A simple law for the motion of a dislocation under force may be given by the power law (Head, 1972)

$$-\frac{dZ}{dt} = A \cdot F^n \qquad (3)$$

where Z is depth, and A is a constant (in the Savage model, A = 0). The power coefficient n must be selected on the basis of either experimental results or additional models. In a rough way, the value of n may be related to laboratory observed power law creep in rocks (e.g., Kirby, 1977). We may consider therefore the range $1 \leq n \leq 5$ for lithospheric behavior. Combining equation (1) and (3) for the screw dislocation, we obtain

$$-\frac{dZ}{dt} = C_o\left(\frac{1}{Z}\right)^n \qquad (4)$$

with the constant $C_o = A\left(\frac{\mu b^2}{4\pi}\right)^n$.

Assuming a dislocation source at Z_o, the initial condition $Z = Z_o$ at $t = 0$ and the final state $Z = 0$ at $t = t_1$ we obtain the position of the dislocation as a function of time

$$\overline{Z} = (1 - T)^{1/(n+1)} \qquad (5)$$

and the dislocation velocity

$$\overline{V} = -\frac{d\overline{Z}}{dt} = \left(\frac{1}{n + 1}\right)(1 - T)^{-n/(n+1)} \qquad (6)$$

where $\overline{Z} = Z/Z_o$ and $T = t/t_1$.

Equation (6) shows that at early time $T \ll 1$, the dislocation moves slowly towards the surface. However, it greatly accelerates as T approaches 1, and becomes infinite as it breaks out. We suggest that this explosive breakout of the dislocation, for $n \geq 1$, may be useful in describing the instability process which is responsible for very large earthquakes. The effect is quite insensitive to details of stress,

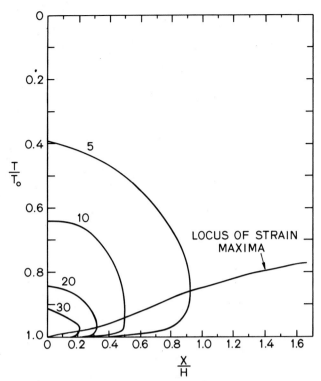

Fig. 3. The time-space distribution of surface strain for an upward moving dislocation with power law coefficient $n = 1$. Note that surface strain for most points away from the fault $(x \neq 0)$, first increase gradually and then decrease _before_ the breakout at $T/T_o = 1$, $X = 0$.

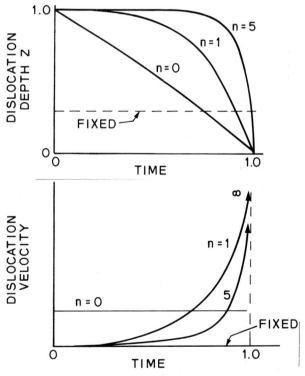

Fig. 2. The depth and velocity of the screw dislocation moving under the attractive surface force. The factor n indicates the motion power law used in the computation. The Savage and Turcotte models correspond to the fixed dislocation depth.

geometry and the exact value of n. It does depend however, on the assumption that the material involved fails by localization of strain, and that the localized zone obeys in some gross sense a power law behavior.

In Figure 2 we show the depth $\overline{Z}$ and the velocity $\overline{V}$ of the screw dislocation for various values for n--invariably showing the breakout instability.

Using the dislocation model we calculate the free surface displacement U_3 and strain fields e_{23} as a function of time.

$$U_3(X,t) = 2b\left[\tan^{-1}\left(\frac{X}{Z(t)}\right)\right]$$

and $\qquad\qquad\qquad\qquad\qquad\qquad (7)$

$$e_{23}(X,t) = 2b\left[\frac{Z(t)}{X^2 + Z^2(t)}\right]$$

where $X = x/Z_o$, is the distance perpendicular to the fault. Figure 3 shows an example of these fields for $n = 1$. The strain field is first very broad becoming more concentrated with time.

In general we find that most of the surface strain accumulation occurs only in the last 30% to 10% of earthquake repeat cycle.

The Length of Surface Ruptures

The two-dimensional model can be extended to three dimensions, using the line tension concept. We consider a situation as shown in Figure 4 in which two types of forces act on the bowed dislocation line shown: the attraction towards the free surface, given approximately by equation (3), and the self attraction of the line, given by

$$T = \alpha\mu b^2 \frac{1}{R} \qquad (8)$$

where R is the radius of curvature, and α is a constant. Expressing R in terms of the depth Z of the closest point of the line to the surface, normalized with respect to the depth of the source Z_o, and the half length of the slip zone L, we obtain the line tension

$$T = \alpha\mu b^2 \frac{2(1 - Z)}{L^2 + (1 - Z)^2} . \qquad (9)$$

The configuration of the dislocation line at any given moment is controlled by the equilibrium between the line tension T which tends to keep the line straight and thus opposes the motion towards the free surface, and the image attraction towards the free surface F. Equating the two forces we obtain the relation between the half length L of the slip zone at depth and the depth of the dislocation line Z as shown in Figure 4.

$$L = [(8\pi\alpha + 2)Z - (8\pi\alpha + 1)Z^2 - 1]^{\frac{1}{2}}. \qquad (10)$$

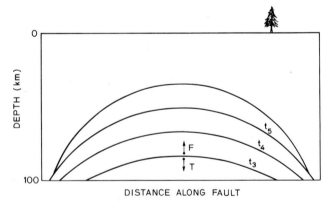

Fig. 4. Equilibrium configurations of bowed dislocation line, subject to line tension T and surface attraction f, at consecutive time t_i. Shortly after t_3 the system becomes unstable and rupture accelerates toward the free surface.

We notice at once that equation (10) yields a half length L which is double valued. This implies that for $Z < Z_c$, where Z_c is some critical depth, $0 < Z_c < Z_o$, the motion of the dislocation line becomes unstable, breaking out at speeds approaching wave speeds. Differentiating equation (10) with respect to Z, we find the critical depth Z_c at which the equilibrium rupture half length L_c is largest

$$Z_c = \frac{4\pi\alpha + 1}{8\pi\alpha + 1} . \qquad (11)$$

Reasonable values are (Cotrell, 1964) $1/\pi \leq \alpha \leq 4/\pi$, leading to $.5 \leq Z_c \leq .6$. This suggests that the unstable motion of the dislocation line begins at a depth of about 1/2 to 2/3 the depth of the lithosphere. Creep or slip below this depth is slow, involving a gradual increase of the length of the zone. At the onset of the instability, the critical half length is

$$L_c = \frac{2\pi\alpha(8\pi\alpha - 2)}{(8\pi\alpha + 1)} . \qquad (12)$$

For $\alpha = 1/\pi$, $2/\pi$, and $4/\pi$, the values of the critical length L_c are 1.15, 1.94 and 2.7 times the thickness of the lithosphere. Taking this thickness as 70 km, computed rupture lengths range between 170 and 500 km. These values are in good agreement with observed values for large earthquakes, such as 400 km for the 1906 San Francisco earthquake.

Assuming again that our lithospheric model behaves elastically outside the slip zone, we may use dislocation theory also to estimate the spatial and temporal pattern of surface deformation--displacement, tilt, and strain--preceeding the breakout of rupture, in the form of a major earthquake. Figure 5 shows the computed surface displacement history for the bowing dislocation line model.

Discussion and Conclusion

I have shown in this paper that a mechanical instability--drastically different from stick-slip type models (e.g., Brace and Byerlee, 1970) may be responsible for very large earthquakes $M \geq 7$ or so. In the model, the beginning of the rupture cycle consists of aseismic yielding at the bottom part of the lithosphere. The yield zone propagates upwards, initially very slowly, with accelerating speed. It breaks out at sonic speeds, with surface rupture length which is determined by the thickness of the lithosphere. Predicted values for surface rupture range from 100 to 500 km or so for strike-slip events, in agreement with observed values.

The instability as outlined in the previous sections is at best only a conceptual outline of the process. A comprehensive development of the model must include much more realistic

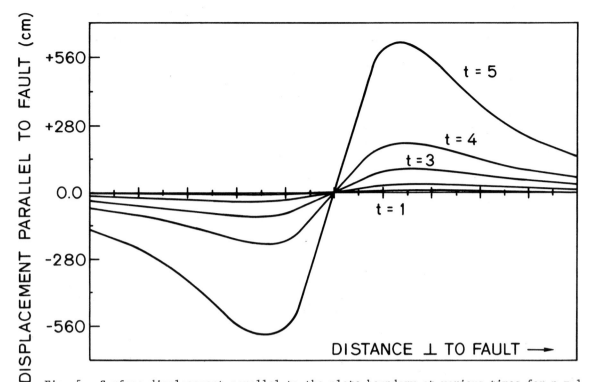

Fig. 5. Surface displacement parallel to the plate boundary at various times for n = 1, and slip history as shown in Figure 4.

and more detailed characteristics of the lithosphere, and comparison with observed inter-, pre-, and co-seismic deformation data such as from California and Japan, including for example, rupture length vs. magnitude. It will also be necessary to analyze the instability in a plate with a lower free boundary, in contrast to the present half space analysis (e.g., Chou, 1965). The net effect of the lower plate boundary is likely to retard the motion of the dislocation at early times in the interseismic cycle, and accelerate the motion in its later part, relative to the half space case.

The entire analysis must also be developed for thrust faulting--using edge dislocations, moving under glide forces, with dependence on the power n in dislocation velocity power law, the tip of the slip plane, and the possible effects of climb forces.

The uniform slip dislocation model can further be generalized to a variable slip model by superposition of dislocations. In particular we need to study the effects of a stiff brittle crust over a more compliant lithosphere. It is well known that additional forces on dislocations are exerted by elastic interfaces. The effect of a stiff crust might be to retard the dislocation motion at depth, but accelerate it once it penetrates the crust, thus enhancing the instability.

The model may also be extended to include

post-seismic deformation in which the dynamic stress drop forces the dislocation to move back down away from the free surface. Because this motion just below the seismic zone is downward, it is stable, with a velocity which decreases with depth or time.

Acknowledgement. This research was funded by grant EAR 76-22501 from the Geophysics Program, National Science Foundation.

References

Brace, W.F., and J.D. Byerlee, California earthquakes: Why only shallow focus, Science, 168, 1573-1575, 1970.

Chou, Y.T., Screw dislocations in and near lamellar inclusions, Phy. Stat. Sol., 17, 509-516, 1966.

Cottrell, A.H., Theory of Crystal Dislocations, Gordon and Breach, New York, pp. 91, 1964.

Head, A.K., Dislocation group dynamics I. Similarity solutions of the n-body problem, Philosophical Magazine, 26(1), 43-53, 1972.

Kirby, S.H., State of stress in the lithosphere: Inferences from the flow laws of olivine, Pure and Appl. Geophys., 115(1,2), 245-258, 1977.

Nur, A., Nonuniform friction as a physical basis for earthquake mechanics, Pageoph. 116, 959-991, 1978.

Nur, A. and M. Israel, The role of heterogeneities in faulting, Physics of the Earth and Planetary Interiors, 21, 225-236, 1980.

Prescott, W.H. and A. Nur, The accommodation of relative motion at depth on the San Andreas fault system in California, submitted, J.G.R., 1980.

Prescott, W.H., J.C. Savage and W.T. Konoshita, Strain accumulation rates in the Western United States between 1970-1978, J. Geophys. Res., 84(B10), 5423-5435, 1979.

Savage, J.C., Comment on 'An analysis of strain accumulation on a strike slip fault' by

D.L. Turcotte and D.A. Spence, J. Geophys. Res., 80, 29, 4111-4114, 1975.

Savage, J.C. and R.O. Burford, Accumulation of tectonic strain in California, Bull. Seism. Soc. Am., 60, 6, 1877-1896, 1970.

Thatcher, W., Strain accumulation and release mechanism of the 1906 San Francisco earthquake, J. Geophys. Res., 80, 4862-4872.

Turcotte, D.L. and D.A. Spence, An analysis of strain accumulation on a strike-slip fault, J. Geophys. Res., 79, 29, 4407-4412, 1974.

EARTHQUAKE PREDICTION PROGRAM IN JAPAN

Kiyoo Mogi

Earthquake Research Institute, Tokyo University, Tokyo 113

Abstracts. The first 5-year earthquake project in Japan was started in 1965. Now, the fourth 5-year project (1979-1983) is in progress. Following the initial principles of earthquake prediction adopted at the start of the project, several fundamental observations have been made. This research includes geodetic, seismic, geomagnetic, geoelectric, and geochemical observations; continuous observations of crustal deformation by strain and tilt meters; and gravimetric surveys. Detailed surveys of active faults and laboratory experiments on rock failure have also been conducted. The density of the observation network has rapidly increased with time. In recent years, telemetering systems for microearthquake observations exist in a major portion of the land area. The first attempts, such as the real-recording-time observations by an array of ocean bottom seismographs in the Tokai region and seismic observations at 3000-meter-deep wells in Tokyo area, were very successful.

During the last few years, several destructive earthquakes including the 1974 Izu-hanto-oki (M 6.9), the 1978 Izu-Oshima-kinkai (M 7.0) and the 1978 Miyagi-oki (near Sendai)(M 7.4) earthquakes occurred successively in Japan. In spite of the recently intensified observations, reliable precursor data are very limited, because the density of the observation network is still insufficient for predicting earthquakes of magnitude ~7. Before the Izu-Oshima-kinkai earthquake, various observations had been intensified in the Izu Peninsula; microearthquake swarms and anomalous crustal uplift in the Izu Peninsula were observed about two years before the earthquake. Several kinds of remarkable precursory phenomena were observed here, but the Izu-Oshima-kinkai earthquake was not predicted. In the Izu Peninsula, we still have no real-time data processing system for various observations related to earthquakes of magnitude ~7. The Yamasaki fault in the Kinki district in western Japan has been investigated as a test field for earthquakes of magnitude ~4-5. Some earthquakes along this fault were forecast on the basis of the regularity in seismic activity and the observations of various kinds of precursors in this region.

On the basis of the recurrence time of historical, large earthquakes along the Nankai trough and the crustal deformation observed by geodetic surveys, the Tokai region has been indicated as a potential region for a great shallow earthquake of magnitude ~8 since 1969. For the purpose of predicting this potentially large earthquake, a dense network of observation instruments has been set up in the Tokai region. The various data are telemetered to the center of the Japan Meteorological Agency (JMA) in Tokyo. When anomalous behaviour is noted, the Earthquake Prediction Council evaluates the data and conveys their findings to the director general of JMA.

The characteristics of precursory earthquake phenomena in Japan are compared with those in different tectonic regions, such as central California and some regions in China.

Introduction

The Japanese islands and vicinity constitute one of the most active seismic regions in the world (Fig. 1). Because the Tokai region has recently been pointed out as a potential region for a future large shallow earthquake of magnitude ~ 8, the problem of earthquake prediction has attracted much scientific and social attention. The Tokai region, an area in central Japan between Tokyo and Nagoya, is a very important region, which has a high population density and heavy industrial activity; hence, the prediction of this earthquake is critical. Due to this need, a very dense network of different kinds of instruments for monitoring various precursory earthquake phenomena has been set up in this region. The various data are telemetered to the center of the Japan Meteorological Agency (JMA) in Tokyo, which is staffed 24 hours a day and 7 days a week. When anomalous behaviour is observed in these data, the Earthquake Prediction Council (chairman, Prof. T. Hagiwara) evaluates the data and conveys its findings to the director general of JMA, who reports to the prime minister. On June 15, the Large-scale Earthquake Counter-measures Act was promulgated as Law No. 74 of 1978. This law will now be applied to the Tokai region.

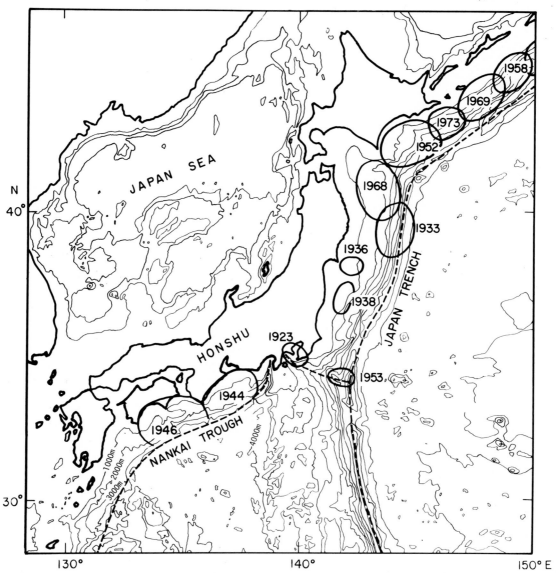

Fig. 1. Submarine topography and recent large shallow earthquakes (M 7.7) around the Japanese islands.

On the basis of various observational data obtained by universities, the Geographic Survey Institute, the Japan Meteorological Agency, the National Research Center for Disaster Prevention, the Geological Survey of Japan, the Hydrographic Department, the International Latitude Observatory, and other Governmental institutions, the long-term earthquake prediction problems throughout Japan are discussed every three months by the Coordinating Committee for Earthquake Prediction (chairman, Prof. T. Hagiwara).

The first 5-year earthquake prediction project was started in 1965, just after the Niigata earthquake of 1964, and the project has been growing at a steady rate (Kanamori, 1970; Hagiwara, 1975; Rikitake, 1976). The yearly budget of the earthquake-prediction project in Japan is shown in Table 1. The yearly budget, exclusive of salaries, was 5,800,000,000 yen for 1979 and 6,000,000,000 yen for 1980. The project consists of three categories. The first is various observations for long-term prediction, such as geodetic surveys and seismic observations. The second is various observations for short-term prediction, such as continuous observations of crustal deformation. The third is various fundamental studies, such as rock fracture experiments. These will be discussed in the following two

TABLE 1. Yearly budget of earthquake
prediction project in Japan

Year	Budget in Yen
1965	212 539 000
1966	290 035 000
1967	334 362 000
1968	328 559 000
1969	496 116 000
1970	596 120 000
1971	805 720 000
1972	898 900 000
1973	761 164 000
1974	1 552 674 000
1975	2 007 637 000
1976	2 312 676 000
1977	2 928 000 000
1978	4 124 000 000
1979	5 847 000 000
1980	6 000 000 000*

* approximate value

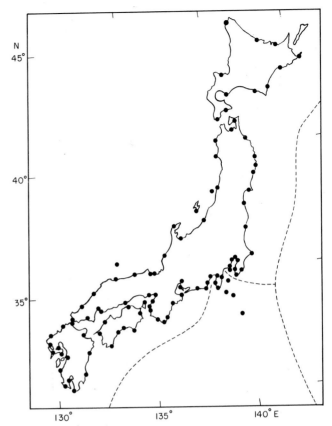

Fig. 3.   Stations of tide-gage observation.

chapters. Major topics of earthquake prediction
in the past 10 years in Japan will be discussed
in some detail in the later chapter.

Outline of Basic Observations
for Earthquake Prediction

Observation of Crustal Deformation

A shallow earthquake occurs by sudden faulting
of the earth's crust when the stress or strain
reaches a critical value. From observations of
crustal deformation, the accumulation of strain
(or the stress state) can be approximately estimated; this estimate is useful in evaluating the
potential for earthquake occurrence in a region.
Observations of the deformation preceding earthquakes are essential for earthquake prediction.

Vertical Displacement Observation. The leveling surveys in Japan have been conducted by the
Geographical Survey Institute (GSI). Figure 2
shows the first-order leveling routes in Japan.
These routes have been surveyed at about 5-year
intervals. In some regions, the survey have been
carried out more frequently, including the survey
along the second-order leveling routes, as
shown in the later chapter.

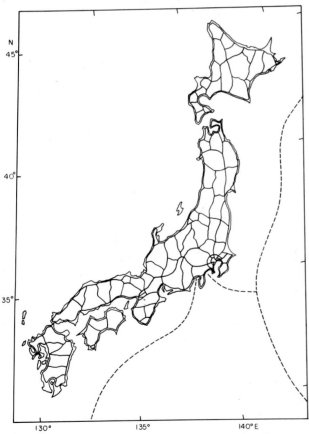

Fig. 2.   First-order leveling routes in Japan.
Broken curves are deep sea trenches or troughs.

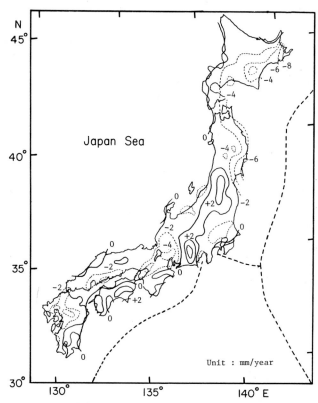

Fig. 4. Average rate of vertical crustal movement in the Japanese islands during the period
from about 1895 to about 1965 (after Dambara,
1971).

The tide-gage observation is useful for detecting the crustal upheaval and subsidence of the
coastal region. The 105 tide-gage stations are
shown in Figure 3. This observation has been
carried out by the Japan Meteorological Agency,
the Geographical Survey Institute, the Hydrographic Department, and Tokyo University. Data
are sent to the central office in the Geographical Survey Institute. In many cases, the measurement of crustal movement was obtained from the
difference between the mean-sea-level readings of
adjacent tide-gage stations. The tide-gage observations are also useful for the continuous observation of crustal movement.

Figure 4 shows the spatial distribution of the
average rate of vertical movement in the Japanese
islands during the period from about 1895 to
about 1965; this distribution was compiled by
Dambara (1971) on the basis of leveling survey
and tidal observation data. In this compilation,
Dambara tried to eliminate local crustal movements caused by major earthquakes. However, effects of coastal uplift due to great earthquakes
in 1923, 1944 and 1946 seem to be still present.
Recently Kato and Tsumura (1979) reported vertical movements deduced from tidal records at the
coast of Japan for the period 1951 to 1978. Both

conclusions note the appreciable subsidence in
the northeastern part of Japan and the Setouchi
between the Chugoku and the Shikoku districts of
western Japan.

Horizontal Displacement Observation. Figure 5
shows the first-order triangulation network
covering Japan. The first-order triangulation
surveys by GSI have been made twice, for 1883-
1909 and 1948-1967. Figure 6 shows the displacement vectors of triangulation points that
were obtained by Harada and Isawa (1969) and
Harada and Kassai (1971). In this figure, it is
assumed that several stations (closed circles)
which are located in relatively quiet regions,
are immovable. It is obvious that the pattern
of the recent horizontal deformation of the
earth's crust in Japan is markedly affected by
the occurrence of large shallow earthquakes during the period from (1883-1909) to (1948-1967).
The relation between the horizontal deformation
of the earth's crust and the occurrence of large
earthquakes along deep sea trenches (subduction
zones) will be discussed in the later chapter.

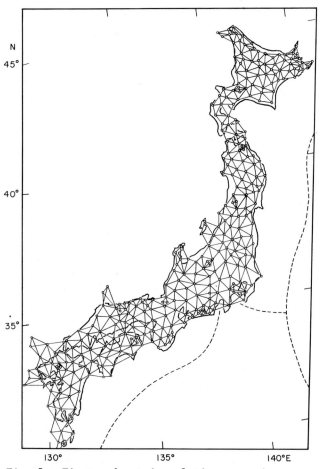

Fig. 5. First-order triangulation network covering the Japanese islands.

Continuous Observation of Crustal Deformation.
Two methods of continuous observation of crustal
deformation are used. One method uses an underground-gallery equipped with strainmeters and
water-tube tiltmeters. This type of observation
is conducted mainly by universities. The locations of these underground-gallery observatories
are shown as large solid circles in Figure 7.
The other continuous observation method uses a
bore-hole-type observatory. The locations of the
strainmeters or tiltmeters of this type of equipment are shown as small solid circles in Figure
7. A dense observation network of the volumetric
strainmeters of the Sacks-Evertson type is operated by JMA in the Kanto-Tokai region. The observation network in the Tokai region will be
explained in the later chapter. A number of
array stations for crustal-deformation measurements along several belts are under construction
by universities in Japan. These planned array
stations are shown as small open circles in
Figure 7.

By making continuous observations of crustal
deformations, scientists have observed, in some
cases, deformations before earthquakes. The
1978 Izu-Oshima-kinkai earthquake will be discussed in the later chapter.

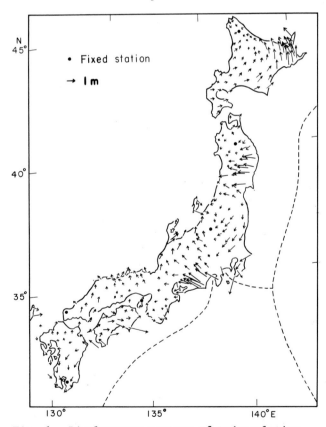

Fig. 6. Displacement vectors of triangulation
points during the period from 1883-1909 to 1948-
1967 (Harada and Isawa, 1969; Harada and Kassai,
1971).

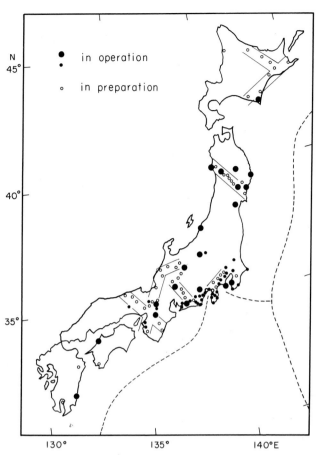

Fig. 7. Locations of continuous crustal movement
observatories.

Gravity Survey

The gravity survey has been conducted repeatedly in some areas, such as the Izu-Tokai region.
Hagiwara et al. (1977) have found marked gravity
changes in the uplift region in the eastern part
of the Izu Peninsula; these changes will be discussed in the later chapter. The gravity survey
is useable to quickly find approximate vertical
crustal movements, and it is also useful for
studying the changes in mechanical states of the
earth's crust by a comparison with leveling data.

Seismic Observation

Large and Small Earthquake Observations. Seismic observations for earthquakes of M 3 and over
in and near Japan have been conducted by JMA.
Figure 8 shows the seismic stations of JMA. Large
circles show stations with equipped with high sensitive seismographs. Small circles show seismographic stations used only to detect strong earthquakes. They include array stations in the Tokai
region having ocean-bottom seismographs.

Figure 9 shows locations of major shallow

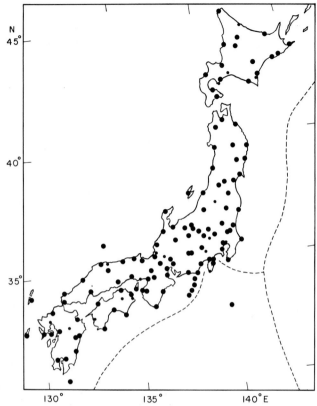

Fig. 8. JMA seismographic stations for small or large earthquake observation.

earthquakes of M 6.0 and over in about the past one hundreds years (1885-1979) determined by instrumental observations. Data in this figure are from the JMA catalogues for the period (1926-1979) and from Utsu (1979a) for the period (1885-1925). On the basis of these data, foreshock activities, seismic gaps, and migration phenomena of seismic activity have been studied in relation to earthquake prediction.

Microearthquake Observation. Figure 10 shows the microearthquake observation networks operating in 1978 (Suzuki et al., 1979). The solid lines in this figure represent telemetering systems. In recent years, these telemetering systems, operated by universities and the National Research Center for Disaster Prevention, have been rapidly increasing. In 1980, the Earthquake Prediction Data Center, a university telemetering system, was established in the Earthquake Research Institute, Tokyo University. By the use of these microearthquake observation networks, microearthquakes M∼(1-2) and over in the major part of the Japanese islands can be observed. The spatio-temporal distributions of microearthquakes are considered important to earthquake prediction. Various features of microearthquakes, such as the b-value in the magnitude-frequency relation and the waveform similar-

ity, which will be explained in the later chapter, may give important clues for discriminating between foreshocks and earthquake swarms. Lineations of microearthquakes have been found along a number of active faults in western Japan (Huzita et al., 1973). Along some active faults, however, no appreciable activity has been observed. This discrepancy suggests either a difference in the mechanical state of the active faults or a difference in stages of fault movements.

Measurement of Seismic Velocity Changes

In the past ten years, precise measurements of temporal variations in seismic velocity, using explosions, have been made by the Geological Survey of Japan (GSJ) in the Kanto-Tokai region (GSJ, 1979). Although a large shallow earthquake occurred within the region, no appreciable changes in seismic velocity have been observed. This fact will be discussed in the later chapter. On the basis of data from natural earthquake observations, a number of previous papers have reported marked changes in seismic wave velocity occurring before earthquakes (Ohtake and Katsumata, 1977). However, most of these reports are questionable, because of the low accuracy of arrival-time measurements.

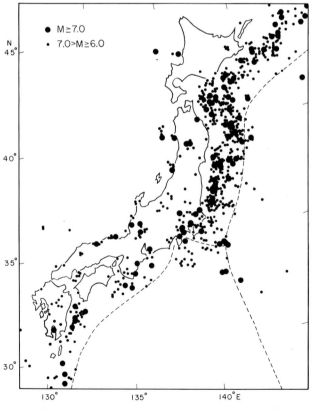

Fig. 9. Epicentral locations of major shallow earthquakes of M 6.0 and over during the period from 1885 to 1979.

Geomagnetic and Geoelectric Observation

Figure 11 shows the sites of various kinds of geomagnetic and geoelectric observations. Proton-procession magnetometers are distributed at 12 localities for the measurement of geomagnetic secular variations. In addition to these permanent stations, proton magnetometers are also densely deployed in the areas where various geophysical observations are concentrated, such as the Izu Peninsula. As will be mentioned in the later chapter, a marked anomalous change in the total geomagnetic intensity was observed before an earthquake in the Izu Peninsula (Sasai and Ishikawa, 1980).

Earth resistivity is being measured in some active fault areas. Continuous observation of variations in resistivity by direct current methods was recently initiated at the Yamasaki fault as a joint project of several universities. Repeated measurements of resistivity have been conducted in the area of the Oshima volcano crater. Anomalous variation in resistivity was observed a few months before the 1978 Izu-Oshima-kinkai earthquake (Yukutake et al., 1978).

At Aburatsubo, resistivity of rocks is being measured continuously by a "resistivity variome-

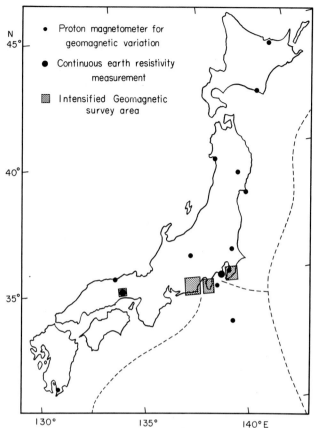

Fig. 11. Sites of geomagnetic and geoelectric observation.

ter" as will be discussed in the later chapter. Precursory change was observed just before the 1974 Izu-hanto-oki earthquake (Rikitake and Yamazaki, 1977).

Geochemical Observation and Groundwater-level Measurement

Geochemical observations for earthquake prediction were started near the end of 1973. Figure 12 shows the sites for continuous or periodical measurements of radon, helium, argon, and other elements. These observatories are mainly operated by Tokyo University, Nagoya University, and the Geological Survey of Japan. Anomalous geochemical changes have been observed before some earthquakes. The most reliable precursory chemical change is the radon concentration, as shown in Figure 30 (Wakita et al., 1980). The temperature and the water level in deep or shallow wells are being observed by universities, the Geological Survey of Japan, and other groups (e.g. the Catfish Club). A marked precursory change in the level of groundwater prior to the 1978 Izu-Oshima-kinkai earthquake is shown in Figure 29 (Yamaguchi and Odaka, 1978). Sites of geochemical observations and groundwater-level measure-

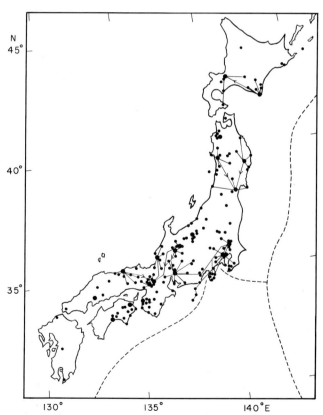

Fig. 10. Microearthquake observation networks in 1978 (Suzuki et al., 1979). Solid lines show telemetering systems.

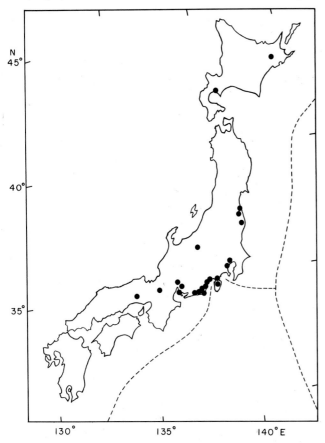

Fig. 12. Sites of geochemical observations in the earthquake-prediction program.

ment concentrate in the Tokai region, which will be mentioned in the later chapter.

Active Fault Survey

Detailed surveys of active faults in and around the Japanese islands have been made by geologists and geomorphologists of universities, the Geological Survey of Japan, the Hydrographic Department, and the Geographical Survey Institute. The active fault maps were compiled by the Research Group for Active Faults (1980). Figure 13 shows a summary of this fault map. Matsuda (1981) found a close relation between the active faults found by geological surveys and recent large shallow earthquakes. Since 1979, a few active faults have been investigated by the trenching method developed in United States (Okada et al., 1979).

Fundamental Research

Laboratory Study

Parallel to the above-mentioned field observations, universities and the Geological Survey of

Japan have performed laboratory experiments to clarify the nature of precursory phenomena and, more fundamentally, the physical processes of earthquakes. Figure 14 shows some experimental results (Mogi, 1980a). Figure 14a shows that the relation between the maximum amplitude of acoustic emission (AE) and the number of AE events is different for different frequency ranges of acoustic waves. The amplitude-number relation of low frequency (30KHz) AE events, which is linear in the logarithmic scale in this case, seems to show the true relation between the magnitude and the number of AE events.

Figure 14b (top) shows the b-value of AE events (low frequency) in a rock sample under a nearly constant load as a function of time. The AE activity and the ratio between the number of AE events counted through the low frequency band-pass filter (30KHz) and those counted through the high frequency band-pass filter (500KHz) are also shown as a function of time. In this figure, it is noted that the b-value decreases before the fracture caused by a constant external stress (even under a decreasing external stress). Scholz (1968) argued that the b-value was dependent on stress level and that it decreased with increasing stress. In the field, it has been frequently observed that the b-value decreases before earthquakes (e.g. Suyehiro et al., 1964). Therefore, on the basis of Scholz's report, a number of investigators argued that the external stress increased before earthquakes. However, the above-mentioned experimental result shows that the decrease in the b-value before an earthquake can be expected without any increase of stress.

The ratio between the number of low frequency acoustic waves and the high frequency ones increases just before the main fracture. The increase of this ratio may be consistent with the decrease in the b-value. After the main fracture, the above-mentioned ratio increases markedly. According to Utsu (1980), the predominant frequency of seismic waves in aftershocks is appreciably lower than that in earthquakes before large earthquakes (including foreshocks). This result in the field seems to be consistent with the laboratory result.

Historical Earthquake Study

The spatio-temporal distribution of large earthquakes for a long period is very useful for long-term prediction of large earthquakes. In Japan, there are abundant historical documents dating from one thousand years ago to the present. Historical earthquakes have been mainly investigated by universities on the basis of these documents. A revised list of historical destructive earthquakes was published by Usami (1975).

Theoretical and Statistical Research

One of the most essential problems in earthquake prediction research is why precursory phe-

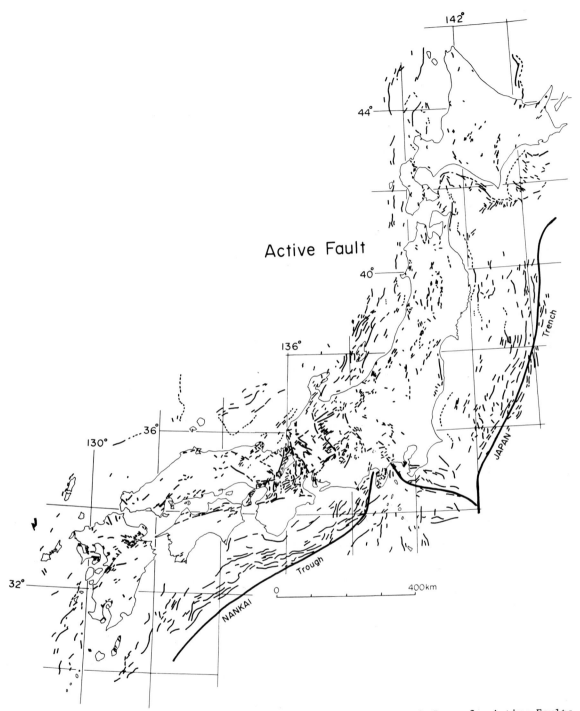

Fig. 13. Active faults in and around the Japanese islands (Research Group for Active Faults, 1980).

nomena occur, Mogi (1963, 1967, 1977a) theorized that remarkable precursory phenomena were mainly caused by stress concentration due to heterogeneous structures. Tsumura (1979) tried to explain the space-time distribution of the precursory phenomena by means of the spatial distribution of the regions with different strengths under increasing tectonic stresses. The problem concerning the order of appearance of precursory phenomena will be discussed in the later chapter.

Rikitake (1975, 1976) summarized various kinds of precursory events reported by a number of in-

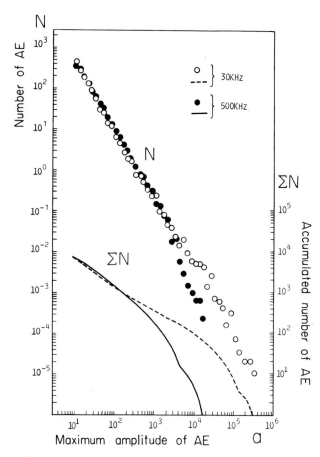

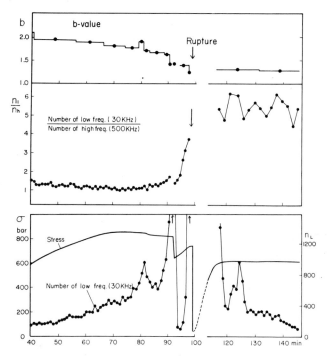

Fig. 14(b). The b-value of AE events in a rock sample under nearly a constant stress (top) and the ratio of the number of AE events of low and high frequencies (middle) are shown as functions of time.

Fig. 14(a). Relation between the maximum amplitude of acoustic emission (AE) and the number of AE events, for different frequency ranges of acoustic events (Mogi, 1980a).

vestigators, although some of these data are open to question. He presented a relation between the duration of some precursory phenomena and the magnitude of the resultant earthquake. Precursory events of other types occurred just before earthquakes.

Utsu (1979c) discussed the probability of the successful prediction of earthquakes. If the probability that an anomalous event is a precursor of an earthquake is given, the probability of successful prediction can be calculated. This probability greatly increases with the increase of independent anomalous events.

Major Topics of Earthquake Prediction in The Past 10 Years

The left side of Figure 15 shows the sites of major earthquake prediction studies in the past 10 years in Japan and the right part of the figure shows the history of studies at these sites. Large solid circles are major destructive earthquakes discussed from the standpoint of earth-

quake prediction. Horizontal broken and solid lines show the degree of scientific or social interest in the possibility of large earthquake occurrence. In this figure, the 1973 Nemuro-oki earthquake of M 7.4, the 1978 Miyagi-oki (near Sendai) earthquake of M 7.4, the activity in the Izu Peninsula with the 1974 Izu-hanto-oki earthquake of M 6.9, the 1978 Izu-Oshima-kinkai earthquake of M 7.0, and the Tokai region, as a potential region of a future great shallow earthquake, are listed. The Yamasaki fault is also listed as a test field for earthquake prediction. It should be noted that there are some interesting topics besides the above-mentioned ones. One is that the prediction of major earthquakes (M~4-5) at Wakayama in the Kinki district of western Japan was discussed from regularities in the space-time distribution of major earthquakes in this region (Mizoue et al., 1978). Another topic is that the four recent major earthquakes in western Japan (the 1975 Aso earthquake of M 6.1, the 1975 Oita earthquake of M 6.4, the 1978 Sanbe-yama earthquake of M 6.1, and the 1979 Suho-nada earthquake of M 6.3) occurred in the region predicted from the recent migration of seismic activity from south to north along the Kyushu-Ryukyu seismic belt (Mogi, 1980b). The location and the magnitude (roughly) of the 1973 Nemuro-oki earthquake were predicted by the seismic gap hypothesis (Utsu, 1972). This earth-

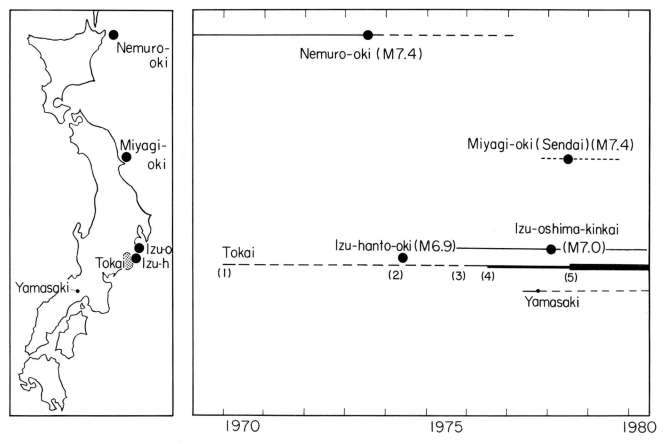

Fig. 15. Major topics of earthquake-prediction studies in the past 10 years in Japan. Large solid circles represent major destructive earthquakes. Horizontal broken and solid lines in the right figure show the degree of scientific or social interest in the probability of large earthquake occurrence.

quake is not discussed in this paper, because the prediction of this earthquake is discussed in detail in a number of previous paper (Utsu, 1972; Shimazaki, 1974; Rikitake, 1974; Mogi, 1977).

Relation between The Crustal Movements and Recent Earthquakes

Mogi (1970) discussed that locations of recent large shallow earthquakes along trenches could be approximately predicted from geodetic data. Figure 16 explains schematically the vertical and the horizontal deformation of the earth's crust along a trench. The right top figure shows the horizontal deformation of the island areas in the stationary stage. Japan is being compressed in the direction of plate motion. The right bottom figure shows the horizontal deformation affected by the occurrence of two great shallow earthquakes along the trench. The dotted region, which remains as the compressed one, is a potential region for a future large earthquake. The horizontal crustal movements of the Japanese islands during the period from (1883-1909) to

(1948-1967), shown in Figure 6, was analyzed from the above-mentioned hypothesis of earthquake prediction. In Figure 17, the compressed regions are shown by dotted areas. In 1969, after the last survey, it was suggested that large shallow earthquakes along deep sea trenches may occur in these compressed regions (Mogi 1970). All large shallow earthquakes along the Pacific coast that

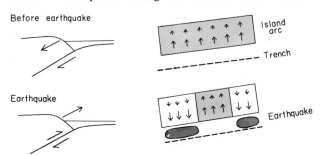

Fig. 16. Relation between the vertical (left) and the horizontal (right) crustal deformations, and great shallow earthquakes along a deep sea trench.

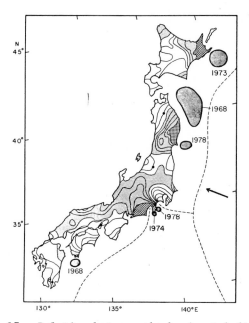

Fig. 17. Relation between the horizontal displacements of triangulation points during the period from 1883-1909 to 1948-1967 and recent large shallow earthquakes along deep sea trenches. A thick arrow shows approximately the direction of motion of oceanic plates in this area. The regions compressed in this direction are shown by dotted areas (Mogi, 1970).

happened after this triangulation survey occurred approximately in these compressed regions. In Mogi's paper, the Tokai region, one of the highly compressed regions, was given as a potential region for a future large shallow earthquake. Although the horizontal displacement vectors in Figure 6 may have some systematic errors (Fujita, 1973; Sato, 1977), the above-mentioned discussion seems to not be affected significantly.

Miyagi-oki Earthquake

Before the 1978 Miyagi-oki earthquake, Seno (1979) suggested the possibility of large shallow earthquake occurring to the east, off the Miyagi prefecture. He suggested a seismic gap of the first and second kind, as shown in Figure 18. This seismic gap was also pointed out by Utsu (1979b). The 1978 Miyagi-oki earthquake occurred near the seismic gap suggested by Seno and Utsu, but was significantly different, as shown in Figure 18. Seno (1979) still suggested a future large earthquake would occur in his seismic gap.

A major earthquake of magnitude 6.7 occurred about four months before the Miyagi-oki earthquake. Probably this earthquake is one of the foreshocks of the Miyagi-oki earthquake. Takagi (1981) noted three dimensional spatio-temporal variations in seismic activity before the main shock. These variations were reported by Takagi at the Ewing Symposium.

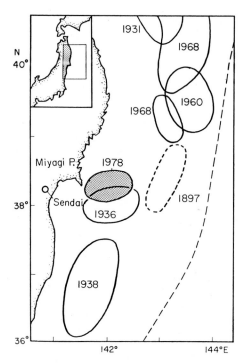

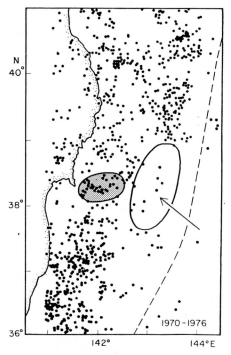

Fig. 18. The 1978 Miyagi-oki earthquake (dotted region) and seismic gap (a solid curve in the right figure), which was suggested by Seno (1979) before earthquake.

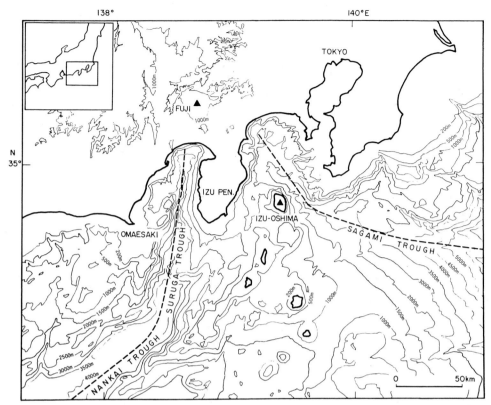

Fig. 19. Topography of the Kanto-Tokai region including the Izu Peninsula.

Earthquakes along The Yamasaki Fault

The Yamasaki fault is one of the major active faults in the Kinki district of western Japan. Along this fault, earthquakes of M~4-5 have occurred frequently. Since 1975, various kinds of observations have been carried out in this area by the cooperation of the Kyoto University and other universities as a test-field program for earthquake prediction. This test-field program includes observations of microearthquakes, crus-

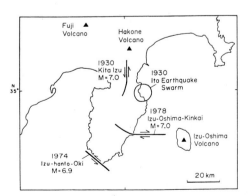

Fig. 20. Major seismic events in the Izu Peninsula in this century. These large earthquakes occurred by means of strike-slip faulting.

tal movements, radioactivities, the level and chemical components of underground water, geomagnetism, earth current, and others. Some earthquakes of M~4-5 along the Yamasaki fault were approximately predicted on the basis of the regularity in seismic activity in this region and by observations of several kinds of precursors. This topic is discussed in detail by Kishimoto (1981) in this volume.

Earthquakes in The Izu Region

Figure 19 shows the submarine topography of the Kanto-Tokai region. Deep sea trenches or troughs are shown by broken curves. They may correspond to the plate boundaries. In this region, the tectonic structures are very complex, and large earthquakes have occurred frequently. In this section, the recent crustal activity in the Izu Peninsula and its surrounding region is reviewed. This activity was also reviewed by Takahashi and Kakimi (1977), Tsumura (1977) and Mogi (1979a).

Figure 20 shows major seismic events in the Izu Peninsula in this century. In 1930, a marked uplift of the earth's crust near Ito and a very active earthquake swarm occurred. Several months after the Ito earthquake swarm, the Kita-Izu earthquake of M 7.0 occurred in the northwestern adjacent region. This activity, limited to the northern part of the Izu Peninsula, have been an

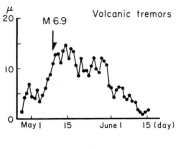

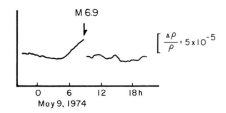

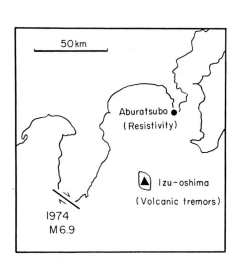

Fig. 21. Precursory phenomena of the 1974 Izu-hanto-oki earthquake of M 6.9. Right top figure: amplitude of volcanic tremors of the Izu-Oshima-volcano (Nakamura and Tazawa, 1975); right bottom figure: electric resistivity of the ground at Aburatsubo (Rikitake and Yamazaki, 1977).

aftereffect of the 1923 Kanto earthquake of M 8.0. Since this activity, the Izu Peninsula has been extremely calm, except for minor activities in 1935 (Abe, 1978) and 1965 (Sekiya, 1977; Yamakawa et al., 1977).

In 1974, the Izu-hanto-oki earthquake of M 6.9 occurred rather suddenly. Following this event, an anomalous uplift occurred in the eastern part of the Izu Peninsula, and the Izu-Oshima-kinkai earthquake of M 7.0 occurred in 1978. Thus, the Izu Peninsula has been very active since 1974. It is suggested that this activity may have originated because of the occurrence of the two Hachijozima-oki earthquakes (M 7.2) of 1972. These Hachijozima-oki earthquakes were located in the southeastern extension of the Izu-hanto-oki earthquake fault and along the Izu-Bonin trench.

1974 Izu-hanto-oki Earthquake. Before the occurrence of the Izu-hanto-oki earthquake, a very few observation stations were distributed near the source region. This earthquake occurred rather suddenly. Ohtake (1976) and Sekiya (1977) pointed out that the region was very quiet before the earthquake. Only a few short-term precursory phenomena were observed at distant places. The right top figure in Figure 21 shows the amplitudes of volcanic tremors at the Izu-Oshima volcano, which is located about 40 km to the NE of the earthquake (Nakamura and Tazawa, 1975). The amplitude began to increase about 10 days before the earthquake and decreased monotonically after the earthquake. The right bottom figure in Figure 21 shows the change in electric resistivity of the ground at Aburatsubo which is located about 100 km in the NE direction; this change was observed by Rikitake and Yamazaki

(1977). In this case, resistivity of rock under the ground increases slightly just before the earthquake.

Crustal Uplift and Earthquake Swarm in The Eastern Izu Peninsula. After the 1974 Izu-hanto-

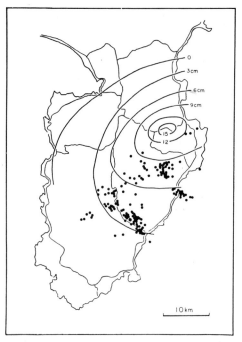

Fig. 22. Crustal uplift during the period 1967-1976 in the eastern part of the Izu Peninsula, obtained using leveling surveys (GSI, 1976) and earthquake swarms (Tsumura et al., 1977).

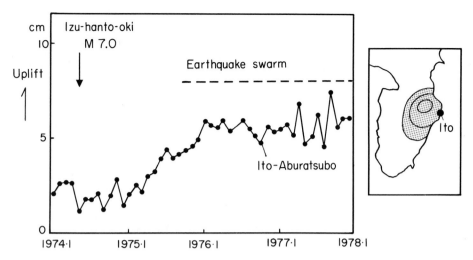

Fig. 23. Changes in the relative mean sea level at the Ito tidal station (GSI, 1979).

oki earthquake, microearthquake swarms were observed in the eastern part of the Izu Peninsula by the Earthquake Research Institute (Tsumura et al., 1977), and a marked uplift of the ground surface was found by GSI (1976). Figure 22 shows the uplift determined by leveling surveys and the locations of microearthquakes. The maximum uplift was about 15 cm.

The temporal variation of the vertical movement was deduced from the tide-gage observation at Ito, as shown in Figure 23 (GSI, 1979a). The left figure shows changes in the relative mean sea level at Ito; this data was obtained from the difference between the sea levels of Ito and Aburatsubo. An arrow shows the 1974 Izu-hanto-oki earthquake, and a horizontal broken line shows the active period of earthquake swarms. This result shows that the uplift started a half year after the 1974 Izu-hanto-oki earthquake and continued about one year. The earthquake swarm began to occur at a later stage of this uplift.

Figure 24 shows the daily number of microearthquakes observed at the Okuno station in the eastern part of the Izu Peninsula (ERI, 1980). A microearthquake swarm, with an interval of a calm period of several months, began one year after the Izu-hanto-oki earthquake. In the later stage of the period shown in this figure, the Co-operative Committee for Earthquake Prediction discussed frequently whether or not these earthquake swarms and the crustal uplift suggested the occurrence of a future large earthquake. In the

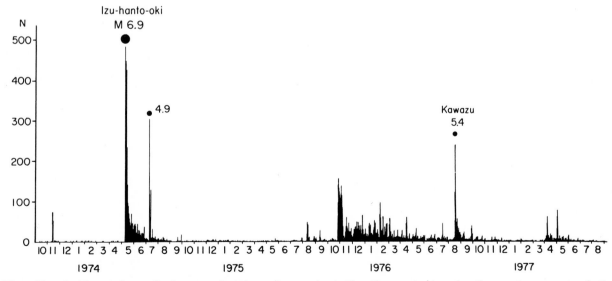

Fig. 24. Daily number of microearthquakes observed at the Okuno station in the eastern part of the Izu Peninsula during the period before the occurrence of the 1978 Izu-Oshima-kinkai earthquake of M 7.0 (from Tsumura et al., 1978; ERI, 1980).

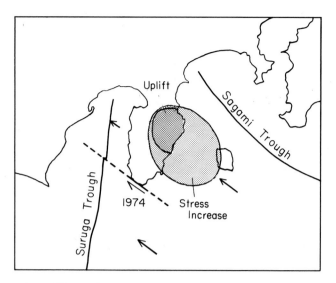

Fig. 25. A hypothesis for explaining the activation of the eastern part of the Izu Peninsula after the 1974 Izu-hanto-oki earthquake (Mogi, 1977b).

Izu region, a famous resort area near Tokyo, a number of earthquake swarms have occurred frequently. Therefore, the scientific evaluation of these events was not easy, and a strong social response to the evaluation was expected. One opinion is that this activity is only an after-effect of the 1974 Izu-hanto-oki earthquake and that it may decrease with time. Another opinion is that this crustal activity indicates stress build up in this area (Mogi, 1977b). Figure 25 shows one model that explains the mechanism of stress build up in the dotted region. In the 1974 Izu-hanto-oki earthquake, the plate motion in the south-west side of the earthquake fault

may have been accelerated via this lateral strike-slip fault, and the stress in the dotted region may have increased. In this model, an active structural zone along the fault of the Izu-hanto-oki earthquake was proposed. Various observations were continued and were rather intensified in the Izu Peninsula.

1978 Izu-Oshima-kinkai Earthquake. On January 14, 1978, the Izu-Oshima-kinkai earthquake of M 7.0 occurred in the southern boundary of the above-mentioned uplift region. The epicenter of the main shock is located between the Izu-Oshima island and the Izu Peninsula. This earthquake was preceded by a marked foreshock activity. The left figure in Figure 26 shows the locations of the regions of foreshocks, a main shock, and aftershocks (Tsumura et al., 1978). The main fault is a right lateral strike-slip fault with an EW direction (Shimazaki and Somerville, 1978). The right figure in Figure 26 shows temporal variations in seismic activity in different regions: A, B, C and D (Tsumura et al., 1978). In region A, remarakable foreshock activity increased several hours before the main shock and decreased suddenly before the main shock. It is noted that the aftershock activity is not high in region B, where the main rupture occurred. The aftershock activity developed rapidly westward. In region D, the secondary fault, which is conjugate to the extension of the main fault, activity occurred several hours after the main shock.

The discrimination of foreshocks from earthquake swarms on the basis of seismic observations is an important problem for earthquake prediction. It has been observed that the b-value of foreshocks is appreciably lower than that of other earthquakes in many cases, but it is a future problem whether or not this is always true. Tsujiura (1979) found that the waveforms

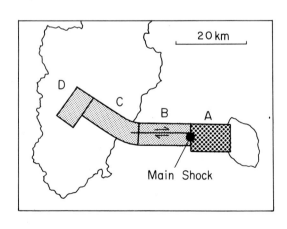

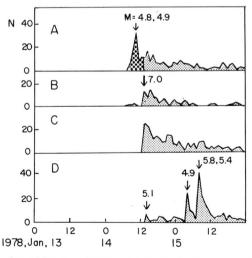

Fig. 26. Foreshocks, main shock, and aftershocks of the 1978 Izu-Oshima-kinkai earthquake of M 7.0 (simplified Tsumura et al., 1978).

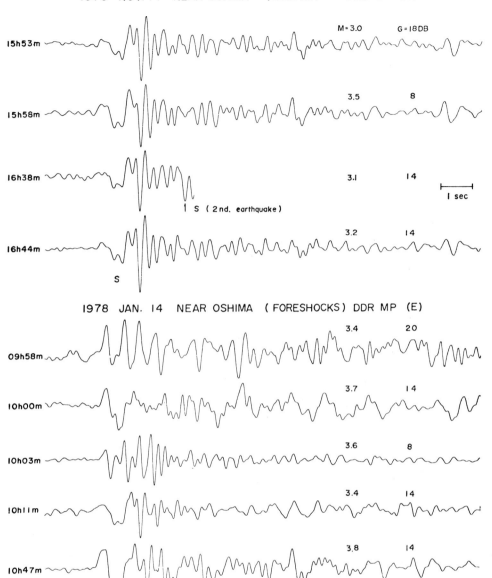

Fig. 27. Difference in the waveform similarity between earthquake swarms and foreshocks in the Izu region (Tsujiura, 1979).

of some earthquakes in an earthquake swarm were completely similar, but the waveforms in the foreshock sequence of the 1978 Izu-Oshima-kinkai earthquake were always different. In Figure 27, the upper group of seismograms represents an earthquake swarm in the Izu region, and the lower group represents foreshocks of the 1978 earthquake. If Tsujiura's result is generally applicable, waveform similarity may be a useful method of discriminating foreshocks from earthquake swarms.

Several other short-term precursory phenomena were observed. Figure 28 shows the records of the Sacks-Evertson-type volumetric strainmeters, which were observed at four stations around the Izu-Oshima-kinkai earthquake by JMA. Curves at the distant observatories do not show any changes, but curves at the Irozaki and the Ajiro observatories, about 20 km distant from the 1978 earthquake fault, showed some anomalous changes (JMA, 1978). The change that occurred a few days before the main shock at Irozaki is particularly noticeable.

Figure 29 shows the change in water level of a

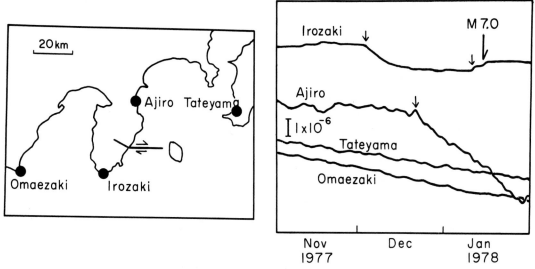

Fig. 28. Observed result made by the volumetric strainmeters of the Sacks-Evertson type at four stations around the 1978 Izu-Oshima-kinkai earthquake (JMA, 1978).

deep well at Funabara in the central part of Izu Peninsula. This well is located near the western end of the extension of the main fault. Clearly, water level became lower about one month before the earthquake (Yamaguchi and Odaka, 1978).

Figure 30 shows the variation in the radon concentration at a station in the central part of the Izu Peninsula (Wakita et al., 1980). The radon concentration changed several days before the 1978 earthquake. Changes in the temperatures of hot springs were also reported, but changes on the hot springs seem not to be so clear.

Geomagnetic surveys in the Izu Peninsula have

been carried out repeatedly and some changes have been found during the anomalous crustal activities in the Izu Peninsula. At some stations, continuous observations have been made by proton magnetometers. Figure 31 shows a remarkable variation in the total geomagnetic intensity at Kawazu before a major aftershock of the Izu-Oshima-kinkai earthquake (Sasai and Ishikawa, 1980). This aftershock occurred along the fault of the main shock and the Kawazu station (K) is located on the fault. At station (S), 10 km distant from the fault, no appreciable geomagnetic changes related to this aftershock were observed.

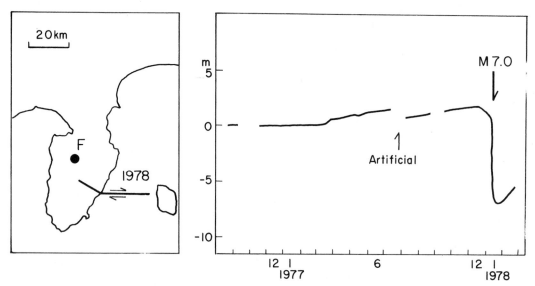

Fig. 29. The change in water level of a deep well at Funabara (F) in the central part of the Izu Peninsula (Yamaguchi and Odaka, 1978).

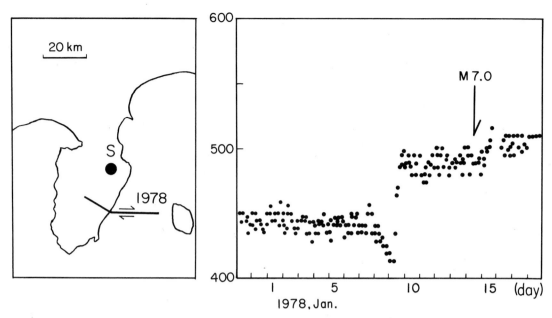

Fig. 30. Temporal variation in the radon concentration at a station (S) in the central part of the Izu Peninsula (Wakita et al., 1980).

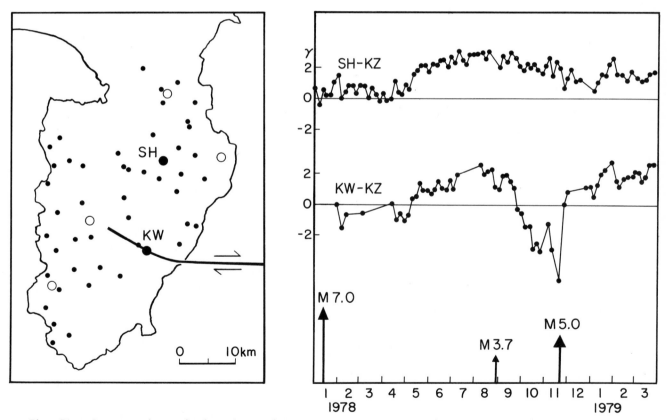

Fig. 31. Geomagnetic variation observed by proton magnetometers (Sasai and Ishikawa, 1980). Left figure shows sites of geomagnetic surveys (large circles: continuous observation stations). A remarkable variation in the total geomagnetic intensity before a major aftershock of the 1978 Izu-Oshima-kinkai earthquake was observed at the Kawazu station (KW).

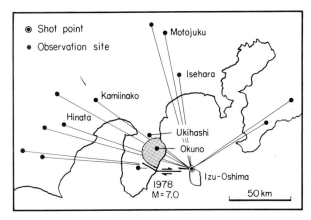

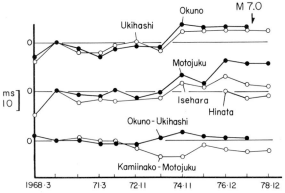

Fig. 32. Observation of seismic velocity changes by the Izu-Oshima explosions in the Kanto-Tokai region (GSJ, 1979).

This result suggests that the magnetic precursory change may be observed only at the station very near the fault.

Since 1968, the Geological Survey of Japan has reported the travel time observation at Izu-Oshima island in the Izu-Tokai region (GSJ, 1979). These observations were made by recording the travel time of waves from explosions. The accuracy of the travel time observation is very high (± 5 milliseconds). The results of these observation for the past 10 years is shown in Figure 32. Although this explosion seismic network covers the uplift area in the Izu Peninsula and the focal region of the 1978 Izu-Oshima-kinkai earthquake of M 7.0, any appreciable change can be recognized before the 1978 earthquake. The coseismic change cannot be recognized either. This observation may be one of the most accurate measurements of velocity variation in relation to the occurrence of large shallow earthquakes. Therefore, this result shows that the likelihood of a large velocity change before a large earthquake is open to question. Because of this fact, the dilatancy model formulated on the basis of a large velocity change occurring before a large earthquake should be reconsidered.

Figure 33 shows the vertical crustal movement

in the Izu Peninsula from 1967 to 1978, including the coseismic movement of the Izu-Oshima-kinkai earthquake (GSI, 1978b). A remarkable uplift in the northeastern part of the Izu Peninsula remained. Since an early stage of anomalous crustal movements, leveling and gravity surveys have been frequently repeated. Figure 34a shows gravity changes in the Izu Peninsula from 1972 to 1978 (Hagiwara et al., 1978). This spatial pattern is quite similar to that of the vertical crustal movement in Figure 33. Figure 34b shows the temporal variation of gravity at station A, which is located at the center of the uplifted area. Gravity decreased markedly in the early stage in this figure, and then decreased gradually. No noticeable coseismic change due to the 1978 Izu-Oshima-kinkai earthquake was observed.

Figure 35 shows the daily number of micro-earthquakes in the past eight years in the Izu region, observed at the Okuno station (Tsumura et al., 1978; ERI, 1980). After the Izu-Oshima-kinkai earthquake, remarkable earthquake swarms and ground uplift occurred near Ito (GSI, 1980a). In Figure 36, the locations of earthquakes in the following successive periods since 1974 are shown:

(a) the 1974 Izu-hanto-oki earthquake of M 6.9 and its aftershocks,

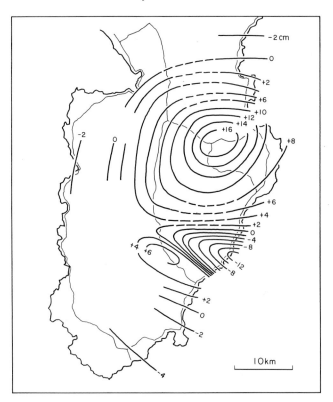

Fig. 33. Vertical crustal movement in the Izu Peninsula during the period from 1967 to 1978 (including the 1978 Izu-Oshima-kinkai earthquake) (GSI, 1978).

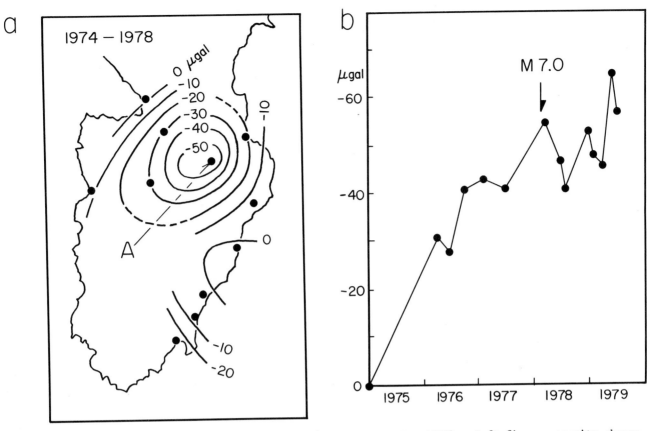

Fig. 34. Gravity changes in the Izu Peninsula (Hagiwara et al., 1978). Left figure: gravity change during the period from 1972 to 1978; right figure: temporal variation of gravity at station A indicated in the left figure.

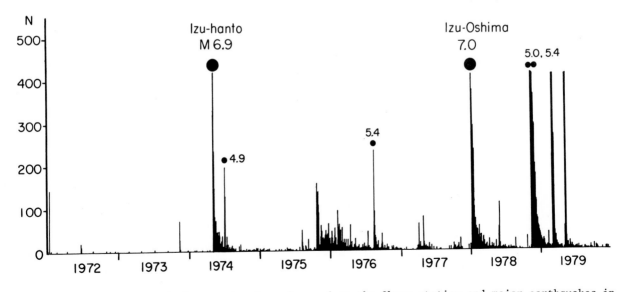

Fig. 35. Daily number of microearthquakes observed at the Okuno station and major earthquakes in the Izu region in the past eight years (from Tsumura et al., 1978; ERI, 1980).

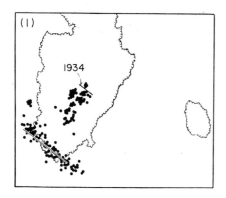 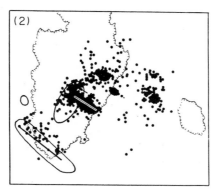 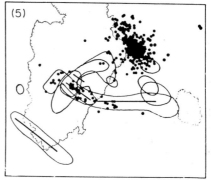

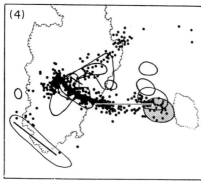

Fig. 36. Locations of earthquakes in the successive periods since 1974 (compiled from Research Group for Aftershocks, 1975; Ishibashi and Yamashina, 1975; Tsumura et al., 1977, 1978; Abe, 1978; Shimazaki and Somerville, 1978; ERI, 1980). (a) The 1974 Izu-hanto-oki earthquake and its aftershocks; (b) 1975 November- 1976 December (including the 1976 Kawazu earthquake of M 5.5); (c) 1977 January-December (earthquake swarms before the 1978 Izu-Oshima-kinkai earthquake); (d) 1978 January 1-November 23 (the 1978 Izu-Oshima-kinkai earthquake and its aftershocks); (e) 1978 November 24-1979 March 31 (earthquake swarms near Ito).

(b) earthquake swarms including the 1976 Kawazu earthquake of M 5.5,

(c) earthquake swarm activities before the 1978 Izu-Oshima-kinkai earthquake,

(d) the 1978 Izu-Oshima-kinkai earthquake of M 7.0, and its foreshocks and aftershocks,

(e) earthquake swarms near Ito until 1979. In each figure of Figure 36, the previous active regions are shown by solid curves.

Figure 37 shows locations of major events since 1974. The active region seems to have migrated from SW to NE. Various observations are continuing in the north-eastern part of the Izu Peninsula.

Since microearthquake swarm activities and anomalous crustal uplift in the eastern part of the Izu Peninsula were found about two or three years before the earthquake, various kinds of observations had been particularly intensified in the Izu Peninsula and then the 1978 Izu-Oshima-kinkai earthquake occurred. As mentioned above, several kinds of precursors were observed. Nevertheless, the occurrence of the Izu-Oshima-kinkai earthquake was not predicted. One of the reasons is that we have no real-time data processing system

for various kinds of observations needed to predict an earthquake of M 7 in this region.

After the Ewing Symposium on Earthquake Prediction (May, 1980), a large earthquake of M 6.7, occurred on June 29, 1980, east of Ito. This

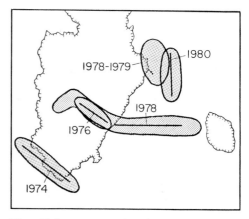

Fig. 37. Major events in the Izu region since 1974.

656 MOGI

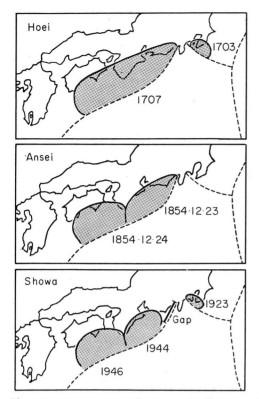

Fig. 38. Rupture zones of great shallow earth-quakes along the Nankai and the Sagami trough in the successive periods. This figure suggests that the Tokai region along the Suruga trough is a seismic gap of the first kind (compiled from Hagiwara, 1970; Usami, 1975; Hatori, 1976; Ishibashi, 1977).

earthquake was preceded by a marked swarm activi-ty. The earthquake fault is a left-lateral strike-slip fault having a NS direction (Shimazaki, personal communication). The preli-minary location of the focal region of this earthquake is shown in Figure 37. This earth-quake will be discussed in detail, on the basis of various observations including high frequency measurements by a hydrophone in the future paper.

Earthquake Prediction Problem in The Tokai Region

The Tokai region was first selected as a poten-tial region for a future great earthquake in 1969, on the basis of horizontal crustal deforma-tion, as mentioned in the preceding chapter (Mogi,1970). The Coordinating Committee for Earthquake Prediction designated this region as one of the areas that should receive specific observation, on the basis of its recent crustal deformation and historical earthquake data. Thereafter, by considering detailed historical data about the preceding great earthquake of 1854, Ishibashi (1977) proposed a model of a future rupture zone, which is an underthrust

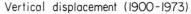
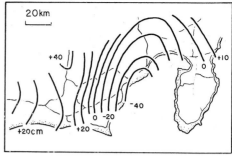

Horizontal deformation (1884-1973)

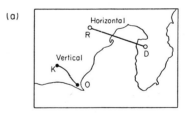

Fig. 39. Recent vertical (upper figure) and horizontal (lower figure) crustal movements in the Tokai region (GSI, 1978a, 1980c).

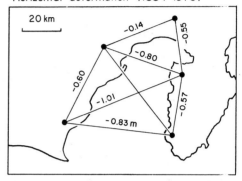

(b) Vertical displacement (O - K)

(c) Horizontal displacement (R - D)

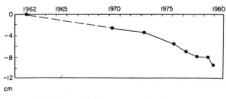
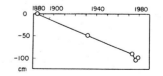

Fig. 40. Temporal variations of subsidence of the east coast of the Suruga bay (GSI, 1980b) and the horizontal contraction between the east coast of the Suruga bay and the Izu Peninsula (GSI, 1978c, 1979b).

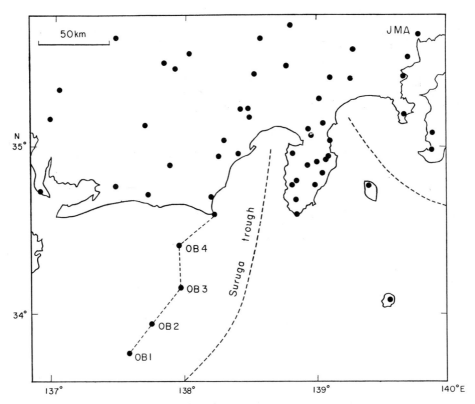

Fig. 41. Seismographic stations in the Tokai region. An array of ocean-bottom seismographs (OB in this figure) on a cable system is being operated for the real-time observation.

◑ Linear strainmeter; Tiltmeter
● Volumetric strainmeter
○ Tide gage

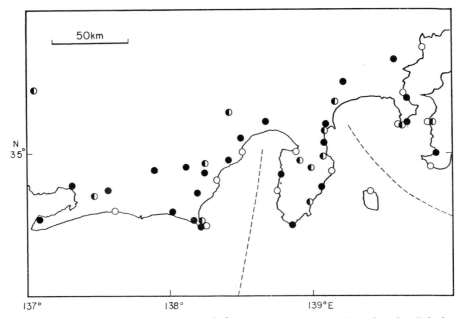

Fig. 42. Sites of continuous crustal deformation observatories in the Tokai region.

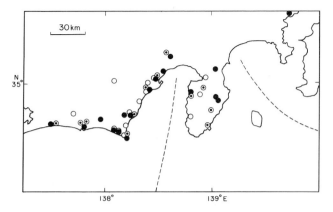

Fig. 43. Sites of various kinds of observations, except for seismic observations and crustal deformation measurements.

fault along the Suruga trough. The underthrust fault mechanism is expected from the recent geodetic data in this region and the submarine topography along the Suruga trough. The possibility of the occurrence of the Tokai earthquake was also discussed by Aoki (1977) and Utsu (1977).

Figure 38 shows that the Tokai region is a seismic gap of the first kind.

Recent Crustal Activity. The upper figure in Figure 39 shows the vertical displacement during the period from 1900 to 1973 (GSI, 1978a). During this time period, the east coast of the Suruga bay subsided markedly. The lower figure shows changes in horizontal distances between triangulation points in this region during the period from 1884 to 1973 (GSI, in preparation). The contraction between the east coast of the Suruga bay and the west coast of the Izu Peninsula was about one meter. These vertical and horizontal crustal movements strongly suggest the possibility of subduction along the Suruga trough.

Figure 40 shows the temporal variations of sub-

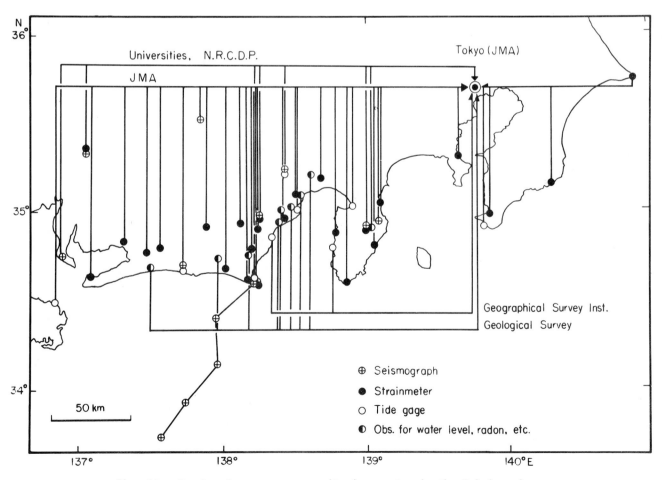

Fig. 44. Earthquake precursor monitoring system in the Tokai region.

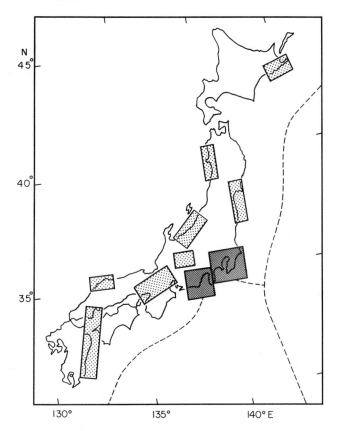

Area of intensified observation
Area of specific observation

Fig. 45. Areas of intensified observation and specific observation, which were designated by the Coordinating Committee for Earthquake Prediction in 1978. The Tokai region and the southern Kanto region including Tokyo area are the areas of specific observation.

sidence of the east coast of the Suruga bay and the horizontal contraction between the east coast of the Suruga bay and the Izu Peninsula. Figure 40b shows the relative vertical movement between Kakegawa (K) and Omaezaki (O) (GSI, 1980b), and Figure 40c shows the contraction between Ryusozan (R) and Darumayama (D) (GSI, 1978b, 1979b). Both subsidence and contraction are continuing in a stationary manner.

Furthermore, the surrounding part of the Tokai region, such as the Izu Peninsula, has been seismically active in recent years (Utsu, 1977; Seno, 1979). This activity may be an indication of a high level of stress in this region.

Observation in The Tokai Region. As mentioned above, the occurrence of a great shallow earthquake along the Suruga trough is expected in the future. However, no short-term precursory pheno-

mena have been observed until now. Perhaps, the occurrence of the earthquake can be predicted accurately by observation of short-term precursors. The following report is very important in relation to the probability of the appearance of short-term precursors of this future earthquake. The 1944 Tonankai earthquake, the source region of which is adjacent to the expected rupture zone in the Tokai region, was preceded by a marked crustal uplift at the coastal region just before the great earthquake. This precursory uplift was observed by the leveling survey, which was carried out near Kakegawa. Therefore, it can be expected that the future Tokai earthquake may be

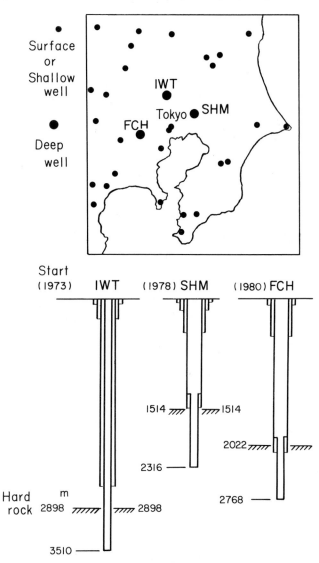

Fig. 46. Upper figure: seismic observation network in the Tokyo area. Large circles are deep-bore-hole observatories with about 3000 m depth. Lower figure: vertical sections of the deep-bore-holes for seismic and tilt observations in the Tokyo area.

preceded by appreciable precursory events.

For the purpose of predicting the earthquake, a very dense network of various instruments has been set up in the Tokai region. The observation network is shown in the following figures. Figure 41 shows seismographic stations in this region; some of them are telemetered to the center of JMA in Tokyo. Along the Nankai-Suruga trough, an array of ocean-bottom seismographs on a cable system was set up, and real-time observation for microearthquakes has been carried out at the ocean bottom (Meteorological Res. Inst., 1980). Figure 42 shows sites of continuous crustal deformation observations made by various instruments, including linear strainmeters, volumetric strainmeters, tiltmeters, and tide gages. Figure 43 shows sites of other various observations, including groundwater level observations, groundwater or hot spring temperature measurements, and radon content observations.

The earthquake precursor monitoring system in the Tokai region is shown in Figure 44. The data, which are obtained by JMA, GSI, the National Research Center for Disaster Prevention, the Geological Survey of Japan, Nagoya University, and Tokyo University, are telemetered to the center of JMA in Tokyo, which is staffed 24 hours a day and 7 days a week.

Intensified Observation in Other Areas

Figure 45 shows the areas of intensified observation and the areas of specific observation that were designated by the Coordinating Committee for Earthquake Prediction in 1978. The above-men-

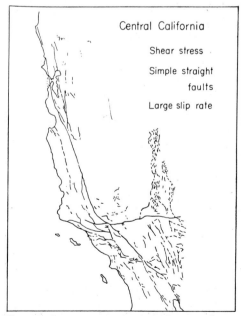

Fig. 47. Fault pattern in California (from Jennings, 1975). Central California is characterized by a simple shear movement along a linear long fault zone.

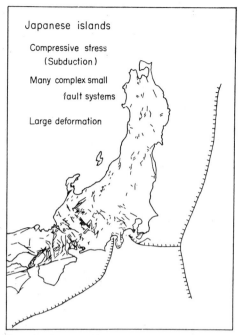

Fig. 48. Fault pattern in the northeastern Honshu, Japan (from Matsuda et al., 1976). This region is adjacent to subduction zones, which are highly compressed and fractured.

tioned Tokai region and the southern Kanto region, including the Tokyo area, are the areas of specific observation. In the Tokyo area, destructive earthquakes have occurred frequently. Because of its dense population and heavy industrial activity, serious damage may be caused even by local earthquakes. The occurrence of large shallow earthquakes along the Sagami trough should also be considered in the future.

Very precise observations of crustal activity are required for earthquake prediction; however, the Tokyo area is covered by a thick soft layer and is very noisy due to traffic and industrial activities. To avoid these measurement difficulties, the National Research Center for Disaster Prevention has constructed three deep-bore-hole observatories, which are shown by solid circles in the upper figure of Figure 46 (Takahashi and Hamada, 1975; Suzuki and Takahashi, 1979). The lower figure of Figure 46 shows schematically the vertical sections of these deep (about 3000m) bore-holes. Using this deep-bore-hole method, very precise observations are obtained by seismographs and tilt meters. For the purpose of local earthquakes in the Tokyo area, a dense network of special observations, such as those in deep wells, are required.

Comparison of Earthquake Prediction in Japan with That in Other Countries

In China, some destructive earthquakes were successfully predicted. On the other hand, pre-

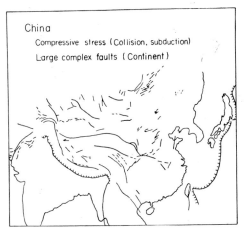

Fig. 49. Simplified fault pattern in China (from Molnar and Tapponier, 1975). Some regions in China are characterized by the complex large-scale continental structure and the highly compressed state.

cursory phenomena have not been observed prior to recent earthquakes in central California. In Japan, some recent earthquakes were preceded by noticeable precursors, as mentioned above. In this chapter, these different regions are compared from the standpoint of earthquake prediction.

Figure 47 shows the fault pattern in California (Jennigs, 1975). Central California is structurally characterized by simple shear movement along a long linear fault zone. This slip movement has continued for a very long time ($\sim 10^7$ years). Therefore, it is deduced that the fault plane may be flat and homogeneous, and so the fault movement may occur without any marked stress concentration. In such a homogeneous fault zone, the degree of appearance of precursory phenomena may be low, as suggested by laboratory experiments (Mogi, 1980c). Figure 48 shows the fault pattern in northeastern Honshu, Japan (Matsuda et al., 1976). This region is adjacent to subduction zones, and is highly compressed and fractured. The dimension of complex structures is small. In this case, the stress is concentrated at many singular points, and appreciable precursors, such as foreshocks and precursory deformations, may occur at these points, as also shown in laboratory experiments. Figure 49 shows a very simplified fault pattern in China (Molnar and Tapponnier, 1975). Some regions in China are mainly characterized by a complex large-scale continental structure and a highly compressed state. In this case, marked stress concentrations are expected at structural singular points before earthquakes and appreciable precursors may occur at these points. It may be explained by the large-scale continental structure that precursory phenomena were sometimes observed at distant places in China. In Figure 50, the simplified tectonics

and the nature of precursory phenomena in these areas are compared. The nature of precursory phenomena is different for different tectonic situations. In Japan, very precise and dense observation networks are required for prediction of earthquakes of M~7.

Conclusion

The first 5-year earthquake prediction project in Japan was started in 1965, and now the fourth 5-year project (1979-1983) is in progress. Geodetic, seismic, geomagnetic, geoelectric, and geochmical observations, continuous observations of crustal deformation, gravimetric surveys, and measurements of seismic velocity changes have been made. The active fault map of all of Japan was published in 1980. Laboratory experiments on rock failure have also been conducted to clarify the physical processes of the precursory phenomena.

The density of various observation networks has rapidly increased with time. Telemetering systems for microearthquake observations have been established in the major part of the land area by universities and Government agencies. The real-time observations made by JMA used an array of ocean-bottom seismographs in the Tokai region; those made by the National Research Center for Disaster Prevention, which used seismic observations at 3000-meter-deep wells in Tokyo area, were very successful.

During the past few years, several destructive earthquakes, including the 1974 Izu-hanto-oki earthquake (M 6.9), the 1978 Izu-Oshima-kinkai earthquake (M 7.0), and the 1978 Miyagi-oki earthquake (M 7.4), occurred successively in Japan. In spite of the recently intensified observation, noticeable precursor data were very limited. To observe precursors of earthquakes of M~7, the density of the present observation network is still insufficient.

In the case of the 1978 Izu-Oshima-kinkai earthquake, various kinds of observations have been particularly intensified in the Izu Peninsula; microearthquake swarms and anomalous crustal uplift were observed in the eastern part of the Izu Peninsula about two years before the

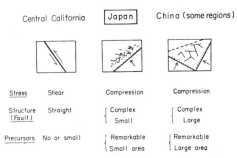

Fig. 50. Simplified tectonics and the nature of precursory phenomena in central California, Japan and China.

large earthquake, and several kinds of remarkable precursors of the Izu-Oshima-kinkai earthquake were observed. This set of data is probably the most accurate one for such a large earthquake. One of the important results in relation to this earthquake is that we could not find any appreciable change in seismic velocity before the earthquake, nor was any seismic change found, although the velocity measurement was very accurate.

The 1978 Miyagi-oki earthquake occurred in the region adjacent to the area that had been pointed out as a seismic gap. Prior to this low-angle-thrust-type earthquake, noticeable changes in seismic activity in and around the fault zone were observed.

It is also noted that the 1975 Aso earthquake of M 6.1, the 1975 Ōita earthquake of M 6.4, the 1978 Sanbe-yama earthquake of M 6.1, and the 1979 Suho-nada earthquake of M 6.3 in western Japan occurred in the region predicted from the recent migration phenomenon from south to north along the Kyushu-Ryukyu seismic zone.

The Yamasaki fault in the Kinki district of western Japan has been investigated as a test field for earthquake prediction. On the basis of the regularity in seismic activity in this region and observations of various kinds of precursors, moderate earthquakes (M~4-5) along the fault were approximately predicted.

The Tokai region has been named as a potential region for a great shallow earthquake since 1969, on the basis of the recurrence time of historical great shallow earthquakes along the Nankai trough and the crustal deformation observed by geodetic surveys. In the case of the 1944 Tonankai earthquake, the source region of which is just adjacent to the expected source region in the Tokai region, a marked uplift just before the great earthquake was observed by the leveling survey. Therefore, we can reasonably expect that the future Tokai earthquake may be preceded by appreciable precursory events.

For the purpose of predicting the earthquake, a very dense observation network of various instruments has been set up in the Tokai region. These data are telemetered to a center at JMA in Tokyo. When anomalous behaviour is observed, the Earthquake Prediction Council eveluates the data and conveys its findings to the director general of JMA.

Accumulated data on earthquake precursors in Japan suggest that the order of appearance of precursors depends on the degree of the structural heterogeneity of the earth's crust or fault system. From this point of view, Japan may be compared with central California and some regions in China.

(This paper is not only the national report on earthquake-prediction program in Japan, but it also contains the author's opinions.)

Acknowledgements. I thank Professors Takahiro Hagiwara and Lynn R. Sykes for their valuable advice on this paper. I thank a number of Japanese colleagues of the earthquake prediction project for their help in preparation of the manuscript , particularly Dr. Norio Yamakawa for his critical reading the paper.

References

Abe, K., Dislocations, source dimensions and stresses associated with earthquakes in the Izu Peninsula, Japan, J. Phys. Earth, 26, 253-274, 1978.

Aoki, H., Possibility of a great earthquake in Tokai district, Symposium on Earthquake Prediction Research (1976), 56-68, 1977 (in Japanese).

Dambara, T., Synthetic vertical movements in Japan during the recent 70 years, J. Geodetic Soc. Japan, 17, 100-108, 1971 (in Japanese).

Earthquake Research Institute, Seismic activity in the Izu Peninsula and its vicinity (November, 1979 - April, 1980). Rep. Coord. Comm. Earthquake Prediction, 24, 108-112, 1980 (in Japanese).

Fujita, N., Horizontal displacement vectors in Kanto and Chubu districts, Rep. Coord. Comm. Earthquake Prediction, 10, 64-67, 1973 (in Japanese).

Geographical Survey Institute, Crustal deformation in the central part of Izu Peninsula, Rep. Coord. Comm. Earthquake Prediction, 16, 82-87, 1976 (in Japanese).

Geographical Survey Institute, Crustal movement in the Tokai district, Rep. Coord. Comm. Earthquake Prediction, 19, 96-98, 1978a (in Japanese).

Geographical Survey Institute, Crustal deformation in Izu Peninsula, Rep. Coord. Comm. Earthquake Prediction, 20, 92-99, 1978b (in Japanese).

Geographical Survey Institute, Horizontal strains in the Tokai district, Rep. Coord. Comm. Earthquake Prediction, 20, 166-171, 1978c (in Japanese).

Geographical Survey Institute, Crustal deformation in the eastern Izu district, Rep. Coord. Comm. Earthquake Prediction, 22, 68-71, 1979a (in Japanese).

Geographical Survey Institute, Horizontal crustal movement in the Tokai district, Rep. Coord. Comm. Earthquake Prediction, 22, 159-162, 1979b (in Japanese).

Geographical Survey Institute, Crustal deformation in the eastern Izu district, Rep. Coord. Comm. Earthquake Prediction, 23, 48-52, 1980a (in Japanese).

Geographical Survey Institute, Crustal deformation in the Tokai district, Rep. Coord. Comm. Earthquake Prediction, 24, 152-158, 1980b (in Japanese).

Geographical Survey Institute, Horizontal strains in the Tokai district, (in preparation), 1980c.

Geological Survey of Japan, Measurements of variation in seismic wave velocity by using explosion seismic method - Preliminary report of the result in 10th (1976) - 12th (1978) Oshima explosion -, Rep. Coord. Comm. Earthquake Prediction, 22, 83-85, 1979 (in Japanese).

Hagiwara, T., Distribution of seismic intensity of the 1854 Tokai earthquake, Rep. Coord. Comm. Earthquake Prediction, 3, 51-52, 1970 (in Japanese).

Hagiwara, T., Earthquake prediction research in Japan, Rep. Coord. Comm. Earthquake Prediction (1975), 30pp, 1975 (in Japanese).

Hagiwara, Y., H. Tajima, S. Izutuya and H. Hanada, Gravity change associated with earthquake swarm activities in the eastern part of Izu Peninsula, Bull. Earthq. Res. Inst., 52, 141-150, 1977 (in Japanese).

Hagiwara, Y., H. Tajima, S. Izutuya, K. Nagasawa, I. Murata and S. Shimada, Gravity change during the Izu-Oshima-kinkai earthquake of 1978, Bull. Earthq. Res. Inst., 53, 875-880, 1978 (in Japanese).

Harada, K. and N. Isawa, Horizontal deformation of the crust in Japan - Result obtained by multiple fixed stations, J. Geodetic Soc. Japan, 14, 101-105, 1969 (in Japanese).

Harada, K. and A. Kassai, Horizontal strain of the crust in Japan for the last 60 years, J. Geodetic Soc. Japan, 17, 4-7, 1971 (in Japanese).

Hatori, T., Documents of tsunami and crustal deformation in Tokai district associated with the Ansei earthquake of Dec. 23, 1854, Bull. Earthq. Res. Inst., 51, 13-28, 1976 (in Japanese).

Huzita, K., Y. Kishimoto and K. Shiono, Neotectonics and seismicity in the Kinki area, southwest Japan, J. Geosciences, Osaka City Univ., 16, 93-124, 1973.

Ishibashi, K., Re-examination of a great earthquake expected in the Tokai district, central Japan - Possibility of the "Suruga Bay Earthquake", Rep. Coord. Comm. Earthquake Prediction, 17, 126-132, 1977 (in Japanese).

Ishibashi, K. and K. Yamashina, The largest induced shock and active faults associated with it, in conjuction with the Off-Izu-Peninsula earthquake of 1974, Rep. 1974 Izu-hanto-oki Earthquake and Damages Due to the Earthquake, 27-32, 1975 (in Japanese).

Japan Meteorological Agency, On the Izu-Oshima-kinkai earthquake, 1978, Rep. Coord. Comm. Earthquake Prediction, 20, 45-50, 1978 (in Japanese).

Jennings, C. W., Fault Map of California, Geological Data Map No. 1, California Division of Mines and Geology, 1975.

Kanamori, H., Recent development in earthquake prediction research in Japan, Tectonophysics, 9, 291-300, 1970.

Kato, T. and K. Tsumura, Vertical land movement in Japan as deduced from tidal record (1951-1978), Bull. Earthq. Res. Inst., 54, 559-628, 1979 (in Japanese).

Kishimoto, Y., On precursory phenomena observed at the Yamasaki fault, southwest Japan, as a test field for earthquakes prediction (in this volume).

Matsuda, T., Active faults and damaging earthquakes in Japan - Macroseismic zoning and precaution fault zones (in this volume).

Matsuda, T., A Okada and K. Huzita, Fault and Earthquakes, Mem. Geol. Soc. Japan, 12, Appendix, 1976.

Meteorological Research Institute (Seismology and Volcanology Research Division), Permanent ocean-bottom seismograph observation systems, Tech. Rep. Meteorological Res. Inst., 4, 1-233, 1980 (in Japanese).

Mizoue, M., M. Nakamura, Y. Ishiketa and N. Seto, Earthquake prediction from micro-earthquake observation in the vicinity of Wakayama City, northwestern part of the Kii Peninsula, central Japan, J. Phys. Earth, 26, 397-416, 1978.

Molnar, P. and P. Tapponnier, Cenozoic tectonics of Asia: Effects of a continental collision, Science, 189, 419-426, 1975.

Mogi, K., Some discussions on aftershocks, foreshocks and earthquake swarms - the fracture of a semi-infinite body caused by an inner stress origin and its relation to earthquake phenomena, 3., Bull. Earthq. Res. Inst., 41, 615-658, 1963.

Mogi, K., Earthquakes and fractures, Tectonophysics, 5, 35-55, 1967.

Mogi, K., Recent horizontal deformation of the earth's crust and tectonic activity in Japan (1), Bull. Earthq. Res. Inst., 48, 413-430, 1970.

Mogi, K., Dilatancy of rocks under general triaxial stress states with special reference to earthquake precursors, J. Phys. Earth, 25, Suppl., S203-S217, 1977a.

Mogi, K., An interpretation of recent tectonic activity in the Izu-Tokai district, Bull. Earthq. Res. Inst., 52, 315-331, 1977b (in Japanese).

Mogi, K., Izu - recent crustal activity -, The Ten-Years Activity of the Coordinating Committee for Earthquake Prediction, Geographical Survey Institute, 121-140, 1979a (in Japanese).

Mogi, K., Two kinds of seismic gaps, Pageoph, 117, 1172-1186, 1979b.

Mogi, K., Amplitude-frequency relationship of acoustic emission events, Programme and Abstracts, Seismol. Soc. Japan, 1980 (1), 176, 1980a (in Japanese).

Mogi, K., Migration of seismic activity in western Japan, Rep. Coord. Comm. Earthquake Prediction, 23, 149-150, 1980b (in Japanese).

Mogi, K., Experimental study on precursory deformations preceding to slips of heterogeneous faults, Programme and Abstracts, Seismol. Soc. Japan, 1980 (2), 140, 1980c.

Nakamura, K. and K. Tazawa, A possible precursor of 1974 Izu-hanto-oki earthquake - Rise of the magma head and amplitude increase of volcanic tremors of Miharayama, Izu-Oshima volcano, Rep.

Coord. Comm. Earthquake Prediction, 13, 75-78, 1975 (in Japanese).

Ohtake, M., Search for precursors of the 1974 Izu-hanto-oki earthquake, Japan, Pure. Appl. Geophys.,114, 1083-1093, 1976.

Ohtake, M. and M. Katsumata, Detection of premonitory change in seismic waves velocity, Symposium on Earthquake Prediction Research (1976), 106-115, 1977 (in Japanese).

Okada, A., M. Ando and T. Tsukuda, Surveys of active faults by trenching method, The Earth Monthly, 8th Symposium (Active fault), 608-615, 1979.

Research Group for Active Faults, Active faults in Japan, University of Tokyo Press, Tokyo, 363pp, 1980 (in Japanese).

Research Group for Aftershocks, Observation of the main and aftershocks of the earthquake off the Izu Peninsula, 1974, Rep. 1974 Izu-hanto-oki Earthquake and Damages Due to the Earthquake, 11-19, 1975 (in Japanese).

Rikitake, T., Possibility of earthquake occurrences as estimated from crustal strain, Tectonophysics, 23, 299-313, 1974.

Rikitake, T., Earthquake precursors, Bull. Seismol. Soc. Am., 65, 1133-1162, 1975.

Rikitake, T., Earthquake Prediction, Elsevier, Amsterdam, 357pp, 1976.

Rikitake, T. and Y. Yamazaki, Precursory and coseismic changes in ground resistivity, J. Phys. Earth, 25, Suppl. S161-S173, 1977.

Sasai, Y. and Y. Ishikawa (Geomagnetic Mobile Survey , ERI), Repeated magnetic survey and observation of total force intensity in the eastern part of the Izu peninsula (5), Rep. Coord. Comm. Earthquake Prediction, 23, 57-59, 1980 (in Japanese).

Sato, H., Crustal movements in the Tokai district by geodetic surveys, Rep. Regional Sub- comm., Coord. Comm. Earthquake Prediction, 1, 19-27, 1977 (in Japanese).

Scholz, C. H., The frequency-magnitude relation of microfracturing in rock and its relation to earthquakes, Bull. Seismol. Soc. Am., 58, 399-415, 1968.

Sekiya, H., Anomalous seismic activity and earthquake prediction, J. Phys. Earth, 25, Suppl., S85-S93, 1977.

Seno. T., Intraplate seismicity in Tohoku and Hokkaido, northern Japan, and a possibility of a large interplate earthquake off the southern Sanriku coast, J. Phys. Earth, 27, 21-51, 1979.

Shimazaki, K., Preseismic crustal deformation caused by an underthrusting oceanic plate in eastern Hokkaido, Japan, Phys. Earth Planet. Int., 8, 148-157, 1974.

Shimazaki, K. and P. Somerville, Summary of the static and dynamic parameters of the Izu-Oshima-kinkai earthquake of January 14, 1978, Bull. Earthq. Res. Inst., 53, 613-628, 1978.

Suyehiro, S., T. Asada and M. Ohtake, Foreshocks and aftershocks accompanying a perceptible earthquake in central Japan, Papers Meteo-rol. Geophys., 15, 71-88, 1964.

Suzuki, H. and H. Takahashi, Fuchu deep borehole observatory of the crustal activity and geological structure in this region, Proceedings of 16th Symposium on Natural Disaster Sciences, 599-602, 1979 (in Japanese).

Suzuki, Z., K. Tsumura, K. Oike and K. Katsumata, Micro-earthquake observatories in Japan (Second Edition), 161pp, 1979 (in Japanese).

Takagi, A., Ewing Symposium presentation, 1980.

Takahashi, H. and K. Hamada, Deep borehole observation of the earth's crust activities around Tokyo - Introduction of the Iwatsuki Observatory, Pageoph, 113, 311-320, 1975.

Takahashi, H. and T. Kakimi, Anomalous crustal activity in the eastern part of Izu peninsula, Geological News, Geological Survey of Japan, 270, 1-15, 1977 (in Japanese).

Tsujiura, M., The difference between foreshocks and earthquake swarms, as inferred from the similarity of seismic waveform (preliminary report), Bull. Earthq. Res. Inst., 54, 309-315, 1979.

Tsumura, K., Anomalous crustal activity in the Izu Peninsula, Central Honshu, J. Phys. Earth, 25, Suppl., S51-S68, 1977.

Tsumura, K., A hypothesis for explanation of the space-time distribution of earthquake precursor phenomena, Programme and Abstracts, Seismol. Soc. Japan, 1979 (1), 159, 1979 (in Japanese).

Tsumura, K., I. Karakama, I. Ogino, K. Sakai and M. Takahashi, Observation of the earthquake swarm in the Izu Peninsula (1975-1977), Bull. Earthq. Res. Inst., 52, 113-140, 1977 (in Japanese).

Tsumura, K., I. Karakama, I. Ogino and M. Takahashi, Seismic activities before and after the Izu-Oshima-kinkai earthquake of 1978, Bull. Earthq. Res. Inst., 53, 675-706, 1978 (in Japanese).

Usami, T., Descriptive Catalogue of Disaster Earthquakes in Japan, University of Tokyo Press, Tokyo, 327pp, 1975 (in Japanese).

Utsu, T., Large earthquakes near Hokkaido and the expectancy of the occurrence of a large earthquake off Nemuro, Rep. Coord. Comm. Earthquake Prediction, 7, 7-13, 1972 (in Japanese).

Utsu, T., Possibility of a great earthquake in the Tokai district, central Japan, J. Phys. Earth, 25, Suppl., S219-S230, 1977.

Utsu, T., Seismicity of Japan from 1885 through 1925 - A new catalogue of earthquakes of $M \geq 6$ felt in Japan and smaller earthquakes which caused damage in Japan -, Bull. Earthq. Res. Inst., 54, 253-308, 1979a (in Japanese).

Utsu, T., Some remarks on a seismic gap off Miyagi prefecture, Rep. Coord. Comm. Earthquake Prediction, 21, 44-46, 1979b (in Japanese).

Utsu, T., Calculation of the probability of success of an earthquake prediction (In the case of the Izu-Oshima-kinkai earthquake of 1978), Rep. Coord. Comm. Earthquake Prediction, 21, 164-166, 1979c (in Japanese).

Utsu, T., Spatial and temporal distribution of low frequency earthquakes in Japan, J. Phys. Earth, 28, 361-384, 1980.

Wakita, H., Y. Nakamura, K. Notsu, M. Moguchi and T. Asada, Radon anomaly: a possible precursor of the 1978 Izu-Oshima-kinkai earthquake, Science, 207, 882-883, 1980.

Yamaguchi, R. and T. Odaka, Precursory changes in water level at Funabara and Kakigi before the Izu-Oshima-kinkai earthquake of 1978, Bull. Earthq. Res. Inst., 53, 841-854, 1978 (in Japanese).

Yamakawa, N., A. Takayanagi and M. Kishio, Earthquakes and tectonics in the southern Izu peninsula (I), Abstracts, Seismol. Soc. Japan, 1977 (2), 64, 1977 (in Japanese).

Yukutake, T., T. Yoshino, H. Utada and T. Shimomura, Time variations observed in the earth resistivity on the Oshima volcano before the Izu-Oshima-kinkai earthquake on January 14, 1978, Bull. Earthq. Res. Inst., 53, 961-972, 1978 (in Japanese).

STRATEGY OF EARTHQUAKE PREDICTION IN CHINA (A PERSONAL VIEW)

Gong-xu Gu

Institute of Geophysics, State Seismological Bureau
Beijing (Peking), The People's Republic of China

Abstract. This report begins with some fundamental guiding philosophy regarding earthquake
prediction and some controversy on opinions regarding: (1) basic research and prediction
practice; (2) observations and theoretical
studies; (3) the question of composite prediction.
Two urgent tasks now confronting all Chinese
seismologists are discussed, namely protection of
the Peking metropolitan area against earthquake
disaster and betterment of the ways of assessing
earthquake risk for a given region. Finally, some
considerations on future scientific approaches
are put forward, involving the discovery of more
empirical precursory relations from actual earthquake occurrences in this country, experimental
work in-situ, and exhaustive and penetrating
studies of some of the more promising precursory
phenomena for earthquake prediction.

Background of Earthquake Prediction Studies in China

The disastrous earthquake in March 1966 at
Hsingtai, about 300 km southwest of Peking, the
capital of China, had a great impact upon this
highly populated agricultural area. The long
sequence of aftershocks, which lasted for many
years, provided an opportunity to predict stronger
aftershocks of the sequence, a few days or weeks
before their occurrences, although such an attempt
did not have much success. However, it initiated,
in our country, a scientific endeavour to predict
earthquakes, a problem which has always been considered an invincible difficulty. Certain phenomena observed in this earthquake area indicated
very vaguely the possibility that before a
stronger aftershock one can see something precursory. Three of them may be mentioned here:
1. Immediately preceding a strong aftershock,
   small earthquakes in clusters usually broke
   out in the vicinity of the coming epicenter.
   This was then followed by a quiet period of
   almost no earthquakes and finally by the
   actual strong aftershock. The entire time
   interval ranges from say a few hours to a few
   days.

2. It was also observed that during the entire
   length of the aftershock sequence, strong
   aftershocks "leaped" back and forth between
   the southwestern end of the aftershock belt
   and its northeastern end, a distance of 40 to
   50 km. In that way, one could sometimes make
   certain guesses as to the possible location
   and the time of occurrence of the next strong
   aftershock based on the preceding one.
3. The time of occurrence of a strong aftershock
   to a certain extent was related to the position of the moon - that is the 1st and 15th
   of a month on the lunar calendar. This, of
   course, was not always so.
These three cases merely gave us something worth
noticing in the search for precursory phenomena of
strong aftershocks.

Our late Premier Chou Enlai, who went to the
earthquake stricken area of Hsingtai several
times, hinted to the seismologists in the field
that they should try their best to learn, from
such a natural calamity, lessons and experiences
so as to be able in the future to tell the people
living in earthquake prone areas some precautionary information prior to an impending large earthquake, and not to leave only records of earthquakes like our ancestors did. Encouraged by his
instruction and support, seismologists in our
country began for the first time to open up their
minds to study earthquakes for the purpose of
their prediction, instead of simply recording or
compiling earthquake bulletins and catalogues.

To predict correctly the time, place and magnitude of an impending earthquake is, of course, an
extremely difficult scientific task. These factors must be correct to a very high degree of
accuracy at the same time, otherwise the prediction would cause great panic and social consequences even if the earthquakes do not eventually
occur.

In the 14 years from 1966 to the present, there
have occurred more than 60 earthquakes of magnitude greater than 6, including aftershocks, in
continental China. Some of them were very disastrous, inflicting heavy losses to life and property. Naturally, the seismologists of this coun-

try have been called upon to take up the responsibility to do something to minimize the losses caused by large earthquakes.

This is the background of attempting prediction of earthquakes in China.

Controversy in Guiding Thoughts on the Earthquake Prediction Problem

As the attempt to predict earthquakes proceeds, the seismologists have been faced with at least three fundamental questions on which opinions are controversial, namely:

1. Prediction practice and basic research

    First opinion - Prediction of earthquakes should not be practiced unless the science actually reaches such an advanced stage that methods for accurate prediction become available through basic research, since false prediction is just as bad as a disastrous earthquake.

    Opposite opinion - Others think that to achieve accurate prediction of earthquakes is a very distant goal. We should not wait but do something, even if scientifically far from maturity, because earthquakes occur so often in China. The government and people are all much concerned with earthquake disasters, which in a few minutes could cause great destruction and confusion. Seismology is required to meet the situation despite all difficulties - "being unable to predict but having to predict" as we sometimes say. Basic research should be carried out intensively along with the practice of prediction.

2. Observation and theoretical studies

    First opinion - Basic research, some people think, should start from theoretical studies of the physics of the earthquake source or of the origin of the earthquake and then explain the observed precursory phenomena based on the theoretical results acquired from such studies.

    Opposite opinion - Other people believe that the studies should be done in the other way, that is to start with observed facts, then deduce from them certain possible empirical relations or rules connecting those phenomena with the earthquake occurrences. In this way one could do certain prediction right now empirically, without deferring till the physical processes involved in the earthquake source are understood. Perhaps, in this way, earthquake prediction could possibly be realized on a more or less scientific basis in quicker steps. As time proceeds, empirical relations may be brought up gradually to a theoretical level. According to this opinion, much more emphasis should be put on observations.

3. Composite prediction

    First opinion - Because of the extreme prematurity in scientific understanding of all the presumably precursory phenomena, it is generally thought that prediction should be done only by compositing all kinds of phenomena observed, as many as possible, then making the prediction decision on the basis of the majority of the anomalous indications.

    Opposite opinion - Others assert that, given the present state of the art of earthquake prediction, it might be necessary to do prediction in the above manner by compositing many phenomena, but basic research should be confined only to the few promising phenomena which appear destined in the future to provide the eventual scientific breakthrough in accurate prediction. Prediction based on composite phenomena is not our final objective.

The present author prefers, in all three cases, the second opinion. That is, (i) basic research should be combined with prediction practice. Research scientists should make prediction themselves on the basis of the results of their investigations, despite all uncertainties, but predictions should not be made by those who do not do basic research. (ii) Start with observations, searching for more empirical relations for prediction first. (iii) Composite prediction is a temporary method; final breakthrough will be made using only a few precursory phenomena.

Two Urgent Tasks

One

Peking is the national capital of China, situated in a seismically active region. The question confronted is exceedingly serious - how can the city area be protected from extreme suffering in a disastrous earthquake? The 1976 disaster of the Tangshan earthquake took place only about 150 km from Peking, causing tremendous destruction and a great calamity. The recurrence of a similar event in this area is thought to be not entirely impossible in the future.

Such a densely populated metropolitan area - a political, cultural and also industrial center - is entirely different from the cities and rural districts of western China, which, although strong earthquakes frequently occur there, are much less vulnerable. Any large earthquake event will certainly bring about an unprecedented calamity to human life and property in the Peking area. Of course, apart from monitoring and the prediction of impending earthquakes, other more realistic and effective protective measures to mitigate earthquake hazards, such as anti-seismic engineering construction and so forth, should be taken, but in this report we shall deal only with monitoring and prediction.

First of all, a complete, highly-reliable and technically-advanced station network for observing a number of phenomena should be installed, so as to "fortify" the area against possible earthquake hazards, even though we do not yet know which of them are the true precursors of an earthquake. Nevertheless, some observations are fundamental and perhaps more promising as far as earthquake prediction is concerned. The network should be set up as quickly as possible and, in terms of data acquisition, as rigorously as possible. It may involve for example: dense seismometric stations; stations for observing crustal deformations; stations for observing deep water wells; etc. In addition, taking the Peking area as an experimental site, portable observation units are necessary to supplement the permanent observations. In addition, certain other kinds of observations (such as measurements of geomagnetic field, gravity, near surface stress by over-coring on rock outcrops and hydraulic fracturing in drill holes, deep earth resistivity, seismic wave velocity ratio, etc.) should be subjected to exhaustive long range scientific experimentation to explore their possibility as true precursors of earthquakes.

Meanwhile, the deep "fine structure" of the crust and upper mantle of the region should be explored by perhaps the COCORP technique in order to provide some more detailed crustal information for the study of the physics of the earthquake source.

Two

The second question confronting us is how to raise the scientific level of assessing earthquake risk of a region or of the entire country. For many years, we have been constantly making such assessments of the possible occurrence of strong earthquakes, purely on phenomenological and intuitive grounds. An assessment was usually made by conjecture and in favour of the majority of the opinions. Most people are now dissatisfied with the results of such assessments and find them rather meaningless. Therefore, the present situation of earthquake risk assessment must be improved. Some progress in the method of forecasting earthquake occurrences must be made.

In endeavouring to make the situation a little better, it is suggested that seismologists should study recent destructive earthquakes which have occurred in China, to "grasp" them and to learn from them, by collecting more reliable data before and after earthquakes in the attempt to expose certain possible empirical relations which might be of help for earthquake risk assessment, even though they would remain empirical. It is believed that if we could discover some such empirics by observations and studies, the situation will be better and more scientific than now. It may constitute a first step in improving the way of earthquake risk analysis. In actual practice, such circumstances do exist. Their governing principles may not be understood, yet they could be employed for the cause of human welfare, like Chinese medicine and other things.

Future Scientific Approaches

Apart from making all possible efforts to do something for the two urgent tasks mentioned above, we must of course do basic research incessantly and rigorously, destined to achieve our final goal: that one day science will be able to identify true precursory phenomena of destructive earthquakes so that their three elements, time, magnitude and location can be accurately predicted simultaneously - a real breakthrough of earthquake prediction. Research work should be principally confined now to the study of earthquake precursors, not other areas, because, as we think, earthquake prediction relies chiefly on observations of precursory signals which indicate an impending earthquake. Other areas, such as the study of the physics of the earthquake source, should be aimed at understanding the true relation between it and the precursors.

It is our opinion that efforts should be directed along the following lines of approach:

(i). To discover more empirical precursory relations - Aside from judging the risk of occurrence of a strong earthquake for a region empirically, research work should be carried out seriously to put empirical relations gradually onto scientific ground. We think, at the present stage, earthquake prediction is principally empirical in nature. Basic research strives to carry it over to the real scientific stage. It should be done in close connection with work in the actual earthquake regions which are widespread in China. The two large earthquakes in recent years (Tangshan and Haicheng in northern China) seem to display very vaguely certain precursory indications before their occurrences. One must carry on this kind of study in an extremely cautious and strict manner. We must "summarize" the large earthquakes scientifically but not administratively, i.e., simply collecting observed facts and compiling data.

(ii). To do experiments and observations in situ - In order to understand the nature of any phenomena which are supposed to be earthquake precursors, it is absolutely necessary to study them in the actual earthquake area in-situ. Without doing any such field experiments, but using only existing data acquired during past earthquakes passively, one could get hardly anything worthwhile in attempting to understand the phenomena. "Practice is the only test of truth." For instance, before some of our recent earthquakes both gravity and earth-resistivity seemed to decrease near and around the epicentral area and resumed after the earthquakes. What caused the variations? Some people think they were the effect of the rise and fall of the underground water level at the time of the earthquake. To resolve such a question, one must resort to field experiments, for correct conclusions can never be arrived at by mere supposition or conjecture. In this case, the experiments may be carried out in a rather simplified manner - just by observing the change of water level in some water wells and at the same time the variations of gravity values at a nearby observation point.

In the case of the 1975 Haicheng earthquake, it was reported that a few hours before the occurrence of the earthquake, at some places within the epicentral area, the local people smelled a distinct, offensive odor in the air. A field experiment should start with examining the reliability of the report in a very strict way. Then soil samples taken from the places where the phenomena were displayed should be chemically analyzed to see whether the material which gave rise to the odor came from underground. If such a phenomena did appear without any doubt, it may be considered as a short range precursory phenomena.

Reports also indicate that before certain large earthquakes in China, the local inhabitants noticed quite a number of kinds of abnormal phenomena which were possibly precursory indications. We must firmly "grasp" them and endeavor to understand them by doing experiments in-situ. It is highly desirable for us to learn from the scientific phenomena that nature has provided for us. In addition, those phenomena observed by the broad mass of people living in earthquake regions of this country must be at first cautiously examined as to their reliability and then placed on scientific ground.

(iii). To promote much more exhaustive and penetrating studies on certain phenomena which are promising as earthquake precursors - A list of such phenomena for the present may be:

a) the appearance of abnormal distribution of small earthquakes and variation of their focal parameters,
b) variations of seismic velocity ratio,
c) anomalous crustal deformation,
d) variation of ground water level,
e) gravity field,
f) earth-resistivity,
g) shallow stress field,
h) chemical content of underground water,
i) geomagnetic field, etc.

These phenomena should be subjected to long, continued research, equipped with sophisticated observational techniques and with unwearied spirit in attacking such a difficult scientific goal.

It must be admitted that all such phenomena are extremely complicated. Without making the utmost effort in our work, any accomplishment, even the slightest, is hardly concievable.

We need a long-range and far-sighted plan for basic research, including all that has been narrated above, although it is by no means complete. It should be a plan consisting of chiefly scientific but not administrative ideas. It will guide us in making our scientific steps forward. It is not a rigid plan but will be adjusted from time to time in order to comply with developments in the actual work. We are faced with enormous difficulties in earthquake prediction. We have come to this Ewing Symposium to learn from scientists from all parts of the world. We welcome your advice and criticism.

DEVELOPMENT AND STRATEGY OF THE EARTHQUAKE PREDICTION
PROGRAM IN THE UNITED STATES

Robert L. Wesson and John R. Filson

U.S. Geological Survey, Reston, VA 22092

Abstract. A review of the strategy of the
current earthquake prediction program in the
United States indicates that it follows a course
charted in the early 1900's. The organization of
the program in the United States involves
participation of Federal and non-Federal
scientists representing many institutions and
disciplines. The structure of the current program
is described in terms of long-range goals and more
immediate objectives and tasks.

Philosophy

Earth sciences, armed with the understanding and
experience gained through the development of the
concepts of plate tectonics, are now embarking on
an even more challenging course--the temporal
prediction of geologic phenomena. Earthquake
prediction is an outstanding example of this new
direction. While the concepts of plate tectonics
enable an understanding of the spatial
distribution of tectonic belts and explain many
features of the long-term behavior of the plates
that cover the Earth's surface, they provide only
a framework for attempting to predict the short-
term behavior of geologic phenomena. While these
concepts provide a geometrical description of the
constraints on the character of tectonic
processes, much more is required before we fully
understand the tectonic processes which shape the
surface of the Earth. Indeed, the goal of
earthquake prediction is implicit in this broader
goal of understanding the physical and geologic
mechanisms of tectonism.

But earthquake prediction is not a scientific
goal alone. The power for destruction that a
large earthquake can visit upon us and our works
requires that we take mitigating actions. The
particular set of mitigating actions appropriate
for a society will depend upon the conditions of
the building inventory and other factors within
that society. Given adequate economic resources,
the technology of earthquake engineering permits
construction to resist the shaking of almost any
conceivable earthquake. However, many existing
structures around the world, including those in
the United States, and indeed many structures
currently under construction, cannot be expected
to resist the strong ground shaking they are apt
to experience. Consequently, earthquake
prediction must be relied upon to play the role of
saving the lives of people who live and work in
these structures. Earthquake predicition has a
role to play not only in the saving of lives but
in the reduction of economic losses and social
disruptions from large earthquakes. For example,
it provides the opportunity to prepare or move
fragile inventories and to reduce the potential
for damage from secondary failures.

These two distinct motivations for pursuing the
goal of earthquake predicition, scientific
understanding and the imperative to reduce human
suffering, are reflected in what sometimes appear
to be two distinctive approaches to the problem of
earthquake predicition, the mechanistic and
empirical approaches. The mechanistic approach is
suggested by the desire to understand the
processes leading to a large earthquake. The
empirical approach is motivated by the immediate
need to protect lives and property. In actual
practice, these two approaches represent possible
emphases in a blend of approaches. Indeed at our
current state of knowledge about the earthquake
process, progress cannot be made without vast new
quantities of observational information. This is
required not only for the empirical predicition of
earthquakes but also for the testing and
development of hypotheses and, therefore, of
understanding. At the same time, reliable
prediction cannot be based simply on empirical
observations but--to address adequately the
uncertainities--must include an understanding of
the processes at work.

These then are the most descriptive
characteristics of an earthquake predicition
program. It is basically a program of scientific
inquiry, but one which is motivated by social,
political, and economic as well as scientific
imperatives. It is a pursuit that cannot rely on
empirical observations alone nor can it be carried
out solely on a blackboard or in the laboratory.
Experiments must by carried out in the real Earth.

History and Development

Development of the earthquake prediction program within the United States has taken place within the context of a multiphased approach to earthquake hazard reduction. While commonly earthquake prediction is the aspect of this program that catches the public eye, earthquake engineering and the investigation of earthquake risk and hazard assessment are also critical elements of the total effort. Indeed, it is often difficult to delineate in detail meaningful boundaries between these elements.

Earthquake prediction was not taken seriously as a respectable avenue of scientific endeavor in the United States until the mid-1960's. Prior to the great Alaskan earthquake of 1964, there were only a few indications that a major program would be in place some 16 years later. First of these was the general increase in the level of effort and scientific thought about earthquake problems brought about by the Department of Defense VELA Uniform Program. Second was the general elevation of the level of geophysics in the United States brought about during the 1950's and 60's as exemplified by the National Academy of Science-National Research Council Report on Solid Earth Geophysics (Panel on Solid-Earth Problems, 1964), which contains one of the first mentions of earthquake prediction as a scientific goal.

Following the impressive display of destructive power in the 1964 Alaskan earthquake, concern in the United States began to focus on the problems of earthquake hazards reduction, particularly in the rapidly developing areas of California and the western United States. The first systematic approach to defining an earthquake program was that of the so-called "Press panel" led by Frank Press which resulted in a report published in 1965 entitled "Earthquake Prediction - A Proposal for a 10 Year Program of Research" (U.S. Office of Science and Technology, 1965). This report continues to make fascinating reading even in 1980 because many of the trends that we are still following were outlined at that time. The Press panel was composed of a cross section of academic, industrial, and government scientists. Following the release of this report, an ad hoc interagency working group of the Federal Council for Science and Technology was established. This group, composed of government officials in scientific agencies, prepared a report entitled "Proposal for a 10 Year National Earthquake Hazards Program" (Federal Council for Science and Technology, 1968). At the time of these reports, total funding for earthquake research, excluding engineering, in the government agencies including NOAA, USGS, and NSF was on the order of $5 million dollars. The organization of an integrated earthquake hazards program did not take place rapidly however. Indeed, bureaucratic scrambling among government agencies and friction among various disciplines including geophysics, geology, and engineering, significantly retarded the

development of the program (U.S. Office of Science and Technology, 1970). However, through the late 60's and into the early 70's a number of events took place that led to the eventual consolidation and development of an integrated earthquake hazards reduction program. Most significant among these events was the San Fernando Earthquake of 1971 which reemphasized the vulnerability of the rapidly developing areas such as southern California to earthquake damage (Joint Panel on the San Fernando earthquake, 1971). Second of these events was the consolidation in 1973 of the solid earth geophysics program within the U.S. Government in the Geological Survey. Third, the increased interaction with foreign earthquake prediction and loss-reduction efforts, particularly in the Soviet Union, Japan, and the Peoples Republic of China stimulated thought in the United States. The fourth event was the discovery in 1975-1976 of the apparently anomalous widespread crustal movements in southern California (Castle and others, 1976). Throughout this period ferment continued within the scientific and engineering communities, as evidenced by a continual flow of panel reports relevant to various aspects of the earthquake hazards reduction problem (Committee on Earthquake Engineering Research, 1969; Committee on the Alaska Earthquake, 1969; Committee on Seismology, 1969; Joint Panel on Problems Concerning Seismology and Rock Mechanics, 1972; Panel on Strong-Motion Seismology, 1973; Joint Committee on Seismic Safety, 1974; Panel on Public Policy Implications of Earthquake Prediction, 1975; Panel on Earthquake Prediction of the Committee on Seismology, 1976.) These events led in 1976 to the formation by the then President's Science Advisor, H. Guyford Stever, of the so-called Newmark-Stever Panel which issued a report titled "Earthquake Prediction and Hazards Reduction Options for U.S. Geological Survey and National Science Foundation Programs" (U.S. National Science Foundation and U.S. Geological Survey, 1976). The recommendations in this report were adopted by the Administration and preparations were made for the initiation of a much expanded program in the fall of 1977. This Administration initiative was strengthened significantly by the passage in the fall of 1977 by the Congress of the Earthquake Hazard Reduction Act which established in law the goals of earthquake prediction and hazards reduction. Upon the completion of these efforts in the fall of 1977, the combined funding for the Geological Survey and National Science Foundation in the areas of earthquake studies, including earthquake engineering, reached a level of about $50 million dollars. However, at least as important as the substantially increased funding for earthquake prediction was the general coalescence of the relevant disciplines within the scientific community toward the overall goal of earthquake hazards reduction--and the goal of earthquake prediction in particular. Indeed, within the community of earthquake scientists, it is now difficult to remember the previous difficulties in communication among engineers, geologist, seismologists, and practitioners of rock mechanics. Today many of these barriers to communication have been removed, resulting in a much stronger and unified approach to the problems posed by earthquakes.

Organization

Responsibility for the management of research within the U.S. National Earthquake Hazards Reduction Program is shared between the National Science Foundation (NSF) and the U.S. Geological Survey (USGS). The NSF is responsible for fundamental studies, earthquake engineering, and research for policy. The USGS is responsible for earthquake hazards and risk assessment, earthquake prediction, global seismology, induced seismicity, and the operation of seismic engineering programs. The distribution of funds for these elements of the overall program for Fiscal Year 1980 is shown in table 1.

This paper--following the theme of the Ewing Symposium--is concerned with earthquake prediction in a slightly broader sense than that used in defining the elements above. Consequently, it will focus not only on the earthquake prediction element listed above but discuss some aspects of the NSF fundamental studies program and of the USGS earthquake hazards and risk assessment elements.

All NSF-funded programs related to earthquake prediction are carried out by university researchers and funded through unsolicited proposals. The elements of the program managed by the USGS are carried out by university and industry researchers, as well as by USGS personnel. Work carried out by non-government researchers in the elements managed by the USGS is funded almost entirely through proposals submitted in response to a request for proposals which lists the kinds of research sought.

Despite the fact that the management of the earthquake prediction program rests with the government agencies, academic and industrial researchers have opportunities to influence the direction of the program through service on advisory committees and panels and through the preparation and submission of proposals for spectific research projects. This process is outlined diagrammatically in figure 1.

Scientific Foundations and Current Understanding

Many of the ideas that form the scientific basis for the United States earthquake predicition program can be traced to those of H. F. Reid (1910) and G. K. Gilbert (1909). Indeed, the elastic rebound hypothesis of Reid, which is well supported by geodetic and geological observations of large earthquakes and by the character of seismic radiation from the earthquake source, forms the core of our ideas about strain

Table 1. Funding for Research in U.S. National Earthquake Hazards Reduction Program Fiscal Year 1980, total $55.3 million.

Organization/Program	Funding
U.S. Geological Survey	
Earthquake Prediction	$15.3 M
Earthquake Hazards and Risk Assessment	$11.1 M
Fundamental Studies (Global Seismology)	$3.6 M
Induced Seismicity	$1.2 M
U.S. Geological Survey, total	$31.2 M
National Science Foundation	
Earthquake Engineering	$17.5 M
Fundamental Studies	$6.6 M
National Science Foundation, total	$24.1 M
National Earthquake Hazards Reduction Program, total	$55.3 M

accumulation. The basic idea of elastic rebound--somewhat abstract when it was first conceived--taken together with powerful notions of plate tectonics, gives us a basic framework for understanding the accumulation and release of strain as episodes in the inexorable process of plate motion. In fact, it can be argued that presently we understand the processes of very long-term strain accumulation, strain release, and elastic radiation relatively well. In contrast, we understand only very poorly, if at all, the actual processes by which the upper crustal rocks are loaded and, perhaps most importantly, the processes leading to failure and slippage along the fault. Understanding of these processes is the key to earthquake prediction.

Reid (1910), from his studies of the geodetic measurements made before and after the 1906 San Francisco earthquake, deduced how elastic strains must be accumulated. He further speculated about the origin of the stresses causing these strains, and argued that they arise from "flows of material at some depth below the surface." Although Reid admitted that he could not demonstrate the existence of the flows, he quoted Willis (1908) who speculated on geologic grounds "that there is a general sub-surface flow towards the north (beneath the Pacific Ocean) which would produce

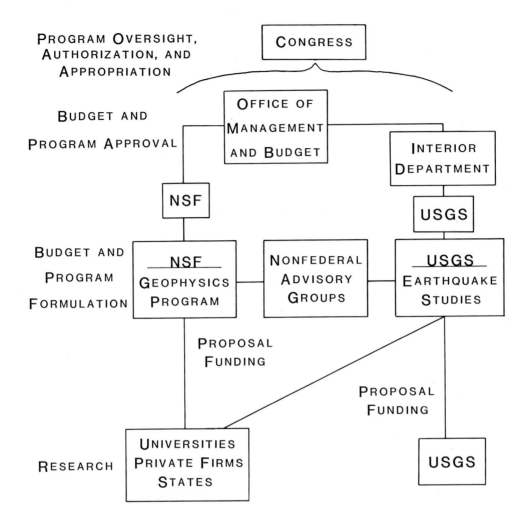

CONGRESS

OFFICE OF
MANAGEMENT
AND BUDGET

INTERIOR
DEPARTMENT

BUDGET AND

PROGRAM APPROVAL

NSF

USGS

BUDGET AND

PROGRAM

FORMULATION

NSF
GEOPHYSICS
PROGRAM

NONFEDERAL
ADVISORY
GROUPS

USGS
EARTHQUAKE
STUDIES

PROPOSAL
FUNDING

PROPOSAL
FUNDING

RESEARCH

UNIVERSITIES
PRIVATE FIRMS
STATES

USGS

ORGANIZATION OF EARTHQUAKE PREDICTION STUDIES

Figure 1. Organization of earthquake prediction research in the United States. The government activities responsible for the management of the relevant elements of the national program reside within the Geophysics Program of the National Science Foundation and the Office of Earthquake Studies of the U.S. Geological Survey. Program decisions by these activities are influenced significantly by advice received from panels of non-federal advisors, and by the nature of specific research proposed for funding.

strains and earthquakes along the western coast of North America". Gilbert (1909), in considering the problem of "earthquake forecasts", described four elements: "rhythm" - now established in our thinking as the concept of recurrence interval and statistics; "alternation" - now established as the concept of the seismic gaps; the "trigger or starter" - which embodies the ideas of a slow accumulation of elastic strain with the question of "what is the straw that breaks the camels back", (i.e. the variation of stress due to tidal, atmospheric and other phenomena), and the "prelude" - a period prior to the earthquake during which precursors are observed. The analyses of Reid and Gilbert provide a useful

framework for evaluating what we know and what we do not know about how to predict earthquakes. Geologic and seismologic studies, such as those described by others in this volume, overwhelmingly demonstrate the validity of the concept of a secular accumulation of elastic strain leading to a series of failures, the occurrence of which in time is reflected by statistics on recurrence intervals, and the distribution of which along the fault zone occurs in a pattern of filling seismic gaps. We currently fail to understand--in a systematic way: (1) the process by which plate motions are translated into strain accumulation and the mechanism and significance of the apparent short-term variations in this process, (2) whether the final initiation of an earthquake is the result of (a) some trigger stress induced by external forces as envisioned by Gilbert, (b) a surge of the internal process responsible for accumulating strain in the upper crust, or (c) some progressive failure mechanism internal to the crustal rocks at the earthquake focus, and (3) how the current collection of precursory observations fits into in any kind of synoptic framework useful for prediction. Indeed, we currently have a poor understanding of which of the precursory observations made to date owe their origins to processes responsible for the initiation of a subsequent large earthquake.

Specific recent accomplishments of the American program in earthquake prediction are elaborated by Ward (1978 and 1979), Kisslinger and Wyss (1975), and elsewhere in this volume; however, it is useful to summarize accomplishments broadly. During the past decade significant technical accomplishments toward the solution of problems related to earthquake prediction have been attained. We now have the experience and ability to operate dense seismometer networks. We understand how to apply the data from these networks in the routine location of earthquakes and the determination of source parameters. We know how to use the data from global seismometer networks to identify gaps in worldwide seismicity patterns. Our kinematic and dynamic source models are becoming more sophisticated and, we hope, more realistic, with the introduction of concepts such as the source moment tensor and stress drop and the study of dynamic crack or rupture propagation models.

Today, the installation and maintenance of seismographs makes up the single greatest instrumentation effort in the American earthquake program. About 400 seismographs are located in California, and nearly 300 are scattered throughout most of the other seismic areas of the United States. Routine and comprehensive analysis of the data from dense seismometer networks is a more difficult problem than data collection; to this end, systems have been developed to monitor by computer data from more than 150 stations at a time and to process these data interactively. A major challenge facing the American program is to consolidate these gains and to move on toward the solution of other problems.

A list of other geophysical instrumentation that has been developed and deployed in the effort to predict earthquakes includes: tiltmeters, creep meters, strainmeters, gravimeters, magnetometers, resistivity monitoring equipment, water-well-level sensors, and soil-gas emission indicators. Anomalies or rapid variations in all of the geophysical fields sensed by these instruments have been reported or suggested as precursors to earthquakes. The major observational advance in the past decade has been the realization that many, if not most, variations in certain fields occur due to circumstances not directly relatable to an imminent earthquake. No one observable

variation has been demonstrated to occur, and only occur, before every type and size of earthquake; thus, it must be assumed that observations of a variety of precursors will be required to predict earthquakes reliably.

The last decade has shown an increase in our ability to conduct more realistic laboratory experiments on rock and other materials and to relate this experience to field situations. Laboratory studies have demonstrated that certain rocks increase in volume prior to fracture and that this dilatancy effect is caused by stress-induced microcracks. It was proposed that this effect would increase the permeability and pore pressure, thus reducing the effective pressure at depth and accelerating the failure process or faulting in the crust. The appeal in the dilatancy-diffusion theory is that, if applicable, it should result in a number of observable variations in seismic velocities, ground level or tilt, electrical resistivity, water well levels, and gas emission. Detailed laboratory studies suggest, however, that dilatancy cannot account for the large size of many precursory signals, nor for their time history. Another important process observed in laboratory studies is preseismic fault slip. During certain experiments designed to model the faulting process ("stick-slip" conditions) there is usually a slow propagation of creep along the fault, followed by a rapid stage of slip that becomes unstable after propagating some distance, and ending with sudden earthquake-like displacement. At this time the relative importance of dilatancy and preseismic slip as the causes of observed precursory phenomena is unclear.

The materials used in laboratory experiments have increased in realism and complexity from solid blocks, to blocks with ground surfaces, to models with material representing fault gouge between two competent blocks. As the realism of the rock models has increased, the ranges of physical conditions reproducible in the laboratory has also increased.

One of the more exciting and puzzling observations during the past decade has been geodetic evidence that the elevation of a 12,000 square kilometer area in southern California had increased by up to 15 to 25 cm in the 1970's. Because the uplift lies astride the section of the San Andreas fault that has been essentially aseismic since a large earthquake in 1857, and uplifts had been reported before other damaging earthquakes, concern developed that this uplift might be precursory to the next great California earthquake. This incident points up the increased use of the geodetic data in earthquake prediction studies in the past few years. Precise triangulation, trilateration, and leveling techniques that allow mapping of horizontal and vertical strain changes over survey lines kilometers long are now being conducted, in some cases routinely, to determine the strain patterns in regions where earthquakes are expected. These efforts have shown, for example, what appears to be a regional strain accumulation pattern of north-south compression and east-west tension in central California, as one would expect for the San Andreas, but that in southern California the strain pattern is north-south compression with a lesser degree of a east-west motion. Variations from this pattern with periods of weeks to months, described elsewhere in this volume, are intriguing but not understood. New methods being developed using laser ranging to satellites and the correlation of signals from satellites and distant quasars offer the possibility of measuring strain changes over distances of hundreds to thousands of kilometers.

A key issue is the absolute level of stress in the crust and the stresses involved in active faulting. A major dispute in seismic source studies is whether the typical 10 to 100 bar stress drop observed during an earthquake occurs in a high ambient stress field of up to kilobar or in a low stress field of tens to hundreds of bars. Interpretation of heat flow data suggest an upper limit on stress of 100-200 bars. Recent results using hydrofracturing techniques borrowed from the oil industry indicate that shallow stress values near the San Andreas fault are relatively low and that the stress decreases toward the fault and toward the free surface. Detailed seismic reflection studies provide evidence for a low velocity zone at the depth of earthquakes in the rocks adjacent to the creeping section of the San Andreas fault in central California. Such a zone may reflect a reduced effective stress in the rock at depth. If the stresses vary significantly with fault type and location, the possiblility is raised that processes that might lead to the prediction of strike-slip faulting are fundamentally different from those acting in thrust faulting or faults not occurring on plate boundaries.

A preponderance of historical and current reports has convinced many that a variation in some, yet unidentified, field affects animals prior to an earthquake. Serious projects are underway to study this phenomena and to determine what phenomena related to earthquakes animals may sense.

Clearly some of the most significant gains in understanding have come through the application of plate tectonic concepts to the investigation of great earthquakes. Foremost among these gains is the development and elaboration of the seismic gap hypothesis detailed elsewhere in the volume. Another key development has been the application and vertification of the Hubbert-Rubey effective stress hypothesis to the occurrence of earthquakes, at least under some special circumstances in which fluids are injected into the ground at relatively high pressures. The application of the hypothesis to natural tectonic earthquakes offers considerable appeal, but the role of pore fluids in natural earthquakes remains to be elucidated.

Strategy

With a program of scientific research at this stage of development and with such a deverse group of participants, including government, university and industrial researchers, it is useful to have a statement of the future broad directions of the program. Meaningful research in earthquake prediction requires substantial resources. These resources are required for long-term observations and for the maintenance of large and expensive operational facilities. For this reason, geologists, geophysicists, and others involved in earthquake prediction must realize that they face some of the same problems that have already been faced by colleagues in a number of scientific fields where similar large capital expenditures are required. These fields include high energy physics, radio astronomy, oceanography, and marine geophysics. Because the tools of these disciplines are so expensive, practitioners have developed mechnisms for jointly planning and coordinating research. It would be false, however, to argue that these disciplines have produced any less creativity or significant new breakthroughs than those disciplines whose practitioners work in isolation. Indeed, the contrary seems to be true. Disciplines that have learned to manage large programs have produced some of the most exciting and stimulating breakthroughs in science in recent years.

To meet the need for a program statement for the earthquake prediction program, the U.S. Geological Survey has prepared a 5 year plan. This plan, and similar plans for earthquake hazards assessment and the other parts of the program for which the USGS is responsible, are the result of internal debate, and review and discussion by the USGS Earthquake Studies Advisory Panel. The plan will be reviewed annually and revised as appropriate.

A 5-year plan for earthquake research must provide the basis for decisions about the expenditure of resources and provide for the scheduling and phasing of expensive facilities, but at the same time, it must provide an environment that allows an individual scientist freedom to pursue the leads arising from his or her own research.

The plan for a research program must be based upon a strategy that takes into account the current state of understanding of the phenomena, the current capability for making observations and available hypotheses. The plan should indicate long-term goals. To measure progress, objectives should established that can be accomplished within a given period of time.

The primary goals of the earthquake prediction program are to:

o   Determine the tectonic framework of seismically active regions as a means for understanding the basic process that result in earthquakes.

o   Provide a physical basis for the short term prediction of earthquakes through understanding the mechanics of faulting.

o   Determine the measurable phenomena that may be useful as earthquake precursors.

o   Develop and evaluate an experimental earthquake prediction system.

These goals, if reached, should lead to a successful earthquake prediction capability and to a reassuring understanding of the physics of the phenomena monitored in the predictive process. To reach these goals, a broad interdisciplinary and multi-institutional strategy must be employed. Given the current level of understanding of the predicition problem, no one institution or discipline can demonstrate that it alone possesses the most direct path to successful attainment of these goals. Indeed, while it may be inappropriate to establish short-term objectives for the fundamental studies element of the program managed by the National Science Foundation, much of that work--and its accomplishments--can be associated with these long-term goals.

To some extent, how quickly we attain these goals is based on chance. Because of limited resources, we must place heavy emphasis on certain geographic regions within which we expect earthquakes of intermediate size or larger. In most areas, we must limit our activities to reconnaissance or partial instrumentation efforts, because--although very large earthquakes are possible--the probability of these earthquakes occurring in the next 5 years must be judged as low.

This problem can be mitigated through international cooperation, where we hope to share expertise and experience with foreign earthquake prediction research activities. This pooling of scientific knowledge and field investigations, which we can promote and even influence through modest allotments of resources, allows us to increase the number of prediction case histories and the number of seismic regions in which precursory phenomena may be observed.

Eventually, in order to direct our resources efficiently, we must winnow the techniques that, although they have enthusiastic and competent spokesmen, do not appear to justify intensive instrumentation efforts. This requirement for evaluation includes the instrumentation itself as well as the analytical techniques employed for analysis. Devices that often show spurious responses or that are affected by extremely local conditions must be identified and replaced with those of more stable characteristics and reliable design.

Objectives

The goals stated above are clear and valid statements of the aims of earthquake prediction.

But how do we measure our progress toward their attainment? We need to find some statement of what we can reasonably expect to accomplish in 5 years. The following statements of 5-year objectives from the earthquake prediction element of the program appear accomplishable within that time period. The allocation of resources among these objectives is shown in Table 2. The objectives are:

o   Observe at a reconnaissance scale or better--and analyze--the pattern of the seismicity and geodetic changes leading to two earthquakes of magnitude 6.5 or larger

o   Observe at a detailed scale, with continuously recording instruments, the local pattern of seismicity, strain, magnetic field, and other possible measurements, the geophysical processes leading to five earthquakes of magnitude 5 or larger.

o   Complete detailed profiles through the crust and conduct analyses of physical parameters including seismic velocities and attenuation, deep geologic structure, and physical state (including stress, temperature, and fluid pressure) in regions of active seismicity with emphasis on identified fault zones.

Table 2.   Earthquake Prediction Program of the U.S. Geological Survey, Fiscal Year 1980.

Program Element	Percentage of Resources
Earthquake Prediction	
Deatailed observations of variations preceeding earthquakes of magnitude 5 or larger	57
Reconnaissance scale observations of earthquakes of magnitude 6.5 or larger	15
Studies of fault zone properties and the Earth's crust	14
Laboratory experiments and measurements	11
Continuous crustal strain measurement development	3
Total	100

o   Complete development of a reliable system for field deployment for the continuous measurement of real crustal strain at the level of 10^{-6} or better.

o   Conduct detailed laboratory measurements to determine the physical properties of rock at the temperatures, pressures, and fluid pressures expected to occur at depths of 5-15 km in the Earth's crust.

Also relevant is one of the objectives of the hazards assessment element of the overall program:

o   Improve capability to evaluate earthquake potential and predict character of surface faulting.

Some of these objectives, if successfully completed, will further our progress toward more than one goal and, thus, the reader must keep in mind the inter-relation of the various efforts within the prediction program. For this reason, the phasing or timing of many of the tasks is difficult to define. For example, an automated data analysis system will be a little use if we cannot demonstrate a reliable prediction technique based on geophysical data. However, it does not seem prudent to wait until the applicability of precursory geophysical phenomena has been exhaustively proved before work on such a system is begun. Indeed such a system should help confirm or deny prediction hypotheses. It is commonly difficult to estimate how long certain tasks should be pursued. If data come to light that reveal the futility of a certain area of investigation, that task, and other dependent on its success, must of course, be discontinued.

Each of the objectives is derived from a strategy and implies a set of tasks discussed below:

o   Observe at reconnaissance scale or better--and analyze--the local patterns of seismicity and geodetic changes leading to two earthquakes of magnitude 6.5 or larger.

The chances of a magnitude 6.5 or larger earthquake occurring in a densely instrumented area in the next 5 years are sufficiently low that reconnaissance scale observations over fairly broad areas of the western U.S., Alaska, and selected foreign areas are warranted. These reconnaissance observations consist primarily of regional-scale networks of seismographs and geodetic survey networks. Sample tasks to be undertaken include: the development and operation of reconnaissance prediction networks; establishment of baseline control for future repeatable geodetic networks and strain-measurement networks in seismic gaps in Alaska and foreign countries; analysis of the usefulness of teleseismically-derived seismicity patterns and

stress estimates as indicators of seismic gaps about to be filled, and evaluation of the usefulness of multi-criteria determination of sites of potential great earthquakes through the use of geologic considerations, routinely available geophysical data, and pattern-recognition techniques.

o   Observe at a detailed scale with continuously recording instruments the local patterns of seismicity, strain, magnetic field, and other possible measurements the geophysical processes leading to five earthquakes of magnitude 5 or larger and analyze the results.

Detailed--and costly--networks of instruments and surveys must be focussed on the most highly seismic areas where observations can be carried out effectively. Emphasis will be placed on development of intensive observational systems in California. It may be possible to obtain detailed observations preceding an earthquake of magnitude 6 or larger, but it is likely that several earthquakes of magnitude 5 will occur in the next 5 years within this region. Sample tasks include: maintenance and operation of prediction networks and data analysis for detection of premonitory signals in central and southern California and other locations; development of automated network data acquisition and analysis systems; determination of the usefulness of local seismic velocity data, seismicity patterns, and foreshock characteristics as indicators of an imminent earthquake; annual surveys of 200 km of level lines in California and Nevada; gravity profiles across active faults in California and elsewhere; observation and analysis of near field creep, strain, and tilt measurements to determine the spatial and temporal characteristics of these phenomena prior to earthquakes; determination of the characteristics of magnetic and electric phenomena prior to earthquakes, and observation and analysis of data on other short-term precursors to earthquakes.

o   Complete detailed profiles of the crust and conduct analyses of physical parameters including seismic velocities and attenuation, deep geologic structure and physical state including stress, and temperature and fluid pressure in regions of active seismicity with emphasis on identified fault zones.

A physical basis for the understanding of the earthquake process requires detailed knowledge of the conditions within the focal zone of earthquakes. Measurements using geological, geophysical, and subsurface exploration techniques will be used to gather information, including: measurements of in-situ properties of fault zones, i.e., stress, temperature, fluid pressure, permeability, elastic properties, seismic attenuation, and sample properties; determination, through geophysical field studies, of the detailed structure of fault zones; geologic studies of exposed or eroded fault zones to infer stresses, strain rates, temperatures, fracture distribution, and porosity, and understanding of the dynamic properties of shallow faulting through well-defined experiments.

o   Complete development of a reliable system for field deployment for the continuous measurement of real crustal strain at the level 10^{-6}.

Current methods for monitoring crustal strain are either suspect because of environmental or siting problems or discontinuous and expensive because they require survey-type methods. Efforts will be made to develop continuously-recording systems that eliminate (or reduce to acceptable levels) contamination from environmental and site effects by averaging over sizeable regions, emplacement in boreholes, or other means. Sample tasks include: evaluation of the reliability and stability of instrumentation used to report previous anomalies, e.g., magnetometers, point tiltmeters, telluric current detectors, long-base tiltmeters, laser strainmeters, and point strain gauges; field testing of multicolor laser strainmeters, and development of an economical technique for continuous monitoring of strain using satellite based positioning techniques along long profiles (10 km or greater) along the San Andreas fault and elsewhere.

o   Conduct detailed laboratory measurements to determine the physical properties of rock at the temperatures, pressures, and fluid pressures expected to occur at depths of 5 to 15 km in the Earth's crust.

Because most observations of the focal zone of earthquakes will—of necessity—be made indirectly, knowledge of the properties of rock at earthquake source conditions must be obtained in the laboratory under conditions of high temperature, pressure, and fluid pressure. These studies should be extended to include dynamic experiments of rock failure under shallow crustal conditions. Laboratory experiments and measurements serve as a guide to what should be looked for in the field and as a confirmation of models and hypotheses based on field observations. Tasks include: laboratory measurements of the physical properties of rocks and fault zone materials; theoretical and experimental studies to relate near-source observations to physical models of the failure or faulting process.

o   Improve capability to evaluate earthquake potential and predict the character of surface faulting.

Because the time periods between large earthquakes in a given region are measured in decades and centuries, the instrumental and historic records of seismicity are often inadequate to establish the recurrence intervals of these damaging events. In many cases the record of recent geologic activity can be studied to understand the sequence of occurrence of great earthquakes. To do this reliably, methods for dating Quaternary geologic phenomena must be developed and refined. Furthermore, our understanding of the nature of surface faulting as related to fault geometry and setting must be improved by studying active faults worldwide. New techniques in in situ stress measurement should be applied near active faults and evaluated as a tool for assessing earthquake potential. Sample tasks include: improvement of techniques and tools for estimating earthquake recurrence intervals, such as new methods for dating prehistoric earthquakes and episodes of faulting; expansion of the data base on the characteristics of surface faulting, including the distribution of slip in time and space, in relation to the size, depth, and tectonic setting of associated earthquakes; investigation of surface faulting associated with earthquakes worldwide and evaluation of direct stress measurement in the vicinity of a fault as a tool for assessing earthquake potential, and measurement of stress in the vicinity of selected active and inactive faults.

Outlook

The attainment of capability for earthquake prediction, like the goals of other directed basic research programs, is no easy task. Earthquake prediction research is similar to a space exploration program in that, if current indications are correct, high technology will be required. But technology alone is not enough. Breakthroughs will be required in our understanding of the natural tectonic processes at work in the Earth. Earthquake prediction provides a particularly sharp challenge for the Earth sciences, because it will require the best of the classical methods of geology as well as emerging technology and analytical techniques from a host of disciplines. In addition, research in earthquake prediction will require cooperative efforts of increasingly large scales, distributed over braod regions of seismically active zones and sustained over periods of years to decades. Local studies will always play a critical role, but the essential experiment of earthquake prediction is carried out in volumes of the Earth's crust involving millions of cubic kilometers and over periods of time as long as centuries or more. We must sustain an environment that fosters individual creativity with experiments of this overwhelming scale, which of necessity will require planning and coordination.

The pressures from society, certainly in the United States, will be great. Society will demand increasingly accurate predictions much sooner than the scientific community will be able to provide them (Wesson, 1979). At the same time, society will experience quite significant stresses in coming to grips with the responsible application of predictions. These stresses are not insurmountable, but the productive use of earthquake prediction will require strenuous efforts on the part of scientists to educate and communicate. There are certain to be disappointments, and it will be vitally important to reduce the loss of credibility from incorrect predictions and to learn as much as possible from mistakes to prevent their repetition. From the research perspective, large budgets will bring high visibility and increased public scrutiny.

International cooperation in earthquake prediction research, so important in the formative stages of the United States program, will play an increasingly important role as we benefit from shared knowledge and experience.

Earthquake prediction, and the ultimate prospect of exploiting this capability to reduce human suffering, is a recognized goal within the United States, as reflected by the budgetary support for this work given by United States Government. But the success of the program will require much more than money alone. It will require the best possible allocation of resources, and most importantly, the creativity and dedication of our scientists.

References

Castle, R. O., Church, J. P., and Elliott, M. R., Aseismic uplift in southern California: Science, v. 192, pp. 251-253, 1976.

Committee on the Alaska Earthquake, Toward reduction of losses from earthquakes: National Academy of Sciences, Washington, D. C., 1969.

Committee on Earthquake Engineering Research, National Academy of Engineering and National Academy of Sciences, Washington, D. C., 1969.

Committee on Seismology, Seismology: Responsibilities and requirements of a growing science. Part I, Summary and recommendations, 37 pages. Part II, Problems and prospects, 59 pages, National Academy of Sciences, Washington, D. C., 1969.

Federal Council for Science and Technology, Ad hoc interagency group for earthquake research, Proposal for a 10 -year National Earthquake Hazards Program--A partnership of science and the community: Washington, D. C., December, 81 pages, 1968.

Gilbert, G. K., 1909, Earthquake forecasts: Science, v. 29, pp. 122-138, 1909.

Joint Committee on Seismic Safety, State of California, Report of the Legislature: Meeting the earthquake challenge, Sacramento, California, 1974.

Joint Panel on Problems Concerning Seismology and

Rock Mechanics, Earthquakes related to reservoir filling: National Academy of Sciences and National Academy of Engineering, Washington, D. C., 1972.

Joint Panel on the San Fernando Earthquake, The San Fernando Earthquake of February 9, 1971, Washington, D. C., National Academy of Sciences and National Academy of Engineering, 24 pages, 1971.

Kisslinger, C. and Wyss, M., Earthquake Prediction: Reviews of Geophysics and Space Physics, v. 13, no. 3, pp. 298-319, 1975.

Panel on Earthquake Prediction of the Committee on Seismology, Predicting earthquakes: A scientific and technical evaluation with implications for society, National Academy of Sciences, Washington, D. C., 62 pages, 1976.

Panel on the Public Policy Implications of Earthquake Prediction, Earthquake prediction and public policy, National Academy of Sciences, Washington, D. C., 142 pages, 1975.

Panel on Solid-Earth Problems, Solid-earth geophysics-survey and outlook: National Academy of Sciences-National Research Council Publication 1231, 198 pages, 1964.

Panel on Strong-Motion Seismology, Strong-motion engineering seismology: The key to understanding and reducing the damaging effects of earthquakes, National Academy of Sciences, Washington, D. C., 1973.

Reid, H. F., The mechanics of the earthquake, vol. II of the California earthquake of April 18, 1906: Washington, D. C., Carnegie Institute of Washington, 1910.

U.S. National Science Foundation and U.S. Geological Survey, Earthquake prediction and hazard mitigation options for USGS and NSF programs, September 15, 1976: Washington, D. C., U.S. Government Printing Office, 76 pages, 1977.

U.S. Office of Science and Technology, Ad Hoc Panel on Earthquake Prediction, Earthquake prediction--A proposal for a ten year program of research: Washington, D. C., variously paged, 1965.

U.S. Office of Science and Technology, Executive Office of the President, Report of the task force on earthquakes, 1970.

Ward, P. L., Earthquake Prediction in Geophysical Predictions: Washington, D. C., National Academy of Sciences, pp. 37-46, 1978.

Ward, P. L., Earthquake prediction, Reviews of Geophysics and Space Physics, v. 17, no. 2, pp. 343-353, 1979.

Wesson, R. L., Procedures for the evaluation and communication of earthquake predictions in the United States: International Symposium on Earthquake Prediction; UNESCO Headquarters, Paris, France, April 2-6; 22 pages, 1979.

Willis, Bailey, The mobility of the lithosphere, summarized in Science, v. 27, pp. 695-697, 1908.